Linear and Nonlinear Functional Analysis with Applications

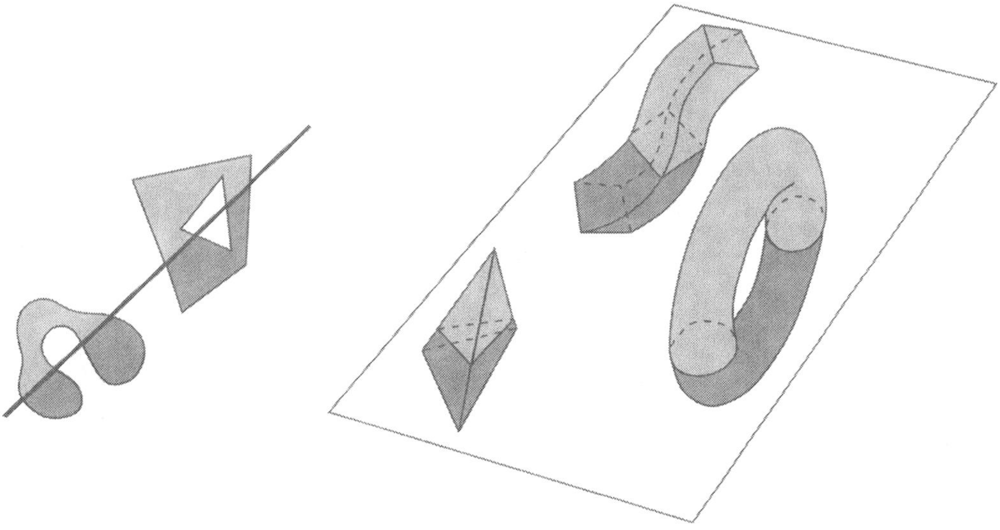

Philippe G. Ciarlet
City University of Hong Kong

Linear and Nonlinear Functional Analysis with Applications

with 401 Problems and 52 Figures

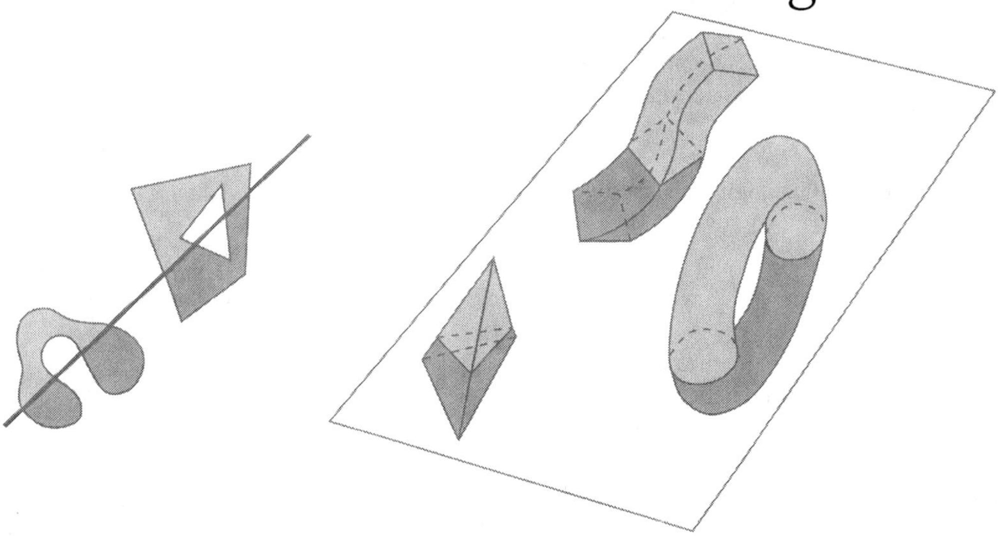

SIAM
Society for Industrial and Applied Mathematics
Philadelphia

Philippe G. Ciarlet
University Distinguished Professor
City University of Hong Kong
Hong Kong
and
Emeritus Professor
Université Pierre et Marie Curie
Paris, France

10 9 8 7 6 5 4 3 2 1

Library of Congress Cataloging-in-Publication Data

Ciarlet, Philippe G., author.
Linear and nonlinear functional analysis with applications / Philippe G. Ciarlet, university distinguished professor, City University of Hong Kong, Hong Kong, emeritus professor, Université Pierre et Marie Curie, Paris, France.
pages cm. -- (Applied mathematics ; 130)
Includes bibliographical references and index.
ISBN 978-1-611972-58-0 (alk. paper)
1. Functional analysis--Textbooks. 2. Nonlinear functional analysis--Textbooks. I. Title.
QA320.C52 2013
515'.7--dc23 2013018736

To the memory of my parents,

Hélène and Gaston

CONTENTS

PREFACE

Why write another textbook on functional analysis and its applications, since there are already many excellent textbooks around?

Apart from the personal pleasure that such an exercise provides to an author, there are other reasons: One, which perhaps constitutes the main originality of this text, was to assemble in a single volume the most basic theorems of linear and of nonlinear functional analysis; another reason was to simultaneously illustrate the wide applicability of these theorems by treating an abundance of applications.

Applications to linear and nonlinear partial differential equations treated here include Korn's inequality and existence theorems in linear elasticity, obstacle problems, the Babuška–Brezzi inf-sup condition, existence theorems for the Stokes and Navier–Stokes equations of fluid mechanics, existence theorems for the von Kármán equations of a nonlinearly elastic plate, and John Ball's existence theorem in nonlinear elasticity. A variety of other applications deals with selected topics from numerical analysis and optimization theory, such as approximation theory, error estimates for polynomial interpolation, numerical linear algebra, basic algorithms of optimization, Newton's method, or finite difference methods.

A special effort has been made to enhance the pedagogical appeal of the book. After Chapter 1, which is essentially a review of results from real analysis and the theory of functions that will be used in the text, self-contained and complete proofs of most of the theorems are provided.[1] These include proofs that are not always easy to locate in the literature, or difficult to reconstitute without an extended knowledge of collateral topics; for instance, self-contained proofs are given of the Poincaré lemma, of the hypoellipticity of the Laplacian, of the existence theorem for Pfaff systems, or of the fundamental theorem of surface theory. Numerous figures and problems (almost 400) have also been included. Historical notes and original references (at least those that I have been able to trace with a reasonable assurance of veracity) have also been included[2] (mostly as footnotes), so as to provide an idea of the genesis of some important results.

It is my belief that this book contains most of the core topics from functional analysis that any analyst interested in linear and nonlinear applications should have encountered at least once in his or her life. More specifically, linear functional analysis and its applications are the subjects of Chapters 2–6, while nonlinear functional analysis and its applications are the subjects of Chapters 7–9.

Of course, choices had to be made, in particular so as to keep the length of the book within reasonable limits. For instance, more specialized topics, such as the Fourier transform,

[1] The symbol $^\flat$ to the left of a theorem indicates one without proof.

[2] With the full knowledge that doing so sometimes constitutes a perilous exercise...

wavelets, spectral theory (save for compact self-adjoint operators), or time-dependent partial differential equations, are not treated.

Several one-semester courses, at the last-year undergraduate or graduate levels, can be taught from this book, such as "Linear Functional Analysis," "Linear and Nonlinear Boundary Value Problems," "Differential Calculus and Applications," "Introduction to Differential Geometry," "Nonlinear Functional Analysis," or "Mathematical Elasticity and Fluid Mechanics." In this respect, it should be easy for an instructor to identify from the table of contents those parts of the book that should be used for any such course. Indeed, I had the pleasure of teaching such courses, primarily at the University Pierre et Marie Curie and at City University of Hong Kong, but also at the University of Texas at Austin, at Cornell University, at Fudan University, at the University of Stuttgart, at l'Ecole Polytechnique Fédérale de Lausanne, at the ETH-Zürich, and at the University of Zürich.

The main prerequisites are a reasonable acquaintance with real analysis, i.e., elementary topology (such as continuity and compactness), the basic properties of metric spaces and Lebesgue integration, and the theory of real-valued functions of one or several real variables. For the reader's convenience, the basic definitions and theorems from these subjects needed in this book are assembled without proofs in the first chapter.

During the writing of this book, I have greatly benefitted from the comments of Liliana Gratie, George Dinca, Cristinel Mardare, Sorin Mardare, and Pascal Azerad, who were kind enough to very carefully read most of the chapters and to suggest numerous significant improvements. Bernard Dacorogna and Vicentiu Radulescu have also provided me with much precious advice. To all of them, my most sincere thanks!

My gratitude is also due to Douglas N. Arnold for his early—and strong—support of the project, and also to Elizabeth Greenspan, Gina Rinelli, and Lisa Briggeman from the Editorial Office of SIAM, with whom it is a real pleasure to cooperate.

Last but not least, I express my deep gratitude and my lasting admiration to my "mathematical heroes" Laurent Schwartz, Richard S. Varga, Jacques-Louis Lions, and Robert Dautray, whose teaching and advice over the years have been invaluable.

I am perfectly aware that, most likely, there are still at places inadequacies, inconsistencies of notations, inadvertently omitted references, or inappropriate attributions of original results. But any adventure (mathematical or otherwise) must come to an end, even if its main protagonist is not fully satisfied with it. Or equivalently, as Paul Halmos said in a much better way, in a pure gem of a paper[3] that any mathematician, pure or applied, should read and reread (I paraphrase him): "The last step for most authors is to stop writing. That's hard."

This is one more reason why I welcome in advance all comments, remarks, criticisms, etc., which should be sent to `mapgc@cityu.edu.hk`, and—who knows—could be used in a second edition.

Hong Kong, November 2012 Philippe G. Ciarlet

[3] P.R. Halmos [1970]: How to write mathematics, *L'Enseignement Mathématique* **16**, 123–152.

Chapter 1

Real Analysis and Theory of Functions: A Quick Review

Introduction

This first chapter constitutes a quick review of *real analysis*, which traditionally comprises: *set theory*, the *axiom of choice*, and the construction of the *sets* $\mathbb{R}$ *and* $\mathbb{R}^n$; the basic properties of *topological and metric spaces*, such as those related to the notions of continuity, compactness, completeness, connectedness, and simple-connectedness; the Tietze–Urysohn extension theorem (a crucial use of which will be made at several places in Chapter 9); and the construction and the main properties of the *Lebesgue measure* and *Lebesgue integral* in $\mathbb{R}^n$: the Radon–Nikodym theorem; Fatou's lemma; the Beppo Levi monotone convergence theorem; the Lebesgue dominated convergence theorem; Tonelli's and Fubini's theorems; volumes, areas, and lengths in $\mathbb{R}^n$; and the change of variable formula in multiple integrals.

This first chapter also includes a quick review of some aspects of the *theory of real-valued functions of several real variables*. More specifically, basic *function spaces*, such as $\mathcal{C}^m(\Omega)$ and $\mathcal{C}^m(\overline{\Omega})$, where Ω is an open subset of $\mathbb{R}^n$, are introduced (other function spaces, such as the Lebesgue spaces $L^p(\Omega)$, or the Sobolev spaces $H^m(\Omega)$ and $W^{m,p}(\Omega)$, will be introduced and studied in later chapters). *Domains* in $\mathbb{R}^n$, that is, open subsets $\Omega \subset \mathbb{R}^n$ that are bounded, connected, and have a Lipschitz-continuous boundary, with Ω being locally on the same side of the boundary, are then singled out among all open subsets of $\mathbb{R}^n$, one reason being that they insure the validity of the *fundamental Green's formula* for functions in the space $\mathcal{C}^1(\overline{\Omega})$ (this Green's formula over domains in $\mathbb{R}^n$ will be later extended to functions in the Sobolev spaces $W^{m,p}(\Omega)$; cf. Chapter 6).

Otherwise the reader is assumed to be already familiar with *linear algebra in finite-dimensional spaces* (bases, linear dependence, matrices, determinants, etc.), as well as with basic notions of *differential calculus for real-valued functions of several real variables* (partial derivatives, Taylor formulas, etc.).

The objective of this chapter is essentially to state in the form of *theorems* the various results from these topics that will be used throughout the book, and to list the various notations and definitions needed for this purpose.

No proofs are given and no exercises are provided, since the reader is assumed to be already reasonably familiar with these results. *References* are provided in the *Bibliographical Notes*.

1.1 Sets

The most commonly adopted set theory is the **Zermelo–Fraenkel set theory**. It starts with *six axioms*, which will not be explicitly stated here; only their consequences will be described.

In this respect, notions such as those of *element* or *set*, or notations such as "$=, \neq, \in, \notin$," or words or an assemblage of words such as "implies," "for all," "there exists," "such that," etc., are not defined; these are assumed instead to be given their intuitive or usual sense (whatever that means).

Let X be a set. The notation $A \subset X$ or $X \supset A$ means that the set A is a **subset** of X, i.e., that $x \in A$ implies $x \in X$. The notation $A \subsetneqq X$ or $X \supsetneqq A$ means that A is a **proper subset** of X, i.e., that $A \subset X$ but $A \neq X$.

Let X be a set. There exists a set, denoted $\mathcal{P}(X)$, whose elements are all the subsets of X. The set $\mathcal{P}(X)$ comprises in particular the **empty set** $\varnothing$ (whose existence is a consequence of the axioms) and the set X itself. If $X \neq \varnothing$ and $x \in X$, the subset of X whose only element is x is denoted $\{x\}$.

Let X be a set. If $A \subset X$, the **complement of** A **relative to** X, or simply the **complement of** A if there is no ambiguity as to what is the set X, is the subset of X defined by

$$X - A := \{x \in X;\ x \notin A\}.$$

If A and B are subsets of a set X, their **union** and **intersection** are respectively denoted, and defined, by

$$A \cup B = \{x \in X;\ x \in A \text{ or } x \in B\},$$
$$A \cap B = \{x \in X;\ x \in A \text{ and } x \in B\}.$$

The sets A and B are **disjoint** if $A \cap B = \varnothing$.

Let X and Y be two sets. The set

$$X \times Y := \{(x, y);\ x \in X \text{ and } y \in Y\},$$

whose elements are all the **ordered pairs** (x, y) with $x \in X$ and $y \in Y$, is called the **product** of X and Y.

A **relation** $\mathcal{R}$ on a set X is any subset $\mathcal{R}$ of the product $X \times X$, i.e., $\mathcal{R}$ consists of specific ordered pairs (x, y), with $x \in X$ and $y \in X$.

An **equivalence relation** on X is a relation $\mathcal{R}$ that satisfies the following properties, where the notation $x \sim y$ means that $(x, y) \in \mathcal{R}$:

reflexivity : $x \sim x$ for all $x \in X$,
symmetry : $x \sim y$ implies $y \sim x$,
transitivity : $x \sim y$ and $y \sim z$ implies $x \sim z$.

Equivalently, $(x, x) \in \mathcal{R}$ for all $x \in X$; if $(x, y) \in \mathcal{R}$, then $(y, x) \in \mathcal{R}$; if $(x, y) \in \mathcal{R}$ and $(y, z) \in \mathcal{R}$, then $(x, z) \in \mathcal{R}$.

Given an element x in a set X endowed with an equivalence relation $\mathcal{R}$, the **equivalence class of** x **modulo** $\mathcal{R}$ is the *subset of* X defined by

$$\dot{x} := \{y \in X;\ y \sim x\}.$$

Two equivalence classes of elements of X are thus either identical or disjoint.

The **quotient set** $X/\mathcal{R}$ is the *subset of* $\mathcal{P}(X)$ consisting of all equivalence classes modulo $\mathcal{R}$ of elements of X.

All the above definitions and properties rely only on the first five axioms of the Zermelo–Fraenkel set theory. The sixth one, called the *axiom of infinity*, is of crucial importance, since it implies both the *existence* of the *set*

$$\mathbb{N} := \{0, 1, 2, \ldots\}$$

formed by all **natural integers**: $0, 1, 2, \ldots$, and the possibility of proving statements by *recursion*: To prove that a property holds for all $n \in \mathbb{N}$, it suffices to prove that it holds for $n = 0$ and that, if it holds for some $n \in \mathbb{N}$, then it also holds for $n + 1$. Specific subsets of $\mathbb{N}$ are designated by self-explanatory notations, such as $\{0, 1, \ldots, n\}$, $\{j \in \mathbb{N};\ 1 \le j \le n\} = \{1, 2, \ldots, n\}$, $\{n \in \mathbb{N};\ n \ge n_0\} = \{n_0, n_0 + 1, \ldots\}$, etc.

1.2 Mappings

Let X and Y be two nonempty sets. A **mapping**, or a **function, of** X **into** Y is a subset f of the product $X \times Y$ such that, for *each* $x \in X$, there exists *one* and *only one* element $y \in Y$ such that (x, y) belongs to f. This element y is then denoted either $f(x)$ or y_x. When the notation y_x is used, x is called an **index**.

The notations

$$f : X \to Y \quad \text{or} \quad X \xrightarrow{f} Y,$$

mean that X and Y are two sets and that f is a mapping of X into Y. The notation

$$f : x \in X \to f(x) \in Y$$

with an explicit expression for $f(x)$ is used to *define* a mapping f.

Let X be a set. The mapping $x \in X \to x \in X$ is called the **identity mapping** of X; it is denoted id or id_X, or I or I_X if X is a vector space.

If A is a subset of a set X, the function $\chi_A : X \to \mathbb{R}$ defined by

$$\chi_A(x) := 1 \ \text{ if } x \in A \quad \text{and} \quad \chi_A(x) := 0 \ \text{ if } x \notin A$$

is called the **characteristic function** of A.

Let $f : X \to Y$ be a mapping. The **direct image under** f **of a subset** A **of** X is the subset $f(A)$ of Y defined by

$$f(A) := \{y \in Y;\ \text{there exists } x \in A \text{ such that } y = f(x)\}.$$

The **inverse image under** f **of a subset** B **of** Y is the subset of X defined by

$$f^{-1}(B) := \{x \in X;\ f(x) \in B\}.$$

If $b \in Y$, the (improper but convenient) notation $f^{-1}(b)$ will be blithely used to designate the inverse image $f^{-1}(\{b\})$.

Care should be exercised when using notations such as $f(A)$ and $f^{-1}(B)$: The notation f designates a mapping of X into Y, *not* a mapping of $\mathcal{P}(X)$ into $\mathcal{P}(Y)$ (as the notation $f(A)$ tends to suggest). Likewise, the notation f^{-1} designates the inverse mapping of f when it exists (see below), in which case f^{-1} is a mapping of Y onto X, *not* a mapping of $\mathcal{P}(Y)$ into $\mathcal{P}(X)$ (as the notation $f^{-1}(B)$ tends to suggest).

The inverse image "preserves all the set operations," in that it satisfies

$$\begin{aligned} f^{-1}(B) &\subset f^{-1}(\widetilde{B}) \text{ if } B \subset \widetilde{B}, \\ f^{-1}(B \cup \widetilde{B}) &= f^{-1}(B) \cup f^{-1}(\widetilde{B}), \\ f^{-1}(B \cap \widetilde{B}) &= f^{-1}(B) \cap f^{-1}(\widetilde{B}), \\ f^{-1}(Y - B) &= X - f^{-1}(B). \end{aligned}$$

By contrast, the direct image only satisfies

$$\begin{aligned} f(A) &\subset f(\widetilde{A}) \text{ if } A \subset \widetilde{A}, \\ f(A \cup \widetilde{A}) &= f(A) \cup f(\widetilde{A}), \\ f(A \cap \widetilde{A}) &\subset f(A) \cap f(\widetilde{A}). \end{aligned}$$

A mapping $f : X \to Y$ is **surjective**, or **onto**, or is a **surjection**, if for each $y \in Y$, there exists *at least one* element $x \in X$ such that $y = f(x)$.

A mapping $f : X \to Y$ is **injective**, or **one-to-one**, or is an **injection**, if for each $y \in Y$, there exists *at most one* element $x \in X$ such that $y = f(x)$. If X is a *subset* of Y, the mapping $\iota : X \to Y$ defined by $\iota(x) = x$ for all $x \in X$ is called the **canonical injection** from X into Y.

A mapping $f : X \to Y$ is **bijective**, or is **one-to-one and onto**, or is a **bijection**, if it is both surjective and injective. In this case, for each $y \in Y$, there thus exists one and only one element $x \in X$ such that $y = f(x)$, and the mapping $f^{-1} : y \in Y \to x \in X$ defined in this fashion is the **inverse mapping of** f.

Let $f : X \to Y$ be a mapping and let A be a subset of X. The mapping $f|_A : A \to Y$ defined by $f|_A(x) := f(x)$ for all $x \in A$ is the **restriction** of f to A.

Let $g : A \to Y$ be a mapping, where A is a subset of X. A mapping $f : X \to Y$ is an **extension** of g if $f|_A = g$.

Let $f : X \to Y$ and $g : Y \to Z$ be two mappings. The mapping $h : X \to Z$ defined by $h(x) = g(f(x))$ for all $x \in X$ is called the **composition of f and g**. It is denoted $h = g \circ f$, or $h = gf$.

Let $f : X \times Y \to Z$ be a mapping and let a be a point in the set X. The mapping $f(a, \cdot) : Y \to Z$ defined by

$$f(a, \cdot) : y \in Y \to f(a, y) \in Z$$

is a **partial mapping**. Given a point $b \in Y$, a similar definition holds for the partial mapping $f(\cdot, b) : X \to Z$.

Given a mapping $f : (x, y) \in X \times Y \to f(x, y) \in Z$, the elements $x \in X$, *resp.* $y \in Y$, are sometimes called *first arguments of f*, *resp. second arguments of f*.

1.3 The axiom of choice and Zorn's lemma

Let $I \neq \varnothing$ and $X \neq \varnothing$ be two sets. A **family of elements of X indexed by I** is a mapping $f : I \to X$ defined as $f : i \in I \to x_i \in X$, i.e., the elements of the set I are regarded as **indices**. Such a family is then denoted by

$$(x_i)_{i \in I},$$

or simply (x_i) if the definition of the set I is unambiguous. Naturally, a family $(x_i)_{i\in I}$ of elements of X is to be carefully distinguished from the subset $\bigcup_{i\in I}\{x_i\}$ of X (which for instance consists of a single point $a \in X$ if $x_i = a$ for all $i \in I$).

A **subfamily** $(x_i)_{i\in J}$ of the family $(x_i)_{i\in I}$ is a mapping $g : J \to X$ such that $J \subset I$ and $f|_J = g$.

If $I = \{1, \ldots, n\}$ for some $n \geq 1$, the family $(x_i)_{i\in I}$ is called an **n-tuple** and is denoted

$$(x_j)_{j=1}^n \quad \text{or} \quad (x_1, \ldots, x_n).$$

If $I = \mathbb{N}$, the family $(x_i)_{i\in I}$ is called a **sequence** and is denoted

$$(x_n)_{n=0}^\infty, \quad \text{or} \quad (x_0, x_1, \ldots, x_n, \ldots), \quad \text{or} \quad (x_n)_{n\geq 0}, \quad \text{or simply} \quad (x_n).$$

Other self-explanatory notations are also used, such as

$$(x_n)_{n\geq 0}, \quad (x_n)_{n=n_0}^\infty \quad \text{or} \quad (x_n)_{n\geq n_0} \quad \text{if } I = \{n_0, n_0+1, \ldots\}, \text{ etc.}$$

A **subsequence** of a sequence is a subfamily that is also a sequence. For instance, given any *strictly increasing* mapping $\sigma : \mathbb{N} \to \mathbb{N}$ (i.e., such that $\sigma(n) < \sigma(n+1)$ for all $n \in \mathbb{N}$), the sequence $(x_{\sigma(n)})_{n=0}^\infty$ is a subsequence of the sequence $(x_n)_{n=0}^\infty$. *This notation will be often used to denote subsequences in the sequel.*

Let I and X be two sets. A **family $(A_i)_{i\in I}$ of subsets of X indexed by I** is a family $i \in I \to A_i \in \mathcal{P}(X)$ (i.e., the mapping appearing in the definition of a family now takes its values in the set $\mathcal{P}(X)$, instead of the set X for a family of elements of X).

Given a family $(A_i)_{i\in I}$ of subsets of a set X, the **union** $\bigcup_{i\in I} A_i$ and **intersection** $\bigcap_{i\in I} A_i$ are respectively defined by

$$\bigcup_{i\in I} A_i := \{x \in X;\ \text{there exists } i \in I \text{ such that } x \in A_i\},$$

$$\bigcap_{i\in I} A_i := \{x \in X;\ x \in A_i \text{ for all } i \in I\}.$$

Other self-explanatory notations are also used, such as

$$\bigcup_{i=0}^n A_i \text{ and } \bigcap_{i=0}^n A_i \quad \text{if } I = \{0, 1, \ldots, n\}, \qquad \bigcup_{i=0}^\infty A_i \text{ and } \bigcap_{i=0}^\infty A_i \quad \text{if } I = \mathbb{N}, \text{ etc.}$$

Given a family $(A_i)_{i\in I}$ of subsets of a set X, the **disjoint union** $\bigsqcup_{i\in I} A_i$ is defined by

$$\bigsqcup_{i\in I} A_i := \bigcup_{i\in I}\{(x,i); x \in A_i\}.$$

The disjoint union $\bigsqcup_{i\in I} A_i$ is thus a subset of the product $(\bigcup_{i\in I} A_i) \times I$, itself a subset of the product $X \times I$.

If A and B are two subsets of a set X, the union $A \cup B$ and the intersection $A \cap B$ coincide with the union $\bigcup_{i\in I} A_i$ and the intersection $\bigcap_{i\in I} A_i$ with $A_1 := A$, $A_2 := B$, and $I := \{1, 2\}$.

The following identities are constantly used:

$$A \cup \left(\bigcap_{i\in I} A_i\right) = \bigcap_{i\in I}(A \cup A_i) \quad \text{and} \quad A \cap \left(\bigcup_{i\in I} A_i\right) = \bigcup_{i\in I}(A \cap A_i),$$

$$X - \bigcup_{i\in I} A_i = \bigcap_{i\in I}(X - A_i) \quad \text{and} \quad X - \bigcap_{i\in I} A_i = \bigcup_{i\in I}(X - A_i),$$

the last two identities constituting **de Morgan's laws**.

Let $(A_i)_{i\in I}$ be a family of subsets of a set X where $I \neq \varnothing$. Then the **product** $\prod_{i\in I} A_i$ is, by definition, the set of all mappings $f : I \to X$ such that $f(i) \in A_i$ for all $i \in I$. Any such mapping f is called a **choice function**, as it asserts that it is possible to "choose" one element $f(i) \in A_i$ for each $i \in I$.

Whereas the definitions of the union $\bigcup_{i\in I} A_i$ and intersection $\bigcap_{i\in I} A_i$ as subsets of the set X do not pose specific difficulties, the definition of the product $\prod_{i\in I} A_i$ raises an immediate question, as nothing guarantees the *existence* of at least one such choice function f. This is why the following axiom, called the **axiom of choice**, was introduced in 1904 by Ernest Zermelo:

Axiom of choice *Let $(A_i)_{i\in I}$ be a family of subsets of a set. If $I \neq \varnothing$ and $A_i \neq \varnothing$ for all $i \in I$, then $\prod_{i\in I} A_i \neq \varnothing$.* □

In 1963, Paul J. Cohen established in a landmark paper[1] that *the axiom of choice is independent of the six axioms of the Zermelo–Fraenkel set theory.*

Other notations for the product $\prod_{i\in I} A_i$ are also used, viz.,

$$A_1 \times A_2 \text{ if } I = \{1, 2\} \text{ (as in Section 1.1)}, \qquad \prod_{i=I}^{n} A_i \text{ if } I = \{1, 2, \dots, n\},$$

$$A^n \text{ if } I = \{1, 2, \dots, n\} \quad \text{and} \quad A_i = A \text{ for all } 1 \le i \le n, \text{ etc.}$$

The element of the product $\prod_{i\in I} A_i$ corresponding to a choice function f will be denoted $x = (x_i)_{i\in I}$, where $x_i := f(i)$, each element $i \in I$ being thus regarded as an *index* (Section 1.2). Each element $x_i \in A_i$, $i \in I$, is called the **ith coordinate** of x. This notation is coherent with those used for a finite sequence $(x_i)_{i=1}^n$ or for a sequence $(x_i)_{i=0}^\infty$ of scalars, which are simply special cases of elements in a product, viz., $\mathbb{K}^n$, or $\prod_{i=0}^\infty A_i$ with $A_i = \mathbb{K}$ for all $i \in \mathbb{N}$, respectively.

The axiom of choice is in fact often used in disguise in proofs, by means of one of its different, but *equivalent*, forms, each one of which then taking the form of a *theorem*. Zorn's lemma (Theorem 1.3-1 below) provides such an example. Note, however, that while the statement of the axiom of choice is intuitively clear, the same cannot be said of Zorn's lemma.

---

[1] P.J. Cohen [1963]: The independence of the continuum hypothesis, *Proceedings of the National Academy of Sciences, USA* **50**, 1143–1148.

In order to state this lemma, we need several definitions.

A set X is **partially ordered** by a *relation* $\mathcal{R}$ (Section 1.1), or equivalently, a relation $\mathcal{R}$ is a **partial ordering on** X, if $\mathcal{R}$ satisfies the following properties, where the notation $x \preccurlyeq y$ means that $(x, y) \in \mathcal{R}$:

reflexivity :	$x \preccurlyeq x$ for all $x \in X$,
antisymmetry :	$x \preccurlyeq y$ and $y \preccurlyeq x$ implies $x = y$,
transitivity :	$x \preccurlyeq y$ and $y \preccurlyeq z$ implies $x \preccurlyeq z$.

Equivalently, $(x, x) \in \mathcal{R}$ for all $x \in X$; if $(x, y) \in \mathcal{R}$ and $(y, x) \in \mathcal{R}$, then $x = y$; if $(x, y) \in \mathcal{R}$ and $(y, z) \in \mathcal{R}$, then $(x, z) \in \mathcal{R}$.

For instance, the relation "$x = (x_i)_{i=1}^n \preccurlyeq y = (y_i)_{i=1}^n$ if and only if $x_i \leq y_i$ for all $1 \leq i \leq n$" defines a partial ordering on the set $\mathbb{R}^n$; the relation "$A \preccurlyeq B$ if and only if $A \subset B$" defines a partial ordering on the set $\mathcal{P}(X)$ formed by all subsets of a set X.

The notation $y \succcurlyeq x$ means that $x \preccurlyeq y$. The notation $x \prec y$, or $y \succ x$, means that $x \preccurlyeq y$ and $x \neq y$.

A subset A of a partially ordered set X is **totally ordered** if any two elements $a \in A$ and $b \in A$ are **comparable**, in the sense that either $a \preccurlyeq b$ or $b \preccurlyeq a$ (if $a \preccurlyeq b$ and $b \preccurlyeq a$, then $a = b$). Clearly, if such a set A is *finite*, i.e., of the form $A = \bigcup_{i=1}^m \{a_i\}$, there exists $1 \leq i_0 \leq m$ such that $a_i \preccurlyeq a_{i_0}$ for all $1 \leq i \leq m$. For instance, if a finite subset $\{A_1, A_2, \ldots, A_m\}$ of $\mathcal{P}(X)$ is totally ordered for the inclusion, then there exists $1 \leq m_0 \leq m$ such that $A_i \subset A_{m_0}$ for all $1 \leq i \leq m$, so that $A_{m_0} = \bigcup_{i=1}^m A_i$; this observation is often used.

Let A be a subset of a partially ordered set X. Then an element $b \in X$ is an **upper bound** for A if $a \preccurlyeq b$ for all $a \in A$. Note that *all* elements of A must then be comparable to b, but that b need not belong to A.

Let X be a partially ordered set. An element $m \in X$ is **maximal** if any element $x \in X$ that is *comparable* to m satisfies $x \preccurlyeq m$; or equivalently, if $x \in X$ satisfies $m \preccurlyeq x$, then $x = m$. Note that m need not be comparable to all elements $x \in X$.

Then *the following result is equivalent to the axiom of choice.*

Theorem 1.3-1 (Zorn's lemma) *Let X be a nonempty partially ordered set with the property that every totally ordered subset has an upper bound in X. Then X has at least one maximal element.* □

Zorn's lemma constitutes an extremely powerful tool for establishing the *existence* of certain mathematical objects. For instance, it is used for proving that there exist *non-Lebesgue measurable subsets of* $\mathbb{R}$ (Section 1.14); for proving that any vector space possesses a *Hamel basis* (Theorem 2.1-1); for proving that any vector space can be *normed* (Theorem 2.2-8); for proving that any inner-product space possesses a maximal orthonormal family (Theorem 4.8-4); or for proving the fundamental *Hahn–Banach theorems*, which assert the existence of *extensions of linear functionals* (Theorems 5.8-1 and 5.9-1).

Note that, each time that Zorn's lemma is applied in a set X, particular care should be given to verifying that X is *nonempty*.

1.4 Construction of the sets $\mathbb{R}$ and $\mathbb{C}$

The set $\mathbb{N}$, whose existence is implied by the axiom of infinity (Section 1.1), is used for constructing the set

$$\mathbb{Z} := \{\ldots, -2, -1, 0, 1, 2, \ldots\}$$

of all **integers**, as the quotient set $(\mathbb{N} \times \mathbb{N})/\mathcal{R}$, where $\mathcal{R}$ denotes the equivalence relation on $\mathbb{N} \times \mathbb{N}$ defined by

$$((m,n),(m',n')) \in \mathcal{R} \quad \text{if and only if } m + n' = m' + n.$$

Equipped with the addition and multiplication, $\mathbb{Z}$ becomes a *commutative ring* that is also an *integral domain* (i.e., if $mp = mq$ and $m \neq 0$, then $p = q$) and *totally ordered* by $\leq$ (total ordering is defined in Section 1.3).

The set $\mathbb{Z}$ is then used for constructing the set $\mathbb{Q}$ of all **rational numbers**, as the set formed by all the *equivalence classes* modulo the following equivalence relation in the set $\mathbb{Z} \times (\mathbb{Z} - \{0\}) : (m,n) \sim (p,q)$ if and only if $mq = np$. Equipped with the operations $+$ and $\times$, the set $\mathbb{Q}$ then becomes a *totally ordered commutative field.* The field $\mathbb{Q}$ is **Archimedean**, that is, given any rational numbers $r > 0$ and $s > 0$, there exists an integer $n \geq 0$ such that $nr > s$. The *absolute value* of $r \in \mathbb{Q}$ is defined by $|r| := r$ if $r \geq 0$, or by $|r| = -r$ if $r \leq 0$.

A sequence $(r_n)_{n=1}^{\infty}$ of *rational numbers* is said to be a **Cauchy sequence** if, given any $\varepsilon > 0$, there exists an integer $m_0 = m_0(\varepsilon) \geq 1$ such that $|r_m - r_n| \leq \varepsilon$ for all $m, n \geq m_0$. The set $\mathbb{Q}$ is then used for constructing the set $\mathbb{R}$ of all **real numbers**, as the set formed by all *equivalence classes* modulo the following *equivalence relation* $\mathcal{R}$ in the *set formed by all Cauchy sequences of rational numbers*: $((r_n)_{n=1}^{\infty}, (s_n)_{n=1}^{\infty}) \in \mathcal{R}$ if, given any $\varepsilon > 0$, there exists an integer $n_0 = n_0(\varepsilon) \geq 1$ such that $|r_n - s_n| \leq \varepsilon$ for all $n \geq n_0$. Equipped with the operations $+$ and $\times$ and the total ordering $\leq$, the set $\mathbb{R}$ is also a *totally ordered, Archimedean, commutative field*, and the *absolute value* of $x \in \mathbb{R}$ is likewise defined by $|x| := x$ if $x \geq 0$, or by $|x| := -x$ if $x \leq 0$. Alternatively, the set $\mathbb{R}$ may be also constructed by means of *Dedekind cuts.*

The set $\{-\infty\} \cup \mathbb{R} \cup \{\infty\}$ of **extended real numbers** is defined by adjoining to the set $\mathbb{R}$ two elements, denoted $-\infty$ and ∞, which obey the usual rules; for instance, $-\infty < x$ for all $x \in \mathbb{R}$, $x + \infty = \infty$ for all $x \in \mathbb{R}$, etc. Naturally, $-\infty + \infty$ is not defined.

Finally, the *commutative field* $\mathbb{C}$ of **complex numbers** is constructed in the usual way from the set $\mathbb{R}$. If $z \in \mathbb{C}$, then $\operatorname{Re} z$ and $\operatorname{Im} z$ respectively denote the *real* and *imaginary parts* of z; in other words, $z = \operatorname{Re} z + i \operatorname{Im} z$. The *absolute value* of $z \in \mathbb{C}$ is defined by $|z| := \sqrt{|\operatorname{Re} z|^2 + |\operatorname{Im} z|^2}$.

It is often convenient to designate by the same letter $\mathbb{K}$ either the field $\mathbb{R}$ or the field $\mathbb{C}$, in which case the elements of the field $\mathbb{K}$ are called **scalars**.

Once sets $\mathbb{R}$ and $\mathbb{C}$ have been constructed as outlined above, various properties of $\mathbb{R}$ and $\mathbb{C}$ can then be *established.* The next theorem gathers the most important ones. **Cauchy sequences** *of real or complex numbers* are defined like Cauchy sequences of rational numbers.

Theorem 1.4-1 (a) *The set* $\mathbb{R}$, resp. $\mathbb{C}$, *is* **complete**, *i.e., any Cauchy sequence* $(x_n)_{n=1}^{\infty}$ *of real,* resp. *complex, numbers* **converges** *to a real,* resp. *complex, number; this means that there exists* $x \in \mathbb{R}$, resp. $x \in \mathbb{C}$, *and, given any* $\varepsilon > 0$, *there exists an integer* $n_0 = n_0(\varepsilon) \geq 1$, *such that*

$$|x_n - x| \le \varepsilon \quad \textit{for all } n \ge n_0.$$

(b) **Bolzano–Weierstraß property for $\mathbb{R}$ and $\mathbb{C}$:** *Any sequence* $(x_n)_{n=1}^{\infty}$ *of real,* resp. *complex, numbers that is* **bounded**, *i.e., such that there exists* $M \in \mathbb{R}$ *with the property that* $|x_n| \le M$ *for all* $n \ge 1$, *contains a convergent subsequence.*

(c) *Let* A *be a nonempty subset of* $\mathbb{R}$ *that has an upper bound in* $\mathbb{R}$ (Section 1.3). *Then there exists* $a \in \mathbb{R}$ *that is the* **least upper bound**, *or* **supremum**, *of* A*; this means that* a *is an upper bound for* A *and any upper bound* $b \in \mathbb{R}$ *for* A *necessarily satisfies* $a \le b$.

Likewise, any nonempty subset of $\mathbb{R}$ *that has a lower bound in* $\mathbb{R}$ *admits a* **greatest lower bound**, *or* **infimum**, *in* $\mathbb{R}$. □

That a sequence $(x_n)_{n=1}^{\infty}$ of real, or complex, numbers converges to x (according to the definition in Theorem 1.4-1(a)) is also denoted:

$$x = \lim_{n\to\infty} x_n, \quad \text{or} \quad x_n \underset{n\to\infty}{\longrightarrow} x, \quad \text{or} \quad x_n \to x \text{ as } n \to \infty.$$

A mapping of a set X into $\mathbb{R}$, *resp.* $\mathbb{C}$, is called a **real-valued**, *resp.* **complex-valued, function.**

A mapping of a set X into $\mathbb{R}^n$, *resp.* the set of all $m \times n$ matrices, is called a **vector field**, *resp.* a **matrix field.**

1.5 Cardinal numbers; finite and infinite sets

It is immediately seen that the relation "there exists a bijection of A onto B," where A and B are subsets of a set X, defines an *equivalence relation* $\mathcal{R}$ (Section 1.1) *on the set* $\mathcal{P}(X)$. The elements of the *quotient set* $\mathcal{P}(X)/\mathcal{R} \subset \mathcal{P}(\mathcal{P}(X))$ are then called the **cardinal numbers** of the subsets of X. If A is a subset of X, its cardinal number, denoted

$$\text{card}\, A,$$

is thus the *equivalence class of* A *modulo* $\mathcal{R}$, and as such, is an element of the set $\mathcal{P}(\mathcal{P}(X))$.

Remarkably, *the set* $\mathcal{P}(X)/\mathcal{R}$ *can be totally ordered* (the definition of a totally ordered set is given in Section 1.3), according to the following fundamental theorem:

Theorem 1.5-1 *The set* $\mathcal{P}(X)/\mathcal{R}$ *of all the cardinal numbers of the subsets of a set* X *is totally ordered by the relation* $\boldsymbol{R}$, *where* $(\text{card}\, A, \text{card}\, B) \in \boldsymbol{R}$ *means that there exists an injection of* A *into* B.

Equivalently, $(\text{card}\, A, \text{card}\, B) \in \boldsymbol{R}$ *if and only if there exists a surjection of* B *onto* A. □

Let us give some indications about the proof of this result (the proof may seem innocuous at first glance, but is in effect anything but trivial). First, it is clear that the definition of the relation $\boldsymbol{R}$ is unambiguous. For, if $\text{card}\, A = \text{card}\, \widetilde{A}$ and $\text{card}\, B = \text{card}\, \widetilde{B}$ and if there exists an injection of A into B, then there exists an injection of $\widetilde{A}$ into $\widetilde{B}$.

Second, one has to show that $\boldsymbol{R}$ is a *partial ordering* on the set $\mathcal{P}(X)$. While the reflexivity and transitivity are straightforward to verify, the antisymmetry is not, since it amounts to showing that, *if there exist an injection of A into B and an injection of B into A, then there exists a bijection of A onto B.*[2]

Third, one has to show that $\mathcal{P}(X)/\mathcal{R}$ is *totally ordered* by the relation $\boldsymbol{R}$, i.e., that, *given any two subsets $A \subset X$ and $B \subset X$, either there exists an injection of A into B, or there exists an injection of B into A,* these two properties being not exclusive. This part of the proof[3] requires the *axiom of choice.*

Finally, one has to show that there exists an injection of A into B if and only if there exists a surjection of B onto A. The proof of the "if" part again requires the *axiom of choice.*

As in Section 1.3, we will henceforth use the more "transparent" notations

$$\operatorname{card} A \preccurlyeq \operatorname{card} B, \quad \textit{resp.} \quad \operatorname{card} A \prec \operatorname{card} B,$$

to express that $(\operatorname{card} A, \operatorname{card} B) \in \boldsymbol{R}$, *resp.* $(\operatorname{card} A, \operatorname{card} B) \in \boldsymbol{R}$ and $\operatorname{card} A \neq \operatorname{card} B$.

From now on, we shall compare cardinals of sets, even if these are not *a priori* given as subsets of a given set. This entails no difficulty, however, since the union of such sets can always be defined, within the Zermelo–Fraenkel set theory (Section 1.1).

The next result[4] implies that, loosely speaking, there is no cardinal number that would be the "largest" (with respect to the relation $\boldsymbol{R}$).

Theorem 1.5-2 *Let X be any set. Then there does not exist a bijection of X onto the set $\mathcal{P}(X)$. Consequently, the relation*

$$\operatorname{card} X \prec \operatorname{card} \mathcal{P}(X)$$

always holds. □

A set X is **finite** if either $X = \varnothing$ or there exists an integer $n \geq 1$ such that $\operatorname{card} X = \operatorname{card}\{1, \ldots, n\} = n$. A set is **infinite** if it is not finite. In particular, a set X is **countably infinite** if $\operatorname{card} X = \operatorname{card} \mathbb{N}$, and a set is **uncountably infinite** if it is neither finite nor countably infinite. Note that some authors call *countable* a set that is either finite or countably infinite and call *denumerable* a countably infinite set.

The next theorem gathers some important properties of infinite sets. The proofs of both (a) and (b) require the *axiom of choice*. Property (a) asserts that $\operatorname{card} \mathbb{N}$ is the "smallest" of all "infinite" cardinals.

Theorem 1.5-3 (a) *Let X be an infinite set. Then*

$$\operatorname{card} \mathbb{N} \preccurlyeq \operatorname{card} X.$$

---

[2]This is the content of the famous theorem, proved in 1897 by Felix Bernstein (1878–1956).

[3]Due to:

E. ZERMELO [1904]: Beweis dass jede Menge wohlgeordnet werden kann, *Mathematische Annalen* **LIX**, 514–516.

[4]The theory of cardinal numbers is due to Georg Cantor (1845–1918), who expounded it in a highly influential (and for a long time highly controversial among mathematicians, some very famous) book:

G. CANTOR [1899]: *Beiträge zur Begründung der transfiniten Mengenlehre*, Georg Olms Verlag (English translation: *Contributions to the Founding of Transfinite Numbers*, Dover, New York, 1955).

(b) *Let X be an infinite set. Then*

$$\operatorname{card}(X \times X) = \operatorname{card} X.$$

(c) *The cardinal of the set $\mathbb{R}$ satisfies*

$$\operatorname{card} \mathbb{R} = \operatorname{card} \mathcal{P}(\mathbb{N}). \qquad \square$$

Note, however, that the important special case

$$\operatorname{card}(\mathbb{N} \times \mathbb{N}) = \operatorname{card} \mathbb{N}$$

of property (b) can be proved directly by means of a simple counting argument, i.e., without using the axiom of choice. This special case in turn easily implies that

$$\operatorname{card} \mathbb{Q} = \operatorname{card} \mathbb{N},$$

and that *a finite or countably infinite union of countably infinite sets is also countably infinite.*

Combined with Theorem 1.5-2, property (c) implies that

$$\operatorname{card} \mathbb{N} \prec \operatorname{card} \mathbb{R}.$$

The **continuum hypothesis** asserts that *there does not exist any infinite set X whose cardinal would satisfy* $\operatorname{card} \mathbb{N} \prec \operatorname{card} X \prec \operatorname{card} \mathbb{R}$. The long-standing question of whether the continuum hypothesis is true was beautifully settled in 1963 and 1964, when Paul J. Cohen showed in two landmark papers[5] that *the continuum hypothesis is independent of the six axioms of the Zermelo–Fraenkel set theory.* In an equally famous monograph, Kurt Gödel had already shown in 1940[6] that, if the Zermelo–Fraenkel set theory is noncontradictory, it remains so under the addition of the continuum hypothesis.

1.6 Topological spaces

A **topological space** is a pair $(X, \mathcal{O})$, where X is a set, and $\mathcal{O}$ is a subset of $\mathcal{P}(X)$ with the following properties:

Given *any* family $(O_i)_{i \in I}$ of subsets $O_i \in \mathcal{O}$, their union $\bigcup_{i \in I} O_i$ belongs to $\mathcal{O}$ (the set I may thus be finite, countably infinite, or uncountably infinite); given any *finite* family $(O_j)_{j=1}^n$ of subsets $O_j \in \mathcal{O}$, their intersection $\bigcap_{j=1}^n O_j$ belongs to $\mathcal{O}$; and the set X and the empty set $\varnothing$ belong to $\mathcal{O}$.

If $(X, \mathcal{O})$ is a topological space, the set X is said to be equipped with a **topology** (corresponding to the subset $\mathcal{O}$ of $\mathcal{P}(X)$), a subset of X that belongs to $\mathcal{O}$ is **open** (for this topology), and a subset F of X is **closed** (for this topology) if the set $X - F$ is open.

[5] P.J. COHEN [1963, 1964]: The independence of the continuum hypothesis, *Proceedings of the National Academy of Sciences, USA* **50**, 1143–1148, and *Proceedings of the National Academy of Sciences, USA* **51**, 105–110.

[6] K. GÖDEL [1940]: *The Consistency of the Axiom of Choice and of the Generalized Continuum Hypothesis with the Axioms of Set Theory*, Princeton University Press, Princeton, NJ.

Clearly, given *any* family $(F_i)_{i\in I}$ of closed subsets F_i, their intersection $\bigcap_{i\in I} F_i$ is closed; given any *finite* family $(F_j)_{j=1}^n$ of closed subsets F_j, their union $\bigcup_{i=1}^n F_i$ is closed; and the set X and the empty set are closed (these properties simply follow from de Morgan's laws; cf. Section 1.3).

In a topological space $(X, \mathcal{O})$, a **neighborhood of a point** $x \in X$ is any subset of X that contains an open set containing x. The set formed by all the neighborhoods of point $x \in X$ is denoted $\mathcal{V}(x)$.

Given two points $a, b \in \mathbb{R}$ such that $a < b$, let $]a, b[:= \{y \in \mathbb{R};\ a < y < b\}$. A fundamental example of topological space is $(\mathbb{R}, \mathcal{O})$, where a nonempty subset O of $\mathbb{R}$ belongs to $\mathcal{O}$ if and only if, for each $x \in O$, there exist $a < b$ such that $x \in]a, b[$ and $]a, b[\subset O$.

Let $a, b \in \mathbb{R}$ be two points that satisfy $a < b$. The set $]a, b[$, which is open for this topology, is called the *open interval with end-points a and b*. The unbounded sets $]-\infty, a[:= \{y \in \mathbb{R};\ y < a\}$, $]b, \infty[:= \{y \in \mathbb{R};\ b < y\}$, and $\mathbb{R}$ itself, which are likewise open for this topology, are also called *open intervals*.

Unless explicitly stated otherwise, the set $\mathbb{R}$ will be always considered as equipped with this topology, called its **usual topology**. The following characterization of the open sets for this topology is very useful.

Theorem 1.6-1 *Let $\mathbb{R}$ be equipped with its usual topology. Then any nonempty open subset of $\mathbb{R}$ can be written as a finite or countably infinite union of disjoint, bounded or unbounded, open intervals.* □

Let $(X, \mathcal{O})$ be a topological space and let A be a subset of X. The **interior** of A, denoted $\mathring{A}$ or $\operatorname{int} A$, is the union of all the open sets contained in A; equivalently,

$$\mathring{A} = \operatorname{int} A := \{x \in X;\ A \in \mathcal{V}(x)\}.$$

The **closure** of A, denoted $\overline{A}$, is the intersection of all the closed sets containing A; equivalently,

$$\overline{A} := \{x \in X;\ V \cap A \neq \varnothing \text{ for all } V \in \mathcal{V}(x)\} = X - \{\operatorname{int}(X - A)\}.$$

The **boundary** of A, denoted ∂A, is defined as the intersection of $\overline{A}$ and $\overline{X - A}$; equivalently,

$$\partial A := \{x \in X;\ V \cap A \neq \varnothing \text{ and } V \cap (X - A) \neq \varnothing \text{ for all } V \in \mathcal{V}(x)\}.$$

Note that $\partial A = \overline{A} - \mathring{A}$.

Let $(X, \mathcal{O})$ be a topological space. The **support** of a real-valued or complex-valued function $f : X \to \mathbb{K}$ is the set

$$\operatorname{supp} f := \overline{\{x \in X;\ f(x) \neq 0\}}.$$

A subset A of a topological space $(X, \mathcal{O})$ is **dense in** X if $\overline{A} = X$.

A topological space $(X, \mathcal{O})$ is **separable** if it contains a finite or countably infinite dense subset, i.e., there exist elements $x_n \in X$, $n \geq 0$, such that $\overline{\bigcup_{n=0}^{\infty} \{x_n\}} = X$.

A topological space $(X, \mathcal{O})$ is said to be **Hausdorff**, or equivalently, to be equipped with a **Hausdorff topology**, if, given any two distinct points $x \in X$ and $y \in X$, there exist a neighborhood V of x and a neighborhood W of y such that $V \cap W = \varnothing$.

A topological space $(X, \mathcal{O})$ is said to be **normal** if, given any two disjoint closed subsets F_1 and F_2 of X, there exist disjoint open subsets O_1 and O_2 such that $F_1 \subset O_1$ and $F_2 \subset O_2$.

Let X be a topological space. A sequence $(x_n)_{n=0}^{\infty}$ of points $x_n \in X$ is **convergent in** X if there exists a point $x \in X$ such that, given any neighborhood V of x, there exists an integer $n_0 = n_0(V) \geq 0$ such that $x_n \in V$ for all $n \geq n_0$.

If X is Hausdorff, such a point x is unique and is called the **limit** of the sequence $(x_n)_{n=1}^{\infty}$. *In this case*, the notations

$$x = \lim_{n\to\infty} x_n, \quad \text{or} \quad x_n \underset{n\to\infty}{\longrightarrow} x, \quad \text{or} \quad x_n \to x \quad \text{as } n \to \infty,$$

are equivalently used to express that *x is the limit of the convergent sequence $(x_n)_{n=0}^{\infty}$*.

Let X be a set and let $(Y, \mathcal{O})$ be a Hausdorff topological space. A sequence $(f_n)_{n=0}^{\infty}$ of mappings $f_n : X \to Y$ is said to be **pointwise convergent** to a mapping $f : X \to Y$ if

$$\text{for each } x \in X, \quad f_n(x) \to f(x) \ \text{ as } n \to \infty.$$

Let $(X, \mathcal{O})$ be a topological space, let A be a subset of X, and let $\mathcal{O}_A$ denote the subset of $\mathcal{P}(A)$ consisting of all the subsets O_A of A that can be written as $O_A = O \cap A$ for some $O \in \mathcal{O}$. Then $(A, \mathcal{O}_A)$ is also a topological space, and A is said to be equipped with the **topology induced on A by the topology of $(X, \mathcal{O})$**, or simply the **induced topology** if there is no ambiguity regarding the nature of the set $\mathcal{O}$. By definition, the subsets of A that belong to $\mathcal{O}_A$ are thus the *open sets for the induced topology*.

Then a subset F_A of A is closed in the topological space $(A, \mathcal{O}_A)$ if and only if there exists a subset F of X that is closed in $(X, \mathcal{O})$ such that $F_A = F \cap A$; likewise, a subset V_A of A is a neighborhood of a point $x \in A$ in $(A, \mathcal{O}_A)$ if and only if there exists a neighborhood V of x in $(X, \mathcal{O})$ such that $V_A = V \cap A$.

Naturally, if A is a subset of X and B is a subset of A, the topological properties of B in $(X, \mathcal{O})$ and the topological properties of B in $(A, \mathcal{O}_A)$ have to be carefully distinguished. For instance, A is always open in $(A, \mathcal{O}_A)$ but evidently not necessarily so in $(X, \mathcal{O})$.

Let $(X_j, \mathcal{O}_j)$, $1 \leq j \leq n$, be topological spaces, let $X := \prod_{j=1}^{n} X_j$ denote the (finite) *product* of the sets X_j, $1 \leq j \leq n$, and let

$$\begin{aligned} \mathcal{O} := \{O \in \mathcal{P}(X);\ &\text{for each } x \in O, \text{ there exist } O_j \in \mathcal{O}_j, 1 \leq j \leq n, \\ &\text{such that } x \in O_1 \times \cdots \times O_n \text{ and } O_1 \times \cdots \times O_n \subset O\}. \end{aligned}$$

Then $(X, \mathcal{O})$ is a topological space, and X is said to be equipped with the **product topology**, corresponding to the subsets $\mathcal{O}_j$ of $\mathcal{P}(X_j)$, $1 \leq j \leq n$.

More generally, given *any* family of topological spaces $(X_i, \mathcal{O}_i)$, $i \in I$, the **product topology** in the product $X = \prod_{i\in I} X_i$ is defined as follows: A subset $O \subset X$ is *open* in this topology if, for each $x \in O$, there exists a *finite* family $(O_i)_{i\in J(x)}$ of open sets $O_i \in \mathcal{O}_i$ such that

$$x \in \Big(\prod_{i\in J(x)} O_i\Big) \times \Big(\prod_{i\in I(x)} X_i\Big) \quad \text{and} \quad \Big(\prod_{i\in J(x)} O_i\Big) \times \Big(\prod_{i\in I(x)} X_i\Big) \subset O,$$

where $I(x) := I - J(x)$.

For the sake of notational brevity, we shall no longer mention $\mathcal{O}$ in the notation so far used for a topological space $(X, \mathcal{O})$, whenever no ambiguity should arise about the nature of the subset $\mathcal{O} \subset \mathcal{P}(X)$ that is considered.

1.7 Continuity in topological spaces

A mapping $f : X \to Y$ from a topological space X into a topological space Y is **continuous at a point** $x \in X$ if, given any neighborhood V of $f(x)$ in Y, there exists a neighborhood U of x in X such that the direct image $f(U)$ of U under f (Section 1.2) is contained in V.

A basic property of continuous mappings between Hausdorff spaces is that "they map convergent sequences into convergent sequences":

Theorem 1.7-1 *Let X and Y be two Hausdorff topological spaces and let $f : X \to Y$ be a mapping that is continuous at a point $x \in X$. Then, given any sequence $(x_n)_{n=0}^{\infty}$ of points $x_n \in X$ that converges to x in X, the sequence $(f(x_n))_{n=0}^{\infty}$ converges to $f(x)$ in Y.* □

The converse holds in the important special case where the topology of X is that of a metric space (Theorem 1.11-1).

The following criterion of continuity at a point of a composite mapping is constantly used (and immediate to prove):

Theorem 1.7-2 *Let X, Y, Z be three topological spaces, let $f : X \to Y$ be a mapping that is continuous at a point $x \in X$, and let $g : Y \to Z$ be a mapping that is continuous at the point $f(x) \in Y$. Then the composite mapping $g \circ f : X \to Z$ is continuous at x.* □

A mapping $f : X \to Y$ is **continuous** if it is continuous at all points of X. The following characterization of continuous mappings (also immediate to prove) is fundamental:

Theorem 1.7-3 *Let X and Y be two topological spaces. A mapping $f : X \to Y$ is continuous if and only if the inverse image under f of any open set in Y is open in X; or equivalently, if and only if the inverse image under f of any closed set in Y is closed in X.* □

The set formed by all the continuous mappings from X to Y is denoted

$$\mathcal{C}(X;Y), \quad \text{or} \quad \mathcal{C}(X) \quad \text{if } Y = \mathbb{R}.$$

Let X and Y be two topological spaces. A mapping $f : X \to Y$ is said to be a **homeomorphism** *of X onto Y* if f is a bijection, $f \in \mathcal{C}(X;Y)$, and $f^{-1} \in \mathcal{C}(Y;X)$.

The following characterization of homeomorphisms immediately follows from the definition and Theorem 1.7-3:

Theorem 1.7-4 *Let X and Y be two topological spaces and let $f \in \mathcal{C}(X;Y)$ be a bijection. Then f is a homeomorphism of X onto Y if and only if the direct image under f of any open subset of X is an open subset of Y; or equivalently, if and only if the direct image under f of any closed subset of X is a closed subset of Y.* □

Two topological spaces X and Y are said to be **homeomorphic** if there exists a homeomorphism of X onto Y.

Let us now examine the special cases where the set X, or the set Y, is a finite product.

Theorem 1.7-5 *Let X_j, $1 \le j \le n$, and Y be topological spaces, let the product $X := \prod_{j=1}^{n} X_j$ be equipped with the product topology (Section 1.6), and let $f : X \to Y$ be a mapping* *that is continuous at a point* $a = (a_j)_{j=1}^n \in X$. *Then for each* $1 \leq j \leq n$, *the mapping*

$$x_j \in X_j \to f(a_1, \ldots, a_{j-1}, x_j, a_{j+1}, \ldots, a_n) \in Y$$

is continuous at the point a_j. □

Note that the converse does *not* necessarily hold; consider for instance the special case where $n = 2$, $X_1 = X_2 = Y = \mathbb{R}$, $f(x_1, x_2) = \dfrac{x_1 x_2}{x_1^2 + x_2^2}$ if $(x_1, x_2) \neq (0, 0)$ and $f(x_1, x_2) = 0$ if $(x_1, x_2) = (0, 0)$, and $(a_1, a_2) = (0, 0)$.

Theorem 1.7-6 *Let* X *and* Y_i, $1 \leq i \leq m$, *be topological spaces, and let the product* $Y := \prod_{i=1}^m Y_i$ *be equipped with the product topology. Then a mapping* $f = (f_i)_{i=1}^m : X \to Y$ *is continuous at a point* $a \in X$ *if and only if each mapping* $f_i : X \to Y_i$, $1 \leq i \leq m$, *is continuous at* $a \in X$. □

The following *extension theorem* for continuous functions is fundamental. In particular, it will be abundantly used in Chapter 9, for defining the *Brouwer topological degree in* $\mathbb{R}^n$, or for establishing the *hairy ball theorem* and the *Borsuk–Ulam theorem*.

Theorem 1.7-7 (Tietze–Urysohn extension theorem) *Let* X *be a normal topological space,* F *a closed subset of* X, *and* $f : F \to \mathbb{R}$ *a continuous function. Then there exists a continuous function* $\widetilde{f} : X \to \mathbb{R}$ *such that*

$$\widetilde{f}(x) = f(x) \quad \textit{for all } x \in F. \qquad \square$$

Finally, we mention a fundamental way to construct a specific topology on a set, by means of given mappings from this set into topological spaces.

Theorem 1.7-8 *Let there be given a set* X *and a family* $(\varphi_i)_{i \in I}$ *of mappings* φ_i *from* X *into topological spaces* Y_i. *Then there exists a topology on* X *with the following two properties:*

First, all the mappings $\varphi_i : X \to Y_i$, $i \in I$, *are continuous.*

Second, any subset of X *that is open for this topology is necessarily open for any topology on* X *for which all the mappings* $\varphi_i : X \to Y_i$, $i \in I$, *are continuous.* □

In view of these two properties, the (clearly unique) topology defined in Theorem 1.7-8 is aptly called the **weakest topology** *on* X *that renders all the mappings* $\varphi_i : X \to Y_i$, $i \in I$, *continuous.* As we shall see, the *weak* and *weak* * *topologies*, on a normed vector space and its dual space, constitute fundamental examples of weakest topologies (Section 5.12).

1.8 Compactness in topological spaces

Let $(X, \mathcal{O})$ be a *topological space*. A subset K of X is **compact** if, given *any* family $(O_i)_{i \in I}$ of open sets $O_i \in \mathcal{O}$ such that $K \subset \bigcup_{i \in I} O_i$, there exists a *finite* subfamily $(O_j)_{j \in J}$ of the family $(O_i)_{i \in I}$ such that $K \subset \bigcup_{j \in J} O_j$.

This property, which constitutes the **Heine–Borel–Lebesgue property**, is often expressed as follows: A subset K of X is compact if *any open covering of* K *admits a finite subcovering.*

That a subset K of X is compact does *not* depend on whether K is considered as a subset of X, or as a topological space *per se*, equipped with the induced topology. In other words, *a subset K of a topological space $(X, \mathcal{O})$ is compact if and only if the topological space $(K, \mathcal{O}_K)$ equipped with the induced topology* (Section 1.6) *is compact.*

The following theorems assemble some basic (and elementary to prove) properties involving compactness.

Theorem 1.8-1 *A topological space X is compact if and only if, given any family $(F_i)_{i\in I}$ of closed subsets F_i of X with the property that $\bigcap_{j\in J} F_j \neq \varnothing$ for any finite subfamily $(F_j)_{j\in J}$ of the family $(F_i)_{i\in I}$, we have $\bigcap_{i\in I} F_i \neq \varnothing$.* □

Theorem 1.8-2 (a) *Any compact subset of a Hausdorff topological space is closed.*
(b) *A closed subset of a compact topological space is compact.* □

Theorem 1.8-3 *Let X and Y be two topological spaces and let $f : X \to Y$ be a continuous mapping. Then the direct image $f(K)$ of any compact subset K of X is a compact subset of Y.* □

Theorem 1.8-4 *Any continuous bijection from a compact topological space X onto a topological space Y is a homeomorphism of X onto Y (then Y is also compact by Theorem* 1.8-3). □

Theorem 1.8-5 *Let X_j, $1 \le j \le n$, be compact topological spaces. Then the product $\prod_{j=1}^n X_j$ equipped with the product topology (Section* 1.6) *is compact.* □

Theorem 1.8-5 is a special case of **Tychonoff's theorem**, one of the most important results in general topology. This theorem asserts that, given *any* family $(X_i)_{i\in I}$ of compact topological spaces, the product $\prod_{i\in I} X_i$ equipped with the product topology is compact.[7] But, by contrast with that of Theorem 1.8-5, its proof requires the *axiom of choice*.

A subset A of a topological space X is **relatively compact** if its closure $\overline{A}$ is a compact subset of X.

1.9 Connectedness and simple-connectedness in topological spaces

A topological space $(X, \mathcal{O})$ is **connected** if the only subsets of X that are *both* open and closed are X and $\varnothing$. A subset A of X is **connected** if it is a connected topological space when it is equipped with the topology induced by that of X (Section 1.6).

That a subset A of X is connected does *not* depend on whether A is considered as a subset of X, or as a topological space by itself, i.e., when it is equipped with the induced topology.

[7] This theorem was first proved in the special case where $X_i = [0, 1]$ for all $i \in I$ in:
A. Tychonoff [1930]: Über die topologische Erweiterung von Räumen, *Mathematische Annalen* **102**, 544–561.
The general case was then proved in:
E. Čech [1937]: On bicompact spaces, *Annals of Mathematics* **38**, 823–844.

The following theorems gather some basic properties involving connectedness.

Theorem 1.9-1 *Let A be a connected subset of a topological space X. Then any subset B of X that satisfies $A \subset B \subset \overline{A}$, hence $B = \overline{A}$ in particular, is also connected.* □

Theorem 1.9-2 *Let X be a connected topological space, let Y be a topological space, and let $f : X \to Y$ be a locally constant function, i.e., such that each point $x \in X$ possesses a neighborhood V_x such that the restriction $f|_{V_x}$ is a constant function. Then f is a constant function.* □

Like compactness, connectedness is a property that is "preserved by continuous mappings":

Theorem 1.9-3 *Let X and Y be two topological spaces and let $f : X \to Y$ be a continuous mapping. Then the direct image $f(A)$ of any connected subset A of X is a connected subset of Y.* □

The next result characterizes the connected subsets of $\mathbb{R}$.

Theorem 1.9-4 *Let $\mathbb{R}$ be equipped with its usual topology. Then a subset of $\mathbb{R}$ is connected if and only if it is an interval, bounded or unbounded.* □

An immediate corollary of Theorems 1.9-3 and 1.9-4 then follows:

Theorem 1.9-5 (Bolzano intermediate value theorem) *Let X be a connected topological space, let $f : X \to \mathbb{R}$ be a continuous function, and let $a, b \in X$ be such that $f(a) < f(b)$ (to fix ideas). Then, given any $y \in]f(a), f(b)[$, there exists $x \in X$ such that $f(x) = y$.* □

The next three theorems provide useful sufficient conditions for connectedness.

Theorem 1.9-6 *Let X be a topological space and let $(A_i)_{i\in I}$ be any family of connected subsets A_i of X. If the intersection $\bigcap_{i\in I} A_i$ is nonempty, then the union $\bigcup_{i\in I} A_i$ is connected.* □

Theorem 1.9-7 *Let X_j, $1 \leq j \leq n$, be connected topological spaces and let their product $X = \prod_{j=1}^{n} X_j$ be equipped with the product topology (Section 1.6). Then X is connected.* □

Let X be a topological space. The relation $\mathcal{R}$ defined by "$(x, y) \in \mathcal{R}$ if and only if there exists a connected subset of X that contains both x and y," is an equivalence relation on X. The equivalence classes modulo this relation, which are thus *subsets of* X, are called the **connected components** of X.

Given any $x \in X$, the connected component of X that contains x is called the **connected component of** x; it is also the largest connected subset of X that contains x, according to the following result.

Theorem 1.9-8 *Let X be a topological space and let $x \in X$. Then the connected component of x is the union of all the connected subsets of X that contain x.* □

Let x and y be two points in a topological space X. A **path joining** x **to** y is a continuous mapping $\gamma : [0, 1] \to X$ such that $\gamma(0) = x$ and $\gamma(1) = y$.

A topological space X is **arcwise-connected** if, given any two distinct points x, y in X, there exists a path joining x to y.

Theorem 1.9-9 *An arcwise-connected topological space is connected.* □

The converse implication does not necessarily hold. For instance, let

$$A := \{(x, y) \in \mathbb{R}^2;\ x = 0, |y| \leq 1\} \quad \text{and} \quad B := \{(x, y) \in \mathbb{R}^2;\ x \neq 0,\ y = \sin(x^{-1})\}.$$

Then $A \cup B$ is a connected subset of $\mathbb{R}^2$ that is not arcwise-connected.

Let x and y be two points in a topological space X. Two paths $\gamma_0 : [0, 1] \to X$ and $\gamma_1 : [0, 1] \to X$ joining x to y are **homotopic** if there exists a continuous mapping $H : [0, 1] \times [0, 1] \to X$, called a **homotopy joining** γ_0 **to** γ_1, such that $H(\cdot, 0) = \gamma_0$ and $H(\cdot, 1) = \gamma_1$, and $H(0, \cdot) = x$ and $H(1, \cdot) = y$.

A *topological space* X is said to be **simply connected** if it is arcwise-connected, hence connected (Theorem 1.9-9), *and* if any two paths $\gamma_0 : [0, 1] \to X$ and $\gamma_1 : [0, 1] \to X$ such that $\gamma_0(0) = \gamma_1(0)$ and $\gamma_0(1) = \gamma_1(1)$ are homotopic.

1.10 Metric spaces

Let X be a set. A **distance** on X is a function $d : X \times X \to \mathbb{R}$ that satisfies the following properties for all $x, y, z \in X$:

$$\begin{aligned}
&d(x, x) = 0 \quad \text{and} \quad d(x, y) > 0 \quad \text{if } x \neq y,\\
&d(x, y) = d(y, x),\\
&d(x, z) \leq d(x, y) + d(y, z).
\end{aligned}$$

The last property is called the **triangle inequality**. A **metric space** is a pair (X, d) where X is a set and d is a distance on X.

In what follows, (X, d) is a metric space. Given a point $x \in X$ and a number $r > 0$, the **ball** with **center** x, or **centered at** x, and **radius** $r > 0$ is the subset of X defined by

$$B(x; r) = \{y \in X;\ d(y, x) < r\}.$$

A subset A of X is **bounded** if there exists a ball $B(x; r) \subset X$ such that $A \subset B(x; r)$. It is **unbounded** otherwise.

The **diameter** of a nonempty subset A of X is defined as the extended real number

$$\operatorname{diam} A := \sup\{d(x, y);\ x \in A, y \in A\} \in [0, \infty].$$

Clearly, a subset A of X is bounded if and only if $\operatorname{diam} A < \infty$.

The *distance from a point* $x \in X$ *to a nonempty subset* A of X is defined as the real number

$$\operatorname{dist}(x, A) := \inf\{d(x, y);\ y \in A\}.$$

Unless otherwise mentioned, *a metric space* (X, d) *will be always viewed as a topological space* $(X, \mathcal{O})$, whose *open sets*, i.e., the subsets of X that belong to $\mathcal{O}$, are those described in the next theorem. This "canonical" topology is called the **topology induced on** X **by the distance** d.

Theorem 1.10-1 *Let (X, d) be a metric space. Let $\mathcal{O}$ denote the subset of $\mathcal{P}(X)$ consisting of the empty set and of all the subsets O of X with the following property: Given any $x \in O$, there exists $r > 0$ such that the ball $B(x; r)$ is contained in O. Then the pair $(X, \mathcal{O})$ is a topological space, which is Hausdorff and normal.*

Besides, any ball $B(x; r)$, with $x \in X$ and $r > 0$, is an open set for this topology. □

For instance, the **usual distance on** $\mathbb{R}$, *defined by $d(x, y) = |x - y|$ for all $x, y \in \mathbb{R}$, induces the usual topology of* $\mathbb{R}$ (Section 1.6). Likewise, the **usual distance on** $\mathbb{C}$, defined by $d(x, y) = |x - y|$ for all $x, y \in \mathbb{C}$, induces a topology on $\mathbb{C}$, called the **usual topology of** $\mathbb{C}$.

When viewed as a metric space, the set $\mathbb{R}$, or a subset of $\mathbb{R}$, will be always implicitly considered as endowed with this distance d, called the *usual distance on* $\mathbb{R}$.

Of course, d is far from being the only distance on $\mathbb{R}$ that induces its usual topology. For instance, the distance $\rho : \mathbb{R} \times \mathbb{R} \to \mathbb{R}$ defined by $\rho(x, y) = \dfrac{|x - y|}{1 + |x - y|}$ for all $x, y \in \mathbb{R}$ also induces the usual topology on $\mathbb{R}$. Note, however, that the metric space $(\mathbb{R}, d)$ is unbounded while the metric space $(\mathbb{R}, \rho)$ is bounded; incidentally, this simple example shows that boundedness is a metric notion, not a topological one.

The topology of a topological space $(X, \mathcal{O})$ is said to be **metrizable** if it can be induced by a metric on X, and any such metric is said to be **compatible with the topology**. For instance, the usual topology of $\mathbb{R}$ is metrizable, and the above metrics d and ρ are both compatible with it.

Another fundamental example of metric space is that of $\mathbb{K}^n$, where $\mathbb{K} = \mathbb{R}$ or $\mathbb{K} = \mathbb{C}$, or of a subset X of $\mathbb{K}^n$, equipped with one of the distances d_p, $1 \leq p \leq \infty$, defined for any n-tuples $x = (x_i)_{i=1}^n \in \mathbb{K}^n$ and $y = (y_i)_{i=1}^n \in \mathbb{K}^n$ by

$$d_p(x, y) = \left(\sum_{i=1}^n |x_i - y_i|^p \right)^{1/p} \quad \text{if } 1 \leq p < \infty,$$

$$d_\infty(x, y) = \max_{1 \leq i \leq n} |x_i - y_i|.$$

All the axioms of a distance are immediately verified, save the triangle inequality for $1 < p < \infty$, for which we refer the reader to Theorem 2.4-1. The distance d_2 is called the **Euclidean distance**.

Given any $1 \leq p, q \leq \infty$, any ball corresponding to the distance d_p is contained in a ball corresponding to the distance d_q and centered at the same point. Hence the topology induced on $\mathbb{K}^n$, or on a subset of $\mathbb{K}^n$, by any one of the distances d_p, $1 \leq p \leq \infty$, is the same, and is called the **usual topology of** $\mathbb{K}^n$. It is also easily verified that this topology coincides with the product topology (Section 1.6) on $\mathbb{K}^n = \prod_{j=1}^n X_j$, where each topological space X_j, $1 \leq j \leq n$, is the set $\mathbb{K}$ equipped with its usual topology, and that $\mathbb{K}^n$ equipped with this topology is a *separable* topological space.

More generally, any subset X of a finite-dimensional vector space over $\mathbb{K}$ with a basis $(e_i)_{i=1}^n$ becomes a metric space when it is equipped with one of the above distances d_p, $1 \leq p \leq \infty$, with x and y in $d_p(x, y)$ now replaced by $x = \sum_{i=1}^n x_i e_i$ and $y = \sum_{i=1}^n y_i e_i$.

In what follows, a metric space (X, d) will be often simply denoted X for notational brevity, in which case it is implicitly understood that its metric is denoted d when needed.

The various definitions given in Section 1.6 then have the following equivalent "metric" counterparts in a metric space X:

A *neighborhood of a point* $x \in X$ is any subset of X that contains a ball centered at x.

The *interior* $\mathring{A}$ of a subset A of X is the set of all points $x \in A$ such that there is a ball centered at x and contained in A.

The *closure* $\overline{A}$ of a subset A of X is the set of all points $x \in X$ such that any ball centered at x has a nonempty intersection with A.

The *boundary* ∂A of a subset A of X is the set of all points $x \in X$ such that any ball centered at x has a nonempty intersection with both A and $X - A$.

A metric space (X, d) is *separable* if there exist elements $x_n \in X$, $n \geq 1$, such that, given any $x \in X$ and any $\varepsilon > 0$, there exists $n = n(x, \varepsilon) \geq 1$ such that $x \in B(x_n; \varepsilon)$.

A sequence $(x_n)_{n=0}^{\infty}$ of points $x_n \in X$ is *convergent* if there exists a point x such that, given any $\varepsilon > 0$, there exists $n_0 \geq 0$ such that $x_n \in B(x; \varepsilon)$ for all $n \geq n_0$, or equivalently, such that $d(x_n, x) \to 0$ as $n \to \infty$. Such a point x, which is necessarily *unique* because the associated topological space is Hausdorff (Theorem 1.10-1), is thus the *limit* of the sequence.

Thanks to this characterization of limits in terms of distance, the closure of a subset in a metric space can be given a simple characterization by means of convergent sequences.

Theorem 1.10-2 *Let A be a subset of a metric space X. Then a point $x \in X$ belongs to $\overline{A}$ if and only if there exists a sequence $(x_n)_{n=0}^{\infty}$ of points $x_n \in A$ that converges to x as $n \to \infty$.*

Consequently, a subset A of X is closed if and only if

$$x_n \in A, \quad n \geq 0, \quad \textit{and} \quad x_n \to x \textit{ in } X \textit{ as } n \to \infty \textit{ implies } x \in A. \qquad \square$$

Incidentally, note that $\operatorname{dist}(x, A) > 0$ *if $x \notin A$ and A is closed.*

Let (X, d) be a metric space, let A be a subset of X, and let $d_A : A \times A \to \mathbb{R}$ denote the restriction of the distance d to $A \times A$. Clearly, d_A is a distance on A, called the **distance induced by** d **on** A, and thus (A, d_A) is also a metric space. Furthermore, the following properties hold.

Theorem 1.10-3 *Let (X, d) be a metric space and let A be a subset of X. The topology induced on A by the metric d_A coincides with the topology induced on A by the topology induced on X by the metric d (Section 1.6).*

Furthermore, (A, d_A) is separable if (X, d) is separable. $\square$

Finally, let (X_j, d_j), $1 \leq j \leq n$, be metric spaces, and let $X := \prod_{j=1}^{n} X_j$. Then a subset $O \subset X$ is open for the product topology on the product X if and only if, given any point $x = (x_j)_{j=1}^{n} \in X$, there exist $r_j > 0$, $1 \leq j \leq n$, such that $\prod_{j=1}^{n} B(x_j, r_j) \subset O$. Any distance d on the product space X that induces the product topology on X is then said to be **compatible with the product topology**. Examples of such compatible distances are provided by the functions $d : X \times X \to \mathbb{R}$ and $\rho : X \times X \to \mathbb{R}$ defined by

$$d(x, y) := \sum_{j=1}^{n} d_j(x_j, y_j) \quad \text{and} \quad \rho(x, y) := \max_{1 \leq j \leq n} d_j(x_j, y_j)$$

for all $x = (x_j)_{j=1}^{n} \in X$ and $y = (y_j)_{j=1}^{n} \in X$.

1.11 Continuity and uniform continuity in metric spaces

A mapping from a topological space X into a topological space Y that is continuous at a point $x \in X$ maps sequences converging to x in X into sequences converging to $f(x)$ in Y (Theorem 1.7-1). The converse holds if the topologies of both X and Y are those of a metric space (in fact, Theorem 1.11-1 still holds if X is a metric space and Y is a Hausdorff topological space):

Theorem 1.11-1 *Let X and Y be metric spaces. Then a mapping $f : X \to Y$ is continuous at a point $x \in X$ if and only if, given any sequence $(x_n)_{n=0}^{\infty}$ of points $x_n \in X$ that converges to x in X, the sequence $(f(x_n))_{n=0}^{\infty}$ converges to $f(x)$ in Y.* □

The next theorem gives a simple, and often used, property of continuous mappings in metric spaces:

Theorem 1.11-2 *Let X be a dense subset of a metric space $\widetilde{X}$, let Y be a Hausdorff topological space, and let $f : \widetilde{X} \to Y$ and $g : \widetilde{X} \to Y$ be two continuous mappings that coincide on X, i.e., $f(x) = g(x)$ for all $x \in X$. Then $f = g$.* □

If X and Y are both metric spaces, the continuity at a point can be also expressed in terms of balls, or in terms of distances: Let (X, d) and (Y, ρ) be two metric spaces. Then a mapping $f : X \to Y$ is *continuous at a point* $x \in X$ if the inverse image of any ball in Y centered at $f(x)$ contains a ball in X centered at x, or equivalently, if, given any $\varepsilon > 0$, there exists $\delta = \delta(\varepsilon, x) > 0$ such that $\rho(f(x), f(\widetilde{x})) < \varepsilon$ for all $\widetilde{x} \in X$ such that $d(x, \widetilde{x}) < \delta$. This equivalent definition, specific to metric spaces, is sometimes referred to as the "*ε-δ definition of continuity.*"

If a mapping $f : X \to Y$ is *continuous*, i.e., if it is continuous at *all* points $x \in X$, it may happen that, given any $\varepsilon > 0$, the above number $\delta(\varepsilon, x) > 0$ can be chosen *independently of* $x \in X$. This possibility leads to the following definition: Let (X, d) and (Y, ρ) be two metric spaces. A mapping $f : X \to Y$ is **uniformly continuous** if, given any $\varepsilon > 0$, there exists $\delta(\varepsilon) > 0$ such that $\rho(f(x), f(\widetilde{x})) < \varepsilon$ for all $x, \widetilde{x} \in X$ that satisfy $d(x, \widetilde{x}) < \delta(\varepsilon)$.

An important *example of a uniformly continuous mapping* is provided by a **Lipschitz-continuous mapping**, i.e., a mapping $f : (X, d) \to (Y, \rho)$ with the property that there exists a constant k such that

$$\rho(f(x), f(\widetilde{x})) \le k d(x, \widetilde{x}) \quad \text{for all } x, \widetilde{x} \in X.$$

Such a mapping is then said to satisfy a *Lipschitz condition*, with *Lipschitz constant k*.

Another *example* is provided by a **Hölder-continuous mapping**, i.e., a mapping $f : (X, d) \to (Y, \rho)$ with the property that there exist constants C and $0 < \lambda < 1$ such that

$$\rho(f(x), f(\widetilde{x})) \le C(d(x, \widetilde{x}))^{\lambda} \quad \text{for all } x, \widetilde{x} \in X.$$

Such a mapping is then said to satisfy a *Hölder condition of exponent λ*.

Let (X, d) be a metric space and let the product $X \times X$ be equipped with the distance D defined by

$$D((x, \widetilde{x}), (y, \widetilde{y})) := d(x, y) + d(\widetilde{x}, \widetilde{y}) \quad \text{for all } (x, \widetilde{x}) \in X \times X \text{ and all } (y, \widetilde{y}) \in X \times X.$$

Then the function $d : (X\times X, D) \to \mathbb{R}$ provides an *example of a Lipschitz-continuous function, with Lipschitz constant one*. To see this, simply note that, by the triangular inequality,

$$|d(x,\widetilde{x}) - d(y,\widetilde{y})| \leq d(x,y) + d(\widetilde{x},\widetilde{y}) = D((x,\widetilde{x}),(y,\widetilde{y})).$$

Another similar example is provided by the distance to a subset:

Theorem 1.11-3 *Let (X,d) be a metric space and let A be a nonempty subset of X. Then*

$$|\text{dist}\,(x,A) - \text{dist}\,(y,A)| \leq d(x,y) \quad \text{for all } x,y \in X. \qquad \square$$

Given two metric spaces (X,d) and (Y,ρ), a mapping $f : X \to Y$ is an **isometry from X into Y** if f "preserves the distances," i.e., $\rho(f(x),f(y)) = d(x,y)$ for all $x,y \in X$. An isometry thus provides another *example of a uniformly continuous mapping*.

Otherwise, a general, and very useful, sufficient condition for uniform continuity will be given in Theorem 1.13-2.

1.12 Complete metric spaces

In a *metric space* (X,d), a sequence $(x_n)_{n=0}^{\infty}$ of points $x_n \in X$ is a **Cauchy sequence** if the diameter (Section 1.10) of the set $\bigcup_{m=n}^{\infty}\{x_m\}$ converges to zero as $n \to \infty$, or equivalently, if, for each $\varepsilon > 0$, there exists an integer $n_0(\varepsilon) \geq 0$ such that $d(x_m,x_n) < \varepsilon$ for all $m \geq n_0(\varepsilon)$ and $n \geq n_0(\varepsilon)$.

The following theorem gathers elementary properties of Cauchy sequences:

Theorem 1.12-1 (a) *A Cauchy sequence is bounded.*

(b) *A convergent sequence is a Cauchy sequence.*

(c) *A Cauchy sequence that contains a convergent subsequence is convergent, and its limit is the limit of the subsequence.* $\square$

A *metric space* (X,d) is **complete** if every Cauchy sequence of points of X converges in X. A *subset* A of a metric space is **complete** if the metric space (X,d_A), where d_A denotes the distance induced by d on A (Section 1.10), is complete. Consequently, the property "X is a complete metric space" is independent of whether X is a subset of a larger metric space.

The following theorem gathers elementary properties of complete metric spaces:

Theorem 1.12-2 *Let A be a subset of a metric space X.*

(a) *If A is complete, A is closed in X.*

(b) *If X is complete and A is closed in X, A is complete.*

(c) *If X is complete, a subset A of X is complete if and only if A is closed in X.* $\square$

Fundamental examples of complete metric spaces are $\mathbb{R}$ and $\mathbb{C}$, each equipped with its usual distance, and $\mathbb{R}^n$ and $\mathbb{C}^n$, $n \geq 2$, each equipped with any one of the distances d_p, $1 \leq p \leq \infty$ (Section 1.10): That $\mathbb{R}$ and $\mathbb{C}$ are complete follows from their construction (Section 1.4); that $(\mathbb{R}^n, d_p)$ and $(\mathbb{C}^n, d_p)$ are complete in turn easily follows from the completeness of $\mathbb{R}$ and $\mathbb{C}$.

For a given integer $n \geq 2$, the distances d_p, $1 \leq p \leq \infty$, thus provide examples of distances that induce the same topology on $\mathbb{R}^n$ and *simultaneously* render the metric spaces

$(\mathbb{R}^n, d_p)$ complete. This is not a general circumstance, however. For example, let $\mathbb{R}_+ := \{x \in \mathbb{R};\ x \geq 0\}$ and let the distances d and ρ on $\mathbb{R}_+$ be defined by $d(x,y) = |x-y|$ and $\rho(x,y) = \left|\frac{x}{1+x} - \frac{y}{1+y}\right|$ for all $x, y \in \mathbb{R}_+$. Then these distances induce the same topology on $\mathbb{R}_+$, but $(\mathbb{R}_+, d)$ is complete while $(\mathbb{R}_+, \rho)$ is not (the sequence $(x_n)_{n=0}^{\infty}$ with $x_n = n$ is a Cauchy sequence in $(\mathbb{R}_+, \rho)$ but does not converge in $(\mathbb{R}_+, \rho)$).

The next theorem is fundamental. It provides sufficient conditions insuring that a mapping defined and continuous on a dense subset of a metric space can be extended to a continuous mapping on the whole space (this result will be proved later in Theorem 3.1-1 for normed vector spaces).

Theorem 1.12-3 (unique continuous extension) *Let X be a dense subset of a metric space $\widetilde{X}$, let Y be a complete metric space, and let $f : X \to Y$ be a uniformly continuous mapping.*

Then there exists one and only one continuous extension $\widetilde{f} : \widetilde{X} \to Y$ of f to the space $\widetilde{X}$. The mapping $\widetilde{f}$ is also uniformly continuous on $\widetilde{X}$. □

The next theorem is also fundamental. It asserts that any metric space that is not complete can be always identified with a dense subset of a *complete* metric space by means of an *isometry* (this result will be proved later in Theorem 3.1-2, again for normed vector spaces).

Theorem 1.12-4 (completion of a metric space) (a) *Let (X, d) be a metric space. There exists a complete metric space $(\widetilde{X}, \widetilde{d})$ and an isometry $\sigma : X \to \widetilde{X}$ such that $\sigma(X)$ is dense in $\widetilde{X}$.*

(b) *The space $\widetilde{X}$ is separable if the space X is separable.*

(c) *If $(\widehat{X}, \widehat{d})$ is any complete metric space such that there exists an isometry from X onto a dense subset of $\widehat{X}$, then there exists an isometry from $(\widetilde{X}, \widetilde{d})$ onto $(\widehat{X}, \widehat{d})$.* □

The space $(\widetilde{X}, \widetilde{d})$, which is thus "essentially unique" as a metric space, in the sense that it is *unique up to bijective isometries* thanks to property (c), is called the **completion** of the metric space (X, d).

Two other fundamental theorems about complete metric spaces, viz., the *Banach fixed point theorem* and *Baire's theorem*, will be proved in the next chapters (Theorems 3.7-1 and 5.1-2).

1.13 Compactness in metric spaces

Let (X, d) be a metric space and let K be a subset of X, equipped with the topology induced by the metric d. Then K is **compact** if, as a topological space, it satisfies the Heine–Borel–Lebesgue property (Section 1.8).

Noting that any compact subset is closed in a Hausdorff topological space (Theorem 1.8-2(a)), and that any covering by balls of radius one (to fix ideas) admits a finite subcovering in a compact metric space (by the Heine–Borel–Lebesgue property), we immediately obtain two necessary conditions for compactness in a *metric* space:

Theorem 1.13-1 *A compact subset of a metric space is closed and bounded.* □

Another simple consequence of the Heine–Borel–Lebesgue property in a metric space is a sufficient condition of uniform continuity:

Theorem 1.13-2 *Let X be a compact metric space and let Y be a metric space. Then a continuous mapping from X into Y is uniformly continuous.* □

A subset A of a metric space X is **precompact** if, given any $\varepsilon > 0$, there exists a finite number $n = n(\varepsilon)$ of points $x_j = x_j(\varepsilon) \in A$, $1 \leq j \leq n$, such that

$$A \subset \bigcup_{j=1}^{n} B(x_j; \varepsilon).$$

Note that *$A \subset X$ is precompact if and only if $\overline{A}$ is also precompact.*

The following characterizations of compact and precompact subsets of a metric space are fundamental.

Theorem 1.13-3 *Let X be a metric space and let K be a subset of X. The following three assertions are equivalent:*

(a) *K is a compact subset of X.*

(b) *Given any sequence $(x_n)_{n=0}^{\infty}$ of points $x_n \in K$, there exists a subsequence $(x_{\sigma(n)})_{n=0}^{\infty}$ that converges to a point in K.*

(c) *K is precompact and complete.* □

A topological space A that satisfies the above property (b) is said to satisfy the **Bolzano–Weierstraß property**, to reflect that it generalizes Theorem 1.4-1(b).

Theorem 1.13-4 *A subset A of a metric space is relatively compact if and only if any sequence $(x_n)_{n=0}^{\infty}$ of points $x_n \in A$ contains a subsequence $(x_{\sigma(n)})_{n=0}^{\infty}$ that converges to a point in $\overline{A}$.* □

While the converse of Theorem 1.13-1 "seldom holds," an easy application of Theorem 1.13-3(c) shows that it does hold in the following fundamental special case:

Theorem 1.13-5 *Let the space $\mathbb{K}^n$, where $\mathbb{K} = \mathbb{R}$ or $\mathbb{K} = \mathbb{C}$, be equipped with any one of the distances d_p, $1 \leq p \leq \infty$ (Section 1.10). Then a subset of $\mathbb{K}^n$ is compact if and only if it is closed and bounded.* □

We shall prove later that this property is in effect a characterization of finite dimensionality. This is the essence of the fundamental *F. Riesz theorem* (Theorem 2.7-3).

While the characterization of compact subsets in a *finite-dimensional* space is thus settled by Theorem 1.13-5, the characterization of compact subsets in *infinite-dimensional* spaces is often a delicate issue. An important instance[8] of such a characterization is the *Ascoli–Arzelà theorem* in the space of functions that are continuous on a compact set (Theorem 3.10-1).

By Theorem 1.13-5, a subset of $\mathbb{R}$ is compact if and only if it is closed and bounded. Combining this observation with Theorem 1.8-3 yields another basic result, asserting that

[8] Another important instance is *Kolmogorov's theorem* in the spaces $L^p(\Omega)$; for a proof, see, e.g., Brezis [2011, Theorem 4.26].

continuous functions on compact sets attain their infimum and supremum:

Theorem 1.13-6 *Let K be a compact topological space and let $f : K \to \mathbb{R}$ be a continuous function. Then the direct image $f(K)$ is a compact subset of $\mathbb{R}$. Therefore there exist $x_0 \in K$ and $x_1 \in K$ such that*

$$f(x_0) = \inf_{x\in K} f(x) \quad \textit{and} \quad f(x_1) = \sup_{x\in K} f(x). \qquad \square$$

A spectacular application of Theorem 1.13-6 will be given in the next chapter, where it will be shown to provide a simple proof of the *fundamental theorem of algebra* (Theorem 2.8-1).

1.14 The Lebesgue measure in $\mathbb{R}^n$; measurable functions

In what follows, the notations $[0, \infty]$ and $[-\infty, \infty]$ respectively denote the sets $[0, \infty[\cup \{\infty\}$ and $\{-\infty\} \cup \mathbb{R} \cup \{\infty\}$ (Section 1.4).

Let X be a set. A **σ-algebra of subsets of** X is a subset $\mathcal{A}$ of $\mathcal{P}(X)$ that satisfies the following properties:

$$X \in \mathcal{A},$$
$$A \in \mathcal{A} \text{ implies } (X - A) \in \mathcal{A},$$
$$\bigcup_{i=1}^{\infty} A_i \in \mathcal{A} \text{ if } A_i \in \mathcal{A} \text{ for all } i \geq 1.$$

Given a set X and a σ-algebra $\mathcal{A}$ of subsets of X, a **measure** is a function $\mu : \mathcal{A} \to [0, \infty]$ that satisfies the following properties:

$$\mu(\varnothing) = 0,$$
$$\mu\Big(\bigcup_{i=1}^{\infty} A_i\Big) = \sum_{i=1}^{\infty} \mu(A_i) \quad \text{if } A_i \in \mathcal{A} \text{ for all } i \geq 1 \text{ and } A_i \cap A_j = \varnothing \text{ for all } i \neq j.$$

The last property is called the **σ-additivity** of the measure μ. The triple $(X, \mathcal{A}, \mu)$ is then called a **measure space**.

Of fundamental importance is the set $X = \mathbb{R}^n$, where n is any integer ≥ 1, equipped with its usual topology (Section 1.10), when the σ-algebra is the **Borel σ-algebra $\widetilde{\mathcal{A}}$ in $\mathbb{R}^n$**, defined as the *smallest σ-algebra that contains all the open subsets of* $\mathbb{R}^n$ (the σ-algebra $\widetilde{\mathcal{A}}$ is then uniquely defined, as the intersection of all the σ-algebras of subsets of $\mathbb{R}^n$ that possess this property), and the measure $\widetilde{\mu}$ is defined by

$$\widetilde{\mu}(\widetilde{A}) = \inf\left\{ \sum_{k=1}^{\infty} \Big(\prod_{j=1}^{n} (b_j^k - a_j^k) \Big);\ \widetilde{A} \subset \bigcup_{k=1}^{\infty} \Big(\prod_{j=1}^{n}]a_j^k, b_j^k[\Big) \right\} \quad \text{for all } \widetilde{A} \in \widetilde{\mathcal{A}}.$$

For a given set $\widetilde{A} \in \widetilde{\mathcal{A}}$, the infimum is meant here to be taken over all the countably infinite families of products $\prod_{j=1}^{n}]a_j^k, b_j^k[$ of open intervals, $k \geq 1$, the union of which covers the set $\widetilde{A}$. The elements of the σ-algebra are called the **Borel-measurable subsets of $\mathbb{R}^n$**.

The measure space $(\mathbb{R}^n, \widetilde{\mathcal{A}}, \widetilde{\mu})$ constructed in this fashion lacks one desirable property, namely that any subset of a set $\widetilde{A} \in \widetilde{\mathcal{A}}$ that satisfies $\widetilde{\mu}(\widetilde{A}) = 0$ is also in the σ-algebra $\widetilde{\mathcal{A}}$. To obviate this difficulty, let

$$\mathcal{A} := \{A \in \mathcal{P}(\mathbb{R}^n);\ \text{there exist } \widetilde{A} \in \widetilde{\mathcal{A}} \text{ and } \widetilde{A}' \in \widetilde{\mathcal{A}} \text{ with } \widetilde{\mu}(\widetilde{A}') = 0 \\ \text{such that } A = \widetilde{A} \cup \widetilde{B} \text{ with } \widetilde{B} \subset \widetilde{A}'\},$$

and let, for any $A \in \mathcal{A}$,

$$\mu(A) := \widetilde{\mu}(\widetilde{A})$$
for any $\widetilde{A} \in \widetilde{\mathcal{A}}$ such that $A = \widetilde{A} \cup \widetilde{B}$ with $\widetilde{B} \subset \widetilde{A}'$ for some $\widetilde{A}' \in \widetilde{\mathcal{A}}$ with $\mu(\widetilde{A}') = 0$.

One can then show that $\mathcal{A}$ *is again a* σ*-algebra of subsets of* $\mathbb{R}^n$ (which clearly contains the Borel σ-algebra $\widetilde{\mathcal{A}}$), that the above definition of $\mu(A)$ makes sense (i.e., that it is independent of the particular set $\widetilde{A} \in \widetilde{\mathcal{A}}$ chosen as above), and that *the function* $\mu : \mathcal{A} \to [0, \infty]$ *defined in this fashion is again a measure.*

The σ-algebra $\mathcal{A}$ is called the **Lebesgue σ-algebra in** $\mathbb{R}^n$, the elements of $\mathcal{A}$ are called the **Lebesgue-measurable subsets** of $\mathbb{R}^n$, and μ is called the **Lebesgue measure in** $\mathbb{R}^n$, or the **n-dimensional Lebesgue measure**. Evidently, $\mu(\widetilde{A}) = \widetilde{\mu}(\widetilde{A})$ for all $\widetilde{A} \in \widetilde{\mathcal{A}}$.

The Lebesgue measure in $\mathbb{R}^n$ is denoted

$$dx, \quad \text{or} \quad \text{meas}, \quad \text{or} \quad dx\text{-meas},$$

according to the context.

Cardinality arguments show that

$$\operatorname{card} \widetilde{\mathcal{A}} = \operatorname{card} \mathbb{R} \quad \text{and} \quad \operatorname{card} \mathcal{A} = \operatorname{card} \mathcal{P}(\mathbb{R}).$$

Hence there are "many more" Lebesgue-measurable subsets of $\mathbb{R}^n$ than Borel-measurable subsets of $\mathbb{R}^n$, since $\operatorname{card} \mathbb{R} \prec \operatorname{card} \mathcal{P}(\mathbb{R})$ (Theorem 1.5-2).

The next theorem recapitulates four basic properties of the resulting measure space $(\mathbb{R}^n, \mathcal{A}, \mu)$. The first three are direct consequences of the above construction. The fourth one expresses that the Lebesgue measure is *translation invariant.*

Theorem 1.14-1 *The* σ*-algebra* $\mathcal{A}$ *of Lebesgue-measurable subsets of* $\mathbb{R}^n$ *and the Lebesgue measure* $\mu : \mathcal{A} \to [0, \infty]$ *satisfy the following properties:*

(a) *Every open subset of* $\mathbb{R}^n$ *belongs to* $\mathcal{A}$*; hence every closed subset of* $\mathbb{R}^n$*, and any countably infinite intersection or union of open or closed subsets of* $\mathbb{R}^n$*, also belong to* $\mathcal{A}$*.*

(b) *The Lebesgue measure of any subset of* $\mathbb{R}^n$ *of the form* $\prod_{j=1}^n]a_j, b_j[$*, which belongs to* $\mathcal{A}$ *by* (a)*, is given by*

$$\mu\Big(\prod_{j=1}^{n}]a_j, b_j[\Big) = \prod_{j=1}^{n} (b_j - a_j).$$

(c) *If* $A \in \mathcal{A}$ *and* $\mu(A) = 0$*, then every subset of* A *is also Lebesgue-measurable and its Lebesgue-measure is zero.*

(d) *Given any point* $x \in \mathbb{R}^n$ *and any set* $A \in \mathcal{A}$*, the set*

$$x + A := \{(x + y) \in \mathbb{R}^n;\ y \in A\}$$

also belongs to $\mathcal{A}$ *and* $\mu(x + A) = \mu(A)$*.* □

A noteworthy consequence of the translation invariance (d) of the Lebesgue measure is that, when appropriately combined with the *axiom of choice*, it implies the existence of subsets of $\mathbb{R}^n$ that are *not* Lebesgue-measurable.

We next examine how a *product* of measure spaces $(X, \mathcal{A}, \mu)$ and $(Y, \mathcal{B}, \nu)$ can be also made a measure space. First, let $\mathcal{A} \otimes \mathcal{B}$ denote the *smallest* σ-algebra *that contains all the sets* $A \times B \in \mathcal{P}(X \times Y)$, *with* $A \in \mathcal{A}$ *and* $B \in \mathcal{B}$ (hence $\mathcal{A} \otimes \mathcal{B}$ is uniquely defined by these conditions). Then one can show that there exists one and only one **product measure**

$$\mu \otimes \nu : \mathcal{A} \otimes \mathcal{B} \to [0, \infty]$$

with the (expected) property that

$$(\mu \otimes \nu)(A \times B) = \mu(A)\nu(B) \quad \text{for all } A \in \mathcal{A} \text{ and } B \in \mathcal{B}.$$

When μ is the Lebesgue measure on $\mathbb{R}^m$ and ν is the Lebesgue measure on $\mathbb{R}^n$, one can show that the product measure $\mu \otimes \nu$ is (again as expected) precisely the Lebesgue measure on $\mathbb{R}^{m+n}$.

Let $\mathcal{A}$ and μ respectively designate the Lebesgue σ-algebra in $\mathbb{R}^n$ and the Lebesgue measure in $\mathbb{R}^n$. For brevity, the elements of $\mathcal{A}$ will be simply called **measurable subsets of** $\mathbb{R}^n$.

Let A be a measurable subset of $\mathbb{R}^n$. A property is said to hold **almost everywhere (a.e.) in** A, or equivalently to hold **for almost all** $x \in A$, if the set of points in A where it does *not* hold is measurable and of measure zero. For instance, two functions $f, g : A \to \mathbb{R}$ are *equal almost everywhere* if the set $\{x \in A;\ f(x) \neq g(x)\}$ is measurable and of measure zero; a sequence $(f_n)_{n=1}^{\infty}$ of functions $f_n : A \to [-\infty, \infty]$ *converges almost everywhere in* A *as* $n \to \infty$ if the complement of the set $\{x \in A;\ \lim_{n\to\infty} f_n(x) \text{ exists in } [-\infty, \infty]\}$ is measurable and of measure zero, etc.

A much less trivial example of a property that holds almost everywhere is provided by the following fundamental result. In what follows, the spaces $\mathbb{R}^n$ and $\mathbb{R}^m$ are equipped with any one of the distances d_p, $1 \leq p \leq \infty$, defined in Section 1.10.

Theorem 1.14-2 (Rademacher's theorem[9]) *Let* Ω *be an open subset of* $\mathbb{R}^n$ *and let* $f : \Omega \to \mathbb{R}^m$ *be a Lipschitz-continuous function* (*Section* 1.11). *Then* f *is differentiable almost everywhere in* Ω. □

Given any Lebesgue-measurable subset $A \in \mathcal{A}$, a function $f : A \to [-\infty, \infty]$ is said to be **Lebesgue-measurable**, or simply **measurable**, if

$$f^{-1}([-\infty, \alpha[) = \{x \in A;\ f(x) < \alpha\} \in \mathcal{A} \quad \text{for all } \alpha \in \mathbb{R}.$$

The next theorem recapitulates a first series of basic properties of measurable functions. Note that property (c) is restricted to *real-valued* functions.

Theorem 1.14-3 *Let* A *be a measurable subset of* $\mathbb{R}^n$.

(a) *Let* $f : A \to [-\infty, \infty]$ *be a measurable function. Then the function* $|f| : A \to [0, \infty]$ *is also measurable.*

[9] So named after Hans Adolph Rademacher (1892–1969).

(b) *Let* $f_n : A \to [-\infty, \infty]$, $n \geq 1$, *be measurable functions. Then the functions*

$$\sup_{n\geq 1} f_n, \quad \inf_{n\geq 1} f_n, \quad \limsup_{n\to\infty} f_n, \quad \liminf_{n\to\infty} f_n : A \to [-\infty, \infty]$$

are also measurable.

(c) *Let* $f, g : A \to \mathbb{R}$ *be measurable functions. Then the functions* $f + g : A \to \mathbb{R}$ *and* $fg : A \to \mathbb{R}$ *are also measurable.* □

The next theorem recapitulates three other basic properties, this time *linking measurability and continuity.* Property (c) constitutes **Lusin's property**.

Theorem 1.14-4 *Let A be a measurable subset of* $\mathbb{R}^n$.

(a) *Let* $f : A \to \mathbb{R}$ *be a continuous function. Then f is measurable.*

(b) *Let* $f : \mathbb{R} \to \mathbb{R}$ *be a continuous function and let* $g : A \to \mathbb{R}$ *be a measurable function. Then the composite function* $f \circ g : A \to \mathbb{R}$ *is measurable.*

(c) *Let* $f : \mathbb{R}^n \to \mathbb{R}$ *be a measurable function with the property that* $\mu(A) < \infty$, *where* $A := \{x \in \mathbb{R}^n;\ f(x) \neq 0\}$. *Then, given any* $\varepsilon > 0$, *there exists a function* $f_\varepsilon \in \mathcal{C}(\mathbb{R}^n)$ *whose support is a compact subset of A and such that*

$$\sup_{x\in\mathbb{R}^n} |f_\varepsilon(x)| \leq \sup_{x\in\mathbb{R}^n} |f(x)| \quad \text{and} \quad \mu(\{x \in \mathbb{R}^n;\ f(x) \neq f_\varepsilon(x)\}) \leq \varepsilon. \qquad \square$$

Let A be any measurable subset of $\mathbb{R}^n$. A **simple function** on A is a function $s : A \to \mathbb{R}$ whose image is a finite subset of $\mathbb{R}$; equivalently, there exists a finite number of pairwise disjoint (i.e., $A_i \cap A_j = \varnothing$ if $i \neq j$) subsets A_i of A, $1 \leq i \leq m$, and real numbers α_i, $1 \leq i \leq m$, such that

$$s = \sum_{i=1}^{m} \alpha_i \chi_{A_i},$$

where $\chi_{A_i} : A \to \mathbb{R}$ denotes the characteristic function of each set A_i (Section 1.2). Clearly, a simple function s is *measurable* if and only if each set A_i, $1 \leq i \leq m$, is measurable.

Important links between measurability and simple functions are given in the next theorem.

Theorem 1.14-5 *Let A be a measurable subset of* $\mathbb{R}^n$.

(a) *Let* $f : A \to [-\infty, \infty]$ *be a measurable function. Then there exists a sequence of measurable simple functions* $s_n : A \to \mathbb{R}$, $n \geq 1$, *with the following properties:*

$$|s_n| \leq |s_{n+1}| \leq |f| \text{ for all } n \geq 1 \text{ and, for each } x \in A,\ s_n(x) \to f(x) \text{ as } n \to \infty.$$

(b) *Let* $f : A \to [0, \infty]$ *be a measurable function. Then there exists a sequence of measurable simple functions* $s_n : A \to \mathbb{R}$, $n \geq 1$, *with the following properties:*

$$0 \leq s_n \leq s_{n+1} \leq f \text{ for all } n \geq 1 \text{ and, for each } x \in A,\ s_n(x) \to f(x) \text{ as } n \to \infty. \qquad \square$$

1.15 The Lebesgue integral in $\mathbb{R}^n$; the basic theorems

Let A be any measurable subset of $\mathbb{R}^n$. Given a measurable simple function $s = \sum_{i=1}^m \alpha_i \chi_{A_i}$ (Section 1.14) that is ≥ 0 (equivalently, such that $\alpha_i \geq 0$, $1 \leq i \leq m$), let the extended real number $\int_A s(x)\,\mathrm{d}x \in [0,\infty]$ be *defined* as

$$\int_A s(x)\,\mathrm{d}x := \sum_{i=1}^m \alpha_i \mu(A_i).$$

The **Lebesgue integral** of any *measurable function* $f : A \to [0,\infty]$ is then *defined* as

$$\int_A f(x)\,\mathrm{d}x := \sup\left\{\int_A s(x)\,\mathrm{d}x;\ s \text{ is a measurable simple function and } 0 \le s \le f \text{ in } A\right\}.$$

Hence $\int_A f(x)\,\mathrm{d}x$ is again either ≥ 0 or equal to ∞. If the Lebesgue integral $\int_A f(x)\,\mathrm{d}x$ of a measurable function $f : A \to [0,\infty]$ is finite, then f is necessarily finite almost everywhere.

Finally, a function $f : A \to [-\infty,\infty]$ is said to be **Lebesgue-integrable**, or in short **integrable**, if it is *measurable* and is such that

$$\int_A \max\{f(x);\ 0\}\,\mathrm{d}x < \infty \quad \text{and} \quad \int_A \max\{-f(x);\ 0\}\,\mathrm{d}x < \infty.$$

If $f : A \to [-\infty,\infty]$ is Lebesgue-integrable, its **Lebesgue integral** is *defined* by

$$\int_A f(x)\,\mathrm{d}x := \int_A \max\{f(x);\ 0\}\,\mathrm{d}x - \int_A \max\{-f(x);\ 0\}\,\mathrm{d}x.$$

The following immediate properties of Lebesgue-integrable functions and their Lebesgue integrals are constantly used: *The Lebesgue integral of an integrable function is a real number*, i.e., it is not equal to ∞ or $-\infty$. The Lebesgue measure of a measurable subset of $\mathbb{R}^n$ is also given by $\int_{\mathbb{R}^n} \chi_A\,\mathrm{d}x = \int_A \mathrm{d}x$; a Lebesgue-integrable function is *finite almost everywhere*; *a measurable function $f : A \to [-\infty,\infty]$ is Lebesgue-integrable if and only if the function $|f| : A \to [0,\infty]$ is Lebesgue-integrable, and in this case,*

$$\left|\int_A f(x)\,\mathrm{d}x\right| \le \int_A |f(x)|\,\mathrm{d}x.$$

The set, denoted

$$\mathcal{L}^1(A),$$

of all Lebesgue-integrable functions $f : A \to \mathbb{R}$ is clearly a *vector space over* $\mathbb{R}$ (vector spaces over $\mathbb{R}$ or $\mathbb{C}$ are defined in Section 2.1).

The relation "$f = g$ almost everywhere in A" defines an equivalence relation $\mathcal{R}$ over the space $\mathcal{L}^1(A)$, and the quotient set

$$L^1(A) := \mathcal{L}^1(A)/\mathcal{R}$$

is also a *vector space over* $\mathbb{R}$. Besides,

$$\int_A f(x)\,\mathrm{d}x = \int_A g(x)\,\mathrm{d}x \quad \text{if } f, g \in \mathcal{L}^1(A) \text{ are such that } f = g \text{ a.e. in } A.$$

As a consequence, the *Lebesgue integral of any equivalence class in $L^1(A)$* is unambiguously defined, as the Lebesgue integral of any function in the class.

As is customary, we shall also refer to elements in $L^1(A)$ as *integrable functions*, even though they are in effect *equivalence classes* of integrable functions modulo $\mathcal{R}$.

Clearly, the identification of functions in $\mathcal{L}^1(A)$ with their equivalence classes in $L^1(A)$ constitutes a flagrant *abuse of language*, but it avoids many cumbersome statements and what is meant should be always unambiguous. For instance, "$f \in L^1(A)$ is a continuous function" means that, in the equivalence class of f, there is a (unique) continuous function in $\mathcal{L}^1(A)$; likewise, "$f \in L^1(A)$ is finite almost everywhere in A" means that in the equivalence class of f, there is a function which is finite everywhere in A, etc.

There are other ways of defining the Lebesgue integral and the space $L^1(A)$. For instance, let $\mathcal{S}(A)$ denote the set formed by all *integrable simple functions* $s : A \to \mathbb{R}$, i.e., those that satisfy

$$\mu(\{x \in A;\ s(x) \neq 0\}) < \infty.$$

It is then immediately seen that the *quotient set* $S(A) := \mathcal{S}(A)/\mathcal{R}$ is a *vector space*, and that the mapping

$$\|\cdot\|_{L^1(A)} : s \in S(A) \to \int_A |s(x)|\,\mathrm{d}x,$$

where the integral of a simple function on A is defined as earlier, is a *norm* on $S(A)$.

Then the space $L^1(A)$ may be equivalently defined as the completion (Theorem 1.12-4) of the space $(S(A), \|\cdot\|_{L^1(A)})$. In this case, the *Lebesgue integral of functions* $f \in L^1(A)$ is then simply defined as the *unique continuous extension* (Theorem 1.12-3) of the linear functional

$$s \in S(A) \to \int_A s(x)\,\mathrm{d}x,$$

which is defined and continuous over the dense subset $S(A)$ of $L^1(A)$.

When the set A is an *open subset of* $\mathbb{R}^n$, yet another definition is possible: Let $\mathcal{C}_c(A)$ denote the space of all *continuous functions* $f : A \to \mathbb{R}$ *with compact support in* A, and let

$$\|f\|_{L^1(A)} := \int_A |f(x)|\,\mathrm{d}x,$$

the symbol $\int_A g(x)\,\mathrm{d}x$ denoting here the *Riemann integral* of a function $g \in \mathcal{C}_c(A)$. Then the space $L^1(A)$ may be also equivalently defined as the *completion of the space* $(\mathcal{C}_c(A), \|\cdot\|_{L^1(A)})$, and by construction, the space $\mathcal{C}_c(A)$ is then *dense* in $L^1(A)$ (when the space $L^1(A)$ is defined as earlier in this section, the denseness of $\mathcal{C}_c(A)$ in $L^1(A)$ becomes a theorem, which therefore needs to be proved; cf. Theorem 2.5-3). In this case, the Lebesgue integral is then again defined as the *unique continuous extension* of a linear functional defined and continuous on a dense subset.

The notion of the Lebesgue-integrable functions can be easily extended to complex-valued functions: Let A be any measurable subset of $\mathbb{R}^n$. Then a *complex-valued function* $f : A \to \mathbb{C}$ is said to be **Lebesgue-integrable** if

$$\operatorname{Re} f \in \mathcal{L}^1(A) \quad \text{and} \quad \operatorname{Im} f \in \mathcal{L}^1(A).$$

If this is the case, the **Lebesgue integral** of f is *defined* by

$$\int_A f(x)\,\mathrm{d}x := \int_A \operatorname{Re} f(x)\,\mathrm{d}x + i \int_A \operatorname{Im} f(x)\,\mathrm{d}x.$$

It is easily seen that it again satisfies the inequality

$$\left|\int_A f(x)\,\mathrm{d}x\right| \le \int_A |f(x)|\,\mathrm{d}x.$$

The set, denoted

$$\mathcal{L}^1(A;\mathbb{C}),$$

of all Lebesgue-integrable functions $f : A \to \mathbb{C}$ is clearly a *vector space over* $\mathbb{C}$. The relation "$f = g$ almost everywhere in A" again defines an equivalence relation $\mathcal{R}$ over the space $\mathcal{L}^1(A;\mathbb{C})$, and the quotient set

$$L^1(A;\mathbb{C}) := \mathcal{L}^1(A;\mathbb{C})/\mathcal{R}$$

is also a *vector space over* $\mathbb{C}$.

Finally, note that, for brevity, we shall often omit the dependence on $x \in A$, by simply letting

$$\int_A f\,\mathrm{d}x := \int_A f(x)\,\mathrm{d}x \quad \text{if } f \in L^1(A) \text{ or if } f \in L^1(A;\mathbb{C}).$$

The next theorems recapitulate the *most fundamental properties of the Lebesgue integral.* Note in this respect that the *order* in which these properties can be established may in effect depend on the way the Lebesgue integral has been defined. The first three theorems list basic *convergence properties of sequences of integrable functions.*

Theorem 1.15-1 (Beppo Levi monotone convergence theorem) *Let A be a measurable subset of $\mathbb{R}^n$ and let $(f_k)_{k=1}^\infty$ be a sequence of functions $f_k \in \mathcal{L}^1(A)$ with the property that*

$$0 \le f_1 \le \cdots \le f_k \le f_{k+1} \le \cdots \text{ a.e. in } A \quad \text{and} \quad \lim_{k\to\infty} \int_A f_k(x)\,\mathrm{d}x < \infty.$$

Then there exists a function $f \in \mathcal{L}^1(A)$ such that

$$f_k(x) \to f(x) \text{ for almost all } x \in A \quad \text{and} \quad \int_A |f_k(x) - f(x)|\,\mathrm{d}x \to 0 \text{ as } k \to \infty.$$

In particular then, $\lim_{k\to\infty} f_k(x) < \infty$ *for almost all* $x \in A$. □

Theorem 1.15-2 (Fatou's lemma) *Let A be a measurable subset of $\mathbb{R}^n$ and let $(f_k)_{k=1}^\infty$ be a sequence of measurable functions $f_k : A \to \mathbb{R}$ with the property that*

$$f_k \ge 0 \text{ a.e. in } A.$$

Then

$$\int_A \left(\liminf_{k\to\infty} f_k(x)\right) \mathrm{d}x \le \liminf_{k\to\infty} \int_A f_k(x)\,\mathrm{d}x,$$

where the right-hand side, or both sides, of this inequality may be equal to ∞. □

Theorem 1.15-3 (Lebesgue dominated convergence theorem) *Let A be a measurable subset of $\mathbb{R}^n$ and let $(f_k)_{k=1}^\infty$ be a sequence of functions $f_k \in \mathcal{L}^1(A)$,* resp. *$f_k \in \mathcal{L}^1(A;\mathbb{C})$, such that*

$$f(x) := \lim_{k\to\infty} f_k(x) \quad \text{exists for almost all } x \in A,$$

and such that there exists a function $g \in \mathcal{L}^1(A)$ with the property that

$$|f_k(x)| \le g(x) \quad \textit{for all } k \ge 1 \textit{ and almost all } x \in A.$$

Then $f \in \mathcal{L}^1(A)$, resp. $f \in \mathcal{L}^1(A; \mathbb{C})$, and

$$\int_A |f_k(x) - f(x)| \, dx \to 0 \quad \textit{as } k \to \infty.$$

In particular then,

$$\int_A f(x) \, dx = \lim_{k \to \infty} \int_A f_k(x) \, dx. \qquad \square$$

Let B be a measurable subset of $\mathbb{R}^n$ and let $\mathcal{A}$ denote the σ-algebra formed by all the Lebesgue measurable subsets of B. Given a function $f \in \mathcal{L}^1(B)$, let

$$\nu(A) := \int_A f(x) \, dx \quad \text{for each } A \in \mathcal{A}.$$

Hence $|\nu(A)| \le \int_A |f(x)| \, dx \le \int_B |f(x)| \, dx < \infty$ for each $A \in \mathcal{A}$. Then it is clear that the function $\nu : \mathcal{A} \to \mathbb{R}$ defined in this fashion possesses the following properties: *first*, it is a **signed measure**, in the sense that

$$\nu(\varnothing) = 0,$$
$$\nu\Big(\bigcup_{i=1}^{\infty} A_i\Big) = \sum_{i=1}^{\infty} \nu(A_i) \text{ if } A_i \in \mathcal{A},\ i \ge 1, \text{ are such that } A_i \cap A_j = \varnothing \text{ if } i \neq j$$

(this countably additive property easily follows from the Lebesgue dominated convergence theorem); *second*, it is **absolutely continuous** *with respect to the Lebesgue measure* dx, in the sense that

$$A \in \mathcal{A} \quad \text{and} \quad dx\text{-}\operatorname{meas} A = 0 \quad \text{imply} \quad \nu(A) = 0.$$

Remarkably, the converse property holds:

Theorem 1.15-4 (Radon–Nikodym theorem) *Let B be a measurable subset of $\mathbb{R}^n$, let $\mathcal{A}$ denote the σ-algebra formed by all the measurable subsets of B, and let $\nu : \mathcal{A} \to \mathbb{R}$ be a signed measure that is absolutely continuous with respect to the Lebesgue measure. Then there exists a function $f \in \mathcal{L}^1(B)$ such that*

$$\nu(A) = \int_A f(x) \, dx \quad \textit{for each } A \in \mathcal{A}. \qquad \square$$

The next theorem gives a fundamental *criterion of Lebesgue-integrability* of a function defined over a *product* of measurable sets in $\mathbb{R}^m \times \mathbb{R}^n$, as well as a *way of computing the Lebesgue integral*

$$\iint_{A \times B} f(x, y) \, dx \, dy$$

of a Lebesgue-integrable function $f : A \times B \to [-\infty, \infty]$, where $dx \, dy$ denotes the Lebesgue measure on $\mathbb{R}^m \times \mathbb{R}^n$.

Theorem 1.15-5 *Let A be a measurable subset of $\mathbb{R}^m$, let B be a measurable subset of $\mathbb{R}^n$, and let $f : A \times B \subset \mathbb{R}^m \times \mathbb{R}^n \to [-\infty, \infty]$ be a measurable function.*

(a) **(Tonelli's theorem)** *For each $x \in A$, the function $f(x, \cdot) : y \in B \to f(x, y) \in [-\infty, \infty]$ is measurable and, for each $y \in B$, the function $f(\cdot, y) : x \in A \to f(x, y) \in [-\infty, \infty]$ is measurable. Besides, the function f is integrable over $A \times B$, i.e.,*

$$\iint_{A\times B} |f(x,y)|\,\mathrm{d}x\,\mathrm{d}y < \infty,$$

if and only if one of the following two conditions is satisfied:

$$\int_A \left(\int_B |f(x,y)|\,\mathrm{d}y \right) \mathrm{d}x < \infty,$$

$$\int_B \left(\int_A |f(x,y)|\,\mathrm{d}x \right) \mathrm{d}y < \infty.$$

(b) **(Fubini's theorem)** *If the function f is Lebesgue-integrable on $A \times B$, the function $f(\cdot, y) : A \to [-\infty, \infty]$ is integrable for almost all $y \in B$, the function $f(x, \cdot) : B \to [-\infty, \infty]$ is integrable for almost all $x \in A$, and the Lebesgue integral of f on $A \times B$ is given by*

$$\iint_{A\times B} f(x,y)\,\mathrm{d}x\,\mathrm{d}y = \int_A \left(\int_B f(x,y)\,\mathrm{d}y \right) \mathrm{d}x = \int_B \left(\int_A f(x,y)\,\mathrm{d}x \right) \mathrm{d}y. \qquad \square$$

1.16 Change of variable in Lebesgue integrals in $\mathbb{R}^n$

In this section, we examine how a *Lebesgue integral* defined over an *open subset* $\mathbb{R}^n$ is transformed under a *change of variable*; this means that the open set is the image $\varphi(\Omega)$ of another open subset Ω of $\mathbb{R}^n$ under a mapping $\varphi = (\varphi_i)_{i=1}^n : \Omega \to \mathbb{R}^n$, the variable $y \in \varphi(\Omega)$ being replaced by the variable $x \in \Omega$ in the process.

In what follows, the notation $\nabla\varphi$ designates the $n \times n$ matrix field defined by $(\nabla\varphi)_{ij} = \partial_j \varphi_i$, $1 \le i, j \le n$, where ∂_j denotes the partial derivative operator with respect to the jth variable.

Theorem 1.16-1 (injective change of variable in Lebesgue integrals in $\mathbb{R}^n$) *Let Ω be an open subset of $\mathbb{R}^n$ and let $\varphi : \Omega \to \mathbb{R}^n$ be a continuously differentiable injective mapping.*

Then a function $f : \varphi(\Omega) \to \mathbb{R}$ is Lebesgue-integrable on $\varphi(\Omega)$ if and only if the function

$$x \in \Omega \to f(\varphi(x))\,|\det \nabla\varphi(x)| \in \mathbb{R}$$

is Lebesgue-integrable on Ω. If this is the case, then

$$\int_{\varphi(\Omega)} f(y)\,\mathrm{d}y = \int_\Omega f(\varphi(x))\,|\det \nabla\varphi(x)|\,\mathrm{d}x. \qquad \square$$

Remarks (1) Under the assumptions of Theorem 1.16-1, the set $\varphi(\Omega)$ is automatically *open* (hence Lebesgue-integrability on $\varphi(\Omega)$ makes sense): this is a consequence of the deep *Brouwer invariance of domain theorem* in $\mathbb{R}^n$ (Theorem 9.17-3), which in fact holds even if $\varphi : \Omega \to \mathbb{R}^n$ is only assumed to be continuous.

(2) That f be real-valued is not a restrictive assumption since a Lebesgue-integrable function is necessarily finite almost everywhere. □

While the case where the mapping φ is injective (as in Theorem 1.16-1) is considered in many texts, the case where φ is *not* injective (as in the next theorem) is not often treated.[10]

Theorem 1.16-2 (noninjective change of variable in Lebesgue integrals in $\mathbb{R}^n$) *Let Ω be an open subset of $\mathbb{R}^n$ and let $\varphi : \Omega \to \mathbb{R}^n$ be a continuously differentiable mapping such that the image $\varphi(\Omega)$ is open. For each $y \in \varphi(\Omega)$, let*

$$\begin{aligned} \operatorname{card} \varphi^{-1}(y) &:= \textit{cardinal of the set } \varphi^{-1}(y) \textit{ if } \varphi^{-1}(y) \textit{ is finite,} \\ \operatorname{card} \varphi^{-1}(y) &:= \infty \quad \textit{if } \varphi^{-1}(y) \textit{ is infinite.} \end{aligned}$$

Then, given a function $f : \varphi(\Omega) \to \mathbb{R}$, the function $f \operatorname{card} \varphi^{-1} : \varphi(\Omega) \to \mathbb{R}$ is Lebesgue-integrable on $\varphi(\Omega)$ if and only if the function

$$x \in \Omega \to f(\varphi(x))\, |\det \nabla\varphi(x)| \in \mathbb{R}$$

is Lebesgue-integrable on Ω. If this is the case, then

$$\int_{\varphi(\Omega)} f(y) \operatorname{card} \varphi^{-1}(y)\, \mathrm{d}y = \int_\Omega f(\varphi(x))\, |\det \nabla\varphi(x)|\, \mathrm{d}x. \qquad \square$$

Remarks (1) As expected, Theorem 1.16-1 is a special case of Theorem 1.16-2.

(2) By contrast with Theorem 1.16-1, it must now be *assumed* that $\varphi(\Omega)$ is open in Theorem 1.16-2. □

1.17 Volumes, areas, and lengths in $\mathbb{R}^n$

The n-**volume**, or simply, the **volume**, of a measurable subset A of $\mathbb{R}^n$, denoted $\mathrm{d}x$-meas A, or simply meas A, is *by definition* the Lebesgue measure of A; in other words,

$$\mathrm{d}x\text{-}\operatorname{meas} A = \operatorname{meas} A := \int_A \mathrm{d}x,$$

where $\mathrm{d}x$ denotes the n-dimensional Lebesgue measure.

Thanks to the *formula for change of variables in Lebesgue integrals* (Theorem 1.16-1 or 1.16-2), one can compute the *volume of n-parallelepipeds* (these particular subsets of $\mathbb{R}^n$ are defined in the next theorem):

Theorem 1.17-1 (volume of an n-parallelepiped) *The volume of an* n-**parallelepiped** *in $\mathbb{R}^n$, i.e., a subset P of $\mathbb{R}^n$ of the form*

$$P = \left\{ a + \sum_{i=1}^{n} \lambda_i b_i;\ 0 \le \lambda_i \le 1,\ 1 \le i \le n \right\},$$

[10] See, however, Rado & Reichelderfer [1955], Schwartz [1993b, Corollary 6.2.14], Federer [1969], or Smith [1983, Chapter 16].

where $a \in \mathbb{R}^n$ *and* $b_i \in \mathbb{R}^n$, $1 \le i \le n$, *is given by*

$$dx\text{- meas } P = |\det B| = \sqrt{\det(b_i \cdot b_j)},$$

where B *denotes the* $n \times n$ *matrix whose* i*th column is the vector* b_i *(identified here with an* $n \times 1$ *matrix), and* $(b_i \cdot b_j)$ *denotes the* $n \times n$ *matrix whose coefficient at the* i*th row and* j*th column is the Euclidean inner product of the vectors* b_i *and* b_j. □

Remarks (1) The second formula giving dx-meas P is an immediate consequence of the first one (since $(\det B)^2 = \det(B^T B)$ for any square matrix in B).

(2) The volume of the n-parallelepiped P is thus zero if the n vectors b_i are *linearly dependent.* □

Note that while the coefficients of the above matrix B vary in general under a change of orthogonal basis in $\mathbb{R}^n$, those of the matrix $(b_i \cdot b_j)$ do *not* vary under such a change, since the Euclidean inner product is invariant under a change of orthogonal basis. Consequently, the second formula can be still used for defining the *n-dimensional volume of an n-parallelepiped, now defined as a subset of* $\mathbb{R}^m$ *with* $m \ge n$. This observation is the basis for the next definition, that of *n-dimensional area.*

Let Ω be an open subset in $\mathbb{R}^n$, let $m \ge n$, and let $\Theta = (\Theta_j)_{j=1}^m : \Omega \to \mathbb{R}^m$ be a continuously differentiable injective mapping. At each point $x \in \Omega$, the matrix $\nabla\Theta(x) \in \mathbb{M}^{m\times n}$, where $(\nabla\Theta)_{ij} := \partial_j \Theta_i$, maps the n basis vectors of $\mathbb{R}^n$ into the n vectors $\partial_i\Theta(x) := (\partial_i \Theta_j(x))_{j=1}^m \in \mathbb{R}^m$, $1 \le i \le n$, which in turn are used for defining an *n-parallelepiped in* $\mathbb{R}^m$, of the form

$$\left\{\Theta(x) + \sum_{i=1}^n \lambda_i \partial_i \Theta(x);\ 0 \le \lambda_i \le 1,\ 1 \le i \le n\right\}.$$

Since by Theorem 1.17-1 the n-dimensional volume of this parallelepiped is

$$\sqrt{\det(\partial_i\Theta(x) \cdot \partial_j\Theta(x))},$$

it is thus natural to *define* the n-**dimensional area**, or simply the **area**, area $\Theta(\Omega)$, of the set $\Theta(\Omega)$ as the "infinite sum of the elementary n-dimensional volumes $\sqrt{\det(\partial_i\Theta(x) \cdot \partial_j\Theta(x))}\,dx$," i.e., by

$$\text{area } \Theta(\Omega) := \int_\Omega \sqrt{\det(\partial_i\Theta(x) \cdot \partial_j\Theta(x))}\,dx.$$

Remarks (1) If $m = n$ and $\Theta = \mathrm{id}_\Omega$, the area of $\Theta(\Omega) = \Omega$ is thus (as expected) none other than the *n-volume* of Ω, as defined above.

(2) If $m = n$ and Θ is in addition an *immersion*, i.e., the matrix $(\nabla\Theta)(x)$ is invertible at each point $x \in \Omega$, the matrix $(\partial_i\Theta(x) \cdot \partial_j\Theta(x)) \in \mathbb{M}^n$ is the *metric tensor* at $x \in \Omega$ of the set $\Theta(\Omega)$; cf. Section 8.2.

(3) If $n = 2$ and $m = 3$ and Θ is in addition an *immersion*, i.e., the matrix $(\partial_i\Theta(x) \cdot \partial_j\Theta(x))$ is of rank two at each point $x \in \Omega$, the matrix $(\partial_i\Theta(x) \cdot \partial_j\Theta(x)) \in \mathbb{M}^2$ is the *first fundamental form* $x \in \Omega$ of the set $\Theta(\Omega)$, which is then called a *surface in* $\mathbb{R}^3$; cf. Section 8.9. □

Finally, consider the case where the set Ω is an open interval I of $\mathbb{R}$ (hence $n = 1$) and $\Theta = (\Theta_j)_{j=1}^m$ is an injective mapping from I into $\mathbb{R}^m$, $m \ge 1$. Then the image $\Theta(I)$

of the interval I under Θ is said to be a **curve** in $\mathbb{R}^m$ and the variable $t \in I$ is said to **parametrize** the curve $\Theta(I)$. The **length** of the curve $\Theta(I)$ is then naturally *defined* as the *one-dimensional area* of the set $\Theta(I)$, viz., by

$$\operatorname{length}\Theta(I) := \int_I \sqrt{\Theta'(t)\cdot\Theta'(t)}\,dt,$$

where $\Theta'(t) = (\Theta_j'(t))_{j=1}^m \in \mathbb{R}^m$, $t \in I$. Note that the integrand $\sqrt{\Theta'(t)\cdot\Theta'(t)}$ is nothing but the *Euclidean norm* of the vector $\Theta'(t) \in \mathbb{R}^m$; cf. Section 2.2.

If $\Theta'(t) \neq 0$ for all $t \in I$ and $t_0 \in I$, the **arc length** along the curve $\Theta(I)$, measured from the point $\Theta(t_0)$, is defined by

$$s := \sigma(t) = \int_{t_0}^t \sqrt{\Theta'(t)\cdot\Theta'(t)}\,dt.$$

The function $\sigma : I \to \mathbb{R}$ defined in this fashion is then invertible, and the derivative of its inverse function $\tau : \sigma(I) \to I$ is given by

$$\tau'(s) = \frac{1}{\sqrt{\Theta'(t)\cdot\Theta'(t)}} \quad \text{for all } s = \sigma(t),\ \ t \in I.$$

1.18 The spaces $\mathcal{C}^m(\Omega)$ and $\mathcal{C}^m(\overline{\Omega})$; domains in $\mathbb{R}^n$

All the functions considered in this section are *real-valued.*

The coordinates of a point $x \in \mathbb{R}^n$ are denoted x_i, $1 \le i \le n$, and the corresponding partial derivative operators are denoted $\partial_i := \partial/\partial x_i$, $\partial_{ij} := \partial^2/\partial x_i \partial x_j$, $\partial_{ijk} := \partial^3/\partial x_i\partial x_j\partial x_k$, etc. Partial derivative operators of any order are also denoted with the *multi-index notation* as

$$\partial^{\boldsymbol{\alpha}} := \partial^{|\boldsymbol{\alpha}|}/\partial_{x_1}^{\alpha_1}\partial_{x_2}^{\alpha_2}\cdots\partial_{x_n}^{\alpha_n},$$

where $\boldsymbol{\alpha} = (\alpha_1, \alpha_2, \dots, \alpha_n)$ with $\alpha_i \in \mathbb{N}$, $1 \le i \le n$, is a *multi-index*, and $|\boldsymbol{\alpha}| := \sum_{i=1}^n \alpha_i \ge 0$; note that $\mathbf{0} = (0, 0, \dots, 0)$ is allowed, with the convention that $\partial^{\mathbf{0}} v := v$. Finally, if $x = (x_i) \in \mathbb{R}^n$, we let $|x| := (\sum_{i=1}^n |x_i|^2)^{1/2}$ (the function $|\cdot| : \mathbb{R}^n \to \mathbb{R}$ defined in this fashion is the Euclidean norm; cf. Section 2.2).

To begin with, let Ω be an *arbitrary open subset* of $\mathbb{R}^n$ (later on in this section additional assumptions will be made on the set Ω).

For any integer $m \ge 1$, the space of all functions that are m times, *resp.* infinitely, continuously differentiable over Ω is denoted

$$\mathcal{C}^m(\Omega), \quad \text{resp.} \quad \mathcal{C}^\infty(\Omega) := \bigcap_{m=0}^{\infty} \mathcal{C}^m(\Omega).$$

For $m = 0$, we let

$$\mathcal{C}^0(\Omega) := \mathcal{C}(\Omega) \quad \text{and} \quad \mathcal{C}^0(\overline{\Omega}) := \mathcal{C}(\overline{\Omega}).$$

For any integer $m \ge 1$, we also define the spaces

$$\mathcal{C}^m(\overline{\Omega}) := \{f \in \mathcal{C}^m(\Omega);\ \text{ for each } |\boldsymbol{\alpha}| \le m, \text{ there exists } g^{\boldsymbol{\alpha}} \in \mathcal{C}(\overline{\Omega}) \text{ such that } \partial^{\boldsymbol{\alpha}} f = g^{\boldsymbol{\alpha}}|_\Omega\}.$$

In other words, $\mathcal{C}^m(\overline{\Omega})$ consists of all functions $f \in \mathcal{C}^m(\Omega)$ that, together with all their partial derivatives $\partial^{\boldsymbol{\alpha}} f$, $1 \leq |\boldsymbol{\alpha}| \leq m$, possess continuous extensions to $\overline{\Omega}$, or equivalently, such that, at each point $x_0 \in \partial\Omega$, $\lim_{x \to x_0} \partial^{\boldsymbol{\alpha}} f(x)$ exists in $\mathbb{R}$ for all $0 \leq |\alpha| \leq m$, or equivalently, when Ω is *bounded*, if each function $\partial^{\boldsymbol{\alpha}} f$, $0 \leq |\alpha| \leq m$, is uniformly continuous in Ω.

The subspace of $\mathcal{C}^m(\overline{\Omega})$ that consists of functions whose partial derivatives of order m satisfy a *Hölder condition of exponent* λ if $0 < \lambda < 1$ in Ω, or are *Lipschitz-continuous* in Ω if $\lambda = 1$ (Section 1.11), is denoted

$$\mathcal{C}^{m,\lambda}(\overline{\Omega}) := \{f \in \mathcal{C}^m(\overline{\Omega});\ \text{there exists } L \text{ such that } |\partial^{\boldsymbol{\alpha}} f(x) - \partial^{\boldsymbol{\alpha}} f(y)| \leq L\,|x - y|^{\lambda} \\ \text{for all } |\boldsymbol{\alpha}| = m \text{ and for all } x, y \in \Omega\}.$$

The *boundary* Γ of an open subset Ω of $\mathbb{R}^n$ is said to be **Lipschitz-continuous** if the following conditions are satisfied (see Figure 1.18-1 when $n = 2$): There exist constants $\alpha > 0$ and $L > 0$ and a *finite* number of *local coordinate systems*, with coordinates $\boldsymbol{\zeta}_r' = (\zeta_1^r, \zeta_2^r, \ldots, \zeta_{n-1}^r) \in \mathbb{R}^{n-1}$ and $\zeta_r = \zeta_n^r$, and corresponding functions $\theta_r : \omega_r := \{\boldsymbol{\zeta}_r \in \mathbb{R}^{n-1};\ |\boldsymbol{\zeta}_r| < \alpha\} \to \mathbb{R}$, $1 \leq r \leq s$, such that

$$\Gamma = \bigcup_{r=1}^{s} \{(\boldsymbol{\zeta}_r', \zeta_r);\ \boldsymbol{\zeta}_r' \in \omega_r \text{ and } \zeta_r = \theta_r(\boldsymbol{\zeta}_r')\},$$

$$|\theta_r(\boldsymbol{\zeta}_r') - \theta_r(\boldsymbol{\eta}_r')| \leq L\,|\boldsymbol{\zeta}_r' - \boldsymbol{\eta}_r'| \quad \text{for all } \boldsymbol{\zeta}_r', \boldsymbol{\eta}_r' \in \omega_r,\ \ 1 \leq r \leq s,$$

the last inequalities expressing the *Lipschitz-continuity of the mappings* θ_r. Note that, by a convenient abuse of notation, $\{(\boldsymbol{\zeta}_r', \zeta_r);\ \boldsymbol{\zeta}_r' \in \omega_r \text{ and } \zeta_r = \theta_r(\boldsymbol{\zeta}_r')\}$ designates the set formed by those points whose coordinates ζ_i^r, $1 \leq i \leq n$, *in the* rth *local coordinate system* satisfy $|(\zeta_1^r, \zeta_2^r, \ldots, \zeta_{n-1}^r)| < \alpha$ and $\zeta_n^r = \theta_r(\zeta_1^r, \zeta_2^r, \ldots, \zeta_{n-1}^r)$.

Remark While a Lipschitz-continuous boundary Γ is thus necessarily *bounded*, this is not necessarily true of the set Ω, which can be interchanged with the set $\mathbb{R}^n - \overline{\Omega}$ in the definition. □

Likewise, the boundary Γ is said to be **of class** $\mathcal{C}^m$, $m \geq 1$, if the mappings θ_r, $1 \leq r \leq s$, are in the space $\mathcal{C}^m(\omega_r)$.

More generally, a *subset* Γ_0 of Γ is said to be *Lipschitz-continuous*, resp. *of class* $\mathcal{C}^m$, if the same definitions apply with Γ replaced with Γ_0.

The *open set* Ω is said to be **locally on the same side of its boundary** Γ if in addition there exists a constant $\beta > 0$ such that

$$\{(\boldsymbol{\zeta}_r', \zeta_r);\ \boldsymbol{\zeta}_r' \in \omega_r \text{ and } \theta_r(\boldsymbol{\zeta}_r') < \zeta_r < \theta_r(\boldsymbol{\zeta}_r') + \beta\} \subset \Omega, \quad 1 \leq r \leq s,$$
$$\{(\boldsymbol{\zeta}_r', \zeta_r);\ \boldsymbol{\zeta}_r' \in \omega_r \text{ and } \theta_r(\boldsymbol{\zeta}_r') - \beta < \zeta_r < \theta_r(\boldsymbol{\zeta}_r')\} \subset \mathbb{R}^n - \overline{\Omega}, \quad 1 \leq r \leq s.$$

A **domain** Ω **in** $\mathbb{R}^n$ is a bounded connected open subset of $\mathbb{R}^n$ with a Lipschitz-continuous boundary Γ, the set Ω being locally on the same side of Γ (see Figure 1.18-1 and the counter-examples of Figure 1.18-2, in the case $n = 2$).

The possibility of giving another equivalent definition (Theorem 1.18-1) of the spaces $\mathcal{C}^m(\overline{\Omega})$ *when* Ω *is a domain* (instead of an arbitrary open subset in $\mathbb{R}^n$ as until now in this

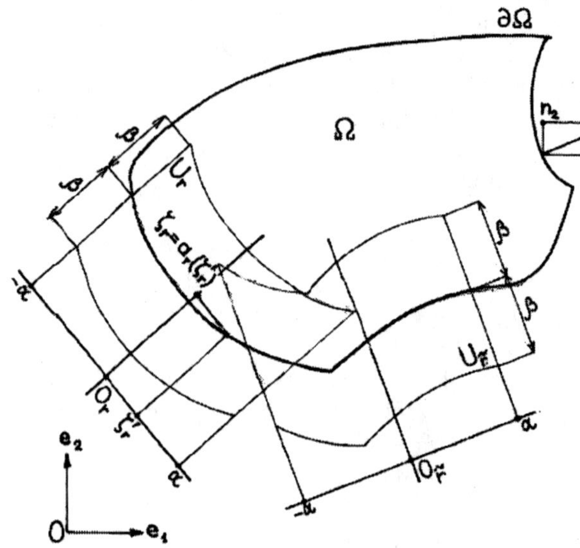

Figure 1.18-1 A domain in $\mathbb{R}^2$. This figure originally appeared in P.G. CIARLET [1988]: *Mathematical Elasticity, Volume* I: *Three-Dimensional Elasticity*, North–Holland, Amsterdam.

section) constitutes a crucial property of domains. Note that the next theorem[11] may be viewed as a generalization of the *Tietze–Urysohn extension theorem* (Theorem 1.7-7) for continuous functions ($m = 0$) to continuously differentiable functions ($m \geq 1$).

Theorem 1.18-1 *Let Ω be a domain in $\mathbb{R}^n$. Then, for any integer $m \geq 1$ and for $m = \infty$, the space $\mathcal{C}^m(\overline{\Omega})$ can be also defined as*

$$\mathcal{C}^m(\overline{\Omega}) = \{f|_\Omega;\ f \in \mathcal{C}^m(\mathbb{R}^n)\}. \qquad \square$$

The interest of Lipschitz-continuous boundaries is that, even though they are not too smooth, *surface integrals* can still be defined along them and *Green's formula* holds, as we now briefly indicate. We do not discuss the *measurability* of the function involved.

A function $f : \Gamma \to \mathbb{R}$ is $d\Gamma$-**almost everywhere defined** if each function $\zeta'_r \in \omega_r \to f(\zeta'_r, \theta_r(\zeta'_r))$, $1 \leq r \leq s$, is defined almost everywhere (in the sense of the $(n-1)$-dimensional Lebesgue measure) on the set ω_r. If in addition each function $\zeta'_r \in \omega_r \to f(\zeta'_r, \theta_r(\zeta'_r))$ is

[11]For a proof, see, e.g., STEIN [1970, Chapter 6], or:

P.G. CIARLET; C. MARDARE [2004]: Recovery of a manifold with boundary and its continuity as a function of its metric tensor, *Journal de Mathématiques Pures et Appliquées* **83**, 811–843.

As expected, the proof is somewhat delicate; in particular, it relies on a deep *extension theorem*, due to:

H. WHITNEY [1934]: Analytic extensions of differentiable functions defined in closed sets, *Transactions of the American Mathematical Society* **36**, 63–89.

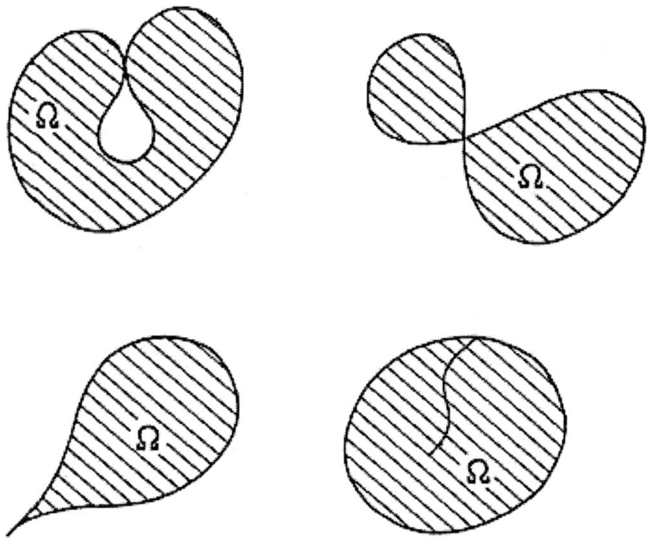

Figure 1.18-2 Examples of bounded connected open subsets $\Omega \subset \mathbb{R}^2$ that are not domains. This figure originally appeared in P.G. Ciarlet [1988]: *Mathematical Elasticity, Volume* I: *Three-Dimensional Elasticity*, North–Holland, Amsterdam.

Lebesgue integrable, i.e., if

$$\int_{\omega_r} \left| f(\zeta_r', \theta_r(\zeta_r')) \right| \mathrm{d}\zeta_r' < \infty,$$

the function f is said to be **integrable on** Γ, and the vector space formed by such functions is denoted

$$\mathcal{L}^1(\Gamma).$$

In order to define the *integral of a function* $f \in \mathcal{L}^1(\Gamma)$, we need a *partition of unity* associated with the covering of the boundary Γ by the open sets (Figure 1.18-3)

$$U_r := \{(\zeta_r', \zeta_r);\ \zeta_r' \in \omega_r \text{ and } \theta_r(\zeta_r') - \beta < \zeta_r < \theta_r(\zeta_r') + \beta\},$$

that is, a family of functions $\psi_r \in \mathcal{C}^\infty(\mathbb{R}^n)$, $1 \le r \le s$, that satisfy

$$\operatorname{supp} \psi_r \subset U_r \text{ and } 0 \le \psi_r \le 1,\ \ 1 \le r \le s, \quad \text{and} \quad \sum_{r=1}^{s} \psi_r(x) = 1 \text{ for all } x \in \Gamma.$$

Then the **surface integral** of a function $f \in \mathcal{L}^1(\Gamma)$ is *defined* as

$$\int_\Gamma f\,\mathrm{d}\Gamma := \sum_{r=1}^{s} \int_{\omega_r} f\left(\zeta_r', \theta_r(\zeta_r')\right) \psi_r\left(\zeta_r', \theta_r(\zeta_r')\right) \left(1 + \sum_{i=1}^{n-1} \left| \frac{\partial \theta_r}{\partial \xi_i^r} \right|^2 \right)^{1/2} \mathrm{d}\zeta_r',$$

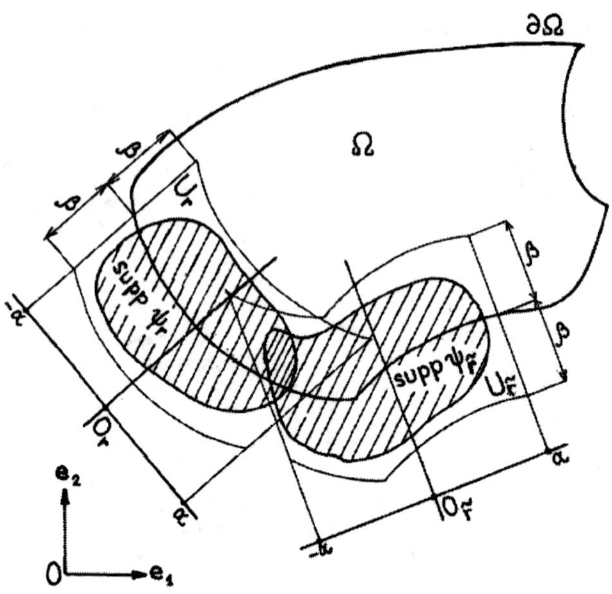

Figure 1.18-3 The supports of two functions ψ_r and $\psi_{\tilde{r}}$ in a partition of unity associated with the covering $\Gamma \subset \bigcup_{r=1}^{s} U_r$ of the boundary Γ of a domain in $\mathbb{R}^2$. This figure originally appeared in P.G. CIARLET [1988]: *Mathematical Elasticity, Volume* I: *Three-Dimensional Elasticity,* North–Holland, Amsterdam.

and $\mathrm{d}\Gamma$ is said to be the **area element** along Γ. This definition makes sense: *First,* by *Rademacher's theorem* (Theorem 1.14-2), the functions θ_r are almost everywhere (in the sense of the $(n-1)$-dimensional Lebesgue measure) differentiable since they are Lipschitz-continuous, and their partial derivatives satisfy

$$\left|\frac{\partial \theta_r}{\partial \zeta_i^r}(\boldsymbol{\zeta}_r')\right| \leq L \quad \text{for almost all } \boldsymbol{\zeta}_r' \in \omega_r, \;\; 1 \leq i \leq n-1, \;\; 1 \leq r \leq s.$$

Second, it is a simple exercise to verify that the $(n-1)$-*area,* according to the definition given in Section 1.17, of each surface $\Theta(\omega_r) \subset \mathbb{R}^n$, where $\Theta(\boldsymbol{\zeta}_r') := (\boldsymbol{\zeta}_r', \theta_r(\boldsymbol{\zeta}_r'))$ for each $\boldsymbol{\zeta}_r' \in \omega_r$, is given in this special case by $\int_{\omega_r} \left(1 + \sum_{i=1}^{n-1} \left|\frac{\partial \theta_r}{\partial \zeta_i^r}\right|^2\right)^{1/2} \mathrm{d}\boldsymbol{\zeta}_r'$, thus justifying the expression used for defining $\int_\Gamma f \,\mathrm{d}\Gamma$.

Third, it can be shown that the number $\int_\Gamma f \,\mathrm{d}\Gamma$ defined in this fashion is *independent of the local coordinate systems considered* and *independent of the partition of unity considered.*

The **area** of a Lipschitz-continuous subset Γ_0 of Γ is denoted and defined by

$$\mathrm{d}\Gamma\text{-meas}\,\Gamma_0 := \int_\Gamma \chi_{\Gamma_0} \,\mathrm{d}\Gamma, \quad \text{or} \quad \text{area}\,\Gamma_0 = \int_\Gamma \chi_{\Gamma_0} \,\mathrm{d}\Gamma \;\text{ if } n = 3,$$

where $\chi_{\Gamma_0} : \Gamma \to \mathbb{R}$ denotes the characteristic function of the set Γ_0.

Another important consequence of the almost everywhere differentiability of the functions θ_r is that a **unit outer normal vector field** $\boldsymbol{\nu} = (\nu_i)_{i=1}^n$ *exists* $\mathrm{d}\Gamma$*-almost everywhere along* Γ. Here, "unit" and "outer" respectively mean that for $\mathrm{d}\Gamma$-almost all $x \in \Gamma$, $|\boldsymbol{\nu}(x)| = 1$ and $\{x + t\boldsymbol{\nu}(x);\ 0 \le t < \varepsilon(x)\} \cap \Omega = \varnothing$ for some $\varepsilon(x) > 0$ and "normal" means that $\boldsymbol{\nu}(x)$ is normal to the tangent hyperplane to Γ, which, for the same reason, exists $\mathrm{d}\Gamma$-almost everywhere.

Another crucial property of *domains* is the validity of the following *fundamental Green's formula*, which is nothing but the multidimensional extension of the well-known integration by parts formula $\int_a^b f'(t)g(t)\,\mathrm{d}t = -\int_a^b f(t)g'(t)\,\mathrm{d}t + f(b)g(b) - f(a)g(a)$.

Theorem 1.18-2 (fundamental Green's formula) *Let Ω be a domain in $\mathbb{R}^n$ and let $\boldsymbol{\nu} = (\nu_i)_{i=1}^n$ denote the unit outer normal vector field along the boundary Γ of Ω. Then, given any functions $u, v \in \mathcal{C}^1(\overline{\Omega})$,*

$$\int_\Omega (\partial_i f) g\,\mathrm{d}x = -\int_\Omega f \partial_i g\,\mathrm{d}x + \int_\Gamma f g \nu_i\,\mathrm{d}\Gamma, \quad \textit{for each } 1 \le i \le n. \qquad \square$$

Using the fundamental Green's formula, one can prove other **Green's formulas** where, in essence, *a particular combination of integrals over Ω is written as a combination of surface integrals over Γ*. For example, let there be given a vector field $\boldsymbol{v} = (v_i) \in \mathcal{C}^1(\overline{\Omega}; \mathbb{R}^n)$; then the fundamental Green's formula shows that $\int_\Omega \partial_i v_i\,\mathrm{d}x = \int_\Gamma v_i \nu_i\,\mathrm{d}\Gamma$ for each $1 \le i \le n$. Consequently,

$$\int_\Omega \operatorname{div} \boldsymbol{v}\,\mathrm{d}x = \int_\Gamma \boldsymbol{v} \cdot \boldsymbol{\nu}\,\mathrm{d}\Gamma, \quad \text{where } \operatorname{div} \boldsymbol{v} := \sum_{i=1}^n \partial_i v_i.$$

This Green's formula constitutes the **divergence theorem for vector fields**.

CHAPTER 2

NORMED VECTOR SPACES

Introduction

Linear functional analysis constitutes the subject of Chapters 2–5.

More specifically, the aim of the present chapter is to establish basic properties that hold in any *normed vector space, complete or not.* Then Chapter 3 will be devoted to *complete* normed vector spaces and Chapter 4 to normed vector spaces, complete or not, whose norm is derived from an *inner product.* Finally, Chapter 5 will address more elaborate properties of these spaces, assembled under the appellation "great theorems."

Among the main notions introduced in this chapter are those of *continuous linear or multilinear operators* (Sections 2.9 and 2.11) and of *compact linear operators* (Section 2.10). Another key notion is that of *compactness*, which in particular characterizes *finite dimensionality*, as shown by the beautiful *F. Riesz theorem* (Theorem 2.7-3); compactness also lies at the heart of the proof of the *fundamental theorem of algebra* (Theorem 2.8-1).

Basic *examples of infinite-dimensional normed vector spaces* are introduced in this chapter, such as the space $\mathcal{C}(K;Y)$ of all continuous functions from a compact set K into a normed vector space Y (Section 2.3), the spaces ℓ^p, $1 \leq p \leq \infty$ (Section 2.4) and $L^p(\Omega)$, $1 \leq p \leq \infty$, with Ω an arbitrary open subset of $\mathbb{R}^n$ (Section 2.5), and the space $\mathcal{L}(X;Y)$ of all continuous linear operators from a normed vector space X into a normed vector space Y (Section 2.9). A detailed treatment is given in particular of the *approximation of functions in* $L^p(\Omega)$, $1 \leq p < \infty$, *by smooth functions*, by way of *mollifiers* (Section 2.6).

Applications include some basic results in *approximation theory*, such as the *Weierstraß approximation theorems* for continuous functions, either by means of usual polynomials (Theorems 2.13-3 and 2.15-2) or by means of trigonometric polynomials (Theorem 2.14-3): these theorems are given constructive proofs by means of *Korovkin's theorem* (Theorem 2.12-1) applied to *Bernstein polynomials* (Theorem 2.13-2) or *Fejér's trigonometric polynomials* (Theorem 2.14-2). It is shown how such results can be also derived from the more general, but more abstract, *Stone–Weierstraß theorems* (Theorems 2.15-1 and 2.15-3).

This chapter also includes an introduction to convexity (Sections 2.16 and 2.17), a notion that plays a crucial role in the projection theorem (Chapter 4), in the Banach–Saks–Mazur theorem (Chapter 5), in the characterization of minima (Chapter 7), or in the calculus of variations (Chapter 9).

2.1 Vector spaces; Hamel bases; dimension of a vector space

In what follows, $\mathbb{K}$ denotes either the field $\mathbb{R}$ or the field $\mathbb{C}$, and the elements of $\mathbb{K}$ are called *scalars*. A set X is a **vector space over** $\mathbb{K}$ if there exist two mappings:

$$(x, y) \in X \times X \to (x + y) \in X \quad \text{and} \quad (\alpha, x) \in \mathbb{K} \times X \to \alpha x \in X,$$

called respectively **addition** and **scalar multiplication**, that together satisfy the following properties:

$$x + y = y + x \quad \text{and} \quad x + (y + z) = (x + y) + z \quad \text{for all } x, y, z \in X;$$

there exists an element of X, denoted 0, such that $x + 0 = x$ for all $x \in X$; given any $x \in X$, there exists an element of X, denoted $(-x)$, such that $x + (-x) = 0$ (equipped with the addition, the set X is thus an Abelian group); and

$$\begin{aligned} \alpha(x + y) = \alpha x + \alpha y \quad &\text{and} \quad (\alpha + \beta)x = \alpha x + \beta x \qquad \text{for all } \alpha, \beta \in \mathbb{K} \text{ and } x, y \in X, \\ \alpha(\beta x) = (\alpha\beta)x \quad &\text{and} \quad 1x = x \qquad \text{for all } \alpha, \beta \in \mathbb{K} \text{ and } x \in X. \end{aligned}$$

These properties immediately imply the following consequences: The element 0 is unique; given any $x \in X$, the element $(-x)$ is unique; $-(-x) = x$ and $-(x + y) = (-x) + (-y)$ for all $x, y \in X$; $\lambda 0 = 0$ and $0x = 0$ and $(-\lambda)x = -(\lambda x)$ for all $\lambda \in \mathbb{K}$ and $x \in X$; if $x \neq 0$, then $\lambda x = 0$ implies $\lambda = 0$; a vector space is nonempty, since $0 \in X$; since the addition is associative, the notation $x+y+z := x+(y+z)$ is justified. The shorter notations $-x := (-x)$ and $x - y := x + (-y)$ are also used.

A **real vector space** is a vector space over $\mathbb{K} = \mathbb{R}$. A **complex vector space** is a vector space over $\mathbb{K} = \mathbb{C}$. A **vector space** is either a real vector space or a complex vector space.

The elements of X and $\mathbb{K}$ are respectively called **vectors** and **scalars**. The element $0 \in X$ is called the **origin**, or the **zero vector**, of X; in this respect, note that the same symbol 0 denotes both the zero vector of X and the zero of $\mathbb{K}$. If $X \neq \{0\}$, any vector $x \in X$ such that $x \neq 0$ is called a **nonzero vector** of X.

A **subspace** of a vector space X over $\mathbb{K}$ is any subset of X that is also a vector space over $\mathbb{K}$. In particular, $\{0\}$ is a subspace of X. A **proper subspace** Y of X is a subspace Y of X that satisfies $Y \subsetneq X$.

Let Y and Z be two subspaces of a vector space X. Then X is said to be the **direct sum** of Y and Z if any element $x \in X$ can be written as

$$x = y + z \quad \text{with } y \in Y \text{ and } z \in Z,$$

and such a decomposition is *unique*.

Another example of a subspace is the **subspace spanned by a subset** A of X, consisting of all **finite linear combinations** of vectors of A, i.e., vectors $x \in X$ of the form $x = \sum_{j \in J} \alpha_j a_j$, where the set J of indices is *finite*, and $\alpha_j \in \mathbb{K}$ and $a_j \in A$ for all $j \in J$. This subspace is denoted

$$\operatorname{Span} A.$$

If the subset A of X is of the form $A = \bigcup_{i=1}^{n} \{x_i\}$ or $A = \bigcup_{i \in I} \{x_i\}$, the subspace $\operatorname{Span} A$ is also denoted

$$\operatorname{Span}(x_i)_{i=1}^{n} \quad \text{or} \quad \operatorname{Span}(x_i)_{i \in I}.$$

The following notion was introduced by G. Hamel[1] (for the purpose of solving a particular functional equation; cf. Problem 2.1-1). Let $X \neq \{0\}$ be a vector space. Then a **Hamel basis** in X is any family $(e_i)_{i\in I}$ of vectors $e_i \in X$ (Section 1.3) that satisfies the following two properties:

First, the family is **linearly independent**, in the sense that, given any *finite* subfamily $(e_j)_{j\in J}$ of the family $(e_i)_{i\in I}$ and given any scalars $\alpha_j \in \mathbb{K}$, $j \in J$, such that $\sum_{j\in J} \alpha_j e_j = 0$, then $\alpha_j = 0$, $j \in J$. *Second*, $\text{Span}(e_i)_{i\in I} = X$, i.e., given any vector $x \in X$, there exists a *finite* subfamily $(e_j)_{j\in J(x)}$ of the family $(e_i)_{i\in I}$ and there exist scalars $x_j \in \mathbb{K}$, $j \in J(x)$, such that $x = \sum_{j\in J(x)} x_j e_j$. Note that the first property implies that all the vectors e_i, $i \in I$, of a Hamel basis are necessarily *nonzero* and *distinct*, and that, given any $x \in X$, the scalars x_j, $j \in J(x)$, are *uniquely determined*.

For instance, the family $(e_n)_{n=0}^{\infty}$, where $e_n(x) = x^n$, $x \in \mathbb{R}$, constitutes a Hamel basis in the space of all polynomials of one real variable.

As a first application of the *axiom of choice* (used here in the form of Zorn's lemma), we now establish the *existence of Hamel bases* in *any* vector space, together with a crucial property of their *cardinals*. Another related property, which extends to any vector space a well-known property of finite-dimensional spaces, is the object of Problem 2.1-2.

Theorem 2.1-1 *Let $X \neq \{0\}$ be a vector space.*

(a) *There exists a Hamel basis of X.*

(b) *Let E and F be two Hamel bases of X. Then* $\text{card}\, E = \text{card}\, F$.

Proof (i) Let $\mathcal{F}$ denote the set formed by all linearly independent families of vectors of X. Hence $\mathcal{F}$ is *nonempty*, since $\mathcal{F}$ contains $\{e\}$, where e is any nonzero vector of X. Furthermore, $\mathcal{F}$ is *partially ordered* by the relation $\preccurlyeq$, where $E = (e_i)_{i\in I} \preccurlyeq F = (e_j)_{j\in J}$ means that $\bigcup_{i\in I}\{e_i\} \subset \bigcup_{j\in J}\{e_j\}$. Since a family $E = (e_i)_{i\in I}$ can be identified with the subset $\bigcup_{i\in I}\{e_i\}$ (the elements of a linearly independent family are all distinct), the relation $E \preccurlyeq F$ is thus simply the *inclusion relation* $E \subset F$.

Let $\mathcal{E}$ be a *totally ordered* subset of $\mathcal{F}$. Then the family $G := \bigcup_{E\in\mathcal{E}} E$ is an element of $\mathcal{F}$, since any finite subfamily $(e_i)_{i=1}^m$ of G is a subfamily of some family $E \in \mathcal{E}$, because the set $\mathcal{E}$ is assumed to be totally ordered. Therefore, the vectors e_i, $1 \leq i \leq m$, are linearly independent. Besides, G is clearly an *upper bound* of $\mathcal{E}$, since $E \subset G$ for all $E \in \mathcal{E}$, by the very construction of G.

By *Zorn's lemma* (Theorem 1.3-1), the set $\mathcal{F}$ thus possesses a *maximal element* M, which is a *Hamel basis of* X. For otherwise, there would exist a nonzero vector $e \in X$ that cannot be written as a linear combination of elements of M. In this case, $M \cup \{e\}$ would be an element of $\mathcal{F}$ (clearly, $M \cup \{e\}$ is a linearly independent family) that satisfies $M \prec M \cup \{e\}$, in contradiction with the maximal character of M. This proves (a).

(ii) Let next $E = \bigcup_{i\in I}\{e_i\}$ and $F = \bigcup_{j\in J}\{f_j\}$ be two Hamel bases of X. In particular then, each vector e_i of the basis E can be written as a finite linear combination of elements f_j, $j \in J(i)$, i.e., where $J(i)$ is a finite subset of the set J.

Then we claim that $F = \bigcup_{i\in I} F_i$, where $F_i := \bigcup_{j\in J(i)}\{f_j\}$. To see this, assume that there exists $j_0 \in J$ such that $f_{j_0} \notin \bigcup_{i\in I} F_i$. Then $F - \{f_{j_0}\}$ would be a basis since E is a basis, in

[1]G. Hamel [1905]: Eine Basis aller Zahlen und die unstetigen Lösungen der Funktionalgleichung $f(x+y) = f(x) + f(y)$, *Mathematische Annalen* **60**, 459–462.

contradiction with the assumption that F is a basis. Hence $F = \bigcup_{i\in I} F_i$.

Assume first that one of the bases, say E, is *finite*. Then the relation $F = \bigcup_{i\in I} F_i$ shows that the basis F is also finite (the sets I and $J(i)$, $i \in I$, are all finite). Hence $\operatorname{card} E = \operatorname{card} F$ in this case.[2]

Assume next that the basis E, or equivalently the set I, is *infinite*. Then, for each $i \in I$, there exists a surjection $f_i : \mathbb{N} \to F_i$ (since the set F_i is finite), and thus the mapping $(i,n) \in I \times \mathbb{N} \to f_i(n) \in \bigcup_{i\in I} F_i = F$ is also a surjection. This implies that $\operatorname{card} F \preccurlyeq \operatorname{card}(I \times \mathbb{N})$ (Theorem 1.5-1). But $\operatorname{card} \mathbb{N} \preccurlyeq \operatorname{card} I$ since the set I is infinite (Theorem 1.5-3(a)), so that

$$\operatorname{card}(I \times \mathbb{N}) \preccurlyeq \operatorname{card}(I \times I) = \operatorname{card} I$$

(Theorem 1.5-3(b)). Therefore $\operatorname{card} F \preccurlyeq \operatorname{card} I = \operatorname{card} E$. A similar argument shows that $\operatorname{card} E \preccurlyeq \operatorname{card} F$. Hence $\operatorname{card} E = \operatorname{card} F$ also in this case. □

A vector space X is **finite-dimensional**, *resp.* **infinite-dimensional**, if there exists a finite, *resp.* infinite, Hamel basis of X, and its **dimension**, denoted

$$\dim X,$$

is then the cardinal of any one of its Hamel bases (this definition makes sense since any two Hamel bases of a given vector space have the same cardinal by Theorem 2.1-1(b)). A Hamel basis of a *finite-dimensional* vector space X is simply called a **basis**.

A Hamel basis thus generalizes to arbitrary vector spaces the notion of a basis in a finite-dimensional vector space.

The space $\mathcal{P}$ of real polynomials $p : x \in \mathbb{R} \to p(x) = \sum_{j=0}^{n} c_j x^j$ of *arbitrary* degree $n \geq 0$ provides an example of an *infinite-dimensional* vector space, since the family $\mathcal{H} := (e_j)_{j=0}^{\infty}$, where e_j denotes the polynomial $x \in \mathbb{R} \to x^j$, $j \geq 0$, is a *Hamel basis of* $\mathcal{P}$, called the **canonical basis** of $\mathcal{P}$. Besides, $\dim \mathcal{P} = \operatorname{card} \mathcal{H} = \operatorname{card} \mathbb{N}$ in this case.

Remark We will show (Theorem 5.1-4) that, by contrast, the cardinal of a Hamel basis H of any infinite-dimensional *complete* normed vector space always satisfies $\operatorname{card} \mathbb{N} \prec \operatorname{card} H$. □

Problems

2.1-1 (1) Describe the set $\mathcal{F}$ of all functions $f : \mathbb{R} \to \mathbb{R}$ that satisfies the functional equation $f(x+y) = f(x) + f(y)$ for all $x, y \in \mathbb{R}$.

Hint: Use a Hamel basis of $\mathbb{R}$, considered as a vector space over the field $\mathbb{Q}$.

(2) What is the cardinal of the set $\mathcal{F}$?

2.1-2 Let $X \neq \{0\}$ be a vector space and let $(e_j)_{j\in J}$ be any linearly independent family of elements $e_i \in X$. Show that there exists a Hamel basis of X that contains the family $(e_j)_{j\in J}$ as a subfamily.

---

[2]The reader is assumed to be already familiar with the basic properties of finite-dimensional spaces, such as this one.

2.2 Normed vector spaces; first properties and examples; quotient spaces

Let X be a vector space over $\mathbb{K}$, where either $\mathbb{K} = \mathbb{R}$ or $\mathbb{K} = \mathbb{C}$. A **norm** on X is any mapping $\|\cdot\| : X \to \mathbb{R}$ that satisfies the following properties:

$$\begin{aligned} \|x\| &\geq 0 && \text{for all } x \in X \text{ and } \|x\| = 0 \text{ if and only if } x = 0, \\ \|\alpha x\| &= |\alpha|\|x\| && \text{for all } \alpha \in \mathbb{K} \text{ and } x \in X, \\ \|x + y\| &\leq \|x\| + \|y\| && \text{for all } x, y \in X, \end{aligned}$$

the last property constituting the **triangle inequality**. A **normed vector space** is a pair $(X, \|\cdot\|)$, where X is a vector space and $\|\cdot\|$ is a norm on X.

Occasionally, we shall also need the following weaker definition. Let X be a vector space over $\mathbb{K}$. A **seminorm** on X is any mapping $|\cdot| : X \to \mathbb{R}$ that satisfies the following properties:

$$\begin{aligned} |x| &\geq 0 && \text{for all } x \in X, \\ |\alpha x| &= |\alpha|\,|x| && \text{for all } \alpha \in \mathbb{K} \text{ and } x \in X, \\ |x + y| &\leq |x| + |y| && \text{for all } x \in X. \end{aligned}$$

Let $(X, \|\cdot\|)$ be a normed vector space. The inequalities

$$\begin{aligned} \big|\|x\| - \|y\|\big| &\leq \|x - y\| && \text{for all } x, y \in X, \\ \Big\|\sum_{i=1}^{n} x_i\Big\| &\leq \sum_{i=1}^{n} \|x_i\| && \text{for all } x_i \in X,\ 1 \leq i \leq n, \end{aligned}$$

and the following property are immediate consequences of the definition of a norm.

Theorem 2.2-1 *Let $(X, \|\cdot\|)$ be a normed vector space. Then the mapping $d : X \times X \to \mathbb{R}$ defined by $d(x, y) = \|x - y\|$ for all $x, y \in X$ is a distance on X.* □

Equipped with the above distance d, a normed vector space $(X, \|\cdot\|)$ thus becomes a *metric space* (X, d). The topology induced on X by this distance (Section 1.10) is then called the **topology induced on X by the norm $\|\cdot\|$**, or the **norm topology of X**, or the **strong topology**.

Unless otherwise stated, a normed vector space will be always considered as equipped with its norm topology.

Remark Later on (Section 5.12), we shall see that any *infinite-dimensional* vector space can be also equipped with an equally important, but *different*, topology, called the *weak topology*. □

Let X be a vector space equipped with a topology. Then its topology is said to be **normable** if it can be induced by a norm on X. Examples of topologies on a vector space that are *not* normable are provided in Problems 2.3-2 and 2.3-3.

The norms defined in the next theorem are the most commonly used in *finite-dimensional vector spaces*, which thus provide our first *examples* of normed vector spaces. The notation $\|\cdot\|_\infty$ is justified in Problem 2.4-1.

Theorem 2.2-2 *Let X be a finite-dimensional vector space over $\mathbb{K} = \mathbb{R}$ or $\mathbb{K} = \mathbb{C}$, and let $(e_i)_{i=1}^n$ denote a basis of X.*

(a) *For each extended real number $1 \le p \le \infty$, the mapping $\|\cdot\|_p$ defined by*

$$x = \sum_{i=1}^n x_i e_i \in X \to \|x\|_p = \left(\sum_{i=1}^n |x_i|^p\right)^{1/p} \quad \text{if } 1 \le p < \infty,$$

$$x = \sum_{i=1}^n x_i e_i \in X \to \|x\|_\infty = \max_{1 \le i \le n} |x_i| \quad \text{if } p = \infty,$$

is a norm on X.

(b) *For each $1 \le p \le \infty$, the space $(X, \|\cdot\|_p)$ is separable.*

Proof In the proof of (a), the only nontrivial property is the triangle inequality when $1 < p < \infty$, itself a special case of the more general *Minkowski inequality for sequences* established in the proof of Theorem 2.4-1 below (to which the reader is therefore referred).

To prove (b) for all $1 \le p \le \infty$, it suffices to notice that the countably infinite set $\{\sum_{i=1}^n y_i e_i \in X;\ y_i \in \mathbb{Q}, 1 \le i \le n\}$ if $\mathbb{K} = \mathbb{R}$, or the countably infinite set $\{\sum_{i=1}^n y_i e_i \in X;\ \operatorname{Re} y_i \in \mathbb{Q}$ and $\operatorname{Im} y_i \in \mathbb{Q}$, $1 \le i \le n\}$ if $\mathbb{K} = \mathbb{C}$, is dense in the space $(X, \|\cdot\|_p)$, which is thus separable. □

Remark We shall see later that, in fact, *any* finite-dimensional normed vector space is separable (Theorem 2.7-1). □

Note that the distances $d_p : X \times X \to \mathbb{R}, 1 \le p \le \infty$, associated with these norms, i.e., defined by $d_p(x, y) = \|x - y\|_p$ for all $(x, y) \in X \times X$, are none other than the distances introduced in Section 1.10. The norm $\|\cdot\|_2$ is called the **Euclidean norm**; for brevity, it will be simply denoted

$$|\cdot| := \|\cdot\|_2,$$

whenever no confusion should arise.

For each integer $n \ge 2$, the vector space $\mathbb{K}^n$, which consists of all n-tuples $(x_i)_{i=1}^n$ of scalars $x_i \in \mathbb{K}$, thus becomes a *normed vector space* when it is equipped with one of the norms $\|\cdot\|_p$, $1 \le p \le \infty$, and the topology induced on $\mathbb{K}^n$ by any one of these norms is the *usual topology of* $\mathbb{K}^n$ (Section 1.10), which is thus *normable* (an analogous topology can be defined in the vector space of $n \times n$ *matrices*, once it is identified with the space $\mathbb{K}^{n^2}$; cf. Problem 2.2-1).

Another *example* of normed vector space is provided by a *product* $X = X_1 \times X_2 \times \cdots \times X_n$ of normed vector spaces on the *same* field $\mathbb{K}$, when X is equipped with any one of the following norms:

$$x = (x_j)_{j=1}^n \to \left(\sum_{j=1}^n \|x_j\|_{X_j}^p\right)^{1/p} \quad \text{for any } 1 \le p < \infty,$$

$$x = (x_j)_{j=1}^n \to \max_{1 \le j \le n} \|x_j\|_{X_j},$$

each of which induces the *product topology* on X.

Let next X be a vector space over $\mathbb{K} = \mathbb{R}$ or $\mathbb{K} = \mathbb{C}$ and let Z be a subspace of X. It is immediately verified that the relation

$$x \sim y \quad \text{if and only if} \quad (x - y) \in Z$$

is an *equivalence relation* (Section 1.1) on X. Let

$$[x] = \{y \in X;\ (x - y) \in Z\} = \{(x - z) \in X;\ z \in Z\} \subset \mathcal{P}(X)$$

denote the equivalence class of x *modulo* this relation.

It is then readily seen that the *quotient set* X/Z (the set formed by all the above equivalence classes; cf. again Section 1.1) becomes also a vector space over $\mathbb{K}$, called the **quotient space** X/Z, if the addition and scalar multiplication are respectively defined by

$$[x] + [y] = [x + y] \quad \text{and} \quad \alpha\,[x] = [\alpha x] \quad \text{for all } x, y \in X \text{ and } \alpha \in \mathbb{K}$$

and the zero vector in X/Z is $[0] = Z$. When there is no ambiguity about the definition of the space Z, we will also use the notation

$$[X] := X/Z.$$

Remark The equivalence class $[x]$ of $x \in X$ and the quotient space $[X]$ will be also denoted $\dot{x}$ and $\dot{X}$ at other places. □

For instance, let e_1, e_2, e_3 denote the canonical basis in $\mathbb{R}^3$. Then the quotient space $X/\operatorname{span} e_1$ is the (real) vector space formed by all straight lines parallel to the line $\operatorname{Span} e_1$; the quotient space $X/\operatorname{span}(e_1, e_2)$ is the (real) vector space formed by all planes parallel to the plane $\operatorname{Span}(e_1, e_2)$, etc.

If the space X is a normed vector space and Z is a *closed* subspace of X, the quotient space X/Z provides a *basic example* of a normed vector space:

Theorem 2.2-3 *Let $(X, \|\cdot\|_X)$ be a normed vector space and let Z be a closed subspace of X. Then the mapping $\|\cdot\| : X/Z \to \mathbb{R}$ defined by*

$$\|[x]\| := \inf_{y \in [x]} \|y\|_X = \inf_{z \in Z} \|x - z\|_X$$

is a norm over the quotient space X/Z, called the **quotient norm**.

Proof That $\|[x]\| \geq 0$ for all $[x] \in X/Z$ and $\|[0]\| = 0$ is clear. If $[x] \in X/Z$ satisfies $\|[x]\| = \inf_{z \in Z} \|x - z\|_X = 0$, then $x \in \overline{Z}$; but Z is closed, so that $x \in Z$. Hence $[x] = Z$, which is the zero vector in X/Z. Besides,

$$\|\alpha[x]\| = \|[\alpha x]\| = \inf_{z \in Z} \|\alpha x - z\|_X = \inf_{u \in Z} \|\alpha(x - u)\|_X = |\alpha| \inf_{z \in Z} \||x - z\|_X = |\alpha|\, \|[x]\|$$

$$\begin{aligned} \|[x] + [y]\| = \|[x + y]\| &= \inf_{z \in Z} \|x + y - z\|_X = \inf_{u, v \in Z} \|(x - u) + (y - v)\|_X \\ &\leq \inf_{u, v \in Z} (\|x - u\|_X + \|y - v\|_X) = \inf_{u \in Z} \|x - u\|_X + \inf_{v \in Z} \|y - v\|_X = \|[x]\| + \|[y]\| \end{aligned}$$

for all $\alpha \in \mathbb{K}$ and $x, y \in X$. Thus $\|\cdot\|$ indeed defines a norm on the quotient space. □

Since it is both a topological and a metric space, a normed vector space $(X, \|\cdot\|)$ inherits all the definitions and properties of metric spaces that were recalled in Chapter 1. In particular:

A **ball** with **center** $x \in X$ and **radius** $r > 0$ in X is any subset of X of the form

$$B(x; r) = \{y \in X;\ \|y - x\| < r\}$$

for some $x \in X$ and $r > 0$, which is thus *open* in X (Theorem 1.10-1); the **unit ball** is the particular ball $B(0; 1) = \{x \in X;\ \|x\| < 1\}$.

A subset A of X is open if and only if, given any point $x \in A$, there exists a ball $B(x; r)$ contained in A.

A sequence $(x_n)_{n=0}^{\infty}$ of vectors x_n in X **converges to** $x \in X$ if

$$\|x_n - x\| \to 0 \quad \text{as } n \to \infty.$$

Note in passing that a convergent sequence (as defined above) is also said to **strongly converge**, especially when this "norm-convergence" is to be distinguished from the *weak convergence* (Section 5.12).

A subset A of $(X, \|\cdot\|)$ is **bounded** if there exists M such that $\|x\| \leq M$ for all $x \in A$.

By contrast, the following notions are *specific to normed vector spaces.* Given any $x \in X$ and any $r > 0$, the closure

$$\overline{B(x; r)} = \{y \in X;\ \|y - x\| \leq r\}$$

of the (open) ball $B(x; r)$ is called the **closed ball** with **center** x and **radius** r, or simply the **closed unit ball** if $x = 0$ and $r = 1$; the boundary

$$\partial B(x; r) = \{y \in X;\ \|y - x\| = r\}$$

of the ball $B(x; r)$ is called the **sphere** with **center** x and **radius** r, or simply the **unit sphere** if $x = 0$ and $r = 1$.

But, because a normed vector space X is endowed with two specific operations and its distance d (constructed as in Theorem 2.2-1) is "compatible with these operations," in the sense that it satisfies $d(x + z, y + z) = d(x, y)$ and $d(\lambda x, \lambda y) = |\lambda| d(x, y)$ for all $x, y, z \in X$ and all $\lambda \in \mathbb{K}$, the space X is "much more" than an arbitrary metric space, and *a fortiori* than an arbitrary topological space. Accordingly, our main objective in this and the following chapters will be to study the *additional* topological, or metric, or otherwise, properties that are specific to normed vector spaces.

In this direction, we begin with a definition: Two norms $\|\cdot\|$ and $\|\cdot\|'$ on a given vector space X are said to be **equivalent** if the topologies induced on X by $\|\cdot\|$ and $\|\cdot\|'$ are identical. The next theorem then provides a simple, yet basic, criterion for the equivalence of two norms.

Theorem 2.2-4 *Two norms $\|\cdot\|$ and $\|\cdot\|'$ on a vector space X are equivalent if and only if there exist constants C and C' such that*

$$\|x\|' \leq C\|x\| \quad \textit{and} \quad \|x\| \leq C'\|x\|' \quad \textit{for all } x \in X.$$

Proof (i) Assume that $\|\cdot\|$ and $\|\cdot\|'$ are equivalent norms. Hence the identity mapping $\mathrm{id} : (X, \|\cdot\|) \to (X, \|\cdot\|')$ is continuous (since the open sets are the same; cf. Theorem 1.7-3). Then in particular the inverse image $\mathrm{id}^{-1}(B')$ of the set $B' := \{y \in X;\ \|y\|' < 1\}$, which is open in $(X, \|\cdot\|')$, is an open set of $(X, \|\cdot\|)$ that contains 0 (since $I(0) = 0 \in B'$). There thus exists a constant $C > 0$ such that the closure of the set $\{y \in X;\ \|y\| < \frac{1}{C}\}$ is contained in $\mathrm{id}^{-1}(B')$. Therefore,

$$\|y\| \leq \frac{1}{C} \quad \text{implies} \quad \|y\|' \leq 1.$$

Given any nonzero vector $x \in X$, the vector $y := \dfrac{1}{C\|x\|}x$ satisfies $\|y\| = \dfrac{1}{C}$, and hence $\|y\|' = \dfrac{1}{C\|x\|}\|x\|' \leq 1$. The inequality $\|x\|' \leq C\|x\|$ thus holds for all $x \in X$. The other inequality follows by the same argument.

(ii) Assume that $\|x\|' \leq C\|x\|$ for all $x \in X$. Then this inequality implies that the closure of any ball centered at any point $y \in X$ and of radius r in the metric space $(X, \|\cdot\|')$ contains a ball centered at $y \in X$ and of radius r/C in the metric space $(X, \|\cdot\|)$. Hence any open set for the topology induced by $\|\cdot\|'$ is open for the topology induced by $\|\cdot\|$. The other implication follows by the same argument. □

The next theorem gathers other elementary, and constantly used, properties of a normed vector space (understood as equipped with its norm topology).

Theorem 2.2-5 *Let $(X, \|\cdot\|)$ be a normed vector space over $\mathbb{K}$. Then the mappings*

$$\|\cdot\| : x \in X \to \|x\| \in \mathbb{R},$$
$$(x, y) \in X \times X \to (x + y) \in X,$$
$$(\alpha, x) \in \mathbb{K} \times X \to X$$

are continuous.

Proof The continuity of the mapping $x \in X \to \|x\| \in \mathbb{R}$ follows from the inequality

$$\big|\|x\| - \|\widetilde{x}\|\big| \leq \|x - \widetilde{x}\|.$$

The continuity of the last two mappings follows from the inequality

$$\|(x + y) - (\widetilde{x} + \widetilde{y})\| \leq \|x - \widetilde{x}\| + \|y - \widetilde{y}\|,$$
$$\|\alpha x - \widetilde{\alpha}\widetilde{x}\| \leq |\widetilde{\alpha}|\, \|x - \widetilde{x}\| + |\alpha - \widetilde{\alpha}|\, \|\widetilde{x}\| + |\alpha - \widetilde{\alpha}|\, \|x - \widetilde{x}\|,$$

combined with the definition of the product topology (Section 1.6) and the boundedness of a convergent sequence (for the last mapping). □

A **topological vector space** is a vector space equipped with a topology that makes both the addition and scalar multiplication continuous mappings. Theorem 2.2-5 thus shows that *a normed vector space is a topological vector space.*

As a first application of Theorem 2.2-5, we establish an interesting property of *open* subsets in a *normed vector space* (this property does not necessarily hold in an arbitrary topological space); the notions of connectedness used here are found in Section 1.9.

Theorem 2.2-6 *Let X be a normed vector space and let A be an open subset of X. Then the connected components of A are open in X.*

Proof Let C be a connected component of A and let $x \in C$. Since $C \subset A$ and A is open, there exists a ball $B(x;r)$ contained in A. Given $y, z \in B(x;r)$, define a mapping $\gamma : [0,1] \to X$ by

$$\gamma(\lambda) := (1-\lambda)y + \lambda z, \quad 0 \le \lambda \le 1.$$

Then γ maps the interval $[0,1]$ into $B(x;r)$, since

$$\begin{aligned}\|\gamma(\lambda) - x\| &= \|(1-\lambda)(y-x) + \lambda(z-x)\| \\ &\le (1-\lambda)\,\|y-x\| + \lambda\,\|z-x\| < r \quad \text{for all } \lambda \in [0,1],\end{aligned}$$

and $\gamma : [0,1] \to B(x;r)$ is continuous by Theorem 2.2-5.

Hence γ *is a path joining* y *to* z, which implies that $B(x;r)$ is arcwise-connected, and hence connected. As the largest connected set containing x, the set C thus contains $B(x;r)$. Consequently, C is open. □

Normed vector spaces that are *separable* possess an interesting property (often used later):

Theorem 2.2-7 *Let X denote a separable normed vector space. Then there exists a countably infinite family $(X_n)_{n=1}^{\infty}$ of finite-dimensional subspaces of X such that*

$$\dim X_n = n \quad \textit{and} \quad X_n \subset X_{n+1}, \quad n \ge 1, \quad \textit{and} \quad \overline{\bigcup_{n=1}^{\infty} X_n} = X.$$

Proof Let $x_k \in X$, $k \ge 1$, be such that

$$\overline{\bigcup_{k=1}^{\infty} \{x_k\}} = X.$$

First, we note that there is no loss of generality in assuming that $x_k \ne 0$ for all $k \ge 1$ (otherwise let $K := \{k \ge 1;\ x_k \ne 0\}$, and let $\widetilde{x}_k \in X$, $k \ge 1$, be such that $\widetilde{x}_k \ne 0$ for all $k \ge 1$ and $\widetilde{x}_k \to 0$ as $k \to \infty$; then the countable family $\Big(\bigcup_{k\in K}\{x_k\}\Big) \cup \Big(\bigcup_{k=1}^{\infty}\{\widetilde{x}_k\}\Big)$ is dense in X). This being the case, let the vectors $e_k \in X$, $k \ge 1$, be recursively defined by

$$\begin{aligned} e_1 &:= x_{\sigma(1)} \quad \text{with } \sigma(1) := 1, \\ e_k &:= x_{\sigma(k)} \quad \text{with } \sigma(k) := \min\{m \ge \sigma(k-1)+1;\ x_m \notin \operatorname{Span}(e_\ell)_{\ell=1}^{k-1}\}, \quad k \ge 2. \end{aligned}$$

Then the subspaces defined by

$$X_n := \operatorname{Span}(e_k)_{k=1}^{n}$$

clearly possess all the required properties (that $\overline{\bigcup_{n=1}^{\infty} X_n} = X$ follows from the inclusion $\bigcup_{k=1}^{\infty}\{x_k\} \subset \bigcup_{n=1}^{\infty} X_n$). □

We conclude this section by showing that it is always possible to endow *any* vector space with a norm. Not unexpectedly, the remarkable generality of this result has its price, viz., the inevitable recourse to the *axiom of choice* (by way of Theorem 2.1-1).

Theorem 2.2-8 *Any vector space can be normed.*

Proof Given any vector space X over $\mathbb{K}$, let $(e_i)_{i\in I}$ be a Hamel basis of X (Theorem 2.1-1). Given any vector $x \in X$, there thus exist a unique *finite* subset $I(x)$ of I and uniquely determined scalars $x_i \in \mathbb{K}$, $i \in I(x)$, such that $x = \sum_{i\in I(x)} x_i e_i$. It is then immediately verified that (for instance) the mapping

$$x = \sum_{i\in I(x)} x_i e_i \in X \to \sum_{i\in I(x)} |x_i|$$

is a norm on X. □

Problems

2.2-1 (1) Show that the set of all invertible real matrices of order n is open in the set $\mathbb{M}^n$ of all real matrices of order n, identified here with $\mathbb{R}^{n^2}$ equipped with its usual topology.

(2) Show that the set $\mathbb{S}^n$ of all real symmetric matrices of order n is closed in $\mathbb{M}^n$.

(3) Show that the set $\mathbb{S}^n_>$ of all real symmetric and positive-definite matrices of order n is open in $\mathbb{S}^n$ equipped with the topology induced by that of $\mathbb{M}^n$.

(4) What can be said of $\mathbb{S}^n_>$ as a subset of $\mathbb{M}^n$?

(5) Show that $\{\boldsymbol{A} \in \mathbb{M}^n;\ \det \boldsymbol{A} > 0\}$ is a connected subset of $\mathbb{M}^n$.

2.2-2 Show that any connected open subset of a normed vector space is arcwise-connected (Section 1.9).

2.2-3 Is the following proposition true or false? Let $\|\cdot\|$ and $\|\cdot\|'$ be two norms on the same vector space X, and let $(x_n)_{n=1}^{\infty}$ be a sequence of elements $x_n \in X$ such that $\lim_{n\to\infty} x_n = x$ in $(X, \|\cdot\|)$ and $\lim_{n\to\infty} x_n = x'$ in $(X, \|\cdot\|')$. Then $x = x'$.

2.2-4 Let K be a compact subset of a normed vector space $(X, \|\cdot\|)$.

(1) Show that, given any $x \in X$, there exists $y \in K$ such that $\|x - y\| = \inf_{z\in K} \|x - z\|$.

(2) Show that, if in addition y is unique for each $x \in X$, the mapping $P : X \to K$ defined by $\|x - Px\| = \inf_{z\in K} \|x - z\|$ for each $x \in X$ is continuous.

2.3 The space $\mathcal{C}(K;Y)$ with K compact; uniform convergence and local uniform convergence

We now define another *basic example* of a normed vector space, viz., that formed by continuous functions on a compact set. Further basic examples will be given later, such as the spaces ℓ^p (Section 2.4) and $L^p(\Omega)$ (Section 2.5), $1 \le p \le \infty$.

Notations such as $\mathcal{C}(K;Y)$ and $\mathcal{C}(K)$ have been defined in Section 1.7.

Theorem 2.3-1 *Let K be a compact topological space and let $(Y; \|\cdot\|)$ be a normed vector space. Then $\mathcal{C}(K;Y)$ is a vector space, and the function $|||\cdot||| : \mathcal{C}(K;Y) \to \mathbb{R}$ defined by*

$$|||f||| := \sup_{x\in K} \|f(x)\| \quad \textit{for each } f \in \mathcal{C}(K;Y),$$

is a norm on $\mathcal{C}(K;Y)$.

Proof That $\mathcal{C}(K;Y)$ is a vector space is clear. That $\sup_{x\in K}\|f(x)\| < \infty$ follows from Theorem 1.13-6, which can be applied since K is compact and the function $x \in K \to \|f(x)\|$ is continuous, as a composite mapping of continuous functions (Theorems 1.7-2 and 2.2-5). Finally, that $|||\cdot|||$ is a norm is immediately verified. □

The norm $|||\cdot|||$ on the space $\mathcal{C}(K;Y)$ introduced in Theorem 2.3-1 is called the **sup-norm**. A sequence $(f_n)_{n=1}^{\infty}$ of functions $f_n \in \mathcal{C}(K;Y)$ is said to **converge uniformly** as $n \to \infty$ to a function $f \in \mathcal{C}(K;Y)$ if $\lim_{n\to\infty} |||f_n - f||| = 0$, i.e., if

$$\lim_{n\to\infty}\left(\sup_{x\in K}\|f_n(x) - f(x)\|\right) = 0.$$

In the important special cases where $Y = \mathbb{R}$ or $Y = \mathbb{C}$, the sup-norm on the space $\mathcal{C}(K)$, or the space $\mathcal{C}(K;\mathbb{C})$, is denoted $\|\cdot\|$. In other words,

$$\|f\| := \sup_{x\in K}|f(x)| \quad \text{for all } f \in \mathcal{C}(K), \text{ or for all } f \in \mathcal{C}(K;\mathbb{C}).$$

The space $(\mathcal{C}(\overline{\Omega}), \|\cdot\|)$, where Ω is a *bounded* open subset of $\mathbb{R}^n$ and $\|\cdot\|$ denotes the *sup-norm*, thus defined in this case by

$$\|f\| := \sup_{x\in\overline{\Omega}}|f(x)| \quad \text{for all } f \in \mathcal{C}(\overline{\Omega}),$$

provides a fundamental example of such a space, including when $n = 1$ and $\Omega =]a,b[\subset \mathbb{R}$ (as will be abundantly illustrated later in this chapter). For notational brevity, we shall let in this case

$$\mathcal{C}[a,b] := \mathcal{C}([a,b]).$$

Remark By contrast, the "seemingly similar" space $\mathcal{C}(\Omega)$, where Ω is any open subset of $\mathbb{R}^n$ (bounded or not), is in effect quite different, since its "natural" topology is *not normable*, although it is *metrizable* (Problem 2.3-2). □

It is likewise clear that, for each integer $m \geq 1$, the space

$$\mathcal{C}^m(\overline{\Omega}) = \{f \in \mathcal{C}^m(\Omega);\ \text{for each } |\boldsymbol{\alpha}| \leq m, \text{ there exists } g^{\boldsymbol{\alpha}} \in \mathcal{C}(\overline{\Omega}) \text{ such that } \partial^{\boldsymbol{\alpha}} f = g^{\boldsymbol{\alpha}}|_{\Omega}\}$$

(Section 1.18), where Ω is again a *bounded* open subset of $\mathbb{R}^n$, becomes a *normed vector space* when it is equipped with the norm $\|\cdot\|_{\mathcal{C}^m(\overline{\Omega})}$ defined by

$$\|f\|_{\mathcal{C}^m(\overline{\Omega})} := \max_{0\leq|\boldsymbol{\alpha}|\leq m}\sup_{x\in\overline{\Omega}}|g^{\boldsymbol{\alpha}}(x)| = \max_{0\leq|\boldsymbol{\alpha}|\leq m}\sup_{x\in\Omega}|\partial^{\boldsymbol{\alpha}} f(x)| \quad \text{for each } f \in \mathcal{C}^m(\overline{\Omega}).$$

The notion of uniform convergence is in fact not restricted to *continuous* mappings defined on a compact space and taking their values in a normed vector space (the situation described in Theorem 2.3-1). More specifically, let X be *any set* and let Y be a *normed vector space*. Then a sequence $(f_n)_{n=1}^{\infty}$ of mappings $f_n : X \to Y$ is said to **converge uniformly on** X to a mapping $f : X \to Y$ as $n \to \infty$ if

$$\lim_{n\to\infty}\left(\sup_{x\in X}\|f_n(x) - f(x)\|\right) = 0.$$

Note that, in this definition, the functions f_n, $n \geq 1$, and f may be *unbounded*. Consider for instance the functions $f_n : x \in]0, \infty[\to \frac{1}{x} + \frac{1}{n}$, $n \geq 1$, and $f : x \in]0, \infty[\to \frac{1}{x}$.

This more general notion of uniform convergence can be viewed as a convergence with respect to a *norm topology* if the functions f_n, $n \geq 1$, and f are *bounded*, according to the following result (whose proof is straightforward and for this reason omitted):

Theorem 2.3-2 *Let X be any set and let $(Y, \|\cdot\|)$ be a normed vector space. Then the set*

$$\mathcal{B}(X;Y)$$

of all **bounded mappings** *$f : X \to Y$, i.e., such that the direct image $f(X)$ is a bounded subset of Y, is a vector space. Besides, the function $|||\cdot||| : \mathcal{B}(X;Y) \to \mathbb{R}$ defined by*

$$|||f||| := \sup_{x \in X} \|f(x)\| \quad \textit{for each } f \in \mathcal{B}(X;Y)$$

is a norm on $\mathcal{B}(X;Y)$. □

This notion can be in turn further extended as follows: Let X be a *topological space* and let Y be a *normed vector space*. Then a sequence $(f_n)_{n=1}^\infty$ of mappings $f_n : X \to Y$ is said to **converge locally uniformly** to a mapping $f : X \to Y$ as $n \to \infty$ if, given any $x_0 \in X$, there exists a neighborhood $V(x_0)$ of x_0 such that

$$\lim_{n \to \infty} \Big(\sup_{x \in V(x_0)} \|f_n(x) - f(x)\| \Big) = 0.$$

Again, the functions f_n, $n \geq 1$, and f may be unbounded. Consider for instance the functions $f_n : x \in]0, \infty[\to \frac{1}{x} + \max\{0, x - n\}$; then the sequence $(f_n)_{n=1}^\infty$ converges locally uniformly (but not uniformly) to the function $f : x \in]0, \infty[\to \frac{1}{x}$.

Naturally, each one of the above uniform convergences implies the **pointwise convergence** of the sequence $(f_n)_{n=1}^\infty$ to f as $n \to \infty$, i.e., that

$$\text{for each } x \in X, \quad f_n(x) \to f(x) \text{ as } n \to \infty.$$

A key property is that *continuity is preserved by local uniform convergence*:

Theorem 2.3-3 *Let X be a topological space, let Y be a normed vector space, and let $(f_n)_{n=1}^\infty$ be a sequence of mappings $f_n : X \to Y$ that converges locally uniformly to a mapping $f : X \to Y$ as $n \to \infty$. Then, if the mappings f_n, $n \geq 1$, are continuous at a point $x_0 \in X$,* resp. *continuous in X, the mapping f is continuous at x_0,* resp. *continuous in X.*

Proof Assume that the mappings f_n, $n \geq 1$, are continuous at a point $x_0 \in X$, and let $\varepsilon > 0$ be given. Since the sequence $(f_n)_{n=1}^\infty$ converges locally uniformly, there exists a neighborhood $V(x_0)$ of x_0 such that $\lim_{n\to\infty}(\sup_{x \in V(x_0)} \|f_n(x) - f(x)\|) = 0$. Let then $n_0 \geq 1$ be so chosen that

$$\sup_{x \in V(x_0)} \|f_{n_0}(x) - f(x)\| \leq \frac{\varepsilon}{3}.$$

Since the mapping f_{n_0} is continuous at x_0, there exists a neighborhood $W(x_0) \subset V(x_0)$ of x_0 such that

$$\|f_{n_0}(x) - f_{n_0}(x_0)\| \leq \frac{\varepsilon}{3} \quad \text{for all } x \in W(x_0).$$

The mapping f is thus continuous at x_0 since

$$\begin{aligned}\|f(x) - f(x_0)\| &\leq \|f(x) - f_{n_0}(x)\| + \|f_{n_0}(x) - f_{n_0}(x_0)\| + \|f_{n_0}(x_0) - f(x_0)\| \\ &\leq \varepsilon \quad \text{for all } x \in W(x_0).\end{aligned}$$

If the mappings f_n, $n \geq 1$, are continuous at all points in X, the same argument shows that f is continuous at all points in X. □

Problems

2.3-1 (**Dini's theorem**[3]) Given a compact metric space K, let $(f_n)_{n=1}^{\infty}$ be an increasing ($f_n(x) \leq f_m(x)$ for all $x \in K$ if $n \leq m$) sequence of functions $f_n \in \mathcal{C}(K)$ that pointwise converge to a function $f \in \mathcal{C}(K)$. Show that $(f_n)_{n=1}^{\infty}$ converges uniformly to f.

2.3-2 In what follows, Ω is an open subset of $\mathbb{R}^n$. Given any function $f \in \mathcal{C}(\Omega)$ and any compact subset K of Ω, let

$$|f|_K := \sup_{x \in K} |f(x)|.$$

Then the mapping $|\cdot|_K : \mathcal{C}(\Omega) \to \mathbb{R}$ defined in this fashion is clearly a *seminorm*, but *not a norm*, on the space $\mathcal{C}(\Omega)$.

(1) Show that there exists a sequence $(K_i)_{i=1}^{\infty}$ of compact subsets K_i of Ω such that

$$K_i \subset \operatorname{int} K_{i+1} \quad \text{for all } i \geq 1 \quad \text{and} \quad \Omega = \bigcup_{i=1}^{\infty} K_i.$$

(2) Let $\sum_{i=1}^{\infty} \alpha_i$ with $\alpha_i > 0$ for all $i \geq 1$ be a convergent series. Given two functions $f, g \in \mathcal{C}(\Omega)$, let

$$d(f,g) := \sum_{i=1}^{\infty} \alpha_i \frac{|f-g|_{K_i}}{1 + |f-g|_{K_i}}.$$

Show that the mapping $d : \mathcal{C}(\Omega) \times \mathcal{C}(\Omega) \to \mathbb{R}$ defined in this fashion is a distance on the vector space $\mathcal{C}(\Omega)$.

(3) Show that a sequence $(f_n)_{n=1}^{\infty}$ of functions $f_n \in \mathcal{C}(\Omega)$ converges to a function $f \in \mathcal{C}(\Omega)$ in the metric space $(\mathcal{C}(\Omega), d)$ if and only if

$$\text{for any compact subset } K \subset \Omega, \quad \lim_{n \to \infty} |f_n - f|_K = 0.$$

(4) Is the metric space $(\mathcal{C}(\Omega), d)$ complete?

(5) Show that the topology induced on the space $\mathcal{C}(\Omega)$ by the distance d of (2) is not normable.

The topology induced by the above distance d on the space $\mathcal{C}(\Omega)$ is called the *Fréchet topology associated with the family* $(|\cdot|_K)_{K \in \mathcal{K}}$ *of seminorms* $|\cdot|_K$, where $\mathcal{K}$ denotes the family of all the compact subsets of Ω.

Remark A similar Fréchet topology can be defined on the space of functions that are m times continuously differentiable in Ω; cf. Problem 7.8-3. □

[3]U. Dini [1878]: *Fondamenti per la Teoria delle Funzioni di Variabili Reali*, T. Nistri, Pisa.

2.3-3 Given any function $f \in \mathcal{C}^\infty[0,1]$ and any integer $n \geq 0$, let

$$\|f\|_n := \max_{0\leq m\leq n} \sup_{0\leq x\leq 1} |f^{(m)}(x)|.$$

(1) Let $\sum_{n=0}^\infty \alpha_n$ with $\alpha_n > 0$ for all $n \geq 1$ be a convergent series. Given two functions $f, g \in \mathcal{C}^\infty[0,1]$, let

$$d(f,g) := \sum_{n=1}^\infty \alpha_n \frac{\|f-g\|_n}{1+\|f-g\|_n}.$$

Show that the mapping $d : \mathcal{C}^\infty[0,1] \times \mathcal{C}^\infty[0,1] \to \mathbb{R}$ defined in this fashion is a distance on the space $\mathcal{C}^\infty[0,1]$.

(2) Show that the metric space $(\mathcal{C}^\infty[0,1], d)$ is complete.

(3) Show that the topology induced on the space $\mathcal{C}^\infty[0,1]$ by the distance d is not normable.

2.3-4 Let X be a topological space, let $(Y, \|\cdot\|)$ be a normed vector space, and let $(\mathcal{B}(X;Y); |||\cdot|||)$ be the normed vector space defined in Theorem 2.3-2. Show that $\mathcal{B}(X;Y) \cap \mathcal{C}(X;Y)$ is a closed subspace of $(\mathcal{B}(X;Y); |||\cdot|||)$.

2.4 The spaces ℓ^p, $1 \leq p \leq \infty$

We saw in Theorem 2.2-2 that the mappings $\|\cdot\|_p$ defined for all $x = (x_i)_{i=1}^n \in \mathbb{K}^n$ by $\|x\|_p = (\sum_{i=1}^n |x_i|^p)^{1/p}$ if $1 \leq p < \infty$ and $\|x\|_\infty = \max_{1\leq i\leq n} |x_i|$ if $p = \infty$ are norms on $\mathbb{K}^n$. The next theorem paves the way for extending these norms to vector spaces consisting of *infinite* sequences $(x_i)_{i=1}^\infty$ of scalars $x_i \in \mathbb{K}$.

Theorem 2.4-1 (Hölder's and Minkowski's inequalities for sequences) (a) *Given a real number $p > 1$, let the real number q be defined by*

$$\frac{1}{p} + \frac{1}{q} = 1 \quad (\textit{hence } q > 1),$$

and let $x = (x_i)_{i=1}^\infty$ and $y = (y_i)_{i=1}^\infty$ be two sequences of scalars that satisfy

$$\sum_{i=1}^\infty |x_i|^p < \infty \quad \textit{and} \quad \sum_{i=1}^\infty |y_i|^q < \infty.$$

Then the series $\sum_{i=1}^\infty |x_i y_i|$ converges and **Hölder's inequality**[4] *holds:*

$$\sum_{i=1}^\infty |x_i y_i| \leq \left(\sum_{i=1}^\infty |x_i|^p\right)^{1/p} \left(\sum_{i=1}^\infty |y_i|^q\right)^{1/q}.$$

(b) *Given a real number $p \geq 1$, let $x = (x_i)_{i=1}^\infty$ and $y = (y_i)_{i=1}^\infty$ be two sequences of scalars that satisfy*

$$\sum_{i=1}^\infty |x_i|^p < \infty \quad \textit{and} \quad \sum_{i=1}^\infty |y_i|^p < \infty.$$

[4] O. Hölder [1889]: *Über einen Mittelwertsatz*, Göttinger Nachrichten, 38–47.

Then the series $\sum_{i=1}^{\infty}|x_i+y_i|^p$ *converges and* **Minkowski's inequality**[5] *holds:*

$$\left(\sum_{i=1}^{\infty}|x_i+y_i|^p\right)^{1/p} \le \left(\sum_{i=1}^{\infty}|x_i|^p\right)^{1/p} + \left(\sum_{i=1}^{\infty}|y_i|^p\right)^{1/p}.$$

Proof (i) *A simple inequality: If* $p>1$ *and* $\frac{1}{p}+\frac{1}{q}=1$, *then*

$$\alpha\beta \le \frac{\alpha^p}{p}+\frac{\beta^q}{q} \quad \text{for all } \alpha>0 \text{ and } \beta>0.$$

To see this, note that the convexity of the exponential function implies that

$$e^{\theta r+(1-\theta)s} \le \theta e^r + (1-\theta)e^s \quad \text{for all } 0<\theta<1,\ r\in\mathbb{R},\ s\in\mathbb{R}.$$

The announced inequality follows by letting $\theta = \frac{1}{p}$, $r = p\,\mathrm{Log}\,\alpha$, and $s = q\,\mathrm{Log}\,\beta$ in this inequality.

(ii) *Hölder's inequality.* Assume that $x \neq 0$ and $y \neq 0$ (otherwise, Hölder's inequality clearly holds), and let $\|x\|_p := \left(\sum_{i=1}^{\infty}|x_i|^p\right)^{1/p}$ and $\|y\|_p := \left(\sum_{i=1}^{\infty}|y_i|^p\right)^{1/p}$ (at this stage, $\|x\|_p$ and $\|y\|_p$ should be simply regarded as convenient notations). Letting $\alpha = \frac{|x_i|}{\|x\|_p}$ and $\beta = \frac{|y_i|}{\|y\|_p}$ in the inequality of (i) gives

$$\frac{|x_iy_i|}{\|x\|_p\|y\|_q} \le \frac{|x_i|^p}{p(\|x\|_p)^p}+\frac{|y_i|^q}{q(\|y\|_q)^q} \quad \text{for each integer } i\ge 1,$$

so that

$$\frac{\sum_{i=1}^{n}|x_iy_i|}{\|x\|_p\|y\|_q} \le \frac{\sum_{i=1}^{n}|x_i|^p}{p(\|x\|_p)^p}+\frac{\sum_{i=1}^{n}|y_i|^q}{q(\|y\|_q)^q} \le \frac{1}{p}+\frac{1}{q}=1 \quad \text{for any integer } n\ge 1.$$

Passing to the limit as $n\to\infty$ then shows that the series $\sum_{i=1}^{\infty}|x_iy_i|$ converges and that Hölder's inequality holds.

(iii) *Minkowski's inequality.* Assume $p>1$ (Minkowski's inequality clearly holds if $p=1$) and let q be again defined by $\frac{1}{p}+\frac{1}{q}=1$, so that $pq-q=p$. Hölder's inequality (part (ii)) then gives, for any integer $n\ge 1$,

$$\begin{aligned}
\sum_{i=1}^{n}(|x_i|+|y_i|)^p &= \sum_{i=1}^{n}|x_i|(|x_i|+|y_i|)^{p-1}+\sum_{i=1}^{n}|y_i|(|x_i|+|y_i|)^{p-1}\\
&\le \left(\sum_{i=1}^{n}|x_i|^p\right)^{1/p}\left(\sum_{i=1}^{n}(|x_i|+|y_i|)^p\right)^{1/q}+\left(\sum_{i=1}^{n}|y_i|^p\right)^{1/p}\left(\sum_{i=1}^{n}(|x_i|+|y_i|)^p\right)^{1/q}\\
&= \left(\left(\sum_{i=1}^{n}|x_i|^p\right)^{1/p}+\left(\sum_{i=1}^{n}|y_i|^p\right)^{1/p}\right)\left(\sum_{i=1}^{n}(|x_i|+|y_i|)^p\right)^{1/q}.
\end{aligned}$$

[5] H. Minkowski [1896]: *Geometrie der Zahlen*, Leipzig.

Since $1 - \dfrac{1}{q} = \dfrac{1}{p}$, the above inequality gives

$$\left(\sum_{i=1}^{n} |x_i + y_i|^p\right)^{1/p} \le \left(\sum_{i=1}^{n} (|x_i| + |y_i|)^p\right)^{1/p} \le \left(\sum_{i=1}^{n} (x_i^p)\right)^{1/p} + \left(\sum_{i=1}^{n} (y_i^p)\right)^{1/p}.$$

Letting first $n \to \infty$ in the right-hand side, then $n \to \infty$ in the left-hand side, shows that the series $\sum_{i=1}^{\infty} |x_i + y_i|^p$ converges and that Minkowski's inequality holds. □

We are now in a position to define the real or complex normed vector spaces

$$(\ell^p, \|\cdot\|_p), \quad 1 \le p \le \infty,$$

which constitute the announced generalization of the spaces $(\mathbb{K}^n, \|\cdot\|_p)$ (Theorem 2.2-2) to spaces of *infinite sequences*.[6] We also show that, *except for* $p = \infty$, these spaces are *separable*.

Theorem 2.4-2 *For each extended real number $1 \le p \le \infty$, let ℓ^p denote the set of all infinite sequences $x = (x_i)_{i=1}^{\infty}$ of scalars $x_i \in \mathbb{K}$ that satisfy*

$$\sum_{i=1}^{\infty} |x_i|^p < \infty \textit{ if } 1 \le p < \infty, \quad \textit{or} \quad \sup_{i \ge 1} |x_i| < \infty \textit{ if } p = \infty.$$

(a) *For each $1 \le p \le \infty$, the set ℓ^p is a vector space, and the mapping $\|\cdot\|_p$ defined by*

$$\begin{aligned} x = (x_i)_{i=1}^{\infty} \in \ell^p \quad &\to \quad \|x\|_p = \left(\sum_{i=1}^{\infty} |x_i|^p\right)^{1/p} \qquad \textit{if } 1 \le p < \infty, \\ x = (x_i)_{i=1}^{\infty} \in \ell^\infty \quad &\to \quad \|x\|_\infty = \sup\nolimits_{i \ge 1} |x_i| \qquad \textit{if } p = \infty, \end{aligned}$$

is a norm on ℓ^p.

(b) *The normed vector spaces $(\ell^p, \|\cdot\|_p)$, $1 \le p < \infty$, are separable.*

(c) *The normed vector space $(\ell^\infty, \|\cdot\|_\infty)$ is not separable.*

Proof That ℓ^p is a vector space follows from Minkowski's inequality (Theorem 2.4-1) for $p \ge 1$ and is clear for $p = \infty$. That $\|\cdot\|_p$ is a norm on ℓ^p likewise follows from Minkowski's inequality (which is precisely the triangle inequality for this norm; the other properties of a norm are immediately verified) for $p \ge 1$ and is clear for $p = \infty$. Hence (a) is proved.

Given $1 \le p < 1$, let

$$A := \bigcup_{n=1}^{\infty} \{(y_i)_{i=1}^{\infty} \in \ell^p;\ y_i \in \mathbb{Q} \text{ for } i \le n,\ y_i = 0 \text{ for } i \ge n+1\} \text{ if } \mathbb{K} = \mathbb{R},$$

$$A := \bigcup_{n=1}^{\infty} \{(y_i)_{i=1}^{\infty} \in \ell^p;\ \operatorname{Re} y_i \in \mathbb{Q} \text{ and } \operatorname{Im} y_i \in \mathbb{Q} \text{ for } i \le n,\ y_i = 0 \text{ for } i \ge n+1\} \text{ if } \mathbb{K} = \mathbb{C}.$$

[6]The spaces ℓ^p and $L^p(\Omega)$ (Section 2.5) were introduced in 1910 by Frigyes Riesz (1880–1956), who made many landmark contributions to functional analysis, which accordingly bear his name (as will be abundantly illustrated in this and the next chapters). Together with his student Béla Szökefalvi Nagy (1913–1998), he coauthored Riesz & Nagy [1955], a masterpiece justly considered as one of the most influential texts in functional analysis.

Frigyes Riesz had a brother, Marcel Riesz (1886–1969), also a famous mathematician.

Then the set A *is countably infinite*, as a countably infinite union of countably infinite sets (Section 1.5). Furthermore, A *is dense in* ℓ^p: Given any $x = (x_i)_{i=1}^\infty \in \ell^p$ and any $\varepsilon > 0$, there exists $n_0 = n_0(x, \varepsilon) \geq 1$ such that $\sum_{i=n_0+1}^\infty |x_i|^p \leq \frac{\varepsilon^p}{2}$. Then there exist $y_i \in \mathbb{Q}, 1 \leq i \leq n_0$, if $\mathbb{K} = \mathbb{R}$, or there exist $y_i \in \mathbb{C}$ with $\operatorname{Re} y_i \in \mathbb{Q}$ and $\operatorname{Im} y_i \in \mathbb{Q}$, $1 \leq i \leq n_0$, if $\mathbb{K} = \mathbb{C}$, such that $\sum_{i=1}^{n_0} |x_i - y_i|^p \leq \frac{\varepsilon^p}{2}$. Then the vector $y := (y_1, \ldots, y_{n_0}, 0, \ldots)$ belongs to A and satisfies $\|y - x\|_p \leq \varepsilon$. This proves (b).

The set
$$B := \{(x_i)_{i=1}^\infty \in \ell^\infty;\ x_i = 0 \text{ or } x_i = 1,\ i \geq 1\}$$
is an *uncountably infinite subset of* ℓ^∞ since $(x_i)_{i=1}^\infty \in B \to \sum_{i=1}^\infty \frac{1}{2^i} x_i$ is a surjection of B onto $[0, 1[$ (cf. Section 1.5; we use here the property that every real number in the interval $[0, 1[$ has such a binary expansion).

Let then C be any dense subset of ℓ^∞. Given any $x \in B$, there thus exists $y(x) \in C$ such that $\|x - y(x)\|_\infty < \frac{1}{2}$, and the mapping $x \in B \to y(x) \in C$ defined in this fashion is an injection (since $x, \widetilde{x} \in B$ with $x \neq \widetilde{x}$ implies $\|x - \widetilde{x}\|_\infty = 1$, which in turn implies $y(x) \neq y(\widetilde{x})$). Therefore C is necessarily *uncountably infinite* (Section 1.5). This proves (c). □

Interesting complements to Theorems 2.4-1 and 2.4-2 are given in the next exercises.

Problems

2.4-1 (1) Given any vector $x \in \mathbb{K}^n$, show that $\|x\|_\infty = \lim_{p \to \infty} \|x\|_p$.
(2) Given any $x = (x_i)_{i=1}^\infty \in \ell^\infty$, show that $\|x\|_\infty = \lim_{n \to \infty} \{\lim_{p \to \infty} (\sum_{i=1}^n |x_i|^p)^{1/p}\}$.

2.4-2 (1) Show that equality holds in Hölder's inequality (Theorem 2.4-1) if and only if there exist constants $\alpha \geq 0$ and $\beta \geq 0$ with $\alpha + \beta > 0$ such that $\alpha |x_i|^p = \beta |y_i|^q$ for all $i \geq 1$.

(2) Show that equality holds in Minkowski's inequality (Theorem 2.4-1) if and only if there exist constants $\alpha \geq 0$ and $\beta \geq 0$ with $\alpha + \beta > 0$ such that $\alpha x_i = \beta y_i$ for all $i \geq 1$.

2.4-3 Given a real number $0 < p < 1$, let X denote the set of all sequences $(x_i)_{i=1}^\infty$ of scalars that satisfy $\sum_{i=1}^\infty |x_i|^p < \infty$.
(1) Show that X is a vector space.
(2) Show that the mapping $(x_i)_{i=1}^\infty \in X \to (\sum_{i=1}^\infty |x_i|^p)^{1/p}$ is not a norm on X.
(3) Show that the mapping $d : X \times X \to \mathbb{R}$ defined by $d(x, y) = \sum_{i=1}^\infty |x_i - y_i|^p$ for all $x = (x_i)_{i=1}^\infty \in X$ and $y = (y_i)_{i=1}^\infty \in X$ is a distance on X.

2.4-4 Let p and q be two real numbers satisfying $0 < p < q$ and let $(x_i)_{i=1}^\infty$ be an infinite sequence of scalars $x_i \in \mathbb{K}$ such that $\sum_{i=1}^\infty |x_i|^p < \infty$. Show that the series $\sum_{i=1}^\infty |x_i|^q$ is convergent and that **Jensen's inequality**[7] **in** ℓ^p holds:
$$\left(\sum_{i=1}^\infty |x_i|^q\right)^{1/q} \leq \left(\sum_{i=1}^\infty |x_i|^p\right)^{1/p}.$$

Note that Jensen's inequality implies that, for each $p \geq 1$, the space ℓ^p, $p \geq 1$, is contained in all the spaces ℓ^q, $p < q \leq \infty$, and that $\|x\|_q \leq \|x\|_p$ for all $x \in \ell^p$.

[7] J.L.W.V. Jensen [1906]: Sur les fonctions convexes et les inégalités entre les valeurs moyennes, *Acta Mathematica* **30**, 175–193.

2.5 The Lebesgue spaces $L^p(\Omega), 1 \leq p \leq \infty$

Let Ω be an *open* (thus measurable) *subset of* $\mathbb{R}^n$. The corresponding space $L^1(\Omega)$ (we refer to Section 1.15 for the definition and the main properties of the vector space $L^1(A)$, where A is any measurable subset of $\mathbb{R}^n$) thus consists of all (equivalence classes of) real *Lebesgue-integrable functions*, i.e., those *measurable* functions $f : \Omega \to [-\infty, \infty]$ that satisfy

$$\int_\Omega |f(x)| \, dx < \infty.$$

We now extend this definition. Given any $1 < p < \infty$, we let $L^p(\Omega)$ denote the set formed by all (equivalence classes of) *measurable* functions $f : \Omega \to [-\infty, \infty]$ such that $|f|^p \in L^1(\Omega)$, or equivalently, that satisfy

$$\int_\Omega |f(x)|^p \, dx < \infty \quad \text{for some } 1 < p < \infty.$$

The first objective of this section is to show that the sets $L^p(\Omega)$, $1 \leq p < \infty$, defined in this fashion, and also a set $L^\infty(\Omega)$ that will be defined below (Theorem 2.5-2), are (real) *normed vector spaces*. To this end, we shall proceed along lines reminiscent of those followed for the normed vector spaces $(\ell^p, \|\cdot\|_p)$, $1 \leq p \leq \infty$ (Section 2.4): compare the statements and proofs of Theorems 2.5-1 and 2.5-2 with those of Theorems 2.4-1 and 2.4-2(a), respectively.

Theorem 2.5-1 (Hölder's and Minkowski's inequalities for functions) *Let Ω be an open subset of $\mathbb{R}^n$.*

(a) *Given a real number $p > 1$, let the real number q be defined by*

$$\frac{1}{p} + \frac{1}{q} = 1 \quad (\textit{hence } q > 1),$$

and let $f : \Omega \to [-\infty, \infty]$ and $g : \Omega \to [-\infty, \infty]$ be two measurable functions that satisfy

$$\int_\Omega |f(x)|^p \, dx < \infty \quad \textit{and} \quad \int_\Omega |g(x)|^q \, dx < \infty.$$

Then $fg \in L^1(\Omega)$ and **Hölder's inequality** *holds:*

$$\int_\Omega |f(x)g(x)| \, dx \leq \left(\int_\Omega |f(x)|^p \, dx \right)^{1/p} \left(\int_\Omega |g(x)|^q \, dx \right)^{1/q}.$$

(b) *Given a real number $p \geq 1$, let $f : \Omega \to [-\infty, \infty]$ and $g : \Omega \to [-\infty, \infty]$ be two measurable functions that satisfy*

$$\int_\Omega |f(x)|^p \, dx < \infty \quad \textit{and} \quad \int_\Omega |g(x)|^p \, dx < \infty.$$

Then $(f + g) \in L^p(\Omega)$ and **Minkowski's inequality** *holds:*

$$\left(\int_\Omega |f(x) + g(x)|^p \, dx \right)^{1/p} \leq \left(\int_\Omega |f(x)|^p \, dx \right)^{1/p} + \left(\int_\Omega |g(x)|^p \, dx \right)^{1/p}.$$

Proof (i) *Hölder's inequality.* Assume $f \neq 0$ and $g \neq 0$ (otherwise, Hölder's inequality clearly holds), and let $\|f\|_p := (\int_\Omega |f(x)|^p \, dx)^{1/p}$ and $\|g\|_q := (\int_\Omega |g(x)|^q \, dx)^{1/q}$ (at this stage, $\|f\|_p$ and $\|g\|_q$ should be simply regarded as convenient notations). Letting $\alpha := \dfrac{|f(x)|}{\|f\|_p}$ and $\beta := \dfrac{|g(x)|}{\|g\|_q}$ in the inequality $\alpha\beta \leq \dfrac{\alpha^p}{p} + \dfrac{\beta^q}{q}$ (cf. part (i) of the proof of Theorem 2.4-1) shows that

$$\frac{|f(x)g(x)|}{\|f\|_p \|g\|_q} \leq \frac{1}{p} \frac{|f(x)|^p}{(\|f\|_p)^p} + \frac{1}{q} \frac{|g(x)|^q}{(\|g\|_q)^q} \quad \text{for all } x \in \Omega,$$

and therefore that

$$\begin{aligned} \frac{1}{\|f\|_p \|g\|_q} \int_\Omega |f(x)g(x)| \, dx &\leq \frac{1}{p(\|f\|_p)^p} \int_\Omega |f(x)|^p \, dx + \frac{1}{q(\|g\|_q)^q} \int_\Omega |g(x)|^q \, dx \\ &= \frac{1}{p} + \frac{1}{q} = 1. \end{aligned}$$

Hence $fg \in L^1(\Omega)$ and Hölder's inequality holds. This proves (a).

(ii) *Minkowski's inequality.* Assume $p > 1$ (Minkowski's inequality clearly holds for $p = 1$), and let q be again defined by $\dfrac{1}{p} + \dfrac{1}{q} = 1$, so that $pq - q = p$. Hölder's inequality (part (ii)) then gives

$$\begin{aligned} \int_\Omega (|f(x)| + |g(x)|)^p \, dx &= \int_\Omega |f(x)|(|f(x)| + |g(x)|)^{p-1} \, dx \\ &\quad + \int_\Omega |g(x)|(|f(x)| + |g(x)|)^{p-1} \, dx \\ &\leq \left(\int_\Omega |f(x)|^p \, dx\right)^{1/p} \left(\int_\Omega (|f(x)| + |g(x)|)^p \, dx\right)^{1/q} \\ &\quad + \left(\int_\Omega |g(x)|^p \, dx\right)^{1/p} \left(\int_\Omega (|f(x)| + |g(x)|)^p \, dx\right)^{1/q} \\ &= \left\{\left(\int_\Omega |f(x)|^p \, dx\right)^{1/p} + \left(\int_\Omega |g(x)|^p \, dx\right)^{1/p}\right\} \left(\int_\Omega (|f(x)| + |g(x)|)^p \, dx\right)^{1/q}. \end{aligned}$$

Since $1 - \dfrac{1}{q} = \dfrac{1}{p}$, the above inequality implies that

$$\begin{aligned} \left(\int_\Omega (|f(x) + g(x)|)^p \, dx\right)^{1/p} &\leq \left(\int_\Omega (|f(x)| + |g(x)|)^p\right)^{1/p} \\ &\leq \left(\int_\Omega |f(x)|^p \, dx\right)^{1/p} + \left(\int_\Omega |g(x)|^p \, dx\right)^{1/p}. \end{aligned}$$

Hence $(f + g) \in L^p(\Omega)$ and Minkowski's inequality holds. Thus (b) is proved. □

We are now in a position to define the real normed vector spaces

$$(L^p(\Omega), \|\cdot\|_{L^p(\Omega)}), \quad 1 \leq p \leq \infty,$$

which are called the **Lebesgue spaces:**[8]

Theorem 2.5-2 *Let Ω be an open subset of $\mathbb{R}^n$. For each extended real number $1 \leq p \leq \infty$, let $L^p(\Omega)$ denote the set of all measurable functions $f : \Omega \to [-\infty, \infty]$ that satisfy*

$$\int_\Omega |f(x)|^p \, dx < \infty \quad \text{if } 1 \leq p < \infty,$$
$$\inf\{C \geq 0;\ |f| \leq C \text{ a.e. in } \Omega\} < \infty \quad \text{if } p = \infty.$$

Then, for each $1 \leq p \leq \infty$, the set $L^p(\Omega)$ is a vector space, and the mapping $\|\cdot\|_{L^p(\Omega)}$ defined by

$$f \in L^p(\Omega) \to \|f\|_{L^p(\Omega)} := \left(\int_\Omega |f(x)|^p \, dx \right)^{1/p} \quad \text{if } 1 \leq p < \infty,$$
$$f \in L^\infty(\Omega) \to \|f\|_{L^\infty(\Omega)} := \inf\{C \geq 0;\ |f| \leq C \text{ a.e. in } \Omega\},$$

is a norm on $L^p(\Omega)$.

Proof For $1 \leq p < \infty$, Minkowski's inequality (Theorem 2.5-1) shows that $L^p(\Omega)$ is a vector space and that $\|\cdot\|_{L^p(\Omega)}$ is a norm on $L^p(\Omega)$ (Minkowski's inequality is nothing but the triangle inequality for $\|\cdot\|_{L^p(\Omega)}$, and the other properties of a norm are immediately verified). It is clear that $L^\infty(\Omega)$ is a vector space and that $\|\cdot\|_{L^\infty(\Omega)}$ is a norm on $L^\infty(\Omega)$. □

Remark For each $1 \leq p < \infty$, one can similarly define the *complex* spaces

$$L^p(\Omega; \mathbb{C}) := \{f : \Omega \to \mathbb{C};\ \operatorname{Re} f \text{ and } \operatorname{Im} f \text{ are measurable and } |f|^p \in L^1(\Omega)\},$$

which share most of the properties of the spaces $L^p(\Omega)$.[9] □

Given a measurable function $f : \Omega \to [-\infty, \infty]$, the extended real number

$$\inf\{C \geq 0;\ |f| \leq C \text{ a.e. in } \Omega\} \in [0, \infty]$$

is called the **essential supremum** of f. The space $L^\infty(\Omega)$ thus consists of all (equivalence classes of) measurable functions whose essential supremum is finite.

While the issue of *separability* was fairly easy to settle for the spaces ℓ^p (Theorem 2.4-2(b)), it is no longer so for the Lebesgue spaces $L^p(\Omega)$. As a preliminary for the case where $1 \leq p < \infty$, we first prove in Theorem 2.5-3 that any function in the Lebesgue space $(L^p(\Omega), \|\cdot\|_{L^p(\Omega)})$, $1 \leq p < \infty$, can be approximated as close as we please by *continuous functions with compact support in* Ω. This result, which is already of interest *per se*, will be considerably refined in Theorem 2.6-2, where we will show that any function in $L^p(\Omega)$, $1 \leq p < \infty$, can be in fact approximated as close as we please by *infinitely differentiable* functions with compact support in Ω.

Note that the next result does *not* hold for $p = \infty$; cf. Problem 2.5-5.

[8] So named after the founder of the Lebesgue measure and of the Lebesgue integral, Henri Lebesgue (1875–1941), and his seminal *Note aux Comptes Rendus*:

H. LEBESGUE [1901]: Sur une généralisation de l'intégrale définie, *Comptes Rendus des Séances de l'Académie des Sciences* **132**, 1025–1027.

[9] The spaces $L^p(\Omega; \mathbb{C})$ are analyzed in detail in HEWITT & STROMBERG [1965, Section 13].

Theorem 2.5-3 *Let Ω be an open subset of $\mathbb{R}^n$. Define the space*

$$\mathcal{C}_c(\Omega) := \{g \in \mathcal{C}(\Omega);\ \operatorname{supp} g \textit{ is a compact subset of } \Omega\}.$$

Then, for each $1 \le p < \infty$, the subspace $\mathcal{C}_c(\Omega)$ is dense in the space $L^p(\Omega)$.

Proof Let a function $f \in L^p(\Omega)$ and $\varepsilon > 0$ be given. Our objective is to find a function $g \in \mathcal{C}_c(\Omega)$ that satisfies $\|f - g\|_{L^p(\Omega)} \le \varepsilon$.

(i) *There exists a measurable simple function $s = s(f, \varepsilon)$ such that*

$$\mu(\{x \in \Omega;\ s(x) \neq 0\}) < \infty \quad \text{and} \quad \|f - s\|_{L^p(\Omega)} \le \frac{\varepsilon}{2},$$

where μ denotes the Lebesgue measure in $\mathbb{R}^n$; measurable simple functions are defined in Section 1.14.

Assume first that $f \ge 0$. Then, by Theorem 1.14-5, there exists a sequence $(s_k)_{k=1}^{\infty}$ of measurable simple functions with the following properties:

$$0 \le s_k \le f \text{ for all } k \ge 1 \quad \text{and} \quad s_k(x) \to f(x) \text{ as } k \to \infty \text{ for each } x \in \Omega.$$

Consequently,

$$\begin{aligned} s_k \in L^p(\Omega) \quad &\text{and thus} \quad \mu(\{x \in \Omega;\ s_k(x) \neq 0\}) < \infty \text{ for all } k \ge 1,\\ |(f - s_k)(x)|^p &\le |f(x)|^p \quad \text{for all } x \in \Omega \text{ and all } k \ge 1,\\ |(f - s_k)(x)|^p &\to 0 \quad \text{as } k \to \infty \text{ for each } x \in \Omega. \end{aligned}$$

Lebesgue's dominated convergence theorem (Theorem 1.15-3) applied to the functions $|f - s_k|^p \in L^1(\Omega)$, $k \ge 1$, which are all dominated by the same function $|f|^p \in L^1(\Omega)$, therefore shows that

$$\int_\Omega |f(x) - s_k(x)|^p \, dx \to 0 \quad \text{as } k \to \infty.$$

Returning to the general case, let

$$\Omega^+ := \{x \in \Omega;\ f(x) > 0\} \quad \text{and} \quad \Omega^- := \{x \in \Omega;\ f(x) < 0\}.$$

The above argument then shows that there exist measurable simple functions $s_k^+ : \Omega^+ \to [0, \infty[$ and $s_k^- : \Omega_- \to [0, \infty[$, $k \ge 1$, such that

$$\int_{\Omega^+} |f(x) - s_k^+(x)|^p \, dx \to 0 \quad \text{and} \quad \int_{\Omega^-} |-f(x) - s_k^-(x)|^p \, dx \to 0 \quad \text{as } k \to \infty.$$

The measurable simple functions $s_k : \Omega \to \mathbb{R}$ defined for each $k \ge 1$ by $s_k := s_k^+$ on Ω^+, $s_k := -s_k^-$ on Ω^-, and $s_k := 0$ on $\Omega - (\Omega^+ \cup \Omega^-)$, therefore satisfy

$$\int_\Omega |f(x) - s_k(x)|^p \, dx = \int_{\Omega^+} |f(x) - s_k^+(x)|^p \, dx + \int_{\Omega^-} |-f(x) - s_k^-(x)|^p \, dx \to 0 \quad \text{as } k \to \infty.$$

Hence (i) is proved.

(ii) *Let $s = s(f, \varepsilon)$ be the measurable simple function constructed in* (i). *Then there exists a function $g = g(s, \varepsilon) = g(f, \varepsilon) \in \mathcal{C}_c(\Omega)$ such that*

$$\|s - g\|_{L^p(\Omega)} \leq \frac{\varepsilon}{2}.$$

Since $\mu(\{x \in \Omega;\ s(x) \neq 0\}) < \infty$, *Lusin's property* (Theorem 1.14-4(c)) implies that there exists a function $g \in \mathcal{C}_c(\Omega)$ that satisfies

$$\sup_{x\in\Omega} |g(x)| \leq \|s\|_{L^\infty(\Omega)} \quad \text{and} \quad \mu\left(\{x \in \Omega;\ g(x) \neq s(x)\}\right) \leq \left(\frac{\varepsilon}{4\,\|s\|_{L^\infty(\Omega)}}\right)^p.$$

Consequently,

$$\|s - g\|_{L^p(\Omega)} = \left(\int_{\{x\in\Omega;\ g(x)\neq s(x)\}} |s(x) - g(x)|^p \,\mathrm{d}x\right)^{1/p} \leq \frac{\varepsilon}{2},$$

since $|s(x) - g(x)| \leq 2\,\|s\|_{L^\infty}$ for all $x \in \Omega$. Hence (ii) is proved. □

Note in passing that Theorem 2.5-3 and part (i) of its proof provide two other ways of defining each Lebesgue space $L^p(\Omega)$, $1 \leq p < \infty$, either as the *completion of the space $\mathcal{C}_c(\Omega)$ with respect to the norm $\|\cdot\|_{L^p(\Omega)}$* (in the definition of which the Riemann integral is used), or as the *completion with respect to the norm $\|\cdot\|_{L^p(\Omega)}$ of the space formed by all measurable simple functions $s : \Omega \to \mathbb{R}$ that satisfy $\int_\Omega |s|^p \,\mathrm{d}s < \infty$.*

Note also that Theorem 2.5-3 also implies that, *if Ω is bounded, the space $\mathcal{C}(\overline{\Omega})$ is dense in $L^p(\Omega)$, $1 \leq p < \infty$.*

We are now in a position to settle the issue of *separability* for the spaces $L^p(\Omega)$.

Theorem 2.5-4 *Let Ω be an open subset of $\mathbb{R}^n$.*

(a) *The normed vector spaces $L^p(\Omega)$, $1 \leq p < \infty$, are separable.*

(b) *The space $L^\infty(\Omega)$ is not separable.*

Proof In what follows, χ_A denotes the characteristic function of a set A.

(i) First, let a function $f \in L^p(\Omega)$, $1 \leq p < \infty$, and $\varepsilon > 0$ be given. By Theorem 2.5-3, there exists a function $g = g(f, \varepsilon) \in \mathcal{C}_c(\Omega)$ such that

$$\|f - g\|_{L^p(\Omega)} \leq \frac{\varepsilon}{2}.$$

Since the set $K := \operatorname{supp} g$ is a compact subset of Ω, there exists a bounded open set U such that $K \subset U \subset \Omega$ (to see this, consider any covering of K by open balls centered at points of K and contained in Ω, and let U be a finite subcovering of K).

Since the continuous function g is uniformly continuous on the compact set $\overline{U}$ (Theorem 1.13-2), there exists $\delta_0 > 0$ such that

$$|g(x) - g(y)| \leq \widetilde{\varepsilon} := \frac{\varepsilon}{2(\mu(U))^{1/p}} \quad \text{for all } x, y \in \overline{U} \text{ such that } \|y - x\|_\infty < \delta_0,$$

where μ denotes the Lebesgue measure in $\mathbb{R}^n$. Besides, the continuity of the function $x \in K \to \inf_{y \in \mathbb{R}^n - U} \|x - y\|_\infty$ (Theorem 1.11-3) and the compactness of K together imply that there exists $\delta_1 > 0$ such that

$$\{y \in \mathbb{R}^n;\ \|y - x\|_\infty < \delta_1\} \subset U \quad \text{for all } x \in K.$$

Let then $\delta \in \mathbb{Q}$ be such that $0 < \delta \leq \min\{\delta_0, \delta_1\}$.

Let $(B_i)_{i\in I}$ denote the countably infinite family formed by all the open balls of the form

$$\left\{y \in \mathbb{R}^n;\ \|y - x\|_\infty < \frac{\delta}{2} \text{ with } x_j = p_j\delta \text{ for some } p_j \in \mathbb{Z},\ 1 \leq j \leq n\right\},$$

and let $(B_i)_{i\in I(K)}$ denote the subfamily formed by those balls B_i, $i \in I$, that satisfy $B_i \cap K \neq \varnothing$. Then, for each $i \in I(K)$, there exists $\alpha_i \in \mathbb{Q}$ such that

$$|g(y) - \alpha_i| \leq \widetilde{\varepsilon} \quad \text{for all } y \in B_i$$

(if the function $g|_{\overline{B}_i}$ is not a constant, choose any $\alpha_i \in \mathbb{Q}$ between its minimum and its maximum; if $g|_{\overline{B}_i}$ is equal to a constant β_i, choose any $\alpha_i \in \mathbb{Q}$ that satisfies $|\alpha_i - \beta_i| \leq \widetilde{\varepsilon}$). Consequently, the function

$$h := \sum_{i\in I(K)} \alpha_i \chi_{B_i},$$

which by construction satisfies $|h(x) - g(x)| \leq \widetilde{\varepsilon}$ for almost all $x \in U$, is such that

$$\|h - g\|_{L^p(\Omega)} = \left(\int_U |h(x) - g(x)|^p \,\mathrm{d}x\right)^{1/p} \leq (\mu(U))^{1/p}\widetilde{\varepsilon} = \frac{\varepsilon}{2}.$$

The resulting inequality $\|f - h\|_{L^p(\Omega)} \leq \frac{\varepsilon}{2}$, combined with the observation that such functions h form a countably infinite family (since $\alpha_i \in \mathbb{Q}$ for all $i \in I(K)$ and the set $I(K)$ is countably infinite), then shows that each space $L^p(\Omega)$, $1 \leq p < \infty$, is separable.

(ii) Second, let $p = \infty$. Given any $x \in \Omega$, let $B(x)$ be any open ball centered at x and contained in Ω, and let

$$O(x) := \left\{f \in L^\infty(\Omega);\ \|f - \chi_{B(x)}\|_{L^\infty(\Omega)} < \frac{1}{2}\right\}.$$

Then $(O(x))_{x\in\Omega}$ is an uncountably infinite family of nonempty open subsets of $L^\infty(\Omega)$ that satisfies

$$O(x) \cap O(y) = \varnothing \quad \text{if } x \neq y$$

(if $x \neq y$, there exists an open ball B such that, e.g., $B \subset B(x)$ and $B \cap B(y) = \varnothing$; but the inequalities $|f(b) - 1| < \frac{1}{2}$, $b \in B$, and $|f(b)| < \frac{1}{2}$, $b \in B$, cannot hold simultaneously).

Assume that a countable subset $\bigcup_{n\in\mathbb{N}}\{f_n\}$ of $L^\infty(\Omega)$ is dense in $L^\infty(\Omega)$. For each $x \in \Omega$, there thus exists an integer $n(x) \geq 0$ such that $f_{n(x)} \in O(x)$. But the mapping $x \in \Omega \to n(x) \in \mathbb{N}$ defined in this fashion is necessarily injective ($f_{n(x)} \in O(x)$ and $f_{n(y)} \in O(y)$, and $O(x) \cap O(y) = \varnothing$ if $x \neq y$), which constitutes a contradiction (Section 1.5). $\square$

Remark Another, and in a sense simpler, proof of Theorem 2.5-4(a) relies on the *Weierstraß approximation theorem in several variables* (Theorem 2.15-2); cf. Problem 2.15-2. □

Problems

In the following problems, Ω denotes an open subset of $\mathbb{R}^n$.

2.5-1 (1) Show that equality occurs in Hölder's inequality (Theorem 2.5-1(a)) if and only if there exist constants $\alpha \geq 0$ and $\beta \geq 0$ with $\alpha + \beta > 0$ such that $\alpha\,|f(x)|^p = \beta\,|g(x)|^q$ for almost all $x \in \Omega$.

(2) Show that equality occurs in Minkowski's inequality (Theorem 2.5-1(b)) if and only if there exists a measurable function $h : \Omega \to [0, \infty[$ such that $f = gh$ almost everywhere on the set $\{x \in \Omega, f(x) \neq 0 \text{ and } g(x) \neq 0\}$ if $p = 1$, and if and only if there exist constants $\alpha \geq 0$ and $\beta \geq 0$ with $\alpha + \beta > 0$ such that $\alpha f(x) = \beta g(x)$ for almost all $x \in \Omega$ if $1 < p < \infty$.

2.5-2 Given $1 \leq p < \infty$, let functions $f_k \in L^p(\Omega)$, $k \geq 1$, and $f \in L^p(\Omega)$ be such that

$$\|f_k\|_{L^p(\Omega)} \to \|f\|_{L^p(\Omega)} \quad \text{and} \quad (f_k)_{k=1}^\infty \text{ converges a.e. in } \Omega \text{ to } f \text{ as } k \to \infty.$$

Show that $\|f_k - f\|_{L^p(\Omega)} \to 0$ as $k \to \infty$.

2.5-3 (1) Given $0 < p < 1$, let again q be defined by $\dfrac{1}{p} + \dfrac{1}{q} = 1$ (hence $q < 0$). Let $f : \Omega \to [-\infty, \infty]$ and $g : \Omega \to [-\infty, \infty]$ be two measurable functions that satisfy

$$\int_\Omega |f(x)|^p \, dx < \infty \quad \text{and} \quad 0 < \int_\Omega |g(x)|^q \, dx < \infty.$$

Show that the following *reverse Hölder's inequality* holds:

$$\int_\Omega |f(x)g(x)| \, dx \geq \left(\int_\Omega |f(x)|^p \, dx\right)^{1/p} \left(\int_\Omega |g(x)|^q \, dx\right)^{1/q},$$

where the left-hand side of this inequality is possibly equal to ∞.

(2) Given $0 < p < 1$, let again $f : \Omega \to [-\infty, \infty]$ and $g : \Omega \to [-\infty, \infty]$ be two measurable functions that satisfy

$$\int_\Omega |f(x)|^p \, dx < \infty \quad \text{and} \quad \int_\Omega |g(x)|^p \, dx < \infty.$$

Show that the following *reverse Minkowski's inequality* holds:

$$\left(\int_\Omega (|f(x)| + |g(x)|)^p \, dx\right)^{1/p} \geq \left(\int_\Omega |f(x)|^p \, dx\right)^{1/p} + \left(\int_\Omega |g(x)|^p \, dx\right)^{1/p}.$$

2.5-4 Given $0 < p < 1$, let $L^p(\Omega)$ denote the set of all measurable functions $f : \Omega \to [-\infty, \infty]$ that satisfy $\int_\Omega |f(x)|^p \, dx < \infty$.

(1) Show that $L^p(\Omega)$ is a vector space.

Hint: Show that, for all $f, g \in L^p(\Omega)$,

$$\int_\Omega |f(x) + g(x)|^p \, dx \leq 2^{1-p}\left\{\left(\int_\Omega |f(x)|^p \, dx\right)^{1/p} + \left(\int_\Omega |g(x)|^p\right)^{1/p}\right\}^p.$$

(2) Show that the mapping $d_p : L^p(\Omega) \times L^p(\Omega) \to [0, \infty[$ defined by

$$d_p(f, g) = \int_\Omega |f(x) - g(x)|^p \, dx \quad \text{for all } f, g \in L^p(\Omega),$$

is a distance on $L^p(\Omega)$.

2.5-5 Show that the subspace $\mathcal{C}_c(\Omega)$ is not dense in $L^\infty(\Omega)$.

2.5-6 Let $1 < p < \infty$.
(1) Show that any function $f \in L^p(0,\infty)$ that is ≥ 0 almost everywhere in $(0,\infty)$ satisfies[10]

$$\int_0^\infty \left(\frac{1}{x}\int_0^x f(t)\,\mathrm{d}t\right)^p \mathrm{d}x \leq \left(\frac{p}{p-1}\right)^p \int_0^\infty f(x)^p \,\mathrm{d}x.$$

(2) Show that the constant $\left(\frac{p}{p-1}\right)^p$ is the best possible in this inequality.
(3) Show that there is no nonzero function f for which this inequality becomes an equality.

2.6 Regularization and approximation in the spaces $L^p(\Omega)$, $1 \leq p < \infty$

Let Ω be an open subset of $\mathbb{R}^n$. A function $f : \Omega \to [-\infty, \infty]$ is said to be **locally integrable** in Ω if f is measurable and the restriction $f|_K$ of f to any *compact* subset K of Ω belongs to the space $\mathcal{L}^1(K)$. Since any compact subset of Ω admits a finite covering by open subsets with compact closures in Ω (balls for instance), a measurable function $f : \Omega \to [-\infty, \infty]$ is thus locally integrable if (and only if), given any *open* subset U of Ω such that $\overline{U}$ *is a compact subset of* Ω, its restriction $f|_U$ is in the space $L^1(U)$.

Such locally integrable functions clearly form a vector space, denoted $\mathcal{L}^1_{\text{loc}}(\Omega)$. The quotient set

$$L^1_{\text{loc}}(\Omega) := \mathcal{L}^1_{\text{loc}}(\Omega)/\mathcal{R},$$

where the equivalence relation $\mathcal{R}$ is that of equality almost everywhere in Ω, also clearly forms a *vector space*. As is customary, functions in $\mathcal{L}^1_{\text{loc}}(\Omega)$ will be identified with their equivalence classes in $L^1_{\text{loc}}(\Omega)$.

Remark The space $L^1_{\text{loc}}(\Omega)$ can be equipped with a *metrizable topology*; cf. Problem 2.6-1. □

More generally, one can define the space

$$L^p_{\text{loc}}(\Omega), \quad 1 \leq p \leq \infty,$$

as the set of all measurable functions $f : \Omega \to [-\infty, \infty]$ with the property that $f|_U \in L^p(U)$ for any open subset U of Ω such that $\overline{U}$ is a compact subset of Ω.

Any function $f \in L^p(\Omega)$, $1 \leq p \leq \infty$, *is locally integrable in* Ω, since for any compact subset K of Ω,

$$\int_K |f(x)|\,\mathrm{d}x \leq \|f\|_{L^1(\Omega)} < \infty$$

[10]This is the famed *Hardy inequality*, due to:
G.H. Hardy [1925]: Notes on some points in the integral calculus. LX. An inequality between integrals, *Messengers of Mathematics* **54**, 150–156.
Since its inception, this inequality has generated numerous developments, well documented in the survey:
A. Kufner; L. Maligranda; L.E. Persson [2007]: *The Hardy Inequality: About Its History and Some Related Results*, Vydavatelský Servis, Pilsen.

if $p = 1$, and

$$\int_K |f(x)|\,\mathrm{d}x \le \Big(\int_K \mathrm{d}x\Big)^{1/q}\Big(\int_K |f(x)|^p\,\mathrm{d}x\Big)^{1/p} \le \Big(\int_K \mathrm{d}x\Big)^{1/q}\|f\|_{L^p(\Omega)} < \infty$$

with $q = \dfrac{p}{p-1}$ if $1 < p < \infty$ (by Hölder's inequality, cf. Theorem 2.5-1(a)), or with $q = 1$ if $p = \infty$.

Clearly, *any function in the space $\mathcal{C}(\Omega)$ is locally integrable in* Ω, since for any compact subset K of Ω,

$$\int_K |f(x)|\,\mathrm{d}x \le \Big(\int_K \mathrm{d}x\Big)\sup_{x\in K}|f(x)| < \infty.$$

A **family of mollifiers in** $\mathbb{R}^n$ is a family $(\omega_\varepsilon)_{\varepsilon>0}$ of functions $\omega_\varepsilon : \mathbb{R}^n \to \mathbb{R}$ of the form

$$\omega_\varepsilon(x) = \frac{1}{\varepsilon^n}\omega\Big(\frac{x}{\varepsilon}\Big), \quad x \in \mathbb{R}^n,$$

where $\omega : \mathbb{R}^n \to \mathbb{R}$ is any function that possesses the following properties:

$$\omega \in \mathcal{C}^\infty(\mathbb{R}^n), \quad \omega(x) \ge 0 \ \text{ for all } x \in \mathbb{R}^n, \quad \operatorname{supp}\omega \subset \overline{B(0;1)}, \quad \text{and} \quad \int_{\mathbb{R}^n}\omega(x)\,\mathrm{d}x = 1.$$

Hence, for each $\varepsilon > 0$,

$$\omega_\varepsilon \in \mathcal{C}^\infty(\mathbb{R}^n), \quad \omega_\varepsilon(x) \ge 0 \text{ for all } x \in \mathbb{R}^n, \quad \operatorname{supp}\omega_\varepsilon \subset \overline{B(0;\varepsilon)}, \quad \text{and} \quad \int_{\mathbb{R}^n}\omega_\varepsilon(x)\mathrm{d}x = 1.$$

An example of a function ω with the above properties is given by

$$\omega(x) := ce^{\frac{1}{|x|^2-1}} \text{ for } |x| < 1 \quad \text{and} \quad \omega(x) = 0 \text{ for } |x| \ge 1,$$

where the constant $c > 0$ is such that $\int_{B(0;1)}\omega(y)\,\mathrm{d}y = 1$ (Problem 2.6-2).

Let Ω be an open subset of $\mathbb{R}^n$. Given a function $f \in L^1_{\mathrm{loc}}(\Omega)$ and a family $(\omega_\varepsilon)_{\varepsilon>0}$ of mollifiers, let the set Ω_ε and the function $f_\varepsilon : \Omega_\varepsilon \to \mathbb{R}$ be defined for each $\varepsilon > 0$ by

$$\Omega_\varepsilon := \{x \in \Omega;\ \operatorname{dist}(x, \mathbb{R}^n - \Omega) > \varepsilon\},$$
$$f_\varepsilon(x) := \int_\Omega \omega_\varepsilon(x-y)f(y)\,\mathrm{d}y \quad \text{for all } x \in \Omega_\varepsilon.$$

The family $(f_\varepsilon)_{\varepsilon>0}$ is then called a **regularizing family of** f.

For each $\varepsilon > 0$, the set Ω_ε is clearly <u>open</u> (the function $x \in \Omega \to \operatorname{dist}(x, \mathbb{R}^n - \Omega)$ is continuous; cf. Theorem 1.11-3), the ball $\overline{B(x;\varepsilon)}$ is contained in Ω (so that the above definition of the function f_ε makes sense), and $f_\varepsilon(x)$ is equivalently given for each $x \in \Omega_\varepsilon$ by

$$\begin{aligned} f_\varepsilon(x) &= \int_{B(x;\varepsilon)}\omega_\varepsilon(x-y)f(y)\,\mathrm{d}y = \int_{B(0;\varepsilon)}\omega_\varepsilon(z)f(x-z)\,\mathrm{d}z \\ &= \frac{1}{\varepsilon^n}\int_{B(x;1)}\omega\Big(\frac{x-y}{\varepsilon}\Big)f(y)\,\mathrm{d}y. \end{aligned}$$

Note also that, unless $\Omega = \mathbb{R}^n$, in which case $\Omega_\varepsilon = \Omega$ for all $\varepsilon > 0$, each function $f_\varepsilon : \Omega_\varepsilon \to \mathbb{R}$ is only defined on the *proper* subset Ω_ε of Ω.

The next theorem establishes two important properties of such a regularizing family, namely that the functions f_ε are *infinitely differentiable* (a "regularization" property) and that, if $f \in \mathcal{C}(\Omega)$, *the functions f_ε converge uniformly to f on compact subsets of Ω as* $\varepsilon \to 0$ (an "approximation" property). Other, equally important, approximation properties of regularizing families will be established in Theorems 2.6-3 and 2.6-4.

Theorem 2.6-1 (a) *Let Ω be an open subset of $\mathbb{R}^n$, and let a function $f \in L^1_{\mathrm{loc}}(\Omega)$ and a regularizing family $(f_\varepsilon)_{\varepsilon>0}$ of f be given. Then*

$$f_\varepsilon \in \mathcal{C}^\infty(\Omega_\varepsilon) \quad \textit{for all } \varepsilon > 0.$$

Besides,

$$\partial^{\boldsymbol{\alpha}} f_\varepsilon(x) = \int_\Omega \partial^{\boldsymbol{\alpha}} \omega_\varepsilon(x-y) f(y)\,\mathrm{d}y = \int_{B(x;\varepsilon)} \partial^{\boldsymbol{\alpha}} \omega_\varepsilon(x-y) f(y)\,\mathrm{d}y \quad \textit{at each } x \in \Omega_\varepsilon,$$

for any multi-index $\boldsymbol{\alpha} = (\alpha_1, \alpha_2, \ldots, \alpha_n)$ with $|\boldsymbol{\alpha}| = \sum_{i=1}^n \alpha_i \geq 1$.

(b) *Assume in addition that $f \in \mathcal{C}^m(\Omega)$ for some integer $m \geq 1$. Then, given any compact subset K of Ω, there exists $\varepsilon_0 = \varepsilon_0(K) > 0$ such that $K \subset \Omega_\varepsilon$ for all $0 < \varepsilon \leq \varepsilon_0$, $f_\varepsilon(x)$ is well defined for all $x \in K$ and all $0 < \varepsilon \leq \varepsilon_0$, and*

$$\sup_{x \in K} |\partial^{\boldsymbol{\alpha}} f_\varepsilon(x) - \partial^{\boldsymbol{\alpha}} f(x)| \to 0 \quad \textit{for all } |\boldsymbol{\alpha}| \leq m \textit{ as } \varepsilon \to 0.$$

Proof (i) In this part, $\varepsilon > 0$ is fixed. Let $x \in \Omega_\varepsilon$ and, for some $1 \leq i \leq n$, let $\boldsymbol{e}_i$ be a vector of the canonical basis in $\mathbb{R}^n$. Since Ω_ε is open, there exists $h_0 > 0$ such that $(x + h\boldsymbol{e}_i) \in \Omega_\varepsilon$ for all $|h| \leq h_0$. Consequently, we can write

$$\frac{1}{h}\{f_\varepsilon(x + h\boldsymbol{e}_i) - f_\varepsilon(x)\} = \frac{1}{\varepsilon^n}\int_\Omega \frac{1}{h}\Big\{\omega\Big(\frac{x + h\boldsymbol{e}_i - y}{\varepsilon}\Big) - \omega\Big(\frac{x-y}{\varepsilon}\Big)\Big\} f(y)\,\mathrm{d}y \quad \text{for all } |h| \leq h_0.$$

Since $\omega \in \mathcal{C}^\infty(\mathbb{R}^n)$ by assumption and since the set $\Big\{\Big(\frac{x + h\boldsymbol{e}_i - y}{\varepsilon}\Big) \in \mathbb{R}^n;\ |h| \leq h_0\Big\}$ is compact, there exists a constant M such that

$$\Big|\frac{1}{h}\Big\{\omega\Big(\frac{x + h\boldsymbol{e}_i - y}{\varepsilon}\Big) - \omega\Big(\frac{x-y}{\varepsilon}\Big)\Big\} - \frac{1}{\varepsilon}\partial_i\omega\Big(\frac{x-y}{\varepsilon}\Big)\Big| \leq \frac{h}{2\varepsilon^2}M \quad \text{for all } |h| \leq h_0.$$

Noting that $\partial_i\omega\Big(\frac{x-y}{\varepsilon}\Big) = \varepsilon^{n+1}\partial_i\omega_\varepsilon(x-y)$, we thus infer that

$$\Big|\frac{1}{\varepsilon^n}\int_\Omega \frac{1}{h}\Big\{\omega\Big(\frac{x + h\boldsymbol{e}_i - y}{\varepsilon}\Big) - \omega\Big(\frac{x-y}{\varepsilon}\Big)\Big\} f(y)\,\mathrm{d}y - \int_\Omega \partial_i\omega_\varepsilon(x-y) f(y)\,\mathrm{d}y\Big| \leq \frac{hM}{2\varepsilon^2}\int_{B(x;\varepsilon)} |f(y)|\,\mathrm{d}y \quad \text{for all } |h| \leq h_0.$$

Letting $h \to 0$ then shows that the partial derivative $\partial_i f_\varepsilon(x)$ exists and is given by

$$\partial_i f_\varepsilon(x) = \int_\Omega \partial_i\omega_\varepsilon(x-y) f(y)\,\mathrm{d}y = \int_{B(x;\varepsilon)} \partial_i\omega_\varepsilon(x-y) f(y)\,\mathrm{d}y \quad \text{at each } x \in \Omega_\varepsilon.$$

An analogous argument clearly applies to any partial derivative $\partial^{\boldsymbol{\alpha}} f_\varepsilon(x)$ with $|\boldsymbol{\alpha}| \ge 2$.

(ii) Assume that $\Omega \neq \mathbb{R}^n$ and that $f \in \mathcal{C}(\Omega)$. Given any compact subset K of Ω, the set

$$K_0 := \{x \in \Omega; \operatorname{dist}(x, K) \le \delta\}, \quad \text{where } 2\delta := \inf_{x \in K} \operatorname{dist}(x; \mathbb{R}^n - \Omega) > 0,$$

is a compact subset of Ω. Since $\bigcup_{\varepsilon>0} \Omega_\varepsilon = \Omega$ constitutes an open covering of K_0 and $\Omega_{\varepsilon'} \subset \Omega_\varepsilon$ if $\varepsilon < \varepsilon'$, there exists $\varepsilon_0 = \varepsilon_0(K) > 0$ such that $K_0 \subset \Omega_\varepsilon$ for all $\varepsilon \le \varepsilon_0$. Hence $f_\varepsilon(x) = \int_{B(x;\varepsilon)} \omega_\varepsilon(x-y) f(y)\,\mathrm{d}y$ is well defined for all $x \in K$ and all $0 < \varepsilon \le \varepsilon_0$. Recalling that $\omega_\varepsilon(z) \ge 0$ for all $z \in \mathbb{R}^n$ and $\int_{B(0;\varepsilon)} \omega_\varepsilon(z)\,\mathrm{d}z = 1$, we infer that, for all $x \in K$ and all $\varepsilon \le \varepsilon_0$,

$$\begin{aligned} |f_\varepsilon(x) - f(x)| &= \Big| \int_{B(0;\varepsilon)} \omega_\varepsilon(z)(f(x-z) - f(x))\,\mathrm{d}z \Big| \\ &\le \sup_{\substack{x \in K \\ z \in B(0;\varepsilon)}} |f(x-z) - f(x)|. \end{aligned}$$

The uniform continuity of the function f on the compact set K_0 therefore implies that

$$\sup_{x \in K} |f_\varepsilon(x) - f(x)| \to 0 \quad \text{as } \varepsilon \to 0.$$

(iii) Assume again that $\Omega \neq \mathbb{R}^n$ and that $f \in \mathcal{C}^m(\Omega)$ for some $m \ge 1$. Then, for all $x \in K$ and all $0 < \varepsilon \le \varepsilon_0$,

$$\partial^{\boldsymbol{\alpha}} f_\varepsilon(x) = \int_{B(x;\varepsilon)} \partial_x^{\boldsymbol{\alpha}} \omega_\varepsilon(x-y) f(y) \mathrm{d}y = (-1)^{|\boldsymbol{\alpha}|} \int_{B(x;\varepsilon)} \partial_y^{\boldsymbol{\alpha}} \omega_\varepsilon(x-y) f(y)\,\mathrm{d}y,$$

where $\partial_x^{\boldsymbol{\alpha}}$ and $\partial_y^{\boldsymbol{\alpha}}$ respectively denote partial differentiation with respect to the x and y variables. Then m successive integration by parts (these do not require any regularity on $\partial\Omega$ since $\operatorname{supp} \omega_\varepsilon(x - \cdot) \subset \overline{B(x;\varepsilon)} \subset \Omega$) give

$$\int_{B(x;\varepsilon)} \partial_y^{\boldsymbol{\alpha}} \omega_\varepsilon(x-y) f(y)\,\mathrm{d}y = (-1)^{|\boldsymbol{\alpha}|} \int_{B(x;\varepsilon)} \omega_\varepsilon(x-y) \partial^{\boldsymbol{\alpha}} f(y)\,\mathrm{d}y.$$

Consequently,

$$\begin{aligned} |\partial^{\boldsymbol{\alpha}} f_\varepsilon(x) - \partial^{\boldsymbol{\alpha}} f(x)| &= \Big| \int_{B(0;\varepsilon)} \omega_\varepsilon(z)(\partial^{\boldsymbol{\alpha}} f(x-z) - \partial^{\boldsymbol{\alpha}} f(x))\,\mathrm{d}z \Big| \\ &\le \sup_{\substack{x \in K \\ z \in B(0;\varepsilon)}} |\partial^{\boldsymbol{\alpha}} f(x-z) - \partial^{\boldsymbol{\alpha}} f(x)|, \end{aligned}$$

and the conclusion follows from the uniform continuity of $\partial^{\boldsymbol{\alpha}} f$ on K_0.

(iv) If $\Omega = \mathbb{R}^n$, the same argument holds, with $\delta > 0$ and $\varepsilon_0 > 0$ being now arbitrarily chosen, since $f_\varepsilon(x)$ is well defined for all $x \in \mathbb{R}^n$ and $\varepsilon > 0$ in this case. ☐

Remark The formula $\partial_i f_\varepsilon(x) = \int_\Omega \partial_i \omega_\varepsilon(x-y) f(y)\,\mathrm{d}y$ at each $x \in \Omega_\varepsilon$ established in part (i) also follows from a general criterion of differentiability for functions defined by an integral (this criterion will be established later; cf. Theorem 7.4-1). □

As we shall see later, the subspace

$$\mathcal{D}(\Omega) := \{f \in \mathcal{C}^\infty(\Omega);\ \operatorname{supp} f \text{ is a compact subset of } \Omega\}$$

of the space $\mathcal{C}^\infty(\Omega)$ plays a role of paramount importance in the definition of *weak*, or *distributional, derivatives* as found in, e.g., Sobolev spaces (Chapter 6); note that the space $\mathcal{D}(\Omega)$ contains *nonzero* functions (such as the functions denoted $\widetilde{g}_\varepsilon$ in the next proof). At this stage, we only prove one, but very important, property of this space, which considerably extends that proved in Theorem 2.5-3.

Remark The space $\mathcal{D}(\Omega)$ is sometimes denoted $\mathcal{C}_c^\infty(\Omega)$ in the literature. The letter $\mathcal{D}$ reflects that its elements play a key role in the definition of *distributions* over Ω (Section 6.3). □

Theorem 2.6-2 *Let Ω be an open subset of $\mathbb{R}^n$. For each $1 \le p < \infty$, the space $\mathcal{D}(\Omega)$ is dense in the space $L^p(\Omega)$.*

Proof Let a function $f \in L^p(\Omega)$ and $\eta > 0$ be given. By Theorem 2.5-3, there exists a function $g = g(f, \eta) \in \mathcal{C}_c(\Omega)$ such that

$$\|f - g\|_{L^p(\Omega)} \le \frac{\eta}{2}.$$

Let $(g_\varepsilon)_{\varepsilon>0}$ be a regularizing family of g. Since $\operatorname{supp} g$ is a compact subset of Ω, the same argument as in part (ii) of the proof of Theorem 2.6-1 can be repeated, showing that there exist a compact subset K_0 of Ω and $\varepsilon_1 > 0$ such that

$$\operatorname{supp} g \subset \operatorname{supp} g_\varepsilon \subset K_0 \subset \Omega_\varepsilon \quad \text{for all } \varepsilon \le \varepsilon_1.$$

Besides,

$$\sup_{x \in K_0} |g_\varepsilon(x) - g(x)| \to 0 \quad \text{as } \varepsilon \to 0$$

since K_0 is a compact subset of Ω (Theorem 2.6-1(b)). Therefore,

$$\|g - g_\varepsilon\|_{L^p(\Omega_\varepsilon)} = \left(\int_{K_0} |g_\varepsilon(x) - g(x)|^p\,\mathrm{d}x\right)^{1/p} \le \left(\int_{K_0} \mathrm{d}x\right)^{1/p} \sup_{x\in K_0} |g_\varepsilon(x) - g(x)| \le \frac{\eta}{2}$$

if $\varepsilon > 0$ is small enough. Let $\widetilde{g}_\varepsilon$ denote the extension by 0 of g_ε on $\Omega - \Omega_\varepsilon$. Then $\widetilde{g}_\varepsilon \in \mathcal{D}(\Omega)$ since $g_\varepsilon \in \mathcal{D}(\Omega_\varepsilon)$, and

$$\|f - \widetilde{g}_\varepsilon\|_{L^p(\Omega)} = \|f - g_\varepsilon\|_{L^p(\Omega_\varepsilon)} \le \|f - g\|_{L^p(\Omega)} + \|g - g_\varepsilon\|_{L^p(\Omega_\varepsilon)} \le \eta$$

if $\varepsilon > 0$ is small enough. Since $\eta > 0$ is arbitrary, the conclusion follows. □

Note that Theorem 2.6-2 does *not* hold for $p = \infty$ (Problem 2.6-3). Note also that this theorem provides another way of defining each Lebesgue space $L^p(\Omega)$, $1 \le p < \infty$, as the *completion of the space* $\mathcal{D}(\Omega)$ *with respect to the norm* $\|\cdot\|_{L^p(\Omega)}$ (in the definition of which the Riemann integral is used).

In the remainder of this section, we assume that $\Omega = \mathbb{R}^n$, in which case, given a function $f \in L^1_{\text{loc}}(\mathbb{R}^n)$, the function f_ε found in any regularizing family $(f_\varepsilon)_{\varepsilon>0}$ of f is also defined on $\mathbb{R}^n$ (since $\Omega_\varepsilon = \mathbb{R}^n$ for all $\varepsilon > 0$ if $\Omega = \mathbb{R}^n$). More specifically, we now have

$$f_\varepsilon(x) = \int_{\mathbb{R}^n} \omega_\varepsilon(x-y) f(y)\,\mathrm{d}y = \int_{\mathbb{R}^n} \omega_\varepsilon(y) f(x-y)\,\mathrm{d}y \quad \text{for all } x \in \mathbb{R}^n$$

(the equality of the two integrals over $\mathbb{R}^n$ holds thanks to Theorem 1.16-1).

Remark The function f_ε is in effect the *convolution product* of the functions ω_ε and f; see Problem 2.6-4 for some details about this important notion. □

The next result establishes a fundamental property of any regularizing family of a function $f \in L^p(\mathbb{R}^n)$, $1 \le p < \infty$. Note that it also provides another, more constructive, proof of Theorem 2.6-2 when $\Omega = \mathbb{R}^n$.

Theorem 2.6-3 (regularization and approximation in $L^p(\mathbb{R}^n)$, $1 \le p < \infty$**)** *Let a function* $f \in L^p(\mathbb{R}^n)$, $1 \le p < \infty$, *be given, and let* $(f_\varepsilon)_{\varepsilon>0}$ *be a regularizing family of* f. *Then*

$$f_\varepsilon \in \mathcal{C}^\infty(\mathbb{R}^n) \cap L^p(\mathbb{R}^n) \quad \textit{for all } \varepsilon > 0,$$

and

$$\|f_\varepsilon - f\|_{L^p(\mathbb{R}^n)} \to 0 \quad \textit{as } \varepsilon \to 0.$$

Proof (i) *First, we show that* $f_\varepsilon \in L^p(\mathbb{R}^n)$ *for each* $\varepsilon > 0$ (we already know from Theorem 2.6-1 that $f_\varepsilon \in \mathcal{C}^\infty(\mathbb{R}^n)$) *and that*

$$\|f_\varepsilon\|_{L^p(\mathbb{R}^n)} \le \|f\|_{L^p(\mathbb{R}^n)} \quad \text{for all } \varepsilon > 0.$$

If $p = 1$, Fubini's theorem (Theorem 1.15-5(b)), combined with the relations $\omega_\varepsilon(x) \ge 0$ for all $x \in \mathbb{R}^n$ and $\int_{\mathbb{R}^n} \omega_\varepsilon(x)\,\mathrm{d}x = 1$, gives

$$\begin{aligned}\int_{\mathbb{R}^n} |f_\varepsilon(x)|\,\mathrm{d}x &\le \int_{\mathbb{R}^n} \left(\int_{\mathbb{R}^n} \omega_\varepsilon(x-y)\,|f(y)|\,\mathrm{d}y \right) \mathrm{d}x \\ &= \int_{\mathbb{R}^n} |f(y)| \left(\int_{\mathbb{R}^n} \omega_\varepsilon(x-y)\,\mathrm{d}x \right) \mathrm{d}y = \|f\|_{L^1(\mathbb{R}^n)}.\end{aligned}$$

If $1 < p < \infty$, let q be defined by $\dfrac{1}{p} + \dfrac{1}{q} = 1$. Then Hölder's inequality (Theorem 2.5-1) gives

$$\begin{aligned}|f_\varepsilon(x)| &\le \int_{B(x;\varepsilon)} \omega_\varepsilon(x-y)\,|f(y)|\,\mathrm{d}y \\ &\le \left(\int_{B(x;\varepsilon)} \omega_\varepsilon(x-y)\,\mathrm{d}y \right)^{1/q} \left(\int_{B(x;\varepsilon)} \omega_\varepsilon(x-y)\,|f(y)|^p\,\mathrm{d}y \right)^{1/p} \\ &= \left(\int_{\mathbb{R}^n} \omega_\varepsilon(x-y)\,|f(y)|^p\,\mathrm{d}y \right)^{1/p} \quad \text{for all } x \in \mathbb{R}^n,\end{aligned}$$

and thus, again by Fubini's theorem,

$$\begin{aligned}\int_{\mathbb{R}^n} |f_\varepsilon(x)|^p \,\mathrm{d}x &\le \int_{\mathbb{R}^n} \left(\int_{\mathbb{R}^n} \omega_\varepsilon(x-y)\,|f(y)|^p \,\mathrm{d}y \right) \mathrm{d}x \\ &= \int_{\mathbb{R}^n} |f(y)|^p \left(\int_{\mathbb{R}^n} \omega_\varepsilon(x-y)\,\mathrm{d}x \right) \mathrm{d}y = \left(\|f\|_{L^p(\mathbb{R}^n)}\right)^p .\end{aligned}$$

(ii) *Second, we show that* $\|f_\varepsilon - f\|_{L^p(\mathbb{R}^n)} \to 0$ *as* $\varepsilon \to 0$.

Let $\eta > 0$ be given. By Theorem 2.5-3, there exists a function $g = g(f,\eta) \in \mathcal{C}_c(\Omega)$ such that

$$\|f - g\|_{L^p(\mathbb{R}^n)} \le \frac{\eta}{3}.$$

Since $\operatorname{supp} g$ is a compact subset of Ω, there exists a compact subset K_0 of $\mathbb{R}^n$ and $\varepsilon_1 > 0$ such that

$$\operatorname{supp} g \subset \operatorname{supp} g_\varepsilon \subset K_0 \quad \text{for all } \varepsilon \le \varepsilon_1.$$

Hence

$$\|g_\varepsilon - g\|_{L^p(\mathbb{R}^n)} = \|g_\varepsilon - g\|_{L^p(K_0)} \le \left(\int_{K_0} \mathrm{d}x \right)^{1/p} \sup_{x \in K_0} |g_\varepsilon(x) - g(x)| .$$

Besides,

$$\sup_{x \in K_0} |g_\varepsilon(x) - g(x)| \to 0 \quad \text{as } \varepsilon \to 0,$$

by Theorem 2.6-1(b). Therefore there exists $\varepsilon_0 > 0$ such that

$$\|g_\varepsilon - g\|_{L^p(\mathbb{R}^n)} \le \frac{\eta}{3} \quad \text{for all } \varepsilon \le \varepsilon_0.$$

Noting that $f_\varepsilon - g_\varepsilon = (f-g)_\varepsilon$ and that $\|(f-g)_\varepsilon\|_{L^p(\mathbb{R}^n)} \le \|f-g\|_{L^p(\mathbb{R}^n)}$ by (i), we finally obtain

$$\|f_\varepsilon - f\|_{L^p(\mathbb{R}^n)} \le \|(f-g)_\varepsilon\|_{L^p(\mathbb{R}^n)} + \|g_\varepsilon - g\|_{L^p(\mathbb{R}^n)} + \|g - f\|_{L^p(\mathbb{R}^n)} \le \eta \quad \text{for all } \varepsilon \le \varepsilon_0.$$

Since $\eta > 0$ is arbitrary, the conclusion follows. □

The next result provides a useful complement to both Theorems 2.6-1 and 2.6-3.

Theorem 2.6-4 *Let Ω be an open subset of $\mathbb{R}^n$. Let there be given a function $f \in L^p_{\mathrm{loc}}(\Omega)$, $1 \le p < \infty$, and a regularizing family $(f_\varepsilon)_{\varepsilon>0}$ of f. Then, given any open subset U of Ω such that $\overline{U}$ is a compact subset of Ω, there exists $\varepsilon_0 = \varepsilon_0(U) > 0$ such that $\overline{U} \subset \Omega_\varepsilon$ for all $0 < \varepsilon \le \varepsilon_0$, and*

$$\|f_\varepsilon - f\|_{L^p(U)} \to 0 \quad \textit{as } \varepsilon \to 0.$$

Proof Let V be an open subset of Ω such that

$$\overline{U} \subset V \text{ and } \overline{V} \text{ is a compact subset of } \Omega.$$

Then $f|_V \in L^p(V)$ by assumption, and the function $\widetilde{f}$ defined on $\mathbb{R}^n$ by $\widetilde{f}|_V = f|_V$ and $\widetilde{f}|_{\mathbb{R}^n - V} = 0$ belongs to $L^p(\mathbb{R}^n)$. Besides, there exists $\varepsilon_0 > 0$ such that the regularizing families $(f_\varepsilon)_{\varepsilon>0}$ and $(\widetilde{f}_\varepsilon)_{\varepsilon>0}$ coincide over U for $\varepsilon \le \varepsilon_0$. Since then

$$\|f_\varepsilon - f\|_{L^p(U)} = \|\widetilde{f}_\varepsilon - \widetilde{f}\|_{L^p(U)} \le \|\widetilde{f}_\varepsilon - \widetilde{f}\|_{L^p(\mathbb{R}^n)} \quad \text{for all } \varepsilon \le \varepsilon_0,$$

Theorem 2.6-3 shows that $\|f_\varepsilon - f\|_{L^p(U)} \to 0$ as $\varepsilon \to 0$. □

Problems

2.6-1 A sequence $(f_n)_{n=1}^\infty$ of functions $f_n \in L^1_{\text{loc}}(\Omega)$ is said to *converge in* $L^1_{\text{loc}}(\Omega)$ *to a function* $f \in L^1_{\text{loc}}(\Omega)$ if

$$\lim_{n\to\infty} \|f_n - f\|_{L^1(K)} = 0 \quad \text{for all compact subsets } K \subset \Omega.$$

Show that there exists a distance d on the vector space $L^1_{\text{loc}}(\Omega)$ such that $(f_n)_{n=1}^\infty$ converges to f in $L^1_{\text{loc}}(\Omega)$ if and only if

$$\lim_{n\to\infty} d(f_n, f) = 0.$$

This property shows that the above notion of convergence defines a *metrizable topology*, which is called the *Fréchet topology associated with the family of seminorms* $(\|\cdot\|_{L^1(K)})_{K\in\mathcal{K}}$, where $\mathcal{K}$ denotes the family of all compact subsets of Ω.

Hint: Mimic Problem 2.3-2.

2.6-2 (1) Show that the function $\theta : \mathbb{R}^n \to \mathbb{R}$ defined by

$$\theta(x) = e^{\frac{1}{|x|^2-1}} \quad \text{for } |x| < 1 \quad \text{and} \quad \theta(x) = 0 \text{ for } |x| \ge 1,$$

is infinitely differentiable in $\mathbb{R}^n$.

(2) What does the Taylor formula with integral remainder (to fix ideas) at a point $x_0 \in \mathbb{R}^n$ such that $|x_0| = 1$ look like for the function θ?

2.6-3 Show that the subspace $\mathcal{D}(\Omega)$ is not dense in $L^\infty(\Omega)$.

2.6-4 Let there be given two functions $f \in L^1(\mathbb{R}^n)$ and $g \in L^p(\mathbb{R}^n)$, $1 \le p \le \infty$.

(1) Show that the function $y \in \mathbb{R}^n \to f(x-y)g(y)$ is integrable in $\mathbb{R}^n$ for almost all $x \in \mathbb{R}^n$. Hence

$$(f * g)(x) := \int_{\mathbb{R}^n} f(x-y)g(y)\,\mathrm{d}y$$

is a well-defined real number for almost all $x \in \mathbb{R}^n$, which thus defines a function $f * g : \mathbb{R}^n \to \mathbb{R}$, called the **convolution product** of f and g.

(2) Show that

$$f * g \in L^p(\mathbb{R}^n) \quad \text{and} \quad \|f * g\|_{L^p(\mathbb{R}^n)} \le \|f\|_{L^1(\mathbb{R}^n)} \|g\|_{L^p(\mathbb{R}^n)}.$$

Remark By a result that will be proved later (Theorem 2.11-1), this inequality implies that the bilinear mapping

$$(f, g) \in L^1(\mathbb{R}^n) \times L^p(\mathbb{R}^n) \to f * g \in L^p(\mathbb{R}^n)$$

defined in this fashion is *continuous*. □

2.6-5 (1) Let $\varphi \in L^\infty(\mathbb{R}^n)$ be a function that is ≥ 0 almost everywhere and has a compact support. Show that there exist a bounded open subset U of $\mathbb{R}^n$ and functions $\varphi_k : U \to \mathbb{R}$, $k \ge 1$, with the following properties:

$$\varphi_k \in \mathcal{D}(U), \quad \varphi_k \ge 0 \text{ in } U, \quad \text{and} \quad \|\varphi_k\|_{L^\infty(U)} \le \|\varphi\|_{L^\infty(U)} \quad \text{for all } k \ge 1,$$
$$\text{for almost all } x \in U, \quad \varphi_k(x) \to \varphi(x) \quad \text{as } k \to \infty.$$

(2) Let Ω be an open subset of $\mathbb{R}^n$ and let a function $f \in L^1_{\text{loc}}(\Omega)$ be such that

$$\int_\Omega f\varphi\,\mathrm{d}x \ge 0 \quad \text{for all } \varphi \in \mathcal{D}(\Omega) \text{ that are } \ge 0 \text{ in } \Omega.$$

Show that $f \geq 0$ almost everywhere in Ω.

Hint: First show, using (1), that $\int_\Omega f\varphi \mathrm{d}x \geq 0$ for all $\varphi \in L^\infty(\Omega)$ that are ≥ 0 almost everywhere in Ω and have compact support. Then show that, given any open subset V of Ω such that $\overline{V}$ is compact, $\mathrm{d}x$-meas$\{x \in V;\ f(x) < 0\} = 0$.

2.7 Compactness and finite-dimensional normed vector spaces; F. Riesz theorem

The objective of this section is to review basic properties of *finite-dimensional* normed vector spaces, most of them related to the notion of *compactness*.

To begin with, property (a) in the next theorem essentially asserts that *there is only one norm topology in a finite-dimensional vector space*, which may thus be defined as that defined by one of the norms $\|\cdot\|_p$, $1 \leq p \leq \infty$ (Theorem 2.2-2). Properties (b) and (c) extend to arbitrary finite-dimensional vector spaces properties of finite-dimensional spaces equipped with one of the norms $\|\cdot\|_p$, $1 \leq p \leq \infty$ (Theorems 1.13-5 and 2.2-2(b)). Property (d) is an important topological property of finite-dimensional subspaces. Note that the proofs of properties (b), (c), and (d) all rely on property (a).

Theorem 2.7-1 (a) *Any two norms $\|\cdot\|$ and $\|\cdot\|'$ in a finite-dimensional vector space X are equivalent, i.e., the topologies induced on X by $\|\cdot\|$ and $\|\cdot\|'$ are identical.*

(b) *Any finite-dimensional normed vector space is separable.*

(c) *A subset of a finite-dimensional normed vector space is compact if and only if it is closed and bounded.*

(d) *A finite-dimensional subspace of a normed vector space X is closed in X.*

Proof (i) Let $(e_i)_{i=1}^n$ be a basis of X. It clearly suffices to prove that any norm on X is equivalent to the particular norm $\|\cdot\|_1 : x = \sum_{i=1}^n x_i e_i \to \sum_{i=1}^n |x_i|$ (Theorem 2.2-4). To this end, first notice that

$$\|x\| = \Big\|\sum_{i=1}^n x_i e_i\Big\| \leq C_1 \|x\|_1 \quad \text{for all } x \in X,$$

with $C_1 := \max_{1 \leq i \leq n} \|e_i\|$.

Consider next the function

$$f : x \in (X, \|\cdot\|_1) \to f(x) := \|x\| \in \mathbb{R},$$

and the set

$$K := \{y \in X;\ \|y\|_1 = 1\}.$$

Then f is a *continuous function on* X, since

$$|f(x) - f(y)| = \big|\|x\| - \|y\|\big| \leq \|x - y\| \leq C_1 \|x - y\|_1 \quad \text{for all } x, y \in X,$$

and K is a *compact subset of* X, as a bounded and closed subset of the metric space (X, d_1) (Theorem 1.13-5). Hence there exists $y_0 \in K$ such that $f(y_0) = \inf_{y \in K} f(y)$ (Theorem 1.13-6) and $\frac{1}{C} := f(y_0) = \|y_0\| > 0$ since $y_0 \neq 0$. Therefore,

$$\|y\|_1 = 1 \quad \text{implies} \quad \|y\| \geq \frac{1}{C}.$$

Given any nonzero vector $x \in X$, the vector $y := \frac{x}{\|x\|_1}$ satisfies $\|y\|_1 = 1$, and hence $\|y\| = \frac{\|x\|}{\|x\|_1} \geq \frac{1}{C}$. We thus have

$$\|x\|_1 \leq C\|x\| \quad \text{for all } x \in X.$$

By Theorem 2.2-4, the topologies induced by $\|\cdot\|$ and $\|\cdot\|'$ are thus identical. This proves (a), which, combined with Theorem 2.2-2(b), in turn proves (b).

(ii) Let K be a closed and bounded subset in a finite-dimensional normed vector space $(X, \|\cdot\|)$. Then K is closed and bounded in, e.g., $(X, \|\cdot\|_1)$ by (a), hence compact in $(X, \|\cdot\|_1)$ by Theorem 1.13-5, and hence compact in $(X, \|\cdot\|)$ since the topologies of $(X, \|\cdot\|)$ and $(X, \|\cdot\|_1)$ are the same, again by (a). The "only if" part holds in any metric space (Theorem 1.13-1); therefore it also holds in the present situation (in fact irrespective of whether X is finite-dimensional or not). This proves (c).

(iii) Let Y be a finite-dimensional subspace of a normed vector space $(X, \|\cdot\|)$, let $(e_i)_{i=1}^n$ be a basis of Y, and let $(y^k)_{k=1}^\infty$ be a sequence of vectors $y^k = \sum_{i=1}^n y_i^k e_i \in Y$ that converges in the space X.

The convergence of $(y^k)_{k=1}^\infty$ then implies that each sequence $(y_i^k)_{k=1}^\infty$ of scalars $y_i^k \in \mathbb{K}$, $1 \leq i \leq n$, is a Cauchy sequence, since by (a), there exists a constant C such that

$$\sum_{i=1}^n |y_i^k - y_i^\ell| = \|y^k - y^\ell\|_1 \leq C\|y^k - y^\ell\| \quad \text{for all } k, \ell \geq 1,$$

and $(y^k)_{k=1}^\infty$ is a Cauchy sequence (Theorem 1.12-1(b)). The scalar field $\mathbb{K}$ being complete, there exist $y_i \in \mathbb{K}$, $1 \leq i \leq n$, such that $y_i^k \to y_i$ as $k \to \infty$, and thus $\lim_{k\to\infty} \|y^k - y\|_1 = 0$ where $y := \sum_{i=1}^n y_i e_i$.

This implies that $\lim_{k\to\infty} \|y^k - y\| = 0$, since, again by (a), there exists a constant C_1 such that $\|y^k - y\| \leq C_1\|y^k - y\|_1$ for all $k \geq 1$ (the vectors y^k, $k \geq 1$, and y all belong to Y). Hence the sequence $(y^k)_{k=1}^\infty$ converges to the vector $y \in Y$, and thus Y is closed. This proves (d). $\square$

As an application, let $\mathcal{P}_n$ denote the space of all real polynomials $p : x \in \mathbb{R} \to p(x) = \sum_{j=0}^n c_j(p)x^j$ of degree $\leq n$. Then Theorem 2.7-1(a) shows that there exist constants C and C_1 (depending on n) such that

$$\sum_{j=0}^n |c_j(p)| \leq C \sup_{0\leq x\leq 1} |p(x)| \quad \text{and} \quad \sup_{0\leq x\leq 1} |p(x)| \leq C_1 \sum_{j=0}^n |c_j(p)| \quad \text{for all } p \in \mathcal{P}_n.$$

Incidentally, notice that, while the latter inequality is trivial to establish directly (with $C_1 = 1$), the former is not.

We can also prove that the equivalence of norms established in Theorem 2.7-1(a) in fact *characterizes finite-dimensional vector spaces.* Not surprisingly, the *axiom of choice* is again needed for this purpose, as the existence of a Hamel basis (used in the proof) depends on this axiom.

Theorem 2.7-2 *Let X be any infinite-dimensional vector space. Then there exist norms on X that are not equivalent.*

Proof Let $(e_i)_{i\in I}$ be a Hamel basis of X (Section 2.1); this means that any vector $x \in X$ can be written in a unique fashion as $x = \sum_{j\in J(x)} x_j e_j$, where $J(x)$ is a finite subset of I. It is then immediately verified that the mappings $\|\cdot\|_1 : X \to \mathbb{R}$ and $\|\cdot\|_\infty : X \to \mathbb{R}$ defined by

$$\|x\|_1 := \sum_{j\in J(x)} |x_j| \quad \text{and} \quad \|x\|_\infty := \max_{j\in J(x)} |x_j|,$$

are both norms on X. Since the set I is infinite (X is infinite-dimensional by assumption), the Hamel basis $(e_i)_{i\in I}$ contains a countably infinite subfamily $(e_j)_{j=1}^\infty$ (Theorem 1.5-3(a)). Then the sequence $(x_n)_{n=1}^\infty$ with $x_n := \sum_{j=1}^n \frac{1}{n} e_j$ is such that $\|x_n\|_1 = 1$ and $\|x_n\|_\infty = \frac{1}{n}$, $n \geq 1$. Hence there is no constant C such that $\|x\|_1 \leq C\|x\|_\infty$ for all $x \in X$. $\square$

Theorem 2.7-1(c) shows that, in a finite-dimensional normed vector space $(X, \|\cdot\|)$, the unit sphere $\{x \in X;\ \|x\| = 1\}$ is compact, as a particular closed and bounded subset of X. It is remarkable that this property also *characterizes finite-dimensional vector spaces*, according to the following fundamental theorem.

Theorem 2.7-3 (F. Riesz theorem) *A normed vector space $(X, \|\cdot\|)$ is finite-dimensional if and only if the unit sphere of X is compact.*

Proof Assume that the unit sphere

$$K := \{x \in X;\ \|x\| = 1\}$$

is compact in $(X, \|\cdot\|)$. There thus exist a finite number of points $x_i \in X$, $1 \leq i \leq n$, such that $K \subset \bigcup_{i=1}^n B\Big(x_i; \frac{1}{2}\Big)$ (Section 1.13).

The idea of the proof then simply consists in showing that X coincides with the finite-dimensional vector space

$$Y := \operatorname{Span}(x_i)_{i=1}^n.$$

To this end, it is enough to prove that, given any $x \in X$,

$$\inf_{y\in Y} \|x - y\| = 0,$$

since this will imply that $x \in \overline{Y}$, and $\overline{Y} = Y$ since Y is finite-dimensional (Theorem 2.7-1(d)).

So let $x \in X$ be given. If $x \in Y$, there is nothing to prove. Otherwise, given any $y \in Y$, let $\widetilde{x} := \dfrac{x}{\|x - y\|}$ and $\widetilde{y} := \dfrac{y}{\|x - y\|}$. Since $(\widetilde{x} - \widetilde{y}) \in K$, there exists $1 \leq i_0 \leq n$ such that $(\widetilde{x} - \widetilde{y}) \in B(x_{i_0}; 1/2)$, and thus

$$\big\|x - \|x - y\|(\widetilde{y} + x_{i_0})\big\| = \|x - y\| \left(\|(\widetilde{x} - \widetilde{y}) - x_{i_0}\|\right) < \frac{1}{2}\|x - y\|.$$

But the vector $y_1 := \|x - y\|(\widetilde{y} + x_{i_0})$ belongs to Y since both $\widetilde{y}$ and x_{i_0} belong to Y.

To sum up, if $x \notin Y$, then given any $y \in Y$, there exists $y_1 \in Y$ such that $\|x - y_1\| < \frac{1}{2}\|x - y\|$. By induction, there exist vectors $y_n \in Y$ such that

$$\|x - y_n\| < \frac{1}{2^n}\|x - y\|.$$

Hence $\inf_{y \in Y} \|x - y\| = 0$, which proves the "if" part (evidently, $\frac{1}{2}$ may be replaced in this argument by any number in the open interval $]0, 1[$).

The "only if" part was proved in Theorem 2.7-1(c). □

Note that the F. Riesz theorem may be equivalently stated as follows: *A normed vector space is finite-dimensional if and only if the closed unit ball is compact* (since in this case, the unit sphere is also compact as a closed subset of the closed unit ball).

Problems

2.7-1 Let Y be a finite-dimensional subspace of a normed vector space $(X, \|\cdot\|)$.

(1) Show that, given any vector $x \in X$, there exists a (not necessarily unique) vector $\widetilde{y} \in Y$ such that $\|x - \widetilde{y}\| = \inf_{y \in Y} \|x - y\|$.

(2) Assume that, in addition, the space $(X, \|\cdot\|)$ is *strictly convex*, in the sense that $\|x\| = \|y\| = 1$ and $x \neq y$ implies that $\left\|\frac{x+y}{2}\right\| < 1$. Show that the vector $\widetilde{y} \in Y$ found in (1) is unique.

(3) Show that the space $(\mathbb{K}^n, \|\cdot\|_p)$ (Section 2.2) is strictly convex for any $1 < p < \infty$, but not for $p = 1$ and $p = \infty$.

2.7-2 Show that the interior of any compact subset of an infinite-dimensional normed vector space is empty.

2.7-3 In what follows, the space $\mathcal{C}[0, 2\pi]$ is equipped with the sup-norm. Let the functions $g_n \in \mathcal{C}[0, 2\pi]$, $n \geq 1$, be defined by $g_n(\theta) := \sin n\theta$, $0 \leq \theta \leq 2\pi$. Show *directly*, i.e., without recourse to the F. Riesz theorem, that the sequence $(g_n)_{n=1}^{\infty}$ (which is clearly bounded) does not contain any convergent subsequence.

2.8 Application of compactness in finite-dimensional normed vector spaces: The fundamental theorem of algebra

The *fundamental theorem of algebra* states that any real or complex polynomial, i.e., with real or complex coefficients, of degree $n \geq 1$ has at least one complex root. The formula $z^k - z_0^k = (z - z_0)(z^{k-1} + z^{k-2}z_0 + \cdots + z_0^{k-1})$ then shows that such a polynomial has exactly n complex roots, counting multiplicities. The quest for this elusive result has fascinated mathematicians for a very long time.

The Greeks already knew the formula for computing the real roots (when they exist) of a real polynomial of degree 2. But it was only in 1545 that Girolamo Cardano published the formulas, in effect due to Scipione del Ferro and Nicolo Tartaglia, for computing the roots of a polynomial of degree 3, while Lodovico Ferrari had already found in 1540 (but did not publish until much later) the formulas for computing the roots of a polynomial of degree 4.

Nevertheless, such formulas[11] were shrouded in uncertainty as their full understanding would have required the theory of complex numbers, then only at a nascent stage.

After Niels Henrick Abel[12] proved in 1823, at the incredibly young age of 21, that no such formula exists for a general polynomial of degree 5 ("formula" means any finite expression involving only the elementary operations and extraction of pth roots, $p \geq 2$), the final blow was struck in 1832 by Evariste Galois who, at the equally incredibly young age of 20, established that no such formula exists for a general polynomial of arbitrary degree $n \geq 5$. The discovery of Galois[13] stands as one of the greatest achievements in the history of mathematics.

Meanwhile, many attempts were made to establish the fundamental theorem of algebra without trying to find *ad hoc* formulas, i.e., by using instead the full power of *analysis*.

This approach, likewise rendered all the more difficult by necessary manipulations of the ever mysterious complex numbers, was pursued by many mathematicians. Among them, Jean Le Rond d'Alembert is generally credited as having produced in 1746 the first "serious attempt" at a proof of the fundamental theorem of algebra, although a flaw remained in his argument; incidentally, this explains why the fundamental theorem of algebra is sometimes called *d'Alembert's theorem*. The first correct proof ("correct" according to our current standards) for real polynomials is due to Carl-Friedrich Gauß, who published it in 1816 (after a first, but still incomplete, attempt in his Doctoral Thesis of 1799). He also gave the first correct proof for complex polynomials in 1849.

Incidentally, notice the irony: The "fundamental theorem of algebra," as it is commonly called, is in effect essentially a theorem of *analysis*!

The remarkably simple proof[14] given below relies on two basic *compactness* properties, viz., the characterization of compact subsets in the finite-dimensional normed vector space $(\mathbb{R}^2; \|\cdot\|_2)$ (Theorem 2.7-1(c)) and the property that a function that is continuous on a compact set attains its minimum (Theorem 1.13-6).

Theorem 2.8-1 (fundamental theorem of algebra) *Any complex polynomial of degree* ≥ 1 *has at least one root in* $\mathbb{C}$.

Proof Let $p : \mathbb{C} \to \mathbb{C}$ be a complex polynomial of degree $n \geq 1$, given by

$$p(z) := a_n z^n + \cdots + a_1 z + a_0, \quad z \in \mathbb{C},$$

where $a_0, a_1, \ldots, a_n$ are complex numbers and $a_n \neq 0$.

[11] A clever way to find $z \in \mathbb{C}$ such that $p(z) = 0$, where p is a polynomial of degree n, consists (after the monomial of degree $n-1$ has been eliminated) in finding an $n \times n$ *circulant matrix* (then with trace zero) whose characteristic polynomial is precisely p. For $n = 3$ and $n = 4$, this procedure leads to explicit formulas for the roots of p; see the illuminating account given in:

I. Kra; S.R. Simanca [2012]: On circulant matrices, *Notices of the American Mathematical Society* **59**, 368–377.

[12] A fascinating biography has been devoted to Abel (1802–1829):

A. Stubhaug [2000]: *Niels Henrik Abel and his Times—Called Too Soon by Flames Afar*, Springer, Heidelberg (translated from the Norwegian).

[13] There does not seem to be any authoritative biography (like Stubhaug's about Abel) about Galois (1811–1832), an equally prodigious mathematician. A scholarly account of all of Galois' mathematical contributions, re-examined from a modern perspective, is given in:

P.M. Neumann [2011]: *The Mathematical Writings of Évariste Galois*, European Mathematical Society, Zürich.

[14] This proof is found in Schwartz [1991, Theorem 2.7.10].

(i) *Using compactness arguments, we first show that there exists $z_0 \in \mathbb{C}$ such that*

$$|p(z_0)| = \inf_{z \in \mathbb{C}} |p(z)|.$$

To this end, we note that there exists $r > 0$ such that

$$|p(z)| \geq |p(0)| \quad \text{if } |z| \geq r,$$

since

$$\lim_{|z| \to \infty} |p(z)| = \lim_{|z| \to \infty} \left(|z^n| \left| a_n + \frac{a_{n-1}}{z} + \cdots + \frac{a_0}{z^n} \right| \right) = \infty.$$

Identifying $(\mathbb{C}, |\cdot|)$ with $(\mathbb{R}^2, \|\cdot\|_2)$ in the obvious way, i.e., such that $|z| = \|(x, y)\|_2$ if $z = x+iy$, we next note that the set

$$K := \{z \in \mathbb{C};\ |z| \leq r\}$$

is *compact* by Theorem 2.7-1(c) and that the function $z \in \mathbb{C} \to |p(z)| \in \mathbb{R}$ is *continuous* since

$$\big| |p(z_1)| - |p(z_2)| \big| \leq \sum_{k=0}^{n} |a_k|\, |z_1^k - z_2^k|.$$

We thus infer (Theorem 1.13-6) that there exists $z_0 \in K$ such that

$$\inf_{z \in K} |p(z)| = |p(z_0)|.$$

Since

$$|p(z)| \geq |p(0)| \geq |p(z_0)| \quad \text{if } |z| \geq r,$$

it therefore follows that $|p(z)| \geq |p(z_0)|$ for all $z \in \mathbb{C}$.

If $p(z_0) = 0$, the theorem is proved. It thus remains to consider the case where $p(z_0) \neq 0$.

(ii) *Using elementary algebra of complex numbers, we next show that, if $z_0 \in \mathbb{C}$ is such that $p(z_0) \neq 0$, then there exists $z_1 \in \mathbb{C}$ such that $|p(z_1)| < |p(z_0)|$.* Taylor's expansion around z_0 shows that there exists an integer k with $1 \leq k \leq n$ and there exist complex numbers $c_k, c_{k+1}, \ldots, c_n$ with $c_k \neq 0$ and $c_n = a_n \neq 0$ such that

$$p(z) = p(z_0) + c_k(z - z_0)^k + c_{k+1}(z - z_0)^{k+1} + \cdots + c_n(z - z_0)^n.$$

The idea then consists in showing that $|p(z)|$ *becomes strictly less than* $|p(z_0)|$ *when z describes a circle with a small enough radius ε centered at z_0.* More specifically, let $\varepsilon > 0$ be such that

$$|c_{k+1}|\varepsilon + \cdots + |c_n|\varepsilon^{n-k} < |c_k| \quad \text{and} \quad |c_k|\varepsilon^k < |p(z_0)|,$$

with the convention that the left-hand side of this inequality is equal to zero if $k = n$. When z describes the circle $\Gamma := \{z \in \mathbb{C};\ |z - z_0| = \varepsilon\}$, the point $\{p(z_0) + c_k(z - z_0)^k\}$ describes k times the circle of radius $|c_k|\varepsilon^k$ centered at $p(z_0)$ (Figure 2.8-1). Because $|c_k|\varepsilon^k < |p(z_0)|$, there exists $z_1 \in \Gamma$ such that the point $\{p(z_0) + c_k(z_1 - z_0)^k\}$ is on the segment joining the origin of $\mathbb{C}$ to $p(z_0)$ (it suffices to solve the equation $(z - z_0)^k = -\varepsilon^k \frac{|c_k|}{c_k} \frac{p(z_0)}{|p(z_0|}$), so that

$$|p(z_0) + c_k(z_1 - z_0)^k| = |p(z_0)| - |c_k|\varepsilon^k.$$

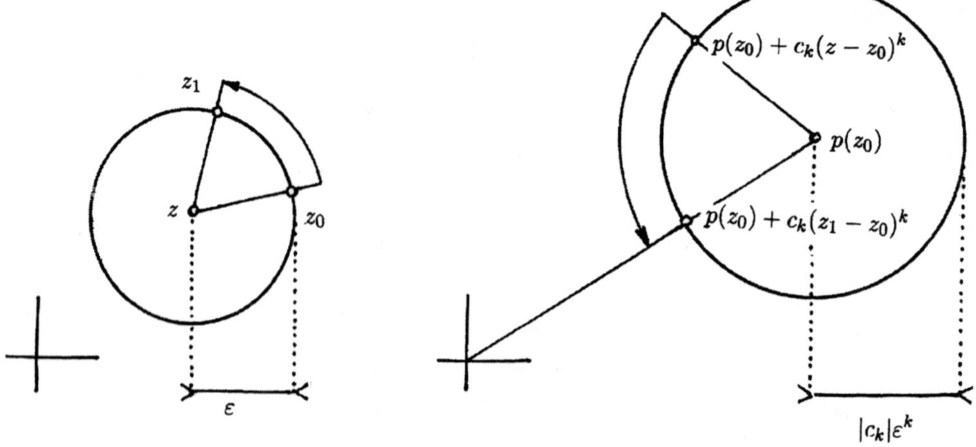

Figure 2.8-1 Geometric interpretation of the construction of z_1 in the proof of Theorem 2.8-1.

Consequently,

$$|p(z_1)| \le |p(z_0) + c_k(z_1 - z_0)^k| + |c_{k+1}(z_1 - z_0)^{k+1} + \cdots + c_n(z_1 - z_0)^n| < |p(z_0)|.$$

(iii) The fundamental theorem of algebra immediately follows from the conjunction of (i) and (ii). □

Remark In many texts, part (ii) in the above proof is replaced by a recourse to *Liouville's theorem*, a fundamental result from the theory of functions of a complex variable. This theorem asserts that an analytic function that is bounded on the whole complex plane $\mathbb{C}$ is constant. Hence if a polynomial p of degree ≥ 1 had no root in $\mathbb{C}$, the function $\frac{1}{p}$ would be analytic in $\mathbb{C}$ and bounded, since $\frac{1}{|p(z)|} \le \frac{1}{|p(z_0)|}$ for all $z \in \mathbb{C}$ by part (i). Consequently, p would be a constant function by Liouville's theorem, a contradiction. □

2.9 Continuous linear operators in normed vector spaces; the spaces $\mathcal{L}(X;Y)$, $\mathcal{L}(X)$, and X'

In what follows, X and Y are two vector spaces over the *same* field $\mathbb{K} = \mathbb{R}$ or $\mathbb{K} = \mathbb{C}$; the same notation 0 stands for both the zero vector of X and the zero vector of Y.

A mapping $A : X \to Y$ is a **linear operator** from X into Y, or a **linear functional** or **linear form** if $Y = \mathbb{K}$, if

$$A(x + y) = A(x) + A(y) \quad \text{and} \quad A(\alpha x) = \alpha A(x) \quad \text{for all } x, y \in X \text{ and } \alpha \in \mathbb{K}$$

(hence $A(0) = 0$). When no confusion should arise, it is a common practice to simply write Ax in lieu of $A(x)$ if A is a linear operator; AB in lieu of $A \circ B$ for the composition of two linear operators A and B; A^2, A^3, etc., in lieu of $A \circ A, A \circ A \circ A$, etc., if $X = Y$, with the convention that $A^0 = I_X$.

If $\mathbb{K} = \mathbb{C}$, a mapping $A : X \to Y$ is **semilinear** if

$$A(x) + A(y) = A(x) + A(y) \quad \text{and} \quad A(\alpha x) = \overline{\alpha} A(x) \quad \text{for all } x, y \in X \text{ and } \alpha \in \mathbb{K},$$

where $\overline{\alpha}$ denotes the complex conjugate of α. This related notion arises naturally in the definition of an inner product in a complex vector space (Section 4.1).

Let $A : X \to Y$ be a linear operator. Then the **kernel** of A is the subset of X defined by

$$\operatorname{Ker} A := \{x \in X;\ Ax = 0\},$$

and the **direct image** $A(X)$ of X under A (Section 1.2), also called the **range** of A, is also denoted $\operatorname{Im} A$ in this case; in other words,

$$\operatorname{Im} A := A(X) = \{y \in Y;\ \text{there exists } x \in X \text{ such that } y = Ax\}.$$

Clearly, $\operatorname{Ker} A$ is *a subspace of* X, and $\operatorname{Im} A$ *is a subspace of* Y.

Remark The *same* notation Im is also used to denote the imaginary part $\operatorname{Im} z$ of a complex number z; however, the risk of confusion is admittedly low... □

The following elementary properties of linear operators are constantly used.

Theorem 2.9-1 (a) *A linear operator* $A : X \to Y$ *is injective if and only if* $\operatorname{Ker} A = \{0\}$.

(b) *If a linear operator* $A : X \to Y$ *is injective, the inverse mapping* $B : \operatorname{Im} A \to X$ *of* $A : X \to \operatorname{Im} A$ *is a linear operator from* $\operatorname{Im} A$ *onto* X.

Proof A mapping $A : X \to Y$ is injective if and only if $Ax = A\widetilde{x}$ implies $x = \widetilde{x}$; hence if and only if $Ax = 0$ implies $x = 0$ if A is a linear operator.

If A is injective, then $BA = I_X$, where I_X denotes the identity mapping of X. Given any two vectors $y, \widetilde{y} \in \operatorname{Im} A$, there thus exist uniquely defined vectors $x, \widetilde{x} \in X$ such that $y = Ax$ and $\widetilde{y} = A\widetilde{x}$. Therefore, for any scalars $\beta, \widetilde{\beta} \in \mathbb{K}$,

$$B(\beta y + \widetilde{\beta}\widetilde{y}) = B(\beta Ax + \widetilde{\beta} A\widetilde{x}) = B(A(\beta x + \widetilde{\beta}\widetilde{x})) = \beta x + \widetilde{\beta}\widetilde{x} = \beta By + \widetilde{\beta} B\widetilde{y}. \qquad \square$$

Endowed with an addition and a scalar multiplication defined by

$$(A + B) : x \in X \to (Ax + Bx) \in Y \quad \text{and} \quad \alpha A : x \in X \to \alpha(Ax) \in Y,$$

the set formed by all the linear operators from X *into* Y *becomes itself a vector space, over the same field* $\mathbb{K}$. Its zero vector is $0 : x \in X \to 0 \in Y$ (this zero vector is thus denoted like the zero vectors of X and Y).

Let X be a vector space over $\mathbb{K}$ and let $A : X \to X$ be a *linear operator*. Then a scalar $\lambda \in \mathbb{K}$ is an **eigenvalue** of A if there exists a vector $p \in X$ such that

$$Ap = \lambda p \quad \text{and} \quad p \neq 0.$$

Such a *nonzero* vector is then called an **eigenvector** of A, *corresponding to the eigenvalue* λ, and the subspace

$$\{p \in X;\ Ap = \lambda p\} \neq \{0\}$$

of X is called the **eigenspace** *corresponding to the eigenvalue* λ.

Note that A *is injective if and only if* 0 *is not an eigenvalue of* A.

When both X and Y are *normed* vector spaces and are equipped as such with their *norm topologies* (Section 2.2), *continuous* linear operators from X into Y, or *continuous* linear functionals if $Y = \mathbb{K}$, possess specific properties. The next theorems list the most elementary, yet basic, of these properties. For notational brevity, the same notation $\|\cdot\|$ designates in what follows the norm in vector spaces that are not necessarily the same. The context should always prevent any confusion, however.

Theorem 2.9-2 *Let X and Y be two normed vector spaces and let $A : X \to Y$ be a linear operator. Then the following properties are equivalent:*

(a) *The linear operator A is continuous on X.*

(b) *The linear operator A is continuous at the origin of X.*

(c) *There exists a constant $C > 0$ such that*

$$\|Ax\| \leq C\|x\| \quad \textit{for all } x \in X.$$

(d) *The direct image under A of any bounded subset of X is a bounded subset of Y.*

Proof Clearly, (a) implies (b).

If (b) holds, the inverse image under A of the closed unit ball of Y contains a closed ball centered at the origin of X; let $\frac{1}{C} > 0$ denote its radius. Consequently, any nonzero vector $x \in X$ satisfies $\left\|A\left(\frac{x}{C\|x\|}\right)\right\| \leq 1$, i.e., $\|Ax\| \leq C\|x\|$. Hence (b) implies (c).

Assume that (c) holds. Since any bounded subset B of X is contained in a ball with center at the origin of X and radius $r = r(B) > 0$, the direct image $A(B)$ is contained in a ball with center at the origin of Y and radius Cr; consequently, $A(B)$ is bounded. Hence (c) implies (d).

If (d) holds, the direct image of the closed unit ball of X is bounded in Y, i.e., there exists $M > 0$ such that $\|x\| \leq 1$ implies $\|Ax\| \leq M$. Given any $x_0 \in X$ and any $\varepsilon > 0$, let $\delta := \frac{\varepsilon}{M}$. Then $\|x - x_0\| \leq \delta$ implies $\frac{1}{\delta}\|A(x - x_0)\| = \left\|A\left(\frac{x - x_0}{\delta}\right)\right\| \leq M$, and thus $\|Ax - Ax_0\| \leq \varepsilon$. This proves (a). $\square$

Property (d) explains why continuous linear operators in normed vector spaces are also called **bounded linear operators**.

If X is a subspace of Y, the notation

$$X \hookrightarrow Y$$

means that the canonical injection from X into Y (Section 1.2), which is clearly linear, is continuous, or equivalently (Theorem 2.9-2) that there exists a constant C such that

$$\|x\|_Y \leq C\,\|x\|_X \quad \text{for all } x \in X.$$

Theorem 2.9-3 *Let X and Y be two normed vector spaces.*

(a) *Any continuous linear operator from X into Y is uniformly continuous.*

(b) *If X is finite-dimensional, any linear operator from X into Y is continuous.*

Proof The uniform continuity of a continuous linear operator $A : X \to Y$ follows from the relation $\|Ax - A\widetilde{x}\| \leq C\|x - \widetilde{x}\|$ for all $x, \widetilde{x} \in X$ (Theorem 2.9-2(c)).

Assume next that X is finite-dimensional and let $(e_i)_{i=1}^n$ be a basis of X. Then, for any vector $x = \sum_{i=1}^n x_i e_i \in X$,

$$\|Ax\| = \Big\|A\Big(\sum_{i=1}^{n} x_i e_i\Big)\Big\| \leq C_1\|x\|_1 \quad \text{with } C_1 := \max_{1\leq i\leq n} \|Ae_i\|,$$

where $\|x\|_1 = \sum_{i=1}^n |x_i|$ (Theorem 2.2-2). The continuity of A follows from Theorem 2.9-2(c) again, combined with the property that any two norms in a finite-dimensional space are equivalent (Theorem 2.7-1(a)). □

As an application, let $\mathcal{P}_n$ denote the space of all real polynomials of degree $\leq n$, equipped with the norm $p \in \mathcal{P}_n \to \sup_{-1\leq x\leq 1} |p(x)|$, and let the linear operator $A : \mathcal{P}_n \to \mathcal{P}_n$ be defined by $Ap = p'$. Then Theorems 2.9-2(c) and 2.9-3(b) show that there exists a constant $C(n)$ such that

$$\sup_{-1\leq x\leq 1} |p'(x)| \leq C(n) \sup_{-1\leq x\leq 1} |p(x)| \quad \text{for all } p \in \mathcal{P}_n.$$

One can further show that the "best" (i.e., the smallest) constant $C(n)$ in this inequality is n^2 (the equality corresponding to, e.g., the particular polynomial $x \in \mathbb{R} \to \cos(n \operatorname{Arcos} x)$). This result constitutes the famed **Markoff inequality**[15] whose proof is, incidentally, anything but trivial.[16] Similar inequalities, but corresponding to other norms, follow from Theorem 2.9-2(c). For instance, for any integer $n \geq 0$ and any $r \geq 1$, there exists a constant $C(n,r)$ such that[17]

$$\Big(\int_{-1}^{1} |p'(x)|^r\Big)^{1/r} \leq C(n,r)\Big(\int_{-1}^{1} |p(x)|^r\Big)^{1/r} \quad \text{for all } p \in \mathcal{P}_n.$$

By contrast with property (b) in Theorem 2.9-3, the continuity of a linear operator $A : X \to Y$ has no relation to the dimension of the space Y when the space X is infinite-dimensional.

To illustrate how continuity may fail even if $\dim Y = 1$, consider for instance the space $\mathcal{P}$ of all real polynomials of arbitrary degree, equipped with the norm $\|\cdot\| : p \in \mathcal{P} \to \|p\| = \sup_{-1\leq x\leq 1} |p(x)|$, and let the linear functional $f : \mathcal{P} \to \mathbb{R}$ be defined by $f(p) = p(3)$. For each integer $k \geq 0$, let the polynomial $p_k \in \mathcal{P}$ be defined by $p_k(x) = \left(\frac{x}{2}\right)^k$ for all $x \in \mathbb{R}$. Then $\|p_k\| \to 0$ as $k \to \infty$ while $|f(p_k)| \to \infty$ as $k \to \infty$, which shows that f is not continuous.

We next examine the continuity of the *inverse* (when it is defined) of a linear operator.

[15] A.A. Markoff [1889]: Sur une question posée par Mendeleieff, *Izvestia Akademii Nauk SSSR* **62**, 1–24.

[16] For such a proof, see, e.g., Cheney [1966, Chapter 3, Section 7].

[17] For an estimate of the best constant $C(n,r)$, see:

E. Hille; C. Szegö; J.D. Tamarkin [1937]: On some generalizations of a theorem of A. Markoff, *Duke Mathematical Journal* **8**, 729–739.

Theorem 2.9-4 *Let X and Y be two normed vector spaces and let $A : X \to Y$ be a linear operator. Then the following two properties are equivalent:*

(a) *The linear operator A is injective and the inverse mapping $B : \operatorname{Im} A \to X$ of $A : X \to \operatorname{Im} A$ is a continuous linear operator;*

(b) *There exists a constant $C > 0$ such that*

$$\|x\| \le C\|Ax\| \quad \textit{for all } x \in X.$$

Proof If (a) holds, then $A : X \to \operatorname{Im} A$ is a linear bijection (Theorem 2.9-1). Therefore, given any $x \in X$, there exists a unique vector $y \in \operatorname{Im} A$ such that $Ax = y$, or equivalently such that $By = x$. Furthermore, the continuity of B implies that there exists a constant $C > 0$ such that $\|By\| \le C\|y\|$ for all $y \in \operatorname{Im} A$ (Theorem 2.9-2). Hence (b) holds.

If (b) holds, then A is injective since $\operatorname{Ker} A = \{0\}$ (Theorem 2.9-1), and the inequality $\|x\| \le C\|Ax\|$ for all $x \in X$ implies that $\|By\| \le C\|y\|$ for all $y \in \operatorname{Im} A$. Hence B is continuous, again by Theorem 2.9-2. □

Let X and Y be two normed vector spaces over the same field $\mathbb{K}$. Then the vector space, also over $\mathbb{K}$, formed by all the *continuous* linear operators from X into Y, which is denoted

$$\mathcal{L}(X;Y), \quad \text{or} \quad \mathcal{L}(X) \quad \text{if } Y = X,$$

can be also endowed with a *norm.* The definition and some elementary properties of this norm are given in the next theorem.

Although this is not mentioned for notational brevity, it should be clear that the vector x appearing in each supremum belongs to the space X (the same observation can be made at many other places in the sequel).

Theorem 2.9-5 *Let X and Y be two normed vector spaces.*

(a) *The mapping defined by*

$$\|\cdot\| : A \in \mathcal{L}(X;Y) \to \|A\| := \sup_{x \ne 0} \frac{\|Ax\|}{\|x\|}$$

is a norm on the vector space $\mathcal{L}(X;Y)$. This definition implies that

$$\|Ax\| \le \|A\|\,\|x\| \quad \textit{for all } x \in X.$$

(b) *The norm of any $A \in \mathcal{L}(X;Y)$ may be equivalently defined as*

$$\begin{aligned}\|A\| &= \sup_{\|x\|\le 1} \|Ax\| = \sup_{\|x\|<1} \|Ax\| = \sup_{\|x\|=1} \|Ax\| \\ &= \frac{1}{r} \sup_{\|x\|\le r} \|Ax\| = \frac{1}{r} \sup_{\|x\|=r} \|Ax\| \quad \textit{for any } \alpha > 0 \\ &= \inf\{C > 0;\ \|Ax\| \le C\|x\| \textit{ for all } x \in X\}.\end{aligned}$$

(c) *If X is finite-dimensional, there exists $x_0 \in X$ such that*

$$x_0 \ne 0 \quad \textit{and} \quad \|A\|\,\|x_0\| = \|Ax_0\|.$$

(d) *Let Z be a normed vector space. If $A \in \mathcal{L}(X;Y)$ and $B \in \mathcal{L}(Y;Z)$, then $BA \in \mathcal{L}(X;Y)$ and*

$$\|BA\| \leq \|A\| \, \|B\| .$$

Consequently, if $A \in \mathcal{L}(X)$, then

$$\|A^n\| \leq \|A\|^n \quad \textit{for any integer } n \geq 0.$$

(e) *If $A \in \mathcal{L}(X)$, any eigenvalue λ of A satisfies $|\lambda| \leq \|A\|$.*

Proof Properties (a), (b), and (d) are immediately verified (to prove that $\sup_{\|x\|\leq 1} \|Ax\| = \sup_{\|x\|<1} \|Ax\|$, note that the unit ball is dense in the closed unit ball).

If X is finite-dimensional, then the unit sphere $K = \{x \in X; \|x\| = 1\}$ is compact (Theorem 2.7-1). Therefore the function $x \in X \to \|Ax\| \in \mathbb{R}$, which is continuous (as the composition of two continuous mappings, viz., $x \in X \to Ax \in Y$ and $y \in Y \to \|y\| \in \mathbb{R}$), attains its supremum over K; this proves (c).

If $A \in \mathcal{L}(X)$, $Ap = \lambda p$, and $p \neq 0$, then $\|Ap\| = |\lambda| \, \|p\| \leq \|A\| \, \|p\|$; this proves (e). □

The norm $\|\cdot\|$ over the space $\mathcal{L}(X;Y)$ defined in Theorem 2.9-5 is called the **operator norm**. It is also denoted

$$\|\cdot\|_{\mathcal{L}(X;Y)}$$

whenever any confusion could arise.

In the *fundamental* special case where $Y = \mathbb{K}$, the space

$$X' := \mathcal{L}(X;\mathbb{K})$$

is called the **dual space** of X, or simply the **dual** of X. The norm of any $\ell \in X'$ is thus given by

$$\|\ell\| = \sup_{x \neq 0} \frac{|\ell(x)|}{\|x\|}.$$

Note that, in this case, the notations

$${}_{X'}\langle \ell, x \rangle_X := \ell(x), \quad \text{or simply} \quad \langle \ell, x \rangle := \ell(x), \quad \text{for any } \ell \in X' \text{ and } x \in X,$$

will be also used.

In the special case where $X = \mathbb{K}$, *the space $\mathcal{L}(\mathbb{K};Y)$ can be identified with the space Y*, by means of the linear bijection $A \in \mathcal{L}(\mathbb{K};Y) \to A(1) \in Y$.

Examples and properties of operator norms in finite-dimensional spaces are given in Problems 2.9-1 and 2.9-2. A counterexample to property (c) in Theorem 2.9-5 when X is an infinite-dimensional space is provided in Problem 2.9-4. A characterization of finite dimensionality in terms of continuous linear functionals is provided in Problem 2.9-5. Examples of dual spaces will be given in Chapter 3.

Problems

2.9-1 The norms $\|\cdot\|_p$, $1 \leq p \leq \infty$, on $\mathbb{K}^n$ are those defined in Theorem 2.2-2. A linear operator in $\mathcal{L}(\mathbb{K}^n)$ is identified here with an $n \times n$ matrix $\boldsymbol{A} = (a_{ij})$ with coefficients in $\mathbb{K}$.

(1) Show that

$$\|A\|_1 := \sup_{x \neq 0} \frac{\|Ax\|_1}{\|x\|_1} = \max_{1 \leq j \leq n} \sum_{i=1}^{n} |a_{ij}|,$$

$$\|A\|_2 := \sup_{x \neq 0} \frac{\|Ax\|_2}{\|x\|_2} = \{\rho(A^* A)\}^{1/2},$$

$$\|A\|_\infty := \sup_{x \neq 0} \frac{\|Ax\|_\infty}{\|x\|_\infty} = \max_{1 \leq i \leq n} \sum_{j=1}^{n} |a_{ij}|,$$

where A^* designates the adjoint matrix of A and $\rho(B)$ designates the largest modulus of the eigenvalues of a matrix B.

Remark Since the vector norm $\|\cdot\|_2$ is also denoted $|\cdot|$, the above matrix norm $\|\cdot\|_2$ is also denoted $|\cdot|$ whenever no confusion should arise. □

(2) Find a formula for the operator norms defined by

$$\sup_{x \neq 0} \frac{\|Ax\|_p}{\|x\|_1}, \ 1 < p \leq \infty, \quad \text{and} \quad \sup_{x \neq 0} \frac{\|Ax\|_\infty}{\|x\|_p}, \ 1 \leq p < \infty.$$

2.9-2 The notations are the same as in Problem 2.9-1.

(1) Let A be a real $n \times n$ matrix. Show that, for $p = 1$, $p = 2$, and $p = \infty$,

$$\sup\left\{\frac{\|Ax\|_p}{\|x\|_p};\ x \in \mathbb{R}^n,\ x \neq 0\right\} = \sup\left\{\frac{\|Ax\|_p}{\|x\|_p};\ x \in \mathbb{C}^n,\ x \neq 0\right\}.$$

(2) Is the equality of (1) still valid for $1 < p < 2$ and $2 < p < \infty$?

(3) Find a real $n \times n$ matrix A and a norm $\|\cdot\|$ on $\mathbb{C}^n$ such that

$$\sup\left\{\frac{\|Ax\|}{\|x\|};\ x \in \mathbb{R}^n,\ x \neq 0\right\} < \sup\left\{\frac{\|Ax\|}{\|x\|};\ x \in \mathbb{C}^n,\ x \neq 0\right\}.$$

2.9-3 Given any vector norm $\|\cdot\|$ on $\mathbb{K}^n$, the corresponding operator norm on $\mathcal{L}(\mathbb{K}^n)$ is defined by $\|A\| = \sup_{x \neq 0} \dfrac{\|Ax\|}{\|x\|}$ for any $A \in \mathcal{L}(\mathbb{K}^n)$. This operator norm is called a *subordinate matrix norm* (to reflect that it is "subordinate" to the given vector norm) when A is identified with an $n \times n$ matrix, as in this problem (for examples of subordinate matrix norms, see Problems 2.9-1 and 2.9-2). Clearly, $\rho(A) \leq \|A\|$ for any subordinate matrix norm, where $\rho(A)$ designates the largest modulus of the eigenvalues of A (since $Ap = \lambda p$ implies $|\lambda|\,\|p\| \leq \|A\|\,\|p\|$). In what follows, A is a given $n \times n$ matrix with coefficients in $\mathbb{K}$.

(1) Show that, given any $\varepsilon > 0$, there exists a subordinate matrix norm $\|\cdot\|$ such that $\|A\| \leq \rho(A) + \varepsilon$.

(2) Show that $A^k \to 0$ as $k \to \infty$ if and only if $\rho(A) < 1$.

(3) Show that, given any subordinate matrix norm,

$$\|A^k\|^{1/k} \to \rho(A) \quad \text{as } k \to \infty.$$

(4) Show that

$$\limsup_{k \to \infty} \left(|\operatorname{tr}(A^k)|^{1/k} \right) = \rho(A).$$

2.9-4 The space ℓ^∞ and its norm $\|\cdot\|_\infty$ are defined in Section 2.4. Let c_0 denote the set of all infinite sequences $(x_i)_{i=1}^\infty$ of scalars $x_i \in \mathbb{K}$ such that $\lim x_i = 0$ as $i \to \infty$. Hence it is clear that c_0 is a subspace of ℓ^∞ (a convergent sequence is bounded).

(1) Let $\sum_{i=1}^\infty \alpha_i$ be a convergent series with $\alpha_i > 0$ for all $i \geq 1$. Show that

$$f : x = (x_i)_{i=1}^\infty \in (c_0; \|\cdot\|_\infty) \to f(x) := \sum_{i=1}^\infty \alpha_i x_i \in \mathbb{K}$$

is a continuous linear functional on c_0, with $\|f\| := \sup_{x \neq 0} \dfrac{|f(x)|}{\|x\|_\infty} = \sum_{i=1}^\infty \alpha_i$.

(2) Show that there does not exist a nonzero vector $x \in c_0$ such that $|f(x)| = \|f\| \, \|x\|_\infty$.

2.9-5 Show that a normed vector space X is finite-dimensional if and only if all the linear functionals on X are continuous.

2.9-6 (1) Let X and Y be two vector spaces over the same field, and let $A : X \to Y$ be a linear operator. Show that there exists a linear bijection from the quotient space $X/\operatorname{Ker} A$ onto the subspace $\operatorname{Im} A$ of Y.

(2) Given a linear operator $A : X \to Y$, define the mapping $[A] : X/\operatorname{Ker} A \to Y$ by $[A]\,[x] := Ax$ for all $[x] \in X/\operatorname{Ker} A$. Show that this definition makes sense, that $[A]$ is a linear operator, and that $[A]$ is a bijection from $X/\operatorname{Ker} A$ onto $\operatorname{Im} A$.

(3) Assume that $\operatorname{Im} A$ is closed in Y. Show that $[A]$ is continuous if and only if A is continuous, and that $\|[A]\| = \|A\|$ in this case.

2.9-7 (1) Let Y be a real vector space, and let $A : \mathcal{L}(\mathbb{R}^3) \to Y$ be a linear operator with the property that its restriction to the set of all 3×3 proper orthogonal matrices (identified here with a subset of $\mathcal{L}(\mathbb{R}^3)$) is a constant mapping. Show that $A = 0$.

(2) Does the same property hold in any dimension $n \neq 3$?

2.10 Compact linear operators in normed vector spaces

In this section, we introduce an important class of continuous linear operators.

Let X and Y be two normed vector spaces over the same field. A *linear operator* $A : X \to Y$ is said to be **compact** if the image under A of any bounded subset of X is a relatively compact subset of Y; in other words, whenever B is bounded in X, then $\overline{A(B)}$ is compact.

We now prove some elementary, but important, properties of compact operators. The first one (which is immediate) asserts that *any compact linear operator is continuous*, thus showing that the set formed by all compact operators from X to Y is a subset of the space $\mathcal{L}(X;Y)$ (this set is clearly a subspace of $\mathcal{L}(X;Y)$). The "if" part of the second property is often used to prove that a linear operator is compact. The last two (again immediate) properties give sufficient conditions for a linear operator to be compact.

Theorem 2.10-1 *Let X and Y be two normed vector spaces over the same field, and let $A : X \to Y$ be a linear operator.*

(a) *If A is compact, then A is continuous.*

(b) *The operator A is compact if and only if, given any bounded sequence $(x_n)_{n=1}^\infty$ in X, the sequence $(Ax_n)_{n=1}^\infty$ contains a subsequence that converges in Y.*

(c) *If X is finite-dimensional, A is compact.*

(d) *If A is continuous and the direct image $A(X)$ of X under A is finite-dimensional, A is compact.*

Proof (i) If a linear operator $A : X \to Y$ is compact, the direct image $A(B)$ of any bounded subset B of X is bounded in Y (as a subset of the compact subset $\overline{A(B)}$ of Y; cf. Theorem 1.13-1). Hence A is continuous (Theorem 2.9-2(d)). This proves (a).

(ii) Assume that, given any bounded sequence $(x_n)_{n=1}^\infty$ in X, the sequence $(Ax_n)_{n=1}^\infty$ contains a subsequence that converges in Y, and let B be any bounded subset of X.

Given any sequence $(y_n)_{n=1}^\infty$ in the set $A(B)$, let $x_n \in B$ be such that $y_n = Ax_n$. The sequence $(x_n)_{n=1}^\infty$ being thus bounded, there exists by assumption a subsequence $(Ax_{\sigma(n)})_{n=1}^\infty$ that converges to a limit y in Y, and $y \in \overline{A(B)}$ since $y_{\sigma(n)} = Ax_{\sigma(n)} \in A(B)$ for all $n \geq 1$. Hence

$$\lim_{n\to\infty} y_{\sigma(n)} = y \in \overline{A(B)},$$

which shows that the set $A(B)$ is relatively compact, by Theorem 1.13-4. This proves the "if" part of (b).

(iii) To prove the "only if" part of (b), assume that A is compact, and let $(x_n)_{n=1}^\infty$ be any bounded sequence in X.

The set $B := \bigcup_{n=1}^\infty \{x_n\}$ being thus bounded, the set $\overline{A(B)}$ is compact by assumption. Since $Ax_n \in A(B) \subset \overline{A(B)}$ for all $n \geq 1$, there thus exists a subsequence $(Ax_{\sigma(n)})_{n=1}^\infty$ that converges in $\overline{A(B)} \subset Y$.

(iv) If X is finite-dimensional, any linear operator $A : X \to Y$ is continuous (Theorem 2.9-3) and thus maps any bounded subset B of X into a bounded subset $A(B)$ of Y (Theorem 2.9-2(d)). Since $A(B) \subset A(X)$ and the direct image $A(X)$ of the space X under A is a finite-dimensional subspace of Y, the set $\overline{A(B)}$ is compact (Theorem 2.7-1(c)). This proves (c).

(v) If $A \in \mathcal{L}(X;Y)$, the direct image $A(X)$ is finite-dimensional, and B is any bounded subset of X, the set $\overline{A(B)}$ is compact as a closed and bounded (by the assumed continuity of A) subset of the finite-dimensional normed vector space $A(X)$ (Theorem 2.7-1(c)). $\square$

If X is a subspace of Y, the notation

$$X \Subset Y$$

means that the canonical injection from X into Y (Section 1.2) is a compact linear operator, or equivalently, that *any bounded sequence in X contains a subsequence that converges in Y* (Theorem 2.10-1(b)).

Remark The above definition of compactness is specific to *linear* operators; for *nonlinear* operators, it needs to be complemented by the assumption of *continuity* (which here is automatic; cf. Theorem 2.10-1(a)); see Section 9.12. $\square$

Simple *examples* of compact linear operators acting from the space $\mathcal{C}[0,1]$ into itself, or from the space $L^2(0,1)$ into itself, will be given in Problems 3.10-4 and 4.9-5 (such examples need to be postponed, because their analysis rests on notions not yet introduced). Another, and particularly important, example is provided by the canonical injection from the Sobolev space $H^1(\Omega)$ into $L^2(\Omega)$ (Theorem 6.6-3).

Compact linear operators acting in an *inner-product space*, and that are in addition *self-adjoint* (Section 4.10), play a key role in the spectral theory of linear second-order elliptic operators (Section 6.10), thanks to the fundamental *spectral theorem* for such operators (Theorem 4.11-1).

Problems

2.10-1 Let X and Y be two normed vector spaces over the same field and let $A : X \to Y$ be a compact linear operator. Show that the space $\operatorname{Im} A$ is separable.

2.10-2 Let X be a normed vector space over $\mathbb{K} = \mathbb{R}$ or $\mathbb{K} = \mathbb{C}$, let $A : X \to X$ be a compact linear operator, and let $\lambda \in \mathbb{K}$, $\lambda \neq 0$. Show that $(\lambda I - A)$ is injective[18] if and only if $\|(\lambda I - A)x\| \to \infty$ as $\|x\| \to \infty$.

2.11 Continuous multilinear mappings in normed vector spaces; the space $\mathcal{L}_k(X_1, X_2, \dots, X_k; Y)$

Throughout this section, k is an integer ≥ 2 and $X_\ell, 1 \leq \ell \leq k$, and Y denote vector spaces over the same field $\mathbb{K} = \mathbb{R}$ or $\mathbb{K} = \mathbb{C}$.

The *product space*

$$X_1 \times X_2 \times \cdots \times X_k$$

is the set of all elements of the form $(x_1, x_2, \dots, x_k)$ where $x_\ell \in X_\ell$, $1 \leq \ell \leq k$. Equipped with the addition and scalar multiplication defined by

$$\begin{aligned}(x_1, x_2, \dots, x_k) + (y_1, y_2, \dots, y_k) &:= (x_1 + y_1, x_2 + y_2, \dots, x_k + y_k),\\ \alpha(x_1, x_2, \dots, x_k) &:= (\alpha x_1, \alpha x_2, \dots, \alpha x_k),\end{aligned}$$

the product $X_1 \times X_2, \dots \times X_k$ becomes also a *vector space over* $\mathbb{K}$.

A mapping $A : X_1 \times X_2 \times \cdots \times X_k \to Y$ is a **multilinear**, or k-**linear**, **mapping** from $X_1 \times X_2 \times \cdots \times X_k$ into Y if it is linear with respect to each variable $x_\ell \in X_\ell$, $1 \leq \ell \leq k$, when the $(k-1)$ other variables are kept fixed. If $k = 2$, *resp.* $k = 3$, a multilinear mapping is also said to be **bilinear**, *resp.* **trilinear**. If $Y = \mathbb{K}$, a multilinear mapping is also called a **multilinear functional** or a **multilinear form**.

Naturally, this notion is to be carefully distinguished from that of a *linear* mapping from the product space $X_1 \times X_2 \times \cdots \times X_k$ into Y. If for instance $k = 2$, a *linear* mapping from $X_1 \times X_2$ into Y satisfies (with self-explanatory notations)

$$\begin{aligned}A\left((x_1, x_2) + (y_1, y_2)\right) &= A(x_1, x_2) + A(y_1, y_2),\\ A\left(\alpha(x_1, x_2)\right) &= \alpha A(x_1, x_2),\end{aligned}$$

while a *bilinear* mapping from $X_1 \times X_2$ into Y satisfies

$$\begin{aligned}A\left((x_1, x_2) + (y_1, y_2)\right) &= A(x_1, x_2) + A(y_1, y_2) + A(x_1, y_2) + A(x_2, y_2),\\ A\left(\alpha(x_1, x_2)\right) &= \alpha^2 A(x_1, x_2).\end{aligned}$$

---

[18] This observation is due to:

G. Dinca [2001]: A Fredholm-type result for a couple of nonlinear operators, *Comptes Rendus de l'Académie des Sciences de Paris, Série* 1, **333**, 4015–4019.

Endowed with the addition and scalar multiplication defined (again with self-explanatory notations) by

$$(A+B)(x_1,x_2,\ldots,x_k) := A(x_1,x_2,\ldots,x_k) + B(x_1,x_2,\ldots,x_k),$$
$$\alpha A(x_1,x_2,\ldots,x_k) := \alpha\left(A(x_1,x_2,\ldots,x_k)\right),$$

the set formed by all multilinear mappings from $X_1 \times X_2 \times \cdots \times X_k$ into Y becomes also a vector space over $\mathbb{K}$.

Let $\mathfrak{S}_k$ denote the set of all permutations of the set $\{1,2,\ldots,k\}$. If $X_1 = X_2 = \cdots = X_k = X$, a multilinear mapping from $X \times X \times \cdots \times X$ (k factors) into Y is said to be **symmetric** if, for all $x_\ell \in X_\ell$, $1 \le \ell \le k$, and all $\sigma \in \mathfrak{S}_k$,

$$A(x_{\sigma(1)}, x_{\sigma(2)}, \ldots, x_{\sigma(k)}) = A(x_1, x_2, \ldots, x_k),$$

and **alternate** if

$$A(x_{\sigma(1)}, x_{\sigma(2)}, \ldots, x_{\sigma(k)}) = \varepsilon(\sigma) A(x_1, x_2, \ldots, x_k),$$

where $\varepsilon(\sigma)$ denotes the signature of σ. The *determinant* of a $k \times k$ matrix with coefficients over $\mathbb{K}$, viewed as a function of the column vectors of the matrix, provides an example of an alternate multilinear functional, with $X = \mathbb{K}^k$.

When the spaces $X_1, X_2, \ldots, X_k$, and Y are *normed* vector spaces and the product space $X_1 \times X_2 \times \cdots \times X_k$ is equipped with the product topology (Section 2.2), *continuous* multilinear mappings possess specific properties. The next theorem gathers various characterizations of such operators, which may be viewed as the "multilinear analogues" of the characterizations of continuous *linear* operators established in Theorem 2.9-2.

Theorem 2.11-1 *Let X_ℓ, $1 \le \ell \le k$, and Y be normed vector spaces and let*

$$A : X := X_1 \times X_2 \times \cdots X_k \to Y$$

be a multilinear mapping. Then the following properties are equivalent:

(a) *The mapping $A : X \to Y$ is continuous.*

(b) *The mapping A is continuous at the origin.*

(c) *There exists a constant $C > 0$ such that*

$$\|Ax\|_Y \le C \|x_1\|_{X_1} \|x_2\|_{X_2} \cdots \|x_k\|_{X_k} \quad \textit{for all } x = (x_1, x_2, \ldots, x_k) \in X.$$

(d) *The direct image of any bounded subset of X is a bounded subset of Y.*

Proof Throughout this proof, we assume without loss of generality that the product topology on the space $X_1 \times X_2 \times \cdots \times X_k$ is induced by the norm $|||\cdot|||_\infty$ defined by

$$|||x|||_\infty = \max_{1\le \ell \le k} \|x_\ell\|_{X_\ell} \quad \text{for all } x = (x_1, x_2, \ldots, x_k) \in X.$$

It is clear that (a) implies (b). If (b) holds, the inverse image under A of the closed unit ball of Y contains a closed ball centered at the origin of X; let $\alpha > 0$ denote its radius. By definition of the norm $|||\cdot|||_\infty$, this means that, if a vector $\widetilde{x} = (\widetilde{x}_1, \widetilde{x}_2, \ldots, \widetilde{x}_k) \in X$ is such that $\|\widetilde{x}_\ell\|_{X_\ell} \le \alpha$, $1 \le \ell \le k$, then $\|A\widetilde{x}\|_Y \le 1$.

Given any vector $x = (x_1, x_2, \dots, x_k) \in X$ such that $x_\ell \neq 0$, $1 \leq \ell \leq k$ (otherwise, the inequality of (c) is surely satisfied), let $\widetilde{x} = (\widetilde{x}_1, \widetilde{x}_2, \dots, \widetilde{x}_\ell)$, with $\widetilde{x}_\ell := \alpha(\|x_\ell\|_{X_\ell})^{-1} x_\ell$, $1 \leq \ell \leq k$. Then $\|\widetilde{x}_\ell\|_{X_\ell} = \alpha$, $1 \leq \ell \leq k$, and thus $\|A\widetilde{x}\|_Y \leq 1$. Since

$$Ax = \frac{1}{\alpha^k} \|x_1\|_{X_1} \|x_2\|_{X_2} \cdots \|x_\ell\|_{X_\ell} A\widetilde{x},$$

by the assumed multilinearity of A, the inequality of (c) holds with $C := \dfrac{1}{\alpha^k}$.

Assume that (c) holds. Since any bounded subset B of X is contained in a ball with center at the origin of X and radius $r = r(B) > 0$, the direct image $A(B)$ is therefore contained in a ball with center at the origin of Y and radius Cr^k. This shows that $A(B)$ is bounded. Hence (c) implies (d).

If (d) holds, the direct image of the closed unit ball of X is bounded in Y. By definition of the norm $|||\cdot|||_\infty$, there thus exists $C > 0$ such that, if $\|x_\ell\|_{X_\ell} \leq 1$, $1 \leq \ell \leq k$, then $\|A(x_1, x_2, \dots, x_k)\|_Y \leq C$. Therefore, by the assumed multilinearity of A,

$$\|Ax\|_Y \leq C \|x_1\|_{X_1} \|x_2\|_{X_2} \cdots \|x_k\|_{X_k} \quad \text{for all } x = (x_1, x_2, \dots, x_k) \in X.$$

Given $x = (x_1, x_2, \dots, x_k) \in X$ and $a = (a_1, a_2, \dots, a_k) \in X$, the difference $A(x) - A(a)$ may be written as

$$\begin{aligned} A(x) - A(a) \;=\; & A(x_1 - a_1, x_2, x_3, \dots, x_k) \\ & + A(a_1, x_2 - a_2, x_3, \dots, x_k) \\ & \vdots \\ & + A(a_1, a_2, \dots, a_{k-1}, x_k - a_k), \end{aligned}$$

thanks again to the multilinearity of A. Consequently,

$$\begin{aligned} \|A(x) - A(a)\|_Y \leq C\big(& \|x_1 - a_1\|_{X_1} \|x_2\|_{X_2} \cdots \|x_k\|_{X_k} \\ & + \|a_1\|_{X_1} \|x_2 - a_2\|_{X_2} \cdots \|x_k\|_{X_k} \\ & \vdots \\ & + \|a_1\|_{X_1} \|a_2\|_{X_2} \cdots \|x_k - a_k\|_{X_k} \big). \end{aligned}$$

Let $M := |||a|||_\infty$ and $\delta := |||x - a|||_\infty$. Therefore, by the above inequality,

$$\|A(x) - A(a)\|_Y \leq C\delta\big\{(M + \delta)^{k-1} + M(M + \delta)^{k-2} + \cdots + M^{k-1}\big\},$$

since $|||x|||_\infty \leq M + \delta$. If the point a is fixed, the right-hand side of this inequality approaches zero as $\delta = |||x - a|||_\infty$ approaches zero, and therefore the mapping A is continuous. Hence (d) implies (a). □

It is instructive to compare the inequality

$$\|Ax\|_Y \leq C \|x_1\|_{X_1} \|x_2\|_{X_2} \cdots \|x_k\|_{X_k} \quad \text{for all } x = (x_1, x_2, \dots, x_k) \in X,$$

which characterizes a continuous *multilinear* mapping $A : X_1 \times X_2 \times \cdots \times X_k \to Y$, with the inequality

$$\|Ax\|_Y \leq C\left(\|x_1\|_{X_1} + \|x_2\|_{X_2} + \cdots + \|x_k\|_{X_k}\right) \quad \text{for all } x = (x_1, x_2, \ldots, x_k) \in X$$

(or the inequality $\|Ax\|_Y \leq C \max_{1\leq \ell\leq k} \|x_\ell\|_{X_\ell}$, etc.) which characterizes a continuous *linear* mapping $A : X_1 \times X_2 \times \cdots \times X_k \to Y$.

Given a vector space X over $\mathbb{K}$, the mapping $(\alpha, x) \in \mathbb{K} \times X \to \alpha x \in X$ provides an instance of a continuous bilinear mapping, since $\|\lambda x\| = |\lambda|\,\|x\|$.

Given $1 \leq p \leq \infty$, let q denote the conjugate exponent of p. The bilinear mappings

$$((x_i)_{i=1}^\infty, (y_i)_{i=1}^\infty) \in \ell^p \times \ell^q \to (x_i y_i)_{i=1}^\infty \in \ell^\infty$$

and

$$((x_i)_{i=1}^\infty, (y_i)_{i=1}^\infty) \in \ell^p \times \ell^q \to \sum_{i=1}^\infty x_i y_i \in \mathbb{K}$$

likewise provide examples of continuous bilinear mappings, thanks to Hölder's inequality (Theorem 2.4-1).

Remark For any $1 \leq p \leq \infty$, the *convolution product* $(f, g) \in L^1(\mathbb{R}^n) \times L^p(\mathbb{R}^n) \to f * g \in L^p(\mathbb{R}^n)$ provides another example of a continuous bilinear mapping (Problem 2.6-4). □

The following result, which extends a property of linear mappings (Theorem 2.9-3(b)), also provides other examples of continuous multilinear mappings.

Theorem 2.11-2 *If all the spaces $X_1, X_2, \ldots, X_k$ are finite-dimensional and Y is a normed vector space, any multilinear mapping $A : X_1 \times X_2 \times \cdots \times X_k \to Y$ is continuous.*

Proof For each $1 \leq \ell \leq k$, let $(e^\ell_{i(\ell)})_{i(\ell)=1}^{m(\ell)}$ be a basis of X_ℓ. Given any vector $x = (x_1, \ldots, x_k) \in X_1 \times X_2 \times \cdots \times X_k$ with $x_\ell = \sum_{i(\ell)=1}^{m(\ell)} x^\ell_{i(\ell)} e^\ell_{i(\ell)}, 1 \leq k \leq \ell$, the multilinearity of A gives

$$A(x) = \sum_{i(1)=1}^{m(1)} \sum_{i(2)=1}^{m(2)} \cdots \sum_{i(k)=1}^{m(k)} x^1_{i(1)} x^2_{i(2)} \cdots x^k_{i(k)} A\big(e^1_{i(1)}, e^2_{i(2)}, \ldots, e^k_{i(k)}\big).$$

Since the sum in the right-hand side of this relation is finite, there exists a constant C such that (here assume for definiteness that each space X_j, $1 \leq j \leq n$, is equipped with the norm $\|\cdot\|_\infty$)

$$\|A(x)\|_Y \leq C\, \|x_1\|_\infty \|x_2\|_\infty \cdots \|x_k\|_\infty\,.$$

Since all norms are equivalent in a finite-dimensional space (Theorem 2.7-1(a)), the conclusion follows by Theorem 2.11-1(c). □

For instance, Theorem 2.11-2 shows that *the determinant of a $k\times k$ matrix with coefficients in $\mathbb{K}$ is a continuous multilinear functional from $\mathbb{K}^k$ into $\mathbb{K}$.*

By contrast, the property that any continuous linear operator is uniformly continuous (Theorem 2.9-3(a)) does *not* hold for multilinear mappings:

Theorem 2.11-3 *A nonzero continuous multilinear mapping is not uniformly continuous.*

Proof Given a nonzero continuous multilinear mapping $A : X = X_1 \times X_2 \times \cdots \times X_k \to Y$, let $a \in X$ be such that $A(a) \neq 0$. For each integer $n \geq 1$, let

$$x_n := na \quad \text{and} \quad y_n := \left(n - \frac{1}{n^{k-1}}\right) a,$$

so that

$$\|x_n - y_n\|_X = \frac{1}{n^{k-1}} \|a\|_X \to 0 \quad \text{as } n \to \infty$$

(recall that $k \geq 2$ by assumption). Since

$$\|A(x_n) - A(y_n)\|_Y = \left(n^k - \left(n - \frac{1}{n^{k-1}}\right)^k\right) \|A(a)\|_Y$$

by the assumed multilinearity of A, and $\lim_{n\to\infty} \left(n^k - \left(n + \frac{1}{n^{k-1}}\right)^k\right) = k$, there exists $n_0 \geq 1$ such that

$$\|A(x_n) - A(y_n)\|_Y \geq \frac{k}{2} \|A(a)\|_Y \quad \text{for all } n \geq n_0,$$

which shows that A is not uniformly continuous. □

Let $X_1, X_2, \ldots, X_k$ and Y be normed vector spaces over the same field $\mathbb{K}$. Then the vector space, also over $\mathbb{K}$, formed by all the *continuous* multilinear mappings from $X_1 \times X_2 \times \cdots \times X_k$ into Y, which is denoted

$$\mathcal{L}_k(X_1, X_2, \ldots, X_k; Y) \quad \text{or} \quad \mathcal{L}_k(X_1 \times X_2 \times \cdots \times X_k; Y),$$
$$\text{or} \quad \mathcal{L}_k(X; Y) \ \text{if } X_\ell = X,\ 1 \leq \ell \leq k,$$

can be also endowed with a *norm.* The definition and some elementary properties of this norm are given in the next theorem, which constitutes the "multilinear analogue" of parts (a) and (b) in Theorem 2.9-5.

Theorem 2.11-4 *Let $X_1, X_2, \ldots, X_k$ and Y be normed vector spaces.*

(a) *The mapping defined by*

$$\|\cdot\| : A \in \mathcal{L}_k(X_1, X_2, \ldots, X_k; Y) \to \|A\| := \sup_{\left\{\substack{x_\ell \neq 0,\\ 1 \leq \ell \leq k}\right.} \frac{\|A(x_1, x_2, \ldots, x_k)\|_Y}{\|x_1\|_{X_1} \|x_2\|_{X_2} \cdots \|x_k\|_{X_k}}$$

is a norm on the vector space $\mathcal{L}_k(X_1, X_2, \ldots, X_k; Y)$. This definition implies that

$$\|A(x_1, x_2, \ldots, x_k)\|_Y \leq \|A\| \, \|x_1\|_{X_1} \|x_2\|_{X_2} \cdots \|x_k\|_{X_k}$$
$$\text{for all } (x_1, x_2, \ldots, x_k) \in X_1 \times X_2 \times \cdots \times X_k.$$

(b) *The norm of any $A \in \mathcal{L}_k(X_1, X_2, \ldots, X_k; Y)$ may be equivalently defined as*

$$\|A\| = \sup_{\left\{\substack{\|x_\ell\|_{X_\ell} \leq 1,\\ 1 \leq \ell \leq k}\right.} \|A(x_1, x_2, \ldots, x_k)\|_Y = \sup_{\left\{\substack{\|x_\ell\|_{X_\ell} = 1,\\ 1 \leq \ell \leq k}\right.} \|A(x_1, x_2, \ldots, x_k)\|_Y,$$

or as

$$\|A\| = \inf\{C > 0;\ \|A(x_1, x_2, \dots, x_k)\|_Y \le C\,\|x_1\|_{X_1}\,\|x_2\|_{X_2} \cdots \|x_k\|_{X_k}$$
$$\text{for all } (x_1, x_2, \dots, x_k) \in X_1 \times X_2 \times \cdots \times X_k\}.$$

Proof Properties (a) and (b) immediately follow from Theorem 2.11-1(c) and from the definition of a multilinear mapping. □

We will see in Chapter 7 that fundamental examples of continuous multilinear mappings are provided by *Fréchet derivatives of order* ≥ 2. The following result, which shows that "any space of continuous *multilinear* mappings can be obtained by iteration of spaces of continuous *linear* mappings," will then be particularly useful for the study of such derivatives.

Theorem 2.11-5 *Let $X_1, X_2, \dots, X_k$ and Y be normed vector spaces. Then there exists a linear and bijective isometry*

$$\iota : \mathcal{L}(X_1; \mathcal{L}(X_2; \dots; \mathcal{L}(X_k; Y)) \dots) \to \mathcal{L}_k(X_1, X_2, \dots, X_k; Y).$$

Proof We give the proof for $k = 2$, i.e., we prove that, given normed vector spaces X_1, X_2, and Y, there exists a linear and bijective isometry

$$\iota : \mathcal{L}(X_1; \mathcal{L}(X_2; Y)) \to \mathcal{L}_2(X_1, X_2; Y).$$

For notational brevity, all norms are designated by the same notation $\|\cdot\|$.

Given any element $A \in \mathcal{L}(X_1; \mathcal{L}(X_2; Y))$, the mapping $\widetilde{A} : X_1 \times X_2 \to Y$ defined by

$$\widetilde{A}(x_1, x_2) := (Ax_1)x_2 \quad \text{for all } (x_1, x_2) \in X_1 \times X_2$$

is clearly bilinear. Besides, $\widetilde{A}$ is continuous since

$$\|\widetilde{A}(x_1, x_2)\| = \|(Ax_1)x_2\| \le \|Ax_1\|\,\|x_2\| \le \|A\|\,\|x_1\|\,\|x_2\| \quad \text{for all } (x_1, x_2) \in X_1 \times X_2,$$

which shows that $\|\widetilde{A}\| \le \|A\|$. To show that $\|\widetilde{A}\| \ge \|A\|$, let $\varepsilon > 0$ be given. By definition of $\|A\|$, there exists $\widetilde{x}_1 \in X_1$ such that $\|\widetilde{x}_1\| = 1$ and $\|A\widetilde{x}_1\| \ge \|A\|\,(1 - \varepsilon)$. With the element $\widetilde{x}_1 \in X_1$ so determined, there likewise exists $\widetilde{x}_2 \in X_2$ such that $\|\widetilde{x}_2\| = 1$ and $\|(A\widetilde{x}_1)\widetilde{x}_2\| \ge \|A\widetilde{x}_1\|\,(1 - \varepsilon)$, by definition of $\|A\widetilde{x}_1\|$.

Consequently,

$$\|\widetilde{A}\| = \sup_{\left\{\begin{smallmatrix}\|x_1\|=1\\ \|x_2\|=1\end{smallmatrix}\right.} \|\widetilde{A}(x_1, x_2)\| \ge \|\widetilde{A}(\widetilde{x}_1, \widetilde{x}_2)\| = \|(A\widetilde{x}_1)\widetilde{x}_2\| \ge \|A\|\,(1 - \varepsilon)^2.$$

Hence $\|\widetilde{A}\| \ge \|A\|$ since $\varepsilon > 0$ is arbitrary. We have thus shown that

$$\|\widetilde{A}\| = \|A\|.$$

It remains to show that the linear isometry defined by

$$\iota : A \in \mathcal{L}(X_1; \mathcal{L}(X_2; Y)) \to \iota(A) := \widetilde{A} \in \mathcal{L}_2(X_1, X_2; Y)$$

is surjective. Given any element $\widetilde{B} \in \mathcal{L}_2(X_1, X_2; Y)$, let the mapping $Bx_1 : X_2 \to Y$ be defined for each $x_1 \in X_1$ by

$$(Bx_1)x_2 := \widetilde{B}(x_1, x_2) \quad \text{for all } x_2 \in X_2.$$

Then for each $x_1 \in X_1$, the mapping $Bx_1 : X_2 \to Y$, which is clearly linear, is also continuous since

$$\|(Bx_1)x_2\| = \|\widetilde{B}(x_1, x_2)\| \leq \left(\|\widetilde{B}\| \, \|x_1\|\right) \|x_2\| \quad \text{for all } x_2 \in X_2,$$

which shows that $\|Bx_1\| \leq \|\widetilde{B}\| \, \|x_1\|$. Hence $Bx_1 \in \mathcal{L}(X_2; Y)$ for each $x_1 \in X_1$. The mapping $B : x_1 \in X_1 \to \mathcal{L}(X_2; Y)$ defined in this fashion, which is clearly linear, is also continuous since $\|Bx_1\| \leq \|\widetilde{B}\| \, \|x_1\|$. Hence $B \in \mathcal{L}(X_1; \mathcal{L}(X_2; Y))$. That $\iota(B) = \widetilde{B}$ shows that ι is surjective.

The proof for $k \geq 3$ is similar and, for this reason, is omitted. □

For instance, given two normed vector spaces X and Y, Theorem 2.11-5 allows us to *identify* the spaces $\mathcal{L}(X; \mathcal{L}(X; Y))$ and $\mathcal{L}_2(X; Y)$ and, more generally, the spaces $\mathcal{L}(X; \mathcal{L}_{k-1}(X; Y))$ and $\mathcal{L}_k(X; Y)$ for any integer $k \geq 2$.

Problems

2.11-1 Let Ω be an open subset of $\mathbb{R}^n$ and let $1 < p_j < \infty$, $1 \leq j \leq m$, be such that $\frac{1}{s} := \sum_{j=1}^m \frac{1}{p_j} \leq 1$. Show that the multilinear mapping

$$(f_1, f_2, \ldots, f_m) \in L^{p_1}(\Omega) \times L^{p_2}(\Omega) \times \cdots \times L^{p_m}(\Omega) \to \prod_{j=1}^m f_j \in L^s(\Omega)$$

is well defined and continuous.

2.11-2 Let X and Y be two normed vector spaces over the same field, and let A be a multilinear, symmetric, and continuous mapping from $X \times X \times \cdots \times X$ (k factors) into Y. Show that there exists a constant $C(k)$ such that[19]

$$\|A\| \leq C(k) \sup_{\|x\| \leq 1} \|A(x, x, \ldots, x)\|.$$

2.12 Korovkin's theorem

The following "abstract" theorem will be put to use in the next sections to give disarmingly simple proofs of some of the most basic results concerning the approximation of a continuous function on a compact interval of $\mathbb{R}$ by polynomials or by trigonometric polynomials.

Recall that, given a compact metric space (K, d), the space of all continuous functions $f : K \to \mathbb{R}$ is denoted $\mathcal{C}(K)$ and is equipped with the sup-norm, defined by $\|f\| = \sup_{x \in K} |f(x)|$ (Section 2.3).

[19] One can show that the best constant $C(k)$ is $\frac{k^k}{k!}$; see, e.g.:
L. Nachbin [1969]: *Topology on Spaces of Holomorphic Mappings*, Springer, Berlin.

Theorem 2.12-1 (Korovkin's theorem[20]) *Let (K,d) be a compact metric space, let $\phi \in \mathcal{C}\,[0,\infty[$ be a function that satisfies $\phi(t) > 0$ for all $t > 0$, and let the function $\psi_x \in \mathcal{C}(K)$ be defined for each $x \in K$ by*

$$\psi_x(y) := \phi(d(x,y)) \quad \textit{for all } y \in K.$$

Let $(A_n)_{n=0}^{\infty}$ be a sequence of linear operators $A_n : \mathcal{C}(K) \to \mathcal{C}(K)$ that possess the following three properties. First, each A_n, $n \geq 0$, is **nonnegativity-preserving**, *in the sense that*

$$f \in \mathcal{C}(K) \textit{ and } f(x) \geq 0 \textit{ for all } x \in K \textit{ implies } A_n f(x) \geq 0 \textit{ for all } x \in K.$$

Second,

$$\lim_{n\to\infty} \|f_0 - A_n f_0\| = 0,$$

where the function $f_0 \in \mathcal{C}(K)$ is defined by $f_0(x) = 1$ for all $x \in K$. Third,

$$\lim_{n\to\infty} \left(\sup_{x\in K} |(A_n\psi_x)(x)| \right) = 0.$$

Then

$$\textit{for each } f \in \mathcal{C}(K), \quad \lim_{n\to\infty} \|f - A_n f\| = 0.$$

Proof The reader is referred to Chapter 1 for those properties of continuity and compactness used in this proof. The objective is to show that, given any function $f \in \mathcal{C}(K)$ and given any $\varepsilon > 0$, there exists $n_0 = n_0(f,\varepsilon) \geq 0$ such that

$$\sup_{x\in K} |(A_n f)(x) - f(x)| \leq \varepsilon \quad \text{for all } n \geq n_0.$$

So, let $f \in \mathcal{C}(K)$ and $\varepsilon > 0$ be given in what follows.

(i) *A technical inequality: There exists a constant $C = C(f,\varepsilon)$ such that*

$$|f(y) - f(x)| \leq \widetilde{\varepsilon} + 2C\|f\|\psi_x(y) \quad \text{for all } x, y \in K,$$

where (note that $\sup_{n\geq 0} \|A_n f_0\| < \infty$ since the sequence $(A_n f_0)_{n\geq 0}$ converges in $\mathcal{C}(K)$)

$$\widetilde{\varepsilon} := \frac{\varepsilon}{2\sup_{n\geq 0}\|A_n f_0\|} > 0.$$

Since the function $f \in \mathcal{C}(K)$ is uniformly continuous (the set K is compact), there exists $\delta = \delta(f,\varepsilon) > 0$ such that

$$|f(y) - f(x)| \leq \widetilde{\varepsilon} \quad \text{for all } x, y \in K \text{ that satisfy } d(x,y) \leq \delta.$$

It thus remains to consider points $x, y \in K$ that satisfy $d(x,y) \geq \delta$. The function $d : K \times K \to \mathbb{R}$ is continuous (since $|d(\widetilde{x},\widetilde{y}) - d(x,y)| \leq d(\widetilde{x},x) + d(\widetilde{y},y)$ and the right-hand side of this inequality defines a distance that induces the topology of the product $K \times K$), and

[20] P.P. Korovkin [1959]: On convergence of linear positive operators in the space of continuous functions, *Doklady Akademii Nauk SSR* **90**, 961–964 (in Russian).

the product space $K \times K$ is compact. As the inverse image under the continuous function d of the closed interval $[\delta, \infty[$, the set

$$\mathcal{K} := \{(x, y) \in K \times K;\ d(x, y) \geq \delta\}$$

is closed, and thus compact, in $K \times K$.

The composite function $\phi \circ d : \mathcal{K} \to \mathbb{R}$, which is continuous and > 0 on $\mathcal{K}$ (the function ϕ is assumed to be continuous and > 0 on $]0, \infty[$), attains its infimum on $\mathcal{K}$. So, let $C = C(\phi, \delta) = C(\phi, f, \varepsilon) > 0$ be defined as

$$C := \frac{1}{\inf_{(x,y)\in\mathcal{K}} \phi(d(x, y))}.$$

Since $\psi_x(y) := \phi(d(x, y))$ for all $x, y \in K$, it follows that

$$C\psi_x(y) \geq 1 \quad \text{for all } x, y \in K \text{ that satisfy } d(x, y) \geq \delta.$$

As a result,

$$|f(y) - f(x)| \leq 2\|f\| \leq 2C\|f\|\psi_x(y) \quad \text{for all } x, y \in K \text{ that satisfy } d(x, y) \geq \delta.$$

The announced technical inequality then follows by combining the two cases considered above, viz., $d(x, y) \leq \delta$ and $d(x, y) \geq \delta$.

(ii) For each $x \in K$, the technical inequality of (i) can be recast as an inequality between functions, viz.,

$$-\tilde{\varepsilon} f_0 - 2C\|f\|\psi_x \leq f - f(x) f_0 \leq \tilde{\varepsilon} f_0 + 2C\|f\|\psi_x.$$

The assumption that the linear operators A_n are nonnegativity-preserving therefore implies that, for all $n \geq 0$,

$$-\tilde{\varepsilon} A_n f_0 - 2C\|f\|A_n\psi_x \leq A_n f - f(x) A_n f_0 \leq \tilde{\varepsilon} A_n f_0 + 2C\|f\|A_n\psi_x,$$

or equivalently that

$$|(A_n f)(y) - f(x)\,[(A_n f_0)(y)]\,| \leq \tilde{\varepsilon}(A_n f_0)(y) + 2C\|f\|A_n\psi_x(y) \quad \text{for all } y \in K.$$

Letting $y = x$ in this inequality gives

$$|(A_n f)(x) - f(x)\,[(A_n f_0)(x)]\,| \leq \tilde{\varepsilon}(A_n f_0)(x) + 2C\|f\|A_n\psi_x(x) \quad \text{for all } x \in K \text{ and all } n \geq 0.$$

(iii) The last inequality implies that

$$\begin{aligned}|(A_n f)(x) - f(x)| &\leq |(A_n f)(x) - f(x)\,[(A_n f_0)(x)]\,| + |f(x)\,[(A_n f_0 - f_0)(x)]\,| \\ &\leq \tilde{\varepsilon}(A_n f_0)(x) + 2C\|f\|A_n\psi_x(x) + \|f\|\,|(A_n f_0 - f_0)(x)| \quad \text{for all } x \in K \text{ and all } n \geq 0.\end{aligned}$$

The definition of $\tilde{\varepsilon}$ in (i) shows that

$$\tilde{\varepsilon}(A_n f_0)(x) \leq \tilde{\varepsilon}\|A_n f_0\| \leq \frac{\varepsilon}{2} \quad \text{for all } x \in K \text{ and all } n \geq 0,$$

on the one hand. On the other hand,

$$\begin{aligned}&2C\,\|f\|\,(A_n\psi_x)(x)+\|f\|\,|(A_nf_0-f_0)(x)|\\&\le 2C\,\|f\|\sup_{x\in K}|(A_n\psi_x)(x)|+\|f\|\,\|A_nf_0-f_0\|\quad\text{for all } x\in K \text{ and all } n\ge 0.\end{aligned}$$

Therefore the assumptions made on the operators A_n imply that there exists $n_0 = n_0(f, \varepsilon)$ such that

$$2C\,\|f\|\sup_{x\in K}|(A_n\psi_x)(x)|+\|f\|\,\|A_nf_0-f_0\|\le\frac{\varepsilon}{2}\quad\text{for all } n\ge n_0,$$

as desired. □

Remarks (1) The linear operators $A_n : \mathcal{C}(K) \to \mathcal{C}(K)$ are not assumed to be continuous.

(2) The function $\phi \in \mathcal{C}[0, \infty[$ necessarily satisfies $\phi(0) = 0$. For, Theorem 2.12-1 shows that in particular $\lim_{n\to\infty} \|A_n\psi_x - \psi_x\| = 0$ for each $x \in K$. Then, again in particular, $\lim_{n\to\infty}(A_n\psi_x)(x) = \psi_x(x) = \phi(d(x, x)) = \phi(0)$, but $\lim_{n\to\infty} |(A_n\psi_x)(x)| = 0$ by assumption.

(3) *Linear* operators that are nonnegativity-preserving are sometimes called *monotone operators*, a potentially misleading terminology, since matrices whose *inverse* has nonnegative coefficients are called *monotone matrices*. In fact, "*monotone operators*" most often refer to a special class of *nonlinear* operators, which will be introduced later (Section 9.13). □

2.13 Application of Korovkin's theorem to polynomial approximation; Bohman's, Bernstein's, and Weierstraß' theorems

A first, and remarkable, application of Korovkin's theorem is to the space $\mathcal{C}[0, 1]$, equipped with the sup-norm $\|\cdot\|$.

Theorem 2.13-1 (Bohman's theorem[21]) *Let $(A_n)_{n=0}^{\infty}$ be a sequence of linear operators $A_n : \mathcal{C}[0,1] \to \mathcal{C}[0,1]$ that possess the following two properties. First, each A_n, $n \ge 0$, is non-negativity preserving:*

$$f\in\mathcal{C}[0,1] \text{ and } f(x)\ge 0,\ 0\le x\le 1,\quad\text{implies } (A_nf)(x)\ge 0,\ 0\le x\le 1.$$

Second,

$$\lim_{n\to\infty}\|f_p - A_nf_p\| = 0\quad\text{for}\ \ p=0,1,2,$$

where the functions $f_p \in \mathcal{C}[0,1]$, $p = 0, 1, 2$, are defined by

$$f_0(x)=1,\quad f_1(x)=x,\quad f_2(x)=x^2,\quad 0\le x\le 1.$$

Then

$$\text{for each } f\in\mathcal{C}[0,1],\quad \lim_{n\to\infty}\|f-A_nf\|=0.$$

[21] H. Bohman [1952]: On approximation of continuous and of analytic functions, *Arkiv för Mathematik* **2**, 43–56.

Proof We show that Korovkin's theorem (Theorem 2.12-1) can be applied, with the particular function $\phi \in \mathcal{C}\,[0,\infty[$ defined by $\phi(t) = t^2$ for all $t \geq 0$. The only assumption that remains to be checked is thus that

$$\lim_{n\to\infty} \Big(\sup_{0\leq x\leq 1} |(A_n\psi_x)(x)| \Big) = 0,$$

where each function $\psi_x \in \mathcal{C}[0,1]$, $0 \leq x \leq 1$, is defined in this case by

$$\psi_x(y) := \phi(|x-y|) = |x-y|^2 = x^2 f_0(y) - 2x f_1(y) + f_2(y), \quad 0 \leq y \leq 1,$$

or equivalently, by $\psi_x := x^2 f_0 - 2x f_1 + f_2$. Combined together, the relations

$$A_n\psi_x = x^2 A_n f_0 - 2x A_n f_1 + A_n f_2 \quad \text{and} \quad x^2 f_0(x) - 2x f_1(x) + f_2(x) = 0, \quad 0 \leq x \leq 1,$$

then imply that

$$(A_n\psi_x)(x) = x^2 (A_n f_0 - f_0)(x) - 2x(A_n f_1 - f_1)(x) + (A_n f_2 - f_2)(x), \quad 0 \leq x \leq 1.$$

Consequently,

$$\sup_{0\leq x\leq 1} |(A_n\psi_x)(x)| \leq \|A_n f_0 - f_0\| + 2\|A_n f_1 - f_1\| + \|A_n f_2 - f_2\| \quad \text{for all } n \geq 0,$$

and thus $\lim_{n\to\infty}(\sup_{0\leq x\leq 1} |(A_n\psi_x)(x)|) = 0$. The conclusion then follows from Korovkin's theorem. □

The next theorem provides an important instance of a sequence of linear operators from $\mathcal{C}[0,1]$ into $\mathcal{C}[0,1]$ (now denoted B_n, $n \geq 0$) that satisfy the assumptions of Theorem 2.13-1.

Theorem 2.13-2 (Bernstein's theorem[22]) *Let the* **Bernstein operators** $B_n : \mathcal{C}[0,1] \to \mathcal{C}[0,1]$, $n \geq 1$, *be defined for each function* $f \in \mathcal{C}[0,1] \to \mathbb{R}$ *by*

$$(B_n f)(x) = \sum_{k=0}^{n} \frac{n!}{(n-k)!k!} f\Big(\frac{k}{n}\Big) x^k (1-x)^{n-k}, \quad 0 \leq x \leq 1.$$

Then

$$\textit{for each } f \in \mathcal{C}[0,1], \quad \lim_{n\to\infty} \|f - B_n f\| = 0.$$

Proof The operators $B_n : \mathcal{C}[0,1] \to \mathcal{C}[0,1]$ defining the Bernstein polynomials are clearly linear and nonnegativity-preserving. Besides, simple computations (Problem 2.13-1) show that

$$\lim_{n\to\infty} \|B_n f_p - f_p\| = 0 \quad \text{for } p = 0, 1, 2,$$

where the functions f_p are those defined in Theorem 2.13-1. The conclusion then follows from this theorem, the assumptions of which are thus all satisfied. □

[22] S.N. Bernstein [1912]: Démonstration du théorème de Weierstrass fondée sur le calcul de probabilités, *Communications of the Kharkov Mathematical Society* **13**, 1–2.

The functions $B_n f$, $n \geq 0$, defined in Theorem 2.13-2 are called the **Bernstein polynomials** of f.[23] Clearly, each $B_n f$ is of degree $\leq n$.

Remarks (1) The operators $B_n : \mathcal{C}[0,1] \to \mathcal{C}[0,1]$ defining the Bernstein polynomials are *continuous*, since $\|B_n f\| \leq \|f\| \sup_{0\leq x\leq 1}(B_n f_0)(x) = \|f\|$ for all $f \in \mathcal{C}[0,1]$ and $B_n f_0(x) = 1$, $0 \leq x \leq 1$. Hence $\|B_n\| \leq 1$, and in fact, $\|B_n\| = 1$ since $\|B_n f_0\| = \|f_0\| = 1$.

(2) An estimate of the rate of convergence to 0 of $\|f - B_n f\|$ can be obtained under the additional assumption that $f \in \mathcal{C}^2\,[0,1]$; cf. Problem 2.13-3. □

Bernstein's theorem provides as an immediate corollary a constructive proof of *one of the most basic theorems in analysis.* Let

$$\mathcal{P}[0,1], \quad \textit{resp.} \quad \mathcal{P}([0,1];\mathbb{C}),$$

denote the real, *resp.* complex, vector space formed by the restrictions to $[0,1]$ of all polynomials of the real variable with real, *resp.* complex, coefficients.

Theorem 2.13-3 (Weierstraß polynomial approximation theorem[24]) *The space $\mathcal{P}[0,1]$ is dense in the space $\mathcal{C}[0,1]$ equipped with the sup-norm.*

Likewise, the space $\mathcal{P}([0,1];\mathbb{C})$ is dense in the space $\mathcal{C}([0,1];\mathbb{C})$ equipped with the sup-norm.

Proof Given any function $f \in \mathcal{C}[0,1]$, Bernstein's theorem shows that the sequence $(B_n f)_{n=1}^{\infty}$ formed by its associated Bernstein polynomials uniformly converges to f as $n \to \infty$. Hence $\mathcal{P}[0,1]$ is dense in $\mathcal{C}[0,1]$.

The complex case follows from the same argument applied to the real and imaginary parts of any function in $\mathcal{C}([0,1];\mathbb{C})$. □

Weierstraß' theorem provides in turn an interesting corollary:

Theorem 2.13-4 *The spaces $\mathcal{C}[0,1]$ and $\mathcal{C}([0,1];\mathbb{C})$ equipped with the sup-norm are separable.*

Proof Let a function $f \in \mathcal{C}[0,1]$ and $\varepsilon > 0$ be given. By the Weierstraß approximation theorem, there exists a polynomial $p : x \in [0,1] \to \sum_{k=0}^{n} c_k x^k$ with real coefficients c_k such that

$$\|f - p\| \leq \frac{\varepsilon}{2}.$$

Since $\overline{\mathbb{Q}} = \mathbb{R}$, there exist $r_k \in \mathbb{Q}$ such that $|c_k - r_k| \leq \dfrac{\varepsilon}{2(n+1)}$, $0 \leq k \leq n$. Let $q(x) := \sum_{k=0}^{n} r_k x^k$, $0 \leq x \leq 1$. Then

$$\|p - q\| \leq \sup_{0\leq x\leq 1}\left(\sum_{k=0}^{n} |c_k - r_k| x^k\right) \leq \frac{\varepsilon}{2},$$

[23] Extensive studies of the Bernstein polynomials are found in:
G.G. Lorentz [1986]: *Bernstein Polynomials*, Chelsea, New York.
R. DeVore, G.G. Lorentz [1993]: *Constructive Approximation*, Springer, Heidelberg.

[24] K. Weierstraß [1885]: Über die analytische Darstellbarkeit sogenannter willkürlicher Funktionen einer reellen Veränderlichen, *Sitzungsberichte der Akademie zu Berlin*, 633–639 and 789–805.

and thus $\|f - q\| \leq \varepsilon$. It then suffices to observe that the set formed by all polynomials with rational coefficients is countably infinite (Section 1.5).

The complex case is treated in an analogous manner. □

To conclude our analysis of polynomial approximations in the space $\mathcal{C}[0,1]$, we mention a more specialized, but spectacular, result. To this end, we first need a general definition: A subset A of a normed vector space X is *total* if $\overline{\operatorname{Span} A} = X$, i.e., if the subspace of X formed by all *finite* linear combinations of elements of A is dense in X.

The Weierstraß polynomial approximation theorem thus asserts that *the set* $A = \bigcup_{n=0}^{\infty}\{p_n\}$, *where* $p_n(x) := x^n$, $0 \leq x \leq 1$, *is total in* $\mathcal{C}[0,1]$. A natural question then arises: Are there other subsets of $\mathcal{C}[0,1]$ formed again by *powers of* x that are also total in $\mathcal{C}[0,1]$? The next theorem gives a beautiful answer to this question.

$^\flat$**Theorem 2.13-5 (Müntz theorem[25])** *Let* $\alpha_0 = 0 < \alpha_1 < \alpha_2 < \cdots < \alpha_n < \cdots$ *be such that* $\lim_{n\to\infty} \alpha_n = \infty$ (the numbers α_n, $n \geq 1$, are not necessarily integers), *and let* $e_n(x) := x^{\alpha_n}$, $0 \leq x \leq 1$. *Then the set* $A = \bigcup_{n=0}^{\infty}\{e_n\}$ *is total in* $\mathcal{C}[0,1]$ *if and only if the series* $\sum_{n=1}^{\infty} \dfrac{1}{\alpha_n}$ *diverges.* □

Problems

2.13-1 Show that the Bernstein polynomials of the functions f_p, $p = 0, 1, 2$, defined in Theorem 2.13-1 are given for $n \geq 2$ by

$$(B_n f_0)(x) = 1, \quad (B_n f_1)(x) = x, \quad (B_n f_2)(x) = x^2 + \frac{x - x^2}{n}, \quad 0 \leq x \leq 1,$$

which shows that $\lim_{n\to\infty} \|B_n f_p - f_p\| = 0$ for $p = 0, 1, 2$ (a property used in the proof of Theorem 2.13-2).

2.13-2 Let the function $f \in \mathcal{C}[0,1]$ be defined by $f(x) = \sqrt{x}$, $0 \leq x \leq 1$, and let the polynomials $p_n \in \mathcal{P}[0,1]$, $n \geq 0$, be recursively defined by

$$p_n(x) = 0 \quad \text{and} \quad p_n(x) = p_{n-1}(x) + \frac{1}{2}\left(x - [p_{n-1}(x)]^2\right), \quad n \geq 1,\ 0 \leq x \leq 1.$$

Show that $\lim_{n\to\infty} \|p_n - f\| = 0$ (this exercise thus provides an explicit construction of polynomials that converge uniformly to this particular function f; naturally, the Bernstein polynomials provide another example).

Hint: Apply Dini's theorem (Problem 2.3-1) to the sequence $(p_n)_{n=0}^{\infty}$.

2.13-3 Show that the Bernstein polynomials $B_n f$, $n \geq 0$, of a function $f \in \mathcal{C}^2[0,1]$ satisfy

$$f(x) - B_n f(x) = \frac{1}{n}\frac{x(x-1)}{2} f''(x) + \frac{1}{n}\varepsilon(n,x), \quad 0 \leq x \leq 1,$$

where $\lim_{n\to\infty}(\sup_{0\leq x\leq 1} |\varepsilon(n,x)|) = 0$. This property constitutes **Voronovskaja's theorem.**[26]

[25]C. Müntz [1914]: Über den Approximationssatz von Weierstraß, in *H.A. Schwarz Festschrift*, pp. 303–312, Mathematische Abhandlungen, Springer, Berlin.

Proofs are also found in, e.g., Goffman & Pedrick [1965, Chapter 4, Section 4.6], or in Cheney [1966, Chapter 6, Section 2].

[26]E.V. Voronovskaja [1932]: Détermination de la forme asymptotique de l'approximation des fonctions

2.14 Application of Korovkin's theorem to trigonometric polynomial approximation; Fejér's theorem

A second, and equally remarkable, application of Korovkin's theorem is to the space

$$\mathcal{C}_{\text{per}}[0, 2\pi],$$

formed by all 2π-*periodic* continuous functions $g : [0, 2\pi] \to \mathbb{R}$, equipped with the sup-norm $\|\cdot\|$ defined by $\|g\| := \sup_{0\leq\theta\leq 2\pi} |g(\theta)|$. Notice the analogies with Theorem 2.13-1.

Theorem 2.14-1[27] *Let $(A_n)_{n=0}^{\infty}$ be a sequence of linear operators $A_n : \mathcal{C}_{\text{per}}[0, 2\pi] \to \mathcal{C}_{\text{per}}[0, 2\pi]$ that possess the following two properties: First, each A_n, $n \geq 0$, is nonnegativity-preserving:*

$$g \in \mathcal{C}_{per}[0, 2\pi] \text{ and } g(\theta) \geq 0,\ 0 \leq \theta \leq 2\pi, \quad \text{implies } (A_n g)(\theta) \geq 0,\ 0 \leq \theta \leq 2\pi.$$

Second,

$$\lim_{n\to\infty} \|g_p - A_n g_p\| = 0 \quad \text{for } p = 0, 1, 2,$$

where the functions $g_p \in \mathcal{C}_{\text{per}}[0, 2\pi]$, $p = 0, 1, 2$, are defined by

$$g_0(\theta) = 1, \quad g_1(\theta) = \cos\theta, \quad g_2(\theta) = \sin\theta, \quad 0 \leq \theta \leq 2\pi.$$

Then

$$\text{for each } g \in \mathcal{C}_{\text{per}}[0, 2\pi], \quad \lim_{n\to\infty} \|g - A_n g\| = 0.$$

Proof Let the set

$$K := \{x = (x_1, x_2) \in \mathbb{R}^2;\ x_1^2 + x_2^2 = 1\}$$

be equipped with the distance d induced by the Euclidean norm in $\mathbb{R}^2$ and, given any function $g \in \mathcal{C}_{\text{per}}[0, 2\pi]$, let the function $g^\sharp : K \to \mathbb{R}$ be defined by

$$g^\sharp(x) := g(\theta), \quad x = (\cos\theta, \sin\theta), \quad 0 \leq \theta < 2\pi.$$

Then the function $g^\sharp$ belongs to $\mathcal{C}(K)$, because

$$\frac{1}{\sqrt{2}}|\theta - \varphi| \leq d((\cos\theta, \sin\theta), (\cos\varphi, \sin\varphi)) \leq |\theta - \varphi| \quad \text{for } |\theta - \varphi| \leq \frac{\pi}{2}$$

(continuity is a local property), and because $g(\theta)$ converges to $g(0)$ as $\theta \in [0, 2\pi[$ approaches 2π since the function g is 2π-periodic. Clearly, the mapping

$$g \in \mathcal{C}_{\text{per}}[0, 2\pi] \to g^\sharp \in \mathcal{C}(K)$$

par les polynômes de M. Bernstein, *Doklady Akademii Nauk SSSR* **4**, 79–85.

This result was then immediately extended to functions $f \in \mathcal{C}^{2m}[0, 1]$, $m \geq 1$, by:

S.N. Bernstein [1932]: Complément à l'article de E. Voronovskaya "Détermination de la forme asymptotique de l'approximation des fonctions par les polynômes de M. Bernstein," *Doklady Akademii Nauk SSSR* **4**, 86–92.

[27] P.P. Korovkin [1959]: *Linear Operators and Approximation Theory*, Fitzmatgiz, Moscow (in Russian) [English translation, Hindustan Publishing Corporation, Delhi, 1960].

defined in this fashion is a bijection.

We then show that Korovkin's theorem (Theorem 2.12-1) can be applied to the space $\mathcal{C}(K)$ equipped with the sup-norm, also denoted $\|\cdot\|$, with the particular function $\phi \in \mathcal{C}\,[0, \infty[$ defined by $\phi(t) = t^2$ for all $t \geq 0$ (as in the proof of Theorem 2.13-1), and with the linear operators $A_n^\sharp : \mathcal{C}(K) \to \mathcal{C}(K)$, $n \geq 0$, defined by

$$A_n^\sharp g^\sharp := (A_n g)^\sharp \quad \text{for all } g \in \mathcal{C}_{\text{per}}[0, 2\pi].$$

To this end, we first note that (K, d) is a compact metric space (as a closed and bounded subset of $\mathbb{R}^2$), that the linear operators $A_n^\sharp$ are also nonnegativity-preserving, and that $\lim_{n\to\infty} \|A_n^\sharp f_0 - f_0\| = 0$ if $f_0(x) := 1$, $x \in K$, since $f_0 = g_0^\sharp$. The only assumption that remains to be checked is that

$$\lim_{n\to\infty} \left(\sup_{x\in K} |(A_n^\sharp \psi_x^\sharp)(x)| \right) = 0,$$

where, for each $x = (\cos\theta, \sin\theta) \in K$, the function $\psi_x^\sharp \in \mathcal{C}(K)$ is defined by

$$\begin{aligned} \psi_x^\sharp(y) &= \phi(d(x, y)) = |d(x, y)|^2 = 4\sin^2\left(\frac{\theta - \varphi}{2}\right) \\ &= 2g_0(\varphi) - 2\cos\theta g_1(\varphi) - 2\sin\theta g_2(\varphi), \quad y = (\cos\varphi, \sin\varphi) \in K, \end{aligned}$$

or equivalently, by $\psi_x^\sharp = 2g_0^\sharp - 2\cos\theta g_1^\sharp - 2\sin\theta g_2^\sharp$. Consequently,

$$A_n^\sharp \psi_x^\sharp = 2\left(A_n g_0 - \cos\theta(A_n g_1) - \sin\theta(A_n g_2)\right)^\sharp,$$

which in particular implies that

$$\begin{aligned} (A_n^\sharp \psi_x^\sharp)(x) &= 2\left(A_n g_0 - \cos\theta(A_n g_1) - \sin\theta(A_n g_2)\right)^\sharp(x) \\ &= 2A_n g_0(\theta) - 2\cos\theta(A_n g_1)(\theta) - 2\sin\theta(A_n g_2)(\theta), \quad \text{for all } x = (\cos\theta, \sin\theta) \in K. \end{aligned}$$

Since $g_0(\theta) - \cos\theta g_1(\theta) - \sin\theta g_2(\theta) = 0$ for all $0 \leq \theta \leq 2\pi$, the last relation may be also rewritten as

$$\begin{aligned} (A_n^\sharp \psi_x^\sharp)(x) = \; & 2(A_n g_0 - g_0)(\theta) - 2\cos\theta(A_n g_1 - g_1)(\theta) \\ & - 2\sin\theta(A_n g_2 - g_2)(\theta) \quad \text{for all } x = (\cos\theta, \sin\theta) \in K. \end{aligned}$$

Consequently,

$$\sup_{x\in K} |(A_n^\sharp \psi_x^\sharp)(x)| \leq 2\left(\|A_n g_0 - g_0\| + \|A_n g_1 - g_1\| + \|A_n g_2 - g_2\|\right),$$

and thus $\lim_{n\to\infty} \sup_{x\in K} |(A_n^\sharp \psi_x^\sharp)(x)| = 0$. The conclusion then follows from Korovkin's theorem. □

The next theorem will provide an important instance of a sequence of linear operators from $\mathcal{C}_{\text{per}}[0, 2\pi]$ into $\mathcal{C}_{\text{per}}[0, 2\pi]$ (now denoted F_n, $n \geq 0$) that satisfy the assumptions of Theorem 2.14-1.

But first, a few definitions are in order. For each integer $n \geq 0$, let

$$\mathcal{Q}_n[0, 2\pi]$$

denote the space formed by all **real 2π-periodic trigonometric polynomials of degree** $\leq n$, i.e., functions in $\mathcal{C}_{\text{per}}[0, 2\pi]$ of the form

$$\theta \in [0, 2\pi] \to \sum_{k=0}^{n} c_k \cos k\theta + \sum_{k=1}^{n} d_k \sin k\theta \quad \text{with real coefficients } c_k \text{ and } d_k,$$

and let

$$\mathcal{Q}[0, 2\pi] := \bigcup_{n=0}^{\infty} \mathcal{Q}_n[0, 2\pi] \subset \mathcal{C}_{\text{per}}[0, 2\pi]$$

denote the space formed by all **real 2π-periodic trigonometric polynomials**. Clearly, $\dim \mathcal{Q}_n[0, 2\pi] = 2n + 1$. The functions $S_n g$ and $F_{n+1} g$ defined in the next theorem provide *examples* of such trigonometric polynomials of degree $\leq n$. The functions $S_n g$ draw their name from the fact that they are the nth *partial sum of the Fourier series* of the function g, as we will see later (Theorem 4.9-2).

Theorem 2.14-2 (Fejér's theorem[28]**)** *Let the* **Fourier partial sum operators** $S_n : g \in \mathcal{C}_{\text{per}}[0, 2\pi] \to S_n g \in \mathcal{C}_{\text{per}}[0, 2\pi]$ *be defined for any integer* $n \geq 0$ *by*

$$(S_0 g)(\theta) := \frac{a_0}{2} \quad \textit{and} \quad (S_n g)(\theta) := \frac{a_0}{2} + \sum_{k=1}^{n} (a_k \cos k\theta + b_k \sin k\theta) \;\; \textit{for } n \geq 1,\ 0 \leq \theta \leq 2\pi,$$

where

$$a_k := \frac{1}{\pi} \int_0^{2\pi} g(\varphi) \cos k\varphi \, \mathrm{d}\varphi, \;\; k \geq 0, \quad \textit{and} \quad b_k := \frac{1}{\pi} \int_0^{2\pi} g(\varphi) \sin k\varphi \, \mathrm{d}\varphi, \;\; k \geq 1,$$

and let the **Fejér operators** $F_n : \mathcal{C}_{\text{per}}[0, 2\pi] \to \mathcal{C}_{\text{per}}[0, 2\pi]$ *be defined for any integer* $n \geq 1$ *by*

$$F_n : g \in \mathcal{C}_{\text{per}}[0, 2\pi] \to F_n g := \frac{1}{n}(S_0 g + S_1 g + \cdots + S_{n-1} g).$$

Then

$$\textit{for each } g \in \mathcal{C}_{\text{per}}[0, 2\pi], \quad \lim_{n \to \infty} \|g - F_n g\| = 0.$$

Proof The Fejér operators A_n are clearly linear. Besides, straightforward computations (Problem 2.14-1) show that, for any $n \geq 1$,

$$F_n g(\theta) = \frac{1}{2n\pi} \int_0^{2\pi} g(\theta + \varphi) \left(\frac{\sin \frac{1}{2} n\varphi}{\sin \frac{1}{2}\varphi} \right)^2 \mathrm{d}\varphi, \quad 0 \leq \theta \leq 2\pi,$$

$$\lim_{n \to \infty} \|F_n g_p - g_p\| = 0, \quad p = 0, 1, 2,$$

[28]L. Fejér [1900]: Sur les fonctions bornées et intégrables, *Comptes Rendus de l'Académie des Sciences, Paris* **131**, 984–987.

where the functions g_p, $p = 0, 1, 2$, are defined as in Theorem 2.14-1. The operators F_n, which are therefore nonnegativity-preserving by the first formula above, thus satisfy all the assumptions of Theorem 2.14-1. □

Remark The Fejér operators $F_n : \mathcal{C}_{\rm per}[0, 2\pi] \to \mathcal{C}_{\rm per}[0, 2\pi]$, $n \geq 1$, are *continuous*, since

$$\|F_n g\| \leq \|g\| \left(\frac{1}{2n\pi} \int_0^{2\pi} \left(\frac{\sin \frac{1}{2} n\varphi}{\sin \frac{1}{2}\varphi} \right)^2 \right) \mathrm{d}\varphi \text{ and } F_n g_0(\theta) = \frac{1}{2n\pi} \int_0^{2\pi} \left(\frac{\sin \frac{1}{2} n\varphi}{\sin \frac{1}{2}\varphi} \right)^2 \mathrm{d}\varphi = 1,\ 0 \leq \theta \leq 2\pi$$

(Problem 2.14-1). Hence $\|F_n\| \leq 1$, and in fact, $\|F_n\| = 1$ since $\|F_n g_0\| = \|g_0\| = 1$. □

The functions $F_n g$, $n \geq 0$, are called the **Fejér trigonometric polynomials** of g.

While the above theorem thus asserts that, given any function $g \in \mathcal{C}_{\rm per}[0, 2\pi]$, its Fejér trigonometric polynomials $F_n g$ uniformly converge on $[0, 2\pi]$ to g as $n \to \infty$, nothing can be said *in general* about the pointwise convergence,[29] let alone about the uniform convergence, of the nth Fourier partial sums $S_n g$ to g as $n \to \infty$ (unless additional assumptions are made on g; cf. Problem 2.14-2). As we shall see later (Section 4.9), what *can* be proved is that $\lim_{n\to\infty} \|S_n g - g\|_{L^2(0,2\pi)} = 0$, where $\|\cdot\|_{L^2(0,2\pi)}$ denotes the norm of the space $L^2(0, 2\pi)$ (Section 2.5), in fact not only for any function $g \in \mathcal{C}_{\rm per}[0, 2\pi]$, but for any function $g \in L^2(0, 2\pi)$.

Remark The operators F_n are the *Cesàro means*[30] $F_n := \dfrac{1}{n}(S_0 + S_1 + \cdots + S_{n-1})$ of the operators $S_n : \mathcal{C}_{\rm per}[0, 2\pi] \to \mathcal{C}_{\rm per}[0, 2\pi]$, an "averaging procedure" that often improves convergence properties, as in the present case. □

Fejér's theorem immediately provides a constructive proof of another *basic result of analysis*, which constitutes the "trigonometric polynomial" equivalent of the Weierstraß approximation theorem (Theorem 2.13-3). This result applies to the real space $\mathcal{C}_{\rm per}[0, 2\pi]$ (defined earlier) as well as to the *complex* space

$$\mathcal{C}_{\rm per}\left([0, 2\pi]\,; \mathbb{C}\right),$$

formed by all 2π-*periodic* continuous functions $g : [0, 2\pi] \to \mathbb{C}$, equipped with the sup-norm $\|\cdot\|$ defined by $\|g\| := \sup_{0\leq\theta\leq 2\pi} |g(\theta)|$. For each integer $n \geq 0$, let

$$\mathcal{Q}_n\left([0, 2\pi]\,; \mathbb{C}\right)$$

denote the space formed by all **complex 2π-periodic trigonometric polynomials of degree $\leq n$**, i.e., functions in $\mathcal{C}_{\rm per}([0, 2\pi]\,; \mathbb{C})$ of the form

$$\theta \in [0, 2\pi] \to \sum_{k=-n}^{n} c_k e^{ik\theta} \quad \text{with complex coefficients } c_k.$$

[29]The first example of a continuous periodic function whose Fourier series diverges at one point was given in:

P. Du Bois-Raymond [1876]: Untersuchungen über die Convergenz und Divergenz der Fourierschen Darstellungsformeln, *Abhandlungen der Mathematisch-Physikalischen Klasse der Königlich Bayerischen Akademie der Wissenschaften* **12**, 1–103.

[30]So named after Ernesto Cesàro (1859–1906).

Finally, let

$$\mathcal{Q}([0,2\pi];\mathbb{C}) := \bigcup_{n=0}^{\infty} \mathcal{Q}_n([0,2\pi];\mathbb{C}) \subset \mathcal{C}_{\text{per}}([0,2\pi];\mathbb{C})$$

denote the space formed by all **complex 2π-periodic trigonometric polynomials.**

Theorem 2.14-3 (Weierstraß trigonometric polynomial approximation theorem) *The space $\mathcal{Q}[0,2\pi]$ formed by all real 2π-periodic trigonometric polynomials is dense in the space $\mathcal{C}_{\text{per}}[0,2\pi]$.*

Likewise, the space $\mathcal{Q}([0,2\pi];\mathbb{C})$ formed by all complex 2π-periodic trigonometric polynomials is dense in the space $\mathcal{C}_{\text{per}}([0,2\pi];\mathbb{C})$.

Proof Given any function $g \in \mathcal{C}_{\text{per}}[0,2\pi]$, the sequence $(F_n g)_{n=1}^{\infty}$, where F_n denote the Fejér operators (Theorem 2.14-2), uniformly converges to g as $n \to \infty$. Hence $\mathcal{Q}[0,2\pi]$ is dense in $\mathcal{C}_{\text{per}}[0,2\pi]$.

The same argument, applied to both the real and imaginary parts of any complex-valued function $g \in \mathcal{C}_{\text{per}}([0,2\pi],\mathbb{C})$, shows that g can be uniformly approximated by trigonometric polynomials of the specific form

$$\theta \in [0,2\pi] \to \sum_{k=0}^{n} a_k \cos k\theta + \sum_{k=1}^{n} b_k \sin k\theta \quad \text{with complex coefficients } a_k \text{ and } b_k.$$

But such a trigonometric polynomial can be immediately rewritten as a complex trigonometric polynomial in the space $\mathcal{C}_{\text{per}}([0,2\pi],\mathbb{C})$, i.e., of the form

$$\theta \in [0,2\pi] \to \sum_{k=-n}^{k=n} c_k e^{ik\theta} \quad \text{with complex coefficients } c_k$$

(to see this, let $c_0 := a_0$, $c_k = \dfrac{1}{2}(a_k - ib_k)$ for $1 \le k \le n$, and $c_k := \dfrac{1}{2}(a_{-k} + ib_{-k})$ for $-n \le k \le -1$). □

As we shall see (Theorem 2.15-4), the same conclusion in the complex case can be also reached from a stronger version of the Weierstraß approximation theorem, which constitutes the object of the next section.

Problems

2.14-1 (1) Given a function $g \in \mathcal{C}_{\text{per}}[0,2\pi]$, show that the nth Fourier partial sum of g (Theorem 2.14-2) is also given for any $n \ge 0$ by

$$(S_n g)(\theta) = \frac{1}{2\pi}\int_0^{2\pi} g(\theta+\varphi)\frac{\sin(n+\frac{1}{2})\varphi}{\sin\frac{1}{2}\varphi}\,d\varphi.$$

The function $\varphi \in [0,2\pi] \to \dfrac{1}{2\pi}\dfrac{\sin(n+\frac{1}{2})\varphi}{\sin\frac{1}{2}\varphi}\,d\varphi$ appearing in this formula is called the *Dirichlet kernel* (naturally, its value at $\varphi = 0$ is defined as $n + \dfrac{1}{2}$).

(2) Show that the function $F_n g$, where F_n denotes the Fejér operator (Theorem 2.14-2), is also given for any $n \geq 1$ by

$$(F_n g)(\theta) = \frac{1}{2n\pi} \int_0^{2\pi} g(\theta + \varphi) \left(\frac{\sin \frac{1}{2} n\varphi}{\sin \frac{1}{2}\varphi} \right)^2 \mathrm{d}\varphi.$$

The function $\varphi \in [0, 2\pi] \to \left(\frac{\sin \frac{1}{2} n\varphi}{\sin \frac{1}{2}\varphi} \right)^2$ appearing in this formula is called the *Fejér kernel* (naturally, its value at $\varphi = 0$ is defined as n^2).

(3) The functions g_p, $p = 0, 1, 2$, being defined as in Theorem 2.14-1, show that, for any $n \geq 1$,

$$(F_n g_0)(\theta) = 1, \quad (F_n g_1)(\theta) = \frac{n-1}{n} \cos \theta, \quad (F_n g_2)(\theta) = \frac{n-1}{n} \sin \theta, \quad 0 \leq \theta \leq 2\pi,$$

which shows that $\lim_{n\to\infty} \|F_n g_p - g_p\| = 0$ (a property used in the proof of Theorem 2.14-2).

2.14-2 Let $g \in \mathcal{C}_{\text{per}}[0, 2\pi]$ be differentiable at a point $\theta_0 \in [0, 2\pi]$. Show that $\lim_{n\to\infty}(S_n g)(\theta_0) = g(\theta_0)$, where $S_n g$ denotes the nth Fourier partial sum of g (Theorem 2.14-2).

Hint: Use Problem 2.14-1(1).

2.15 The Stone–Weierstraß theorem

An **algebra** is a vector space X over $\mathbb{K} = \mathbb{R}$ or $\mathbb{K} = \mathbb{C}$ that is endowed with an additional mapping

$$(x, y) \in X \times X \to xy \in X,$$

called **multiplication**, satisfying the following properties for all $x, y, z \in X$ and all $\alpha, \beta \in \mathbb{K}$:

$$(xy)z = x(yz), \quad x(y + z) = xy + xz, \quad (x + y)z = xz + yz, \quad (\alpha x)(\beta y) = (\alpha\beta)(xy).$$

If $\mathbb{K} = \mathbb{R}$, *resp.* $\mathbb{K} = \mathbb{C}$, X is called a **real**, *resp.* **complex**, **algebra**.

A **subalgebra** of an algebra X is a subspace of X that is also an algebra. For instance, *the space* $\mathcal{C}[0, 1]$ *is a real algebra and its subspace* $\mathcal{P}[0, 1]$ *is a subalgebra of* $\mathcal{C}[0, 1]$. The Weierstraß polynomial approximation theorem (Theorem 2.13-3) thus asserts that *the subalgebra* $\mathcal{P}[0, 1]$ *is dense in the algebra* $\mathcal{C}[0, 1]$ (equipped as usual with the sup-norm).

While the Weierstraß polynomial approximation theorem was proved in Section 2.13 as a corollary of Korovkin's theorem applied to Bernstein's polynomials, it can also be given an entirely different proof, which simply uses that $\mathcal{P}[0, 1]$ is a subalgebra of the algebra $\mathcal{C}[0, 1]$ that possesses two specific properties: first, it contains the constant functions, and second, it *separates the elements of* $\mathcal{C}[0, 1]$, in the sense that, given any two distinct points $\xi, \eta \in [0, 1]$, there exists a function $g \in \mathcal{P}[0, 1]$ that satisfies $g(\xi) \neq g(\eta)$ (for instance, that defined by $g(x) := x$, $0 \leq x \leq 1$).

It is remarkable that, given *any* compact metric space K and *any* subalgebra of the (real) space $\mathcal{C}(K)$ that satisfies the same simple assumptions, the same density property holds. This is the essence of the next result, *one of the most basic theorems in functional analysis.*

Theorem 2.15-1 (Stone–Weierstraß theorem[31]) *Let K be a compact metric space, and let $\mathcal{A}$ be a subalgebra of the* (real) *space $\mathcal{C}(K)$ that possesses the following two properties:*

[31] M.H. Stone [1948]: The generalized Weierstrass approximation theorem, *Mathematics Magazine* **21**, 167–183 and 237–254.

(a) *The constant functions belong to* $\mathcal{A}$.

(b) *Given any two distinct points* $\xi, \eta \in K$, *there exists a function* $g = g(\xi, \eta) \in \mathcal{A}$ *that satisfies* $g(\xi) \neq g(\eta)$.

Then $\mathcal{A}$ *is dense in* $\mathcal{C}(K)$.

Proof (i) *The closure* $\overline{\mathcal{A}}$ *of* $\mathcal{A}$ *is also a subalgebra of* $\mathcal{C}(K)$.

This property holds simply because addition and scalar multiplication are continuous (Theorem 2.2-5) and the multiplication is also a continuous mapping from $\mathcal{C}(K) \times \mathcal{C}(K)$ into $\mathcal{C}(K)$. To see this, apply the triangle inequality to the identity $\widetilde{f}\widetilde{g} - fg = (\widetilde{f} - f)g + (\widetilde{g} - g)f + (\widetilde{f} - f)(\widetilde{g} - g)$ and use the inequality $\|fg\| \leq \|f\|\,\|g\|$; the commutativity of the multiplication in the algebra $\mathcal{C}(K)$ is also used here.

(ii) *If* $f \in \overline{\mathcal{A}}$, *then* $|f| \in \overline{\mathcal{A}}$.

First, note that $f \in \mathcal{C}(K)$ implies $|f| \in \mathcal{C}(K)$ (since $||f(x)| - |f(y)|| \leq |f(x) - f(y)|$ for all $x, y \in K$).

Next, let a function $f \in \overline{\mathcal{A}}$ and $\varepsilon > 0$ be given. Without recourse to the Bernstein polynomials (Theorem 2.13-2) or to the Weierstraß polynomial approximation theorem (Theorem 2.13-3), it is easily seen (Problem 2.15-3) that there exists a polynomial $p \in \mathcal{P}$ such that

$$\sup_{-\|f\| \leq t \leq \|f\|} ||t| - p(t)| \leq \varepsilon.$$

Consequently,

$$\sup_{x \in K} ||f(x)| - p(f(x))| = \||f| - p \circ f\| \leq \varepsilon.$$

But the function $p \circ f$ also belongs to $\overline{\mathcal{A}}$ because p is a *polynomial* and $\overline{\mathcal{A}}$ is a *subalgebra* by (i). Hence the function $|f| \in \mathcal{C}(K)$ belongs to $\overline{\mathcal{A}}$ since $\varepsilon > 0$ is arbitrary.

(iii) *If* $f, g \in \overline{\mathcal{A}}$, *then* $\max\{f, g\} \in \overline{\mathcal{A}}$ *and* $\min\{f, g\} \in \overline{\mathcal{A}}$.

To see this, it suffices to combine (ii) and the relations

$$\max\{f, g\} = \frac{1}{2}(f + g + |f - g|) \quad \text{and} \quad \min\{f, g\} = \frac{1}{2}(f + g - |f - g|).$$

(iv) *Given any points* $\xi, \eta \in K$ *and any* $\alpha, \beta \in \mathbb{R}$, *there exists a function* $g \in \mathcal{A}$ *such that* $g(\xi) = \alpha$ *and* $g(\eta) = \beta$.

By assumption, there exists a function $g_0 \in \mathcal{A}$ such that $g_0(\xi) \neq g_0(\eta)$. Then the function $g \in \mathcal{C}(K)$ defined by

$$g(x) := \frac{\alpha g_0(\eta) - \beta g_0(\xi)}{g_0(\eta) - g_0(\xi)} + \frac{\beta - \alpha}{g_0(\eta) - g_0(\xi)} g_0(x), \quad x \in K,$$

belongs to $\mathcal{A}$ (the constant functions belong to $\mathcal{A}$ by assumption and $g_0 \in \mathcal{A}$) and satisfies $g(\xi) = \alpha$ and $g(\eta) = \beta$.

(v) *Let a function* $f \in \mathcal{C}(K)$ *and* $\varepsilon > 0$ *be given. Then there exists a function* $g \in \overline{\mathcal{A}}$ *that satisfies* $\|f - g\| \leq \varepsilon$.

Given any points $\xi, \eta \in K$, there exists by (iv) a function $g(\xi, \eta) \in \mathcal{A}$ such that

$$g(\xi, \eta)(\xi) = f(\xi) \quad \text{and} \quad g(\xi, \eta)(\eta) = f(\eta).$$

Each set

$$U(\xi,\eta) := \{x \in K;\ g(\xi,\eta)(x) < f(x) + \varepsilon\}$$

is open in K (both functions $g(\xi,\eta)$ and f are continuous), and $K = \bigcup_{\xi\in K} U(\xi,\eta)$ for all $\eta \in K$ since $\xi \in U(\xi,\eta)$ for all $\xi,\eta \in K$. Since the set K is compact, the above open covering of K admits a finite subcovering, thus of the form

$$K = \bigcup_{i=1}^{m(\eta)} U(\xi_i,\eta).$$

For each $\eta \in K$, define the function

$$g(\eta) := \min_{1\le i\le m(\eta)} \{g(\xi_i,\eta)\},$$

which belongs to $\overline{\mathcal{A}}$ by (iii) (this is why a "finite minimum" is needed here). Given any $x \in K$, there exists $i = i(x,\eta) \in \{1,2,\dots,m(\eta)\}$ such that $x \in U(\xi_i,\eta)$, which implies that $g(\xi_i,\eta)(x) < f(x) + \varepsilon$. Consequently,

$$g(\eta)(x) \le g(\xi_{i(x,\eta)},\eta)(x) < f(x) + \varepsilon \quad \text{for all } x \in K.$$

Each set

$$V(\eta) := \{x \in K;\ g(\eta)(x) > f(x) - \varepsilon\}$$

is open in K (both functions $g(\eta)$ and f are continuous), and $K = \bigcup_{\eta\in K} V(\eta)$ since $\eta \in V(\eta)$ for all $\eta \in K$. Hence there exists a finite subcovering of K, thus of the form

$$K = \bigcup_{j=1}^{n} V(\eta_j).$$

Define the function

$$g := \max_{1\le j\le n} \{g(\eta_j)\},$$

which belongs to $\overline{\mathcal{A}}$ by (iii) (this is why a "finite maximum" is needed here). Given any $x \in K$, there exists $j = j(x) \in \{1,2,\dots,n\}$ such that $x \in V(\eta_j)$, which implies that $g(\eta_j)(x) > f(x) - \varepsilon$. Consequently,

$$g(x) \ge g(\eta_{j(x)})(x) > f(x) + \varepsilon \quad \text{for all } x \in K,$$

on the one hand. On the other hand, there exists $k = k(x) \in \{1,2,\dots,n\}$ such that $g(x) = g(\eta_{k(x)})(x)$; consequently,

$$g(x) = g(\eta_{k(x)})(x) \le g(\xi_{i(x,\eta_{k(x)})},\eta_{k(x)})(x) < f(x) + \varepsilon \quad \text{for all } x \in K.$$

We have thus found a function $g \in \overline{\mathcal{A}}$ that satisfies

$$\|f - g\| = \sup_{x\in K} |f(x) - g(x)| < \varepsilon.$$

Since $\varepsilon > 0$ is arbitrary, this shows that f belongs to the closure of $\overline{\mathcal{A}}$, which coincides with $\overline{\mathcal{A}}$ since $\overline{\mathcal{A}}$ is closed. This completes the proof. $\square$

A first corollary of the Stone–Weierstraß theorem is the following generalization of the classical Weierstraß polynomial approximation theorem in the real case (Theorem 2.13-3). Another interesting consequence is proposed in Problem 2.15-1.

Theorem 2.15-2 (Weierstraß polynomial approximation theorem in several variables) *Let K be a compact subset of $\mathbb{R}^n$, and let $\mathcal{P}(K)$ denote the space formed by the restrictions to K of all the real polynomials in n variables. Then $\mathcal{P}(K)$ is dense in $\mathcal{C}(K)$.*

Proof It is clear that $\mathcal{P}(K)$ is a subalgebra that contains the constant functions. If $\xi = (\xi_i)_{i=1}^n$ and $\eta = (\eta_i)_{i=1}^n$ are two distinct points in K, there necessarily exists $i \in \{1, 2, \dots, n\}$ such that $\xi_i \neq \eta_i$; therefore the polynomial g defined by $g(x) := x_i$ for all $x = (x_j)_{j=1}^n$ satisfies $g(\xi) \neq g(\eta)$. The assertion then follows from the Stone–Weierstraß theorem. $\square$

Another noticeable feature of the Stone–Weierstraß theorem is its following extension to *complex-valued* functions. Note that an extra assumption (cf. (c) below) is needed in this case, however.

Theorem 2.15-3 (complex Stone–Weierstraß theorem) *Let K be a compact metric space, and let $\mathcal{A}$ be a complex subalgebra of the* (complex) *space $\mathcal{C}(K; \mathbb{C})$ that possesses the following three properties:*

(a) *The constant functions belong to $\mathcal{A}$.*

(b) *Given any two distinct points $\zeta, \eta \in K$, there exists a function $g \in \mathcal{A}$ that satisfies $g(\xi) \neq g(\eta)$.*

(c) *If $f \in \mathcal{A}$, then the conjugate function $\overline{f}$ also belongs to $\mathcal{A}$.*

Then $\mathcal{A}$ is dense in $\mathcal{C}(K; \mathbb{C})$.

Proof Thanks to (c), the real and imaginary parts of any function $f \in \mathcal{A}$, viz.,

$$\operatorname{Re} f := \frac{1}{2}(f + \overline{f}) \quad \text{and} \quad \operatorname{Im} f := \frac{1}{2i}(f - \overline{f}),$$

belong to the subalgebra $\mathcal{A}_{\mathbb{R}}$ of $\mathcal{C}(K; \mathbb{C})$ formed by all the *real-valued* continuous functions in $\mathcal{A}$. It thus suffices to show that $\mathcal{A}_{\mathbb{R}}$ satisfies the assumptions of the (real) Stone–Weierstraß theorem (Theorem 2.15-1), and to apply this theorem to the real and imaginary parts of any function $f \in \mathcal{C}(K; \mathbb{C})$.

First, by (a), $\mathcal{A}_{\mathbb{R}}$ clearly contains the real constant functions. Second, given any two distinct points $\xi, \eta \in K$, there exists a function $g \in \mathcal{A}$ such that $g(\xi) = 0$ and $g(\xi) = 1$ (use (b) and part (iv) of the proof of Theorem 2.15-1). Hence the real-valued function $\operatorname{Re} g$ belongs to $\mathcal{A}$ and satisfies $\operatorname{Re} g(\xi) = 0$ and $\operatorname{Re} g(\eta) = 1$. $\square$

An immediate corollary of the complex Stone–Weierstraß theorem is the following basic result in approximation theory, already encountered in Theorem 2.14-3, where it was derived from the Weierstraß trigonometric approximation theorem in the real case.

Theorem 2.15-4 (complex trigonometric polynomial approximation theorem) *The space $\mathcal{Q}([0,2\pi];\mathbb{C})$ formed by all complex 2π-periodic trigonometric polynomials, i.e., functions in $\mathcal{C}_{\rm per}([0,2\pi];\mathbb{C})$ of the form (Section 2.14)*

$$\theta \in [0,2\pi] \to \sum_{k=-n}^{n} c_k e^{ik\theta} \quad \textit{with complex coefficients } c_k,$$

is a subalgebra of the space $\mathcal{C}_{\rm per}([0,2\pi]);\mathbb{C})$, which is dense in the space $\mathcal{C}_{\rm per}([0,2\pi];\mathbb{C})$.

Proof As in the proof of Theorem 2.14-1, the functions $g \in \mathcal{C}_{\rm per}([0,2\pi];\mathcal{C})$ are first identified with functions $g^\sharp \in \mathcal{C}(K;\mathbb{C})$, where

$$K := \{x = (x_1,x_2) \in \mathbb{R}^2;\ x_1^2 + x_2^2 = 1\}.$$

Given two distinct points $\xi, \eta \in K$, the function $g^\sharp$, where $g(\theta) := e^{i\theta}$, $0 \le \theta \le 2\pi$, is such that $g^\sharp(\xi) \ne g^\sharp(\eta)$. The proof is thus an immediate application of the complex Stone–Weierstraß theorem (the other assumptions of which are clearly satisfied). □

Problems

2.15-1 Let K be a compact metric space. Using the Stone–Weierstraß theorem, show that the space $\mathcal{C}(K)$ is separable[32] (the special case $K = [0,1]$ has been established in Theorem 2.13-4 as a corollary to the Weierstraß polynomial approximation theorem).

2.15-2 This exercise provides another proof of the *separability of the Lebesgue spaces* $L^p(\Omega)$, $1 \le p < \infty$ (Theorem 2.5-4(a)), this time based on the *Weierstraß polynomial approximation theorem in several variables* (Theorem 2.15-2).

(1) Let Ω be any open subset of $\mathbb{R}^n$. Show that, for each integer $\ell \ge 1$, the set

$$K_\ell := \left\{x \in \Omega;\ \mathrm{dist}(x,\partial\Omega) \ge \frac{1}{\ell}\right\} \cap \overline{B(0,\ell)}$$

is a compact subset of Ω and that, given any compact subset K of Ω, there exists $\ell = \ell(K) \ge 1$ such that $K \subset K_\ell$.

(2) For each integer $\ell \ge 1$, define the set

$$\Pi_\ell := \{q : \Omega \to \mathbb{R};\ q|_{K_\ell} \text{ is a polynomial in } n \text{ variables with rational coefficients, and } q|_{\Omega - K_\ell} = 0\}.$$

Show that the set $\Pi = \bigcup_{\ell=1}^\infty \Pi_\ell$ is a countably infinite subset of $L^p(\Omega)$, $1 \le p \le \infty$.

(3) Let $f \in L^p(\Omega)$ for some $1 \le p < \infty$, and $\varepsilon > 0$, be given. By Theorem 2.5-3, there exists a function $g = g(f,\varepsilon) \in \mathcal{C}_c(\Omega)$ such that $\|f - g\|_{L^p(\Omega)} \le \dfrac{\varepsilon}{2}$. Show that there exists a function $h = h(f,\varepsilon) \in \Pi$ such that $\|g - h\|_{L^p(\Omega)} \le \dfrac{\varepsilon}{2}$, thus showing that $L^p(\Omega)$, $1 \le p < \infty$, is separable.

(4) Show likewise that each Lebesgue space $L^p(\Omega;\mathbb{C})$, $1 \le p < \infty$, is separable.

2.15-3 Using Problem 2.13-2, show that, given any $a > 0$, there exists a sequence of polynomials $p_n \in \mathcal{P}$ such that $\lim_{n\to\infty} \sup_{-a \le t \le a} ||t| - p_n(t)| = 0$.

[32] See, e.g., DIEUDONNÉ [1960, Theorem 7.4.4].

2.16 Convex sets

Given a vector space X and two points $a \in X$ and $b \in X$, the subset

$$[a, b] := \{x \in X;\ x = \lambda a + (1 - \lambda)b,\ 0 \leq \lambda \leq 1\}$$

of X is called a **segment**, or a **closed segment**, and the points a and b are called its *end-points*.

A subset A of a vector space X is **convex** if, whenever it contains two points a and b, it contains the segment $[a, b]$. Note that the empty set and a subset of X consisting of a single element are convex subsets of X and that the intersection $\bigcap_{i \in I} A_i$ of any family of convex subsets $A_i \subset X$ is also convex.

In a normed vector space, the closure $\overline{A}$ *of a convex subset* A *is convex* (given any $a, b \in \overline{A}$ and any $0 \leq \lambda \leq 1$, let $a_k, b_k \in A$ be such that $\lim_{k \to \infty} a_k = a$ and $\lim_{k \to \infty} b_k = b$ and note that $(\lambda a_k + (1 - \lambda)b_k) \in A$ for all k). Likewise, *the* (open) *balls, and thus their closures, are convex subsets in a normed vector space.*

The interior of a convex subset is convex (how to prove this assertion is the object of Problem 2.16-2). Note in passing that in *infinite-dimensional* spaces, interiors of convex sets have an unfortunate tendency to be empty; cf. Problem 2.16-7.

Given a subset A of a vector space X, the **convex hull** of A, denoted

$$\operatorname{co} A,$$

is the intersection of all the *convex* subsets of X that contain A, or equivalently, the smallest convex subset of X that contains A. The following result gives a useful characterization of convex hulls. Other useful properties of convex hulls are proposed in Problems 2.16-4–2.16-6.

Theorem 2.16-1 *Let A be a subset of a vector space X. Then the convex hull of A is also the subset of X formed by all* **convex combinations** *of elements of A, i.e., those finite linear combinations $\sum_{i \in I} \lambda_i a_i$ of elements $a_i \in A$* (Section 2.1) *that satisfy*

$$\lambda_i \geq 0 \text{ for all } i \in I \text{ and } \sum_{i \in I} \lambda_i = 1.$$

Proof (i) *Let C be a convex subset of X. Then any point of the form*

$$a = \sum_{i=1}^{n} \lambda_i a_i, \text{ where } \lambda_i \geq 0 \text{ and } a_i \in C,\ 1 \leq i \leq n, \text{ and } \sum_{i=1}^{n} \lambda_i = 1,$$

belongs to C.

Assume that this property holds for $1, 2, \ldots, n-1$ (it clearly holds for $n = 1, 2$), and let a point $a \in X$ of the above form be given. Since $a_n \in C$, we may assume that $\lambda := \sum_{i=1}^{n-1} \lambda_i > 0$. Then it suffices to write a as

$$a = \sum_{i=1}^{n} \lambda a_i = \lambda \Big(\sum_{i=1}^{n-1} \frac{\lambda_i}{\lambda} a_i \Big) + (1 - \lambda) a_n$$

and to observe that $0 < \lambda \leq 1$ and $\sum_{i=1}^{n-1} \dfrac{\lambda_i}{\lambda} a_i \in C$ by the induction hypothesis. Hence $a \in C$ since C is convex.

(ii) Let

$$T := \left\{ \sum_{i \in I} \lambda_i a_i \in X;\ I \text{ finite, } \lambda_i \geq 0 \text{ and } a_i \in A \text{ for all } i \in I, \sum_{i \in I} \lambda_i = 1 \right\}.$$

By (i), any point in T belongs to any convex set that contains A. Hence $T \subset \operatorname{co} A$.

(iii) The set T as defined in (ii) is convex since, given any $a = \sum_{i \in I} \lambda_i a_i \in T$ and $b = \sum_{j \in J} \mu_j b_j \in T$ and any $0 \leq \nu \leq 1$, we can write

$$\nu a + (1-\nu) b = \sum_{i \in I} (\nu \lambda_i) a_i + \sum_{j \in J} ((1-\nu)\mu_j) b_j,$$

with $\nu \lambda_i \geq 0$, $(1-\nu)\mu_j \geq 0$, and $\sum_{i \in I} (\nu \lambda_i) + \sum_{j \in J} ((1-\nu)\mu_j) = 1$.
Hence $\operatorname{co} A \subset T$ since T is convex. $\square$

A finite linear combination $a = \sum_{i \in I} \lambda_i a_i$ with $\lambda_i \geq 0$ for all $i \in I$ and $\sum_{i \in I} \lambda_i = 1$ (such as those encountered in Theorem 2.16-1) is called a *convex combination* of the points a_i, $i \in I$, and the point a is called the *barycenter* of the points a_i with *weights* λ_i.

For instance, let there be given $(n+1)$ points $a_j = (a_{ij})_{i=1}^n \in \mathbb{R}^n$, $1 \leq j \leq n+1$, that are affinely independent, in the sense that they are not contained in a hyperplane of $\mathbb{R}^n$; equivalently, the $(n+1) \times (n+1)$ matrix (a_{ij}), where $a_{n+1,j} := 1$, $1 \leq j \leq n+1$, is invertible. Then the *convex hull of the set* $\bigcup_{j=1}^{n+1} \{a_j\}$, which by Theorem 2.16-1 is thus given by

$$T = \left\{ x \in \mathbb{R}^n;\ x = \sum_{j=1}^{n+1} \lambda_j a_j,\ \lambda_j \geq 0,\ 1 \leq j \leq n+1,\ \sum_{j=1}^{n+1} \lambda_j = 1 \right\},$$

is called an *n-simplex*, and the points a_j are called its *vertices* (a 2-simplex is a triangle and a 3-simplex is a tetrahedron). A simple compactness argument then shows that T is *closed* (a special case of a general property; cf. Problem 2.16-5).

Such convex hulls of *finite* sets share the following property:

Theorem 2.16-2 *Let A be a finite subset of a normed vector space X. Then $\operatorname{co} A$ is a compact subset of X.*

Proof Let $A = \bigcup_{j=1}^m \{x_j\} \subset X$. Then, by Theorem 2.16-1,

$$\operatorname{co} A = \left\{ \sum_{j=1}^m \lambda_j x_j;\ \lambda_j \geq 0, 1 \leq j \leq m, \text{ and } \sum_{j=1}^m \lambda_j = 1 \right\}.$$

Given any infinite sequence $(x^k)_{k=1}^\infty$ with $x^k = \sum_{j=1}^m \lambda_j^k x_j \in \operatorname{co} A$ for each $k \geq 1$, the corresponding sequence $((\lambda_1^k, \lambda_2^k, \ldots, \lambda_m^k))_{k=1}^\infty$ is bounded in $\mathbb{R}^m$. Hence there exist a subsequence

$((\lambda_1^{\sigma(k)}, \lambda_2^{\sigma(k)}, \ldots, \lambda_m^{\sigma(k)}))_{k=1}^{\infty}$ and an element $(\lambda_1, \lambda_2, \ldots, \lambda_m) \in \mathbb{R}^m$ such that $\lambda_j^{\sigma(k)} \to \lambda_j \geq 0$ as $k \to \infty$, $1 \leq j \leq m$, and $\sum_{j=1}^m \lambda_j = \lim_{k\to\infty} \sum_{j=1}^m \lambda_j^{\sigma(k)} = 1$. Therefore,

$$x^{\sigma(k)} = \sum_{j=1}^m \lambda_j^{\sigma(k)} x_j \to \sum_{j=1}^m \lambda_j x_j \in \operatorname{co} A \quad \text{as } k \to \infty,$$

which shows that $\operatorname{co} A$ is compact. □

Another equally important notion is that of the **closed convex hull**

$$\overline{\operatorname{co}} A$$

of a subset A of a *normed* vector space X, defined as the intersection of all the *closed* and *convex* subsets of X that contain A, or equivalently, as the smallest closed convex subset of X that contains A.

Theorem 2.16-3 *Let A be a subset of a normed vector space X. Then $\overline{\operatorname{co}}\, A = \overline{\operatorname{co} A}$.*

Proof On the one hand, $\overline{\operatorname{co} A}$ is closed (as a closure) and convex (as the closure of the convex set $\operatorname{co} A$).

On the other hand, let C be a closed convex subset of X that contains A. Then C contains $\operatorname{co} A$ since C is convex and any convex set containing A contains $\operatorname{co} A$; therefore C also contains $\overline{\operatorname{co} A}$ since C is closed.

Hence $\overline{\operatorname{co} A}$ is a closed convex subset contained in any closed convex subset that contains A; therefore $\overline{\operatorname{co} A} = \overline{\operatorname{co}} A$. □

Remarks (1) An important property of closed convex hulls in *complete* normed vector space will be established later on (Theorem 3.1-5).

(2) Since $\overline{\operatorname{co} A}$ is a closed convex set that contains $\overline{A}$ (as a closed set containing A), it is clear that $\operatorname{co} \overline{A} \subset \overline{\operatorname{co} A}$. However, this inclusion may be *strict*; consider for example the subset $A := \{(x_1, x_2) \in \mathbb{R}^2;\ x_2 \geq (1 + x_1^2)^{-1}\} = \overline{A}$ of $\mathbb{R}^2$. □

Problems

2.16-1 Let A be a convex subset of $\mathbb{R}^2$ containing the origin O and possessing the following property: given any constants $\alpha_1, \alpha_2 \in \mathbb{R}^2$ such that $|\alpha_1| + |\alpha_2| > 0$, the subset $\{(\alpha_1 t, \alpha_2 t) \in \mathbb{R}^2; t \geq 0\}$ of $\mathbb{R}^2$ (i.e., a half-line originating at O) contains at least one point that does not belong to A. Show that the set A is bounded.

2.16-2 Let A be a convex subset of a normed vector space X.
(1) Show that, if $a \in \operatorname{int} A$ and $b \in \overline{A}$, then $\{x \in X;\ x = \lambda a + (1 - \lambda) b,\ 0 < \lambda \leq 1\} \subset \operatorname{int} A$.
(2) Show that $\operatorname{int} A$ is convex.
(3) Show that $\overline{A} = \overline{\operatorname{int} A}$ if $\operatorname{int} A \neq \varnothing$ (clearly, this property need not hold if A is an arbitrary subset of X).

2.16-3 For any integer $n \geq 2$, let $\mathbb{M}^n$ denote the vector space formed by all $n \times n$ real matrices.
(1) Show directly that the set $\mathbb{M}_+^n := \{\boldsymbol{A} \in \mathbb{M}^n;\ \det \boldsymbol{A} > 0\}$ is not convex.
(2) Show that $\operatorname{co} \mathbb{M}_+^n = \mathbb{M}^n$.

2.16-4 Show that the convex hull of an open set is also open.

2.16-5 (1) (**Carathéodory theorem**[33]) Let A be a subset in $\mathbb{R}^n$. Show that any point $x \in \operatorname{co} A$ can be written as

$$x = \sum_{i=1}^{n+1} \lambda_i a_i, \text{ where } \lambda_i \geq 0 \text{ and } a_i \in A,\ 1 \leq i \leq n+1, \text{ and } \sum_{i=1}^{n+1} \lambda_i = 1,$$

i.e., as a convex combination of at most $(n+1)$ points of A.

(2) Using (1), show that the convex hull of a compact subset of $\mathbb{R}^n$ is also compact.

Remark Property (2) is a special case of a more general one, which holds in any *Banach space*; cf. Theorem 3.1-5. □

2.16-6 (**Birkhoff's theorem**[34]) Given a permutation τ of the set $\{1, 2, \ldots, n\}$, the associated $n \times n$ *permutation matrix* $\boldsymbol{P}_\tau$ is defined by $(\boldsymbol{P}_\tau)_{ij} = \delta_{i\tau(j)}$. An $n \times n$ matrix (a_{ij}) is a *doubly stochastic matrix* if

$$a_{ij} \geq 0,\ 1 \leq i,\ j \leq n;\quad \sum_{j=1}^{n} a_{ij} = 1,\ 1 \leq i \leq n;\quad \sum_{i=1}^{n} a_{ij} = 1,\ 1 \leq j \leq n.$$

Show that the convex hull of the set of all $n \times n$ permutation matrices is the set of all $n \times n$ doubly stochastic matrices.

2.16-7 (1) Show that the set $A_n := \{x = (x_i)_{i=1}^n \in \mathbb{R}^n;\ x_i \geq 0,\ 1 \leq i \leq n\}$ is convex in $\mathbb{R}^n$ and that $\mathring{A}_n = \{x = (x_i)_{i=1}^n \in \mathbb{R}^n;\ x_i > 0,\ 1 \leq i \leq n\}$.

(2) Show that the set $A := \{x = (x_i)_{i=1}^\infty \in \ell^2;\ x_i \geq 0,\ i \geq 1\}$ is convex in ℓ^2 and identify its interior.

2.16-8 Show that the **field of values** $F(\boldsymbol{A})$ of an $n \times n$ complex matrix $\boldsymbol{A}$, defined by[35]

$$F(\boldsymbol{A}) := \{\boldsymbol{x}^* \boldsymbol{A} \boldsymbol{x} \in \mathbb{C};\ \boldsymbol{x} \in \mathbb{C}^n,\ \|\boldsymbol{x}\|_2 = 1\},$$

is a convex subset of $\mathbb{C}$.

This (not easy to prove) result constitutes the **Toeplitz–Hausdorff theorem**[36]; it also applies to linear operators acting in infinite-dimensional inner-product spaces[37].

---

[33] C. Carathéodory [1907]: Über den Variabilitätsbereich der Fourier'schen Konstanten von positiven harmonischen Funktionen, *Rendiconti del Circolo Matematico di Palermo* **32**, 193–217.

[34] G. Birkhoff [1946]: Tres observaciones sobre el algebra lineal, *Universidad Nacional de Tucumán Revista A*, **5**, 147–151.

[35] The field of values of a matrix plays an important role in *matrix theory*; see Horn & Johnson [1991, Chapter 1].

[36] O. Toeplitz [1918]: Das algebraische Analogon zu einem Satze von Fejér, *Mathematische Zeitschrift* **2**, 187–197.

F. Hausdorff [1919]: Der Wertvorrat einer Bilinearform, *Mathematische Zeitschrift* **3**, 314–316.

[37] See, e.g., Halmos [1982, Chapter 22], or:

C. Davis [1971]: The Toeplitz-Hausdorff theorem explained, *Canadian Mathematical Bulletin* **14**, 245–246.

A. McIntosch [1978]: The Toeplitz-Hausdorff theorem and ellipticity conditions, *The American Mathematical Monthly* **85**, 475–477.

2.17 Convex functions

Let X be a vector space and let A be a *convex* subset of X. A function $f : A \to \mathbb{R}$ is said to be **convex** over A if, given any two points $a, b \in A$,

$$f(\lambda a + (1-\lambda)b) \leq \lambda f(a) + (1-\lambda)f(b) \quad \text{for all } 0 \leq \lambda \leq 1,$$

A simple induction then shows that, given any points $a_i \in A$, $1 \leq i \leq n$,

$$f\Big(\sum_{i=1}^{n} \lambda_i a_i\Big) \leq \sum_{i=1}^{n} \lambda_i f(a_i) \quad \text{for all } 0 \leq \lambda_i \leq 1,\ 1 \leq i \leq n, \text{ such that } \sum_{i=1}^{n} \lambda_i = 1.$$

A function $f : A \to \mathbb{R}$ is **strictly convex** over A if, given any two *distinct* points $a, b \in A$,

$$f(\lambda a + (1-\lambda)b) < \lambda f(a) + (1-\lambda)f(b) \quad \text{for all } 0 < \lambda < 1.$$

A function $g : A \to \mathbb{R}$ is **concave**, *resp.* **strictly concave**, over A if the function $-g : A \to \mathbb{R}$ is convex, *resp.* strictly convex.

For instance, if $f_i : A \to \mathbb{R}$, $1 \leq i \leq n$, are convex functions, then clearly so are the functions $\max_{1 \leq i \leq n}\{f_i\}$ and $\sum_{i=1}^n f_i$. If X is a real vector space, a linear functional $\ell : X \to \mathbb{R}$ is a convex, but not strictly convex, function. If $(X; \|\cdot\|)$ is a normed vector space, *the norm* $\|\cdot\| : X \to \mathbb{R}$ *is a convex function* since, for each $x, y \in X$,

$$\|\lambda x + (1-\lambda)y\| \leq \lambda\|x\| + (1-\lambda)\|y\| \quad \text{for all } 0 \leq \lambda \leq 1.$$

The notion of convexity can be extended in the following natural way to functions that are not defined on convex sets (this extension will be needed in particular in the definition of polyconvex functions; cf. Section 9.7): Let X be a vector space and let A be a subset of X. A function $f : A \to \mathbb{R}$ is said to be **convex** over A if there exists a convex function (in the previous sense) $\widetilde{f} : \operatorname{co} A \to \mathbb{R}$ such that $\widetilde{f}|_A = f$.

Convex functions defined over *open* subsets of *finite-dimensional* vector spaces possess a remarkable property:

Theorem 2.17-1 *Let Ω be an open convex subset of a finite-dimensional space X. Then any convex function $f : X \to \mathbb{R}$ is continuous.*

Proof (i) We first show that *the function f is locally bounded from above*, i.e., that, given any point $a \in \Omega$, there exist a neighborhood B of a contained in Ω and a constant M such that $f(x) \leq M$ for all $x \in B$.

Let $(e_i)_{i=1}^n$ denote a basis in the space X, and let X be equipped with the norm $\|\cdot\|_1 : x = \sum_{i=1}^n x_i e_i \to \sum_{i=1}^n |x_i|$ (this is no loss of generality, since all norms are equivalent in a finite-dimensional space). Since the set Ω is open, there exists $r > 0$ such that

$$B := \{x \in X;\ \|x - a\|_1 \leq r\} \subset \Omega.$$

Let $a_i := a + re_i$, $1 \leq i \leq n$, and $a_i := a - re_{i-n}$, $n+1 \leq i \leq 2n$ (Figure 2.17-1). The definition of the norm $\|\cdot\|_1$ then shows that any point $x \in B$ can be written as

$$x = \sum_{i=1}^{2n} \lambda_i a_i \text{ with } 0 \leq \lambda_i \leq 1,\ 1 \leq i \leq 2n, \quad \text{and} \quad \sum_{i=1}^{2n} \lambda_i = 1,$$

i.e., as a convex combination of the points a_i, $1 \leq i \leq 2n$. The assumed convexity of the function f then implies that

$$f(x) \leq \sum_{i=1}^{2n} \lambda_i f(a_i) \leq M := \max_{1 \leq i \leq 2n} f(a_i) \quad \text{for all } x \in B.$$

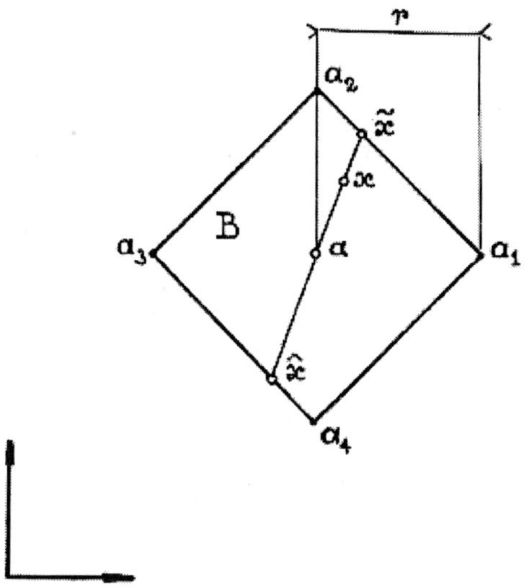

Figure 2.17-1 The points $x, \widetilde{x}, \widehat{x}$, and a_i appearing in the proof of Theorem 2.17-1 in the special case of $\mathbb{R}^2$.

(ii) *The function f is continuous.*

Let the point a and the closed ball B be as in (i). Any point $x \in B$ can be written as (Figure 2.17-1)

$$x = \lambda \widetilde{x} + (1 - \lambda)a, \text{ where } 0 < \lambda := \frac{\|x - a\|_1}{r} \leq 1 \quad \text{and} \quad \widetilde{x} = a + \frac{1}{\lambda}(x - a).$$

Consequently, $f(x) \leq \lambda f(\widetilde{x}) + (1 - \lambda) f(a)$, which in turn implies that

$$f(x) - f(a) \leq \lambda(f(\widetilde{x}) - f(a)) \leq \frac{2M}{r}\|x - a\|_1 \quad \text{for all } x \in B,$$

on the one hand. On the other hand, the definition of λ also shows that (Figure 2.17-1)

$$a = \frac{1}{1+\lambda} x + \frac{\lambda}{1+\lambda} \widehat{x}, \quad \text{where } \widehat{x} := a - (\widetilde{x} - a).$$

Consequently, $f(a) \leq \dfrac{1}{1+\lambda} f(x) + \dfrac{\lambda}{1+\lambda} f(\widehat{x})$, which in turn implies that

$$f(a) - f(x) \leq \lambda(f(\widehat{x}) - f(a)) \leq \frac{2M}{r}\|x - a\|_1 \quad \text{for all } x \in B.$$

□

Remark One can even prove that the function f is *locally Lipschitz-continuous*; cf. Problem 2.17-11. □

It is worth pointing out that Theorem 2.17-1 does *not* hold in an infinite-dimensional normed vector space. Consider for example the space $\mathcal{P}$ equipped with the norm $\|p\| = \sup_{0\le x\le 1}|p(x)|$ and the linear functional $\ell : p \in \mathcal{P} \to p(3)$, which is a convex function. Then ℓ is *not* continuous: the sequence $(p_n)_{n=1}^{\infty}$, where $p_n(x) = \left(\frac{x}{2}\right)^n$, $x \in \mathbb{R}$, is such that $\|p_n\| \underset{n\to\infty}{\longrightarrow} 0$; yet, $\ell(p_n) \underset{n\to\infty}{\longrightarrow} \infty$.

Other examples of convex functions are given in Problems 2.17-1 and 2.17-2.

In any normed vector space, $\|x\| = \|y\| = 1$ implies $\left\|\frac{x+y}{2}\right\| \le 1$. A *normed* vector space $(X, \|\cdot\|)$ is said to be **strictly convex**, or **rotund**, if the following stronger property holds:

$$\|x\| = \|y\| = 1 \quad \text{and} \quad x \neq y \text{ implies } \left\|\frac{x+y}{2}\right\| < 1.$$

A *normed* vector space $(X, \|\cdot\|)$ is said to be **uniformly convex** if the following even stronger property holds: for each $\varepsilon > 0$, there exists $\delta(\varepsilon) > 0$ such that

$$\|x\| = \|y\| = 1 \quad \text{and} \quad \|x - y\| \ge \varepsilon \quad \text{implies } \left\|\frac{x+y}{2}\right\| \le 1 - \delta(\varepsilon).$$

As we shall see later, the notion of uniform convexity is particularly important in the analysis of *weak convergence* (Theorem 5.12-3) and of *reflexive spaces* (Theorem 5.14-3).

The spaces ℓ^p and $L^p(\Omega)$ for $1 < p < \infty$ (Sections 2.4 and 2.5) *provide basic examples of uniformly convex spaces* (Problems 2.17-8 and 2.17-9).

Problems

2.17-1 Let $(X, \|\cdot\|)$ be a normed vector space. Show that, for any $p \ge 1$, the function $f : x \in X \to \|x\|^p$ is convex.

2.17-2 Let X be a real vector space and let A be a convex subset of X.

(1) Show that a function $f : A \to \mathbb{R}$ is convex if and only if its *epigraph*, defined by

$$\operatorname{Epi} f := \{(x, y) \in X \times \mathbb{R};\ x \in A, y \ge f(x)\},$$

is a convex subset of the vector space $X \times \mathbb{R}$.

Remark This property also holds for convex functions that take their values in $\mathbb{R} \cup \{\infty\}$; cf. Theorem 9.2-1. □

(2) Let $(f_i)_{i\in I}$ be any family of convex functions $f_i : A \to \mathbb{R}$ with the property that $\sup_{i\in I} f_i(x) < \infty$ for all $x \in A$. Show that the function $\sup_{i\in I} f_i : A \to \mathbb{R}$ is also convex.

2.17-3 Let $(X, \|\cdot\|)$ be a normed vector space and let $f : X \to \mathbb{R}$ be a convex function with a *local minimum* $x_0 \in X$, i.e., such that there exists $r > 0$ such that $f(x) \ge f(x_0)$ for all $x_0 \in B(x; r)$. Show that x_0 is a *global minimum* of f, i.e., that $f(x) \ge f(x_0)$ for all $x \in X$.

2.17-4 Let X be a vector space and let $f : X \to \mathbb{R}$ be a strictly convex function. Show that f has *at most one minimum* and that, if f has one minimum x_0, it is a *strict minimum*, i.e., $f(x) > f(x_0)$ for all $x \in X, x \neq x_0$.

2.17-5 Let X be a finite-dimensional normed vector space, and let $f : X \to \mathbb{R}$ be a convex function with a *strict global minimum* $x_0 \in X$, i.e., such that $f(x) > f(x_0)$ for all $x \in X$, $x \neq x_0$. Show that $f(x) \to \infty$ uniformly as $\|x\| \to \infty$, i.e., $\lim_{r\to\infty}(\inf_{\|x\|\geq r} f(x)) = \infty$.

2.17-6 Show that a finite-dimensional strictly convex space is uniformly convex.

2.17-7 Show that the spaces ℓ^1 and ℓ^∞, and the spaces $L^1(\Omega)$ and $L^\infty(\Omega)$, are not strictly convex.

2.17-8 In this problem, a number $1 < p \leq 2$ is given, and $q > 1$ is defined by $\dfrac{1}{p} + \dfrac{1}{q} = 1$.

(1) Show that $(1+t)^q + (1-t)^q \leq 2(1+t^p)^{q/p}$ for all $0 \leq t \leq 1$.

(2) Using (1), show that $|\alpha+\beta|^q + |\alpha-\beta|^q \leq 2(|\alpha|^p + |\beta|^p)^{q/p}$ for all $\alpha, \beta \in \mathbb{K}$.

(3) Deduce from (2) that the following **Clarkson's inequalities**[38] hold for $1 < p \leq 2$: For all $x, y \in \ell^p$,

$$(\|x+y\|_{\ell^p})^q + (\|x-y\|_{\ell^p})^q \leq 2\Big((\|x\|_{\ell^p})^p + (\|y\|_{\ell^p})^p\Big)^{q/p};$$

for all $f, g \in L^p(\Omega)$,

$$\big(\|f+g\|_{L^p(\Omega)}\big)^q + \big(\|f-g\|_{L^p(\Omega)}\big)^q \leq 2\Big(\big(\|f\|_{L^p(\Omega)}\big)^p + \big(\|g\|_{L^p(\Omega)}\big)^p\Big)^{q/p}.$$

(4) Show that, for $p = 2$ (in which case $q = 2$), the Clarkson inequalities become an *equality*; this equality is in fact a special case of the *parallelogram law*, which holds in any inner-product space (Theorem 4.1-2).

(5) Conclude that, for $1 < p \leq 2$, the spaces ℓ^p and $L^p(\Omega)$ are uniformly convex.

2.17-9 In this problem, a number $p \geq 2$ is given.

(1) Show that $\Big(\dfrac{1+t}{2}\Big)^p + \Big(\dfrac{1-t}{2}\Big)^p \leq \dfrac{1}{2}(1+t^p)$ for all $0 \leq t \leq 1$.

(2) Using (1), show that $\Big|\dfrac{\alpha+\beta}{2}\Big|^p + \Big|\dfrac{\alpha-\beta}{2}\Big|^p \leq \dfrac{|\alpha|^p}{2} + \dfrac{|\beta|^p}{2}$ for all $\alpha, \beta \in \mathbb{K}$.

(3) Deduce from (2) that the following **Clarkson's inequalities** hold for $p \geq 2$ (the observation made in Problem 2.17-8(4) applies as well to these inequalities): For all $x, y \in \ell^p$,

$$(\|x+y\|_{\ell^p})^p + (\|x-y\|_{\ell^p})^p \leq 2^{p-1}\big((\|x\|_{\ell^p})^p + (\|y\|_{\ell^p})^p\big);$$

for all $f, g \in L^p(\Omega)$,

$$\big(\|f+g\|_{L^p(\Omega)}\big)^p + \big(\|f-g\|_{L^p(\Omega)}\big)^p \leq 2^{p-1}\Big(\big(\|f\|_{L^p(\Omega)}\big)^p + \big(\|g\|_{L^p(\Omega)}\big)^p\Big).$$

(4) Conclude that, for $p \geq 2$, the spaces ℓ^p and $L^p(\Omega)$ are uniformly convex.

2.17-10 Let $\varphi : \mathbb{R} \to \mathbb{R}$ be a convex function (hence continuous by Theorem 2.17-1).

(1) Show that, for any integer $m \geq 1$ and any $\zeta_i \in \mathbb{R}$ and $\lambda_i > 0$, $1 \leq i \leq m$,

$$\varphi\Big(\frac{\sum_{i=1}^m \lambda_i\zeta_i}{\sum_{i=1}^m \lambda_i}\Big) \leq \Big(\frac{\sum_{i=1}^m \lambda_i\varphi(\zeta_i)}{\sum_{i=1}^m \lambda_i}\Big).$$

(2) Using (1) and the convexity of the function $x \in]0,\infty[\to -\log x$, show that the *arithmetic mean-geometric inequality* holds, viz.,

$$\Big(\prod_{i=1}^m \zeta_i\Big)^{1/m} \leq \frac{1}{m}\sum_{i=1}^m \zeta_i \quad \text{for any } \zeta_i > 0,\ 1 \leq i \leq m.$$

[38] J.A. Clarkson [1936]: Uniformly convex spaces, *Transactions of the American Mathematical Society* **40**, 396–414.

(3) Show that, for any bounded open subset Ω of $\mathbb{R}^n$ and any nonnegative function $f \in L^1(\Omega)$,

$$\varphi\left(\frac{1}{\operatorname{meas}\Omega}\int_\Omega f(x)\,\mathrm{d}x\right) \le \frac{1}{\operatorname{meas}\Omega}\int_\Omega \varphi(f(x))\,\mathrm{d}x.$$

The inequalities of (1) and (3) constitute the **Jensen inequalities;**[39] there is also a *Jensen inequality in* ℓ^p (Problem 2.4-4).

2.17-11 The assumptions and notations are those of Theorem 2.17-1. Show that, given any point $a \in \Omega$, there exists a neighborhood $V(a)$ and a constant $C = C(a, V(a)) > 0$ such that $|f(x) - f(y)| \le C\|x - y\|_1$ for all $x, y \in V(a)$.

[39] J.L.W.V. Jensen [1906]: Sur les fonctions convexes et les inégalités entre les valeurs moyennes, *Acta Mathematica* **30**, 175–193.

CHAPTER 3

BANACH SPACES

Introduction

Banach spaces, i.e., *complete normed vector spaces*, play a central role in linear and nonlinear functional analysis. The aim of this chapter is to establish their most immediate basic properties.

To begin with, basic examples of Banach spaces, which will pervade the rest of the book, are given and studied, such as the space $\mathcal{C}(K;Y)$ of all continuous functions on a compact set K into a Banach space Y equipped with the sup-norm (Section 3.2), the spaces ℓ^p and $L^p(\Omega)$, $1 \leq p \leq \infty$ (Section 3.4), or the spaces $\mathcal{L}(X;Y)$ when Y is a Banach space (Section 3.2), which include the all-important *dual spaces* as special cases (Section 3.5). In particular, a complete proof is given of the fundamental *F. Riesz representation theorem in the spaces* $L^p(\Omega)$, $1 \leq p < \infty$ (Theorem 3.5-3), which fully identifies their dual spaces.

The assumption that a normed vector space is complete allows us to prove numerous far-reaching results, among which three are established and applied in this chapter (further far-reaching results, but of a more elaborate nature, will be established in Chapter 5).

The first result is the possibility of defining *convergent series in a Banach space* (Section 3.6). For instance, one can compute the inverse of a linear operator of the form $(I - A)$ when A acts in a Banach space and $\|A\| < 1$, by means of the *Neumann series* (Theorem 3.6-2).

The second result, perhaps the most basic result of Banach space theory, is *Banach fixed point theorem* (Theorem 3.7-1). Its importance is already highlighted in this chapter by two applications, the first one by means of the *Cauchy–Lipschitz theorem* (Theorem 3.8-1) to *nonlinear ordinary differential equations* such as the *pendulum equation*, and the second one to *nonlinear two-point boundary value problems* (Theorem 3.9-1). But the Banach fixed point theorem will be also put to use later on, for instance as the keystone to the fundamental *implicit function theorem* (Chapter 7). Such applications highlight that the Banach fixed point theorem is also a *basic theorem of nonlinear functional analysis* (in effect the first one encountered in this book), again perhaps the most basic one.

The third result, also a basic one, is the *Ascoli–Arzelà theorem*, which characterizes compact subsets of the space $\mathcal{C}(K;\mathbb{R})$ when K is compact (Theorem 3.10-1). Its importance is illustrated by means of the *Cauchy–Peano theorem* (Theorem 3.11-1) for *nonlinear ordinary differential equations.*

3.1 Banach spaces; first properties

A normed vector space $(X, \|\cdot\|)$ is a **Banach space**[1] if the *metric space* (X, d), where d is the distance on X defined by $d(x, y) := \|x - y\|$ (Theorem 2.2-1), is *complete.*

Banach spaces thus inherit all the properties of *complete metric spaces*, such as those that were recalled in Section 1.12. Besides, some of these properties can be further refined when the richer structure inherent to a normed vector space is taken into account. For instance, *the unique continuous extension to the whole space of a uniformly continuous mapping that is defined and continuous on a dense subset and takes its values in a complete space* (Theorem 1.12-3) now takes the specific form stated in Theorem 3.1-1 when it is applied to a *linear* mapping between *normed vector spaces.* In view of its importance, we give a self-contained proof of this result (i.e., without appealing to Theorem 1.12-3 for the first parts of the proof).

Theorem 3.1-1 (unique continuous linear extension) *Let X be a dense subspace of a normed vector space $\widetilde{X}$, let Y be a Banach space, and let $A : X \to Y$ be a continuous linear operator.*

Then there exists one and only one continuous linear operator $\widetilde{A} : \widetilde{X} \to Y$ that is an extension of A, i.e., such that $\widetilde{A}x = Ax$ for all $x \in X$. This extension is defined for any $\widetilde{x} \in \widetilde{X}$ by

$$\widetilde{A}\widetilde{x} := \lim_{n\to\infty} Ax_n,$$

where $(x_n)_{n=1}^{\infty}$ is any sequence of elements $x_n \in X$ such that $\lim_{n\to\infty} x_n = \widetilde{x}$ in $\widetilde{X}$. Besides,

$$\|\widetilde{A}\|_{\mathcal{L}(\widetilde{X};Y)} = \|A\|_{\mathcal{L}(X;Y)}.$$

Proof (i) *First, we need to define such an extension.*

So, given any $\widetilde{x} \in \widetilde{X}$, let $(x_n)_{n=1}^{\infty}$ be a sequence of vectors $x_n \in X$ that converges to x. Since

$$\|Ax_m - Ax_n\| \leq \|A\| \, \|x_m - x_n\| \quad \text{for all } m, n \geq 1,$$

and Y is complete, there exists $y \in Y$ such that $Ax_n \to y$ as $n \to \infty$ (what is used here is in effect the uniform continuity of A; cf. Theorem 2.9-3(a)). Besides, y does not depend on the particular sequence of vectors $x_n \in X$ that converges to x. To see this, consider another such sequence $(x'_n)_{n=1}^{\infty}$; then both sequences $(Ax_n)_{n=1}^{\infty}$ and $(Ax'_n)_{n=1}^{\infty}$ must have the same limit, since they are both subsequences of the same Cauchy sequence $(Ax_1, Ax'_1, Ax_2, Ax'_2, \ldots)$.

Letting $\widetilde{A}\widetilde{x} := y$ thus defines a mapping $\widetilde{A} : \widetilde{X} \to Y$ that is clearly an *extension of* A (if $x \in X$, consider the particular sequence $(x, x, \ldots)$ for defining $\widetilde{A}x$).

(ii) *The extension $\widetilde{A} : \widetilde{X} \to Y$ of $A \in \mathcal{L}(X; Y)$ defined in* (i) *is a continuous mapping, and $\widetilde{A}$ is the only continuous extension of A to $\widetilde{X}$.*

[1]Banach spaces are so named after the Polish mathematician Stefan Banach (1892–1945), who essentially created their theory and then expounded it at length in BANACH [1932], one of the most influential books in the history of mathematics (for biographical and historical accounts, see PIETSCH [2007] and JAKIMOWICZ & MIRANOVICZ [2011]). Together with other mathematicians, such as Karol Borsuk (1905–1982), Stanislaw Saks (1897–1942), Juliusz Schauder (1899–1943), or Hugo Steinhaus (1887–1972) (all of those names will be encountered later in this book), he often worked at the "Scottish Café" in Lwów (at that time in eastern Poland, now Lviv in western Ukraine), a legendary emblem of this bygone era of mathematics; see MAUDLIN [1981].

Without loss of generality, we may assume $A \neq 0$. Given $\varepsilon > 0$, there exists $\delta := \frac{\varepsilon}{2\|A\|}$ such that, if $x, x' \in X$ satisfy $\|x - x'\| \leq 2\delta$, then $\|Ax - Ax'\| \leq \varepsilon$. Let then $\widetilde{x}, \widetilde{x}' \in \widetilde{X}$ satisfy $\|\widetilde{x} - \widetilde{x}'\| \leq \delta$, and let $x_n, x'_n \in X$, $n \geq 1$, be such that $x_n \to \widetilde{x}$ and $x'_n \to \widetilde{x}'$ as $n \to \infty$. Then there exists $n_0 = n_0(\widetilde{x}, \widetilde{x}')$ such that $\|x_n - x'_n\| \leq 2\delta$ for all $n \geq n_0$. Consequently, $\|Ax_n - Ax'_n\| \leq \varepsilon$ for all $n \geq n_0$, and thus

$$\|\widetilde{A}\widetilde{x} - \widetilde{A}\widetilde{x}'\| = \lim_{n\to\infty} \|Ax_n - Ax'_n\| \leq \varepsilon$$

for all $\widetilde{x}, \widetilde{x}' \in \widetilde{X}$ satisfying $\|x - x'\| \leq \delta$. This shows that the extension $\widetilde{A} : \widetilde{X} \to Y$ of A is continuous (in fact, even uniformly continuous).

Let $\widetilde{A}' : \widetilde{X} \to Y$ be another continuous extension of A. Given $\widetilde{x} \in \widetilde{X} - X$, let $x_n \in X$, $n \geq 1$, be such that $x_n \to \widetilde{x}$ as $n \to \infty$. Then

$$\widetilde{A}x = \lim_{n\to\infty} Ax_n \quad \text{and} \quad \widetilde{A}'x = \lim_{n\to\infty} \widetilde{A}'x_n = \lim_{n\to\infty} Ax_n$$

by definition of $\widetilde{A}x$ and by continuity of $\widetilde{A}'$. Hence $\widetilde{A}x = \widetilde{A}'x$ since the limit of a sequence in a normed vector space is unique (Theorem 1.10-1). Hence $\widetilde{A}$ is the unique continuous extension of A to $\widetilde{X}$.

(iii) *The continuous mapping $\widetilde{A}$ is also a linear operator; besides,* $\|\widetilde{A}\|_{\mathcal{L}(\widetilde{X};Y)} = \|A\|_{\mathcal{L}(X;Y)}$.

Given any $\widetilde{x}, \widetilde{x}' \in \widetilde{X}$, let $x_n \in X$ and $x'_n \in X$, $n \geq 1$, be such that $x_n \to \widetilde{x}$ and $x'_n \to \widetilde{x}'$ as $n \to \infty$. Then, given any scalars $\alpha, \alpha' \in \mathbb{K}$,

$$\widetilde{A}(\alpha\widetilde{x} + \alpha'\widetilde{x}') = \lim_{n\to\infty} \big(A(\alpha x_n + \alpha' x'_n)\big) = \lim(\alpha A x_n + \alpha' A x'_n) = \alpha\widetilde{A}\widetilde{x} + \alpha'\widetilde{A}\widetilde{x}',$$

since the addition and scalar multiplication in a normed vector space are continuous mappings (Theorem 2.2-5).

Clearly $\|A\|_{\mathcal{L}(X;Y)} \leq \|\widetilde{A}\|_{\mathcal{L}(\widetilde{X};Y)}$ since $X \subset \widetilde{X}$. Given any $\widetilde{x} \in \widetilde{X}$, let $x_n \in X$, $n \geq 1$, be such that $x_n \to \widetilde{x}$ as $n \to \infty$. Then

$$\|Ax_n\| \leq \|A\|\,\|x_n\| \quad \text{for all } n \geq 1 \quad \text{and} \quad \|\widetilde{A}\widetilde{x}\| = \lim_{n\to\infty} \|Ax_n\|,$$

and thus $\|\widetilde{A}\widetilde{x}\| \leq \|A\|\,\|\widetilde{x}\|$ since $\|x_n\| \to \|\widetilde{x}\|$ as $n \to \infty$. Hence $\|\widetilde{A}\|_{\mathcal{L}(\widetilde{X};Y)} \leq \|A\|_{\mathcal{L}(X;Y)}$. □

An important example of such a unique continuous linear extension is provided by the *trace operator*, which allows us to define "boundary values" for functions in Sobolev spaces (Section 6.6).

The *completion procedure* in metric spaces (Theorem 1.12-4) provides another example of a result that can be refined in normed vector spaces. Again in view of its importance, we give a self-contained proof of this result. See also Problem 3.1-1 for an interesting complement.

First, we need a definition: A linear operator σ from a normed vector space $(X, \|\cdot\|_X)$ into, *resp.* onto, a normed vector space $(Y, \|\cdot\|_Y)$ that satisfies

$$\|\sigma x\|_Y = \|x\|_X \quad \text{for all } x \in X$$

is called a **linear isometry from** X **into** Y, *resp.* **onto** Y. A linear isometry is evidently injective and continuous.

Theorem 3.1-2 (completion of a normed vector space) *Let $(X, \|\cdot\|_X)$ be a normed vector space over $\mathbb{K}$. Then there exist a Banach space $(\widetilde{X}, \|\cdot\|_{\widetilde{X}})$ over $\mathbb{K}$ and a linear isometry $\sigma : X \to \widetilde{X}$ such that $\sigma(X)$ is dense in $(\widetilde{X}, \|\cdot\|_{\widetilde{X}})$.*

Besides, if $(\widehat{X}, \|\cdot\|_{\widehat{X}})$ is any Banach space over $\mathbb{K}$ such that there also exists a linear isometry from X onto a dense subset of $\widehat{X}$, then there exists a linear isometry from $(\widetilde{X}, \|\cdot\|_{\widetilde{X}})$ onto $(\widehat{X}, \|\cdot\|_{\widehat{X}})$.

The space $(\widetilde{X}, \|\cdot\|_{\widetilde{X}})$, which is called the **completion of the space** *$(X, \|\cdot\|)$, is thus unique up to bijective linear isometries. As a normed vector space, the space X may thus be identified with a dense subset of its completion $\widetilde{X}$.*

Proof (i) *Construction of the would-be completion $\widetilde{X}$.*

For notational conciseness, a sequence $(x_n)_{n=1}^{\infty}$ is abbreviated as (x_n) in what follows. It is readily verified that the relation $(x_n) \sim (y_n)$ if and only if $\|x_n - y_n\| \to 0$ as $n \to \infty$ defines an *equivalence relation* $\mathcal{R}$ on the set $\mathcal{C}$ formed by all *Cauchy sequences* (x_n) of *vectors* $x_n \in X$.

First, we show that the *quotient set* $\widetilde{X} := \mathcal{C}/\mathcal{R}$ can be naturally equipped with an addition, a scalar multiplication, and a norm that make it a *normed vector space over the same field* $\mathbb{K}$. We denote by $[(x_n)]$ the *equivalence class* of (x_n) (see Section 1.1 for the notions of equivalence relation, equivalence class, and quotient set), and we let

$$\begin{aligned} 0 &:= [(x_n)] \text{ where } x_n = 0 \text{ for all } n \geq 1, \\ [(x_n)] + [(y_n)] &:= [(x_n + y_n)], \\ \alpha[(x_n)] &:= [(\alpha x_n)] \text{ for all } \alpha \in \mathbb{K}, \\ \|[(x_n)]\|_{\widetilde{X}} &:= \lim_{n\to\infty} \|x_n\|. \end{aligned}$$

To verify that these definitions of addition and scalar multiplication make sense, i.e., that they are independent of the particular Cauchy sequence chosen in a given equivalence class, we note that, if $(x_n) \sim (x_n')$ and $(y_n) \sim (y_n')$, then both $(x_n + y_n)$ and $(x_n' + y_n')$ are evidently again Cauchy sequences and $(x_n + y_n) \sim (x_n' + y_n')$ since

$$\|(x_n + y_n) - (x_n' + y_n')\| \leq \|x_n - x_n'\| + \|y_n - y_n'\|.$$

Likewise, if $(x_n) \sim (x_n')$, then (αx_n) is evidently again a Cauchy sequence and $(\alpha x_n) \sim (\alpha x_n')$ since $\|\alpha x_n - \alpha x_n'\| = |\alpha|\,\|x_n - x_n'\|$.

The inequality $|\|x_n\| - \|x_m\|| \leq \|x_n - x_m\|$ shows that if (x_n) is a Cauchy sequence of vectors of X, then $(\|x_n\|)$ *is a Cauchy sequence of real numbers*. Hence $\lim_{n\to\infty} \|x_n\|$ is a well-defined real number *because* $(\mathbb{R}, |\cdot|)$ *is complete*. Besides, if $(x_n) \sim (x_n')$, then the inequality $|\|x_n\| - \|x_n'\|| \leq \|x_n - x_n'\|$ shows that $\lim_{n\to\infty} \|x_n\| = \lim_{n\to\infty} \|x_n'\|$; consequently, the number $\|[(x_n)]\|_{\widetilde{X}}$ is indeed independent of the particular Cauchy sequence chosen in $[(x_n)]$.

That the mapping $\|\cdot\|_{\widetilde{X}} : \widetilde{X} \to \mathbb{R}$ as defined above is indeed a *norm* on $\widetilde{X}$ is likewise immediately verified.

(ii) Given any $x \in X$, let $\sigma(x) \in \widetilde{X}$ denote the *equivalence class* of the particular Cauchy sequence (x_n) with $x_n = x$ for all $n \geq 1$. It is then immediate to verify that the mapping $\sigma : X \to \widetilde{X}$ defined in this fashion is a *linear isometry from X into $\widetilde{X}$*.

Besides, *the direct image* $\sigma(X)$ *is dense in* $(\widetilde{X}, \|\cdot\|_{\widetilde{X}})$, as we now show. Given any $\widetilde{x} = [(x_n)] \in \widetilde{X}$ and any $\varepsilon > 0$, there exists an integer $n_0 = n_0(\widetilde{x}, \varepsilon) \geq 1$ such that $\|x_n - x_{n_0}\| \leq \varepsilon$ for all $n \geq n_0$, since (x_n) is a Cauchy sequence. Let then $\widetilde{x}_0 := [(y_n)]$ where $y_n = x_{n_0}$ for all $n \geq 1$. Then clearly $\widetilde{x}_0 \in \sigma(X)$ since $\widetilde{x}_0 = \sigma(x_{n_0})$ and

$$\|\widetilde{x} - \widetilde{x}_0\|_{\widetilde{X}} = \lim_{n\to\infty} \|\widetilde{x}_n - \widetilde{x}_{n_0}\|_{\widetilde{X}} \leq \varepsilon.$$

Hence $\sigma(X)$ is dense in $(\widetilde{X}, \|\cdot\|_{\widetilde{X}})$.

(iii) *The normed vector space* $(\widetilde{X}, \|\cdot\|_{\widetilde{X}})$ *is complete.* Let $(\widetilde{x}^k)_{k=1}^\infty$ be a Cauchy sequence in $(\widetilde{X}, \|\cdot\|_{\widetilde{X}})$. For each $k \geq 1$, there exists $x^k \in X$ such that $\|\widetilde{x}^k - \sigma(x^k)\|_{\widetilde{X}} \leq \dfrac{1}{k}$ since $\sigma(X)$ is dense in $(\widetilde{X}, \|\cdot\|_{\widetilde{X}})$ by (ii). Then $(x^k)_{k=1}^\infty$ is a Cauchy sequence in X since, for all $k, \ell \geq 1$,

$$\|x^k - x^\ell\|_X = \|\sigma(x^k) - \sigma(x^\ell)\|_{\widetilde{X}}$$

because $\sigma : X \to \widetilde{X}$ is an isometry by (ii), and thus

$$\begin{aligned}\|x^k - x^\ell\|_X &\leq \|\widetilde{x}^k - \sigma(x^k)\|_{\widetilde{X}} + \|\widetilde{x}^\ell - \sigma(x^\ell)\|_X + \|\widetilde{x}^k - \widetilde{x}^\ell\|_{\widetilde{X}} \\ &\leq \frac{1}{k} + \frac{1}{\ell} + \|\widetilde{x}^k - \widetilde{x}^\ell\|_{\widetilde{X}}.\end{aligned}$$

Let $\widetilde{x} := [(x^k)]$. Then we claim that $\|\widetilde{x}^k - \widetilde{x}\|_{\widetilde{X}} \to 0$ as $k \to \infty$. To see this, note that

$$\|\widetilde{x}^k - \widetilde{x}\|_{\widetilde{X}} \leq \|\widetilde{x}^k - \sigma(x^k)\|_{\widetilde{X}} + \|\sigma(x^k) - \widetilde{x}\|_{\widetilde{X}} \leq \frac{1}{k} + \|\sigma(x^k) - \widetilde{x}\|_{\widetilde{X}}$$

and that, by definition of the norm $\|\cdot\|_{\widetilde{X}}$,

$$\|\sigma(x^k) - \widetilde{x}\|_{\widetilde{X}} = \lim_{n\to\infty} \|x^k - x^n\| = 0,$$

since $\sigma(x^k) = [(x^k, x^k, \dots, x^k, \dots)]$ and $\widetilde{x} = [(x^1, x^2, \dots, x^k, x^{k+1}, \dots)]$. Therefore,

$$\lim_{k\to\infty} \|\sigma(x^k) - \widetilde{x}\|_{\widetilde{X}} = 0,$$

which shows that the space $(\widetilde{X}, \|\cdot\|_{\widetilde{X}})$ is complete.

(iv) Assume that there also exists a linear isometry $\tau : X \to \widehat{X}$ into a Banach space $(\widehat{X}, \|\cdot\|_{\widehat{X}})$ such that $\tau(X)$ is a dense subset of $\widehat{X}$.

Then the mapping $\tau \circ \sigma^{-1} : \sigma(X) \to \widehat{X}$ is a linear isometry from $(\sigma(X), \|\cdot\|_{\widetilde{X}})$ into $(\widehat{X}, \|\cdot\|_{\widehat{X}})$. Since $\sigma(X)$ is *dense in* $\widetilde{X}$ and $\widehat{X}$ is *complete*, the *unique continuous linear extension theorem* (Theorem 3.1-1) shows that $\tau \circ \sigma^{-1}$ has a unique continuous linear extension $\varphi : \widetilde{X} \to \widehat{X}$, and φ is clearly also a linear isometry (to see this, consider sequences in $\sigma(X)$ and use the continuity of the norm).

Likewise, the linear isometry $\sigma \circ \tau^{-1} : \tau(X) \to \widetilde{X}$ has a unique continuous linear extension $\psi : \widehat{X} \to \widetilde{X}$, which is a linear isometry from $\widehat{X}$ into $\widetilde{X}$.

By construction, the restriction to $\tau(X)$ of the linear isometry $\varphi \circ \psi : \widehat{X} \to \widehat{X}$ is the identity mapping $I_{\tau(X)}$. Another application of the *unique continuous linear extension theorem* thus shows that $\varphi \circ \psi = I_{\widehat{X}}$.

Therefore the linear isometry $\varphi : \widetilde{X} \to \widehat{X}$, which as such is already injective, is also surjective (since, given any $\widehat{x} \in \widehat{X}$, $\varphi(\psi(\widehat{x})) = \widehat{x}$). Hence φ *is a linear isometry from* $(\widetilde{X}, \|\cdot\|_{\widetilde{X}})$ *onto* $(\widehat{X}, \|\cdot\|_{\widehat{X}})$. This completes the proof. $\square$

Since the *Lebesgue spaces* $L^p(\Omega)$, $1 \leq p < \infty$, where Ω is an open subset in $\mathbb{R}^n$, are complete (as will be shown in Section 3.4), they provide fundamental *examples of completions*, viz., as the *completion of the space* $(\mathcal{C}(\overline{\Omega}), \|\cdot\|_{L^p(\Omega)})$ if Ω is *bounded* (which insures that $\|f\|_{L^p(\Omega)} < \infty$ if $f \in \mathcal{C}(\overline{\Omega})$); or in general, as the *completion of the space* $(\mathcal{C}_c(\Omega), \|\cdot\|_{L^p(\Omega)})$, where $\mathcal{C}_c(\Omega)$ denotes the space of all functions that are continuous in Ω and have compact support in Ω; or, again in general, as the *completion of the space* $(\mathcal{D}(\Omega), \|\cdot\|_{L^p(\Omega)})$, where $\mathcal{D}(\Omega)$ denotes the space of all functions that are infinitely differentiable in Ω and have compact support in Ω (Theorems 2.5-3 and 2.6-2). Naturally, this last denseness property implies the two other ones.

Remark The construction of the completion given in the proof of Theorem 3.1-2 is reminiscent of the construction of the complete normed vector space $(\mathbb{R}, |\cdot|)$ from the set $\mathbb{Q}$, i.e., by means of equivalence classes of Cauchy sequences of rational numbers. Note, however, that *the completeness of the normed vector space* $(\mathbb{R}, |\cdot|)$ *was used in an essential way in the above proof* (see part (i)). $\square$

We conclude this section by three general properties of Banach spaces. Although almost obvious, the first two properties are nevertheless worth recording.

Theorem 3.1-3 *Let X be a Banach space, let Y be a normed vector space, and let $A \in \mathcal{L}(X;Y)$ be a bijection such that $A^{-1} \in \mathcal{L}(Y;X)$. Then Y is a Banach space.*

Proof Let (y_n) be a Cauchy sequence in Y. Then $(A^{-1}y_n)$ is a Cauchy sequence in X (since $A^{-1} \in \mathcal{L}(Y;X)$), which converges to a limit $x \in X$ (since X is complete). Hence $y_n = A(A^{-1}y_n)$ converges in Y, to $Ax \in Y$ (since $A \in \mathcal{L}(X;Y)$). $\square$

Theorem 3.1-4 *Let X be a Banach space, let Y be a normed vector space, and let $A \in \mathcal{L}(X;Y)$. Assume that there exists a constant C such that*

$$\|x\| \leq C\,\|Ax\| \quad \textit{for all } x \in X.$$

Then $\operatorname{Im} A$ *is also a Banach space, and hence in particular a closed subspace of* Y.

Proof Let $y_n = Ax_n \in \operatorname{Im} A$, $n \geq 1$, be such that (y_n) is a Cauchy sequence in Y. Then the assumed inequality implies that (x_n) is a Cauchy sequence in X, which thus converges to a limit $x \in X$ since X is complete. Hence

$$y_n = Ax_n \to y := Ax \quad \text{as } n \to \infty$$

since A is continuous, and therefore $y = Ax \in \operatorname{Im} A$. This shows that the subspace $\operatorname{Im} A$ of Y is complete. In particular then, $\operatorname{Im} A$ is necessarily closed in Y (Theorem 1.12-2(a)). $\square$

Remark The assumed inequality in Theorem 3.1-4 also implies that A is injective and that the inverse mapping of $A : X \to \operatorname{Im} A$ is a continuous linear operator from $\operatorname{Im} A$ onto X (Theorem 2.9-4). □

The third result (which by contrast is not as easy to establish) constitutes an important property of Banach spaces. It will for instance play a crucial role in the proof of *Schauder's fixed point theorem* (Theorem 9.12-1). Closed convex hulls have been defined in Section 2.16.

Theorem 3.1-5 *The closed convex hull of a compact subset of a Banach space is also compact.*

Proof Let A be a compact subset of a Banach space X. Hence $\overline{\mathrm{co}}A$ is a complete metric space, as a closed subset of a complete metric space; this is why the assumption that X is a *Banach space* is needed. By Theorem 1.13-3, it therefore suffices to show that $\overline{\mathrm{co}}A = \overline{\mathrm{co}\, A}$ (Theorem 2.16-3) is precompact, or equivalently, that $\mathrm{co}\, A$ *is precompact*.

So, let any $\varepsilon > 0$ be given. Since A is *compact* by assumption, there exists a finite subset $A(\varepsilon)$ of A such that

$$A \subset \bigcup_{x \in A(\varepsilon)} B\Big(x; \frac{\varepsilon}{2}\Big).$$

Since $A(\varepsilon)$ is finite, its convex hull $\mathrm{co}\, A(\varepsilon)$ is compact (Theorem 2.16-2), and hence precompact. So, there exists a finite number $m = m(\varepsilon)$ of points $y_i = y_i(\varepsilon) \in \mathrm{co}\, A(\varepsilon)$, $1 \le i \le m$, such that

$$\mathrm{co}\, A(\varepsilon) \subset \bigcup_{i=1}^{m} B\Big(y_i; \frac{\varepsilon}{2}\Big).$$

Given any point $y \in \mathrm{co}\, A$, there exists a *finite* set $J = J(y)$ of indices such that (Theorem 2.16-1)

$$y = \sum_{j \in J} \lambda_j x_j \quad \text{with } x_j \in A \text{ and } \lambda_j \ge 0,\ j \in J, \text{ and } \sum_{j \in J} \lambda_j = 1.$$

Then, for each $j \in J$, there exists a point $x(j) \in A(\varepsilon)$ such that $x_j \in B\Big(x(j); \frac{\varepsilon}{2}\Big)$, i.e., such that $\|x_j - x(j)\| < \frac{\varepsilon}{2}$. The point

$$z := \sum_{j \in J} \lambda_j x(j)$$

therefore satisfies

$$z \in \mathrm{co}\, A(\varepsilon) \quad \text{and} \quad \|y - z\| = \Big\| \sum_{j \in J} \lambda_j (x_j - x(j)) \Big\| < \frac{\varepsilon}{2}.$$

Since $z \in \mathrm{co}\, A(\varepsilon)$, there exists an integer $i = i(z)$ with $1 \le i \le m$ such that $z \in B\Big(y_i; \frac{\varepsilon}{2}\Big)$, i.e., such that

$$\|z - y_i\| < \frac{\varepsilon}{2}.$$

The resulting inequality $\|y - y_i\| \le \|y - z\| + \|z - y_i\| < \varepsilon$ then shows that $y \in B(y_i; \varepsilon)$, and hence that

$$\mathrm{co}\, A \subset \bigcup_{i=1}^{m} B(y_i; \varepsilon).$$

Therefore $\operatorname{co} A$ is precompact since $\varepsilon > 0$ is arbitrary. □

Problems

3.1-1 The notations and assumptions are as in Theorem 3.1-2. Show that, if the space $(X, \|\cdot\|)$ is separable, its completion $(\widetilde{X}, \|\cdot\|_{\widetilde{X}})$ is also separable.

3.1-2 Let $(X, \|\cdot\|)$ be a uniformly convex Banach space (uniformly convex spaces are defined in Section 2.17) and let Z be a nonempty, closed, convex subset of X. Show that given any point $x \in X$, there exists one and only one point $Px \in Z$ such that $\|x - Px\| = \inf_{z \in Z} \|x - z\|$.

Remark In a *Hilbert space* (a special case of a uniformly convex Banach space), this result is part of the fundamental *projection theorem* (Theorem 4.3-1(a)). □

3.2 First examples of Banach spaces; the spaces $\mathcal{C}(K;Y)$ with K compact and Y complete, and $\mathcal{L}(X;Y)$ with Y complete

We begin by the simplest example of Banach space.

Theorem 3.2-1 *Any finite-dimensional normed vector space is a Banach space.*

Proof Let $(X, \|\cdot\|)$ be a finite-dimensional normed vector space over $\mathbb{K}$, and let $(e_i)_{i=1}^n$ be a basis of X. Equipped with the norm

$$\|\cdot\|_1 : x = \sum_{i=1}^n x_i e_i \to \|x\|_1 = \sum_{i=1}^n |x_i|,$$

the space $(X, \|\cdot\|_1)$ is complete, since

$$\sum_{i=1}^n |x_i^k - x_i^\ell| \le \|x^k - x^\ell\| \quad \text{for all } k, \ell \ge 1$$

for any Cauchy sequence $(x^k)_{k=1}^\infty$ of vectors $x^k = \sum_{i=1}^n x_i^k e_i \in X$, and the field $\mathbb{K}$ is complete.

The space $(X, \|\cdot\|)$ is thus also complete, since all norms are equivalent in a finite-dimensional vector space (Theorem 2.7-1). □

The next example is fundamental. Recall that the notation $\mathcal{C}(X;Y)$, or simply $\mathcal{C}(X)$ if $Y = \mathbb{R}$, designates the set of all continuous mappings of a topological space X into a topological space Y.

Theorem 3.2-2 *Let K be a compact topological space and let $(Y; \|\cdot\|)$ be a Banach space. Then the space $\mathcal{C}(K;Y)$, equipped with the sup-norm $|||\cdot|||$ defined by*

$$|||f||| := \sup_{x \in K} \|f(x)\| \quad \textit{for each } f \in \mathcal{C}(K;Y)$$

(Theorem 2.3-1), *is a Banach space.*

Proof Let $(f_n)_{n=1}^{\infty}$ be a Cauchy sequence in the space $(\mathcal{C}(K;Y), |||\cdot|||)$. Given any $x \in K$, the inequality

$$\|f_m(x) - f_n(x)\| \leq |||f_m - f_n||| \quad \text{for all } m, n \geq 1,$$

shows that $(f_n(x))_{n=1}^{\infty}$ is a Cauchy sequence in the complete space $(Y, \|\cdot\|)$. Hence this sequence converges. Let then the mapping $f : K \to Y$ be defined by

$$f(x) := \lim_{n\to\infty} f_n(x) \quad \text{for all } x \in K.$$

Given $\varepsilon > 0$, there exists $n_0 = n_0(\varepsilon) \geq 1$ such that

$$\|f_m(x) - f_n(x)\| \leq |||f_m - f_n||| \leq \varepsilon \quad \text{for all } x \in K \text{ and all } m, n \geq n_0.$$

Letting $m \to \infty$ in this relation, we obtain

$$\|f(x) - f_n(x)\| \leq \varepsilon \quad \text{for all } x \in K \text{ and all } n \geq n_0,$$

or equivalently,

$$\sup_{x\in K} \|f(x) - f_n(x)\| \leq \varepsilon \quad \text{for all } n \geq n_0.$$

It thus remains to show that the mapping $f : K \to Y$ is continuous, a property which, in particular, will allow us to rewrite the left-hand side of the last inequality as $|||f - f_n|||$.

So, let x_0 be any point in K. Given $\varepsilon > 0$, let $n_0 = n_0(\varepsilon)$ be chosen as above. The mapping $f_{n_0} : K \to Y$ being continuous at x_0, there exists a neighborhood $V(x_0) \subset K$ of x_0 such that

$$\|f_{n_0}(x) - f_{n_0}(x_0)\| \leq \varepsilon \quad \text{for all } x \in V(x_0).$$

Consequently,

$$\|f(x) - f(x_0)\| \leq \|f(x) - f_{n_0}(x)\| + \|f_{n_0}(x) - f_{n_0}(x_0)\| + \|f_{n_0}(x_0) - f(x_0)\| \leq 3\varepsilon,$$

and thus $f : K \to Y$ is continuous at x_0 since $3\varepsilon > 0$ is arbitrary. This completes the proof. □

In particular, the space $(\mathcal{C}(\overline{\Omega}); \|\cdot\|)$, where Ω is a *bounded* open subset of $\mathbb{R}^n$ and $\|\cdot\|$ denotes the *sup-norm*, defined by

$$\|f\| := \sup_{x\in\overline{\Omega}} |f(x)| \quad \text{for each } x \in \overline{\Omega},$$

provides a fundamental example of a *Banach space*.

Remarks (1) For any integer $m \geq 1$ and any bounded open subset Ω of $\mathbb{R}^n$, the space $\mathcal{C}^m(\overline{\Omega})$, which consists of the restrictions to $\overline{\Omega}$ of all the functions that are m times continuously differentiable in $\mathbb{R}^n$, provides another example of a Banach space (Problem 3.2-1).

(2) The space $\mathcal{C}(\overline{\Omega})$, where Ω is again a bounded open subset of $\mathbb{R}^n$, is *not* complete when it is equipped with any one of the norms $\|\cdot\|_{L^p(\Omega)}$, $1 \leq p < \infty$ (Problem 3.2-2). □

With a proof similar to that of Theorem 3.2-2 (and for this reason omitted), we also have:

Theorem 3.2-3 *Let X be any set and let Y be a Banach space. Then the space $\mathcal{B}(X;Y)$ of all bounded mappings from X into Y, equipped with the sup-norm $|||\cdot|||$ defined by*

$$|||f||| := \sup_{x\in X} \|f(x)\| \quad \textit{for each } f \in \mathcal{B}(X;Y)$$

(Theorem 2.3-2), is a Banach space. □

We conclude this section by another fundamental example of a Banach space. Recall that, given two normed vector spaces X and Y over the same field, $\mathcal{L}(X;Y)$ denotes the normed vector space formed by all the continuous linear operators $A : X \to Y$, with $\|A\|_{\mathcal{L}(X;Y)} = \sup_{x\neq 0} \dfrac{\|Ax\|_Y}{\|x\|_X}$ (Theorem 2.9-5).

Theorem 3.2-4 *Let X be a normed vector space and let Y be a Banach space. Then $(\mathcal{L}(X;Y), \|\cdot\|_{\mathcal{L}(X;Y)})$ is a Banach space.*

In particular, the dual space $X' = \mathcal{L}(X;\mathbb{K})$ of a normed vector space X over $\mathbb{K}$, equipped with the norm

$$x' \in X' \to \|x'\| = \sup_{x\neq 0} \frac{|x'(x)|}{\|x\|_X},$$

is a Banach space.

Proof For brevity, the same notation $\|\cdot\|$ denotes the norms in the spaces X, Y, and $\mathcal{L}(X;Y)$. Let $(A_n)_{n=1}^\infty$ be a Cauchy sequence in the space $\mathcal{L}(X;Y)$. Given any $x \in X$, the inequality

$$\|A_m x - A_n x\| \leq \|A_m - A_n\| \, \|x\| \quad \text{for all } m, n \geq 1$$

shows that $(A_n x)_{n=1}^\infty$ is a Cauchy sequence in the Banach space Y. Hence this sequence converges. Let then the mapping $A : X \to Y$ be defined by

$$Ax := \lim_{n\to\infty} A_n x \quad \text{for all } x \in X.$$

Then A is a linear operator, since, for any scalars $\alpha, \widetilde{\alpha} \in \mathbb{K}$ and any vectors $x, \widetilde{x} \in X$,

$$\begin{aligned} A(\alpha x + \widetilde{\alpha}\widetilde{x}) &= \lim_{n\to\infty} A_n(\alpha x + \widetilde{\alpha}\widetilde{x}) = \lim_{n\to\infty} (\alpha A_n x + \widetilde{\alpha} A_n \widetilde{x}) \\ &= \alpha \lim_{n\to\infty} A_n x + \widetilde{\alpha} \lim_{n\to\infty} A_n \widetilde{x} = \alpha A x + \widetilde{\alpha} A \widetilde{x} \end{aligned}$$

(the continuity of the addition and scalar multiplication is used here).

Let $C := \sup_{n\geq 1} \|A_n\| < \infty$ (recall that a Cauchy sequence is bounded; cf. Theorem 1.12-1(a)). Then, for any $x \in X$, the relations

$$\|Ax\| = \lim_{n\to\infty} \|A_n x\| \text{ and } \|A_n x\| \leq C\|x\| \quad \text{for all } n \geq 1$$

show that $\|Ax\| \leq C\|x\|$. Hence the linear operator $A : X \to Y$ is continuous.

It remains to show that $\|A_n - A\| \to 0$ as $n \to \infty$. Given any $\varepsilon > 0$, there exists $n_0 = n_0(\varepsilon)$ such that $\|A_m - A_n\| \leq \varepsilon$ for all $m, n \geq n_0$, and hence such that

$$\|A_m x - A_n x\| \leq \varepsilon\|x\| \quad \text{for all } x \in X \text{ and all } m, n \geq n_0.$$

Letting $m \to \infty$ in the above inequality, we obtain

$$\|Ax - A_n x\| \leq \varepsilon \|x\| \quad \text{for all } x \in X \text{ and all } n \geq n_0.$$

Consequently

$$\|A_n - A\| = \sup_{x \neq 0} \frac{\|A_n x - Ax\|}{\|x\|} \leq \varepsilon \quad \text{for all } n \geq n_0,$$

which completes the proof. □

We will show later (Theorem 3.6-5) that another important example of a Banach space is provided by the *quotient space* X/Z (Section 2.2) when X is itself a *Banach space.* See also Problems 3.2-4–3.2-5 for other examples.

Problems

3.2-1 Let Ω be a domain in $\mathbb{R}^n$. Show that, for any integer $m \geq 1$, the space $\mathcal{C}^m(\overline{\Omega})$ (Theorem 1.18-1) becomes a Banach space when it is equipped with the norm $\|\cdot\|_{\mathcal{C}^m(\overline{\Omega})}$ defined by

$$\|f\|_{\mathcal{C}^m(\overline{\Omega})} := \max_{|\alpha| \leq m} \sup_{x \in \Omega} |\partial^\alpha f(x)| \quad \text{for each } f \in \mathcal{C}^m(\overline{\Omega}).$$

3.2-2 Let Ω be a bounded open subset of $\mathbb{R}^n$ and let $1 \leq p < \infty$. Show that the space $(\mathcal{C}(\overline{\Omega}), \|\cdot\|_{L^p(\Omega)})$ is not complete.
Hint: Construct a Cauchy sequence that does not converge.

3.2-3 Show that the space $\mathcal{P}$ of all polynomials $p : \mathbb{R} \to \mathbb{R}$ equipped with the norm defined by $\|p\| = \sup_{0 \leq x \leq 1} |p(x)|$ is not complete.

Remark In fact, there is no norm that can make $\mathcal{P}$ a Banach space. As we shall see (Theorem 5.1-4), this nontrivial result is a consequence of *Baire's theorem* (Section 5.1). □

3.2-4 Let X and Y be two normed vector spaces over the same field. Show that, if Y is complete, the subspace $\mathcal{K}(X;Y)$ of $\mathcal{L}(X;Y)$ formed by all *compact* operators from X into Y is closed in $\mathcal{L}(X;Y)$. Hence $\mathcal{K}(X;Y)$ *is a Banach space when* Y *is a Banach space*, as a closed subset of a Banach space.

3.2-5 Let $X_1, X_2, \ldots, X_k$, $k \geq 2$, and Y be normed vector spaces. Show that the space $\mathcal{L}_k(X_1, X_2, \ldots, X_k; Y)$ formed by all continuous multilinear mappings from $X_1 \times X_2 \times \cdots \times X_k$ into Y (Section 2.11) is complete if Y is complete.

3.3 Integral of a continuous function of a real variable with values in a Banach space

An interesting application of the *unique continuous linear extension theorem* (Theorem 3.1-1) and of the *completion of a normed vector space* (Theorem 3.1-2) is the construction of the integral $\int_a^b f(x)\,\mathrm{d}x$ when the function $f : [a,b] \to Y$ takes its values in a *Banach space* Y and is *continuous.* This type of integral will be needed in the proof of the *mean value theorem in a Banach space* (Theorem 7.6-1), which in turn will be used for establishing the *Newton–Kantorovich theorem in a Banach space* (Theorem 7.7-3). In addition, the construction of

$\int_a^b f(x)\,\mathrm{d}x$ naturally leads to the definition of a *Banach space*, denoted $\mathcal{R}([a,b]\,;Y)$ below, which contains the space $\mathcal{C}([a,b]\,;Y)$.

Remark More generally, a *Lebesgue integral* can be constructed for functions defined on a *measure space* (Section 1.14) and taking their values in a *Banach space* (once the notion of measurability has been appropriately defined for such functions).[2] □

The definition of $\int_a^b f(x)\,\mathrm{d}x$ is carried out in *two stages*. *First*, let $f : [a,b] \subset \mathbb{R} \to Y$ be a *step function over* $[a,b]$: This means that there exist finitely many points $x_i \in [a,b]$, $0 \leq i \leq n$, and vectors $c_i \in Y$, $1 \leq i \leq n$, such that

$$a = x_0 < x_1 < \cdots < x_i < \cdots < x_{n-1} < x_n = b,$$
$$f(x) = c_i \quad \text{for all } x_{i-1} < x < x_i,\ 1 \leq i \leq n,$$
$$\max_{0\leq i\leq n} \|f(x_i)\|_Y \leq \max_{1\leq j\leq n} \|c_j\|_Y \,.$$

We then *define* the integral of such a step function in the most natural way, i.e., by

$$\ell(f) := \sum_{i=1}^{n} (x_i - x_{i-1})c_i \in Y.$$

It is easily seen that the set $\mathcal{S}([a,b]\,;Y)$ formed by all step functions over $[a,b]$ with values in Y is a vector space and that, equipped with the *sup-norm*, defined by

$$\|f\| := \sup_{a\leq x\leq b} \|f(x)\|_Y \,,$$

the space $\mathcal{S}([a,b]\,;Y)$ becomes a *normed vector space*. Then the above mapping $\ell : \mathcal{S}([a,b]\,;Y) \to Y$, which is clearly linear, is continuous over this space since

$$\|\ell(f)\|_Y \leq (b-a)\,\|f\| \quad \text{for all } f \in \mathcal{S}([a,b]\,;Y),$$

as the definition of $\ell(f)$ immediately shows.

Second, let $\mathcal{R}([a,b]\,;Y)$ denote the *completion* of the space $\mathcal{S}([a,b]\,;Y)$ with respect to the sup-norm $\|\cdot\|$. The Banach space $\mathcal{R}([a,b]\,;Y)$ is thus a *closed* subspace of the Banach space $\mathcal{B}([a,b]\,;Y)$ of all *bounded* functions from $[a,b]$ into Y, *equipped with the same sup-norm* $\|\cdot\|$ (Theorem 3.2-3). Therefore the continuous linear mapping $\ell : \mathcal{S}([a,b]\,;Y) \to Y$ admits a unique continuous extension to the space $\mathcal{R}([a,b]\,;Y)$, since $\mathcal{S}([a,b]\,;Y)$ is by construction *dense* in $\mathcal{R}([a,b]\,;Y)$ and Y is *complete* (Theorems 3.1-1 and 3.1-2); this is why the assumed completeness of Y is essential. This observation thus provides a natural *definition of the integral of any function* $f \in \mathcal{R}([a,b]\,;Y)$ *over* $[a,b]$, as

$$\int_a^b f(x)\,\mathrm{d}x := \lim_{n\to\infty} \ell(f_n)$$

for any sequence $(f_n)_{n=1}^\infty$ of step functions $f_n \in \mathcal{S}([a,b]\,;Y)$ such that $\|f_n - f\| \to 0$ as $n \to \infty$, the norm $\|\cdot\|$ being thus again given by

$$\|f\| = \sup_{a\leq x\leq b} \|f(x)\|_Y \quad \text{for each } f \in \mathcal{R}([a,b]\,;Y).$$

[2] See, e.g., SCHWARTZ [1993a] or LANG [1993].

The Banach space $\mathcal{R}([a,b]\,;Y)$ and the integral $\int_a^b f(x)\,\mathrm{d}x$ so constructed then possess the following properties:

Theorem 3.3-1 (a) *For each function $f \in \mathcal{R}([a,b]\,;Y)$,*

$$\left\| \int_a^b f(x)\,\mathrm{d}x \right\|_Y \le \int_a^b \|f(x)\|_Y \,\mathrm{d}x \le (b-a)\,\|f\|\,.$$

(b) *The space $\mathcal{R}([a,b]\,;Y)$ contains the space $\mathcal{C}([a,b]\,;Y)$.*

Proof First, the inequalities

$$\|\ell(f)\|_Y \le \int_a^b \|f(x)\|_Y \,\mathrm{d}x \le (b-a)\,\|f\|\,,$$

which clearly hold for all step functions $f \in \mathcal{S}([a,b]\,;Y)$, also hold for all functions in the closure $\mathcal{R}([a,b]\,;Y)$ of $\mathcal{S}([a,b]\,;Y)$, since each term in these inequalities is a continuous function of $f \in \mathcal{S}([a,b]\,;Y)$.

Let next $f : [a,b] \to Y$ be any *continuous* function. Since the interval $[a,b]$ is compact, the function f is *uniformly continuous* over $[a,b]$, which easily implies that *f is a uniform limit of step functions $f_n \in \mathcal{S}([a,b]\,;Y)$, $n \ge 1$*. Hence the integral $\int_a^b f(x)\,\mathrm{d}x \in Y$ is well defined as $\lim_{n\to\infty} \ell(f_n)$, since this limit is independent of the sequence of step functions chosen to approximate f. ☐

Remark The functions in the space $\mathcal{R}([a,b]\,;Y)$ are called **regulated functions**. The above considerations thus show that the space $\mathcal{C}([a,b]\,;Y)$ of all continuous functions with values in Y satisfy the inclusion $\mathcal{C}([a,b]\,;Y) \subset \mathcal{R}([a,b]\,;Y)$. Besides, one can show that this inclusion is always *strict*; for instance, the space $\mathcal{R}([a,b]\,;Y)$ contains all monotone functions, which may be discontinuous at a countably infinite number of points of $[a,b]$.[3] ☐

3.4 Further examples of Banach spaces: the spaces ℓ^p and $L^p(\Omega), 1 \le p \le \infty$

We begin by considering the normed vector spaces $(\ell^p, \|\cdot\|_p)$, $1 \le p \le \infty$, introduced in Section 2.4. Recall that the norm of a sequence $x = (x_i)_{x=1}^{\infty} \in \ell^p$ of scalars x_i is defined by

$$\|x\|_p := \left(\sum_{i=1}^{\infty} |x_i|^p \right)^{1/p} \text{ if } 1 \le p < \infty, \quad \text{or} \quad \|x\|_\infty = \sup_{i\ge1} |x_i| \text{ if } p = \infty.$$

Theorem 3.4-1 *The spaces $(\ell^p, \|\cdot\|_p), 1 \le p \le \infty$, are Banach spaces.*

Proof Let $(x^n)_{n=1}^{\infty}$ be a Cauchy sequence of elements $x^n = (x_i^n)_{i=1}^{\infty} \in \ell^p$. Since

$$|x_i^m - x_i^n| \le \|x^m - x^n\|_p \quad \text{for each } i \ge 1,$$

[3]For more details about such notions, see, e.g., DIEUDONNÉ [1960] or LANG [1993].

each sequence $(x_i^n)_{n=1}^\infty$ converges, as a Cauchy sequence of scalars. Let then

$$x := (x_i)_{i=1}^\infty, \quad \text{where } x_i := \lim_{n\to\infty} x_i^n \quad \text{for each } i \geq 1.$$

First, we show that $x \in \ell^p$. Let M be such that $\|x^n\|_p \leq M$ for all $n \geq 1$ (a Cauchy sequence is bounded). Consequently, for any integer $k \geq 1$,

$$\left(\sum_{i=1}^k |x_i|^p\right)^{1/p} = \lim_{n\to\infty} \left(\sum_{i=1}^k |x_i^n|^p\right)^{1/p} \leq M \quad \text{if } 1 \leq p < \infty,$$
$$\sup_{1\leq i\leq k} |x_i| = \lim_{n\to\infty} \sup_{1\leq i\leq k} |x_i^n| \leq M \quad \text{if } p = \infty.$$

Letting $k \to \infty$ in the left-hand sides thus shows that $x \in \ell^p$, since the upper bound M is independent of the integer k.

Second, we show that $\|x^n - x\|_p \to 0$ as $n \to \infty$. Let then $\varepsilon > 0$ be given. Since $(x^n)_{n=1}^\infty$ is a Cauchy sequence in ℓ^p, there exists $n_0 = n_0(\varepsilon)$ such that, for all $m, n \geq n_0$,

$$\left(\sum_{i=1}^\infty |x_i^m - x_i^n|^p\right)^{1/p} \leq \varepsilon \text{ if } 1 \leq p < \infty, \quad \text{or} \quad \sup_{i\geq 1} |x_i^m - x_i^n| \leq \varepsilon \text{ if } p = \infty,$$

hence *a fortiori* such that, for any given integer $k \geq 1$ and again for all $m, n \geq n_0$,

$$\left(\sum_{i=1}^k |x_i^m - x_i^n|^p\right)^{1/p} \leq \varepsilon \text{ if } 1 \leq p < \infty, \quad \text{or} \quad \sup_{1\leq i\leq k} |x_i^m - x_i^n| \leq \varepsilon \text{ if } p = \infty.$$

Keeping the integer k fixed and letting $m \to \infty$ thus implies that, for all $n \geq n_0$,

$$\left(\sum_{i=1}^k |x_i - x_i^n|^p\right)^{1/p} \leq \varepsilon \text{ if } 1 \leq p < \infty, \quad \text{or} \quad \sup_{1\leq i\leq k} |x_i - x_i^n| \leq \varepsilon \text{ if } p = \infty.$$

It thus remains to let $k \to \infty$ in the left-hand sides, which shows that

$$\|x - x^n\|_p \leq \varepsilon \quad \text{for all } n \geq n_0.$$

This completes the proof. □

We now turn our attention to the (real) Lebesgue spaces $(L^p(\Omega), \|\cdot\|_{L^p(\Omega)})$, $1 \leq p \leq \infty$, where Ω is an open subset of $\mathbb{R}^n$, introduced in Section 2.5. Recall that a function $f \in L^p(\Omega)$ is finite almost everywhere in Ω, and that its norm is defined by

$$\|f\|_{L^p(\Omega)} = \left(\int_\Omega |f(x)|^p \, \mathrm{d}x\right)^{1/p} \quad \text{if } 1 \leq p < \infty,$$
$$\|f\|_{L^\infty(\Omega)} = \inf\{C \geq 0;\ |f| \leq C \text{ a.e. in } \Omega\} \quad \text{if } p = \infty.$$

As expected, the completeness of these spaces is not as simple to establish as that of the spaces ℓ^p.

Theorem 3.4-2 *The spaces $(L^p(\Omega), \|\cdot\|_{L^p(\Omega)})$, $1 \le p \le \infty$, where Ω is an open subset of $\mathbb{R}^n$, are Banach spaces.*

Proof For brevity, we let $\|\cdot\|_p = \|\cdot\|_{L^p(\Omega)}$ in this proof. We begin by considering the case where $1 \le p < \infty$. So, let $(f_m)_{m=1}^{\infty}$ be a Cauchy sequence of functions $f_m \in L^p(\Omega)$. There thus exists a subsequence $(f_{\sigma(m)})_{m=1}^{\infty}$ such that

$$\|f_{\sigma(m+1)} - f_{\sigma(m)}\|_p \le \frac{1}{2^m} \quad \text{for all } m \ge 1.$$

The functions g_k defined for each integer $k \ge 1$ by

$$g_k := \sum_{m=1}^{k} |f_{\sigma(m+1)} - f_{\sigma(m)}|$$

clearly belong to the space $L^p(\Omega)$ and, in addition, they satisfy

$$0 \le g_1 \le \cdots \le g_k \le g_{k+1} \le \cdots \quad \text{and} \quad \|g_k\|_p \le \sum_{m=1}^{k} \frac{1}{2^m} \le 1.$$

Therefore,

$$g(x) := \lim_{k\to\infty} g_k(x) \quad \text{exists in } [0,\infty] \text{ for all } x \in \Omega.$$

Then *Fatou's lemma* (Theorem 1.15-2) applied to the function $g : \Omega \to [0,\infty]$ so defined shows that

$$\int_\Omega |g(x)|^p \,\mathrm{d}x = \int_\Omega \lim_{k\to\infty} |g_k(x)|^p \le \liminf_{m\to\infty} \int_\Omega |g_k(x)|^p \,\mathrm{d}x \le 1.$$

Hence $g \in L^p(\Omega)$ and, consequently, $0 \le g(x) < \infty$ for almost all $x \in \Omega$. Since

$$\sum_{m=1}^{k} |f_{\sigma(m+1)}(x) - f_{\sigma(m)}(x)| = g_k(x) \le g(x) \quad \text{for all } x \in \Omega,$$

it next follows that

$$f(x) := \lim_{k\to\infty} f_{\sigma(k+1)}(x) = \lim_{k\to\infty} \Big(f_{\sigma(1)}(x) + \sum_{m=1}^{k} (f_{\sigma(m+1)}(x) - f_{\sigma(m)}(x))\Big)$$

exists in $\mathbb{R}$ for almost all $x \in \Omega$ on the one hand, and that, on the other hand, *the function f defined in this fashion is in $L^p(\Omega)$*, since

$$\begin{aligned}|f(x)| &\le |f_{\sigma(1)}(x)| + \sum_{m=1}^{k} |f_{\sigma(m+1)}(x) - f_{\sigma(m)}(x)| \\ &= |f_{\sigma(1)}(x)| + g_k(x) \le |f_{\sigma(1)}(x)| + g(x) \quad \text{for almost all } x \in \Omega,\end{aligned}$$

and both functions $f_{\sigma(1)}$ and g are in $L^p(\Omega)$ (recall that g is ≥ 0).

It remains to show that $\|f_m - f\|_p \to 0$ as $m \to \infty$. Given $\varepsilon > 0$, let $m_0 = m_0(\varepsilon)$ be such that $\|f_\ell - f_m\|_p \le \varepsilon$ for all $\ell, m \ge m_0$. Another application of *Fatou's lemma* then implies that, for all $m \ge m_0$,

$$\begin{aligned}\int_\Omega |f(x) - f_m(x)|^p \, dx &= \int_\Omega \lim_{k\to\infty} |f_{\sigma(k)}(x) - f_m(x)|^p \, dx \\ &\le \liminf_{k\to\infty} \int_\Omega |f_{\sigma(k)}(x) - f_m(x)|^p \, dx \le \varepsilon^p,\end{aligned}$$

i.e., that $\|f - f_m\|_p \le \varepsilon$ for all $m \ge m_0$. This completes the proof for $1 \le p < \infty$.

Given a Cauchy sequence $(f^m)_{m=1}^\infty$ in $L^\infty(\Omega)$, let M be such that $\|f^m\|_\infty \le M$ for all $m \ge 1$. Consequently,

$$|f^m(x)| \le M \quad \text{and} \quad |f^\ell(x) - f^m(x)| \le \|f^\ell - f^m\| \quad \text{for almost all } x \in \Omega.$$

An argument similar to that used in the proof of Theorem 3.2-2 then shows that $f(x) := \lim_{m\to\infty} f_m(x)$ exists for almost all $x \in \Omega$, that the function f defined in this fashion is in $L^\infty(\Omega)$, and finally, that

$$\|f^m - f\|_\infty \to 0 \quad \text{as } m \to \infty. \qquad \square$$

The first part of the above proof has shown in passing a *remarkable property of convergent sequences in* $L^p(\Omega)$ (since the same proof has also shown that such sequences coincide with the Cauchy sequences in $L^p(\Omega)$):

Theorem 3.4-3 *Let* $(f_m)_{m=1}^\infty$ *be a convergent sequence in* $L^p(\Omega)$, $1 \le p < \infty$, *and let* $f \in L^p(\Omega)$ *be its limit. Then there exists a subsequence* $(f_{\sigma(m)})_{m=1}^\infty$ *that* **pointwise converges to f almost everywhere in Ω**, *i.e., such that*

$$\lim_{m\to\infty} f_{\sigma(m)}(x) = f(x) \quad \textit{for almost all } x \in \Omega. \qquad \square$$

Problem

3.4-1 Show that, for any $0 < p < 1$, the metric space $(L^p(\Omega), d_p)$ defined in Problem 2.5-4 is complete.

3.5 Dual of a normed vector space; first examples; F. Riesz representation theorem in $L^p(\Omega)$, $1 \le p < \infty$

Let X be a normed vector field over $\mathbb{K} = \mathbb{R}$ or over $\mathbb{K} = \mathbb{C}$. Recall that the space

$$X' := \mathcal{L}(X; \mathbb{K}),$$

which is called the **dual space** of X, or simply the **dual** of X (Section 2.9), thus consists of all the *linear* functionals $x' : X \to \mathbb{K}$ that are *continuous* on X. Since the field $\mathbb{K}$ is complete, *the space X' equipped with the operator norm, defined in this case by*

$$\|x'\| = \sup_{x\ne 0} \frac{|x'(x)|}{\|x\|} \quad \text{for all } x' \in X',$$

is always a Banach space (Theorem 3.2-4), i.e., irrespective of whether the space X is complete or not.

Dual spaces play a central role in linear functional analysis, as will be abundantly illustrated in Chapter 5, where their basic properties will be studied at length. The more modest purpose of the present section is simply to describe some basic *examples* of dual spaces.

Given any extended real number $1 \le p \le \infty$, the extended real number $1 \le q \le \infty$ defined by

$$q = \infty \text{ if } p = 1, \quad \frac{1}{p} + \frac{1}{q} = 1 \text{ if } 1 < p < \infty, \quad \text{and} \quad q = 1 \text{ if } p = \infty,$$

is called the **conjugate exponent** of p.

To begin with, we consider the spaces $(\ell_p, \|\cdot\|_p)$, $1 \le p \le \infty$, introduced in Section 2.4. As shown in the next theorem, it is remarkable that, if $1 \le p < \infty$, then *the dual of ℓ^p can be identified with the space ℓ^q*, where q denotes the conjugate exponent of p. Note that this result does *not* hold for $p = \infty$, however (Theorem 3.5-2).

Theorem 3.5-1 (dual of $\ell^p, 1 \le p < \infty$) *Given a real number $1 \le p < \infty$, let $1 < q \le \infty$ denote its conjugate exponent. Then, given any element $a = (a_i)_{i=1}^\infty \in \ell^q$, the relation*

$$x'(x) = \sum_{i=1}^{\infty} a_i x_i \quad \text{for all } x = (x_i)_{i=1}^\infty \in \ell^p$$

defines a continuous linear functional x' on ℓ^p. Besides,

$$\|x'\|_{(\ell^p)'} = \|a\|_q .$$

The linear isometry $a \in \ell^q \to x' \in (\ell^p)'$ defined in this fashion is bijective, i.e., given any continuous linear functional x' on ℓ^p, there exists one and only one element $a = (a_i)_{i=1}^\infty \in \ell^q$ such $x'(x) = \sum_{i=1}^\infty a_i x_i$ for all $x = (x_i)_{i=1}^\infty \in \ell^p$.

Consequently, for any $1 \le p < \infty$, the dual space of ℓ^p can be identified as a normed vector space with space ℓ^q.

Proof (i) By Hölder's inequality for sequences (Theorem 2.4-1),

$$\Big|\sum_{i=1}^{\infty} a_i x_i\Big| \le \|a\|_q \|x\|_p \quad \text{for all } a = (a_i)_{i=1}^\infty \in \ell^q \text{ and all } x = (x_i)_{i=1}^\infty \in \ell^p$$

if $1 < p < \infty$; otherwise it is clear that this inequality holds with $q = \infty$ if $p = 1$. Given $a = (a_i)_{i=1}^\infty \in \ell^q$, the relation $x'(x) = \sum_{i=1}^\infty a_i x_i$ for all $x = (x_i)_{i=1}^\infty \in \ell^p$ therefore defines a continuous linear functional x' on ℓ^p, the norm of which satisfies $\|x'\| \le \|a\|_q$.

To establish the opposite inequality, we distinguish two cases. First, assume that $p = 1$. For any integer $n \ge 1$, let

$$x^n = (x_i^n)_{i=1}^\infty \in \ell^p, \quad \text{where } x_n^n := \operatorname{sgn} \overline{a}_n \text{ and } x_i^n := 0 \text{ if } i \ne n.$$

Then $x'(x^n) = a_n x_n^n = |a_n|$, and $\|x^n\|_1 \le 1$ (since $\|x^n\|_1 = 1$ if $a_n \ne 0$, or $\|x^n\|_1 = 0$ if $a_n = 0$), so that

$$|a_n| = |x'(x_n)| \le \|x'\| \, \|x^n\|_1 \le \|x'\|,$$

which implies that $\|a\|_\infty = \sup_{n\ge 1}|a_n| \le \|x'\|$. Hence $\|x'\| = \|a\|_\infty$ when $p = 1$.

Second, assume that $1 < p < \infty$. For any integer $m \ge 1$, let

$$x^n = (x_i^n)_{i=1}^\infty, \quad \text{where } x_i^n := |a_i|^{q-1}\operatorname{sgn}\bar{a}_i \text{ if } 1 \le i \le n \text{ and } x_i^n = 0 \text{ if } n < i.$$

Then $x'(x^n) = \sum_{i=1}^n a_i x_i = \sum_{i=1}^n |a_i|^q$, and $\|x^n\|_p = (\sum_{i=1}^n |a_i|^q)^{1/p}$, so that

$$\sum_{i=1}^n |a_i|^q = |x'(x^n)| \le \|x'\|\,\|x^n\|_p = \|x'\|\left(\sum_{i=1}^n |a_i|^q\right)^{1/p},$$

which implies that $\|a\|_q = \lim_{n\to\infty}(\sum_{i=1}^n |a_i|^q)^{1/q} \le \|x'\|$. Hence $\|x'\| = \|a\|_p$ when $1 < p < \infty$.

(ii) It remains to show that, for any $1 \le p < \infty$, the isometry $a \in \ell^q \to x' \in (\ell^p)'$ defined in (i) is *surjective* (that it is linear and injective is clear). So, let $x' \in (\ell^p)'$ be given.

Given any integer $i \ge 1$, define the element $e_i \in \ell^p$ and the scalar $a_i \in \mathbb{K}$ by

$$e_i := (\delta_{ij})_{j=1}^\infty \quad \text{and} \quad a_i := x'(e_i).$$

Given any $x = (x_i)_{i=1}^\infty \in \ell^p$, the relation $\|x - \sum_{i=1}^n x_i e_i\|_p = (\sum_{i=n+1}^\infty |x_i|^p)^{1/p}$ shows that $\lim_{n\to\infty}\sum_{i=1}^n x_i e_i = x$ in the space ℓ^p. The assumed continuity of x' therefore implies that

$$x'\left(\sum_{i=1}^n x_i e_i\right) = \sum_{i=1}^n a_i x_i \to x'(x) \quad \text{as } n \to \infty.$$

Consequently, the series $\sum_{i=1}^\infty a_i x_i$ converges in $\mathbb{K}$ and its sum is $x'(x)$, as desired. □

We now consider the case where $p = \infty$.

Theorem 3.5-2 *Given any element $a = (a_i)_{i=1}^\infty \in \ell^1$, the relation*

$$x'(x) = \sum_{i=1}^\infty a_i x_i \quad \textit{for each } x = (x_i)_{i=1}^\infty \in \ell^\infty$$

defines a continuous linear functional x' on ℓ^∞. Besides,

$$\|x'\|_{(\ell^\infty)'} = \|a\|_1.$$

The isometry $a \in \ell^1 \to x' \in (\ell^\infty)'$ defined in this fashion is linear and injective, but is not surjective. This means that, as a normed vector space, ℓ^1 can be only identified with a proper subspace of the dual space of ℓ^∞.

Proof Since

$$\left|\sum_{i=1}^\infty a_i x_i\right| \le \|a\|_1\,\|x\|_\infty \quad \text{for all } a = (a_i)_{i=1}^\infty \in \ell^1 \text{ and all } x = (x_i)_{i=1}^\infty \in \ell^\infty,$$

the relation $x'(x) = \sum_{i=1}^\infty a_i x_i$ for each $x = (x_i)_{i=1}^\infty \in \ell^\infty$ defines a continuous linear functional x' on ℓ^∞, the norm of which satisfies $\|x'\| \le \|a\|_1$.

In view of establishing the opposite inequality, let $x = (\operatorname{sgn} \overline{a}_i)_{i=1}^{\infty}$. Then $x'(x) = \sum_{i=1}^{\infty} |a_i|$ and $\|x\|_\infty \leq 1$, so that

$$\|a\|_1 = \sum_{i=1}^{\infty} |a_i| = |x'(x)| \leq \|x'\| \, \|x\|_\infty \leq \|x'\|.$$

Hence $\|x'\| = \|a\|_1$. Therefore the mapping $a \in \ell^1 \to x' \in (\ell^\infty)'$ defined in this fashion, which is clearly linear, is an isometry (hence injective).

The quickest way to prove that this isometry, or more generally any linear isometry from ℓ^1 into $(\ell^\infty)'$, is not surjective, is to resort to the following result, whose proof (fortunately independent of the present one) has to be postponed, as it relies on the Hahn–Banach theorem (Theorem 5.9-5): If the dual space of a normed vector space X is separable, then X is also separable. So, if $(\ell^\infty)'$ could be identified as a normed vector space with ℓ^1 by means of a linear isometry, $(\ell^\infty)'$ would be separable, like ℓ^1 (Theorem 2.4-2(b)), since separability is a property that only involves the norm. But then ℓ^∞ itself would be separable, which is not the case (Theorem 2.4-2(c)). □

We now turn our attention to the (real) Lebesgue spaces $(L^p(\Omega), \|\cdot\|_{L^p(\Omega)})$, $1 \leq p \leq \infty$, introduced in Section 2.5. Although the conclusions are analogous to those for the spaces ℓ^p, $1 \leq p \leq \infty$ (compare Theorems 3.5-1 and 3.5-2 with Theorems 3.5-3 and 3.5-4), the proofs are not unexpectedly slightly more delicate. Note that the notation ℓ, rather than x', will be henceforth preferred in the rest of this section for designating a generic element of the dual space of $L^p(\Omega)$ (so as to avoid confusion, since x will designate as usual a generic point in the set Ω). The next result is fundamental.

Theorem 3.5-3 (F. Riesz representation theorem[4] in $L^p(\Omega)$, $1 \leq p < \infty$) *Let Ω be an open subset of $\mathbb{R}^n$ and, given a real number $1 \leq p < \infty$, let $1 < q \leq \infty$ denote the conjugate exponent of p. Then, given any function $g \in L^q(\Omega)$, the relation*

$$\ell(f) = \int_\Omega f(x) g(x) \, dx \quad \textit{for all } f \in L^p(\Omega)$$

defines a continuous linear functional ℓ on $L^p(\Omega)$. Besides,

$$\|\ell\|_{(L^p(\Omega))'} = \|g\|_{L^q(\Omega)}.$$

The linear isometry $g \in L^q(\Omega) \to \ell \in (L^p(\Omega))'$ defined in this fashion is bijective, i.e., given any continuous linear functional ℓ on $L^p(\Omega)$, there exists one and only one function $g \in L^q(\Omega)$ such that $\ell(f) = \int_\Omega f(x) g(x) \, dx$ for all $f \in L^p(\Omega)$.

Consequently, the dual space of $L^p(\Omega)$, $1 \leq p < \infty$, can be identified as a normed vector space with the space $L^q(\Omega)$.

Proof For notational brevity, the norms $\|\cdot\|_{L^p(\Omega)}$ or $\|\cdot\|_{(L^q(\Omega))'}$ will be abbreviated as $\|\cdot\|_{L^p}$ or $\|\cdot\|_{(L^q)'}$ throughout this proof.

---

[4]So named in honor of F. Riesz, who began to study the spaces ℓ^p and $L^p(\Omega)$ in 1910 and proved this representation theorem (with Ω an open interval of $\mathbb{R}$) in 1913; the genesis of his ideas is beautifully described in DIEUDONNÉ [1981, Chapter 6, Section 2].

(i) By Hölder's inequality for functions (Theorem 2.5-1),

$$\left|\int_\Omega fg\,\mathrm{d}x\right| \le \|f\|_{L^p}\,\|g\|_{L^q} \quad \text{for all } f\in L^p(\Omega) \text{ and all } g\in L^q(\Omega)$$

if $1<p<\infty$; otherwise, this inequality clearly holds with $q=\infty$ if $p=1$. Given $g\in L^q(\Omega)$, the relation $\ell(f)=\int_\Omega fg\,\mathrm{d}x$ for all $f\in L^p(\Omega)$ therefore defines a continuous linear functional on $L^p(\Omega)$, the norm of which satisfies $\|\ell\|_{(L^p)'}\le\|g\|_{L^q}$.

It remains to show that the continuous linear operator $g\in L^q(\Omega)\to\ell\in(L^p(\Omega))'$ defined in this fashion is isometric and surjective. To this end, we first consider the case where $\mu(\Omega)<\infty$, where μ denotes the Lebesgue measure in $\mathbb{R}^n$; cf. parts (ii)–(vi).

In the remainder of this proof, ℓ denotes a given continuous linear functional on $L^p(\Omega)$.

(ii) *Assume that $\mu(\Omega)<\infty$. Then there exists a function $g\in L^1(\Omega)$ such that*

$$\ell(s)=\int_\Omega sg\,\mathrm{d}x$$

for all measurable simple functions $s:\Omega\to\mathbb{R}$.

Let $\mathcal{A}$ denote the σ-algebra formed by all the Lebesgue-measurable subsets of Ω. Since $\mu(\Omega)<\infty$, the characteristic function χ_A of any $A\in\mathcal{A}$ is in the space $L^p(\Omega)$. Our objective then consists in showing that the function $\nu:\mathcal{A}\to\mathbb{R}$ defined by

$$\nu(A):=\ell(\chi_A)\quad\text{for all } A\in\mathcal{A},$$

is a *signed measure*, which is *absolutely continuous* with respect to the Lebesgue measure μ (Section 1.15).

First, it is clear that $\nu(\varnothing)=0$ since $\chi_\varnothing=0$, and that ν is *finitely additive*: If $A_1\cap A_2=\varnothing$, then $\chi_{A_1\cup A_2}=\chi_{A_1}+\chi_{A_2}$, so that $\nu(A_1\cup A_2)=\nu(A_1)+\nu(A_2)$ since ℓ is linear.

Second, given a countably infinite family of pairwise disjoint sets $A_i\in\mathcal{A}$, $i\ge1$, let

$$A:=\bigcup_{i=1}^\infty A_i \quad\text{and}\quad B_m:=A-\bigcup_{i=1}^m A_i \ \text{ for all } m\ge1.$$

Therefore, by the finite additivity of ν,

$$\nu(A)=\nu\Big(\bigcup_{i=1}^m A_i\Big)+\nu(B_m)=\sum_{i=1}^m\nu(A_i)+\nu(B_m)\quad\text{for all } m\ge1.$$

Since the Lebesgue measure μ is countably additive,

$$\mu(A)=\sum_{i=1}^m\mu(A_i)+\mu(B_m)=\sum_{i=1}^\infty\mu(A_i)<\infty,$$

and thus $\mu(B_m)\to0$ as $m\to\infty$. The continuity of ℓ then implies that $\nu(B_m)\to0$ as $m\to\infty$, since

$$|\nu(B_m)|=|\ell(\chi_{B_m})|\le\|\ell\|\,\|\chi_{B_m}\|_{L^p}=\|\ell\|\,(\mu(B_m))^{1/p}\quad\text{for all } m\ge1.$$

This shows that ν is *countably additive.* Besides,

$$|\nu(A)| < \infty \text{ for all } A \in \mathcal{A}, \quad \text{and} \quad \mu(A) = 0 \text{ implies } \nu(A) = 0,$$

since $|\nu(A)| = |\ell(\chi_A)| \leq \|\ell\| \, (\mu(A))^{1/p}$ for all $A \in \mathcal{A}$.

By the *Radon–Nikodym theorem* (Theorem 1.15-4), there thus exists a function $g \in L^1(\Omega)$ such that

$$\nu(A) = \int_A g \, \mathrm{d}x \quad \text{for each } A \in \mathcal{A},$$

or equivalently, such that

$$\ell(\chi_A) = \int_\Omega \chi_A g \, \mathrm{d}x \quad \text{for each } A \in \mathcal{A}.$$

Since any simple measurable function is of the form $s = \sum_{i=1}^m \alpha_i \chi_{A_i}$ with $\alpha_i \in \mathbb{R}$ and $A_i \in \mathcal{A}$, $1 \leq i \leq m$, the linearity of ℓ implies that $\ell(s) = \int_\Omega s g \, \mathrm{d}x$ for all such functions s.

(iii) *Assume that* $\mu(\Omega) < \infty$, *and let* $g \in L^1(\Omega)$ *be the function found in* (ii). *For each integer* $k \geq 1$, *define the measurable set*

$$B_k := \{x \in \Omega; \; |g(x)| \leq k\}.$$

Then

$$\ell(f) = \int_\Omega f g \, \mathrm{d}x \quad \text{for all } f \in L^p(\Omega) \text{ such that } f|_{\Omega - B_k} = 0.$$

In what follows, the integer $k \geq 1$ is fixed and a function $f \in L^p(\Omega)$ that satisfies $f|_{\Omega - B_k} = 0$ is given. By Theorem 1.14-5, there exist measurable simple functions s_m, $m \geq 1$, such that

$$|s_m(x)| \leq |f(x)| \quad \text{and} \quad s_m(x) \to f(x) \text{ for almost all } x \in \Omega.$$

Consequently

$$(s_m(x) - f(x))g(x) \underset{m \to \infty}{\longrightarrow} 0 \quad \text{and} \quad |(s_m(x) - f(x))g(x)| \leq 2k \, |f(x)| \quad \text{for almost all } x \in \Omega.$$

Since $\int_\Omega |f| \, \mathrm{d}x \leq \|f\|_{L^p} (\mu(\Omega))^{1/q} < \infty$, the function f belongs to the space $L^1(\Omega)$. Therefore, the *Lebesgue dominated convergence theorem* (Theorem 1.15-3) can be applied to the sequence $((s_m - f)g)_{m=1}^\infty$, showing that $\int_\Omega |(s_m(x) - f(x))g(x)| \, \mathrm{d}x \to 0$ as $m \to \infty$. Hence

$$\ell(s_m) \to \int_\Omega f g \, \mathrm{d}x \quad \text{as } m \to \infty,$$

since $|\ell(s_m) - \int_\Omega f g \, \mathrm{d}x| \leq \int_\Omega |(s_m - f)g| \, \mathrm{d}x$ on the one hand. On the other hand,

$$|s_m(x) - f(x)|^p \underset{m \to \infty}{\longrightarrow} 0 \quad \text{and} \quad |s_m(x) - f(x)|^p \leq 2^p \, |f(x)|^p \quad \text{for almost all } x \in \Omega,$$

and the function $|f|^p$ belongs to the space $L^1(\Omega)$. Therefore, $\|s_m - f\|_{L^p}^p \to 0$ as $m \to \infty$, again by Lebesgue's dominated convergence theorem, thus showing that

$$\ell(s_m) \to \ell(f) \quad \text{as } m \to \infty,$$

by the continuity of ℓ. Hence $\ell(f) = \int_\Omega fg\,\mathrm{d}x$.

(iv) *Assume that* $\mu(\Omega) < \infty$ *and that* $p = 1$. *Then the function* $g \in L^1(\Omega)$ *found in* (ii) *satisfies*

$$g \in L^\infty(\Omega) \quad \text{and} \quad \|g\|_{L^\infty} \leq \|\ell\|_{(L^1)'}.$$

For brevity, let $\|\ell\| := \|\ell\|_{(L^1)'}$. Given $\varepsilon > 0$, define the set

$$A_\varepsilon := \{x \in \Omega;\ |g(x)| \geq \|\ell\| + \varepsilon\},$$

and, given any integer $k \geq 1$, define the function $f_k^\varepsilon \in L^1(\Omega)$ by

$$f_k^\varepsilon(x) := \operatorname{sgn} g(x) \text{ if } x \in A_\varepsilon \cap B_k \quad \text{and} \quad f_k^\varepsilon(x) := 0 \text{ if } x \in \Omega - (A_\varepsilon \cap B_k),$$

where the set B_k is defined as in (iii). Since $|g(x)| \geq \|\ell\| + \varepsilon$ for all $x \in A_\varepsilon \cap B_k$, it follows that

$$\mu(A_\varepsilon \cap B_k)(\|\ell\| + \varepsilon) \leq \int_{A_\varepsilon \cap B_k} |g|\,\mathrm{d}x = \int_\Omega f_k^\varepsilon g\,\mathrm{d}x,$$

by definition of f_k^ε. Besides,

$$\int_\Omega f_k^\varepsilon g\,\mathrm{d}x = \ell(f_k^\varepsilon) \leq \|\ell\|\,\|f_k^\varepsilon\|_{L^1(\Omega)} = \|\ell\|\,\mu(A_\varepsilon \cap B_k),$$

by (iii). The conjunction of these inequalities thus implies that

$$\mu(A_\varepsilon \cap B_k) = 0 \quad \text{for all } k \geq 1.$$

The set Ω can be written as $\Omega = (\bigcup_{k=1}^\infty B_k) \cup B$ with $\mu(B) = 0$, since the function g is finite almost everywhere (recall that $g \in L^1(\Omega)$). The relation $A_\varepsilon = (\bigcup_{k=1}^\infty (A_\varepsilon \cap B_k)) \cup (A \cap B)$ then implies that $\mu(A_\varepsilon) = 0$. But $\varepsilon > 0$ is arbitrary; hence $\mu(\{x \in \Omega;\ |g(x)| \geq \ell\}) = \mu(\bigcup_{m=1}^\infty A_{1/m}) = 0$, so that $|g(x)| \leq \|\ell\|$ for almost all $x \in \Omega$.

(v) *Assume that* $\mu(\Omega) < \infty$ *and that* $1 < p < \infty$. *Then the function* $g \in L^1(\Omega)$ *found in* (ii) *satisfies*

$$g \in L^q(\Omega) \quad \text{and} \quad \|g\|_{L^q} \leq \|\ell\|_{(L^p)'}.$$

Define the set $A := \{x \in \Omega;\ g(x) \neq 0\}$ and, given any integer $k \geq 1$, define the function $f_k : \Omega \to \mathbb{R}$ by

$$f_k(x) := \operatorname{sgn} g(x)\,|g(x)|^{q-1} \text{ if } x \in A \cap B_k \quad \text{and} \quad f_k(x) := 0 \text{ if } x \in \Omega - (A \cap B_k),$$

where the set B_k is defined as in (iii). Then

$$\int_\Omega |f_k|^p\,\mathrm{d}x = \int_{A \cap B_k} |g|^q\,\mathrm{d}x \leq \int_{B_k} |g|^q\,\mathrm{d}x,$$

by definition of f_k; hence

$$\int_{B_k} |g^q|\,\mathrm{d}x = \int_{B_k} f_k g\,\mathrm{d}x = \int_\Omega f_k g\,\mathrm{d}x = \ell(f_k) \leq \|\ell\|_{(L^p)'}\,\|f_k\|_{L^p} \leq \|\ell\|_{(L^p)'} \left(\int_{B_k} |g|^q\,\mathrm{d}x\right)^{1/p},$$

by (iii). Consequently,

$$\left(\int_{B_k} |g|^q \,\mathrm{d}x\right)^{1-1/p} = \left(\int_{B_k} |g|^q \,\mathrm{d}x\right)^{1/q} \le \|\ell\|_{(L^p)'}.$$

Since this inequality holds for any integer $k \ge 1$ and since $\Omega = (\bigcup_{k=1}^{\infty} B_k) \cup B$ with $\mu(B) = 0$, letting $k \to \infty$ gives

$$\|g\|_{L^q} = \lim_{k\to\infty} \left(\int_{B_k} |g|^q \,\mathrm{d}x\right)^{1/q} \le \|\ell\|_{(L^p)'}.$$

(vi) *Assume that $\mu(\Omega) < \infty$ and that $1 \le p < \infty$. Then, given any $\ell \in (L^p(\Omega))'$, there exists a function $g \in L^q(\Omega)$ such that*

$$\ell(f) = \int_\Omega fg \,\mathrm{d}x \ \text{ for all } f \in L^p(\Omega) \quad \text{and} \quad \|g\|_{L^q} = \|\ell\|_{(L^p)'}.$$

It follows from parts (ii), (iv), and (v) that, for any $1 \le p < \infty$, there exists a function $g \in L^q(\Omega)$ such that $\ell(s) = \int_\Omega sg \,\mathrm{d}x$ for all measurable simple functions s.

Let then a function $f \in L^p(\Omega)$ be given. By Theorem 1.14-5, there exist measurable simple functions $s_m : \Omega \to \mathbb{R}$, $m \ge 1$, such that $|s_m| \le |f|$ for all $m \ge 1$ and $f(x) = \lim_{m\to\infty} s_m(x)$ for almost all $x \in \Omega$. Then the Lebesgue dominated convergence theorem applied to the functions $|s_m - f|^p$, $m \ge 1$, which converge to zero almost everywhere in Ω and are bounded above by the function $2^p |f|^p \in L^1(\Omega)$, shows that $\|s_m - f\|_{L^p(\Omega)} \to 0$ as $m \to \infty$. Therefore,

$$\ell(s_m) \to \ell(f) \quad \text{as } m \to \infty,$$

since ℓ is continuous. Besides,

$$\left|\ell(s_m) - \int_\Omega fg \,\mathrm{d}x\right| = \left|\int_\Omega (s_m - f) g \,\mathrm{d}x\right| \le \|s_m - f\|_{L^p(\Omega)}| \, \|g\|_{L^q(\Omega)}.$$

Hence

$$\ell(s_m) \to \int_\Omega fg \,\mathrm{d}x \quad \text{as } m \to \infty.$$

Consequently, $\ell(f) = \int_\Omega fg \,\mathrm{d}x$ for all $f \in L^p(\Omega)$.

That $\|g\|_{L^q} = \|\ell\|_{(L^p)'}$ follows from the inequality $\|\ell\|_{(L^p)'} \le \|g\|_{L^q}$ established in part (i) and from the inequality $\|g\|_{L^q} \le \|\ell\|_{(L^p)'}$ established in parts (iv) and (v).

(vii) *Finally, assume that $\mu(\Omega) = \infty$.*

For each integer m, let

$$\Omega_m := \left\{x \in \Omega;\ \operatorname{dist}(x, \mathbb{R}^n - \Omega) > \frac{1}{m}\right\} \cap B(0, m).$$

Hence

$$\Omega = \bigcup_{m=1}^{\infty} \Omega_m, \quad \mu(\Omega_m) < \infty \quad \text{and} \quad \Omega_m \subset \Omega_{m+1} \ \text{for all } m \ge 1.$$

Given any function $f \in L^p(\Omega_m)$, the function $f^\sharp : \Omega \to [-\infty, \infty]$ defined by $f^\sharp|_{\Omega_m} := f$ and $f^\sharp|_{\Omega - \Omega_m} := 0$ belongs to $L^p(\Omega)$. Therefore, for each $m \geq 1$, the relation

$$\ell_m(f) := \ell(f^\sharp) \quad \text{for all } f \in L^p(\Omega_m)$$

defines a continuous linear functional on $L^p(\Omega_m)$, which clearly satisfies $\|\ell_m\|_{(L^p(\Omega_m))'} \leq \|\ell\|_{(L^p)'}$. Besides, since $\mu(\Omega_m) < \infty$, the result established in part (vi) shows that there exists a function $g_m \in L^q(\Omega_m)$ such that

$$\ell_m(f) = \int_{\Omega_m} g_m f \, dx \quad \text{for all } f \in L^p(\Omega_m) \quad \text{and} \quad \|g_m\|_{L^q} = \|\ell_m\|_{(L^p(\Omega_m))'}.$$

Given any function $f \in L^p(\Omega_m)$, the function $\widetilde{f} : \Omega_{m+1} \to [-\infty, \infty]$ defined by $\widetilde{f}|_{\Omega_m} := f$ and $\widetilde{f}|_{\Omega_{m+1} - \Omega_m} := 0$ is such that $\widetilde{f}^\sharp = f^\sharp$, $\widetilde{f}^\sharp|_{\Omega_{m+1}} = \widetilde{f}$, and $\widetilde{f}^\sharp|_{\Omega - \Omega_{m+1}} = 0$. Therefore, for all $f \in L^p(\Omega_m)$,

$$\int_{\Omega_m} g_m f \, dx = \ell_m(f) = \ell(f^\sharp) = \ell(\widetilde{f}^\sharp) = \ell_{m+1}(\widetilde{f}) = \int_{\Omega_{m+1}} g_{m+1} \widetilde{f} \, dx = \int_{\Omega_m} g_{m+1} f \, dx.$$

Besides, both functions g_m and $g_{m+1}|_{\Omega_m}$ belong to the space $L^q(\Omega_m)$. Consequently, the relation

$$\int_{\Omega_m} (g_{m+1} - g_m) f \, dx = 0 \quad \text{for all } f \in L^p(\Omega_m),$$

which *a fortiori* holds for all $\varphi \in \mathcal{D}(\Omega_m)$, implies that

$$g_{m+1} - g_m = 0 \quad \text{a.e. in } \Omega.$$

So we can unambiguously define a function $g : \Omega \to [-\infty, \infty]$ by letting $g(x) := g_{m(x)}(x)$ for each $x \in \Omega$, where $m(x) := \min\{m \geq 1;\ x \in \Omega_m\}$; besides, this function clearly satisfies

$$g|_{\Omega_m} = g_m \in L^q(\Omega_m) \quad \text{for each } m \geq 1.$$

We now show that $g \in L^q(\Omega)$. If $p = 1$, in which case $q = \infty$, this is clear since

$$\|g\|_{L^\infty} = \lim_{m \to \infty} \|g_m\|_{L^\infty} = \lim_{m \to \infty} \|\ell_m\|_{(L^1(\Omega_m))'} \leq \|\ell\|_{(L^1)'}.$$

If $1 < p < \infty$, consider the functions $|g|^q \chi_{\Omega_m}$, $m \geq 1$, which satisfy

$$|g(x)|^q \chi_{\Omega_m}(x) \underset{m \to \infty}{\longrightarrow} |g(x)|^q \quad \text{for almost all } x \in \Omega,$$

$$\int_\Omega |g|^q \chi_{\Omega_m} \, dx = \int_{\Omega_m} |g|^q \, dx = \int_{\Omega_m} |g_m|^q \, dx = \left(\|\ell_m\|_{(L^p(\Omega_m))'}\right)^q \leq \left(\|\ell\|_{(L^p)'}\right)^q$$

for all $m \geq 1$. Hence *Fatou's lemma* (Theorem 1.15-2) applied to the sequence $(|g|^q \chi_{\Omega_m})_{m \geq 1}$ shows that

$$\int_\Omega |g|^q \, dx \leq \liminf_{m \to \infty} \int_\Omega |g_m|^q \chi_{\Omega_m} \, dx \leq \left(\|\ell\|_{(L^p)'}\right)^q.$$

Consequently, for all $1 \leq p < \infty$, $g \in L^q(\Omega)$ and $\|g\|_{L^q} \leq \|\ell\|_{(L^p)'}$.

Finally, let a function $f \in L^p(\Omega)$ be given. Then

$$\ell(f\chi_{\Omega_m}) = \ell_m(f|_{\Omega_m}) = \int_{\Omega_m} f|_{\Omega_m} g_m \, \mathrm{d}x = \int_\Omega f\chi_{\Omega_m} g \mathrm{d}x \quad \text{for each } m \geq 1.$$

Since $\|f\chi_{\Omega_m} - f\|_{L^p(\Omega)} \to 0$ as $m \to \infty$ (to see this, use Lebesgue's dominated convergence theorem), it follows that

$$\ell(f\chi_{\Omega_m}) \to \ell(f) \quad \text{as } m \to \infty,$$

on the one hand; since $g \in L^q(\Omega)$, it follows that

$$\int_\Omega (f\chi_{\Omega_m}) g \mathrm{d}x \to \int_\Omega f g \mathrm{d}x \quad \text{as } m \to \infty,$$

on the other hand. Consequently, $\ell(f) = \int_\Omega fg \mathrm{d}x$ for all $f \in L^p(\Omega)$.

That $\|g\|_{L^q} = \|\ell\|_{(L^p)'}$ follows from the inequality $\|\ell\|_{(L^p)'} \leq \|g\|_{L^q}$ established in part (i) and from the inequality $\|g\|_{L^q} \leq \|\ell\|_{(L^p)'}$ established above. □

When $p = 2$, Theorems 3.5-1 and 3.5-3 become special cases of a general result, the *F. Riesz representation theorem in a Hilbert space* (Theorem 4.6-1), which is valid for *any Hilbert space* (hence in particular for the spaces ℓ^2 and $L^2(\Omega)$).

Remark A particularly elegant proof of Theorem 3.5-3 for $1 < p < \infty$ can be also given,[5] based on the *reflexivity* of the spaces $L^p(\Omega)$ (reflexive spaces are defined in Section 5.14). □

Finally, we consider the case where $p = \infty$.

Theorem 3.5-4 *Given any function $g \in L^1(\Omega)$, the relation*

$$\ell(f) = \int_\Omega f(x) g(x) \, \mathrm{d}x \quad \textit{for all } f \in L^\infty(\Omega)$$

defines a continuous linear functional on $L^\infty(\Omega)$. Besides,

$$\|\ell\|_{(L^\infty(\Omega))'} = \|g\|_{L^1(\Omega)}.$$

The isometry $g \in L^1(\Omega) \to \ell \in (L^\infty(\Omega))'$ defined in this fashion is linear and injective, but is not surjective. This means that, as a normed vector space, $L^1(\Omega)$ can be only identified with a proper subspace[6] *of the space $(L^\infty(\Omega))'$.*

Proof The same shortened notations as in the proof of Theorem 3.5-3 are used here. Since

$$\left| \int_\Omega fg \mathrm{d}x \right| \leq \|f\|_{L^\infty} \|g\|_{L^1} \quad \text{for all } f \in L^\infty(\Omega) \text{ and all } f \in L^1(\Omega),$$

the relation $\ell(f) = \int_\Omega fg \, \mathrm{d}x$ for all $f \in L^\infty(\Omega)$ defines a continuous linear functional ℓ on $L^\infty(\Omega)$, the norm of which satisfies $\|\ell\|_{(L^\infty)'} \leq \|g\|_{L^1}$.

[5]See Brezis [2011, Theorem 4.11].

[6]A complete description of the space $(L^\infty(\Omega))'$ is given in Yosida [1965, Chapter 4, Section 9].

In view of establishing the opposite inequality, let the function $f \in L^\infty(\Omega)$ be defined by $f(x) = \operatorname{sgn} g(x)$, $x \in \Omega$, so that $\|f\|_{L^\infty} \leq 1$. Then

$$\|g\|_{L^1} = \int_\Omega |g| \,\mathrm{d}x = \int_\Omega fg \,\mathrm{d}x = \ell(f) \leq \|\ell\|_{(L^\infty)'} \|f\|_{L^\infty} \leq \|\ell\|_{(L^\infty)'}.$$

Hence $\|\ell\|_{(L^\infty)'} = \|g\|_{L^1(\Omega)}$. So the mapping $g \in L^1(\Omega) \to \ell \in (L^\infty(\Omega))'$ defined in this fashion, which is clearly linear, is an isometry (hence injective).

The quickest way to prove that this isometry is not surjective is to notice that the space $L^\infty(\Omega)$ is not separable (Theorem 2.5-4), and then to mimic the argument given at the end of the proof of Theorem 3.5-2. □

3.6 Series in Banach spaces

We now turn our attention to another remarkable feature of Banach spaces, viz., a very simple sufficient condition for the *convergence of a series* in such spaces. But first, we need a few definitions.

Let $(X, \|\cdot\|)$ be a normed vector space, and let $(x_n)_{n=1}^\infty$ be a sequence of vectors $x_n \in X$. Then the *notation* $\sum_{n=1}^\infty x_n$ is called a **series**, and for each integer $k \geq 1$,

$$s_k := \sum_{n=1}^k x_n$$

designates the **kth partial sum** of the series $\sum_{n=1}^\infty x_n$. The series $\sum_{n=0}^\infty x_n$ is said to be **convergent** if the *sequence* $(s_k)_{k=1}^\infty$ *is convergent* in X. In this case, we write

$$\sum_{n=1}^\infty x_n = s, \quad \text{where } s := \lim_{k\to\infty} s_k,$$

and s is called the **sum** of the series. Note that, when such a series is convergent, the *same* notation $\sum_{n=1}^\infty x_n$ thus denotes both the series itself and its sum.

The following sufficient condition for the convergence of a series is fundamental.

Theorem 3.6-1 (convergence of a series in a Banach space) *Let $(X, \|\cdot\|)$ be a Banach space, and let $\sum_{n=1}^\infty x_n$ be a series of vectors $x_n \in X$ such that*[7]

$$\sum_{n=1}^\infty \|x_n\| < \infty.$$

Then the series $\sum_{n=1}^\infty x_n$ converges and its sum satisfies

$$\left\|\sum_{n=1}^\infty x_n\right\| \leq \sum_{n=1}^\infty \|x_n\|.$$

[7]The reader is assumed to be familiar with the basic properties of series of real or complex numbers. Given numbers $\alpha_n \geq 0$, $n \geq 1$, the notation $\sum_{n=1}^\infty \alpha_n < \infty$ means that the series $\sum_{n=1}^\infty \alpha_n$ is convergent in $\mathbb{R}$.

Proof Because the series $\sum_{n=1}^{\infty} \|x_n\|$ converges by assumption, the sequence $(\sigma_k)_{k=1}^{\infty}$, where $\sigma_k := \sum_{n=1}^{k} \|x_n\|$, is a Cauchy sequence of real numbers. Since the partial sums $s_k = \sum_{n=1}^{k} x_n$, $k \geq 1$, satisfy

$$\|s_k - s_\ell\| = \left\| \sum_{n=\ell+1}^{k} x_n \right\| \leq \sum_{n=\ell+1}^{k} \|x_n\| = \sigma_k - \sigma_\ell \quad \text{for all } k \geq \ell + 1,$$

the sequence $(s_k)_{k=1}^{\infty}$ is thus a Cauchy sequence in the Banach space $(X, \|\cdot\|)$ and, as such, converges in X. Besides, its limit s satisfies $\|s\| \leq \sum_{n=1}^{\infty} \|x_n\|$, since

$$\|s\| = \lim_{k\to\infty} \|s_k\| \quad \text{and} \quad \|s_k\| \leq \sum_{n=1}^{k} \|x_n\| \leq \sum_{n=1}^{\infty} \|x_n\| \quad \text{for all } k \geq 1. \qquad \square$$

A first application of this result is a simple sufficient condition allowing us to define the *inverse* of a linear operator of the specific form $(I - A)$, by means of the **Neumann series**[8] $\sum_{n=0}^{\infty} A^n$, where $A^0 := I$. Note that the next theorem extends to general Banach spaces the well-known formula $\dfrac{1}{1-z} = \sum_{n=0}^{\infty} z^n$ for $|z| < 1$, $z \in \mathbb{C}$.

Theorem 3.6-2 (convergence of the Neumann series) *Let $(X, \|\cdot\|)$ be a Banach space and let $A \in \mathcal{L}(X)$ be such that*

$$\|A\| < 1,$$

where $\|A\|$ denotes the operator norm of A (Section 2.9). Then the continuous linear operator $(I - A) : X \to X$ is bijective and its inverse $(I - A)^{-1} : X \to X$ is also a continuous linear operator. Besides,

$$(I - A)^{-1} = \sum_{n=0}^{\infty} A^n \quad \textit{and} \quad \|(I - A)^{-1}\| \leq \frac{1}{1 - \|A\|}.$$

Proof The assumed inequality $\|A\| < 1$ and the inequalities $\|A^n\| \leq \|A\|^n$ for each $n \geq 0$ (Theorem 2.9-5(d)) together imply that

$$\sum_{n=0}^{\infty} \|A^n\| \leq \sum_{n=0}^{\infty} \|A\|^n < \infty.$$

Since $\mathcal{L}(X)$ is a *Banach space* (Theorem 3.2-4), *the series $\sum_{n=0}^{\infty} A^n$ converges in $\mathcal{L}(X)$* by Theorem 3.6-1. Let $B \in \mathcal{L}(X)$ denote its sum, i.e.,

$$B = \sum_{n=0}^{\infty} A^n := \lim_{k\to\infty} B_k, \quad \text{where } B_k := \sum_{n=0}^{k} A^n.$$

Then

$$AB = \lim_{k\to\infty} AB_k = \lim_{k\to\infty} (B_{k+1} - I) = B - I,$$
$$BA = \lim_{k\to\infty} B_k A = \lim_{k\to\infty} (B_{k+1} - I) = B - I,$$

[8]So named after Carl Neumann (1832–1925).

so that

$$I = B(I - A) = (I - A)B.$$

Hence $(I - A) \in \mathcal{L}(X)$ is bijective (since $(I - A)$ has a left and a right inverse), and

$$(I - A)^{-1} = B = \sum_{n=0}^{\infty} A^n.$$

Besides, again by Theorem 3.6-1,

$$\|(I - A)^{-1}\| \leq \sum_{n=0}^{\infty} \|A^n\| \leq \sum_{n=0}^{\infty} \|A\|^n = \frac{1}{1 - \|A\|}. \qquad \square$$

As a first application of Theorem 3.6-2, we establish an important property of continuous linear operators with a continuous inverse, acting from a Banach space into a normed vector space.

Theorem 3.6-3 *Let X be a Banach space and let Y be a normed vector space. Then the set*

$$\mathcal{U} := \{A \in \mathcal{L}(X;Y);\ A : X \to Y \text{ is a bijection and } A^{-1} \in \mathcal{L}(Y;X)\}$$

is open in the normed vector space $\mathcal{L}(X;Y)$, $\|\cdot\|_{\mathcal{L}(X;Y)}$.

More specifically, let $A \in \mathcal{U}$. Then $B \in \mathcal{U}$ if

$$\|B - A\| < \frac{1}{\|A^{-1}\|},$$

and in this case,

$$\|B^{-1}\| \leq \frac{\|A^{-1}\|}{1 - \|A^{-1}(B - A)\|} \leq \frac{\|A^{-1}\|}{1 - \|A^{-1}\|\,\|B - A\|},$$

$$\|B^{-1} - A^{-1}\| \leq \frac{\|A^{-1}\|^2\,\|B - A\|}{1 - \|A^{-1}(B - A)\|} \leq \frac{\|A^{-1}\|^2\,\|B - A\|}{1 - \|A^{-1}\|\,\|B - A\|}.$$

Consequently, the mapping $A \in \mathcal{U} \to A^{-1} \in \mathcal{U}$ is continuous.

Proof Let $A \in \mathcal{U}$. Since $\mathcal{L}(X)$ is a Banach space (because X is a Banach space; cf. Theorem 3.2-4), Theorem 3.6-2 can be applied, showing that $(I_X + A^{-1}(B - A)) \in \mathcal{L}(X)$ is a bijection with a continuous inverse if

$$\|B - A\|_{\mathcal{L}(X;Y)} < \frac{1}{\|A^{-1}\|_{\mathcal{L}(Y;X)}},$$

since this condition implies that $\|A^{-1}(B - A)\|_{\mathcal{L}(X)} < 1$. Therefore,

$$B = A(I + A^{-1}(B - A)) \in \mathcal{L}(X;Y)$$

is also a bijection with a continuous inverse if $\|B - A\| < (\|A^{-1}\|)^{-1}$, with an inverse given by

$$B^{-1} = (I_X + A^{-1}(B - A))^{-1}A^{-1} \in \mathcal{L}(Y;X).$$

Hence the *set $\mathcal{U}$ is open in $\mathcal{L}(X;Y)$*. Besides, the above expression of B^{-1} shows that

$$\|B^{-1}\| \le \frac{\|A^{-1}\|}{1-\|A^{-1}(B-A)\|} \le \frac{\|A^{-1}\|}{1-\|A^{-1}\|\,\|B-A\|} \quad \text{if } \|B-A\| < \frac{1}{\|A^{-1}\|},$$

again by Theorem 3.6-2. The identity $B^{-1}-A^{-1} = B^{-1}(A-B)A^{-1}$ then implies that

$$\|B^{-1}-A^{-1}\| \le \frac{\|A^{-1}\|^2\,\|B-A\|}{1-\|A^{-1}\|\,\|B-A\|} \quad \text{if } \|B-A\| < \frac{1}{\|A^{-1}\|}.$$

□

Remarks (1) It will be proved later that the mapping $A \in \mathcal{U} \subset \mathcal{L}(X;X) \to A^{-1} \in \mathcal{L}(Y;X)$ is not only continuous, but in effect *infinitely differentiable* (Theorem 7.12-2).

(2) If the set $\mathcal{U}$ is nonempty, the space Y is thus necessarily a *Banach space*. □

A series $\sum_{n=1}^{\infty} x_n$ in a normed vector space $(X, \|\cdot\|)$ is said to be **absolutely convergent** if

$$\sum_{n=1}^{\infty} \|x_n\| < \infty.$$

Theorem 3.6-1 thus asserts that *any absolutely convergent series in a Banach space is convergent.* Remarkably, *the converse also holds*, thus providing a useful criterion for showing that a normed vector space is a Banach space (as immediately illustrated by Theorem 3.6-5):

Theorem 3.6-4 *Let X be a normed vector space in which every absolutely convergent sequence is convergent. Then X is a Banach space.*

Proof Let $(x_n)_{n=1}^{\infty}$ be a Cauchy sequence in X. Hence there exists a subsequence $(x_{\sigma(n)})_{n=1}^{\infty}$ such that $\|x_{\sigma(n+1)} - x_{\sigma(n)}\| \le \frac{1}{2^n}$ for all $n \ge 1$, so that $\sum_{n=1}^{\infty} \|x_{\sigma(n+1)} - x_{\sigma(n)}\| < \infty$. By assumption, there thus exists an element $x \in X$ such that

$$x = \lim_{k\to\infty} \sum_{n=1}^{k} (x_{\sigma(n+1)} - x_{\sigma(n)}) = \lim_{k\to\infty} (x_{\sigma(k+1)} - x_{\sigma(1)}).$$

Therefore the subsequence $(x_{\sigma(n)})_{n=1}^{\infty}$ is convergent. But a Cauchy sequence that contains a convergent subsequence is also convergent (Theorem 1.12-1(c)). □

We now apply the above criterion to *quotient spaces* (Section 2.2).

Theorem 3.6-5 *Let X be a Banach space and let Z be a closed subspace of X. Then the quotient space X/Z equipped with the quotient norm (Theorem 2.2-3) is also a Banach space.*

Proof Let $\sum_{n=1}^{\infty}[x_n]$ be an absolutely convergent series in the quotient space X/Z. By definition of the quotient norm (Theorem 2.2-3), for any $n \ge 1$, there exists $y_n \in [x_n]$ such that $\|y_n\| \le \|[x_n]\| + \frac{1}{2^n}$. Hence $\sum_{n=1}^{\infty} \|y_n\| \le 1 + \sum_{n=1}^{\infty} \|[x_n]\| < \infty$. Since X is complete, there exists $x \in X$ such that $x = \lim_{k\to\infty} \sum_{n=1}^{k} y_n$. Now,

$$[x - \sum_{n=1}^{k} y_n] = [x] - \sum_{n=1}^{k} [y_n] = [x] - \sum_{n=1}^{k} [x_n],$$

and thus

$$\left\|[x] - \sum_{n=1}^{k}[x_n]\right\| = \left\|[x - \sum_{n=1}^{k} y_n]\right\| \leq \left\|x - \sum_{n=1}^{k} y_n\right\|.$$

This shows that the series $\sum_{n=1}^{\infty}[x_n]$ converges to $[x]$. Hence the space X/Z is a Banach space by Theorem 3.6-4. □

Problems

3.6-1 Let $\boldsymbol{A}$ be a real $N \times N$ matrix.

(1) Show that the series $\sum_{n=0}^{\infty} \frac{1}{n!}\boldsymbol{A}^n$ is convergent in the vector space $\mathbb{M}^N$ formed by all real $N \times N$ matrices. Its sum is denoted $e^{\boldsymbol{A}} := \sum_{n=0}^{\infty} \frac{1}{n!}\boldsymbol{A}^n$ and is called the *matrix exponential* of the matrix $\boldsymbol{A}$.

(2) Show that $e^{\boldsymbol{A}} = \lim_{k\to\infty}\left(\boldsymbol{I} + \frac{\boldsymbol{A}}{k}\right)^k$.

(3) Show that $\det(e^{\boldsymbol{A}}) = e^{\operatorname{tr}\boldsymbol{A}}$ (this implies that the matrix $e^{\boldsymbol{A}}$ is always invertible).

(4) Let $\boldsymbol{B}$ be a real $N \times N$ matrix. Show that, if $\boldsymbol{A}$ and $\boldsymbol{B}$ commute, then $e^{(\boldsymbol{A}+\boldsymbol{B})} = e^{\boldsymbol{A}}e^{\boldsymbol{B}}$, which shows in particular that the matrices $e^{\boldsymbol{A}}$ and $e^{\boldsymbol{B}}$ also commute in this case.

3.6-2 (1) Given a real $N \times N$ matrix $\boldsymbol{A}$ and a vector $\boldsymbol{u}_0 \in \mathbb{R}^N$ and any $t \geq 0$, let $\boldsymbol{u}(t) = (u_i(t))_{i=1}^N := e^{t\boldsymbol{A}}\boldsymbol{u}_0 \in \mathbb{R}^N$, where $e^{t\boldsymbol{A}}$ denotes the matrix exponential of the matrix $t\boldsymbol{A}$ (Problem 3.6-1). Show that each function $t \in [0,\infty[\to u_i(t)$, $1 \leq i \leq N$, is differentiable, and that

$$\boldsymbol{u}'(t) = \boldsymbol{A}\boldsymbol{u}(t), \quad t \geq 0, \quad \text{and} \quad \boldsymbol{u}(0) = \boldsymbol{u}_0,$$

where $\boldsymbol{u}'(t) := (u_i'(t))_{i=1}^N$.

(2) Let $\boldsymbol{b} \in \mathcal{C}([0,\infty[\,;\mathbb{R}^N)$ be a given vector field. Find an explicit expression for the solution $t \in [0,\infty[\to \boldsymbol{u}(t) = (u_i(t))_{i=1}^N$ of

$$\boldsymbol{u}'(t) = \boldsymbol{A}\boldsymbol{u}(t) + \boldsymbol{b}(t), \quad t \geq 0, \quad \text{and} \quad \boldsymbol{u}(0) = \boldsymbol{0}.$$

Questions (1) and (2) thus provide explicit solutions to *Cauchy problems* for specific *linear ordinary differential equations*. The existence of solutions to Cauchy problems for *nonlinear* ordinary differential equations is established in Sections 3.8 and 3.11.

3.6-3 Let $(X, \|\cdot\|)$ be a Banach space and let $A \in \mathcal{L}(X)$ be such that $\|A^p\| < 1$ for some power $p \geq 2$. Show that $(I - A) \in \mathcal{L}(X)$ is bijective and that its inverse $(I - A)^{-1} : X \to X$ is also continuous.

3.7 Banach fixed point theorem

Let $f : X \to X$ be a mapping from a set X into itself. A **fixed point** of f is any point $x \in X$ that satisfies

$$f(x) = x.$$

Let (X, d) be a metric space. A mapping $f : X \to X$ is a **contraction** if there exists a constant k such that

$$0 < k < 1 \text{ and } d(f(x), f(y)) \leq kd(x, y) \quad \text{for all } x, y \in X.$$

The next theorem, which is due to Stefan Banach,[9] is *one of the most important results from analysis.* Its proof is simple, but it has numerous crucial applications, such as the convergence of iterative methods for solving linear equations (Problem 3.7-6), the existence of solutions to ordinary differential equations (Theorem 3.8-1) and to two-point boundary value problems (Theorem 3.9-1), the Lax–Milgram lemma (Theorem 6.2-1), or the implicit function theorem (Theorem 7.12-1), to name a few.

Even though this chapter is devoted to Banach spaces, we prove this theorem for complete *metric spaces* (in fact, the proof in this more general case is the same). Various interesting complements are provided in Problems 3.7-1–3.7-5.

Theorem 3.7-1 (Banach fixed point theorem) *Let (X, d) be a complete metric space. Then any contraction $f : X \to X$ has one and only one fixed point $x \in X$.*

Besides, given any point $x_0 \in X$, the sequence $(x_n)_{n=0}^{\infty}$ defined by

$$x_{n+1} = f(x_n), \quad n \geq 0,$$

converges to x as $n \to \infty$, and the following estimate holds:

$$\|x_n - x\| \leq Ck^n, \quad n \geq 0, \qquad \textit{with } C := \frac{d(f(x_0), x_0)}{1-k}.$$

Proof Let d denote the distance in X. Given any point $x_0 \in X$, the sequence $(x_n)_{n=0}^{\infty}$ defined by $x_{n+1} = f(x_n)$, $n \geq 0$, is a Cauchy sequence since, for any $p \geq 1$,

$$d(x_{p+1}, x_p) \leq kd(x_p, x_{p-1}) \leq \cdots \leq k^p d(x_1, x_0)$$

so that, for any $m > n \geq 0$,

$$\begin{aligned} d(x_m, x_n) &\leq \sum_{p=n}^{m-1} d(x_{p+1}, x_p) \leq \left(\sum_{p=n}^{m-1} k^p \right) d(x_1, x_0) \\ &\leq k^n \left(\sum_{p=0}^{m-n-1} k^p \right) d(x_1, x_0) \leq \frac{k^n}{1-k} d(x_1, x_0). \end{aligned}$$

The space (X, d) being complete, there exists $x \in X$ such that $\lim_{n\to\infty} x_n = x$. Since a contraction is clearly continuous,

$$f(x) = \lim_{n\to\infty} f(x_n) = \lim_{n\to\infty} x_{n+1} = x.$$

Hence x is a fixed point of f. Let $y \in X$ be also a fixed point of f. Then

$$d(f(x), f(y)) = d(x, y) \leq kd(x, y).$$

Hence $y = x$, and thus the fixed point of f is unique. □

[9] S. Banach [1922]: Sur les opérations dans les ensembles abstraits et leurs applications aux équations intégrales, *Fundamenta Mathematicae* **3**, 133–181.

The approximation of the fixed point of f by means of the sequence $(x_n)_{n=0}^{\infty}$, where $x_{n+1} = f(x_n)$, $n \geq 0$, and x_0 is any point in X, is called the **method of successive approximations,** or **Picard's method.**[10]

As illustrated by Problem 3.7-6, Picard's method constitutes in effect the essence of some of the most basic *iterative methods for solving linear systems.*

Remark All the assumptions in Theorem 3.7-1 are essential, as shown by the following simple counterexamples:

The contraction $x \to \frac{x}{2}$ has no fixed point in the noncomplete metric space $]0, \infty[$.

The mapping $f : x \to x + \frac{1}{x}$ from the complete metric space $X := [1, \infty[$ into itself satisfies $d(f(x), f(y)) < d(x, y)$ for all $x, y \in X$, $x \neq y$; yet, f has no fixed point. Note, however, that such a mapping *does* have a fixed point if it is defined over a *compact* metric space; cf. Problem 3.7-3.

In any metric space X (complete or not) with at least two elements, the identity mapping f of X satisfies $d(f(x), f(y)) \leq d(x, y)$ for all $x, y \in X$, but has more than one fixed point. □

Problems

3.7-1 Let (X, d) be a complete metric space, let T be a topological space, and let $(f_t)_{t \in T}$ be a family of mappings $f_t : X \to X$ with the following properties: for each $x \in X$, the mapping $t \in T \to f_t(x) \in X$ is continuous, and there exists a constant k such that

$$0 < k < 1 \quad \text{and} \quad d(f_t(x), f_t(y)) \leq k d(x, y) \quad \text{for all } x, y \in X \text{ and all } t \in T.$$

Let $x_t \in X$ denote for each $t \in T$ the unique fixed point of f_t. Show that the mapping $t \in T \to x_t \in X$ is continuous.

3.7-2 Let (X, d) be a complete metric space and let $f : X \to X$ be a mapping such that, for some $p \geq 2$, the composite mapping $f \circ f \circ \cdots \circ f$ with p factors is a contraction. Note that the mapping f is not assumed to be continuous.

(1) Show that f has one and only one fixed point x.

(2) Show that, given any $x_0 \in X$, the sequence $(x_n)_{n=0}^{\infty}$ defined by $x_{n+1} = f(x_n)$, $n \geq 0$, converges to x.

3.7-3 Let (X, d) be a compact metric space (hence (X, d) is complete; cf. Theorem 1.13-3) and let $f : X \to X$ be a mapping that satisfies $d(f(x), f(y)) < d(x, y)$ for all $x, y \in X$, $x \neq y$.

(1) Show that f has one and only one fixed point.

(2) Find an example showing that f is not necessarily a contraction.

3.7-4 Let (X, d) be a compact metric space and let $f : X \to X$ be a continuous mapping that satisfies $d(f(x), f(y)) \geq d(x, y)$ for all $x, y \in X$. Show that $d(f(x), f(y)) = d(x, y)$ for all $x, y \in X$.

3.7-5 Let $k > 0$, let A be a subset of a metric space (X, d), and let $f : A \to \mathbb{R}$ be a function that satisfies $|f(x) - f(y)| \leq k d(x, y)$ for all $x, y \in A$. Show that there exists a mapping $\widetilde{f} : X \to \mathbb{R}$ that satisfies $\widetilde{f}(x) = f(x)$ for all $x \in A$ and $|\widetilde{f}(x) - \widetilde{f}(y)| \leq k d(x, y)$ for all $x, y \in X$. This result

[10]This method was introduced for solving two-point boundary value problems (of the form considered in Problem 3.9-1) in:

E. PICARD [1893]: Sur l'application des méthodes d'approximations successives à l'étude de certaines équations différentielles ordinaires, *Journal de Mathématiques Pures et Appliquées* **9**, 217–271.

constitutes the **MacShane lemma.**[11]

3.7-6 (1) Consider a *linear system* of the form $\boldsymbol{u} = \boldsymbol{B}\boldsymbol{u} + \boldsymbol{c}$, where $\boldsymbol{B}$ is a real $N \times N$ matrix and $\boldsymbol{c} \in \mathbb{R}^n$, and assume that $\|B\| < 1$ for some matrix norm $\|\cdot\|$. Show that this linear system has a unique solution $\boldsymbol{u}$, and that the sequence $(\boldsymbol{u}^n)_{n=0}^{\infty}$ of vectors $\boldsymbol{u}^n \in \mathbb{R}^N$ defined by

$$\boldsymbol{u}^{n+1} = \boldsymbol{B}\boldsymbol{u}^n + \boldsymbol{c}, \quad n \geq 0, \quad \text{where } \boldsymbol{u}^0 \in \mathbb{R}^N \text{ is an arbitrary vector,}$$

converges as $n \to \infty$ to $\boldsymbol{u}$.

In the rest of this problem, the following notations are used: Given an $N \times N$ matrix $\boldsymbol{A} = (a_{ij})$, the matrices $\boldsymbol{D}, \boldsymbol{E}$, and $\boldsymbol{F}$ are defined by $(\boldsymbol{D})_{ij} := a_{ij}\delta_{ij}$, $(-\boldsymbol{E})_{ij} := a_{ij}$ if $i > j$ and $(-\boldsymbol{E})_{ij} := 0$ if $i \leq j$, and $(-\boldsymbol{F})_{ij} := a_{ij}$ if $i < j$ and $(-\boldsymbol{F})_{ij} := 0$ if $i \geq j$, $1 \leq i, j \leq N$. Note that $\boldsymbol{A} = \boldsymbol{D} - \boldsymbol{E} - \boldsymbol{F}$.

(2) Consider the linear system $\boldsymbol{A}\boldsymbol{u} = \boldsymbol{b}$, where $\boldsymbol{A} = (a_{ij})$ is an $N \times N$ real matrix such that $a_{ii} \neq 0$, $1 \leq i \leq N$, and $\boldsymbol{b} = (b_i)$ is a vector in $\mathbb{R}^N$. The *Jacobi method* is the simplest *iterative method* for computing a solution $\boldsymbol{u} \in \mathbb{R}^N$ to such a system: Given any vector $\boldsymbol{u}^0 \in \mathbb{R}^N$, it consists in defining a sequence $(\boldsymbol{u}^n)_{n=0}^{\infty}$ of vectors $\boldsymbol{u}^n \in \mathbb{R}^N$, $n \geq 0$, by

$$\boldsymbol{D}\boldsymbol{u}^{n+1} = (\boldsymbol{E} + \boldsymbol{F})\boldsymbol{u}^n + \boldsymbol{b}, \quad n \geq 0.$$

Show that, if the matrix $\boldsymbol{A} = (a_{ij})$ is *strictly diagonally dominant*, in the sense that $|a_{ii}| > \sum_{\substack{j=1 \\ j \neq i}}^{N} |a_{ij}|$, $1 \leq i \leq N$, the matrix A is invertible and, given any vector $\boldsymbol{u}^0 \in \mathbb{R}^N$, the above sequence $(\boldsymbol{u}^n)_{n=0}^{\infty}$ converges to $\boldsymbol{u} = \boldsymbol{A}^{-1}\boldsymbol{b}$ as $n \to \infty$.

Hint: Use the matrix norm $\|\cdot\|_{\infty}$ (Problem 2.9-1).

(3) The *Gauß–Seidel method* for solving the linear system $\boldsymbol{A}\boldsymbol{u} = \boldsymbol{b}$, where $\boldsymbol{b}$ is a vector in $\mathbb{R}^N$, consists in defining a sequence $(\boldsymbol{u}^n)_{n=0}^{\infty}$ of vectors $\boldsymbol{u}^n \in \mathbb{R}^n$ by

$$(\boldsymbol{D} - \boldsymbol{E})\boldsymbol{u}^{n+1} = \boldsymbol{F}\boldsymbol{u}^n + \boldsymbol{b}, \quad n \geq 0,$$

where $\boldsymbol{u}^0$ is an arbitrary vector in $\mathbb{R}^N$. Show that the Gauß–Seidel method is convergent if the matrix A is strictly diagonally dominant.

Hint: Show that any eigenvalue λ of the matrix $(\boldsymbol{D} - \boldsymbol{E})^{-1}\boldsymbol{F}$ satisfies $|\lambda| < 1$, and use Problem 2.9-3(1).

(4) Given an $N \times N$ real matrix $\boldsymbol{A} = (a_{ij})$ with $a_{ii} \neq 0$, $1 \leq i \leq N$, and given a parameter $\omega \neq 0$, the *relaxation method* for solving the linear system $\boldsymbol{A}\boldsymbol{u} = \boldsymbol{b}$ consists in defining a sequence $(\boldsymbol{u}^n)_{n=0}^{\infty}$ of vectors $\boldsymbol{u}^n \in \mathbb{R}^N$ by

$$\left(\frac{1}{\omega}\boldsymbol{D} - \boldsymbol{E}\right)\boldsymbol{u}^{n+1} = \left(\frac{1-\omega}{\omega}\boldsymbol{D} + \boldsymbol{F}\right)\boldsymbol{u}^n + \boldsymbol{b}, \quad n \geq 0,$$

where $\boldsymbol{u}^0$ is an arbitrary vector in $\mathbb{R}^N$ (the Gauß–Seidel method thus corresponds to the special case $\omega = 1$). Show that the relaxation method is convergent if $0 < \omega < 2$ and the matrix A is symmetric and positive-definite; this result constitutes the *Ostrowski–Reich*[12] *theorem*.

Hint: Show that $\|\left(\frac{1}{\omega}\boldsymbol{D} - \boldsymbol{E}\right)^{-1}\left(\frac{1-\omega}{\omega}\boldsymbol{D} + \boldsymbol{F}\right)\| < 1$, where $\|\cdot\|$ is the matrix norm subordinate to the vector norm $\boldsymbol{v} \in \mathbb{R}^N \to (\boldsymbol{v}^T\boldsymbol{A}\boldsymbol{v})^{1/2}$.

[11] E.J. MacShane [1934]: Extension of range of functions, *Bulletin of the American Mathematical Society* **40**, 837–842.

[12] A.M. Ostrowski [1954]: On the linear iteration procedures for symmetric matrices, *Rendiconti Lincei – Matematica e Applicazioni* **14**, 140–163.

E. Reich [1949]: On the convergence of the classical iterative method of solving linear simultaneous equations, *Annals of Mathematical Statistics* **20**, 448–451.

Note that one iteration of Jacobi's method involves the solution of a linear system whose matrix is *diagonal*, while one iteration of the Gauß–Seidel or relaxation method involves the solution of a linear system whose matrix is *lower triangular*.[13]

3.8 Application of Banach fixed point theorem: Existence of solutions to nonlinear ordinary differential equations; Cauchy–Lipschitz theorem; the pendulum equation

As a first application of the Banach fixed point theorem in a Banach space of the form $\mathcal{C}(K;Y)$ with K compact and Y a Banach space (Theorem 3.2-2), we establish the existence and uniqueness of a solution to the *initial value problem*, or *Cauchy problem*, for a specific class of *systems of ordinary differential equations*.[14]

Since the variable often stands for the time in applications, it will be denoted t. In this respect, note that there is no loss of generality in assuming that the "initial time" is $t_0 = 0$ (should it be $t_0 \neq 0$, then use $(t - t_0)$ as the new "time variable"). The space of all continuously differentiable mappings $\boldsymbol{v} := (v_i)_{i=1}^N : [0,T] \to \mathbb{R}^N$ is denoted $\mathcal{C}^1([0,T];\mathbb{R}^N)$ and, for each $t \in [0,T]$, the notation $\boldsymbol{v}'(t)$ denotes the vector $(v_i'(t))_{i=1}^n$.

Theorem 3.8-1 (Cauchy–Lipschitz theorem) *Let $\|\cdot\|$ denote any norm in $\mathbb{R}^N$. Given $T > 0$, let $\boldsymbol{g} \in \mathcal{C}([0,T] \times \mathbb{R}^N;\mathbb{R}^N)$ be a mapping with the property that there exists a constant $\gamma > 0$ such that*

$$\|\boldsymbol{g}(t,\boldsymbol{w}) - g(t,\boldsymbol{v})\| \leq \gamma\|\boldsymbol{w} - \boldsymbol{v}\| \quad \textit{for all } t \in [0,T] \textit{ and all } \boldsymbol{w},\boldsymbol{v} \in \mathbb{R}^N.$$

Let also $\boldsymbol{u}_0 \in \mathbb{R}^N$ be a given vector. Then the **initial value problem**, *or* **Cauchy problem**,

$$\boldsymbol{u}'(t) = \boldsymbol{g}(t,\boldsymbol{u}(t)), \quad 0 \leq t \leq T, \quad \textit{and} \quad \boldsymbol{u}(0) = \boldsymbol{u}_0,$$

has one and only one solution $\boldsymbol{u} \in \mathcal{C}^1([0,T];\mathbb{R}^N)$.

Proof (i) It is immediately verified that, if $\boldsymbol{u} \in \mathcal{C}([0,T];\mathbb{R}^N)$ is a solution to the integral equation

$$\boldsymbol{u}(t) = \boldsymbol{u}_0 + \int_0^t \boldsymbol{g}(s,\boldsymbol{u}(s))\,\mathrm{d}s, \quad 0 \leq t \leq T,$$

then $\boldsymbol{u} \in \mathcal{C}^1([0,1];\mathbb{R}^N)$ and $\boldsymbol{u}$ is a solution to the initial value problem, and conversely, if $\boldsymbol{u} \in \mathcal{C}^1([0,1];\mathbb{R}^N)$ is a solution to the initial value problem, then $\boldsymbol{u}$ is a solution to the integral equation.

(ii) Equipped with the norm

$$|||\cdot||| : \boldsymbol{v} \in \mathcal{C}([0,T];\mathbb{R}^N) \to \sup_{0 \leq t \leq T} (e^{-\gamma t}\|\boldsymbol{v}(t)\|),$$

[13] A particularly illuminating treatment of iterative methods for solving linear systems, together with bibliographical references to the iterative methods described here, is found in Varga [1962].

[14] Extensive treatments of ordinary differential equations, which include bibliographical and historical references, are found in two great classics, Coddington & Levinson [1955] and Hartman [2002].

the space $\mathcal{C}([0,T];\mathbb{R}^N)$ is a Banach space (since this norm is clearly equivalent to the usual sup-norm over this space; cf. Theorem 3.2-2). Then *the mapping* $\boldsymbol{F} : \mathcal{C}([0,T];\mathbb{R}^N) \to \mathcal{C}([0,T];\mathbb{R}^N)$ *defined by*

$$\boldsymbol{F}(\boldsymbol{v})(t) = \boldsymbol{u}_0 + \int_0^t \boldsymbol{g}(s,\boldsymbol{v}(s))\,\mathrm{d}s, \quad 0 \le t \le T,$$

is a contraction with respect to this norm.

To see this, observe that, for all $\boldsymbol{v}, \boldsymbol{w} \in \mathcal{C}([0,T];\mathbb{R}^N)$, we can write

$$(\boldsymbol{F}(\boldsymbol{w}) - \boldsymbol{F}(\boldsymbol{v}))(t) = \int_0^t e^{\gamma s} e^{-\gamma s}(\boldsymbol{g}(s,\boldsymbol{w}(s)) - \boldsymbol{g}(s,\boldsymbol{v}(s)))\,\mathrm{d}s, \quad 0 \le t \le T.$$

From this relation, we deduce that

$$\begin{aligned}
\|(\boldsymbol{F}(\boldsymbol{w}) - \boldsymbol{F}(\boldsymbol{v}))(t)\| &\le \left(\int_0^t e^{\gamma s}\,\mathrm{d}s\right) \sup_{0\le s\le T} \left(e^{-\gamma s}\|\boldsymbol{g}(s,\boldsymbol{w}(s)) - \boldsymbol{g}(s,\boldsymbol{v}(s))\|\right) \\
&\le \gamma\left(\int_0^t e^{\gamma s}\,\mathrm{d}s\right) |||\boldsymbol{w} - \boldsymbol{v}||| \\
&\le e^{\gamma t}(1 - e^{-\gamma T})|||\boldsymbol{w} - \boldsymbol{v}|||, \quad 0 \le t \le T,
\end{aligned}$$

since $\gamma \int_0^t e^{\gamma s}\,\mathrm{d}s = e^{\gamma t} - 1 = e^{\gamma t}(1 - e^{-\gamma t}) \le e^{\gamma t}(1 - e^{\gamma T})$. Consequently,

$$\begin{aligned}
|||\boldsymbol{F}(\boldsymbol{w}) - \boldsymbol{F}(\boldsymbol{v})||| &= \sup_{0\le t\le T} \left(e^{-\gamma t}\|(\boldsymbol{F}(\boldsymbol{w}) - \boldsymbol{F}(\boldsymbol{v}))(t)\|\right), \\
&\le (1 - e^{-\gamma T})|||\boldsymbol{w} - \boldsymbol{v}|||.
\end{aligned}$$

Therefore f is a contraction. By the Banach fixed point theorem (Theorem 3.7-1), this contraction has one and only one fixed point $\boldsymbol{u}$, i.e., a function $\boldsymbol{u} \in \mathcal{C}([0,T];\mathbb{R}^N)$ that satisfies

$$\boldsymbol{u}(t) = \boldsymbol{u}_0 + \int_0^t \boldsymbol{g}(s,\boldsymbol{u}(s))\,\mathrm{d}s, \quad 0 \le t \le T.$$

We then conclude from (i) that $u \in \mathcal{C}^1([0,T];\mathbb{R}^N)$ and that $\boldsymbol{u}$ is the unique solution to the initial value problem. □

We shall see later (Theorem 3.11-1) that the existence of a solution to the initial value problem considered in Theorem 3.8-1 can still be established, but then only for "small enough times," under a substantially weaker assumption on the mapping $\boldsymbol{g}$, thanks this time to the *Ascoli–Arzelà theorem* (Theorem 3.10-1).

Remarks (1) Unless $\gamma < \dfrac{1}{T}$ (i.e., for $T > 0$ given, γ should be small enough), the mapping $\boldsymbol{F}$ introduced in the above proof is *not* necessarily a contraction in the space $\mathcal{C}([0,T];\mathbb{R}^N)$ equipped with its "usual" sup-norm, viz., $\boldsymbol{v} \to \sup_{0\le t\le T}\|\boldsymbol{v}(t)\|$, where $\|\cdot\|$ is any norm in $\mathbb{R}^N$. It can be established, however, that the composite mapping $f \circ f \circ \cdots \circ f$ *is* a contraction with respect to this norm, provided the number of factors is large enough. The existence of a solution then follows by resorting to Problem 3.7-2.

(2) A "local" version of the Cauchy–Lipschitz theorem is given in Problem 3.8-1.

(3) When $N = 1$ and $u(0) = 0$, the integral equation that the function $u \in \mathcal{C}([0, T])$ satisfies (cf. part (i) of the above proof) is a special case of a *nonlinear Volterra integral equation of the first kind*, an equation that takes the general form

$$u(t) = \int_0^t h(t, s, u(s))\,\mathrm{d}s, \quad 0 \leq t \leq T,$$

where $h \in \mathcal{C}([0, T] \times \mathbb{R} \times \mathbb{R})$ is a given function.

(4) The existence result of Theorem 3.8-1 does not depend on the norm chosen on $\mathbb{R}^N$ (changing the norm in $\mathbb{R}^N$ may only affect the constant γ). □

The following existence and uniqueness result for a *linear* system of ordinary differential equations is an immediate corollary of Theorem 3.8-1. The notation $\mathbb{M}^N$ stands for the space of all real $N \times N$ matrices.

Theorem 3.8-2 *For some $T > 0$, let there be given a matrix field $\boldsymbol{A} \in \mathcal{C}([0, T]; \mathbb{M}^N)$ and a vector field $\boldsymbol{b} \in \mathcal{C}([0, T]; \mathbb{R}^N)$. Let also $\boldsymbol{u}_0 \in \mathbb{R}^N$ be a given vector. Then the initial value problem*

$$\boldsymbol{u}'(t) = \boldsymbol{A}(t)\boldsymbol{u}(t) + \boldsymbol{b}(t), \quad 0 \leq t \leq T, \quad \text{and} \quad \boldsymbol{u}(0) = \boldsymbol{u}_0,$$

has one and only one solution $\boldsymbol{u} \in \mathcal{C}^1([0, T]; \mathbb{R}^N)$. □

Remark When the matrix field $\boldsymbol{A}$ is constant and $\boldsymbol{u}(0) = \boldsymbol{0}$, an *explicit solution* is provided by means of the *matrix exponential* (Problem 3.6-1). □

A noteworthy application of the Cauchy–Lipschitz theorem is to the vertical motion of a pendulum. A *pendulum*, or more accurately, an "ideal pendulum," is a rigid weightless rod of length ℓ, one end of which rotates freely around a point O, while a mass m is concentrated at the other end. Under the additional assumption that the pendulum moves in a *vertical* plane, its position at a time t is thus entirely determined by the angle $\theta(t)$ between a vertical axis with origin O and directed downward and the pendulum itself (Figure 3.8-1).

The equation of motion is then obtained by projecting at any time t Newton's law on the tangent vector to the oriented circle with center O and radius ℓ. This immediately gives $-mg \sin\theta(t) = m\ell\theta''(t)$, where the constant $g > 0$ denotes the earth gravity. The motion of the pendulum is thus governed by the nonlinear second-order differential equation

$$\theta''(t) = -\frac{g}{\ell}\sin\theta(t) \quad \text{for all time } t,$$

which is called the **pendulum equation**.

We now show that, once supplemented by initial conditions (that simply specify the initial angle θ_0 and initial velocity ω_0), *the pendulum equation has a unique solution for all times.*

Theorem 3.8-3 *Given any constants θ_0 and ω_0, the initial value problem*

$$\theta''(t) = -\frac{g}{\ell}\sin\theta(t), \quad 0 \leq t, \quad \text{and} \quad \theta(0) = \theta_0, \quad \theta'(0) = \omega_0,$$

has one and only one solution $\theta \in \mathcal{C}^\infty([0, \infty[)$.

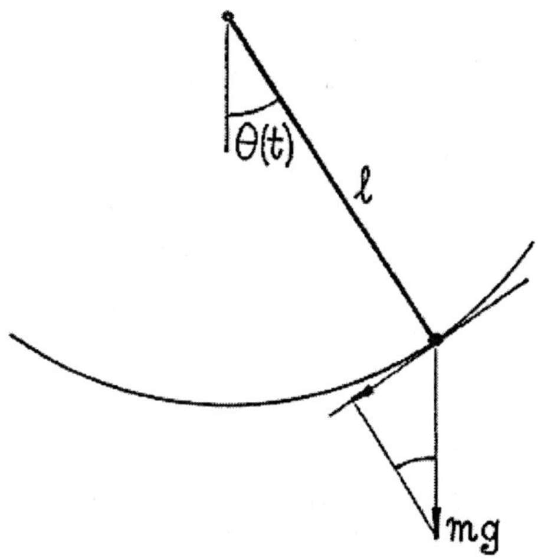

Figure 3.8-1 A pendulum.

Proof Define the vector field $\boldsymbol{u} : [0, \infty[\to \mathbb{R}^2$ by $\boldsymbol{u}(t) = (u_i(t))_{i=1}^2$ with $u_1(t) := \theta(t)$ and $u_2(t) := \theta'(t)$ for all $t \geq 0$. Hence this vector field satisfies

$$\boldsymbol{u}'(t) = \boldsymbol{g}(t, \boldsymbol{u}(t)), \quad 0 \leq t, \quad \text{and} \quad \boldsymbol{u}'(0) = \boldsymbol{u}_0,$$

where the vector-valued function $\boldsymbol{g} : [0, \infty[\times \mathbb{R}^2 \to \mathbb{R}^2$ is defined by

$$\boldsymbol{g}(t, \boldsymbol{v}) := \left(v_2, -\frac{g}{\ell} \sin v_1 \right) \quad \text{for all } (t, \boldsymbol{v}) \in [0, \infty[\times \mathbb{R}^2,$$

and $\boldsymbol{u}_0 := (\theta_0, \omega_0)$. The above system of two first-order ordinary differential equations is now of the form considered in the Cauchy–Lipschitz theorem (Theorem 3.8-1). Besides,

$$\begin{aligned} \|\boldsymbol{g}(t, \boldsymbol{w}) - \boldsymbol{g}(t, \boldsymbol{v})\|_1 &= |w_2 - v_2| + \frac{g}{\ell} |\sin w_1 - \sin v_1| \\ &\leq \max \left\{ 1, \frac{g}{\ell} \right\} \|\boldsymbol{w} - \boldsymbol{v}\|_1 \quad \text{for all } \boldsymbol{w}, \boldsymbol{v} \in \mathbb{R}^2. \end{aligned}$$

Hence this theorem can be applied for arbitrarily large times T, thus establishing the existence and uniqueness of a solution $\boldsymbol{u} \in \mathcal{C}^1([0, \infty[; \mathbb{R}^2)$ (for any $t > 0$, $\boldsymbol{u}(t)$ is defined as the value at t of the solution corresponding to any $T \geq t$; therefore $\boldsymbol{u}(t)$ is indeed uniquely defined by Theorem 3.8-1), hence also the existence and uniqueness of a solution $\theta \in \mathcal{C}^2([0, \infty[)$ to the pendulum equation. That $\theta \in \mathcal{C}^\infty([0, \infty[)$ immediately follows by differentiating the pendulum equation. □

Interesting complements to Theorem 3.8-3 are proposed in Problem 3.8-2.

Problems

3.8-1 Using a proof similar to that of Theorem 3.8-1, establish the following *local version of the Cauchy–Lipschitz theorem*: Let $\|\cdot\|$ denote any norm in $\mathbb{R}^N$. Given $T > 0$, $r > 0$, and $\boldsymbol{u}_0 \in \mathbb{R}^N$, let $\boldsymbol{g} \in \mathcal{C}([0,T] \times \overline{B(\boldsymbol{u}_0; r)}; \mathbb{R}^N)$, where $B(\boldsymbol{u}_0; r) := \{\boldsymbol{v} \in \mathbb{R}^N;\ \|\boldsymbol{v} - \boldsymbol{u}_0\| < r\}$, be a mapping with the property that there exists a constant $\gamma > 0$ such that

$$\|\boldsymbol{g}(t, \boldsymbol{w}) - \boldsymbol{g}(t, \boldsymbol{v})\| \leq \gamma \|\boldsymbol{w} - \boldsymbol{v}\| \quad \text{for all } t \in [0,T] \text{ and all } \boldsymbol{w}, \boldsymbol{v} \in \overline{B(\boldsymbol{u}_0; r)}.$$

Then there exists $0 < \tau \leq T$ such that the initial value problem

$$\boldsymbol{u}'(t) = \boldsymbol{g}(t, \boldsymbol{u}(t)), \quad 0 \leq t \leq \tau, \quad \text{and} \quad \boldsymbol{u}(0) = \boldsymbol{u}_0$$

has one and only one solution $\boldsymbol{u} \in \mathcal{C}^1([0,\tau]; \mathbb{R}^N)$.

Note that it follows from the *mean value theorem* (Theorem 7.2-1) that any mapping $\boldsymbol{g}$ of class $\mathcal{C}^1$ in a neighborhood of a point $(0, \boldsymbol{u}_0) \in \mathbb{R} \times \mathbb{R}^N$ satisfies the above assumptions for some $T > 0$ and $r > 0$.

3.8-2 Let $\theta \in \mathcal{C}^\infty([0, \infty[)$ be the solution to the pendulum equation corresponding to the initial conditions $\theta(0) = 0$ and $\theta'(0) = \omega_0 > 0$ (Theorem 3.8-3).

(1) Show that any $t > 0$ such that $\theta'(\tau) > 0$, $0 \leq \tau \leq t$, satisfies

$$t = \frac{1}{\omega_0} \int_0^{\theta(t)} \frac{\mathrm{d}\psi}{\sqrt{1 - \frac{4g}{\omega_0^2 \ell} \sin^2 \frac{\psi}{2}}}.$$

(2) Deduce from (1) that the pendulum can undergo three possible types of motions: if $\omega_0 > 2\sqrt{\frac{g}{\ell}}$, the pendulum rotates *ad infinitum* with periodically varying velocity (i.e., $\theta'(t) > 0$ for all $t > 0$ and $\lim_{t\to\infty} \theta(t) = \infty$); if $\omega_0 = 2\sqrt{\frac{g}{\ell}}$, then $0 < \theta(t) < \pi$ and $\theta'(t) > 0$ for all $t > 0$, and $\lim_{t\to\infty} \theta(t) = \pi$; if $\omega_0 < 2\sqrt{\frac{g}{\ell}}$, the pendulum oscillates periodically between two angles $-\alpha$ and α, where the angle $0 < \alpha < \pi$ and the *period* $T > 0$ are given by[15]

$$\frac{4g}{\omega_0^2 \ell} \sin^2 \frac{\alpha}{2} = 1 \quad \text{and} \quad \frac{T}{4} = \sqrt{\frac{\ell}{g}} \int_0^{\frac{\pi}{2}} \frac{\mathrm{d}\varphi}{\sqrt{1 - \sin^2 \frac{\alpha}{2} \sin^2 \varphi}}.$$

(3) Show that, in the third case considered in (2), the period can be expanded as a series of the form

$$T = 2\pi \sqrt{\frac{\ell}{g}} \left(1 + \frac{\alpha^2}{16} + \cdots\right).$$

---

[15] Letting $k = \sin \frac{\alpha}{2}$ and $\sin \varphi = t$ in the last integral shows that it is of the form $\displaystyle\int_0^{t_0} \frac{\mathrm{d}t}{\sqrt{(1-t^2)(1-k^2t^2)}}$, which provides an example of an *elliptic integral of the first kind*. Such integrals, together with the *elliptic integrals of the second kind*, which are of the form $\displaystyle\int_0^{t_0} \frac{\sqrt{1-k^2t^2}}{\sqrt{1-t^2}} \mathrm{d}t$, have been the object of extensive studies (neither type of integrals can be computed by means of elementary functions), notably by such luminaries as Leonhard Euler (1707–1783), Adrien-Marie Legendre (1752–1833), and Carl Gustav Jacob Jacobi (1804–1851). The adjective "elliptic" reflects that the length on an arc along an *ellipse* is precisely given by such an integral (of the second kind in this case).

A detailed study of elliptic integrals is found in, e.g.:

D.F. Lawden [1989]: *Elliptic Functions and Applications*, Applied Mathematical Sciences Series, Volume 98, Springer, Heidelberg.

3.9 Application of Banach fixed point theorem: Existence of solutions to nonlinear two-point boundary value problems

As an application of the completeness of the space $\mathcal{C}[a,b]$ equipped with the sup-norm and of the Banach fixed point theorem, we now establish the existence and uniqueness of a *classical solution* to a specific class of *nonlinear boundary value problems posed over a bounded open interval* $I = \,]a,b[\, \subset \mathbb{R}$ (see also Problem 3.9-1 for an extension to more general boundary value problems). A "classical" solution is one that is twice continuously differentiable over I and continuous over $\overline{I}$, as opposed to a "weak" solution, which is in $L^2(I)$ with a derivative in the sense of distributions also in $L^2(I)$ (weak solutions will be introduced and studied in Chapter 6). Without loss of generality, we assume that $I = \,]0,1[$.

Theorem 3.9-1 *Let* $I = \,]0,1[$, *let* $f \in \mathcal{C}(\overline{I} \times \mathbb{R})$ *be a function with the property that there exists a constant* γ *such that*

$$0 \le \gamma < 8 \quad \text{and} \quad |f(x,u) - f(x,v)| \le \gamma |u - v| \quad \text{for all } 0 \le x \le 1 \text{ and all } u, v \in \mathbb{R},$$

and let $\alpha, \beta \in \mathbb{R}$ *be two constants. Then the* **two-point boundary value problem**

$$-u''(x) = f(x, u(x)), \quad 0 < x < 1, \quad \text{and} \quad u(0) = \alpha, \quad u(1) = \beta,$$

has one and only one solution $u \in \mathcal{C}(\overline{I}) \cap \mathcal{C}^2(I)$.

Proof For clarity, the proof is divided in three parts. Note that only the assumption that $f \in \mathcal{C}(\overline{I} \times \mathbb{R})$ is needed in parts (i) and (ii).

(i) *If* $u \in \mathcal{C}(\overline{I}) \cap \mathcal{C}^2(I)$ *is a solution to the boundary value problem, then* $u \in \mathcal{C}^2(\overline{I})$. Since $f \in \mathcal{C}(\overline{I} \times \mathbb{R})$ and $u \in \mathcal{C}(\overline{I})$, the relation

$$u'(x) = u'\Big(\frac{1}{2}\Big) - \int_{1/2}^{x} f(t, u(t))\, \mathrm{d}t, \quad 0 < x < 1,$$

shows that the function $u' \in \mathcal{C}(I)$ can be extended to a continuous function over $\overline{I}$. Besides, Rolle's theorem shows that, for each $0 < x < 1$, there exists $\xi \in \,]0,x[$ such that

$$\frac{u(x) - u(0)}{x} = u'(\xi).$$

This shows that u is differentiable at 0 and that $u'(0) = \lim_{\xi \to 0} u'(\xi)$; a similar argument shows that u is differentiable at 1 and that $u'(1) = \lim_{\xi \to 1} u'(\xi)$. Hence $u \in \mathcal{C}^1(\overline{I})$. The relation $-u''(x) = f(x, u(x))$, $0 < x < 1$, similarly implies that $u \in \mathcal{C}^2(\overline{I})$.

(ii) *If* $u \in \mathcal{C}^2(\overline{I})$ *is a solution to the boundary value problem, then* u *is a solution to the integral equation*

$$u(x) = \alpha(1 - x) + \beta x + \int_0^1 G(x,\xi) f(\xi, u(\xi))\, \mathrm{d}\xi, \quad 0 \le x \le 1,$$

where the function $G \in \mathcal{C}(\overline{I} \times \overline{I})$ *is defined by*

$$G(x,\xi) := \xi(1 - x) \text{ if } 0 \le \xi \le x \le 1 \quad \text{and} \quad G(x,\xi) := x(1 - \xi) \text{ if } 0 \le x < \xi \le 1.$$

Conversely, if $u \in \mathcal{C}(\overline{I})$ is a solution to the integral equation above, then $u \in \mathcal{C}^2(\overline{I})$ and u is a solution to the boundary value problem.

Assume that $u \in \mathcal{C}^2(\overline{I})$ is a solution to the boundary value problem. Then

$$\begin{aligned}\int_0^1 G(x,\xi)f(\xi,u(\xi))\,\mathrm{d}t &= -(1-x)\int_0^x \xi u''(\xi)\,\mathrm{d}\xi - x\int_x^1 (1-\xi)u''(\xi)\,\mathrm{d}\xi \\ &= u(x) - \alpha(1-x) - \beta x, \quad 0 \le x \le 1,\end{aligned}$$

by definition of the functions G and u. Conversely, assume that $u \in \mathcal{C}(\overline{I})$ is a solution to the integral equation. First, it is clear that $u(0) = \alpha$ and $u(1) = \beta$. Second, two successive differentiations show that u is twice continuously differentiable in $[0,1]$ and that

$$\begin{aligned}u'(x) &= -\alpha + \beta - \int_0^x \xi f(\xi,u(\xi)) + \int_x^1 (1-\xi)f(\xi,u(\xi))\,\mathrm{d}\xi, \quad 0 \le x \le 1, \\ -u''(x) &= xf(x,u(x)) + (1-x)f(x,u(x)) = f(x,u(x)), \quad 0 \le x \le 1.\end{aligned}$$

(iii) Let the space $\mathcal{C}(\overline{I})$ be equipped with the sup-norm $\|\cdot\|$, which makes it a *Banach space* (Theorem 3.2-2). Then *the mapping $F : \mathcal{C}(\overline{I}) \to \mathcal{C}(\overline{I})$ defined by*

$$F(v)(x) = \alpha(1-x) + \beta x + \int_0^1 G(x,\xi)f(\xi,u(\xi))\,\mathrm{d}\xi, \quad 0 \le x \le 1,$$

is a contraction. This follows from the inequalities

$$0 \le G(x,\xi) \text{ for all } 0 \le x,\ \xi \le 1 \quad \text{and} \quad \int_0^1 G(x,\xi)\,\mathrm{d}\xi \le \frac{1}{8} \text{ for all } 0 \le x \le 1,$$

and from the ensuing inequality

$$\begin{aligned}|(F(w) - F(v))(x)| &\le \int_0^1 G(x,\xi)|f(\xi,w(\xi)) - f(\xi,v(\xi))|\,\mathrm{d}\xi \\ &\le \left(\sup_{0\le x\le 1}\int_0^1 G(x,\xi)\,\mathrm{d}\xi\right)\sup_{0\le\xi\le 1}|f(\xi,w(\xi)) - f(\xi,v(\xi))| \\ &\le \frac{\gamma}{8}\sup_{0\le\xi\le 1}|w(\xi) - v(\xi)|, \quad 0 \le x \le 1, \quad \text{for all } v,w \in \mathcal{C}(\overline{I}),\end{aligned}$$

combined with the assumption $\gamma < 8$. Hence the contraction F has one and only one fixed point $u \in \mathcal{C}(\overline{I})$ and, by part (ii), $u \in \mathcal{C}^2(\overline{I})$ and u is the unique solution to the boundary value problem. □

The function G appearing in (ii) is the *Green's function* associated with the differential operator $u \in \{v \in \mathcal{C}^2(\overline{I});\, v(0) = v(1) = 0\} \to -u'' \in \mathcal{C}(\overline{I})$. This means that, given any function $f \in \mathcal{C}(\overline{I})$, the unique solution $u \in \mathcal{C}^2(\overline{I})$ to the boundary value problem

$$-u''(x) = f(x), \ 0 \le x \le 1, \quad \text{and} \quad u(0) = u(1) = 0,$$

is given by $u(x) = \int_0^1 G(x,\xi)f(\xi)\,\mathrm{d}\xi$, $0 \le x \le 1$.

Remark The integral equation that the function $u \in \mathcal{C}(\overline{I})$ satisfies (cf. part (ii) of the above proof) is a special case of a *nonlinear Fredholm integral equation of the first kind*, an equation that takes the general form

$$u(x) = \int_0^1 h(x, \xi, u(\xi))\,\mathrm{d}\xi, \quad 0 \le x \le 1,$$

where $h \in \mathcal{C}(\overline{I} \times \mathbb{R} \times \mathbb{R})$ is a given function. □

The key to the above proof consists in replacing the requirement that $u \in \mathcal{C}^2(\overline{I})$ by the considerably milder requirement that $u \in \mathcal{C}(\overline{I})$, thanks to the equivalence between the boundary value problem and the integral equation (cf. part (ii)). This replacement in turn allows us to use the Banach fixed point theorem in the space $\mathcal{C}(\overline{I})$. Unfortunately, this welcome circumstance is restricted to dimension one.

If the function f is differentiable with respect to its second argument, the second assumption takes the equivalent form

$$\left|\frac{\partial f}{\partial u}(x, v)\right| \le \gamma < 8 \quad \text{for all } (x, v) \in \overline{I} \times \mathbb{R}.$$

In fact, the existence and uniqueness of a solution to the two-point boundary value problem of Theorem 3.9-1 can still be established by means of the *theory of monotone operators* (Problem 9.14-3), under the much less stringent assumption that there exists a constant γ such that

$$\frac{\partial f}{\partial u}(x, v) \le \gamma < \pi^2 \quad \text{for all } (x, v) \in \overline{I} \times \mathbb{R}.$$

It is no surprise that π^2 appears here. For, consider the boundary problem

$$-u''(x) = \pi^2 u(x), \quad 0 < x < 1, \quad \text{and} \quad u(0) = 0, \quad u(1) = \beta,$$

which has infinitely many solutions $u : x \in [0,1] \to u(x) = C \sin \pi x$ for any constant C if $\beta = 0$ and no solution if $\beta \neq 0$. The reason is that π^2 is an *eigenvalue* (in fact the smallest) of the operator $u \in \{v \in \mathcal{C}^2(\overline{I});\ v(0) = v(1) = 0\} \to -u'' \in \mathcal{C}^0(\overline{I})$, with these functions u for $C \neq 0$ as the associated *eigenfunctions*.

Remark When $\dfrac{\partial f}{\partial u}(x, v) \le 0$ for all $(x, v) \in \overline{I} \times \mathbb{R}$, an existence theorem can be obtained by means of the *Ascoli–Arzelà theorem*; cf. Problem 3.10-3 in the linear case.[16] □

Problem

3.9-1 Let $I = \,]0, 1[$ and let $f \in \mathcal{C}(\overline{I} \times \mathbb{R} \times \mathbb{R})$ be a function with the property that there exist constants $\gamma > 0$ and $\delta > 0$ such that

$$|f(x, u, p) - f(x, v, q)| \le \gamma |u - v| + \delta |p - q| \quad \text{for all } x \in \overline{I} \text{ and all } u, v, p, q \in \mathbb{R}.$$

[16] In the nonlinear case, the problem needs first to be reduced to one with a bounded right-hand side, thanks to an *a priori* bound on the solution; see:

P.G. Ciarlet; M.H. Schultz; R.S. Varga [1969]: Numerical methods of high-order accuracy for nonlinear boundary value problems V: Monotone operator theory, *Numerische Mathematik* **13**, 51–79.

Using the same method as in the proof of Theorem 3.9-1, show that, if $(\gamma+\delta)$ is small enough, the two-point boundary value problem

$$-u''(x) = f(x, u(x), u'(x)), \quad 0 < x < 1, \quad \text{and} \quad u(0) = \alpha, \quad u(1) = \beta,$$

has one and only one solution $u \in \mathcal{C}^0(\overline{I}) \cap \mathcal{C}^2(I)$.

3.10 Ascoli–Arzelà's theorem

Let K be a compact metric space and let $\mathcal{C}(K)$ denote as usual the space formed by all continuous functions $f : K \to \mathbb{R}$. The space $\mathcal{C}(K)$ is endowed with the sup-norm $\|\cdot\|$ defined by

$$\|f\| = \sup_{x \in K} |f(x)| \quad \text{for all } f \in \mathcal{C}(K),$$

which makes it a Banach space (Theorem 3.2-2).

The next result provides a fundamental *characterization of the compact subsets of the space* $(\mathcal{C}(K), \|\cdot\|)$.

Theorem 3.10-1 (Ascoli–Arzelà theorem[17]) *Let (K, d) be a compact metric space. Then a subset $\mathcal{F} \subset \mathcal{C}(K)$ is relatively compact in $(\mathcal{C}(K), \|\cdot\|)$ if and only if the following two properties are simultaneously satisfied:*

(a) *There exists M such that*

$$\|f\| \le M \quad \textit{for all } f \in \mathcal{F}.$$

(b) *Given any $\varepsilon > 0$, there exists $\delta(\varepsilon) > 0$ such that*

$$|f(x) - f(y)| < \varepsilon \textit{ for all } x, y \in K \textit{ such that } d(x,y) < \delta(\varepsilon) \textit{ and for all } f \in \mathcal{F}.$$

Proof Recall that, in a metric space, $B(a; r)$ denotes the (open) ball of center a and radius $r > 0$ and diam A denotes the diameter of a subset A.

(i) Let a subset $\mathcal{F}$ of $\mathcal{C}(K)$ be such that $\overline{\mathcal{F}}$ is a compact subset of $\mathcal{C}(K)$. Then $\overline{\mathcal{F}}$ is bounded in $\mathcal{C}(K)$ (Theorem 1.13-1) and thus property (a) is satisfied.

Given any $\varepsilon > 0$, the union $\bigcup_{f \in \overline{\mathcal{F}}} \left(f; \frac{\varepsilon}{3}\right)$ constitutes an open covering of the compact set $\overline{\mathcal{F}}$. Hence (Section 1.8) there exists a finite number of functions $f_j \in \overline{\mathcal{F}} \subset \mathcal{C}(K), 1 \le j \le n$, such that

$$\mathcal{F} \subset \overline{\mathcal{F}} \subset \bigcup_{j=1}^{n} B\left(f_j; \frac{\varepsilon}{3}\right).$$

Since the functions f_j are uniformly continuous (Theorem 1.13-2) and their number is finite, there exists $\delta(\varepsilon) > 0$ such that

$$|f_j(x) - f_j(y)| < \frac{\varepsilon}{3} \text{ for all } x, y \in K \text{ such that } d(x,y) < \delta(\varepsilon) \text{ and for all } 1 \le j \le n.$$

[17]C. Arzelà [1883]: Un' osservazione intorno alle serie di funzioni, *Rendiconti delle Sessioni dell' Accademia Reale delle Scienze dell' Istituto di Bologna*, 142–159.

C. Ascoli [1883]: Le curve limiti di una varietà data di curve, *Atti della Accademia Nazionale dei Lincei, Classe di Scienze Fisiche, Matematiche e Naturali* **18**, 521–586.

Given any function $f \in \mathcal{F}$, let j_0 be such that $f \in B\left(f_{j_0}; \frac{\varepsilon}{3}\right)$. Then

$$|f(x) - f(y)| \leq |f(x) - f_{j_0}(x)| + |f_{j_0}(x) - f_{j_0}(y)| + |f_{j_0}(y) - f(y)| < \varepsilon$$

for all $x, y \in K$ such that $d(x,y) < \delta(\varepsilon)$. Hence property (b) is satisfied.

(ii) Conversely, let $\mathcal{F}$ be a subset of $\mathcal{C}(K)$ that satisfies properties (a) and (b). To show that $\overline{\mathcal{F}}$ is compact in $\mathcal{C}(K)$, it suffices to show that $\overline{\mathcal{F}}$ is complete and precompact (Theorem 1.13-3). Since $\overline{\mathcal{F}}$ is complete as a closed subset of a complete metric space, viz., the Banach space $\mathcal{C}(K)$ (Theorem 1.12-2(b)), it remains to show that $\overline{\mathcal{F}}$ is precompact, or equivalently, that $\mathcal{F}$ itself is precompact.

So let $\varepsilon > 0$ be given. First, by property (b), there exists $\delta(\varepsilon) > 0$ such that

$$|f(x) - f(y)| \leq \frac{\varepsilon}{3} \text{ for all } (x,y) \in K \text{ such that } d(x,y) < \delta(\varepsilon) \text{ and for all } f \in \mathcal{F}.$$

Besides, since $K \subset \bigcup_{x \in K} B(x; \delta(\varepsilon))$ and K is compact, there exist a finite number of points $x_\ell \in K$, $1 \leq \ell \leq p$, such that

$$K \subset \bigcup_{\ell=1}^{p} B(x_\ell; \delta(\varepsilon)).$$

Second, there exist a finite number of points $y_m \in \mathbb{R}$, $1 \leq m \leq q$, such that

$$-M = y_1 < y_2 < \cdots < y_q = M \quad \text{and} \quad y_{m+1} - y_m < \frac{\varepsilon}{2}, \; 1 \leq m \leq q-1,$$

where M is the constant appearing in property (b).

Let then $\{\sigma_j;\ 1 \leq j \leq n\}$ denote the finite set formed by all the mappings from the set $\{1, 2, \ldots, p\}$ into the set $\{1, 2, \ldots, q\}$, and let the subsets A_j, $1 \leq j \leq n$, of $\mathcal{F}$ (some possibly empty) be defined by

$$A_j := \left\{ f \in \mathcal{F};\ |f(x_\ell) - y_{\sigma_j(\ell)}| < \frac{\varepsilon}{4},\ 1 \leq \ell \leq p \right\}, \quad 1 \leq j \leq n.$$

Then we claim that

$$\mathcal{F} \subset \bigcup_{j=1}^{n} A_j \quad \text{and} \quad \operatorname{diam} A_j \leq \varepsilon, \; 1 \leq j \leq n,$$

which will show that $\mathcal{F}$ is precompact.

To prove our assertion, let f be any function in $\mathcal{F}$. Since $|f(x_\ell)| \leq M$, $1 \leq \ell \leq p$, by property (a), there exists for each $\ell \in \{1, 2, \ldots, p\}$ an integer $k = k(f, \ell) \in \{1, \ldots, q\}$ such that $|f(x_\ell) - y_{k(f,\ell)}| \leq \frac{\varepsilon}{4}$ (by construction, $y_{m+1} - y_m < \frac{\varepsilon}{2}$, $1 \leq m \leq q-1$). Then, by definition of the mappings σ_j, $1 \leq j \leq n$, there exists an integer $j(f) \in \{1, \ldots, n\}$ such that the mapping $\sigma_{j(f)}$ satisfies $\sigma_{j(f)}(\ell) = k(f, \ell)$, $1 \leq \ell \leq p$, or equivalently such that $f \in A_{\sigma_j(f)}$. Hence $\mathcal{F} \subset \bigcup_{j=1}^{n} A_1$.

Given any integer $j \in \{1, 2, \ldots, n\}$, let $f, g \in A_j$ and $x \in K$. Since $K \subset \bigcup_{\ell=1}^{p} B(x_\ell; \delta(\varepsilon))$, there exists $\ell = \ell(x) \in \{1, 2, \ldots, p\}$ such that $x \in B(x_\ell; \delta(\varepsilon))$. Then, by definition of $\delta(\varepsilon)$ and by definition of the set A_j,

$$\begin{aligned} |f(x) - g(x)| &\leq |f(x) - f(x_\ell)| + |f(x_\ell) - y_{\sigma_{j(f)}\ell}| \\ &\quad + |y_{\sigma_{j(f)}\ell} - g(x_\ell)| + |g(x_\ell) - g(x)| < \varepsilon. \end{aligned}$$

Hence $\operatorname{diam} A_j \leq \varepsilon$, and the proof is complete. □

A subset $\mathcal{F} \subset \mathcal{C}(K)$ that satisfies property (b) in Theorem 3.10-1 is said to be **equicontinuous**. The prefix "equi" reflects that $\delta(\varepsilon)$ can be chosen not only independently of $x, y \in K$ (each function $f \in \mathcal{F}$ is uniformly continuous since K is compact), but also independently of $f \in \mathcal{F}$.

Thanks to these definitions, Ascoli–Arzelà's theorem takes the shorter form: *The closure of a subset $\mathcal{F}$ of $\mathcal{C}(K)$ is compact if and only if $\mathcal{F}$ is bounded and equicontinuous.*

In applications (such as those treated in Problem 3.10-3 and in the next section), the following corollary of Ascoli–Arzelà's theorem is frequently used (its proof is an immediate consequence of Theorems 3.10-1 and 1.13-3):

Theorem 3.10-2 (corollary to Ascoli–Arzelà's theorem) *Let K be a compact metric space and let $(f_n)_{n=0}^{\infty}$ be a sequence of functions $f_n \in \mathcal{C}(K)$ that satisfies the following properties:*

(a) *There exists M such that*

$$\|f_n\| \leq M \quad \textit{for all } n \geq 0.$$

(b) *Given any $\varepsilon > 0$, there exists $\delta(\varepsilon) > 0$ such that*

$$|f_n(x) - f_n(y)| < \varepsilon \textit{ for all } x, y \in K \textit{ such that } d(x,y) < \delta(\varepsilon) \textit{ and for all } n \geq 0.$$

Then there exist a subsequence $(f_{\sigma(n)})_{n=0}^{\infty}$ and a function $f \in \mathcal{C}(K)$ such that

$$\lim_{n \to \infty} \|f_{\sigma(n)} - f\| = 0.$$ □

It should be clear that *both* Theorems 3.10-1 *and* 3.10-2 *hold as well if the space $\mathcal{C}(K)$ is replaced by the space $\mathcal{C}(K; \mathbb{R}^N)$*, the only modifications being that $|\cdot|$ is to be replaced by some norm in $\mathbb{R}^N$ and $\|\cdot\|$ is to be replaced by the corresponding sup-norm (it suffices to argue componentwise and to extract N successive subsequences). In fact, it is easy to establish that *Ascoli–Arzelà's theorem holds as well in the space $\mathcal{C}(K; Y)$, where Y is any Banach space*; cf. Problem 3.10-1.

Ascoli–Arzelà's theorem provides in particular a powerful tool for proving existence theorems for *two-point boundary value problems* (Problem 3.10-3), as well as for *ordinary differential equations* (Section 3.11).

Problems

3.10-1 Show that the following extension of *Ascoli–Arzelà's theorem* (Theorem 3.10-1) holds. Let (K, d) be a compact metric space, let $(Y, \|\cdot\|)$ be a Banach space, and let the space $\mathcal{C}(K; Y)$ be equipped with the sup-norm $|||\cdot|||$ (Section 3.2). Then the closure $\mathcal{F}$ of a subset $\mathcal{F} \subset \mathcal{C}(K; Y)$ is compact if and only if the following two properties are satisfied:

(a) For each $x \in X$, the closure of the set $\{f(x);\ x \in X\}$ is a compact subset of Y.

(b) Given any $\varepsilon > 0$, there exists $\delta(\varepsilon) > 0$ such that $\|f(x) - f(y)\| < \varepsilon$ for all $x, y \in K$ such that $d(x,y) < \delta(\varepsilon)$ and for all $f \in \mathcal{F}$.

3.10-2 Let K be a compact metric space, and let $(f_n)_{n=1}^{\infty}$ be an equicontinuous family of functions $f_n \in \mathcal{C}(K)$ that pointwise converges to a function $f : K \to \mathbb{R}$. Show that $f \in \mathcal{C}(K)$ and that $(f_n)_{n=1}^{\infty}$ converges uniformly to f.

3.10-3 Let there be given two functions $c \in \mathcal{C}[0,1]$ and $f \in \mathcal{C}[0,1]$. The aim of this problem is to establish the existence of a solution $u \in \mathcal{C}^2[0,1]$ to the two-point boundary value problem

$$-u''(x) + c(x)u(x) = f(x),\ \ 0 < x < 1, \quad \text{and} \quad u(0) = u(1) = 0,$$

under the assumption that $c(x) \geq 0$, $0 \leq x \leq 1$ (there is no loss of generality in assuming that $u(0) = u(1) = 0$; if instead $u(0) = \alpha$ and $u(1) = \beta$ with $|\alpha| + |\beta| > 0$, then introduce the new unknown $x \in [0,1] \to u(x) - \alpha(1-x) - \beta x$). The method consists in applying *Ascoli–Arzelà's theorem* to a sequence of functions (denoted $\widehat{u}_n$ below; cf. question (7)) that are constructed from a natural finite-difference approximation to this boundary value problem. Note that Theorem 3.9-1, which uses the *Banach fixed point theorem*, also establishes the existence of a solution to this problem, but under the *different* assumption that $|c(x)| \leq \gamma$ for some constant $\gamma < 8$.

Given any integer $n \geq 1$, let $h := \dfrac{1}{n+1}$. Then the *finite-difference method* for approximating the above boundary value problem consists in finding a vector $\boldsymbol{u}_h \in \mathbb{R}^n$ that satisfies the linear system $\boldsymbol{A}_h \boldsymbol{u}_h = \boldsymbol{f}_h$, where the $n \times n$ matrix $\boldsymbol{A}_h$ and the vector $\boldsymbol{f}_h \in \mathbb{R}^n$ are defined by

$$\boldsymbol{A}_h := \frac{1}{h^2}\begin{pmatrix} 2+c_1h^2 & -1 & & & \bigcirc \\ -1 & 2+c_2h^2 & -1 & & \\ & \ddots & \ddots & \ddots & \\ & & -1 & 2+c_{n-1}h^2 & -1 \\ \bigcirc & & & -1 & 2+c_nh^2 \end{pmatrix} \quad \text{and} \quad \boldsymbol{f}_h := \begin{pmatrix} f_1 \\ f_2 \\ \vdots \\ f_{n-1} \\ f_n \end{pmatrix}.$$

where $c_i := c(ih)$ and $f_i := f(ih)$, $1 \leq i \leq n$. This approximation thus amounts to replacing $-u''(x_i)$ by its *finite-difference approximation* $\dfrac{-u_{i-1} + 2u_i - u_{i+1}}{h^2}$, $1 \leq i \leq n$, with $u_0 = u_{n+1} = 0$.

(1) Show that, for each $h = \dfrac{1}{n+1}$, the matrix $\boldsymbol{A}_h$ has the following property: Whenever a vector $\boldsymbol{v} = (v_i) \in \mathbb{R}^n$ is such that $(\boldsymbol{A}_h\boldsymbol{v})_i \geq 0$, $1 \leq i \leq n$, then $v_i \geq 0$, $1 \leq i \leq n$.

(2) Deduce from (1) that, for each $h = \dfrac{1}{n+1}$, the matrix A_h is *monotone*, i.e., that $\boldsymbol{A}_h$ is invertible and $(\boldsymbol{A}_h^{-1})_{ij} \geq 0$, $1 \leq i, j \leq n$.

(3) Let $\boldsymbol{A}_{oh}$ denote the matrix $\boldsymbol{A}_h$ corresponding to $c(x) = 0$, $0 \leq x \leq 1$. Show that $\|\boldsymbol{A}_{oh}^{-1}\|_\infty \leq \dfrac{1}{8}$ for all $h = \dfrac{1}{n+1}$ (recall that $\|\boldsymbol{B}\|_\infty = \max_{1\leq i\leq n}\sum_{j=1}^n |(\boldsymbol{B})_{ij}|$; cf. Problem 2.9-1).

(4) Using (2), show that $\|\boldsymbol{A}_h^{-1}\|_\infty \leq \|\boldsymbol{A}_{oh}^{-1}\|_\infty$ for all $h = \dfrac{1}{n+1}$.

(5) Show that a function $u \in \mathcal{C}^2[0,1]$ is a solution to the two-point boundary value problem if and only if $u \in \mathcal{C}[0,1]$ and u is a solution of the *integral equation*

$$u(x) = \int_0^1 G(x,\xi)(-c(\xi)u(\xi) + f(\xi))\,\mathrm{d}\xi, \quad 0 \leq x \leq 1,$$

where the function $G \in \mathcal{C}([0,1] \times [0,1])$ is defined by $G(x,\xi) := \xi(1-x)$ if $0 \leq \xi \leq x \leq 1$ and $G(x,\xi) := x(1-\xi)$ if $0 \leq x < \xi \leq 1$.

(6) Show that the vector $\boldsymbol{u}_h$ is a solution of the equation $\boldsymbol{A}_h\boldsymbol{u}_h = \boldsymbol{b}_h$ if and only if its components u_i, $1 \leq i \leq n$, satisfy the *summation equation* (which is the discrete analogue of the integral equation of question (5))

$$u_i = h\sum_{j=1}^n G(ih, j)(-c_j u_j + f_j), \quad 1 \leq i \leq n.$$

(7) For each $h = \dfrac{1}{n+1}$, let the continuous function $\widehat{u}_n : [0,1] \to \mathbb{R}$ be defined by the following conditions: $\widehat{u}_n(0) = \widehat{u}_n(1) = 0$; $\widehat{u}_n(ih) = (\boldsymbol{u}_h)_i$, $1 \le i \le n$; and $\widehat{u}_n$ is affine over each interval $[ih, i+1)h]$, $0 \le i \le n$. Show that there exists a constant M independent of n such that

$$\sup_{0 \le x \le 1} |\widehat{u}_n(x)| \le M \quad \text{for all } n \ge 1,$$

and that the sequence $(\widehat{u}_n)_{n=1}^{\infty}$ is *equicontinuous*, i.e., that, given $\varepsilon > 0$, there exists $\delta(\varepsilon) > 0$ such that

$$|\widehat{u}_n(x) - \widehat{u}_n(y)| \le \varepsilon \text{ for all } x, y \in [0,1] \text{ satisfying } |x-y| \le \delta(\varepsilon) \text{ and for all } n \ge 1.$$

Hint: To establish this last property, use the discrete analogue of the inequality

$$|\varphi'(x)| \le |\varphi(1) - \varphi(0)| + \frac{1}{2} \sup_{0 \le \xi \le 1} |\varphi''(\xi)|, \quad 0 \le x \le 1,$$

which holds for every function $\varphi \in \mathcal{C}^2[0,1]$.

(8) Deduce from *Ascoli–Arzelà's theorem* that there exists a subsequence of the sequence $\widehat{u}_n$ that converges uniformly to a function $u \in \mathcal{C}[0,1]$. Show that $u \in \mathcal{C}^2[0,1]$ and that u is a solution to the two-point boundary value problem.

(9) Show that, in fact, the full sequence $(\widehat{u}_n)_{n=1}^{\infty}$ converges uniformly to this function u.

(10) It follows from (9) that the finite difference method considered in this problem is *convergent*, in the sense that

$$\max_{1 \le i \le n} |u(ih) - u_i| \to 0 \quad \text{as } n+1 = \frac{1}{h} \to \infty.$$

Show that this convergence can be improved if the solution u has a certain smoothness property. More specifically, show that

$$\max_{1 \le i \le n} |u(ih) - u_i| \le \frac{2}{3} h \sup_{0 \le \xi \le 1} \left| \frac{\mathrm{d}^3 u}{\mathrm{d}x^3}(\xi) \right| \quad \text{if } u \in \mathcal{C}^3[0,1],$$

$$\max_{1 \le i \le n} |u(ih) - u_i| \le \frac{h^2}{92} \sup_{0 \le \xi \le 1} \left| \frac{\mathrm{d}^4 u}{\mathrm{d}x^4}(\xi) \right| \quad \text{if } u \in \mathcal{C}^4[0,1].$$

(11) Show that the *order of convergence* of this finite-difference method, viz., $O(h^2)$ if $u \in \mathcal{C}^4[0,1]$, cannot be improved in general, i.e., even if the solution u exhibits additional smoothness.

3.10-4 In what follows, the space $\mathcal{C}[0,1]$ is equipped with the sup-norm, the space $L^2(0,1)$ is equipped with the norm $\|\cdot\|_{L^2(0,1)}$, and G is a given function in the space $\mathcal{C}([0,1] \times [0,1])$.

(1) Given any function $v \in \mathcal{C}[0,1]$, let

$$Av(x) := \int_0^1 G(x,\xi) v(\xi) \,\mathrm{d}\xi, \quad 0 \le x \le 1.$$

Show that this relation defines a function $Av \in \mathcal{C}[0,1]$ and that the linear operator $A : \mathcal{C}[0,1] \to \mathcal{C}[0,1]$ defined in this fashion is compact (Section 2.10).

(2) Given any function $v \in L^2(0,1)$, let $Av(x)$, $0 \le x \le 1$, be defined as in (1). Show that the function $Av : [0,1] \to \mathbb{R}$ is continuous, and that the linear operators (still denoted) $A : L^2(0,1) \to \mathcal{C}[0,1]$ and $A : L^2(0,1) \to L^2(0,1)$ defined in this fashion are both compact.

Hint: Apply *Ascoli–Arzelà's theorem.*

(3) Show that, if $G(x,\xi) = G(\xi,x)$ for all $(x,\xi) \in [0,1] \times [0,1]$, the operator A satisfies

$$\int_0^1 (Av(x)) w(x) \,\mathrm{d}x = \int_0^1 v(x) Aw(x) \,\mathrm{d}x \quad \text{for all } v, w \in \mathcal{C}[0,1].$$

Remarks (1) The analysis of the two-point boundary value problem $-u''(x) = f(x)$, $0 \le x \le 1$, and $u(0) = u(1)$, provides an example of such an operator A, since its unique solution u is given by $u(x) = \int_0^1 G(x,\xi)f(\xi)\,\mathrm{d}\xi$, where $G(x,\xi) := \xi(1-x)$ if $0 \le \xi \le x \le 1$ and $G(x,\xi) := x(1-\xi)$ if $0 \le x < \xi \le 1$ (as is immediately verified).

(2) The case where the function G is only in the space $L^2(]0,1[\times]0,1[)$ will be the object of Problem 4.9-5. □

3.11 Application of Ascoli–Arzelà's theorem: Existence of solutions to nonlinear ordinary differential equations; Cauchy–Peano theorem; Euler's method

Using the *Banach fixed point theorem*, we have established (Theorem 3.8-1) the existence and uniqueness of a solution $\boldsymbol{u} \in \mathcal{C}^1([0,T];\mathbb{R}^N)$ to the *initial value problem* for a *system of ordinary differential equations* of the form

$$\boldsymbol{u}'(t) = \boldsymbol{g}(t,\boldsymbol{u}(t)), \quad 0 \le t \le T, \quad \text{and} \quad \boldsymbol{u}(0) = \boldsymbol{u}_0,$$

where the mapping $\boldsymbol{g} : (t,\boldsymbol{v}) \in [0,T]\times\mathbb{R} \to \boldsymbol{g}(t,\boldsymbol{v}) \in \mathbb{R}^N$ appearing in the right-hand side is continuous on $[0,T]\times\mathbb{R}^N$ and satisfies a Lipschitz condition with respect to its second argument $\boldsymbol{v}$, uniformly with respect to its first argument $t \in [0,T]$.

Using *Ascoli–Arzelà's theorem*, we now show that there still exists a solution to such a system under the much weaker assumption that the mapping $\boldsymbol{g}$ is continuous on a set of the form $\{(t,\boldsymbol{v}) \in \mathbb{R}\times\mathbb{R}^N;\ 0 \le t \le T,\ \|\boldsymbol{v}-\boldsymbol{u}_0\| \le r\}$ for some $T > 0$ and some $r > 0$. Of course, there is a "price to pay" for this increased generality.

First, this result will provide only *local* existence, in the sense that the solution may exist only for $t \in [0,\tau]$, with $\tau > 0$ but arbitrarily small, even if the right-hand side is smooth and is defined for all $(t,\boldsymbol{v}) \in \mathbb{R}\times\mathbb{R}^N$. Consider for instance the initial value problem

$$u'(t) = (u(t))^2, \quad 0 \le t, \quad \text{and} \quad u(0) = u_0.$$

Then the (unique) solution, which is given by $u(t) = \dfrac{u_0}{1-u_0 t}$, is defined for all $t \ge 0$ if $u_0 \le 0$, but only for $t \in [0,\tau]$ where $\tau > 0$ is any number that satisfies $\tau < \dfrac{1}{u_0}$ if $u_0 > 0$. The solution is thus only defined on an interval $[0,\tau]$ that becomes arbitrarily small as $u_0 \to +\infty$ (because the solution "blows up" when t approaches $\dfrac{1}{u_0}$ from the left).

Second, *nonuniqueness* may occur. Consider for instance the initial value problem

$$u'(t) = 3(u(t))^{3/2}, \quad 0 \le t, \quad \text{and} \quad u(0) = u_0,$$

where the function $v \in \mathbb{R} \to v^{3/2} \in \mathbb{R}$ appearing in the right-hand side is continuous but does not satisfy a Lipschitz condition at $v = 0$. Hence the existence and uniqueness of such a solution cannot be deduced from the Cauchy–Lipschitz theorem, while, by contrast, Theorem 3.11-1 below will always provide local existence. More specifically, this problem has a unique solution given by $u(t) = (t + u_0^{1/3})^{1/3}$ for all $t \geq 0$ if $u_0 \neq 0$ while, if $u_0 = 0$, it has infinitely many solutions, given by

$$\begin{aligned} u(t) &= 0 \quad \text{for all } t \geq 0, \\ u(t) &= t^3 \quad \text{for all } t \geq 0, \\ u(t) &= 0 \quad \text{for all } 0 \leq t \leq t_0 \quad \text{and} \quad u(t) = (t - t_0)^3 \ \text{ for } t_0 \leq t, \end{aligned}$$

where $t_0 > 0$ is arbitrarily chosen (note that, when $u_0 \neq 0$, the local existence and uniqueness of a solution could also be deduced from the "local" version of the Cauchy–Lipschitz theorem, proposed in Problem 3.8-1).

Theorem 3.11-1 (Cauchy–Peano theorem) *Let $\|\cdot\|$ denote any norm in $\mathbb{R}^N$. Given $T > 0$, $r > 0$, and $\boldsymbol{u}_0 \in \mathbb{R}^N$, let there be given a mapping $\boldsymbol{g} \in \mathcal{C}([0, T] \times \overline{B(\boldsymbol{u}_0; r)}; \mathbb{R}^N)$, where $B(\boldsymbol{u}_0; r) := \{\boldsymbol{v} \in \mathbb{R}^N; \|\boldsymbol{v} - \boldsymbol{u}_0\| \leq r\}$. Then there exists $0 < \tau \leq T$ such that the initial value problem*

$$\boldsymbol{u}'(t) = \boldsymbol{g}(t, \boldsymbol{u}(t)), \quad 0 \leq t \leq \tau, \quad \text{and} \quad \boldsymbol{u}(0) = \boldsymbol{u}_0,$$

has at least one solution $\boldsymbol{u} \in \mathcal{C}^1([0, \tau]; \mathbb{R}^N)$.

Proof (i) Let

$$M := \sup \left\{ \|\boldsymbol{g}(t, \boldsymbol{v})\|; \ (t, \boldsymbol{v}) \in [0, T] \times \overline{B(\boldsymbol{u}_0; r)} \right\} \text{ and } \tau := \min \left\{ \frac{r}{M}, T \right\}.$$

As already observed in the proof of Theorem 3.8-1, it is enough to establish the existence and uniqueness of a solution $\boldsymbol{u} \in \mathcal{C}^0([0, \tau]; \mathbb{R}^N)$ to the *integral equation*

$$\boldsymbol{u}(t) = \boldsymbol{u}_0 + \int_0^t \boldsymbol{g}(s, \boldsymbol{u}(s)) \mathrm{d}s, \quad 0 \leq t \leq \tau.$$

(ii) Given any integer $n \geq 1$, let $h := \dfrac{\tau}{n}$ and $t_i := ih$, $0 \leq i \leq m$, so that $0 = t_0 \leq t_i \leq \tau \leq T$, $0 \leq i \leq n$. Then the simplest *finite-difference method* for approximating this initial value problem consists in recursively defining vectors $\boldsymbol{u}_i \in \mathbb{R}^N$, $1 \leq i \leq n$, by

$$\frac{\boldsymbol{u}_{i+1} - \boldsymbol{u}_i}{h} = \boldsymbol{g}(t_i, \boldsymbol{u}_i), \quad 0 \leq i \leq n - 1.$$

Of course, we must first check that $\boldsymbol{u}_i \in \overline{B(\boldsymbol{u}_0; r)}$, $1 \leq i \leq n - 1$ (otherwise, $(t_i, \boldsymbol{u}_i)$ would fall outside the domain of definition of the mapping $\boldsymbol{g}$, and thus $\boldsymbol{u}_{i+1}$ could not be defined). To this end, we note that

$$\|\boldsymbol{u}_1 - \boldsymbol{u}_0\| = h\|\boldsymbol{g}(t_0, u_0)\| \leq hM \leq \tau M \leq r.$$

So, assume that $\|\boldsymbol{u}_{i-1} - \boldsymbol{u}_0\| \leq (i - 1)hM \leq \tau M \leq r$ for an integer $i \in \{2, \ldots, m - 1\}$. Then

$$\|\boldsymbol{u}_i - \boldsymbol{u}_0\| \leq \|\boldsymbol{u}_i - \boldsymbol{u}_{i-1}\| + \|\boldsymbol{u}_{i-1} - \boldsymbol{u}_0\| \leq hM + (i - 1)hM = ihM \leq \tau M \leq r.$$

Hence the successive iterates $\boldsymbol{u}_1, \boldsymbol{u}_2, \ldots, \boldsymbol{u}_n$ are well defined.

(iii) For each integer $n \geq 1$, define the vector-valued function $\widehat{\boldsymbol{u}}_n : [0,\tau] \to \mathbb{R}^N$ by

$$\widehat{\boldsymbol{u}}_n(t) := \boldsymbol{u}_i + \frac{(t-t_i)}{h}(\boldsymbol{u}_{i+1} - \boldsymbol{u}), \quad t_i \leq t \leq t_{i+1}, \quad 0 \leq i \leq n-1.$$

In other words, $\widehat{\boldsymbol{u}}_n(t_i) = \boldsymbol{u}_i$, $0 \leq i \leq n$, and $\widehat{\boldsymbol{u}}_n$ is affine over $[t_i, t_{i+1}]$, $0 \leq i \leq n-1$. Hence $\widehat{\boldsymbol{u}}_n \in \mathcal{C}^0([0,\tau];\mathbb{R}^N)$ for each $n \geq 1$.

The sequence $(\widehat{\boldsymbol{u}}_n)_{n=1}^{\infty}$ is *bounded* in the space $\mathcal{C}^0([0,\tau];\mathbb{R}^N)$, equipped as usual with the sup-norm $|||\cdot|||$, since

$$|||\widehat{\boldsymbol{u}}_n||| = \sup_{0\leq t\leq \tau} \|\widehat{\boldsymbol{u}}_n(t)\| = \max_{0\leq i\leq n} \|\boldsymbol{u}_i\|$$

and

$$\|\boldsymbol{u}_i\| \leq \|\boldsymbol{u}_0\| + \|\boldsymbol{u}_i - \boldsymbol{u}_0\| \leq \|\boldsymbol{u}_0\| + r.$$

The sequence $(\widehat{\boldsymbol{u}}_n)_{n=1}^{\infty}$ is also *equicontinuous*, since, for each $i \in \{0,1,\ldots,n-1\}$,

$$\|\widehat{\boldsymbol{u}}_n(t) - \widehat{\boldsymbol{u}}_n(t_i)\| = \|\widehat{\boldsymbol{u}}_n(t) - \boldsymbol{u}_i\| \leq (t-t_i)\left\|\frac{\boldsymbol{u}_{i+1} - \boldsymbol{u}_i}{h}\right\| \leq (t-t_i)M, \quad t_i \leq t \leq t_{i+1},$$

so that

$$\|\widehat{\boldsymbol{u}}_n(t) - \widehat{\boldsymbol{u}}_n(\widetilde{t})\| \leq |t-\widetilde{t}|M \quad \text{for all } t, \widetilde{t} \in [0,\tau].$$

Ascoli–Arzelà's theorem then shows that *there exist a subsequence* $(\widehat{\boldsymbol{u}}_{\sigma(n)})_{n=1}^{\infty}$ *of the sequence* $(\widehat{\boldsymbol{u}}_n)_{n=1}^{\infty}$ and a mapping $\boldsymbol{u} \in \mathcal{C}^0([0,\tau];\mathbb{R}^n)$ such that

$$\sup_{0\leq t\leq \tau} \|\widehat{\boldsymbol{u}}_{\sigma(t)}(t) - \boldsymbol{u}(t)\| \to 0 \text{ as } n \to \infty.$$

(iv) It remains to show that $\boldsymbol{u}$ is a solution to the integral equation of (i). To this end, we first note that

$$\boldsymbol{u}_{i+1} = \boldsymbol{u}_0 + h(\boldsymbol{g}(t_0, u_0) + \boldsymbol{g}(t_1, u_1) + \cdots + \boldsymbol{g}(t_i, u_i)) = \boldsymbol{u}_0 + \int_0^{t_i} \boldsymbol{g}_n(s)\,\mathrm{d}s, \quad 0 \leq i \leq n-1,$$

where the piecewise constant mapping $\boldsymbol{g}_n : [0,\tau] \to \mathbb{R}^N$ is defined by

$$\boldsymbol{g}_n(s) := \boldsymbol{g}(t_i, \boldsymbol{u}_i), \quad t_i \leq s \leq t_{i+1}, \quad 0 \leq i \leq n.$$

Observing that integrating a constant mapping over each interval $[t_i, t_{i+1}]$ produces an affine mapping, we infer that, for each integer $n \geq 1$, the mapping $\widehat{\boldsymbol{u}}_n \in \mathcal{C}^0([0,T];\mathbb{R}^n)$ is also given by

$$\widehat{\boldsymbol{u}}(t) = \boldsymbol{u} + \int_0^t \boldsymbol{g}_n(s)\,\mathrm{d}s, \quad 0 \leq t \leq \tau.$$

Combining straightforward "(ε,δ)-arguments" with the uniform continuity of the limit $\boldsymbol{u}$ found in (iv) and of the mapping $s \in [0,\tau] \to \boldsymbol{g}(s, \boldsymbol{u}(s))$, we then easily deduce from the convergence $|||\widehat{\boldsymbol{u}}_{\sigma(n)} - \boldsymbol{u}||| \to 0$ as $n \to \infty$ that

$$\sup_{0\leq s\leq \tau} \|\boldsymbol{g}_{\sigma(n)}(s) - \boldsymbol{f}(s, \boldsymbol{u}(s))\| \to 0 \quad \text{as } n \to \infty,$$

which in turn implies that

$$\sup_{0\le t\le\tau}\left\|\int_0^t \boldsymbol{g}_{\sigma(n)}(s)\,\mathrm{d}s - \int_0^t \boldsymbol{f}(s,\boldsymbol{u}(s))\,\mathrm{d}s\right\| \to 0 \quad \text{as } n\to\infty.$$

We thus conclude that

$$\boldsymbol{u}(t) = \boldsymbol{u}_0 + \int_0^t \boldsymbol{g}(s,\boldsymbol{u}(s))\,\mathrm{d}s, \quad 0\le t\le\tau,$$

which completes the proof. □

The finite-difference method described in (ii) constitutes **Euler's method** for approximating initial value problems for ordinary differential equations.

If the *uniqueness* of the solution to the initial value problem can be established by some means, the above proof shows that the *whole* sequence $(\boldsymbol{u}_n)_{n=1}^{\infty}$ converges to $\boldsymbol{u}$ in the space $(\mathcal{C}^0([0,\tau];\mathbb{R}^N), |||\cdot|||)$, thus providing as a bonus the *convergence of Euler's method*.

Remark While it can be shown without much further ado that the Cauchy–Lipschitz theorem (Theorem 3.8-1) holds *verbatim* with $\mathbb{R}^N$ replaced by an arbitrary *Banach space* X (once the integral of a continuous mapping $[0,T]\to X$ has been defined as in Section 3.3), the Cauchy–Peano theorem does *not* necessarily hold in this more general situation.[18] □

[18] J. Dieudonné [1950]: Deux exemples singuliers d'équations différentielles, *Acta Scientiarum Mathematicarum B (Szeged)* **12**, 38–40.

CHAPTER 4

INNER-PRODUCT SPACES AND HILBERT SPACES

Introduction

Among infinite-dimensional normed vector spaces, *inner-product spaces*, and especially *Hilbert spaces*, i.e., complete inner-product spaces, such as their archetypes, the spaces ℓ^2 and $L^2(\Omega)$ (Section 4.2), are by far "the best."

A basic reason for their attractiveness is that their norm shares many properties of the Euclidean norm in $\mathbb{R}^n$, because it is defined by means of an *inner product* (the natural generalization of the well-known scalar product in $\mathbb{R}^n$). As a result, most of the "geometry" of the n-dimensional Euclidean space carries over to such spaces, such as the *Cauchy–Schwarz–Bunyakovskiĭ inequality* and the *parallelogram law* (Section 4.1), the fundamental *projection theorem* (Theorem 4.3-1), the *orthogonality of vectors* (Section 4.5), or the possibility of representing any element by means of a *Fourier series* over an *orthonormal basis* if the space is complete (Theorem 4.9-1); this possibility is illustrated in the text by way of fundamental examples, such as the *classical* (i.e., trigonometric) *Fourier series*, or the *Legendre*, *Laguerre*, or *Hermite polynomials* (Section 4.8).

We also show in passing that the projection theorem provides a transparent proof of the existence of a *least-squares solution to a linear system* (Theorem 4.4-1).

Another basic reason for the attractiveness of a Hilbert space is that any such space can be *identified with its dual space*, by means of a specific linear isometry: this is the content of the fundamental *F. Riesz representation theorem in a Hilbert space* (Theorem 4.6-1). This theorem has many far-reaching applications, such as a simple proof of the *Hahn–Banach theorem in a Hilbert space*, or a straightforward definition of the *adjoint* of a continuous linear operator (Section 4.7); note that, by contrast, the analysis of the analogous notion of a *dual* operator in an arbitrary normed vector space requires the axiom of choice (via the Hahn–Banach extension theorem; cf. Chapter 5).

This chapter concludes with a detailed treatment of the *spectral theory of compact self-adjoint operators* (Sections 4.10 and 4.11); in particular, the *spectral theorem* (Theorem 4.11-1) will be the basis for analyzing eigenvalue problems for second-order elliptic boundary value problems in Chapter 6. Note that this will be our only incursion into spectral theory, as its treatment in arbitrary normed vector spaces is beyond the scope of this book.

4.1 Inner-product spaces and Hilbert spaces; first properties; Cauchy–Schwarz–Bunyakovskiĭ inequality; parallelogram law

Let first X be a *real* vector space. An **inner product** on X is a function $(\cdot,\cdot) : X \times X \to \mathbb{R}$ with the following properties: for all $x, y, z \in X$ and all $\alpha, \beta \in \mathbb{R}$,

$$\begin{aligned}(\alpha x + \beta y, z) &= \alpha(x,z) + \beta(y,z),\\ (x, \alpha y + \beta z) &= \alpha(x,y) + \beta(x,z),\\ (x,y) &= (y,x),\\ (x,x) &\geq 0 \text{ and } (x,x) = 0 \text{ implies } y = 0.\end{aligned}$$

In other words, an inner product on a *real* vector space is a *bilinear form* (i.e., a function that is linear with respect to each one of its two arguments; cf. Section 2.11) that is *symmetric* (third property; note that the second property evidently follows from the first and third ones) and *positive-definite* (fourth property).

A **real inner-product space** is a pair $(X, (\cdot,\cdot))$, where X is a real vector space and $(\cdot,\cdot)$ is an inner product on X.

Let next X be a *complex* vector space. An **inner product** on X is a complex-valued function $(\cdot,\cdot) : X \times X \to \mathbb{C}$ with the following properties: for all $x, y, z \in X$ and all $\alpha, \beta \in \mathbb{C}$ (the notation $\overline{\alpha}$ designates the complex conjugate of $\alpha \in \mathbb{C}$),

$$\begin{aligned}(\alpha x + \beta y, z) &= \alpha(x,z) + \beta(y,z),\\ (x, \alpha y + \beta z) &= \overline{\alpha}(x,y) + \overline{\beta}(x,z),\\ (x,y) &= \overline{(y,x)},\\ (x,x) &\geq 0 \text{ and } (x,x) = 0 \text{ implies } y = 0.\end{aligned}$$

In other words, an inner product on a *complex* vector space is a *Hermitian form* (first, second, and third properties; note that the second property again evidently follows from the first and third ones) that is *positive-definite* (fourth property). An inner product on a complex vector space is thus *linear* with respect to its first argument (first property) and *semilinear* with respect to its second argument (second property); for this reason, the inner product in a complex vector space is sometimes said to be *sesquilinear* (the prefix "sesqui" means "one and a half").

A **complex inner-product space** is a pair $(X, (\cdot,\cdot))$, where X is a complex vector space and $(\cdot,\cdot)$ is an inner product on X.

Let $\mathbb{K}$ denote either the field $\mathbb{R}$ or the field $\mathbb{C}$. An **inner-product space** is either a real inner-product space ($\mathbb{K} = \mathbb{R}$) or a complex inner-product space ($\mathbb{K} = \mathbb{C}$).

The fundamental inequality established in the next theorem (cf. (a)) pervades the theory of inner-product spaces. As its first consequences, it implies that *an inner-product space is also a normed vector space* (cf. (b)), and that *the inner product is a continuous function of its two arguments* (cf. (c)).

Theorem 4.1-1 *Let $(X, (\cdot,\cdot))$ be a real or complex inner-product space* ($\mathbb{K} = \mathbb{R}$ or $\mathbb{K} = \mathbb{C}$).

(a) *The* **Cauchy–Schwarz–Bunyakovskiĭ inequality**[1] *holds:*

$$|(x,y)| \leq \sqrt{(x,x)}\sqrt{(y,y)} \quad \textit{for all } x,y \in X.$$

(b) *The function*

$$\|\cdot\| : x \in X \to \|x\| := \sqrt{(x,x)} \in \mathbb{R}$$

is a norm on X. *Besides,*

$$\|x\| = \sup_{y \neq 0} \frac{|(x,y)|}{\|y\|} \quad \textit{for all } x \in X.$$

(c) *The mapping*

$$(\cdot,\cdot) : X \times X \to \mathbb{K}$$

is continuous, the topology on X *being that induced by the norm* $\|\cdot\|$ *of* (b) *and the topology of* $X \times X$ *being the corresponding product topology.*

Proof Assume first that $\mathbb{K} = \mathbb{R}$. Given two vectors $x, y \in X$ with $y \neq 0$ (so that $(y,y) > 0$), the real quadratic polynomial

$$p : t \in \mathbb{R} \to p(t) := (x+ty, x+ty) = (x,x) + 2t(x,y) + t^2(y,y)$$

satisfies $p(t) \geq 0$ for all $t \in \mathbb{R}$. In particular then,

$$p\left(-\frac{(x,y)}{(y,y)}\right) = (x,x) - \frac{|(x,y)|^2}{(y,y)} \geq 0,$$

and thus the Cauchy–Schwarz–Bunyakovskiĭ inequality holds when $\mathbb{K} = \mathbb{R}$ (if $y = 0$, this inequality also holds, since it then reduces to $0 = 0$).

Assume next that $\mathbb{K} = \mathbb{C}$. Given again two vectors $x, y \in X$ with $y \neq 0$, consider this time the complex-valued function

$$p : z \in \mathbb{C} \to p(z) := (x+zy, x+zy) = (x,x) + z(y,x) + \overline{z}(y,x) + z\overline{z}(y,y),$$

which thus satisfies $p(z) \geq 0$ for all $z \in \mathbb{C}$. In particular then,

$$p\left(-\frac{(x,y)}{(y,y)}\right) = (x,x) - \frac{(x,y)\overline{(x,y)}}{(y,y)} \geq 0,$$

[1]This inequality was first established for vectors in a finite-dimensional space in:

A.L. Cauchy [1821]: *Cours d'Analyse de l'École Royale Polytechnique*, de Bure, Paris.

See Corollary to Theorem XVI, in Note II of:

R.E. Bradley; C.E. Sandifer [2009]: *Cauchy's Cours d'Analyse—An Annotated Translation*, Springer, Heidelberg.

This inequality was then extended to integrals by:

V. Bunyakovskiĭ: [1859]: Sur quelques inégalités concernant les intégrales aux différences finies, *Mémoires de l'Académie des Sciences de Saint-Peterbourg, 7ème Série, Tome* 1, No. 9, 1–18.

The extension to general inner-product spaces, as stated here, is due to:

H.A. Schwarz [1885]: Über ein Flächen kleinsten Flächeninhalts betreffendes Problem der Variationsrechnung, *Acta Societatis Scientiarum Fennicae* **15**, 315–362.

and thus the Cauchy–Schwarz–Bunyakovskiĭ inequality also holds when $\mathbb{K} = \mathbb{C}$. This proves (a).

The triangle inequality for the function $\|\cdot\| : X \to \mathbb{R}$ defined in (b) follows from the identities

$$\|x+y\|^2 = \|x\|^2 + 2(x,y) + \|y\|^2 \quad \text{if } \mathbb{K} = \mathbb{R},$$
$$\|x+y\|^2 = \|x\|^2 + 2\,\mathrm{Re}(x,y) + \|y\|^2 \quad \text{if } \mathbb{K} = \mathbb{C},$$

which, combined with the Cauchy–Schwarz–Bunyakovskiĭ inequality, imply that

$$\|x+y\|^2 \leq \|x\|^2 + 2\,\|x\|\,\|y\| + \|y\|^2 = (\|x\| + \|y\|)^2$$

in each case. Hence the function $\|\cdot\|$ is a norm (the other properties of a norm clearly hold).

The Cauchy–Schwarz–Bunyakovskiĭ inequality also shows that, for any $x \in X$,

$$\|x\| = \frac{(x,x)}{\|x\|} \leq \sup_{y \neq 0} \frac{|(x,y)|}{\|y\|} \leq \|x\|,$$

and thus $\|x\| = \sup_{y\neq 0} \dfrac{|(x,y)|}{\|y\|}$.

The continuity of the inner product follows from the identity

$$(x,y) - (x_0,y_0) = (x - x_0, y_0) + (x_0, y - y_0) + (x - x_0, y - y_0),$$

which holds for all $x, y, x_0, y_0 \in X$, combined with another application of the Cauchy–Schwarz–Bunyakovskiĭ inequality. □

Remarks (1) The Cauchy–Schwarz–Bunyakovskiĭ inequality still holds if the function $(\cdot,\cdot) : X \times X \to \mathbb{R}$ satisfies all the properties of an inner product, save that $(x,x) = 0$ implies $y = 0$ (i.e., the fourth property reduces to $(x,x) \geq 0$ for all $y \in X$). To see this, observe that the above proof covers the case where $(y,y) > 0$, and thus also the case where $(x,x) > 0$. In the remaining case where $(x,x) = (y,y) = 0$, we are left with

$$p(-(x,y)) = -2(x,y)^2 \geq 0 \quad \text{if } \mathbb{K} = \mathbb{R}, \quad \text{or} \quad p(-(x,y)) = -2|(x,y)|^2 \geq 0 \quad \text{if } \mathbb{K} = \mathbb{C},$$

so that $(x,y) = 0$. Hence the Cauchy–Schwarz–Bunyakovskiĭ inequality also holds in this case (it reduces to $0 = 0$).

(2) The proof of Theorem 4.1-1 shows that *equality holds in the Cauchy–Schwarz–Bunyakovskiĭ inequality if and only if the two vectors x and y are linearly independent.* □

It will be always implicitly understood in the sequel that, *when viewed as a normed vector space, an inner-product space $(X, (\cdot,\cdot))$ is equipped with the norm defined in* Theorem 4.1-1(b), which is called the **norm induced by the inner product** $(\cdot,\cdot)$.

Two illustrations of this implicit understanding are provided by the next two theorems, which give *two basic properties* of this norm, which are *specific to inner-product spaces*.

Theorem 4.1-2 (parallelogram law) *Let $(X, (\cdot,\cdot))$ be a real or complex inner-product space. Then*

$$\|x+y\|^2 + \|x-y\|^2 = 2\|x\|^2 + 2\|y\|^2 \quad \textit{for all } x, y \in X.$$

The above parallelogram law implies that an inner-product space is uniformly convex (Section 2.17).

Proof The parallelogram law immediately follows from the identities

$$\|x \pm y\|^2 = \|x\|^2 \pm 2(x, y) + \|y\|^2 \quad \text{if } \mathbb{K} = \mathbb{R},$$
$$\|x \pm y\|^2 = \|x\|^2 \pm 2\,\mathrm{Re}(x, y) + \|y\|^2 \quad \text{if } \mathbb{K} = \mathbb{C}.$$

The parallelogram law may be rewritten as

$$\left\|\frac{x+y}{2}\right\|^2 = \frac{1}{2}\|x\|^2 + \frac{1}{2}\|y\|^2 - \frac{1}{4}\|x-y\|^2 \quad \text{for all } x, y \in X.$$

Consequently,

$$\|x\| = \|y\| = 1 \quad \text{and} \quad \|x - y\| \geq \varepsilon > 0 \quad \text{implies} \quad \left\|\frac{x+y}{2}\right\| \leq 1 - \delta(\varepsilon),$$

with $\delta(\varepsilon) := 1 - \sqrt{1 - \frac{\varepsilon^2}{4}} > 0$. Hence a real or complex inner-product space is uniformly convex. □

If two vectors x and y in an inner-product space satisfy $(x, y) = 0$, in which case the two vectors x and y are said to be *orthogonal* (Section 4.5), the identities used at the beginning of the above proof reduce to

$$\|x + y\|^2 = \|x\|^2 + \|y\|^2 \quad \text{if } (x, y) = 0.$$

To reflect that it likewise extends to arbitrary inner-product spaces a well-known property of a right-angled triangle, this identity is often called **Pythagoras theorem.**[2]

The identity established in Theorem 4.1-2 is called the "*parallelogram law*" to reflect that it extends to arbitrary inner-product spaces a well-known property of parallelograms in $\mathbb{R}^2$ (Figure 4.1-1). Note that the parallelogram law admits an interesting converse, showing that *it in fact characterizes inner-product spaces* (Problem 4.1-3).

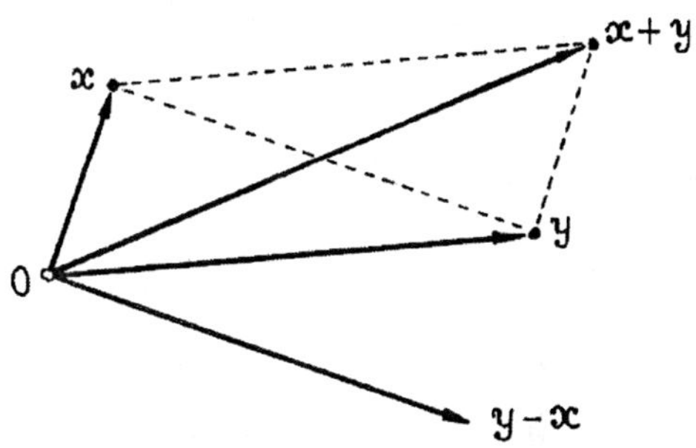

Figure 4.1-1 (parallelogram law) The sum of the squares of the lengths of the two diagonals of a parallelogram in $\mathbb{R}^2$ is equal to the sum of the squares of the lengths of its four edges.

[2]So named after the famed Greek philosopher Pythagoras of Samos, who gave the first proof of this identity for a triangle ca. 520 BC (in fact this identity had been known to the Babylonians since around 1500 BC).

We now show that the *operator norm* of a continuous linear operator acting in an *inner-product space* has another characterization than that defined in Theorem 2.9-5, again as a supremum. Recall that $\mathcal{L}(X)$ denotes the space of all continuous linear operators from a normed vector space X into itself (Section 2.9).

Theorem 4.1-3 *Let $(X, (\cdot,\cdot))$ be a real or complex inner-product space. Then the operator norm $\|A\| = \sup_{x\neq 0} \dfrac{\|Ax\|}{\|x\|}$ of any $A \in \mathcal{L}(X)$ is also given by*

$$\|A\| = \sup_{\begin{cases} x\neq 0 \\ y\neq 0\end{cases}} \frac{|(Ax, y)|}{\|x\|\,\|y\|}.$$

Proof Let $A \in \mathcal{L}(X)$ be given. By the Cauchy–Schwarz–Bunyakovskiĭ inequality,

$$|(Ax, y)| \leq \|Ax\|\,\|y\| \leq \|A\|\,\|x\|\,\|y\|,$$

and thus

$$\sup_{\begin{cases} x\neq 0 \\ y\neq 0\end{cases}} \frac{|(Ax, y)|}{\|x\|\,\|y\|} \leq \|A\|.$$

Let $x \in X$ be such that $Ax \neq 0$ (hence $x \neq 0$). Then

$$\frac{\|Ax\|^2}{\|x\|^2} = \frac{(Ax, Ax)}{\|x\|\,\|x\|} = \frac{\|Ax\|}{\|x\|}\frac{(Ax, Ax)}{\|x\|\,\|Ax\|} \leq \frac{\|Ax\|}{\|x\|} \sup_{\begin{cases} x\neq 0 \\ y\neq 0\end{cases}} \frac{|(Ax, y)|}{\|x\|\,\|y\|},$$

and thus

$$\frac{\|Ax\|}{\|x\|} \leq \sup_{\begin{cases} x\neq 0 \\ y\neq 0\end{cases}} \frac{|(Ax, y)|}{\|x\|\,\|y\|}.$$

Clearly the last inequality remains true if $Ax = 0$ and $x \neq 0$. Hence

$$\|A\| = \sup_{x\neq 0} \frac{\|Ax\|}{\|x\|} \leq \sup_{\begin{cases} x\neq 0 \\ y\neq 0\end{cases}} \frac{|(Ax, y)|}{\|x\|\,\|y\|}. \qquad \square$$

An inner-product space $(X, (\cdot,\cdot))$ is a **Hilbert space**[3] if, as a normed vector space, it is a *Banach space*, i.e., if X is complete with respect to the norm $\|\cdot\|$ defined by $\|x\| = \sqrt{(x,x)}$ for all $x \in X$ (Theorem 4.1-1(b)).

Any noncomplete inner-product space X can be identified with a dense subset of a Banach space $\widetilde{X}$, by means of the completion of the associated normed vector space (Theorem 3.1-2). As expected, $\widetilde{X}$ is also a Hilbert space and its inner product is an extension, *modulo* a linear isometry, of the inner product of X:

[3]So named as a tribute to David Hilbert (1862–1943), who extensively studied special cases of Hilbert spaces at the beginning of the twentieth century (see in particular the chapter by Hermann Weyl in Hilbert's biography by REID [1970], and DIEUDONNÉ [1981, Chapter 5, Section 2]). But the idea of an "abstract" Hilbert space (i.e., not a particular one such as ℓ^2 or $L^2(\Omega)$) is in effect due to John von Neumann (1903–1957), who was the first to coin the term "*Hilbert space*" in 1929.

Theorem 4.1-4 (completion of an inner-product space) *Let $(X, (\cdot,\cdot))$ be an inner-product space over $\mathbb{K} = \mathbb{R}$ or over $\mathbb{K} = \mathbb{C}$. Then the completion $(\widetilde{X}, \|\cdot\|_{\widetilde{X}})$ of its associated normed space $(X, \|\cdot\|)$ (Theorem 3.1-2) is a Hilbert space over $\mathbb{K}$, whose inner product $(\cdot,\cdot)_{\widetilde{X}}$ satisfies*

$$(\sigma x, \sigma y)_{\widetilde{X}} = (x, y) \quad \textit{for all } x, y \in X,$$

where σ is the linear isometry from X onto a dense subspace of $\widetilde{X}$ given by Theorem 3.1-2.

Proof For any $\widetilde{x} = \sigma x \in \sigma(X)$ and $\widetilde{y} = \sigma y \in \sigma(X)$, let

$$(\widetilde{x}, \widetilde{y})_{\sigma(X)} := (x, y).$$

Clearly, the mapping from $\sigma(X) \times \sigma(X)$ into $\mathbb{K}$ defined in this fashion is an inner product on the vector space $\sigma(X)$, since σ is a linear isometry. For each $\widetilde{x} \in \widetilde{X}$ and $\widetilde{y} \in \widetilde{X}$, let $\widetilde{x}_n \in \sigma(X)$, $n \geq 0$, and $\widetilde{y}_n \in \sigma(X)$, $n \geq 0$, be such that

$$\|\widetilde{x}_n - \widetilde{x}\|_{\widetilde{X}} \to 0 \quad \text{and} \quad \|\widetilde{y}_n - \widetilde{y}\|_{\widetilde{X}} \to 0 \quad \text{as } n \to \infty$$

(by construction, $\sigma(X)$ is a dense subspace in $(\widetilde{X}, \|\cdot\|_{\widetilde{X}})$; cf. Theorem 3.1-2), and let

$$\begin{aligned}
(\widetilde{x}, \widetilde{y})_{\widetilde{X}} &:= \lim_{n\to\infty} (\widetilde{x}_n, \widetilde{y}_n)_{\sigma(X)} \\
&= \frac{1}{4} \lim_{n\to\infty} \left(\|\widetilde{x}_n + \widetilde{y}_n\|^2 - \|\widetilde{x}_n - \widetilde{y}_n\|^2\right) \quad \text{if } \mathbb{K} = \mathbb{R}, \\
&= \frac{1}{4} \lim_{n\to\infty} \left(\|\widetilde{x}_n + \widetilde{y}_n\|^2 - \|\widetilde{x}_n - \widetilde{y}_n\|^2 + i\|\widetilde{x}_n + i\widetilde{y}_n\|^2 - i\|\widetilde{x}_n - i\widetilde{y}_n\|^2\right) \quad \text{if } \mathbb{K} = \mathbb{C}
\end{aligned}$$

(hence $\lim_{n\to\infty}(\widetilde{x}_n, \widetilde{y}_n)_{\sigma(X)}$ only depends on $\widetilde{x}$ and $\widetilde{y}$). It is then easily verified that the mapping from $\widetilde{X} \times \widetilde{X}$ into $\mathbb{K}$ defined in this fashion is an inner product on $\widetilde{X}$ and that it satisfies

$$(\sigma x, \sigma y)_{\widetilde{X}} = (x, y) \quad \text{for all } x, y \in X. \qquad \square$$

Remark Another proof uses the converse to the parallelogram law (Problem 4.1-3). $\square$

Most results presented in this chapter apply *verbatim* to *both* real and complex inner-product spaces, even if for clarity the two cases are sometimes separated (see, e.g., the proof of the Cauchy–Schwarz–Bunyakovskiĭ inequality in Theorem 4.1-1). Particular caution should be exercised, however, as some results hold only in *one* case (see, e.g., Problem 4.1-1, which provides an example of a property that holds in a complex inner-product space, but not in a real one), or require *different proofs* for each case (see, e.g., the converse to the parallelogram law proposed in Problem 4.1-3).

Problems

4.1-1 (1) Let $(X, (\cdot,\cdot))$ be a *complex* inner-product space, and let $A : X \to X$ be a linear operator that satisfies $(Ax, x) = 0$ for all $x \in X$. Show that $A = 0$.

(2) Show that the implication of (1) does not necessarily hold if $(X, (\cdot,\cdot))$ is a *real* inner-product space.

4.1-2 Let X be a vector space over $\mathbb{K} = \mathbb{R}$ or $\mathbb{K} = \mathbb{C}$ and let $(\cdot,\cdot) : X \times X \to \mathbb{K}$ be a mapping that satisfies all the properties of an inner-product, save possibly the fourth one (positive-definiteness). Show that the mapping $(\cdot,\cdot)$ is entirely determined by its restriction to the *diagonal* of the product $X \times X$, i.e., the subset $\{(x,y) \in X \times X;\ x = y\}$ of $X \times X$.

4.1-3 Let $(X, \|\cdot\|)$ be a normed vector space over $\mathbb{K} = \mathbb{R}$ or $\mathbb{K} = \mathbb{C}$, whose norm satisfies the parallelogram law:

$$\|x+y\|^2 + \|x-y\|^2 = 2\|x\|^2 + 2\|y\|^2 \quad \text{for all } x, y \in X.$$

(1) Show that X is also an inner-product space, whose inner-product $(\cdot,\cdot)$ satisfies $\|x\| = \sqrt{(x,x)}$ for all $x \in X$.

Hint: Verify that the sought inner product is given by

$$(x,y) := \frac{1}{4}\left(\|x+y\|^2 - \|x-y\|^2\right) \quad \text{if } \mathbb{K} = \mathbb{R},$$
$$(x,y) := \frac{1}{4}\left(\|x+y\|^2 - \|x-y\|^2 + i\|x+iy\|^2 - i\|x-iy\|^2\right) \quad \text{if } \mathbb{K} = \mathbb{C}.$$

(2) Use (1) to give another proof of Theorem 4.1-4.

4.1-4 (1) Let X and Y be two real inner-product spaces and let $A : X \to Y$ be a mapping that satisfies

$$A(0) = 0 \quad \text{and} \quad \|Ax - A\tilde{x}\|_Y = \|x - \tilde{x}\|_X \quad \text{for all } x, \tilde{x} \in X.$$

Show that A is a linear operator (which is clearly continuous).

(2) Does this result still hold if X and Y are complex inner-product spaces?

Remark The special case $X = Y = \mathbb{R}^n$ constitutes the well-known *Mazur–Ulam theorem*; cf. Problem 8.7-1. □

4.1-5 Let X be a Hilbert space and let Y be a closed subspace of X. Show that the quotient space X/Y (which is a Banach space; cf. Theorem 3.6-5), is also a Hilbert space.

4.1-6 (1) Show that the *Cauchy–Schwarz–Bunyakovskiĭ* inequality in $\mathbb{R}^n$, viz.,

$$\sum_{i=1}^{n} |x_i y_i| \le \left(\sum_{i=1}^{n} |x_i|^2\right)^{1/2} \left(\sum_{i=1}^{n} |y_i|^2\right)^{1/2} \quad \text{for any } x_i, y_i \in \mathbb{R}, 1 \le i \le n,$$

is equivalent[4] to the *arithmetic mean-geometric inequality* (Problem 2.17-10), viz.,

$$\left(\prod_{i=1}^{n} x_i\right)^{1/n} \le \frac{1}{n}\sum_{i=1}^{n} x_i \quad \text{for any } x_i > 0,\ 1 \le i \le n.$$

(2) Show that the *arithmetic mean-geometric inequality* is equivalent[5] to the *Bernoulli inequality*, viz.,

$$1 + n(x-1) \le x^n \quad \text{for all } x > 0 \text{ and } n \ge 1.$$

[4]Minghua Lin [2012]: The AM-GM inequality and CBS inequality are equivalent, *The Mathematical Intelligencer* **34**, 6.

[5]L. Maligranda [2012]: The AM-GM inequality is equivalent to the Bernoulli inequality, *The Mathematical Intelligencer* **34**, 1–2.

4.2 First examples of inner-product spaces and Hilbert spaces; the spaces ℓ^2 and $L^2(\Omega)$

The space $\mathbb{R}^n$ equipped with the **Euclidean inner product**, also called **scalar product**, defined by

$$\boldsymbol{x}\cdot\boldsymbol{y} := \sum_{i=1}^{n} x_i y_i \quad \text{for all } \boldsymbol{x} = (x_i)_{i=1}^n,\ \boldsymbol{y} = (y_i)_{i=1}^n \in \mathbb{R}^n,$$

and the space $\mathbb{C}^n$ equipped with the **Hermitian inner product** defined by

$$\boldsymbol{x}\cdot\boldsymbol{y} := \sum_{i=1}^{n} x_i \overline{y}_i \quad \text{for all } \boldsymbol{x} = (x_i)_{i=1}^n,\ \boldsymbol{y} = (y_i)_{i=1}^n \in \mathbb{C}^n,$$

provide the simplest examples of real and complex Hilbert spaces. The norm induced by this inner product is thus given by

$$|\boldsymbol{x}| := (\boldsymbol{x}\cdot\boldsymbol{x})^{1/2} = \left(\sum_{i=1}^{n} |x_i|^2\right)^{1/2} \quad \text{for any vector } \boldsymbol{x} = (x_i)_{i=1}^n \text{ in } \mathbb{R}^n \text{ or } \mathbb{C}^n.$$

The space $\mathbb{R}^n$ equipped with the Euclidean inner product is called the **n-dimensional Euclidean space.**

More generally, the space $\mathbb{R}^n$, *resp.* $\mathbb{C}^n$, likewise becomes a Hilbert space if it is equipped with the inner product defined by

$$(\boldsymbol{x},\boldsymbol{y})_{\boldsymbol{A}} := \sum_{i,j=1}^{n} a_{ij} x_i y_j, \quad \text{resp.} \quad (\boldsymbol{x},\boldsymbol{y})_{\boldsymbol{A}} := \sum_{i,j=1}^{n} a_{ij} x_i \overline{y}_j,$$

where $\boldsymbol{A} = (a_{ij})$ is a given positive-definite symmetric, *resp.* Hermitian, matrix of order n.

Analogous inner products can be evidently defined over any *finite-dimensional vector space.*

Another example of a real, *resp.* complex, finite-dimensional Hilbert space is provided by the vector space consisting of all real, *resp.* complex, $m \times n$ matrices, equipped with the **matrix inner product** defined by

$$\boldsymbol{A} : \boldsymbol{B} := \sum_{i=1}^{m}\sum_{j=1}^{n} a_{ij} b_{ij} \text{ if } \mathbb{K} = \mathbb{R}, \quad \textit{resp.} \quad \boldsymbol{A} : \boldsymbol{B} := \sum_{i=1}^{m}\sum_{j=1}^{n} a_{ij} \overline{b}_{ij} \text{ if } \mathbb{K} = \mathbb{C},$$

for all $m \times n$ matrices $A = (a_{ij})$ and $B = (b_{ij})$. The norm $\|\cdot\|_F$ induced by this inner product, thus defined by

$$\|\boldsymbol{A}\|_F := (\boldsymbol{A} : \boldsymbol{A})^{1/2} = \left(\sum_{i=1}^{m}\sum_{j=1}^{n} |a_{ij}|^2\right)^{1/2} \quad \text{for any } m \times n \text{ matrix } \boldsymbol{A} = (a_{ij}),$$

is called the *Frobenius norm.*

The above Hilbert spaces are all *separable* (since they are finite-dimensional; cf. Theorem 2.7-1(b)).

The *real or complex space* ℓ^2 (the special case $p = 2$ of the spaces ℓ^p, $1 \le p \le \infty$, introduced in Section 2.4) consists of all infinite sequences $x = (x_i)_{i=1}^{\infty}$ of scalars $x_i \in \mathbb{K}$ that satisfy $\sum_{i=1}^{\infty} |x_i|^2 < \infty$. Equipped with the inner product defined by

$$(x, y) := \sum_{i=1}^{\infty} x_i y_i \quad \text{for all } x = (x_i)_{i=1}^{\infty},\ y = (y_i)_{i=1}^{\infty} \in \ell^2 \text{ if } \mathbb{K} = \mathbb{R},$$

$$(x, y) := \sum_{i=1}^{\infty} x_i \overline{y}_i \quad \text{for all } x = (x_i)_{i=1}^{\infty},\ y = (y_i)_{i=1}^{\infty} \in \ell^2 \text{ if } \mathbb{K} = \mathbb{C},$$

the space ℓ^2 provides an example of an *infinite-dimensional*, real or complex, *separable Hilbert space*, since it is separable (Theorem 2.4-2(b)), and complete when it is equipped with the induced norm, defined by

$$\|x\| := \sqrt{(x, x)} = \left(\sum_{i=1}^{\infty} |x_i|^2 \right)^{1/2} \quad \text{for all } x = (x_i)_{i=1}^{\infty} \in \ell^2$$

(Theorem 3.4-1). Note that the corresponding Cauchy–Schwarz–Bunyakovskiĭ inequality is the special case $p = q = 2$ of Hölder's inequality for sequences (Theorem 2.4-1(a)) and that the corresponding triangle inequality is the special case $p = 2$ of Minkovski's inequality for sequences (Theorem 2.4-1(b)).

The *real space* $L^2(\Omega)$ (the special case $p = 2$ of the spaces $L^p(\Omega)$, $1 \le p \le \infty$, introduced in Section 2.5) consists of all the equivalence classes of measurable functions $f : \Omega \to [-\infty, \infty]$ that satisfy $\int_\Omega |f(x)|^2 \, dx < \infty$, where Ω is any open subset of $\mathbb{R}^n$. Equipped with the inner product defined by

$$(f, g) := \int_\Omega f(x) g(x) \, dx \quad \text{for all } f, g \in L^2(\Omega),$$

the space $L^2(\Omega)$ provides another example of an *infinite-dimensional separable real Hilbert space*, since it is separable (Theorem 2.5-4(a)), and complete when it is equipped with the associated norm, defined by

$$\|f\|_{L^2(\Omega)} := \left(\int_\Omega |f(x)|^2 \, dx \right)^{1/2} \quad \text{for all } f \in L^2(\Omega)$$

(Theorem 3.4-2). Note that the corresponding Cauchy–Schwarz–Bunyakovskiĭ inequality is the special case $p = q = 2$ of Hölder's inequality for functions (Theorem 2.5-1(a)) and that the corresponding triangle inequality is the special case $p = 2$ of Minkovski's inequality for functions (Theorem 2.5-1(b)).

One can similarly define the *complex* space

$$L^2(\Omega; \mathbb{C}) := \{ f : \Omega \to \mathbb{C};\ \operatorname{Re} f \text{ and } \operatorname{Im} f \text{ are measurable and } |f|^2 \in L^1(\Omega) \},$$

which is easily seen to provide an example of an infinite-dimensional separable complex Hilbert space when it is equipped with the inner product defined by

$$(f, g) := \int_\Omega f(x) \overline{g(x)} \, dx \quad \text{for all } f, g \in L^2(\Omega; \mathbb{C}).$$

An example of a *noncomplete real inner-product space* (thus necessarily infinite-dimensional) is provided by the space $\mathcal{C}(\overline{\Omega})$, where Ω is a bounded open subset of $\mathbb{R}^N$, and the inner product is that of the space $L^2(\Omega)$ (Problem 3.2-2). Clearly, the completion of the space $\mathcal{C}(\overline{\Omega})$ with respect to the norm $\|\cdot\|_{L^2(\Omega)}$ is precisely the larger space $L^2(\Omega)$ (Theorem 2.5-3 or 2.6-2).

The spaces ℓ^2 and $L^2(\Omega)$ constitute basic examples of infinite-dimensional separable Hilbert spaces. Other basic examples will be provided later by the *Sobolev spaces* $H^m(\Omega)$ and $H_0^m(\Omega)$ (Chapter 6).

As we shall later see, ℓ^2 is in effect the paradigm of such spaces, in the sense that *any infinite-dimensional separable Hilbert space can be identified with ℓ^2* by means of a linear bijection that preserves the inner product (Theorem 4.9-4).

Problem

4.2-1 The *angle* between two nonzero vectors $\boldsymbol{x}, \boldsymbol{y} \in \mathbb{C}^n$ can be defined either as the unique solution $\theta(\boldsymbol{x}, \boldsymbol{y}) \in [0, \pi]$ of the equation

$$\cos\theta(\boldsymbol{x}, \boldsymbol{y}) = \frac{\mathrm{Re}(\boldsymbol{x} \cdot \boldsymbol{y})}{|\boldsymbol{x}|\,|\boldsymbol{y}|}$$

(a definition that extends that of the angle between two nonzero vectors $\boldsymbol{x}, \boldsymbol{y} \in \mathbb{R}^n$), or as the unique solution $\varphi(\boldsymbol{x}, \boldsymbol{y}) \in [0, \frac{\pi}{2}]$ of the equation

$$\cos\varphi(\boldsymbol{x}, \boldsymbol{y}) = \frac{|\boldsymbol{x} \cdot \boldsymbol{y}|}{|\boldsymbol{x}|\,|\boldsymbol{y}|}.$$

Remark If $\boldsymbol{x} = i\boldsymbol{y}$, then $\theta(\boldsymbol{x}, \boldsymbol{y}) = \frac{\pi}{2}$ while $\varphi(\boldsymbol{x}, \boldsymbol{y}) = 0$. □

(1) Show that $\theta(\boldsymbol{x}, \boldsymbol{z}) \le \theta(\boldsymbol{x}, \boldsymbol{y}) + \theta(\boldsymbol{y}, \boldsymbol{z})$ for all nonzero vectors $\boldsymbol{x}, \boldsymbol{y}, \boldsymbol{z} \in \mathbb{C}^n$.

(2) Using (1), show that $\varphi(\boldsymbol{x}, \boldsymbol{z}) \le \varphi(\boldsymbol{x}, \boldsymbol{y}) + \varphi(\boldsymbol{y}, \boldsymbol{z})$ for all nonzero vectors $\boldsymbol{x}, \boldsymbol{y}, \boldsymbol{z} \in \mathbb{C}^n$.

4.3 The projection theorem

The next result, which pervades the theory of Hilbert spaces, is fundamental. It is in particular the keystone for several other basic results, such as the direct sum theorem (Theorem 4.5-2), the F. Riesz representation theorem in a Hilbert space (Theorem 4.6-1), or the minimization of quadratic functionals over convex sets (Theorem 6.1-1). Its illuminating geometrical interpretation (which in particular justifies its name) is discussed after the proof.

Theorem 4.3-1 (projection theorem) *Let Z be a nonempty, convex, and complete, subset of a real ($\mathbb{K} = \mathbb{R}$) or complex ($\mathbb{K} = \mathbb{C}$) inner-product space $(X, (\cdot,\cdot))$.*

(a) *Given any element $x \in X$, there exists a unique element $Px \in Z$ that satisfies*

$$\|x - Px\| = \inf_{z \in Z} \|x - z\|,$$

where $\|\cdot\|$ is the norm induced by the inner product $(\cdot,\cdot)$ (Theorem 4.1-1(c)).

(b) *The unique element $Px \in Z$ found in* (a) *satisfies*

$$\begin{aligned} (Px - x, z - Px) \geq 0 &\quad \textit{for all } z \in Z \textit{ if } \mathbb{K} = \mathbb{R}, \\ \operatorname{Re}(Px - x, z - Px) \geq 0 &\quad \textit{for all } z \in Z \textit{ if } \mathbb{K} = \mathbb{C}. \end{aligned}$$

Conversely, if an element $y \in Z$ satisfies

$$\begin{aligned} (y - x, z - y) \geq 0 &\quad \textit{for all } z \in Z \textit{ if } \mathbb{K} = \mathbb{R}, \\ \operatorname{Re}(y - x, z - y) \geq 0 &\quad \textit{for all } z \in Z \textit{ if } \mathbb{K} = \mathbb{C}, \end{aligned}$$

then $y = Px$.

(c) *The mapping $P : X \to Z$ defined in* (a) *satisfies*

$$\|Px_1 - Px_2\| \leq \|x_1 - x_2\| \quad \textit{for all } x_1, x_2 \in X.$$

Hence P is a Lipschitz-continuous mapping with Lipschitz constant one.

(d) *Assume that the subset Z is a complete subspace of X. Then the element Px found in* (a) *satisfies*

$$(Px - x, z) = 0 \quad \textit{for all } z \in Z.$$

Conversely, if an element $y \in Z$ satisfies

$$(y - x, z) = 0 \quad \textit{for all } z \in Z,$$

then $y = Px$.

(e) *The mapping $P : X \to Z$ is linear if and only if the subset Z is a subspace of X. In this case,*

$$\|P\|_{\mathcal{L}(X;Z)} = 1 \quad \textit{if } Z \neq \{0\}.$$

Proof (i) If $x \in Z$, then $Px = x$. If $x \notin Z$, then $\delta := \inf_{z \in Z} \|x - z\|$ is a well-defined ≥ 0 number since the set Z is *nonempty* by assumption. In fact δ is > 0, since $\delta = 0$ would imply that $x \in \overline{Z}$, and hence that $x \in Z$ since Z is *closed* by assumption (a complete subset is closed; cf. Theorem 1.12-2(a)), a contradiction. Let then $y_n \in Z$, $n \geq 0$, be such that

$$\|x - y_n\| \underset{n \to \infty}{\longrightarrow} \delta = \inf_{z \in Z} \|x - z\| > 0.$$

The parallelogram law (Theorem 4.1-2) implies that, for all $m, n \geq 0$,

$$\begin{aligned} \|y_m - y_n\|^2 &= \|(x - y_m) - (x - y_n)\|^2 \\ &= 2\|x - y_m\|^2 + 2\|x - y_n\|^2 - \|2x - (y_m + y_n)\|^2 \\ &= 2\|x - y_m\|^2 + 2\|x - y_n\|^2 - 4\left\|x - \frac{y_m + y_n}{2}\right\|^2. \end{aligned}$$

The assumed *convexity* of the set Z implies that $\dfrac{y_m + y_n}{2} \in Z$; therefore

$$\left\|x - \frac{y_m + y_n}{2}\right\|^2 \geq \delta^2,$$

which in turn implies that

$$0 \leq \|y_m - y_n\|^2 \leq 2\|x - y_m\|^2 + 2\|x - y_n\|^2 - 4\delta^2 \quad \text{for all } m, n \geq 0.$$

The sequence $(y_n)_{n=0}^{\infty}$ is thus a *Cauchy sequence*, since $\|x - y_m\| \underset{m\to\infty}{\longrightarrow} \delta$ and $\|x - y_n\| \underset{n\to\infty}{\longrightarrow} \delta$. The set Z being *complete* by assumption, there exists $y \in Z$ such that $y_n \underset{n\to\infty}{\longrightarrow} y$. Besides, the continuity of the norm (Theorem 2.2-5) implies that

$$\|x - y\| = \lim_{n\to\infty} \|x - y_n\| = \delta = \inf_{z \in Z} \|x - z\|.$$

To show that such an element $y \in Z$ is *unique*, let $y_0 \in Z$ and $y_1 \in Z$ be such that

$$\delta = \|x - y_0\| = \|x - y_1\|.$$

Then the sequence $(y_n)_{n=0}^{\infty}$ defined by $y_{2k} := y_0$ and $y_{2k+1} := y_1$ for all $k \geq 0$ evidently satisfies $\|x - y_n\| \underset{n\to\infty}{\longrightarrow} \delta$. The same argument as above therefore shows that this sequence converges. Hence $y_0 = y_1$ since the limit of a convergent sequence is unique in a normed vector space. This proves (a).

(ii) Assume first that $\mathbb{K} = \mathbb{R}$. If $x \in Z$, the announced inequalities hold since $Px - x = 0$ in this case. If $x \notin Z$, let $y := Px \in Z$, and let $z \in Z$ be given. Since $(y + \theta(z - y)) \in Z$ for all $0 \leq \theta \leq 1$ (the set Z is convex by assumption), the definition of $y = Px$ (cf. (a)) implies that

$$\begin{aligned} \|x - y\|^2 &\leq \|x - (y + \theta(z - y))\|^2 \\ &= \|x - y\|^2 - 2\theta(x - y, z - y) + \theta^2\|z - y\|^2 \quad \text{for all } 0 \leq \theta \leq 1. \end{aligned}$$

Consequently,

$$0 \leq 2\theta(y - x, z - y) + \theta^2\|z - y\|^2 \quad \text{for all } 0 \leq \theta \leq 1,$$

which implies that $(y - x, z - y) \geq 0$.

Conversely, assume that an element $y \in Z$ satisfies $(y - x, z - y) \geq 0$ for all $z \in Z$. Then

$$\begin{aligned} \|x - z\|^2 = \|x - y + y - z\|^2 &= \|x - y\|^2 + 2(y - x, z - y) + \|z - y\|^2 \\ &\geq \|x - y\|^2 \quad \text{for all } z \in Z, \end{aligned}$$

which shows that $y = Px$. If $\mathbb{K} = \mathbb{C}$, the corresponding conclusions hold, thanks this time to the relations

$$\begin{aligned} \|x - (y + \theta(z - y))\|^2 &= \|x - y\|^2 - 2\theta \operatorname{Re}(x - y, z - y) + \theta^2\|z - y\|^2, \\ \|x - z\|^2 &= \|x - y\|^2 + 2 \operatorname{Re}(y - x, z - y) + \|z - y\|^2. \end{aligned}$$

This proves (b).

(iii) Assume first that $\mathbb{K} = \mathbb{R}$. Part (b) implies that, for all $x_1, x_2 \in X$,

$$\begin{aligned} (Px_1 - x_1, Px_2 - Px_1) \geq 0 &\quad \text{since } Px_2 \in Z, \\ (x_2 - Px_2, Px_2 - Px_1) \geq 0 &\quad \text{since } Px_1 \in Z, \end{aligned}$$

so that

$$(Px_1 - Px_2, Px_2 - Px_1) + (x_2 - x_1, Px_2 - Px_1) \geq 0.$$

This inequality, combined with the Cauchy–Schwarz–Bunyakovskiĭ inequality, in turn implies that

$$\|Px_1 - Px_2\|^2 \leq (x_2 - x_1, Px_2 - Px_1) \leq \|x_2 - x_1\| \, \|Px_2 - Px_1\| .$$

Therefore the announced inequality holds.

If $\mathbb{K} = \mathbb{C}$, the same conclusion holds, thanks this time to the inequalities

$$\operatorname{Re}(Px_1 - x_1, Px_2 - Px_1) \geq 0, \quad \operatorname{Re}(x_2 - Px_2, Px_2 - Px_1) \geq 0,$$

which in turn imply that

$$\|Px_1 - Px_2\|^2 = \operatorname{Re}(Px_1 - Px_2, Px_1 - Px_2) \leq \operatorname{Re}(x_2 - x_1, Px_2 - Px_1).$$

This proves (c).

(iv) Let now Z be a complete subspace of X, and assume first that $\mathbb{K} = \mathbb{R}$. If $x \in Z$, the announced equalities hold since $Px - x = 0$ in this case. If $x \notin Z$, let $z \in Z$ be given. Since $(Px + \theta z) \in Z$ for all $\theta \in \mathbb{R}$ (the set Z is assumed here to be subspace), the inequalities of (b) show that

$$(Px - x, Px + \theta z - Px) = \theta(Px - x, z) \geq 0 \quad \text{for all } \theta \in \mathbb{R}.$$

Hence $(Px - x, z) = 0$.

Conversely, assume that $y \in Z$ satisfies $(y - x, z) = 0$ for all $z \in Z$, so that $(y - x, y) = 0$ in particular. Consequently,

$$(y - x, z - y) = 0 \geq 0 \quad \text{for all } z \in Z,$$

and thus $y = Px$ by (b).

Assume next that $\mathbb{K} = \mathbb{C}$ and let $z \in Z$ be given. Since $(Px + \theta z) \in Z$ for all $\theta \in \mathbb{R}$, the inequalities of (b) show that

$$\operatorname{Re}(Px - x, Px + \theta z - Px) = \theta \operatorname{Re}(Px - x, z) \geq 0 \quad \text{for all } \theta \in \mathbb{R},$$

and hence that $\operatorname{Re}(Px - x, z) = 0$. Since $(Px + i\theta z) \in Z$ for all $\theta \in \mathbb{R}$, the same inequalities of (b) show that

$$\operatorname{Re}(Px - x, Px + i\theta z - Px) = \theta \operatorname{Im}(Px - x, z) \geq 0 \quad \text{for all } \theta \in \mathbb{R},$$

which implies that $\operatorname{Im}(Px - x, z) = 0$. Consequently, $(Px - x, z) = 0$ also holds in the complex case. The converse property likewise holds if $\mathbb{K} = \mathbb{C}$, since

$$(y - x, z - y) = 0 = \operatorname{Re}(y - x, z - y) \geq 0$$

in this case. This proves (d).

(v) Assume first that Z is a subspace. Let $x_1, x_2 \in X$ and $\alpha_1, \alpha_2 \in \mathbb{K}$ be given. Then

$$(Px_1 - x_1, z) = (Px_2 - x_2, z) = 0 \quad \text{for all } z \in Z,$$

by (d). Consequently,

$$((\alpha_1 Px_1 + \alpha_2 Px_2) - (\alpha_1 x_1 + \alpha_2 x_2), z) = 0 \quad \text{for all } z \in Z.$$

Since $(\alpha_1 Px_1 + \alpha_2 Px_2) \in Z$ in this case, the characterization established in (d) shows that

$$\alpha_1 Px_1 + \alpha_2 Px_2 = P(\alpha_1 x_1 + \alpha_2 x_2).$$

Hence the mapping $P : X \to Z$ is linear.

Conversely, assume that $P : X \to Z$ is *linear.* Since the direct image of X under P is Z (clearly, $Px = x$ if $x \in Z$) and since the direct image of a linear mapping is necessarily a vector space, the set Z is a subspace.

Finally, letting $x_2 = 0$ in the inequality of (c) shows that

$$\|Px\| \leq \|x\| \quad \text{for all } x \in X,$$

since $Px_2 = x_2 = 0 \in Z$ if Z is a subspace. Hence $\|P\|_{\mathcal{L}(X;Z)} = 1$, unless $Z = \{0\}$. This proves (e). □

Remark It is immediately realized from the proof that the converses to properties (b) and (d) hold if Z is any nonempty subset of the space X. □

Several comments are in order about the projection theorem (see also Problems 4.3-1 to 4.3-3 for various complements).

If $(X, (\cdot, \cdot))$ is a Hilbert space, the assumption "Z is complete" is of course equivalent to "Z is closed in X."

The geometrical interpretation of the element $Px \in Z$ defined in Theorem 4.3-1(a) is clear in the special case where $X = \mathbb{R}^2$ and $(\cdot, \cdot)$ is the Euclidean inner product (Figure 4.3-1): Px is that element in Z that is the "nearest" to x. Besides, the absolute value of the angle between the two vectors $(Px - x)$ and $(z - Px)$ should be $\leq \pi/2$ for all $z \in Z$ (cf. Theorem 4.3-1(b)), while the vector $(Px - x)$ should be orthogonal to any vector $z \in Z$, or equivalently the absolute value of the angle between the vector $(Px - x)$ and any vector $z \in Z$ should be equal to $\pi/2$, if Z is a subspace (cf. Theorem 4.3-1(d)).

For these reasons, the element $Px \in Z$ is called the **projection** of $x \in X$ on the set Z, and the operator $P : X \to Z$ is called the **projection operator** of X onto Z.

The Lipschitz-continuity with constant one of the projection operator P established in (c) expresses another intuitively clear property, viz., that "the projection does not increase the distances" (Figure 4.3-1).

It should be also emphasized that the *linearity* of the projection operator $P : X \to Z$ when Z is a subspace (Theorem 4.3-1(e)) crucially hinges on the fact that *the norm* (in the space X) *is derived from an inner product*; in this respect, see Problem 4.3-4.

Let us give some *examples of projection operators.* In the space $\mathbb{R}^n$ equipped with its Euclidean inner product (Section 4.2) defined by $(\boldsymbol{x}, \boldsymbol{y}) := \boldsymbol{x}^T \boldsymbol{y}$ (the matrix notation is used here; this means that vectors are identified with column vectors, i.e., $n \times 1$ matrices), consider a *hyperplane*

$$\boldsymbol{Z} := \{\boldsymbol{z} \in \mathbb{R}^n;\ \boldsymbol{a}^T \boldsymbol{z} = 0\},$$

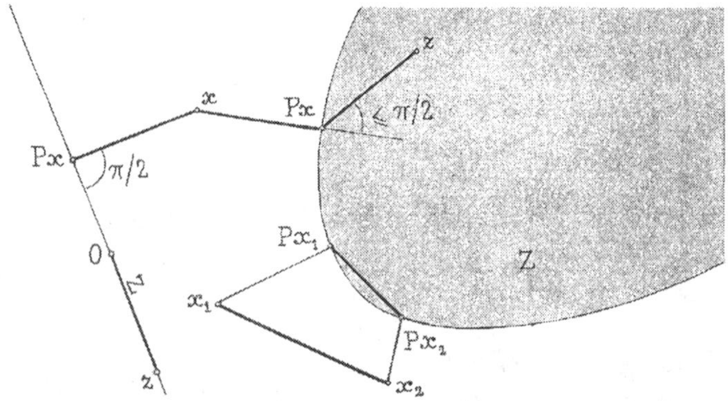

Figure 4.3-1 Geometrical interpretation of the projection $Px \in Z$ of an element $x \in X$ and of the properties established in Theorem 4.3-1 when X is the space $\mathbb{R}^2$ equipped with the Euclidean inner product.

i.e., the subspace formed by all the vectors of $\mathbb{R}^n$ orthogonal to a unit ($\boldsymbol{a}^T\boldsymbol{a} = 1$) vector $\boldsymbol{a} \in \mathbb{R}^n$. Then the mapping

$$\boldsymbol{P} := \boldsymbol{I} - \boldsymbol{a}\boldsymbol{a}^T$$

(thus identified here with an $n \times n$ matrix) is the *projection operator, parallel to the vector* $\boldsymbol{a}$, *from* $\mathbb{R}^n$ *onto the hyperplane* $\boldsymbol{Z}$ (Figure 4.3-2). To see this, observe that $\boldsymbol{Px} \in \boldsymbol{Z}$ for all $\boldsymbol{x} \in \mathbb{R}^n$, since

$$\boldsymbol{a}^T\boldsymbol{Px} = \boldsymbol{a}^T\boldsymbol{x} - \boldsymbol{a}^T\boldsymbol{a}\boldsymbol{a}^T\boldsymbol{x} = 0 \quad \text{for all } \boldsymbol{x} \in \mathbb{R}^n,$$

and that

$$(\boldsymbol{Px} - \boldsymbol{x})^T\boldsymbol{z} = -\boldsymbol{x}^T\boldsymbol{a}\boldsymbol{a}^T\boldsymbol{z} = 0 \quad \text{for all } \boldsymbol{z} \in \boldsymbol{Z}.$$

Hence the conclusion follows from Theorem 4.3-1(d).

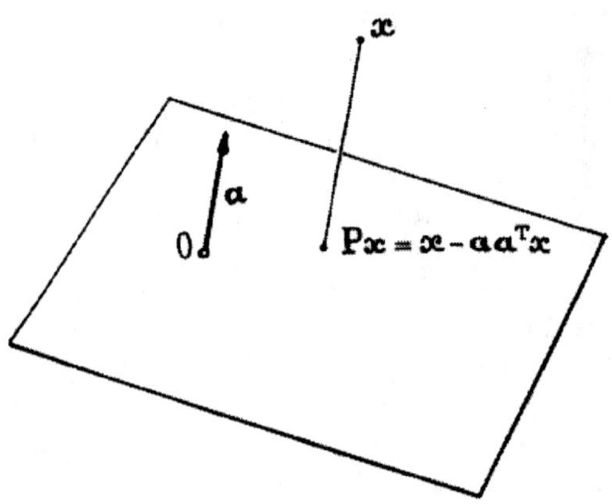

Figure 4.3-2 Projection, parallel to a unit vector $\boldsymbol{a}$, from $\mathbb{R}^3$ onto a hyperplane in $\mathbb{R}^3$.

This example can be immediately extended to any Hilbert space $(X, (\cdot,\cdot))$ and *hyperplane*

$$Z := \{z \in X;\ (a, z) = 0\},$$

where a is an element of X that satisfies $(a, a) = 1$ (the set Z is clearly a closed subspace of X). The *projection operator, parallel to a, from X onto Z* is now given by

$$Px = x - (a, x)a \quad \text{for all } x \in X.$$

Consider next the real Hilbert space $L^2(\Omega)$, where Ω is an open subset of $\mathbb{R}^n$. Let

$$Z := \{g \in L^2(\Omega);\ g = 0 \text{ a.e. on } A\},$$

where A is a measurable subset of Ω. The set Z is a subspace of $L^2(\Omega)$ that is closed in $L^2(\Omega)$, since any sequence converging in any $L^p(\Omega)$, $1 \leq p < \infty$, contains a subsequence that converges almost everywhere to the same limit (Theorem 3.4-3). Then the *projection operator* $P : L^2(\Omega) \to Z$ is given by

$$Pf = f\chi_{\Omega - A} \quad \text{for all } f \in L^2(\Omega),$$

where $\chi_{\Omega-A}$ denotes the characteristic function of the set $\Omega - A$ (Figure 4.3-3). To see this, it suffices to note that $Pf \in Z$ for all $f \in L^2(\Omega)$ and that

$$\int_\Omega (Pf - f)g\,\mathrm{d}x = 0 \quad \text{for all } g \in Z,$$

since $Pf - f = 0$ almost everywhere on $\Omega - A$ and $g = 0$ almost everywhere on A. Hence the conclusion again follows from Theorem 4.3-1(d).

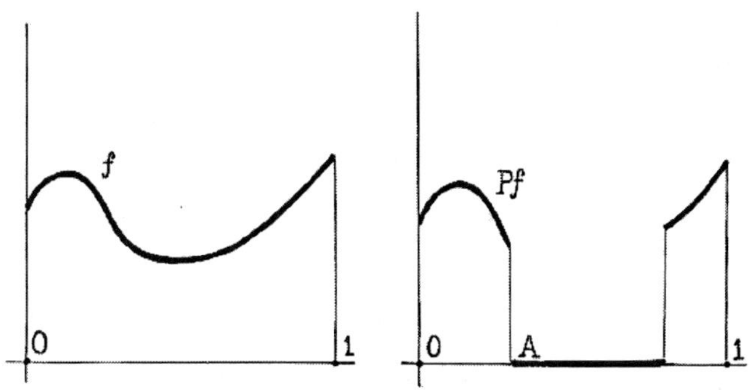

Figure 4.3-3 Projection from $L^2(\Omega)$ onto $Z = \{g \in L^2(\Omega);\ g = 0 \text{ a.e. on } A\}$, when $\Omega =]0, 1[$.

While the projection operators described in the two above examples are *linear* (in each case the set Z is a subspace; cf. Theorem 4.3-1(e)), the next one is *nonlinear*. As in the first example, the space X is $\mathbb{R}^n$ equipped with the Euclidean inner product $(\cdot,\cdot)$, but the subset Z is now defined as

$$\mathbb{R}^n_+ := \{(z_i)_{i=1}^n \in \mathbb{R}^n;\ z_i \geq 0, 1 \leq i \leq n\}.$$

The set $\mathbb{R}^n_+$, which is clearly a nonempty closed convex subset, but not a subspace, of $\mathbb{R}^n$ is sometimes called the *nonnegative hyperoctant.*

As suggested by an inspection of all possible cases in two dimensions (Figure 4.3-4), it is intuitively clear that the ith component $(\boldsymbol{Px})_i$ of the projection $\boldsymbol{Px} \in \mathbb{R}^n_+$ of an arbitrary vector $\boldsymbol{x} = (x_i) \in \mathbb{R}^n$ should be given by

$$(\boldsymbol{Px})_i = \max\{0, x_i\}, \quad 1 \le i \le n.$$

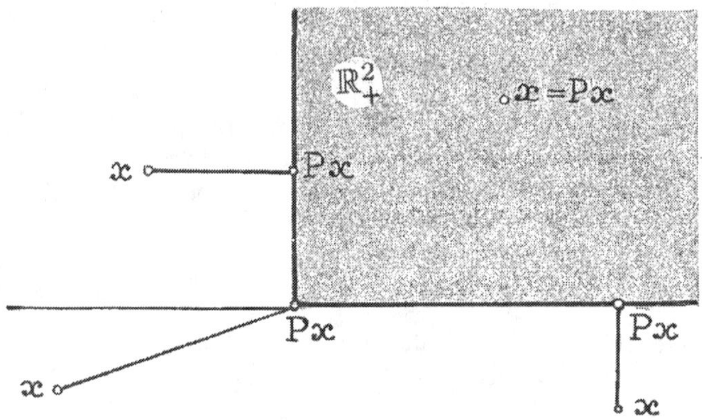

Figure 4.3-4 Projection from $\mathbb{R}^2$ onto the set $\mathbb{R}^2_+ := \{(z_i)^2_{i=1} \in \mathbb{R}^2;\ z_i \ge 0,\ i = 1, 2\}$. This figure originally appeared in P.G. CIARLET [2007]: *Introduction a l'Analyse Numerique Matricelle et a l'Optimisation*, Dunod, Paris.

In order to check that this is indeed the case, it suffices (according to Theorem 2.4-1(b)) to verify that $\boldsymbol{Px} \in \mathbb{R}^n_+$, which clearly holds, and that $(\boldsymbol{Px} - \boldsymbol{x}, \boldsymbol{z} - \boldsymbol{Px}) \ge 0$ for all $\boldsymbol{z} \in \mathbb{R}^n_+$, which also holds since

$$(\boldsymbol{Px} - \boldsymbol{x}, \boldsymbol{z} - \boldsymbol{Px}) = \sum_{i=1}^n ((\boldsymbol{Px})_i - x_i)(z_i - (\boldsymbol{Px})_i) \ge 0 \quad \text{for all } \boldsymbol{z} = (z_i) \in \mathbb{R}^n_+$$

(if $x_i \ge 0$, $(\boldsymbol{Px})_i = x_i$; if $x_i < 0$, $(\boldsymbol{Px})_i - x_i = -x_i > 0$ and $z_i - (\boldsymbol{Px})_i = z_i \ge 0$).

This example can be easily extended to subsets of $\mathbb{R}^n$ of the form

$$\boldsymbol{Z} := \{(z_i) \in \mathbb{R}^n;\ a_i \le z_i \le b_i,\ 1 \le i \le n\},$$

in which case the components of the projection $\boldsymbol{Px} \in \boldsymbol{Z}$ of an arbitrary element $\boldsymbol{x} \in \mathbb{R}^n$ are given by

$$(\boldsymbol{Px})_i = \min\{\max\{x_i, a_i\}, b_i\}, \quad 1 \le i \le n,$$

with obvious modifications if some inequalities $a_i \le x_i \le b_i$ are replaced by either $a_i \le x_i$ or $x_i \le b_i$, or no longer appear in the definition of the set $\boldsymbol{Z}$.

The familiar *polar factorization* of an invertible matrix provides an interesting example of a projection operator from a finite-dimensional inner-product space onto a nonempty closed subset that is *nonconvex*; cf. Problem 4.3-5.

We conclude this section by a first application of the projection theorem, viz., an interesting *characterization of a dense subspace in a Hilbert space*, which asserts that the *only vector orthogonal to all its elements is the zero vector.*

Theorem 4.3-2 *Let $(X, (\cdot,\cdot))$ be a Hilbert space and let Y be a subspace of X. Then*

$$\overline{Y} = X$$

if and only if the only element $x \in X$ that satisfies $(x, y) = 0$ for all $y \in Y$ is $x = 0$.

Proof Assume that $\overline{Y} \neq X$ and pick any element $\widetilde{x} \in (X - \overline{Y})$. Then $x := \widetilde{x} - P\widetilde{x}$, where P is the projection operator of X onto $\overline{Y}$, is not the zero vector; yet it satisfies $(x, y) = 0$ for all $y \in \overline{Y}$, hence a fortiori for all $y \in Y$, by the projection theorem (Theorem 4.3-1(d)). This proves the "if" part.

Assume that $\overline{Y} = X$ and let a vector $x \in X$ be given that satisfies $(x, y) = 0$ for all $y \in Y$. Since $\overline{Y} = X$, there exist $y_n \in Y$, $n \geq 0$, such that $\lim_{n\to\infty} y_n = x$. Hence $(x, x) = \lim_{n\to\infty}(x, y_n) = 0$. Note that this "only if" part holds irrespective of whether X is complete. □

Remark A similar property holds in fact in *any* normed vector space X, with the inner product replaced by the duality between X' and X (but then its proof requires the Hahn–Banach theorem; cf. Theorem 5.9-4). □

Problems

4.3-1 Assume that all the assumptions of Theorem 4.3-1 are satisfied, save that the set Z is not convex.

(1) Give a counterexample to the uniqueness of the projection.

(2) Give a counterexample to the existence of the projection.

4.3-2 Let X be a Hilbert space and let Z_n, $n \geq 1$, be nonempty closed convex subsets of X that satisfy $Z_1 \supset Z_2 \supset \cdots \supset Z_n \supset \cdots$. Given an element $x \in X$, let y_n denote the projection of x onto Z_n, $n \geq 1$.

(1) Show that, if $Z := \bigcap_{n=1}^{\infty} Z_n \neq \varnothing$, then $y_n \to y$ as $n \to \infty$, where y is the projection of x onto the set Z.

(2) Show that, if $Z = \varnothing$, then $\|x - y_n\| \to \infty$ as $n \to \infty$.

4.3-3 Let X be a Hilbert space and let Z_n, $n \geq 1$, be nonempty closed convex subsets of X that satisfy $Z_1 \subset Z_2 \subset \cdots \subset Z_n \subset \cdots$. Given an element $x \in X$, let y_n denote the projection of x onto Z_n, $n \geq 1$.

Show that $y_n \to y$ as $n \to \infty$, where y denotes the projection of x onto the set $\overline{\bigcup_{n=1}^{\infty} Z_n}$.

4.3-4 Let $\mathcal{P}_n[0, 1] = \{p|_{[0,1]};\ p \in \mathcal{P}_n\}$, where $\mathcal{P}_n$ denotes the space of all polynomials $p : \mathbb{R} \to \mathbb{R}$ of degree $\leq n$, and let a number $q > 1$ be given.

(1) Show that, given any function $f \in \mathcal{C}[0, 1]$, there exists a unique polynomial $Pf \in \mathcal{P}_n[0, 1]$ such that

$$\|f - Pf\|_{L^q(0,1)} = \inf_{p \in \mathcal{P}_n[0,1]} \|f - p\|_{L^q(0,1)}.$$

(2) Show that the mapping $P : \mathcal{C}[0, 1] \to \mathcal{P}_n[0, 1]$ defined in this fashion is linear if and only if $q = 2$ (the proof of the "if" part is similar to that of Theorem 4.3-1(e)).

4.3-5 Let $\mathbb{M}^n$, $\mathbb{S}^n_>$, $\mathbb{O}^n$, and $\mathbb{O}^n_+$, respectively denote the set of all square, positive-definite symmetric, orthogonal, and proper orthogonal, real matrices of order n.

(1) Show that, given any matrix $\boldsymbol{A} \in \mathbb{S}^n_>$, there exists one, and only one, matrix $\boldsymbol{B} \in \mathbb{S}^n_>$ such that $\boldsymbol{B}^2 = \boldsymbol{A}$. The matrix $\boldsymbol{B}$, which is called the *square root* of $\boldsymbol{A}$, is often denoted $\boldsymbol{A}^{1/2}$.

(2) Show that any *invertible* matrix $\boldsymbol{F} \in \mathbb{M}^n$ can be factored as $\boldsymbol{F} = \boldsymbol{RU}$ where $\boldsymbol{R} \in \mathbb{O}^n$ and $\boldsymbol{U} \in \mathbb{S}^n_>$ and that both matrices $\boldsymbol{R}$ and $\boldsymbol{U}$ are unique. The relation $\boldsymbol{F} = \boldsymbol{RU}$ constitutes the *polar factorization* of the invertible matrix $\boldsymbol{F}$.

(3) Let $\mathbb{U}^n := \{\boldsymbol{F} \in \mathbb{M}^n; \det \boldsymbol{F} \neq 0\}$. Show that both mappings $\boldsymbol{F} \in \mathbb{U}^n \to \boldsymbol{R} \in \mathbb{O}^n$ and $\boldsymbol{F} \in \mathbb{U}^n \to \boldsymbol{U} \in \mathbb{S}^n_>$ defined in this fashion are infinitely differentiable (to this end, notions from Chapter 7 are needed).

(4) Show that $\mathbb{O}^n_+$ is a nonempty closed subset of $\mathbb{M}^n$ that is not convex.

(5) Assume that $\det \boldsymbol{F} > 0$, so that $\boldsymbol{R} \in \mathbb{O}^n_+$. Show that

$$\|\boldsymbol{F} - \boldsymbol{R}\|_F = \inf_{\boldsymbol{S} \in \mathbb{O}^n_+} \|\boldsymbol{F} - \boldsymbol{S}\|_F,$$

where $\|\cdot\|_F$ denotes the Frobenius matrix norm (Section 4.2).

Remark In the same manner as in (2), one can show that any invertible *complex* matrix can be factored in a unique fashion as a product of a unitary matrix by a positive-definite Hermitian matrix. In this case, the terminology "polar factorization" reflects that such a factorization is an extension of the factorization $z = |z|e^{i \arg z}$ of a nonzero complex number z. □

4.3-6 Let $|\cdot|$ denote the matrix norm subordinate to the Euclidean vector norm (Problem 2.9-1). Show that

$$\inf_{\boldsymbol{S} \in \mathbb{O}^n_+} |\boldsymbol{F} - \boldsymbol{S}| = |(\boldsymbol{F}^T\boldsymbol{F})^{1/2} - \boldsymbol{I}| \leq |\boldsymbol{F}^T\boldsymbol{F} - \boldsymbol{I}|^{1/2}.$$

4.3-7 Let $(X, (\cdot,\cdot))$ be a Hilbert space.

(1) Let Z be a closed subspace of X. Show that the associated continuous linear projection operator $P : X \to Z$ (Theorem 4.3-1) possesses the following three properties: $\|P\| = 1$ (except if $Z = \{0\}$), P is *idempotent*, in the sense that $P^2 = P$, and P is *symmetric*, in the sense that $(Px, y) = (x, Py)$ for all $x, y \in X$.

(2) Let $Q : X \to X$ be a continuous linear operator that is idempotent and symmetric. Show that $Q(X)$ is a closed subspace of X and that Q is the projection operator of X onto $Q(X)$.

(3) Let $Q : X \to X$ be a continuous linear operator that is idempotent and satisfies $\|Q\| \leq 1$. Show that $Q(X)$ is a closed subspace of X and that Q is the projection operator of X onto $Q(X)$.

4.3-8 Let X be a Hilbert space, let Z be a closed subspace of X, and let $P : X \to Z$ be the associated projection operator. Show that, if $\lambda \neq 0$ and $\lambda \neq 1$, the continuous linear operator $(\lambda I - P) : X \to X$ is bijective and that its inverse is also continuous.

4.3-9 Let X be a Hilbert space and let an operator $A \in \mathcal{L}(X)$ be given such that $\|A\| \leq 1$. Show that, for any $x \in X$, the sequence $(y_n)_{n=1}^\infty$ defined by $y_n := \dfrac{1}{n}(x + Ax + \cdots + A^{n-1}x)$, $n \geq 1$, converges in X.

Hint: Show that $\lim_{n\to\infty} y_n$ is the projection of x onto the closure of $\mathrm{Span}(A^k f)_{k=0}^\infty$.

4.3-10 Let Y be a nonempty convex and closed subset of a real Hilbert space $(X, (\cdot,\cdot))$ and let $b \in (X - Y)$. Show that there exist $a \in X$ and $\alpha \in \mathbb{R}$ such that

$$(b, y) < \alpha < (y, a) \quad \text{for all } y \in Y.$$

Remark This property expresses that *the hyperplane* $\{x \in X;\ (x-a)=\alpha\}$ *strictly separates the convex sets* Y *and* $\{b\}$, a property that holds in fact in arbitrary normed vector spaces (but then is substantially harder to prove at this level of generality; cf. Theorem 5.10-2). □

4.3-11 The objective of this problem is to establish the **Farkaš lemma.**[6] *Let* $(X, (\cdot,\cdot))$ *be a real Hilbert space, and let* b *and* c_i, $1 \le i \le m$, *be vectors in* X. *Then the inclusion*

$$\{x \in X;\ (c_i, x) \ge 0,\ 1 \le i < m\} \subset \{x \in X;\ (b, x) \ge 0\}$$

holds if and only if there exist real numbers λ_i, $1 \le i \le m$, *such that*

$$\lambda_i \ge 0,\quad 1 \le i \le m, \quad \text{and} \quad b = \sum_{i=1}^{m} \lambda_i c_i.$$

(1) Show that the set

$$Y := \left\{ \sum_{i=1}^{m} \lambda_i c_i \in X;\ \lambda_i \ge 0,\ 1 \le i \le m \right\}$$

(which is clearly a cone with vertex 0) is a convex and closed subset of X.

(2) Using question (1) and Problem 4.3-10, show that, if a point $b \in X$ does not belong to the set Y, there exists a vector $a \in X$ such that $(c_i, a) \ge 0$, $1 \le i \le m$, and $(b, a) < 0$.

(3) Deduce from (2) the "only if" part of the Farkaš–Minkowski lemma (the "if" part is clear).

Remark The Farkaš lemma plays a key role in proving the existence of *Kuhn–Tucker multipliers* found in constrained optimization problems when the constraints take the form of *inequalities* (Problem 7.15-3). □

4.4 Application of the projection theorem: Least-squares solution of a linear system

Given an arbitrary $m \times n$ real matrix $\boldsymbol{A}$ and an arbitrary vector $\boldsymbol{c} \in \mathbb{R}^m$, there is generally no vector $\boldsymbol{x} \in \mathbb{R}^n$ that satisfies $\boldsymbol{Ax} = \boldsymbol{c}$. Finding a *least-squares solution*[7] to this linear system consists instead in finding a vector $\boldsymbol{x} \in \mathbb{R}^n$ that minimizes the Euclidean distance in $\mathbb{R}^m$ between the vectors $\boldsymbol{Ax}$ and $\boldsymbol{c}$ (hence the terminology "least-squares" solution). The following simple corollary to the *projection theorem* shows that, by contrast with the former problem, the latter always has at least one solution.

Theorem 4.4-1 (least-squares solution of a linear system) *Let* $|\cdot|$ *denote the Euclidean norm in* $\mathbb{R}^m$.

(a) *Let there be given an* $m \times n$ *matrix* $\boldsymbol{A}$ *and a vector* $\boldsymbol{c} \in \mathbb{R}^m$. *Then the following minimization problem: Find* $\boldsymbol{x} \in \mathbb{R}^n$ *such that*

$$|\boldsymbol{Ax} - \boldsymbol{c}| = \inf_{\boldsymbol{y} \in \mathbb{R}^n} |\boldsymbol{Ay} - \boldsymbol{c}|$$

[6] J. Farkaš [1901]: Theorie der einfachen Ungleichungen, *Journal für die Reine und Angewandte Mathematik* **124**, 1–27.

[7] This method was discovered, for the purpose of computing (by hand, of course) orbits of celestial bodies, by:
A.M. Legendre [1805]: *Nouvelle Méthode pour la Détermination des Orbites des Comètes*, Chez Didot, Paris.
C.F. Gauß [1809]: *Theoria Motus Corporum Coelestium in Sectionibus Conicis Solum Ambientium*, Perthes und Besser, Hamburg.

has at least one solution.

(b) *A vector* $\boldsymbol{x} \in \mathbb{R}^n$ *satisfies the above minimization problem if and only if* $\boldsymbol{x}$ *is a solution of the linear system*

$$\boldsymbol{A}^T\boldsymbol{A}\boldsymbol{x} = \boldsymbol{A}^T\boldsymbol{c}.$$

Proof Since $\operatorname{Im}\boldsymbol{A}$ is a closed subspace of $\mathbb{R}^m$ (as a finite-dimensional subspace; cf. Theorem 2.7-1(c)), the projection theorem (Theorem 4.3-1) asserts that there exists a unique element $\widetilde{\boldsymbol{x}} \in \operatorname{Im}\boldsymbol{A}$ that satisfies

$$|\widetilde{\boldsymbol{x}} - \boldsymbol{c}| = \inf_{\widetilde{\boldsymbol{y}} \in \operatorname{Im}\boldsymbol{A}} |\widetilde{\boldsymbol{y}} - \boldsymbol{c}|,$$

or equivalently, that satisfies

$$(\widetilde{\boldsymbol{x}} - \boldsymbol{c}, \widetilde{\boldsymbol{y}})_m = 0 \quad \text{for all } \widetilde{\boldsymbol{y}} \in \operatorname{Im}\boldsymbol{A},$$

where $(\cdot,\cdot)_m$ denotes the Euclidean inner product in $\mathbb{R}^m$. By definition of the space $\operatorname{Im}\boldsymbol{A}$, there thus exists at least one vector $\boldsymbol{x} \in \mathbb{R}^n$ that satisfies

$$|\boldsymbol{A}\boldsymbol{x} - \boldsymbol{c}| = \inf_{\boldsymbol{y} \in \mathbb{R}^n} |\boldsymbol{A}\boldsymbol{y} - \boldsymbol{c}|,$$

or equivalently, that satisfies

$$(\boldsymbol{A}\boldsymbol{x} - \boldsymbol{c}, \boldsymbol{A}\boldsymbol{y})_m = (\boldsymbol{A}^T\boldsymbol{A}\boldsymbol{x} - \boldsymbol{A}^T\boldsymbol{c}, \boldsymbol{y})_n = 0 \quad \text{for all } \boldsymbol{y} \in \mathbb{R}^n,$$

where $(\cdot,\cdot)_n$ denotes the Euclidean inner product in $\mathbb{R}^n$ and the matrix $\boldsymbol{A}^T$ denotes the transpose of $\boldsymbol{A}$.

Both assertions (a) and (b) are thus proved. □

The linear system $\boldsymbol{A}^T\boldsymbol{A}\boldsymbol{x} = \boldsymbol{A}^T\boldsymbol{c}$, which therefore always has *at least one solution* by the above theorem, constitutes the **normal equations**[8] associated to the least-squares solution of the linear system $\boldsymbol{A}\boldsymbol{x} = \boldsymbol{c}$. Naturally, if the set $\{\boldsymbol{x} \in \mathbb{R}^n;\ \boldsymbol{A}\boldsymbol{x} = \boldsymbol{c}\}$ is nonempty, it coincides with the set of solutions to the normal equations.

It should be again emphasized that finding the least-squares solution to a linear system gives rise to a *linear* problem (namely, the normal equations) only because the norm used for that purpose is *induced by an inner product* (a similar observation was made at the end of Section 4.3 about the linearity of the projection operator onto a subspace). This observation explains why, from a numerical standpoint, least-squares solutions are overwhelmingly preferred to "least-$\|\cdot\|_p$ norm solutions" with $p \neq 2$.

Remarks (1) The above considerations can be immediately extended to the *complex* case, in which case the normal equations become $\boldsymbol{A}^*\boldsymbol{A}\boldsymbol{x} = \boldsymbol{A}^*\boldsymbol{c}$, where $\boldsymbol{A}^*$denotes the adjoint matrix of $\boldsymbol{A}$.

(2) A criterion for the *uniqueness* of the solution to the normal equations is given in Problem 4.7-2. □

[8]Discovered, and so called, in:

C.F. GAUß [1822]: Anwendung der Wahrscheinlichkeitsrechnung auf eine Aufgabe der practischen Geometrie, *Astronomische Nachrichten* **1**, 81–86.

4.5 Orthogonality; direct sum theorem

Let $(X, (\cdot,\cdot))$ be a real or complex inner-product space. Two vectors $x \in X$ and $y \in X$ are said to be **orthogonal** if

$$(x, y) = 0,$$

and the **orthogonal complement** of any nonempty subset Z of X is the subset of X defined as

$$Z^{\perp} := \{x \in X;\ (x, z) = 0 \text{ for all } z \in Z\}.$$

The next result lists some elementary properties of orthogonal complements.

Theorem 4.5-1 *Let Z be a nonempty subset of an inner-product space X. Then the set $Z^{\perp}$ is a closed subspace of X. Besides, $(\overline{Z})^{\perp} = Z^{\perp}$, and $Z \cap Z^{\perp} = \{0\}$ if $0 \in Z$ and $Z \cap Z^{\perp} = \varnothing$ if $0 \notin Z$.*

Proof That $Z^{\perp}$ is a subspace follows from the linearity of the inner product with respect to its first argument; that $Z^{\perp}$ is closed follows from the continuity of the inner product with respect to its first argument (Theorem 4.1-1(c)).

The definition of the orthogonal complement implies that $(\overline{Z})^{\perp} \subset Z^{\perp}$. To show that $Z^{\perp} \subset (\overline{Z})^{\perp}$, let $x \in Z^{\perp}$; since $(x, z) = 0$ for all $z \in Z$, the continuity of the inner product with respect to its second argument implies that $(x, z) = 0$ for all $z \in \overline{Z}$, and hence that $x \in (\overline{Z})^{\perp}$.

The relation $Z \cap Z^{\perp} = \{0\}$, *resp.* $Z \cap Z^{\perp} = \varnothing$, clearly holds if $0 \in Z$, *resp.* $0 \notin Z$. □

When X is a *Hilbert space* and Y is a *closed subspace* of X, it turns out that the space X can be written as the *direct sum* (Section 2.1) of its subspaces Y and $Y^{\perp}$, which is also a closed subspace of X (Theorem 4.5-1). As shown below, this remarkable property is in effect a simple corollary of the *projection theorem.*

Theorem 4.5-2 (direct sum theorem) *Let X be a real or complex Hilbert space and let Y be a closed subspace of X. Then the space X is the direct sum*

$$X = Y \oplus Y^{\perp},$$

i.e., any element $x \in X$ can be written as

$$x = y + y^{\perp} \quad \textit{with } y \in Y \textit{ and } y^{\perp} \in Y^{\perp},$$

and such a decomposition is unique. In fact,

$$y = Px \quad \textit{and} \quad y^{\perp} = P^{\perp}x,$$

where $P : X \to Y$ denotes the projection operator from X onto Y, and

$$P^{\perp} := I - P$$

is the projection operator from X onto $Y^{\perp}$.

Proof Any element $x \in X$ can be written as

$$x = Px + (I - P)x.$$

Then $Px \in Y$ by definition of the projection operator. Besides, $(I - P)x \in Y^\perp$, since $((I - P)x, z) = 0$ for all $z \in Y$ by the characterization of the projection onto a closed subspace (Theorem 4.3-1(d)). Hence

$$x = y + y^\perp \quad \text{with } y := Px \in Y \text{ and } y^\perp := (I - P)x \in Y^\perp.$$

To verify that such a decomposition is unique, let

$$x = y + y^\perp = \widehat{y} + \widehat{y}^\perp \quad \text{with } y, \widehat{y} \in Y \text{ and } y^\perp, \widehat{y}^\perp \in Y^\perp.$$

Since $(y - \widehat{y}) \in Y$ and $(y^\perp - \widehat{y}^\perp) \in Y^\perp$ (the set $Y^\perp$ is also a subspace; cf. Theorem 4.5-1), it follows that $y - \widehat{y} = y^\perp - \widehat{y}^\perp = 0$ since $Y \cap Y^\perp = \{0\}$.

That $P^\perp := I - P$ is indeed the projection operator from X onto the subspace $Y^\perp$ follows from the characterization of the projection: for any element $x \in X$,

$$(x - P^\perp x, y^\perp) = (Px, y^\perp) = 0 \quad \text{for all } y^\perp \in Y^\perp,$$

since $Px \in Y$. □

Remarks (1) Theorem 4.5-2 implies that $Y = (Y^\perp)^\perp$ if Y is a *closed subspace*, since $X = Y \oplus Y^\perp = (Y^\perp)^\perp \oplus Y^\perp$.

(2) If the subspace Y is not necessarily closed, then X can still be written as a direct sum, viz., $X = \overline{Y} \oplus Y^\perp$, since $(\overline{Y})^\perp = Y^\perp$ (Theorem 4.5-1). □

Problems

4.5-1 Let the space $\mathcal{C}^1[0,1]$ be equipped with the inner product $(\cdot,\cdot)$ defined by

$$(f, g) := \int_0^1 (f'g' + fg)\,\mathrm{d}x,$$

and let the subset Y of $\mathcal{C}^1[0,1]$ be defined by

$$Y := \{g \in \mathcal{C}^1[0,1];\ g(0) = g(1) = 0\}.$$

(1) Show that Y is a closed subspace of $(\mathcal{C}^1[0,1], (\cdot,\cdot))$, and also of $(\mathcal{C}^2[0,1], (\cdot,\cdot))$.
Hint: Show that there exists a constant C such that $\sup_{0 \le x \le 1} |f(x)| \le C(f,f)^{1/2}$ for all $f \in \mathcal{C}^1[0,1]$.
(2) Identify the orthogonal complement $Y^\perp$ of Y in $(\mathcal{C}^2[0,1], (\cdot,\cdot))$. What is the dimension of $Y^\perp$?

4.5-2 Let the subset Y of the Hilbert space ℓ^2 (Section 4.2) be defined by

$$Y := \{x = (x_i)_{i=1}^\infty;\ x_{2k-1} = x_{2k} \text{ for all integers } k \ge 1\}.$$

(1) Show that Y is a closed subspace of ℓ^2.
(2) Identify the orthogonal complement of Y in ℓ^2.
(3) Identify the projection operators $P : \ell^2 \to Y$ and $P^\perp : \ell^2 \to Y^\perp$.

4.6 F. Riesz representation theorem in a Hilbert space

Let $(X, (\cdot,\cdot))$ be an inner-product space over $\mathbb{K} = \mathbb{R}$ or $\mathbb{K} = \mathbb{C}$, and let X' designate its dual space. Then, given any vector $y \in X$, the linear functional $\ell_y : X \to \mathbb{K}$ defined by

$$\ell_y(x) := (x, y) \in \mathbb{K} \quad \text{for all } x \in X,$$

is continuous and

$$\|\ell_y\|_{X'} = \|y\|,$$

since, if $y \neq 0$,

$$\|\ell_y\|_{X'} = \sup_{x\neq 0} \frac{|\ell_y(x)|}{\|x\|} = \sup_{x\neq 0} \frac{|(x,y)|}{\|x\|} = \|y\|,$$

by Theorem 4.1-1.

It is remarkable, and of paramount importance, that *the converse holds if X is a Hilbert space*, thanks to the *direct sum theorem* (itself a corollary to the projection theorem).

Theorem 4.6-1 (F. Riesz representation theorem in a Hilbert space[9]**)** *Let $(X, (\cdot,\cdot))$ be a Hilbert space over $\mathbb{K} = \mathbb{R}$ or $\mathbb{K} = \mathbb{C}$. Then, given any continuous linear functional $\ell \in X'$, there exists one and only one vector $y_\ell \in X$ such that*

$$\ell(x) = (x, y_\ell) \quad \textit{for all } x \in X.$$

Besides,

$$\|\ell\|_{X'} = \|y_\ell\|_X,$$

and the **F. Riesz isometry**

$$\sigma : \ell \in X' \to \sigma(\ell) := y_\ell \in X$$

defined in this fashion is a bijection, which is linear if $\mathbb{K} = \mathbb{R}$, or semilinear if $\mathbb{K} = \mathbb{C}$.

Consequently, any Hilbert space can be identified with its dual space X' by means of the F. Riesz isometry $\sigma : X' \to X$. Besides, the dual space X' becomes a Hilbert space when it is equipped with the inner product $(\cdot,\cdot)_{X'} : X' \times X' \to \mathbb{K}$ defined by

$$(x', y')_{X'} := \overline{(\sigma x', \sigma y')} \quad \textit{for each } x', y' \in X'.$$

Proof If $\ell = 0$, it suffices to let $y_\ell = 0$. If $\ell \neq 0$, let

$$Y := \{x \in X;\ \ell(x) = 0\}.$$

Then Y is a closed subspace of X since $\ell : X \to \mathbb{K}$ is linear and continuous; besides, $Y \subsetneq X$ since $\ell \neq 0$. Hence

$$X = Y \oplus Y^\perp$$

by the *direct sum theorem* (Theorem 4.5-2), and $Y^\perp$ contains nonzero vectors ($Y^\perp = \{0\}$ would imply $Y = X$). So, let $y_0 \in Y^\perp$ with $y_0 \neq 0$; hence $\ell(y_0) \neq 0$ (otherwise $\ell(y_0) = 0$

[9] F. Riesz [1907]: Sur une espèce de géométrie analytique des systèmes de fonctions sommables, *Comptes Rendus de l'Académie des Sciences de Paris* **144**, 1409–1411.

would imply that $y_0 \in Y$; but $Y \cap Y^\perp = \{0\}$) and thus there is no loss of generality in assuming that

$$\ell(y_0) = 1.$$

The characterization of the projection onto a subspace (Theorem 4.3-1(d)) then shows that the projection operator $P : X \to Y$ is given by

$$Px = x - \ell(x)y_0 \quad \text{for all } x \in X,$$

since $Px \in Y$ and $(Px - x, y) = -\ell(x)(y_0, y) = 0$ for all $y \in Y$.

Consequently, the projection operator $P^\perp : X \to Y^\perp$ is given by (Theorem 4.5-2)

$$P^\perp x = (I - P)x = \ell(x)y_0 \quad \text{for all } x \in X.$$

The vector

$$y_\ell := \frac{1}{\|y_0\|^2} y_0$$

thus satisfies the announced property, since

$$(x, y_\ell) = \frac{1}{\|y_0\|^2}(Px + P^\perp x, y_0) = \frac{1}{\|y_0\|^2}(P^\perp x, y_0) = \ell(x) \quad \text{for all } x \in X.$$

That y_ℓ is uniquely defined is clear since $(x, y) = (x, \widetilde{y})$ for all $x \in X$ implies $y = \widetilde{y}$ (take $x = y - \widetilde{y}$). That the mapping $\ell \in X' \to y_\ell \in X$ defined in this fashion is a bijection, which is linear if $\mathbb{K} = \mathbb{R}$ or semilinear if $\mathbb{K} = \mathbb{C}$, is equally clear. Besides,

$$\|\ell\|_{X'} = \sup_{x \neq 0} \frac{|\ell(x)|}{\|x\|} = \sup_{x \neq 0} \frac{|(x, y_\ell)|}{\|x\|} = \|y_\ell\|.$$

Finally, it is immediately verified that the function $(\cdot, \cdot)_{X'} : X' \times X' \to \mathbb{K}$ as defined in the statement of the theorem is an inner product on X', as a consequence of the sesquilinearity of the inner product $(\cdot, \cdot)$ on X. It is also clear that the inner-product space $(X', (\cdot, \cdot)_{X'})$ is complete since

$$\|x'\|_{X'} = \left((x', x')_{X'}\right)^{1/2} = \left((\sigma x', \sigma x')\right)^{1/2} = \|\sigma x'\|_X \quad \text{for any } x' \in X'. \qquad \square$$

Remark The relation $P^\perp x = \ell(x)y_0$ for all $x \in X$ established in the above proof shows that

$$Y^\perp = P^\perp(X) = \{\alpha y_0 \in X;\ \alpha \in \mathbb{K}\} = \mathrm{Span}(y_0)$$

is a *one-dimensional* subspace of the space X. $\square$

For example, let Ω be an open subset of $\mathbb{R}^N$ and let A be a measurable subset of Ω that satisfies $\int_A \mathrm{d}x < \infty$. Hence the characteristic function χ_A of the set A belongs to the (real) Hilbert space $L^2(\Omega)$. Then the functional $\ell : f \in L^2(\Omega) \to \int_A f(x)\,\mathrm{d}x$, which is clearly continuous by the Cauchy–Schwarz–Bunyakovskiĭ inequality, is also given by $f \in L^2(\Omega) \to \int_\Omega \chi_A(x) f(x)\,\mathrm{d}x$.

More generally, Theorem 4.6-1 shows that, given any continuous linear functional ℓ over the space $L^2(\Omega)$, there exists a function $g_\ell \in L^2(\Omega)$ such that

$$\ell(f) = \int_\Omega f(x) g_\ell(x)\,\mathrm{d}x \quad \text{for all } f \in L^2(\Omega).$$

While this remarkable result is thus an effortless application of the F. Riesz representation theorem in a *Hilbert space*, its extension to the space $L^p(\Omega)$ for any $1 < p < \infty$, $p \neq 2$, requires by contrast a specific, and substantially more delicate, proof (in this case the function g_ℓ belongs to the space $L^q(\Omega)$ with $\dfrac{1}{p} + \dfrac{1}{q} = 1$; cf. Theorem 3.5-3).

4.7 First applications of the F. Riesz representation theorem: Hahn–Banach theorem in a Hilbert space; adjoint operators; reproducing kernels

Together with the direct sum theorem, the F. Riesz representation theorem provides a remarkably simple proof of the *"Hilbert space version" of one of the most basic results from linear functional analysis*, whose proof in an arbitrary normed vector space otherwise requires the *axiom of choice* (Theorem 5.9-1). Recall that X' denotes the dual space of a normed vector space X.

Theorem 4.7-1 (Hahn–Banach theorem in a Hilbert space) *Let X be a Hilbert space over $\mathbb{K} = \mathbb{R}$ or $\mathbb{K} = \mathbb{C}$, let Y be a subspace of X, and let $\ell : Y \to \mathbb{K}$ be a continuous linear form on Y. Then there exists a continuous linear form $\widetilde{\ell} : X \to \mathbb{K}$ that satisfies*

$$\widetilde{\ell}(y) = \ell(y) \text{ for all } y \in Y \quad \text{and} \quad \|\widetilde{\ell}\|_{X'} = \|\ell\|_{Y'}.$$

Besides, such an extension is unique.

Proof Let $\overline{Y}$ denote the closure of Y in X. Since the field $\mathbb{K}$ is complete, there exists a unique continuous linear form $\widehat{\ell} : \overline{Y} \to \mathbb{K}$ that satisfies

$$\widehat{\ell}(y) = \ell(y) \text{ for all } y \in Y \quad \text{and} \quad \|\widehat{\ell}\|_{(\overline{Y})'} = \|\ell\|_{Y'}$$

(Theorem 3.1-1). By the *direct sum theorem* (Theorem 4.5-2), any element $x \in X$ can be written in a unique fashion as

$$x = Px + P^\perp x,$$

where P and $P^\perp$ respectively denote the projection operators from the Hilbert space X onto its closed subspaces $\overline{Y}$ and $(\overline{Y})^\perp$. Let then the linear form $\widetilde{\ell} : X \to \mathbb{K}$ be defined by

$$\widetilde{\ell}(x) := \widehat{\ell}(Px) \quad \text{for all } x \in X.$$

Then $\widetilde{\ell}$ is an extension of ℓ since

$$\widetilde{\ell}(y) = \widehat{\ell}(Py) = \widehat{\ell}(y) = \ell(y) \quad \text{for all } y \in Y,$$

and

$$\|\ell\|_{Y'} = \sup_{\begin{cases} y\in Y \\ y\neq 0\end{cases}} \frac{|(\ell(y)|}{\|y\|} \le \sup_{\begin{cases} x\in X \\ x\neq 0\end{cases}} \frac{|\widetilde{\ell}(x)|}{\|x\|} = \|\widetilde{\ell}\|_{X'} = \sup_{\begin{cases} x\in X \\ x\neq 0\end{cases}} \frac{|\widehat{\ell}(Px)|}{\|x\|} \le \|\widehat{\ell}\|_{\overline{Y}'} = \|\ell\|_{Y'},$$

since $\|Px\| \le \|x\|$ for all $x \in X$; hence $\|\widetilde{\ell}\|_{X'} = \|\ell\|_{Y'}$.

To verify that such an extension is *unique*, there is no loss of generality in assuming that Y is closed (since the extension from Y to $\overline{Y}$ is unique). This being the case, let $\ell^\sharp \in X'$ be an extension of $\ell \in Y'$ that satisfies $\|\ell^\sharp\|_{X'} = \|\ell\|_{Y'}$. Hence, by the *F. Riesz representation theorem in a Hilbert space*, there exists a unique vector $z \in X$ such that

$$\ell^\sharp(x) = (x, z) \text{ for all } x \in X \quad \text{and} \quad \|\ell^\sharp\|_{X'} = \|z\|\,.$$

Since then

$$\ell^\sharp(y) = \ell(y) = (y, z) = (y, Pz) \quad \text{for all } y \in Y,$$

it also follows that $\|\ell\|_{Y'} = \|Pz\|$. Hence $\|\ell^\sharp\|_{X'} = \|\ell\|_{Y'}$ implies that $\|Pz\| = \|z\|$, and hence that $z = Pz$. Consequently,

$$\ell^\sharp(x) = (x, Pz) = (Px, z) = \ell^\sharp(Px) = \ell(Px) \quad \text{for all } x \in X,$$

which shows that $\ell^\sharp = \widetilde{\ell}$. □

As a preparation to another application of the F. Riesz representation theorem in a Hilbert space, consider the spaces $\mathbb{R}^n$ and $\mathbb{R}^m$, both equipped with their Euclidean inner product (Section 4.2), respectively denoted $(\cdot,\cdot)_n$ and $(\cdot,\cdot)_m$. Then the $n \times m$ *transpose matrix* $\boldsymbol{A}^T$ of any real $m \times n$ matrix $\boldsymbol{A} = (a_{ij})$, which is defined by $(\boldsymbol{A}^T)_{ij} = a_{ji}$, can be also defined as the unique $n \times m$ real matrix that satisfies

$$(\boldsymbol{Ax}, \boldsymbol{y})_m = (\boldsymbol{x}, \boldsymbol{A}^T\boldsymbol{y})_n \quad \text{for all } \boldsymbol{x} \in \mathbb{R}^n, \boldsymbol{y} \in \mathbb{R}^m.$$

Similarly, the $n \times m$ *adjoint matrix* $\boldsymbol{A}^*$ of any complex $m \times n$ matrix $\boldsymbol{A} = (a_{ij})$ can be also defined as the unique $n \times m$ complex matrix that satisfies

$$(\boldsymbol{Ax}, \boldsymbol{y})_m = (\boldsymbol{x}, \boldsymbol{A}^*\boldsymbol{y})_n \quad \text{for all } \boldsymbol{x} \in \mathbb{C}^n, \boldsymbol{y} \in \mathbb{C}^m,$$

where $(\cdot,\cdot)_n$ and $(\cdot,\cdot)_m$ now denote the Hermitian inner product on $\mathbb{C}^n$ and $\mathbb{C}^m$ (Section 4.2).

It is remarkable that, thanks to the *F. Riesz representation theorem*, the *transpose* in the real case, or the *adjoint* in the complex case, of any *continuous linear operator* between two *Hilbert spaces* can be similarly defined. For brevity, only the *complex* case is considered in the next theorem; the modifications in the real case are indicated after the proof. Various complements are proposed in Problem 4.7-1.

Theorem 4.7-2 (adjoint operator) *Let $(X, (\cdot,\cdot)_X)$ and $(Y, (\cdot,\cdot)_Y)$ be two complex Hilbert spaces and let an operator $A \in \mathcal{L}(X;Y)$ be given.*

(a) *There exists a unique operator $A^* \in \mathcal{L}(Y;X)$, called the* **adjoint** *of A, that satisfies*

$$(Ax, y)_Y = (x, A^*y)_X \quad \textit{for all } x \in X,\ y \in Y.$$

The mapping $A \in \mathcal{L}(X;Y) \to A^* \in \mathcal{L}(Y;X)$ *defined in this fashion is semilinear. Besides,*

$$\|A^*\|_{\mathcal{L}(Y;X)} = \|A\|_{\mathcal{L}(X;Y)}.$$

(b) *The following relations hold:*

$$(\operatorname{Im} A)^\perp = \operatorname{Ker} A^* \quad \textit{and} \quad (\operatorname{Im} A^*)^\perp = \operatorname{Ker} A,$$
$$Y = \operatorname{Ker} A^* \oplus \overline{\operatorname{Im} A} \quad \textit{and} \quad X = \operatorname{Ker} A \oplus \overline{\operatorname{Im} A^*}.$$

Proof For each element $y \in Y$, the mapping $x \in X \to (Ax, y)_Y \in \mathbb{K}$ is a continuous linear functional since $|(Ax,y)_Y| \le \|A\|\,\|x\|\,\|y\|$ for all $x \in X$. Hence the *F. Riesz representation theorem* (Theorem 4.6-1) applied in the Hilbert space X shows that there exists a uniquely defined element $A^*y \in X$ such that

$$(Ax, y)_Y = (x, A^*y)_X \quad \text{for all } x \in X.$$

The mapping $A^* : Y \to X$ defined in this fashion is linear since, for all $\alpha, \beta \in \mathbb{C}$, $x \in X$, and $y, z \in Y$,

$$\begin{aligned}(x, A^*(\alpha y + \beta z)) &= (Ax, \alpha y + \beta z) = \overline{\alpha}(Ax, y) + \overline{\beta}(Ax, z) \\ &= \overline{\alpha}(x, A^*y) + \overline{\beta}(x, A^*z) = (x, \alpha A^*y + \beta A^*z).\end{aligned}$$

That $(\alpha A + \beta B)^* = \overline{\alpha}A^* + \overline{\beta}B^*$ for all $A, B \in \mathcal{L}(X;Y)$ is clear.

The linear operator $A^* : Y \to X$ is continuous, since

$$\|A^*y\|^2 = (A^*y, A^*y)_X = (AA^*y, y)_Y \le \|A\|\,\|A^*y\|\,\|y\| \quad \text{for all } y \in Y,$$

so that

$$\|A^*\|_{\mathcal{L}(Y;X)} = \sup_{y \ne 0} \frac{\|A^*y\|}{\|y\|} \le \|A\|_{\mathcal{L}(X;Y)}.$$

Likewise,

$$\|Ax\|^2 = (Ax, Ax)_Y = (x, A^*Ax)_X \le \|A^*\|\,\|Ax\|\,\|x\| \quad \text{for all } x \in X,$$

so that

$$\|A\|_{\mathcal{L}(X;Y)} = \sup_{x \ne 0} \frac{\|Ax\|}{\|x\|} \le \|A^*\|_{\mathcal{L}(Y;X)}.$$

Hence $\|A^*\|_{\mathcal{L}(Y;X)} = \|A\|_{\mathcal{L}(X;Y)}$. This proves (a).

To prove (b), simply note that

$$\begin{aligned}(\operatorname{Im} A)^\perp &= \{y \in Y;\ (y,z)_Y = 0 \text{ for all } z \in \operatorname{Im} A\}, \\ &= \{y \in Y;\ (y, Ax)_Y = 0 \text{ for all } x \in X\}, \\ &= \{y \in Y;\ (A^*y, x)_X = 0 \text{ for all } x \in X\} = \operatorname{Ker} A^*.\end{aligned}$$

Since $(\overline{\operatorname{Im} A})^\perp = \operatorname{Im} A^\perp$ (Theorem 4.5-1), it follows from the direct sum theorem (Theorem 4.5-2) that

$$Y = \overline{\operatorname{Im} A} \oplus \left(\overline{\operatorname{Im} A}\right)^\perp = \overline{\operatorname{Im} A} \oplus (\operatorname{Im} A)^\perp = \overline{\operatorname{Im} A} \oplus \operatorname{Ker} A^*.$$

The other relations in (b) are established in an analogous manner. □

Remarks (1) The completeness of the space Y is not used for establishing the existence of the adjoint A^*. It is only needed for concluding that Y can be written as the direct sum $Y = \operatorname{Ker} A^* \oplus \overline{\operatorname{Im} A}$.

(2) Naturally, $\overline{\operatorname{Im} A} = \operatorname{Im} A$ if Y is finite-dimensional, and $\overline{\operatorname{Im} A^*} = \operatorname{Im} A^*$ if X is finite-dimensional. □

If $(X, (\cdot,\cdot)_X)$ and $(Y, (\cdot,\cdot)_Y)$ are *real* Hilbert spaces, one similarly establishes that, given any operator $A \in \mathcal{L}(X;Y)$, there exists a unique operator $A^T \in \mathcal{L}(Y;X)$, called the **transpose** of A, that satisfies

$$(Ax, y)_Y = (x, A^T y) \quad \text{for all } x \in X,\ y \in Y.$$

Save that the mapping $A \in \mathcal{L}(X;Y) \to A^T \in \mathcal{L}(Y;X)$ defined in this fashion is now *linear*, all the other properties established in Theorem 4.7-2 hold *verbatim* with A^T in lieu of A^*.

A simple corollary of the above theorem is the following classical result in matrix theory.

Theorem 4.7-3 (Fredholm alternative in finite-dimensional spaces) *Let there be given a real* ($\mathbb{K} = \mathbb{R}$) *or complex* ($\mathbb{K} = \mathbb{C}$) $m \times n$ *matrix* $\boldsymbol{A}$ *and a vector* $\boldsymbol{b} \in \mathbb{K}^m$.

Then either the linear system $\boldsymbol{Ax} = \boldsymbol{b}$ *has at least one solution* $\boldsymbol{x} \in \mathbb{K}^n$, *or it has no solution and then there exists at least one vector* $\boldsymbol{y} \in \mathbb{K}^m$ *such that*

$$\boldsymbol{A}^T\boldsymbol{y} = \boldsymbol{0} \text{ and } \boldsymbol{y}^T\boldsymbol{b} \neq 0 \text{ if } \mathbb{K} = \mathbb{R}; \quad \text{or} \quad \boldsymbol{A}^*\boldsymbol{y} = \boldsymbol{0} \text{ and } \boldsymbol{y}^*\boldsymbol{b} \neq 0 \text{ if } \mathbb{K} = \mathbb{C}.$$

Proof To fix ideas, assume that $\mathbb{K} = \mathbb{C}$ (the proof is analogous if $\mathbb{K} = \mathbb{R}$) and let $\mathbb{C}^n$ be equipped with its Hermitian inner product (Section 4.2). Noting that the finite-dimensional space $\operatorname{Im} A$ is closed (Theorem 2.7-1(c)), we infer from Theorem 4.7-2(b) that

$$\mathbb{C}^m = \operatorname{Ker} \boldsymbol{A}^* \oplus \operatorname{Im} \boldsymbol{A}.$$

Therefore, either $\boldsymbol{b} \in \operatorname{Im} \boldsymbol{A}$, in which case the linear system $\boldsymbol{Ax} = \boldsymbol{b}$ has at least one solution $\boldsymbol{x} \in \mathbb{C}^n$, or $\boldsymbol{b} \notin \operatorname{Im} \boldsymbol{A}$, in which case the linear system has no solution and the projection $\boldsymbol{w}$ of $\boldsymbol{b}$ on the space $\operatorname{Ker} \boldsymbol{A}^*$, which cannot be the zero vector of $\mathbb{C}^m$ since $\boldsymbol{b} \notin \operatorname{Im} \boldsymbol{A}$, satisfies $\boldsymbol{A}^*\boldsymbol{y} = \boldsymbol{0}$ and $\boldsymbol{y}^*\boldsymbol{b} = \boldsymbol{y}^*\boldsymbol{y} \neq 0$. □

As a preparation for another application, consider the space ℓ^2 whose elements $x = (x_i)_{i=0}^\infty$ are in effect *functions* $x : i \in \mathbb{N} \to \mathbb{K}$. For each integer $j \in \mathbb{N}$, let $e_j := (\delta_{ij})_{i=0}^\infty$. It is then clear that $e_j \in \ell^2$ for each $j \in \mathbb{N}$ and that

$$x_j = (x, e_j)_{\ell^2} \quad \text{for all } j \geq 0 \text{ and all } x = (x_i) \in \ell^2.$$

A simple criterion, insuring that this property may be shared by more general Hilbert spaces whose elements are also *functions*, is provided by another corollary of the F. Riesz representation theorem.

Theorem 4.7-4 (reproducing kernel) *Let A be a nonempty set and let $(X, (\cdot,\cdot))$ be a Hilbert space over $\mathbb{K} = \mathbb{R}$ or $\mathbb{K} = \mathbb{C}$ whose elements are functions $x : A \to \mathbb{K}$. Assume that, for each $a \in A$, there exists a constant $C(a) > 0$ such that*

$$|x(a)| \leq C(a)\|x\| \quad \textit{for all } x \in X.$$

Then there exists a function

$$K : A \times A \to \mathbb{K}$$

called a **reproducing kernel** *of* X, *such that, for each* $a \in A$, *the function* $K(\cdot, a) : A \to \mathbb{K}$ *is an element of the space* X, *and*

$$x(a) = (x, K(\cdot, a)) \quad \textit{for all } x \in X.$$

Proof For each $a \in A$, the linear functional $x \in X \to x(a) \in \mathbb{K}$ is continuous by assumption. The F. Riesz representation theorem thus shows that there exists an element $K(\cdot, a)$ in the space X, which is therefore a function $K(\cdot, a) : A \to \mathbb{K}$, that satisfies $x(a) = (x, K(\cdot, a))$ for all $x \in X$. □

This seemingly innocuous corollary of the F. Riesz representation theorem in a Hilbert space has important consequences, regarding in particular the *existence of nonnegative Green's functions* for certain classes of boundary value problems.[10]

Another important application of the F. Riesz representation theorem is proposed in Problem 4.7-3.

Problems

4.7-1 Let the assumptions and notations be those of Theorem 4.7-2.

(1) Show that $(A^*)^* = A^*$.

(2) Show that $\operatorname{Ker} A^* = \operatorname{Ker}(AA^*)$ and $\overline{\operatorname{Im} A} = \overline{\operatorname{Im}(AA^*)}$.

(3) Show that $\|A^*A\|_{\mathcal{L}(X)} = \|A\|^2_{\mathcal{L}(X;Y)} = \|AA^*\|_{\mathcal{L}(Y)}$.

(4) Show that, if $A \in \mathcal{L}(X;Y)$ is bijective and $A^{-1} \in \mathcal{L}(Y;X)$, then $A^* \in \mathcal{L}(Y;X)$ is also bijective with $(A^*)^{-1} \in \mathcal{L}(X;Y)$; besides, $(A^*)^{-1} = (A^{-1})^*$.

(5) Let $(Z, (\cdot,\cdot)_Z)$ be another complex Hilbert space and let $B \in \mathcal{L}(Y;Z)$ be given. Show that $(AB)^* = B^*A^*$.

4.7-2 Using Theorem 4.7-2, show that the solution to the normal equations $\boldsymbol{A}^T\boldsymbol{A}\boldsymbol{x} = \boldsymbol{A}^T\boldsymbol{c}$ (Section 4.4) is unique if and only if the rank of the matrix $\boldsymbol{A}$ is n (which of course implies that $n \le m$), or equivalently, if and only if the symmetric matrix $\boldsymbol{A}^T\boldsymbol{A}$ is positive-definite.

Naturally, an analogous result holds in the complex case.

4.7-3 **(Lax–Milgram lemma)** Let $(X, (\cdot,\cdot))$ be a real or complex Hilbert space and let $a : X \times X \to \mathbb{K}$ be a bilinear form if $\mathbb{K} = \mathbb{R}$, or a function linear with respect to its first argument and semilinear with respect to its second argument if $\mathbb{K} = \mathbb{C}$, such that there exist constants $M > 0$ and $\alpha > 0$ such that

$$\begin{aligned} |a(x,y)| &\le M\,\|x\|\,\|y\| && \text{for all } x, y \in X, \\ |a(x,x)| &\ge \alpha\,\|x\|^2 && \text{for all } x \in X, \end{aligned}$$

where $\|\cdot\|$ denotes the norm associated with the inner product $(\cdot,\cdot)$.

(1) Show that there exists a mapping $A \in \mathcal{L}(X)$ that satisfies

$$a(x,y) = (Ax, y) \quad \text{for all } x, y \in X.$$

[10] S. Bergman; M. Schiffer [1948]: Kernel functions in the theory of partial differential equations of elliptic type, *Duke Mathematical Journal* **15**, 535–566.

N. Aronszajn; K.T. Smith [1957]: Characterization of positive reproducing kernels. Applications to Green's functions, *American Journal of Mathematics* **79**, 611–622.

Show that the mapping $A : X \to X$ defined in this fashion is injective and that the inverse operator from $A(X)$ onto X is continuous.

(2) Show that $A(X)$ is a closed subspace of X.

(3) Show that $A(X) = X$. Conclude that, *given any element* $c \in X$, *there exists a unique element* $x \in X$ *that satisfies*

$$a(x, y) = (c, y) \quad \text{for all } y \in X,$$

and that the linear mapping $b \in X \to x \in X$ *defined in this fashion is continuous.*

The result of (3) constitutes the *Lax–Milgram lemma* (another proof of the Lax–Milgram lemma will be proposed in Theorem 6.2-1).

4.7-4 (1) Let X and Y be two complex Hilbert spaces (the real case is similar) and let $A \in \mathcal{L}(X; Y)$ be such that $\operatorname{Im} A$ is a closed subspace of Y. Show that there exists a unique $A^\dagger \in \mathcal{L}(Y; X)$, called the *Moore–Penrose inverse*[11] of A, that satisfies the following four properties:

$$AA^\dagger A = A, \quad A^\dagger AA^\dagger = A^\dagger, \quad (AA^\dagger)^* = AA^\dagger, \quad (A^\dagger A)^* = A^\dagger A.$$

(2) Assume that $X = \mathbb{C}^n$ and $Y = \mathbb{C}^m$, in which case A and $A^\dagger$ (which always exists since $\operatorname{Im} A$ is finite-dimensional) can be respectively identified with an $m \times n$ complex matrix $\boldsymbol{A}$ and an $n \times m$ complex matrix $\boldsymbol{A}^\dagger$. Show that

$$\boldsymbol{A}^\dagger = \lim_{\varepsilon \to 0} \left((\boldsymbol{A}^*\boldsymbol{A} + \varepsilon \boldsymbol{I})^{-1} \boldsymbol{A}^* \right) = \lim_{\varepsilon \to 0} \left(\boldsymbol{A}^* (\boldsymbol{A}\boldsymbol{A}^* + \varepsilon \boldsymbol{I})^{-1} \right).$$

(3) Show that, if $\boldsymbol{A}$ is an $n \times n$ complex invertible matrix, then $\boldsymbol{A}^\dagger = \boldsymbol{A}^{-1}$. This observation explains why the Moore–Penrose inverse $\boldsymbol{A}^\dagger$ is also called the *generalized inverse* of the matrix $\boldsymbol{A}$.

(4) Let A be an $m \times n$ complex matrix. Given any vector $\boldsymbol{c} \in \mathbb{C}^m$, there then exists at least one *least-squares solution* $\boldsymbol{x}$ *to the linear system* $\boldsymbol{A}x = \boldsymbol{c}$, i.e., a vector $\boldsymbol{x} \in \mathbb{C}^n$ that satisfies $|\boldsymbol{A}\boldsymbol{x} - \boldsymbol{c}| = \inf_{\boldsymbol{y} \in \mathbb{C}^n} |\boldsymbol{A}\boldsymbol{y} - \boldsymbol{c}|$, where $|\cdot|$ denotes the Euclidean norm in $\mathbb{C}^m$ (Theorem 4.4-1). Show that there exists a unique vector

$$\boldsymbol{x}^\dagger \in \mathbb{C}^n \quad \text{such that } |\boldsymbol{x}^\dagger| = \inf \left\{ |\boldsymbol{x}|;\ x \in \mathbb{C}^n \text{ and } |\boldsymbol{A}\boldsymbol{x} - \boldsymbol{c}| = \inf_{\boldsymbol{y} \in \mathbb{C}^n} |\boldsymbol{A}\boldsymbol{y} - \boldsymbol{c}| \right\},$$

and that the mapping $\boldsymbol{c} \in \mathbb{C}^m \to \boldsymbol{x}^\dagger \in \mathbb{C}^n$ defined in this fashion is precisely given by $\boldsymbol{x}^\dagger = \boldsymbol{A}^\dagger \boldsymbol{c}$.

This observation thus provides another definition of the matrix $\boldsymbol{A}^\dagger$.

(5) Let $\boldsymbol{A}(\varepsilon) = \begin{pmatrix} 1 & 0 \\ 0 & \varepsilon \end{pmatrix}$. Show that $\lim_{\varepsilon \to 0} \boldsymbol{A}(\varepsilon)^\dagger$ does not exist[12], thus showing that *the Moore–Penrose inverse of a matrix* $\boldsymbol{A}$ *is not necessarily a continuous function of the elements of* $\boldsymbol{A}$ (clearly, $\lim_{\varepsilon \to 0} \boldsymbol{A}(\varepsilon)$ exists).

---

[11] E.H. Moore [1920]: On the reciprocal of the general algebraic matrix, *Bulletin of the American Mathematical Society* **26**, 394–395.

R. Penrose [1955]: A generalized inverse for matrices, *Proceedings of the Cambridge Philosophical Society* **51**, 406–413.

These two authors independently proposed two different definitions of such an operator (in the finite-dimensional case), the equivalence of which was established by:

R. Rado [1956]: Note on generalized inverses of matrices, *Proceedings of the Cambridge Philosophical Society* **52**, 600–601.

For various extensions (such as the infinite-dimensional case), see, e.g.:

A. Ben-Israel; T.N.E. Greville [2003]: *Generalized Inverses: Theory and Applications*, Second Edition, Springer.

[12] This example is due to:

G.W. Stewart [1969]: On the continuity of the generalized inverse, *SIAM Journal on Applied Mathematics* **17**, 33–45.

4.8 Maximal orthonormal families in an inner-product space

As we shall see in the next section, *maximal orthonormal families* in a *Hilbert space* play a fundamental role, because any element in such a space can be expanded as a *Fourier series* over the elements of such a family.

Recall that, given any family $(e_i)_{i\in I}$ of vectors $e_i \in X$, where X is a real ($\mathbb{K} = \mathbb{R}$) or complex ($\mathbb{K} = \mathbb{C}$) vector space, $\mathrm{Span}(e_i)_{i\in I}$ designates the subspace of X formed by all *finite linear combinations* of vectors of the family, i.e., vectors of X of the form $\sum_{j\in J} \alpha_j e_j$, where J is a finite subset of I and $\alpha_j \in \mathbb{K}$, $j \in J$ (Section 2.1).

Let $(X, (\cdot,\cdot))$ be a real or complex inner-product space. A family $(e_i)_{i\in I}$ of elements $e_i \in X$ is called an **orthonormal family** if

$$(e_i, e_j) = \delta_{ij} \quad \text{for all } i, j \in I.$$

Any orthonormal family is necessarily a linearly independent family since, given any finite subset J of I, the relation $\sum_{j\in J} \alpha_j e_j = 0$ implies that $(\sum_{j\in J} \alpha_j e_j, e_i) = \alpha_i = 0$ for all $i \in J$.

The next theorem provides a simple way of constructing orthonormal families. For definiteness, it is stated and proved in the infinite-dimensional case; its finite-dimensional version should be clear.

Theorem 4.8-1 (Gram–Schmidt[13] orthonormalization) *Let $(X, (\cdot,\cdot))$ be a real or complex infinite-dimensional inner-product space, and let $(f_n)_{n=0}^{\infty}$ be a countably infinite linearly independent family of vectors $f_n \in X$. Let*

$$\widetilde{e}_0 := f_0 \quad \textit{and} \quad \widetilde{e}_k := f_k - P_k f_k \quad \textit{for } k = 1, 2, \dots,$$

where P_k denotes the projection operator from X onto $\mathrm{Span}(f_n)_{n=0}^{k-1}$. Then $\widetilde{e}_k \neq 0$ for all $k \geq 1$, and the family $(e_n)_{n=0}^{\infty}$ where

$$e_n := \frac{\widetilde{e}_n}{\|\widetilde{e}_n\|}, \quad n \geq 0,$$

is an orthonormal family of vectors $e_n \in X$ that satisfies

$$\mathrm{Span}\,(e_n)_{n=0}^{k} = \mathrm{Span}\,(f_n)_{n=0}^{k} \quad \textit{for all } k \geq 0, \quad \textit{and} \quad \mathrm{Span}\,(e_n)_{n=0}^{\infty} = \mathrm{Span}\,(f_n)_{n=0}^{\infty}.$$

Proof Let $\widetilde{e}_0 := f_0$; hence $\mathrm{Span}(\widetilde{e}_0) = \mathrm{Span}(f_0)$. Assume that, for some integer $k \geq 1$, nonzero vectors $\widetilde{e}_0, \dots, \widetilde{e}_{k-1}$ have been found that satisfy

$$(\widetilde{e}_m, \widetilde{e}_n) = 0 \quad \text{for all } m \neq n,\ 0 \leq m, n \leq k-1, \quad \text{and} \quad \mathrm{Span}(\widetilde{e}_n)_{n=0}^{k-1} = \mathrm{Span}(f_n)_{n=0}^{k-1}.$$

[13] So named after:

J.P. Gram [1883]: Über die Entwicklung reeller Funktionen in Reihen mittelst der Methode der kleinsten Quadrate, *Journal für die Reine und Angewandte Mathematik* **94**, 41–73.

E. Schmidt [1907]: Zur Theorie der linearen und nichtlinearen Integralgleichungen. 1. Teil: Entwicklung willkürlicher Funktionen nach Systemen vorgeschriebener, *Mathematische Annalen* **63**, 433–476.

But in fact, this orthonormalization procedure is already found in:

P.S. Laplace [1820]: *Théorie Analytique des Probabilités, Troisième Edition, Premier Supplément: Sur l'Application du Calcul des Probabilités à la Philosophie Naturelle*, Courcier, Paris.

Let P_k designate the projection operator from X onto $X_k := \mathrm{Span}(f_n)_{n=0}^{k-1}$ (as a finite-dimensional subspace, X_k is closed in X; cf. Theorem 2.7-1(c)). Then the vector $\widetilde{e}_k := f_k - P_k f_k$, which is nonzero since the vectors f_n, $0 \le n \le k$, are linearly independent, is orthogonal to the subspace X_k (Theorem 4.3-1(d)), and hence to all the vectors $\widetilde{e}_n$, $0 \le n \le k-1$.

It is clear that

$$\mathrm{Span}(\widetilde{e}_n)_{n=0}^{k} = \mathrm{Span}(f_n)_{n=0}^{k} \text{ for all } k \ge 0; \text{ hence } \mathrm{Span}(\widetilde{e}_n)_{n=0}^{\infty} = \mathrm{Span}(f_n)_{n=0}^{\infty}.$$

Consequently, the family $(e_n)_{n=0}^{\infty}$ defined by $e_n := \dfrac{\widetilde{e}_n}{\|\widetilde{e}_n\|}$ for all $n \ge 0$ possesses all the required properties. □

Remark An explicit expression of the vectors $\widetilde{e}_n$ in terms of the vectors f_n is provided in Problem 4.8-1. □

We now describe several *basic examples of orthonormal families*. To begin with, consider the (real) space $\mathcal{C}[-1,1]$ equipped with the inner product of the space $L^2(-1,1)$, viz., $(f,g) = \int_{-1}^{1} f(x)g(x)\,\mathrm{d}x$. For each integer $n \ge 0$, let the function $f_n \in \mathcal{C}[-1,1]$ be defined by

$$f_n(x) := x^n, \quad -1 \le x \le 1.$$

Then the orthonormal family $(e_n)_{n=0}^{\infty}$ constructed as in Theorem 4.8-1 from the family $(f_n)_{n\ge 0}$ (which is clearly linearly independent) consists of the **Legendre polynomials**,[14] which are defined by

$$e_n(x) := \frac{\sqrt{n+\frac{1}{2}}}{2^n n!}\frac{\mathrm{d}^n}{\mathrm{d}x^n}\left[(x^2-1)^n\right], \quad -1 \le x \le 1$$

(Problem 4.8-2). Note that the same Legendre polynomials also form an orthonormal family in the complex space $\mathcal{C}([-1,1];\mathbb{C})$ equipped with the inner product of the space $L^2(-1,1;\mathbb{C})$, viz., $(f,g) = \int_{-1}^{1} f(x)\overline{g(x)}\,\mathrm{d}x$. Naturally, the Legendre polynomials *a fortiori* constitute an orthonormal family in the larger Hilbert spaces $L^2(-1,1)$ or $L^2((-1,1);\mathbb{C})$.

More generally, one can construct real polynomials on a compact interval $[a,b]$ that are orthogonal with respect to an inner product of the form $(f,g) = \int_a^b f(x)g(x)\omega(x)\,\mathrm{d}x$, where ω is a given *weight function*. Such polynomials possess remarkable properties;[15] cf. Problem 4.8-3.

Consider next the (real) space $\mathcal{C}_{\mathrm{per}}[0,2\pi]$ equipped with the inner product of the space $L^2(0,2\pi)$, viz., $(f,g) = \int_0^{2\pi} f(\theta)g(\theta)\,\mathrm{d}\theta$. Elementary trigonometry calculations then show that *the functions defined by*

$$\frac{1}{\sqrt{2\pi}}, \quad \frac{1}{\sqrt{\pi}}\cos n\theta \ \text{ for all } n \ge 1, \quad \frac{1}{\sqrt{\pi}}\sin n\theta \ \text{ for all } n \ge 1, \quad 0 \le \theta \le 2\pi,$$

form an orthonormal family in the space $\mathcal{C}_{\mathrm{per}}[0,2\pi]$, *and hence also in the Hilbert space* $L^2(0,2\pi)$. Consider likewise the space $\mathcal{C}_{\mathrm{per}}([0,2\pi];\mathbb{C})$, equipped with the inner product of the

[14] So named after Adrien-Marie Legendre (1752–1833).

[15] Clear and highly readable introductions to orthonormal families of polynomials are found in Wong [2010] and Beals & Wong [2010]. The great classic on the subject is Szegő [1975].

space $L^2(0,2\pi;\mathbb{C})$, viz., $(f,g)=\int_0^{2\pi} f(\theta)\overline{g(\theta)}\,\mathrm{d}\theta$. Then *the functions defined by*

$$\frac{1}{\sqrt{2\pi}}e^{in\theta} \quad \text{for all } n\in\mathbb{Z},\ \ 0\le\theta\le 2\pi,$$

form an orthonormal family in $C_{\text{per}}([0,2\pi];\mathbb{C})$, *and also in the Hilbert space* $L^2(0,2\pi;\mathbb{C})$ (Section 4.2). To see this, it suffices to observe that

$$\int_0^{2\pi} e^{i(m-n)\theta}\,\mathrm{d}\theta = \frac{e^{i(m-n)2\pi}-1}{i(m-n)} = 0 \ \text{ if } m\ne n \quad\text{and}\quad \int_0^{2\pi} e^{i(m-n)\theta}\,\mathrm{d}\theta = 2\pi \ \text{ if } m=n.$$

Consider next the (real) Hilbert space $L^2(0,\infty)$ equipped with the inner product $(f,g)=\int_0^\infty f(x)g(x)\,\mathrm{d}x$. Then *the* **Laguerre**[16] **functions** L_n, $n\ge 0$, *defined by*

$$L_n(x) := \frac{1}{n!}e^{x/2}\frac{\mathrm{d}^n}{\mathrm{d}x^n}[x^n e^{-x}], \quad x\in(0,\infty),$$

form an orthonormal family in $L^2(0,\infty)$ (Problem 4.8-4).

Consider finally the (real) Hilbert space $L^2(\mathbb{R})$ equipped with the inner product $(f,g)=\int_{-\infty}^\infty f(x)g(x)\,\mathrm{d}x$. Then *the* **Hermite**[17] **functions** H_n, $n\ge 0$, *defined by*

$$H_n(x) := \frac{(-1)^n}{2^n n!\sqrt{2\pi}}e^{x^2/2}\frac{\mathrm{d}^n}{\mathrm{d}x^n}[e^{-x^2}], \quad x\in\mathbb{R},$$

form an orthonormal family in $L^2(\mathbb{R})$ (Problem 4.8-5).

An *orthonormal family* $(e_i)_{i\in I}$ in a real or complex inner-product space $(X,(\cdot,\cdot))$ is said to be **maximal** if *the only vector* $x\in X$ *that satisfies* $(x,e_i)=0$ *for all* $i\in I$ *is* $x=0$. The following simple *sufficient* condition of maximality is often used.

Theorem 4.8-2 *An orthonormal family* $(e_i)_{i\in I}$ *in an inner-product space* X *is maximal if*[18]

$$\overline{\text{Span}\,(e_i)_{i\in I}} = X.$$

Proof Since $(x,e_i)=0$ for all $i\in I$ if and only if $x\in(\text{Span}(e_i)_{i\in I})^\perp$, and since $(\overline{W})^\perp = W^\perp$ in general (Theorem 4.5-1), *an orthonormal family is thus maximal if and only if*

$$(\overline{\text{Span}(e_i)_{i\in I}})^\perp = \{0\}.$$

If $\overline{\text{Span}(e_i)_{i\in I}}=X$, then $(\overline{\text{Span}(e_i)_{i\in I}})^\perp=\{0\}$, and thus the family $(e_i)_{i\in I}$ is maximal in this case. □

[16] So named after Edmond Laguerre (1834–1886).

[17] So named after Charles Hermite (1822–1901).

[18] But, unless the space X is *complete* (in which case the converse implication clearly follows from the direct sum theorem), *the converse implication does not necessarily hold*: there exist (necessarily noncomplete) inner-product spaces X in which there does *not* exist any orthonormal family $(e_i)_{i\in I}$ such that $\overline{\text{Span}(e_i)_{i\in I}}=X$; see:

J. Dixmier [1953]: Sur les bases orthonormales dans les espaces préhilbertiens, *Acta Scientiarum Mathematicarum Szeged* **15**, 29–30.

Remarkably, *any* inner-product space (complete or not) possesses maximal orthonormal families, as shown in the next theorems. While the existence of such maximal orthonormal families can be established by means of a simple recursion argument if the space is *separable* (Theorem 4.8-3(a)), its proof otherwise requires *Zorn's lemma*, or equivalently the *axiom of choice* (Section 1.3), in the general case (Theorem 4.8-4). For definiteness we consider only the infinite-dimensional case in Theorem 4.8-3 (its finite-dimensional version, which holds *a fortiori*, should be clear). We also establish (cf. (b)) an interesting property of orthonormal families in such a space. Note also that a converse to Theorem 4.8-3(a) holds in a *Hilbert space*; cf. Problem 4.8-6.

Remark The usually encountered Hilbert spaces are indeed separable (such as $\ell^2, L^2(\Omega)$, or the Sobolev spaces $H^m(\Omega), m \geq 1$; cf. Chapter 6). But it is easy to construct an example of a *nonseparable Hilbert space*; cf. Problem 4.8-7. □

Theorem 4.8-3 (maximal orthonormal families in a separable inner-product space) *Let $(X, (\cdot,\cdot))$ be a separable, infinite-dimensional, inner-product space.*

(a) *There exists a countably infinite maximal orthonormal family $(e_n)_{n=0}^{\infty}$ of vectors $e_n \in X$, i.e., such that*

$$(e_m, e_n) = \delta_{mn} \quad \textit{for all } m, n \geq 0,$$
$$x \in X \quad \textit{and} \quad (x, e_n) = 0 \ \textit{ for all } n \geq 0 \textit{ implies } x = 0.$$

(b) *Any orthonormal family (maximal or not) is either finite or countably infinite.*

Proof Since X is separable, there exists a countably infinite family of linearly independent vectors $f_n \in X$, $n \geq 0$, such that

$$\overline{\text{Span}(f_n)_{n=0}^{\infty}} = X$$

(Theorem 2.2-7). Then the orthonormal family $(e_n)_{n=0}^{\infty}$ constructed from the linearly independent family $(f_n)_{n=0}^{\infty}$ by the Gram–Schmidt orthonormalization (Theorem 4.8-1) satisfies $\text{Span}(e_n)_{n=0}^{\infty} = \text{Span}(f_n)_{n=0}^{\infty}$. Therefore,

$$\overline{\text{Span}(e_n)_{n=0}^{\infty}} = \overline{\text{Span}(f_n)_{n=0}^{\infty}} = X.$$

Hence, the family $(e_n)_{n=0}^{\infty}$ is maximal by Theorem 4.8-2. This proves (a).

To prove (b), note first that the elements of any orthonormal family $(e_i)_{i\in I}$ necessarily satisfy

$$\|e_i - e_j\| = \sqrt{\|e_i\|^2 + \|e_j\|^2} = \sqrt{2} \quad \text{if } i \neq j.$$

Let vectors $g_k \in X$, $k \geq 0$, be such that $\overline{\bigcup_{k=0}^{\infty}\{g_k\}} = X$. Then, for each $i \in I$, there exists an integer $k(i) \geq 0$ such that $\|e_i - g_{k(i)}\| \leq \dfrac{\sqrt{2}}{3}$. The relation

$$\sqrt{2} = \|e_i - e_j\| \leq \|e_i - g_{k(i)}\| + \|g_{k(i)} - g_{k(j)}\| + \|e_j - g_{k(j)}\| \quad \text{if } i \neq j$$

then implies that $\|g_{k(i)} - g_{k(j)}\| \geq \dfrac{\sqrt{2}}{3}$, hence that $k(i) \neq k(j)$, since

$$g_{k(i)} \neq g_{k(j)} \quad \text{if } i \neq j.$$

Because the mapping $i \in I \to k(i) \in \mathbb{N}$ established in this fashion is therefore an injection, the set I is either finite or countably infinite (Section 1.5). □

Theorem 4.8-4 (existence of maximal orthonormal families in any inner-product space) *Let $(X, (\cdot,\cdot))$ be an inner-product space. Then there exists a family $(e_i)_{i\in I}$ of vectors $e_i \in X$ such that*

$$(e_i, e_j) = \delta_{ij} \quad \textit{for all } i, j \in I,$$
$$x \in X \textit{ and } (x, e_i) = 0 \quad \textit{for all } i \in I \textit{ implies that } x = 0.$$

Proof Assume that $\dim X \geq 2$, and let $e_1, e_2 \in X$ be such that $\|e_1\| = \|e_2\| = 1$ and $(e_1, e_2) = 0$ (for instance, e_1 and e_2 are constructed as in Theorem 4.8-3 from two linearly independent vectors $f_1, f_2 \in X$).

Let $\mathcal{F}$ denote the subset of $\mathcal{P}(X)$ consisting of all orthonormal families of vectors of X that contain $(e_i)_{i=1,2}$ (hence $\mathcal{F} \neq \varnothing$), and let $\mathcal{F}$ be partially ordered by the set-inclusion relation (since its elements e_i are all distinct, an orthonormal family $(e_i)_{i\in I}$ is identified here with the set $\bigcup_{i\in I}\{e_i\}$).

Given any totally ordered subset $\mathcal{E}$ of $\mathcal{F}$, the set $G = \bigcup_{E\in\mathcal{E}} E$ belongs to $\mathcal{F}$. For, if $e, \widetilde{e} \in G$, then $e \in E$ for some $E \in \mathcal{E}$ and $\widetilde{e} \in \widetilde{E}$ for some $\widetilde{E} \in \mathcal{E}$; since $\mathcal{E}$ is totally ordered, either $\widetilde{E} \subset E$ or $E \subset \widetilde{E}$, so that $(e, \widetilde{e}) = 0$ if $e \neq \widetilde{e}$ or $(e, \widetilde{e}) = 1$ if $e = \widetilde{e}$ (both E and $\widetilde{E}$ are orthonormal families). Since $E \subset G$ for all $E \in \mathcal{E}$, the set G is thus an upper bound of $\mathcal{E}$.

By the *axiom of choice* (Theorem 1.3-1), the set $\mathcal{F}$ has a maximal element $M = (e_i)_{i\in I}$. First, $(e_i, e_j) = \delta_{ij}$ for all $i, j \in I$ since $(e_i)_{i\in I} \in \mathcal{F}$. Second, let $x \in X$ be such that $(x, e_i) = 0$ for all $i \in I$; then necessarily $x = 0$, for otherwise the set $M \cup \left\{\frac{x}{\|x\|}\right\}$, which clearly belongs to $\mathcal{F}$, would satisfy $M \subsetneqq M \cup \left\{\frac{x}{\|x\|}\right\}$, contradicting the maximal character of M. □

Remarkably, *all the orthonormal families described earlier in this section are also maximal.* We now give a proof of this assertion for the first three examples, leaving the proof for the last two examples as problems (Problems 4.8-4 and 4.8-5).

Theorem 4.8-5 (examples of maximal orthonormal families) (a) *The Legendre polynomials, defined by*

$$e_n(x) = \frac{\sqrt{n+\frac{1}{2}}}{2^n n!}\frac{\mathrm{d}^n}{\mathrm{d}x^n}[(x^2-1)^n] \quad \textit{for all } n \geq 0, \quad -1 \leq x \leq 1,$$

form a maximal orthonormal family in the Hilbert spaces $L^2(-1,1)$ and $L^2(-1,1;\mathbb{C})$.

(b) *The functions defined by*

$$\frac{1}{\sqrt{2\pi}}, \quad \frac{1}{\sqrt{\pi}}\cos m\theta \quad \textit{for all } m \geq 1, \quad \frac{1}{\sqrt{\pi}} \in \sin n\theta \quad \textit{for all } n \geq 1, \quad 0 \leq \theta \leq 2\pi,$$

form a maximal orthonormal family in the Hilbert spaces $L^2(0,2\pi)$ and $L^2(0,2\pi;\mathbb{C})$.

(c) *The functions defined by*

$$\frac{1}{\sqrt{2\pi}}e^{in\theta} \quad \textit{for all } n \in \mathbb{Z}, \quad 0 \leq \theta \leq 2\pi,$$

form a maximal orthonormal family in the Hilbert space $L^2(0, 2\pi; \mathbb{C})$.

Proof (i) Recall that, by construction, all the above functions form orthonormal families in the corresponding Hilbert spaces; so it remains to show that these families are maximal.

(ii) Let a function $f \in L^2(-1, 1)$ and $\varepsilon > 0$ be given. Since *the space* $\mathcal{C}[-1, 1]$ *is dense in* $L^2(-1, 1)$ (Theorem 2.5-3), there exists a function $\widetilde{f} \in \mathcal{C}[-1, 1]$ such that

$$\|f - \widetilde{f}\|_{L^2(-1,1)} \le \frac{\varepsilon}{2},$$

and, by the *Weierstraß approximation theorem* (Theorem 2.13-3), there exists a polynomial p such that

$$\|\widetilde{f} - p\|_{L^2(-1,1)} \le \sqrt{2} \sup_{-1 \le x \le 1} |\widetilde{f}(x) - p(x)| \le \frac{\varepsilon}{2}.$$

The last two inequalities combined imply that

$$\overline{\mathrm{Span}(f_n)_{n=0}^{\infty}} = L^2(-1, 1),$$

where $f_n(x) := x^n$, $-1 \le x \le 1$. Since, by construction, the Legendre polynomials satisfy (Theorem 4.8-1)

$$\mathrm{Span}(e_n)_{n=0}^{\infty} = \mathrm{Span}(f_n)_{n=0}^{\infty},$$

Theorem 4.8-2 shows that they form a maximal orthonormal family in the space $L^2(-1, 1)$.

The same argument applied to both the real and imaginary parts of a function in the space $L^2(-1, 1; \mathbb{C})$ shows that the Legendre polynomials also form a maximal orthonormal family in the space $L^2(-1, 1; \mathbb{C})$. This proves (a).

(iii) Let next a function $g \in L^2(0, 2\pi)$ and $\varepsilon > 0$ be given. Since the space $\mathcal{D}(0, 2\pi)$ is dense in $L^2(0, 2\pi)$ (Theorem 2.6-2), there exists a function $\widetilde{g} \in \mathcal{D}(0, 2\pi)$ such that

$$\|g - \widetilde{g}\|_{L^2(0,2\pi)} \le \frac{\varepsilon}{2}.$$

Since $\widetilde{g} \in \mathcal{C}_{\mathrm{per}}[0, 2\pi]$, the *Weierstraß trigonometric polynomial approximation theorem* (Theorem 2.14-3) can be applied, showing that there exists a trigonometric polynomial q such that

$$\|\widetilde{g} - q\|_{L^2(0,2\pi)} \le \sqrt{2\pi} \sup_{0 \le \theta \le 2\pi} |\widetilde{g}(\theta) - q(\theta)| \le \frac{\varepsilon}{2}.$$

This proves (b).

(iv) The proof of (c) is similar to that of (b), save that the Weierstraß trigonometric polynomial approximation theorem is now replaced by its complex version (Theorem 2.15-4), which asserts that, given any function $\widehat{g} \in \mathcal{C}_{\mathrm{per}}([0, 2\pi]; \mathbb{C})$, there exists a complex trigonometric polynomial q, i.e., of the form

$$q(\theta) = \sum_{k=-n}^{n} c_k e^{ik\theta}, \quad 0 \le \theta \le 2\pi,$$

with $c_k \in \mathbb{C}$, $-n \le k \le n$, and $n \ge 0$, that is arbitrary close to $\widehat{g}$ with respect to the sup-norm over the interval $[0, 2\pi]$. $\square$

Remark Naturally, the Legendre polynomials also form a maximal orthonormal family in any subspace of $L^2(-1,1)$ that contains them, such as the space $\mathcal{C}[-1,1]$ (equipped with the inner product of $L^2(-1,1)$). □

We shall see that the *spectral theory of compact self-adjoint operators in a separable Hilbert space* (Section 4.11) provides another, and powerful, way of constructing maximal orthonormal families in such a space (Theorem 4.11-3). Fundamental *specific examples*, found when solving *eigenvalue problems for second-order elliptic operators*, will be also given later (Theorem 6.10-2).

Problems

4.8-1 Show that the vectors $\widetilde{e}_n$, $n \geq 1$, found by the *Gram–Schmidt orthonormalization* (Theorem 4.8-1) may be also recursively defined by $\widetilde{e}_0 = f_0$ and $\widetilde{e}_n := f_n - \sum_{k=0}^{n-1} \frac{(f_n, \widetilde{e}_k)}{\|\widetilde{e}_k\|^2}\widetilde{e}_k$ for $n \geq 1$.

4.8-2 (1) For each integer $n \geq 0$, let the functions $f_n : [-1,1] \to \mathbb{R}$ be defined by $f_n(x) := x^n$, $-1 \leq x \leq 1$. Show that the orthonormal family $(e_n)_{n=0}^{\infty}$ constructed as in Theorem 4.8-1 from the family $(f_n)_{n=0}^{\infty}$ consists of the *Legendre polynomials of degree* n, which are defined for all $n \geq 0$ by

$$e_n(x) := \frac{\sqrt{n+\frac{1}{2}}}{2^n n!} \frac{\mathrm{d}^n}{\mathrm{d}x^n} \left[(x^2-1)^n\right], \quad -1 \leq x \leq 1.$$

(2) Show directly that $\int_{-1}^{1} e_m(x)e_n(x)\,\mathrm{d}x = \delta_{mn}$ for all $m, n \geq 0$.
(3) Show directly that, for each $n \geq 0$,

$$e_n(x) = \sqrt{n+\frac{1}{2}} \sum_{0 \leq k \leq \frac{n}{2}} (-1)^k \frac{(2(n-k))!}{2^k (n-k)! k! (n-2k)!} x^{n-2k}, \quad -1 \leq x \leq 1,$$

which shows in particular that e_n is a polynomial of degree n (this also follows from Theorem 4.8-1).
(4) Show that conversely, for each $n \geq 0$,

$$x^n = n! \sum_{0 \leq k \leq \frac{n}{2}} \left(\frac{1}{(2k)!\sqrt{n-2k+\frac{1}{2}}}\right)\left(\frac{2n-4k+1}{(2k+1)(2k+3)\cdots(2n-2k+1)}\right) e_{n-2k}(x), \quad -1 \leq x \leq 1.$$

(5) Let the second-order differential operator $\mathcal{L}$ be defined by

$$\mathcal{L}u(x) := -\frac{\mathrm{d}}{\mathrm{d}x}\left[(1-x^2)\frac{\mathrm{d}u}{\mathrm{d}x}\right], \quad -1 \leq x \leq 1.$$

Show that, for each $n \geq 0$, the *Legendre polynomial* e_n is an *eigenfunction of the operator* $\mathcal{L}$, in the sense that e_n satisfies

$$\mathcal{L}e_n(x) = \lambda_n e_n(x), \quad -1 \leq x \leq 1, \quad \text{with } \lambda_n = n(n+1).$$

4.8-3 (orthogonal polynomials with respect to a weight function) Let ω be a *weight function over the interval* $[0,1]$, i.e., a function $\omega \in L^1(0,1)$ that satisfies $\omega > 0$ almost everywhere in $[0,1]$. Then

$$(f,g) := \int_0^1 f(x)g(x)\omega(x)\,\mathrm{d}x$$

clearly defines an inner product over the space $\mathcal{C}[0,1]$.

In this problem, "orthogonal" or "orthonormal" means with respect to this inner product. For each $n \geq 0$, let $f_n(x) := x^n$, $0 \leq x \leq 1$, and let $(e_n)_{n=0}^{\infty}$ be the orthonormal family of polynomials constructed from the family $(f_n)_{n=0}^{\infty}$ as in Theorem 4.8-1. Since $\mathrm{Span}(e_n)_{n=0}^{k} = \mathrm{Span}(f_n)_{n=0}^{k}$ for all $k \geq 0$, each polynomial e_n, $n \geq 0$, is of the form $e_n(x) = c_n x^n + c_{n-1}x^{n-1} + \cdots + c_0$, $0 \leq x \leq 1$, with $c_n \neq 0$. The polynomials $p_n \in \mathcal{P}_n[0,1]$, $n \geq 0$, defined by

$$p_n(x) := \frac{1}{c_n} e_n(x) = x^n + \frac{c_{n-1}}{c_n} x^{n-1} + \cdots + \frac{c_0}{c_n}, \quad 0 \leq x \leq 1,$$

are called the *monic orthogonal polynomials with respect to the weight function* ω ("monic" means that their leading coefficient is one). Note that these polynomials still satisfy $(p_m, p_n) = 0$ if $m \neq n$, but may no longer satisfy $(p_n, p_n) = 1$ for all $n \geq 0$. The object of this problem is to establish two basic properties of these polynomials.

(1) Show that the polynomials p_n satisfy a *three-term recursion formula* of the form

$$p_n(x) = (x + b_n)p_{n-1}(x) + c_n p_{n-2}(x), \quad 0 \leq x \leq 1, \text{ for all } n \geq 2,$$

where the constants $b_n, c_n \in \mathbb{R}$ are functions of the coefficients of the polynomials p_n, p_{n-1}, and p_{n-2}.

(2) Show that, for each $n \geq 1$, all the roots of the polynomial p_n, now viewed as a polynomial on $\mathbb{R}$, are real, simple, and lie in the open interval $]0,1[$.

4.8-4 (1) For each $n \geq 0$, let the function $f_n : [0,\infty[\to \mathbb{R}$ be defined by $f_n(x) := e^{-\frac{x}{2}} x^n$ $x \in [0,\infty[$. Show that the functions f_n belong to the space $L^2(0,\infty)$ and that the orthonormal family constructed as in Theorem 4.8-1 from the family $(f_n)_{n=0}^{\infty}$ (which is clearly linearly independent) consists of the *Laguerre functions* L_n, $n \geq 0$, defined by

$$L_n(x) := \frac{1}{n!} e^{x/2} \frac{\mathrm{d}^n}{\mathrm{d}x^n} [x^n e^{-x}], \quad x \in [0,\infty[.$$

(2) Show that the orthonormal family $(L_n)_{n=0}^{\infty}$ is maximal in the Hilbert space $L^2(0,\infty)$.[19]

4.8-5 (1) For each $n \geq 0$, let the function $f_n : \mathbb{R} \to \mathbb{R}$ be defined by $f_n(x) := e^{-\frac{x^2}{2}} x^n$, $x \in \mathbb{R}$. Show that the functions f_n belong to the space $L^2(\mathbb{R})$ and that the orthonormal family constructed as in Theorem 4.8-1 from the family $(f_n)_{n=0}^{\infty}$ (which is clearly linearly independent) consists of the *Hermite functions* H_n, $n \geq 0$, defined by

$$H_n(x) := \frac{(-1)^n}{2^n n! \sqrt{2\pi}} e^{x^2/2} \frac{\mathrm{d}^n}{\mathrm{d}x^n} [e^{-x^2}], \quad x \in \mathbb{R}.$$

(2) Show that the orthonormal family $(H_n)_{n=0}^{\infty}$ is maximal in the Hilbert space $L^2(\mathbb{R})$.[20]

4.8-6 Let X be a real or complex Hilbert space that has a finite or countably infinite maximal orthonormal basis. Show that X is separable.

4.8-7 This problem provides an example of a *nonseparable Hilbert space* and of an *uncountably infinite orthonormal family*.

(1) Let the subspace $Y$of the complex vector space $\mathcal{C}(\mathbb{R};\mathbb{C})$ be defined as

$$Y := \mathrm{Span}(e_\lambda)_{\lambda \in \mathbb{R}}, \quad \text{where } e_\lambda(x) := e^{i\lambda x}, \quad x \in \mathbb{R}.$$

Show that $(f,g) := \lim_{T \to \infty} \frac{1}{2T} \int_{-T}^{T} f(x)\overline{g(x)}\,\mathrm{d}x$ defines an inner product over Y.

[19] For a proof, see, e.g., GOFFMAN & PEDRICK [1965, Section 4.10].

[20] For a proof, see, e.g., AKHIEZER & GLAZMAN [1961, Section 11].

(2) Show that $(e_\lambda)_{\lambda\in\mathbb{R}}$ is an orthonormal family in the space $(Y, (\cdot,\cdot))$.

(3) Show that $(e_\lambda)_{\lambda\in\mathbb{R}}$ is a maximal orthonormal family in $(Y, (\cdot,\cdot))$.

(4) Show that the completion X of Y is a Hilbert space (Theorem 4.1-4) that is not separable.

4.9 Hilbert bases and Fourier series in a Hilbert space

We saw in the previous section that maximal orthonormal families $(e_i)_{i\in I}$ always exist in any infinite-dimensional inner-product space X and that they are countably infinite if X is separable (Theorems 4.8-3 and 4.8-4). We now show that, if X is *complete*, such families possess the fundamental property that any element $x \in X$ can be expanded as a *series* of the form $\sum_{i\in I}(x, e_i)e_i$. For this reason, a maximal orthonormal family *in a Hilbert space X* is called a **Hilbert basis** of X.

The following result is *one of the most basic results of linear functional analysis.* We consider here only the separable case, leaving some complements in the separable case and the nonseparable case as problems (Problems 4.9-1 and 4.9-2).

Theorem 4.9-1 (Fourier series in a separable Hilbert space) *Let $(X, (\cdot,\cdot))$ be an infinite-dimensional separable Hilbert space, and let $(e_n)_{n=1}^\infty$ be a Hilbert basis in X.*

(a) *Any element $x \in X$ can be expanded as the convergent series*

$$x = \sum_{n=1}^{\infty}(x, e_n)e_n,$$

which is called the **Fourier series**[21] *of x.*

(b) *The scalars $(x, e_n) \in \mathbb{K}$, $n \geq 1$, which are called the* **Fourier coefficients** *of x* (relative to the basis $(e_n)_{n=1}^\infty$), *satisfy* **Parseval's formula**:[22]

$$\|x\|^2 = \sum_{n=1}^{\infty} |(x, e_n)|^2.$$

Proof Let $x \in X$ be given.

(i) *We first show that* $\sum_{n=1}^\infty |(x, e_n)|^2 < \infty$. Using the assumption that $(e_n)_{n=1}^\infty$ is in

[21] So named after Jean Baptiste Joseph Fourier (1768–1830) and his seminal book on the theory of heat: *Théorie Analytique de la Chaleur*, published in 1822. In this masterpiece, Fourier established the convergence of the "classical" Fourier series (i.e., in terms of sines and cosines; these series are defined later in this section) in some specific cases, and showed how Fourier series could be used for solving partial differential equations, such as the heat equation.

[22] So named after Marc-Antoine Parseval, who in 1799 inferred from direct computations, but without a proof of convergence, that the coefficients a_k and b_k of the classical Fourier series should satisfy $\frac{1}{\pi}\int_0^{2\pi} |g(\theta)|^2 \,d\theta = \frac{a_0^2}{2} + \sum_{k=1}^\infty (a_k^2 + b_k^2)$.

particular an orthonormal family, we obtain

$$\begin{aligned}
0 &\le \Big\| x - \sum_{n=1}^{k} (x, e_n) e_n \Big\|^2 \\
&= \Big(x - \sum_{n=1}^{k} (x, e_n) e_n, x - \sum_{m=1}^{k} (x, e_m) e_m \Big) \\
&= \|x\|^2 - \sum_{n=1}^{k} |(x, e_n)|^2 - \sum_{m=1}^{k} |(x, e_m)|^2 + \sum_{m,n=1}^{k} (x, e_n)\overline{(x, e_m)}(e_n, e_m) \\
&= \|x\|^2 - \sum_{n=1}^{k} |(x, e_n)|^2 \quad \text{for any integer } k \ge 1.
\end{aligned}$$

We thus infer from this inequality that

$$\sum_{n=1}^{k} |(x, e_n)|^2 \le \|x\|^2 \quad \text{for any } k \ge 1,$$

which in turn implies that the series $\sum_{n=1}^{\infty} |(x, e_n)|^2$ is convergent.

(ii) *We next show that* $\sum_{n=1}^{\infty} (x, e_n) e_n$ *is a convergent series* (Section 3.6) *in the space* X. Since X is complete, it suffices to show that the sequence $(x_k)_{k=1}^{\infty}$ defined by

$$x_k := \sum_{n=1}^{k} (x, e_n) e_n$$

is a Cauchy sequence. To this end, using again that the family $(e_n)_{n=1}^{\infty}$ is orthonormal, we note that, for any integers $\ell \ge 1$ and $k \ge \ell + 1$,

$$\|x_k - x_\ell\|^2 = \Big(\sum_{n=\ell+1}^{k} (x, e_n) e_n, \sum_{m=\ell+1}^{k} (x, e_m) e_m \Big) = \sum_{n=\ell+1}^{k} |(x, e_n)|^2.$$

Hence $(x_k)_{k=1}^{\infty}$ is a Cauchy sequence, since the series $\sum_{n=1}^{\infty} |(x, e_n)|^2$ converges by (i).

(iii) *Let*

$$y := \lim_{k\to\infty} x_k = \sum_{n=1}^{\infty} (x, e_n) e_n.$$

It remains to show that $x = y$, or equivalently that

$$(x - y, e_n) = 0 \quad \text{for all } n \ge 1,$$

since the orthonormal family $(e_n)_{n=1}^{\infty}$ is maximal by assumption. The definition of y and the continuity of the inner product together imply that

$$(x - y, e_n) = \lim_{k\to\infty} \Big(x - \sum_{m=1}^{k} (x, e_m) e_m, e_n \Big).$$

But

$$\Big(x - \sum_{m=1}^{k} (x, e_m)e_m, e_n\Big) = 0 \quad \text{if } k \geq n.$$

Hence $x = y$.

(iv) The relations used in (i) and the relation $x = \lim_{k\to\infty} \sum_{n=1}^{k} (x, e_n)e_n$ established in (ii) and (iii) together imply that

$$0 = \lim_{k\to\infty} \Big\|x - \sum_{n=1}^{k} (x, e_n)e_n\Big\|^2 = \lim_{k\to\infty} \Big(\|x\|^2 - \sum_{n=1}^{k} |(x, e_n)|^2\Big).$$

This proves *Parseval's formula.* □

Part (i) of the above proof shows that, *in any inner-product space* X (complete or not), *the inequality*

$$\sum_{n=1}^{\infty} |(x, e_n)|^2 \leq \|x\|^2,$$

holds for any $x \in X$ *and any orthonormal family* $(e_n)_{n=1}^{\infty}$ (maximal or not). This inequality is called **Bessel's inequality.**[23]

Note that the convergence of the series $\sum_{n=1}^{\infty} |(x, e_n)|^2$ (itself a consequence of Bessel's inequality) evidently implies that, *given any orthonormal family* $(e_n)_{n=1}^{\infty}$ *in any inner-product space* X,

$$\lim_{n\to\infty} (x, e_n) = 0 \quad \text{for each } x \in X.$$

Theorem 4.9-1 has many important consequences. For instance, when it is applied to the space $L^2(0, 2\pi)$ it implies that any (real) function $g \in L^2(0, 2\pi)$ can be expanded as a *"classical" Fourier series* over the Hilbert basis defined in Theorem 4.8-5(b) ("classical" as opposed to the "general" Fourier series over arbitrary Hilbert bases considered in Theorem 4.9-1):

Theorem 4.9-2 (classical Fourier series) *Given any function* $g \in L^2(0, 2\pi)$, *let the* n*th Fourier partial sum* $S_n g \in \mathcal{C}_{\text{per}}[0, 2\pi]$ *of* g *be defined for all* $n \geq 0$ *by*

$$(S_n g)(\theta) := \frac{a_0}{2} + \sum_{k=1}^{n} (a_k \cos k\theta + b_k \sin k\theta), \quad 0 \leq \theta \leq 2\pi,$$

where

$$a_k := \frac{1}{\pi} \int_0^{2\pi} g(\varphi) \cos k\varphi \, \mathrm{d}\varphi, \ k \geq 0, \quad \textit{and} \quad b_k := \frac{1}{\pi} \int_0^{2\pi} g(\varphi) \sin k\varphi \, \mathrm{d}\varphi, \ k \geq 1.$$

Then

$$\|S_n g - g\|_{L^2(0,2\pi)} \to 0 \quad \textit{as } n \to \infty,$$

[23] So named after Friedrich Wilhelm Bessel (1784–1846), who established in 1828 this inequality for the coefficients of the "classical" Fourier series.

and the corresponding **Parseval formula for classical Fourier series** *holds:*

$$\|g\|^2_{L^2(0,2\pi)} = \pi\left(\frac{|a_0|^2}{2} + \sum_{k=1}^{\infty}|a_k|^2 + \sum_{k=1}^{\infty}|b_k|^2\right). \qquad \square$$

There thus also exists a subsequence $(S_{\sigma(n)}g)_{n=1}^{\infty}$ that pointwise converges to g almost everywhere in the interval $[0, 2\pi]$ (Theorem 3.4-3). The *Lusin conjecture*,[24] enunciated in 1913, asserted that in fact the *whole* sequence $(S_n g)_{n=1}^{\infty}$ *pointwise converges to* g *almost everywhere in* $[0, 2\pi]$. This seemingly innocuous statement remained one of the most challenging open problems for several decades, until it was finally shown to be true in a landmark paper by Lennart Carleson[25] in 1966.

Remark Even if the function g is *continuous* and *periodic* over $[0, 2\pi]$, its Fourier series does *not* necessarily converge uniformly on $[0, 2\pi]$: we shall establish later this assertion (Theorem 5.5-1), as a consequence of the *Banach–Steinhaus theorem*. Recall that, by contrast, the trigonometric polynomials $F_n g$, where F_n denotes the *Fejér operators*, do converge uniformly to g (Theorem 2.14-2). $\square$

Likewise, any *complex-valued* function $g \in L^2(0, 2\pi; \mathbb{C})$ can be expanded as a Fourier series over the Hilbert basis defined in Theorem 4.8-5(c):

Theorem 4.9-3 (classical Fourier series in the complex case) *Given any function* $g \in L^2(0, 2\pi; \mathbb{C})$, *let the* n*th Fourier partial sums* $g_n \in \mathcal{C}_{\text{per}}([0, 2\pi]; \mathbb{C})$ *be defined for all* $n \geq 0$ *by*

$$g_n(\theta) := \sum_{k=-n}^{n} c_k e^{ik\theta}, \quad 0 \leq \theta \leq 2\pi, \quad \textit{where} \quad c_k := \frac{1}{2\pi}\int_0^{2\pi} g(\varphi)e^{-ik\varphi}\,\mathrm{d}\varphi, \quad k \geq 0.$$

Then

$$\|g_n - g\|_{L^2(0,2\pi;\mathbb{C})} \to 0 \quad \textit{as } n \to \infty,$$

and the corresponding **Parseval formula for classical complex Fourier series** *holds:*

$$\|g\|^2_{L^2(0,2\pi;\mathbb{C})} = 2\pi \sum_{k=-\infty}^{\infty} |c_k|^2. \qquad \square$$

Remark The coefficients a_k and b_k defined in Theorem 4.9-2, *resp.* c_k defined in Theorem 4.9-3, are *not* genuine Fourier coefficients according to the definition given in Theorem 4.9-1; instead these are $\sqrt{\pi/2}a_0$, $\sqrt{\pi}a_k$ and $\sqrt{\pi}b_k$, $k \geq 1$, *resp.* $\sqrt{2\pi}c_k$, $k \in \mathbb{Z}$. This observation also explains why the factors π, *resp.* 2π, appear in the corresponding Parseval formulas. $\square$

Note in passing that the convergence to zero of the Fourier coefficients applied to the

---

[24] N. Lusin [1913]: Sur la convergence des séries trigonométriques de Fourier, *Comptes Rendus de l'Académie des Sciences de Paris* **156**, 1655–1658.

[25] L. Carleson [1966]: On convergence and growth of partial sums of Fourier series, *Acta Mathematica* **116**, 135–157.
For this and other mathematical feats, Carleson was awarded the Abel Prize in 2006.

above instances asserts that, *for any function* $g \in L^2(0, 2\pi)$,

$$\lim_{n\to\infty} \int_0^{2\pi} g(\varphi) \cos n\varphi \, \mathrm{d}\varphi = 0 \quad \text{and} \quad \lim_{n\to\infty} \int_0^{2\pi} g(\varphi) \sin n\varphi \, \mathrm{d}\varphi = 0,$$

and that, for any function $g \in L^2(0, 2\pi; \mathbb{C})$,

$$\lim_{n\to\infty} \int_0^{2\pi} g(\theta) e^{-in\theta} \, \mathrm{d}\theta = 0 \quad \text{and} \quad \lim_{n\to-\infty} \int_0^{2\pi} g(\theta) e^{-in\theta} \, \mathrm{d}\theta = 0.$$

These relations, which constitute the **Riemann–Lebesgue lemma**, provide examples of *weak convergence*, a fundamental notion that will be studied in Chapter 5.

Another important consequence of Theorem 4.9-1 applied to the space $L^2(0, 2\pi; \mathbb{C})$ is the *F. Riesz–Fischer theorem* (see Problem 4.9-4).

Naturally, similar Fourier series expansions and Parseval formulas hold in the spaces $L^2(-1, 1)$ or $L^2(-1, 1; \mathbb{C})$, $L^2(0, \infty)$, and $L^2(\mathbb{R})$, in terms respectively of the Legendre polynomials, Laguerre functions, and Hermite functions.

Returning to the general case, we next use Theorem 4.9-1 to show that there exists a *Hilbert space isomorphism* between *any separable Hilbert space* X and the *space* ℓ^2: this means that there exists a linear bijective mapping (denoted σ in the next theorem) between X and ℓ^2 that preserves the inner product (and is thus an isometry); hence the Hilbert space structures of the two spaces are identical.

Theorem 4.9-4 *Let* $(X, (\cdot, \cdot))$ *be a real,* resp. *complex, infinite-dimensional, separable Hilbert space. Then there exists a linear bijective mapping* σ *from* X *onto the real,* resp. *complex, space* ℓ^2*, such that*

$$(x, y)_X = (\sigma x, \sigma y)_{\ell^2} \quad \textit{for all } x, y \in X.$$

Consequently, any infinite-dimensional separable Hilbert space can be identified with the space ℓ^2*, by means of a linear isometry that preserves the inner product.*

Proof Since X is an infinite-dimensional separable Hilbert space, there exists a countably infinite Hilbert basis $(e_n)_{n=1}^\infty$ in X (Theorem 4.8-3). Hence any $x \in X$ can be expanded as the Fourier series $x = \sum_{n=1}^\infty (x, e_n) e_n$ (Theorem 4.9-1). Let then

$$\sigma(x) := ((x, e_n))_{n=1}^\infty \quad \text{for each } x \in X.$$

First, $\sigma(x) \in \ell^2$ for each $x \in X$ since $\|\sigma(x)\|_{\ell^2} = \sum_{n=1}^\infty |(x, e_n)|^2 = \|x\|^2 < \infty$ by Parseval's formula (Theorem 4.9-1).

Second, the mapping $\sigma : X \to \ell^2$ defined in this fashion is linear (the inner product is linear with respect to its first argument), isometric (again by Parseval's formula), and preserves the inner product, since

$$\begin{aligned}(x, y)_X &= \lim_{k\to\infty} \Big(\sum_{n=1}^{k} (x, e_m) e_m, \sum_{n=1}^{k} (y, e_n) e_n \Big) \\ &= \lim_{k\to\infty} \sum_{n=1}^{k} (x, e_n) \overline{(y, e_n)} = (\sigma(x), \sigma(y))_{\ell^2}\end{aligned}$$

(by continuity of the inner product; cf. Theorem 4.1-1(c)).

Third, given any element $\xi = (\xi_n)_{n=1}^{\infty} \in \ell^2$ and any integer $k \geq 1$, let $x_k := \sum_{n=1}^{k} \xi_n e_n$. Then the sequence $(x_k)_{k=1}^{\infty}$ is a Cauchy sequence in X, since

$$\|x_k - x_\ell\|^2 = \sum_{n=\ell+1}^{k} |\xi_n|^2 \quad \text{for all } k-1 \geq \ell \geq 1,$$

and $\sum_{n=1}^{\infty} |\xi_i|^2 < \infty$ by assumption. Let $x := \lim_{k\to\infty} x_k$ (the space X is complete by assumption). Then,

$$\text{for each } x \geq 1, \quad (x, e_n) = \lim_{k\to\infty} (x_k, e_n) = \xi_n.$$

Hence $\sigma(x) = \xi$, which shows that $\sigma : X \to \ell^2$ is surjective. The mapping σ thus possesses all the announced properties. □

Problems

4.9-1 Let $(X, (\cdot,\cdot))$ be an infinite-dimensional separable Hilbert space, and let $(e_n)_{n=1}^{\infty}$ be an orthonormal family in X. Show that the following properties are equivalent:

(a) The family $(e_n)_{n=1}^{\infty}$ is maximal.

(b) Any element $x \in X$ can be expanded as $x = \sum_{n=1}^{\infty}(x, e_n)e_n$.

(c) For any $x \in X$, $\|x\|^2 = \sum_{n=1}^{\infty} |(x, e_n)|^2$.

(d) For any $x, y \in X$, $(x, y) = \sum_{n=1}^{\infty}(x, e_n)\overline{(y, e_n)}$.

(e) $\overline{\text{Span}(e_n)_{n=1}^{\infty}} = X$.

Remark That (a) implies (b) and (c) has been established in Theorem 4.9-1. □

4.9-2 (Fourier series in a nonseparable Hilbert space) This problem constitutes the "nonseparable version" of Theorem 4.9-1.

(1) Let $(X, (\cdot,\cdot))$ be an inner-product space and let $(e_i)_{i\in I}$ be an uncountably infinite orthonormal family of elements $x_i \in X$. Show that given any $x \in X$, $(x, e_i) = 0$ for at most a countably infinite number of indices $i \in I$.

(2) Let $(X, (\cdot,\cdot))$ be a nonseparable Hilbert space, and let $(e_i)_{i\in I}$ be a Hilbert basis in X (such a Hilbert basis always exists by Theorem 4.8-4, and is necessarily uncountably infinite by Problem 4.8-6). Given any $x \in X$, let the nonzero scalars $(x, e_i)_{i\in I}$ be arranged as a sequence $(x, e_n)_{n=0}^{\infty}$ (the case where there are only a finite number of nonzero scalars $(x, e_i)_{i\in I}$ is left to the reader). Show that $x = \sum_{n=0}^{\infty}(x, e_n)e_n$ and that this series is *commutatively convergent*, in the sense that $x = \sum_{n=0}^{\infty}(x, e_{\tau(n)})e_{\tau(n)}$ for any bijection $\tau : \mathbb{N} \to \mathbb{N}$.

(3) Show that $\|x\|^2 = \sum_{n=0}^{\infty} |(x, e_n)|^2$ and that the series $\sum_{n=0}^{\infty} |(x, e_n)|^2$ is likewise commutatively convergent.

4.9-3 Let X be a Hilbert space. Show that any two Hilbert bases of X have the same cardinal number (Section 1.5).

4.9-4 (F. Riesz–Fischer[26] theorem) Let scalars $c_k \in \mathbb{C}$, $k \in \mathbb{Z}$, be given such that $\sum_{k=-\infty}^{\infty} |c_k|^2 < \infty$. Show that there exists a function $g \in L^2(0, 2\pi; \mathbb{C})$ such that $c_k = \frac{1}{2\pi}\int_0^{2\pi} g(\varphi)e^{-ik\varphi}\,d\varphi$ for all $k \geq 0$.

[26] This theorem was first established in:

F. RIESZ [1907]: Sur les systèmes orthogonaux de fonctions, *Comptes Rendus de l'Académie des Sciences* **144**, 615–619.

4.9-5 Let G be a function in the space $L^2(]0,1[\times]0,1[)$.
(1) Given any function $f \in L^2(0,1)$, let

$$Af(x) := \int_0^1 G(x,\xi)f(\xi)\mathrm{d}\xi, \quad 0 \le x \le 1.$$

Show that this relation defines a function $Af \in L^2(0,1)$ and that the linear operator $A : L^2(0,1) \to L^2(0,1)$ defined in this fashion is compact (Section 2.10).

Hint: Let $(e_n)_{n=1}^\infty$ be a Hilbert basis in the space $L^2(0,1)$, and, for each $n \ge 1$, define a linear operator $A_n : L^2(0,1) \to \mathrm{Span}(e_k)_{k=1}^n$ by $A_n f := \sum_{\ell,k=1}^n (Ae_\ell, e_k)(f, e_\ell)e_k$ for any $f \in L^2(0,1)$. Show that $\|A_n - A\| \to 0$ as $n \to \infty$, and use Problem 3.2-4.

(2) Show that, if $G(x,\xi) = G(\xi,x)$ for almost all $(x,\xi) \in [0,1] \times [0,1]$, the operator A satisfies $(Af,g) = (f,Ag)$ for all $f,g \in L^2(0,1)$, where $(\cdot,\cdot)$ denotes the inner product of the space $L^2(0,1)$.

4.10 Eigenvalues and eigenvectors of self-adjoint operators in inner-product spaces

Let $(X,(\cdot,\cdot))$ be an inner-product space over $\mathbb{K}$. A *linear operator* $A : X \to X$ is **self-adjoint** if it coincides with its adjoint A^* (Section 4.7), i.e., if it satisfies

$$(Ax,y) = (x,Ay) \quad \text{for all } x,y \in X.$$

A self-adjoint operator is also said to be **symmetric** if $\mathbb{K} = \mathbb{R}$, or **Hermitian** if $\mathbb{K} = \mathbb{C}$.

Remark We shall see later (Theorem 5.7-2) that, *if X is a Hilbert space, any self-adjoint operator from X into X is continuous*; this remarkable, and somewhat surprising, property is a simple corollary of the *Banach closed graph theorem.* □

For instance, let $\mathbb{R}^n$ be equipped with the Euclidean inner product (Section 4.2). Since any linear operator from $\mathbb{R}^n$ into $\mathbb{R}^n$ can be identified with an $n \times n$ real matrix $\boldsymbol{A}$, it is clear that such a linear operator is symmetric if and only if the associated matrix $\boldsymbol{A} = (a_{ij})$ is symmetric in the matrix sense, i.e., if $a_{ij} = a_{ji}$ for all $1 \le i, j \le n$.

Similarly, let $\mathbb{C}^n$ be equipped with the Hermitian inner product (Section 4.2). Since any linear operator from $\mathbb{C}^n$ into $\mathbb{C}^n$ can be identified with an $n \times n$ complex matrix $\boldsymbol{A}$, it is likewise clear that such a linear operator is Hermitian if and only if the associated matrix $\boldsymbol{A} = (a_{ij})$ is Hermitian in the matrix sense, i.e., if $a_{ij} = \overline{a_{ji}}$ for all $1 \le i, j \le n$.

A self-adjoint linear operator $A : X \to X$ is **nonnegative-definite** if $(Ax,x) \ge 0$ for all $x \in X$, or **positive-definite** if $(Ax,x) > 0$ for all nonzero $x \in X$. Note that, *if A is positive-definite, then* $\mathrm{Ker}\, A = \{0\}$.

The notions of nonnegative-definiteness and positive-definiteness as defined above for general self-adjoint operators thus extend well-known properties of real symmetric, or complex Hermitian, matrices.

Examples of symmetric linear operators acting from the space $(\mathcal{C}[0,1],(\cdot,\cdot))$ into itself, or from the space $(L^2(0,1),(\cdot,\cdot))$ into itself, where $(\cdot,\cdot)$ denotes in both cases the inner product of the space $L^2(0,1)$, are provided in Problems 3.10-4 and 4.9-5.

Another proof was almost immediately thereafter given by:

E. Fischer [1907]: Sur la convergence en moyenne, *Comptes Rendus de l'Académie des Sciences* **144**, 1022–1024.

The next theorem gathers some elementary, yet constantly used, properties of self-adjoint linear operators and of their eigenvalues and eigenvectors, which generalize well-known properties of real symmetric, or complex Hermitian, matrices.

Theorem 4.10-1 *Let $(X, (\cdot,\cdot))$ be an inner-product space, and let $A : X \to X$ be a self-adjoint linear operator.*

(a) *For any $x \in X$, the scalar (Ax, x) is real.*

(b) *Let λ be any eigenvalue of A. Then λ is real. Moreover, $\lambda \geq 0$ if A is nonnegative-definite, and $\lambda > 0$ if A is positive-definite.*

(c) *Eigenvectors corresponding to distinct eigenvalues are orthogonal.*

(d) *If $A \in \mathcal{L}(X)$, the operator norm of A, viz., $\|A\| := \sup_{x\neq 0} \dfrac{\|Ax\|}{\|x\|}$, is also given by*

$$\|A\| = \sup_{x\neq 0} \frac{|(Ax,x)|}{\|x\|^2}.$$

Proof If $\mathbb{K} = \mathbb{C}$, $(Ax, x) = (x, Ax) = \overline{(Ax, x)}$ for any $x \in X$. Consequently, $(Ax, x) \in \mathbb{R}$. This proves (a) (if $\mathbb{K} = \mathbb{R}$, there is nothing to prove).

If $Ap = \lambda p$ and $p \neq 0$, then $(Ap, p) = \lambda(p, p)$ and thus $\lambda = \dfrac{(Ap,p)}{(p,p)} \in \mathbb{R}$ by (a). That $\lambda \geq 0$ if A is nonnegative-definite and $\lambda > 0$ if A is positive-definite is clear. This proves (b).

If $Ap_1 = \lambda_1 p_1$ and $Ap_2 = \lambda_2 p_2$, then

$$(Ap_1, p_2) = \lambda_1(p_1, p_2) = (p_1, Ap_2) = \lambda_2(p_1, p_2).$$

Hence $(\lambda_1 - \lambda_2)(p_1, p_2) = 0$, which implies that $(p_1, p_2) = 0$ if $\lambda_1 \neq \lambda_2$. This proves (c).

If $A \in \mathcal{L}(X)$, the Cauchy–Schwarz–Bunyakovskiĭ inequality immediately gives

$$\nu(A) := \sup_{x\neq 0} \frac{|(Ax,x)|}{\|x\|^2} \leq \|A\|.$$

To prove (d), it thus remains to show that $\|A\| \leq \nu(A)$. To this end, recall that the operator norm $\|A\| = \sup_{x\neq 0} \dfrac{\|Ax\|}{\|x\|}$ in an inner-product space is also given by (Theorem 4.1-3)

$$\|A\| = \sup_{\begin{cases} x\neq 0 \\ y\neq 0 \end{cases}} \frac{|(Ax,y)|}{\|x\|\,\|y\|} = \sup_{\|x\|=\|y\|=1} |(Ax,y)|.$$

So, let x and y be such that $(Ax, y) \neq 0$. Then $|(Ax, y)|$ can be rewritten as

$$|(Ax,y)| = \frac{(Ax,y)\overline{(Ax,y)}}{|(Ax,y)|} = (A\widetilde{x}, y), \quad \text{with } \widetilde{x} := \frac{\overline{(Ax,y)}}{|(Ax,y)|}\, x.$$

Since $(A\widetilde{x}, y) = (y, A\widetilde{x})$ because $(A\widetilde{x}, y) = |(Ax, y)| \in \mathbb{R}$, and $(y, A\widetilde{x}) = (Ay, \widetilde{x})$ by the assumed self-adjointness of A, $|(Ax, y)|$ can be further rewritten as

$$\begin{aligned}|(Ax,y)| = (A\widetilde{x},y) &= \frac{1}{2}\{(A\widetilde{x},y) + (Ay,\widetilde{x})\}\\ &= \frac{1}{4}\{(A(\widetilde{x}+y),\widetilde{x}+y) - (A(\widetilde{x}-y,\widetilde{x}-y)\}.\end{aligned}$$

Using successively the definition of $\nu(A)$, the parallelogram law, and the relation $\|\widetilde{x}\| = \|x\|$, we obtain

$$\begin{aligned}|(Ax,y)| &\leq \frac{1}{4}\nu(A)\{\|\widetilde{x}+y\|^2 + \|\widetilde{x}-y\|^2\}\\ &= \frac{1}{2}\nu(A)\{\|\widetilde{x}\|^2 + \|y\|^2\} = \frac{1}{2}\nu(A)\{\|x\|^2 + \|y\|^2\}.\end{aligned}$$

Hence

$$\|A\| = \sup_{\|x\|=\|y\|=1} |(Ax,y)| \leq \nu(A),$$

as desired. □

4.11 The spectral theorem for compact self-adjoint operators

It is well known that any $n \times n$ real symmetric, or $n \times n$ complex Hermitian, matrix possesses exactly n real eigenvalues (counting multiplicities), which can be computed by means of its *Rayleigh quotient*,[27] and that there exist exactly n corresponding eigenvectors that form an orthonormal basis in $\mathbb{R}^n$, or in $\mathbb{C}^n$. It is remarkable that any *compact* (Section 2.10) and *self-adjoint* (Section 4.10) linear operator acting in an *infinite-dimensional* inner-product space possesses similar properties. Moreover, such an operator possesses an at most *countably infinite number of nonzero real eigenvalues*, each one of *finite multiplicity* (i.e., whose corresponding eigenspace is finite-dimensional), and *the corresponding eigenvectors form a maximal orthonormal family* if the operator is in addition *injective*. This is the essence of the next theorem, which constitutes the *spectral theorem for such operators*.

This result is all the more remarkable, since the existence of eigenvalues for such operators is established therein without any recourse to the notions of determinants or characteristic polynomials as in the finite-dimensional case.

Remark By contrast, the study of eigenvalues and eigenvectors of "general" linear operators acting in "general" infinite-dimensional normed vector spaces is much more delicate[28] (as already suggested by the finite-dimensional case; think of the Jordan canonical form *versus* the diagonalization theorem for real symmetric, or complex Hermitian, matrices). □

Theorem 4.11-1 (spectral theorem for compact self-adjoint operators with infinite-dimensional range) *Let $(X, (\cdot,\cdot))$ be an infinite-dimensional inner-product space and let $A : X \to X$ be a compact and self-adjoint linear operator with an infinite-dimensional range. Then:*

[27] See, e.g., CIARLET [1987, Section 1.3].

[28] For a short introduction, see, e.g., TAYLOR [1958, Chapter 5], or TAYLOR & LAY [1980, Chapter 5]. For an extensive treatment, see DUNFORD & SCHWARTZ [1963].

(a) *There exist an infinite sequence* $(\lambda_n)_{n=1}^{\infty}$ *of eigenvalues of* A *and an infinite sequence* $(p_n)_{n=1}^{\infty}$ *of corresponding eigenvectors that satisfy*

$$|\lambda_1| = \|A\|, \quad |\lambda_1| \geq |\lambda_2| \geq \cdots \geq |\lambda_n| \geq \cdots, \quad \lambda_n \neq 0 \text{ for all } n \geq 1, \quad \lim_{n\to\infty} \lambda_n = 0,$$

$$Ap_n = \lambda_n p_n \text{ for all } n \geq 1 \quad \text{and} \quad (p_k, p_\ell) = \delta_{k\ell} \text{ for all } k, \ell \geq 1.$$

$$|\lambda_1| = \frac{|(Ap_1, p_1)|}{\|p_1\|^2} = \sup_{x\neq 0} \frac{|(Ax, x)|}{\|x\|^2},$$

$$|\lambda_n| = \frac{|(Ap_n, p_n)|}{\|p_n\|^2} = \sup_{\left\{\substack{x\neq 0 \\ (x,p_k)=0,\, 1\leq k\leq n-1}\right.} \frac{|(Ax, x)|}{\|x\|^2} \text{ for all } n \geq 2.$$

(b) *For any vector* $x \in X$,

$$Ax = \sum_{n=1}^{\infty} \lambda_n (x, p_n) p_n.$$

(c) *Let* λ *be any nonzero eigenvalue of* A. *Then, there exists* $n \geq 1$ *such that* $\lambda_n = \lambda$. *Besides, the set* $I(\lambda) := \{n \geq 1;\ \lambda_n = \lambda\}$ *is finite, and*

$$\{p \in X;\ Ap = \lambda p\} = \text{Span}(p_n)_{n\in I(\lambda)}.$$

(d) *The kernel of* A *is also given by*

$$\text{Ker}\, A = (\text{Span}\,(p_n)_{n=1}^{\infty})^{\perp}.$$

Proof For convenience, the proof is broken into several parts, numbered from (i) to (vii). Recall that all the eigenvalues of a self-adjoint operator are real (Theorem 4.10-1(b)).

(i) *There exist an eigenvalue* λ_1 *and a corresponding eigenvector* p_1 *that satisfy*

$$Ap_1 = \lambda_1 p_1, \quad \|p_1\| = 1, \quad \text{and} \quad 0 < |\lambda_1| = \|A\| = \sup_{x\neq 0} \frac{|(Ax, x)|}{\|x\|^2}. = \frac{|(Ap_1, p_1)|}{\|p_1\|^2}.$$

The self-adjointness of A implies that $\sup_{\|x\|=1} |(Ax, x)| = \|A\|$ (Theorem 4.10-1(d)), and $\|A\| > 0$ since $A \neq 0$ (by assumption, the direct image $A(X)$ is infinite-dimensional); hence there exists a sequence $(x_n)_{n=1}^{\infty}$ such that $|(Ax_n, x_n)| \to \|A\|$ as $n \to \infty$. Consequently, there exist a subsequence $(x_m)_{m=1}^{\infty}$ and $\lambda_1 \in \mathbb{R}$ such that

$$\|x_m\| = 1 \text{ for all } m \geq 1, \quad (Ax_m, x_m) \underset{m\to\infty}{\longrightarrow} \lambda_1, \quad \text{and} \quad |\lambda_1| = \sup_{\|x\|=1} |(Ax, x)| = \|A\| > 0.$$

The sequence $(x_m)_{m=1}^{\infty}$ being thus bounded, the assumed compactness of A implies that there exists a subsequence $(x_\ell)_{\ell=1}^{\infty}$ of the sequence $(x_m)_{m=1}^{\infty}$ such that the sequence $(Ax_\ell)_{\ell=1}^{\infty}$ converges in X. Noting that

$$\|Ax_\ell - \lambda_1 x_\ell\|^2 = \|Ax_\ell\|^2 - 2\lambda_1 (Ax_\ell, x_\ell) + \lambda_1^2 \leq \|A\|^2 - 2\lambda_1 (Ax_\ell, x_\ell) + \lambda_1^2$$

(recall that $(Ax_\ell, x_\ell) \in \mathbb{R}$ even in the complex case; cf. Theorem 4.10-1(a)), and that

$$(\|A\|^2 - 2\lambda_1 (Ax_\ell, x_\ell) + \lambda_1^2) \underset{\ell\to\infty}{\longrightarrow} 0,$$

we infer that

$$(Ax_\ell - \lambda_1 x_\ell) \underset{\ell\to\infty}{\longrightarrow} 0,$$

and therefore that

$$x_\ell = \left\{ -\frac{1}{\lambda_1}(Ax_\ell - \lambda_1 x_\ell) + \frac{1}{\lambda_1} Ax_\ell \right\} \underset{\ell\to\infty}{\longrightarrow} p_1 := \frac{1}{\lambda_1} \lim_{\ell\to\infty} Ax_\ell.$$

Besides,

$$\|p_1\| = 1,$$

since $\|x_\ell\| = 1$ for all $\ell \geq 1$. Finally,

$$Ap_1 = A\Big(\lim_{\ell\to\infty} x_\ell\Big) = \lim_{\ell\to\infty} Ax_\ell = \lambda_1 p_1,$$

since A is continuous. Hence either $\lambda_1 = \|A\|$ or $\lambda_1 = -\|A\|$ is an eigenvalue of A.

(ii) *There exist an infinite sequence $(\lambda_n)_{n=1}^\infty$ of eigenvalues and an infinite sequence $(p_n)_{n=1}^\infty$ of corresponding eigenvectors that together satisfy*

$$Ap_n = \lambda_n p_n \quad \textit{for all } n \geq 1 \quad \textit{and} \quad (p_k, p_\ell) = \delta_{k\ell} \quad \textit{for all } k, \ell \geq 1,$$

$$0 < |\lambda_n| = \sup_{\substack{x \neq 0 \\ (x,p_k)=0,\, 1\leq k\leq n-1}} \frac{|(Ax,x)|}{\|x\|^2} = \frac{|(Ap_n,p_n)|}{\|p_n\|^2} \leq |\lambda_{n-1}| \leq \cdots \leq |\lambda_1| \quad \text{for all } n \geq 2.$$

Define the subspace

$$X_2 := \{x \in X;\ (x, p_1) = 0\},$$

where p_1 is the eigenvector found in (i). Then the direct image $A(X_2)$ of X_2 under A is contained in X_2, since

$$(Ax, p_1) = (x, Ap_1) = \lambda_1 (x, p_1) = 0 \quad \text{for all } x \in X_2.$$

Clearly, the restriction A_2 of A to X_2 is again a compact and self-adjoint linear operator. Besides, $A_2 \neq 0$; otherwise $A_2 = 0$ would imply that

$$A(x - (x,p_1)p_1) = 0 \quad \text{for all } x \in X,$$

since $(x - (x,p_1)p_1) \in X_2$ for all $x \in X$, and hence that

$$Ax = \lambda_1 (x, p_1) p_1 \quad \text{for all } x \in X,$$

which would contradict the assumption that $A(X)$ is infinite-dimensional. The argument of part (i) can thus be applied *verbatim* to $A_2 : X_2 \to X_2$, showing that there exist $\lambda_2 \in \mathbb{R}$ and a vector $p_2 \in X_2$ that together satisfy

$$A_2 p_2 = Ap_2 = \lambda_2 p_2, \quad (p_2, p_1) = 0, \quad \text{and} \quad \|p_2\| = 1,$$

$$0 < |\lambda_2| = \|A_2\| \sup_{\substack{x\neq 0 \\ x \in X_2}} \frac{|(A_2 x, x)|}{\|x\|^2} = \sup_{\substack{x\neq 0 \\ x \in X_2}} \frac{|(Ax, x)|}{\|x\|^2} \leq \sup_{x\neq 0} \frac{|(Ax, x)|}{\|x\|^2} = |\lambda_1|.$$

We then iterate the above procedure: Assume that, for some integer $n \geq 2$, we have found eigenvalues λ_k, $1 \leq k \leq n$, and corresponding eigenvectors p_k, $1 \leq k \leq n$, that together satisfy

$$Ap_k = \lambda_k p_k \quad \text{and} \quad (p_k, p_\ell) = \delta_{k\ell} \quad \text{for all } 1 \leq k, \ell \leq n,$$

$$0 < |\lambda_n| = \sup_{\left\{\substack{x \neq 0 \\ (x, p_k)=0,\, 1 \leq k \leq n-1}\right.} \frac{|(Ax, x)|}{\|x\|^2} \leq |\lambda_{n-1}| \leq \cdots \leq |\lambda_1|.$$

Define the subspace

$$X_{n+1} := \{x \in X;\ (x, p_k) = 0,\ 1 \leq k \leq n\}.$$

Then the direct image $A(X_{n+1})$ of X_{n+1} under A is contained in X_{n+1} since

$$(Ax, p_k) = (x, Ap_k) = \lambda_k (x, p_k) = 0 \quad \text{for all } x \in X_{n+1} \text{ and all } 1 \leq k \leq n.$$

Clearly, the restriction A_{n+1} of A to X_{n+1} is again a compact and self-adjoint linear operator. Besides, $A_{n+1} \neq 0$; otherwise $A_{n+1} = 0$ would imply that

$$A\left(x - \sum_{k=1}^{n} (x, p_k) p_k\right) = 0 \quad \text{for all } x \in X,$$

since $(x - \sum_{k=1}^{n} (x, p_k) p_k) \in X_{n+1}$ for any $x \in X$, hence that

$$Ax = \sum_{k=1}^{n} \lambda_k (x, p_k) p_k \quad \text{for all } x \in X,$$

which would contradict the assumption that $A(X)$ is infinite-dimensional. The argument of part (i) can thus be again applied *verbatim* to A_{n+1}, showing that there exist $\lambda_{n+1} \in \mathbb{R}$ and a vector $p_{n+1} \in X_{n+1}$ that satisfy

$$A_{n+1} p_{n+1} = Ap_{n+1} = \lambda_{n+1} p_{n+1}, \quad (p_{n+1}, p_k) = 0, \quad 1 \leq k \leq n, \quad \text{and} \quad \|p_{n+1}\| = 1,$$

$$0 < |\lambda_{n+1}| = \sup_{\left\{\substack{x \neq 0 \\ x \in X_{n+1}}\right.} \frac{|(A_{n+1}x, x)|}{\|x\|^2} = \sup_{\left\{\substack{x \neq 0 \\ (x, p_k)=0,\, 1 \leq k \leq n}\right.} \frac{|(Ax, x)|}{\|x\|^2}$$

$$\leq \sup_{\left\{\substack{x \neq 0 \\ (x, p_k)=0,\, 1 \leq s \leq n-1}\right.} \frac{A(x, x)}{\|x\|^2} = |\lambda_n| \leq \cdots \leq |\lambda_1|.$$

Hence the announced property indeed holds for all $n \geq 2$.

(iii) *The eigenvalues* λ_n, $n \geq 1$, *found in* (ii) *satisfy* $\lim_{n \to \infty} \lambda_n = 0$.
Assume otherwise that there exists $\delta > 0$ such that

$$|\lambda_n| \geq \delta > 0 \quad \text{for all } n \geq 1$$

(recall that $(|\lambda_n|)_{n=1}^{\infty}$ is a decreasing sequence; cf. part (ii)). The sequence $\left(\frac{1}{\lambda_n}p_n\right)_{n=1}^{\infty}$ being then bounded in X, there would exist a subsequence $\left(\frac{1}{\lambda_{\sigma(n)}}p_{\sigma(n)}\right)_{n=1}^{\infty}$ such that the sequence

$$\left(A\left(\frac{1}{\lambda_{\sigma(n)}}p_{\sigma(n)}\right)\right)_{n=1}^{\infty} = \left(p_{\sigma(n)}\right)_{n=1}^{\infty}$$

converges, by the compactness of A (Theorem 2.10-1(b)). But this is impossible, since the orthonormality of the eigenvectors established in part (ii) implies that

$$\|p_k - p_\ell\|^2 = \|p_k\|^2 + \|p_\ell\|^2 = 2 \quad \text{for all } k \neq \ell.$$

All the properties announced in part (a) of Theorem 4.11-1 have thus been established.

(iv) *For any vector $x \in X$, the vector $Ax \in X$ is given as the sum of the convergent series $Ax = \sum_{k=1}^{\infty} \lambda_k (x, p_k) p_k$.*

Given any $x \in X$ and any integer $n \geq 1$, define the vector

$$x_n := x - \sum_{k=1}^{n} (x, p_k) p_k,$$

which belongs to the subspace

$$X_{n+1} = \{x \in X;\ (x, p_k) = 0,\ 1 \leq k \leq n\} = \left(\mathrm{Span}(p_k)_{k=1}^{n}\right)^{\perp},$$

already encountered in part (ii). Since, for each $n \geq 1$, the vector x_n is orthogonal to the vector $\sum_{k=1}^{n}(x, p_k)p_k$, the Pythagoras theorem implies that

$$\|x_n\| \leq \|x\|.$$

Since $Ax_n = A_{n+1}x_n$ (because $x_n \in X_{n+1}$ and $A_{n+1} = A|_{X_{n+1}}$), and

$$\|A_{n+1}\| = \sup_{\substack{x \neq 0 \\ x \in X_{n+1}}} \frac{|(A_{n+1}x, x)|}{\|x\|^2} = |\lambda_{n+1}|,$$

by part (ii), it thus follows that

$$\begin{aligned} \|Ax - \sum_{k=1}^{n} \lambda_k (x, p_k) p_k\| &= \|Ax_n\| = \|A_{n+1}x_n\| \\ &\leq \|A_{n+1}\|\,\|x_n\| = |\lambda_{n+1}|\,\|x_n\| \leq |\lambda_{n+1}|\,\|x\|. \end{aligned}$$

Hence $\lim_{n\to\infty}\{Ax - \sum_{k=1}^{n} \lambda_k (x, p_k) p_k\} = 0$ since $\lim_{n\to\infty} |\lambda_{n+1}| = 0$ by part (iii). This proves (b).

(v) *All the nonzero eigenvalues of A have been found by the iterative procedure described in* part (ii).

Let $\lambda \neq 0$ and $p \neq 0$ be such that $Ap = \lambda p$; hence $Ap \neq 0$. If $\lambda \neq \lambda_k$ for all $k \geq 1$, then $(p, p_k) = 0$ for all $k \geq 1$ (Theorem 4.10-1(c)). Hence, by part (iv),

$$Ap = \lim_{n\to\infty} \left(\sum_{k=1}^{n} \lambda_k (p, p_k) p_k \right) = 0,$$

a contradiction.

(vi) *Given any nonzero eigenvalue λ of A, define the set*

$$I(\lambda) := \{n \geq 1,\ \lambda_n = \lambda\},$$

which is nonempty (by (v)) *and finite* (since $\lim_{n\to\infty} \lambda_n = 0$ by (iii)). *Then*

$$\{p \in X;\ Ap = \lambda p\} = \mathrm{Span}(p_n)_{n\in I(\lambda)}.$$

In other words, each nonzero eigenvalue of A is of finite multiplicity, and all the eigensubspaces of A corresponding to the nonzero eigenvalues of A have been found by the iterative procedure described in (ii).

By part (ii), there exist eigenvectors p_n, $n \in I(\lambda)$, such that

$$Ap_n = \lambda p_n \quad \text{for all } n \in I(\lambda) \quad \text{and} \quad (p_k, p_\ell) = \delta_{k\ell} \quad \text{for all } k, \ell \in I(\lambda).$$

Hence $\mathrm{Span}(p_n)_{n\in I(\lambda)} \subset \{p \in X;\ Ap = \lambda p\}$. To prove that this inclusion is an equality, assume otherwise that there exists a vector $\widetilde{p} \in X$ such that $A\widetilde{p} = \lambda\widetilde{p}$, $\widetilde{p} \neq 0$, and $\widetilde{p} \notin \mathrm{Span}(p_n)_{n\in I(\lambda)}$. The Gram–Schmidt orthonormalization (Theorem 4.8-1) applied to the vectors $\widetilde{p}$ and p_n, $n \in I(\lambda)$, would then provide a vector $p \in \mathrm{Span}\{(p_n)_{n\in I(\lambda)}, \widetilde{p}\}$ that satisfies

$$(p, p_n) = 0 \quad \text{for all } n \in I(\lambda) \quad \text{and} \quad \|p\| = 1.$$

Besides,

$$Ap = \lambda p,$$

since p is a linear combination of the vectors $\widetilde{p}$ and p_n, $n \in I(\lambda)$. Therefore, by Theorem 4.10-1(c),

$$(p, p_n) = 0 \quad \text{for all } n \notin I(\lambda),$$

since $\lambda \neq \lambda_n$ for all $n \notin I(\lambda)$. Hence the nonzero vector p satisfies $(p, p_n) = 0$ for all $n \geq 1$. The same argument as in part (v) above then leads to a contradiction.

All the properties announced in (c) have thus been established.

(vii) *It remains to show that* (orthogonal complements are defined in Section 4.5)

$$\mathrm{Ker}\, A = (\mathrm{Span}(p_n)_{n=1}^{\infty})^{\perp}.$$

Let $x \in X$ be such that $Ax = 0$. Then, for any $n \geq 1$,

$$(x, p_n) = \frac{1}{\lambda_n}(x, \lambda_n p_n) = \frac{1}{\lambda_n}(x, Ap_n) = \frac{1}{\lambda_n}(Ax, p_n) = 0,$$

which means that $x \in (\mathrm{Span}(p_n)_{n=1}^{\infty})^{\perp}$.

If, conversely, $x \in (\mathrm{Span}(p_n)_{n=1}^{\infty})^{\perp}$, then $(x, p_n) = 0$ for all $n \geq 1$, and thus $Ax = 0$ by (iv). Hence (d) is proved. □

Naturally, if A is *nonnegative-definite* or *positive-definite*, the eigenvalues $\lambda_n, n \geq 1$, found in Theorem 4.11-1 satisfy

$$\lambda_n > 0 \quad \text{for all } n \geq 1 \text{ and } \lambda_1 \geq \lambda_2 \geq \cdots \geq \lambda_n \geq \cdots,$$

and there is no need to use absolute values in their characterization, which now takes the form

$$\lambda_1 = \sup_{x \neq 0} \frac{(Ax, x)}{\|x\|^2} \quad \text{and} \quad \lambda_n = \sup_{\left\{\substack{x \neq 0 \\ (x, p_k) = 0,\ 1 \leq k \leq n-1}\right.} \frac{(Ax, x)}{\|x\|^2}.$$

Remark A converse to Theorem 4.11-1 holds; cf. Problem 4.11-1. □

For completeness, we also consider the simpler case where $A : X \to X$ is a continuous linear operator with a *finite-dimensional range* (in which case A is necessarily compact; cf. Theorem 2.10-1(d)). Such operators thus include those in finite-dimensional spaces that are represented by *real symmetric*, or *complex Hermitian, matrices.*

Theorem 4.11-2 (spectral theorem for continuous self-adjoint operators with finite-dimensional range) *Let $(X, (\cdot, \cdot))$ be an inner-product space and let $A : X \to X$ be a continuous and self-adjoint linear operator with a range of finite dimension $N \geq 1$. Then:*

(a) *There exist exactly N nonzero eigenvalues λ_n of A and N corresponding eigenvectors p_n, $1 \leq n \leq N$, that satisfy*

$$|\lambda_1| \geq |\lambda_2| \geq \cdots \geq |\lambda_N| > 0,$$

and

$$Ap_n = \lambda_n p_n, \quad 1 \leq n \leq N, \quad \textit{and} \quad (p_k, p_\ell) = \delta_{k\ell} \quad \textit{for } 1 \leq k, \ell \leq N,$$

$$|\lambda_1| = \sup_{x \neq 0} \frac{|(Ax, x)|}{\|x\|^2} \quad \textit{and} \quad |\lambda_n| = \sup_{\left\{\substack{x \neq 0 \\ (x, p_k) = 0,\ 1 \leq k \leq n-1}\right.} \frac{|(Ax, x)|}{\|x\|^2}, \quad 2 \leq n \leq N \quad \textit{if } N \geq 2.$$

(b) *Let λ be any nonzero eigenvalue of A. Then there exists $n \in \{1, \ldots, N\}$ such that $\lambda_n = \lambda$, and*

$$\{p \in X;\ Ap = \lambda p\} = \mathrm{Span}(p_n)_{n \in I(\lambda)}, \quad \textit{where } I(\lambda) := \{n \in \{1, \ldots, N\};\ \lambda_n = \lambda\}.$$

(c) *For any vector $x \in X$,*

$$Ax = \sum_{n=1}^{N} \lambda_n (x, p_n) p_n.$$

(d) *The kernel of A is given by*

$$\mathrm{Ker} A = \left(\mathrm{Span}(p_n)_{n=1}^{N}\right)^{\perp}.$$

Proof The proof is an easy adaptation of parts (i)–(vii) of that of Theorem 4.11-2, the notations of which are reused below.

Since A is compact and $A \neq 0$ (the range of A is of dimension $N \geq 1$ by assumption), part (i) holds *verbatim.*

If $N = 1$, the range of A then necessarily coincides with the subspace $\mathrm{Span}(p_1)$. If $N \geq 2$, then $A_2 \neq 0$ and the iterative procedure of part (ii) can thus be initialized. The essential difference is that this procedure must now terminate in N iterations, since necessarily $A_{N+1} = 0$, where A_{N+1} denotes the restriction of A to the subspace

$$X_{N+1} = \{x \in X;\ (x, p_k) = 0,\ 1 \leq k \leq N\} = \left(\mathrm{Span}(p_k)_{k=1}^N\right)^\perp.$$

To see this, it suffices to recall that $A_{N+1} = 0$ implies that

$$Ax = \sum_{k=1}^{N} \lambda_k (x, p_k) p_k \quad \text{for all } x \in X,$$

which shows that the range of A is of dimension N (the eigenvalues λ_k, $1 \leq k \leq N$, are nonzero; the eigenvectors p_k, $1 \leq k \leq N$, are linearly independent; and $Ap_k = \lambda_k p_k$, $1 \leq k \leq N$). This proves (a) and (c).

The arguments of parts (v)–(vii) hold almost *verbatim*, thus proving (b) and (d). □

To conclude this analysis, we show that Theorem 4.11-1 provides as a simple corollary an important means of constructing *maximal orthonormal families* and *Hilbert bases* when the operator A is in addition assumed to be *injective.*

Theorem 4.11-3 (a) *Let $(X, (\cdot,\cdot))$ be an infinite-dimensional inner-product space and let $A : X \to X$ be an injective, compact, and self-adjoint, linear operator. Then the eigenvectors $(p_n)_{n=1}^\infty$ found in Theorem* 4.11-1 *form a maximal orthonormal family in X.*

(b) *If in addition X is a separable Hilbert space, the eigenvectors $(p_n)_{n=1}^\infty$ found in Theorem* 4.11-1 *form a Hilbert basis in X.*

Proof First, we note that the range of A is necessarily infinite-dimensional, since A is injective and X is infinite-dimensional. Hence all the assumptions of Theorem 4.11-1 are satisfied. By the same theorem, the assumption $\mathrm{Ker}\, A = \{0\}$ implies that

$$(\mathrm{Span}(p_n)_{n=1}^\infty)^\perp = \{0\},$$

i.e., that $(p_n)_{n=1}^\infty$ is a maximal orthonormal family in X, or a Hilbert basis if X is a separable Hilbert space. □

Under the assumptions of Theorem 4.11-3(b), two remarkable formulas thus simultaneously hold, viz.,

$$x = \sum_{n=1}^{\infty} (x, p_n) p_n \quad \text{and} \quad Ax = \sum_{n=1}^{\infty} \lambda_n (x, p_n) p_n \quad \text{for any } x \in X$$

thanks to Theorems 4.9-1 and 4.11-1.

Problem

4.11-1 Let X be a separable Hilbert space, let $(e_n)_{n=1}^{\infty}$ be a Hilbert basis of X (Section 4.9), and let $(\lambda_n)_{n=1}^{\infty}$ be a bounded sequence of real numbers.

(1) Show that, for any $x \in X$, the series $\sum_{n=1}^{\infty} \lambda_n (x, e_n) e_n$ is convergent in the space X.

(2) For any $x \in X$, let $Ax := \sum_{n=1}^{\infty} \lambda_n (x, e_n) e_n$. Show that the mapping $A : X \to X$ defined in this fashion is a continuous and self-adjoint linear operator.

(3) Show that for each $n \geq 1$, λ_n is an eigenvalue of A and e_n is a corresponding eigenvector.

(4) Show that, if $\lambda_n \neq 0$ for all $n \geq 1$, the operator A is injective.

(5) Show that, if $\lim_{n \to \infty} \lambda_n = 0$, the operator A is compact.

Chapter 5

The "Great Theorems" of Linear Functional Analysis

Introduction

This chapter is devoted to the proofs of most of the *"great theorems" of linear functional analysis.* Their common characteristic is that they hinge on one, or on both, of *two fundamental results*: *Baire's theorem* (Theorem 5.1-2) and the *Hahn–Banach theorem in a normed vector space* (Theorem 5.9-1).

Baire's theorem asserts that a countably infinite intersection of dense open subsets of a *Banach space* (or more generally of a complete metric space) is still dense.

Direct consequences of Baire's theorem include the noncompleteness of the space of all polynomials (whatever the norm it is equipped with; cf. Theorem 5.1-4) and the existence of "many" continuous functions that are nowhere differentiable (Theorem 5.2-1).

Another consequence of Baire's theorem is the *Banach–Steinhaus theorem*, alias the *uniform boundedness principle* (Theorem 5.3-1), one of the cornerstones of linear functional analysis. This theorem implies for instance the existence of continuous functions whose Lagrange interpolation by polynomials does not uniformly converge (Theorem 5.4-2), or the existence of continuous functions whose Fourier series does not uniformly converge (Theorem 5.5-1).

Two other such cornerstones, also consequences of Baire's theorem, are the *Banach open mapping theorem* (Theorem 5.6-1) and the *Banach closed graph theorem* (Theorem 5.7-1). Their efficiency is illustrated by two remarkable applications, the first one to the continuity of the inverse of a differential operator under minimal assumptions (Theorem 5.6-3), and the second one to the surprising *Hellinger–Toeplitz theorem* (Theorem 5.7-2), which asserts that any self-adjoint operator in a Hilbert space is automatically continuous.

The *Hahn–Banach theorem* (Theorem 5.9-1) is of a different nature, if only because its proof requires the *axiom of choice.* This theorem asserts that, in *any* normed vector space X, *any* continuous linear form on *any* subspace of X can be extended to the whole space X by a continuous linear form with the same norm.

The list of its consequences is also impressive: it includes for instance basic theorems of linear functional analysis, such as the *"geometric form" of the Hahn–Banach theorem* (Theorems 5.10-1 and 5.10-2), which shows how to *"separate convex sets"* in *any* normed vector spaces, or, together with the Banach open mapping theorem, the deep *Banach closed range theorem* (Theorems 5.11-5 and 5.11-6), which provides in particular a strikingly simple *sufficient condition for the surjectivity of a linear operator* or, more generally, a *characterization of its image*, in terms of its *dual operator.*

Note that, in a sense, the Hahn–Banach theorem allows us to extend to an *arbitrary* normed vector space X properties that hold in a *Hilbert space*, such as for instance the notion of *dual operator* (Section 5.11), which extends the notion of adjoint operator, or the *projection theorem* onto a closed subspace (Theorem 5.9-7).

This chapter concludes with two fundamental notions, those of *weak convergence* (Section 5.12) and of a *reflexive Banach space* (Section 5.14), whose analysis often relies on *both* Baire's theorem and the Hahn–Banach theorem. This analysis culminates with two major results: the *Banach–Saks–Mazur theorem* (Theorem 5.13-1) and the *Banach–Eberlein–Šmulian theorem* (Theorem 5.14-4), which will both play a major role in the sequel.

Perhaps paradoxically, the "great theorems" of linear functional analysis established in this chapter will not be of much use in the next chapter, which is devoted to *linear* partial differential equations; but, by contrast, they will be used at many places in the last chapter for establishing, together with basic theorems of nonlinear functional analysis, the existence of solutions to *nonlinear* partial differential equations.

5.1 Baire's theorem; a first application: Noncompleteness of the space of all polynomials

Baire's theorem (Theorem 5.1-2) is one of the two keystones of Linear Functional Analysis, the other one being the *Hahn–Banach theorems* (Sections 5.8–5.10). Baire's theorem's far-reaching consequences include such basic theorems as the Banach–Steinhaus theorem (Theorem 5.3-1), the Banach open mapping theorem (Theorem 5.6-1), or the Banach closed graph theorem (Theorem 5.7-1).

Although Baire's theorem will be applied in the remainder of this chapter to Banach spaces, its proof is given in the more general setting of *complete metric spaces*, as this greater generality involves no extra cost.

Baire's theorem rests on the following interesting property of complete metric spaces (recall that $\operatorname{diam} A = \sup\{d(x,y);\ x \in A,\ y \in A\} \in [0,\infty]$ denotes the diameter of a subset A of a metric space (X,d); cf. Section 1.10).

Theorem 5.1-1 (Cantor's intersection theorem) *Let X be a complete metric space, and let $(A_n)_{n=0}^{\infty}$ be a sequence of nonempty closed subsets A_n of X that satisfy*

$$A_0 \supset A_1 \supset \cdots \supset A_n \supset A_{n+1} \supset \cdots \quad \textit{and} \quad \operatorname{diam} A_n \to 0 \textit{ as } n \to \infty.$$

Then there exists $x \in X$ such that

$$\bigcap_{n=0}^{\infty} A_n = \{x\}.$$

Proof For each $n \geq 0$, pick an element $x_n \in A_n$ (each subset A_n is nonempty). Then the sequence $(x_n)_{n=0}^{\infty}$ is a Cauchy sequence, since the inclusions $A_m \subset A_n$ for all $m \geq n$ imply that

$$d(x_m, x_n) \leq \operatorname{diam} A_n \quad \text{for all } m \geq n,$$

and $\operatorname{diam} A_n \to 0$ as $n \to \infty$. Let $x := \lim_{n\to\infty} x_n$.

Given any integer $n \geq 0$, $x_m \in A_n$ for all $m \geq n$. Hence $x = \lim_{n\to\infty} x_m \in A_n$ since A_n is closed. Consequently, $x \in \bigcap_{n=0}^{\infty} A_n$.

Assume that the intersection $\bigcap_{n=0}^{\infty} A_n$ contains a point $y \neq x$. Then there exists $n_0 \geq 0$ such that $\operatorname{diam} A_{n_0} < d(x,y)$. But, since both x and y belong to A_{n_0}, there also holds $d(x,y) \leq \operatorname{diam} A_{n_0}$, a contradiction. Hence $\bigcap_{n=0}^{\infty} A_n = \{x\}$. □

Remarks (1) The assumption $\operatorname{diam} A_n \to 0$ as $n \to \infty$ is essential. Consider for example the special case $X = \mathbb{R}$ and $A_n = [n, \infty[$, $n \geq 1$.

(2) The property established in Theorem 5.1-1 in fact *characterizes* complete metric spaces (Problem 5.1-1).

(3) Cantor's intersection theorem will be also put to an essential use later on for establishing *Ekeland's variational principle* (Theorem 9.8-1). □

Theorem 5.1-2 (Baire's theorem[1]) *Let X be a complete metric space. Then the following two equivalent properties hold:*

(a) *Let $(F_n)_{n=0}^{\infty}$ be a sequence of closed subsets of X such that* $\operatorname{int} F_n = \varnothing$ *for all* $n \geq 0$. *Then* $\operatorname{int}(\bigcup_{n=0}^{\infty} F_n) = \varnothing$.

(b) *Let $(O_n)_{n=0}^{\infty}$ be a sequence of open subsets of X such that $\overline{O_n} = X$ for all $n \geq 0$. Then* $\overline{\bigcap_{n=0}^{\infty} O_n} = X$.

Proof For typographical reasons, the notation $\operatorname{int} A$ is preferred here to $\mathring{A}$.

(i) To begin with, we show that (a) *and* (b) *are equivalent properties.* Assume for instance that property (a) holds.

First, we note that the relation $\overline{A} = X - \operatorname{int}(X - A)$ for any $A \subset X$ implies that

$$\overline{A} = X \quad \text{if and only if } \operatorname{int}(X - A) = \varnothing.$$

Given open sets $O_n \in X$, $n \geq 0$, such that $\overline{O_n} = X$ for all $n \geq 0$, let $F_n := X - O_n$. Then the closed sets F_n satisfy $\operatorname{int} F_n = \varnothing$ for all $n \geq 0$ and thus $\operatorname{int}(\bigcup_{n=0}^{\infty} F_n) = \varnothing$ by (a). But

$$\bigcup_{n=0}^{\infty} F_n = \bigcup_{n=0}^{\infty} (X - O_n) = X - \bigcap_{n=0}^{\infty} O_n,$$

by de Morgan's laws (Section 1.3). Hence $\operatorname{int}(X - \bigcap_{n=0}^{\infty} O_n) = \varnothing$, so that $\overline{\bigcap_{n=0}^{\infty} O_n} = X$. Therefore property (b) holds.

That (b) implies (a) is proved analogously, this time by noting that the relation $\overline{X - B} = X - \operatorname{int} B$ for any $B \subset X$ implies that

$$\operatorname{int} B = \varnothing \quad \text{if and only if} \quad \overline{X - B} = X.$$

(ii) *Let us prove* (a). To begin with, observe that a subset A of X has a nonempty interior if and only if there exists a nonempty open subset $O \subset X$ contained in A, or equivalently such that $O \cap (X - A) = \varnothing$. Consequently,

$$\operatorname{int} A = \varnothing \quad \text{if and only if } O \cap (X - A) \neq \varnothing \text{ for all nonempty open subsets } O \subset X.$$

[1]This theorem was first proved for $X = [a,b]$ in:
R. Baire [1899]: Sur les fonctions de variables réelles, *Annali di Matematica Pura ed Applicata* **3**, 1–123.

Let then X be a complete metric space and let F_n, $n \neq 0$, be closed subsets of X such that $\operatorname{int} F_n = \varnothing$ for all $n \geq 0$. We thus wish to prove that $\operatorname{int} \bigcup_{n=0}^{\infty} F_n = \varnothing$, or equivalently that, given any nonempty open subset $O \subset X$,

$$O \cap \left(X - \bigcup_{n=0}^{\infty} F_n \right) \neq \varnothing.$$

Given a nonempty open subset $O \subset X$, let $O_0 := O$. Since $\operatorname{int} F_0 = \varnothing$ and O_0 is open, $O_0 \cap (X - F_0)$ is a nonempty open subset of X. There thus exists a nonempty open subset $O_1 \subset X$ such that

$$\overline{O}_1 \subset O_0 \cap (X - F_0) \quad \text{and} \quad \operatorname{diam} \overline{O}_1 < 1$$

(such as a ball O_1 centered at any point in $O_0 \cap (X - F_0)$ with a small enough radius). Since $\operatorname{int} F_1 = \varnothing$ and O_1 is open, so that $O_1 \cap (X - F_1)$ is a nonempty open subset of X, there likewise exists a nonempty open subset $O_2 \subset X$ such that

$$\overline{O}_2 \subset O_1 \cap (X - F_1) \quad \text{and} \quad \operatorname{diam} \overline{O}_2 < \frac{1}{2},$$

and so on. In this fashion, we construct a sequence of nonempty open subsets $O_n \subset X$, $n \geq 0$, such that

$$\overline{O}_{n+1} \subset O_n \cap (X - F_n) \quad \text{and} \quad \operatorname{diam} \overline{O}_{n+1} < \frac{1}{n+1}, \qquad n \geq 0.$$

The nonempty closed subsets $\overline{O}_n$, $n \geq 0$, clearly satisfy all the assumptions of Cantor's intersection theorem (Theorem 5.1-1). Hence there exists $x \in X$ such that $\{x\} = \bigcap_{n=0}^{\infty} \overline{O}_n$.

On the one hand, $x \in O$ since $x \in \overline{O}_1 \subset O_0 = O$. On the other hand, the relations $x \in \overline{O}_{n+1}$ and $\overline{O}_{n+1} \subset X - F_n$ for all $n \geq 0$ imply that $x \in \bigcap_{n=0}^{\infty} (X - F_n)$. Noting that $\bigcap_{n=0}^{\infty} (X - F_n) = X - \bigcup_{n=0}^{\infty} F_n$, we thus have

$$x \in O \cap \left(X - \bigcup_{n=0}^{\infty} F_n \right).$$

Consequently, the set $O \cap (X - \bigcup_{n=0}^{\infty} F_n)$ is nonempty, as was to be proven. □

Baire's theorem is often put to use in the form of property (a) or (b) in the next theorem (both properties follow immediately from Theorem 5.1-2(a)).

Theorem 5.1-3 *Let X be a metric space, and let F_n, $n \geq 0$, be closed subsets of X such that $X = \bigcup_{n=0}^{\infty} F_n$.*

(a) *If $\operatorname{int} F_n = \varnothing$ for all $n \geq 0$, then X is not complete.*

(b) *If X is complete, there exists $n_0 \geq 0$ such that $\operatorname{int} F_{n_0} \neq \varnothing$.* □

Consider for instance the countably infinite set $\mathbb{Q} = \bigcup_{n=0}^{\infty} \{q_n\}$ of all rational numbers (ordered in any fashion), equipped with its usual distance d defined by $d(p, q) = |p - q|$ for all $p, q \in \mathbb{Q}$. Since each subset $\{q_n\}$ of $\mathbb{Q}$ is closed and has an empty interior, Theorem 5.1-3(a) implies that $(\mathbb{Q}, d)$ is not complete (the same conclusion can of course be reached directly; for instance, the sequence $(r_n)_{n=0}^{\infty}$ of rational numbers defined by $r_0 = 0$ and $r_n = \frac{1}{4} + r_{n-1} - \frac{1}{2}(r_{n-1})^2$, $n \geq 1$, is a Cauchy sequence that does not converge in $\mathbb{Q}$).

A likewise immediate, but nevertheless worthwhile, consequence of Theorem 5.1-3(b) is that the plane $\mathbb{R}^2$ cannot be written as a countably infinite union of lines, or more generally, that the space $\mathbb{R}^n$ cannot be written as a countably infinite union of hyperplanes (note that this conclusion cannot be reached by cardinality arguments, since $\operatorname{card}\mathbb{R}^n = \operatorname{card}\mathbb{R}$; cf. Theorem 1.5-3).

A more striking application of Theorem 5.1-3(a) is the following result.

Theorem 5.1-4 *An infinite-dimensional Banach space cannot have a countably infinite Hamel basis* (*Section* 2.1).

In particular, the space of all polynomials of one, or several, variables cannot be equipped with a norm that would make it a Banach space.

Proof Given a countably infinite Hamel basis $(e_j)_{j=0}^{\infty}$ in a normed vector space $(X, \|\cdot\|)$, define the subsets F_n of X by

$$F_n := \operatorname{Span}(e_j)_{j=0}^{n} \quad \text{for each } n \geq 0.$$

We first note that $X = \bigcup_{n=0}^{\infty} F_n$ and that each set F_n, $n \geq 0$, is closed in X (as a finite-dimensional subspace of X; cf. Theorem 2.7-1(d)).

We next show that $\operatorname{int} F_n = \varnothing$ for all $n \geq 0$. For, if otherwise $\operatorname{int} F_n \neq \varnothing$ for some $n \geq 0$, there exist $x = \sum_{j=0}^{n} x_j e_j \in F_n$ and $r > 0$ such that $\overline{B(x;r)} \subset F_n$. Then the point

$$y := \frac{r}{\|e_{n+1}\|} e_{n+1} + x$$

belongs to $\overline{B(x;r)}$, but y cannot belong to $F_n = \operatorname{Span}(e_j)_{j=0}^{n}$ since the family $(e_j)_{j=0}^{n+1}$ is linearly independent.

That the space X cannot be complete then follows from Theorem 5.1-3(a).

The application to the space $\mathcal{P}$ of all polynomials in n variables is immediate, since the polynomials

$$x = (x_1, x_2, \ldots, x_n) \in \mathbb{R}^n \to x_1^{k_1} x_2^{k_2} \cdots x_n^{k_n}, \quad \text{with } k_i \in \mathbb{N}, \quad 1 \leq i \leq n,$$

form a countably infinite Hamel basis of the space $\mathcal{P}$. □

Problems

5.1-1 The object of this exercise is to establish a *converse to Cantor's intersection theorem* (Theorem 5.1-1). Let (X, d) be a metric space that possesses the following property: Given any sequence $(A_n)_{n=0}^{\infty}$ of nonempty closed subsets of X that satisfy $A_n \supset A_{n+1}$ for all $n \geq 0$ and $\lim_{n\to\infty} \operatorname{diam} A_n = 0$, then the intersection $\bigcap_{n=0}^{\infty} A_n$ is nonempty (since $\lim_{n\to\infty} \operatorname{diam} A_n = 0$, this intersection consists of a single element).

Show that (X, d) is complete.

5.1-2 Let $f \in \mathcal{C}^\infty(\mathbb{R})$ be a function with the following property: At each $x \in \mathbb{R}$, there exists an integer $n(x) \geq 1$ such that $f^{(n(x))} = 0$. Show that f is a polynomial.[2]

5.2 Application of Baire's theorem: Existence of nowhere differentiable continuous functions

In 1872, Karl Weierstraß[3] published a startling example of a continuous function on $[0,1]$ that is nowhere differentiable on $[0,1]$ (Problem 5.2-1). It is remarkable that the *existence* of such functions can be in fact deduced from Baire's theorem, without having to explicitly produce their expressions. This approach, which is the object of the next theorem, even shows that such functions constitute the rule rather than the exception.

Theorem 5.2-1 *There exist continuous functions on* $[0,1]$ *that are nowhere differentiable on* $[0,1]$.

Proof (i) *Let* $f \in \mathcal{C}[0,1]$ *be differentiable at at least one point* $a \in [0,1]$. *Then*

$$f \in \bigcup_{n=1}^{\infty} F_n,$$

where

$$F_n := \{f \in \mathcal{C}[0,1];\ \text{there exists } a \in [0,1] \text{ such that } \sup_{h \neq 0} \left|\frac{f(a+h)-f(a)}{h}\right| \leq n\}.$$

Given such a function f, there exists $h_0 > 0$ such that

$$\begin{aligned}\left|\frac{f(a+h)-f(a)}{h}\right| &\leq \left|\frac{f(a+h)-f(a)}{h} - f'(a)\right| + |f'(a)| \\ &\leq 1 + |f'(a)| \quad \text{for all } 0 < |h| \leq h_0\end{aligned}$$

on the one hand. Since, on the other hand,

$$\left|\frac{f(a+h)-f(a)}{h}\right| \leq \frac{2}{h_0} \sup_{0 \leq x \leq 1} |f(x)| \quad \text{for all } |h| \geq h_0,$$

it follows that

$$\sup_{h \neq 0} \left|\frac{f(a+h)-f(a)}{h}\right| < \infty$$

(above and below it is tacitly understood that such inequalities are restricted to those points $(a+h)$ that belong to the interval $[0,1]$). Consequently, there exists $n_0 \geq 1$ such that

---

[2]This spectacular result is due to:

E. Corominas; F.S. Balaguer [1954]: Condiciones para que una función infinitamente derivable sea un polinomio, *Revista Matemática Hispano-Americana* **14**, 26–43.

It was then extended to functions of several variables by:

A.B. Boghossian; P.D. Johnson, Jr. [1990]: A pointwise condition for an infinitely differentiable function of several variables to be a polynomial, *Journal of Mathematical Analysis and Applications* **151**, 17–19.

[3]K. Weierstraß [1872]: Über continuirliche Functionen eines reellen Arguments, die für keinen Werth des letzteren einen bestimmten Differentialquotienten besitzen, *Königliche Akademie der Wissenschaften*.

$$\sup_{h\neq 0}\left|\frac{f(a+h)-f(a)}{h}\right| \leq n_0.$$

Hence $f \in F_{n_0} \subset \bigcup_{n=1}^{\infty} F_n$.

(ii) *Each set* F_n, $n \geq 1$, *is closed in* $\mathcal{C}[0,1]$. Let functions $f_k \in F_n$, $k \geq 0$, and $f \in \mathcal{C}[0,1]$ be such that $\lim_{k\to\infty} \|f_k - f\| = 0$, where $\|\cdot\|$ denotes the sup-norm in the space $\mathcal{C}[0,1]$ (the integer $n \geq 1$ is fixed here).

For each integer $k \geq 0$, there exists a point $a_k \in [0,1]$ such that

$$\sup_{h\neq 0}\left|\frac{f_k(a_k+h)-f_k(a_k)}{h}\right| \leq n,$$

since $f_k \in F_n$. By the Bolzano–Weierstraß property (Theorem 1.4-1(b)), a subsequence $(a_\ell)_{\ell=0}^{\infty}$ of the sequence $(a_k)_{k=0}^{\infty}$ converges in the interval $[0,1]$; let $a := \lim_{\ell\to\infty} a_\ell$.

Given any $h \neq 0$, let h_ℓ be defined by $a_\ell + h_\ell = a + h$; hence there exists an integer $\ell_0 = \ell_0(h) \geq 0$ such that $h_\ell \neq 0$ for all $\ell \geq \ell_0$. The relations

$$\begin{aligned}
|f_\ell(a_\ell+h_\ell) - f(a+h)| &= |f_\ell(a_\ell+h_\ell) - f(a_\ell+h_\ell)| \leq \|f_\ell - f\|,\\
|f_\ell(a_\ell) - f(a)| &\leq |f_\ell(a_\ell) - f(a_\ell)| + |f(a_\ell) - f(a)| \leq \|f_\ell - f\| + |f(a_\ell) - f(a)|,\\
\lim_{\ell\to\infty} h_\ell &= h,
\end{aligned}$$

combined with the continuity of the function f, then imply that

$$\left|\frac{f(a+h)-f(a)}{h}\right| = \lim_{\substack{\ell\to\infty\\ \ell\geq \ell_0}} \left|\frac{f_\ell(a_\ell+h_\ell)-f_\ell(a_\ell)}{h_\ell}\right| \leq n,$$

since $f_\ell \in F_n$ for all $\ell \geq \ell_0$. This shows that $f \in F_n$; hence F_n is closed.

(iii) *Each set* F_n, $n \geq 1$, *has an empty interior.* This amounts to proving that, given any function $f \in F_n$ and given any $\varepsilon > 0$, there exists a function $g \in \mathcal{C}[0,1]$ such that $\|g - f\| \leq \varepsilon$ and $g \notin F_n$ (the integer $n \geq 1$ is again fixed here).

To this end, we first note that, by the *Weierstraß approximation theorem* (Theorem 2.13-3), there exists a polynomial $p = p(f,\varepsilon)$ such that $\|f - p\| \leq \frac{\varepsilon}{2}$. Given such a polynomial p, we then construct (starting from, e.g., the point $(0, p(0))$ a piecewise affine function $g = g(p) = g(f,\varepsilon) \in \mathcal{C}[0,1]$ that satisfies $\|g - p\| \leq \frac{\varepsilon}{2}$ and $|g'(x)| > n$ at all the points $x \in [0,1]$ where its derivative $g'(x)$ is defined (Figure 5.2-1). Note in passing that the existence of such a function g crucially hinges on the fact that $\sup_{0\leq x\leq 1} |p'(x)| < \infty$, *because* p *is a polynomial.*

The function g so constructed clearly possesses the required properties.

(iv) *Baire's theorem* (Theorem 5.1-2) then implies that

$$\operatorname{int}\left(\bigcup_{n=1}^{\infty} F_n\right) = \varnothing,$$

since the space $\mathcal{C}[0,1]$ is *complete* (Theorem 3.2-2). Consequently, $\mathcal{C}[0,1] - (\bigcup_{n=1}^{\infty} F_n) \neq \varnothing$, i.e., there exist functions that are continuous on $[0,1]$, but nowhere differentiable on $[0,1]$. □

In fact, the relation $\operatorname{int}(\bigcup_{n=1}^{\infty} F_n) = \varnothing$ established at the end of the above proof shows much more, namely that, given any function $f \in \mathcal{C}[0,1]$ that is differentiable at at least one

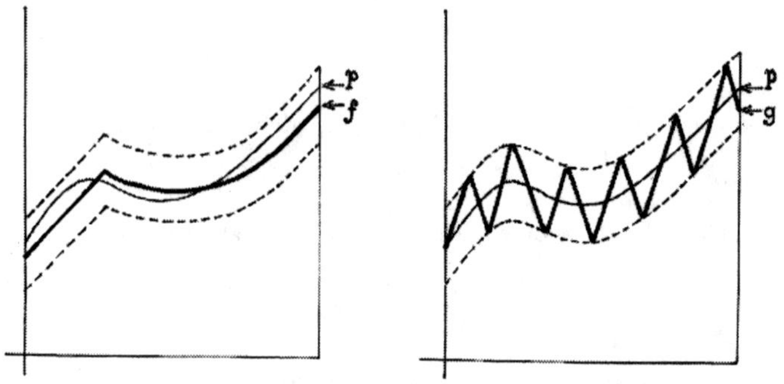

Figure 5.2-1 Construction of the piecewise affine function g in the proof of Theorem 5.2-1.

point, so that $f \in \bigcup_{n=1}^{\infty} F_n$, and given any $\varepsilon > 0$, there exists a function $g \in \mathcal{C}[0,1]$ that is nowhere differentiable on $[0,1]$ and such that $\|f - g\| \le \varepsilon$. This shows that *any function that is continuous on* $[0,1]$ *and differentiable at at least one point in* $[0,1]$ *is the uniform limit of a sequence of nowhere differentiable continuous functions.* In other words, there are indeed "many" continuous functions that are nowhere differentiable!

Problems

5.2-1 Show that the *Weierstraß function* $f : \mathbb{R} \to \mathbb{R}$, given by

$$f : x \in \mathbb{R} \to f(x) := \sum_{n=0}^{\infty} \frac{1}{2^n} \sin(3^n x),$$

is well-defined and continuous, but nowhere differentiable, on $\mathbb{R}$.

5.2-2 Show that the *Hardy function*[4] $f : \mathbb{R} \to \mathbb{R}$, defined by

$$f : x \in \mathbb{R} \to f(x) := \sum_{n=1}^{\infty} \frac{1}{n^2} \sin(n^2 \pi x),$$

is well defined and continuous, but nowhere differentiable, on $\mathbb{R}$.

5.3 Banach–Steinhaus theorem, *alias* the uniform boundedness principle; application to numerical quadrature formulas

Given two normed vector spaces X and Y, consider a family $(A_i)_{i \in I}$ of continuous linear operators $A_i \in \mathcal{L}(X;Y)$ that are *"uniformly bounded"* in the sense that

$$\sup_{i \in I} \|A_i\|_{\mathcal{L}(X;Y)} < \infty.$$

[4] G.H. Hardy [1916]: Weierstraß's non-differentiable function, *Transactions, American Mathematical Society* **17**, 301–325.

Since $\|A_i x\| \leq \|A_i\| \, \|x\|$ for all $i \in I$ and all $x \in X$, it immediately follows that, necessarily,

$$\text{for each } x \in X, \quad \sup_{i \in I} \|A_i x\|_Y < \infty.$$

It is remarkable that, *if the space X is complete* (this assumption is essential; cf. Problem 5.3-1), this necessary condition becomes *sufficient* for the *uniform boundedness* (in the above sense) of the mappings A_i, $i \in I$. This is the content of the following *"uniform boundedness principle,"* itself a consequence of *Baire's theorem*.

Theorem 5.3-1 (Banach–Steinhaus theorem,[5] *alias* the uniform boundedness principle) *Let X be a Banach space, let Y be a normed vector space, and let $(A_i)_{i \in I}$ be a family of mappings $A_i \in \mathcal{L}(X; Y)$ that satisfy*

$$\textit{for each } x \in X, \quad \sup_{i \in I} \|A_i x\|_Y < \infty.$$

Then

$$\sup_{i \in I} \|A_i\|_{\mathcal{L}(X;Y)} < \infty.$$

Proof The same notation $\|\cdot\|$ designates the various norms encountered throughout this proof. For each integer $n \geq 0$, define the set

$$F_n := \left\{ x \in X; \ \sup_{i \in I} \|A_i x\| \leq n \right\}.$$

Given any $x \in X$, $\sup_{i \in I} \|A_i x\| < \infty$ by assumption; hence there exists an integer $n(x) \geq 0$ such that $\sup_{i \in I} \|A_i x\| \leq n(x)$, which means that $x \in F_{n(x)}$. Consequently,

$$X = \bigcup_{n=0}^{\infty} F_n.$$

By definition, $x \in F_n$ if and only if $\|A_i x\| \leq n$ for all $i \in I$, or equivalently, if and only if $x \in \{z \in X; \ \|A_i z\| \leq n\}$ for all $i \in I$. Hence the set F_n is also given by

$$F_n = \bigcap_{i \in I} \{z \in X; \ \|A_i z\| \leq n\},$$

which shows that F_n is closed in X as an intersection of closed subsets of X (each linear operator A_i is continuous by assumption).

Since X is complete, Baire's theorem can be applied, showing that there exists an integer $n_0 \geq 0$ such that $\operatorname{int} F_{n_0} \neq \varnothing$ (Theorem 5.1-3(b)). Hence there exist $x_0 \in F_{n_0}$ and $r > 0$ such that $\overline{B(x_0; r)} \subset F_{n_0}$; by definition of F_{n_0}, this means that

$$\|A_i z\| \leq n_0 \quad \text{for all } z \in \overline{B(x_0; r)} \text{ and all } i \in I.$$

[5] S. Banach; H. Steinhaus [1927]: Sur le principe de la condensation de singularités, *Fundamenta Mathematicae* **9**, 50–61.

Since any nonzero vector $x \in X$ can be written as

$$x = \frac{\|x\|}{r}(z - x_0) \quad \text{with } z := \left(x_0 + \frac{r}{\|x\|}x\right) \in \overline{B(x_0; r)},$$

it follows that

$$\begin{aligned}\|A_i x\| &\leq \frac{\|x\|}{r}\Big(\|A_i z\| + \|A_i x_0\| \Big) \leq \frac{1}{r}(n_0 + \|A_i x_0\|)\, \|x\| \\ &\leq \frac{1}{r}\Big(n_0 + \sup_{i \in I} \|A_i x_0\|\Big)\|x\| \quad \text{for all } i \in I \text{ and all } x \in X.\end{aligned}$$

Therefore,

$$\sup_{i \in I} \|A_i\| \leq \frac{1}{r}\Big(n_0 + \sup_{i \in I} \|A_i x_0\|\Big) < \infty,$$

since $\sup_{i \in I} \|A_i x_0\| < \infty$ by assumption. □

The Banach–Steinhaus theorem is often used in the form of its following consequence, referred to in the sequel as "the" *corollary to the Banach–Steinhaus theorem.*

Theorem 5.3-2 (corollary to the Banach–Steinhaus theorem) *Let X be a Banach space, let Y be a normed vector space, and let $(A_n)_{n=1}^{\infty}$ be a family of mappings $A_n \in \mathcal{L}(X;Y)$ such that, for each $x \in X$, the sequence $(A_n x)_{n=1}^{\infty}$ converges in Y. Then*

$$\sup_{n \geq 1} \|A_n\| < \infty.$$

Furthermore, let the mapping $A : X \to Y$ be defined by

$$Ax := \lim_{n \to \infty} A_n x \quad \textit{for each } x \in X.$$

Then

$$A \in \mathcal{L}(X;Y) \quad \textit{and} \quad \|A\| \leq \liminf_{n \to \infty} \|A_n\|.$$

Proof The convergence of each sequence $(A_n x)_{n=1}^{\infty}$ implies that

$$\text{for each } x \in X, \quad \sup_{n \geq 1} \|A_n x\| < \infty.$$

The Banach–Steinhaus theorem (Theorem 5.3-1) thus shows that

$$\sup_{n \geq 1} \|A_n\| < \infty.$$

The linearity of each mapping A_n, combined with the continuity of the addition and scalar multiplication (Theorem 2.2-5) shows that the mapping $A : X \to Y$ defined by $Ax := \lim_{n \to \infty} A_n x$ for each $x \in X$ is linear. Besides, given any nonzero vector $x \in X$,

$$\frac{\|Ax\|}{\|x\|} = \lim_{n \to \infty} \frac{\|A_n x\|}{\|x\|} = \liminf_{n \to \infty} \frac{\|A_n x\|}{\|x\|} \leq \liminf_{n \to \infty} \|A_n\| < \infty,$$

since $\dfrac{\|A_n x\|}{\|x\|} \leq \|A_n\|$ for all $n \geq 1$. Consequently, $A \in \mathcal{L}(X;Y)$ and

$$\|A\| = \sup_{x \neq 0} \frac{\|Ax\|}{\|x\|} \leq \liminf_{n \to \infty} \|A_n\|. \qquad \square$$

Remarks (1) Theorem 5.3-2 is sometimes also referred to as the *Banach–Steinhaus theorem.*

(2) Under the sole assumptions of Theorem 5.3-2, no conclusion can be reached regarding the possible convergence of the sequence $(A_n)_{n=1}^{\infty}$ to A *in the space* $\mathcal{L}(X;Y)$. $\square$

As a first illustration of the power of the Banach–Steinhaus theorem, we show how it yields a beautiful *criterion* of convergence for a large class of *numerical quadrature formulas* (Theorem 5.3-3).

More specifically, given a *weight function* $\omega \in L^1(0,1)$, the objective consists in approximating "as well as possible" for any function $f \in \mathcal{C}[0,1]$, the integral

$$\ell(f) := \int_0^1 f(x)\omega(x)\,\mathrm{d}x,$$

by means of an "easily computable" finite sum (the interval $[0,1]$ is of course chosen here only for definiteness). One natural way of achieving this goal consists in appropriately choosing $(n+1)$ distinct *nodes* $0 \leq x_0^n < x_1^n < \cdots < x_n^n \leq 1$ and $(n+1)$ *weights* $\omega_j^n \in \mathbb{R}$, $0 \leq j \leq n$, for each integer $n \geq 0$ and then in approximating the integral $\ell(f)$ by the *numerical quadrature formula*

$$\ell_n(f) := \sum_{j=0}^{n} \omega_j^n f(x_j^n).$$

Note that the mappings $\ell : \mathcal{C}[0,1] \to \mathbb{R}$ and $\ell_n : \mathcal{C}[0,1] \to \mathbb{R}$ defined above are clearly *continuous linear functionals*, the space $\mathcal{C}[0,1]$ being equipped with the sup-norm.

There are many ways of constructing numerical quadrature formulas.[6] For instance, let $\sum_{j=0}^{n} f(x_j^n) p_j^n \in \mathcal{P}_n[0,1]$ denote the *Lagrange interpolating polynomial of degree* $\leq n$ of a function $f \in \mathcal{C}[0,1]$ associated with *equally spaced* nodes $x_j^n = \dfrac{j}{n}$, $0 \leq j \leq n$ (such polynomials are defined in Section 5.4). Then one way of approximating the integral $\ell(f)$ is by means of the *Newton–Cotes quadrature formula*:

$$\ell_n(f) := \sum_{j=0}^{n} \left(\int_0^1 p_j^n(x)\omega(x)\,\mathrm{d}x \right) f(x_j^n),$$

which, by construction, is thus *exact for all polynomials of degree* $\leq n$.

A (presumably more efficient) procedure consists in seeking whether $(n+1)$ nodes x_j^n and $(n+1)$ weights ω_j^n, $0 \leq j \leq n$, can be simultaneously chosen in such a way that the resulting numerical quadrature formula is *exact for all polynomials of degree* $\leq 2n+1$.

Whether this is possible or not is not *a priori* obvious, since this involves solving a system of $(2n+2)$ equations that are *nonlinear* with respect to the unknowns x_j^n and ω_j^n, $0 \leq j \leq n$.

[6] An in-depth treatment of numerical quadrature is found in Davis & Rabinowitz [1975].

It nevertheless turns out that such a nonlinear system can be solved, thanks in particular to an unsuspected relation between this problem and properties of polynomials that are orthogonal with respect to a strictly positive weight function (Problem 5.3-3). The corresponding formula $\ell_n(f)$ is the *Gauß–Jacobi quadrature formula.*

The following theorem shows that there exists a surprisingly simple *necessary and sufficient* condition for a sequence of numerical quadrature formulas $\ell_n(f)$ of the general form considered here to converge to the integral $\ell(f)$ as $n \to \infty$ for any $f \in \mathcal{C}[0,1]$. Note that, save that they are *distinct* for each integer $n \geq 0$, *no other assumption is made on the nodes* x_j^n, $0 \leq j \leq n$.

Theorem 5.3-3 (Polya's theorem[7]) *Given a weight function $\omega \in L^1(0,1)$, let there be given a sequence of continuous linear functionals $\ell_n : \mathcal{C}[0,1] \to \mathbb{R}$, $n \geq 0$, of the form*

$$\ell_n : f \in \mathcal{C}[0,1] \to \ell_n(f) := \sum_{j=0}^{n} \omega_j^n f(x_j^n) \in \mathbb{R}, \quad \textit{where } 0 \leq x_0^n < x_1^n < \cdots < x_n^n \leq 1,$$

with the following property:

$$\lim_{n\to\infty} \left| \int_0^1 p(x)\omega(x)\,\mathrm{d}x - \ell_n(p) \right| = 0 \quad \textit{for any } p \in \mathcal{P}[0,1].$$

Then

$$\lim_{n\to\infty} \left| \int_0^1 f(x)\omega(x)\,\mathrm{d}x - \ell_n(f) \right| = 0 \quad \textit{for any } f \in \mathcal{C}[0,1]$$

if and only if

$$\sup_{n\geq 0} \left(\sum_{j=0}^{n} |\omega_j^n| \right) < \infty.$$

Proof Clearly, $|\ell_n(f)| \leq (\sum_{j=0}^n |\omega_j^n|)\,\|f\|$ for all $f \in \mathcal{C}[0,1]$, so that

$$\|\ell_n\| = \sup_{f\neq 0} \frac{|\ell_n(f)|}{\|f\|} \leq \sum_{j=0}^{n} |\omega_j^n|.$$

Let $f_0 \in \mathcal{C}[0,1]$ denote the piecewise affine continuous function defined by

$$f_0(0) = \operatorname{sgn}\omega_0^n, \quad f_0(x_j^n) = \operatorname{sgn}\omega_j^n, \ \ 0 \leq j \leq n, \quad \text{and} \quad f_0(1) = \operatorname{sgn}\omega_n^n.$$

Then $|\ell_n(f_0)| = \sum_{j=0}^n |\omega_j^n|$ and $\|f_0\| = 1$. Consequently,

$$\|\ell_n\| \geq \frac{|\ell_n(f_0)|}{\|f_0\|} = \sum_{j=0}^{n} |\omega_j^n| \quad \text{for each } n \geq 0.$$

Hence

$$\|\ell_n\| = \sum_{j=0}^{n} |\omega_j^n|.$$

[7] G. Pólya [1933]: Über die Konvergenz von Quadraturverfahren, *Mathematische Zeitschrift* **37**, 264–286.

The "only if" part thus follows from the *corollary to the Banach–Steinhaus theorem* (Theorem 5.3-2), which implies that $\sup_{n\geq 0}\|\ell_n\| < \infty$.

Conversely, assume that $\sup_{n\geq 0}(\sum_{j=0}^{n}|\omega_j^n|) = \sup_{n\geq 0}\|\ell_n\| < \infty$. For any $f \in \mathcal{C}[0,1]$ and any $p \in \mathcal{P}[0,1]$, we may write

$$\begin{aligned}\left|\ell_n(f) - \int_0^1 f(x)\omega(x)\,\mathrm{d}x\right| &\leq |\ell_n(f-p)| + \left|\ell_n(p) - \int_0^1 p(x)\omega(x)\,\mathrm{d}x\right| \\ &\quad + \left|\int_0^1 (f(x)-p(x))\omega(x)\,\mathrm{d}x\right| \\ &\leq \Big(\sup_{n\geq 0}\|\ell_n\| + \|\omega\|_{L^1(0,1)}\Big)\|f-p\| + \left|\ell_n(p) - \int_0^1 p(x)\omega(x)\,\mathrm{d}x\right|.\end{aligned}$$

Given any $f \in \mathcal{C}[0,1]$ and any $\varepsilon > 0$, the *Weierstraß approximation theorem* (Theorem 2.13-3) shows that there exists a polynomial $p = p(f;\varepsilon) \in \mathcal{P}[0,1]$ such that

$$\Big(\sup_{n\geq 0}\|\ell_n\| + \|\omega\|_{L^1(0,1)}\Big)\|f-p\| \leq \frac{\varepsilon}{2}.$$

By assumption, there then exists $n_0 = n_0(p) = n_0(f;\varepsilon)$ such that

$$\left|\ell_n(p) - \int_0^1 p(x)\omega(x)\,\mathrm{d}x\right| \leq \frac{\varepsilon}{2} \quad \text{for all } n \geq n_0,$$

and hence such that

$$\left|\ell_n(f) - \int_0^1 f(x)\omega(x)\,\mathrm{d}x\right| \leq \varepsilon \quad \text{for all } n \geq n_0.$$

This proves the "if" part. □

Remarks (1) By contrast with the "only if" part, the "if" part does not use the Banach–Steinhaus theorem.

(2) The Newton–Cotes and Gauß–Jacobi quadrature formulas clearly satisfy

$$\lim_{n\to\infty}\left|\ell_n(p) - \int_0^1 p(x)\omega(x)\,\mathrm{d}x\right| = 0 \text{ for any polynomial } p \in \mathcal{P}[0,1],$$

since, given any polynomial $p \in \mathcal{P}[0,1]$, there exists an integer $n = n(p)$ such that $\ell_n(p) = \int_0^1 p(x)\omega(x)\,\mathrm{d}x$. □

Problems

5.3-1 This problem provides a counterexample to the Banach–Steinhaus theorem when the space is not complete. The notation $\mathcal{P}$ designates the space of all real polynomials.

(1) Given a polynomial $p : x \in \mathbb{R} \to \sum_{k=0}^{m} c_k x^k$, let $\|p\| := \max_{0\leq k\leq m}|c_k|$. Show that $\|\cdot\|$ defines a norm on the space $\mathcal{P}$.

(2) Show directly that $(\mathcal{P}, \|\cdot\|)$ is not complete (i.e., without a recourse to Theorem 5.1-4).

(3) For each $n \geq 0$, define the linear operator $A_n : \mathcal{P} \to \mathbb{R}$ by $A_n p := \sum_{k=0}^{\min\{m,n\}} c_k$. Show that each operator A_n is continuous.

(4) Show that $\sup_{n\geq 0}|A_n p| < \infty$ for each $p \in \mathcal{P}$, but that $\sup_{n\geq 0}\|A_n\| = \infty$.

5.3-2 (1) Let X be a Banach space, let Y and Z be normed vector spaces, and let $B : X\times Y \to Z$ be a bilinear mapping that is "separately continuous" in the sense that

$$\text{for each } y \in Y, \quad \lim_{n\to\infty} x_n = x \text{ in } X \quad \text{implies} \quad \lim_{n\to\infty} B(x_n, y) = B(x, y) \text{ in } Z,$$
$$\text{for each } x \in X, \quad \lim_{n\to\infty} y_n = y \text{ in } Y \quad \text{implies} \quad \lim_{n\to\infty} B(x, y_n) = B(x, y) \text{ in } Z.$$

Using the Banach–Steinhaus theorem, show that B is continuous; i.e., that, for each $(x, y) \in X \times Y$,

$$\lim_{n\to\infty} x_n = x \text{ in } X \quad \text{and} \quad \lim_{n\to\infty} y_n = y \text{ in } Y \ \text{ implies } \lim_{n\to\infty} B(x_n, y_n) = B(x, y) \text{ in } Z.$$

(2) Give an example of normed vector spaces X, Y, Z and of a separately continuous bilinear mapping $B : X \times Y \to Z$ that is not continuous.

5.3-3 Given a weight function $\omega \in L^1(0,1)$ that satisfies $\omega > 0$ almost everywhere in $[0,1]$, let p_n, $n \geq 0$, denote the *orthogonal polynomials with respect to the weight function* ω (Problem 4.8-3).

(1) For each integer $n \geq 1$, let x_j^n, $0 \leq j \leq n$, designate the zeros of p_n (these zeros are all real and simple and they all lie in the open interval $]0,1[$; cf. *ibid.*). Show that there exist constants ω_j^n, $0 \leq j \leq n$, that satisfy

$$\omega_j^n > 0, \quad 0 \leq j \leq n, \quad \text{and} \quad \sum_{j=0}^{n} \omega_j^n p(x_j^n) = 0 \quad \text{for all } p \in \mathcal{P}_{2n+1}[0,1].$$

(2) Show that, if $f \in \mathcal{C}^{2n+2}[0,1]$, there exists a point $\xi = \xi(f) \in]0,1[$ such that

$$\int_0^1 f(x)\omega(x)\,\mathrm{d}x - \sum_{j=0}^{n} \omega_j^n f(x_j^n) = \frac{1}{(2n+2)!} f^{(2n+2)}(\xi).$$

Hence

$$\lim_{n\to\infty} \left(\sum_{j=0}^{n} \omega_j^n f(x_j^n) \right) = \int_0^1 f(x)\omega(x)\,\mathrm{d}x$$

if $f \in \mathcal{C}^{\infty}[0,1]$ and $\lim_{n\to\infty} \left(\frac{1}{n!} \sup_{0\leq x\leq 1} |f^{(n)}(x)| \right) = 0$.

5.3-4 Given a weight function $\omega \in L^1(0,1)$, let there be given a sequence of continuous linear functionals $\ell_n : \mathcal{C}[0,1] \to \mathbb{R}$, $n \geq 0$, of the form

$$\ell_n : f \in \mathcal{C}[0,1] \to \ell_n(f) := \sum_{j=0}^{n} \omega_j^n f(x_j^n) \in \mathbb{R}, \quad \text{where } 0 \leq x_0^n < x_1^n < \cdots < x_n^n \leq 1,$$

with the following property:

$$\lim_{n\to\infty} \left| \int_0^1 p(x)\omega(x)\,\mathrm{d}x - \ell_n(p) \right| = 0 \quad \text{for any } p \in \mathcal{P}[0,1].$$

Show that, if $\omega_j^n \geq 0$ for all $n \geq 0$ and all $0 \leq j \leq n$, then

$$\lim_{n\to\infty} \left| \int_0^1 f(x)\omega(x)\,\mathrm{d}x - \ell_n(f) \right| = 0 \quad \text{for any } f \in \mathcal{C}[0,1].$$

This result constitutes **Steklov's theorem.**[8]

[8] So named after Vladimir Andreevich Steklov (1864–1926).

5.4 Application of the Banach–Steinhaus theorem: Divergence of Lagrange interpolation

In what follows, $[a,b]$ is a compact interval $[a,b] \subset \mathbb{R}$ with $a < b$, and $\mathcal{C}[a,b]$ denotes the Banach space formed by all continuous functions $f : [a,b] \to \mathbb{R}$, equipped with the sup-norm defined by $\|f\| = \sup_{a \le x \le b} |f(x)|$. For each $n \ge 0$, $\mathcal{P}_n$ denotes the space of all polynomials of degree $\le n$ of one real variable, and $\mathcal{P}_n[a,b]$ denotes the subspace of $\mathcal{C}[a,b]$ formed by the restrictions to $[a,b]$ of all $p \in \mathcal{P}_n$.

The next theorem describes the well-known **Lagrange interpolation**,[9] which consists in interpolating a given function at a finite number of points by a polynomial. The more general **Hermite interpolation**[10] consists in interpolating in addition some *derivatives* of the function, again at a finite number of points (examples of Hermite interpolation are provided in Problems 5.4-2 and 5.4-3). Lagrange interpolation in *several* variables will be studied in Section 7.11.

Theorem 5.4-1 *For each integer $n \ge 0$, let there be given any $(n+1)$ distinct points $a \le x_0 < x_1 < \cdots < x_n \le b$. Then, given any function $f \in \mathcal{C}[a,b]$, there exists one and only one polynomial $L_n f \in \mathcal{P}_n[a,b]$ that satisfies*

$$L_n f(x_i) = f(x_i), \quad 0 \le i \le n.$$

This polynomial, which is called the **Lagrange interpolating polynomial** *of f of degree $\le n$ associated with the $(n+1)$* **nodes** *$x_i \in [a,b]$ is given by*

$$L_n f(x) = \sum_{j=0}^{n} f(x_j) p_j(x), \quad a \le x \le b,$$

where the $(n+1)$ polynomials $p_j \in \mathcal{P}_n[a,b]$, $0 \le j \le n$, are defined by

$$p_j(x) := \prod_{\begin{cases} i=0 \\ i \ne j \end{cases}}^{n} \frac{(x - x_i)}{(x_j - x_i)}, \quad a \le x \le b.$$

The operator $L_n : \mathcal{C}[a,b] \to \mathcal{C}[a,b]$ defined in this fashion is linear and continuous, with

$$\|L_n\| = \sup_{a \le x \le b} \left(\sum_{j=0}^{n} |p_j(x)| \right),$$

and it satisfies

$$L_n p = p \quad \textit{for all } p \in \mathcal{P}_n[a,b].$$

[9] So named after:

J.-L. Lagrange [1812]: Leçons élémentaires de mathématiques données à l'Ecole Normale en 1795, *Journal de l'Ecole Polytechnique*, VIIe et VIIIe cahiers, t-II.

[10] So named after:

C. Hermite [1878]: Sur la formule d'interpolation de Lagrange, *Journal für die reine und angewandte Mathematik* **84**, 70–79.

Proof The relations $p_j(x_i) = \delta_{ij}$, $0 \le i, j \le n$, show that, given any function $f \in \mathcal{C}[a, b]$, the particular polynomial $p := \sum_{j=0}^n f(x_j)p_j \in \mathcal{P}_n[a, b]$ satisfies $p(x_i) = f(x_i)$, $0 \le i \le n$.

Finding such an interpolating polynomial amounts to solving a linear system of $(n+1)$ equations (the number of nodes) with the same number of unknowns (the $(n+1)$ coefficients of the unknown polynomial over the canonical basis of $\mathcal{P}_n[a, b]$). But a well-known property of linear systems with square matrices asserts that existence (as shown above) implies uniqueness. Hence the unique interpolating polynomial of degree $\le n$ of f is given by

$$L_n f = \sum_{j=0}^n f(x_j)p_j.$$

This also shows that the $(n+1)$ polynomials p_j, $0 \le j \le n$, form a *basis* of $\mathcal{P}_n$.

It is clear that the operator $L_n : \mathcal{C}[a, b] \to \mathcal{C}[a, b]$ defined in this fashion is linear and that $L_n p = p$ for all $p \in \mathcal{P}_n[a, b]$ (the interpolating polynomial is unique). That

$$\|L_n\| \le \sup_{a \le x \le b} \left(\sum_{j=0}^n |p_j(x)| \right)$$

follows immediately from the formula $L_n f = \sum_{j=0}^n f(x_j)p_j$.

Let $\zeta \in [a, b]$ be such that

$$\sum_{j=0}^n |p_j(\zeta)| = \sup_{a \le x \le b} \left(\sum_{j=0}^n |p_j(x)| \right),$$

and let $\widetilde{f} \in \mathcal{C}[a, b]$ denote the piecewise affine continuous function defined by

$$\widetilde{f}(a) = \operatorname{sgn} p_0(\zeta), \quad \widetilde{f}(x_j) = \operatorname{sgn} p_j(\zeta), \quad 0 \le j \le n, \quad \widetilde{f}(b) = \operatorname{sgn} p_n(\zeta).$$

Let the function $f_0 \in \mathcal{C}[a, b]$ be defined by $f_0(x) := 1$, $a \le x \le b$. The relation $L_n f_0 = f_0$ then implies that

$$\sum_{j=0}^n p_j(x) = 1, \quad a \le x \le b.$$

Then $\|\widetilde{f}\| = 1$ because the definition of $\widetilde{f}$ shows that either $\|\widetilde{f}\| = 0$ or $\|\widetilde{f}\| = 1$; but $\|\widetilde{f}\| = 0$ is impossible since $\sum_{j=0}^n p_j(\zeta) = 1$ (so the numbers $p_j(\zeta)$, $0 \le j \le n$, cannot all vanish simultaneously). Besides,

$$\|L_n \widetilde{f}\| \ge |L_n \widetilde{f}(\zeta)| = \left| \sum_{j=0}^n \widetilde{f}(x_j)p_j(\zeta) \right| = \sum_{j=0}^n |p_j(\zeta)|.$$

These relations, combined with

$$\|L_n\| = \sup_{f \ne 0} \frac{\|L_n f\|}{\|f\|} \ge \frac{\|L_n \widetilde{f}\|}{\|\widetilde{f}\|},$$

imply that $\|L_n\| \ge \sup_{a \le x \le b} (\sum_{j=0}^n |p_j(x)|)$. Hence

$$\|L_n\| = \sup_{a\le x\le b}\left(\sum_{j=0}^{n}|p_j(x)|\right). \qquad \square$$

A natural question immediately arises: Under what kind of assumptions on a function $f \in \mathcal{C}[a,b]$ is the Lagrange interpolation a *convergent* approximation scheme, in the sense that $\lim_{n\to\infty}\|L_n f - f\| = 0$? While it is easily established that this is so if the function f is infinitely differentiable and its derivatives "do not grow too fast" (Problem 5.4-1), it has been known, since a famous counterexample given by Sergei Natanovich Bernstein in 1918, that the Lagrange interpolating polynomials of the function $f : x \in [-1,1] \to |x|$ at equally spaced nodes on $[-1,1]$, and with $-1, 0$, and 1 as particular nodes, *do not even pointwise converge* to f, save at the points $-1, 0$, and 1.

Remark One can show[11] that $\lim_{n\to\infty}\left|\dfrac{L_n f(x) - |x|}{\prod_{i=0}^{n}(x - x_i)}\right|^{1/2} = e$ for all $x \in \mathbb{R}$. $\square$

Remarkably, the divergence of the Lagrange interpolation for some continuous functions can be established without providing any explicit expression of such functions, thanks to the *Banach–Steinhaus theorem.* Note that, save that they are *distinct* for each $n \ge 0$, *no other assumption is made on the nodes* x_i^n, $0 \le i \le n$.

Theorem 5.4-2 *For each integer $n \ge 0$, let there be given $(n+1)$ distinct nodes $a \le x_0^n < x_1^n < \cdots < x_n^n \le b$. Given any function $f \in \mathcal{C}[a,b]$, let its Lagrange interpolating polynomial $L_n f \in \mathcal{P}_n[a,b]$ be defined for any $n \ge 0$ by $L_n f(x_i^n) = f(x_i^n)$, $0 \le i \le n$. Then*

$$\sup_{n\ge 0}\|L_n f\| = \infty \quad \textit{for some } f \in \mathcal{C}[a,b],$$

a property that a fortiori *prevents the uniform convergence of $L_n f$ to f for such a function.*

Proof It was established in Theorem 5.4-1 that

$$\|L_n\| = \sup_{a\le x\le b}\left(\sum_{j=0}^{n}|p_j^n(x)|\right), \quad \text{where } p_j^n(x) := \prod_{\substack{i=0\\ i\ne j}}^{n}\left(\frac{x - x_i^n}{x_j^n - x_i^n}\right), \quad a \le x \le b,$$

and it can be shown that

$$\lim_{n\to\infty}\|L_n\| = \infty$$

(Problem 5.4-4).

If we had $\sup_{n\ge 0}\|L_n f\| < \infty$ for each $f \in \mathcal{C}[a,b]$, the Banach–Steinhaus theorem (Theorem 5.3-1), would imply that $\sup_{n\ge 0}\|L_n\| < \infty$, in contradiction with $\lim_{n\to\infty}\|L_n\| = \infty$. Hence $\sup_{n\ge 0}\|L_n f\| = \infty$ for at least one function $f \in \mathcal{C}[a,b]$. $\square$

Remark If, instead of the Banach–Steinhaus theorem, we had used its corollary (Theorem 5.3-2), we could still conclude that there exist continuous functions f whose Lagrange interpolating polynomials do not uniformly converge to f. But we could not conclude that $\sup_{n\ge 1}\|L_n f\| = \infty$. $\square$

---

[11] X. Li; R.N. Mohapatra [1993]: On the convergence of Lagrange interpolation with equidistant nodes, *Proceedings of the American Mathematical Society* **118**, 1205–1212.

The norms $\|L_n\|$ that appeared in the above proof are called the **Lebesgue constants.**[12]

We conclude this section with some general considerations about the approximation by *polynomials* of continuous functions over compact intervals.

The "ideal objective" in this respect consists in finding a sequence $(A_n)_{n=0}^{\infty}$ of mappings $A_n : \mathcal{C}[a,b] \to \mathcal{P}_n[a,b]$ that possess the following *four properties*: The operators A_n are *linear* and *continuous*, they *preserve all polynomials of degree* $\leq n$, i.e., $A_n p = p$ for all $p_n \in \mathcal{P}_n[a,b]$ and all $n \geq 1$ (the operator L_n associated with Lagrange interpolation satisfy these three properties), *and* (most importantly!) *they satisfy*:

$$\text{for each } f \in \mathcal{C}[a,b], \quad \lim_{n\to\infty} \|A_n f - f\| = 0.$$

But this objective is unattainable, according to the following beautiful result (whose proof again depends on the Banach–Steinhaus theorem, as expected), which shows that the divergence phenomenon established in Theorem 5.4-2 is not restricted to Lagrange interpolation.

$^\flat$**Theorem 5.4-3 (Kharshiladze–Lozinski approximation theorem)**[13] *Any sequence* $(A_n)_{n=0}^{\infty}$ *of mappings* $A_n : \mathcal{C}[a,b] \to \mathcal{P}_n[a,b] \subset \mathcal{C}[a,b]$ *that are linear, continuous, and preserve all polynomials of degree* $\leq n$*, is such that*

$$\sup_{n\geq 0} \|A_n f\| = \infty \quad \textit{for some } f \in \mathcal{C}[a,b],$$

a property that a fortiori *prevents the uniform convergence of* $A_n f$ *to* f *for such a function* f. □

In light of this negative result, it is worthwhile to briefly review other types of polynomial approximation (over the interval $[a,b] = [0,1]$ for definiteness).

First, consider the *Bernstein polynomials* $B_n f \in \mathcal{P}_n[0,1]$, $n \geq 0$, which are defined for any function $f \in \mathcal{C}[0,1]$ by (Theorem 2.13-2)

$$(B_n f)(x) = \sum_{k=0}^{n} \frac{n!}{(n-k)!k!} f\left(\frac{k}{n}\right) x^k (1-x)^{n-k}, \quad 0 \leq x \leq 1.$$

Then the associated *Bernstein operators* $B_n : \mathcal{C}[0,1] \to \mathcal{P}_n[0,1] \subset \mathcal{C}[0,1]$ satisfy (see *ibid.*)

$$\text{for each } f \in \mathcal{C}[0,1], \quad \lim_{n\to\infty} \|B_n f - f\| = 0.$$

[12] So named after:

H. Lebesgue [1909]: Sur les intégrales singulières, *Annales de la Faculté des Sciences de l'Université de Toulouse* **1**, 25–117.

The asymptotic behavior of the Lebesgue constants $\|L_n\|$ as $n \to \infty$ has generated considerable scrutiny; see, e.g.:

L. Brutman [1997]: Lebesgue functions for polynomial interpolations – a survey, *Annals of Numerical Mathematics* **4**, 111–127.

A. Eisinberg; G. Fedele; G. Franzè [2004]: Lebesgue constant for Lagrange interpolation on equidistant nodes, *Analysis in Theory and Applications* **20**, 323–331.

S.J. Smith [2006]: Lebesgue constants in polynomial interpolation, *Annales Mathematicae et Informaticae* **33**, 109–123.

R.B. Platte; L.N. Trefethen; A.B.J. Kuijlaars [2011]: Impossibility of fast stable approximation of analytic functions from equispaced samples, *SIAM Review* **53**, 308–318.

[13] S. Lozinski [1948]: On a class of linear operators, *Doklady Akademii Nauk SSSR* **61**, 193–196 (in Russian). A proof is found in Cheney [1966, Chapter 6, Section 5].

Hence, by Theorem 5.4-3, at least one of the four above properties must fail. Since each operator B_n is linear and continuous ($\|B_n\| = 1$ for all $n \geq 2$), it thus follows that B_n does *not* preserve all polynomials of degree $\leq n$ (see Problem 2.13-1, which provides an indication in this direction), even though the range of B_n is the space $\mathcal{P}_n[0,1]$ (Problem 5.4-5).

Second, consider the mappings $A_n : \mathcal{C}[0,1] \to \mathcal{C}[0,1]$ defined for any $n \geq 0$ by

$$A_n f \in \mathcal{P}_n[0,1] \quad \text{and} \quad \|f - A_n f\| = \inf_{p \in \mathcal{P}_n[0,1]} \|f - p\| \quad \text{for each } f \in \mathcal{C}[0,1]$$

(while establishing the existence of $A_n f$ is straightforward, establishing its uniqueness is not as easy; cf. Problem 5.4-6). Then each mapping $A_n : \mathcal{C}[0,1] \to \mathcal{C}[0,1]$, $n \geq 0$, is *continuous* (Problem 5.4-6) and clearly preserves the space $\mathcal{P}_n[0,1]$, i.e., $A_n p = p$ for all $p \in \mathcal{P}_n[0,1]$. Besides, the Weierstraß approximation theorem (Theorem 2.13-3) clearly implies that

$$\text{for each } f \in \mathcal{C}[0,1], \quad \lim_{n \to \infty} \|A_n f - f\| = 0.$$

Since at least one of the four above properties must fail, again by Theorem 5.4-3, there remains only the *linearity* as a candidate for the missing property: indeed, each mapping A_n, $n \geq 0$, is *nonlinear* (Problem 5.4-6).

By contrast, consider the mappings $P_n : \mathcal{C}[0,1] \to \mathcal{C}[0,1]$ defined for any $n \geq 0$ by

$$P_n f \in \mathcal{P}_n[0,1] \quad \text{and} \quad \|f - P_n f\|_{L^2(0,1)} = \inf_{p \in \mathcal{P}_n[0,1]} \|f - p\|_{L^2(0,1)} \,.$$

By the projection theorem (Theorem 4.3-1), which can be applied since, as a finite-dimensional subspace, $\mathcal{P}_n[0,1]$ is a complete subset of $(\mathcal{C}[0,1], \|\cdot\|_{L^2(0,1)})$, each $P_n : \mathcal{C}[0,1] \to \mathcal{C}[0,1]$ is a linear and continuous operator, which, in addition, clearly preserves the space $\mathcal{P}_n[0,1]$. Furthermore, the Weierstraß approximation theorem implies that

$$\text{for each } f \in \mathcal{C}[0,1], \quad \lim_{n \to \infty} \|f - P_n f\|_{L^2(0,1)} = 0$$

(since $\inf_{p \in \mathcal{P}_n[0,1]} \|f - p\|_{L^2(0,1)} \leq \inf_{p \in \mathcal{P}_n[0,1]} \|f - p\|$). This type of polynomial approximation thus satisfies all four properties of the above "ideal objective."

But of course this is not in contradiction with the Kharshiladze–Lozinski theorem, which applies when the space $\mathcal{C}[0,1]$ is equipped with the *sup-norm* $\|\cdot\|$.

Problems

5.4-1 In this problem $L_n f$ denotes the Lagrange interpolation polynomial of a function $f \in \mathcal{C}[a,b]$, as defined in Theorem 5.4-1.

(1) Assume that $f \in \mathcal{C}^{n+1}[a,b]$. Show that, given any $x \in [a,b]$, there exists a point $\xi_x \in \,]a,b[$ such that

$$f(x) - L_n f(x) = \frac{1}{(n+1)!} f^{(n+1)}(\xi_x) \prod_{j=0}^{n} (x - x_j),$$

so that

$$\|L_n f - f\| \leq \frac{1}{(n+1)!} \|f^{(n+1)}\| \sup_{x \in [a,b]} \left| \prod_{j=0}^{n} (x - x_j) \right|.$$

Remark The formula

$$f(x) = L_n f(x) + \frac{1}{(n+1)!} f^{(n+1)}(\xi_x) \prod_{j=0}^{n} (x - x_j), \quad a \le x \le b,$$

provides an example of a one-dimensional *multipoint Taylor formula*. As we shall see in Section 7.11, similar multipoint Taylor formulas hold as well in $\mathbb{R}^n$. □

Hint: Given any point $y \in [a, b]$ with $y \neq x_i$, $0 \le i \le n$, apply Rolle's theorem to the auxiliary function $h_y \in \mathcal{C}^{n+1}[a, b]$ defined by

$$h_y(x) := (f - L_n f)(x) - (f - L_n f)(y) \prod_{j=0}^{n} \left(\frac{x - x_j}{y - x_j} \right).$$

(2) Show that, if $f \in \mathcal{C}^\infty[a, b]$ and there exists a constant C such that $\|f^{(n)}\| \le C^n$ for all $n \ge 0$, then $\lim_{n\to\infty} \|L_n f - f\| = 0$.

(3) Show that $\lim_{n\to\infty} \|L_n f - f\| = 0$ if the function f can be extended to an analytic function in an open subset of $\mathbb{C}$ that contains the closed set $\{z \in \mathbb{C}; \operatorname{dist}(z; [a, b]) \le 1\}$.

5.4-2 (example of Hermite interpolation) For each integer $n \ge 0$, let there be given $(n + 1)$ distinct points $a \le x_0 < x_1 < \cdots < x_n \le b$.

(1) Show that, given any function $f \in \mathcal{C}^1[a, b]$, there exists one and only one polynomial $p_n = p_n(f) \in \mathcal{P}_{2n+1}[a, b]$ such that

$$p_n(x_i) = f(x_i) \quad \text{and} \quad p_n'(x_i) = f'(x_i), \quad 0 \le i \le n.$$

(2) Let the space $\mathcal{C}^1[a, b]$ be equipped with the norm $f \to \max\{\|f\|, \|f'\|\}$. Show that the mapping $f \in \mathcal{C}^1[a, b] \to p_n(f) \in \mathcal{C}^1[a, b]$ defined in (1) is linear, continuous, and is such that $p_n(f) = f$ for all $f \in \mathcal{P}_{2n+1}[a, b]$.

(3) Assume that $f \in \mathcal{C}^{2n+2}[a, b]$. Show that

$$\|p_n(f) - f\| \le \frac{1}{(2n+2)!} \|f^{(2n+2)}\| \sup_{a \le x \le b} \left| \prod_{j=0}^{n} (x - x_j)^2 \right|.$$

Hint: Use an argument similar to that proposed in Problem 5.4-1(1).

5.4-3 (example of Hermite interpolation) Let an integer $n \ge 1$ and a compact interval $[a, b]$ with $a < b$ be given.

(1) Show that, given any function $f \in \mathcal{C}^n[a, b]$, there exists one and only one polynomial $p_n = p_n(f) \in \mathcal{P}_{2n+1}[a, b]$ that satisfies

$$p_n^{(k)}(a) = f^{(k)}(a) \quad \text{and} \quad p_n^{(k)}(b) = f^{(k)}(b), \quad 0 \le k \le n.$$

(2) Let the space $\mathcal{C}^n[a, b]$ be equipped with the norm $f \to \max_{0 \le k \le n} \|f^{(k)}\|$. Show that the mapping $f \in \mathcal{C}^n[a, b] \to p_n(f) \in \mathcal{C}^n[a, b]$ defined in (1) is linear, continuous, and is such that $p_n(f) = f$ for all $f \in \mathcal{P}_{2n+1}[a, b]$.

(3) Assume that $f \in \mathcal{C}^{2n+2}[a, b]$. Show that

$$\|(p_n(f) - f)^{(k)}\| \le \frac{(b-a)^k}{k!(2n+2-2k)!} \|f^{2n+2}\| \big((x-a)(b-x)\big)^{n+1-k}$$

for all $x \in [a, b]$ and all $0 \le k \le n + 1$.

Hint: Given any point $y \in]a,b[$ and any $0 \le k \le n+1$, apply Rolle's theorem to the auxiliary function $h_y^k \in \mathcal{C}^{2n+1-k}[a,b]$ defined by

$$h_y(x) = (f - p_n f)^{(k)}(x) - (f - p_n f)^k(y)\left(\frac{(x-a)(b-x)}{(y-a)(b-y)}\right)^{n+1-k}, \quad a \le x \le b.$$

5.4-4 For each integer $n \ge 1$. let there be given $(n+1)$ distinct points $0 \le x_0 < x_1 < \cdots < x_n^n \le 1$, and let $p_j^n(x) := \prod_{\substack{i=0 \\ i \ne j}}^n \left(\frac{x - x_i^n}{x_j^n - x_i^n}\right)$, $0 \le x \le 1$. Show that there exists a constant $C > 0$, which depends on the points x_j^n, $1 \le j \le n$, $n \ge 0$, such that

$$\sup_{0 \le x \le 1} \left(\sum_{j=0}^n |p_j^n(x)|\right) \ge C \log n \quad \text{for all } n \ge 1.$$

Consequently, the Lebesgue constants $\|L_n\| = \sup_{0 \le x \le 1}\left(\sum_{j=0}^n |p_j^n(x)|\right)$ satisfy $\lim_{n \to \infty} \|L_n\| = \infty$.

5.4-5 Show that, for each integer $n \ge 1$, the image of $\mathcal{C}[0,1]$ by the Bernstein operator B_n is the space $\mathcal{P}_n[0,1]$.

5.4-6 (1) Given any function $f \in \mathcal{C}[0,1]$, show that, for each $n \ge 0$, there exists one and only one polynomial $A_n f \in \mathcal{P}_n[0,1]$ that satisfies

$$\|f - A_n f\| = \inf_{p \in \mathcal{P}_n[0,1]} \|f - p\|.$$

Hence $A_n p = p$ for all $p \in \mathcal{P}_n[0,1]$.

(2) Let $A_n : \mathcal{C}[0,1] \to \mathcal{C}[0,1]$ denote for each $n \ge 0$ the mapping defined in (1). Show that, for each $f \in \mathcal{C}[0,1]$, there exists a constant $C(f,n)$ such that

$$\|A_n f - A_n \widetilde{f}\| \le C(f,n)\|f - \widetilde{f}\| \quad \text{for all } \widetilde{f} \in \mathcal{C}[0,1].$$

Hence each mapping A_n is continuous.

(3) Show that, for each $n \ge 0$, the mapping $A_n : \mathcal{C}[0,1] \to \mathcal{C}[0,1]$ is nonlinear.

Hints: This problem is by no means trivial. While the existence of $A_n f$ in (1) follows from a simple compactness argument (as in Problem 2.7-1(1)), the uniqueness of $A_n f$ relies on the following beautiful (but not simple to prove) *de la Vallée Poussin alternation theorem*:[14] A polynomial $p_n \in \mathcal{P}_n[0,1]$ is such that $\|f - p_n\| = \inf_{p \in \mathcal{P}_n[0,1]} \|f - p\|$ if and only if there exists at least $(n+2)$ distinct points $0 \le x_0 < x_1 < \cdots < x_{n+1} \le 1$ such that

$$\begin{aligned} |f(x_i) - p_n(x_i)| &= \|f - p_n\|, \quad 0 \le i \le n+1, \\ \operatorname{sgn}(f(x_i) - p_n(x_i)) &= -\operatorname{sgn}(f(x_{i+1}) - f(x_i)), \quad 0 \le i \le n. \end{aligned}$$

In fact, this alternation theorem holds not only for the subspace $\mathcal{P}_n[0,1]$, but more generally for any subspace $V_n = \mathrm{Span}(e_i)_{i=0}^n$, of $\mathcal{C}[0,1]$ that satisfies the *Haar condition*.[15] This condition is satisfied if, given any $(n+1)$ distinct points $x_j \in [0,1]$, $0 \le j \le n$, one has $\det(e_i(x_j)) \ne 0$. If $V_n = \mathcal{P}_n[0,1]$, $\det(e_i(x_j))$ is nothing but the familiar *Vandermonde determinant*, and thus the Haar condition is satisfied by the space $\mathcal{P}_n[0,1]$.[16]

[14] C.J. DE LA VALLÉE POUSSIN [1910]: Sur les polynômes d'approximation et la représentation approchée d'un angle, *Académie Royale de Belgique, Bulletins de la Classe des Sciences* **12**.

[15] A. HAAR [1918]: Die Minkowskische Geometrie und die Annäherung an stetige Funktionen, *Mathematische Annalen* **78**, 294–311.

[16] For proofs of (1) and (2), see, e.g., CHENEY [1966, Chapter 3, Sections 4 and 5].

5.5 Application of the Banach–Steinhaus theorem: Divergence of Fourier series

Recall that $\mathcal{C}_{\text{per}}[0,2\pi]$ denotes the Banach space formed by all continuous, 2π-periodic functions $g:\mathbb{R}\to\mathbb{R}$ equipped with the sup-norm $\|\cdot\|$, i.e., defined by $\|g\| := \sup_{0\le\theta\le 2\pi}|g(\theta)|$, that $\mathcal{Q}_n[0,2\pi]$ denotes the space formed by all real 2π-periodic trigonometric polynomials of degree $\le n$, and that the nth *Fourier partial sum of* g is defined by

$$(S_n g)(\theta) := \frac{a_0}{2} + \sum_{k=1}^{n} a_k \cos k\theta + \sum_{k=1}^{n} b_k \sin k\theta, \quad 0 \le \theta \le 2\pi,$$

where

$$a_k := \frac{1}{\pi}\int_0^{2\pi} g(\theta)\cos k\theta \, \mathrm{d}\theta, \ \ k\ge 0, \quad \text{and} \quad b_k := \frac{1}{\pi}\int_0^{2\pi} g(\theta)\sin k\theta\,\mathrm{d}\theta, \quad k \ge 1.$$

We then showed (Theorem 2.14-2) that the *Fejér operators* $F_n : \mathcal{C}_{\text{per}}[0,2\pi] \to \mathcal{Q}_{n-1}[0,2\pi] \subset \mathcal{C}_{\text{per}}[0,2\pi]$, defined by

$$F_n : g \in \mathcal{C}_{\text{per}}[0,2\pi] \to F_n g := \frac{1}{n}(S_0 g + S_1 g + \cdots + S_{n-1} g) \quad \text{for each } n \ge 1,$$

have the property that

$$\lim_{n\to\infty} \|F_n g - g\| = 0 \quad \text{for any function } g \in \mathcal{C}_{\text{per}}[0,2\pi].$$

We now show that, by contrast, there exist functions $g \in \mathcal{C}_{\text{per}}[0,2\pi]$ whose nth Fourier partial sums $S_n g$ do *not* uniformly converge to g (it is in this sense that "divergence of Fourier series" in the title of this section is to be understood). This is a consequence of the next theorem, where the Banach–Steinhaus theorem is put to use in the same manner as in the preceding section (compare with Theorem 5.4-2), this time for establishing the existence of functions $g \in \mathcal{C}_{\text{per}}[0,2\pi]$ such that $\sup_{n\ge 0}\|S_n g\| = \infty$.

Theorem 5.5-1 *There exist functions $g \in \mathcal{C}_{\text{per}}[0,2\pi]$ whose* n*th Fourier partial sums $S_n g$ satisfy*

$$\sup_{n\ge 0} \|S_n g\| = \infty,$$

a property that a fortiori *prevents the uniform convergence of $S_n g$ to g for such functions g.*

Proof (i) *The linear operator $S_n : \mathcal{C}_{\text{per}}[0,2\pi] \to \mathcal{C}_{\text{per}}[0,2\pi]$ defining for each $n \ge 0$ the* n*th Fourier partial sum series is continuous, and its norm is given by*

$$\|S_n\| = \frac{1}{2\pi}\int_{-\pi}^{\pi} \left| \frac{\sin\left(\frac{n+1}{2}\right)\varphi}{\sin\frac{1}{2}\varphi} \right| \mathrm{d}\varphi.$$

Let the *Dirichlet kernel* $D_n \in \mathcal{C}_{\text{per}}[0,2\pi]$ be defined by

$$D_n(\varphi) := \frac{1}{2\pi}\frac{\sin(\frac{n+1}{2}\varphi)}{\sin\frac{1}{2}\varphi}, \quad 0 \le \varphi \le 2\pi.$$

Then the nth Fourier partial sum $S_n g$ of any function $g \in \mathcal{C}_{\text{per}}[0, 2\pi]$ is also given by (Problem 2.14-1)

$$S_n g(\theta) = \int_{-\pi}^{\pi} g(\theta + \varphi) D_n(\varphi)\, \mathrm{d}\varphi,$$

and thus

$$\|S_n\| = \sup_{g \neq 0} \frac{\|S_n g\|}{\|g\|} \leq \int_{-\pi}^{\pi} |D_n(\varphi)|\, \mathrm{d}\varphi.$$

Let the function $g_n : [-\pi, \pi] \to \mathbb{R}$ be defined by

$$g_n(\varphi) := \operatorname{sgn} D_n(\varphi), \quad -\pi \leq \varphi \leq \pi,$$

and, for $\varepsilon > 0$ small enough, let $g_n^\varepsilon : [-\pi, \pi] \to \mathbb{R}$ denote the continuous piecewise affine function that is equal to g_n on $[-\pi, \pi] - I_n^\varepsilon$, where I_n^ε denotes the intersection of $[-\pi, \pi]$ with the union of the open intervals of length ε centered at those zeros of the Dirichlet kernel D_n that belong to the interval $[-\pi, \pi]$ (Figure 5.5-1). Then

$$\|g_n^\varepsilon\| = 1 \quad \text{and} \quad \|S_n g_n^\varepsilon\| \geq |S_n g_n^\varepsilon(0)| = \left| \int_{-\pi}^{\pi} g_n^\varepsilon(\varphi) D_n(\varphi)\, \mathrm{d}\varphi \right|.$$

Since

$$\lim_{\varepsilon \to 0} \int_{-\pi}^{\pi} g_n^\varepsilon(\varphi) D_n(\varphi)\, \mathrm{d}\varphi = \int_{-\pi}^{\pi} g_n(\varphi) D_n(\varphi)\, \mathrm{d}\varphi = \int_{-\pi}^{\pi} |D_n(\varphi)|\, \mathrm{d}\varphi$$

(as is easily verified), and

$$\|S_n\| = \sup_{g \neq 0} \frac{\|S_n g\|}{\|g\|} \geq \frac{\|S_n g_n^\varepsilon\|}{\|g_n^\varepsilon\|} = \|S_n g_n^\varepsilon\|,$$

it thus follows that $\|S_n\| \geq \int_{-\pi}^{\pi} |D_n(\varphi)|\, \mathrm{d}\varphi$. Hence $\|S_n\| = \int_{-\pi}^{\pi} |D_n(\varphi)|\, \mathrm{d}\varphi$ as announced.

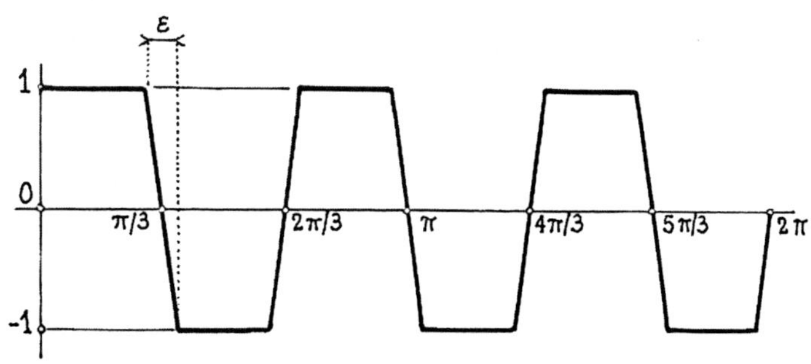

Figure 5.5-1 The function g_n^ε appearing in the proof of Theorem 5.5-1, drawn here for $n = 5$.

(ii) *The norms* $\|S_n\|$ *satisfy*

$$\|S_n\| \geq \frac{4}{\pi^2} \log n \quad \text{for all } n \geq 1.$$

This inequality, which can be established by means of elementary computations, is left as a problem (Problem 5.5-1).

(iii) *Existence of a function* $g \in \mathcal{C}_{\text{per}}[0, 2\pi]$ *whose* nth *Fourier partial sums* $S_n g$ *satisfy* $\sup_{n\geq 0} \|S_n g\| = \infty$.

If we had $\sup_{n\geq 1} \|S_n g\| < \infty$ for each $g \in \mathcal{C}_{\text{per}}[0, 2\pi]$, the *Banach–Steinhaus theorem* (Theorem 5.3-1) would imply that $\sup_{n\geq 1} \|S_n\| < \infty$. But this would contradict the relation $\lim_{n\to\infty} \|S_n\| = \infty$, which follows from (ii). Hence $\sup_{n\geq 0} \|S_n g\| = \infty$ for at least one function $g \in \mathcal{C}_{\text{per}}[0, 2\pi]$. □

Remark If, instead of the Banach–Steinhaus theorem, we had used its corollary (Theorem 5.3-2), we could still conclude that there exist functions $g \in \mathcal{C}_{\text{per}}[0, 2\pi]$ whose nth Fourier partial sums $S_n g$ do not converge to g in the sup-norm. But we could not conclude that $\sup_{n\geq 1} \|S_n g\| = \infty$. □

The norms $\|S_n\| = \dfrac{1}{2\pi} \displaystyle\int_{-\pi}^{\pi} \left| \frac{\sin\left(\frac{n+1}{2}\right)\varphi}{\sin \frac{1}{2}\varphi} \right| \mathrm{d}\varphi$, $n \geq 0$, which naturally appeared in the proof of Theorem 5.5-1, are called the **Lebesgue constants**[17] (to be distinguished from the "other" Lebesgue constants $\|L_n\|$, which appeared in the proof of Theorem 5.4-2).

Just like the divergence phenomenon for polynomial interpolation is not limited to Lagrange interpolation (Section 5.4), *the divergence phenomenon for trigonometric polynomial approximation is not specific to Fourier series*, according to the following beautiful result (whose proof, not surprisingly, again depends on the Banach–Steinhaus theorem).

$^\flat$**Theorem 5.5-2 (Kharshiladze–Lozinski trigonometric approximation theorem**[18]**)** *Any sequence* $(B_n)_{n=0}^{\infty}$ *of mappings* $B_n : \mathcal{C}_{\text{per}}[0, 2\pi] \to \mathcal{Q}_n[0, 2\pi] \subset \mathcal{C}_{\text{per}}[0, 2\pi]$ *that are linear, continuous, and preserve all trigonometric polynomials of degree* $\leq n$, *i.e.,* $B_n q = q$ *for all* $q \in \mathcal{Q}_n[0, 2\pi]$ *and all* $n \geq 0$, *is such that*

$$\|B_n\| \geq \|S_n\| \quad \textit{for all } n \geq 0,$$

where S_n *denotes for each* $n \geq 0$ *the operator associated with the* nth *Fourier partial sums.* □

Since the Banach–Steinhaus theorem again implies that

$$\sup_{n\geq 0} \|B_n g\| = \infty \quad \text{for some } g \in \mathcal{C}_{\text{per}}[0, 2\pi],$$

Theorem 5.5-2 thus implies that there does *not* exist any such sequence $(B_n)_{n=0}^{\infty}$ that would in addition satisfy $\lim_{n\to\infty} \|B_n g - g\| = 0$ for all $g \in \mathcal{C}_{\text{per}}[0, 2\pi]$.

By contrast, the *Fejér operators* F_n satisfy $\lim_{n\to\infty} \|F_n g - g\| = 0$ for all $g \in \mathcal{C}_{\text{per}}[0, 2\pi]$ (Theorem 2.14-2); so they must lack at least one of the above four properties: indeed, they

---

[17] So named after:
H. LEBESGUE [1909]: Sur les intégrales singulières, *Annales de la Faculté des Sciences de l'Université de Toulouse* **1**, 25–117.

[18] S. LOZINSKI [1948]: On a class of linear operators, *Doklady Akademii Nauk SSSR* **61**, 193–196 (in Russian). A proof is found in CHENEY [1966, Chapter 6, Section 5].

are linear and continuous ($\|F_n\| = 1$ for all $n \geq 1$), but they do *not* preserve all trigonometric polynomials of degree $\leq n$ (Problem 2.14-1(3)).

Problem

5.5-1 Show that the Lebesgue constants $\|S_n\| := \frac{1}{2\pi}\int_{-\pi}^{\pi}\left|\frac{\sin\left(\frac{n+1}{2}\right)\varphi}{\sin\frac{1}{2}\varphi}\right| \mathrm{d}\varphi$ (which play a key role in the proof of Theorem 5.5-1) satisfy $\|S_n\| \geq \frac{4}{\pi^2}\log n$ for all $n \geq 1$.

5.6 Banach open mapping theorem; a first application: Well-posedness of two-point boundary value problems

Another fundamental consequence of Baire's theorem is the following sufficient condition for a *continuous linear operator* from one *Banach space* into another *Banach space* to be an **open mapping**, i.e., one that maps open sets into open sets:

Theorem 5.6-1 (Banach open mapping theorem[19]) *Let X and Y be Banach spaces and let $A \in \mathcal{L}(X;Y)$ be surjective.*

Then the direct image $A(U)$ under A of any open subset U of X is an open subset of Y.

Proof Throughout this proof, the following notations are respectively used for denoting the open balls in the space X centered at the origin of X and the open balls in the space Y:

$$B_r := \{x \in X;\ \|x\| < r\} \quad \text{and} \quad B(y;s) := \{\widetilde{y} \in Y;\ \|\widetilde{y} - y\| < s\}.$$

Also, given a vector space Z, a vector $z_0 \in Z$, a scalar α, and a subset $A \subset Z$, we let

$$\{z_0\} + A := \{(z_0 + z) \in Z;\ z \in A\} \quad \text{and} \quad \alpha A = \{(\alpha z) \in Z;\ \ z \in A\}.$$

Hence $\{z_0\} + A \subset 2A$ if $z_0 \in A$.

(i) *The set $\overline{A(B_1)}$ contains an open ball.*

Given any $y \in Y$, there exists $x \in X$ such that $y = Ax$ since A is *surjective* (this is the only place where this assumption is used). Since $x \in B_n$ for some integer $n \geq 1$, this shows that $Y = \bigcup_{n=1}^{\infty} A(B_n)$; hence *a fortiori*

$$Y = \bigcup_{n=1}^{\infty} \overline{A(B_n)}.$$

The space Y being complete, Baire's theorem (used here in the form of its corollary, Theorem 5.1-3(b)) shows that there exists an integer $n_0 \geq 1$ such that

$$\operatorname{int} \overline{A(B_{n_0})} \neq \varnothing.$$

Therefore $\operatorname{int} \overline{A(B_1)} \neq \varnothing$, since $\overline{A(B_1)} = \frac{1}{n_0}\overline{A(B_{n_0})}$ by the linearity of A. Hence the set $\overline{A(B_1)}$ contains an open ball.

[19] S. Banach [1932]: *Théorie des Opérateurs Linéaires*, Monografje Matematyczne, Warsaw.

(ii) *The set* $\overline{A(B_1)}$ *contains an open ball centered at the origin of* Y.

By (i), there exist $y \in Y$ and $s > 0$ such that $B(y; 2s) \subset \overline{A(B_1)}$, and hence such that

$$B(0; 2s) = \{-y\} + B(y; 2s) \subset \{-y\} + \overline{A(B_1)}.$$

Since $-y \in \overline{A(B_1)}$ (because $y \in \overline{A(B_1)}$ and A is linear), it follows that

$$\{-y\} + \overline{A(B_1)} \subset 2\overline{A(B_1)}.$$

The resulting inclusion $B(0; 2s) \subset 2\overline{A(B_1)}$, combined with the linearity of A, then implies that

$$B(0; s) \subset \overline{A(B_1)}.$$

(iii) *The set* $A(B_1)$ *contains an open ball centered at the origin of* Y.

To prove this assertion, we will show that $B\left(0; \frac{s}{2}\right) \subset A(B_1)$, where $s > 0$ is the radius of the ball $B(0; s)$ found in part (ii) above. This means that, given any $y \in B\left(0; \frac{s}{2}\right)$, we need to find $x \in B_1$ such that $y = Ax$. So, let $y \in B\left(0; \frac{s}{2}\right)$ be given.

Since $y \in B\left(0; \frac{s}{2}\right) \subset \overline{A(B_{1/2})}$ by (ii), there exists $x_1 \in B_{1/2}$ such that $\|y - Ax_1\| < \frac{s}{2^2}$; since

$$(y - Ax_1) \in B\left(0; \frac{s}{2^2}\right) \subset \overline{A(B_{1/2^2})},$$

again by (ii), there exists $x_2 \in B_{1/2^2}$ such that $\|y - Ax_1 - Ax_2\| < \frac{s}{2^3}$; and so on. In this fashion we construct a sequence $(x_n)_{n=1}^{\infty}$ of points $x_n \in X$ with the following properties:

$$x_n \in B_{1/2^n} \quad \text{and} \quad \left\| y - A\left(\sum_{k=1}^{n} x_k\right) \right\| < \frac{s}{2^{n+1}} \quad \text{for all } n \geq 1.$$

Since the series $\sum_{n=1}^{\infty} x_n$ is thus *uniformly convergent* (because $\|x_n\| < \frac{1}{2^n}$ for each $n \geq 1$) and *the space* X *is complete*, there exists $x \in X$ such that

$$\lim_{n\to\infty} \sum_{k=1}^{n} x_k = x \quad \text{and} \quad \|x\| \leq \sum_{k=1}^{\infty} \|x_n\| < \frac{1}{2} + \frac{1}{2^2} + \cdots + \frac{1}{2^{n+1}} + \cdots = 1$$

(Theorem 3.6-1), so that $x \in B_1$. Furthermore,

$$y = \lim_{n\to\infty} A\left(\sum_{k=1}^{n} x_k\right) = Ax,$$

since A is *continuous*. Hence the assertion is proved.

(iv) *The mapping* A *is open.*

Given any open subset U of X and given any $y \in A(U)$, we must find $\sigma > 0$ such that $B(y; \sigma) \subset A(U)$. So let $x \in U$ be such that $y = Ax$.

Since U is open, there exists $r > 0$ such that $B(x; r) = \{x\} + B_r \subset U$, and by (iii) there exists $\sigma > 0$ such that $B(0; \sigma) \subset A(B_r)$. Hence

$$B(y;\sigma) = \{y\} + B(0;\sigma) \subset \{y\} + A(B_r) = A(\{x\} + B_r) \subset A(U). \qquad \square$$

The following easy consequence of the Banach open mapping theorem, which shall be referred to in the sequel as *"the" corollary to the Banach open mapping theorem*, is a frequently used sufficient condition for the *continuity of the inverse of a linear operator.*

Theorem 5.6-2 (corollary to the Banach open mapping theorem) *Let X and Y be Banach spaces and let $A \in \mathcal{L}(X;Y)$ be bijective. Then $A^{-1} \in \mathcal{L}(Y;X)$.*

Proof Open balls in X and Y are denoted as in the proof of Theorem 5.6-1. Since the mapping $A^{-1} : Y \to X$ is also *linear* (Theorem 2.9-1(b)), it suffices to show that it is *continuous at the origin* (Theorem 2.9-2(b)), i.e., that, given any open ball $B_r \subset X$, there exists an open ball $B(0;\sigma) \subset Y$ such that

$$A^{-1}(B(0;\sigma)) \subset B_r,$$

by definition of continuity at a point; cf. Section 1.11. But the inclusion $A^{-1}(B(0;\sigma)) \subset B_r$ is equivalent to the inclusion

$$B(0;\sigma) \subset A(B_r),$$

since the mapping A is bijective; and the Banach open mapping theorem precisely shows that this last inclusion holds for some $\sigma > 0$. $\square$

The following application to *two-point boundary value problems* provides a first indication of the power of the corollary to the Banach open mapping theorem. Under the *sole* assumptions that its solution $u \in C^2[0,1]$ exists and is unique for *all* right-hand sides $f \in \mathcal{C}[0,1]$, this theorem shows that this problem is *well-posed*, in the sense that "small" perturbations of f in the sup-norm induce "small" variations of u, u', and u'', also in the sup-norm. It is indeed remarkable that such a powerful continuity result can be derived from such minimal assumptions, satisfied for instance if $a(x) = -1$ and $\|b\| + \|c\|$ is small enough (Problem 3.9-1), or if $a(x) = -1$, $b(x) = 0$, $c(x) \geq 0$, $0 \leq x \leq 1$ (Problem 3.10-3; in fact, the inequality $c(x) \geq 0$, $0 \leq x \leq 1$, can be relaxed to $c(x) \geq \gamma > -\pi^2$, $0 \leq x \leq 1$; cf. Problem 9.14-3).

Theorem 5.6-3 *Let functions $a, b, c \in \mathcal{C}[0,1]$ be given such that the two-point boundary value problem*

$$a(x)u''(x) + b(x)u'(x) + c(x)u(x) = f(x), \quad 0 \leq x \leq 1, \quad \textit{and} \quad u(0) = u(1) = 0,$$

has one and only one solution $u \in \mathcal{C}^2[0,1]$ for each $f \in \mathcal{C}[0,1]$. Then there exists a constant C such that

$$\|u\| + \|u'\| + \|u''\| \leq C\|f\| \quad \textit{for all } f \in \mathcal{C}[0,1],$$

where $\|\cdot\|$ denotes the sup-norm of the space $\mathcal{C}[0,1]$.

Proof The space

$$X := \{v \in \mathcal{C}^2[0,1];\ v(0) = v(1) = 0\}$$

equipped with the norm $v \to (\|v\| + \|v'\| + \|v''\|)$ is a Banach space (Problem 5.6-2), and the linear operator $L : v \in X \to Lv \in Y := \mathcal{C}[0,1]$, where

$$Lv(x) := a(x)v''(x) + b(x)v'(x) + c(x)v(x), \quad 0 \leq x \leq 1,$$

is continuous, since

$$\|Lv\| \le \max\{\|a\|, \|b\|, \|c\|\}(\|v\| + \|v'\| + \|v''\|) \quad \text{for all } v \in X.$$

The conclusion then immediately follows from the corollary to the Banach open mapping theorem (Theorem 5.6-2). □

Another consequence of the Banach open mapping theorem is the following sufficient condition for *two norms to be equivalent in an infinite-dimensional space*. In particular, it will provide a quick proof of the Banach closed graph theorem (Theorem 5.7-1).

Theorem 5.6-4 *Let $\|\cdot\|$ and $\|\cdot\|'$ be two norms on the same vector space X, with the following properties: both spaces $(X, \|\cdot\|)$ and $(X, \|\cdot\|')$ are complete, and there exists a constant C such that*

$$\|x\|' \le C\|x\| \quad \textit{for all } x \in X.$$

Then the two norms $\|\cdot\|$ and $\|\cdot\|'$ are equivalent.

Proof The bijective and linear identity mapping $\iota : (X, \|\cdot\|) \to (X, \|\cdot\|')$ is continuous by assumption. Theorem 5.6-2 therefore shows that the inverse mapping $\iota^{-1} : (X, \|\cdot\|') \to (X, \|\cdot\|)$ is also continuous: this means that there exists a constant C' such that $\|x\| \le C'\|x\|'$ for all $x \in X$. Hence the two norms are equivalent (Theorem 2.2-4). □

Another interesting application of the Banach open mapping theorem is given in Problem 5.6-3.

Problems

5.6-1 Do there exist a vector space X and two norms $\|\cdot\|$ and $\|\cdot\|'$ on X such that both spaces $(X, \|\cdot\|)$ and $(X, \|\cdot\|')$ are complete, but the two norms $\|\cdot\|$ and $\|\cdot\|'$ are not equivalent?

5.6-2 In this problem $\|\cdot\|$ denotes the sup-norm of the space $\mathcal{C}[0,1]$, and m is an integer ≥ 1.

(1) Show that the function $v \in \mathcal{C}^m[0,1] \to (\|v\| + \|v'\| + \cdots + \|v^{(m)}\|)$ defines a norm on the space $\mathcal{C}^m[0,1]$, which makes it a Banach space.

(2) Show that the space $(\mathcal{C}^m[0,1], \|\cdot\|)$ is not complete.

Hint: Rather than exhibiting a Cauchy sequence that does not converge, use (1) with $m = 1$ and Theorem 5.6-4.

5.6-3 Let X and Y be Banach spaces. Using the Banach open mapping theorem, show that the set $\{A \in \mathcal{L}(X;Y);\ A \text{ is surjective}\}$ is open in the space $\mathcal{L}(X;Y)$ equipped with the operator norm.

In other words, if $A \in \mathcal{L}(X;Y)$ is thus that the equation $Ax = y$ has at least one solution $x \in Y$ for any $y \in Y$, then the equation $\widetilde{A}x = y$ has again at least one solution $x \in X$ for any $y \in Y$ if $\widetilde{A}$ is close enough to A.

5.7 Banach closed graph theorem; a first application: Hellinger–Toeplitz theorem

Given two sets X and Y, the **graph** $\mathrm{Gr}\, A$ of a mapping $A : X \to Y$ is the subset of the product $X \times Y$ defined by

$$\mathrm{Gr}\, A := \{(x, Ax) \in X \times Y;\ x \in X\}.$$

If X and Y are *topological spaces*, a mapping $A : X \to Y$ is said to be **closed** *if its graph* $\mathrm{Gr}\, A$ *is closed in the product* $X \times Y$ (equipped with the product topology; cf. Section 1.6).

Therefore, *if X and Y are metric spaces, a mapping $A : X \to Y$ is closed if and only if*

$$\lim_{n\to\infty} x_n = x \text{ in } X \quad \text{and} \quad \lim_{n\to\infty} Ax_n = y \text{ in } Y \quad \text{implies } y = Ax$$

(Theorem 1.11-1), and *any continuous mapping $A : X \to Y$ has a closed graph* (since $\lim_{n\to\infty} x_n = x$ in X implies that $\lim_{n\to\infty} Ax_n = Ax$ and the limit of a convergent sequence is unique in a metric space). However, *the converse need not hold in general:* if a mapping is closed, it may happen that the convergence of a sequence $(x_n)_{n=1}^\infty$ in X does *not* imply the convergence of the sequence $(Ax_n)_{n=1}^\infty$ in Y (see Problem 5.7-1 for such an example).

But remarkably, if *both* X and Y are *Banach spaces*, a simple corollary of the *Banach open mapping theorem* shows that any *closed* and *linear* mapping $A : X \to Y$ is *continuous*:

Theorem 5.7-1 (Banach closed graph theorem[20]) *Let X and Y be Banach spaces, and let $A : X \to Y$ be a closed linear operator. Then $A \in \mathcal{L}(X; Y)$.*

Proof Define another norm on X by

$$\|x\|' := \|x\|_X + \|Ax\|_Y \quad \text{for all } x \in X.$$

By definition of the norm $\|\cdot\|'$, any Cauchy sequence $(x_n)_{n=1}^\infty$ with respect to $\|\cdot\|'$ is such that $(x_n)_{n=1}^\infty$ and $(Ax_n)_{n=1}^\infty$ are Cauchy sequences in the spaces X and Y, respectively. Since both spaces are complete, there exist $x \in X$ and $y \in Y$ such that

$$\lim_{n\to\infty} x_n = x \text{ in } X \quad \text{and} \quad \lim_{n\to\infty} Ax_n = y \text{ in } Y,$$

and thus $y = Ax$ since A is closed by assumption. Therefore,

$$\|x_n - x\|' = \|x_n - x\|_X + \|Ax_n - Ax\|_Y = \|x_n - x\|_X + \|Ax_n - y\| \to 0 \quad \text{as } n \to \infty,$$

which shows that $(X, \|\cdot\|')$ is also complete.

Since $\|x\| \le \|x\|'$ for all $x \in X$, Theorem 5.6-4 (itself a corollary of the Banach open mapping theorem) shows that there exists a constant C such that

$$\|Ax\| \le \|x\| + \|Ax\| = \|x\|' \le C\|x\| \quad \text{for all } x \in X.$$

Hence the linear operator A is continuous. □

The following spectacular result ("spectacular" in that a strong conclusion is derived from a seemingly innocuous assumption) constitutes a first application of the Banach closed graph theorem (other applications are proposed in Problems 5.7-2 and 5.7-3).

[20] S. Banach [1932]: *Théorie des Opérateurs Linéaires*, Monografje Matematyczne, Volume 1, Warsaw.

Theorem 5.7-2 (Hellinger–Toeplitz theorem[21]) *Let $(X, (\cdot,\cdot))$ be a Hilbert space and let $A : X \to X$ be a self-adjoint linear operator (Section 4.10), i.e., that satisfies*

$$(Ax, y) = (x, Ay) \quad \textit{for all } x, y \in X.$$

Then A is continuous.

Proof By the Banach closed graph theorem, it suffices to show that A is closed. So, let $(x_n)_{n=1}^{\infty}$ be a sequence of elements $x_n \in X$ such that $x_n \to x \in X$ and $Ax_n \to y \in X$ as $n \to \infty$. Then the continuity of the inner product (Theorem 4.1-1) implies that, for any $z \in X$,

$$(Ax_n, z) \to (y, z) \quad \text{and} \quad (Ax_n, z) = (x_n, Az) \to (x, Az) = (Ax, z) \quad \text{as } n \to \infty.$$

Consequently, $(y, z) = (Ax, z)$ for all $z \in X$, and thus $y = Ax$, which shows that the linear operator A is *closed*. The *Banach closed graph theorem* then implies that A is continuous (the space X is complete). □

Remarks (1) If X is only an inner-product space, the above proof shows that the linear operator A is *closed*.

(2) In fact, it is easily seen that *any* mapping $A : X \to X$ that satisfies $(Ax, y) = (x, Ay)$ for all $x, y \in X$ is automatically *linear*. □

Recall that a self-adjoint linear operator on a Hilbert space on $\mathbb{K} = \mathbb{R}$ is said to be *symmetric*. Examples of such symmetric operators arise in particular in the weak formulation of linear elliptic boundary value problems, the central theme of the next chapter.

Problems

5.7-1 In this problem, $\|\cdot\|$ denotes the sup-norm of the space $\mathcal{C}[0,1]$.

(1) Show that the linear operator $A : (\mathcal{C}^1[0,1]; \|\cdot\|) \to (\mathcal{C}[0,1], \|\cdot\|)$ defined by $(Av)(x) = v'(x)$, $0 \le x \le 1$, for all $v \in \mathcal{C}^1[0,1]$, is not continuous.

(2) Show that A is closed (an application of the closed graph theorem thus provides a further proof that $(\mathcal{C}^1[0,1]; \|\cdot\|)$ is not a Banach space; cf. Problem 5.6-2(2)).

5.7-2 Given $1 < p < \infty$, let $q > 1$ be defined by $\frac{1}{p} + \frac{1}{q} = 1$. Then, given any $a = (a_i)_{i=1}^{\infty} \in \ell^q$, the series $\sum_{i=1}^{\infty} a_i x_i$ converges in $\mathbb{K}$ for all $x = (x_i)_{i=1}^{\infty} \in \ell^p$, since

$$\left|\sum_{i=1}^{\infty} a_i x_i\right| \le \|a\|_q \|x\|_p \quad \text{for all } x = (x_i)_{i=1}^{\infty} \in \ell^p,$$

by Hölder's inequality (Theorem 2.4-1).

Show that, conversely, if a sequence $a = (a_i)_{i=1}^{\infty}$ of scalars a_i is such that the series $\sum_{i=1}^{\infty} a_i x_i$ converges for all $(x_i)_{i=1}^{\infty} \in \ell^p$, then $a \in \ell^q$.

Hint: Show that the linear operator $A : \ell^p \to \ell^\infty$ defined by $A : x = (x_i)_{i=1}^{\infty} \in \ell^p \to Ax = (\sum_{i=1}^{j} a_i x_i)_{j=1}^{\infty}$ is closed, and apply the closed graph theorem.

[21] E. Hellinger; O. Toeplitz [1910]: Grundlagen für eine Theorie der unendlichen Matrizen, *Mathematische Annalen* **69**, 281–330.

5.7-3 Given $1 < p < \infty$, let $q > 1$ be defined by $\frac{1}{p} + \frac{1}{q} = 1$. Let $(a_{ij})_{i,j=1}^{\infty}$ be an "infinite matrix" of scalars with the following properties: Given any $x = (x_j)_{j=1}^{\infty} \in \ell^p$, each series $\sum_{j=1}^{\infty} a_{ij}x_j$, $i \geq 1$, converges. Besides, $y = (y_i)_{i=1}^{\infty} \in \ell^q$, where $y_i := \sum_{j=1}^{\infty} a_{ij}x_j$, $i \geq 1$.

Show that the linear operator $A : \ell^p \to \ell^q$ defined by $Ax := y$ for all $x \in \ell^p$ is continuous.

Hint: Using Problem 5.7-2, show that $A : \ell^p \to \ell^q$ is closed, and apply the closed graph theorem.

5.8 The Hahn–Banach theorem in a vector space

The *Hahn–Banach theorem in a vector space* (Theorem 5.8-1 below) is one of the two keystones of linear functional analysis, the other one being *Baire's theorem* (Theorem 5.1-2). Indeed, this theorem, which is also often referred to as the **analytic form of the Hahn–Banach theorem**, will pervade the rest of this chapter, mostly through its corollaries proved in the next two sections. Among these corollaries, two stand out owing to their importance: the *Hahn–Banach theorem in a normed vector space* (Theorem 5.9-1) and the *geometric forms of the Hahn–Banach theorem* (Theorems 5.10-1 and 5.10-2).

The proof is given in the real case only, as it is the only one that will be encountered in the remainder of this book; the proof in the complex case, where the assumptions are more restrictive, is left as a problem (Problem 5.8-1).

Note that the proof of the Hahn–Banach theorem requires the *axiom of choice* (by way of Zorn's lemma), while that of Baire's theorem does not.

Theorem 5.8-1 (Hahn–Banach theorem[22] in a real vector space) *Let X be a real vector space and let p be a* **sublinear functional on** X, *i.e., a function $p : X \to \mathbb{R}$ that satisfies*

$$
\begin{aligned}
p(\alpha x) &= \alpha p(x) && \textit{for all } \alpha > 0 \textit{ and all } x \in X, \\
p(x + x) &\leq p(x) + p(y) && \textit{for all } x, y \in X.
\end{aligned}
$$

Let Y be a subspace of X and let $\ell : Y \to \mathbb{R}$ be a linear functional on Y that satisfies

$$\ell(y) \leq p(y) \quad \textit{for all } y \in Y.$$

Then there exists a linear functional $\widetilde{\ell} : X \to \mathbb{R}$ that satisfies

$$\widetilde{\ell}(y) = \ell(y) \;\; \textit{for all } y \in Y \quad \textit{and} \quad \widetilde{\ell}(x) \leq p(x) \;\; \textit{for all } x \in X.$$

Proof (i) Assume that $Y \subsetneq X$, *pick any element* $x_0 \in X - Y$ (so that $x_0 \neq 0$), *and define the subspace*

$$\operatorname{Dom} f := \{(\alpha x_0 + y) \in X;\ \alpha \in \mathbb{R},\ y \in Y\}$$

of X, which clearly contains Y. We then show that *there exists a linear functional f : $\operatorname{Dom} f \to \mathbb{R}$ that satisfies*

$$f(y) = \ell(y) \;\text{ for all } y \in Y \quad \text{and} \quad f(x) \leq p(x) \;\text{ for all } x \in \operatorname{Dom} f.$$

[22]This result, which was independently rediscovered by Stefan Banach in 1929, first appeared (in effect in its normed vector space version of Theorem 5.9-1) in:

H. Hahn [1927]: Über lineare Gleichungssysteme in linearen Räumen, *Journal de Crelle* **157**, 214–229.

Finding f amounts to finding a real number $\lambda := f(x_0)$ such that

$$f(\alpha x_0 + y) = \alpha\lambda + f(y) = \alpha\lambda + \ell(y) \le p(\alpha x_0 + y) \quad \text{for all } \alpha \in \mathbb{R} \text{ and all } y \in Y.$$

Since this inequality holds for $\alpha = 0$, it remains to find $\lambda \in \mathbb{R}$ that satisfies

$$\begin{aligned}\lambda &\le \alpha^{-1}\Big(p(\alpha x_0 + y) - \ell(y)\Big)\\ &= p\left(x_0 + \alpha^{-1}y\right) - \ell\left(\alpha^{-1}y\right) \quad \text{for all } \alpha > 0 \text{ and all } y \in \operatorname{Dom} f\end{aligned}$$

and

$$\begin{aligned}\lambda &\ge \alpha^{-1}(p(\alpha x_0 + y) - \ell(y)) = \alpha^{-1}\Big(p\big(-\alpha(-x_0 - \alpha^{-1}y)\big) - \ell(y)\Big)\\ &= -p\left(-x_0 - \alpha^{-1}y\right) + \ell\left(-\alpha^{-1}y\right) \quad \text{for all } \alpha < 0 \text{ and all } y \in Y.\end{aligned}$$

The linearity of $\ell : Y \to \mathbb{R}$ and the sublinearity of $p : X \to \mathbb{R}$ together imply that, for all $u, v \in Y$,

$$\ell(u) + \ell(v) = \ell(u+v) \le p(u+v) = p(-x_0 + u + x_0 + v) \le p(-x_0 + u) + p(x_0 + v),$$

and hence that

$$-p(-x_0 + u) + \ell(u) \le p(x_0 + v) - \ell(v) \quad \text{for all } u, v \in Y.$$

Since then

$$a := \sup_{u \in Y}\{-p(-x_0 + u) + \ell(u)\} \le b := \inf_{v \in Y}\{p(x_0 + v) - \ell(v)\},$$

it thus suffices to pick any λ that satisfies $a \le \lambda \le b$.

(ii) Let $\mathcal{F}$ denote the set of all linear functionals $f : \operatorname{Dom} f \to \mathbb{R}$ that are defined on a subspace $\operatorname{Dom} f$ of X containing Y and that satisfy

$$f(y) = \ell(y) \text{ for all } y \in Y \quad \text{and} \quad f(x) \le p(x) \text{ for all } x \in \operatorname{Dom} f.$$

The set $\mathcal{F}$ is *nonempty*, since $\ell \in \mathcal{F}$. Besides, $\mathcal{F}$ is *partially ordered* (Section 1.3) by the relation $\preccurlyeq$, where $f_1 \preccurlyeq f_2$ means that

$$\operatorname{Dom} f_1 \subset \operatorname{Dom} f_2 \quad \text{and} \quad f_2(x) = f_1(x) \quad \text{for all } x \in \operatorname{Dom} f_1.$$

Given a *totally ordered* (cf. *ibid.*) subset $\mathcal{E}$ of $\mathcal{F}$, let

$$\operatorname{Dom} g := \bigcup_{f \in \mathcal{E}} \operatorname{Dom} f,$$

which is clearly a subspace of X, since $\mathcal{E}$ is totally ordered. We then show that, for any $x \in \operatorname{Dom} g$, the relation

$$g(x) := f(x) \quad \text{for all } f \in \mathcal{E} \text{ such that } x \in \operatorname{Dom} f,$$

unambiguously defines a linear functional $g : \operatorname{Dom} g \to \mathbb{R}$ that satisfies $g(x) \le p(x)$ for all $x \in \operatorname{Dom} g$.

To see this, let $x \in \operatorname{Dom} g$ be such that $x \in \operatorname{Dom} f_1$ and $x \in \operatorname{Dom} f_2$ with $f_1, f_2 \in \mathcal{E}$, and assume for instance that $f_1 \preccurlyeq f_2$ (the subset $\mathcal{E}$ is totally ordered). Therefore

$$g(x) = f_1(x) = f_2(x) \leq p(x).$$

If $x_1 \in \operatorname{Dom} f_1$ and $x_2 \in \operatorname{Dom} f_2$, then $(x_1 + x_2) \in \operatorname{Dom} f_2$ if $f_1 \preccurlyeq f_2$ (to fix ideas) and thus $g(x_1 + x_2) = f_2(x_1 + x_2) = f_2(x_1) + f_2(x_2) = g(x_1) + g(x_2)$. Likewise, $g(\alpha x_1) = f_1(\alpha x_1) = \alpha f_1(x_1) = \alpha g(x_1)$ for any $\alpha \in \mathbb{R}$.

Furthermore, g is clearly an *upper bound* of $\mathcal{E}$, since, by construction, $\operatorname{Dom} f \subset \operatorname{Dom} g$ for all $f \in \mathcal{E}$. By *Zorn's lemma* (Theorem 1.3-1), the set $\mathcal{F}$ thus possesses a *maximal element* $\widetilde{\ell}$, defined on a subspace $\operatorname{Dom} \widetilde{\ell}$ of X.

It then follows that

$$\operatorname{Dom} \widetilde{\ell} = X,$$

which implies that the linear functional $\widetilde{\ell} \in \mathcal{F}$ possesses all the desired properties. For, if $\operatorname{Dom} \widetilde{\ell} \subsetneqq X$, the same construction as in (i) would produce a linear functional $\widetilde{f} : \operatorname{Dom} \widetilde{f} \to \mathbb{R}$ that satisfies

$$\operatorname{Dom} \widetilde{\ell} \subsetneqq \operatorname{Dom} \widetilde{f} \subset X, \quad \widetilde{f}(y) = \ell(y) \quad \text{for all } y \in Y, \quad \text{and} \quad \widetilde{f}(x) \leq p(x) \quad \text{for all } x \in \operatorname{Dom} \widetilde{f},$$

in contradiction with the maximal character of $\widetilde{\ell}$. Hence $\operatorname{Dom} \widetilde{\ell} = X$. □

Clearly, *sublinear functionals* (as defined in Theorem 5.8-1) include *norms* and *seminorms* as examples. But they are more general, since the scalar multiplication property $p(\alpha x) = |\alpha|\, p(x)$ need only hold for $\alpha > 0$. An example of a sublinear functional that is not necessarily a seminorm is provided by the *Minkowski functional* encountered in the proof of the geometric form of the Hahn–Banach theorem (Theorem 5.10-1).

Problem

5.8-1 (Hahn–Banach theorem in a complex vector space[23]) Let X be a *complex* vector space and let $p : X \to \mathbb{R}$ be a *seminorm* on X (hence a more restrictive assumption than in Theorem 5.8-1, where $p : X \to \mathbb{R}$ was only assumed to be a sublinear functional). Let Y be a subspace of X and let $\ell : Y \to \mathbb{C}$ be a linear functional on Y that satisfies $|\ell(y)| \leq p(y)$ for all $y \in Y$.

Show that there exists a linear functional $\widetilde{\ell} : X \to \mathbb{C}$ on X that satisfies $\widetilde{\ell}(y) = \ell(y)$ for all $y \in Y$ and $|\widetilde{\ell}(x)| \leq p(x)$ for all $x \in X$.

Hint: For each $y \in Y$, write $\ell(y)$ as $\ell(y) = \operatorname{Re}(\ell(y)) - i \operatorname{Re}(\ell(iy))$, and observe that a complex vector space is *a fortiori* a real vector space. Then apply Theorem 5.8-1 to the real linear functionals $y \in Y \to \operatorname{Re}(\ell(y)) \in \mathbb{R}$ and $y \in Y \to \operatorname{Re}(\ell(iy)) \in \mathbb{R}$.

---

[23] This theorem is due to:

H.F. Bohnenblust; A. Sobczyk [1938]: Extensions of functionals on complex linear spaces. *Bulletin of the American Mathematical Society* **44**, 91–93.

G.A. Soukhomlinoff [1938]: Über Fortsetzung von linearen Funktionalen in linearen komplexen Räumen und linearen Quaternionräumen, *Mathematiceskii Sbornik* **3**, 353–358.

5.9 The Hahn–Banach theorem in a normed vector space; first consequences

Recall (Section 3.5) that the notation X' designates the *dual space* of a normed vector space X, i.e., X' is the Banach space formed by all the *continuous linear functionals* $\ell : X \to \mathbb{K}$ defined on X, and that the norm of any $\ell \in X'$ is given by

$$\|\ell\|_{X'} = \sup_{\left\{\substack{x\in X\\ x\neq 0}\right.} \frac{|\ell(x)|}{\|x\|_X}.$$

Like its "vector space counterpart" (Theorem 5.8-1), the Hahn–Banach theorem in a *normed* vector space X (Theorem 5.9-1) asserts that, given a linear functional ℓ defined on a subspace Y of X, there exists a linear functional $\widetilde{\ell}$ that is an *extension* of ℓ to the whole space X and that shares with ℓ a specific property. This property took the form of the inequalities $\ell(y) \leq p(y)$ for all $y \in Y$ and $\widetilde{\ell}(x) \leq p(x)$ for all $x \in X$ in Theorem 5.8-1; it will now take the form of the relation $\|\ell\|_{Y'} = \|\widetilde{\ell}\|_{X'}$.

It is to be emphasized that *all the theorems in this section hold* verbatim *in the real as well as in the complex cases.*

In what follows, notations such as ℓ or ℓ_x will be usually preferred for designating a particular element of the dual space X', while the notation x' will be usually preferred for designating a generic element of X'.

Theorem 5.9-1 (Hahn–Banach theorem in a normed vector space) *Let X be a normed vector space, let Y be a subspace of X, and let $\ell : Y \to \mathbb{K}$ be a continuous linear functional.*

Then there exists a continuous linear functional $\widetilde{\ell} : X \to \mathbb{K}$ that satisfies

$$\widetilde{\ell}(y) = \ell(y) \text{ for all } y \in Y \quad \text{and} \quad \|\widetilde{\ell}\|_{X'} = \|\ell\|_{Y'}.$$

Proof Let X be a *real* normed vector space. The function $p : X \to \mathbb{R}$ defined by $p(x) := \|\ell\|\,\|x\|$ for all $x \in X$, where $\|\ell\| := \|\ell\|_{Y'}$, is a norm (unless $\ell = 0$), hence a sublinear functional, on X; besides,

$$\ell(y) \leq \|\ell\|\,\|y\| = p(y) \quad \text{for all } y \in Y.$$

By the Hahn–Banach theorem in a real vector space (Theorem 5.8-1), there thus exists a linear functional $\widetilde{\ell} : X \to \mathbb{R}$ that satisfies

$$\widetilde{\ell}(y) = \ell(y) \text{ for all } y \in Y,$$
$$\widetilde{\ell}(x) \leq p(x) = \|\ell\|\,\|x\| \quad \text{and} \quad -\widetilde{\ell}(x) = \widetilde{\ell}(-x) \leq p(-x) = \|\ell\|\,\|x\| \quad \text{for all } x \in X.$$

Hence $|\widetilde{\ell}(x)| \leq \|\ell\|\,\|x\|$ for all $x \in X$. Consequently, the linear functional $\widetilde{\ell}$ is continuous; besides

$$\|\ell\| = \sup_{\left\{\substack{y\in Y\\ y\neq 0}\right.} \frac{|\ell(y)|}{\|y\|} \leq \sup_{\left\{\substack{x\in X\\ x\neq 0}\right.} \frac{|\widetilde{\ell}(x)|}{\|x\|} = \|\widetilde{\ell}\|_{X'} \leq \|\ell\|.$$

Hence $\|\widetilde{\ell}\|_{X'} = \|\ell\|_{Y'}$.

The proof in the *complex* case is left as a problem (Problem 5.9-1). □

Remark The Hahn–Banach theorem in a *Hilbert space* can be proved in a much simpler way, by means of the direct sum theorem and of the F. Riesz representation theorem, which provides in addition the uniqueness of the extension (Theorem 4.7-1). In particular, a recourse to the axiom of choice is no longer needed in this case. □

It should be emphasized that such a norm-preserving extension $\widetilde{\ell}$ is *not necessarily unique*. For instance, let $X = \mathcal{P}[0,1]$ equipped with the sup-norm $\|\cdot\|$, let $Y = \mathcal{P}_3[0,1]$, and let the linear functional $\ell : Y \to \mathbb{R}$ be defined by

$$\ell(p) = \frac{1}{6}\Big(p(0) + 4p\Big(\frac{1}{2}\Big) + p(1)\Big) \quad \text{for all } p \in \mathcal{P}_3[0,1].$$

Then ℓ is continuous, with $\|\ell\| = \sup_{\left\{\substack{p\in\mathcal{P}_3[0,1]\\ p\neq 0}\right.} \dfrac{|\ell(p)|}{\|p\|} = 1.$

It is then immediately verified that the *distinct* linear forms $\widetilde{\ell}_1 : X \to \mathbb{R}$ and $\widetilde{\ell}_2 : X \to \mathbb{R}$ defined by

$$\widetilde{\ell}_1(f) = \frac{1}{6}\Big(f(0) + 4f\Big(\frac{1}{2}\Big) + f(1)\Big) \quad \text{and} \quad \widetilde{\ell}_2(f) = \int_0^1 f(t)\,\mathrm{d}t \ \text{ for all } f \in \mathcal{P}[0,1],$$

satisfy

$$\widetilde{\ell}_\alpha(p) = \ell(p) \ \text{ for all } p \in \mathcal{P}_3[0,1] \quad \text{and} \quad \|\widetilde{\ell}_\alpha\| = \sup_{\left\{\substack{f\in\mathcal{C}[0,1]\\ f\neq 0}\right.} \frac{|\widetilde{\ell}_\alpha(f)|}{\|f\|} = 1 \ \text{ for } \alpha = 1, 2.$$

There is, however, a large class of normed vector spaces in which norm-preserving extensions of continuous linear functionals are *unique*, according to the following result. Recall that a real or complex normed vector space is *strictly convex* (Section 2.17) if

$$x \neq y \quad \text{and} \quad \|x\| = \|y\| = 1 \quad \text{implies} \ \Big\|\frac{x+y}{2}\Big\| < 1.$$

Theorem 5.9-2 (Taylor–Foguel theorem[24]) *Let X be a normed vector space. Then all the continuous linear functionals defined on subspaces of X have a unique norm-preserving extension to X if and only if the dual space X' of X is strictly convex.*

[24] A.E. TAYLOR [1939]: The extension of linear functionals, *Duke Mathematical Journal* **5**, 538–547.

S.R. FOGUEL [1958]: On a theorem of A.E. Taylor, *Proceedings of the American Mathematical Society* **9**, 325.

The simple proof given here is adapted from:

P.R. BEESACK; E. HUGHES; M. ORTEL [1979]: Rotund complex linear spaces, *Proceedings of the American Mathematical Society* **75**, 42–44.

The Taylor–Foguel theorem can be also derived as a consequence of the more general **Phelps theorem**, due to:

R. PHELPS [1960]: Uniqueness of Hahn–Banach extensions and unique best approximation, *Transactions of the American Mathematical Society* **95**, 238–255.

Proof We give the proof in the real case, leaving the complex one as a problem (Problem 5.9-3).

(i) *Assume that there exists a subspace Y of X and continuous linear functionals $\ell \in Y'$ and $\ell_1, \ell_2 \in X'$ such that*

$$\ell_1(y) = \ell_2(y) = \ell(y) \text{ for all } y \in Y, \quad \|\ell_1\|_{X'} = \|\ell_2\|_{X'} = \|\ell\|_{Y'} = 1, \quad \text{and } \ell_1 \neq \ell_2$$

(there is clearly no loss of generality in assuming that $\|\ell\|_{Y'} = 1$). Since then $\frac{1}{2}(\ell_1 + \ell_2)(y) = \ell(y)$ for all $y \in Y$, it follows that

$$1 = \|\ell\|_{Y'} = \left\|\frac{\ell_1 + \ell_2}{2}\right\|_{Y'} \leq \left\|\frac{\ell_1 + \ell_2}{2}\right\|_{X'} \leq 1,$$

which shows that X' *is not strictly convex.*

(ii) *Assume that all norm-preserving extensions to X of continuous linear functionals defined on subspaces of X are unique.*

Let $\ell_1, \ell_2 \in X'$ be such that

$$\|\ell_1\|_{X'} = \|\ell_2\|_{X'} = 1 \quad \text{and} \quad \ell_1 \neq \ell_2.$$

Then

$$Y := \{y \in X;\ \ell_1(y) = \ell_2(y)\}$$

is a proper subspace of X, and the continuous linear functional $\ell \in Y'$ defined by $\ell(y) := \ell_1(y) = \ell_2(y)$ for all $y \in Y$ is such that

$$\|\ell\|_{Y'} < 1.$$

To see this, note that $\|\ell\|_{Y'} \leq \|\ell_1\|_{X'} = 1$ and that, if $\|\ell\|_{Y'} = 1$, then ℓ_1 and ℓ_2 would be equal, a contradiction.

By assumption, there exists a unique $\widetilde{\ell} \in X'$ such that

$$\widetilde{\ell}(y) = \ell(y) = \ell_1(y) = \ell_2(y) \text{ for all } y \in Y \quad \text{and} \quad \|\widetilde{\ell}\|_{X'} = \|\ell\|_{Y'} < 1.$$

Let $x_0 \in X - Y$. Since then $\ell_1(x_0) \neq \ell_2(x_0)$, there exist $\lambda = \lambda(x_0) \in \mathbb{R}$ and $\mu = \mu(x_0) \in \mathbb{R}$ such that

$$\lambda\ell_1(x_0) + \mu\ell_2(x_0) = \widetilde{\ell}(x_0) \quad \text{and} \quad \lambda + \mu = 1.$$

Consequently, $\widetilde{\ell}(x_0) = \lambda\ell_1(x_0) + (1 - \lambda)\ell_2(x_0)$, and thus

$$\widetilde{\ell} = \lambda\ell_1 + (1 - \lambda)\ell_2,$$

since each $x \in X$ can be written as $x = y + \alpha x_0$ for some $y \in Y$ and $\alpha \in \mathbb{R}$.

We then claim that $\lambda \in \,]0,1[$. For, if $\lambda \geq 1$, then $\ell_1 = \frac{1}{\lambda}\widetilde{\ell} + \frac{\lambda - 1}{\lambda}\ell_2$ would imply $\|\ell_1\|_{X'} < 1$ a contradiction; while, if $\lambda \leq 0$, then $\ell_2 = \frac{1}{1-\lambda}\widetilde{\ell} - \frac{\lambda}{1-\lambda}\ell_1$ would imply $\|\ell_2\|_{X'} < 1$, also a contradiction. Hence, $\lambda \in \,]0,1[$ as claimed.

If $0 < \lambda < \frac{1}{2}$, then $\frac{\ell_1 + \ell_2}{2} = \frac{1}{2(1-\lambda)}\widetilde{\ell} + \frac{1-2\lambda}{2(1-\lambda)}\ell_1$ implies $\left\|\frac{\ell_1+\ell_2}{2}\right\|_{X'} < 1$; if $\frac{1}{2} < \lambda < 1$, then $\frac{\ell_1+\ell_2}{2} = \frac{1}{2\lambda}\widetilde{\ell} + \left(1 - \frac{1}{2\lambda}\right)\ell_2$ implies $\left\|\frac{\ell_1+\ell_2}{2}\right\|_{X'} < 1$; finally, if $\lambda = \frac{1}{2}$, then $\left\|\frac{\ell_1+\ell_2}{2}\right\|_{X'} = \|\widetilde{\ell}\| < 1$. Hence X *is strictly convex.* □

Remark If X is a *Hilbert space*, the corresponding "if" part of the Taylor–Foguel theorem has already been established, by means of the *F. Riesz representation theorem* (Theorem 4.7-1). □

Thanks to the Hahn–Banach theorem in a normed vector space, we can now answer (positively) the question regarding the *existence of nonzero continuous linear functionals defined on an arbitrary normed vector space* X.

Theorem 5.9-3 *Let $X \neq \{0\}$ be a normed vector space. Given any nonzero vector $x \in X$, there exists $\ell_x \in X'$ such that*

$$\ell_x(x) = \|x\| \quad \textit{and} \quad \|\ell_x\|_{X'} = 1.$$

Consequently, the dual space X' contains nonzero elements.

Proof Let

$$Y := \{\alpha x \in X;\ \alpha \in \mathbb{K}\} = \operatorname{Span}(x) \quad \text{and} \quad \ell(\alpha x) := \alpha\|x\| \text{ for all } \alpha \in \mathbb{K},$$

so that the function $\ell : Y \to \mathbb{R}$ defined in this fashion is a continuous linear functional on the subspace Y of X, with

$$\|\ell\|_{Y'} = \sup_{\alpha \neq 0} \frac{|\ell(\alpha x)|}{\|\alpha x\|} = 1.$$

Then Theorem 5.9-1 shows that there exists a continuous linear functional $\ell_x \in X'$ that satisfies

$$\ell_x(\alpha x) = \ell(\alpha x) = \alpha\|x\| \quad \text{for all } \alpha \in \mathbb{R} \text{ and } \|\ell_x\| = \|\ell\| = 1.$$

Since then $\ell(x) = \|x\|$, the functional ℓ_x thus possesses the announced properties. □

By Theorem 5.9-3, given any $x_0 \in X$ with $\|x_0\| = 1$, there exists $x' \in X'$ such that $x'(x_0) = 1$ and $\|x'\| = 1$. Hence, for such $x' \in X'$,

$$\|x'\| = \sup_{\|x\|=1} |x'(x)| = |x'(x_0)|,$$

i.e., *the supremum defining the norm of x' is attained.* That there are in fact "many" such continuous linear functionals is the content of another *basic theorem of linear functional analysis.*

$^\flat$**Theorem 5.9-4 (Bishop–Phelps theorem[25])** *Let X be a real Banach space, and let*

$$Y' := \{x' \in X';\ \textit{there exists } x_0 \textit{ such that } \|x_0\| = 1 \textit{ and } \sup_{\|x\|=1} |x'(x)| = |x'(x_0)|\}.$$

[25] E. Bishop; R.R. Phelps [1961]: A proof that every Banach space is subreflexive, *Bulletin of the American Mathematical Society* **67**, 97–98.

Then Y' is dense in X'. □

Another remarkable consequence of the Hahn–Banach theorem in a normed vector space X (through its corollary, Theorem 5.9-3) is that the norm $\|x\|$ of any element $x \in X$ is given by a formula that is reciprocal to the formula

$$\|x'\| = \sup_{\left\{\substack{x \in X \\ x \neq 0}\right.} \frac{|x'(x)|}{\|x\|} \quad \text{for any } x' \in X'$$

("reciprocal" in the sense that the roles of X and X' are interchanged). This formula in turn answers the question regarding whether *the dual space X' contains "enough" elements x' to guarantee that $x'(x) = 0$ for all $x' \in X'$ implies $x = 0$.* Note that the existence of nonzero elements $x' \in X'$ (Theorem 5.9-3) insures that the supremum appearing in the next theorem is well defined.

Theorem 5.9-5 *Let $X \neq \{0\}$ be a normed vector space. Then the norm of any vector $x \in X$ is given by*

$$\|x\| = \sup_{\left\{\substack{x' \in X' \\ x' \neq 0}\right.} \frac{|x'(x)|}{\|x'\|} = \sup_{\left\{\substack{x' \in X' \\ \|x'\|=1}\right.} |x'(x)|.$$

Consequently, if $x \in X$ is such that $x'(x) = 0$ for all $x' \in X'$, then $x = 0$.

Proof Given any $x \in X$, let $\ell_x \in X'$ be determined as in Theorem 5.9-3, so that $\ell_x(x) = \|x\|$ and $\|\ell_x\| = 1$. Therefore,

$$\|x\| = \ell_x(x) \leq \sup_{\left\{\substack{x' \in X' \\ \|x'\|=1}\right.} |x'(x)| \leq \|x\|.$$ □

Remark In a *Hilbert space* $(X, (\cdot,\cdot))$, any element of X can be identified with an element of the dual space X' by means of the F. Riesz isometry (Theorem 4.6-1). This observation shows that the relation $\|x\| = \sup_{\left\{\substack{x' \in X' \\ x' \neq 0}\right.} \frac{|x'(x)|}{\|x'\|}$ established in Theorem 5.9-5 is equivalent in this case to the relation $\|x\| = \sup_{y \neq 0} \frac{|(x,y)|}{\|y\|}$ (which holds in fact in *any* inner-product space, complete or not; cf. Theorem 4.1-1). □

Another important consequence of the Hahn–Banach theorem in a normed vector space is a simple sufficient condition for the *separability* of a normed vector space, which will be

This result can be extended to *complex* Banach spaces that possess the "Radon–Nikodym property" as shown in:

J. Bourgain [1977]: On dentability and the Bishop-Phelps property, *Israel Journal of Mathematics* **28**, 268–271.

In this direction, another deep theorem asserts that, if a real Banach space X is such that $Y' = X'$ where Y' is defined as in Theorem 5.9-4, then X is *reflexive* (this notion will be defined in Section 5.14); this result is due to:

R.C. James [1972]: Reflexivity and the sup of linear functionals, *Israel Journal of Mathematics* **13**, 289–301.

established in Theorem 5.9-8. To this end, we first need to prove an interesting *per se* generalization of Theorem 5.9-3 (which can be regarded as the special case $Y = \{0\}$ in the next theorem), which shows in particular that, if Y is a *closed* and *strict* subspace of a normed vector space X, there exists a nonzero continuous functional on X that vanishes on Y.

Theorem 5.9-6 *Let X be a normed vector space and let Y be a closed subspace of X such that $Y \subsetneqq X$. Given any element $x \in X - Y$, there exists $\ell_x \in X'$ such that*

$$\ell_x(y) = 0 \textit{ for all } y \in Y, \quad \ell_x(x) = \inf_{y\in Y} \|x - y\| > 0, \quad \textit{and} \quad \|\ell_x\| = 1.$$

Proof Define the subspace Z of X and the function $\ell : Z \to \mathbb{K}$ by

$$Z := \{(\alpha x + y) \in X;\ \alpha \in \mathbb{K},\ y \in Y\},$$
$$\ell(\alpha x + y) := \alpha\delta \quad \text{for all } \alpha \in \mathbb{K} \text{ and } y \in Y, \text{ where } \delta := \inf_{y\in Y} \|x - y\| > 0.$$

Clearly, ℓ is a linear functional on Z that satisfies $\ell(x) = \delta$ and $\ell(y) = 0$ for all $y \in Y$. We claim that, furthermore, ℓ is continuous and $\|\ell\|_{Z'} = 1$. To see this, we first note that, by definition of δ, any element $(\alpha x + y) \in Z$ with $\alpha \neq 0$ satisfies

$$\|\alpha x + y\| = |\alpha|\, \|x + \alpha^{-1}y\| \geq |\alpha|\delta,$$

since $\alpha^{-1}y \in Y$, so that $|\ell(\alpha x + y)| = |\alpha|\delta \leq \|\alpha x + y\|$. Noting that this inequality also holds for $\alpha = 0$, we thus have

$$\|\ell\|_{Z'} = \sup_{\left\{\substack{\alpha\in\mathbb{K}, y\in Y\\ \alpha x+y\neq 0}\right.} \frac{|\ell(\alpha x + y)|}{\|\alpha x + y\|} \leq 1.$$

Second, we note that, for any $\varepsilon > 0$, there exists $y_\varepsilon \in Y$ such that $\delta \leq \|x - y_\varepsilon\| \leq \delta + \varepsilon$, again by definition of δ. Consequently, since $(x - y_\varepsilon) \in Z$ and $\ell(x - y_\varepsilon) = \ell(x) = \delta$,

$$1 \geq \|\ell\|_{Z'} = \sup_{\left\{\substack{z\in Z\\ z\neq 0}\right.} \frac{|\ell(z)|}{\|z\|} \geq \frac{|\ell(x - y_\varepsilon)|}{\|x - y_\varepsilon\|} \geq \frac{\delta}{\delta + \varepsilon}.$$

Hence $\|\ell\| = 1$ since $\varepsilon > 0$ is arbitrary.

The Hahn–Banach theorem in a normed vector space (Theorem 5.9-1) then shows that there exists a continuous linear functional $\ell_x \in X'$ that satisfies

$$\ell_x(z) = \ell(z) \text{ for all } z \in Z \quad \text{and} \quad \|\ell_x\| = \|\ell\| = 1.$$

In particular then, $\ell_x(x) = \ell(x) = \delta$ and $\ell_x(y) = \ell(y) = 0$ for all $y \in Y$. □

Remark A quick glance at its proof shows that Theorem 5.9-6 still holds for any $x \in (X - Y)$ such that $\inf_{y\in Y} \|x - y\| > 0$, even if the subspace Y is not closed. □

Theorem 5.9-6 is nothing but the *generalization of the projection theorem in an inner-product space* X (Theorem 4.3-1) *to an arbitrary normed vector space.* For, if Z is a *complete subspace* of X and $x \in X - Z$, the projection theorem asserts the existence of an element $Px \in Z$ such that $\|x - Px\| = \inf_{z \in Z} \|x - z\|$, furthermore characterized by the property that $(x - Px, z) = 0$ for all $z \in Z$. Then the element

$$\ell_x := \frac{x - Px}{\|x - Px\|} \in X,$$

identified here with an element $\ell_x \in X'$, thanks again to the F. Riesz representation theorem, satisfies exactly the same properties as those found in Theorem 5.9-6, viz.,

$$\|\ell_x\| = 1,$$
$$\ell_x(z) = \frac{1}{\|x - Px\|}(x - Px, z) = 0 \quad \text{for all } z \in Z,$$
$$\ell_x(x) = \frac{1}{\|x - Px\|}(x - Px, x) = \frac{1}{\|x - Px\|}(x - Px, x - Px + Px) = \|x - Px\|.$$

A first consequence of Theorem 5.9-6 is the following useful criterion for a *subspace* to be *dense* in *any* normed vector space.

Theorem 5.9-7 *Let Y be a subspace of a normed vector space X. Then $\overline{Y} = X$ if and only if*

$$\{x' \in X';\ x'(y) = 0 \text{ for all } y \in Y\} = \{0\}.$$

Proof If $\overline{Y} = X$ and $x' \in X'$ is such that $x'(y) = 0$ for all $y \in Y$, then $x'(y) = 0$ for all $y \in \overline{Y} = X$ (since $x' : X \to \mathbb{K}$ is continuous). Hence $y = 0$.

If $\overline{Y} \subsetneqq X$, then by Theorem 5.9-6, there exists a *nonzero* continuous linear form $x' \in X'$ such that $x'(y) = 0$ for all $y \in \overline{Y}$, and hence for all $y \in Y$. Therefore, $\{x' \in X';\ x'(y) = 0 \text{ for all } y \in Y\} \supsetneqq \{0\}$ in this case. □

Remark If X is a *Hilbert space*, the corresponding "if" part is an immediate consequence of the F. Riesz representation theorem combined with the *projection theorem* (Theorem 4.3-2). □

As exemplified by the spaces ℓ^1 and $L^1(\Omega)$, *the dual of a separable normed vector space is not necessarily separable*, since the duals of these spaces, which can be respectively identified with the spaces ℓ^∞ and $L^\infty(\Omega)$, are for this reason not separable (Theorems 2.4-2, 2.5-4, 3.5-1, and 3.5-3). *But the converse holds*, thanks to Theorem 5.9-6:

Theorem 5.9-8 (sufficient condition for separability) *If the dual space X' of a normed vector space X is separable, then X is separable.*

Proof (i) *Let $S' := \{x' \in X';\ \|x'\| = 1\}$. Then there exist elements $x'_n \in S'$, $n \geq 1$, such that $\overline{\bigcup_{n=1}^{\infty}\{x'_n\}} = S'$.*

Let $\varepsilon > 0$ and $x' \in S'$ be given. Since X' is *separable* by assumption, there exist elements $\widetilde{x}'_n \in X'$, $n \geq 1$, such that $\overline{\bigcup_{n=1}^{\infty}\{\widetilde{x}'_n\}} = X'$. Given any $0 < \varepsilon < 2$ and any $x' \in S'$, there thus exists an integer $n \geq 1$ such that $\|\widetilde{x}'_n - x'\| \leq \frac{\varepsilon}{2}$. Consequently,

$$\big|\|\widetilde{x}'_n\| - 1\big| = \big|\|\widetilde{x}'_n\| - \|x'\|\big| \leq \|\widetilde{x}'_n - x'\| \leq \frac{\varepsilon}{2},$$

and thus $0 < 1 - \frac{\varepsilon}{2} \leq \|\widetilde{x}'_n\|$. Let $x'_n := \frac{\widetilde{x}'_n}{\|\widetilde{x}'_n\|}$; then

$$\|x' - x'_n\| \leq \|x' - \widetilde{x}'_n\| + \|\widetilde{x}'_n - x'_n\| = \|x' - \widetilde{x}'_n\| + \left(\|\widetilde{x}'_n\| - 1\right)\|x'_n\| \leq \varepsilon.$$

This shows that $\overline{\bigcup_{n=1}^{\infty}\{x'_n\}} = S'$.

(ii) Let the functionals $x'_n \in S' \subset X'$, $n \geq 1$, be those found in (i). Since $\|x'_n\| = \sup_{\|x\|=1}|x'_n(x)|$, there exists $x_n \in X$ such that

$$\|x_n\| = 1 \quad \text{and} \quad \frac{1}{2} \leq |x'_n(x_n)| \quad \text{for each } n \geq 1.$$

We will then show that

$$\overline{Y} = X, \quad \text{where } Y := \operatorname{Span}(x_n)_{n=1}^{\infty},$$

thus proving that X is separable, since this relation implies that finite linear combinations with rational coefficients of the vectors x_n, $n \geq 1$, already form a dense subset of X.

To this end, we proceed by contradiction. If $Y \subsetneq X$, let $x \in X - Y$. Then, by Theorem 5.9-6, there exists $\ell_x \in X'$ such that

$$\ell_x(y) = 0 \text{ for all } y \in Y \quad \text{and} \quad \|\ell_x\| = 1$$

(the property $\ell_x(x) = \inf_{y \in Y}\|x - y\|$ is not needed here). Hence $\ell_x \in S'$ and

$$\frac{1}{2} \leq |x'_n(x_n)| = |x'_n(x_n) - \ell_x(x_n)| \leq \|x'_n - \ell_x\| \quad \text{for all } n \geq 1,$$

contradicting the denseness of $\bigcup_{n=1}^{\infty}\{x'_n\}$ in S' established in (i). □

Remark Naturally, a property analogous to (i) holds in *any* separable normed vector space (i.e., whether or not it is a dual space). □

Problems

5.9-1 Prove Theorem 5.9-1 when X is a *complex* normed vector space.
Hint: Apply the Hahn–Banach theorem in a *complex* vector space (Problem 5.8-1).

5.9-2 Let X be a normed vector space, let Y be a strict subspace of X, and let $\ell : Y \to \mathbb{K}$ be a continuous linear functional. Show that there exist continuous linear functionals $\widetilde{\ell} : X \to \mathbb{K}$ that satisfy

$$\widetilde{\ell}(y) = \ell(y) \quad \text{for all } y \in Y \text{ and } \|\widetilde{\ell}\|_{X'} > \|\ell\|_{Y'}.$$

5.9-3 Prove the Taylor–Foguel theorem (Theorem 5.9-2) in the complex case.
Hint: First, show that a complex normed vector space X is strictly convex if and only if, for each $x, y \in X$ such that $x \neq y$ and $\|x\| = \|y\| = 1$, there exists $\lambda \in \mathbb{C}$ such that $\|\lambda x + (1 - \lambda)y\| < 1$ (use Theorem 5.9-2 for the "if" part). Then adapt the proof given in the text to the complex case.

5.10 Geometric forms of the Hahn–Banach theorem; separation of convex sets

Given a *real* normed vector space X, a nonzero continuous linear functional $\ell : X \to \mathbb{R}$, and a number $\gamma \in \mathbb{R}$, the set

$$\{x \in X;\ \ell(x) = \gamma\}$$

is called a **closed affine hyperplane** and the set $\{x \in X;\ \ell(x) \geq \gamma\}$, *resp.* $\{x \in X;\ \ell(x) > \gamma\}$, is called a **closed**, *resp.* **open, half-space** (complements on hyperplanes and their relation to linear functionals are given in Problems 5.10-1 and 5.10-2).

Two subsets A and B of X are said to be **separated by a hyperplane** if there exist a nonzero $\ell \in X'$ and $\gamma \in \mathbb{R}$ such that

$$A \subset \{x \in X;\ \ell(x) \leq \gamma\} \quad \text{and} \quad B \subset \{y \in X;\ \gamma \leq \ell(y)\},$$

i.e., if they are separately contained in two *closed half-spaces*, the intersection of which is the closed affine hyperplane $\{y \in X;\ \ell(y) = \gamma\}$ (Figure 5.10-1).

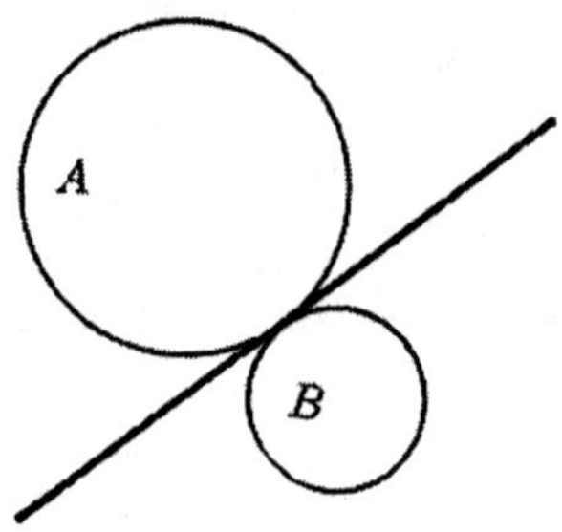

Figure 5.10-1 Two subsets A and B of $\mathbb{R}^2$ separated by a hyperplane (a straight line in this case).

The next theorem, which crucially hinges on the *Hahn–Banach theorem in a vector space* (Theorem 5.8-1), gives sufficient conditions for two subsets of a *real* normed vector space to be separated by a hyperplane. For the extension to the *complex* case, see Problem 5.10-3(1).

Theorem 5.10-1 (first geometric form of the Hahn–Banach theorem: Separation of convex sets) *Let A and B be two nonempty subsets of a real normed vector space X, with the following properties:*

$$A \text{ is convex and open}; \quad B \text{ is convex}; \quad A \cap B = \phi.$$

Then there exist a nonzero $\ell \in X'$ and $\gamma \in \mathbb{R}$ such that (*Figure* 5.10-1)

$$\ell(x) \leq \gamma \leq \ell(y) \quad \text{for all } x \in A,\ y \in B.$$

Proof (i) *Let C be a nonempty, convex, open subset of X containing the origin. Then the function $p : X \to [0, \infty[$ defined by*

$$p(x) := \inf \left\{ \beta > 0;\ \frac{x}{\beta} \in C \right\} \quad \text{for each } x \in X$$

possesses the following four properties:

$$\text{there exists a constant } M \text{ such that } 0 \leq p(x) \leq M\|x\| \quad \text{for all } x \in X,$$
$$C = \{x \in X;\ p(x) < 1\},$$
$$p(\alpha x) = \alpha p(x) \quad \text{for all } \alpha \geq 0 \text{ and all } x \in X,$$
$$p(x + y) \leq p(x) + p(y) \quad \text{for all } x, y \in X.$$

Since C is open and contains the origin, there exists $r > 0$ such that $B(0; r) \subset C$. Given any $x \in X$ and any $\beta > \frac{\|x\|}{r}$, the point $\frac{x}{\beta}$ belongs to $B(0; r)$ (since then $\|\frac{x}{\beta}\| < r$), and thus $\frac{x}{\beta} \in C$. Hence

$$p(x) = \inf \left\{ \beta > 0;\ \frac{x}{\beta} \in C \right\} \leq \inf \left\{ \beta > \frac{\|x\|}{r} \right\} = \frac{\|x\|}{r}.$$

Therefore $0 \leq p(x) \leq M\|x\|$ for all $x \in X$, with $M := \frac{1}{r}$.

Given any $x \in C$, there exists $\delta > 0$ such that $(1 + \delta)x \in C$ since C is open. So,

$$p(x) = \inf \left\{ \beta > 0;\ \frac{x}{\beta} \in C \right\} \leq \frac{1}{1 + \delta} < 1.$$

Conversely, let $x \in X$ be such that $p(x) < 1$, which means that there exists $0 < \beta < 1$ such that $\frac{x}{\beta} \in C$. Since C is convex and contains the origin, $x = \left(\beta \frac{x}{\beta} + (1 - \beta)0\right) \in C$. Therefore $C = \{x \in X;\ p(x) < 1\}$.

Given any $\alpha > 0$ and any $x \in X$,

$$p(\alpha x) = \inf \left\{ \widetilde{\beta} > 0;\ \frac{\alpha x}{\widetilde{\beta}} \in C \right\} = \inf \left\{ \alpha\beta > 0;\ \frac{\alpha x}{\alpha\beta} = \frac{x}{\beta} \in C \right\} = \alpha p(x).$$

Besides, $p(0) = 0$. Hence $p(\alpha x) = \alpha p(x)$ for all $\alpha \geq 0$ and all $x \in X$.

Finally, let two points $x, y \in X$ and $\varepsilon > 0$ be given. By the above properties,

$$p\left(\frac{x}{p(x) + \varepsilon}\right) = \frac{p(x)}{p(x) + \varepsilon} < 1.$$

Hence $\left(\frac{x}{p(x) + \varepsilon}\right) \in C$; and likewise $\left(\frac{y}{p(y) + \varepsilon}\right) \in C$. The convexity of C then implies that

$$\left(\frac{\mu}{p(x) + \varepsilon} x + \frac{1 - \mu}{p(y) + \varepsilon} y\right) \in C \quad \text{for all } 0 < \mu < 1.$$

Noting that the particular choice $\mu := \frac{p(x) + \varepsilon}{p(x) + p(y) + \varepsilon}$ implies that

$$\frac{1}{p(x) + p(y) + 2\varepsilon}(x + y) \in C,$$

we conclude that

$$p(x+y) < p(x) + p(y) + 2\varepsilon.$$

Consequently, $p(x+y) \leq p(x) + p(y)$ since $\varepsilon > 0$ is arbitrary.

The function p thus possesses all the announced properties. In particular, the last two properties show that p is a *sublinear functional* (Section 5.8).

(ii) *Let C be a nonempty, convex, open subset of X, and let $y_0 \notin C$. Then there exists $\ell \in X'$ such that*

$$\ell(x) < \ell(y_0) \quad \text{for all } x \in C$$

(note in passing that (ii) is in effect a special case of Theorem 5.10-1, with $A := \{y_0\}$ and $B := C$).

Assume first that $O \in C$. Let then the function $p : X \to [0, \infty[$ be defined as in (i), let $Y := \text{Span}\{y_0\}$ (the vector y_0 is nonzero since $O \in C$ and $y_0 \notin C$), and let $\ell_0 : Y \to \mathbb{R}$ be the linear functional defined by $\ell_0(\alpha y_0) = \alpha$ for each $\alpha \in \mathbb{R}$. Then

$$\ell_0(y) \leq p(y) \quad \text{for all } y \in Y,$$

since

$$\ell_0(\alpha y_0) = \alpha \leq \alpha p(y_0) = p(\alpha y_0) \quad \text{for all } \alpha \geq 0$$

(by (i), $p(y_0) \geq 1$ since $y_0 \notin C$), and

$$\ell(\alpha y_0) = \alpha \leq 0 \leq p(\alpha_0 y) \quad \text{for all } \alpha \leq 0$$

(by (i), $p(y) \geq 0$ for all $y \in X$).

Since p is a sublinear functional by (i), the *Hahn–Banach theorem in a real vector space* (Theorem 5.8-1) shows that there exists a *nonzero linear functional* $\ell : X \to \mathbb{R}$ such that

$$\ell(y_0) = \ell_0(y_0) = 1 \quad \text{and} \quad \ell(x) \leq p(x) \quad \text{for all } x \in X.$$

Furthermore, the inequality $p(x) \leq M\|x\|$ for all $x \in X$ established in (i) implies that

$$\ell(x) \leq p(x) \leq M\|x\| \quad \text{and} \quad -\ell(x) \leq p(-x) \leq M\|x\| \quad \text{for all } x \in X.$$

Hence the linear functional ℓ is continuous, i.e., $\ell \in X'$. Besides,

$$\ell(x) \leq p(x) < 1 = \ell_0(y_0) = \ell(y) \quad \text{for all } x \in C$$

(by (i), $p(x) < 1$ for all $x \in C$). Hence the assertion is proved if $0 \in C$.

Assume next that $0 \notin C$. Chose any point $x_0 \in C$ and let

$$\widetilde{C} := \{(x - x_0) \in X;\ x \in C\} \quad \text{and} \quad \widetilde{y}_0 := y_0 - x_0.$$

Since $0 \in \widetilde{C}$ and $\widetilde{y}_0 \notin \widetilde{C}$, the above argument shows that there exists $\ell \in X'$ such that

$$\ell(\widetilde{x}) < \ell(\widetilde{y}_0) \quad \text{for all } \widetilde{x} \in \widetilde{C},$$

and hence such that $\ell(x) < \ell(y_0)$ for all $x \in C$, since ℓ is linear.

(iii) Finally, *let A be a nonempty, convex, open subset of X and let B be a nonempty convex subset of X such that $A \cap B = \varnothing$* (as in the statement of the theorem).

Define the set

$$C := \bigcup_{y \in B} \{(x - y) \in X;\ x \in A\},$$

which is open as a union of open sets. It is also convex: let $z_i = (x_i - y_i) \in C$ with $x_i \in A$ and $y_i \in B$ for $i = 1, 2$. Since

$$(\mu x_1 + (1-\mu)x_2) \in A \quad \text{and} \quad (\mu y_1 + (1-\mu)y_2) \in B \quad \text{for all } 0 < \mu < 1$$

(both sets A and B are convex), it follows that

$$\mu z_1 + (1-\mu)z_2 = ((\mu x_1 + (1-\mu)x_2) - (\mu y_1 + (1-\mu)y_2)) \in C \quad \text{for all } 0 < \mu < 1.$$

Finally, $0 \notin C$ since $A \cap B = \varnothing$.

By (ii), there thus exists a nonzero $\ell \in X'$ such that $\ell(z) < \ell(0) = 0$ for all $z \in C$, or equivalently for any $z = (x - y)$ with $x \in A$, $y \in B$. In other words,

$$\ell(x) < \ell(y) \quad \text{for all } x \in A \text{ and all } y \in B.$$

Therefore there exists $\gamma \in \mathbb{R}$ such that

$$\ell(x) \le \gamma \le \inf_{y \in B} \ell(y) \quad \text{for all } x \in A,$$

which concludes the proof. □

The sublinear functional $p : X \to [0, \infty[$ defined in part (i) of the above proof is called the **Minkowski functional**, or the **gauge**, or the **support function**, of the convex set C.

Two subsets A and B of a *real* normed vector space X are said to be **strictly separated by a hyperplane** if there exist a nonzero continuous linear functional $\ell \in X'$ and numbers $\gamma \in \mathbb{R}$ and $\delta > 0$ such that (Figure 5.10-2)

$$\ell(x) \le \gamma - \delta \text{ for all } x \in A \quad \text{and} \quad \gamma + \delta \le \ell(y) \text{ for all } y \in B.$$

The next theorem gives sufficient conditions for two subsets of a *real* normed vector space to be *strictly* separated by a hyperplane. For the extension to the *complex* case, see Problem 5.10-3(2).

Theorem 5.10-2 (second geometric form of the Hahn–Banach theorem: Strict separation of convex sets) *Let A and K be two nonempty subsets of a real normed vector space X, with the following properties:*

$$A \text{ is convex and closed;} \quad K \text{ is convex and compact;} \quad A \cap K = \varnothing.$$

Then there exist a nonzero $\ell \in X'$, $\gamma \in \mathbb{R}$, and $\delta > 0$ such that

$$\ell(x) \le \gamma - \delta < \gamma + \delta \le \ell(y) \quad \text{for all } x \in A,\ y \in K.$$

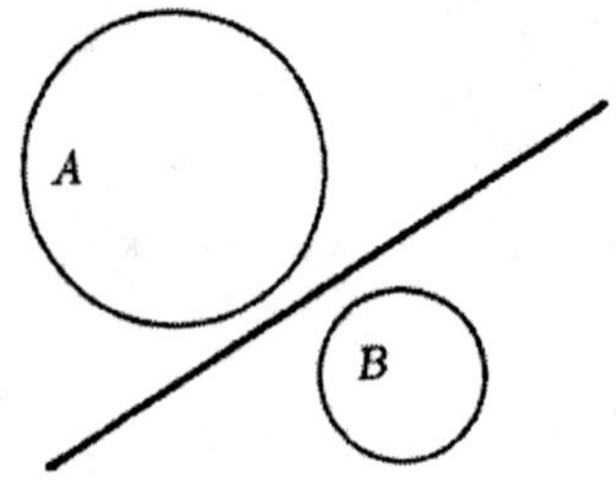

Figure 5.10-2 Two subsets A and B of $\mathbb{R}^2$ strictly separated by a hyperplane (a straight line in this case).

Proof For any $r > 0$, let

$$A(r) := \bigcup_{x \in A} B(x; r) \quad \text{and} \quad K(r) := \bigcup_{y \in K} B(y; r).$$

Then both sets $A(r)$ and $K(r)$ are nonempty, convex, and open. Besides, there exists $r_0 > 0$ such that $A(r) \cap K(r) = \varnothing$ for all $r \le r_0$. To see this, assume on the contrary that there exist $x_n \in A$, $\widetilde{x}_n \in X$, $y_n \in K$, $\widetilde{y}_n \in X$, $n \ge 1$, such that

$$x_n + v_n = y_n + w_n \quad \text{for all } n \ge 1, \text{ and } v_n \to 0 \text{ and } w_n \to 0 \text{ as } n \to \infty.$$

The set K being compact, there exists a subsequence $(y_{\sigma(n)})_{n=1}^{\infty}$ that converges in K. Therefore the subsequence $(x_{\sigma(n)})_{n=1}^{\infty}$ also converges in X, and $\lim_{n\to\infty} x_{\sigma(n)} \in A$ since A is closed. Hence $\lim_{n\to\infty} x_{\sigma(n)} = \lim_{n\to\infty} y_{\sigma(n)} \in A \cap K$, a contradiction.

Therefore, by Theorem 5.10-1, the two sets $A(r_0)$ and $B(r_0)$ are separated by a hyperplane, i.e., there exist a nonzero $\ell \in X'$ and $\gamma \in \mathbb{R}$ such that

$$\ell(x+v) \le \gamma \le \ell(y+w) \quad \text{for all } x \in A,\ y \in K \text{ and all } v, w \in X \text{ with } \|w\| = r_0.$$

Hence

$$\ell(x)+r_0\|\ell\| = \ell(x) + \sup_{\|v\|=r_0} \ell(v) \le \gamma \le \ell(y) + \inf_{\|w\|=r_0} \ell(w) = \ell(y) - r_0\|\ell\| \quad \text{for all } x \in A,\ y \in B,$$

and the proof is complete ($\delta := r_0\|\ell\| > 0$ since $\ell \ne 0$). □

Problems

5.10-1 Let X be a vector space of dimension ≥ 2. A *hyperplane* in X is any subset of X of the form $\{x \in X;\ \ell(x) = 0\}$, where $\ell : X \to \mathbb{K}$ is a *nonzero* linear functional.

(1) Show that a subspace H of X is a hyperplane if and only if the quotient space H/X has dimension one.

(2) Let $\ell : X \to \mathbb{K}$ and $\widetilde{\ell} : X \to \mathbb{K}$ be two nonzero linear functionals. Show that $\{x \in X;\ \ell(x) = 0\} = \{x \in X;\ \widetilde{\ell}(x) = 0\}$ if and only if there exists $\alpha \ne 0$ such that $\ell = \alpha\widetilde{\ell}$.

5.10-2 Let X be a normed vector space. Show that a linear functional $\ell : X \to \mathbb{K}$ is continuous if and only if the hyperplane $\{x \in X;\ \ell(x) = 0\}$ is closed.

5.10-3 (geometric forms of the Hahn–Banach theorem in a complex vector space) (1) Let the assumptions be those of Theorem 5.10-1, save that X is now a *complex* normed vector space. Show that there exists a nonzero $\ell \in X'$ and $\gamma \in \mathbb{R}$ such that

$$\operatorname{Re}\ell(x) < \gamma \le \operatorname{Re}\ell(y) \quad \text{for all } x \in A,\ y \in B.$$

(2) Let the assumptions be those of Theorem 5.10-2, save that X is now a *complex* normed vector space. Show that there exists a nonzero $\ell \in X'$, $\gamma \in \mathbb{R}$, and $\delta > 0$ such that

$$\operatorname{Re}\ell(x) \le \gamma - \delta < \gamma + \delta \le \operatorname{Re}\ell(y) \quad \text{for all } x \in A,\ y \in K.$$

5.10-4 Let X be a normed vector space and Y a subspace of X. Show that $\overline{Y} = X$ if and only if the only continuous linear functional ℓ that satisfies $\ell(y) = 0$ for all $y \in Y$ is $\ell = 0$.
Hint: Use Theorem 5.10-2.

5.10-5 Let X be a normed vector space and let $f : X \to \mathbb{R}$ be a convex and continuous function. Show that there exists $\ell \in X'$ and $c \in \mathbb{R}$ such that $f(x) > \ell(x) + c$ for all $x \in X$.

5.11 Dual operators; Banach closed range theorem

Suppose that, given two infinite-dimensional normed linear spaces X and Y and a mapping $A \in \mathcal{L}(X; Y)$, we wish to decide whether, given any vector $y \in Y$, there exists a vector $x \in X$ that solves the linear equation

$$Ax = y.$$

While the issue of *uniqueness*, i.e., to decide whether $\operatorname{Ker} A = \{0\}$ or not, is usually easy to resolve, that of *existence*, i.e., to decide whether $\operatorname{Im} A = Y$ or to characterize $\operatorname{Im} A$ when $\operatorname{Im} A$ is a strict subspace of Y, is often one of considerable difficulty.

It turns out that remarkably simple, and very useful, necessary and sufficient conditions guaranteeing that $\operatorname{Im} A = Y$ or characterizing $\operatorname{Im} A$ can be found, not in terms of the operator A itself, but instead in terms of the *dual operator* A' of A, the continuous linear operator from Y' into X' defined in Theorem 5.11-1 below.

Such an operator is the natural generalization to arbitrary normed vector spaces of the *adjoint operator* of a continuous linear operator between two *Hilbert spaces*: let X and Y be two Hilbert spaces and let $\sigma : X' \to X$ and $\tau : Y' \to Y$ be the corresponding F. Riesz isometries (Theorem 4.6-1). Then it is immediately verified that the adjoint operator $A^* \in \mathcal{L}(Y; X)$ (as defined in Theorem 4.7-2) and the dual operator $A' \in \mathcal{L}(Y'; X')$ (as defined in Theorem 5.11-1 below) of $A \in \mathcal{L}(X; Y)$ are related by

$$A' = \sigma^{-1} A^* \tau.$$

These necessary and sufficient conditions together constitute the beautiful *Banach closed range theorem*, which derives its name from the fact that the direct image of a vector space under a linear operator is also called the *range* of this linear operator.

This theorem comprises two parts. The *first part* provides in particular a useful *characterization of the subspace* $\operatorname{Im} A$ *of* Y *in terms of the subspace* $\operatorname{Ker} A'$ of Y', under the assumption that $\operatorname{Im} A$ is *closed*; cf. Theorem 5.11-5(a) and (c). This characterization will be later the key to the proofs of the *Babuška–Brezzi theorem* (Theorem 6.12-1), of the existence of a solution to the *Stokes equations* (Theorem 6.14-1), and of the sufficiency of the *Donati conditions* (Theorems 6.19-4–6.19-6).

The *second part* provides in particular an equally useful necessary and sufficient condition that $\operatorname{Im} A = Y$, again in terms of the dual operator A', namely that A' be *injective* with a *closed image* $\operatorname{Im} A'$; cf. Theorem 5.11-6.

Incidentally, while indeed simple to *state*, this theorem is not simple to *prove*. In particular, its proof crucially hinges on the *Banach open mapping theorem*, the *Hahn–Banach theorem in a normed vector space*, and the *geometric form of the Hahn–Banach theorem.*

In the remainder of this section, elements in X' will be typically denoted x' (rather than ℓ) and, for brevity, shorter notations such as $A'y'(x)$ will be preferred to $(A'y')(x)$ whenever no confusion should arise. Recall that the dual space $X' := \mathcal{L}(X;\mathbb{K})$ of a normed vector space X, equipped with the norm defined by $\|x'\| := \sup_{x\neq 0} \dfrac{|x'(x)|}{\|x\|}$, is a Banach space.

Theorem 5.11-1 (dual operator) *Let X and Y be two normed vector spaces over the same field $\mathbb{K}$. Given any operator $A \in \mathcal{L}(X;Y)$, there exists one and only one operator $A' \in \mathcal{L}(Y';X')$, called the* **dual operator** *of A, or simply the* **dual** *of A, such that*

$$A'y'(x) = y'(Ax) \quad \textit{for all } x \in X \textit{ and all } y' \in Y'.$$

Besides,

$$\|A'\|_{\mathcal{L}(Y';X')} = \|A\|_{\mathcal{L}(X;Y)}.$$

Proof (i) Given $y' \in Y' = \mathcal{L}(Y;\mathbb{K})$, the mapping

$$A'y' : x \in X \to A'y'(x) := y'(Ax) \in \mathbb{K}$$

is a continuous linear functional as a composition of continuous linear mappings.

(ii) The mapping $A' : Y' \to X'$ defined in this fashion is *linear*, since for any $y', \widetilde{y}' \in Y'$,

$$A'(y' + \widetilde{y}')(x) = (y' + \widetilde{y}')(Ax) = y'(Ax) + \widetilde{y}'(Ax) = A'y'(x) + A'\widetilde{y}'(x)$$

for all $x \in X$; and for any $\alpha \in \mathbb{K}$ and $y' \in Y'$,

$$A'(\alpha y')(x) = (\alpha y')(Ax) = \alpha(y'(Ax)) = \alpha(A'y'(x)) = (\alpha A'y')(x)$$

for all $x \in X$.

(iii) The linear operator $A' : Y' \to X'$ is *continuous*, since

$$\|A'y'\| = \sup_{x\neq 0} \frac{|A'y'(x)|}{\|x\|} = \sup_{x\neq 0} \frac{|y'(Ax)|}{\|x\|} \leq \|A\|\,\|y'\| \quad \text{for all } y' \in Y'.$$

Hence $\|A'\| \leq \|A\|$ on the one hand. By Theorem 5.9-5,

$$\|Ax\| = \sup_{y'\neq 0} \frac{|y'(Ax)|}{\|y'\|} = \sup_{y'\neq 0} \frac{|A'y'(x)|}{\|y'\|} \leq \left(\sup_{y'\neq 0} \frac{\|A'y'\|}{\|y'\|}\right) \|x\| = \|A'\|\,\|x\| \quad \text{for all } x \in X.$$

Hence $\|A\| \le \|A'\|$ on the other hand. Therefore, $\|A'\| = \|A\|$. □

Remark An interesting example in infinite-dimensional normed vector spaces will be given in Theorem 6.14-1, where it will be shown that the dual operator of

$$A : \mu \in L_0^2(\Omega) := \left\{ \mu \in L^2(\Omega);\ \int_\Omega \mu \, dx = 0 \right\} \to A\mu := \mathbf{grad}\, \mu \in \boldsymbol{H}^{-1}(\Omega)$$

is

$$A' : \boldsymbol{v} \in \boldsymbol{H}_0^1(\Omega) \to A'\boldsymbol{v} := -\operatorname{div} \boldsymbol{v} \in L_0^2(\Omega). \qquad \square$$

Although we will not immediately use the following *sufficient condition for a dual operator to be compact*, we nevertheless record it now for convenience (a converse property also holds; cf. Problem 5.11-1). Notice that, perhaps unexpectedly, the *Ascoli–Arzelà theorem* plays an essential role in the next proof.

Theorem 5.11-2 *Let X and Y be two real normed vector spaces and let $A : X \to Y$ be a compact linear operator. Then the dual operator $A' : Y' \to X'$ is also compact.*

Proof Let $B := \overline{B_X(0,1)}$, so that $K := \overline{A(B)}$ is a compact subset of Y (since A is compact). In particular then, there exists M such that $K \subset \overline{B_Y(0; M)}$.

Let $(y_n')_{n=1}^\infty$ be any bounded sequence in Y', assumed without loss of generality to satisfy $\|y_n'\|_{Y'} \le 1$, $n \ge 1$. Then the functions

$$f_n : y \in K \to f_n(y) := y_n'(y), \quad n \ge 1,$$

form an *equicontinuous* and *bounded* family in the space $\mathcal{C}(K)$, since, for each $n \ge 1$,

$$|f_n(y) - f_n(\widetilde{y})| = |y_n'(y - \widetilde{y})| \le \|y - \widetilde{y}\| \quad \text{for all } y, \widetilde{y} \in K,$$
$$\|f_n\|_{\mathcal{C}(K)} = \sup_{y \in K} |f_n(y)| \le M.$$

Therefore, by the *Ascoli–Arzelà theorem* (Theorem 3.10-1), there exists a subsequence $(f_{\sigma(n)})_{n=1}^\infty$ that converges in the space $\mathcal{C}(K)$. The relations

$$\begin{aligned}
\|A'y_{\sigma(m)}' - A'y_{\sigma(n)}'\|_{X'} &= \sup_{\|x\|_X \le 1} |(A'(y_{\sigma(m)}' - y_{\sigma(n)}'))(x)| \\
&= \sup_{\|x\|_X \le 1} |(y_{\sigma(m)}' - y_{\sigma(n)}')(Ax)| \le \sup_{y \in K} |(y_{\sigma(m)}' - y_{\sigma(n)}')(y)| \\
&= \sup_{y \in K} |f_{\sigma(m)}(y) - f_{\sigma(n)}(y)| = \|f_{\sigma(m)} - f_{\sigma(n)}\|_{\mathcal{C}(K)}
\end{aligned}$$

then show that the sequence $(A'y_{\sigma(n)}')_{n=1}^\infty$ is a Cauchy sequence in X', which thus converges in X' since X' is complete (as a dual space). Consequently, $A' : Y' \to X'$ is compact. □

As a first step towards finding necessary and sufficient conditions guaranteeing that $\operatorname{Im} A = Y$, we show that a surprisingly simple condition involving the *dual* A' of A is equivalent to the more modest requirement that $\overline{\operatorname{Im} A} = Y$. Note that, if both X and Y are Hilbert spaces, this condition follows immediately from the relation $Y = \operatorname{Ker} A^* \oplus \overline{\operatorname{Im} A}$ established in Theorem 4.7-2(b).

Theorem 5.11-3 *Let X and Y be two normed vector spaces over the same field, and let $A \in \mathcal{L}(X;Y)$. Then the following two conditions are equivalent:*

(a) *The operator A has a dense range, i.e., $\overline{\operatorname{Im} A} = Y$.*

(b) *The dual $A' : Y' \to X'$ is injective, i.e., $\operatorname{Ker} A' = \{0\}$.*

Proof By Theorem 5.9-7, $\overline{\operatorname{Im} A} = Y$ if and only if

$$\{y' \in Y';\ y'(z) = 0 \text{ for all } z \in \operatorname{Im} A\} = \{0\}.$$

Hence the announced equivalence simply follows from the relations

$$\begin{aligned}\{y' \in Y';\ y'(z) = 0 \text{ for all } z \in \operatorname{Im} A\} &= \{y' \in Y';\ y'(Ax) = 0 \text{ for all } x \in X\} \\ &= \{y' \in Y';\ A'y'(x) = 0 \text{ for all } x \in X\} \\ &= \{y' \in Y';\ A'y' = 0\} = \operatorname{Ker} A'.\end{aligned}$$

□

We now give a *first fundamental necessary and sufficient condition for an operator* $A \in \mathcal{L}(X;Y)$ *to be such that* $\operatorname{Im} A = Y$, thus providing a *first answer* to the question raised at the beginning of this section regarding the *existence* of a solution to the equation $Ax = y$. Actually, this result will be eventually incorporated into the second part of the Banach closed range theorem (Theorem 5.11-6), but it is needed now as it will be used for proving the first part of the Banach closed range theorem (Theorem 5.11-5).

Theorem 5.11-4 *Let X and Y be two Banach spaces over the same field and let $A \in \mathcal{L}(X;Y)$. Then the following two conditions are equivalent:*

(a) *The operator $A : X \to Y$ is surjective, i.e., $\operatorname{Im} A = Y$.*

(b) *There exists a constant $C > 0$ such that the dual operator $A' : Y' \to X'$ satisfies*

$$\|y'\| \leq C\|A'y'\| \quad \textit{for all } y' \in Y'.$$

Proof The proof is given when $\mathbb{K} = \mathbb{R}$; the proof when $\mathbb{K} = \mathbb{C}$ is left as a problem (Problem 5.11-3). That (a) implies (b) is proved in (i). That (b) implies (a) is proved in parts (ii) and (iii).

The notation $B_Z(z;r)$ designates an open ball with center z and radius r in a space Z.

(i) As a surjective mapping, A is open by the *Banach open mapping theorem* (Theorem 5.6-1). In particular then, the image of $B_X(0;1)$ is an open subset of Y that contains $0 = A0$, i.e., there exists $s > 0$ such that

$$B_Y(0;s) \subset A(B_X(0;1)).$$

Consequently, for any $y' \in Y'$,

$$\begin{aligned}\|A'y'\| &= \sup_{x \in B_X(0;1)} |A'y'(x)| = \sup_{x \in B_X(0;1)} |y'(Ax)| \\ &= \sup_{y \in A(B_X(0;1))} |y'(y)| \geq \sup_{y \in B_Y(0;s)} |y'(y)| = s\|y'\|,\end{aligned}$$

and thus the inequality of (b) holds with $C = s^{-1}$.

(ii) *Assume that* (b) *holds. Then*

$$B_Y\left(0; C^{-1}\right) \subset \overline{A(B_X(0;1))}.$$

Pick any point $y_0 \in Y$ that does not belong to the set

$$Z := \overline{A(B_X(0;1))}.$$

Since Z is a closed convex subset of Y and $\{y_0\}$ is a compact convex subset of Y whose intersection with Z is empty, the *second geometric form of the Hahn–Banach theorem* (Theorem 5.10-2) shows that Z and $\{y_0\}$ can be strictly separated by a hyperplane. This means that there exist a nonzero $\widetilde{y}' \in Y'$, $\gamma \in \mathbb{R}$, and $\delta > 0$, such that

$$\widetilde{y}'(y) \leq \gamma - \delta < \gamma + \delta \leq \widetilde{y}'(y_0) \quad \text{for all } y \in Z,$$

and $\widetilde{y}'(y_0) > 0$ since $0 \in Z$ and $\widetilde{y}'(0) = 0$. The above inequalities thus show that

$$\sup_{y \in Z} \widetilde{y}'(y) \leq \gamma - \delta < \widetilde{y}'(y_0).$$

Noting that $y \in A(B_X(0,1))$ implies $-y \in A(B_X(0,1))$, we infer that

$$\sup_{y \in Z} |\widetilde{y}'(y)| = \sup_{y \in Z} \widetilde{y}'(y) \leq \gamma - \delta < \widetilde{y}'(y_0).$$

Letting

$$y_0' := \left(\sup_{y \in Z} |\widetilde{y}'(y)|\right)^{-1} \widetilde{y}' \quad \text{if } \sup_{y \in Z} |\widetilde{y}'(y)| > 0,$$

or

$$y_0' := \frac{\delta}{\widetilde{y}'(y_0)} \widetilde{y}' \quad \text{for any } \delta > 1 \text{ if } \sup_{y \in Z} |\widetilde{y}'(y)| = 0,$$

we have therefore found a nonzero $y_0' \in Y'$ such that

$$|y_0'(y)| \leq 1 \quad \text{for all } y \in Z \text{ and } y_0'(y_0) > 1.$$

Consequently,

$$\|A' y_0'\| = \sup_{x \in B_X(0;1)} |A' y_0'(x)| = \sup_{x \in B_X(0;1)} |y_0'(Ax)| \leq \sup_{y \in Z} |y_0'(y)| \leq 1,$$

which in turn implies that

$$1 < y_0'(y_0) \leq \|y_0'\| \, \|y_0\| \leq C \|A' y_0'\| \, \|y_0\| \leq C \|y_0\|.$$

In other words, $y_0 \notin \overline{A(B_X(0,1))}$ implies $\|y_0\| \geq C^{-1}$, which means that

$$B_Y\left(0; C^{-1}\right) \subset \overline{A(B_X(0;1))}.$$

(iii) *Assume that* (b) *holds. Then there exists* $s > 0$ *such that*

$$B_Y(0; s) \subset A(B_X(0; 1)).$$

Therefore, the operator A *is surjective.*

The attentive reader will have undoubtedly noticed that what is proved in part (ii) above is exactly what was proved in part (ii) of the proof of the Banach open mapping theorem (Theorem 5.6-1). It then suffices to reproduce *verbatim* part (iii) of that same proof (where the assumption of surjectivity is fortunately not needed) to conclude that there exists $s > 0$ such that $B_Y(0; s) \subset A(B_X(0; 1))$.

It is then clear that A is surjective. □

Given a subset Z of a normed vector space X, *resp.* a subset Z' of its dual space X', the subspace of X' defined by

$$Z^0 := \{x' \in X';\ x'(z) = 0 \text{ for all } z \in Z\},$$

resp. the subspace of X defined by

$${}^0(Z') := \{x \in X;\ z'(x) = 0 \text{ for all } z' \in Z'\},$$

is called the **polar set** of Z, *resp.* of Z'; it is also sometimes called the *orthogonal* (to reflect that it generalizes the notion of orthogonal complement in an inner-product space; cf. Section 4.5), or the *annihilator* (a somewhat bizarre terminology), of Z, *resp.* of Z'. Such subspaces arise naturally in the characterizations found in the next theorem (see also the proof of Theorem 5.11-3 or Problem 5.11-2).

In view of eventually producing in Theorem 5.11-6(c) a second equivalent condition for the surjectivity of A, a fundamental result is first needed, which answers in particular the question raised at the beginning of this section regarding a characterization of $\text{Im}\,A$; cf. (c) in the next theorem.

Note that relations (c) and (d) in the next theorem constitute natural extensions to general Banach spaces of the relations $\text{Im}\,A = (\text{Ker}\,A^*)^\perp$ and $\text{Im}\,A^* = (\text{Ker}\,A)^\perp$ that hold in *Hilbert spaces* when $\text{Im}\,A$ and $\text{Im}\,A^*$ are closed (Theorem 4.7-2).

Theorem 5.11-5 (Banach closed range theorem;[26] first part) *Let X and Y be two Banach spaces over the same field and let $A \in \mathcal{L}(X; Y)$. Then the following four conditions are equivalent:*

(a) *The operator $A : X \to Y$ has a closed range, i.e.,* $\text{Im}\,A$ *is closed in* Y.

(b) *The dual operator $A' : Y' \to X'$ has a closed range, i.e.,* $\text{Im}\,A'$ *is closed in* X'.

(c) $\text{Im}\,A = {}^0(\text{Ker}\,A') = \{y \in Y;\ y'(y) = 0$ *for all* $y' \in \text{Ker}\,A'\}$.

(d) $\text{Im}\,A' = (\text{Ker}\,A)^0 = \{x' \in X';\ x'(x) = 0$ *for all* $x \in \text{Ker}\,A\}$.

Proof (i) *Relation* (a) *implies relation* (b).
Define the subspace $\widetilde{Y}$ of Y and the mapping $\widetilde{A} \in \mathcal{L}(X; \widetilde{Y})$ by

$$\widetilde{Y} := \text{Im}\,A \quad \text{and} \quad \widetilde{A} : x \in X \to \widetilde{A}x := Ax \in \widetilde{Y} \subset Y,$$

[26] S. BANACH [1932]: *Théorie des Opérateurs Linéaires*, Monografje Matematyczne, Volume 1, Warsaw.

so that $\operatorname{Im} A = \operatorname{Im} \widetilde{A} = \widetilde{Y}$ is a closed subspace of Y by assumption.

Given any $y' \in Y'$, the relation $|y'(y)| \le \|y'\| \, \|y\|$ for all $y \in Y$ combined with the inclusion $\widetilde{Y} \subset Y$ shows that y' defines an element $\widetilde{y}' \in \widetilde{Y}'$ by letting $\widetilde{y}'(\widetilde{y}) := y'(\widetilde{y})$ for all $\widetilde{y} \in \widetilde{Y}$. Hence

$$A'y'(x) = y'(Ax) = \widetilde{y}'(Ax) = \widetilde{y}'(\widetilde{A}x) = \widetilde{A}'\widetilde{y}'(x) \quad \text{for all } x \in X,$$

since $Ax \in \widetilde{Y}$ for all $x \in X$. We have thus found $\widetilde{y}' \in \widetilde{Y}'$ such that $\widetilde{A}'\widetilde{y}' = A'y'$.

Given any $\widetilde{y}' \in \widetilde{Y}' = \mathcal{L}(\widetilde{Y}; \mathbb{K})$, there exists $y' \in Y' = \mathcal{L}(Y; \mathbb{K})$ such that

$$y'(\widetilde{y}) = \widetilde{y}'(\widetilde{y}) \quad \text{for all } \widetilde{y} \in \widetilde{Y},$$

by the *Hahn–Banach theorem in a normed vector space* (Theorem 5.9-1). Consequently,

$$A'y'(x) = y'(Ax) = \widetilde{y}'(Ax) = \widetilde{y}'(\widetilde{A}x) = \widetilde{A}'\widetilde{y}'(x) \quad \text{for all } x \in X.$$

We have thus found $y' \in Y'$ such that $A'y' = \widetilde{A}'\widetilde{y}'$.

Combining the two relations above thus shows that

$$\operatorname{Im} A' = \operatorname{Im} \widetilde{A}' \quad \text{in the space } X'.$$

Since $\widetilde{A} : X \to \widetilde{Y}$ is surjective and $\widetilde{Y}$ is complete (as a closed subset of a Banach space), the *Banach open mapping theorem* (Theorem 5.6-1) can be applied, showing that A maps open sets into open sets. In particular then, there exists $\delta > 0$ such that $B_{\widetilde{Y}}(0; 2\delta) \subset \widetilde{A}(B_X(0;1)) = A(B_X(0;1))$. This inclusion implies that, given any $\widetilde{y} \in \widetilde{Y}$ with $\|\widetilde{y}\| = \delta$, there exists $x \in X$ such that $Ax = \widetilde{y}$ and $\|x\| < 1 = \dfrac{\|Ax\|}{\delta}$. Consequently, for any $\widetilde{y}' \in \widetilde{Y}'$,

$$\|\widetilde{y}'\| = \frac{1}{\delta} \sup_{\|\widetilde{y}\| = \delta} |\widetilde{y}'(\widetilde{y})| \le \frac{1}{\delta} \sup_{\|x\| < 1} |\widetilde{y}'(\widetilde{A}x)| = \frac{1}{\delta} \sup_{\|x\| < 1} |\widetilde{A}'\widetilde{y}'(x)| = \frac{1}{\delta} \|\widetilde{A}'\widetilde{y}'\|.$$

This shows that the inverse operator $(\widetilde{A}')^{-1} : \operatorname{Im} \widetilde{A}' \subset X' \to \widetilde{Y}'$ is well defined and that $\widetilde{A}' : \widetilde{Y}' \to \operatorname{Im} \widetilde{A}'$ is a bijective and continuous linear operator with a continuous inverse.

Hence $\operatorname{Im} \widetilde{A}'$ *is closed in* X'. For, let $\widetilde{x}'_n \in \operatorname{Im} \widetilde{A}'$, $n \ge 0$, be such that $\widetilde{x}'_n \to \widetilde{x}'$ in X'; hence $(\widetilde{A}')^{-1}\widetilde{x}'_n$ converges in $\widetilde{Y}'$. Let $\widetilde{y}' = \lim_{n\to\infty} (\widetilde{A}')^{-1}\widetilde{x}'_n$; then $\widetilde{x}'_n = \widetilde{A}'(\widetilde{A}')^{-1}\widetilde{x}'_n \to \widetilde{A}'\widetilde{y}'$, which by definition belongs to $\operatorname{Im} \widetilde{A}'$.

Consequently, $\operatorname{Im} A' = \operatorname{Im} \widetilde{A}'$ is closed in $\widetilde{x}'$, as was to be proven.

(ii) *Relation* (b) *implies relation* (a).

Define the closed subspace $\widehat{Y}$ of Y and the operator $\widehat{A} \in \mathcal{L}(X; \widehat{Y})$ by letting

$$\widehat{A} : x \in X \to \widehat{A}x := Ax \in \widehat{Y} := \overline{\operatorname{Im} A} \subset Y.$$

Since the space $\operatorname{Im} \widehat{A} = \operatorname{Im} A$ is by construction dense in the space $\widehat{Y}$, the dual operator $\widehat{A}' : \widehat{Y}' \to X'$ is injective (Theorem 5.11-3). The same argument as in (i) about the operator $\widetilde{A}$, but now applied to the operator $\widehat{A}$, then shows that

$$\operatorname{Im} A' = \operatorname{Im} \widehat{A}' \quad \text{in the space } X'.$$

But $\operatorname{Im} A'$ is closed in X' by assumption. Hence $\operatorname{Im} A'$ is complete since X' is complete as a dual space, and thus $\operatorname{Im} \widehat{A}'$ is also complete.

Since the injective operator $\widehat{A}' : \widehat{Y}' \to \operatorname{Im} \widehat{A}'$ is surjective (by construction), the *Banach open mapping theorem* (Theorem 5.6-1) can be again applied, showing that $(\widehat{A}')^{-1} : \operatorname{Im} \widehat{A}' \to \widehat{Y}'$ is continuous. This means that there exists a constant C such that

$$\|\widetilde{y}'\| \leq C\|\widehat{A}'\widetilde{y}'\| \quad \text{for all } \widetilde{y}' \in \widehat{Y}'$$

(Theorem 2.9-4). Theorem 5.11-4 then shows that the operator $\widehat{A} : X \to \widehat{Y}$ is surjective, i.e., that

$$\operatorname{Im} \widehat{A} = \widehat{Y} = \overline{\operatorname{Im} A}.$$

But $\operatorname{Im} \widehat{A} = \operatorname{Im} A$; hence $\operatorname{Im} A$ is closed.

(iii) *Relation* (a) *is equivalent to relation* (c).

If $y \in \operatorname{Im} A$, then, for each $y' \in Y'$, $y'(y) = y'(Ax) = A'y'(x)$ for some $x \in X$; hence $y'(y) = 0$ for all $y' \in \operatorname{Ker} A'$. This shows that the inclusion $\operatorname{Im} A \subset {}^0(\operatorname{Ker} A')$ always holds.

Assume next that $\operatorname{Im} A$ is *closed* but that the inclusion ${}^0(\operatorname{Ker} A') \subset \operatorname{Im} A$ does not hold, i.e., that there exists $y_0 \in {}^0(\operatorname{Ker} A')$ such that $y_0 \notin \operatorname{Im} A$. Then, by Theorem 5.9-6 (a corollary to the Hahn–Banach theorem in a normed vector space that can be applied here because $\operatorname{Im} A$ is a closed subspace of Y), there exists $y_0' \in Y'$ such that

$$y_0'(y_0) \neq 0 \quad \text{and} \quad y_0'(y) = 0 \text{ for all } y \in \operatorname{Im} A.$$

Consequently, $A'y_0'(x) = y_0'(Ax) = 0$ for all $x \in X$, which means that $y_0' \in \operatorname{Ker} A'$; but then we should have $y_0'(y_0) = 0$, a contradiction.

We thus conclude that $\operatorname{Im} A = {}^0(\operatorname{Ker} A')$ if $\operatorname{Im} A$ is closed, i.e., that (a) implies (c). That (c) implies (a) is clear since a polar set is always closed.

(iv) *Relation* (b) *is equivalent to relation* (d).

If $x' \in \operatorname{Im} A'$, then, for each $x \in X$, $x'(x) = A'y'(x) = y'(Ax)$ for some $y' \in Y'$; hence $x'(x) = 0$ for all $x \in \operatorname{Ker} A$. This shows that the inclusion $\operatorname{Im} A' \subset (\operatorname{Ker} A)^0$ always holds.

We next show that $(\operatorname{Ker} A)^0 \subset \operatorname{Im} A'$ if $\operatorname{Im} A'$ is *closed*. To this end, define the quotient space $\dot{X} := X/\operatorname{Ker} A$ and define the bijective continuous linear operator $\dot{A} : \dot{X} \to \operatorname{Im} A \subset Y$ by

$$\dot{A}\dot{x} := A\widetilde{x} \quad \text{for each } \dot{x} \in \dot{X},$$

where $\widetilde{x}$ is any element of $\dot{x}$. Equipped with the quotient norm, the quotient space $\dot{X}$ is a Banach space (because $\operatorname{Ker} A$ is a closed subspace; cf. Theorem 3.6-5), and $\operatorname{Im} A$ is also a Banach space as a closed subspace of Y (if $\operatorname{Im} A'$ is closed, then $\operatorname{Im} A$ is also closed by (ii)). By the corollary to the Banach open mapping theorem (Theorem 5.6-2), the inverse $\dot{A}^{-1} : \operatorname{Im} A \to \dot{X}$ of $\dot{A}$ is therefore also a continuous linear operator.

Given any element $x' \in (\operatorname{Ker} A)^0$, let $\dot{x}' \in \dot{X}'$ be defined by $\dot{x}'(\dot{x}) := x'(\widetilde{x})$ for each $\dot{x} \in \dot{X}$, where $\widetilde{x}$ is any element of $\dot{x}$ (this definition makes sense since $x'(x) = 0$ for all $x \in \operatorname{Ker} A$ if $x' \in (\operatorname{Ker} A)^0$). The function

$$y' := \dot{x}' \circ \dot{A}^{-1} : \operatorname{Im} A \to \mathbb{R}$$

is thus a continuous linear functional, i.e., $y' \in (\operatorname{Im} A)'$. Let $\widetilde{y}' \in Y'$ be an extension of y'. Then

$$A'\widetilde{y}'(x) = \widetilde{y}'(Ax) = y'(Ax) = \dot{x}'(\dot{A}^{-1}Ax) = \dot{x}'(\dot{A}^{-1}\dot{A}\dot{x}) = \dot{x}'(\dot{x}) = x'(x)$$

for any $x \in X$. Consequently,

$$x' = A'\widetilde{y}' \in \operatorname{Im} A',$$

which shows that $(\operatorname{Ker} A)^0 \subset \operatorname{Im} A'$.

We thus conclude that $\operatorname{Im} A' = (\operatorname{Ker} A)^0$ if $\operatorname{Im} A'$ is closed, i.e., that (b) implies (d). That (d) implies (b) is clear since a polar set is always closed. □

Remark Other necessary and sufficient conditions that $\operatorname{Im} A$ be closed in Y are proposed in Problem 5.11-4. □

Thanks to Theorem 5.11-5, we are now in a position to give a *second fundamental necessary and sufficient condition for an operator A to be such that* $\operatorname{Im} A = Y$ (cf. (c) in the next theorem), thus providing a *second answer* to the question raised at the beginning of this section. The *first answer* to the same question is repeated in the next theorem (cf. (b)).

Theorem 5.11-6 (Banach closed range theorem;[27] second part) *Let X and Y be two Banach spaces over the same field and let $A \in \mathcal{L}(X;Y)$. Then the following three conditions are equivalent:*

(a) *The operator $A : X \to Y$ is surjective, i.e.,* $\operatorname{Im} A = Y$.

(b) *There exists a constant C such that the dual operator $A' : Y' \to X'$ satisfies*

$$\|y'\| \le C\|A'y'\| \quad \textit{for all } y' \in Y'.$$

(c) *The dual operator A' is injective and* $\operatorname{Im} A'$ *is closed in* X'.

Proof The equivalence between (a) and (b) has already been established in Theorem 5.11-4.

Assume that (a) holds. Then $\operatorname{Im} A = \overline{\operatorname{Im} A} = Y$, and thus A' is injective by Theorem 5.11-3, and $\operatorname{Im} A'$ is closed by Theorem 5.11-5. Hence (a) implies (c).

Assume that (c) holds. Then the mapping (still denoted) $A' : Y' \to \operatorname{Im} A'$ is bijective and $\operatorname{Im} A'$ is complete as a closed subspace of X'. Hence the *corollary to the Banach open mapping theorem* (Theorem 5.6-2) applied to $A' \in \mathcal{L}(Y'; \operatorname{Im} A')$ shows that $(A')^{-1} : \operatorname{Im} A' \to Y'$ is also continuous, which means that there exists a constant C such that $\|y'\| \le C\|A'y'\|$ for all $y' \in Y'$ (Theorem 2.9-4). Hence (c) implies (b). □

Condition (b) in Theorem 5.11-6 is sometimes put to use in the analysis of *linear boundary value problems*. For, it asserts that in order to establish the *existence* of a solution x to a partial differential equation, written symbolically as $Ax = y$ (here, X and Y are *ad hoc* function spaces with X incorporating some boundary conditions, $A : X \to Y$ is a partial differential operator, and b is the right-hand side of the equation), it suffices to have an *a priori bound* on any given solution y' of the *dual equation* $A'y' = x'$, in the form $\|y'\| \le C\|x'\|$ for some constant C independent of x'. What is particularly remarkable is that *there is no need to verify that the dual equation possesses solutions.* This observation therefore provides a powerful technique for establishing the existence of solutions to such problems.

[27] S. Banach [1932]: *Théorie des Opérateurs Linéaires*, Monografje Matematyczne, Volume 1, Warsaw.

To conclude, we mention that the Banach closed range theorem holds more generally for *closed* and *densely defined* linear operators;[28] however, such an extension will not be needed in the rest of this book.

Problems

5.11-1 Let X be a real normed vector space, let Y be a real Banach space, and let $A \in \mathcal{L}(X;Y)$ be such that its dual operator $A' : Y' \to X'$ is compact. Show that A is compact.

5.11-2 Let X be a normed vector space and let Z be a closed subset of X, so that the quotient space X/Z is also a normed vector space (Theorem 2.2-3).

(1) Show that the subspace $Z^0 := \{x' \in X';\ x'(z) = 0 \text{ for all } z \in Z\}$ is closed in X' and that there exists a linear isometry from Y' onto X'/Z^0.

(2) Show that there exists a linear isometry from $(X/Z)'$ onto Z^0.

5.11-3 Show that Theorem 5.11-4 holds *verbatim* if X and Y are *complex* Banach spaces.

Hint: In part (ii) of the proof of Theorem 5.11-4, use the geometric form of the Hahn–Banach theorem about the strict separation of convex sets in a complex vector space (Problem 5.10-3(2)) to establish the existence of $y_0' \in \widetilde{Y}'$ such that $|y_0'(y)| \leq 1$ for all $y \in C$ and $|y'(y_0)| > 1$.

5.11-4 Let X and Y be Banach spaces and let $A \in \mathcal{L}(X;Y)$.

(1) Assume that A is injective. Show that a necessary and sufficient condition that $\operatorname{Im} A$ be closed in Y is that there exists a constant C such that $\|x\| \leq C\|Ax\|$ for all $x \in X$.

(2) Assume that A is not injective. Show that a necessary and sufficient condition that $\operatorname{Im} A$ be closed in Y is that there exists a constant C such that $\|[x]\| \leq C\|Ax\|$ for all $x \in X$, where $\|[x]\|$ denotes the norm of $[x]$ in the quotient space $X/\operatorname{Ker} A$ (which is also a Banach space; cf. Theorem 3.6-5).

5.11-5 Let $(X, (\cdot,\cdot))$ be a Hilbert space and let $A \in \mathcal{L}(X)$ be a symmetric operator. Show that, if there exists a constant $\alpha > 0$ such that $(Ax, x) \geq \alpha\|x\|^2$ for all $x \in X$, then $A : X \to X$ is injective and $\operatorname{Im} A = X$ (hence, for each $y \in X$, the equation $Ax = y$ has one and only one solution $x \in X$).

5.12 Weak convergence and weak * convergence

While every *bounded* sequence in a *finite-dimensional* normed vector space contains a convergent subsequence (the closure of a bounded subset of a finite-dimensional space is compact; cf. Theorem 2.7-1(c)), this is no longer necessarily true in an *infinite-dimensional* space. For instance, a countably infinite orthonormal family (e_i) in an inner-product space (Section 4.8) is bounded since $\|e_i\| = 1$ for all i; yet it cannot contain any convergent subsequence, since $\|e_i - e_j\| = \sqrt{2}$ if $i \neq j$.

A natural question therefore arises: Is there *another* notion of *convergence* with a similar property in an *infinite-dimensional* normed vector space? It turns out that the proper notion is that of *weak convergence*,[29] which will be defined below. For, it will be shown (Theorem 5.14-4) that *any bounded sequence in a reflexive Banach space contains a weakly convergent subsequence* (a reflexive Banach space is one that can be identified with the dual of its dual by means of a specific linear isometry; cf. Section 5.14), which thus provides a positive answer

[28] See YOSIDA [1965, Chapter 7, Section 5] or BREZIS [2011, Sections 2.6 and 2.7].

[29] This notion was introduced by David Hilbert around 1906.

to the question raised above. When applied in particular to infimizing sequences of coercive functionals, this property plays a key role in the *calculus of variations* (Chapter 9).

Let X be a normed vector space and let X' denote its dual. A sequence $(x_n)_{n=1}^{\infty}$ of elements $x_n \in X$ is said to **converge weakly** in X if there exists $x \in X$ such that

$$\text{for each } x' \in X', \quad x'(x_n) \to x'(x) \quad \text{as } n \to \infty,$$

and such an x is then called a **weak limit** of the sequence $(x_n)_{n=1}^{\infty}$. Weak convergence is denoted with a "half-arrow" $\rightharpoonup$, i.e., by

$$x_n \rightharpoonup x \quad \text{as } n \to \infty,$$

so as to distinguish it from *strong convergence*, which is denoted by a "full arrow" $\to$, i.e., by

$$x_n \to x \quad \text{as } n \to \infty.$$

Recall that "strong convergence" is an *alias* for convergence with respect to the *norm* topology: it means that $\|x_n - x\| \to 0$ as $n \to \infty$.

Observe that, *in a Hilbert space* $(X, (\cdot, \cdot))$, *a sequence* $(x_n)_{n=1}^{\infty}$ *converges weakly to* x *if and only if, for each* $y \in X$, $(x_n, y) \to (x, y)$ *as* $n \to \infty$, since the dual X' of X can be identified with X by means of the F. Riesz isometry.

For instance, the sequence $(f_n)_{n=1}^{\infty}$ defined by

$$f_n(\theta) := \sin n\theta, \quad 0 \le \theta \le 2\pi,$$

converges weakly in the Hilbert space $L^2(0, 2\pi)$ to 0. To see this, simply recall that, given any function $g \in L^2(0, 2\pi)$, the numbers

$$a_n = \frac{1}{\pi} \int_0^{2\pi} g(\varphi) \cos n\varphi \, \mathrm{d}\varphi, \ n \ge 0, \quad \text{and} \quad b_n := \frac{1}{\pi} \int_0^{2\pi} g(\varphi) \sin n\varphi \, \mathrm{d}\varphi, \ n \ge 1,$$

are the coefficients of the "classical" Fourier series of g, so that

$$\frac{|a_0|^2}{2} + \sum_{n=1}^{\infty} |a_n|^2 + \sum_{n=1}^{\infty} |b_n|^2 = \frac{1}{\pi} \|g\|_{L^2(0,2\pi)}^2 < \infty$$

by Parseval's formula (Theorem 4.9-2); hence $\lim_{n\to\infty} a_n = \lim_{n\to\infty} b_n = 0$. This also shows that the sequence $(g_n)_{n=0}^{\infty}$ defined by $g_n(\theta) = \cos n\theta$, $0 \le \theta \le 2\pi$, likewise weakly converges to 0 in $L^2(0, 1)$ as $n \to \infty$.

Yet, the sequence $(f_n)_{n=1}^{\infty}$ (or the sequence $(g_n)_{n=1}^{\infty}$ for that matter) does *not* strongly converge in $L^2(0, 1)$; in fact, it does not even contain any strongly convergent subsequence since the family $\left(\frac{1}{\sqrt{\pi}} f_n\right)_{n=1}^{\infty}$ is an orthonormal family.

The next theorem gives two immediate relations between weak and strong convergences. More elaborate relations will be given later (Theorems 5.12-3 and 5.13-1).

Theorem 5.12-1 (a) *In any normed vector space,* $x_n \to x$ *as* $n \to \infty$ *implies that* $x_n \rightharpoonup x$ *as* $n \to \infty$.

(b) *In a finite-dimensional normed vector space, any weakly convergent sequence is also strongly convergent. Consequently, these two notions of convergence coincide in a finite-dimensional space.*

Proof Let $(X, \|\cdot\|)$ be a normed vector space, and let $x_n \to x$ in X as $n \to \infty$. Then

$$\text{for each } x' \in X', \quad |x'(x_n) - x'(x)| \leq \|x'\| \, \|x_n - x\| \, ,$$

which proves (a).

Let X be finite-dimensional, let $(e_i)_{i=1}^k$ be a basis of X, and let X be equipped with the norm $\|\cdot\|_1$ (to fix ideas). Since each linear functional

$$x'_j : x = \sum_{i=1}^{k} x_i e_i \in X \to x_j \in \mathbb{K}, \quad 1 \leq j \leq k,$$

is clearly continuous and since $|x_j| \leq \|x\|$ for all $x \in X$, the weak convergence $x_n = \sum_{i=1}^k x_i^n e_i \rightharpoonup x = \sum_{i=1}^k x_i e_i$ implies in particular that $x_j^n \to x_j$ for each $1 \leq j \leq k$. Hence $\|x_n - x\|_1 \to 0$ as $n \to \infty$, which proves (b). □

Two natural questions about weak convergence immediately arise: Is the limit of a weakly convergent sequence *unique*? Is a weakly convergent sequence *bounded*? Surprisingly, the (positive) answers to these seemingly innocuous questions are by no means trivial. For, the answer to the first question (which is immediate for a strongly convergent sequence) requires no less than the *Hahn–Banach theorem in a normed vector space* (hence the axiom of choice), while the answer to the second question (again immediate for a strongly convergent sequence) requires no less than both the *Hahn–Banach theorem in a normed vector space* and the *Banach–Steinhaus theorem* (hence both the axiom of choice and Baire's theorem). These properties are established in the next theorem (cf. (a) and (b)), which also provides an upper bound for the norm of a weak limit (cf. (c)).

Theorem 5.12-2 *The following properties hold in any normed vector space:*
(a) *The limit of a weakly convergent sequence is unique.*
(b) *A weakly convergent sequence is bounded.*
(c) *Let $x_n \rightharpoonup x$ as $n \to \infty$. Then*

$$\|x\| \leq \liminf_{n\to\infty} \|x_n\|.$$

Proof Let $(x_n)_{n=1}^\infty$ be a weakly convergent sequence in a normed vector space $(X, \|\cdot\|)$, and let $x, \widetilde{x} \in X$ be such that

$$\text{for each } x' \in X', \quad x'(x) = x'(\widetilde{x}) = \lim_{n\to\infty} x'(x_n).$$

Then $x'(x - \widetilde{x}) = 0$ for all $x' \in X'$, and thus $x = \widetilde{x}$ by Theorem 5.9-5 (a corollary to the Hahn–Banach theorem). This proves (a).

For each $n \geq 1$, define the mapping $J_n : X' \to \mathbb{K}$ by

$$J_n : x' \in X' \to J_n(x') := x'(x_n),$$

which is clearly linear. Then $J_n \in (X')' := \mathcal{L}(X'; \mathbb{K})$ since

$$\|J_n\|_{(X')'} = \sup_{\substack{x' \in X' \\ x' \neq 0}} \frac{|J_n(x')|}{\|x'\|} = \sup_{\substack{x' \in X' \\ x' \neq 0}} \frac{|x'(x_n)|}{\|x'\|} = \|x_n\|,$$

by Theorem 5.9-5. Besides, for each $x' \in X'$, the sequence $(J_n(x'))_{n=1}^{\infty}$ converges in $\mathbb{K}$, since

$$\lim_{n\to\infty} J_n(x') = x'(x),$$

where x denotes the weak limit of the sequence $(x_n)_{n=1}^{\infty}$. Then the linear mapping $J_x : X' \to \mathbb{K}$ defined by

$$J_x : x' \in X' \to J_x(x') := \lim_{n\to\infty} J_n(x') = x'(x),$$

is continuous since

$$\|J_x\|_{(X')'} = \sup_{x'\neq 0} \frac{|x'(x)|}{\|x'\|} = \|x\|,$$

again by Theorem 5.9-5.

Since the space X' is *complete* (as a dual space; cf. Theorem 3.2-3), the corollary to the Banach–Steinhaus theorem (Theorem 5.3-2) can be applied, showing that

$$\sup_{n\geq 1} \|J_n\|_{(X')'} < \infty \quad \text{and} \quad \|J_x\|_{(X')'} \leq \liminf_{n\to\infty} \|J_n\|_{(X')'},$$

which is the same as

$$\sup_{n\geq 1} \|x_n\| < \infty \quad \text{and} \quad \|x\| \leq \liminf_{n\to\infty} \|x_n\|,$$

thus proving (b) and (c). □

Remark In a *Hilbert space* $(X, (\cdot,\cdot))$, the *uniqueness* of the weak limit is much easier to prove: since in this case $x_n \underset{n\to\infty}{\rightharpoonup} x$ and $x_n \underset{n\to\infty}{\rightharpoonup} \widetilde{x}$ implies that $(x - \widetilde{x}, y) = 0$ for all $y \in X$, it immediately follows that $x = \widetilde{x}$. □

Incidentally, the uniqueness of the weak limit provides another reason why the sequence $(f_n)_{n=1}^{\infty}$ defined by $f_n(\theta) = \sin n\theta$, $0 \leq \theta \leq 2\pi$ does not strongly converge in $L^2(0,1)$. For, assume otherwise that there exists $f \in L^2(0,1)$ such that $f_n \underset{n\to\infty}{\to} f$. This would imply that $f_n \underset{n\to\infty}{\rightharpoonup} f$, and hence that $f = 0$ since the limit of a weakly convergent sequence is unique (Theorem 5.12-2(a)). But this is impossible, because an immediate computation shows that $\|f_n\|_{L^2(0,2\pi)} = \sqrt{\pi} \neq 0$ for all $n \geq 1$.

We now give a very useful *sufficient condition for a weakly convergent sequence to be strongly convergent.* Recall (Section 2.17) that a normed vector space $(X, \|\cdot\|)$ is *uniformly convex* if, given any $\varepsilon > 0$, there exists $\delta(\varepsilon) > 0$ such that

$$\|x\| = \|y\| = 1 \quad \text{and} \quad \|x - y\| \geq \varepsilon \quad \text{implies} \quad \left\|\frac{x+y}{2}\right\| \leq 1 - \delta(\varepsilon),$$

and that the spaces ℓ^p and $L^p(\Omega)$, $1 < p < \infty$, and any inner-product space are uniformly convex (Problems 2.17-8 and 2.17-9 and Theorem 4.1-2).

Theorem 5.12-3 *Let X be a uniformly convex normed vector space and let a sequence $(x_n)_{n=1}^{\infty}$ of elements $x_n \in X$ and $x \in X$ be such that*

$$x_n \rightharpoonup x \quad \textit{and} \quad \|x_n\| \to \|x\| \quad \textit{as } n \to \infty.$$

Then $x_n \to x$ as $n \to \infty$.

Proof The result clearly holds if $x = 0$. If $x \neq 0$, there exists $n_0 \geq 1$ such that $x_n \neq 0$ for all $n \geq n_0$ since $\|x_n\| \to \|x\| > 0$ as $n \to \infty$. Let

$$y := \frac{x}{\|x\|}, \quad \text{and} \quad y_n := \frac{x_n}{\|x_n\|} \quad \text{for all } n \geq n_0,$$

so that, for all $x' \in X'$ and $n \geq n_0$,

$$x'(y_n) = \frac{1}{\|x_n\|} x'(x_n) \to \frac{1}{\|x\|} x'(x) = x'(y) \quad \text{as } n \to \infty.$$

Hence $y_n \rightharpoonup y$, and thus $(y_n + y) \rightharpoonup 2y$ as $n \to \infty$.

Theorem 5.12-2 (c) then shows that

$$2 = \|2y\| \leq \liminf_{n\to\infty} \|y_n + y\| \leq \limsup_{n\to\infty} \|y_n + y\| \leq 2,$$

since $\|y_n + y\| \leq \|y_n\| + \|y\| = 2$, which implies that

$$\left\| \frac{y_n + y}{2} \right\| \to 1 \quad \text{as } n \to \infty.$$

This relation in turn implies that

$$\|y_n - y\| \to 0 \quad \text{as } n \to \infty.$$

For otherwise there would exist $\varepsilon > 0$ and a subsequence $(y_{\sigma(n)})_{n=n_0}^{\infty}$ such that $\|y_{\sigma(n)} - y\| \geq \varepsilon$ for all $n \geq n_0$. But the uniform convexity assumption would then imply that $\left\| \frac{y_{\sigma(n)} + y}{2} \right\| \leq 1 - \delta(\varepsilon)$ for some $\delta(\varepsilon) > 0$, a contradiction.

The relation

$$x_n - x = \|x_n\| y_n - \|x\| \, y = \|x_n\|(y_n - y) + (\|x_n\| - \|x\|) y,$$

combined with the boundedness of the sequence $(x_n)_{n=1}^{\infty}$ (Theorem 5.12-2(b)), then shows that $\|x_n - x\| \to 0$ as $n \to \infty$. □

Remark As often, there is a much simpler proof if the uniformly convex space X is a *Hilbert space* $(X, (\cdot, \cdot))$, since it suffices to take the limit as $n \to \infty$ in the relation

$$\|x_n - x\|^2 = \|x_n\|^2 - 2(x, x_n) + \|x\|^2.$$

For, thanks to the F. Riesz representation theorem, the weak convergence $x_n \rightharpoonup x$ as $n \to \infty$ implies in this case that $(x, x_n) \to (x, x) = \|x\|^2$ as $n \to \infty$. □

We next describe the effect of applying *linear* or *bilinear operators* to weakly convergent sequences (bilinear operators and spaces such as $\mathcal{L}_2(X \times Y; \mathbb{K})$ are defined in Section 2.11).

Theorem 5.12-4 *Let X and Y be normed vector spaces over the same field $\mathbb{K}$.*

(a) *Let $A \in \mathcal{L}(X; Y)$. Then*

$$x_n \underset{n\to\infty}{\rightharpoonup} x \text{ in } X \quad \textit{implies} \quad Ax_n \underset{n\to\infty}{\rightharpoonup} Ax \textit{ in } Y.$$

(b) *Let $A \in \mathcal{L}(X;Y)$ be compact. Then*

$$x_n \underset{n\to\infty}{\rightharpoonup} x \text{ in } X \quad \textit{implies} \quad Ax_n \underset{n\to\infty}{\to} Ax \textit{ in } Y.$$

(c) *Let $B \in \mathcal{L}_2(X \times Y;\mathbb{K})$. Then*

$$x_n \underset{n\to\infty}{\rightharpoonup} x \textit{ in } X \quad \textit{and} \quad y_n \underset{n\to\infty}{\to} y \textit{ in } Y \quad \textit{implies} \quad B(x_n, y_n) \to B(x,y) \textit{ in } \mathbb{K}.$$

Proof (i) Let $A' \in \mathcal{L}(Y';X')$ denote the *dual* of $A \in \mathcal{L}(X;Y)$ (Theorem 5.11-1). Then $A'y' \in X'$ for all $y' \in Y'$ and thus, by definition of weak convergence,

$$\text{for each } y' \in Y', \quad y'(Ax_n) = A'y'(x_n) \underset{n\to\infty}{\to} A'y'(x) = y'(Ax),$$

which proves (a).

(ii) Let $A \in \mathcal{L}(X;Y)$ be a compact operator and let $x_n \underset{n\to\infty}{\rightharpoonup} x$ in X. Since a weakly convergent sequence is bounded (Theorem 5.12-2(b)), the sequence $(Ax_n)_{n=1}^\infty$ contains a subsequence $(Ax_{\sigma(n)})_{n=1}^\infty$ that strongly converges in Y (by definition of a compact operator; cf. Section 2.10). Besides, its limit is Ax by (a) (a strongly convergent sequence also weakly converges; cf. Theorem 5.12-1(a)). Since this limit is unique, the whole sequence $(Ax_n)_{n=1}^\infty$ strongly converges to Ax. This proves (b).

(iii) The proof of (c) follows from the relation

$$B(x_n, y_n) - B(x,y) = B(x_n, y_n - y) + B(x_n - x, y),$$

combined with the boundedness of the weakly convergent sequence $(x_n)_{n=1}^\infty$, the continuity of B (which implies that, for each $y \in Y$, the mapping $x \in X \to B(x,y) \in \mathbb{K}$ is a continuous linear functional on X), and the definitions of weak and strong convergence. $\square$

It should be noted that property (c) does *not* necessarily hold if *both* sequences only weakly converge. Consider for instance the continuous bilinear form

$$B : (f,g) \in L^2(0,2\pi) \times L^2(0,2\pi) \to B(f,g) := \int_0^{2\pi} f(\theta)g(\theta)\,\mathrm{d}\theta,$$

and the sequence $(f_k)_{k=1}^\infty$ defined by $f_k(\theta) = \sin k\theta$, $0 \le \theta \le 2\pi$, which weakly converges to 0 in $L^2(0,2\pi)$ as we saw earlier. But the sequence $(B(f_k,f_k))_{k=1}^\infty$ does not converge to $B(0,0) = 0$ since $B(f_k,f_k) = \pi$ for all $k \ge 1$.

We conclude this section by mentioning without proof important *complements on weak convergence.*

Let $(X, \|\cdot\|)$ be a normed vector space. Then its dual space X' consists of all the linear forms $x' : X \to \mathbb{R}$ that are continuous when X is equipped with the topology induced by its norm $\|\cdot\|$, also called the **strong topology** on X.

But the same space X can be also equipped with its **weak topology**, which is *by definition* the *weakest topology on X for which all the elements x' of the dual space X' remain continuous* as functions from X equipped with this topology into $\mathbb{K}$ (the existence of such a topology is guaranteed by Theorem 1.7-8). Recall that "weakest" means that any other topology on X

with the same property contains more open sets. A subset of X that is open for the weak topology is thus open for the strong topology, but the converse does not necessarily hold.

One can then prove the following basic properties of the weak topology.[30]

$^\flat$**Theorem 5.12-5** *Let X be a normed vector space.*

(a) *If X is finite-dimensional, the strong and weak topologies coincide. Consequently, the weak topology is normable in this case.*

(b) *If X is infinite-dimensional, there exist open sets for the strong topology that are not open for the weak topology. Furthermore, the weak topology is not metrizable in this case.*

(c) *The weak topology on X is a Hausdorff topology (Section 1.6).*

(d) *A sequence $(x_n)_{n=1}^{\infty}$ of elements $x_n \in X$ converges to $x \in X$ for the weak topology of X if and only if $x'(x_n) \to x'(x)$ for all $x' \in X'$, i.e., if and only if the sequence $(x_n)_{n=1}^{\infty}$ weakly converges to x.* $\square$

Parts (b) and (d) in the above theorem indicate why an infinite-dimensional space "usually" contains weakly convergent sequences that do not strongly converge. There are, however, "pathological" spaces, such as the space ℓ^1, where every weakly convergent sequence is also strongly convergent![31]

Property (d) thus shows that weak convergence (according to the definition given at the beginning of this section) is thus precisely the convergence corresponding to the weak topology.

Incidentally, note that the issue of deciding whether a topology can be defined by identifying the convergent sequences is a subtle one.[32] For instance, the strongly and weakly convergent sequences coincide in the space ℓ^1 (as mentioned above); yet, they correspond to the strong and weak topologies, which are necessarily *different* because ℓ^1 is an infinite-dimensional space (Theorem 5.12-5(b)).

Another basic notion of "weak" convergence can be defined simply by interchanging the role of X and X' in the definition of weak convergence: Let X be a normed vector space and let X' denote its dual. A sequence $(x'_n)_{n=1}^{\infty}$ of elements $x'_n \in X'$ is said to **weakly * converge** in X' if there exists $x' \in X'$ such that

$$\text{for each } x \in X, \quad x'_n(x) \to x'(x) \text{ as } n \to \infty,$$

and such an x', which is clearly *unique*, is called the **weak * limit** of the sequence $(x'_n)_{n=1}^{\infty}$. Weak * convergence is denoted by a "half-arrow with a star above," i.e., by

$$x'_n \overset{*}{\rightharpoonup} x' \quad \text{as } n \to \infty.$$

[30] For proofs and further properties, see the illuminating Sections 3.2 and 3.3 in Brezis [2011].

[31] A proof of this result, which constitutes **Schur's lemma**, is found in Kesavan [2009, Section 5.1].

[32] In this direction, see for instance:

J. Kisynski [1959]: Convergence du type L, *Colloquium Mathematicum* **7**, 205–211.

S.P. Franklin [1965]: Spaces in which sequences suffice, *Fundamenta Mathematicae* **57**, 107–115.

S.P. Franklin [1967]: Spaces in which sequences suffice, *Fundamenta Mathematicae* **61**, 51–56.

R.M. Dudley [1964]: On sequential convergence, *Transactions of the American Mathematical Society* **112**, 483–507.

B. Kripke [1967]: One more reason why sequences are not enough, *American Mathematical Monthly* **74**, 563–565.

Not unexpectedly, weak * and weak convergence share similar properties; cf. Problems 5.12-4 and 5.14-6.

Three different types of convergence can thus be defined in a *dual space* X': the *strong convergence*:

$$x'_n \to x' \quad \text{as } n \to \infty,$$

which means that $\|x'_n - x'\|_{X'} \to 0$ as $n \to \infty$; the *weak convergence*:

$$x'_n \rightharpoonup x' \quad \text{as } n \to \infty,$$

which means that, for each $x'' \in (X')'$, $x''(x'_n) \to x''(x')$ as $n \to \infty$; and the *weak * convergence* as defined above.

We shall see later (Section 5.14) that *any* normed vector space X can be identified with a subspace of the space $(X')'$ by means of a specific linear isometry $J : X \to J(X) \subset (X')'$. In other words, weak * convergence can be viewed as a *restricted* form of weak convergence (as defined at the beginning of this section), which only involves those elements in the dual space $(X')'$ of X' that belong to the image $J(X) \subset (X')'$. Consequently, *these two notions coincide if the space X is such that the isometry $J : X \to (X')'$ is surjective.* Such spaces, which are called *reflexive*, are studied in Section 5.14.

Just like the weak convergence in a normed vector space X, the weak * convergence corresponds to a topology on X'. More specifically, the **weak * topology** on X' *is by definition* the *weakest topology* (Theorem 1.7-8)[33] *on X' such that all the mappings $\varphi_x : x' \in X' \to \varphi_x(x') := x'(x) \in \mathbb{K}$, $x \in X$, are continuous.*

One can then establish the following results (compare with Theorem 5.12-5).

$^\flat$**Theorem 5.12-6** *Let X be a normed vector space.*

(a) *A subset of X' that is open for the weak * topology of X' is open for the weak topology of X', but the converse need not hold.*

(b) *The weak * topology on X' is a Hausdorff topology.*

(c) *The closed unit ball of X' is compact for the weak * topology of X'.*

(d) *A sequence $(x'_n)_{n=0}^\infty$ of elements $x'_n \in X'$ converge to $x' \in X'$ for the weak * topology of X' if and only if it weakly * converges to x'.* □

Perhaps the most important reason for introducing the weak * topology is property (c) above: while the closed unit ball of X' is *never* compact for the strong topology of X' if X is *infinite-dimensional* (by the F. Riesz theorem; cf. Theorem 2.7-3), the same closed unit ball is *always* compact for the weak * topology of X' (i.e., even if X is infinite-dimensional).

Problems

5.12-1 This problem provides a *sufficient condition for weak convergence.* Let X be a normed vector space and let Y' be a dense subset of its dual X'. Let a sequence $(x_n)_{n=1}^\infty$ of elements $x_n \in X$ and $x \in X$ be such that

$$\sup_{n\geq 1} \|x_n\| < \infty \quad \text{and} \quad y'(x_n) \underset{n\to\infty}{\longrightarrow} y'(x) \ \text{ for each } y' \in Y'.$$

[33] A highly readable account of the basic properties of the weak* topology is given in Brezis [2011, Section 3.4].

Show that $x_n \rightharpoonup x$ as $n \to \infty$ (note that, by Theorem 5.12-2(b), the condition $\sup_{n\geq 1} \|x_n\| < \infty$ is also necessary).

5.12-2 Let Ω be an open subset of $\mathbb{R}^n$, let $1 < p < \infty$, and let $(f_k)_{k=1}^{\infty}$ be a bounded sequence of functions $f_k \in L^p(\Omega)$ that pointwise converges almost everywhere to a function $f \in L^p(\Omega)$. Show that $f_k \rightharpoonup f$ in $L^p(\Omega)$ as $k \to \infty$.

5.12-3 For each integer $n \geq 1$, let the function $f_n \in L^2(0,1)$ be defined as follows:

$$f_n(x) := 0 \quad \text{if} \quad \frac{j}{n} \leq x < \frac{j}{n} + \frac{1}{2n}, \quad 0 \leq j \leq n-1,$$
$$f_n(x) := 1 \quad \text{if} \quad \frac{j}{n} + \frac{1}{2n} \leq x < \frac{j+1}{n}, \quad 0 \leq j \leq n-1.$$

(1) Show that the sequence $(f_n)_{n=1}^{\infty}$ weakly converges in the space $L^2(0,1)$.
Hint: Use Problem 5.12-1.
(2) Does the sequence $(f_n)_{n=1}^{\infty}$ strongly converge in $L^2(0,1)$?

5.12-4 Let X be a Banach space and let $(x'_n)_{n=0}^{\infty}$ be a sequence of elements $x'_n \in X'$ that weakly * converges in X'.
(1) Show that the sequence $(x'_n)_{n=0}^{\infty}$ is bounded in X'.
(2) Show that the weak * limit x' of $(x'_n)_{n=0}^{\infty}$ satisfies

$$\|x'\|_{X'} \leq \liminf_{n\to\infty} \|x'_n\|_{X'}.$$

Remark This result constitutes the "weak * analogue" of Theorem 5.12-2. □

5.13 Banach–Saks–Mazur theorem

We saw in Section 5.12 that the sequence $(f_k)_{k=1}^{\infty}$ defined by $f_k(\theta) = \sin k\theta$, $0 \leq \theta \leq 2\pi$, weakly converges to 0 in the space $L^2(0, 2\pi)$, but does not strongly converge in that space. Yet, there are sequences of *convex combinations* (Section 2.16) of the functions f_k that do strongly converge to the *same* limit 0 in $L^2(0, 2\pi)$ such as, for instance, the sequences $(f_n)_{n=1}^{\infty}$ and $(h_n)_{n=1}^{\infty}$ defined by (Problem 5.13-1)

$$f_n := \frac{1}{n}\sum_{k=1}^{n} f_k \quad \text{or} \quad h_n := \frac{1}{pn+1}\sum_{k=n}^{n+pn} f_k \quad \text{for any fixed integer } p \geq 1.$$

Such a result holds in fact in *any* normed vector space, according to the following beautiful, and very often used, result. It plays in particular a key role in the *calculus of variations* (Chapter 9).

As shown in part (i) of its proof, this theorem crucially hinges on the *separation of convex sets by a hyperplane*, as provided by the *first geometric form of the Hahn–Banach theorem* (Theorem 5.10-1).

Notice that, in parts (a) and (c) of the next theorem, the sequences denoted $(y_n)_{n=1}^{\infty}$ and $(z_n)_{n=1}^{\infty}$ are not just sequences of *completely undetermined* convex combinations that strongly converge to the same weak limit x. But, as in the example above, the nth convex combination y_n is one of precisely the first n terms, while the nth convex combination z_n starts with precisely the nth term of the given weakly convergent sequence $(x_n)_{n=1}^{\infty}$.

Notice also that the proof of (c) rests on another, quite important by itself, relation between weak convergence and the assumed *convexity* of the set C, viz., property (b). By contrast, the conclusion of (b) holds for any *strongly* convergent sequence, irrespective of whether the closed set C is convex or not.

Theorem 5.13-1 (Banach–Saks–Mazur theorem[34]) *Let X be a real normed vector space.*

(a) *Let $(x_k)_{k=1}^{\infty}$ be a sequence in X such that*

$$x_k \rightharpoonup x \quad \text{as } k \to \infty.$$

Then, for each $n \geq 1$, there exist $\lambda_k^n \geq 0$, $1 \leq k \leq n$, with $\sum_{k=1}^{n} \lambda_k^n = 1$, such that

$$y_n := \sum_{k=1}^{n} \lambda_k^n x_k \to x \quad \text{as } n \to \infty.$$

(b) *Let C be a nonempty, convex, and closed subset of X, and let $(x_k)_{k=1}^{\infty}$ be a sequence of points $x_k \in C$ that weakly converges to $x \in X$ as $k \to \infty$. Then the weak limit x belongs to C.*

(c) *Let $(x_k)_{k=1}^{\infty}$ be a sequence in X such that*

$$x_k \rightharpoonup x \quad \text{as } k \to \infty.$$

Then, for each $n \geq 1$, there exist an integer $m(n) \geq 0$ and $\mu_k^n \geq 0$, $n \leq k \leq n + m(n)$, with $\sum_{k=n}^{n+m(n)} \mu_k^n = 1$, such that

$$z_n := \sum_{k=n}^{n+m(n)} \mu_k^n x_k \to x \quad \text{as } n \to \infty.$$

Proof Recall that $\operatorname{co} A$ designates the *convex hull* of a subset A of a vector space (Section 2.16).

(i) *Proof of* (a). Define the convex sets

$$A_n := \operatorname{co}\left(\bigcup_{k=1}^{n} \{x_k\}\right) \quad \text{for each integer } n \geq 1, \quad \text{and} \quad A := \bigcup_{n=1}^{\infty} A_n$$

(each set A_n is convex by construction, and thus A is convex since $A_n \subset A_{n+1}$ for all $n \geq 1$). We then claim that

$$\rho_n := \inf_{w \in A_n} \|x - w\| \to 0 \quad \text{as } n \to \infty.$$

If not, there exists $\rho > 0$ such that $A \cap B(x;\rho) = \varnothing$ (since $\rho_n \geq \rho_{n+1}$ for all $n \geq 1$). Hence by the *first geometric form of the Hahn–Banach theorem* (Theorem 5.10-1) there exists a hyperplane that separates A and $B(x;\rho)$ (the ball $B(x;\rho)$ is open and convex): this means that there exist a nonzero $\ell \in X'$ and $\gamma \in \mathbb{R}$ such that

$$\ell(x + \rho v) = \ell(x) + \rho\ell(v) \leq \gamma \leq \ell(w) \quad \text{for all } \|v\| < 1 \text{ and all } w \in A.$$

[34] S. Banach; S. Saks [1930]: Sur la convergence forte dans le champ L^p, *Studia Mathematica* **2**, 51–57. S. Mazur [1933]: Über konvexe Mengen in linearen normierten Räumen, *Studia Mathematica* **5**, 70–84.

Consequently

$$\ell(x) + \rho\|\ell\| = \ell(x) + \rho \sup_{\|v\|<1} \ell(v) \leq \ell(w) \quad \text{for all } w \in A.$$

Letting $w = x_n$ in this inequality gives

$$\ell(x) + \rho\|\ell\| \leq \ell(x_n) \quad \text{for all } n \geq 1,$$

but this is impossible since $\ell(x_n) \to \ell(x)$ as $n \to \infty$, by definition of weak convergence.

Hence $\inf_{w\in A_n} \|x - w\| \to 0$ as $n \to \infty$. Consequently, for each $n \geq 1$, there exists $y_n \in A_n$ such that $\inf_{w\in A_n} \|x - w\| = \|x - y_n\| \to 0$ as $n \to \infty$ (each set A_n is compact). Equivalently, for each $n \geq 1$, there exist $\lambda_k^n \geq 0$, $1 \leq k \leq n$, with $\sum_{k=1}^n \lambda_k^n = 1$ such that

$$\left\| x - \sum_{k=1}^{n} \lambda_k^n x_k \right\| \to 0 \quad \text{as } n \to \infty$$

(there is evidently no loss of generality in assuming that all $x_1, x_2, \ldots, x_n$ enter in each convex combination y_n).

(ii) *Proof of* (b). The convex combinations $y_n = \sum_{k=1}^n \lambda_k^n x_k$ given by (a) belong to the set C since C is convex. Besides, $y_n \to x$ as $n \to \infty$, again by (a). Hence $x \in C$ since C is closed for the strong topology.

(iii) *Proof of* (c). Define the convex sets

$$C_n := \operatorname{co}\left(\bigcup_{k=n}^{\infty} \{x_k\}\right) \quad \text{for each integer } n \geq 1.$$

For each $n \geq 1$, $x_m \in C_n \subset \overline{C}_n$ for all $m \geq n$, and $(x_m)_{m=n}^{\infty}$ weakly converges to x; hence $x \in \overline{C}_n$ by (ii) (each closure $\overline{C}_n$ is also convex). Consequently, for each $n \geq 1$, there exists $z_n \in C_n$ such that $\|x - z_n\| \leq \dfrac{1}{n}$ (to fix ideas); equivalently, for each $n \geq 1$, there exist an integer $m(n) \geq 0$ and $\mu_k^n \geq 0$, $n \leq k \leq n + m(n)$, with $\sum_{k=n}^{n+m(n)} \mu_k^n = 1$, such that

$$\left\| x - \sum_{k=n}^{n+m(n)} \mu_k^n x_k \right\| \to 0 \quad \text{as } n \to \infty. \qquad \square$$

Remark The proof of (b) becomes remarkably simple if X is a *Hilbert space* (in which case X can be identified with its dual space), in which case it does not rest on the Banach–Saks–Mazur theorem: Let $(\cdot,\cdot)$ denote the inner product in X, and let $P : X \to C$ denote the projection operator from X onto C, which thus satisfies $(Px - x, y - Px) \geq 0$ for all $y \in C$ (Theorem 4.3-1). Then in particular $(Px - x, x_k - Px) \geq 0$ for all $k \geq 1$, so that

$$-\|x - Px\|^2 = (Px - x, x - Px) = \lim_{k\to\infty} (Px - x, x_k - Px) \geq 0.$$

Hence $x = Px \in C$. $\square$

The Banach–Saks–Mazur theorem thus shows that *a convex subset of a normed vector space X that is closed for the strong topology of X is* **sequentially weakly closed**, in the sense that it contains the weak limits of all its weakly convergent sequences. Actually, one can further prove[35] that such a subset is indeed *"weakly closed"* in the sense that *it is closed for the weak topology of* X (Section 5.12).

Problem

5.13-1 Let $f_k(\theta) := \sin k\theta$, $0 \leq \theta \leq 2\pi$, $k \geq 1$. Show that the sequences $\left(\frac{1}{n}\sum_{k=1}^{n} f_k\right)_{n=1}^{\infty}$, and $\left(\frac{1}{pn+1}\sum_{k=n}^{n+pn} f_k\right)$ for any fixed integer $p \geq 1$, strongly converge to 0 in $L^2(0, 2\pi)$.

5.14 Reflexive spaces; the Banach–Eberlein–Šmulian theorem

Let X be a normed vector space. Then

$$X'' := (X')'$$

denotes the **bidual space** of X, or simply the **bidual** of X, i.e., the dual space of the dual space of X. As a dual space, *the space* X'' *is thus a Banach space*, with the norm of any element $x'' \in X''$ being given by

$$\|x''\|_{X''} = \sup_{x' \neq 0} \frac{|x''(x')|}{\|x'\|_{X'}}.$$

As we shall see, a basic result (the Banach–Eberlein–Šmulian theorem; cf. Theorem 5.14-4) asserts that a weakly convergent subsequence can be extracted from *any* bounded sequence in a *Banach space* X if, and only if, X *can be identified with the bidual space* X'' by means of a *specific linear isometry*. In view of properly defining this notion, we first show how *any* normed vector space can be identified in a natural way with a *subspace* of its bidual space.

Theorem 5.14-1 *Let* X *be a normed vector space. Then the mapping*

$$J : x \in X \to Jx \in X'',$$

defined for each $x \in X$ *by*

$$Jx(x') := x'(x) \quad \textit{for all } x' \in X',$$

is a linear isometry, called the **canonical isometry** *from* X *into* X''.

Proof Given any element $x \in X$, the functional

$$Jx : x' \in X' \to Jx(x') := x'(x) \in \mathbb{K}$$

is linear and continuous: first, for all $\alpha, \beta \in \mathbb{K}$ and all $x', y' \in X'$,

$$Jx(\alpha x' + \beta y') = (\alpha x' + \beta y')(x) = \alpha x'(x) + \beta y'(x) = \alpha Jx(x') + \beta Jx(y'),$$

[35] See, e.g., BREZIS [1983, Theorem 3.7].

since X' is a vector space; second, $Jx \in X''$ since

$$|Jx(x')| = |x'(x)| \le \|x'\|\,\|x\| \quad \text{for all } x' \in X'.$$

The mapping $J : X \to X''$ defined in this fashion is an isometry, since, by Theorem 5.9-5 (a corollary to the Hahn–Banach theorem in a normed vector space),

$$\|Jx\|_{X''} = \sup_{x' \neq 0} \frac{|Jx(x')|}{\|x'\|} = \sup_{x' \neq 0} \frac{|x'(x)|}{\|x'\|} = \|x\| \quad \text{for any } x \in X. \qquad \square$$

Remark The mappings J_n, $n \ge 1$, and J_x that appeared in the proof of Theorem 5.12-2 were nothing but special cases of the functionals $Jx \in X''$ appearing in the above proof. □

A normed vector space X is **reflexive** if the canonical isometry $J : X \to X''$ defined in Theorem 5.14-1 is *surjective*, thus allowing us to *identify* X *with its bidual space* X''. In other words, X is reflexive if, given any $x'' \in X''$, there exists (a unique) $x \in X$ such that

$$x''(x') = x'(x) \quad \text{for all } x' \in X'.$$

Essential to this definition is that the identification of X with X'' be achieved by means of the *canonical* isometry. Otherwise, there exist Banach spaces that can be identified with their bidual by means of a linear isometry, yet that are *not* reflexive.[36]

Observe that, as a dual space, *a reflexive space is necessarily complete.*

The next two theorems provide *examples* of Banach spaces that are reflexive; see also Problems 5.14-1 and 5.14-2 for further examples, and Problem 5.14-5 for a crucial *counterexample*, that of the space $\mathcal{C}[0,1]$ equipped with the sup-norm.

Theorem 5.14-2 *The following Banach spaces are reflexive:*

(a) *Any finite-dimensional normed vector space;*

(b) *any Hilbert space;*

(c) *any closed subspace of a reflexive Banach space;*

(d) *the dual space of any reflexive Banach space;*

(e) *the spaces* ℓ^p, $1 < p < \infty$, *and the Lebesgue spaces* $L^p(\Omega)$, $1 < p < \infty$, *with* Ω *any open subset of* $\mathbb{R}^n$.

Proof (i) *Proof of* (a). Let X be a finite-dimensional normed vector space. Given a basis $(e_i)_{i=1}^n$ in X, the relations $e'_j(e_i) = \delta_{ij}$, $1 \le i, j \le n$, define a basis $(e'_j)_{j=1}^n$ in X', which in turn defines a basis $(e''_k)_{k=1}^n$ in X'' by means of the relation $e''_k(e'_j) = \delta_{jk}$, $1 \le j, k \le n$. It is then immediately verified that the canonical isometry $J : X \to X''$ of Theorem 5.14-1 is given in this case by

$$J\Big(\sum_{i=1}^n x_i e_i\Big) = \sum_{i=1}^n x_i e''_i \quad \text{for all } x = \sum_{i=1}^n x_i e_i \in X.$$

Hence J is surjective.

[36] R.C. James [1951]: A non-reflexive Banach space isometric with its second conjugate space, *Proceedings of the National Academy of Sciences, USA* **37**, 174–177.

(ii) *Proof of* (b). Given a Hilbert space $(X, (\cdot,\cdot)_X)$, let $\sigma : X' \to X$ denote the corresponding F. Riesz isometry. Then, equipped with the inner product $(\cdot,\cdot)_{X'}$ defined for each $x', y' \in X'$ by

$$(x', y')_{X'} := \overline{(\sigma x', \sigma y')_X},$$

the space X' is also a Hilbert space (Theorem 4.6-1). Let $\sigma' : X'' \to X'$ be the corresponding F. Riesz isometry. It is then immediately verified that the canonical isometry $J : X \to X''$ is given in this case by

$$J = (\sigma \circ \sigma')^{-1}$$

(the composition of two linear if $\mathbb{K} = \mathbb{R}$, or semilinear if $\mathbb{K} = \mathbb{C}$, isometries is a linear isometry).

(iii) *Proof of* (c). Let Y be a closed subspace of a reflexive Banach space X. So, we need to show that, given any $y'' \in Y''$, there exists $y \in Y$ such that

$$y''(y') = y'(y) \quad \text{for all } y' \in Y'.$$

Let then $y'' \in Y''$ be given. The linear functional

$$x'' : x' \in X' \to x''(x') := y''(x'|_Y) \in \mathbb{K},$$

where $x'|_Y$ denotes the restriction of x' to Y, is continuous since

$$|y''(x'|_Y)| \leq \|y''\|\,\|x'|_Y\| \leq \|y''\|\,\|x'\| \quad \text{for all } x' \in X'.$$

Hence $x'' \in X''$ and, since X is reflexive by assumption, there exists $y \in X$ such that

$$x'(y) = x''(x') = y''(x'|_Y) \quad \text{for all } x' \in X'.$$

In particular then, $x'(y) = 0$ for all those $x' \in X'$ whose restriction $x'|_Y$ vanishes; hence $y \in Y$ since Y is closed by assumption (if $y \notin Y$, there would exist $x' \in X'$ such that $x'|_Y = 0$ and $x'(y) \neq 0$, by Theorem 5.9-6).

Given any $x' \in Y' = \mathcal{L}(Y; \mathbb{K})$, let $x' \in X' = \mathcal{L}(X; \mathbb{K})$ be any extension of y' (such an extension exists by the Hahn–Banach theorem in a normed vector space; cf. Theorem 5.9-1). Then

$$y''(y') = y''(x'|_Y) = x'(y) = y'(y)$$

(since $y' = x'|_Y$ and $y \in Y$), as was to be proved.

(iv) *Proof of* (d). Given a reflexive Banach space X, let $J : X \to X''$ denote the canonical isometry of X onto its bidual space X'' and let $J' : X' \to (X')''$ denote the canonical isometry of X' into its bidual space $(X')''$.

Given any $x''' \in (X')''$, define the mapping

$$x' : x \in X \to x'(x) := x'''(Jx) \in \mathbb{K}.$$

Note that this definition makes sense since $Jx \in X''$ and

$$(X')'' = \mathcal{L}(\mathcal{L}(\mathcal{L}(X; \mathbb{K}); \mathbb{K}); \mathbb{K}) = (X'')'.$$

Then $x' \in X'$ since

$$|x'(x)| \leq \|x'''\|_{(X')''} \|Jx\|_{X''} = \|x'''\|_{(X')''} \|x\|_X \quad \text{for all } x \in X.$$

Besides, the definitions of x' and J combined with the bijectivity of $J : X \to X''$ give

$$x'''(x'') = x'''(Jx) = x'(x) = Jx(x') = x''(x') \quad \text{for all } x'' = Jx \in X'',$$

which shows that

$$x''' = J'x',$$

by definition of J'. Hence J' is surjective, as was to be proved.

(v) *Proof of* (e). The reflexivity of the spaces ℓ^p, $1 < p < \infty$, follows from the characterization of their dual spaces (Theorem 3.5-1). Likewise, the reflexivity of the spaces $L^p(\Omega)$, $1 < p < \infty$, follows from the F. Riesz representation theorem in $L^p(\Omega)$ (Theorem 3.5-3). Naturally, their reflexivity for $p = 2$ also follows from (b). □

A different approach for establishing the reflexivity of the spaces ℓ^p and $L^p(\Omega)$, $1 < p < \infty$, consists first in showing that they *are uniformly convex* (Problems 2.17-8 and 2.17-9), then in using the following fundamental sufficient condition for reflexivity, singled out here in view of its importance.

$^\flat$**Theorem 5.14-3 (Milman–Pettis theorem[37])** *A uniformly convex Banach space is reflexive.* □

Remark There are of course reflexive Banach spaces that are *not* strictly convex, let alone uniformly convex, e.g., the spaces $(\mathbb{R}^n, \|\cdot\|_1)$ and $(\mathbb{R}^n, \|\cdot\|_\infty)$. □

We conclude this chapter with *one of the most basic theorems of linear functional analysis.* In particular, the result of part (a) plays a fundamental role in establishing the existence of minimizers of coercive and sequentially weakly lower semicontinuous functionals (Theorem 9.3-1).

Part (b) provides an efficient way to show that a space is reflexive; for example, it can be put to use for proving that the Sobolev spaces $W^{m,p}(\Omega)$, $1 < p < \infty$, are reflexive (Problem 6.11-2).

Theorem 5.14-4 (Banach–Eberlein–Šmulian theorem[38]) (a) *Any bounded sequence in a reflexive Banach space contains a weakly convergent subsequence.*

---

[37] Independently proved by:

D.P. Milman [1938]: On some criteria for the regularity of spaces of type (B), *Doklady Akademii Nauk SSSR* **20**, 243–246 (in Russian).

B.J. Pettis [1939]: A proof that every uniformly convex space is reflexive, *Duke Mathematical Journal* **5**, 249–253.

For a proof, see also Yosida [1966, Chapter V, Section 2], Diestel [1975], or Brezis [2011, Theorem 3.31].

[38] Part (a) is proved in Banach [1932] (under the additional assumption of separability). Part (b) is due to:

V.L. Šmulian [1940]: Über lineare topologische Räume, *Mathematiceskii Sbornik, N.S.* **49**, 425–448.

W.F. Eberlein [1947]: Weak compactness in Banach spaces I, *Proceedings of the National Academy of Sciences, USA* **33**, 51–53.

$^\flat$(b) *Conversely, a Banach space in which every bounded sequence contains a weakly convergent subsequence is reflexive.*[39]

Proof We prove (a) under the additional assumption that the space X is *separable*[40] (an assumption satisfied by all the function spaces encountered in the sequel).

(i) The assumption that X is reflexive means that there exists a linear isometry from X onto X'' (viz., the canonical isometry); therefore, like X, the space X'' is thus also separable. Since, by definition, $X'' = (X')'$, *the space X' is thus also separable*, by Theorem 5.9-8. Let then $x'_k \in X'$, $k \geq 1$, be such that

$$X' = \overline{\bigcup_{k=1}^{\infty} \{x'_k\}}.$$

(ii) *Let $(x_n)_{n=1}^{\infty}$ be a bounded sequence of elements $x_n \in X$.* Therefore, for each $x' \in X'$,

$$|x'(x_n)| \leq M\|x'\| \quad \text{for all } n \geq 1, \text{ where } M := \sup \|x_n\| < \infty;$$

the sequence $(x'(x_n))_{n=1}^{\infty}$, being thus bounded in $\mathbb{K}$, contains a convergent subsequence.

In particular, the sequence $(x'_1(x_n))_{n=1}^{\infty}$ contains a convergent subsequence $(x'_1(x_{\sigma_1(n)}))_{n=1}^{\infty}$; the sequence $(x'_2(x_{\sigma_1(n)}))_{n=1}^{\infty}$, being likewise bounded in $\mathbb{K}$, likewise contains a convergent subsequence $(x'_2(x_{\sigma_2(n)}))_{n=1}^{\infty}$; and so on. Consider the *"diagonal" sequence*

$$(x_{\sigma(n)})_{n=1}^{\infty}, \text{ where } \sigma(n) := \sigma_n(n), \quad n \geq 1.$$

Then by construction, $(x_{\sigma(n)})_{n=1}^{\infty}$ is a *subsequence of the sequence $(x_n)_{n=1}^{\infty}$; hence, for each integer $k \geq 1$, the sequence $(x'_k(x_{\sigma(n)}))_{n=1}^{\infty}$ converges in $\mathbb{K}$ as $n \to \infty$.*

(iii) We next show that, in fact, *for each $x' \in X'$, the sequence $(x'(x_{\sigma(n)}))_{n=1}^{\infty}$ converges in $\mathbb{K}$ as $n \to \infty$.*

Let $x' \in X'$ and $\varepsilon > 0$ be given. By (i), there exists an integer $k = k(x', \varepsilon) \geq 1$ such that $\|x' - x'_k\| \leq \dfrac{\varepsilon}{4M}$. Then, for any integers $m, n \geq 1$,

$$\begin{aligned}|x'(x_{\sigma(m)}) - x'(x_{\sigma(n)})| &\leq |x'_k(x_{\sigma(m)}) - x'_k(x_{\sigma(n)})| + |(x' - x'_k)(x_{\sigma(m)} - x_{\sigma(n)})| \\ &\leq |x'_k(x_{\sigma(m)}) - x'_k(x_{\sigma(n)})| + \frac{\varepsilon}{2},\end{aligned}$$

since $\|x_{\sigma(m)} - x_{\sigma(n)}\| \leq 2M$. But $|x'_k(x_{\sigma(m)}) - x'_k(x_{\sigma(n)})|$ can be made arbitrarily small for m and n large enough, since, as a convergent sequence, $(x'_k(x_{\sigma(n)}))_{n=1}^{\infty}$ is a Cauchy sequence. Hence there exists an integer $n_0 = n_0(k) = n_0(x', \varepsilon) \geq 1$ such that

$$|x'_k(x_{\sigma(m)}) - x'(x_{\sigma(n)})| \leq \varepsilon \quad \text{for all } m, n \geq n_0,$$

which shows that $(x'(x_{\sigma(n)}))_{n=1}^{\infty}$ is also a Cauchy sequence. Therefore $(x'(x_{\sigma(n)}))_{n=1}^{\infty}$ converges in $\mathbb{K}$.

---

[39] A proof of (b) is found in Yosida [1966, Appendix to Chapter 5] or in Dunford & Schwartz [1958, Chapter 5, Section 6].

[40] A proof in the nonseparable case is found in Yosida [1966, Appendix to Chapter 5]. See also:
R. Whitley [1967]: An elementary proof of the Eberlein-Šmulian theorem, *Mathematische Annalen* **172**, 116–118.

(iv) Let $J : X \to X''$ denote the linear isometry given by Theorem 5.14-1, which is surjective since X is assumed to be reflexive. By (iii), the continuous linear functionals $Jx_{\sigma(n)} \in X'' = \mathcal{L}(X'; \mathbb{K})$, which are thus defined for each $n \geq 1$ by

$$Jx_{\sigma(n)}(x') = x'(x_{\sigma(n)}) \quad \text{for all } x' \in X',$$

have the following property:

$$\lim_{n\to\infty} Jx_{\sigma(n)}(x') \text{ exists in } \mathbb{K} \text{ as } n \to \infty \text{ for each } x' \in X'.$$

Since the space X' is complete, the corollary to the Banach–Steinhaus theorem (Theorem 5.3-2) can be applied, showing that there exists $x'' \in X'' = \mathcal{L}(X', \mathbb{K})$ such that

$$Jx_{\sigma(n)}(x') \to x''(x') \quad \text{as } n \to \infty \text{ for each } x' \in X'.$$

But this is the same as

$$x'(x_{\sigma(n)}) \to x'(x) \text{ as } n \to \infty \text{ for each } x' \in X', \text{ where } x := J^{-1}x''.$$

Hence *the subsequence* $(x_{\sigma(n)})_{n=1}^{\infty}$ *weakly converges to* x *as* $n \to \infty$. □

Problems

5.14-1 Show that, if the dual of a *Banach space* X is reflexive, then X itself is reflexive.

Remark This result, combined with Theorem 5.14-2(d), thus shows that *a Banach space is reflexive if and only if its dual is also reflexive.* □

5.14-2 Let Y be a closed subspace of a reflexive Banach space X. Show that the quotient space X/Y is reflexive.

5.14-3 (1) Let X be a reflexive Banach space. Show that, given any $x' \in X'$, there exists $x_0 \in X$ such that $\|x_0\| = 1$ and $\|x'\| = \sup_{\|x\|=1} |x'(x)| = x'(x_0)$.

(2) Show that, conversely, if a Banach space X is such that, given any $x' \in X'$, there exists $x_0 \in X$ such that $\|x_0\| = 1$ and $\|x'\| = \sup_{\|x\|=1} |x'(x)| = x'(x_0)$, then X is reflexive.[41]

5.14-4 Let $(X, \|\cdot\|)$ be a reflexive Banach space, and let Z be a nonempty closed convex subset of X.

(1) Show that, given any element $x \in X$, there exists $y \in Z$ such that $\|x - y\| = \inf_{z\in Z} \|x - z\|$. *Hint*: Consider an infimizing sequence and use the *Banach–Eberlein–Šmulian theorem*.

(2) Show that y is unique if $(X, \|\cdot\|)$ is strictly convex.

Remark Questions (1) and (2), which thus extend the *projection theorem in a Hilbert space* (Theorem 4.3-1), together constitute the **projection theorem in a reflexive Banach space**. □

5.14-5 Let the subset Z of the space $\mathcal{C}[0,1]$ equipped with the sup norm $\|\cdot\|$ be defined by

$$Z = \left\{ f \in \mathcal{C}[0,1]; \int_0^{1/2} f(x)\,\mathrm{d}x = 1 + \int_{1/2}^1 f(x)\,\mathrm{d}x \right\}.$$

(1) Show that Z is a nonempty closed convex subset of $\mathcal{C}[0,1]$.

[41] R.C. James [1964]: Characterizations of reflexivity, *Studia Mathematica* **23**, 205–216.

(2) Show that $\inf_{f \in Z} \|f\| = 1$ but that there is no $f \in Z$ such that $\|f\| = 1$.
(3) Conclude from Problem 5.14-4 that *the Banach space* $(\mathcal{C}[0,1], \|\cdot\|)$ *is not reflexive.*

5.14-6 Let X be a separable Banach space. Show that any bounded sequence in X' contains a weakly * convergent subsequence.

Remark This result constitutes the "weak * analogue" of Theorem 5.14-4(a). □

5.14-7 Let Ω be an open subset of $\mathbb{R}^n$ and let $(f_k)_{k=0}^{\infty}$ be a bounded sequence in $L^\infty(\Omega)$. Show that there exist a subsequence $(f_{\sigma(k)})_{k=0}^{\infty}$ and a function $f \in L^\infty(\Omega)$ such that

$$\text{for each } g \in L^1(\Omega), \quad \int_\Omega f_{\sigma(k)} g \, dx \to \int_\Omega f g \, dx \text{ as } k \to \infty.$$

Hint: Use Problem 5.14-6.

5.14-8 Let X be a normed vector space. Show that the image $J(X)$ of X under the canonical isometry J is closed in X'' if and only if X is a Banach space.

5.14-9 Let Ω be an open subset of $\mathbb{R}^n$, let $1 < p < \infty$, and let functions $f_k \in L^p(\Omega)$, $k \geq 1$, and $f \in L^p(\Omega)$ be such that the sequence $(f_k)_{k=1}^{\infty}$ is bounded in $L^p(\Omega)$ and f_k converges almost everywhere in Ω to f as $k \to \infty$. Show that

$$f_k \rightharpoonup f \quad \text{in } L^p(\Omega) \text{ as } k \to \infty.$$

CHAPTER 6

LINEAR PARTIAL DIFFERENTIAL EQUATIONS

Introduction

In this chapter, we only consider partial differential equations where all the variables are "space variables," i.e., coordinates of points in an open subset of $\mathbb{R}^N$; we do *not* consider "time-dependent" problems.

Problems in *optimization theory*, or in applications such as *linearized elasticity* or *linearized fluid mechanics*, are often modeled by *minimization problems* of the following form: The *unknown* u satisfies

$$u \in U \quad \text{and} \quad J(u) = \inf_{v \in U} J(v),$$

where U is a nonempty closed convex subset of a Hilbert space V, and $J : V \to \mathbb{R}$ is a *quadratic functional*, i.e., of the form

$$J(v) = \frac{1}{2}a(v,v) - \ell(v) \quad \text{for any } v \in V,$$

where $a(\cdot,\cdot)$ is a symmetric bilinear form and ℓ is a linear form, both defined and continuous over the space V.

We first prove, as a simple consequence of the *projection theorem in a Hilbert space* (Chapter 4), a general *existence result* (Theorem 6.1-1) for such minimization problems, the main assumptions of which are the *completeness* of the space V and the *V-coercivity* of the bilinear form. We also describe other equivalent formulations of the same problem (Theorem 6.1-2), called its *variational formulations*, which take the form of *variational inequalities* in general, or of *variational equations* when U is a subspace. When the bilinear form is not symmetric, these formulations make up *abstract variational problems* on their own. For such problems, we then give an existence theorem when $U = V$ (Theorem 6.2-1), which is the celebrated *Lax–Milgram lemma.*

A candidate for the Hilbert space V should therefore have the following properties: It must be *complete* on the one hand, and it must be such that the expression $J(v)$ is well defined for all functions $v \in V$ on the other hand. For the applications that we have in mind (elasticity and fluid mechanics), the *Sobolev spaces* $H^m(\Omega)$ *and* $H_0^m(\Omega)$ fulfill these requirements. The basic properties of these spaces, as well as (for coherence of exposition) those of the more general *Sobolev spaces* $W^{m,p}(\Omega)$ *and* $W_0^{m,p}(\Omega)$, $1 \leq p \leq \infty$, needed in the last chapter for the analysis of *nonlinear* partial differential equations corresponding to the minimization of *nonquadratic* functionals, are reviewed in Sections 6.5 and 6.6; these two sections are preceded by a brief incursion (Section 6.3) into *distribution theory* (the elements of the Sobolev spaces are themselves distributions), which includes in particular a detailed proof of the *Weyl lemma* regarding the *hypoellipticity of the Laplace operator* (Theorem 6.4-2), a crucial property that will be used later at various places.

We then describe and analyze in Sections 6.7–6.9 specific *examples* of variational problems in linearized elasticity that fit in the above abstract setting, such as the *membrane problem*, *plate problems*, or *obstacle problems*. For each example, the main step consists in establishing the V*-coercivity* of the associated bilinear form. As an application of the spectral theorem for compact self-adjoint operators (Chapter 4), we also give a detailed treatment of *eigenvalue problems for second-order elliptic operators* (Theorem 6.10-2).

We also show that, when solving such variational problems, one solves *in the sense of distributions*, and also in the *classical sense* if the solution possesses *ad hoc* regularity, *boundary value problems of the second*, or *fourth, order*. These problems are *linear* if U is a subspace of V, and *nonlinear* if U is not a subspace of V.

The same approach is used for solving two *systems* of linear partial differential equations of paramount importance in applications, the *Stokes equations* (Section 6.14) and the *equations of three-dimensional linearized elasticity* (Section 6.16); incidentally, note that the corresponding *nonlinear* equations that they approximate will be solved in Chapter 9. Establishing the existence of solutions to such equations requires a more elaborate analysis, as it respectively relies on the *Babuška–Brezzi inf-sup theorem* (Theorem 6.12-1) and the *Korn inequality* (Theorem 6.15-1), which both ultimately rely on the *same* fundamental and deep *lemma of Jacques-Louis Lions* (Theorem 6.11-4).

This chapter is then concluded by an analysis of perhaps less conventional topics, such as the *Poincaré lemma*, both in its *classical* form (Theorem 6.17-2) and in its *weak* form (i.e., in Sobolev spaces with negative exponents; cf. Theorem 6.17-4), the *classical* and the *weak Saint-Venant lemma* (Theorems 6.18-1 and 6.18-3), the *Cesàro–Volterra path integral formula* (Theorem 6.18-2), the *Donati lemmas* (Theorems 6.19-5 and 6.19-6), and finally, an existence and uniqueness theorem for *Pfaff systems* (Theorem 6.20-1), which will play a key role in establishing *existence theorems* in *differential geometry* (Chapter 8).

All functions and vector spaces considered in this chapter are real.

6.1 Quadratic minimization problems; variational equations and variational inequalities

We begin with an *existence* and *uniqueness* result for a *minimization problem* that models a wide variety of problems, as it shall be abundantly illustrated in this chapter.

Note that we will henceforth use notations such as $u, v \in V$, etc., rather that $x, y \in X$, etc. This type of notation is often used when the spaces considered are *function spaces*, i.e., spaces whose elements are themselves *functions* (typically, defined over an open subset of $\mathbb{R}^N$), as will be the case in the examples considered later. Again to comply with the common usage, a function $J : V \to \mathbb{R}$ such as that found in Theorem 6.1-1 below will be called a *functional*.

Theorem 6.1-1 *Let $(V, \|\cdot\|)$ be a Banach space, let $a : V \times V \to \mathbb{R}$ be a symmetric and continuous bilinear form with the property that there exists a constant α such that*

$$\alpha > 0 \quad \text{and} \quad a(v, v) \geq \alpha \|v\|^2 \quad \text{for all } v \in V,$$

let $\ell : V \to \mathbb{R}$ *be a continuous linear form, and let the functional* $J : V \to \mathbb{R}$ *be defined by*

$$J(v) := \frac{1}{2}a(v,v) - \ell(v) \quad \textit{for all } v \in V.$$

Finally, let U *be a nonempty closed convex subset of* V.

Then there exists a unique element u *such that*

$$u \in U \quad \textit{and} \quad J(u) = \inf_{v \in U} J(v).$$

The mapping $\ell \in V' \to u \in U$ *defined in this fashion is Lipschitz-continuous, and is linear if and only if* U *is a subspace of* V.

Proof Since the bilinear form a is continuous, there exists $M > 0$ such that $|a(u,v)| \leq M\,\|u\|\,\|v\|$ for all $u, v \in V$ (Theorem 2.11-1). The symmetric bilinear form $a(\cdot,\cdot)$ is clearly an inner product over the space V, and the associated norm is equivalent to the given norm $\|\cdot\|$, since

$$\sqrt{\alpha}\,\|v\| \leq \sqrt{a(v,v)} \leq \sqrt{M}\,\|v\| \quad \text{for all } v \in V.$$

Therefore the space V becomes a Hilbert space when it is equipped with this inner product.

By the *F. Riesz representation theorem* (Theorem 4.6-1), there thus exists a unique element $c = c(\ell) \in V$ such that

$$\ell(v) = a(c,v) \quad \text{for all } v \in V.$$

Again taking into account the symmetry of the bilinear form, we may therefore rewrite the functional J as

$$J(v) = \frac{1}{2}a(v,v) - a(c,v) = \frac{1}{2}a(v-c, v-c) - \frac{1}{2}a(c,c).$$

Hence finding $u \in U$ such that $J(u) = \inf_{v \in U} J(v)$ amounts to minimizing the distance between the element $c \in V$ and the set U, with respect to the norm $\sqrt{a(\cdot,\cdot)}$. In other words, the solution u *is the projection of* c *onto the set* U *with respect to the inner product* $a(\cdot,\cdot)$. By the *projection theorem* (Theorem 4.3-1), such a projection exists and is unique, since U is a nonempty closed convex subset of the space V.

Since both mappings $\ell \in V' \to c \in V$ and $c \in V \to u \in U$ are Lipschitz-continuous (Theorems 4.6-1 and 4.3-1), the composite mapping $\ell \in V' \to u \in U$ is itself Lipschitz-continuous.

Since the mapping $\ell \in V' \to c \in V$ is linear, the mapping $\ell \in V' \to u \in U$ is linear if and only if the projection operator $c \in V \to u \in U$ is itself linear, i.e., if and only if U is a subspace (Theorem 4.3-1). Consequently, *the mapping* $\ell \in V' \to u \in U$ *is linear if* U *is a subspace and nonlinear if* U *is not a subspace* (all other data being considered as fixed). □

Remark One should not forget, however, that if the resulting problem is *linear* when one minimizes the functional J over a subspace, this is so also because J is *quadratic*. The minimization of a nonquadratic functional over a subspace yields a *nonlinear* problem; see Chapter 9 for such examples. □

Let V be a normed vector space with norm $\|\cdot\|$. A bilinear form $a : V \times V \to \mathbb{R}$ (symmetric or not) with the property that there exists a constant α such that

$$\alpha > 0 \quad \text{and} \quad a(v,v) \geq \alpha \|v\|^2 \quad \text{for all } v \in V$$

is said to be V-**coercive**.

Remark A V-coercive bilinear form is also often called V-*elliptic*. Some caution should then be exercised, as this notion is related, but not equivalent, to those of uniformly *elliptic partial differential operators* and *elliptic boundary value problems of the second order* (introduced later; cf. Section 6.7). □

A functional $J : V \to \mathbb{R}$ is said to be **quadratic** if it is of the form $J(v) = \frac{1}{2}a(v,v) - \ell(v)$ for all $v \in V$, where $a : V \times V \to \mathbb{R}$ is a continuous and *symmetric* bilinear form and $\ell : V \to \mathbb{R}$ is a continuous linear form.

Remark If the bilinear form is V-coercive, the associated quadratic functional is *coercive*, in the sense that it satisfies $\lim_{\|v\| \to \infty} J(v) = \infty$ (since $J(v) \geq \frac{\alpha}{2}\|v\|^2 - \|\ell\|\,\|v\|$ for all $v \in V$). More general coercive functionals will be studied in Chapter 9. □

A **quadratic minimization problem** consists in seeking whether there exists an element that minimizes (the restriction of) a quadratic functional $J : V \to \mathbb{R}$ over a nonempty subset U of V.

We next show that the quadratic minimization problem considered in Theorem 6.1-1 can be given other equivalent formulations.

Theorem 6.1-2 *Let the assumptions and notations be as in Theorem* 6.1-1. *An element $u \in U$ is the solution of the minimization problem of Theorem* 6.1-1 *if and only if it satisfies the relations*

$$a(u, v-u) \geq \ell(v-u) \quad \textit{for all } v \in U$$

in the general case, or

$$a(u,v) = \ell(v) \quad \textit{for all } v \in U$$

if U is a closed subspace of V.

Proof Let $c \in V$ be such that $\ell(v) = a(c,v)$ for all $v \in V$. Then the projection theorem (Theorem 4.3-1) asserts that $u \in U$ is the projection of c onto U (cf. the proof of Theorem 6.1-1) if and only if the relations

$$a(u-c, v-u) \geq 0 \quad \text{for all } v \in U$$

hold. Since these relations may be rewritten as

$$a(u, v-u) \geq a(c, v-u) = \ell(v-u) \quad \text{for all } v \in U,$$

the announced inequalities hold.

If U is a *subspace* of V, the projection theorem asserts that $u \in U$ is the projection of c onto U if and only if

$$a(u-c, v) = 0 \quad \text{for all } v \in U,$$

i.e., if and only if $a(u,v) = \ell(v)$ for all $v \in U$. □

It is particularly illuminating to relate the characterizations of Theorem 6.1-2, which are expressed in terms of the bilinear form a and the linear form ℓ, to the functional J itself. To this end, we will use the identity

$$J(u+w) = \frac{1}{2}a(u+w, u+w) - \ell(u+w) = J(u) + \{a(u,w) - \ell(w)\} + \frac{1}{2}a(w,w),$$

which holds for arbitrary elements $u, w \in V$, thanks to the bilinearity and to the *symmetry*, which is *essential* here, of a and to the linearity of ℓ. This identity shows that, for a fixed element $u \in V$, the real number $\{a(u,w) - \ell(w)\}$ is the *linear part with respect to* w *in the exact difference* $J(u+w) - J(u)$. This linear part is called a *first variation* of the functional J at u.

Assume then that an element $u \in U$ satisfies $a(u, v-u) \geq \ell(v-u)$ for all $v \in U$ as in Theorem 6.1-2. Letting $w = v - u$ in the above identity then gives

$$J(v) - J(u) = \{a(u,w) - \ell(w)\} + \frac{1}{2}a(w,w) \geq \frac{\alpha}{2}\|w\|^2 \quad \text{for all } v \in U.$$

Consequently, $J(u) = \inf_{v \in U} J(v)$. Assume conversely that $J(u) = \inf_{v \in U} J(v)$. Given any element $v = u + w \in U$, we thus have $J(u+\theta w) - J(u) \geq 0$ for all $0 \leq \theta \leq 1$ (recall that the set U is convex). Consequently,

$$\theta\{a(u,w) - \ell(w)\} + \frac{\theta^2}{2}a(w,w) \geq 0 \quad \text{for } 0 \leq \theta \leq 1,$$

which implies that $a(u,w) - \ell(w) = a(u, v-u) - \ell(v-u) \geq 0$ for all $v \in U$.

In other words, an element $u \in U$ is such that $J(u) = \inf_{v \in U} J(v)$ if and only if *the first variation* $\{a(u,w) - \ell(w)\}$ *of the functional* J *at* u *is* ≥ 0 *for all* $w \in V$ *such that* $(u+w) \in U$.

In the special case where U is a *subspace*, let an element $u \in U$ satisfy $a(u,w) = \ell(w)$ for all $w \in U$. The above identity then gives

$$J(u+w) - J(u) = \frac{1}{2}a(w,w) \geq \frac{\alpha}{2}\|w\|^2 \quad \text{for all } w \in U.$$

Consequently, $J(u) = \inf_{v \in U} J(v)$. Assume conversely that $J(u) = \inf_{v \in U} J(v)$. Given any element $v = u + w \in U$, we thus have $J(u+\theta w) - J(u) \geq 0$ for all $\theta \in \mathbb{R}$. Consequently,

$$\theta\{a(u,w) - \ell(w)\} + \frac{\theta^2}{2}a(w,w) \geq 0 \quad \text{for all } \theta \in \mathbb{R},$$

which implies that $a(u,w) - \ell(w) = 0$ for all $w \in U$.

In other words, if U is a subspace, an element $u \in U$ is such that $J(u) = \inf_{v \in U} J(v)$ if and only if *the first variation* $\{a(u,w) - \ell(w)\}$ *of the functional* J *at* u *vanishes for all* $w \in U$.

Remark In Chapter 7, each first variation $\{a(u,w) - \ell(w)\}$ will be put in its proper perspective, namely, as the *Gâteaux derivative* $J'(u)w$ of the functional J at u in the direction w. □

The above considerations explain why the characterizations of Theorem 6.1-2 are called **variational formulations** of the minimization problem of Theorem 6.1-1, the relations $a(u, v-u) \geq \ell(v-u)$ for all $v \in U$ are called **variational inequalities**, and the relations $a(u,v) = \ell(v)$ for all $v \in U$ are called **variational equations**.

Remark In Section 6.12, another class of quadratic minimization problems will be introduced, where U is a closed subspace of V of the form $U = \{v \in V;\ b(v, \mu) = 0 \text{ for all } \mu \in M\}$, where V and M are both Hilbert spaces, $b : V \times M \to \mathbb{R}$ is a bilinear form that satisfies the *Babuška–Brezzi inf-sup condition*, and the bilinear form $a : V \times V \to \mathbb{R}$ is only U-elliptic; see Theorem 6.12-2. □

Problem

6.1-1 Let $(V, \|\cdot\|)$ be a Hilbert space and let $a : V \times V \to \mathbb{R}$ be a symmetric and continuous bilinear form with the following properties: $a(v, v) \geq 0$ for all $v \in V$ and, given *any* continuous linear form $\ell : V \to \mathbb{R}$, the variational equations $a(u, v) = \ell(v)$ for all $v \in V$ have one and only one solution $u \in V$.

Show that there exists a constant $\alpha > 0$ such that $a(v, v) \geq \alpha \|v\|^2$ for all $v \in V$.

Remark This result constitutes a converse to Theorem 6.1-2 when $U = V$. □

6.2 The Lax–Milgram lemma

Given a nonempty subset U of a vector space V, a bilinear form $a(\cdot, \cdot) : V \times V \to \mathbb{R}$ and a linear form ℓ, we can also consider the following **abstract variational problem**, in the formulation of which no functional appears: Find an element $u \in U$ such that

$$a(u, v - u) \geq \ell(v - u) \quad \text{for all } v \in U$$

in the general case, or find an element $u \in U$ such that

$$a(u, v) = \ell(v) \quad \text{for all } v \in V$$

if U is a subspace. By Theorem 6.1-2, each one of these problems has one and only one solution if the space V is complete, the nonempty subset U of V is closed and convex, the linear form ℓ is continuous, and the bilinear form is V-coercive, continuous, and *symmetric.*

One can then prove that, if the assumption of *symmetry* of the bilinear form is dropped and V is a *Hilbert space*, such abstract variational problems still have one and only one solution. Here we shall confine ourselves to the case where $U = V$, leaving the general case, which constitutes *Stampacchia's theorem*, as a problem (Problem 6.2-1).

Theorem 6.2-1 (Lax–Milgram lemma[1]) *Let V be a Hilbert space, let $a(\cdot, \cdot) : V \times V \to \mathbb{R}$ be a continuous and V-coercive bilinear form, and let $\ell : V \to \mathbb{R}$ be a continuous linear form.*

Then the following abstract variational problem: Find an element $u \in V$ such that

$$a(u, v) = \ell(v) \quad \textit{for all } v \in V,$$

has one and only one solution, and the mapping $\ell \in V' \to u \in V$ defined in this fashion is linear and continuous.

---

[1]P.D. Lax; A.M. Milgram [1954]: Parabolic equations, in *Contributions to the Theory of Partial Differential Equations, Annals of Mathematics Studies, No.* 33, pp. 167–190, Princeton University Press, Princeton, NJ.

The proof given here is that of:

J.L. Lions; G. Stampacchia [1967]: Variational inequalities, *Communications on Pure and Applied Mathematics* **20**, 493–519.

For his landmark contributions to the theory and approximation of partial differential equations, Peter D. Lax was awarded the Abel Prize in 2005.

Proof Let $(\cdot,\cdot)$ and $\|\cdot\|$ denote the inner product and the norm in V and let M be a constant such that

$$|a(u,v)| \leq M\,\|u\|\,\|v\| \quad \text{for all } u,v \in V.$$

This relation shows that, for each $u \in V$, the linear form $v \in V \to a(u,v) \in \mathbb{R}$ is continuous. Hence, for each $u \in V$, there exists a unique element $Au \in V'$ such that

$$a(u,v) = Au(v) \quad \text{for all } v \in V.$$

The mapping $A : V \to V'$ defined in this fashion is linear since $a(\cdot,\cdot)$ is linear with respect to its first argument, and continuous since

$$\|Au\|_{V'} = \sup_{v \neq 0} \frac{|Au(v)|}{\|v\|} = \sup_{v \neq 0} \frac{|a(u,v)|}{\|v\|} \leq M\,\|u\| \quad \text{for all } u \in V.$$

Hence

$$\|A\|_{\mathcal{L}(V;V')} \leq M.$$

Solving the abstract variational problem is therefore equivalent to solving the equation

$$Au = \ell \text{ in } V', \quad \text{or equivalently} \quad \tau(Au - \ell) = 0 \text{ in } V,$$

where $\tau : V' \to V$ denotes the F. Riesz mapping (Theorem 4.6-1).

We now show that, for each $\ell \in V'$, this equation has one and only one solution $u \in V$ by showing that, for appropriate values of $\rho > 0$, the affine mapping

$$f_\rho : v \in V \to v - \rho\tau(Av - \ell) \in V$$

is a contraction. Let $\alpha > 0$ be such that $a(v,v) \geq \alpha\,\|v\|^2$ for all $v \in V$ Then, for any $\rho > 0$,

$$\|v - \rho\tau Av\|^2 = \|v\|^2 - 2\rho(\tau Av, v) + \rho^2\,\|\tau Av\|^2 \leq \left(1 - 2\rho\alpha + \rho^2 M^2\right)\|v\|^2,$$

since $(\tau Av, v) = a(v,v) \geq \alpha\,\|v\|^2$ and $\|\tau Av\| = \|Av\|_{V'} \leq M\,\|v\|$. Therefore the affine mapping f_ρ is a contraction whenever the number ρ belongs to the interval $]0, \frac{2\alpha}{M^2}[$. The conclusion then follows from the *Banach fixed point theorem* (Theorem 3.7-1), which shows that f_ρ has a unique fixed point $u \in V$, which therefore satisfies $\tau(Au - \ell) = 0$.

The linear operator $A \in \mathcal{L}(V;V')$ is thus a bijection from V onto V'. The inverse mapping $A^{-1} : \ell \in V' \to u = A^{-1}\ell \in V$ is then also linear (Theorem 2.9-1), and it is continuous since the inequality

$$\alpha\,\|u\|^2 \leq a(u,u) = \ell(u) \leq \|\ell\|\,\|u\|$$

implies that

$$\|A^{-1}\ell\| = \|u\| \leq \alpha^{-1}\,\|\ell\| \quad \text{for all } \ell \in V'. \qquad \square$$

Remark In fact, $\|A\|_{\mathcal{L}(V;V')} = \|a\|$, where $\|a\| := \sup_{\{u \neq 0 \atop v \neq 0} \frac{|a(u,v)|}{\|u\|\,\|v\|}$ denotes the norm of the continuous bilinear form $a : V \times V \to \mathbb{R}$ (Theorem 2.11-4). To see this, observe first that, since $|a(u,v)| \leq \|a\|\,\|u\|\,\|v\|$ for all $u,v \in V$, the above proof shows that $\|A\|_{\mathcal{L}(V;V')} \leq \|a\|$. Next, let

$u_n, v_n \in V$, $n \geq 1$, be such that $\|u_n\| = \|v_n\| = 1$ for all $n \geq 1$ and $\|a\| = \lim_{n\to\infty} a(u_n, v_n)$; since, for all $n \geq 1$, $a(u_n, v_n) = Au_n(v_n) \leq \|Au_n\|_{V'} \leq \|A\|_{\mathcal{L}(V;V')}$, it follows that $\|a\| \leq \|A\|_{\mathcal{L}(V;V')}$. □

Using notions that will be introduced in Chapter 7, one can further show that, *if the bilinear form a is not symmetric, there is no longer a functional associated with the abstract variational problem considered in* Theorem 6.2-1; more specifically, the expressions $\{a(u,v) - \ell(v)\}$ for $v \in V$ are no longer the first variations (Section 6.1) at u of an *ad hoc* functional J. For, if they were, they could be written as $a(u,v) - \ell(v) = J'(u)v$ for all $v \in V$, where $J'(u) \in V'$ is the *Fréchet derivative* of a functional $J : V \to \mathbb{R}$ at u; then this relation would in turn imply that the second-order Fréchet derivative $J''(u) \in \mathcal{L}_2(V;\mathbb{R})$ of J is given by $J''(u)(v,w) = a(v,w)$ for all $(v,w) \in V \times V$. But one can show (Theorem 7.8-1) that any second-order Fréchet derivative is necessarily symmetric, a contradiction.

Problems

6.2-1 Questions (1) and (2) of this exercise constitute **Stampacchia's theorem.**[2]

Let V be a Hilbert space, let $a(\cdot,\cdot) : V \times V \to \mathbb{R}$ be a continuous and V-coercive bilinear form, let $\ell : V \to \mathbb{R}$ be a continuous linear form, and finally, let U be a nonempty closed convex subset of V.

(1) Show that the following *abstract variational problem*: Find an element $u \in U$ such that

$$a(u, v-u) \geq \ell(v-u) \quad \text{for all } v \in U,$$

has one and only one solution.

(2) Show that the mapping $\ell \in V' \to u \in V$ defined in this fashion is Lipschitz-continuous.

Hints: For (1), mimic the proof of Theorem 6.2-1; for (2), show that $\|u_1 - u_2\| \leq \alpha^{-1} \|\ell_1 - \ell_2\|$ if $u_1, u_2 \in U$ are solutions corresponding to $\ell_1, \ell_2 \in V'$.

6.2-2 Let $(V, \|\cdot\|)$ be a Hilbert space and let $a(\cdot,\cdot) : V \times V \to \mathbb{R}$ be a continuous bilinear form with the following property: Given *any* continuous linear form $\ell : V \to \mathbb{R}$ and *any* closed subspace U of V, the variational equations $a(u,v) = \ell(v)$ for all $v \in U$ have one and only one solution $u \in U$.

Show that there exists a constant $\alpha > 0$ such that, either $a(v,v) \geq \alpha \|v\|^2$ for all $v \in V$, or $-a(v,v) \geq \alpha \|v\|^2$ for all $v \in V$.

Remark This result[3] constitutes a *converse to the Lax–Milgram lemma* (Theorem 6.2-1). □

6.3 Weak partial derivatives in $L^1_{\text{loc}}(\Omega)$; a brief incursion into distribution theory

The objective of this section is to introduce the notion of *weak partial derivatives*, which play a crucial role in the definition of the *Sobolev spaces* (Section 6.5). We also explain why such weak derivatives are in effect special cases of *derivatives in the sense of distributions.*

Let Ω be an open subset of $\mathbb{R}^N$. Recall that $\mathcal{D}(\Omega)$ denotes the space of infinitely differentiable functions $\varphi : \Omega \to \mathbb{R}$ such that $\operatorname{supp} \varphi$ is a compact subset of Ω (Section 2.6), and that $L^1_{\text{loc}}(\Omega)$ denotes the space of all measurable functions $v : \Omega \to \mathbb{R}$ such that $v|_K \in L^1(K)$ for any compact subset K of Ω (Section 2.6).

[2] G. STAMPACCHIA [1964]: Formes bilinéaires coercitives sur les ensembles convexes, *Comptes Rendus de l'Académie des Sciences de Paris Série A*, **258**, 4413–4416.

[3] Due to Luc Tartar (personal communication).

To begin with, we prove a simple, but very useful, formula, satisfied by the usual partial derivatives of functions of class $\mathcal{C}^m$ on an open subset of $\mathbb{R}^N$. This formula, which can be viewed as an *integration by parts formula without boundary terms* (as it involves functions with compact supports), will in turn provide the basis for defining *weak* partial derivatives of order m.

Theorem 6.3-1 *Let Ω be an open subset of $\mathbb{R}^N$, let $m \geq 1$ be an integer, and let a function $v \in \mathcal{C}^m(\Omega)$ be given. Then*

$$\int_\Omega (\partial^{\boldsymbol{\alpha}} v)\varphi \,\mathrm{d}x = (-1)^{|\boldsymbol{\alpha}|} \int_\Omega v \partial^{\boldsymbol{\alpha}} \varphi \,\mathrm{d}x \quad \textit{for all } \varphi \in \mathcal{D}(\Omega),$$

for each multi-index $\boldsymbol{\alpha}$ such that $|\boldsymbol{\alpha}| \leq m$.

Proof Let a function $v \in \mathcal{C}^1(\Omega)$ be given. Then each integral $\int_\Omega (\partial_i v)\varphi \,\mathrm{d}x$, $1 \leq i \leq N$, is well defined for any function $\varphi \in \mathcal{D}(\Omega)$.

The function $w := v\varphi \in \mathcal{C}^1(\Omega)$ has a compact support in Ω. Let then $\widehat{w}$ denote the extension of w by zero in $\mathbb{R}^N - \Omega$, so that $\widehat{w} \in \mathcal{C}^1(\mathbb{R}^N)$. Since $\operatorname{supp} \widehat{w} = \operatorname{supp} w \subset \operatorname{supp} \varphi$, there exists $a > 0$ such that $\operatorname{supp} \widehat{w} \subset]-a, a[^N$. Therefore,

$$\begin{aligned}
\int_\Omega \partial_i w \,\mathrm{d}x &= \int_{[-a,\,a]^N} \partial_i \widehat{w} \,\mathrm{d}x \\
&= \int_{[-a,\,a]^{N-1}} \left(\int_{-a}^{a} \partial_i \widehat{w}(x_1, \ldots, x_{i-1}, t, x_{i+1}, \ldots, x_N) \,\mathrm{d}t \right) \mathrm{d}x_1 \cdots \mathrm{d}x_{i-1} \mathrm{d}x_{i+1} \cdots \mathrm{d}x_N,
\end{aligned}$$

by Fubini's theorem (Theorem 1.15-5), and thus

$$\int_\Omega \partial_i w \,\mathrm{d}x = \int_\Omega (\partial_i v)\varphi \,\mathrm{d}x + \int_\Omega v \partial_i \varphi \,\mathrm{d}x = 0,$$

since $\widehat{w}(x_i, \ldots, x_{i-1}, t, x_{i+1}, \ldots, x_N) = 0$ for $t = -a$ and $t = a$. Hence

$$\int_\Omega (\partial_i v)\varphi \,\mathrm{d}x = -\int_\Omega v \partial_i \varphi \,\mathrm{d}x \quad \text{for all } \varphi \in \mathcal{D}(\Omega),$$

which proves the theorem for $|\boldsymbol{\alpha}| = 1$.

The proof is similar for any partial derivative operator $\partial^{\boldsymbol{\alpha}}$ of order $|\boldsymbol{\alpha}| \geq 2$. $\square$

The formula established in Theorem 6.3-1 motivates the following fundamental definition: Given a function $v \in L^1_{\text{loc}}(\Omega)$ and any multi-index $\boldsymbol{\alpha}$ with $|\boldsymbol{\alpha}| \geq 1$, a function $v^{\boldsymbol{\alpha}} \in L^1_{\text{loc}}(\Omega)$ is said to be a **weak partial derivative** *in $L^1_{\text{loc}}(\Omega)$ of v, of order $|\boldsymbol{\alpha}|$*, if

$$\int_\Omega v^{\boldsymbol{\alpha}} \varphi \,\mathrm{d}x = (-1)^{|\boldsymbol{\alpha}|} \int_\Omega v \partial^{\boldsymbol{\alpha}} \varphi \,\mathrm{d}x \quad \text{for all } \varphi \in \mathcal{D}(\Omega).$$

For instance, given a function $v \in L^1_{\text{loc}}(\Omega)$, a function $v_i \in L^1_{\text{loc}}(\Omega)$ is a **weak partial derivative** *in $L^1_{\text{loc}}(\Omega)$ of v, of the first order with respect to the ith variable*, if

$$\int_\Omega v_i \varphi \,\mathrm{d}x = -\int_\Omega v \partial_i \varphi \,\mathrm{d}x \quad \text{for all } \varphi \in \mathcal{D}(\Omega).$$

As we will see in Theorem 6.3-3, the justification of the above definition of weak partial derivatives hinges on the following, important *per se*, property of functions in the space $L^1_{\rm loc}(\Omega)$. This property also plays a fundamental role in the *calculus of variations* (as will be shown in Section 9.1), as reflected by its name.

Theorem 6.3-2 (fundamental lemma of the calculus of variations) *Let Ω be an open subset of $\mathbb{R}^N$. Let a function $v \in L^1_{\rm loc}(\Omega)$ be such that*

$$\int_\Omega v\varphi \,\mathrm{d}x = 0 \quad \textit{for all } \varphi \in \mathcal{D}(\Omega).$$

Then $v = 0$.

Proof Define the open sets

$$\Omega_k := \left\{x \in \Omega;\ \mathrm{dist}(x, \mathbb{R}^N - \Omega) > \frac{1}{k}\right\} \cap B(0;k) \quad \text{for each integer } k \geq 1.$$

Then $\Omega = \bigcup_{k=1}^\infty \Omega_k$ and, for each $k \geq 1$, $\overline{\Omega_k}$ is a compact subset of Ω. The assumption that $v \in L^1_{\rm loc}(\Omega)$ then implies that $v|_{\Omega_k} \in L^1(\Omega_k)$ for each $k \geq 1$.

For each integer $k \geq 1$, let the function $v_k \in L^1(\Omega)$ be defined by

$$v^k := v|_{\Omega_{2k}} \text{ on } \Omega_{2k} \quad \text{and} \quad v^k := 0 \text{ on } \Omega - \Omega_{2k},$$

let $\varepsilon_0(k) > 0$ be such that

$$\overline{\Omega_{2k}} \subset \Omega_\varepsilon := \{x \in \Omega;\ \mathrm{dist}(x, \mathbb{R}^n - \Omega) > \varepsilon\} \quad \text{for all } 0 < \varepsilon \leq \varepsilon_0,$$

and let $(v^k_\varepsilon)_{\varepsilon>0}$ denote a *regularizing family* of the function v^k (Section 2.6). Then, by Theorem 2.6-4,

$$\|v^k_\varepsilon - v^k\|_{L^1(\Omega_k)} \to 0 \quad \text{as } \varepsilon \to 0.$$

Since the mollifiers $\omega_\varepsilon : y \in \Omega \to \omega_\varepsilon(x - y)$ used for defining a regularizing family belong to the space $\mathcal{D}(\Omega)$ if $x \in \Omega_k$, the assumption made on the function v implies that, if $\varepsilon > 0$ is small enough,

$$v^k_\varepsilon(x) = \int_{B(x;\varepsilon)} v^k(y)\omega_\varepsilon(x-y)\,\mathrm{d}y = \int_\Omega v(y)\omega_\varepsilon(x-y)\,\mathrm{d}y = 0 \quad \text{at each } x \in \Omega_k.$$

Consequently, $v|_{\Omega_k} = 0$ since

$$\|v\|_{L^1(\Omega_k)} = \lim_{\varepsilon\to 0} \|v^k_\varepsilon\|_{L^1(\Omega_k)} = 0.$$

The relation $\Omega = \bigcup_{k=1}^\infty \Omega_k$ and the relations $v|_{\Omega_k} = 0$, $k \geq 1$, then imply that $v = 0$ in Ω (to see this, note that $\int_\Omega |v|\,\mathrm{d}x \leq \liminf_{k\to\infty} \int_{\Omega_k} |v|\,\mathrm{d}x = 0$ by Fatou's lemma). $\square$

Remark In the special case where $v \in L^p(\Omega)$, $1 < p < \infty$, Theorem 6.3-2 follows from the density of the space $\mathcal{D}(\Omega)$ in the space $L^q(\Omega)$ (Theorem 2.6-2), where q denotes the conjugate exponent of p, and the F. Riesz representation theorem in $L^q(\Omega)$ (Theorem 3.5-3). $\square$

Thanks to Theorem 6.3-2, we can now prove two expected properties of weak partial derivatives, viz., that they are unambiguously defined and that they indeed generalize the usual partial derivatives.

Theorem 6.3-3 *Let Ω be an open subset of $\mathbb{R}^N$. Given a function $v \in L^1_{\rm loc}(\Omega)$ and a multi-index $\boldsymbol{\alpha}$ with $|\boldsymbol{\alpha}| \geq 1$, let a function $v^{\boldsymbol{\alpha}} \in L^1_{\rm loc}(\Omega)$ be a weak partial derivative of v of order $|\boldsymbol{\alpha}|$, i.e., that satisfies*

$$\int_\Omega v^{\boldsymbol{\alpha}} \varphi \, {\rm d}x = (-1)^{|\boldsymbol{\alpha}|} \int_\Omega v \partial^{\boldsymbol{\alpha}} \varphi \, {\rm d}x \quad \textit{for all } \varphi \in \mathcal{D}(\Omega).$$

Then such a weak partial derivative is unique, and $v^{\boldsymbol{\alpha}} = \partial^{\boldsymbol{\alpha}} v$ if $v \in \mathcal{C}^{|\boldsymbol{\alpha}|}(\Omega)$.

Proof Let $v^{\boldsymbol{\alpha}} \in L^1_{\rm loc}(\Omega)$ and $w^{\boldsymbol{\alpha}} \in L^1_{\rm loc}(\Omega)$ be such that

$$\int_\Omega v^{\boldsymbol{\alpha}} \varphi \, {\rm d}x = (-1)^{|\boldsymbol{\alpha}|} \int_\Omega v \partial^{\boldsymbol{\alpha}} \varphi \, {\rm d}x = \int_\Omega w^{\boldsymbol{\alpha}} \varphi \, {\rm d}x \quad \text{for all } \varphi \in \mathcal{D}(\Omega).$$

Then $v^{\boldsymbol{\alpha}} = w^{\boldsymbol{\alpha}}$ by Theorem 6.3-2.

Since

$$\int_\Omega (\partial^{\boldsymbol{\alpha}} v) \varphi \, {\rm d}x = (-1)^{|\boldsymbol{\alpha}|} \int_\Omega v \partial^{\boldsymbol{\alpha}} \varphi \, {\rm d}x = \int_\Omega v^{\boldsymbol{\alpha}} \varphi \, {\rm d}x \quad \text{for all } \varphi \in \mathcal{D}(\Omega)$$

if $v \in \mathcal{C}^{|\boldsymbol{\alpha}|}(\Omega)$ (Theorem 6.3-1), it follows that $v^{\boldsymbol{\alpha}} = \partial^{\boldsymbol{\alpha}} v$ in this case, again by Theorem 6.3-2. □

We now prove an important property of functions in the space $L^1_{\rm loc}(\Omega)$. This property generalizes a well-known property of continuously differentiable functions, namely that a function $v \in \mathcal{C}^1(\Omega)$ such that $\partial_i v = 0$, $1 \leq i \leq N$, in a connected subset Ω of $\mathbb{R}^N$ is a constant function. Indeed, the assumptions $\int_\Omega v \partial_i \varphi \, {\rm d}x = 0$ for all $\varphi \in \mathcal{D}(\Omega)$, $1 \leq i \leq N$, simply mean that all the *weak* partial derivatives of v of the first order vanish in $L^1_{\rm loc}(\Omega)$ (for a generalization, see Problem 6.3-1).

Another important property of functions in $L^1_{\rm loc}(\Omega)$ will be established in the next section (Theorem 6.4-2).

Theorem 6.3-4 *Let Ω be a connected open subset of $\mathbb{R}^N$ and let a function $v \in L^1_{\rm loc}(\Omega)$ be such that*

$$\int_\Omega v \partial_i \varphi \, {\rm d}x = 0 \quad \textit{for all } \varphi \in \mathcal{D}(\Omega), \ 1 \leq i \leq N.$$

Then v is a constant function.

Proof By Theorem 1.9-2, it suffices to show that *the function v is locally constant in Ω* (i.e., that, given any point $x \in \Omega$, there exists a neighborhood of x in which v is a constant), since Ω is connected.

Given any point $x \in \Omega$, there exists $r > 0$ such that $\overline{U} \subset \Omega$, where $U := B(x; r)$. Let then $(v_\varepsilon)_{\varepsilon > 0}$ be a regularizing family (Section 2.6) of the given function $v \in L^1_{\rm loc}(\Omega)$. Since $\overline{U}$ is a compact subset of Ω, Theorems 2.6-1 and 2.6-4 show that there exists $\varepsilon_1 = \varepsilon_1(U) > 0$ such that, for all $0 < \varepsilon \leq \varepsilon_1$,

$$\overline{U} \subset \Omega_\varepsilon := \{x \in \Omega; \ {\rm dist}(x, \mathbb{R}^N - \Omega) > \varepsilon\}, \quad v_\varepsilon \in \mathcal{D}(\Omega_\varepsilon), \quad \|v_\varepsilon - v\|_{L^1(U)} \underset{\varepsilon \to 0}{\longrightarrow} 0,$$

$$\partial_i v_\varepsilon(x) = \int_\Omega \partial_i \omega_\varepsilon(x - y) v(y) \, {\rm d}y \quad \text{for all } x \in \Omega_\varepsilon, \ 1 \leq i \leq N.$$

Since, for each $x \in \Omega_\varepsilon$, each function $y \in \Omega \to \partial_i \omega_\varepsilon(x-y)$, $1 \le i \le N$, belongs to the space $\mathcal{D}(\Omega)$, the assumption made on the function v implies that, for all $0 < \varepsilon \le \varepsilon_1$,

$$\partial_i v_\varepsilon(x) = 0 \quad \text{for all } x \in B(x;r), \quad 1 \le i \le N.$$

By a classical result from calculus (which will be substantially generalized later; cf. Theorem 7.2-4), each restriction $v_\varepsilon|_{B(x;r)}$, $0 < \varepsilon \le \varepsilon_1$, is thus a constant function over the connected open set $B(x;r)$. Hence, the restriction $v|_U$ is also a constant function, since the constant functions $v_\varepsilon|_U$ converge to $v|_U$ in $L^1(U)$ as $\varepsilon \to 0$ (the functions $v_\varepsilon|_U$, $0 < \varepsilon \le \varepsilon_1$ belong to the one-dimensional, hence closed, subspace $P_0(U)$ of the space $L^1(U)$, and they converge in $L^1(U)$). □

Following the common usage, from now on we will denote by the *same symbols*, viz., $\partial_i v$, $\partial_{ij} v$, etc., or $\partial^{\boldsymbol{\alpha}} v$ if we use the multi-index notation, classical and weak partial derivatives. Particular care should be therefore exercised in not blithely attributing to weak partial derivative properties of classical partial derivatives. It may happen that some properties are preserved, but to establish that this is the case usually requires a specific proof. Theorem 6.3-4 and Problem 6.3-1 provide such instances.

We conclude this section by a (very brief) incursion into the fundamental *theory of distributions*, which pervades the modern theory of partial differential equations and of the Laplace and Fourier transforms. The reader interested in further developments (such as the precise definition of the topologies of the spaces $\mathcal{D}(\Omega)$ and $\mathcal{D}'(\Omega)$) should consult the references suggested in the Bibliographical Notes.

Let Ω be an open subset of $\mathbb{R}^N$. A **Schwartz distribution**[4] on Ω is a linear functional $T : \mathcal{D}(\Omega) \to \mathbb{R}$ with the following property: Given any compact subset K of Ω, there exist a constant $C(K)$ and an integer $m(K) \ge 0$ such that

$$|T(\varphi)| \le C(K) \sup_{\left\{\substack{|\boldsymbol{\alpha}| \le m(K) \\ x \in K}\right.} |\partial^{\boldsymbol{\alpha}} \varphi(x)| \quad \text{for all } \varphi \in \mathcal{D}(\Omega) \text{ with } \operatorname{supp} \varphi \subset K.$$

The space formed by all distributions on Ω is denoted

$$\mathcal{D}'(\Omega).$$

The space $\mathcal{D}(\Omega)$ is equipped in a natural way with an "inductive limit" *topology*, which makes it a "locally convex topological vector space." In this topology, a sequence $(\varphi_k)_{k=1}^\infty$ of functions $\varphi_k \in \mathcal{D}(\Omega)$ converges to a function $\varphi \in \mathcal{D}(\Omega)$ if and only if there exists a compact subset K of Ω such that

$$\operatorname{supp} \varphi_k \subset K \text{ for all } k \ge 1 \quad \text{and} \quad \operatorname{supp} \varphi \subset K,$$

and

$$\sup_{x \in K} |\partial^{\boldsymbol{\alpha}} \varphi_k(x) - \partial^{\boldsymbol{\alpha}} \varphi(x)| \to 0 \text{ as } k \to \infty \quad \text{for all multi-indices } \boldsymbol{\alpha} \text{ with } |\boldsymbol{\alpha}| \ge 0.$$

[4] So named after Laurent Schwartz (1915–2002) and his landmark treatise: SCHWARTZ [1966]. A beautiful account of his impressive achievements, as a mathematician (he was awarded the Fields Medal in 1950), a professor, and a very caring person, is found in his autobiography: SCHWARTZ [2001].

Note, however, that the topology of the space $\mathcal{D}(\Omega)$ is *not* metrizable, hence *a fortiori not* normable.

It can then be shown that *the space* $\mathcal{D}'(\Omega)$ (as defined above) *is the dual space of* $\mathcal{D}(\Omega)$, in the sense that $\mathcal{D}'(\Omega)$ consists of all the linear functionals on $\mathcal{D}(\Omega)$ that are continuous with respect to the inductive limit topology of $\mathcal{D}(\Omega)$. As a dual space, $\mathcal{D}'(\Omega)$ is equipped in a natural fashion with a "weak * -like" topology (Section 5.12), which is again *not* metrizable. In this topology, a sequence $(T^k)_{k=1}^\infty$ of distributions $T^k \in \mathcal{D}'(\Omega)$ converges to a distribution $T \in \mathcal{D}(\Omega)$ if and only if

$$T^k(\varphi) \underset{k\to\infty}{\longrightarrow} T(\varphi) \quad \text{for all } \varphi \in \mathcal{D}(\Omega).$$

If this is the case, the sequence (T^k) is said to **converge to T in the sense of distributions**, and such a convergence is denoted

$$T^k \underset{k\to\infty}{\longrightarrow} T \quad \text{in } \mathcal{D}'(\Omega).$$

Given any $T \in \mathcal{D}'(\Omega)$ and any $\varphi \in \mathcal{D}(\Omega)$, we shall also frequently use the notations

$${}_{\mathcal{D}'(\Omega)}\langle T, \varphi\rangle_{\mathcal{D}(\Omega)} := T(\varphi), \quad \text{or simply} \quad \langle T, \varphi\rangle := T(\varphi).$$

Given any function $v \in L^1_{\text{loc}}(\Omega)$, the linear functional

$$T_v : \varphi \in \mathcal{D}(\Omega) \to T_v(\varphi) := \int_\Omega v\varphi \, dx$$

defines a distribution on Ω, since for any compact subset K of Ω and for any function $\varphi \in \mathcal{D}(\Omega)$ with $\operatorname{supp}\varphi \subset K$,

$$|T_v(\varphi)| \le \|v\|_{L^1(K)} \sup_{x\in K} |\varphi(x)|.$$

The distribution T_v is called the **distribution associated with the locally integrable function** v.

There are distributions that are *not* associated with any locally integrable functions, however. Consider for instance the linear functional

$$\delta_a : \varphi \in \mathcal{D}(\Omega) \to \delta_a(\varphi) := \varphi(a),$$

where a is any point in Ω. Since

$$|\delta_a(\varphi)| \le \sup_{x\in K} |\varphi(x)|$$

for any compact subset K of Ω and for any function $\varphi \in \mathcal{D}(\Omega)$ with $\operatorname{supp}\varphi \subset K$, δ_a is a distribution, called the **Dirac distribution**[5] *at* a, or simply the **Dirac distribution** if $a = 0$.

[5]So named after the distinguished physicist Paul Dirac (1902–1984), who was awarded the Nobel Prize in Physics in 1933.

But there does *not* exist any function $v \in L^1_{\text{loc}}(\Omega)$ such that $\varphi(a) = \int_\Omega v\varphi\, dx$ for all $\varphi \in \mathcal{D}(\Omega)$. To see this, consider the functions $\varphi_k \in \mathcal{D}(\mathbb{R}^N)$, $k \geq 1$, defined by

$$\varphi_k(x) = e^{\frac{1}{|k(x-a)|^2-1}} \text{ if } |x-a| < \frac{1}{k} \quad \text{and} \quad \varphi_k(x) = 0 \text{ if } |x-a| \geq \frac{1}{k},$$

so that, for some integer $k_0 \geq 1$, $\varphi_k|_\Omega \in \mathcal{D}(\Omega)$ for all $k \geq k_0$. Let

$$U := \{x \in \Omega;\ \varphi_{k_0}(x) \neq 0\}.$$

Then $v \in L^1(U)$, $|v\varphi_k| \leq |v|$ almost everywhere in Ω for all $k \geq k_0$, and $v\varphi_k \to 0$ almost everywhere in U as $k \to \infty$. Hence $\int_\Omega v\varphi_k\, dx \to 0$ as $k \to \infty$ by Lebesgue's dominated convergence theorem (Theorem 1.15-3), while $\varphi_k(a) = e^{-1}$ for all $k \geq 1$, a contradiction.

A wide source of distributions is provided by the *differentiation in the sense of distributions*: Let T be a distribution on Ω, and let $\boldsymbol{\alpha}$ be any multi-index with $|\boldsymbol{\alpha}| \geq 1$. Then the linear functional defined by

$$\partial^{\boldsymbol{\alpha}} T : \varphi \in \mathcal{D}(\Omega) \to \partial^{\boldsymbol{\alpha}} T(\varphi) := (-1)^{|\boldsymbol{\alpha}|} T(\partial^{\boldsymbol{\alpha}} \varphi)$$

is again a distribution on Ω (this is a simple consequence of the definition), called the **partial derivative of order $\boldsymbol{\alpha}$ of T in the sense of distributions**. For instance, the Dirac distribution on $\mathbb{R}$ (which cannot be associated with any locally integrable function on $\mathbb{R}$, as shown above) is the derivative in the sense of distributions of the locally integrable function $v : \mathbb{R} \to \mathbb{R}$ defined by $v(x) = 0$ if $x < 0$ and $v(x) = 1$ if $x \geq 0$ (Problem 6.3-3).

More generally, given any finite set $\boldsymbol{A}$ of multi-indices and coefficients $a_{\boldsymbol{\alpha}} \in \mathbb{R}$, $\boldsymbol{\alpha} \in \boldsymbol{A}$, the linear functional defined by

$$\mathcal{L}T : \varphi \in \mathcal{D}(\Omega) \to \mathcal{L}T(\varphi) := \sum_{\boldsymbol{\alpha} \in \boldsymbol{A}} (-1)^{|\boldsymbol{\alpha}|} a_{\boldsymbol{\alpha}} T(\partial^{\boldsymbol{\alpha}} \varphi) \quad \text{for all } \varphi \in \mathcal{D}(\Omega),$$

defines a **linear partial differential operator in the sense of distribution**, viz.,

$$\mathcal{L} := \sum_{\boldsymbol{\alpha} \in \boldsymbol{A}} a_{\boldsymbol{\alpha}} \partial^{\boldsymbol{\alpha}} : T \in \mathcal{D}'(\Omega) \to \mathcal{L}T \in \mathcal{D}'(\Omega)$$

(for simplicity, only partial differential operators with constant coefficients are considered here).

Given any distribution $f \in \mathcal{D}'(\Omega)$, one may then seek whether there exists a distribution $T \in \mathcal{D}'(\Omega)$ that satisfies

$$\mathcal{L}T = f \quad \text{in } \mathcal{D}'(\Omega).$$

If this is the case, T is said to be a *solution of $\mathcal{L}T = f$ in the sense of distributions.* An example of such a solution T when $\mathcal{L} = -\Delta$ and $f = \delta$ is provided in Problem 6.3-4 (in this example, $T = T_v$ with $v \in \mathcal{L}^1_{\text{loc}}(\Omega)$).

Problems

6.3-1 Let Ω be a connected open subset of $\mathbb{R}^N$, and let a function $v \in L^1_{\text{loc}}(\Omega)$ be such that, for some integer $m \geq 1$,

$$\int_\Omega v\partial^{\boldsymbol{\alpha}}\varphi \mathrm{d}x = 0 \quad \text{for all } \varphi \in \mathcal{D}(\Omega) \quad \text{and all multi-indices } \boldsymbol{\alpha} \text{ with } |\boldsymbol{\alpha}| = m.$$

In other words, *all the weak partial derivatives of* v *of order* m *vanish.* Then show that v is a polynomial in N variables of degree $\leq m-1$ (the special case $m=1$ was proved in Theorem 6.3-4).

6.3-2 Let Ω be an open subset of $\mathbb{R}^N$ and let a function $v \in L^1_{\text{loc}}(\Omega)$ be such that

$$\int_\Omega v\varphi \mathrm{d}x = 0 \quad \text{for all } \varphi \in \mathcal{D}(\Omega) \text{ that satisfy } \int_\Omega \varphi \mathrm{d}x = 0.$$

Show that v is a constant function.

6.3-3 Let I be any open interval of $\mathbb{R}$ that contains the origin.

(1) Show that the function $v : x \in I \to v(x) := \max\{0, x\}$ (which is clearly in $L^1_{\text{loc}}(I)$) has a weak derivative in $L^1_{\text{loc}}(I)$.

(2) Show that the second derivative of T_v in the sense of distributions is the Dirac distribution (hence v does not possess a weak second derivative in $L^1_{\text{loc}}(I)$).

6.3-4 Let Ω be any open subset of $\mathbb{R}^N$ that contains the origin.

(1) Let ω_N denote the volume of the unit ball in $\mathbb{R}^N$. Show that the function $v : \Omega \to \mathbb{R}$ defined almost everywhere in Ω by

$$v(x) := \frac{1}{2\omega_2} \ln|x| \text{ if } x \neq 0 \text{ and } N = 2, \quad \text{or} \quad v(x) := \frac{1}{N(2-N)\omega_N} |x|^{2-N} \text{ if } x \neq 0 \text{ and } N \geq 3,$$

is in the space $L^1_{\text{loc}}(\Omega)$.

(2) Show that, for any $N \geq 2$,

$$\int_\Omega v\Delta\varphi \mathrm{d}x = \varphi(0) \quad \text{for all } \varphi \in \mathcal{D}(\Omega),$$

i.e., that

$$\Delta v = \delta_0 \quad \text{in the sense of distributions.}$$

For this reason, the function v is called the *fundamental solution to the Laplace equation.*

6.4 Hypoellipticity of Δ

The following result is well known from the theory of analytic functions: Let Ω be an open subset of $\mathbb{R}^2$; then any function $v \in \mathcal{C}^2(\Omega)$ that satisfies the Laplace equation $\Delta v = 0$ in Ω is in effect analytic, hence in particular of class $\mathcal{C}^\infty$, in Ω.

It is remarkable that this result admits the following far-reaching generalization, which holds in any dimension, in the more general sense of distributions, and for the more general Poisson's equation: Let Ω be an open subset of $\mathbb{R}^N$; then any *distribution* $T \in \mathcal{D}'(\Omega)$ that satisfies

$$\Delta T = f \quad \text{in } \mathcal{D}'(\Omega), \text{ where } f \in \mathcal{C}^\infty(\Omega),$$

i.e., that satisfies (Section 6.3),

$$T(\Delta\varphi) = \int_\Omega f\varphi \mathrm{d}x \quad \text{for all } \varphi \in \mathcal{D}(\Omega),$$

is in effect a *function, also in the space* $\mathcal{C}^\infty(\Omega)$. This property, which is called the **hypoellipticity of** Δ, is not easy to establish at this level of generality.[6,7]

Here we shall give a proof in the (important) special case where the distribution T is a *locally integrable function* in Ω; the hypoellipticity of Δ for such functions will be put to use later for proving a "weak" Poincaré lemma (Theorem 6.17-4).

To this end, we first prove interesting *per se* results, which in a sense are reminiscent of earlier results, proved in Theorems 2.6-1(b) and 2.6-3. Note, however, that the functions ρ_ε considered in the next theorem no longer need to have compact supports, in contrast with the mollifiers introduced there.

In what follows, the notation B_δ designates the open ball in $\mathbb{R}^N$ with center at the origin and radius δ and, given two open subsets U and V of $\mathbb{R}^N$, the notation $U \subset\subset V$ means that $\overline{U}$ is a compact subset of V.

Theorem 6.4-1 *Let $(\rho_\varepsilon)_{\varepsilon>0}$ be a family of functions $\rho_\varepsilon \in L^1(\mathbb{R}^N)$ with the following properties:*

$$\rho_\varepsilon(x) \geq 0 \text{ for all } x \in \mathbb{R}^N, \quad \int_{\mathbb{R}^N} \rho_\varepsilon(y)\,\mathrm{d}y = 1,$$

$$\text{for each } \delta > 0, \quad \int_{\mathbb{R}^N - B_\delta} \rho_\varepsilon(y)\,\mathrm{d}y \to 0 \text{ as } \varepsilon \to 0.$$

(a) *Let $w : \mathbb{R}^N \to \mathbb{R}$ be a bounded and uniformly continuous function. Then, for each $\varepsilon > 0$, the function $w \star \rho_\varepsilon : \mathbb{R}^N \to \mathbb{R}$ defined by*

$$(w \star \rho_\varepsilon)(x) := \int_{\mathbb{R}^N} w(x-y)\rho_\varepsilon(y)\,\mathrm{d}y \quad \text{for each } x \in \mathbb{R}^N$$

is also bounded, and

$$\sup_{x\in\mathbb{R}^N} |(w \star \rho_\varepsilon)(x) - w(x)| \to 0 \quad \text{as } \varepsilon \to 0.$$

(b) *Let a function $v \in L^1(\mathbb{R}^N)$ be given. Then $v \star \rho_\varepsilon \in L^1(\mathbb{R}^N)$ for each $\varepsilon > 0$. Besides,*

$$\text{given any open set } V \subset\subset \mathbb{R}^N, \quad \|v \star \rho_\varepsilon - v\|_{L^1(V)} \to 0 \text{ as } \varepsilon \to 0.$$

Proof (i) Let $w : \mathbb{R}^N \to \mathbb{R}$ be a function that satisfies the assumptions of (a). Since $\rho_\varepsilon \geq 0$ and $\int_{\mathbb{R}^N} \rho_\varepsilon\,\mathrm{d}x = 1$,

$$|(w \star \rho_\varepsilon)(x)| \leq \int_{\mathbb{R}^N} |w(x-y)|\,\rho_\varepsilon(y)\,\mathrm{d}y \leq \sup_{z\in\mathbb{R}^N} |w(z)| \quad \text{for each } x \in \mathbb{R}^N.$$

The function $w \star \rho_\varepsilon$ is thus bounded.

[6]For a proof, see, e.g., VO-KHAC [1972b, Chapter DB, Section 3].

[7]More generally, a linear partial differential operator $\mathcal{L}$ with constant coefficients is said to be *hypoelliptic* if every function $v \in L^1_{\text{loc}}(\Omega)$ such that $\mathcal{L}v \in \mathcal{C}^\infty(\Omega)$ is itself of class $\mathcal{C}^\infty$. A necessary and sufficient condition of hypoellipticity for such an operator was given in:

L. HÖRMANDER [1955]: On the theory of general partial differential operators, *Acta Mathematica* **94**, 161–248.

Another proof, which relies on the *closed graph theorem* (Section 5.7), is found in YOSIDA [1966, Chapter 2, Section 7].

Given any $\eta > 0$, there exist $\delta = \delta(\eta) > 0$ and $\varepsilon_0 = \varepsilon_0(\delta) = \varepsilon_0(\eta) > 0$ such that

$$|w(x-y) - w(x)| \leq \frac{1}{2}\eta \quad \text{for all } x \in \mathbb{R}^N \text{ and all } y \in B_\delta,$$
$$\Big(\sup_{z\in\mathbb{R}^N} |w(z)|\Big) \int_{\mathbb{R}^N - B_\delta} \rho_\varepsilon(y)\,\mathrm{d}y \leq \frac{1}{4}\eta \quad \text{for all } \varepsilon \leq \varepsilon_0.$$

Therefore, for any $x \in \mathbb{R}^N$,

$$\begin{aligned}
|(w \star \rho_\varepsilon)(x) - w(x)| &= \Big| \int_{\mathbb{R}^N} \big(w(x-y) - w(x)\big)\rho_\varepsilon(y)\,\mathrm{d}y \Big| \\
&\leq \int_{B_\delta} |w(x-y) - w(x)|\,\rho_\varepsilon(y)\,\mathrm{d}y + \int_{\mathbb{R}^N - B_\delta} |w(x-y) - w(x)|\,\rho_\varepsilon(y)\,\mathrm{d}y \\
&\leq \int_{B_\delta} |w(x-y) - w(x)|\,\rho_\varepsilon(y)\,\mathrm{d}y + 2 \sup_{z\in\mathbb{R}^N} |w(z)| \int_{\mathbb{R}^N - B_\delta} \rho_\varepsilon(y)\,\mathrm{d}y \leq \eta \quad \text{for all } \varepsilon \leq \varepsilon_0.
\end{aligned}$$

Consequently,

$$\sup_{x\in\mathbb{R}^N} |(w \star \rho_\varepsilon)(x) - w(x)| \to 0 \quad \text{as } \varepsilon \to 0,$$

since $\eta > 0$ is arbitrary. This proves (a).

(ii) Let a function $v \in L^1(\mathbb{R}^N)$ be given. Then, for each $\varepsilon > 0$,

$$\int_{\mathbb{R}^N} |(v \star \rho_\varepsilon)|\,(x)\,\mathrm{d}x \leq \int_{\mathbb{R}^N} \Big(\int_{\mathbb{R}^N} |v(x-y)|\,\rho_\varepsilon(y)\,\mathrm{d}y \Big)\,\mathrm{d}x = \int_{\mathbb{R}^N} \Big(\int_{\mathbb{R}^N} |v(x-y)|\,\mathrm{d}x \Big) \rho_\varepsilon(y)\,\mathrm{d}y,$$

by Fubini's theorem (Theorem 1.15-5). Therefore, $v \star \rho_\varepsilon \in L^1(\mathbb{R}^N)$ and

$$\|v \star \rho_\varepsilon\|_{L^1(\mathbb{R}^N)} \leq \|v\|_{L^1(\mathbb{R}^N)} \quad \text{for each } \varepsilon > 0.$$

Since $\overline{\mathcal{D}(\mathbb{R}^N)} = L^1(\mathbb{R}^N)$ (Theorem 2.6-2), there exists a sequence $(v_k)_{k=1}^\infty$ of functions $v_k \in \mathcal{D}(\mathbb{R}^N) \subset L^1(\mathbb{R}^N)$ such that $\|v_k - v\|_{L^1(\mathbb{R}^N)} \to 0$ as $k \to \infty$. Besides, given any open subset $V \subset\subset \mathbb{R}^N$ and any integer $k \geq 1$,

$$\begin{aligned}
\|v \star \rho_\varepsilon - v\|_{L^1(V)} &\leq \|(v - v_k) \star \rho_\varepsilon\|_{L^1(V)} + \|v_k \star \rho_\varepsilon - v_k\|_{L^1(V)} + \|v - v_k\|_{L^1(V)} \\
&\leq 2\,\|v - v_k\|_{L^1(\mathbb{R}^N)} + \Big(\int_V \mathrm{d}x \Big) \sup_{x\in\mathbb{R}^N} |(v_k \star \rho_\varepsilon)(x) - v_k(x)|\,.
\end{aligned}$$

Therefore, for each $k \geq 1$,

$$\limsup_{\varepsilon\to 0} \|v \star \rho_\varepsilon - v\|_{L^1(V)} \leq 2\,\|v - v_k\|_{L^1(\mathbb{R}^N)}\,,$$

since $\lim_{\varepsilon\to 0} \sup_{x\in\mathbb{R}^N} |(v_k \star \rho_\varepsilon)(x) - v_k(x)| = 0$ by (a) (as a function in the space $\mathcal{D}(\mathbb{R}^N)$, each function v_k is bounded and uniformly continuous). Hence

$$\lim_{\varepsilon\to 0} \|v \star \rho_\varepsilon - v\|_{L^1(V)} = 0,$$

since $\|v - v_k i\|_{L^1(\mathbb{R})}$ can be made arbitrarily small by choosing k sufficiently large. This proves (b). $\square$

We are now in a position to prove the hypoellipticity of Δ for functions in $L^1_{\rm loc}(\Omega)$. Note that the function $E \in L^1_{\rm loc}(\mathbb{R}^N)$ introduced in part (iii) of the next proof is nothing but the *fundamental solution to the Laplace equation* (Problem 6.3-4) and that the functions E_ε, $\varepsilon > 0$, introduced in part (ii) converge almost everywhere in $\mathbb{R}^N$ to E as $\varepsilon \to 0$, but, contrary to E, they no longer have a singularity at the origin.

Theorem 6.4-2 (Weyl's lemma:[8] hypoellipticity of Δ) *Let Ω be an open subset of $\mathbb{R}^N$ and let $v \in L^1_{\rm loc}(\Omega)$ and $f \in \mathcal{C}^m(\Omega)$ for some integer $m \geq 0$ be two functions that satisfy*

$$\int_\Omega v\Delta\varphi \,\mathrm{d}x = \int_\Omega f\varphi \,\mathrm{d}x \quad \textit{for all } \varphi \in \mathcal{D}(\Omega).$$

Then $v \in \mathcal{C}^m(\Omega)$. Consequently, $v \in \mathcal{C}^\infty(\Omega)$ if $f \in \mathcal{C}^\infty(\Omega)$.

Proof Let ω_N denote the volume of the unit ball in $\mathbb{R}^N$. For each $\varepsilon > 0$, define the function $E_\varepsilon \in \mathcal{C}^\infty(\mathbb{R}^N)$ by

$$E_\varepsilon(x) := \frac{1}{4\omega_2} \ln\left(|x|^2 + \varepsilon\right) \quad \text{for each } x \in \mathbb{R}^2 \text{ if } N = 2,$$
$$E_\varepsilon(x) := \frac{1}{N(2-N)\omega_N}\left(|x|^2 + \varepsilon\right)^{\frac{2-N}{2}} \quad \text{for each } x \in \mathbb{R}^N \text{ if } N \geq 3.$$

(i) *The functions $\rho_\varepsilon \in \mathcal{C}^\infty(\mathbb{R}^N)$ defined for each $\varepsilon > 0$ by*

$$\rho_\varepsilon := \Delta E_\varepsilon$$

satisfy

$$\rho_\varepsilon(x) \geq 0 \text{ for all } x \in \mathbb{R}^N, \quad \int_\Omega \rho_\varepsilon(y)\,\mathrm{d}y = 1,$$
$$\text{for each } \delta > 0, \quad \int_{\mathbb{R}^N - B_\delta} \rho_\varepsilon(y)\,\mathrm{d}y \to 0 \text{ as } \varepsilon \to 0, \quad \text{where } B_\delta := B(0;\delta).$$

First, the relation

$$\Delta E_\varepsilon(x) = \frac{\varepsilon}{\omega_N(|x|^2 + \varepsilon)^{\frac{N}{2}+1}} \quad \text{at each } x \in \mathbb{R}^N,$$

shows that $\rho_\varepsilon(x) \geq 0$ for all $x \in \mathbb{R}^N$. Next, the well-known formula

$$\int_{\mathbb{R}^N} F(|x|)\,\mathrm{d}x = N\omega_N \int_0^\infty F(r) r^{N-1}\,\mathrm{d}r,$$

[8] H. Weyl [1940]: The method of orthogonal projection in potential theory, *Duke Mathematical Journal* **7**, 414–444.

which holds for any measurable function $F : [0, \infty[\to [0, \infty[$, gives in particular

$$\int_{\mathbb{R}^N} \Delta E_\varepsilon(y)\,\mathrm{d}y = N\varepsilon \int_0^\infty \left(r^2+\varepsilon\right)^{-\frac{N}{2}-1} r^{N-1}\,\mathrm{d}r = \int_0^\infty \frac{\mathrm{d}}{\mathrm{d}r}\left[\left(1+\varepsilon r^{-2}\right)^{-\frac{N}{2}}\right]\mathrm{d}r = 1.$$

Hence $\int_\Omega \rho_\varepsilon(y)\,\mathrm{d}y = 1$. Finally, the formula

$$\int_{\mathbb{R}^N - B_\delta} F(|x|)\,\mathrm{d}x = N\omega_N \int_\delta^\infty F(r) r^{N-1}\,\mathrm{d}r,$$

which holds for each $\delta > 0$, similarly gives

$$\begin{aligned}\int_{\mathbb{R}^N - B_\delta} F(|x|)\,\mathrm{d}x &= N\varepsilon \int_\delta^\infty \left(r^2+\varepsilon\right)^{-\frac{N}{2}-1} r^{N-1}\,\mathrm{d}r \\ &= \int_\delta^\infty \frac{\mathrm{d}}{\mathrm{d}r}\left[\left(1+\varepsilon r^{-2}\right)^{-\frac{N}{2}}\right]\mathrm{d}r = 1 - \left(1+\varepsilon\delta^{-2}\right)^{-N/2},\end{aligned}$$

thus showing that, for each $\delta > 0$, $\int_{\mathbb{R}^N - B_\delta} \rho_\varepsilon(y)\,\mathrm{d}y \to 0$ as $\varepsilon \to 0$.

(ii) *Given a function $v \in L^1_{\mathrm{loc}}(\Omega)$ and an open set $V \subset\subset \Omega$, define the function $\widetilde{v} \in L^1(\mathbb{R}^N)$ by*

$$\widetilde{v}(x) := v(x) \text{ if } x \in V \quad \text{and} \quad \widetilde{v}(x) := 0 \text{ if } x \in \mathbb{R}^N - V.$$

Then there exists a sequence $(\varepsilon(k))_{k=1}^\infty$ such that $\varepsilon(k) > 0$ for all $k \geq 1$, $\varepsilon(k) \to 0$ as $k \to \infty$, and, for almost all $x \in V$,

$$v(x) = \lim_{k\to\infty} \int_{\mathbb{R}^N} \widetilde{v}(x-y)\Delta E_{\varepsilon(k)}(y)\,\mathrm{d}y = \lim_{k\to\infty} \int_{\mathbb{R}^N} \widetilde{v}(y)\Delta E_{\varepsilon(k)}(x-y)\,\mathrm{d}y.$$

Since the functions $\rho_\varepsilon = \Delta E_\varepsilon$, $\varepsilon > 0$, satisfy all the assumptions of Theorem 6.4-1, part (b) of this theorem applied to the function $\widetilde{v} \in L^1(\mathbb{R}^N)$ gives

$$\|\widetilde{v} \star \Delta E_\varepsilon - \widetilde{v}\|_{L^1(V)} \to 0 \quad \text{as } \varepsilon \to 0.$$

Hence there exists a subsequence $(\widetilde{v} \star \Delta E_{\varepsilon(k)})_{k=1}^\infty$ of the family $(\widetilde{v} \star \Delta E_\varepsilon)_{\varepsilon>0}$ that converges almost everywhere to the function $\widetilde{v}|_V = v|_V \in L^1(V)$.

(iii) *Let U and V be two open subsets of $\mathbb{R}^N$ that satisfy $U \subset\subset V \subset\subset \Omega$, let $\delta = \inf_{x\in U} d(x, \mathbb{R}^N - V) > 0$, and let a function $\alpha \in \mathcal{D}(\mathbb{R}^N)$ be so chosen that $\alpha = 1$ in B_{δ_1} for some $0 < \delta_1 < \delta$ and $\alpha|_{B_\delta} \in \mathcal{D}(B_\delta)$. Finally, let the function $E \in L^1_{\mathrm{loc}}(\mathbb{R}^N)$ be defined almost everywhere in $\mathbb{R}^N$ by*

$$\begin{aligned} E(x) &:= \frac{1}{2\omega_2} \ln|x| && \text{for all } x \neq 0 \text{ if } N = 2, \\ E(x) &:= \frac{1}{N(2-N)\omega_N} |x|^{2-N} && \text{for all } x \neq 0 \text{ if } N \geq 3. \end{aligned}$$

Let a function $v \in L^1_{\mathrm{loc}}(\Omega)$ and a function $f \in \mathcal{C}^m(\Omega)$ for some integer $m \geq 0$ be given that satisfy

$$\int_\Omega v\Delta\varphi\,\mathrm{d}x = \int_\Omega f\varphi\,\mathrm{d}x \quad \text{for all } \varphi \in \mathcal{D}(\Omega).$$

Then

$$v(x) = \int_{B_\delta} f(x-y)(\alpha E)(y)\,\mathrm{d}y + \int_V v(y)\big[\Delta((1-\alpha)E)\big]\,(x-y)\,\mathrm{d}y \quad \text{for almost all } x \in U.$$

For notational brevity, let $E_k := E_{\varepsilon(k)}$, $k \geq 1$. In (ii), we showed that

$$v(x) = \lim_{k\to\infty} \int_{\mathbb{R}^N} \widetilde{v}(y)\Delta E_k(x-y)\,\mathrm{d}y \quad \text{for almost all } x \in V,$$

where $\widetilde{v}(x) := v(x)$ if $x \in V$ and $\widetilde{v}(x) = 0$ if $x \in \mathbb{R}^N - V$. We now show that, for almost all $x \in U$, this limit as $k \to \infty$ is indeed given by the announced expression. So, let the functions $\alpha \in \mathcal{D}(\mathbb{R}^N)$ and $E \in L^1_{\mathrm{loc}}(\mathbb{R}^N)$ be defined as above, and let x be a point in the open set V. For each integer $k \geq 1$, we can write

$$\begin{aligned}\int_{\mathbb{R}^N} \widetilde{v}(y)\Delta E_k(x-y)\,\mathrm{d}y &= \int_{\mathbb{R}^N} \widetilde{v}(y)\big[\Delta(\alpha + (1-\alpha)E_k)\big](x-y)\,\mathrm{d}y \\ &= \int_{\mathbb{R}^N} \widetilde{v}(y)\,[\Delta(\alpha E_k)]\,(x-y)\,\mathrm{d}y + \int_{\mathbb{R}^N} \widetilde{v}(y)\big[\Delta((1-\alpha)E_k)\big](x-y)\,\mathrm{d}y.\end{aligned}$$

Let us then study separately the behavior as $k \to \infty$ of each one of the last two integrals.

First, we note that, for each $k \geq 1$, the function

$$\varphi_k : y \in \mathbb{R}^N \to \varphi_k(y) := (\alpha E_k)(x-y)$$

appearing in the first integral is of class $\mathcal{C}^\infty$ (both functions α and E_k are of class $\mathcal{C}^\infty$) and $\operatorname{supp}\varphi_k \subset B_\delta(x) := \{y \in \mathbb{R}^N;\ |x-y| < \delta\}$ since $\alpha|_{B_\delta} \in \mathcal{D}(B_\delta)$. Besides, $B_\delta(x) \subset V \subset \Omega$ by definition of δ, so that $\varphi_k|_\Omega \in \mathcal{D}(\Omega)$. By assumption then,

$$\begin{aligned}\int_{\mathbb{R}^N} \widetilde{v}(y)\,[\Delta(\alpha E_k)]\,(x-y)\,\mathrm{d}y &= \int_\Omega v(y)\Delta\varphi_k(y)\,\mathrm{d}y \\ &= \int_\Omega f(y)\varphi_k(y)\,\mathrm{d}y = \int_\Omega f(y)\,[\alpha E_k]\,(x-y)\,\mathrm{d}y \\ &= \int_{B_{\delta(x)}} f(y)\,[\alpha E_k]\,(x-y)\,\mathrm{d}y = \int_{B_\delta} f(x-y)\,[\alpha E_k]\,(y)\,\mathrm{d}y.\end{aligned}$$

Since $|E_k(y)| \leq |E(y)|$ for all $y \in B_\delta - \{0\}$ (if $N = 2$, we may assume without loss of generality that k is large enough and δ small enough to insure that $0 < |x|^2 + \varepsilon(k) \leq 1$ for all $x \in B_\delta$) and since $E \in L^1(B_\delta)$, the Lebesgue dominated convergence theorem (Theorem 1.15-3) can be applied, showing that

$$\begin{aligned}\lim_{k\to\infty} \int_{\mathbb{R}^N} \widetilde{v}(y)\,[\Delta(\alpha E_k)]\,(x-y)\,\mathrm{d}y &= \lim_{k\to\infty} \int_{B_\delta} f(x-y)\,[\alpha E_k]\,(y)\,\mathrm{d}y \\ &= \int_{B_\delta} f(x-y)\,[\alpha E]\,(y)\,\mathrm{d}y.\end{aligned}$$

Second, we note that the function $\Delta((1-\alpha)E_k)$ appearing in the second integral can be expanded as

$$\Delta((1-\alpha)E_k) = -(\Delta\alpha)E_k - 2\nabla\alpha \cdot \nabla E_k + (1-\alpha)\Delta E_k,$$

where $\nabla v := (\partial_i v)_{i=1}^N$ for any smooth enough function $v : \Omega \to \mathbb{R}^N$. Taking into account that

$$\operatorname{supp}\Delta\alpha \subset B_\delta - B_{\delta_1}, \quad \operatorname{supp}\nabla\alpha \subset B_\delta - B_{\delta_1}, \quad \text{and} \quad \operatorname{supp}(1-\alpha) \subset \mathbb{R}^N - B_{\delta_1},$$

we thus infer that

$$\begin{aligned}
&\int_{\mathbb{R}^N} \widetilde{v}(y)\big[\Delta((1-\alpha)E_k)\big](x-y)\,\mathrm{d}y = \int_{\mathbb{R}^N} \widetilde{v}(x-y)\big[\Delta((1-\alpha)E_k)\big](x-y)\,\mathrm{d}y \\
&= -\int_{B_\delta - B_{\delta_1}} \widetilde{v}(x-y)\big[(\Delta\alpha)E_k\big](y)\,\mathrm{d}y - 2\int_{B_\delta - B_{\delta_1}} \widetilde{v}(x-y)\,[\nabla\alpha\cdot\nabla E_k]\,(y)\,\mathrm{d}y \\
&\qquad + \int_{A_{\delta_1}(x)} \widetilde{v}(x-y)\Delta E_k(y)\,\mathrm{d}y
\end{aligned}$$

where the set $A_{\delta_1}(x) := (\mathbb{R}^N - B_{\delta_1}) \cap \{y \in \mathbb{R}^N;\ (y-x) \in V\}$ is bounded. Since

$$|E_k(y)| \le |E(y)| = \frac{1}{2\pi}\big|\ln|y|\,\big| \le \frac{1}{2\pi}\,|\ln\delta_1| \quad \text{for all } y \in B_\delta - B_{\delta_1} \text{ if } N = 2$$

(with the same *caveat* as above),

$$\begin{aligned}
|E_k(y)| &\le |E(y)| \le \frac{1}{N(N-2)\omega_N\delta_1^{N-2}} \quad \text{for all } y \in B_\delta - B_{\delta_1} \text{ if } N \ge 3, \\
\|\nabla E_k(y)\| &\le \frac{1}{N\omega_N}\frac{\|y\|}{(|y|^2+\varepsilon)^{\frac{N}{2}+1}} \le \frac{1}{N\omega_N}\frac{1}{(|y|^2+\varepsilon)^{\frac{N}{2}}} \le \frac{1}{N\omega_N\delta_1^N} \quad \text{for all } y \in B_\delta - B_{\delta_1}, \\
|\Delta E_k(y)| &= \frac{\varepsilon}{\omega_N(|y|^2+\varepsilon)^{\frac{N}{2}+1}} \le \frac{1}{\omega_N(|y|^2+\varepsilon)^{\frac{N}{2}}} \le \frac{1}{\delta_1^N} \quad \text{for all } y \in \mathbb{R}^N - B_{\delta_1},
\end{aligned}$$

Lebesgue's dominated convergence theorem can be again applied, showing that

$$\begin{aligned}
&\lim_{k\to\infty}\int_{\mathbb{R}^N} \widetilde{v}(y)\big[\Delta((1-\alpha)E_k)\big](y)\,\mathrm{d}y \\
&= -\int_{B_\delta - B_{\delta_1}} \widetilde{v}(x-y)\,[(\Delta\alpha)E]\,(y)\,\mathrm{d}y - 2\int_{B_\delta - B_{\delta_1}} \widetilde{v}(x-y)\,[\nabla\alpha\cdot\nabla E]\,(y)\,\mathrm{d}y \\
&\quad + \int_{A_{\delta_1}(x)} \widetilde{v}(x-y)\Delta E(y)\,\mathrm{d}y \\
&= \int_{\mathbb{R}^N} \widetilde{v}(x-y)\big[\Delta((1-\alpha)E)\big](y)\,\mathrm{d}y = \int_{\mathbb{R}^N} \widetilde{v}(y)\big[\Delta((1-\alpha)E)\big](x-y)\,\mathrm{d}y \\
&= \int_V v(y)\big[\Delta((1-\alpha)E)\big](x-y)\,\mathrm{d}y.
\end{aligned}$$

(iv) *Conclusion.* Because the property to be established, viz., that $v \in \mathcal{C}^m(\Omega)$ if $f \in \mathcal{C}^m(\Omega)$, is local, it is enough to show that $v \in \mathcal{C}^m(U)$ for any open set $U \subset\subset \Omega$ (according to the usual abuse of language, this means that, as an equivalence class of functions almost everywhere equal, $v \in L^1_{\mathrm{loc}}(U)$ contains a function that is of class $\mathcal{C}^m$ in U).

The assumption that $f \in \mathcal{C}^m(\Omega)$ implies that the function

$$x \in U \to \int_{B_\delta} f(x-y)(\alpha E)(y)\,\mathrm{d}y$$

is itself of class $\mathcal{C}^m$ on the open set U; besides, the function

$$x \in U \to \int_V v(y)\big[\Delta((1-\alpha)E)\big](x-y)\,\mathrm{d}y$$

is of class $\mathcal{C}^\infty$ on U (such differentiability properties will be established in Theorem 7.4-1, by independent arguments). The relation

$$v(x) = \int_{B_\delta} f(x-y)(\alpha E)(y)\,\mathrm{d}y + \int_V v(y)\big[\Delta((1-\alpha)E)\big](x-y)\,\mathrm{d}y \quad \text{for almost all } x \in U$$

established in (iii) then shows that $v \in \mathcal{C}^m(U)$. □

Problem

6.4-1 Give a proof of Theorem 6.4-2 when $N = 1$ and Ω is an open interval of $\mathbb{R}$ (the proof provided in the text applies to a dimension $N \geq 2$).

6.5 The Sobolev spaces $W^{m,p}(\Omega)$ and $H^m(\Omega)$: First properties

As will be amply demonstrated in this chapter, the Sobolev spaces $H^m(\Omega)$ and $H_0^m(\Omega)$, where $m \geq 1$ is an integer, play a key role in the analysis of *linear* elliptic boundary value problems and of some "mildly nonlinear" ones, such as *obstacle problems*. As the special cases $p = 2$ of the more general Sobolev spaces $W^{m,p}(\Omega)$, where p is any extended real number satisfying $1 \leq p \leq \infty$, the spaces $H^m(\Omega)$ possess the distinctive feature of being *Hilbert spaces* (Theorem 6.5-1).

In order to avoid repetitions in later chapters, the basic properties of the more general spaces $W^{m,p}(\Omega)$ will be in fact presented in this section and the next one, even if only their special case $p = 2$ is needed in this chapter (as said above).

The properties discussed in this section hold under mild assumptions on the set Ω, which is either an arbitrary open subset of $\mathbb{R}^N$ or one that is "of finite width," while in the next one, the open set Ω will be assumed to be *bounded* and to have a *Lipschitz-continuous boundary* (Section 1.18).

Their *dual spaces* will be studied later in this chapter (Section 6.11).

Let Ω be an open subset of $\mathbb{R}^N$. For each integer $m \geq 1$ and each extended real number $1 \leq p \leq \infty$, the (real) **Sobolev space**[9]

$$W^{m,p}(\Omega), \quad \text{or} \quad H^m(\Omega) \quad \text{if } p = 2,$$

[9] So named after:

S.L. Sobolev [1938]: On a theorem of functional analysis, *Matematicheskii Sbornik* **46**, 471–496 (in Russian).

S.L. Sobolev [1950]: *Applications of Functional Analysis in Mathematical Physics*, Leningrad (in Russian; English translation: American Mathematical Society, Providence, RI, 1963).

The definition of the space $H^1(\Omega)$ (with weak derivatives called "quasi-dérivées"), together with some of its basic properties, is also found on page 205 of:

J. Leray [1933]: Sur le mouvement d'un liquide visqueux emplissant l'espace, *Acta Mathematica* **63**, 193–248.

consists of those functions $v \in L^p(\Omega)$ that possess weak partial derivatives $\partial^{\boldsymbol{\alpha}} v$ also in $L^p(\Omega)$ for all multi-indices $\boldsymbol{\alpha}$ with $1 \leq |\boldsymbol{\alpha}| \leq m$. According to the definition of weak partial derivatives (cf. Section 6.3; note in this respect that $L^p(\Omega) \subset L^1_{\mathrm{loc}}(\Omega)$ for any $1 \leq p \leq \infty$), a function $v \in L^p(\Omega)$ is thus in $W^{m,p}(\Omega)$ if, for each multi-index $\boldsymbol{\alpha}$ with $1 \leq |\boldsymbol{\alpha}| \leq m$, there exists a function $\partial^{\boldsymbol{\alpha}} v \in L^p(\Omega)$ such that

$$\int_\Omega (\partial^{\boldsymbol{\alpha}} v)\varphi \, \mathrm{d}x = (-1)^{|\boldsymbol{\alpha}|} \int_\Omega v \partial^{\boldsymbol{\alpha}} \varphi \, \mathrm{d}x \quad \text{for all } \varphi \in \mathcal{D}(\Omega).$$

Recall that such a function $\partial^{\boldsymbol{\alpha}} v \in L^p(\Omega)$ is then uniquely defined by the above relations, and that $\partial^{\boldsymbol{\alpha}} v$ coincides with the usual partial derivative if in addition $v \in \mathcal{C}^m(\Omega)$ (Theorem 6.3-3).

Note that each space $W^{m,p}(\Omega)$, $m \geq 1$, which is thus defined as a subspace of $L^p(\Omega)$, is *strictly* contained in $L^p(\Omega)$ (Problem 6.5-1).

We now begin to list various fundamental properties of Sobolev spaces that we shall need later on; observe that there are more and more assumptions on the open set Ω as we proceed in this section and the next one (for the sake of simplicity, however, we shall not necessarily state the "weakest" possible assumptions under which each theorem holds). Recall that a normed vector space is *reflexive* if it can be identified with the dual space of its dual space by means of a *specific* isometry (Section 5.14).

Theorem 6.5-1 *Let Ω be an open subset of $\mathbb{R}^N$ and let $m \geq 1$ be an integer. Equipped with the norm*

$$v \to \|v\|_{m,p,\Omega} := \Big(\int_\Omega \sum_{|\boldsymbol{\alpha}| \leq m} |\partial^{\boldsymbol{\alpha}} v|^p \, \mathrm{d}x \Big)^{1/p} = \Big(\sum_{0 \leq |\boldsymbol{\alpha}| \leq m} \|\partial^{\boldsymbol{\alpha}} v\|^p_{L^p(\Omega)} \Big)^{1/p} \quad \text{if } 1 \leq p < \infty,$$

$$v \to \|v\|_{m,\infty,\Omega} := \max_{|\boldsymbol{\alpha}| \leq m} \|\partial^{\boldsymbol{\alpha}} v\|_{L^\infty(\Omega)} \quad \text{if } p = \infty,$$

$$v \to \|v\|_{m,\Omega} := \Big(\int_\Omega \sum_{|\boldsymbol{\alpha}| \leq m} |\partial^{\boldsymbol{\alpha}} v|^2 \, \mathrm{d}x \Big)^{1/2} = \Big(\sum_{0 \leq |\boldsymbol{\alpha}| \leq m} \|\partial^{\boldsymbol{\alpha}} v\|^2_{L^2(\Omega)} \Big)^{1/2} \quad \text{if } p = 2,$$

the Sobolev space $W^{m,p}(\Omega)$ is a Banach space.

The space $W^{m,p}(\Omega)$ is separable if $1 \leq p < \infty$, and reflexive if $1 < p < \infty$.

The space $H^m(\Omega) = W^{m,2}(\Omega)$ is a Hilbert space.

Proof It is easily verified that, for each $1 \leq p \leq \infty$, the mapping $\|\cdot\|_{m,p,\Omega}$ is a norm over $W^{m,p}(\Omega)$.

Let $1 \leq p \leq \infty$, and let $(v_k)_{k=1}^\infty$ be a Cauchy sequence in $W^{m,p}(\Omega)$ equipped with this norm. Since, for each $0 \leq |\boldsymbol{\alpha}| \leq m$,

$$\|\partial^{\boldsymbol{\alpha}} v_k - \partial^{\boldsymbol{\alpha}} v_\ell\|_{L^p(\Omega)} \leq \|v_k - v_\ell\|_{m,p,\Omega} \quad \text{for all } k, \ell \geq 1,$$

and since $L^p(\Omega)$ is complete (Theorem 3.4-2), there exist for each $1 \leq |\boldsymbol{\alpha}| \leq m$ a function $v^{\boldsymbol{\alpha}} \in L^p(\Omega)$ such that $\|\partial^{\boldsymbol{\alpha}} v_k - v^{\boldsymbol{\alpha}}\|_{L^p(\Omega)} \to 0$ as $k \to \infty$, and a function $v \in L^p(\Omega)$ such that $\|v_k - v\|_{L^p(\Omega)} \to 0$ as $k \to \infty$.

Let a function $\varphi \in \mathcal{D}(\Omega)$ be given, so that

$$\int_\Omega (\partial^{\boldsymbol{\alpha}} v_k)\varphi \, dx = (-1)^{|\boldsymbol{\alpha}|} \int_\Omega v_k \partial^{\boldsymbol{\alpha}} \varphi \, dx \quad \text{for all } 1 \leq |\boldsymbol{\alpha}| \leq m \text{ and all } k \geq 1,$$

since $v_k \in W^{m,p}(\Omega)$. Passing to the limit as $k \to \infty$ in the inequalities

$$\Big| \int_\Omega (\partial^{\boldsymbol{\alpha}} v_k)\varphi \, dx - \int_\Omega v^{\boldsymbol{\alpha}} \varphi \, dx \Big| \leq \|\partial^{\boldsymbol{\alpha}} v_k - v^{\boldsymbol{\alpha}}\|_{L^p(\Omega)} \, \|\varphi\|_{L^q(\Omega)} \quad \text{for all } 1 \leq |\alpha| \leq m,$$

$$\Big| \int_\Omega v_k \partial^{\boldsymbol{\alpha}} \varphi \, dx - \int_\Omega v \partial^{\boldsymbol{\alpha}} \varphi \, dx \Big| \leq \|v_k - v\|_{L^p(\Omega)} \, \|\partial^{\boldsymbol{\alpha}} \varphi\|_{L^q(\Omega)},$$

where q denotes the conjugate exponent of p, then shows that

$$\int_\Omega (v^{\boldsymbol{\alpha}})\varphi \, dx = (-1)^{|\boldsymbol{\alpha}|} \int_\Omega v \partial^{\boldsymbol{\alpha}} \varphi \, dx \quad \text{for each } \varphi \in \mathcal{D}(\Omega).$$

Consequently, for each $|\boldsymbol{\alpha}| \leq m$, $v^{\boldsymbol{\alpha}} \in L^p(\Omega)$ is the weak partial derivative of order $\boldsymbol{\alpha}$ of $v \in L^p(\Omega)$; therefore, $v \in W^{m,p}(\Omega)$. Besides, the definitions of the functions $v^{\boldsymbol{\alpha}} \in L^p(\Omega)$ and of the norm $\|\cdot\|_{m,p,\Omega}$ together show that $\|v_k - v\|_{m,p,\Omega} \to 0$ as $k \to \infty$. The space $(W^{m,p}(\Omega), \|\cdot\|_{m,p,\Omega})$ is thus complete.

To verify that the space $W^{m,p}(\Omega)$ is separable if $1 \leq p < \infty$, it is enough to consider the case $m = 1$, since the case $m \geq 2$ is similarly treated. Clearly, the space $W^{1,p}(\Omega)$ can be identified as a normed vector space with the subspace

$$\Big\{ (v_0, v_1, \ldots, v_N) \in (L^p(\Omega))^{N+1}; \int_\Omega v_i \varphi \, dx = -\int_\Omega v_0 \partial_i \varphi \, dx \text{ for all } \varphi \in \mathcal{D}(\Omega), \, 1 \leq i \leq N \Big\}$$

of the *product space* $(L^p(\Omega))^{N+1}$ equipped with the *product norm*. Hence the separability of $W^{m,p}(\Omega)$ follows from that of $(L^p(\Omega))^{N+1}$ (which itself follows from that of $L^p(\Omega)$; cf. Theorem 2.5-4) and the property that any subset of a separable metric space is also separable (Theorem 1.10-3).

Since the above subspace is clearly *closed* in $(L^p(\Omega))^{N+1}$, the reflexivity of $W^{m,p}(\Omega)$ for $1 < p < \infty$ follows from Theorem 5.14-2(c) (which asserts that any closed subspace of a reflexive space is itself reflexive) and from the reflexivity of $L^p(\Omega)$ for $1 < p < \infty$ (Theorem 5.14-2(e)), which clearly implies that of $(L^p(\Omega)^N$.

It is immediately verified that the bilinear mapping $(\cdot, \cdot)_{m,\Omega} : H^m(\Omega) \times H^m(\Omega) \to \mathbb{R}$ defined by

$$(u, v)_{m,\Omega} := \sum_{|\boldsymbol{\alpha}| \leq m} \int_\Omega \partial^{\boldsymbol{\alpha}} u \partial^{\boldsymbol{\alpha}} v \, dx \quad \text{for all } u, v \in H^m(\Omega)$$

possesses all the properties of an inner product on the space $H^m(\Omega)$ and that $\|v\|_{m,\Omega} = \sqrt{(v, v)_{m,\Omega}}$ for all $v \in H^m(\Omega)$. Hence $(H^m(\Omega), \|\cdot\|_{m,\Omega})$ is a Hilbert space. □

Remark A different proof of the reflexivity of the space $W^{m,p}(\Omega)$, $1 < p < \infty$, is proposed in Problem 6.11-2. □

For convenience, we will henceforth also allow $m = 0$ in the above definitions of the spaces $W^{m,p}(\Omega)$ and of the norms $\|\cdot\|_{m,p,\Omega}$, by letting

$$W^{0,p}(\Omega) := L^p(\Omega) \quad \text{and} \quad \|\cdot\|_{0,p,\Omega} := \|\cdot\|_{L^p(\Omega)} \ \text{if } 1 \le p \le \infty,$$
$$H^0(\Omega) := L^2(\Omega) \quad \text{and} \quad \|\cdot\|_{0,\Omega} := \|\cdot\|_{L^2(\Omega)} \ \text{if } p = 2.$$

While the space $\mathcal{D}(\Omega)$ is dense in the space $L^p(\Omega)$ if $1 \le p < \infty$ (Theorem 2.6-2), the space $\mathcal{D}(\Omega)$ is no longer dense in the space $W^{m,p}(\Omega)$, $m \ge 1$, unless the set $R^N - \Omega$ is "very small." For instance, one can show that a necessary (but not sufficient) condition for $\overline{\mathcal{D}(\Omega)} = W^{m,p}(\Omega)$, if $1 < p < \infty$, is that the Lebesgue measure of $\mathbb{R}^N - \Omega$ be zero[10] (in this direction, see also Problems 6.5-2 and 6.5-3). This observation motivates the following definition.

Let Ω be an open subset of $\mathbb{R}^N$. For each integer $m \ge 1$ and each real number $1 \le p < \infty$, the **Sobolev space**

$$W_0^{m,p}(\Omega), \quad \text{or} \quad H_0^m(\Omega) \ \text{ if } p = 2,$$

is defined as the closure of the space $\mathcal{D}(\Omega)$ in the space $(W^{m,p}(\Omega), \|\cdot\|_{m,p,\Omega})$. It then immediately follows from this definition and from Theorems 6.5-1 and 5.14-2(c) that, *for each integer $m \ge 1$, the space $W_0^{m,p}(\Omega)$ is a separable Banach space for each $1 \le p < \infty$, which is reflexive if $1 < p < \infty$, and the space $H_0^m(\Omega)$ is a Hilbert space.*

Other basic properties of the space $W_0^{m,p}(\Omega)$ are proved in the next theorem (see also Problems 6.5-3–6.5-5 for complements) where, for each integer $m \ge 1$ and each real number $1 \le p < \infty$, the following *seminorms* will be used:

$$v \to |v|_{m,p,\Omega} := \Big(\int_\Omega \sum_{|\alpha|=m} |\partial^\alpha v|^p \, \mathrm{d}x \Big) = \Big(\sum_{|\alpha|=m} \|\partial^\alpha v\|^p_{L^p(\Omega)} \Big)^{1/p},$$
$$v \to |v|_{m,\Omega} := \Big(\int_\Omega \sum_{|\alpha|=m} |\partial^\alpha v|^2 \, \mathrm{d}x \Big)^{1/2} = \Big(\sum_{|\alpha|=m} \|\partial^\alpha v\|^2_{L^2(\Omega)} \Big)^{1/2} \quad \text{if } p = 2.$$

A subset of $\mathbb{R}^N$ is said to be *of finite width* if it lies between two parallel hyperplanes in $\mathbb{R}^N$.

Theorem 6.5-2 *Let Ω be an open subset of $\mathbb{R}^N$ of finite width.*

(a) *For each $1 \le p < \infty$, the following* **Poincaré–Friedrichs inequality**[11] *holds: There exists a constant $c = c(\Omega, p)$ such that*

$$\|v\|_{0,p,\Omega} \le c\, |v|_{1,p,\Omega} \quad \textit{for all } v \in W_0^{1,p}(\Omega).$$

(b) *For each $m \ge 1$ and $1 \le p < \infty$, the seminorm $|\cdot|_{m,p,\Omega}$ is a norm over the space $W_0^{m,p}(\Omega)$, equivalent to the norm $\|\cdot\|_{m,p,\Omega}$, i.e., there exists a constant $C = C(\Omega, m, p)$ such that*

$$|v|_{m,p,\Omega} \le \|v\|_{m,p,\Omega} \le C\, |v|_{m,p,\Omega} \quad \textit{for all } v \in W_0^{m,p}(\Omega).$$

[10] J.L. Lions [1965]: *Problèmes aux Limites dans les Equations aux Dérivées Partielles*, Presses de l'Université de Montréal, Montréal, Que.

[11] So named after Henri Poincaré (1854–1912), who established a related inequality for smooth functions, and Kurt Otto Friedrichs (1901–1982), who extended it to the spaces $W_0^{1,p}(\Omega)$.

Proof It is enough to establish the Poincaré inequality for functions in the dense subspace $\mathcal{D}(\Omega)$ of $W_0^{1,p}(\Omega)$, since both the norm $\|\cdot\|_{0,p,\Omega}$ and the seminorm $|\cdot|_{1,p,\Omega}$ are continuous functions on the space $W^{1,p}(\Omega)$, as it immediately follows from the inequalities

$$\big|\|v\|_{0,p,\Omega} - \|w\|_{0,p,\Omega}\big| \le \|v - w\|_{0,p,\Omega} \le \|v - w\|_{1,p,\Omega},$$
$$\big||v|_{1,p,\Omega} - |w|_{1,p,\Omega}\big| \le |v - w|_{1,p,\Omega} \le \|v - w\|_{1,p,\Omega}.$$

Assume first that Ω lies between two parallel hyperplanes that are orthogonal to the vector $(1, 0, \dots, 0)$, and let $a > 0$ be such that $\Omega \subset [-a, a] \times \mathbb{R}^{N-1}$. Given any function $v \in \mathcal{D}(\Omega)$, identified here with its extension by 0 in $]-a, a[\, \times \mathbb{R}^{N-1}$, we have

$$v(x) = \int_{-a}^{x_1} \partial_1 v(t, x_2, \dots, x_N)\,\mathrm{d}t \quad \text{for all } x = (x_1, x_2, \dots, x_N) \in \,]-a, a[\, \times \mathbb{R}^{N-1}.$$

Consequently,

$$\begin{aligned}|v(x)|^p &\le \left(\int_{-a}^{x_1} |\partial_1 v(t, x_2, \dots, x_N)|\,\mathrm{d}t\right)^p \le (a + x_1)^{p-1} \int_{-a}^{x_1} |\partial_1 v(x_1, x_2, \dots, x_N)|^p\,\mathrm{d}x_1 \\ &\le (a + x_1)^{p-1} \int_{-a}^{a} |\partial_1 v(x_1, x_2, \dots, x_N)|^p\,\mathrm{d}x_1.\end{aligned}$$

Since then

$$\int_{-a}^{a} |v(x)|^p\,\mathrm{d}x_1 \le \frac{(2a)^p}{p} \int_{-a}^{a} |\partial_1 v(x_1, x_2, \dots, x_N)|^p\,\mathrm{d}x_1,$$

Fubini's theorem (Theorem 1.15-5) gives

$$\begin{aligned}\|v\|_{0,p,\Omega}^p &= \int_{\mathbb{R}^{N-1}} \left(\int_{-a}^{a} |v(x)|^p\,\mathrm{d}x_1\right)\mathrm{d}x_2 \cdots \mathrm{d}x_N \\ &\le \frac{(2a)^p}{p} \int_{\mathbb{R}^{N-1}} \left(\int_{-a}^{a} |\partial_1 v(x_1, x_2, \dots, x_N)|^p\,\mathrm{d}x_1\right) = \frac{(2a)^p}{p} \|\partial_1 v\|_{0,p,\Omega}^p \le \frac{(2a)^p}{p} |v|_{1,p,\Omega}^p.\end{aligned}$$

This proves (a), with $c = \dfrac{(2a)}{p^{1/p}}$.

The Poincaré inequality immediately implies that

$$|v|_{1,p,\Omega} \le \|v\|_{1,p,\Omega} \le (1 + c^p)^{1/p}\,|v|_{1,p,\Omega} \quad \text{for all } v \in W_0^{1,p}(\Omega),$$

which proves (b) for $m = 1$. The above inequality further implies that

$$\|v\|_{2,p,\Omega}^p = \|v\|_{1,p,\Omega}^p + |v|_{2,p,\Omega}^p \le (1 + c^p)\,|v|_{1,p,\Omega}^p + |v|_{2,p,\Omega}^p \quad \text{for all } v \in \mathcal{D}(\Omega),$$

and another application of Poincaré's inequality shows that

$$|v|_{1,p,\Omega}^p = \sum_{i=1}^N \|\partial_i v\|_{0,p,\Omega}^p \le c^p \sum_{i=1}^N |\partial_i v|_{1,p,\Omega}^p = c^p\,|v|_{2,p,\Omega}^p \quad \text{for all } v \in \mathcal{D}(\Omega).$$

The combination of the last two inequalities therefore proves (b) for $m = 2$. The same type of argument proves (b) for any integer $m \ge 3$.

In the general case, let $x \in \Omega \to \widehat{x} = a + Qx \in \mathbb{R}^N$, with $a \in \mathbb{R}^N$ and Q an orthogonal matrix of order N, be a change of Cartesian coordinates such that the image $\widehat{\Omega}$ of Ω under this transformation lies between two parallel planes orthogonal to the vector $(1, 0, \ldots, 0)$. Then (with self-explanatory notations) $\sum_{i=1}^N |\partial_i v(x)|^2 = \sum_{i=1}^N |\widehat{\partial}_i \widehat{v}(\widehat{x})|^2$ since the matrix Q is orthogonal. The Poincaré inequality over Ω then follows from the Poincaré inequality over $\widehat{\Omega}$, since the Euclidean norm and the norm $\|\cdot\|_p$ are equivalent. □

Problems

6.5-1 Let Ω be an open subset of $\mathbb{R}^N$. Show that, for any $1 \leq p \leq \infty$, $W^{1,p}(\Omega) \subsetneq L^p(\Omega)$.
Hint: If $N = 1$ and $\Omega =]0, 1[$, show that the function $v \in L^p(0, 1)$ defined by $v(x) = 0$ if $0 < x < \frac{1}{2}$ and $v(x) = 1$ if $\frac{1}{2} < x < 1$ does not have a weak derivative of the first order in $L^p(0, 1)$. Then adapt this example to any $N \geq 2$.

6.5-2 Show that, for any integer $m \geq 1$ and any $1 \leq p < \infty$, the space $\mathcal{D}(\mathbb{R}^N)$ is dense in the space $W^{m,p}(\mathbb{R}^N)$[12].

6.5-3 Show that $H_0^1(\Omega) \subsetneq H^1(\Omega)$ if the open subset Ω of $\mathbb{R}^N$ is bounded.
Hint: Identify nonzero functions in the orthogonal complement of $H_0^1(\Omega)$ in $H^1(\Omega)$.

6.5-4 Let $1 \leq p < \infty$.
(1) Show that the seminorm $|\cdot|_{1,p,\mathbb{R}^N}$ is a norm over the space $W^{1,p}(\mathbb{R}^N)$.
(2) Is the norm $|\cdot|_{1,p,\mathbb{R}^N}$ equivalent to the norm $\|\cdot\|_{1,p,\mathbb{R}^N}$ over $W^{1,p}(\mathbb{R}^N)$?

6.5-5 Let Ω be an open subset of $\mathbb{R}^N$. Give a one-line proof that the space $H_0^1(\Omega)$ is infinite-dimensional.

6.5-6 Let Ω be an open subset of $\mathbb{R}^N$, let $1 < p < \infty$, and let q denote the conjugate exponent of p. Show that a function $v \in L^p(\Omega)$ belongs to the space $W^{1,p}(\Omega)$ if and only if there exists a constant C such that

$$\left| \int_\Omega v \partial_i \varphi \, dx \right| \leq C \|\varphi\|_{0,q,\Omega} \quad \text{for all } \varphi \in \mathcal{D}(\Omega).$$

Hint: Use the F. Riesz representation theorem in $L^q(\Omega)$ (Theorem 3.4-3).

6.6 The Sobolev spaces $W^{m,p}(\Omega)$ and $H^m(\Omega)$ with Ω a domain; imbedding theorems, traces, Green's formulas

By contrast with the properties of the Sobolev spaces $W^{m,p}(\Omega)$ described in Section 6.5, which hold if Ω is an arbitrary open subset of $\mathbb{R}^N$ (save the Poincaré–Friedrichs inequality, established under the assumption that Ω is of finite width), those described in this section hold only if some specific assumptions are made on Ω, such as its boundedness and, especially, on the smoothness of its boundary. Their proofs, often long and technical, are not given here. The interested reader should consult the references suggested in the Bibliographical Notes.

[12]For a proof, see, e.g., ADAMS [1975, Corollary 3.19], or ATTOUCH, BUTTAZZO & MICHAILLE [2006, Theorem 5.1-3].

It is easy to prove that, for any integer $m \geq 1$, the space $W^{m,p}(\Omega)$ is strictly contained in the space $L^p(\Omega)$ (Problem 6.5-1). This reflects in effect that some kind of extra "smoothness" is acquired by any function in $L^p(\Omega)$ that possesses weak derivatives in $L^p(\Omega)$. For instance, let Ω be a domain in $\mathbb{R}^2$; then a function in the space $W^{1,p}(\Omega)$ is necessarily continuous on $\overline{\Omega}$ if $p > 2$; or is in any space $L^q(\Omega)$, $1 \leq q < \infty$ if $p = 2$; or is in the space $L^{2p/(2-p)}(\Omega)$ if $1 \leq p < 2$ (such properties are special cases of those given in Theorem 6.6-1 below). Note, however, that if a function is in the space $H^1(\Omega)$ (i.e., if $p = 2$), with $\Omega \subset \mathbb{R}^2$, it is not necessarily continuous, as illustrated in Problem 6.6-1 by a spectacular example of a function in $H^1(\Omega)$ that is even everywhere discontinuous in Ω!

The inclusions $W^{1,p}(\Omega) \subset \mathcal{C}^0(\overline{\Omega})$ or $W^{1,p}(\Omega) \subset L^q(\Omega)$ mentioned above, with Ω as a domain in $\mathbb{R}^2$, are instances of the *imbeddings* stated in the next theorem. There, the notation

$$X \hookrightarrow Y$$

means that a normed vector space X is **continuously imbedded** in a normed vector space Y, in the sense that $X \subset Y$ and, in addition, there exists a constant c such that $\|v\|_Y \leq c\,\|v\|_X$ for all $v \in X$; in other words, the identity mapping $\iota : (X, \|\cdot\|_X) \to (Y, \|\cdot\|_Y)$ is continuous.

Recall that the spaces $\mathcal{C}^{m,\lambda}(\overline{\Omega})$ have been defined in Section 1.18 and that a *domain in* $\mathbb{R}^N$ has also been defined there.

Some care should be taken in interpreting these continuous imbeddings (or, for that matter, the compact imbeddings in Theorem 6.6-3), since an element of a Sobolev space is in effect an *equivalence class* of functions that are almost everywhere equal in Ω. For instance, the imbedding $W^{m,p}(\Omega) \hookrightarrow \mathcal{C}^{0,\lambda}(\overline{\Omega})$ means that there is a constant c such that, in each equivalence class of the space $W^{m,p}(\Omega)$, there is a (unique) representative v that belongs to the space $\mathcal{C}^{0,\lambda}(\overline{\Omega})$ and satisfies $\|v\|_{\mathcal{C}^{0,\lambda}(\overline{\Omega})} \leq c\,\|v\|_{m,p,\Omega}$, etc.

$^\flat$**Theorem 6.6-1 (Sobolev imbedding theorems)** *Let Ω be a domain in $\mathbb{R}^N$, let $m \geq 1$ be an integer, and let $1 \leq p < \infty$. Then the following continuous imbeddings hold:*

$$W^{m,p}(\Omega) \hookrightarrow L^{p^*}(\Omega) \qquad \text{with } \frac{1}{p^*} = \frac{1}{p} - \frac{m}{N} \text{ if } m < \frac{N}{p},$$

$$W^{m,p}(\Omega) \hookrightarrow L^{q}(\Omega) \qquad \text{for all } q \text{ with } 1 \leq q < \infty \text{ if } m = \frac{N}{p},$$

$$W^{m,p}(\Omega) \hookrightarrow \mathcal{C}^{0,m-N/p}(\overline{\Omega}) \qquad \text{if } \frac{N}{p} < m < \frac{N}{p} + 1,$$

$$W^{m,p}(\Omega) \hookrightarrow \mathcal{C}^{0,\lambda}(\overline{\Omega}) \qquad \text{for all } \lambda \text{ with } 0 < \lambda < 1 \text{ if } m = \frac{N}{p} + 1,$$

$$W^{m,p}(\Omega) \hookrightarrow \mathcal{C}^{0,1}(\overline{\Omega}) \qquad \text{if } \frac{N}{p} + 1 < m.$$

□

An important consequence of the Sobolev imbedding theorem is that the same inequality that guarantees that the imbedding $W^{m,p}(\Omega) \hookrightarrow \mathcal{C}^0(\overline{\Omega})$ holds, viz., the inequality $mp > N$, also guarantees that the Sobolev space $W^{m,p}(\Omega)$ is a **Banach algebra**, i.e., a Banach space that is also an algebra according to the definition given in Section 2.15. In this particular case, this means that, if $mp > N$, the product of two functions in $W^{m,p}(\Omega)$ also belongs to $W^{m,p}(\Omega)$, and that the bilinear mapping $(u,v) \in W^{m,p}(\Omega) \times W^{m,p}(\Omega) \to uv \in W^{m,p}(\Omega)$ defined in this fashion is continuous (Section 2.11) with respect to the norm $\|\cdot\|_{m,p,\Omega}$. More specifically, we have:

♭**Theorem 6.6-2** (**$W^{m,p}(\Omega)$ is a Banach algebra if** $mp > N$) *Let Ω be a domain in $\mathbb{R}^N$, let $m \geq 1$ be an integer, and let $1 \leq p < \infty$ be such that $mp > N$. Then*

$$u, v \in W^{m,p}(\Omega) \Rightarrow uv \in W^{m,p}(\Omega),$$

and there exists a constant c such that

$$\|uv\|_{m,p,\Omega} \leq c\,\|u\|_{m,p,\Omega}\,\|v\|_{m,p,\Omega} \quad \textit{for all } u, v \in W^{m,p}(\Omega). \qquad \square$$

A normed vector space X is **compactly imbedded** in a normed vector space Y if $X \hookrightarrow Y$ and the identity mapping $\iota : x \in X \to \iota(x) = x \in Y$ is a *compact* linear operator; equivalently, ι maps each bounded sequence $(x^k)_{k=1}^\infty$ into a sequence $(\iota(x^k))_{k=1}^\infty$ that contains a subsequence converging in Y (Theorem 2.10-1). Such a compact imbedding is denoted

$$X \Subset Y.$$

The next result identifies the continuous imbeddings of Theorem 6.6-1 that are in addition compact; the number p^* is that defined as in Theorem 6.6-1.

♭**Theorem 6.6-3** (**Rellich–Kondrachov compact imbedding theorems**[13]) *Let Ω be a domain in $\mathbb{R}^N$, let $m \geq 1$ be an integer, and let $1 \leq p < \infty$. Then the following compact imbeddings hold:*

$$\begin{aligned} W^{m,p}(\Omega) &\Subset L^q(\Omega) \quad \textit{for all } q \textit{ with } 1 \leq q < p^* \textit{ if } m < \frac{N}{p}, \\ W^{m,p}(\Omega) &\Subset L^q(\Omega) \quad \textit{for all } q \textit{ with } 1 \leq q < \infty \textit{ if } m = \frac{N}{p}, \\ W^{m,p}(\Omega) &\Subset \mathcal{C}(\overline{\Omega}) \quad \textit{if } \frac{N}{p} < m. \end{aligned} \qquad \square$$

Note that the Rellich–Kondrachov theorem implies that *the compact imbedding*

$$W^{1,p}(\Omega) \Subset L^p(\Omega), \quad p \geq 1,$$

always holds, i.e., independently of the dimension N.

Another important property of functions in the Sobolev spaces $W^{m,p}(\Omega)$ when Ω is a domain is that they can be *approximated by smooth functions.* The space $\mathcal{C}^\infty(\overline{\Omega})$ has been defined in Section 1.18.

♭**Theorem 6.6-4** (**approximation by smooth functions**) *Let Ω be a domain in $\mathbb{R}^N$, let $m \geq 0$ be an integer, and let $1 \leq p < \infty$. Then the space $\mathcal{C}^\infty(\overline{\Omega})$ is dense in the space $W^{m,p}(\Omega)$.* $\square$

---

[13] F. Rellich [1930]: Ein Satz über mittlere Konvergenz, *Nachrichten von der Gesellschaft der Wissenschaften zu Göttingen*, 30–35.

V.I. Kondrachov [1945]: Certain properties of functions in the spaces L^p, *Doklady Akademii Nauk SSSR* **48**, 535–538 (in Russian).

The next results are, for simplicity, mostly presented in the special case $m = 1$, i.e., for functions in the Sobolev spaces $W^{1,p}(\Omega)$, although analogous properties hold for the Sobolev spaces $W^{m,p}(\Omega)$, $m \geq 2$. But first, we need to define some function spaces.

Let Ω be a domain in $\mathbb{R}^N$, let Γ denote its boundary, and let $1 \leq p < \infty$. The space $\mathcal{L}^p(\Gamma)$ is defined as that consisting of all functions $f : \Gamma \to \mathbb{R}$ such that $|f|^p \in \mathcal{L}^1(\Gamma)$ (the space $\mathcal{L}^1(\Gamma)$ is defined in Section 1.18). The space formed by the equivalence classes, *modulo* the equality $\mathrm{d}\Gamma$-almost everywhere, of functions in the space $\mathcal{L}^p(\Gamma)$ is denoted

$$L^p(\Gamma).$$

Combining the definition of the integral $\int_\Gamma g\,\mathrm{d}\Gamma$ for a function $g \in \mathcal{L}^1(\Gamma)$ (Section 1.18) with arguments similar to those used in the proofs of Theorems 2.5-2 and 3.4-2, one can then establish that *the function*

$$f \in L^p(\Gamma) \to \|f\|_{L^p(\Gamma)} := \left(\int_\Gamma |f|^p \,\mathrm{d}\Gamma\right)^{1/p}$$

is a norm over the space $L^p(\Gamma)$, *and the space* $(L^p(\Gamma), \|\cdot\|_{L^p(\Gamma)})$ *is a Banach space.*

Let $v : \overline{\Omega} \to \mathbb{R}$ be a continuous function, where Ω is an open subset of $\mathbb{R}^N$. Then its *trace* on the boundary Γ of the set Ω is the continuous function $\operatorname{tr} v : \Gamma \to \mathbb{R}$ defined by $(\operatorname{tr} v)(x) = v(x)$ for all $x \in \Gamma$. A remarkable property of functions in the Sobolev space $W^{1,p}(\Omega)$, where Ω is now a *domain* in $\mathbb{R}^N$, is that generalized "traces" can still be defined on Γ, even when the functions are not continuous on $\overline{\Omega}$. The basis for this extension is the following observation: Let Ω be a domain in $\mathbb{R}^N$ and let $1 \leq p < N$. Then one can show that the mapping

$$\operatorname{tr} : \mathcal{C}^\infty(\overline{\Omega}) \to L^{p^\#}(\Gamma)$$

is well defined and continuous if the space $\mathcal{C}^\infty(\overline{\Omega})$ is endowed with the norm $\|\cdot\|_{1,p,\Omega}$ and the number $p^\sharp > 1$ is defined as in Theorem 6.6-5 below. Since the space $\mathcal{C}^\infty(\overline{\Omega})$ is dense in the space $W^{1,p}(\Omega)$ (Theorem 6.6-4) and since the space $L^{p^\sharp}(\Gamma)$ is complete, there exists a unique continuous linear extension from the space $W^{1,p}(\Omega)$ into the space $L^{p^\sharp}(\Gamma)$ (Theorem 3.1-1) that coincides with the classical trace operator tr on the subspace $\mathcal{C}^\infty(\overline{\Omega})$. This extension, which will still be denoted by the same symbol tr, is called the **trace operator**, and each function $\operatorname{tr} v \in L^{p^\sharp}(\Gamma)$ defined in this fashion is called the **trace** of the function $v \in W^{1,p}(\Omega)$.

We now state various important properties of the trace operator, such as continuity, compactness, and how it is used for providing another equivalent definition of the spaces $W_0^{1,p}(\Omega)$ and $W_0^{2,p}(\Omega)$ when Ω is a domain.

♭**Theorem 6.6-5 (properties of the trace operator)** *Let Ω be a domain in $\mathbb{R}^N$.*

(a) *Let $1 \leq p < \infty$. Then*

$$\operatorname{tr} \in \mathcal{L}(W^{1,p}(\Omega); L^{p^\sharp}(\Gamma)) \quad \text{with } \frac{1}{p^\sharp} := \frac{1}{p} - \frac{p-1}{p(N-1)} \text{ if } 1 \leq p < N,$$

$$\operatorname{tr} \in \mathcal{L}(W^{1,p}(\Omega); L^q(\Gamma)) \quad \text{for all } q \text{ with } 1 \leq q < \infty \text{ if } p = N,$$

$$\operatorname{tr} \in \mathcal{L}(W^{1,p}(\Omega); \mathcal{C}(\Gamma)) \quad \text{if } N < p.$$

(b) *If $1 < p < N$, the trace operator* $\mathrm{tr} : W^{1,p}(\Omega) \to L^q(\Gamma)$ *is compact for all q such that* $1 \le q < p^\sharp$.

(c) *Let $1 \le p < \infty$. Then the space $W_0^{1,p}(\Omega)$, which is by definition the closure of $\mathcal{D}(\Omega)$ in $W^{1,p}(\Omega)$ (Section 6.5), is also given by*

$$W_0^{1,p}(\Omega) = \{v \in W^{1,p}(\Omega);\ \mathrm{tr}\, v = 0\}.$$

(d) *Let $1 \le p < \infty$. Then the space $W_0^{2,p}(\Omega)$, which is by definition the closure of $\mathcal{D}(\Omega)$ in $W^{2,p}(\Omega)$ (Section 6.5), is also given by*[14]

$$W_0^{2,p}(\Omega) = \left\{v \in W^{2,p}(\Omega);\ \mathrm{tr}\, v = 0 \ \text{and} \ \sum_{i=1}^N \nu_i \,\mathrm{tr}\, \partial_i v = 0\right\},$$

where $(\nu_i)_{i=1}^N$ denotes the unit outer normal vector field along Γ (which exists $d\Gamma$-almost everywhere; cf. Section 1.18). □

Note that Theorem 6.6-5(a) implies that *we always have*

$$\mathrm{tr} \in \mathcal{L}(W^{1,p}(\Omega); L^p(\Gamma)), \quad p \ge 1,$$

i.e., independently of the dimension N.

Naturally, tr denotes in (c) the continuous linear operator that corresponds to one of the situations described in (a) (i.e., according to how p compares with N). For instance, if $1 \le p < N$, the relation $\mathrm{tr}\, v = 0$ in (c) means that the function $\mathrm{tr}\, v$ is the zero function of the space $L^{p^\sharp}(\Gamma)$.

The relation $\mathrm{tr}(W^{1,p}(\Omega)) \subsetneqq L^{p^\sharp}(\Gamma)$ is the basis for defining the **trace spaces**

$$W^{1-\frac{1}{p},p}(\Gamma) := \{\mathrm{tr}\, v \in L^{p^\sharp}(\Gamma);\ v \in W^{1,p}(\Omega)\} \quad \text{for } 1 \le p < N,$$
$$H^{1/2}(\Gamma) := \{\mathrm{tr}\, v \in L^2(\Gamma);\ v \in H^1(\Omega)\} \quad \text{if } p = 2,$$

which thus consists of the traces of all the functions in $W^{1,p}(\Omega)$, $1 \le p < N$, or in $H^1(\Omega)$.

As is customary, we shall henceforth omit the symbol "tr" whenever no confusion should arise. For instance, we shall simply rewrite relation (c) in Theorem 6.6-5 as

$$W_0^{1,p}(\Omega) = \{v \in W^{1,p}(\Omega);\ v = 0 \text{ on } \Gamma\}, \quad \text{or} \quad H_0^1(\Omega) = \{v \in H^1(\Omega);\ v = 0 \text{ on } \Gamma\} \ \text{if } p = 2.$$

We shall also encounter spaces such as

$$V := \{v \in W^{1,p}(\Omega);\ \mathrm{tr}\, v = 0 \text{ on } \Gamma_0\}$$

where Γ_0 is a $d\Gamma$-measurable subset of Γ, and the relation "$\mathrm{tr}\, v = 0$ on Γ_0" similarly means that, as a function in the space $L^{p^\sharp}(\Gamma)$, the function $\mathrm{tr}\, v$ vanishes on the subset Γ_0 of Γ. Then we shall likewise rewrite such a space V as

$$V = \{v \in W^{1,p}(\Omega);\ v = 0 \text{ on } \Gamma_0\}.$$

[14]This result is proved in, e.g., NEČAS [1967, Chapter 2, Theorem 4.12]. A similar result holds for $m \ge 3$ if Γ is of class $\mathcal{C}^{m,1}$; cf. Theorem 4.13 in *ibid.*

The *Poincaré–Friedrichs* inequality, which holds for any open subset Ω of $\mathbb{R}^N$ of finite width (Theorem 6.5-2), admits the following generalizations when Ω is a domain (a proof of (a) for $p = 2$ is suggested in Problem 6.7-7(1); a proof of (b) when $p = 2$ will be given in Theorem 6.7-5).

$^\flat$**Theorem 6.6-6 (generalized Poincaré–Friedrichs inequalities)** *Let Ω be a domain in $\mathbb{R}^N$, and let $1 \le p < \infty$.*

(a) *There exists a constant c_0 such that*

$$\int_\Omega |v|^p \,\mathrm{d}x \le c_0 \left\{ \int_\Omega \sum_{i=1}^N |\partial_i v|^p \,\mathrm{d}x + \left| \int_\Omega v \,\mathrm{d}x \right|^p \right\} \quad \textit{for all } v \in W^{1,p}(\Omega).$$

(b) *Let Γ_0 be a $\mathrm{d}\Gamma$-measurable subset of Γ with $\mathrm{d}\Gamma$-meas$\,\Gamma_0 > 0$. Then there exists a constant c_2 such that*

$$\|v\|_{1,p,\Omega} \le c_2 \,|v|_{1,p,\Omega} \quad \textit{for all } v \in W^{1,p}(\Omega) \textit{ that satisfy } v = 0 \textit{ on } \Gamma_0.$$

(c) *Let Γ_0 be a $\mathrm{d}\Gamma$-measurable subset of Γ with $\mathrm{d}\Gamma$-meas$\,\Gamma_0 > 0$. Then there exists a constant c_1 such that*

$$\int_\Omega |v|^p \,\mathrm{d}x \le c_1 \left\{ \int_\Omega \sum_{i=1}^N |\partial_i v|^p \,\mathrm{d}x + \left| \int_{\Gamma_0} v \,\mathrm{d}\Gamma \right|^p \right\} \quad \textit{for all } v \in W^{1,p}(\Omega). \qquad \square$$

We conclude this review by the extension to functions in a Sobolev space of the *fundamental Green's formula* for smooth functions (Section 1.18).

$^\flat$**Theorem 6.6-7 (fundamental Green's formula in Sobolev spaces)** *Let Ω be a domain in $\mathbb{R}^N$, and let $\nu = (\nu_i)_{i=1}^N$ denote the unit outer normal vector field along $\partial\Omega$. Let $1 \le p < \infty$ and $1 \le q < \infty$ be such that*

$$\frac{1}{p} + \frac{1}{q} \le 1 + \frac{1}{N} \quad \textit{if } 1 \le p < N \textit{ and } 1 \le q < N, \textit{ or } 1 < q \textit{ if } N \le p, \textit{ or } 1 < p \textit{ if } N \le q.$$

Then, given functions $u \in W^{1,p}(\Omega)$ and $v \in W^{1,q}(\Omega)$, each function $uv\nu_i$, $1 \le i \le N$, belongs to the space $L^1(\Gamma)$, and

$$\int_\Omega u \partial_i v \,\mathrm{d}x = -\int_\Omega (\partial_i u) v \,\mathrm{d}x + \int_\Gamma uv\nu_i \,\mathrm{d}\Gamma. \qquad \square$$

Note that, *if $u, v \in H^1(\Omega)$, the above fundamental Green's formula holds in any dimension* $N \ge 2$.

More specialized results about Sobolev spaces (such as other Green's formulas or various density theorems in trace spaces) will be also stated in the next sections, at the places where they are needed for the analysis of specific boundary value problems.

Problems

6.6-1 The purpose of this exercise is to show that the inclusion $H^1(\Omega) \subset \mathcal{C}^0(\overline{\Omega})$ does *not* hold in dimension $N \ge 2$.

(1) Given any $0 < \rho < 1$, let $\Omega := \{x \in \mathbb{R}^2, |x| < \rho\}$. Show that, for any $0 < \alpha < 1/2$, the function u defined almost everywhere in Ω by $u(x) := (-\ln|x|)^\alpha$ if $x \neq 0$, belongs to the Sobolev space $H^1(\Omega)$ (as usual, no distinction is made between functions and their equivalence classes). Hence $H^1(\Omega) \not\subset \mathcal{C}^0(\overline{\Omega})$.

(2) Let $B := \bigcup_{k=1}^\infty \{b_k\}$ be a countably infinite dense subset of Ω and let $\beta_k > 0$, $k \geq 1$, be such that $\sum_{k=1}^\infty \beta_k < \infty$. Show that the function v defined almost everywhere in Ω by $v(x) := \sum_{k=1}^\infty \beta_k u(x - b_k)$ if $x \notin B$ belongs to the Sobolev space $H^1(\Omega)$.

(3) Show that any extension $\widetilde{v}$ of the function v to the set Ω is discontinuous everywhere in Ω (i.e., whatever values $\widetilde{v}(b_k)$, $k \geq 1$, may be assigned). Note that any such extension $\widetilde{v}$ also belongs to $H^1(\Omega)$ since $\widetilde{v} = v$ almost everywhere in Ω.

(4) Assume that $N \geq 3$ and let $\Omega := \{x \in \mathbb{R}^N; |x| < 1\}$. Show that, for any $0 < \lambda < (N-2)/2$, the function w defined almost everywhere in Ω by $w(x) := |x|^{-\lambda}$ if $x \neq 0$, belongs to the space $H^1(\Omega)$. Hence $H^1(\Omega) \not\subset \mathcal{C}^0(\overline{\Omega})$.

6.6-2 The purpose of this exercise is to prove a special case of Theorem 6.6-1 when $N = 1$. In what follows, $I :=]0,1[$.

(1) Let v be a function in $H^1(I)$, i.e., $v \in L^2(I)$ and there exists a function $v_1 \in L^2(I)$ such that $\int_0^1 v\varphi' \, dx = -\int_0^1 v_1 \varphi \, dx$ for all $\varphi \in \mathcal{D}(I)$. Show that the function $w : \overline{I} \to \mathbb{R}$ defined by $w(x) = \int_0^x v_1(t) \, dt$, $0 \leq x \leq 1$, belongs to the space $\mathcal{C}(\overline{I})$ and satisfies $\int_0^1 (v - w)\varphi' \, dx = 0$ for all $\varphi \in \mathcal{D}(I)$.

(2) Show that $(v - w)$ is a constant function (so that $v \in \mathcal{C}(\overline{I})$ by (1)) and that $v(y) = v(x) + \int_x^y v_1(t) dt$ for all $0 \leq x \leq y \leq 1$.

(3) Show that $H^1(I) \hookrightarrow \mathcal{C}(\overline{I})$, i.e., that $H^1(I) \subset \mathcal{C}(\overline{I})$ and there exists a constant c such that

$$\sup_{0 \leq x \leq 1} |v(x)| \leq c \, \|v\|_{H^1(I)} \quad \text{for all } v \in H^1(I).$$

6.6-3 Is the following statement true? Let Ω be a bounded open subset of $\mathbb{R}^3$, the boundary of which is a finite union of planar polygons. Then Ω is a domain in $\mathbb{R}^3$.

6.6-4 (1) Show that the N-dimensional Lebesgue measure of the boundary of a domain in $\mathbb{R}^N$ is 0.

(2) Do there exist open subsets of $\mathbb{R}^N$ whose boundary has an N-dimensional Lebesgue measure that is > 0?

6.6-5 Let Ω be a domain in $\mathbb{R}^N$, let $1 \leq p \leq \infty$, and let $\mathcal{P}_m(\Omega)$ denote for each integer $m \geq 1$ the space formed by the restrictions to Ω of all the polynomials of degree $\leq m$ in N variables.

(1) Show that the seminorm $|\cdot|_{m+1,p,\Omega}$ is a norm on the quotient space $W^{m+1,p}(\Omega)/\mathcal{P}_m(\Omega)$, equivalent to the quotient norm over this space.

(2) Let $\ell \in (W^{m+1,p}(\Omega))'$ be such that $\ell(p) = 0$ for all $p \in \mathcal{P}_m(\Omega)$. Show that there exists a constant C *independent of* ℓ such that

$$|\ell(v)| \leq C \, \|\ell\|_{(W^{m+1,p}(\Omega))'} \, |v|_{m+1,p,\Omega} \quad \text{for all } v \in W^{m+1,p}(\Omega).$$

Remark The result of (2) constitutes the **Bramble–Hilbert lemma.**[15] □

[15] J.H. BRAMBLE; S.R. HILBERT [1970]: Estimation of linear functionals on Sobolev spaces with application to Fourier transforms and spline interpolation, *SIAM Journal on Numerical Analysis* **7**, 112–124.

6.7 Examples of second-order linear elliptic boundary value problems; the membrane problem

Now that the needed preliminaries (quadratic minimization problems or abstract variational problems, and Sobolev spaces) have been laid down, we can focus our attention on the description and analysis of various *examples of boundary value problems* posed over a domain Ω in $\mathbb{R}^N$. In each case, we will follow the same three-tier approach:

First, we prescribe a Hilbert space V (either $H^1(\Omega)$ or $H^2(\Omega)$, or a closed subspace thereof, such as $H_0^1(\Omega)$ or $H_0^2(\Omega)$), a nonempty closed convex subset U of V (which in particular may be simply equal to V, as in this section and the next one), a bilinear form $a : V \times V \to \mathbb{R}$, and a linear form $\ell : V \to \mathbb{R}$.

Second, if a is symmetric, we verify that these specific data V, U, a, and ℓ satisfy all the assumptions of the existence result of Theorem 6.1-1. If this is the case, there then exists one and only one function $u \in U$ that satisfies

$$J(u) = \inf_{v \in U} J(v), \quad \text{where } J(v) := \frac{1}{2}a(v,v) - \ell(v) \text{ for all } v \in V,$$

or equivalently, that satisfies the *variational equations*

$$a(u,v) = \ell(v) \quad \text{for all } v \in U$$

if $U = V$, or the *variational inequalities*

$$a(u, v-u) \geq \ell(v-u) \quad \text{for all } v \in U$$

if U is not a subspace of V (Theorem 6.1-2). If a is not symmetric and $U = V$, then we resort to the Lax–Milgram lemma (Theorem 6.2-1), which asserts the existence and uniqueness of $u \in V$ that satisfies the variational equations $a(u,v) = \ell(v)$ for all $v \in V$ (see Problem 6.7-9 for such an example).

Third, under an additional *regularity assumption* on the function $u \in U$ (viz., $u \in H^2(\Omega) \cap V$ if $V \subset H^1(\Omega)$, or $u \in H^4(\Omega) \cap V$ if $V \subset H^2(\Omega)$), we identify a **boundary value problem** that is satisfied over $\overline{\Omega}$ by the solution $u \in U$ of the above variational equations or inequalities. If $U = V$, this problem comprises a *linear partial differential equation* of the form $\mathcal{L}u = f$ that u satisfies in Ω and *linear boundary conditions* that u satisfies on the boundary Γ of Ω. The terminology "boundary value problem" reflects that u satisfies conditions on the *whole* boundary Γ; note that the *type* of such boundary conditions may vary along Γ (see Theorem 6.7-6 for such an example).

In our selection of examples, we proceed "from the simplest to the less simple." This is why we begin by considering linear partial differential operators $\mathcal{L}$ of the particular form $\mathcal{L} : v \to \mathcal{L}v := -\Delta v + cv$ before considering more general *second-order elliptic partial differential operators*; this is why we consider successively *Dirichlet*, then *Neumann*, and finally *mixed*, boundary conditions; this is why we consider partial differential operators of second order before those of *fourth order* (Section 6.8); finally, this is why we consider *linear* problems (when $U = V$) before considering *nonlinear* problems (when U is not a subspace of V; cf. Section 6.9).

As a preparation for the examples treated in this section, we prove two useful *Green's formulas in Sobolev spaces*. Recall that the unit outer normal vector field exists $d\Gamma$-almost everywhere along the boundary Γ of a domain (Section 2.7).

Given any smooth enough vector fields $u, v : \Omega \to \mathbb{R}^N$, we let

$$\nabla v := (\partial_i v)_{i=1}^N, \quad |\nabla v| := \left(\sum_{i=1}^N |\partial_i v|^2\right)^{1/2}, \quad \text{and} \quad \nabla u \cdot \nabla v := \sum_{i=1}^N \partial_i u \partial_i v.$$

Recall in this respect that $|a|$ and $a \cdot b$ respectively denote the norm of $a \in \mathbb{R}^N$ and the Euclidean inner product of $a, b \in \mathbb{R}^N$.

Theorem 6.7-1 *Let Ω be a domain in $\mathbb{R}^N$ and let $(\nu_i)_{i=1}^N$ denote the unit outer normal vector field along $\Gamma := \partial\Omega$.*

(a) *For any $u \in H^2(\Omega)$, let*

$$\Delta u := \sum_{i=1}^N \partial_{ii} u \in L^2(\Omega) \quad \textit{and} \quad \partial_\nu u := \sum_{i=1}^N \nu_i \partial_i u \in L^2(\Gamma),$$

where $\partial_i u \in L^2(\Gamma)$ denotes the trace on Γ of the function $\partial_i u \in H^1(\Omega)$. Then the following **Green's formula** *holds:*

$$\int_\Omega \nabla u \cdot \nabla v \, dx = -\int_\Omega (\Delta u) v \, dx + \int_\Gamma (\partial_\nu u) v \, d\Gamma \quad \textit{for all } u \in H^2(\Omega), v \in H^1(\Omega).$$

(b) *Given functions $a \in \mathcal{C}^1(\overline{\Omega})$ and $u \in H^1(\Omega)$, the function au belongs to the space $H^1(\Omega)$. Besides, the following* **Green's formula** *holds for all $1 \le j \le N$:*

$$\int_\Omega a u \partial_j v \, dx = -\int_\Omega (\partial_j (au)) v \, dx + \int_\Gamma a u v \nu_j \, d\Gamma \quad \textit{for all } u \in H^1(\Omega), v \in H^1(\Omega).$$

Proof If $u \in H^2(\Omega)$, then the definition of the spaces $H^m(\Omega)$ (Section 6.5) implies that each function $\partial_i u$, $1 \le i \le N$, belongs to the space $H^1(\Omega)$. This being the case, the first Green's formula simply follows from the fundamental Green's formula in Sobolev spaces (Theorem 6.6-7).

If $a \in \mathcal{C}^1(\overline{\Omega})$ and $u \in H^1(\Omega)$, then the function au belongs to the space $L^2(\Omega)$. Besides, the functions

$$w_\ell := (\partial_j a) u + a \partial_j u, \quad 1 \le j \le N,$$

which clearly belong to $L^2(\Omega)$, are the weak derivatives of au. To see this, it suffices to remark that, if $u \in \mathcal{C}^\infty(\overline{\Omega})$,

$$\int_\Omega a u \partial_j \varphi \, dx = -\int_\Omega \{(\partial_j a) u \varphi + a (\partial_j u) \varphi\} \, dx = -\int_\Omega w_j \varphi \, dx \quad \text{for all } \varphi \in \mathcal{D}(\Omega).$$

Then the density of $\mathcal{C}^\infty(\overline{\Omega})$ in $H^1(\Omega)$ (Theorem 6.6-4), combined with the continuity of the inner product in $L^2(\Omega)$, shows that the functions $w_j \in L^2(\Omega)$, $1 \le j \le N$, are indeed the weak derivatives of au.

Hence the function au belongs to the space $H^1(\Omega)$. The second Green's formula then again simply follows from another application of the fundamental Green's formula in Sobolev spaces. □

The operator

$$\Delta := \sum_{i=1}^{N} \partial_{ii},$$

which acts on functions defined in Ω, is called the **Laplace operator**, and Δu is called the **Laplacian** of u. The operator

$$\partial_\nu := \sum_{i=1}^{N} \nu_i \partial_i,$$

which acts on functions defined on the boundary Γ, is called the **outer normal derivative operator**, and $\partial_\nu u$ is called the **outer normal derivative** of u.

Remark The outer normal derivative operator was already encountered in Theorem 6.6-5(d). □

We now consider our first example.

Theorem 6.7-2 *Let Ω be an open subset of $\mathbb{R}^N$ of finite width, let functions*

$$c \in L^\infty(\Omega) \text{ such that } c \geq 0 \text{ a.e. in } \Omega \text{ and } f \in L^2(\Omega)$$

be given, and let

$$\begin{aligned} V = U &:= H_0^1(\Omega), \\ a(u,v) &:= \int_\Omega (\nabla u \cdot \nabla v + cuv)\, \mathrm{d}x \quad \text{for all } u, v \in V, \\ \ell(v) &:= \int_\Omega fv\, \mathrm{d}x \quad \text{for all } v \in V. \end{aligned}$$

Then there exists a unique function $u \in H_0^1(\Omega)$ that minimizes over the space $H_0^1(\Omega)$ the functional $J : H_0^1(\Omega) \to \mathbb{R}$ defined by

$$J(v) := \frac{1}{2}a(v,v) - \ell(v) = \frac{1}{2}\int_\Omega \left(|\nabla v|^2 + cv^2\right) \mathrm{d}x - \int_\Omega fv\, \mathrm{d}x \quad \text{for all } v \in H_0^1(\Omega),$$

or equivalently, that satisfies the variational equations:

$$\int_\Omega (\nabla u \cdot \nabla v + cuv)\, \mathrm{d}x = \int_\Omega fv\, \mathrm{d}x \quad \text{for all } v \in H_0^1(\Omega).$$

Besides, the linear mapping

$$f \in L^2(\Omega) \to u \in H_0^1(\Omega)$$

defined in this fashion is continuous.

Finally, the function u satisfies the following boundary value problem:

$$-\Delta u + cu = f \;\; \textit{in}\; \Omega \quad \textit{and} \quad u = 0 \;\; \textit{on}\; \Gamma,$$

where the partial differential equation in Ω is to be understood as an equality in the space $\mathcal{D}'(\Omega)$ and, under the additional assumption that Ω is a domain, the boundary condition on Γ is to be understood as an equality in the space $L^2(\Gamma)$.

Proof The symmetric bilinear form $a(\cdot,\cdot)$ is continuous since, by the Cauchy–Schwarz inequality, first for functions in $L^2(\Omega)$, then for vectors in $\mathbb{R}^{N+1}$,

$$\begin{aligned}
|a(u,v)| &\leq \sum_{i=1}^{N} \|\partial_i u\|_{0,\Omega}\, \|\partial_i v\|_{0,\Omega} + \|c\|_{L^\infty(\Omega)}\, \|u\|_{0,\Omega}\, \|v\|_{0,\Omega} \\
&\leq \max\{1, \|c\|_{L^\infty(\Omega)}\} \Big(\sum_{i=1}^{N} \|\partial_i u\|_{0,\Omega}^2 + \|u\|_{0,\Omega}^2 \Big)^{1/2} \Big(\sum_{i=1}^{N} \|\partial_i v\|_{0,\Omega}^2 + \|v\|_{0,\Omega}^2 \Big)^{1/2} \\
&= \max\{1, \|c\|_{L^\infty(\Omega)}\}\, \|u\|_{1,\Omega}\, \|v\|_{1,\Omega} \quad \text{for all } u, v \in H^1(\Omega).
\end{aligned}$$

Furthermore, the bilinear form a is $H_0^1(\Omega)$-coercive since

$$a(v,v) \geq \int_\Omega |\nabla v|^2 \, dx = |v|_{1,\Omega}^2 \quad \text{for all } v \in H^1(\Omega),$$

and the seminorm $|\cdot|_{1,\Omega}$ is a norm over the space $H_0^1(\Omega)$, equivalent to the norm $\|\cdot\|_{1,\Omega}$ (Theorem 6.5-2). Finally, the linear form ℓ is continuous since

$$|\ell(v)| \leq \|f\|_{0,\Omega}\, \|v\|_{0,\Omega} \leq \|f\|_{0,\Omega}\, \|v\|_{1,\Omega} \quad \text{for all } v \in H^1(\Omega).$$

All the assumptions of Theorem 6.1-1 being therefore satisfied, it follows that there exists one and only one function $u \in H_0^1(\Omega)$ that minimizes the announced functional J over $H_0^1(\Omega)$, or equivalently, that satisfies the announced variational equations (Theorem 6.1-2).

The continuity of the mapping $f \in L^2(\Omega) \to u \in H_0^1(\Omega)$, which is clearly linear, follows from the inequalities

$$|u|_{1,\Omega}^2 \leq a(u,u) = \ell(u) \leq \|f\|_{0,\Omega}\, \|u\|_{0,\Omega} \leq \|f\|_{0,\Omega}\, \|u\|_{1,\Omega},$$

satisfied by the solution u, and from the inequalities

$$\|v\|_{1,\Omega} \leq C(\Omega)\, |v|_{1,\Omega}$$

satisfied by all functions $v \in H_0^1(\Omega)$, which together imply that

$$\|u\|_{1,\Omega} \leq C(\Omega)^2\, \|f\|_{0,\Omega} \quad \text{for all } f \in L^2(\Omega).$$

Since $\int_\Omega \nabla u \cdot \nabla v \, dx = -\langle \Delta u, v \rangle$ for all $v \in \mathcal{D}(\Omega)$, where $\langle \cdot,\cdot \rangle := {}_{\mathcal{D}'(\Omega)}\langle \cdot,\cdot \rangle_{\mathcal{D}(\Omega)}$ (Section 6.3) and Δu is understood as a *distribution*, the equations $a(u,v) = \ell(v)$ for all $v \in V$ imply that

$$\langle -\Delta u + cu - f, v \rangle = 0 \quad \text{for all } v \in \mathcal{D}(\Omega)$$

(since $\mathcal{D}(\Omega) \subset H_0^1(\Omega)$), and hence that

$$-\Delta u + cu = f \quad \text{in } \mathcal{D}'(\Omega).$$

The characterization of the space $H_0^1(\Omega)$ when Ω is a *domain* (Theorem 6.6-5(c)) shows that the function $u \in H_0^1(\Omega)$ satisfies the boundary condition $u = 0$ on Γ, interpreted here as an equality in the space $L^2(\Gamma)$. □

The boundary value problem

$$-\Delta u + cu = f \text{ in } \Omega \quad \text{and} \quad u = 0 \text{ on } \Gamma$$

found in Theorem 6.7-2, or more generally, the boundary value problem

$$-\Delta u + cu = f \ \text{ in } \Omega \quad \text{and} \quad u = u_0 \ \text{ on } \Gamma$$

(which can be likewise derived from a variational problem; cf. Problem 6.7-1) is called a **Dirichlet**[16] **problem** *for the partial differential operator*

$$\mathcal{L} : v \to \mathcal{L}v := -\Delta v + cv.$$

Note that, here and subsequently, it is implicitly understood that either the functions v appearing in the definition of the operator $\mathcal{L}$ are at least defined almost everywhere in Ω and smooth enough for this definition to make sense (e.g., $\mathcal{L}v \in L^2(\Omega)$ if $v \in H^2(\Omega)$); or $\mathcal{L}$ is to be understood as a linear partial differential operator *in the sense of distributions* (Section 6.3).

The boundary condition $u = u_0$ on Γ is called a *homogeneous* if $u_0 = 0$, or *nonhomogeneous* otherwise, **Dirichlet boundary condition**.

The special cases $c = 0$ and $c = f = 0$, of the equation $-\Delta u + cu = f$ in Ω, viz.,

$$-\Delta u = f \text{ in } \Omega \quad \text{and} \quad -\Delta u = 0 \text{ in } \Omega,$$

are respectively called **Poisson's equation**[17] and **Laplace's equation**.[18] These equations are of special importance, because they model an amazingly wide variety of physical phenomena (an example corresponding to $N = 2$ is given below). Besides, in spite of its remarkable simplicity, Laplace's equation is at the root of a whole mathematical theory, viz., that of **harmonic functions**, i.e., those functions $u \in \mathcal{C}^2(\Omega)$ that satisfy $-\Delta u = 0$ in Ω (here Ω may be any open subset of $\mathbb{R}^N$); a brief sample of some of their remarkable properties is proposed in Problems 6.7-3–6.7-6.[19]

Several important comments are in order about this first example. First, it provides an instance of a **well-posed problem**,[20] in the sense that, for any $f \in L^2(\Omega)$, *there exists one*

[16]So named after Gustav Lejeune Dirichlet (1805–1859).

[17]So named after Siméon-Denis Poisson (1781–1840).

[18]So named after Pierre-Simon Laplace (1749–1827).

[19]Illuminating introductions to the theory of harmonic functions are found in EVANS [2010, Section 2.2] or in GILBARG & TRUDINGER [1998, Chapter 2].

[20]The notion of well-posed problem was introduced by Jacques Hadamard (1865–1963), in:

J. HADAMARD [1902]: Sur les problèmes aux dérivées partielles et leur signification physique, *Princeton University Bulletin* **13**, 49–52.

A captivating account of the adventurous life of Jacques Hadamard is given in MAZ'YA & SHAPOSHNIKOVA [1998].

and only one solution $u \in H_0^1(\Omega)$, which *in addition depends continuously on the function* $f \in L^2(\Omega)$.

Remark We shall see in Section 7.10 that, thanks to the *maximum principle*, a continuous dependence of the solution u in terms of the right-hand side f can be also established, but this time with respect to *sup-norms*. □

Second, the last part of the proof of Theorem 6.7-2 shows that the correct way to interpret the partial differential equations $-\Delta u + cu = f$ in Ω is as an *equality in the space* $\mathcal{D}'(\Omega)$. Note that, by contrast, the boundary condition $u = 0$ on Γ always makes sense as an equality of *functions* in the space $L^2(\Gamma)$ *if* Ω *is a domain.*

Remark In fact, *all* the conclusions of Theorem 6.7-2 still hold in the more general case where the *function* $f \in L^2(\Omega)$ is replaced by a *distribution* $f \in H^{-1}(\Omega)$ (the space $H^{-1}(\Omega)$, which denotes the *dual space* of the space $H_0^1(\Omega)$, will be defined in Section 6.11). □

Third, to determine sufficient conditions guaranteeing for instance that $u \in H^2(\Omega)$, in which case each partial derivative $\partial_{ii} u \in \mathcal{D}'(\Omega)$ appearing in the definition of Δu is a function in $L^2(\Omega)$, is a delicate issue, however.[21] For instance, one can show that, if the boundary Γ is of class $\mathcal{C}^2$ (Section 1.18), the solution $u \in H_0^1(\Omega)$ of the variational equations

$$\int_\Omega (\nabla u \cdot \nabla v + uv)\, dx = \int_\Omega fv\, dx \quad \text{for all } v \in H_0^1(\Omega),$$

is in $H^2(\Omega)$ for *any* function $f \in L^2(\Omega)$[22] (in this respect, an interesting complement is proposed in Problem 6.7-2).

Fourth, further assumptions on the data are needed to guarantee that u be a **classical solution** of such a boundary value problem, in the sense that u is in the space $\mathcal{C}(\overline{\Omega}) \cap \mathcal{C}^2(\Omega)$.

In this direction, when $c = f = 0$, a beautiful theorem[23] asserts that, if the open set $\Omega \subset \mathbb{R}^N$ is bounded, the Dirichlet problem $-\Delta u = 0$ in Ω and $u = u_0$ on Γ has a (unique) solution $u \in \mathcal{C}(\overline{\Omega}) \cap \mathcal{C}^2(\Omega)$ for *any* function $u_0 \in \mathcal{C}(\Gamma)$ *if and only if*, given any point $y \in \Gamma$, there exists a *barrier function*, i.e., a function $w_y \in \mathcal{C}(\overline{\Omega}) \cap \mathcal{C}^2(\Omega)$ with the following properties: $-\Delta w_y \geq 0$ in Ω, $w_y(y) = 0$, and $w_y(x) > 0$ for all $x \in (\overline{\Omega} - \{y\})$ (for instance, this is the case if the boundary Γ is of class $\mathcal{C}^2$; however it need not be the case if Γ is only Lipschitz-continuous).

In the general case, a proper functional setting for getting classical solutions is that of the spaces $\mathcal{C}^{m,\lambda}(\overline{\Omega})$ (Section 1.18). In this case, another beautiful theorem[24] asserts that if,

---

[21] If Ω is only a domain, a function $u \in H_0^1(\Omega)$ that satisfies $\Delta u \in L^2(\Omega)$ is not necessarily in $H^2(\Omega)$; see, e.g., the counterexample given in:

D. Jerison; C.E. Kenig [1995]: The inhomogeneous Dirichlet problem in Lipschitz domains, *Journal of Functional Analysis* **130**, 161–219.

[22] See, e.g., Brezis [2011, Theorem 9.25] or Evans [2010, Section 6.3].

[23] Due to:

O. Perron [1923]: Eine neue Behandlung der Randwertaufgabe für $\Delta u = 0$, *Mathematische Zeitschrift* **18**, 42–54.

A modern treatment of Perron's method is found in, e.g., Gilbarg & Trudinger [1998, Chapter 2].

[24] Due to:

O.D. Kellogg [1929]: *Foundations of Potential Theory*, Springer, Berlin.

See also Gilbarg & Trudinger [1998, Theorem 6.14].

for some $0 < \lambda < 1$, the function $c \geq 0$ belongs to the space $\mathcal{C}^{0,\lambda}(\overline{\Omega})$ and the boundary Γ of the domain Ω is of class $\mathcal{C}^{2,\lambda}$, then the Dirichlet problem $-\Delta u + cu = f$ in Ω and $u = u_0$ on Γ has a (unique) solution $u \in \mathcal{C}^{2,\lambda}(\overline{\Omega})$ for *any* functions $f \in \mathcal{C}^{0,\lambda}(\overline{\Omega})$ and $u_0 \in \mathcal{C}^{2,\lambda}(\overline{\Omega})$ (in fact, this existence result holds *verbatim* for general elliptic operators, as defined later in this section, with coefficients in the space $C^{0,\lambda}(\overline{\Omega})$). This theorem hinges on the crucial *Schauder's estimates*,[25] which give a priori bounds on the norms $\|u\|_{\mathcal{C}^{2,\lambda}(\overline{\Omega})}$ of a solution to this problem.

When $c = 0$ and $N = 2$ and Ω lies in the "horizontal" plane, this problem is a mathematical model of the **membrane problem** that arises in linearized elasticity when one considers the problem of finding the equilibrium position of an *elastic membrane*, under the action of a vertical force of density $F = \tau f$ where τ measures the tension of the membrane, and whose vertical displacement $u : \overline{\Omega} \to \mathbb{R}$ is equal to a known function u_0 along the boundary Γ (Figure 6.7-1). By Theorem 6.7-2, when $u_0 = 0$ (to fix ideas), the displacement $u \in H_0^1(\Omega)$ thus minimizes the *membrane energy* $J : H_0^1(\Omega) \to \mathbb{R}$ defined by

$$J(v) := \frac{1}{2}\int_\Omega |\nabla v|^2 \, \mathrm{d}x - \int_\Omega f v \, \mathrm{d}x \quad \text{for all } v \in H_0^1(\Omega)$$

over the space $H_0^1(\Omega)$.

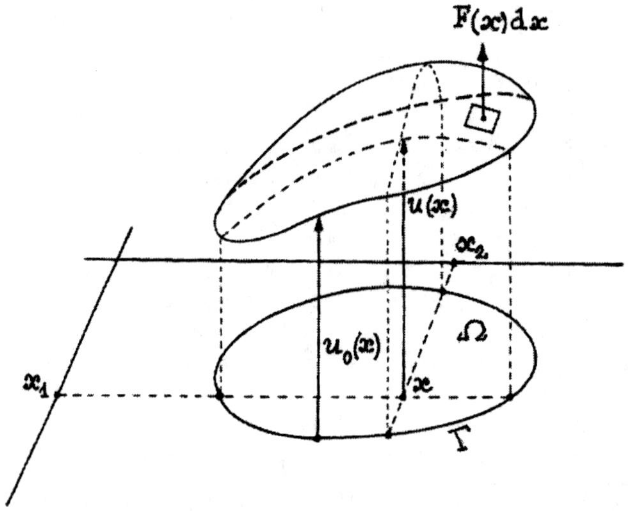

Figure 6.7-1 The membrane problem: The unknown function $u : \overline{\Omega} \subset \mathbb{R}^2 \to \mathbb{R}$ represents the vertical displacement of a membrane subjected to a vertical force of density F per unit area. This figure originally appeared in P.G. CIARLET [1978]: *The Finite Element Method for Elliptic Problems*, North-Holland, Amsterdam.

Remark Interestingly, the data found in the variational formulation of the membrane problem can be rigorously justified by means of an asymptotic analysis applied to the equations of *nonlinear*

[25] J. SCHAUDER [1934]: Über lineare elliptische Differentialgleichungen zweiter Ordnung, *Mathematische Zeitschrift* **38**, 257–282.

three-dimensional elasticity (first, by letting the thickness approach zero;[26] second, by letting the tension τ approach ∞[27]). $\square$

As a preparation for the next examples, we state a result about traces,[28] which will be essential for deriving boundary conditions such as $\partial_\nu u = 0$ along the boundary Γ of a domain (Theorems 6.7-4) or along a subset of Γ (Theorem 6.7-6).

$^\flat$**Theorem 6.7-3** *Let Ω be a domain in $\mathbb{R}^N$, let Γ_1 be a relatively open subset of $\Gamma := \partial\Omega$, and let $w \in L^2(\Gamma_1)$ be a function that satisfies*

$$\int_{\Gamma_1} wv \,\mathrm{d}\Gamma = 0 \textit{ for all } v \in H^1(\Omega) \textit{ such that } v = 0 \textit{ on } \Gamma - \Gamma_1.$$

Then $w = 0$. $\square$

Remark By Theorem 4.3-2, Theorem 6.7-3 is equivalent to stating that *the space* $\{v|_{\Gamma_1} \in L^2(\Gamma_1);\ v \in H^1(\Omega), v = 0 \text{ on } \Gamma - \Gamma_1\}$ *is dense in* $L^2(\Gamma_1)$. $\square$

We now consider our second example, which displays several differences from the first example: an open set Ω that is a domain, a larger space V, a stronger assumption on the function c, and a more general linear form ℓ. For brevity, any argument in the next proofs that is similar to one used in the proof of Theorem 6.7-2 will be omitted.

Theorem 6.7-4 *Let Ω be a domain in $\mathbb{R}^N$, let functions*

$$c \in L^\infty(\Omega) \textit{ such that } c \geq c_0 > 0 \textit{ a.e. in } \Omega, \quad f \in L^2(\Omega), \quad g \in L^2(\Gamma),$$

be given, and let

$$\begin{aligned} V = U &:= H^1(\Omega), \\ a(u,v) &:= \int_\Omega (\nabla u \cdot \nabla v + cuv)\,\mathrm{d}x \quad \textit{for all } u, v \in V, \\ \ell(v) &:= \int_\Omega fv\,\mathrm{d}x + \int_\Gamma gv\,\mathrm{d}\Gamma \quad \textit{for all } v \in V. \end{aligned}$$

Then there exists a unique function $u \in H^1(\Omega)$ that minimizes the functional $J : H^1(\Omega) \to \mathbb{R}$ defined by

$$J(v) := \frac{1}{2}a(v,v) - \ell(v) = \frac{1}{2}\int_\Omega \left(|\nabla v|^2 + cv^2\right)\mathrm{d}x - \left(\int_\Omega fv\,\mathrm{d}x + \int_\Gamma gv\,\mathrm{d}\Gamma\right)$$

[26]P.G. Ciarlet [1980]: A justification of the von Kármán equations, *Archive for Rational Mechanics and Analysis* **73**, 349–389.

G. Friesecke; R.D. James; S. Müller [2006]: A hierarchy of plate models derived from nonlinear elasticity by Gamma-convergence, *Archive for Rational Mechanics and Analysis* **180**, 183–236.

[27]See Ciarlet & Rabier [1980, Section 2.3] or Ciarlet [1997, Section 5.10].

[28]For a proof of this result, which is not usually provided in classical texts about Sobolev spaces, see:

J.M.E. Bernard [2011]: Density results in Sobolev spaces whose elements vanish on a part of the boundary, *Chinese Annals of Mathematics, Series B*, **32**, 823–846.

for all $v \in H^1(\Omega)$, *or, equivalently, that satisfies the variational equations*

$$\int_\Omega (\nabla u \cdot \nabla v + cuv)\, \mathrm{d}x = \int_\Omega fv\, \mathrm{d}x + \int_\Gamma gv\, \mathrm{d}\Gamma \quad \textit{for all } v \in H^1(\Omega).$$

Besides, the linear mapping

$$(f, g) \in L^2(\Omega) \times L^2(\Gamma) \to u \in H^1(\Omega)$$

defined in this fashion is continuous.

Assume in addition that $u \in H^2(\Omega)$. *Then* u *satisfies the following boundary value problem:*

$$-\Delta u + cu = f \ \textit{in } \Omega \quad \textit{and} \quad \partial_\nu u = g \ \text{on } \Gamma,$$

where $\partial_\nu u \in L^2(\Gamma)$ *denotes the outer normal derivative of* u *(Theorem* 6.7-1(a)*).*

Proof The bilinear form is $H^1(\Omega)$-coercive since

$$a(v, v) \geq \min\{1, c_0\} \, \|v\|_{1,\Omega}^2 \quad \text{for all } v \in H^1(\Omega).$$

The linear form $v \in H^1(\Omega) \to \int_\Gamma gv\, \mathrm{d}\Gamma$ is continuous since (Theorem 6.6-5)

$$\left| \int_\Gamma gv\, \mathrm{d}\Gamma \right| \leq \|g\|_{L^2(\Gamma)} \, \|v\|_{L^2(\Gamma)} \leq \|\mathrm{tr}\|_{\mathcal{L}(H^1(\Omega); L^2(\Gamma))} \, \|g\|_{L^2(\Gamma)} \, \|v\|_{1,\Omega} \, .$$

Therefore there exists a unique function $u \in H^1(\Omega)$ that minimizes the announced functional J over the space $H^1(\Omega)$, or equivalently, that satisfies the announced variational equations.

Assume next that $u \in H^2(\Omega)$. Thanks to the Green's formula of Theorem 6.7-1(a), the equations $a(u, v) = \ell(v)$ for all $v \in V$ become

$$\int_\Omega (-\Delta u + cu - f) v\, \mathrm{d}x = \int_\Gamma (g - \partial_\nu u) v\, \mathrm{d}\Gamma \quad \text{for all } v \in H^1(\Omega).$$

In particular then, the function $(-\Delta u + cu - f) \in L^2(\Omega)$ satisfies

$$\int_\Omega (-\Delta u + cu - f) v\, \mathrm{d}x = 0 \quad \text{for all } v \in \mathcal{D}(\Omega),$$

which implies that $-\Delta u + cu - f = 0$ in $L^2(\Omega)$ since $\mathcal{D}(\Omega)$ is dense in $L^2(\Omega)$.

Taking the equation $-\Delta u + cu - f = 0$ into account, we are thus left with

$$\int_\Gamma (g - \partial_\nu u) v\, \mathrm{d}\Gamma = 0 \quad \text{for all } v \in H^1(\Omega),$$

which implies that the function $(g - \partial_\nu u) \in L^2(\Gamma)$ vanishes (apply Theorem 6.7-3 with $\Gamma_1 = \Gamma$). □

The boundary value problem

$$-\Delta u + cu = f \ \text{in } \Omega \quad \text{and} \quad \partial_\nu u = g \ \text{on } \Gamma$$

found in Theorem 6.7-4 is called a **Neumann**[29] **problem** *for the partial differential operator*

$$\mathcal{L} : v \to \mathcal{L}v := -\Delta v + cv.$$

The boundary condition $\partial_\nu u = g$ on Γ is called a *homogeneous* if $g = 0$, or *nonhomogeneous* otherwise, **Neumann boundary condition.**

Notice that, while the Dirichlet boundary condition $u = 0$ on Γ found in the first example makes sense in the space $L^2(\Gamma)$ if u is only in $H^1(\Omega)$, i.e., without assuming additional regularity on the function u, the Neumann boundary condition $\partial_\nu u = g$ on Γ does *not* make sense, at least in any function space over Γ, in this case. Nevertheless, even if u is only in $H^1(\Omega)$, it is a common practice to say that u solves, *at least formally*, the boundary value problem $-\Delta u + cu = f$ in Ω and $\partial_\nu u = g$ on Γ.

Remark Define the space

$$H(\Delta;\Omega) := \{v \in H^1(\Omega);\ \Delta v \in L^2(\Omega)\},$$

where $\Delta v \in L^2(\Omega)$ is to be understood in the sense of distributions. Then one can show[30] that there exists a continuous linear operator $\gamma_1 : H(\Delta;\Omega) \to H^{-1/2}(\Gamma)$, where $H^{-1/2}(\Gamma)$ denotes the dual of the trace space $H^{1/2}(\Gamma) := \{\operatorname{tr} v \in L^2(\Gamma);\ v \in H^1(\Omega)\}$ (Section 6.6), such that $\gamma_1 v = \partial_\nu v|_\Gamma$ for smooth enough functions $v : \overline{\Omega} \to \mathbb{R}$ (the space $H(\Delta;\Omega)$ is equipped here with its natural norm $v \to (\|v\|^2_{1,\Omega} + \|\Delta v\|^2_{0,\Omega})^{1/2}$).

This observation thus provides a well-defined meaning to the boundary condition $\partial_\nu u = g$ on Γ, viz., as an *equality in the dual space* $H^{-1/2}(\Gamma)$, even if the solution u to the variational problem considered in Theorem 6.7-3 is only in $H^1(\Omega)$.

If Γ_1 is a proper, relatively open, subset of Γ, the interpretation of the formal boundary condition $\partial_\nu u = g$ on Γ_1 (such a boundary condition will be found in the next example) is somewhat more delicate, viz., as an *equality in the dual of the space*

$$H^{1/2}_{00}(\Gamma_1) := \{v \in L^2(\Gamma_1);\ \text{there exists } w \in H^1(\Omega) \text{ such that } w = 0 \text{ on } \Gamma - \Gamma_1 \text{ and } w = v \text{ on } \Gamma_1\},$$

which does *not* coincide with the space $H^{1/2}(\Gamma_1) := \{v|_{\Gamma_1};\ v \in H^{1/2}(\Gamma)\}$. □

Note also, while the boundary condition $u = 0$ on Γ found in the first example simply originates from the *definition of the space* V to which u belongs, viz., $V = H^1_0(\Omega)$, the boundary condition $\partial_\nu u = g$ on Γ found in the second example originates instead from an application of a *Green's formula*.

If the assumption $c \geq c_0 > 0$ almost everywhere in Ω is replaced by the assumption $c = 0$ almost everywhere in Ω (a special case of the first example), then the bilinear form a is no longer $H^1(\Omega)$-elliptic. Nevertheless, an existence result still holds, but only if the functions f and g satisfy an appropriate *compatibility condition*; see Problems 6.7-7 and 6.7-8.

Recall that, when Ω is of finite width, there exists a constant $C = C(\Omega)$ such that the *Poincaré-Friedrichs inequality* holds, viz.,

$$\|v\|_{1,\Omega} < C\,|v|_{1,\Omega} \quad \text{for all functions } v \in H^1_0(\Omega)$$

(Theorem 6.5-2). We now show that, if Ω is a domain, this inequality still holds if the functions $v \in H^1(\Omega)$ appearing in it vanish only on a *portion* Γ_0 of the boundary, provided

[29] So named after Carl Neumann (1832–1925).

[30] See, e.g., Dautray & Lions [2000b, Chapter 7, Section 1].

that $d\Gamma$-meas $\Gamma_0 > 0$. The next result, which incidentally provides another proof of Theorem 6.5-2(b) for a domain and when $m = 1$, will be needed for establishing the ellipticity of the bilinear form in the third example.

Theorem 6.7-5 *Let Ω be a domain in $\mathbb{R}^N$, and let Γ_0 be a $d\Gamma$-measurable subset of the boundary Γ that satisfies*

$$d\Gamma\text{-meas}\,\Gamma_0 > 0.$$

Then the space

$$V := \{v \in H^1(\Omega);\ v = 0 \text{ on } \Gamma_0\}$$

is a closed subspace of $H^1(\Omega)$, and there exists a constant $C = C(\Omega)$ such that

$$|v|_{1,\Omega} \le \|v\|_{1,\Omega} \le C\,|v|_{1,\Omega} \quad \textit{for all } v \in V.$$

Proof Let $(v_k)_{k=1}^{\infty}$ be a sequence of functions in the space V that converges to a function $v \in H^1(\Omega)$. Since then the sequence $(\operatorname{tr} v_k)_{k=1}^{\infty}$ converges to $\operatorname{tr} v$ in the space $L^2(\Gamma)$ (Theorem 6.6-5), the sequence $(\operatorname{tr} v_k|_{\Gamma_0})_{k=1}^{\infty}$ converges to $\operatorname{tr} v|_{\Gamma_0}$ in $L^2(\Gamma_0)$. But $\operatorname{tr} v_k|_{\Gamma_0} = 0$ for all $k \ge 1$, and the limit of a sequence in a normed vector space is unique. Hence $\operatorname{tr} v|_{\Gamma_0} = 0$, and thus V is a closed subspace of $H^1(\Omega)$.

Next, let us show that $|\cdot|_{1,\Omega}$ is a norm over the space V. Let v be a function in the space V that satisfies $|v|_{1,\Omega} = 0$. Then v is a constant function by virtue of the connectedness of the set Ω (Theorem 6.3-4). Therefore its trace on Γ is a constant function that takes the same value (recall that tr coincides by construction with the usual trace for functions in $\mathcal{C}^\infty(\overline{\Omega})$; cf. Section 6.6), and this value is zero since the trace vanishes on the set Γ_0, whose $d\Gamma$-measure is > 0.

Finally, assume that the two norms $|\cdot|_{1,\Omega}$ and $\|\cdot\|_{1,\Omega}$ are not equivalent over the space V. Then there exists a sequence $(v_k)_{k=1}^{\infty}$ of functions $v_k \in V$ that satisfy

$$\|v_k\|_{1,\Omega} = 1 \text{ for all } k \quad \text{and} \quad \lim_{k\to\infty} |v_k|_{1,\Omega} = 0.$$

By the Rellich–Kondrachov theorem (Theorem 6.6-3), any bounded sequence in the space $H^1(\Omega)$ contains a subsequence that converges in $L^2(\Omega)$. Hence there exists a subsequence $(v_\ell)_{\ell=1}^{\infty}$ of the sequence $(v_k)_{k=1}^{\infty}$ that converges in the space $L^2(\Omega)$.

Since $\lim_{\ell\to\infty} |v_\ell|_{1,\Omega} = 0$ on the other hand, the sequence $(v_\ell)_{\ell=1}^{\infty}$ is thus a Cauchy sequence in the space V, which is complete as a closed subspace of $H^1(\Omega)$. Therefore the sequence $(v_\ell)_{\ell=1}^{\infty}$ converges with respect to the norm $\|\cdot\|_{1,\Omega}$ to an element $v \in V$.

Since $|v|_{1,\Omega} = \lim_{\ell\to\infty} |v_\ell|_{1,\Omega} = 0$ and $v \in V$, it follows that $v = 0$, which contradicts the equalities $\|v_\ell\|_{1,\Omega} = 1$ for all ℓ. Hence the proof is complete. □

In our third *example*, we extend the previous examples in two directions: First, *boundary conditions of both types*, i.e., Dirichlet *and* Neumann, will appear in the associated boundary value problem; second, the partial differential operator will be more general.

Theorem 6.7-6 *Let Ω be a domain in $\mathbb{R}^N$, let functions $a_{ij} = a_{ji} \in \mathcal{C}^1(\overline{\Omega})$, $1 \le i, j \le N$, be given with the property that there exists a constant μ such that*

$$\mu > 0 \quad \textit{and} \quad \sum_{i,j=1}^{N} a_{ij}(x)\xi_i\xi_j \ge \mu \sum_{i=1}^{N} |\xi_i|^2 \quad \textit{for all } x \in \overline{\Omega} \textit{ and all } (\xi_i)_{i=1}^{N} \in \mathbb{R}^N,$$

let Γ_1 *be a relatively open subset of* $\Gamma := \partial\Omega$ *such that*

$$\mathrm{d}\Gamma\text{-meas}\,\Gamma_0 > 0 \quad \textit{where } \Gamma_0 = \Gamma - \Gamma_1,$$

let functions

$$c \in L^\infty(\Omega) \textit{ such that } c \geq 0 \textit{ a.e. in } \Omega, \quad f \in L^2(\Omega), \quad g \in L^2(\Gamma_1)$$

be given, and finally, let

$$\begin{aligned} V = U &:= \{v \in H^1(\Omega);\ v = 0 \textit{ on } \Gamma_0\}, \\ a(u,v) &:= \int_\Omega \Big(\sum_{i,j=1}^n a_{ij}\partial_i u \partial_j v + cuv \Big) \,\mathrm{d}x \quad \textit{for all } u, v \in V, \\ \ell(v) &:= \int_\Omega fv \,\mathrm{d}x + \int_{\Gamma_1} gv \,\mathrm{d}\Gamma \quad \textit{for all } v \in V. \end{aligned}$$

Then there exists a unique function $u \in V$ *that minimizes over the space* V *the functional* $J : V \to \mathbb{R}$ *defined by*

$$J(v) := \frac{1}{2}a(v,v) - \ell(v) = \frac{1}{2}\int_\Omega \Big(\sum_{i,j=1}^N a_{ij}\partial_i u \partial_j v + cv^2 \Big) \,\mathrm{d}x - \Big(\int_\Omega fv \,\mathrm{d}x + \int_{\Gamma_1} gv \,\mathrm{d}\Gamma \Big)$$

for all $v \in V$, *or equivalently, that satisfies the variational equations*

$$\int_\Omega \Big(\sum_{i,j=1}^N a_{ij}\partial_i u \partial_j v + cuv \Big) \,\mathrm{d}x = \int_\Omega fv \,\mathrm{d}x + \int_{\Gamma_1} gv \,\mathrm{d}\Gamma \quad \textit{for all } v \in V.$$

Besides, the linear mapping

$$(f, g) \in L^2(\Omega) \times L^2(\Gamma_1) \to u \in V \subset H^1(\Omega)$$

defined in this fashion is continuous.

Assume in addition that $u \in H^2(\Omega)$. *Then* u *satisfies the following boundary value problem:*

$$\begin{aligned} -\sum_{i,j=1}^N \partial_j(a_{ij}\partial_i u) + cu = f &\quad \textit{in } \Omega, \\ u = 0 &\quad \textit{on } \Gamma_0, \\ \sum_{i,j=1}^N a_{ij}\nu_j \partial_i u = g &\quad \textit{on } \Gamma_1, \end{aligned}$$

where ν_j, $1 \leq j \leq N$, *denote the components of the unit outer normal vector along* Γ.

Proof As a closed subspace of $H^1(\Omega)$ (Theorem 6.7-5), the space V is a Hilbert space. The bilinear form a is continuous, since

$$|a(u,v)| \leq \max\left\{\|c\|_{L^\infty(\Omega)}, \max_{1\leq i,j\leq N}\|a_{ij}\|_{\mathcal{C}(\overline{\Omega})}\right\}\|u\|_{1,\Omega}\,\|v\|_{1,\Omega} \quad \text{for all } u,v \in H^1(\Omega).$$

The bilinear form is also V-coercive, since

$$a(v,v) = \int_\Omega \left(\sum_{i,j=1}^N a_{ij}\partial_i v \partial_j v + cv^2\right) \mathrm{d}x \geq \mu\,|v|_{1,\Omega}^2 \quad \text{for all } v \in V,$$

and $|\cdot|_{1,\Omega}$ is a norm on V, equivalent to $\|\cdot\|_{1,\Omega}$ (Theorem 6.7-5). The linear form is clearly continuous.

Therefore there exists a unique function u that minimizes the announced functional J over the space V, or equivalently, that satisfies the announced variational equations.

Assume next that $u \in H^2(\Omega)$. Thanks to the Green's formula of Theorem 6.7-1(b) applied to the functions $\partial_i u \in H^1(\Omega)$ (the functions a_{ij} belong to the space $\mathcal{C}^1(\overline{\Omega})$ by assumption) and to the relation $v = 0$ on Γ_0, the variational equations $a(u,v) = \ell(v)$ for all $v \in V$ become

$$\int_\Omega \left(-\sum_{i,j=1}^N \partial_j(a_{ij}\partial_i u) + cu - f\right) v\,\mathrm{d}x = \int_{\Gamma_1}\left(g - \sum_{i,j=1}^n a_{ij}\nu_j\partial_i u\right) v\,\mathrm{d}\Gamma \quad \text{for all } v \in V.$$

In particular then,

$$\int_\Omega \left(-\sum_{i,j=1}^N \partial_j(a_{ij}\partial_i u) + cu - f\right) v\,\mathrm{d}x = 0 \quad \text{for all } v \in \mathcal{D}(\Omega),$$

which implies that $(-\sum_{i,j=1}^N \partial_j(a_{ij}\partial_i u) + cu - f) = 0$ in $L^2(\Omega)$. Taking this equation into account, we are thus left with

$$\int_{\Gamma_1}\left(g - \sum_{i,j=1}^n a_{ij}\nu_j\partial_i u\right) v\,\mathrm{d}\Gamma = 0 \quad \text{for all } v \in V,$$

which implies that $g - \sum_{i,j=1}^n a_{ij}\nu_j\partial_i u = 0$ in $L^2(\Gamma_1)$ by Theorem 6.7-3. Finally, that $u \in V$ implies that $u = 0$ on Γ_0. □

The proof of Theorem 6.7-6 illustrates the three steps needed to recover a second-order boundary value problem from variational equations.

First, apply *ad hoc* Green's formula (assuming sufficient regularity on the solution of the variational equations) and let the function v vary in the space $\mathcal{D}(\Omega)$ in the variational equations (the space V always contains the space $\mathcal{D}(\Omega)$). This provides a *partial differential equation* $\mathcal{L}u = f$ *that always holds at least in the sense of distributions,*[31] i.e., in the space $\mathcal{D}'(\Omega)$.

[31] A particularly illuminating treatment of partial differential equations in the sense of distributions, together with a wealth of physical examples, is found in SCHWARTZ [1965].

Second, taking into account the equation $\mathcal{L}u = f$ in Ω, let the functions v vary in the whole space V. The remaining variational equations, which involve only integrals over Γ_1, then provide a *Neumann boundary condition on* Γ_1 (unless of course $V = H_0^1(\Omega)$, in which case this step has no *raison d'être*).

Third, complete the boundary value problem by the homogeneous Dirichlet boundary condition $u = 0$ on Γ_0 that is contained in the definition of the space V, to which u belongs (unless of course $V = H^1(\Omega)$, in which case this step has no *raison d'être*).

Note also that, like all the other variational problems described in this section, that of Theorem 6.7-6 is *well-posed*, in the sense that the mapping $(f, g) \in L^2(\Omega) \times L^2(\Gamma_1) \to u \in V$ is well defined and continuous.

To reflect that it combines both Dirichlet and Neumann boundary conditions, the boundary value problem

$$\begin{aligned} -\sum_{i,j=1}^N \partial_j(a_{ij}\partial_i u) + cu = f \quad &\text{in } \Omega, \\ u = 0 \quad &\text{on } \Gamma_0, \\ \sum_{i,j=1}^N a_{ij}\nu_j \partial_i u = g \quad &\text{on } \Gamma_1 \end{aligned}$$

found in Theorem 6.7-6 is called a **mixed problem** *for the partial differential operator* $\mathcal{L} : v \to \mathcal{L}v := -\sum_{i,j=1}^N \partial_j(a_{ij}\partial_i v) + bv$.

The *boundary operator*

$$v \to \sum_{i,j=1}^N a_{ij}\nu_j \partial_i v$$

appearing in the boundary condition along Γ_1 is called the **conormal derivative operator** *associated with the partial differential operator* $\mathcal{L}$. Note that it reduces to the normal derivative operator ∂_ν when $\mathcal{L}v := -\Delta v + cv$.

We conclude this section by several *important definitions*.

A linear partial differential operator $\mathcal{L}$ of the second order, which is thus given for all functions $v \in \mathcal{C}^2(\Omega)$ by an expression of the form

$$\mathcal{L}v(x) = -\sum_{i,j=1}^N a_{ij}(x)\partial_{ij}v(x) + \sum_{i=1}^N b_i(x)\partial_i v(x) + c(x)v(x) \quad \text{for all } x \in \Omega,$$

for specific coefficient functions $a_{ij}, b_i, c : \Omega \to \mathbb{R}$, is said to be **elliptic** if the matrices $(a_{ij}(x))$, $x \in \Omega$, which without loss of generality may be assumed to be *symmetric*, is positive-definite at each $x \in \Omega$; equivalently, at each $x \in \Omega$, there exists a constant $\mu(x)$ such that

$$\mu(x) > 0 \quad \text{and} \quad \sum_{i,j=1}^N a_{ij}(x)\xi_i\xi_j \geq \mu(x)\sum_{i=1}^N |\xi_i|^2 \quad \text{for all } (\xi_i)_{i=1}^N \in \mathbb{R}^N.$$

The linear partial differential operator $\mathcal{L}$ is said to be **uniformly elliptic** if the matrices $(a_{ij}(x))$, $x \in \Omega$, are uniformly positive-definite, in the sense that there exists a constant μ

such that

$$\mu > 0 \quad \text{and} \quad \sum_{i,j=1}^{N} a_{ij}(x)\xi_i\xi_j \geq \mu \sum_{i=1}^{N} |\xi_i|^2 \quad \text{for all } x \in \Omega \text{ and all } (\xi_i)_{i=1}^N \in \mathbb{R}^N.$$

A linear boundary value problem is said to be a **second-order elliptic boundary value problem** if the operator $\mathcal{L}$ found in the partial differential equation $\mathcal{L}u = f$ in Ω is an elliptic, or a uniformly elliptic, linear partial differential operator of the second order.

The boundary value problems found in the various examples described in this section thus provide *examples of second-order elliptic boundary value problems* (the matrix $(a_{ij}(x))$ corresponding to the operator $v \to -\Delta v + cv$ is equal to the unit matrix for all $x \in \Omega$), whose corresponding operator $\mathcal{L}$ is uniformly elliptic.

Finally, the variational equations $a(u, v) = \ell(v)$ for all $v \in V$ are said to constitute the **variational formulation** of the associated boundary value problem, and the solution $u \in V$ to these variational equations is said to be a **weak solution** of the associated boundary value problem, as opposed to a *classical solution*, which is typically sought in a space such as $\mathcal{C}(\overline{\Omega}) \cap \mathcal{C}^2(\Omega)$.

Problems

6.7-1 Let Ω be a domain in $\mathbb{R}^N$, let there be given functions c, f, and u_0 that satisfy the following assumptions:

$$c \in L^\infty(\Omega), \quad c \geq 0 \text{ a.e. in } \Omega, \quad f \in L^2(\Omega),\ u_0 \in H^1(\Omega),$$

let V, U, and a be as in Theorem 6.7-2, and let $\ell(v) := \int_\Omega fv\,dx - a(u_0, v)$ for all $v \in H^1(\Omega)$.

(1) Show that the associated variational problem has a unique solution $u \in H_0^1(\Omega)$.

(2) Assuming in addition that $u_0 \in H^2(\Omega)$ and $u \in H^2(\Omega)$, show that $\widetilde{u} := u + u_0$ satisfies the boundary value problem

$$-\Delta\widetilde{u} + c\widetilde{u} = f \ \text{ in } \Omega \quad \text{and} \quad \widetilde{u} = u_0 \ \text{ on } \Gamma.$$

(3) Show that the same boundary value problem, i.e., with a *nonhomogeneous Dirichlet boundary condition*, can be also obtained by minimizing the functional

$$J : v \in H^1(v) \to J(v) := \frac{1}{2}\int_\Omega \left(|\nabla v|^2 + cv^2\right) dx - \int_\Omega fv\,dx$$

over the subset $\{v \in H^1(\Omega);\ v = u_0 \text{ on } \Gamma\}$ of $H^1(\Omega)$.

6.7-2 In Theorem 6.7-2, assume *in addition* that $u \in H^2(\Omega)$ for *any* $f \in L^2(\Omega)$. Then show that there exists a constant C such that $\|u\|_{2,\Omega} \leq C\,\|f\|_{0,\Omega}$ for all $f \in L^2(\Omega)$.

6.7-3 Let Ω be an open subset of $\mathbb{R}^N$ and let ω_N denote the volume of the unit ball of $\mathbb{R}^N$.

(1) Let a function $u \in \mathcal{C}^2(\Omega)$ be such that $-\Delta u = 0$ in Ω. Show that, for any $y \in \Omega$ and any $r > 0$ such that $\overline{B(y;r)} \subset \Omega$, the function u satisfies the *mean value property*, i.e., that

$$u(y) = \frac{1}{N\omega_N r^{N-1}} \int_{\partial B(y;r)} u\,d\Gamma = \frac{1}{\omega_N r^N} \int_{B(y;r)} u\,dx.$$

Show likewise that, if a function $u \in \mathcal{C}^2(\Omega)$ satisfies $-\Delta u \geq 0$ in Ω, then

$$u(y) \geq \frac{1}{N\omega_N r^{N-1}} \int_{\partial B(y;r)} u\,d\Gamma \quad \text{and} \quad u(y) \geq \frac{1}{\omega_N r^N} \int_{B(y;r)} u\,dx.$$

Hints: Use the Green's formula $\int_{B(y;r)} \Delta u \, dx = \int_{\partial B(y;r)} \partial_\nu u \, d\Gamma$ (which itself immediately follows from the divergence theorem for vector fields; cf. Section 1.18), and introduce the variable $|x-y|$ for computing the integrals.

(2) Let a function $u \in \mathcal{C}^2(\Omega)$ be such that $-\Delta u \geq 0$ in Ω. Show that, if there exists a point $y \in \Omega$ such that $u(y) = \inf_{x\in\Omega} u(x)$, then u is a constant function; equivalently, if u is not a constant function, the infimum of u cannot be attained at any point of Ω.

(3) Assume that Ω is bounded and let a function $u \in \mathcal{C}(\overline{\Omega}) \cap \mathcal{C}^2(\Omega)$ satisfy $-\Delta u \geq 0$ in Ω. Show that the minimum of u in $\overline{\Omega}$ is achieved on $\Gamma := \partial\Omega$, i.e., that

$$\inf_{x\in\overline{\Omega}} u(x) = \inf_{x\in\Gamma} u(x).$$

Remark The properties described in (2) and (3) respectively constitute the *strong*, and *weak, minimum principle* for *superharmonic functions*, i.e., those functions $u \in \mathcal{C}^2(\Omega)$ that satisfy $-\Delta u \geq 0$ in Ω. Similar minimum, or maximum, principles for general elliptic operators will be established in greater generality in Section 7.10. □

(4) Assume that Ω is bounded, and let there be given functions $f \in \mathcal{C}(\Omega)$ and $g \in \mathcal{C}(\Gamma)$. Show that the Dirichlet problem

$$-\Delta u = f \text{ in } \Omega \quad \text{and} \quad u = g \text{ on } \Gamma,$$

has at most one solution $u \in \mathcal{C}(\overline{\Omega}) \cap \mathcal{C}^2(\Omega)$.

(5) Assume that Ω is bounded and let $u_\alpha \in \mathcal{C}(\overline{\Omega}) \cap \mathcal{C}^2(\Omega)$, $\alpha \in \{1,2\}$, be solutions to the Dirichlet problems $-\Delta u = f$ in Ω and $u = g_\alpha$, $\alpha \in \{1,2\}$, on Γ. Show that

$$\sup_{x\in\overline{\Omega}} |u_1(x) - u_2(x)| \leq \sup_{x\in\overline{\Omega}} |g_1(x) - g_2(x)| .$$

6.7-4 Let Ω be an open subset of $\mathbb{R}^N$ and let a function $u \in \mathcal{C}(\Omega)$ be such that, for any $y \in \Omega$ and any $r > 0$ such that $\overline{B(y;r)} \subset \Omega$,

$$u(y) = \frac{1}{N\omega_N r^{N-1}} \int_{\partial B(y;r)} u \, d\Gamma.$$

Show that $u \in \mathcal{C}^2(\Omega)$ and that $-\Delta u = 0$ in Ω (in fact, one has even $u \in \mathcal{C}^\infty(\Omega)$; cf. Problem 6.7-5). Combined with Problem 6.7-3(1), this result thus shows that *the mean value property characterizes harmonic functions*.

Hint: Denoting by B the unit ball in $\mathbb{R}^N$, use the classical boundary integral formula[32] that explicitly gives the solution $w \in \mathcal{C}(B) \cap \mathcal{C}^2(B)$ to the Dirichlet problem $-\Delta u = 0$ in B and $u = g$ on ∂B for any function $g \in \mathcal{C}(\partial B)$, and note that the results of questions (2), (3), (4) in Problem 6.7-3 apply as well to the function $u - w$.

6.7-5 Let Ω be an open subset of $\mathbb{R}^N$, let ω_N denote the volume of the unit ball of $\mathbb{R}^N$, and let $u \in \mathcal{C}(\Omega)$ be a function that satisfies the *mean value property* as defined in Problem 6.7-3(1). Then show that $u \in \mathcal{C}^\infty(\Omega)$.

Hint: Let $(u_\varepsilon)_{\varepsilon>0}$ be a *regularizing family* of u, where $u_\varepsilon \in \mathcal{C}^\infty(\Omega^\varepsilon)$ and $\Omega_\varepsilon := \{x \in \Omega; \text{dist}(x, \mathbb{R}^N - \Omega) > \varepsilon\}$ (Section 2.6). Then show that $u = u_\varepsilon$ on Ω_ε for each $\varepsilon > 0$.

6.7-6 Let Ω be an open subset of $\mathbb{R}^N$, and let a function $u \in \mathcal{C}^2(\Omega)$ satisfy $-\Delta u = 0$ in Ω; hence $u \in \mathcal{C}^\infty(\Omega)$ (Problems 6.7-4 and 6.7-5).

(1) Show that, for any $y \in \Omega$ and any $r > 0$ such that $\overline{B(y;r)} \subset \Omega$ and for any integer $k \geq 0$,

$$|\partial^{\boldsymbol{\alpha}} u(y)| \leq \frac{(2^{N+1}Nk)^k}{\omega_N r^{N+k}} \|u\|_{L^1(B(y;r))} \quad \text{for any multi-index } \boldsymbol{\alpha} \text{ with } |\boldsymbol{\alpha}| = k.$$

[32] See, e.g., GILBARG & TRUDINGER [1998, Theorem 2.6].

(2) Deduce from (1) that *any bounded function $u \in \mathcal{C}^2(\mathbb{R}^N)$ that satisfies $-\Delta u = 0$ in $\mathbb{R}^N$ is necessarily a constant function.* This remarkable property constitutes the celebrated *Liouville theorem for harmonic functions.*[33]

(3) Show that Liouville's theorem holds in fact under the weaker assumption that the harmonic function $u \in \mathcal{C}^2(\mathbb{R}^N)$ is bounded above (or below).

Hint: Use Problem 6.7-3(2).

(4) Let Ω be an open subset of $\mathbb{R}^N$ and let a function $u \in \mathcal{C}^2(\Omega)$ be harmonic in Ω. Show that u is *analytic* in Ω, i.e., that, given any $y \in \Omega$, there exists $r > 0$ such that $B(x;r) \subset \Omega$ and u can be expanded as a convergent power series in $B(y;r)$.

Hint: Using the estimates of (1), show that, if r is small enough, the Taylor series of u converges in $B(y;r)$.

6.7-7 Let Ω be a domain in $\mathbb{R}^N$.

(1) Show that there exists a constant C such that the following *generalized Poincaré–Friedrichs inequality* holds:

$$\|v\|_{1,\Omega} \leq C\left\{\int_\Omega |\nabla v|^2 \,\mathrm{d}x + \left|\int_\Omega v\,\mathrm{d}x\right|^2\right\}^{1/2} \quad \text{for all } v \in H^1(\Omega).$$

Hint: First, show that if the right-hand side of this inequality vanishes for some $v \in H^1(\Omega)$, then $v = 0$. Second, proceed by contradiction.

(2) Show that $U := \{v \in H^1(\Omega);\ \int_\Omega v\,\mathrm{d}x = 0\}$ is a closed subspace of $H^1(\Omega)$ and that $|\cdot|_{1,\Omega}$ is a norm on U, equivalent to $\|\cdot\|_{1,\Omega}$ on U.

(3) Let $J(v) := \dfrac{1}{2}\int_\Omega |\nabla v|^2 \,\mathrm{d}x - \ell(v)$, where $\ell(v) := \int_\Omega fv\,\mathrm{d}x$, for all $v \in H^1(\Omega)$. Show that $\inf_{v\in H^1(\Omega)} J(v) > -\infty$ implies that $L(v) = 0$ for all constant functions v.

(4) Show that there exists one and only one function $u \in U$ that satisfies $J(u) = \inf_{v\in U} J(v)$.

(5) Assume that $L(v) = 0$ for all constant functions v, and that $u \in U \cap H^2(\Omega)$. Show that u satisfies the boundary value problem

$$-\Delta u = f \ \text{ in } \Omega, \quad \int_\Omega u\,\mathrm{d}x = 0, \quad \text{and} \quad \partial_\nu u = 0 \ \text{ on } \Gamma.$$

In other words, among all possible solutions u of $-\Delta u = f$ in Ω and $\partial_\nu u = 0$ on Γ (which exist when $\ell(v) = 0$ for all constant functions v and are then defined only up to additive constants), the minimization problem of (3) "selects" the (only) one that satisfies $\int_\Omega u\,\mathrm{d}x = 0$.

Hint: First, show that any function $\widetilde{v} \in H^1(\Omega)$ can be written as $\widetilde{v} = v + C$ with $v \in U$ and $C \in \mathbb{R}$.

6.7-8 Let Ω be a domain in $\mathbb{R}^N$.

(1) Let $\mathcal{P}_0(\Omega)$ denote the space of all constant functions over Ω. Show that the seminorm $|\cdot|_{1,\Omega}$ is a norm over the quotient space $H^1(\Omega)/\mathcal{P}_0(\Omega)$, equivalent to the quotient norm over $H^1(\Omega)/\mathcal{P}_0(\Omega)$.

(2) Letting $\dot{w} \in H^1(\Omega)/\mathcal{P}_0(\Omega)$ denote the equivalence class of a function $w \in H^1(\Omega)$, let

$$\dot{V} := H^1(\Omega)/\mathcal{P}_0(\Omega),$$

$$a(\dot{u},\dot{v}) := \int_\Omega \nabla u\cdot\nabla v\,\mathrm{d}x \ \text{ for all } \dot{u},\dot{v} \in \dot{V} \quad \text{and} \quad \ell(\dot{v}) := \int_\Omega fv\,\mathrm{d}x + \int_\Gamma gv\,\mathrm{d}\Gamma \ \text{ for all } \dot{v} \in \dot{V},$$

where the functions $f \in L^2(\Omega)$ and $g \in L^2(\Gamma)$ satisfy the *compatibility condition*

$$\int_\Omega f\,\mathrm{d}x + \int_\Gamma g\,\mathrm{d}\Gamma = 0.$$

[33] So named by analogy with the "original" *Liouville theorem for holomorphic* (i.e., complex analytic) *functions* of a single complex variable, which Joseph Liouville (1809–1882) presented in a lecture in 1847.

Show that the symmetric bilinear form a is $\dot{V}$-coercive and continuous on $\dot{V} \times \dot{V}$ and that the linear form ℓ is well defined and continuous over $\dot{V}$.

(3) Let $\dot{u} \in \dot{V}$ denote the solution of the variational equations $a(\dot{u}, \dot{v}) = \ell(\dot{v})$ for all $\dot{v} \in V$ (this solution exists and is unique by (2)) and assume that $u \in H^2(\Omega)$. Show that u satisfies the boundary value problem

$$-\Delta u = f \text{ in } \Omega \quad \text{and} \quad \partial_\nu u = g \text{ on } \Gamma,$$

which is thus a *nonhomogeneous Neumann problem for the operator* $-\Delta$.

6.7-9 Let Ω be a domain in $\mathbb{R}^N$, let Γ_1 be a relatively open subset of $\Gamma := \partial\Omega$ such that $\mathrm{d}\Gamma\text{-meas}\,\Gamma_0 > 0$ where $\Gamma_0 := \Gamma - \Gamma_1$, let functions

$$b = (b_i)_{i=1}^N \in L^\infty(\Omega; \mathbb{R}^N), \quad c \in L^\infty(\Omega) \text{ such that } c \geq 0 \text{ a.e. in } \Omega, \quad f \in L^2(\Omega), \quad g \in L^2(\Gamma_1),$$

be given, and let

$$V := \{v \in H^1(\Omega);\ v = 0 \text{ on } \Gamma_0\},$$

$$a(u, v) := \int_\Omega (\nabla u \cdot \nabla v + (b \cdot \nabla u)v + cuv)\,\mathrm{d}x \quad \text{for all } u, v \in V,$$

$$\ell(v) = \int_\Omega f v \,\mathrm{d}x + \int_{\Gamma_1} g v \,\mathrm{d}\Gamma \quad \text{for all } v \in V.$$

(1) Show that, if $\max_{1 \leq i \leq N} \|b_i\|_{L^\infty(\Omega)}$ is small enough, there exists a unique solution $u \in V$ to the variational equations $a(u, v) = \ell(v)$ for all $v \in V$.

(2) If $u \in H^2(\Omega)$, what is the boundary value problem that u satisfies?

6.7-10 The object of this problem is to analyze the behavior as $\varepsilon \to 0$ of the solution of a model **singular perturbation problem,**[34] i.e., a problem parametrized by $\varepsilon > 0$ that becomes "singular" (in some sense) as $\varepsilon \to 0$.

(1) Let a bounded open subset Ω of $\mathbb{R}^N$ and a function $f \in L^2(\Omega)$ be given, and let $u_\varepsilon \in H_0^1(\Omega)$ denote for each $\varepsilon > 0$ the unique solution of

$$-\varepsilon\Delta u_\varepsilon + u_\varepsilon = f \quad \text{in } \Omega \text{ and } u_\varepsilon = 0 \text{ on } \Gamma.$$

Show that $u_\varepsilon \to f$ in $L^2(\Omega)$ as $\varepsilon \to 0$.

Hint: Show that the family $(\sqrt{\varepsilon}u_\varepsilon)_{\varepsilon>0}$ is bounded in $H_0^1(\Omega)$ and that the family $(u_\varepsilon)_{\varepsilon>0}$ is bounded in $L^2(\Omega)$.

(2) Under the additional assumption that $f \in H_0^1(\Omega)$, show that $u_\varepsilon \to f$ in $H_0^1(\Omega)$ as $\varepsilon \to 0$.

6.8 Examples of fourth-order linear boundary value problems; the biharmonic and plate problems

Throughout this section, the boundary of a domain Ω in $\mathbb{R}^N$ is denoted Γ.

Whereas in the preceding section the spaces V were subspaces of the Sobolev space $H^1(\Omega)$, we now consider examples where the spaces V are subspaces of $H^2(\Omega)$. As a preparation for our first such example, we prove two simple preliminary results. To begin with, recall that the seminorm $|\cdot|_{2,\Omega} : H^2(\Omega) \to \mathbb{R}$ becomes a norm equivalent to $\|\cdot\|_{2,\Omega}$ over the space $H_0^2(\Omega)$, if the open set $\Omega \subset \mathbb{R}^N$ is of finite width (Theorem 6.5-2(b)). The next result shows that the space $H_0^2(\Omega)$ can be provided with yet another equivalent norm in this case.

[34] References to singular perturbation problems are provided in the Biographical Notes.

Theorem 6.8-1 (a) *Let Ω be an open subset of $\mathbb{R}^N$. Then*

$$|v|_{2,\Omega} = \|\Delta v\|_{0,\Omega} \quad \textit{for all } v \in H_0^2(\Omega).$$

(b) *If Ω is of finite width, the seminorm $v \to \|\Delta v\|_{0,\Omega}$ becomes a norm over the space $H_0^2(\Omega)$, equivalent to the norm $\|\cdot\|_{2,\Omega}$.*

Proof It suffices to prove the equality $|v|_{2,\Omega} = \|\Delta v\|_{0,\Omega}$ for functions in the dense subspace $\mathcal{D}(\Omega)$ of $H_0^2(\Omega)$. That this equality indeed holds follows by noting that, by definition,

$$|v|_{2,\Omega}^2 = \int_\Omega \left(\sum_{i=1}^N |\partial_{ii} v|^2 + \sum_{\left\{\substack{i,j=1 \\ i \neq j}\right.}^N |\partial_{ij} v|^2 \right) \mathrm{d}x,$$

$$\|\Delta v\|_{0,\Omega} = \int_\Omega \left(\sum_{i=1}^N |\partial_{ii} v|^2 + \sum_{\left\{\substack{i,j=1 \\ i \neq j}\right.}^N \partial_{ii} v \partial_{jj} v \right) \mathrm{d}x,$$

and that, by Theorem 6.3-1,

$$\int_\Omega |\partial_{ij} v|^2 \,\mathrm{d}x = -\int_\Omega \partial_i v \partial_{ijj} v \,\mathrm{d}x = \int_\Omega \partial_{ii} v \partial_{jj} v \,\mathrm{d}x \quad \text{for all } v \in \mathcal{D}(\Omega).$$

Hence (a) is proved. Combining (a) with Theorem 6.5-2(b) with $m = 2$ proves (b). □

The second preliminary result constitutes another *Green's formula in Sobolev spaces*. The operator $\Delta^2 := \sum_{i,j=1}^N \partial_{iijj}$ (which acts on functions defined in Ω) appearing in this formula is called the **biharmonic operator**.

Theorem 6.8-2 *Let Ω be a domain in $\mathbb{R}^N$ and let $\nu = (\nu_i)_{i=1}^N$ denote the unit outer normal vector field along the boundary Γ. Then the following Green's formula holds:*

$$\int_\Omega \Delta u \Delta v \,\mathrm{d}x = \int_\Omega (\Delta^2 u) v \,\mathrm{d}x - \int_\Gamma (\partial_\nu \Delta u) v \,\mathrm{d}\Gamma + \int_\Gamma \Delta u \partial_\nu v \,\mathrm{d}\Gamma \quad \text{for all } u \in H^4(\Omega),\ v \in H^2(\Omega),$$

where

$$\Delta^2 u := \Delta(\Delta u) = \sum_{i,j=1}^N \partial_{iijj} u \in L^2(\Omega),$$

and $\partial_\nu v \in L^2(\Gamma)$ denotes the outer normal derivative of $v \in H^2(\Omega)$ (Theorem 6.7-1).

Proof By Theorem 6.7-1(a),

$$\begin{aligned} \int_\Omega \sum_{i=1}^N \partial_i v \partial_i w &= -\int_\Omega (\Delta v) w \,\mathrm{d}x + \int_\Gamma (\partial_\nu v) w \,\mathrm{d}\Gamma \\ &= -\int_\Omega v \Delta w \,\mathrm{d}x + \int_\Gamma v \partial_\nu w \,\mathrm{d}\Gamma \quad \text{for all } v, w \in H^2(\Omega). \end{aligned}$$

Hence

$$\int_\Omega (v\Delta w - (\Delta v)w)\, dx = \int_\Gamma (v\partial_\nu w - (\partial_\nu v)w)\, dx \quad \text{for all } v, w \in H^2(\Omega).$$

It then suffices to replace w by Δu in this Green's formula to get the announced one. □

We now consider an example of a variational problem posed in the space $H_0^2(\Omega)$.

Theorem 6.8-3 *Let Ω be an open subset of $\mathbb{R}^N$ of finite width, let a function $f \in L^2(\Omega)$ be given, and let*

$$V = U := H_0^2(\Omega),$$
$$a(u, v) := \int_\Omega \Delta u \Delta v\, dx \quad \textit{for all } u, v \in V,$$
$$\ell(v) := \int_\Omega f v\, dx \quad \textit{for all } v \in V.$$

Then there exists a unique function $u \in H_0^2(\Omega)$ that minimizes the functional $J : H_0^2(\Omega) \to \mathbb{R}$ defined by

$$J(v) := \frac{1}{2}a(v, v) - \ell(v) = \frac{1}{2}\int_\Omega |\Delta v|^2\, dx - \int_\Omega f v\, dx \quad \textit{for all } v \in H_0^2(\Omega),$$

or, equivalently, that satisfies the variational equations

$$\int_\Omega \Delta u \Delta v\, dx = \int_\Omega f v\, dx \quad \textit{for all } v \in H_0^2(\Omega).$$

The function u satisfies the following boundary value problem:

$$\Delta^2 u = f \text{ in } \Omega \quad \textit{and} \quad u = \partial_\nu u = 0 \ \textit{ on } \Gamma,$$

where the partial differential equation in Ω is to be understood as an equality in the space $\mathcal{D}'(\Omega)$ and, under the additional assumption that Ω is a domain, the boundary conditions are to be understood as equalities in the space $L^2(\Gamma)$.

Proof The symmetric bilinear form $a : H^2(\Omega) \times H^2(\Omega) \to \mathbb{R}$ is continuous since

$$|a(u, v)| \leq \|\Delta u\|_{0,\Omega} \|\Delta v\|_{0,\Omega} \leq N \|u\|_{2,\Omega} \|v\|_{2,\Omega} \quad \text{for all } u, v \in H^2(\Omega),$$

and $H_0^2(\Omega)$-coercive since

$$a(v, v) = \|\Delta v\|_{0,\Omega}^2 \quad \text{for all } v \in H_0^2(\Omega),$$

and $v \in H^2(\Omega) \to \|\Delta v\|_{0,\Omega}$ is a norm equivalent to $\|\cdot\|_{2,\Omega}$ over $H_0^2(\Omega)$ (Theorem 6.8-1).

All the assumptions of Theorem 6.1-1 being therefore satisfied (the linear form $\ell : H_0^2(\Omega) \to \mathbb{R}$ is clearly continuous), it follows that there exists one and only one function that minimizes the announced functional J, or equivalently, that satisfies the announced variational equations (Theorem 6.1-2).

Since $\int_\Omega \Delta u \Delta v \, dx = \langle \Delta^2 u, v\rangle$ for all $v \in \mathcal{D}(\Omega)$, where $\Delta^2 u$ is now interpreted as a *distribution*, the equations $a(u,v) = \ell(v)$ for all $v \in V$ imply that

$$\langle \Delta^2 u, v\rangle = \int_\Omega f v \, dx \quad \text{for all } v \in \mathcal{D}(\Omega)$$

(since $\mathcal{D}(\Omega) \subset H_0^2(\Omega)$), and hence that

$$\Delta^2 u = f \quad \text{in } \mathcal{D}'(\Omega).$$

When Ω is a *domain*, the characterization

$$H_0^2(\Omega) = \{v \in H^2(\Omega);\ v = \partial_\nu v = 0 \text{ on } \Gamma\}$$

of the space $H_0^2(\Omega)$ (Theorem 6.6-5(d)) shows that the function $u \in H_0^2(\Omega)$ satisfies the boundary conditions $u = 0$ and $\partial_\nu u = 0$ on Γ, interpreted as equalities in the space $L^2(\Gamma)$. □

The boundary value problem

$$\Delta^2 u = f \text{ in } \Omega \quad \text{and} \quad u = \partial_\nu u = 0 \text{ on } \Gamma$$

is called the **biharmonic problem**.

As a preparation for the next variational problem, which is posed over a domain in $\mathbb{R}^2$, we need some preliminary results, which accordingly will be established in dimension two. The first preliminary result (whose proof is similar to that of Theorem 6.7-5 and for this reason is left as a problem; cf. Problem 6.8-1) will be used for establishing the ellipticity of the associated bilinear form.

Theorem 6.8-4 *Let Ω be a domain in $\mathbb{R}^2$, let Γ_0 be a $d\Gamma$-measurable subset of the boundary Γ that satisfies*

$$d\Gamma\text{-}\operatorname{meas} \Gamma_0 > 0,$$

and let

$$V := \{v \in H^2(\Omega);\ v = \partial_\nu v = 0 \quad \text{on } \Gamma_0\}.$$

Then the space V is a closed subspace of $H^2(\Omega)$, and there exists a constant C such that

$$|v|_{2,\Omega} \le \|v\|_{2,\Omega} \le C\, |v|_{2,\Omega} \quad \text{for all } v \in V. \qquad \square$$

The next two results play an essential role in the identification of the boundary conditions appearing in the associated boundary value problem. The first one constitutes the "$H^2(\Omega)$-version" of Theorem 6.7-3.

$^\flat$**Theorem 6.8-5** *Let Ω be a domain in $\mathbb{R}^2$, let Γ_1 be a relatively open subset of class $\mathcal{C}^{1,1}$ of Γ, and let $w_0, w_1 \in L^2(\Gamma_1)$ be two functions that satisfy*

$$\int_{\Gamma_1} w_0 v \, d\Gamma + \int_{\Gamma_1} w_1 \partial_\nu v \, d\Gamma = 0 \quad \text{for all } v \in V := \{v \in H^2(\Omega);\ v = \partial_\nu v = 0 \text{ on } \Gamma - \Gamma_1\}.$$

Then $w_0 = w_1 = 0$. □

Before stating the other result, which constitutes yet another *Green's formula in Sobolev spaces*, we need several definitions and notations *specific to domains in* $\mathbb{R}^2$.

Let Ω denote a domain in $\mathbb{R}^2$ and let $\nu = (\nu_\alpha)_{\alpha=1}^2$ denote the unit outer normal vector field along Γ. A **unit tangential vector field** $\tau = (\tau_\alpha)_{\alpha=1}^2$ along Γ is then defined by

$$\tau_1 = -\nu_2 \quad \text{and} \quad \tau_2 = \nu_1.$$

Like ν, the field τ is thus defined $\mathrm{d}\Gamma$-almost everywhere along Γ.

In addition to the normal derivative operator ∂_ν, we define the boundary differential operators $\partial_\tau, \partial_{\nu\tau}, \partial_{\tau\tau}$ along Γ by

$$\partial_\tau v := \sum_{\alpha=1}^{2} \tau_\alpha \partial_\alpha v, \quad \partial_{\nu\tau} v := \sum_{\alpha,\beta=1}^{2} \nu_\alpha \tau_\beta \partial_{\alpha\beta} v, \quad \partial_{\tau\tau} v := \sum_{\alpha,\beta=1}^{2} \tau_\alpha \tau_\beta \partial_{\alpha\beta} v,$$

for smooth enough functions v. Note in passing that, while $\partial_\tau v$ coincides with the first derivative of the restriction to Γ of the function v considered as a function of the curvilinear abscissa along the boundary at those boundary points where the unit tangential vector is well-defined, $\partial_{\tau\tau} v$ does *not* coincide in general with the second derivative of this restriction.

For brevity, it will be understood in the remainder of this section that *Greek indices* range in the set $\{1, 2\}$ and that the *summation convention* with respect to Greek indices is used; e.g., $m_{\alpha\beta}\partial_{\alpha\beta} v$ stands for $\sum_{\alpha,\beta=1}^2 m_{\alpha\beta}\partial_{\alpha\beta} v$, etc.

Theorem 6.8-6 *Let Ω be a domain in $\mathbb{R}^2$. Then the following* **Green's formula** *holds:*

$$\begin{aligned}\int_\Omega m_{\alpha\beta}\partial_{\alpha\beta} v \,\mathrm{d}x &= \int_\Omega (\partial_{\alpha\beta} m_{\alpha\beta}) v \,\mathrm{d}x \\ &= -\int_\Gamma \big((\partial_\alpha m_{\alpha\beta})\nu_\beta + \partial_\tau (m_{\alpha\beta}\nu_\alpha\tau_\beta)\big)\, v \,\mathrm{d}\Gamma \\ &\quad + \int_\Gamma m_{\alpha\beta}\nu_\alpha\nu_\beta \partial_\nu v \,\mathrm{d}\Gamma \quad \text{for all } m_{\alpha\beta} \in H^2(\Omega),\ v \in H^2(\Omega).\end{aligned}$$

Proof Two successive applications of the fundamental Green's formula in Sobolev spaces (Theorem 6.6-7) give

$$\begin{aligned}\int_\Omega m_{\alpha\beta}\partial_{\alpha\beta} v \,\mathrm{d}x &= -\int_\Omega (\partial_\alpha m_{\alpha\beta})\partial_\beta v \,\mathrm{d}x + \int_\Gamma m_{\alpha\beta}\nu_\alpha \partial_\beta v \,\mathrm{d}\Gamma \\ &= \int_\Omega (\partial_{\alpha\beta} m_{\alpha\beta}) v \,\mathrm{d}x - \int_\Gamma (\partial_\alpha m_{\alpha\beta})\nu_\beta v \,\mathrm{d}\Gamma + \int_\Gamma m_{\alpha\beta}\nu_\alpha\partial_\beta v \,\mathrm{d}\Gamma.\end{aligned}$$

The definition of the boundary operators ∂_ν and ∂_τ imply that each partial derivative of v can be written as $\partial_\beta v = \nu_\beta \partial_\nu v + \tau_\beta \partial_\tau v$ along Γ. Consequently, the last integral on Γ can be rewritten as

$$\int_\Gamma m_{\alpha\beta}\nu_\alpha\partial_\beta v \,\mathrm{d}\Gamma = \int_\Gamma m_{\alpha\beta}\nu_\alpha\nu_\beta\partial_\nu v \,\mathrm{d}\Gamma + \int_\Gamma m_{\alpha\beta}\nu_\alpha\tau_\beta\partial_\tau v \,\mathrm{d}\Gamma,$$

and the announced Green's formula follows by noting that (Problem 6.8-2)

$$\int_\Gamma m_{\alpha\beta}\nu_\alpha\tau_\beta\partial_\tau v\,\mathrm{d}\Gamma = -\int_\Gamma (\partial_\tau(m_{\alpha\beta}\nu_\alpha\tau_\beta))v\,\mathrm{d}\Gamma. \qquad \square$$

We now consider our second example. Recall that the notation $\delta_{\alpha\beta}$ designates the Kronecker symbol.

Theorem 6.8-7 *Let Ω be a domain in $\mathbb{R}^2$, let Γ_1 be a relatively open subset of class $\mathcal{C}^{1,1}$ of Γ such that*

$$\mathrm{d}\Gamma\text{-}\operatorname{meas}\Gamma_0 > 0 \quad \textit{where } \Gamma_0 := \Gamma - \Gamma_1,$$

let

$$0 < \nu < 1 \quad \textit{and} \quad f \in L^2(\Omega)$$

be a given constant and a given function, and finally, let

$$\begin{aligned}
V = U &:= \{v \in H^2(\Omega);\ v = \partial_\nu v = 0 \text{ on } \Gamma_0\},\\
a(u,v) &:= \int_\Omega (\nu\Delta u\Delta v + (1-\nu)\partial_{\alpha\beta}u\partial_{\alpha\beta}v)\,\mathrm{d}x \quad \textit{for all } u, v \in V,\\
\ell(v) &:= \int_\Omega f v\,\mathrm{d}x \quad \textit{for all } v \in V.
\end{aligned}$$

Then there exists a unique function $u \in V$ that minimizes over the space V the functional $J : V \to \mathbb{R}$ defined by

$$J(v) := \frac{1}{2}a(v,v) - \ell(v) = \frac{1}{2}\int_\Omega \left(\nu\,|\Delta v|^2 + (1-\nu)\partial_{\alpha\beta}v\partial_{\alpha\beta}v\right)\mathrm{d}x - \int_\Omega f v\,\mathrm{d}x$$

for all $v \in V$, or equivalently, that satisfies

$$\int_\Omega (\nu\Delta u\Delta v + (1-\nu)\partial_{\alpha\beta}u\partial_{\alpha\beta}v)\,\mathrm{d}x = \int_\Omega f v\,\mathrm{d}x \quad \textit{for all } v \in V.$$

Assume in addition that $u \in H^4(\Omega)$, and let the functions $m_{\alpha\beta}(u) \in H^2(\Omega)$ be defined by

$$m_{\alpha\beta}(u) := \nu\Delta u\delta_{\alpha\beta} + (1-\nu)\partial_{\alpha\beta}u.$$

Then u satisfies the boundary value problem

$$\begin{aligned}
\partial_{\alpha\beta}m_{\alpha\beta}(u) &= f \quad \textit{in } \Omega,\\
u = \partial_\nu u &= 0 \quad \textit{on } \Gamma_0,\\
m_{\alpha\beta}(u)\nu_\alpha\nu_\beta = (\partial_\alpha m_{\alpha\beta}(u))\nu_\beta + \partial_\tau(m_{\alpha\beta}(u)\nu_\alpha\tau_\beta) &= 0 \quad \textit{on } \Gamma_1.
\end{aligned}$$

Proof The symmetric bilinear form $a : V \times V \to \mathbb{R}$ and the linear form $\ell : V \to \mathbb{R}$ are clearly continuous. The bilinear form is V-coercive by Theorem 6.8-4, since $0 < \nu < 1$ and

$$a(v,v) \geq (1-\nu)\,|v|^2_{2,\Omega} \quad \text{for all } v \in V.$$

Hence there exists a unique function u that minimizes the announced functional J over the space V, or equivalently, that satisfies the announced variational equations.

In view of identifying the corresponding boundary value problem, we first note that the left-hand side of the variational equations may be also written as

$$\int_\Omega (\nu\Delta u\Delta v + (1-\nu)\partial_{\alpha\beta}u\partial_{\alpha\beta}v)\,\mathrm{d}x = \int_\Omega m_{\alpha\beta}(u)\partial_{\alpha\beta}v\,\mathrm{d}x,$$

with $m_{\alpha\beta}(u) := \nu\Delta u\delta_{\alpha\beta} + (1-\nu)\partial_{\alpha\beta}u$. Assume then that $u \in H^4(\Omega)$, so that $m_{\alpha\beta}(u) \in H^2(\Omega)$; thanks to the Green's formula of Theorem 6.8-6 and to the relation $v = \partial_\nu v = 0$ on Γ_0, the variational equations $a(u,v) = \ell(v)$ for all $v \in V$ then become

$$\int_\Omega (\partial_{\alpha\beta}m_{\alpha\beta}(u) - f)\,v\,\mathrm{d}x = \int_{\Gamma_1} \{(\partial_\alpha m_{\alpha\beta}(u))\,\nu_\beta + \partial_\tau\,(m_{\alpha\beta}(u)\nu_\alpha\tau_\beta)\}\,v\,\mathrm{d}\Gamma$$
$$- \int_{\Gamma_1} m_{\alpha\beta}(u)\nu_\alpha\nu_\beta\partial_\nu v\,\mathrm{d}\Gamma \quad \text{for all } v \in V.$$

In particular then,

$$\int_\Omega (\partial_{\alpha\beta}m_{\alpha\beta}(u) - f)\,v\,\mathrm{d}x = 0 \quad \text{for all } v \in \mathcal{D}(\Omega),$$

which implies that $\partial_{\alpha\beta}m_{\alpha\beta}(u) = f$ in $L^2(\Omega)$. Taking this equation into account, we are thus left with

$$\int_{\Gamma_1} \{(\partial_\alpha m_{\alpha\beta}(u))\nu_\beta + \partial_\tau\,(m_{\alpha\beta}(u)\nu_\alpha\tau_\beta)\}\,v\,\mathrm{d}\Gamma - \int_{\Gamma_1} m_{\alpha\beta}(u)\nu_\alpha\nu_\beta\partial_\nu v\,\mathrm{d}\Gamma = 0 \quad \text{for all } v \in V,$$

which implies that the announced boundary conditions on Γ_1 are indeed satisfied as equalities in $L^2(\Gamma_1)$, by Theorem 6.8-5. Finally, that $u \in V$ implies that $u = \partial_\nu u = 0$ on Γ_0. □

The data V, $a(\cdot,\cdot)$, and ℓ appearing in Theorem 6.8-6 correspond to the variational formulation of the *flexural equations* of the **Kirchhoff–Love theory of a linearly elastic plate**: The unknown u represents the vertical displacement of a linearly elastic plate of constant thickness e under the action of a transverse force, of density $F = \frac{1}{12}Ee^3f/(1-\nu^2)$ per unit area. The constants $E = \mu(3\lambda+2\mu)/(\lambda+\mu)$ and $\nu = \frac{1}{2}\lambda/(\lambda+\mu)$ are respectively the *Young modulus* and the *Poisson coefficient* of the elastic material constituting the plate, $\lambda \geq 0$ and $\mu > 0$ being the *Lamé constants* of the same material; hence the Poisson coefficient satisfies $0 < \nu < \frac{1}{2}$. When $f = 0$, the plate lies in the "horizontal" plane of coordinates (x_1, x_2) (Figure 6.8-1). The boundary conditions $u = \partial_\nu u = 0$ on Γ_0 contained in the definition of the space V mean that the plate is *clamped* on Γ_0.

The unknown vertical displacement of the plate thus minimizes the *plate energy* $J : V \to \mathbb{R}$, which is defined by

$$J(v) := \frac{1}{2}\int_\Omega \left(\nu\,|\Delta v|^2 + (1-\nu)\partial_{\alpha\beta}v\partial_{\alpha\beta}v\right)\mathrm{d}x - \int_\Gamma fv\,\mathrm{d}x \quad \text{for all } v \in V.$$

Note that, by Theorem 6.8-1(a), the energy of a plate clamped over its *entire* boundary (in which case $V = H_0^2(\Omega)$) takes the simpler form

$$J(v) = \frac{1}{2}\int_\Omega |\Delta v|^2\,\mathrm{d}x - \int_\Gamma fv\,\mathrm{d}x \quad \text{for all } v \in H_0^2(\Omega),$$

i.e., it coincides in this case with the functional corresponding to the *biharmonic problem* (Theorem 6.8-2).

The same biharmonic problem is a mathematical model for a specific class of problems in fluid mechanics: It can be shown that the solution of the *Stokes equations* (Section 6.14) for an incompressible viscous fluid in a simply connected domain $\Omega \subset \mathbb{R}^2$ may be reduced to the solution of the above biharmonic problem, whose unknown u is then an appropriate *stream function*.

Remark The expressions found in the variational formulation of the clamped plate problem can be rigorously justified by means of an *asymptotic analysis* (when the thickness of the plate approaches zero) applied to the variational formulation of the boundary value problem of *three-dimensional linearized elasticity*[35] (which will be studied in Section 6.16). $\square$

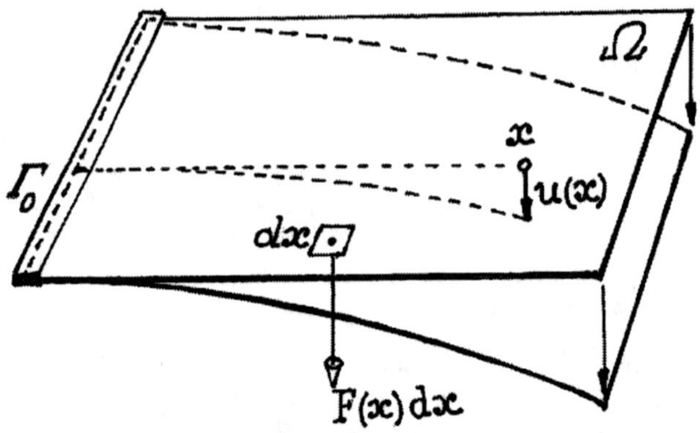

Figure 6.8-1 A plate problem: The unknown $u : \overline{\Omega} \subset \mathbb{R}^2 \to \mathbb{R}$ represents the vertical displacement of a linearly elastic plate occupying the set $\overline{\Omega}$ in the absence of applied forces, subjected to a vertical force of density F per unit area, and clamped along a portion Γ_0 of its boundary Γ.

If $u \in H^4(\Omega)$, the equation $\partial_{\alpha\beta} m_{\alpha\beta}(u) = f$ in Ω may be also written as $\Delta^2 u = f$ in Ω (since $\partial_{\alpha\beta} m_{\alpha\beta}(u) = \Delta^2 u$). The variational problems found in Theorems 6.8-3 and 6.8-7 therefore provide an interesting example of two variational problems with *different* bilinear forms that nevertheless yield the *same partial differential equation* in Ω (in this direction, see also Problem 6.8-3). But *the boundary conditions are different when* $d\Gamma$*-meas* $\Gamma_1 > 0$.

Problems

6.8-1 Let Ω be a domain in $\mathbb{R}^2$ and let $\Gamma_0 \subset \partial\Omega$ be such that $d\Gamma$-meas $\Gamma_0 > 0$.

(1) Show that the space $V := \{v \in H^2(\Omega);\ v = \partial_\nu v = 0 \text{ on } \Gamma_0\}$ is closed in $H^2(\Omega)$ and that $|\cdot|_{2,\Omega}$ is a norm on V.

Hint: Infer from Theorem 6.3-4 that, if a function $v \in H^2(\Omega)$ satisfies $|v|_{2,\Omega} = 0$, then there exist constants a_i, $0 \le i \le 2$, such that $v(x) = a_0 + a_1 x_1 + a_2 x_2$ for all $x = (x_i)_{i=1}^2 \in \Omega$; then show that if,

[35] Such a justification of linear plate models is studied at length in CIARLET [1997, Chapter 1].

in addition, $v = \partial_\nu v = 0$ on Γ_0 and $\mathrm{d}\Gamma\text{-meas}\,\Gamma_0 > 0$, then $v = 0$.

(2) Assume that there exists a sequence $(v_k)_{k=1}^{\infty}$ of functions $v_k \in V$ that satisfy

$$\|v_k\|_{2,\Omega} = 1 \text{ for all } k \quad \text{and} \quad \lim_{k\to\infty} |v_k|_{2,\Omega} = 0.$$

Combining the Rellich–Kondrachov theorem (Theorem 6.6-3) with question (1), show that this assumption leads to a contradiction. Hence there exists a constant C such that $\|v\|_{2,\Omega} \leq C\,|v|_{2,\Omega}$ for all $v \in V$.

6.8-2 Let Ω be a domain in $\mathbb{R}^2$ and let $\Gamma := \partial\Omega$. Show that

$$\int_\Gamma m_{\alpha\beta}\nu_\alpha\tau_\beta\partial_\tau v\,\mathrm{d}\Gamma = -\int_\Gamma (\partial_\tau(m_{\alpha\beta}\nu_\alpha\tau_\beta))v\,\mathrm{d}\Gamma \quad \text{for all functions } m_{\alpha\beta} \in H^2(\Omega),\ v \in H^2(\Omega).$$

6.8-3 Let Ω be a domain in $\mathbb{R}^2$ and let $\Gamma := \partial\Omega$.

(1) Show that the following *Green's formula in Sobolev spaces* holds:

$$\int_\Omega (2\partial_{12}u\partial_{12}v - \partial_{11}u\partial_{22}v - \partial_{22}u\partial_{11}v)\,\mathrm{d}x = \int_\Gamma (-\partial_{\tau\tau}u\partial_\nu v + \partial_{\nu\tau}u\partial_\tau v)\,\mathrm{d}\Gamma$$

for all $u \in H^3(\Omega,\ v \in H^2(\Omega)$.

(2) Show that, for any $\nu \in \mathbb{R}$,

$$\int_\Omega (\nu\Delta u\Delta v + (1-\nu)\partial_{\alpha\beta}u\partial_{\alpha\beta}v)\,\mathrm{d}x = \int_\Omega (\Delta u\Delta v + (1-\nu)(2\partial_{12}u\partial_{12}v - \partial_{11}u\partial_{22}v - \partial_{22}u\partial_{11}v))\,\mathrm{d}x$$

for all $u, v \in H^2(\Omega)$. Combining this observation and the Green's formula of question (1), show that, if the solution u to the variational problem of Theorem 6.8-7 is in the space $H^4(\Omega)$, then u satisfies the partial differential equation $\Delta^2 u = f$ in Ω.

6.8-4 Let ω be a domain in $\mathbb{R}^2$. We established in Theorem 6.8-1 that $\eta \to \|\Delta\eta\|_{0,\omega}$ is a norm over the space $H_0^2(\omega)$, equivalent to $\|\cdot\|_{2,\omega}$.

(1) Assume that ω has smooth boundary γ, and let $\gamma_0 \subset \gamma$ with $0 < \text{length}\,\gamma_0 < \text{length}\,\gamma$. Show that $\eta \to \|\Delta\eta\|_{0,\omega}$ is again a norm over the space $V(\omega) := \{\eta \in H^2(\omega);\ \eta = \partial_\nu\eta = 0 \text{ on } \gamma_0\}$.

(2) Is this norm equivalent to $\|\cdot\|_{2,\omega}$ over $V(\omega)$?

6.8-5 Let Ω be a domain in $\mathbb{R}^N$, let $\Gamma := \partial\Omega$, and let

$$V := \{v \in H^2(\Omega);\ v = 0 \text{ on } \Gamma\}.$$

(1) Show that $v \to \|\Delta v\|_{0,\Omega}$ is a norm on V.

(2) Is this norm equivalent on V to the norm $\|\cdot\|_{2,\Omega}$?

6.9 Examples of nonlinear boundary value problems associated with variational inequalities; obstacle problems

In this section, we study variational problems that are posed in terms of *variational inequalities*, which arise when a quadratic functional is minimized over a set which is *not* a vector space (Theorem 6.1-2). We begin with a specific example, which constitutes an interesting variant of the membrane problem (Section 6.7). Recall that the functional $J : H_0^1(\Omega) \to \mathbb{R}$, defined by

$$J(v) := \frac{1}{2}\int_\Omega |\nabla v|^2 \, dx - \int_\Omega f v \, dx \quad \text{for all } v \in H_0^1(\Omega),$$

represents the energy of an elastic membrane, which passes through the boundary Γ of a domain Ω of the horizontal plane $\mathbb{R}^2$ and is subjected to the action of a vertical force of density $F = \tau f$ with $f \in L^2(\Omega)$, where τ measures the tension of the membrane (Section 6.7).

The **obstacle problem for a membrane** then consists again in finding its equilibrium position under the *additional* assumption that it must lie over an *"obstacle"* represented by a function $\chi : \overline{\Omega} \to \mathbb{R}$, as illustrated in Figure 6.9-1 (the function χ is of course assumed to be ≤ 0 on Γ). The unknown vertical displacement u is thus expected to be a minimizer of the *same* functional J, but now over the set $U := \{v \in H_0^1(\Omega);\ v \geq \chi \text{ almost everywhere in } \Omega\}$, instead of over the whole space $H_0^1(\Omega)$.

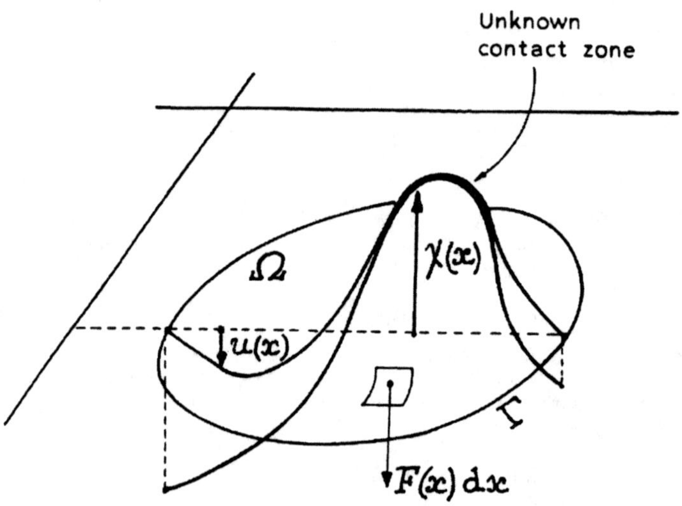

Figure 6.9-1 The obstacle problem: The membrane must lie over an "obstacle," which is represented by a function $\chi : \overline{\Omega} \to \mathbb{R}$. This figure originally appeared in P.G. CIARLET [1978]: *The Finite Element Method for Elliptic Problems*, North-Holland, Amsterdam.

We now establish the existence and uniqueness of such a minimizer $u \in U$, and we also identify the *nonlinear* boundary value problem that u satisfies, as usual under an additional regularity assumption (the justification of which requires special care, however; see the brief discussion after the proof).

Theorem 6.9-1 *Let Ω be a domain in $\mathbb{R}^2$, let functions*

$$\chi \in H^1(\Omega) \cap \mathcal{C}(\overline{\Omega}) \ \textit{with} \ \chi|_\Gamma \leq 0 \quad \textit{and} \quad f \in L^2(\Omega)$$

be given, and let

$$V := H_0^1(\Omega) \quad \textit{and} \quad U := \{v \in H_0^1(\Omega);\ v \geq \chi \textit{ a.e. in } \Omega\},$$

$$a(u,v) := \int_\Omega \nabla u \cdot \nabla v \, \mathrm{d}x \quad \textit{for all } u, v \in V,$$

$$\ell(v) := \int_\Omega f v \, \mathrm{d}x \quad \textit{for all } v \in V.$$

Then there exists a unique function $u \in U$ that minimizes over the set U the functional $J : V \to \mathbb{R}$ defined by

$$J(v) := \frac{1}{2} a(v,v) - \ell(v) = \frac{1}{2} \int_\Omega |\nabla v|^2 \, \mathrm{d}x - \int_\Omega f v \, \mathrm{d}x,$$

or equivalently, that satisfies the variational inequalities

$$\int_\Omega \nabla u \cdot \nabla (v-u) \, \mathrm{d}x \geq \int_\Omega f(v-u) \, \mathrm{d}x \quad \textit{for all } v \in U.$$

Besides, the mapping

$$f \in L^2(\Omega) \to u \in U \subset H_0^1(\Omega)$$

defined in this fashion is nonlinear and Lipschitz-continuous.

Assume in addition that $u \in H^2(\Omega)$. Then u satisfies the following boundary value problem:

$$\begin{aligned} -\Delta u &= f \quad && \textit{in } \Omega^+ := \{y \in \Omega;\ u(y) > \chi(y)\}, \\ -\Delta u &\geq f \quad && \textit{in } \Omega^0 = \{y \in \Omega;\ u(y) = \chi(y)\} := \Omega - \Omega^+, \\ u &\geq \chi \quad && \textit{in } \overline{\Omega}, \\ u &= 0 \quad && \textit{on } \Gamma. \end{aligned}$$

Proof (i) Save those about the set U, all the assumptions of Theorem 6.1-1 have already been verified (see the proof of Theorem 6.7-2).

Given a function $v \in H^1(\Omega)$, the function $\max\{0, v\}$ also belongs to the space $H^1(\Omega)$.[36] It thus follows that, if in addition $\operatorname{tr} v|_\Gamma \leq 0$ dΓ-almost everywhere on Γ (as a function in $L^2(\Gamma)$), the function $\max\{0, v\}$ belongs to the space $H_0^1(\Omega)$. Hence the subset U of $H_0^1(\Omega)$ is nonempty since it contains the function $\max\{0, \chi\}$. It is also convex, since

$$\lambda v + (1-\lambda) w \geq \lambda \chi + (1-\lambda)\chi = \chi \quad \text{a.e. in } \Omega \text{ for all } v, w \in U \text{ and all } 0 < \lambda < 1,$$

and closed: Let functions $v_k \in U$, $k \geq 1$, and $v \in H_0^1(\Omega)$ be such that $\|v_k - v\|_{1,\Omega} \to 0$ as $k \to \infty$, and hence *a fortiori* such that $\|v_k - v\|_{0,\Omega} \to 0$ as $k \to \infty$. Therefore there is a subsequence $(v_{\sigma(k)})_{k=1}^\infty$ that pointwise converges to v almost everywhere in Ω (Theorem 3.4-3). Consequently,

$$v(x) = \lim_{k \to \infty} v_{\sigma(k)}(x) \geq \chi(x) \quad \text{for almost all } x \in \Omega.$$

[36]For a proof of this result (which is nontrivial), see:
G. Stampacchia [1965]: *Equations Elliptiques du Second Ordre à Coefficients Discontinus*, Presses de l'Université de Montréal, Montréal, Que.

There thus exists a unique function $u \in U$ that minimizes the announced functional J over the set U (Theorem 6.1-1), or equivalently, that satisfies the announced variational inequalities (Theorem 6.1-2).

The linear mapping $f \in L^2(\Omega) \to \ell \in V'$ is continuous since

$$\|\ell\|_{V'} = \sup_{\begin{cases} v \in V \\ v \neq 0 \end{cases}} \frac{|\ell(v)|}{\|v\|_{1,\Omega}} \le \|f\|_{0,\Omega} \quad \text{for all } f \in L^2(\Omega),$$

and the *nonlinear* mapping $\ell \in V' \to u \in U \subset H_0^1(\Omega)$ is *Lipschitz-continuous* (Theorem 6.1-1). Hence so is the nonlinear composite mapping $f \in L^2(\Omega) \to u \in U$.

(ii) *We next show that, if $u \in H^2(\Omega)$, then $-\Delta u = f$ in $L^2(\Omega^+)$, where the open set Ω^+ is defined by $\Omega^+ := \{y \in \Omega;\ u(y) > \chi(y)\}$.*

Given any point $x \in \Omega^+$, let $2\delta := u(x) - \chi(x) > 0$. Since $x \in \Omega$ and Ω is open, and since the function $(u - \chi) : \Omega \to \mathbb{R}$ is continuous, there exists $r > 0$ such that

$$B(x;r) \subset \Omega \quad \text{and} \quad u(y) - \chi(y) \ge \delta \text{ for all } y \in B(x;r),$$

which shows that $B(x;r) \subset \Omega^+$; hence the set Ω^+ is open.

Given any nonzero function $\varphi \in \mathcal{D}(\Omega)$ such that $\operatorname{supp}\varphi \subset B(x;r)$, let

$$\alpha_0 = \alpha_0(\varphi) := \frac{\delta}{\sup_{y \in B(x;r)} |\varphi(y)|} > 0.$$

The functions $v_\alpha := u + \alpha\varphi$ therefore belong to the set U for all $|\alpha| \le \alpha_0$, since

$$v_\alpha(y) - \chi(y) = u(y) - \chi(y) + \alpha\varphi(y) \ge \delta - |\alpha\varphi(y)| \ge 0 \quad \text{for all } y \in B(x;r) \text{ and all } |\alpha| \le \alpha_0,$$
$$v_\alpha(y) - \chi(y) = u(y) - \chi(y) \ge 0 \quad \text{for all } y \in (\Omega - B(x;r)).$$

Thanks to the Green's formula of Theorem 6.7-1(a) and to the relation $v - u = 0$ on Γ, the variational inequalities $a(u, v - u) \ge \ell(v - u)$ for all $v \in U$ reduce to

$$\int_\Omega \nabla u \cdot \nabla(v - u)\,dx = -\int_\Omega \Delta u(v - u)\,dx \ge \int_\Omega f(v - u)\,dx \quad \text{for all } v \in U.$$

Letting $v = v_\alpha$ with $|\alpha| \le \alpha_0$ in these inequalities thus gives

$$\alpha \int_{B(x;r)} (-\Delta u - f)\varphi\,dx \ge 0 \quad \text{for all } \varphi \in \mathcal{D}(B(x;r)) \text{ and all } |\alpha| < \alpha_0,$$

which in turn implies that $\int_{B(x;r)}(-\Delta u - f)\varphi\,dx = 0$ for all $\varphi \in \mathcal{D}(B(x;r))$. Hence $-\Delta u = f$ in $L^2(B(x;r))$ and therefore $-\Delta u = f$ in $L^2(\Omega^+)$.

(iii) *It remains to show that, again if $u \in H^2(\Omega)$, then $-\Delta u - f \ge 0$ almost everywhere in $\Omega^0 := \{y \in \Omega;\ u(y) = \chi(y)\}$.*

Given any function $\varphi \in \mathcal{D}(\Omega)$ that satisfies $\varphi \ge 0$ in Ω, the function $v := u + \varphi$ belongs to U. Hence, for such functions v, the variational inequalities combined with the same Green's formula as above imply that

$$\int_\Omega (-\Delta u - f)(v - u)\,dx = \int_\Omega (-\Delta u - f)\varphi\,dx \ge 0 \quad \text{for all } \varphi \in \mathcal{D}(\Omega) \text{ such that } \varphi \ge 0 \text{ in } \Omega.$$

But, if a function $w \in L^1(\Omega)$ satisfies $\int_\Omega w\varphi \, dx \geq 0$ for all $\varphi \in \mathcal{D}(\Omega)$ with $\varphi \geq 0$ in Ω, then $w \geq 0$ almost everywhere in Ω (Problem 2.6-5). Therefore the function $(-\Delta u - f) \in L^2(\Omega)$ satisfies $-\Delta u - f \geq 0$ almost everywhere in Ω, and hence in particular in Ω^0 (in fact, we even have $-\Delta u - f = 0$ almost everywhere in Ω^+ by (ii)). □

Several comments are in order. First, the problem considered in Theorem 6.9-1 provides an instance of a *nonlinear* problem, in the sense that the mapping $f \in L^2(\Omega) \to u \in U$ is nonlinear, which, like the linear problems studied so far, is also *well-posed* since the same mapping $f \in L^2(\Omega) \to u \in U$ is continuous.

Second, by contrast with the solution of the linear membrane problem (Section 6.7), which may be assumed to be as smooth as we please, *the solution of the obstacle problem is not smooth in general, even if the data are very smooth.* To be convinced that this is indeed the case, consider the *one-dimensional analog* of the boundary value problem found in Theorem 6.9-1, with $f = 0$. As shown in Figure 6.9-2, the solution u is then affine in the region where it does not touch the obstacle, and consequently, whatever the smoothness of the function χ, the second derivatives of u will have discontinuities at points such as ξ and η. Therefore *the solution u is "only" in the space $H^2(I)$*, even in this simple case.

These observations carry over to the two-dimensional case, but they are, as expected, not as easy to justify. For example, it is known that if $f = 0$, $\chi \in H^2(\Omega)$, and $\overline{\Omega}$ is a convex polygon, the solution u belongs to the space $H_0^1(\Omega) \cap H^2(\Omega)$; or, if the set $\overline{\Omega}$ is convex with a boundary of class $\mathcal{C}^2$, then again $u \in H_0^1(\Omega) \cap H^2(\Omega)$. Besides, the norm $\|u\|_{2,\Omega}$ can be estimated in both cases in terms of the norms $\|\chi\|_{2,\Omega}$ and $\|f\|_{0,\Omega}$ of the data.[37]

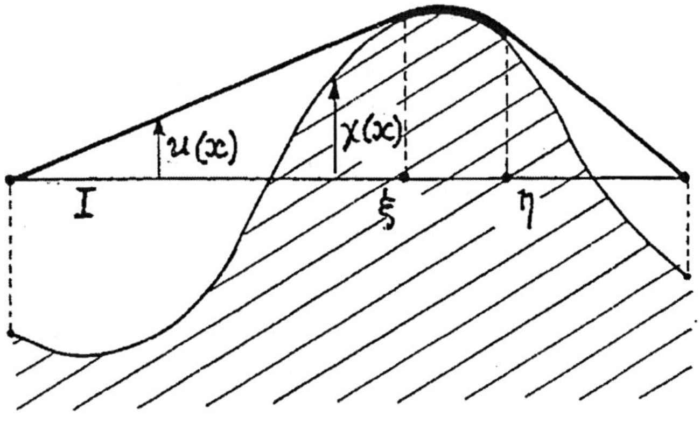

Figure 6.9-2 The one-dimensional analogue of the obstacle problem, with $f = 0$, posed over a bounded open interval I of $\mathbb{R}$. This figure originally appeared in P.G. Ciarlet [1978]: *The Finite Element Method for Elliptic Problems*, North-Holland, Amsterdam.

[37] H. Brezis; G. Stampacchia [1968]: Sur la régularité de la solution d'inéquations elliptiques, *Bulletin de la Société Mathématique de France* **96**, 153–180.
H. Lewy; G. Stampacchia [1969]: On the regularity of the solution of a variational inequality, *Communications on Pure and Applied Mathematics* **22**, 153–188.
These and other similar results are also proved in Kinderlehrer & Stampacchia [1980].

Third, *the region where the membrane touches the obstacle, i.e., the set* Ω^0, *is not known in advance.*

Fourth, the above boundary value problem may be also viewed as an instance of *a free boundary problem*, in the sense that the *"free boundary"* $\Gamma^* := \partial\Omega^+ \cap \partial\Omega^0$ *is one of the unknowns of the problem.* In this perspective, it is customary to adjoin two *transmission conditions* along the unknown free boundary in the formulation of the boundary value problem, viz.,

$$\mathrm{tr}(u|_{\Omega^+}) = \mathrm{tr}(u|_{\Omega^0}) \quad \text{and} \quad \mathrm{tr}\,\partial_\nu(u|_{\Omega^+}) = -\,\mathrm{tr}\,\partial_\nu(u|_{\Omega^0}) \text{ on } \Gamma^*.$$

But these make sense only if Γ^* is smooth enough (e.g., if Γ^* is the boundary of a domain) and u is smooth enough (e.g., if $u \in H^2(\Omega)$).

Other examples of boundary value problems associated with variational inequalities, which include an *obstacle problem for a plate*, are proposed in Problems 6.9-1–6.9-3.

Problems

6.9-1 Let Ω be a domain in $\mathbb{R}^N$, let functions

$$c \in L^\infty(\Omega) \text{ such that } c \geq c_0 > 0 \text{ a.e. in } \Omega, \quad f \in L^2(\Omega), \quad g \in L^2(\Gamma)$$

be given, and let

$$V := H^1(\Omega) \quad \text{and} \quad U := \{v \in H^1(\Omega);\ v \geq 0 \text{ d}\Gamma\text{-a.e. on } \Gamma\},$$

$$a(u,v) := \int_\Omega (\nabla u \cdot \nabla v + cuv)\,\mathrm{d}x \quad \text{and} \quad \ell(v) := \int_\Omega fv\,\mathrm{d}x + \int_\Gamma gv\,\mathrm{d}x \text{ for all } u, v \in H^1(\Omega).$$

(1) Show that the associated variational inequalities have a unique solution $u \in U$.
(2) Show that, if $u \in H^2(\Omega)$,[38] then u satisfies the *nonlinear boundary value problem*:

$$-\Delta u + cu = f \quad \text{in } \Omega,$$

$$u \geq 0 \text{ d}\Gamma\text{-a.e. on } \Gamma, \quad \partial_\nu u \geq g \text{ d}\Gamma\text{-a.e. on } \Gamma, \quad \text{and } u(\partial_\nu u - g) = 0 \text{ d}\Gamma\text{-a.e. on } \Gamma.$$

Remark Such a boundary value problem, where all, or some, boundary conditions take the form of inequalities is called a **Signorini problem.**[39] □

6.9-2 The following variational problem models in particular the *elastoplastic torsion of a thin, cylindrical, linearly elastic rod.* Let Ω be a domain in $\mathbb{R}^2$, and let

$$V := H_0^1(\Omega) \quad \text{and} \quad U := \{v \in H_0^1(\Omega);\ |\nabla v| \leq 1 \text{ a.e. in } \Omega\},$$

$$a(u,v) = \int_\Omega \nabla u \cdot \nabla v\,\mathrm{d}x \quad \text{and} \quad \ell(v) = \tau \int_\Omega v\,\mathrm{d}x,$$

[38] If $g = 0$ and $N = 2$, this regularity assumption is satisfied if Γ is smooth enough, or if Ω is convex and Γ is a polygon; see:

H. BREZIS [1971]: Problèmes unilatéraux, *Journal de Mathématiques Pures et Appliquées* **9**, 1–168.

[39] So named after:

A. SIGNORINI: Sopra alcune questioni di elastostatica, *Atti della Società Italiana per il Progresso della Scienza* (1933).

The first mathematical analysis of a Signorini problem is due to:

G. FICHERA [1964]: Problemi elastostatici con vincoli unilaterali: il problema de Signorini con ambigue condizioni al contorno, *Memorie dell'Accademia Nazionale dei Lincei* **8**, 91–140.

Signorini's problems have been since then extensively studied, notably in FICHERA [1972b], DUVAUT & LIONS [1976], and NEČAS & HLAVÁČEK [1981].

where the constant $\tau \in \mathbb{R}$ measures the torsion of the rod.[40]

(1) Show that the associated variational inequalities have a unique solution $u_\tau \in U$.

(2) Show that, if $u_\tau \in H^2(\Omega)$, then u_τ satisfies

$$-\Delta u_\tau = \tau \quad \text{a.e. in the set } \{x \in \Omega;\ |\nabla v(x)| < 1\}.$$

(3) Show that the set U is a compact subset of $\mathcal{C}(\overline{\Omega})$ and that any function $v \in U$ satisfies $|v(x)| \leq \text{dist}(x, \Gamma)$ for all $x \in \overline{\Omega}$.

(4) Show that $\|u_\tau - u_\infty\|_{1,\Omega} \to 0$ and $\sup_{x\in\overline{\Omega}} |u_\tau(x) - u_\infty(x)| \to 0$ as $\tau \to \infty$, where the function $u_\infty : \overline{\Omega} \to \mathbb{R}$ is defined by $u_\infty(x) := \text{dist}(x, \partial\Omega)$ for all $x \in \overline{\Omega}$.[41]

6.9-3 Let Ω be a domain in $\mathbb{R}^N$ with $N = 2$ or $N = 3$, let x_i, $1 \leq i \leq m$, be distinct points in Ω, let $f \in L^2(\Omega)$, and let

$$V := H_0^2(\Omega) \quad \text{and} \quad U := \{v \in H_0^2(\Omega);\ v(x_i) \geq 0,\ 1 \leq i \leq m\},$$

$$a(u, v) := \int_\Omega \Delta u \Delta v \,\mathrm{d}x \quad \text{and} \quad \ell(v) := \int_\Omega f v \,\mathrm{d}x \quad \text{for all } u, v \in H_0^2(\Omega).$$

(1) Show that the associated variational inequalities have a unique solution $u \in U$.

(2) Show that, if $u \in H^4(\Omega)$, then u satisfies $\Delta^2 u = f$ in the set $\Omega - \bigcup_{i=1}^m \{x_i\}$.

Remarks (1) If $N = 2$, the functional $J : H_0^2(\Omega) \to \mathbb{R}$ defined by $J(v) = \dfrac{1}{2} \int_\Omega |\Delta v|^2 \,\mathrm{d}x - \int_\Omega f v \,\mathrm{d}x$ for all $v \in H_0^2(\Omega)$ represents the energy of a linearly elastic place clamped over its entire boundary (Section 6.8). The above variational problem thus models an **obstacle problem for a clamped plate**, where the unknown vertical displacement $u : \overline{\Omega} \to \mathbb{R}$ is subjected to the inequalities $u(x_i) \geq 0$, $1 \leq i \leq m$.

(2) An interesting complement to question (2) will be provided in Problem 7.15-4, which shows that there exist *"Kuhn–Tucker multipliers"* $\lambda_i \geq 0$, $1 \leq i \leq m$, that satisfy $\Delta^2 u = f + \sum_{i=1}^m \lambda_i \delta_{x_i}$ in $\mathcal{D}'(\Omega)$ (i.e., in the sense of distributions; cf. Section 6.3) and have a remarkable mechanical interpretation. □

6.10 Eigenvalue problems for second-order elliptic operators

Let Ω be a domain in $\mathbb{R}^N$, let $a : H_0^1(\Omega) \times H_0^1(\Omega) \to \mathbb{R}$ be a continuous and $H_0^1(\Omega)$-coercive symmetric bilinear form of the form considered in Theorem 6.7-6, and let $f \in L^2(\Omega)$. There thus exists a unique function $u \in H_0^1(\Omega)$ that satisfies

$$a(u, v) = \langle f, v \rangle \quad \text{for all } v \in H_0^1(\Omega),$$

where we let (for notational brevity throughout this section)

$$\langle f, g \rangle := \int_\Omega f g \,\mathrm{d}x \quad \text{for all } f, g \in L^2(\Omega).$$

[40] Such variational problems have been first analyzed by:

H. Brezis; M. Sibony [1971]: Equivalence de deux inéquations variationnelles, *Archive for Rational Mechanics and Analysis* **41**, 254–265.

R. Glowinski; H. Lanchon [1973]: Torsion élasto-plastique d'une barre cylindrique de section multi-connexe, *Journal de Mécanique* **12**, 151–171.

More general elasto-plastic problems have been studied at length in Duvaut & Lions [1976] and in Nečas & Hlavacek [1981].

[41] This result is proved in Glowinski [1984, Chapter 2, Section 3].

Besides, a smooth enough solution u to these equations satisfies a second-order elliptic boundary value problem of the form

$$\mathcal{L}u = f \text{ in } \Omega \quad \text{and} \quad u = 0 \text{ on } \Gamma := \partial\Omega,$$

where $\mathcal{L}$ is a uniformly elliptic linear partial differential operator of the second order (Section 6.7).

The **eigenvalue problem for the operator** $\mathcal{L}$ consists in seeking whether there exist real numbers μ and *nonzero* functions w that satisfy the boundary value problem

$$\mathcal{L}w = \mu w \text{ in } \Omega \quad \text{and} \quad w = 0 \text{ on } \Gamma.$$

If such a pair (μ, w) exists, μ is called an **eigenvalue** of $\mathcal{L}$ and w is called an **eigenfunction** of $\mathcal{L}$ *associated with the eigenvalue* μ (naturally, each such eigenfunction w should be smooth enough so that the above boundary value problem makes sense). If $w \in H_0^1(\Omega)$, the pair $(\mu, w) \in \mathbb{R} \times H_0^1(\Omega)$ thus satisfies the variational equations

$$a(w, v) = \mu\langle w, v\rangle \quad \text{for all } v \in H_0^1(\Omega),$$

which constitutes the **variational formulation** *of the eigenvalue problem for the operator* $\mathcal{L}$.

Viewed on their own, i.e., without reference to the eigenvalue problem for an elliptic operator $\mathcal{L}$, such variational equations thus provide another example of **abstract variational problems**.

We now show that solving these variational equations is equivalent to finding the *inverses* of the eigenvalues, and the associated eigenvectors, of a *compact, symmetric, positive-definite operator* acting in the Hilbert space $H_0^1(\Omega)$, considered as *equipped with the inner product* $a(\cdot, \cdot)$.

Note that, even though they share the same notation A, this operator is *not* the same as the operator introduced in the proof of the Lax–Milgram lemma (Theorem 6.2-1).

Theorem 6.10-1 *Let Ω be a domain in $\mathbb{R}^N$ and let $a : H_0^1(\Omega) \times H_0^1(\Omega) \to \mathbb{R}$ be a continuous and $H_0^1(\Omega)$-coercive symmetric bilinear form. Given any function $u \in H_0^1(\Omega)$, there thus exists a unique function $Au \in H_0^1(\Omega)$ that satisfies*

$$a(Au, v) = \langle u, v\rangle \quad \textit{for all } v \in H_0^1(\Omega).$$

(a) *The linear operator $A : H_0^1(\Omega) \to H_0^1(\Omega)$ defined in this fashion is compact, symmetric, and positive-definite, hence injective, in the Hilbert space $(H_0^1(\Omega), a(\cdot, \cdot))$; in other words,*

$$\begin{aligned} a(Au, v) &= a(u, Av) && \textit{for all } u, v \in H_0^1(\Omega), \\ a(Av, v) &> 0 && \textit{for all } v \in H_0^1(\Omega),\ v \neq 0. \end{aligned}$$

Finally, A has infinite-dimensional range.

(b) *A pair $(\mu, w) \in \mathbb{R} \times H_0^1(\Omega)$ with $w \neq 0$ satisfies*

$$a(w, v) = \mu\langle w, v\rangle \quad \textit{for all } v \in H_0^1(\Omega)$$

if and only if

$$\mu > 0 \quad \textit{and} \quad Aw = \lambda w \ \textit{ with } \lambda := \frac{1}{\mu}.$$

Proof As already noted (see the proof of Theorem 6.1-1), the bilinear form a is an inner product over the space $H_0^1(\Omega)$ whose associated norm is equivalent to $\|\cdot\|_{1,\Omega}$.

The mapping $A : H_0^1(\Omega) \to H_0^1(\Omega)$ is clearly linear. Besides, the mapping

$$A : (H_0^1(\Omega), \|\cdot\|_{0,\Omega}) \to (H_0^1(\Omega), \|\cdot\|_{1,\Omega})$$

is continuous since there exists $\alpha > 0$ such that

$$\alpha \|Au\|_{1,\Omega}^2 \le a(Au, Au) = \langle u, Au\rangle \le \|u\|_{0,\Omega} \|Au\|_{0,\Omega} \le \|u\|_{0,\Omega} \|Au\|_{1,\Omega}$$

for all $u \in H_0^1(\Omega)$.

Let $(u_n)_{n=1}^\infty$ be a bounded sequence in $H_0^1(\Omega)$; therefore there exists a subsequence $(u_{\sigma(n)})_{n=1}^\infty$ that converges in $L^2(\Omega)$ by the Rellich–Kondrachov theorem (Theorem 6.6-3). The continuity of the mapping $A : (H_0^1(\Omega), \|\cdot\|_{0,\Omega}) \to (H_0^1(\Omega), \|\cdot\|_{1,\Omega})$ then implies that the subsequence $(Au_{\sigma(n)})_{n=1}^\infty$ converges in $H_0^1(\Omega)$. Consequently, A is compact (Theorem 2.10-1).

The symmetry and positive-definiteness of A in the Hilbert space $(H_0^1(\Omega), a(\cdot,\cdot))$ follow from the relations

$$a(Au, v) = \langle u, v\rangle = \langle v, u\rangle = a(Av, u) = a(u, Av) \quad \text{for all } u, v \in H_0^1(\Omega),$$
$$a(Av, v) = \langle v, v\rangle = \|v\|_{0,\Omega}^2 > 0 \quad \text{for all } v \in H_0^1(\Omega),\ v \neq 0.$$

The last relation also shows that $Av = 0$ implies $v = 0$.

Since $A : H_0^1(\Omega) \to H_0^1(\Omega)$ is thus injective and the space $H_0^1(\Omega)$ is infinite-dimensional, so is its range. Hence all the assertions of (a) are proved.

If $(\mu, w) \in \mathbb{R} \times H_0^1(\Omega)$ with $w \neq 0$ satisfies $a(w, v) = \mu\langle w, v\rangle$ for all $v \in H_0^1(\Omega)$, then in particular $\mu\langle w, w\rangle = a(w, w) > 0$, which implies that $\mu > 0$.

Besides, by definition of A,

$$a(w, v) = \mu\langle w, v\rangle = \mu a(Aw, v) \quad \text{for all } v \in H_0^1(\Omega),$$

so that $Aw = \lambda w$ with $\lambda := \frac{1}{\mu}$. Conversely, if $(\mu, w) \in \mathbb{R} \times H_0^1(\Omega)$ with $\mu \neq 0$ and $w \neq 0$ satisfies $Aw = \frac{1}{\mu} w$, then, again by definition of A,

$$\mu\langle w, v\rangle = \mu a(Aw, v) = a(w, v) \quad \text{for all } v \in H_0^1(\Omega).$$

Hence (b) is proved. □

Remark By means of the *same* relation $a(\widetilde{A}f, v) = \langle f, v\rangle$ for all $v \in H_0^1(\Omega)$, one can also define *another* compact, symmetric, and positive-definite operator $\widetilde{A}$, but this time acting from the Hilbert space $L^2(\Omega)$ into itself; cf. Problem 6.10-2. □

When combined with the *spectral theorem for compact symmetric operators* (Section 4.11), Theorem 6.10-1 immediately provides *all* the solutions (μ, w) to the eigenvalue problem $\mathcal{L}w = \mu w$ in Ω and $w = 0$ on Γ (considered at the beginning of this section) for a wide class of *uniformly elliptic operators* $\mathcal{L}$. Besides, the eigenvalues and eigenfunctions of $\mathcal{L}$ have a remarkable characterization in terms of a specific functional, viz., the *Rayleigh quotient* introduced in the next theorem.

Theorem 6.10-2 *Let Ω be a domain in $\mathbb{R}^N$, let functions $a_{ij} = a_{ji} \in \mathcal{C}^1(\overline{\Omega})$, $1 \le i, j \le N$, be given such that, for some constant $\mu > 0$,*

$$\sum_{i,j=1}^{N} a_{ij}(x)\xi_i\xi_j \ge \mu \sum_{i=1}^{N} |\xi_i|^2 \quad \text{for all } x \in \overline{\Omega} \text{ and all } (\xi_i)_{i=1}^N \in \mathbb{R}^N,$$

and let a function $c \in L^\infty(\Omega)$ be given such that $c \ge 0$ almost everywhere in Ω. Define the bilinear form $a : H_0^1(\Omega) \times H_0^1(\Omega) \to \mathbb{R}$ by

$$a(u,v) := \int_\Omega \left(\sum_{i,j=1}^{N} a_{ij}\partial_i u \partial_j v + cuv \right) \mathrm{d}x \quad \text{for all } u, v \in H_0^1(\Omega).$$

(a) *There exist an infinite sequence $(\mu_k)_{k=1}^\infty$ of real numbers and an infinite sequence $(w_k)_{k=1}^\infty$ of nonzero functions $w_k \in H_0^1(\Omega)$ that satisfy*

$$0 < \mu_1 \le \mu_2 \le \cdots \le \mu_k \le \cdots, \quad \lim_{k\to\infty} \mu_k = \infty,$$
$$a(w_k, v) = \mu_k \langle w_k, v\rangle \quad \text{for all } v \in H_0^1(\Omega) \text{ and all } k \ge 1,$$
$$a(w_k, w_\ell) = \delta_{k\ell} \quad \text{and} \quad \langle w_k, w_\ell\rangle = \frac{\delta_{k\ell}}{\mu_k} \quad \text{for all } k, \ell \ge 1.$$

Let $\mu \in \mathbb{R}$ and a nonzero function $w \in H_0^1(\Omega)$ be a solution of the variational equations

$$a(w,v) = \mu\langle w, v\rangle \quad \text{for all } v \in H_0^1(\Omega).$$

Then there exists $k \ge 1$ such that $\mu_k = \mu$. Besides, the set $J(\mu) := \{k \ge 1;\ \mu_k = \mu\}$ is finite, and

$$\{w \in H_0^1(\Omega);\ a(w,v) = \mu\langle w,v\rangle \text{ for all } v \in H_0^1(\Omega)\} = \mathrm{Span}(w_k)_{k\in J(\mu)}.$$

Finally, the family $(w_k)_{k=1}^\infty$ is a Hilbert basis (Section 4.9) in the Hilbert space $(H_0^1(\Omega), a(\cdot,\cdot))$, and the family $(\sqrt{\mu_k}w_k)_{k=1}^\infty$ is a Hilbert basis in the Hilbert space $(L^2(\Omega), \langle\cdot,\cdot\rangle)$.

(b) *Define the* **Rayleigh quotient**[42]

$$R(w) := \frac{a(w,w)}{\langle w,w\rangle} = \frac{\int_\Omega \left(\sum_{i,j=1}^N a_{ij}\partial_i w \partial_j w + c|w|^2\right) \mathrm{d}x}{\int_\Omega |w|^2\, \mathrm{d}x} \quad \text{for all } w \in H_0^1(\Omega),\ w \ne 0.$$

Then

$$\mu_1 = R(w_1) = \inf_{w\in H_0^1(\Omega),\, w\ne 0} R(w),$$
$$\mu_k = R(w_k) = \inf_{\left\{\substack{w\in H_0^1(\Omega),\, w\ne 0\\ \langle w, w_\ell\rangle = 0,\, 1\le \ell\le k-1}\right.} R(w) \quad \text{for all } k \ge 2.$$

[42]So named after John William Strut, third Baron Rayleigh (1842–1919). Lord Rayleigh was awarded the Nobel Prize in Physics in 1904.

(c) *If* $w_k \in H^2(\Omega)$ *for some* $k \geq 1$,[43] *then*

$$\mathcal{L}w_k = \mu_k w_k \text{ in } \Omega \quad \textit{and} \quad w_k = 0 \textit{ on } \Gamma,$$

where the uniformly elliptic operator $\mathcal{L}$ *is defined for all smooth enough functions* v *by*

$$\mathcal{L}v := -\sum_{i,j=1}^{N} \partial_i(a_{ij}\partial_j v) + cv.$$

Proof The linear operator A, considered as acting from the Hilbert space $(H_0^1(\Omega), a(\cdot,\cdot))$ into itself, is compact, symmetric, and positive-definite (Theorem 6.10-1(a)). Hence, by Theorems 4.11-1 and 4.11-3, there exist an infinite sequence $(\lambda_k)_{k=1}^{\infty}$ of eigenvalues of A and an infinite sequence $(w_k)_{k=1}^{\infty}$ of corresponding eigenvectors that satisfy

$$\lambda_1 \geq \lambda_2 \geq \cdots \geq \lambda_k \geq \cdots, \quad \lambda_k > 0 \quad \text{for all } k \geq 1, \lim_{k\to 0} \lambda_k = 0,$$

$$Aw_k = \lambda_k w_k \text{ for all } k \geq 1 \quad \text{and} \quad a(w_k, w_\ell) = \delta_{k\ell} \text{ for all } k, \ell \geq 1,$$

$$\lambda_1 = \frac{a(Aw_1, w_1)}{a(w_1, w_1)} = \sup_{w \in H_0^1(\Omega),\, w \neq 0} \frac{a(Aw, w)}{a(w, w)},$$

$$\lambda_k = \frac{a(Aw_k, w_k)}{a(w_k, w_k)} = \sup_{\left\{\substack{w \in H_0^1(\Omega),\, w \neq 0 \\ a(w, w_\ell) = 0,\, 1 \leq \ell \leq k-1}\right.} \frac{a(Aw, w)}{a(w, w)} \quad \text{for all } k \geq 2.$$

Besides, the family $(w_k)_{k=1}^{\infty}$ is a Hilbert basis in the Hilbert space $(H_0^1(\Omega), a(\cdot,\cdot))$.

The relations $Aw_k = w_k$ and $a(w_k, w_\ell) = \delta_{k\ell}$ then imply that

$$\langle w_k, v_\ell \rangle = \lambda_k a(w_k, w_\ell) = \lambda_k \delta_{k\ell} \quad \text{for all } k, \ell \geq 1,$$

by Theorem 6.10-1(b). To show that the family $(\sqrt{\alpha_k} w_k)_{k=1}^{\infty}$, which is thus orthonormal with respect to $\langle \cdot,\cdot \rangle$, is a Hilbert basis in the space $(L^2(\Omega, \langle \cdot,\cdot \rangle)$, it suffices to show that (Theorem 4.8-2)

$$\overline{\operatorname{Span}(\sqrt{\alpha_k} w_k)_{k=1}^{\infty}} = \overline{\operatorname{Span}(w_k)_{k=1}^{\infty}} = L^2(\Omega),$$

where both closures are meant with respect to the norm $\|\cdot\|_{0,\Omega}$.

So, let a function $u \in L^2(\Omega)$ and $\varepsilon > 0$ be given. Since the space $\mathcal{D}(\Omega)$ is dense in $L^2(\Omega)$, there exists $\varphi = \varphi(u, \varepsilon) \in \mathcal{D}(\Omega) \subset H_0^1(\Omega)$ such that $\|\varphi - u\|_{0,\Omega} \leq \frac{\varepsilon}{2}$. Since $(w_k)_{k=1}^{\infty}$ is a Hilbert basis in $(H_0^1(\Omega); a(\cdot,\cdot))$, there exists $v = v(\varphi) = v(u, \varepsilon) \in \operatorname{Span}(w_k)_{k=1}^{\infty}$ such that $\|v - \varphi\|_{0,\Omega} \leq \|v - \varphi\|_{1,\Omega} \leq \frac{\varepsilon}{2}$. Hence $\|v - u\|_{0,\Omega} \leq \varepsilon$, which shows that $\overline{\operatorname{Span}(w_k)_{k=1}^{\infty}} = L^2(\Omega)$.

All the assertions of (a) are therefore proved, with $\mu_k := \dfrac{1}{\lambda_k}$, $k \geq 1$.

To prove the assertions of (b), we first note that the definition of the operator A shows that the Rayleigh quotient is also given by

$$R(w) := \frac{a(w, w)}{\langle w, w \rangle} = \frac{a(w, w)}{a(Aw, w)} \quad \text{for all } w \in H_0^1(\Omega).$$

[43]This is the case for all $k \geq 1$ if Γ is of class $\mathcal{C}^2$; cf., e.g., Evans [2010, Section 6.3].

We next note that a function $w \in H_0^1(\Omega)$ satisfies $a(w, w_\ell) = 0$ for all $1 \leq \ell \leq k-1$ if and only if $\langle w, w_\ell \rangle = 0$ for all $1 \leq \ell \leq k-1$, since

$$\langle w, w_\ell \rangle = a(Aw_\ell, w) = \lambda_\ell a(w, w_\ell) \quad \text{and} \quad \lambda_\ell > 0, \quad \text{for all } \ell \geq 1.$$

Therefore the characterization of the numbers μ_k, $k \geq 1$, as infimums immediately follows from that of the numbers $\lambda_k = \dfrac{1}{\mu_k}$ as supremums.

The assertion (c) is proved in the usual way (see the proof of Theorem 6.7-6) by means of a Green's formula. □

The power of Theorem 6.10-2 can be already appreciated from the simplest example of eigenvalue problem, viz.,

$$-u''(x) = \mu u(x), \quad 0 < x < 1, \quad \text{and} \quad u(0) = u(1) = 0.$$

In this case, the eigenvalues μ_k, and corresponding eigenvectors w_k, $k \geq 1$, orthonormalized with respect to the inner product $a(\cdot,\cdot)$ defined in this case by $a(u,v) := \int_0^1 u'v' \,\mathrm{d}x$, are given by

$$\mu_k = k^2\pi^2 \quad \text{and} \quad w_k(x) = \frac{\sqrt{2}}{k\pi} \sin k\pi x, \quad 0 \leq x \leq 1.$$

Hence Theorem 6.10-2(a) immediately implies that *the family* $(w_k)_{k=1}^\infty$ *constitutes a Hilbert basis in the Hilbert space* $(H_0^1(0,1); a(\cdot,\cdot))$ *and the family* $(k\pi w_k)_{k=1}^\infty$ *constitutes a Hilbert basis in the space* $L^2(0,1)$. The remarkable *formula*

$$\pi^2 = \inf_{\left\{\substack{w \in H_0^1(0,1) \\ w \neq 0}\right.} \frac{\int_0^1 |w'|^2 \,\mathrm{d}x}{\int_0^1 |w|^2 \,\mathrm{d}x}$$

likewise immediately follows from Theorem 6.10-2(b).

Remark It is easily verified that the space $H_0^1(0,1)$ may be replaced in this infimum by any function space V that satisfies $\mathcal{D}(0,1) \subset V \subset H_0^1(0,1)$. □

Given a subspace W of the space $L^2(\Omega)$, let $W^\perp$ denote its *orthogonal complement* with respect to the inner product $\langle \cdot,\cdot \rangle$ of $L^2(\Omega)$ (Section 4.5). Then the characterization of the kth eigenvalue μ_k in terms of the Rayleigh quotient given in Theorem 6.10-2(b) may be rewritten as

$$\mu_k = R(w_k) = \inf\{R(w);\ w \in W_{k-1}^\perp,\ w \neq 0\},$$

where

$$W_0 = \{0\}, \quad \text{and} \quad W_{k-1} := \operatorname{Span}(w_\ell)_{\ell=1}^{k-1} \text{ if } k \geq 2.$$

It is remarkable that the eigenvalues μ_k, $k \geq 1$, can be also characterized, again in terms of the Rayleigh quotient, but this time *independently of the eigenfunctions:*

Theorem 6.10-3 (**Courant–Fischer theorem**[44]) *Let the assumptions be the same as in Theorem 6.10-2. For each integer $\ell \geq 1$, let $\mathcal{V}_\ell$ denote the set formed by all the subspaces of dimension ℓ of $H_0^1(\Omega)$, and let $\mathcal{V}_0 = \{0\}$. Then, for each integer $k \geq 1$,*

$$\mu_k = \sup_{V \in \mathcal{V}_{k-1}} \left(\inf\{R(w),\ w \in V^\perp,\ w \neq 0\} \right),$$

$$\mu_k = \inf_{V \in \mathcal{V}_k} \left(\sup\{R(w);\ w \in V,\ w \neq 0\} \right).$$

Proof For conciseness, the relation "$w \neq 0$" is omitted throughout the proof.

Assume that $k \geq 2$ (the first relation clearly holds for $k = 1$, since $V^\perp = H_0^1(\Omega)$ if $V \in \mathcal{V}_0$, i.e., if $V = \{0\}$). Since $W_{k-1} = \mathrm{Span}(w_\ell)_{\ell=1}^{k-1} \in \mathcal{V}_{k-1}$ (the eigenfunctions w_ℓ, $1 \leq \ell \leq k-1$, are linearly independent because they are orthogonal), it follows that

$$\mu_k = \inf\{R(w);\ w \in W_{k-1}^\perp\} \leq \sup_{V \in \mathcal{V}_{k-1}} \left(\inf\{R(w);\ w \in V^\perp\} \right).$$

It thus remains to show that, given any subspace $V \in \mathcal{V}_{k-1}$,

$$\inf\{R(w);\ w \in V^\perp\} \leq \mu_k.$$

To this end, we note that there exist functions u that satisfy

$$u \in \mathrm{Span}(w_j)_{j=1}^k, \quad u \neq 0, \quad \text{and} \quad u \in V^\perp,$$

since, given a basis $(v_i)_{i=1}^{k-1}$ in V, the homogeneous linear system

$$\sum_{j=1}^k \alpha_j \langle w_j, v_i \rangle = 0, \quad 1 \leq i \leq k-1,$$

always possesses nonzero solutions. Given such a nonzero solutions, let $u := \sum_{j=1}^k \alpha_j w_j$. The orthogonality relations satisfied by the eigenfunctions then imply that

$$R(u) = R\Bigg(\sum_{j=1}^k \alpha_j w_j \Bigg) = \frac{\sum_{j=1}^k |\alpha_j|^2}{\sum_{j=1}^k \mu_j^{-1} |\alpha_j|^2} \leq \mu_k,$$

since $0 < \mu_1 \leq \cdots \leq \mu_k$. Hence $\inf\{R(w);\ w \in V^\perp\} \leq \mu_k$, and thus

$$\mu_k = \sup_{V \in \mathcal{V}_{k-1}} \left(\inf\{R(w),\ w \in V^\perp\} \right).$$

[44]This theorem was established first for matrices, then for eigenvalue problems of the type considered here, in:

E. Fischer [1905]: Über quadratische Formen mit reellen Koeffizienten, *Monatshefte für Mathematik und Physik* **16**, 234–249.

R. Courant [1920]: Über die Eigenwerte bei den Differentialgleichungen der Mathematischen Physik, *Mathematische Zeitschrift* **7**, 1–57.

To prove the other relation, we first note that

$$\sup_{w \in W_k = \mathrm{Span}(w_\ell)_{\ell=1}^k} R(w) = \sup_{(\alpha_j)_{j=1}^k \in \mathbb{R}^k - \{\mathbf{0}\}} R\Big(\sum_{j=1}^k \alpha_j w_j\Big) = \mu_k = R(w_k), \quad \text{for all } k \geq 1,$$

so that

$$\mu_k = \sup_{w \in W_k} R(w) \geq \inf_{V \in \mathcal{V}_k} \Big(\sup\{R(w);\ w \in V\}\Big),$$

since $W_k \in \mathcal{V}_k$. It thus remains to show that, given any subspace $V \in \mathcal{V}_k$,

$$\mu_k \leq \sup\{R(w);\ w \in V\}.$$

To this end, we note that there exist functions u that satisfy

$$u \in V, \ \ u \neq 0, \quad u \in W_{k-1}^{\perp},$$

since, given a basis $(v_j)_{j=1}^k$ in V that is orthonormal with respect to $\langle \cdot, \cdot \rangle$, the homogeneous linear system

$$\sum_{j=1}^k \beta_j \langle v_j, w_i \rangle = 0, \quad 1 \leq i \leq k-1,$$

always possesses solutions that satisfy $\beta_k \neq 0$. Given such a solution, let $u := \sum_{j=1}^k \beta_j v_j$. Since the relations $\langle u, w_\ell \rangle = 0$, $1 \leq \ell \leq k-1$, imply that $a(u, w_\ell) = 0$, $1 \leq \ell \leq k-1$, it follows from Theorem 6.10-2(b) that $\mu_k = \inf\{R(w);\ w \in W_{k-1}^{\perp}\} \leq R(u)$. Hence

$$\mu_k = \inf_{V \in \mathcal{V}_k} \Big(\sup\{R(w);\ w \in V\}\Big). \qquad \square$$

Remarks (1) The Courant–Fischer theorem plays an essential role in the convergence analysis of the numerical approximation of eigenvalue problems of the type described here by finite element methods.[45]

(2) The Courant–Fischer theorem can be immediately converted into an analogous theorem that holds more generally for any *compact self-adjoint operator* (Section 4.10). □

Problems

6.10-1 Compute explicitly all the eigenvalues and corresponding eigenfunctions for the following eigenvalue problems:

$$\begin{aligned}
-u''(x) + u(x) &= \mu u(x), \ \ 0 < x < 1, \quad \text{and} \quad u(0) = u(1) = 0, \\
-u''(x) + u(x) &= \mu u(x), \ \ 0 < x < 1, \quad \text{and} \quad u'(0) = u'(1) = 0, \\
-u''(x) + u(x) &= \mu u(x), \ \ 0 < x < 1, \quad \text{and} \quad u(0) = u(1) \quad \text{and} \quad u'(0) = u'(1).
\end{aligned}$$

[45] I. Babuška; J.E. Osborn [1991]: Eigenvalue problems, in *Handbook of Numerical Analysis, Volume* II (P.G. Ciarlet & J.L. Lions, editors), pp. 641–787, North-Holland, Amsterdam.

6.10-2 Let Ω be a domain in $\mathbb{R}^N$ and let $a : H_0^1(\Omega) \times H_0^1(\Omega) \to \mathbb{R}$ be a continuous and $H_0^1(\Omega)$-coercive symmetric bilinear form. Given any function $f \in L^2(\Omega)$, there thus exists a unique function $\widetilde{A}f \in H_0^1(\Omega) \subset L^2(\Omega)$ that satisfies

$$a(\widetilde{A}f, v) = \langle f, v\rangle \quad \text{for all } v \in H_0^1(\Omega).$$

Show that the linear operator $\widetilde{A} : L^2(\Omega) \to L^2(\Omega)$ defined in this fashion is compact, symmetric, and positive-definite in the Hilbert space $L^2(\Omega)$ (i.e., $\langle \widetilde{A}f, g\rangle = \langle f, \widetilde{A}g\rangle$ for all $f, g \in L^2(\Omega)$ and $\langle \widetilde{A}f, f\rangle > 0$ for all $f \in L^2(\Omega)$, $f \neq 0$), injective, and has infinite-dimensional range.

6.10-3 Let Ω_1 and Ω_2 be two domains in $\mathbb{R}^N$ such that $\Omega_1 \subset \Omega_2$. Show that the corresponding eigenvalues $\mu_k(\Omega_1)$ and $\mu_k(\Omega_2)$ of the operator $-\Delta$, arranged in increasing order as in Theorem 6.10-2, satisfy $\mu_k(\Omega_2) \leq \mu_k(\Omega_1)$ for all $k \geq 1$.

6.10-4 Let $V := \{v \in H^1(-1,1); v(-1) = v(1) \text{ and } \int_{-1}^1 v(x)\,dx = 0\}$. Show that[46]

$$\pi^2 = \inf_{\left\{\substack{v\in V\\ v\neq 0}\right.} \frac{\int_{-1}^1 |v'|^2\,dx}{\int_{-1}^1 |v|^2\,dx}.$$

6.10-5 Let Ω be a domain in $\mathbb{R}^N$, let $\mu_1 > 0$ denote the smallest eigenvalue of the operator $-\Delta$ on Ω, and let $f : \mathbb{R} \to \mathbb{R}$ be a Lipschitz-continuous function with Lipschitz constant γ. Show that, if $\gamma < \mu_1$, the semilinear boundary value problem

$$-\Delta u = f(u) \text{ in } \Omega, \quad u = 0 \text{ on } \Gamma,$$

has one and only one solution in $H_0^1(\Omega)$.

Hint: Show that u is a fixed point of a Lipschitz-continuous mapping from $L^2(\Omega)$ into itself with Lipschitz constant $\gamma\mu_1^{-1}$.

Remark When $N = 1$ and $\Omega =]0,1[$, this result constitutes an improvement over Theorem 3.9-1 (another application of the Banach fixed point theorem, but in a different Banach space), where it was assumed that $\gamma < 8$ instead of $\gamma < \pi^2$ as here. □

6.11 The spaces $W^{-m,q}(\Omega)$ and $H^{-m}(\Omega)$; J.L. Lions lemma

The Sobolev spaces $W^{m,p}(\Omega)$ and $W_0^{m,p}(\Omega)$, $1 \leq p < \infty$, have been introduced and studied in Sections 6.5 and 6.6. We now identify their *dual spaces*.

Recall that the conjugate exponent q of any $1 \leq p < \infty$ is defined by $q := \dfrac{p}{p-1}$ if $1 < p < \infty$ and $q := \infty$ if $p = 1$. Given an integer $m \geq 1$ and functions $f_{\boldsymbol{\alpha}} \in L^q(\Omega)$ for each multi-index $|\boldsymbol{\alpha}| \leq m$, the linear functional $v \in W^{m,p}(\Omega) \to \sum_{|\alpha|\leq m} \int_\Omega f_{\boldsymbol{\alpha}} \partial^{\boldsymbol{\alpha}} v\,dx$ is clearly continuous over the space $W^{m,p}(\Omega)$. We now show that, *conversely*, any continuous linear functional over $W^{m,p}(\Omega)$ is necessarily of this form. Note that, given such a functional, the functions $f_{\boldsymbol{\alpha}} \in L^q(\Omega)$, $|\boldsymbol{\alpha}| \leq m$, found in the next theorem are *not necessarily unique*, however.

[46]The resulting inequality $\int_{-1}^1 |v'|^2\,dx \geq \pi^2 \int_{-1}^1 |v|^2\,dx$ for all $v \in V$ constitutes **Wirtinger's inequality**, so named after Wilhelm Wirtinger (1865–1945).

Theorem 6.11-1 (dual space of $W^{m,p}(\Omega)$, $1 \le p < \infty$) *Let Ω be an open subset of $\mathbb{R}^N$, let $m \ge 1$ be an integer, let $1 \le p < \infty$, and let q denote the conjugate exponent of p.*

Then $\ell \in (W^{m,p}(\Omega))'$ if and only if there exist functions $f_{\boldsymbol{\alpha}} \in L^q(\Omega)$, defined for each multi-index $\boldsymbol{\alpha}$ with $|\boldsymbol{\alpha}| \le m$, such that (recall that $\partial^{\mathbf{0}} v := v$)

$$\ell(v) = \sum_{|\boldsymbol{\alpha}| \le m} \int_\Omega f_{\boldsymbol{\alpha}} \partial^{\boldsymbol{\alpha}} v \, dx \quad \textit{for all } v \in W^{m,p}(\Omega).$$

Proof As already observed, the "if" part is clear. Let $M := \operatorname{Card}\{\boldsymbol{\alpha}; |\boldsymbol{\alpha}| \le m\}$. As a normed vector space, $W^{m,p}(\Omega)$ can be identified with the subspace

$$Y(\Omega) := \Big\{ (v^{\boldsymbol{\alpha}})_{|\boldsymbol{\alpha}| \le m} \in (L^p(\Omega))^M ; \int_\Omega v^{\boldsymbol{\alpha}} \varphi \, dx = (-1)^{|\boldsymbol{\alpha}|} \int_\Omega v^{\mathbf{0}} \partial^{\boldsymbol{\alpha}} \varphi \, dx \\ \text{for all } \varphi \in \mathcal{D}(\Omega),\ |\boldsymbol{\alpha}| \le m \Big\},$$

of the product space $(L^p(\Omega))^M$ equipped with the norm $(v^{\boldsymbol{\alpha}}) \to \big(\sum_{|\boldsymbol{\alpha}| \le m} \|v^{\boldsymbol{\alpha}}\|^p_{0,p,\Omega} \big)^{1/p}$ (the subspace $Y(\Omega)$ is clearly closed in $(L^p(\Omega))^M$, but this property is not needed in this proof). By the *Hahn–Banach theorem in a normed vector space* (Theorem 5.9-1), any continuous linear functional $\ell : W^{m,p}(\Omega) \to \mathbb{R}$ can thus be extended to a continuous linear functional $\widetilde{\ell} : (L^p(\Omega))^M \to \mathbb{R}$.

By the *F. Riesz representation theorem in $L^p(\Omega)$, $1 \le p < \infty$* (Theorem 3.5-3), there thus exist functions $f_{\boldsymbol{\alpha}} \in L^q(\Omega)$ such that

$$\widetilde{\ell}((v^{\boldsymbol{\alpha}})) = \sum_{|\alpha| \le m} \int_\Omega f_{\boldsymbol{\alpha}} v^{\boldsymbol{\alpha}} \, dx \quad \text{for all } (v^{\boldsymbol{\alpha}}) \in (L^p(\Omega))^M .$$

The "only if" part then follows by restricting $\widetilde{\ell}$ to the subspace $Y(\Omega)$ of $(L^p(\Omega))^M$. □

We next identify the dual space $(W_0^{m,p}(\Omega))'$ of the Sobolev space $W_0^{m,p}(\Omega)$. Recall that $\mathcal{D}'(\Omega)$ denotes the space of all distributions on Ω (Section 6.3). Note that, given a functional $\ell \in (W_0^{m,p}(\Omega))'$, the functions $f_{\boldsymbol{\alpha}} \in L^q(\Omega)$, $|\boldsymbol{\alpha}| \le m$, found in the next theorem, are again *not necessarily unique*.

Theorem 6.11-2 (dual space of $W_0^{m,p}(\Omega)$, $1 \le p < \infty$) *Let Ω be an open subset of $\mathbb{R}^N$, let $m \ge 1$ be an integer, let $1 \le p < \infty$, and let q denote the conjugate exponent of p.*

Then the dual space $(W_0^{m,p}(\Omega))'$ can be identified with the space of all distributions $T \in \mathcal{D}'(\Omega)$ that are of the form

$$T = \sum_{|\boldsymbol{\alpha}| \le m} (-1)^{|\boldsymbol{\alpha}|} \partial^{\boldsymbol{\alpha}} f_{\boldsymbol{\alpha}} \quad \textit{for some } f_{\boldsymbol{\alpha}} \in L^q(\Omega),\ |\boldsymbol{\alpha}| \le m.$$

Proof By the *Hahn–Banach theorem in a normed vector space* (Theorem 5.9-1), any continuous linear functional $\ell : W_0^{m,p}(\Omega) \to \mathbb{R}$ can be extended to a continuous linear functional $\widehat{\ell} : W^{m,p}(\Omega) \to \mathbb{R}$. By Theorem 6.11-1, there thus exist functions $f_{\boldsymbol{\alpha}} \in L^q(\Omega)$ such that

$$\widehat{\ell}(v) = \sum_{|\boldsymbol{\alpha}| \le m} \int_\Omega f_{\boldsymbol{\alpha}} \partial^{\boldsymbol{\alpha}} v \, dx \quad \text{for all } v \in W^{m,p}(\Omega),$$

and hence such that

$$\ell(v) = \sum_{|\alpha|\le m} \int_\Omega f_\alpha \partial^\alpha v \,\mathrm{d}x \quad \text{for all } v \in W_0^{m,p}(\Omega).$$

In particular then,

$$\ell(\varphi) = \Big(\sum_{|\alpha|\le m} (-1)^{|\alpha|}\partial^\alpha f_\alpha\Big)(\varphi) \quad \text{for all } \varphi \in \mathcal{D}(\Omega) \subset W_0^{m,p}(\Omega),$$

by definition of differentiation in the sense of distributions. This shows that the restriction of ℓ to the subspace $\mathcal{D}(\Omega)$ of $W_0^{m,p}(\Omega)$ is the distribution $T \in \mathcal{D}'(\Omega)$ defined by

$$T := \sum_{|\alpha|\le m} (-1)^{|\alpha|}\partial^\alpha f_\alpha.$$

Conversely, let a distribution $T \in \mathcal{D}'(\Omega)$ of this form be given, with functions $f_\alpha \in L^q(\Omega)$, $|\alpha| \le m$. Then T is a continuous linear functional over the space $\mathcal{D}(\Omega)$ equipped with the norm $\|\cdot\|_{m,p,\Omega}$, since

$$\begin{aligned} |T(\varphi)| &= \Big|\Big(\sum_{|\alpha|\le m} (-1)^{|\alpha|}\partial^\alpha f_\alpha\Big)(\varphi)\Big| \le \sum_{|\alpha|\le m} \Big|\int_\Omega f_\alpha \partial^\alpha \varphi \,\mathrm{d}x\Big| \\ &\le \|(f_\alpha)_{|\alpha|\le m}\|_{(L^q(\Omega))^M}\, \|\varphi\|_{m,p,\Omega} \quad \text{for all } \varphi \in \mathcal{D}(\Omega), \end{aligned}$$

where again $M := \mathrm{Card}\{\alpha;\ |\alpha| \le m\}$. Since $\mathcal{D}(\Omega)$ is by definition dense in the space $W_0^{m,p}(\Omega)$, the distribution T possesses a *unique* continuous linear extension to the space $W_0^{m,p}(\Omega)$. This shows that $(W_0^{m,p}(\Omega))'$ can be indeed identified with the space of all such distributions T. □

Let Ω be an open subset of $\mathbb{R}^N$. For each integer $m \ge 1$ and each real number $1 \le p < \infty$, the dual space identified in Theorem 6.11-2 is denoted

$$W^{-m,q}(\Omega) := (W_0^{m,p}(\Omega))',$$

where q denotes the conjugate exponent of p, or

$$H^{-m}(\Omega) := (H_0^m(\Omega))' \quad \text{if } p = 2.$$

For instance, given any integer $m \ge 1$, the distribution T_v associated with a function $v \in L^2(\Omega)$, i.e., that defined by

$$T_v(\varphi) = \int_\Omega v\varphi \,\mathrm{d}x \quad \text{for all } \varphi \in \mathcal{D}(\Omega)$$

(Section 6.3), can be identified with an element $T_v \in H^{-m}(\Omega)$. In this fashion, the space $L^2(\Omega)$ becomes imbedded in the space $H^{-m}(\Omega)$ by means of the *canonical injection from $L^2(\Omega)$ into $H^{-m}(\Omega)$*.

In the remainder of this section, we focus our attention on the space

$$H^{-1}(\Omega) := (H_0^1(\Omega))'$$

when the open set Ω is a *domain* in $\mathbb{R}^N$, as this space will play a key role at various places in this chapter. The first property is a *compactness property* analogous in spirit to that of the Rellich–Kondrachov theorem (Theorem 6.6-3), and thus justifying its name.

Theorem 6.11-3 (Rellich–Kondrachov compact imbedding theorem in $L^2(\Omega)$) *Let Ω be a domain in $\mathbb{R}^N$. Then the canonical injection from $L^2(\Omega)$ into $H^{-1}(\Omega)$ is compact.*

Proof The canonical injection

$$v \in L^2(\Omega) \to T_v \in H^{-1}(\Omega)$$

is nothing but the *dual operator* ι' (Section 5.11) of the canonical injection

$$\iota : H_0^1(\Omega) \to L^2(\Omega),$$

the space $L^2(\Omega)$ being identified here with its dual space. To see this, it suffices to verify that

$$T_v(w) = (v, \iota w) \quad \text{for all } v \in L^2(\Omega) \text{ and all } w \in H_0^1(\Omega),$$

where $(\cdot,\cdot)$ denotes the inner product of $L^2(\Omega)$; equivalently, it suffices to verify that

$$T_v(w) = \int_\Omega vw \, dx \quad \text{for all } v \in L^2(\Omega) \text{ and all } w \in H_0^1(\Omega).$$

But this last relation immediately follows from the definition of T_v and from the denseness of $\mathcal{D}(\Omega)$ in $H_0^1(\Omega)$.

Since ι is compact by the Rellich–Kondrachov imbedding theorem (Theorem 6.6-3), ι' is also compact by Theorem 5.11-2. $\square$

Let Ω be an open subset of $\mathbb{R}^N$. Since a function $v \in L^2(\Omega)$ can be identified with the distribution that it defines (as seen above), it is clear that

$$v \in L^2(\Omega) \text{ implies that } v \in H^{-1}(\Omega) \quad \text{and} \quad \partial_i v \in H^{-1}(\Omega), \ 1 \le i \le N,$$

since

$$|T_v(\varphi)| = \left| \int_\Omega v\varphi \, dx \right| \le \|v\|_{0,\Omega} \|\varphi\|_{1,\Omega} \quad \text{for all } \varphi \in \mathcal{D}(\Omega),$$

$$|\partial_i T_v(\varphi)| = |-T_v(\partial_i\varphi)| = \left| -\int_\Omega v\partial_i\varphi \, dx \right| \le \|v\|_{0,\Omega} \|\varphi\|_{1,\Omega} \quad \text{for all } \varphi \in \mathcal{D}(\Omega).$$

It is *remarkable*, but also *remarkably difficult to prove*, that, if Ω is a domain, the following converse implication holds (note that the assumption $v \in \mathcal{D}'(\Omega)$ is weaker than $v \in H^{-1}(\Omega)$):

$^\flat$**Theorem 6.11-4 (J.L. Lions lemma**[47,48]**)** *Let Ω be a domain in $\mathbb{R}^N$. Then*

$$v \in \mathcal{D}'(\Omega) \quad \textit{and} \quad \partial_i v \in H^{-1}(\Omega),\ 1 \le i \le N, \quad \textit{implies } v \in L^2(\Omega). \qquad \square$$

J.L. Lions lemma is of fundamental importance: As illustrated in the rest of this chapter, it is the key to proving many fundamental results, such as the existence of a solution to the weak formulation of the *Stokes equations* (Section 6.14), the *Korn inequality* (Section 6.15), the *weak Poincaré lemma* (Section 6.17), the *weak Saint-Venant lemma* (Section 6.18), or the *weak Donati lemma* (Section 6.19).

Note that, in *all* the applications that we shall make of J.L. Lions lemma, the distribution $v \in \mathcal{D}'(\Omega)$ such that $\partial_i v \in H^{-1}(\Omega)$, $1 \le i \le N$, belongs to a *strict* subspace of $\mathcal{D}'(\Omega)$, such as $H^{-1}(\Omega)$ or $L^1_{\mathrm{loc}}(\Omega)$.

Remark Although Theorem 6.11-4 shall be referred to as *"the" lemma of J.L. Lions* in this book, there are other results of his that bear the same name in the literature, such as his "compactness lemmas"[49] and "singular perturbation lemma."[50] $\square$

Finally, we mention a useful generalization[51] of J.L. Lions lemma.

$^\flat$**Theorem 6.11-5 (J.L. Lions lemma in $H^m(\Omega)$)** *Let Ω be a domain in $\mathbb{R}^N$, and let $m \in \mathbb{Z}$ be any integer. Then*

$$v \in \mathcal{D}'(\Omega) \quad \textit{and} \quad \partial_i v \in H^{m}(\Omega),\ 1 \le i \le N, \quad \textit{implies } v \in H^{m+1}(\Omega). \qquad \square$$

---

[47]That $v \in H^{-1}(\Omega)$ and $\partial_i v \in H^{-1}(\Omega)$, $1 \le i \le N$, imply $v \in L^2(\Omega)$ was first established, for domains with smooth boundaries, by Jacques-Louis Lions (1928–2001), as stated in footnote [22] of:

E. Magenes; G. Stampacchia [1958]: I problemi al contorno per le equazioni differenziali di tipo ellitico, *Annali della Scuola Normale Superiore di Pisa* **12**, 247–358.

Its first published proof by J.L. Lions appeared in Duvaut & Lions [1976]. Other proofs of this implication have since then been given, some extending it to genuine domains (i.e., with Lipschitz-continuous boundaries, as stated in Theorem 6.11-4), others extending it to the case where the space $H^{-1}(\Omega)$ is replaced by the more general space $W^{-1,q}(\Omega)$, $1 < q < \infty$. See:

L. Tartar [1978]: *Topics in Nonlinear Analysis*, Publications Mathématiques d'Orsay No. 78.13, Université de Paris-Sud, Orsay.

G. Geymonat; P. Suquet [1986]: Functional spaces for Norton-Hoff materials, *Mathematical Methods in the Applied Sciences* **8**, 206–222.

A counterexample to J.L. Lions lemma when Ω is not a domain is given in:

G. Geymonat; G. Gilardi [1998]: Contre-exemple à l'inégalité de Korn et au lemme de Lions dans des domaines irréguliers, in *Equations aux Dérivées Partielles et Applications. Articles Dédiés à Jacques-Louis Lions*, pp. 541–548, Gauthier-Villars, Paris.

[48]That the assumption $v \in H^{-1}(\Omega)$ may be replaced by the weaker assumption $v \in \mathcal{D}'(\Omega)$ has been established by:

W. Borchers; H. Sohr [1990]: On the equations $\operatorname{rot} v = g$ and $\operatorname{div} u = f$ with zero boundary conditions, *Hokkaido Mathematical Journal* **19**, 67–87.

[49]See Lions [1961, Chapter X, Proposition 4.1] or Lions [1969, Chapter X, Section 5.2].

[50]See Lions [1973, Chapter X, Lemma 5.1].

[51]Due to:

C. Amrouche; V. Girault [1994]: Decomposition of vector spaces and application to the Stokes problem in arbitrary dimension, *Czechoslovak Mathematical Journal* **44**, 109–140.

Problems

6.11-1 Under the assumptions and with the notations of Theorem 6.11-1, show that

$$\|\ell\|_{(W^{m,p}(\Omega))'} = \inf\Big\{\|(f^{\boldsymbol{\alpha}})\|_{(L^q(\Omega))^M} := \Big(\sum_{|\boldsymbol{\alpha}|\le m}\|f^{\boldsymbol{\alpha}}\|^q_{L^q(\Omega)}\Big)^{1/q};(f^{\boldsymbol{\alpha}})\in(L^q(\Omega))^M \text{ for each } |\boldsymbol{\alpha}|\le m,$$
$$\text{and } \sum_{|\boldsymbol{\alpha}|\le m}\int_\Omega f_{\boldsymbol{\alpha}}\partial^{\boldsymbol{\alpha}}v\,dx = \ell(v) \text{ for all } v\in W^{m,p}(\Omega)\Big\},$$

and that the above infimum is attained, by a single element in the product space $(L^q(\Omega))^M$.

6.11-2 Let $1 < p < \infty$, and let $(v_k)_{k=1}^\infty$ be a bounded sequence in the space $W^{m,p}(\Omega)$. Using Theorem 6.11-1 and the reflexivity of the space $L^p(\Omega)$ (Theorem 5.14-2), show that there exists a subsequence $(v_{\sigma(k)})_{k=1}^\infty$ that converges weakly in the space $W^{m,p}(\Omega)$.

Combined with part (b) of the *Banach–Eberlein–Šmulian theorem* (Theorem 5.14-4), this property thus provides another proof (i.e., different from that of Theorem 6.5-1) that *the spaces* $W^{m,p}(\Omega)$, $1 < p < \infty$, *are reflexive.*

6.12 The Babuška–Brezzi inf-sup theorem; application to constrained quadratic minimization problems

Our first application of *J.L. Lions lemma* will be to the existence of a weak solution to the *Stokes equations* (see the proof of Theorem 6.14-1). To this end, we first need to develop an abstract functional setting that models various basic linear problems arising in fluid and solid mechanics. This is the object of the present section.

The linear abstract variational problems studied so far are of the form "find $u \in V$ such that $a(u,v) = \ell(v)$ for all $v \in V$" (Theorem 6.2-1). We now consider another class of linear *abstract variational problems*, the definition of which requires a *second space* and a *second bilinear form* (respectively denoted M and b in the next theorem).

To begin with, we establish a fundamental existence and uniqueness result for such problems. The assumption made in the next theorem on the bilinear form b, viz., that there exists a constant β such that

$$\beta > 0 \quad \text{and} \quad \inf_{\substack{\mu\in M\\ \mu\neq 0}}\ \sup_{\substack{v\in V\\ v\neq 0}} \frac{|b(v,\mu)|}{\|v\|_V\,\|\mu\|_M} \ge \beta,$$

constitutes the **Babuška–Brezzi inf-sup condition.**[52]

[52]So named after:

I. Babuška [1971]: Error bound for finite element method, *Numerische Mathematik* **16**, 322–333.

F. Brezzi [1974]: On the existence, uniqueness and approximation of saddle point problems arising from Lagrange multipliers, *Revue Française d'Automatique, Informatique, et Recherche Opérationnelle – Série Rouge* **8**, 129–151.

In these two papers, the "Babuška-Brezzi" condition is in effect stated in two different, but equivalent, forms (the statement of Theorem 6.12-1 is from Brezzi [1974]). Their equivalence is established in, e.g.:

L. Demkowicz [2000]: Babuška ⇔ Brezzi??, Technical Report, Texas Institute for Computational and Applied Mathematics, TICAM Seminar (October 31, 2000).

But Franco Brezzi is to be credited for establishing in addition (again, in Brezzi [1974]) the *necessity* of the inf-sup condition; see Problem 6.12-1. (Footnote continued on next page.)

Remark The next theorem contains the Lax–Milgram theorem as a special case ($M = \{0\}$ and $b = 0$). □

Theorem 6.12-1 (Babuška–Brezzi inf-sup theorem) *Let V and M be two Hilbert spaces, and let $a(\cdot,\cdot) : V \times V \to \mathbb{R}$ and $b : V \times M \to \mathbb{R}$ be two continuous bilinear forms with the following properties: There exists a constant α such that*

$$\alpha > 0 \quad \text{and} \quad a(v,v) \geq \alpha \, \|v\|_V^2 \ \text{ for all } v \in U_0 := \{v \in V;\ b(v,\mu) = 0 \text{ for all } \mu \in M\},$$

i.e., $a(\cdot,\cdot)$ is U_0-coercive, and there exists a constant β such that

$$\beta > 0 \quad \text{and} \quad \inf_{\begin{cases}\mu \in M \\ \mu \neq 0\end{cases}} \sup_{\begin{cases}v \in V \\ v \neq 0\end{cases}} \frac{|b(v,\mu)|}{\|v\|_V \, \|\mu\|_M} \geq \beta.$$

Finally, let $\ell : V \to \mathbb{R}$ and $\chi : M \to \mathbb{R}$ be two continuous linear forms.

Then the variational problem: Find $(u,\lambda) \in V \times M$ such that

$$\begin{aligned} a(u,v) + b(v,\lambda) &= \ell(v) \quad \text{for all } v \in V, \\ b(u,\mu) &= \chi(\mu) \quad \text{for all } \mu \in M, \end{aligned}$$

has one and only one solution, and the linear operator $(\ell,\chi) \in V' \times M' \to (u,\lambda) \in V \times M$ defined in this fashion is continuous.

Proof For each $u \in V$, the linear form $v \in V \to a(u,v) \in \mathbb{R}$ is continuous. Therefore, there exists a unique element $Au \in V'$ such that

$$a(u,v) = Au(v) \quad \text{for all } (u,v) \in V \times V.$$

Besides,

$$\|Au\|_{V'} = \sup_{v \neq 0} \frac{|Au(v)|}{\|v\|_V} = \sup_{v \neq 0} \frac{|a(u,v)|}{\|v\|_V} \leq \|a\|_{\mathcal{L}_2(V;\mathbb{R})} \, \|u\| \quad \text{for all } u \in V,$$

so that the mapping $A : V \to V'$ defined in this fashion, which is clearly linear, is continuous with $\|A\|_{\mathcal{L}(V;V')} \leq \|a\|_{\mathcal{L}_2(V;\mathbb{R})}$. The same argument shows that there exist a mapping $B \in \mathcal{L}(V;M')$ and, consequently (Section 5.11), a dual operator $B' \in \mathcal{L}(M;V')$, such that

$$b(v,\mu) = Bv(\mu) = B'\mu(v) \quad \text{for all } (v,\mu) \in V \times M.$$

This condition already appeared, *albeit* only in the treatment of a specific example (i.e., not in an "abstract" form as in Theorem 6.12-1), in:

O.A. LADYZHENSKAYA [1969]: *The Mathematical Theory of Viscous Flows, Second Edition*, Gordon and Breach, New York.

For this reason, it is also sometimes referred to as the *Ladyzhenskaya-Babuška-Brezzi condition.* In fact, this result, in the form stated in BABUŠKA [1971], is already proved in Theorem 3.1 of:

J. NEČAS [1962]: Sur une méthode pour résoudre les équations aux dérivées partielles du type elliptique, voisine de la variationnelle, *Annali della Scuola Normale Superiore di Pisa, Classe di Scienze, Serie* III, **16**, 305–326.

Ivo Babuška and Franco Brezzi were the first to show that this type of result is also the key to fundamental error estimates for *finite element approximations* of such variational problems.

Solving the abstract variational problem of Theorem 6.12-1 thus amounts to finding a pair $(u, \lambda) \in V \times M$ that satisfies the following system of operator equations:

$$\begin{aligned} Au + B'\lambda &= \ell \quad \text{in } V', \\ Bu &= \chi \quad \text{in } M'. \end{aligned}$$

Expressed in terms of the dual operator B', *the Babuška–Brezzi inf-sup condition is equivalent to*

$$\|B'\mu\|_{V'} = \sup_{v \neq 0} \frac{|B'\mu(v)|}{\|v\|_V} = \sup_{v \neq 0} \frac{|b(v,\mu)|}{\|v\|_V} \geq \beta \|\mu\|_M \quad \text{for all } \mu \in M,$$

a relation that is precisely one of the three equivalent conditions (applied here to the operator $B \in \mathcal{L}(V; M')$) appearing in the *Banach closed range theorem* (second part; cf. Theorem 5.11-6).

We thus infer from this theorem that *the operator* $B : V \to M'$ *is surjective, the operator* $B' : M \to V'$ *is injective*, and *the space* $\operatorname{Im} B'$ is *closed* in V'.

Since B is surjective and $\chi \in M'$, there exists $u_0 \in V$ such that

$$Bu_0 = \chi.$$

Since

$$\begin{aligned} U_0 &= \{v \in V;\ b(v,\mu) = 0 \text{ for all } \mu \in M\} \\ &= \{v \in V;\ Bv(\mu) = 0 \text{ for all } \mu \in M\} \\ &= \{v \in V;\ Bv = 0 \text{ in } M'\} = \operatorname{Ker} B, \end{aligned}$$

the bilinear form $a(\cdot,\cdot)$ is in effect $\operatorname{Ker} B$-coercive. Consequently, there exists a unique $u_1 \in \operatorname{Ker} B$ such that

$$a(u_1, v) = \ell(v) - a(u_0, v) \quad \text{for all } v \in \operatorname{Ker} B$$

by the Lax–Milgram theorem (Theorem 6.2-1; the linear form $v \in V \to \ell(v) - a(u_0, v)$ is clearly continuous). The element

$$u := (u_0 + u_1) \in V$$

therefore satisfies

$$(\sigma(Au - \ell), v)_V = (Au - \ell)(v) = a(u, v) - \ell(v) = 0 \quad \text{for all } v \in \operatorname{Ker} B,$$

where $(\cdot,\cdot)_V$ and $\sigma : V' \to V$ respectively denote the inner product and the F. Riesz isometry of the space V. In other words,

$$\sigma(Au - \ell) \in (\operatorname{Ker} B)^{\perp},$$

where $(\operatorname{Ker} B)^{\perp}$ denotes the orthogonal complement of $\operatorname{Ker} B$ in the Hilbert space $(V, (\cdot,\cdot)_V)$.

Let $(\cdot,\cdot)_M$ and $\tau : M' \to M$ respectively denote the inner product and the F. Riesz isometry of the space M. Then

$$(\tau Bv, \mu)_M = Bv(\mu) = B'\mu(v) = (\sigma B'\mu, v)_V \quad \text{for all } v \in V \text{ and all } \mu \in M,$$

which shows that $\sigma B'$ is the *adjoint operator of* τB *in the Hilbert space sense*, i.e., as defined in Theorem 4.7-2(a).

Hence, by part (b) of the same theorem,

$$\operatorname{Ker}\tau B \oplus \overline{\operatorname{Im}\sigma B'} = V.$$

But $\operatorname{Im}\sigma B'$ is closed in V since $\operatorname{Im} B'$ is closed in V', and $\sigma : V' \to V$ is an isometry; hence

$$\operatorname{Im}\sigma B' = (\operatorname{Ker}\tau B)^{\perp} = (\operatorname{Ker} B)^{\perp}.$$

Since $\sigma(Au - \ell) \in (\operatorname{Ker} B)^{\perp}$, there thus exists $\lambda \in M$ such that $-\sigma B'\lambda = \sigma(Au - \ell)$, or equivalently, such that

$$Au + B'\lambda = \ell,$$

on the one hand. On the other hand, since $u_1 \in \operatorname{Ker} B$,

$$Bu = B(u_0 + u_1) = Bu_0 = \chi,$$

and thus the *existence* of a solution $(u, \lambda) \in V \times M$ to the variational problem of Theorem 6.12-1 is established.

Since this variational problem is linear, establishing the *uniqueness* of the solution amounts to showing that, if $(u, \lambda) \in V \times M$ satisfies

$$\begin{aligned} a(u,v) + b(v,\lambda) &= 0 \quad \text{for all } v \in V, \\ b(u,\mu) &= 0 \quad \text{for all } \mu \in M, \end{aligned}$$

then $(u, \lambda) = (0, 0)$. Letting $v = u$ in the first equations gives

$$a(u,u) + b(u,\lambda) = a(u,u) = 0,$$

since $b(u, \lambda) = 0$. Hence $u = 0$, because the second variational equations mean that $u \in \operatorname{Ker} B$ and the bilinear form $a(\cdot,\cdot)$ is $\operatorname{Ker} B$-coercive by assumption. The first equations then reduce to $b(v, \lambda) = 0$ for all $v \in V$, or equivalently, to

$$Bv(\lambda) = B'\lambda(v) = 0 \quad \text{for all } v \in V,$$

which shows that $B'\lambda = 0$. But B' is injective; hence $\lambda = 0$.

The mapping

$$\mathcal{A} : (v,\mu) \in V \times M \to \mathcal{A}(v,\mu) := (Av + B'\mu, Bv) \in V' \times M',$$

which is clearly continuous, is therefore bijective. The *continuity* of the inverse mapping $\mathcal{A}^{-1} : V' \times M' \to V \times M$ therefore follows from the *Banach open mapping theorem* (Theorem 5.6-1). □

Remarks (1) The Banach closed range theorem shows that the Babuška–Brezzi inf-sup condition holds if and only if the mapping $B \in \mathcal{L}(V; M')$ is surjective, or if and only if the mapping $B' \in \mathcal{L}(M; V')$ is injective and has a closed range in V'.

(2) The Babuška–Brezzi inf-sup condition is also *necessary* for the existence of a solution to the variational problem of Theorem 6.12-1, which means in particular that the equation $Bu = \chi$ in M must have a solution $u \in V$ for any $\chi \in M'$ (as shown in the above proof), i.e., that the mapping

$B \in \mathcal{L}(V; M')$ must be surjective; in this direction, see Problem 6.12-1, where the necessity of the inf-sup condition is established in general. □

Under the additional assumption that the bilinear form $a(\cdot,\cdot)$ is *symmetric*, we next show that Theorem 6.12-1 provides an interesting way to solve *specific quadratic minimization problems* of the form considered in Section 6.1, when the nonempty closed convex subset of a Hilbert space V over which a quadratic functional of the form

$$J : v \in V \to J(v) := \frac{1}{2}a(v,v) - \ell(v)$$

is to be minimized can be written as

$$U_\chi := \{v \in V;\ b(v,\mu) = \chi(\mu) \text{ for all } \mu \in M\},$$

where M is another Hilbert space and $b : V \times M \to \mathbb{R}$ and $\chi : M \to \mathbb{R}$ are continuous bilinear and linear forms that satisfy the assumptions of the Babuška–Brezzi inf-sup theorem.

Such a minimization problem provides an example of a **constrained quadratic minimization problem**, in the sense that any minimizer u (if it exists) should satisfy the **constraint**

$$b(u,\mu) = \chi(\mu) \quad \text{for all } \mu \in M.$$

Theorem 6.12-2 *Let the assumptions on the spaces V and M, on the bilinear forms $a(\cdot,\cdot) : V \times V \to \mathbb{R}$ and $b(\cdot,\cdot) : V \times M \to \mathbb{R}$, and on the linear forms $\ell : V \to \mathbb{R}$ and $\chi : M \to \mathbb{R}$ be as in* Theorem 6.12-1. *Assume in addition that the bilinear form $a(\cdot,\cdot)$ is symmetric.*

Then $(u,\lambda) \in V \times M$ is the unique solution of the variational problem of Theorem 6.12-1, *viz.,*

$$\begin{aligned} a(u,v) + b(v,\lambda) &= \ell(v) \quad \textit{for all } v \in V, \\ b(u,\mu) &= \chi(\mu) \quad \textit{for all } \mu \in M, \end{aligned}$$

if and only if u is the unique solution of the constrained quadratic minimization problem

$$u \in U_\chi \quad \textit{and} \quad J(u) = \inf_{v \in U_\chi} J(v),$$

where the subset U_χ of the space V and the functional $J : V \to \mathbb{R}$ are respectively defined by

$$\begin{aligned} U_\chi &:= \{v \in V;\ b(v,\mu) = \chi(\mu) \quad \textit{for all } \mu \in M\}, \\ J(v) &:= \frac{1}{2}a(v,v) - \ell(v) \quad \textit{for each } v \in V. \end{aligned}$$

Proof Let $(u,\lambda) \in V \times M$ be the unique solution of the variational problem of Theorem 6.12-1. The second variational equations then show that $u \in U_\chi$.

The symmetry of the bilinear form $a(\cdot,\cdot)$ (this assumption is essential here) implies that

$$J(u+w) - J(u) = (a(u,w) - \ell(w)) + \frac{1}{2}a(w,w) \quad \text{for all } w \in V.$$

But the first variational equations imply that

$$a(u,w) - \ell(w) = -b(w,\lambda) = 0 \quad \text{for all } w \in U_0 := \{v \in V; b(v,\mu) = 0 \text{ for all } \mu \in M\}.$$

Hence

$$J(u+w) - J(u) = \frac{1}{2}a(w,w) \geq \frac{\alpha}{2}\|w\|^2 > 0 \quad \text{for all } w \in U_0,\ w \neq 0$$

(since $a(\cdot,\cdot)$ is U_0-elliptic), thus showing that $J(u) = \inf_{v\in U_\chi} J(v)$ and that $u \in U_\chi$ is the unique solution of this constrained quadratic minimization problem.

Conversely, assume that $\widetilde{u} \in U_\chi$ satisfies $J(\widetilde{u}) = \inf_{v\in U_\chi} J(v)$, and let $(u,\lambda) \in V \times M$ be the unique solution to the variational problem of Theorem 6.12-1. Then the above argument shows that $u \in U_\chi$ and $J(u) = \inf_{v\in U_\chi} J(v)$. Hence $\widetilde{u} = u$ since the solution of this minimization problem is unique (we saw above that $J(v) > J(u)$ if $v \in U_\chi$ and $v \neq u$). □

Theorem 6.12-2 thus allows us to find the solution of a specific constrained problem by means of the solution of an unconstrained one. This is perhaps best appreciated by considering the special case where the linear form χ vanishes since, in this case, the following *constrained* variational problem (of the form considered in Section 6.1): Find $u \in U_0 := \{v \in V;\ b(v,\mu) = 0 \text{ for all } \mu \in M\}$ such that

$$a(u,v) = \ell(v) \quad \text{for all } v \in U_0$$

(Theorem 6.1-1) is replaced by the *unconstrained* variational problem: Find $(u,\lambda) \in V \times M$ such that

$$\begin{aligned} a(u,v) + b(v,\lambda) &= \ell(v) \quad \text{for all } v \in V, \\ b(u,\mu) &= 0 \qquad \text{for all } \mu \in M. \end{aligned}$$

Here, "unconstrained" reflects that the variational equations $a(u,v)+b(v,\lambda) = \ell(v)$ are to be satisfied for all v in the whole space V, while the equations $a(u,v) = \ell(v)$ are to be satisfied for all v in the subspace U_0 of V defined by means of the *constraints* $b(v,\mu) = 0$ for all $\mu \in M$.

A first application of both Theorems 6.12-1 and 6.12-2 (to a constrained quadratic minimization problem in $\mathbb{R}^n$) is given in Problem 6.12-2. Other applications are found in the next two sections.

Remark We will show that, under the additional assumption that $a(v,v) \geq 0$ for all $v \in V$, the pair (u,λ) found in Theorem 6.12-2 is a *saddle-point* of an *ad-hoc Lagrangian* $\mathcal{L} : V \times M \to \mathbb{R}$, the second argument $\lambda \in M$ being then the *Lagrange multiplier* associated with the *constraint* $b(u,\mu) = \chi(\mu)$ for all $\mu \in M$ (Section 7.16). □

Problems

6.12-1 Let V and M be two Hilbert spaces and let $a(\cdot,\cdot) : V \times V \to \mathbb{R}$ and $b : V \times M \to \mathbb{R}$ be two continuous bilinear forms.

(1) Assume that, given any two continuous linear forms $\ell : V \to \mathbb{R}$ and $\chi : M \to \mathbb{R}$, there exists one and only one pair $(u,\lambda) \in V \times M$ that satisfies the variational equations

$$a(u,v) + b(v,\lambda) = \ell(v) \ \text{ for all } v \in V, \quad \text{and} \quad b(u,\mu) = \chi(\mu) \ \text{ for all } \mu \in M.$$

Show that the operators $A \in \mathcal{L}(V;V')$ and $B \in \mathcal{L}(V;M')$ defined as in the proof of Theorem 6.12-1 *necessarily* have the following two properties:

First, let the operator $\rho \in \mathcal{L}(V';(\operatorname{Ker} B)')$ be defined by $(\rho v')(v) = v'(v)$ for all $v' \in V'$ and all $v \in \operatorname{Ker} B$; then *the restriction of the operator* $\rho A \in \mathcal{L}(V;(\operatorname{Ker} B)')$ *to* $\operatorname{Ker} B$ *is a bijection with a continuous inverse* (the assumption made in Theorem 6.12-1, viz., that the bilinear form $a(\cdot,\cdot)$ is $\operatorname{Ker} B$-coercive, clearly implies that this property holds).

Second, the inf-sup condition is satisfied (as shown in the proof of Theorem 6.12-1, this is in effect an assumption on the dual operator B' of the operator B).

(2) Assume that, *conversely*, the operators $A \in \mathcal{L}(V; V')$ and $B \in \mathcal{L}(V; M')$ are such that the two properties above are satisfied. Then show that, given any two continuous linear forms $\ell : V \to \mathbb{R}$ and $\chi : M \to \mathbb{R}$, there exists one and only one solution $(u, \lambda) \in V \times M$ to the variational equations of question (1).[53]

Remark Question (2) contains Theorem 6.12-2 as a special case. □

6.12-2 Let $\boldsymbol{A}$ be a real $n \times n$ symmetric matrix and let $\boldsymbol{B}$ be a real $m \times n$ matrix of rank m (hence $m \leq n$) with the property that there exists $\alpha > 0$ such that $\boldsymbol{v}^T\boldsymbol{A}\boldsymbol{v} \geq \alpha \boldsymbol{v}^T\boldsymbol{v}$ for all $\boldsymbol{v} \in \operatorname{Ker} \boldsymbol{B}$ (thus a weaker property than the positive-definiteness of $\boldsymbol{A}$).

Show that, given any vectors $\boldsymbol{c} \in \mathbb{R}^n$ and $\boldsymbol{d} \in \mathbb{R}^m$, the linear system (the matrix of which is symmetric, of order $n + m$)

$$\begin{aligned} \boldsymbol{A}\boldsymbol{u} + \boldsymbol{B}^T\boldsymbol{\lambda} &= \boldsymbol{c} \\ \boldsymbol{B}\boldsymbol{u} \qquad &= \boldsymbol{d} \end{aligned}$$

has a unique solution $(\boldsymbol{u}, \boldsymbol{\lambda}) \in \mathbb{R}^n \times \mathbb{R}^m$ and that $\boldsymbol{u}$ is the unique solution of the following *constrained quadratic minimization problem in* $\mathbb{R}^n$: Find

$$\boldsymbol{u} \in \boldsymbol{U_d} = \{\boldsymbol{v} \in \mathbb{R}^n;\ \boldsymbol{B}\boldsymbol{v} = \boldsymbol{d}\}$$

such that

$$J(\boldsymbol{u}) = \inf_{\boldsymbol{v} \in \boldsymbol{U_d}} J(\boldsymbol{v}), \quad \text{where} \quad J(\boldsymbol{v}) := \frac{1}{2}\boldsymbol{v}^T\boldsymbol{A}\boldsymbol{v} - \boldsymbol{c}^T\boldsymbol{v} \text{ for all } \boldsymbol{v} \in \mathbb{R}^n.$$

Remark If $\boldsymbol{B} = \boldsymbol{0}$ and $\boldsymbol{d} = \boldsymbol{0}$, in which case $\boldsymbol{A}$ is positive-definite and $\boldsymbol{U_d} = \mathbb{R}^n$, the solution $\boldsymbol{u} \in \mathbb{R}^n$ to the above minimization problem is also the solution to the *linear system* $\boldsymbol{A}\boldsymbol{u} = \boldsymbol{c}$ of order n. It is thus remarkable that, in the more general situation considered here, $\boldsymbol{u}$ can still be found by solving again a *linear system*, this time of order $n + m$. As we shall see later (Section 7.15), the *auxiliary unknown* $\boldsymbol{\lambda} \in \mathbb{R}^m$ that appears in this linear system is in effect the *Lagrange multiplier* associated with the *constraint* $\boldsymbol{B}\boldsymbol{v} = \boldsymbol{d}$. □

6.13 Application of the Babuška–Brezzi inf-sup theorem: Primal, mixed, and dual formulations of variational problems

Note that the "*primal formulation*" and "*dual formulations*" defined in this section are to be carefully distinguished from the "*primal problem*" and "*dual problem*" that will be defined in Section 7.16.

In this section, we illustrate the usefulness of Theorems 6.12-1 and 6.12-2, by means of the following *model problem*, which corresponds to a *homogeneous Dirichlet problem for* $-\Delta$ (Section 6.7): Find

$$u \in H_0^1(\Omega) \quad \text{such that} \quad \int_\Omega \sum_{i=1}^N \partial_i u \partial_i v \, dx = \int_\Omega f v \, dx \ \text{ for all } v \in H_0^1(\Omega),$$

[53]The result of this problem is due to Brezzi [1974] *(op. cit.)*.

where Ω is a domain in $\mathbb{R}^N$ and $f \in L^2(\Omega)$ is a given function. As shown earlier (Theorem 6.7-2), this problem has a unique solution, which is also the unique solution of the following quadratic minimization problem: Find $u \in H_0^1(\Omega)$ such that

$$J(u) = \inf_{v \in H_0^1(\Omega)} J(v), \quad \text{where} \quad J(v) := \frac{1}{2}\int_\Omega \sum_{i=1}^N |\partial_i v|^2 \,\mathrm{d}x - \int_\Omega fv \,\mathrm{d}x.$$

In this section, this minimization problem will be regarded as the **primal formulation** (of the model problem).

However, in some applications, it turns out that it is the vector field

$$\mathbf{grad}\, u := (\partial_i u)_{i=1}^N \in \boldsymbol{L}^2(\Omega) := L^2(\Omega; \mathbb{R}^N)$$

that is the unknown of interest, rather than the function u itself. So the question naturally arises as to whether, like the function u, the vector field $\mathbf{grad}\, u$ could be directly characterized as the solution of an *ad hoc* minimization problem. As illustrated in the next theorems, to provide an affirmative answer to this question involves two stages:

First, one constructs a variational problem of the form considered in Theorem 6.12-1 with *both* u and $\mathbf{grad}\, u$ as unknowns; this problem constitutes a **mixed variational formulation** (of the model problem).

Second, Theorem 6.12-2 provides a constrained quadratic minimization problem with $\mathbf{grad}\, u$ as the sole unknown, which constitutes a **dual formulation** (of the model problem).

In what follows, $\boldsymbol{a} \cdot \boldsymbol{b}$ denotes the Euclidean inner product of two vectors $\boldsymbol{a}, \boldsymbol{b} \in \mathbb{R}^N$, $|\boldsymbol{a}| := \sqrt{\boldsymbol{a} \cdot \boldsymbol{a}}$ denotes the Euclidean norm of a vector $\boldsymbol{a} \in \mathbb{R}^N$ (as usual), and $\|\cdot\|_{0,\Omega}$ denotes the product norm in the space $\boldsymbol{L}^2(\Omega)$ defined by

$$\|\boldsymbol{p}\|_{0,\Omega} := \left(\sum_{i=1}^N \|p_i\|_{0,\Omega}^2 \right)^{1/2} \quad \text{for each } \boldsymbol{p} = (p_i)_{i=1}^N \in \boldsymbol{L}^2(\Omega).$$

The next two theorems provide *two different mixed*, and *dual, formulations* (see parts (b) and (c) in Theorems 6.13-1 and 6.13-2) of the *same model problem* (whose primal formulation is for convenience recalled in part (a) of the same theorems).

Theorem 6.13-1 (a first instance of mixed and dual formulation of the homogeneous Dirichlet problem for $-\Delta$) *Let Ω be a domain in $\mathbb{R}^N$ and let a function $f \in L^2(\Omega)$ be given. Then:*

(a) *There exists a unique solution $u \in H_0^1(\Omega)$ to the quadratic minimization problem*

$$J(u) = \inf_{v \in H_0^1(\Omega)} J(v), \quad \textit{where } J(v) := \frac{1}{2}\int_\Omega |\mathbf{grad}\, v|^2 \,\mathrm{d}x - \int_\Omega fv \,\mathrm{d}x \textit{ for each } v \in H_0^1(\Omega).$$

(b) *There exists a unique pair $(\boldsymbol{p}, \lambda) \in \boldsymbol{L}^2(\Omega) \times H_0^1(\Omega)$ that satisfies the variational problem*

$$\int_\Omega \boldsymbol{p} \cdot \boldsymbol{q} \,\mathrm{d}x - \int_\Omega \boldsymbol{q} \cdot \mathbf{grad}\, \lambda \,\mathrm{d}x = 0 \quad \textit{for all } \boldsymbol{q} \in \boldsymbol{L}^2(\Omega),$$

$$\int_\Omega \boldsymbol{p} \cdot \mathbf{grad}\, \mu \,\mathrm{d}x = \int_\Omega f\mu \,\mathrm{d}x \quad \textit{for all } \mu \in H_0^1(\Omega).$$

Besides,

$$\boldsymbol{p} = \mathbf{grad}\, u \quad \text{and} \quad \lambda = u,$$

where $u \in H_0^1(\Omega)$ *is the solution to the minimization problem of* (a).

(c) *The vector field* $\boldsymbol{p} = \mathbf{grad}\, u$ *is the unique solution to the constrained quadratic minimization problem*

$$\boldsymbol{p} \in \boldsymbol{U}_f := \left\{ \boldsymbol{q} \in \boldsymbol{L}^2(\Omega);\ \int_\Omega \boldsymbol{q} \cdot \mathbf{grad}\, \mu \,\mathrm{d}x = \int_\Omega f \mu \,\mathrm{d}x \text{ for all } \mu \in H_0^1(\Omega) \right\},$$

$$I(\boldsymbol{p}) = \inf_{\boldsymbol{q} \in \boldsymbol{U}_f} I(\boldsymbol{q}), \quad \text{where } I(\boldsymbol{q}) := \frac{1}{2} \int_\Omega |\boldsymbol{q}|^2 \,\mathrm{d}x \text{ for each } \boldsymbol{q} \in \boldsymbol{L}^2(\Omega).$$

Proof Let the bilinear forms $a(\cdot,\cdot) : \boldsymbol{L}^2(\Omega) \times \boldsymbol{L}^2(\Omega) \to \mathbb{R}$ and $b(\cdot,\cdot) : \boldsymbol{L}^2(\Omega) \times H_0^1(\Omega) \to \mathbb{R}$, and the linear forms $\ell : \boldsymbol{L}^2(\Omega) \to \mathbb{R}$ and $\chi : H_0^1(\Omega) \to \mathbb{R}$, be respectively defined by

$$a(\boldsymbol{p}, \boldsymbol{q}) := \int_\Omega \boldsymbol{p} \cdot \boldsymbol{q} \,\mathrm{d}x \quad \text{for each } \boldsymbol{p}, \boldsymbol{q} \in \boldsymbol{L}^2(\Omega),$$

$$b(\boldsymbol{q}, \mu) := -\int_\Omega \boldsymbol{q} \cdot \mathbf{grad}\, \mu \,\mathrm{d}x \quad \text{for each } (\boldsymbol{q}, \mu) \in \boldsymbol{L}^2(\Omega) \times H_0^1(\Omega),$$

$$\ell := 0 \quad \text{and} \quad \chi(\mu) := -\int_\Omega f \mu \,\mathrm{d}x \quad \text{for each } \mu \in H_0^1(\Omega),$$

and let the space $H_0^1(\Omega)$ be equipped with the norm $|\cdot|_{1,\Omega}$ (Theorem 6.5-2).

For any $\mu \in H_0^1(\Omega)$, the vector field $\boldsymbol{q}_\mu := \mathbf{grad}\, \mu$ belongs to the space $\boldsymbol{L}^2(\Omega)$, and $\|\boldsymbol{q}_\mu\|_{0,\Omega} = |\mu|_{1,\Omega}$. Consequently, for each nonzero $\mu \in H_0^1(\Omega)$,

$$\sup_{\left\{\substack{\boldsymbol{q} \in \boldsymbol{L}^2(\Omega) \\ \boldsymbol{q} \neq 0}\right.} \frac{\left|\int_\Omega \boldsymbol{q} \cdot \mathbf{grad}\, \mu \,\mathrm{d}x\right|}{\|\boldsymbol{q}\|_{0,\Omega}} \geq \frac{\left|\int_\Omega \boldsymbol{q}_\mu \cdot \mathbf{grad}\, \mu \,\mathrm{d}x\right|}{\|\boldsymbol{q}_\mu\|_{0,\Omega}} = |\mu|_{1,\Omega},$$

which shows that the Babuška–Brezzi inf-sup condition of Theorem 6.12-1 holds, with

$$V := \boldsymbol{L}^2(\Omega) \quad \text{and} \quad M := H_0^1(\Omega).$$

All the other assumptions of Theorem 6.12-1 are clearly satisfied. Hence the variational problem of (b) has a unique solution $(\boldsymbol{p}, \lambda) \in \boldsymbol{L}^2(\Omega) \times H_0^1(\Omega)$.

The first equations in the variational problem of (b) are clearly satisfied with $\boldsymbol{p} = \mathbf{grad}\, u$ and $\lambda = u$. The second equations in the same problem are likewise satisfied with $\boldsymbol{p} = \mathbf{grad}\, u$ since the unique solution $u \in H_0^1(\Omega)$ of the minimization problem of (a) is also a solution to the variational equations

$$\int_\Omega \mathbf{grad}\, u \cdot \mathbf{grad}\, \mu \,\mathrm{d}x = \int_\Omega f \mu \,\mathrm{d}x \quad \text{for all } \mu \in H_0^1(\Omega).$$

Hence (b) is proved.

Finally, (c) follows from Theorem 6.12-2, which can be applied since the bilinear form $a(\cdot,\cdot)$ is symmetric. □

Remark Since the space $\mathcal{D}(\Omega)$ is dense in the space $H_0^1(\Omega)$ and, for each $\boldsymbol{q} \in \boldsymbol{L}^2(\Omega)$, the linear form $\mu \in H_0^1(\Omega) \to \int_\Omega \boldsymbol{q}\cdot\mathbf{grad}\,\mu\,\mathrm{d}x - \int_\Omega f\mu\,\mathrm{d}x$ is continuous, the set $\boldsymbol{U}_f$ appearing in Theorem 6.13-1(c) consists in effect of all the vector fields $\boldsymbol{q} \in \boldsymbol{L}^2(\Omega)$ that satisfy the partial differential equation

$$\operatorname{div}\boldsymbol{q} + f = 0 \quad \text{in } \mathcal{D}'(\Omega),$$

i.e., *in the sense of distributions* (Section 6.3). □

As a preparation for the next theorem, we first need to define a space of vector fields: Given any open subset Ω of $\mathbb{R}^N$, we let

$$\boldsymbol{H}(\mathrm{div};\Omega) := \{\boldsymbol{q} \in \boldsymbol{L}^2(\Omega);\ \operatorname{div}\boldsymbol{q} \in L^2(\Omega)\},$$

where $\operatorname{div}\boldsymbol{q} := \sum_{i=1}^N \partial_i q_i$ for each $\boldsymbol{q} = (q_i)_{i=1}^N \in \boldsymbol{L}^2(\Omega) = L^2(\Omega;\mathbb{R}^N)$. Like the relations $\partial_i v \in L^2(\Omega)$, $1 \le i \le N$, found in the definition of the Sobolev space $H^1(\Omega)$ (Section 6.5), the relation "$\operatorname{div}\boldsymbol{q} \in L^2(\Omega)$" is to be understood as holding *in the sense of distributions.* This means that a vector field $\boldsymbol{q} \in \boldsymbol{L}^2(\Omega)$ belongs to the space $\boldsymbol{H}(\mathrm{div};\Omega)$ if and only if there exists a (uniquely defined) function in $L^2(\Omega)$, *denoted* $\operatorname{div}\boldsymbol{q}$, that satisfies

$$\int_\Omega (\operatorname{div}\boldsymbol{q})\varphi\,\mathrm{d}x = -\int_\Omega \boldsymbol{q}\cdot\mathbf{grad}\,\varphi\,\mathrm{d}x \quad \text{for all } \varphi \in \mathcal{D}(\Omega).$$

It is then easily verified (by means of a proof analogous to that of Theorem 6.5-1) that, *equipped with the norm* $\|\cdot\|_{\boldsymbol{H}(\mathrm{div};\Omega)}$ *defined by*

$$\|\boldsymbol{q}\|_{\boldsymbol{H}(\mathrm{div};\Omega)} := \left(\|\boldsymbol{q}\|_{0,\Omega}^2 + \|\operatorname{div}\boldsymbol{q}\|_{0,\Omega}^2\right)^{1/2} \quad \text{for each } \boldsymbol{q} \in \boldsymbol{H}(\mathrm{div};\Omega),$$

the space $\boldsymbol{H}(\mathrm{div};\Omega)$ *is a Hilbert space.*[54]

Note that, while the bilinear form denoted $a(\cdot,\cdot)$ in the proof of Theorem 6.13-1 is clearly coercive over the *whole* space $\boldsymbol{L}^2(\Omega)$, i.e., not only over the subspace $\{\boldsymbol{q} \in \boldsymbol{L}^2(\Omega);\ \int_\Omega \boldsymbol{q}\cdot\mathbf{grad}\,\mu\,\mathrm{d}x = 0 \text{ for all } \mu \in H_0^1(\Omega)\}$ of $L^2(\Omega)$, the bilinear form denoted $a(\cdot,\cdot)$ in the next proof is coercive only over the *proper* subspace $\{\boldsymbol{q} \in \boldsymbol{H}(\mathrm{div};\Omega);\ \operatorname{div}\boldsymbol{q} = 0 \text{ in } \Omega\}$ of the space $\boldsymbol{H}(\mathrm{div};\Omega)$.

Theorem 6.13-2 (a second instance of mixed and dual formulations of the homogeneous Dirichlet problem for $-\Delta$) *Let Ω be a domain in $\mathbb{R}^N$ and let a function $f \in L^2(\Omega)$ be given. Then:*

(a) *There exists a unique solution $u \in H_0^1(\Omega)$ to the quadratic minimization problem*

$$J(u) = \inf_{v\in H_0^1(\Omega)} J(v), \quad \textit{where } J(v) := \frac{1}{2}\int_\Omega |\mathbf{grad}\,v|^2\,\mathrm{d}x - \int_\Omega fv\,\mathrm{d}x \ \textit{for each } v \in H_0^1(\Omega).$$

[54] The space $\boldsymbol{H}(\mathrm{div};\Omega)$ and other related spaces are of significant importance, as they naturally arise in the mathematical modeling of various problems of physical interest. Further properties of the space $\boldsymbol{H}(\mathrm{div};\Omega)$, such as a specific Green's formula, density of smooth functions, etc., are established in Girault & Raviart [1986, Chapter 1].

(b) *There exists a unique pair* $(\boldsymbol{p}, \lambda) \in \boldsymbol{H}(\mathrm{div};\Omega) \times L^2(\Omega)$ *that satisfies the variational problem*

$$\int_\Omega \boldsymbol{p}\cdot\boldsymbol{q}\,\mathrm{d}x + \int_\Omega (\operatorname{div}\boldsymbol{q})\lambda\,\mathrm{d}x = 0 \quad \textit{for all } \boldsymbol{q} \in \boldsymbol{H}(\mathrm{div};\Omega),$$
$$\int_\Omega (\operatorname{div}\boldsymbol{p})\mu\,\mathrm{d}x = -\int_\Omega f\mu\,\mathrm{d}x \quad \textit{for all } \mu \in L^2(\Omega).$$

Besides,

$$\boldsymbol{p} = \mathbf{grad}\,u \quad \textit{and} \quad \lambda = u,$$

where $u \in H_0^1(\Omega)$ *is the solution to the minimization problem of* (a).

(c) *The vector field* $\boldsymbol{p} = \mathbf{grad}\,u$ *is the unique solution to the constrained quadratic minimization problem*

$$\boldsymbol{p} \in \widetilde{\boldsymbol{U}}_f := \{\boldsymbol{q} \in \boldsymbol{H}(\mathrm{div};\Omega); \operatorname{div}\boldsymbol{q} + f = 0 \textit{ in } L^2(\Omega)\},$$
$$I(\boldsymbol{p}) = \inf_{\boldsymbol{q}\in\widetilde{\boldsymbol{U}}_f} I(\boldsymbol{q}), \quad \textit{where } I(\boldsymbol{q}) := \frac{1}{2}\int_\Omega |\boldsymbol{q}|^2\,\mathrm{d}x \textit{ for each } \boldsymbol{q} \in \boldsymbol{L}^2(\Omega).$$

Proof Let the bilinear forms $a(\cdot,\cdot) : \boldsymbol{H}(\mathrm{div};\Omega)\times\boldsymbol{H}(\mathrm{div};\Omega) \to \mathbb{R}$ and $\widetilde{b}(\cdot,\cdot) : \boldsymbol{H}(\mathrm{div};\Omega)\times L^2(\Omega) \to \mathbb{R}$, and the linear forms $\ell : \boldsymbol{H}(\mathrm{div};\Omega) \to \mathbb{R}$ and $\chi : L^2(\Omega) \to \mathbb{R}$, be respectively defined by

$$a(\boldsymbol{p},\boldsymbol{q}) := \int_\Omega \boldsymbol{p}\cdot\boldsymbol{q}\,\mathrm{d}x \quad \text{for each } \boldsymbol{p},\boldsymbol{q} \in \boldsymbol{H}(\mathrm{div};\Omega),$$
$$\widetilde{b}(\boldsymbol{q},\mu) := \int_\Omega (\operatorname{div}\boldsymbol{q})\mu\,\mathrm{d}x \quad \text{for each } (\boldsymbol{q},\mu) \in \boldsymbol{H}(\mathrm{div};\Omega)\times L^2(\Omega),$$
$$\ell := 0 \quad \text{and} \quad \chi(\mu) := -\int_\Omega f\mu\,\mathrm{d}x \quad \text{for each } \mu \in L^2(\Omega).$$

The bilinear form $a(\cdot,\cdot)$ is coercive over the subspace

$$\widetilde{\boldsymbol{U}}_0 := \left\{\boldsymbol{q} \in \boldsymbol{H}(\mathrm{div};\Omega); \int_\Omega (\operatorname{div}\boldsymbol{q})\mu\,\mathrm{d}x = 0 \text{ for all } \mu \in L^2(\Omega)\right\}$$
$$= \{\boldsymbol{q} \in \boldsymbol{H}(\mathrm{div};\Omega);\ \operatorname{div}\boldsymbol{q} = 0 \text{ in } L^2(\Omega)\}$$

of $\boldsymbol{H}(\mathrm{div};\Omega)$, since

$$a(\boldsymbol{q},\boldsymbol{q}) = \|\boldsymbol{q}\|_{0,\Omega}^2 = \|\boldsymbol{q}\|_{\boldsymbol{H}(\mathrm{div};\Omega)}^2 \quad \text{for all } \boldsymbol{q} \in \widetilde{\boldsymbol{U}}_0.$$

Given any function $\mu \in L^2(\Omega)$, there exists a unique function $w_\mu \in H_0^1(\Omega)$ that satisfies

$$\int_\Omega \mathbf{grad}\,w_\mu\cdot\mathbf{grad}\,v\,\mathrm{d}x = \int_\Omega \mu v\,\mathrm{d}x \quad \text{for all } v \in H_0^1(\Omega).$$

In particular then, $\int_\Omega \mu\varphi\,\mathrm{d}x = -\int_\Omega(-\mathbf{grad}\,w_\mu)\cdot\mathbf{grad}\,\varphi\,\mathrm{d}x$ for all $\varphi \in \mathcal{D}(\Omega)$, which, according to the definition of the space $\boldsymbol{H}(\mathrm{div};\Omega)$, shows that

$$\mathbf{grad}\,w_\mu \in \boldsymbol{H}(\mathrm{div};\Omega) \quad \text{with} \quad -\operatorname{div}\mathbf{grad}\,w_\mu = \mu \in L^2(\Omega).$$

Since there exists a constant C such that $\|\mathbf{grad}\, w_\mu\|_{0,\Omega} \le C\,\|\mu\|_{0,\Omega}$ for all $\mu \in L^2(\Omega)$ (Theorem 6.7-2), it further follows that

$$\begin{aligned}\|\mathbf{grad}\, w_\mu\|_{\boldsymbol{H}(\mathrm{div};\Omega)} &= \left(\|\mathbf{grad}\, w_\mu\|^2_{0,\Omega} + \|\mathrm{div}\,\mathbf{grad}\, w_\mu\|^2_{0,\Omega}\right)^{1/2} \\ &= \left(\|\mathbf{grad}\, w_\mu\|^2_{0,\Omega} + \|\mu\|^2_{0,\Omega}\right)^{1/2} \le \sqrt{C^2+1}\,\|\mu\|_{0,\Omega}\,.\end{aligned}$$

Consequently, for each nonzero $\mu \in L^2(\Omega)$,

$$\sup_{\begin{cases} \boldsymbol{q}\in \boldsymbol{H}(\mathrm{div};\Omega) \\ \boldsymbol{q}\neq 0\end{cases}} \frac{|\int_\Omega (\mathrm{div}\,\boldsymbol{q})\mu\,\mathrm{d}x|}{\|\boldsymbol{q}\|_{\boldsymbol{H}(\mathrm{div};\Omega)}} \ge \frac{|\int_\Omega (\mathrm{div}\,\mathbf{grad}\, w_\mu)\mu\,\mathrm{d}x|}{\|\mathbf{grad}\, w_\mu\|_{\boldsymbol{H}(\mathrm{div};\Omega)}} \ge (C^2+1)^{-1/2}\,\|\mu\|_{0,\Omega}\,,$$

which shows that the Babuška–Brezzi inf-sup condition of Theorem 6.12-1 holds, with

$$V := \boldsymbol{H}(\mathrm{div};\Omega) \quad \text{and} \quad M := L^2(\Omega).$$

All the remaining assumptions of Theorem 6.12-1 are clearly satisfied. Hence the variational problem of (b) has a unique solution $(\boldsymbol{p},\lambda) \in \boldsymbol{H}(\mathrm{div};\ \Omega) \times L^2(\Omega)$.

By definition, any vector field $\boldsymbol{q} \in \boldsymbol{H}(\mathrm{div};\Omega)$ satisfies

$$\int_\Omega \boldsymbol{q}\cdot\mathbf{grad}\,\varphi\,\mathrm{d}x + \int_\Omega (\mathrm{div}\,\boldsymbol{q})\varphi\,\mathrm{d}x = 0 \quad \text{for all } \varphi \in \mathcal{D}(\Omega).$$

Since, for each $\boldsymbol{q} \in \boldsymbol{H}(\mathrm{div};\Omega)$, the linear form $\varphi \in \mathcal{D}(\Omega) \to \int_\Omega \boldsymbol{q}\cdot\mathbf{grad}\,\varphi\,\mathrm{d}x + \int_\Omega (\mathrm{div}\,\boldsymbol{q})\varphi\,\mathrm{d}x$ is continuous and $\overline{\mathcal{D}(\Omega)} = H^1_0(\Omega)$, it follows that

$$\int_\Omega \boldsymbol{q}\cdot\mathbf{grad}\, v\,\mathrm{d}x + \int_\Omega (\mathrm{div}\,\boldsymbol{q})v\,\mathrm{d}x = 0 \quad \text{for all } \boldsymbol{q} \in \boldsymbol{H}(\mathrm{div};\Omega) \text{ and all } v \in H^1_0(\Omega).$$

Letting $v = u$ shows that, in particular,

$$\int_\Omega \mathbf{grad}\, u\cdot\boldsymbol{q}\,\mathrm{d}x + \int (\mathrm{div}\,\boldsymbol{q})u\,\mathrm{d}x = 0 \quad \text{for all } \boldsymbol{q} \in \boldsymbol{H}(\mathrm{div};\Omega).$$

Hence the first equations in the variational problem of (b) are satisfied with $\boldsymbol{p} = \mathbf{grad}\, u$ and $\lambda = u$.

The variational equations satisfied by $u \in H^1_0(\Omega)$, viz., $\int_\Omega \mathbf{grad}\, u\cdot\mathbf{grad}\, v\,\mathrm{d}x = \int_\Omega fv\,\mathrm{d}x$ for all $v \in H^1_0(\Omega)$, hence *a fortiori* for all $v \in \mathcal{D}(\Omega)$, show that $-\mathrm{div}\,\mathbf{grad}\, u = f \in L^2(\Omega)$. Therefore $\mathbf{grad}\, u \in \boldsymbol{H}(\mathrm{div};\Omega)$, and

$$\int_\Omega (\mathrm{div}\,\mathbf{grad}\, u)\mu\,\mathrm{d}x = -\int_\Omega f\mu\,\mathrm{d}x \quad \text{for all } L^2(\Omega).$$

Hence the second equations in the variational problem of (b) are satisfied with $\boldsymbol{p} = \mathbf{grad}\, u$. This proves (b).

Since the set $\widetilde{\boldsymbol{U}}_f$ may be equivalently defined as

$$\widetilde{\boldsymbol{U}}_f = \{\boldsymbol{q} \in \boldsymbol{H}(\mathrm{div};\Omega);\ \widetilde{b}(\boldsymbol{q},\mu) = \chi(\mu) \text{ for all } \mu \in L^2(\Omega)\},$$

Theorem 6.12-2 can be applied, since the bilinear form $a(\cdot,\cdot)$ is symmetric. This proves (c). □

Remark While the set $\widetilde{\boldsymbol{U}}_f$ appearing in Theorem 6.13-2(c) consists of all vector fields $\boldsymbol{q} \in \boldsymbol{H}(\mathrm{div};\Omega)$ that satisfy

$$\mathrm{div}\,\boldsymbol{q} + f = 0 \quad \text{in } L^2(\Omega),$$

it was already remarked that the vector fields $\boldsymbol{q} \in \boldsymbol{L}^2(\Omega)$ appearing in the set $\boldsymbol{U}_f$ found in Theorem 6.13-1(c) satisfy the same partial differential equation in $H^{-1}(\Omega)$, hence only *in the sense of distributions.* □

The analysis carried out in this section on a simple model problem can be clearly extended to the more general elliptic boundary value problems of the second order considered in Theorem 6.7-6, viz.,

$$-\sum_{i,j=1}^{N} \partial_i(a_{ij}\partial_j u) = f \text{ in } \Omega \quad \text{and} \quad u = 0 \text{ on } \partial\Omega$$

(in which case the vector field $(\sum_{j=1}^N a_{ij}\partial_j u)_{i=1}^N \in \boldsymbol{L}^2(\Omega)$ plays the role of $\mathbf{grad}\, u \in \boldsymbol{L}^2(\Omega)$ in the model problem). It can be also extended to linear *systems* of partial differential equations, such as the Stokes equations (Section 6.14), or the equations of linearized elasticity (Problem 6.16-3).

The mixed and dual formulations of such problems have acquired significant importance as the basis of the highly efficient *mixed finite element methods*.[55]

6.14 Application of the Babuška–Brezzi inf-sup theorem and of J.L. Lions lemma: The Stokes equations

The objective of this section is to establish an *existence theorem for the Stokes equations*, which constitutes the most commonly used linear model for incompressible viscous fluids. To this end, we will verify (Theorem 6.14-3) that the weak formulation of these equations constitutes another example of an abstract variational problem of the form described and analyzed in Section 6.12. As expected, the crucial step in the proof of existence will then consist in verifying that the *Babuška–Brezzi inf-sup condition* (Theorem 6.12-1) holds, the verification of which depends on an important *per se* preliminary result, established first in Theorem 6.14-1.

Spaces of vector fields with values in $\mathbb{R}^N$ are again denoted by boldface letters. For instance, $\boldsymbol{H}_0^1(\Omega)$ denotes the space of all vector fields $\boldsymbol{v} = (v_i)_{i=1}^N$ with components v_i in the space $H_0^1(\Omega)$.

Throughout this section, Ω designates a domain in $\mathbb{R}^N$, with $\Gamma := \partial\Omega$. To begin with, we introduce some function spaces and operators. The Hilbert space $\boldsymbol{H}_0^1(\Omega)$ is equipped with

[55]Detailed analyses of mixed finite element methods are found in: Girault & Raviart [1986], Brezzi & Fortin [1991], and Roberts & Thomas [1991].

the inner product and norm defined by

$$(\boldsymbol{u},\boldsymbol{v})_{1,\Omega} := \int_\Omega \boldsymbol{\nabla u} : \boldsymbol{\nabla v}\,\mathrm{d}x \quad \text{for each } \boldsymbol{u},\boldsymbol{v} \in \boldsymbol{H}_0^1(\Omega), \text{ where } \boldsymbol{\nabla u} : \boldsymbol{\nabla v} := \sum_{i,j=1}^N \partial_j u_i \partial_j v_i\,\mathrm{d}x,$$

$$|\boldsymbol{v}|_{1,\Omega} := \sqrt{(\boldsymbol{v},\boldsymbol{v})_{1,\Omega}} \quad \text{for each } \boldsymbol{v} \in \boldsymbol{H}_0^1(\Omega),$$

and the Hilbert space

$$L_0^2(\Omega) := \left\{\mu \in L^2(\Omega);\ \int_\Omega \mu\,\mathrm{d}x = 0\right\}$$

is equipped with the usual inner product and norm of the space $L^2(\Omega)$, respectively denoted $(\cdot,\cdot)_{0,\Omega}$ and $\|\cdot\|_{0,\Omega}$.

First, we note that $\int_\Omega \operatorname{div} \boldsymbol{v} = 0$ for all $\boldsymbol{v} \in \boldsymbol{H}_0^1(\Omega)$ since $\operatorname{div} : \boldsymbol{H}_0^1(\Omega) \to L^2(\Omega)$ is a continuous operator (this relation clearly holds for all vector fields $\boldsymbol{v} = (v_i)_{i=1}^N$ with components v_i in $\mathcal{D}(\Omega)$), and $\mathcal{D}(\Omega)$ is a dense subspace of $H_0^1(\Omega)$. Consequently, the mapping

$$\operatorname{div} : \boldsymbol{v} \in \boldsymbol{H}_0^1(\Omega) \to \operatorname{div} \boldsymbol{v} := \sum_{i=1}^N \partial_i v_i \in L_0^2(\Omega)$$

defined in this fashion is a continuous linear operator.

It turns out that the key to proving that the Babuška–Brezzi inf-sup condition is satisfied by the bilinear form $b(\cdot,\cdot)$ appearing in the variational formulation of the Stokes equations (Theorem 6.14-3) is that *the continuous linear operator*

$$\operatorname{div} : \boldsymbol{H}_0^1(\Omega) \to L_0^2(\Omega)$$

is surjective, a property that is established in Theorem 6.14-1 below.

The proof of this seemingly innocuous property is anything but trivial, however: The proof given below relies in particular on *J.L. Lions lemma* (Theorem 6.11-4) and on the *Banach closed range theorem* (Theorem 5.11-5).[56]

The space

$$\mathbf{Ker}\operatorname{div} := \{\boldsymbol{v} \in \boldsymbol{H}_0^1(\Omega);\ \operatorname{div} \boldsymbol{v} = 0 \text{ in } L^2(\Omega)\}$$

being a closed subspace of $\boldsymbol{H}_0^1(\Omega)$, the direct sum theorem (Theorem 4.5-2) shows that the space $\boldsymbol{H}_0^1(\Omega)$ can be written as

$$\boldsymbol{H}_0^1(\Omega) = \mathbf{Ker}\operatorname{div} \oplus (\mathbf{Ker}\operatorname{div})^\perp,$$

the orthogonality being understood here with respect to the inner product $(\cdot,\cdot)_{1,\Omega}$. It will then follow that the (clearly injective) operator $\operatorname{div} : (\mathbf{Ker}\operatorname{div})^\perp \to L_0^2(\Omega)$ has a continuous inverse, since it is surjective (see part (c) in the next theorem).

The proof rests on the introduction of the mapping

$$\mathbf{grad} : \mu \in L^2(\Omega) \to \mathbf{grad}\,\mu \in \boldsymbol{H}^{-1}(\Omega) := H^{-1}(\Omega;\mathbb{R}^N)$$

[56]The first proof of the surjectivity of $\operatorname{div} : \boldsymbol{H}_0^1(\Omega) \to L_0^2(\Omega)$ is due to LADYZHENSKAYA [1969]; see also TEMAM [1977, Chapter 1] and GIRAULT & RAVIART [1979, Section 3.3, Lemma 3.2].

defined by

$$_{\boldsymbol{H}^{-1}(\Omega)}\langle \mathbf{grad}\,\mu, \boldsymbol{v}\rangle_{\boldsymbol{H}_0^1(\Omega)} := -\int_\Omega \mu \operatorname{div} \boldsymbol{v} \, dx \quad \text{for all } \boldsymbol{v} \in \boldsymbol{H}_0^1(\Omega).$$

Like the definition of each mapping $\partial_i : L^2(\Omega) \to H^{-1}(\Omega)$, $1 \le i \le N$, that of $\mathbf{grad} : L^2(\Omega) \to \boldsymbol{H}^{-1}(\Omega)$ is understood here in the sense of distributions. The norm of the space $\boldsymbol{H}^{-1}(\Omega)$, which is the dual of the space $\boldsymbol{H}_0^1(\Omega)$, will be denoted $\|\cdot\|_{-1,\Omega}$, like that of the space $H^{-1}(\Omega)$.

The mapping $\mathbf{grad} : L^2(\Omega) \to \boldsymbol{H}^{-1}(\Omega)$ thus defined is clearly linear and continuous (each mapping $\mu \in L^2(\Omega) \to \partial_i \mu \in H^{-1}(\Omega)$, $1 \le i \le N$, is continuous). As shown in the next proof, the operator **grad** becomes injective when it is restricted to the subspace $L_0^2(\Omega)$ of $L^2(\Omega)$, and, more importantly, is then nothing but the *dual operator* (Section 5.11) of the operator $-\operatorname{div}$ introduced above.

Note that, in both Theorems 6.14-1 and 6.14-2 and their proofs, *the space* $L_0^2(\Omega)$ *has been implicitly identified with its dual space* (which is licit since $L_0^2(\Omega)$ is a Hilbert space, thanks to the F. Riesz isometry from $(L_0^2(\Omega))'$ onto $L_0^2(\Omega)$), *and the space* $\boldsymbol{H}_0^1(\Omega)$ *has been implicitly identified with its bidual space* (which is licit since $\boldsymbol{H}_0^1(\Omega)$ is reflexive, thanks to the canonical isometry from $\boldsymbol{H}_0^1(\Omega)$ onto $(\boldsymbol{H}_0^1(\Omega))''$; cf. Section 5.14).

Theorem 6.14-1 *Let Ω be a domain in $\mathbb{R}^N$. Then:*

(a) *The continuous linear operator*

$$\mathbf{grad} : L_0^2(\Omega) \to \boldsymbol{H}^{-1}(\Omega)$$

defined for each $\mu \in L_0^2(\Omega)$ by

$$_{\boldsymbol{H}^{-1}(\Omega)}\langle \mathbf{grad}\,\mu, \boldsymbol{v}\rangle_{\boldsymbol{H}_0^1(\Omega)} := -\int_\Omega \mu \operatorname{div} \boldsymbol{v} \, dx \quad \textit{for all } \boldsymbol{v} \in \boldsymbol{H}_0^1(\Omega),$$

is injective and its dual is the continuous linear operator

$$-\operatorname{div} : \boldsymbol{H}_0^1(\Omega) \to L_0^2(\Omega).$$

(b) *The image of the space $L_0^2(\Omega)$ under the operator* **grad** *is closed in $\boldsymbol{H}^{-1}(\Omega)$.*

(c) *The injective continuous linear operator*

$$\operatorname{div} : (\mathbf{Ker}\operatorname{div})^\perp \to L_0^2(\Omega)$$

is surjective and has a continuous inverse.

Proof In what follows, the same letter C designates various constants, which may not be the same at each one of their various occurrences.

(i) *The operator* $\mathbf{grad} : L_0^2(\Omega) \to \boldsymbol{H}^{-1}(\Omega)$ *is injective.*

Let $\mu \in L^2(\Omega)$ be such that $\mathbf{grad}\,\mu = 0$ in $\boldsymbol{H}^{-1}(\Omega)$. By definition of the operator **grad**, this implies that

$$\sum_{i=1}^N \int_\Omega \mu \partial_i \varphi_i \, dx = 0 \quad \text{for all } \varphi_i \in \mathcal{D}(\Omega), \ 1 \le i \le N,$$

so that μ is a constant function by Theorem 6.3-4; hence $\mu = 0$ if $\mu \in L_0^2(\Omega)$.

(ii) *The operator* $-\operatorname{div} : \boldsymbol{H}_0^1(\Omega) \to L_0^2(\Omega)$ *is the dual* (i.e., in the normed vector space sense; cf. Section 5.11) *of* $\mathbf{grad} : L_0^2(\Omega) \to \boldsymbol{H}^{-1}(\Omega)$, *and the operator* $\sigma\,\mathbf{grad} : L_0^2(\Omega) \to \boldsymbol{H}_0^1(\Omega)$, where $\sigma : \boldsymbol{H}^{-1}(\Omega) \to \boldsymbol{H}_0^1(\Omega)$ denotes the F. Riesz isometry of the Hilbert space $\boldsymbol{H}_0^1(\Omega)$, *is the adjoint* (i.e., in the Hilbert space sense; cf. Section 4.7) *of* $-\operatorname{div} : \boldsymbol{H}_0^1(\Omega) \to L_0^2(\Omega)$.

The relation

$$ {}_{\boldsymbol{H}^{-1}(\Omega)}\langle \mathbf{grad}\,\mu, \boldsymbol{v}\rangle_{\boldsymbol{H}_0^1(\Omega)} = -\int_\Omega \mu \operatorname{div} \boldsymbol{v}\,\mathrm{d}x \quad \text{for all } \mu \in L_0^2(\Omega) \text{ and all } \boldsymbol{v} \in \boldsymbol{H}_0^1(\Omega), $$

shows that $-\operatorname{div} : \boldsymbol{H}_0^1(\Omega) \to L_0^2(\Omega)$ is the dual of $\mathbf{grad} : L_0^2(\Omega) \to \boldsymbol{H}^{-1}(\Omega)$ (in the normed vector space sense). Expressed with the F. Riesz isometry $\sigma : \boldsymbol{H}^{-1}(\Omega) \to \boldsymbol{H}_0^1(\Omega)$, the same relation becomes

$$ (\sigma\,\mathbf{grad}\,\mu, \boldsymbol{v})_{1,\Omega} = -(\mu, \operatorname{div} \boldsymbol{v})_{0,\Omega} \quad \text{for all } \mu \in L_0^2(\Omega) \text{ and all } \boldsymbol{v} \in \boldsymbol{H}_0^1(\Omega), $$

thus showing that $-\operatorname{div} : \boldsymbol{H}_0^1(\Omega) \to L_0^2(\Omega)$ is the adjoint of $\sigma\,\mathbf{grad} : L_0^2(\Omega) \to \boldsymbol{H}_0^1(\Omega)$ (in the Hilbert space sense).

(iii) *There exists a constant* C *such that*[57]

$$ \|\mu\|_{0,\Omega} \le C \left(\|\mu\|_{-1,\Omega}^2 + \|\mathbf{grad}\,\mu\|_{-1,\Omega}^2 \right)^{1/2} \quad \text{for all } \mu \in L^2(\Omega). $$

We first claim that *the space*

$$ K(\Omega) := \{\mu \in H^{-1}(\Omega);\ \mathbf{grad}\,\mu \in \boldsymbol{H}^{-1}(\Omega)\} $$

(in this definition, **grad** is again understood in the sense of distributions), *equipped with the norm*

$$ \mu \in K(\Omega) \to \|\mu\|_{K(\Omega)} := \left(\|\mu\|_{-1,\Omega}^2 + \|\mathbf{grad}\,\mu\|_{-1,\Omega}^2 \right)^{1/2} $$

is complete. To see this, let $(\mu_k)_{k=1}^\infty$ be a Cauchy sequence in $K(\Omega)$. Hence $\mu_k \underset{k\to\infty}{\longrightarrow} \mu$ in $H^{-1}(\Omega)$ and $\mathbf{grad}\,\mu_k \underset{k\to\infty}{\longrightarrow} \boldsymbol{w}$ in $\boldsymbol{H}^{-1}(\Omega)$ (as dual spaces, $H^{-1}(\Omega)$ and $\boldsymbol{H}^{-1}(\Omega)$ are complete), and

$$ {}_{\boldsymbol{H}^{-1}(\Omega)}\langle \mathbf{grad}\,\mu_k, \boldsymbol{\varphi}\rangle_{\boldsymbol{H}_0^1(\Omega)} = -\int_\Omega \mu_k \operatorname{div} \boldsymbol{\varphi}\,\mathrm{d}x \quad \text{for all } \boldsymbol{\varphi} \in \mathcal{D}(\Omega; \mathbb{R}^N). $$

Passing to the limit as $k \to \infty$ in this relation yields

$$ {}_{\boldsymbol{H}^{-1}(\Omega)}\langle \boldsymbol{w}, \boldsymbol{\varphi}\rangle_{\boldsymbol{H}_0^1(\Omega)} = -\int_\Omega \mu \operatorname{div} \boldsymbol{\varphi}\,\mathrm{d}x \quad \text{for all } \boldsymbol{\varphi} \in \mathcal{D}(\Omega; \mathbb{R}^N) $$

thus showing that $\boldsymbol{w} = \mathbf{grad}\,\mu$. Hence $(K(\Omega), \|\cdot\|_{K(\Omega)})$ is complete.

The identity mapping $\iota : (L^2(\Omega), \|\cdot\|_{0,\Omega}) \to (K(\Omega), \|\cdot\|_{K(\Omega)})$ is injective, continuous (there clearly exists a constant C such that $\|\mu\|_{K(\Omega)} \le C\,\|\mu\|_{0,\Omega}$ for all $\mu \in L^2(\Omega)$), and surjective since $K(\Omega) = L^2(\Omega)$ by *J.L. Lions lemma* (Theorem 6.11-4).

[57] Another proof of this crucial inequality is found in NEČAS [1965].

Therefore, the *corollary to the Banach open mapping theorem* (Theorem 5.6-2) shows that the inverse mapping ι^{-1} is also continuous, and hence that there exists a constant C such that

$$\|\mu\|_{0,\Omega} \leq C\left(\|\mu\|_{-1,\Omega}^2 + \|\,\mathbf{grad}\,\mu\|_{-1,\Omega}^2\right)^{1/2} \quad \text{for all } \mu \in L^2(\Omega).$$

(iv) *There exists a constant C such that*

$$\|\mu\|_{0,\Omega} \leq C\,\|\mathbf{grad}\,\mu\|_{-1,\Omega} \quad \text{for all } \mu \in L_0^2(\Omega).$$

We proceed by contradiction. If this is not the case, there exists a sequence $(\mu_k)_{k=1}^\infty$ of functions $\mu_k \in L_0^2(\Omega)$ such that

$$\|\mu_k\|_{0,\Omega} = 1 \text{ for all } k \geq 1 \quad \text{and} \quad \|\,\mathbf{grad}\,\mu_k\|_{-1,\Omega} \to 0 \text{ as } k \to \infty.$$

By the *Rellich–Kondrachov compact imbedding theorem in the space* $L^2(\Omega)$ (Theorem 6.11-3), there exists a subsequence $(\mu_{\sigma(k)})_{k=1}^\infty$ that converges in $H^{-1}(\Omega)$. Since the subsequence $(\mathbf{grad}\,\mu_{\sigma(k)})_{k=1}^\infty$ converges in $\boldsymbol{H}^{-1}(\Omega)$ (to $\mathbf{0}$, but this fact is not used at this stage), the subsequence $(\mu_{\sigma(k)})_{k=1}^\infty$ is thus a Cauchy sequence in the space $(K(\Omega), \|\cdot\|_{K(\Omega)})$, hence also a Cauchy sequence in the space $L^2(\Omega)$ by (iii).

Let then $\mu \in L^2(\Omega)$ be such that

$$\mu_{\sigma(k)} \underset{k\to\infty}{\longrightarrow} \mu \quad \text{in } L^2(\Omega).$$

Then $\mu \in L_0^2(\Omega)$ since $\int_\Omega \mu\,\mathrm{d}x = \lim_{k\to\infty} \int_\Omega \mu_{\sigma(k)}\,\mathrm{d}x = 0$, and $\mathbf{grad}\,\mu = \mathbf{0}$ in $\boldsymbol{H}^{-1}(\Omega)$ since

$$\mathbf{grad}\,\mu_{\sigma(k)} \underset{k\to\infty}{\longrightarrow} \mathbf{0} = \mathbf{grad}\,\mu \quad \text{in } \boldsymbol{H}^{-1}(\Omega),$$

and thus $\mu = 0$ by Theorem 6.3-4 ($\mathbf{grad}\,\mu = 0$ in $\boldsymbol{H}^{-1}(\Omega)$ means that $\int_\Omega \mu\partial_i\varphi\,\mathrm{d}x = 0$ for all $\varphi \in \mathcal{D}(\Omega)$, $1 \leq i \leq N$; hence μ is a constant, but this constant is zero since $\mu \in L_0^2(\Omega)$). But this contradicts the relation $\|\mu_{\sigma(k)}\|_{0,\Omega} = 1$ for all $k \geq 1$.

(v) *The image of $L_0^2(\Omega)$ under* **grad** *is closed in $\boldsymbol{H}^{-1}(\Omega)$, and the image of $\boldsymbol{H}_0^1(\Omega)$ under* div *is closed in $L_0^2(\Omega)$.*

It was shown in (iv) that there exists a constant C such that

$$\|\mu\|_{0,\Omega} \leq C\,\|\mathbf{grad}\,\mu\|_{-1,\Omega} \quad \text{for all } \mu \in L_0^2(\Omega).$$

Hence the image of $L_0^2(\Omega)$ under **grad** is *closed* in $\boldsymbol{H}^{-1}(\Omega)$ since $L_0^2(\Omega)$ is complete (Theorem 3.1-4).

Since $-\operatorname{div} : \boldsymbol{H}_0^1(\Omega) \to L_0^2(\Omega)$ is the dual operator of $\mathbf{grad} : L_0^2(\Omega) \to \boldsymbol{H}^{-1}(\Omega)$ (cf. (ii)), the image of $\boldsymbol{H}_0^1(\Omega)$ under div is thus also *closed* in $L_0^2(\Omega)$, by the *Banach closed range theorem* (first part; cf. Theorem 5.11-5).

(vi) *The injective operator* $\operatorname{div} : (\mathbf{Ker}\operatorname{div})^\perp \to L_0^2(\Omega)$ *is surjective and has a continuous inverse.*

Part (v) also shows that the image of $L_0^2(\Omega)$ under the operator $\sigma\,\mathbf{grad} : L_0^2(\Omega) \to \boldsymbol{H}_0^1(\Omega)$ is closed in $\boldsymbol{H}_0^1(\Omega)$ since the F. Riesz map $\sigma : \boldsymbol{H}^{-1}(\Omega) \to \boldsymbol{H}_0^1(\Omega)$ is an isometry, on the one hand. On the other hand, Theorem 4.7-2(b) shows that

$$L_0^2(\Omega) = \operatorname{Ker}(\sigma\,\mathbf{grad}) \oplus \operatorname{Im}(-\operatorname{div}).$$

Consequently, the operator div : $\boldsymbol{H}_0^1(\Omega) \to L_0^2(\Omega)$ is surjective since the operator $\sigma\,\mathbf{grad}$: $L_0^2(\Omega) \to \boldsymbol{H}_0^1(\Omega)$ is injective (like the operator **grad**; cf. (i)) and thus $\mathrm{Ker}(\sigma\,\mathbf{grad}) = \{0\}$. Hence div : $(\mathbf{Ker}\,\mathrm{div})^\perp \to L_0^2(\Omega)$ is a bijection, and its inverse is continuous, by the corollary to the Banach open mapping theorem. Hence (c) is proved. □

Remark Interesting complements to Theorem 6.14-1 are proposed in Problems 6.14-1–6.14-3. □

We now establish as a corollary to Theorem 6.14-1 a first *characterization of vector fields in $\boldsymbol{H}^{-1}(\Omega)$ as gradients of scalar functions in $L^2(\Omega)$*; the *weak Poincaré lemma* (established later; cf. Theorem 6.17-4) constitutes a second characterization of such vector fields (under the additional assumption that Ω be simply connected).

Theorem 6.14-2 *Let Ω be a domain in $\mathbb{R}^N$. Given a vector field $\boldsymbol{h} \in \boldsymbol{H}^{-1}(\Omega)$, there exists a function p such that*

$$p \in L_0^2(\Omega) \quad \textit{and} \quad \mathbf{grad}\, p = \boldsymbol{h} \ \textit{ in } \boldsymbol{H}^{-1}(\Omega)$$

if and only if

$${}_{\boldsymbol{H}^{-1}(\Omega)}\langle \boldsymbol{h}, \boldsymbol{v}\rangle_{\boldsymbol{H}_0^1(\Omega)} = 0 \quad \textit{for all } \boldsymbol{v} \in \boldsymbol{H}_0^1(\Omega) \textit{ that satisfy } \mathrm{div}\,\boldsymbol{v} = 0 \textit{ in } L^2(\Omega).$$

All other solutions $\widetilde{p} \in L^2(\Omega)$ of the equation $\mathbf{grad}\,\widetilde{p} = \boldsymbol{h}$ in $\boldsymbol{H}^{-1}(\Omega)$ are of the form $\widetilde{p} = p + C$ where C is a constant.

Proof Since the dual of **grad** : $L_0^2(\Omega) \to \boldsymbol{H}^{-1}(\Omega)$ is $-\,\mathrm{div}$: $H_0^1(\Omega) \to L_0^2(\Omega)$ and the image $\mathbf{Im}\,\mathbf{grad}$ of $L_0^2(\Omega)$ under **grad** is closed in $\boldsymbol{H}^{-1}(\Omega)$ (Theorem 6.14-1), the *Banach closed range theorem* (second part; cf. Theorem 5.11-6) implies that

$$\mathbf{Im}\,\mathbf{grad} = \{\boldsymbol{h} \in \boldsymbol{H}^{-1}(\Omega);\ {}_{\boldsymbol{H}^{-1}(\Omega)}\langle \boldsymbol{h}, \boldsymbol{v}\rangle_{H_0^1(\Omega)} = 0 \text{ for all } \boldsymbol{v} \in \mathbf{Ker}(-\,\mathrm{div})\}.$$

In other words, given $\boldsymbol{h} \in \boldsymbol{H}^{-1}(\Omega)$, there exists a solution $p \in L_0^2(\Omega) \subset L^2(\Omega)$ to the equation $\mathbf{grad}\,p = \boldsymbol{h}$ if and only if ${}_{\boldsymbol{H}^{-1}(\Omega)}\langle \boldsymbol{h}, \boldsymbol{v}\rangle_{H_0^1(\Omega)} = 0$ for all $\boldsymbol{v} \in \boldsymbol{H}_0^1(\Omega)$ that satisfy $\mathrm{div}\,\boldsymbol{v} = 0$ in $L^2(\Omega)$, as announced in the theorem.

Let $\pi \in L^2(\Omega)$ be such that $\mathbf{grad}\,\pi = \mathbf{0}$ in $\boldsymbol{H}^{-1}(\Omega)$; in particular then,

$${}_{\boldsymbol{H}^{-1}(\Omega)}\langle \partial_i \pi, \varphi\rangle_{H_0^1(\Omega)} := -\int_\Omega \pi \partial_i \varphi \,\mathrm{d}x = 0 \quad \text{for all } \varphi \in \mathcal{D}(\Omega),\ \ 1 \le i \le N.$$

Hence the function π is a constant, by Theorem 6.3-4 (a domain is connected by assumption). This shows that all the other solutions $\widetilde{p} \in L^2(\Omega)$ of $\mathbf{grad}\,\widetilde{p} = \boldsymbol{h}$ are of the form $\widetilde{p} = p + C$ for some constant C. □

We now come to the *Stokes equations*. Like for second-order linear elliptic boundary value problems (Section 6.7), we first give a set of specific variational equations, then show that these have a unique solution, and finally identify the corresponding boundary value problem (assuming as usual that the solution of the variational equations is smooth enough). Since one of the unknowns is a *vector field*, viz., $\boldsymbol{u} = (u_i) \in \boldsymbol{H}_0^1(\Omega)$, this problem comprises, as expected, a *system of partial differential equations* (instead of a single one as in Section 6.7).

Theorem 6.14-3 (existence of a solution to the Stokes equations) *Let Ω be a domain in $\mathbb{R}^N$, and let a constant $\nu > 0$ and a vector field $\boldsymbol{f} \in \boldsymbol{L}^2(\Omega)$ be given. Then:*

(a) *There exists a unique pair $(\boldsymbol{u}, \lambda) \in \boldsymbol{H}_0^1(\Omega) \times L_0^2(\Omega)$ that satisfies the variational problem*

$$\nu \int_\Omega \boldsymbol{\nabla u} : \boldsymbol{\nabla v}\, dx - \int_\Omega (\operatorname{div} \boldsymbol{v})\lambda\, dx = \int_\Omega \boldsymbol{f} \cdot \boldsymbol{v}\, dx \quad \textit{for all } \boldsymbol{v} = \boldsymbol{H}_0^1(\Omega),$$

$$\int_\Omega (\operatorname{div} \boldsymbol{u})\mu\, dx = 0 \quad \textit{for all } \mu \in L_0^2(\Omega),$$

the last equations being equivalent to

$$\operatorname{div} \boldsymbol{u} = 0 \quad \textit{in } L_0^2(\Omega).$$

(b) *The vector field $\boldsymbol{u} \in \boldsymbol{H}_0^1(\Omega)$ is the unique solution to the constrained quadratic minimization problem*

$$\boldsymbol{u} \in \boldsymbol{U}_0 := \{\boldsymbol{v} \in \boldsymbol{H}_0^1(\Omega);\ \operatorname{div} \boldsymbol{v} = 0 \text{ in } \Omega\},$$

$$I(\boldsymbol{u}) = \inf_{\boldsymbol{v} \in \boldsymbol{U}_0} I(\boldsymbol{v}), \quad \textit{where} \quad I(\boldsymbol{v}) := \frac{\nu}{2} \int \boldsymbol{\nabla v} : \boldsymbol{\nabla v}\, dx - \int_\Omega \boldsymbol{f} \cdot \boldsymbol{v}\, dx \ \textit{ for each } \boldsymbol{v} \in \boldsymbol{U}_0.$$

(c) *The pair $(\boldsymbol{u}, \lambda) \in \boldsymbol{H}_0^1(\Omega) \times L_0^2(\Omega)$ satisfies*

$$\begin{aligned} -\nu\Delta \boldsymbol{u} + \mathbf{grad}\, \lambda &= \boldsymbol{f} && \textit{in } \boldsymbol{H}^{-1}\Omega, \\ \operatorname{div} \boldsymbol{u} &= 0 && \textit{in } \Omega, \\ \boldsymbol{u} &= \boldsymbol{0} && \textit{on } \Gamma, \end{aligned}$$

where $\boldsymbol{\Delta u} := (\Delta u_i)_{i=1}^N$.

Proof A vector field $\boldsymbol{u} \in \boldsymbol{H}_0^1(\Omega)$ satisfies the relation $\operatorname{div} \boldsymbol{u} = 0$ in $L_0^2(\Omega)$ appearing in (a) if (and clearly only if)

$$\int_\Omega (\operatorname{div} \boldsymbol{u})\mu\, dx = 0 \quad \text{for all } \mu \in L_0^2(\Omega)$$

(to see this, let $\mu = \operatorname{div} \boldsymbol{u}$ in the above relations).

Let the bilinear forms $a(\cdot,\cdot) : \boldsymbol{H}_0^1(\Omega) \times \boldsymbol{H}_0^1(\Omega) \to \mathbb{R}$ and $b(\cdot,\cdot) : \boldsymbol{H}_0^1(\Omega) \times L_0^2(\Omega) \to \mathbb{R}$, and the linear forms $\ell : \boldsymbol{H}_0^1(\Omega) \to \mathbb{R}$ and $\chi : L_0^2(\Omega) \to \mathbb{R}$, be respectively defined by

$$\begin{aligned} a(\boldsymbol{u}, \boldsymbol{v}) &:= \nu \int_\Omega \boldsymbol{\nabla u} : \boldsymbol{\nabla v}\, dx \quad \text{for each } \boldsymbol{u}, \boldsymbol{v} \in \boldsymbol{H}_0^1(\Omega), \\ b(\boldsymbol{v}, \mu) &:= -\int_\Omega (\operatorname{div} \boldsymbol{v})\mu\, dx \quad \text{for each } (\boldsymbol{v}, \mu) \in \boldsymbol{H}_0^1(\Omega) \times L_0^2(\Omega), \\ \ell(\boldsymbol{v}) &:= \int_\Omega \boldsymbol{f} \cdot \boldsymbol{v}\, dx \quad \text{for each } \boldsymbol{v} \in \boldsymbol{H}_0^1(\Omega) \quad \text{and} \quad \chi := 0. \end{aligned}$$

The symmetric bilinear form $a(\cdot,\cdot)$ is clearly continuous, and is $\boldsymbol{H}_0^1(\Omega)$-coercive since

$$a(\boldsymbol{v}, \boldsymbol{v}) = \nu\, |\boldsymbol{v}|_{1,\Omega}^2 \quad \text{for all } \boldsymbol{v} \in \boldsymbol{H}_0^1(\Omega).$$

Hence $a(\cdot,\cdot)$ is *a fortiori* coercive on

$$\boldsymbol{U}_0 = \{\boldsymbol{v} \in \boldsymbol{H}_0^1(\Omega);\ b(\boldsymbol{v},\mu) = 0 \text{ for all } \mu \in L_0^2(\Omega)\}.$$

By Theorem 6.14-1(c), given any function $\mu \in L_0^2(\Omega)$, there exists a unique vector field $\boldsymbol{w}_\mu \in (\operatorname{Ker}\operatorname{div})^\perp \subset \boldsymbol{H}_0^1(\Omega)$ that satisfies

$$\operatorname{div}\boldsymbol{w}_\mu = \mu \quad \text{in } L_0^2(\Omega),$$

and besides, there exists a constant C such that

$$|\boldsymbol{w}_\mu|_{1,\Omega} \le C\,\|\mu\|_{0,\Omega} \quad \text{for all } \mu \in L_0^2(\Omega).$$

Consequently, for each nonzero $\mu \in L_0^2(\Omega)$,

$$\sup_{\substack{\boldsymbol{v}\in\boldsymbol{H}_0^1(\Omega)\\ \boldsymbol{v}\neq\boldsymbol{0}}} \frac{|\int_\Omega(\operatorname{div}\boldsymbol{v})\mu\,\mathrm{d}x|}{|\boldsymbol{v}|_{1,\Omega}} \ge \frac{\int_\Omega(\operatorname{div}\boldsymbol{w}_\mu)\mu\,\mathrm{d}x}{|\boldsymbol{w}_\mu|_{1,\Omega}} = \frac{\|\mu\|_{0,\Omega}^2}{|\boldsymbol{w}_\mu|_{1,\Omega}} \ge C^{-1}\,\|\mu\|_{0,\Omega},$$

which shows that the *Babuška–Brezzi inf-sup condition* of Theorem 6.12-1 holds, with

$$V := \boldsymbol{H}_0^1(\Omega) \quad \text{and} \quad M := L_0^2(\Omega).$$

The linear form ℓ is clearly continuous on the space $\boldsymbol{H}_0^1(\Omega)$. Hence (a) and (b) respectively follow from Theorems 6.12-1 and 6.12-2.

Finally, the relations

$$\begin{aligned}\nu\int_\Omega \sum_{i,j=1}^N \partial_j u_i \partial_j v_i\,\mathrm{d}x - \int_\Omega \lambda \sum_{i=1}^N \partial_i v_i\,\mathrm{d}x &= \nu\sum_{i=1}^N {}_{H^{-1}(\Omega)}\langle -\Delta u_i + \partial_i\lambda, v_i\rangle_{H_0^1(\Omega)}\\ &= \int_\Omega \sum_{i=1}^N f_i v_i\,\mathrm{d}x \quad \text{for all } (v_i)_{i=1}^N \in \boldsymbol{H}_0^1(\Omega)\end{aligned}$$

show that equations $-\nu\Delta u_i + \partial_i\lambda = f_i$ in $H^{-1}(\Omega)$, $1 \le i \le N$, hold. □

The system

$$\begin{aligned}-\nu\Delta\boldsymbol{u} + \mathbf{grad}\,\lambda &= \boldsymbol{f} \quad \text{in } \Omega,\\ \operatorname{div}\boldsymbol{u} &= 0 \quad \text{in } \Omega,\\ \boldsymbol{u} &= \boldsymbol{0} \quad \text{on } \Gamma,\end{aligned}$$

constitutes the **Stokes equations**[58] in $\mathbb{R}^N$. When $N = 3$, these equations constitute the linearization of the nonlinear *Navier–Stokes equations* (Section 9.11), which model the stationary (i.e., time-independent) flow of an *incompressible viscous fluid* with *kinematic viscosity* $\nu > 0$ filling up a domain $\Omega \subset \mathbb{R}^3$, and subjected to *applied forces* of density $\boldsymbol{f}$ per unit mass. The unknown $\boldsymbol{u}$ is the *velocity* of the fluid, which is subjected to the *incompressibility condition* $\operatorname{div}\boldsymbol{u} = 0$ in Ω and to the boundary condition $\boldsymbol{u} = \boldsymbol{0}$ on Γ, meaning that the velocity of the fluid vanishes on the boundary of the domain; the unknown λ is the *pressure* inside the fluid.

[58] So named after Sir George Gabriel Stokes (1819–1903).

Remarkably, the unknown λ *does not* appear in the formulation of the Stokes problem as a constrained quadratic minimization problem (Theorem 6.14-3(b)), while it *does* appear in its mixed formulation (Theorem 6.14-3(a)). We shall see later (Section 7.16) that the unknown λ also appears in its formulation as a *saddle-point problem*, as the *Lagrange multiplier associated with the constraint* $\operatorname{div} \boldsymbol{v} = 0$ in Ω.

Note that the mathematical analysis of the Stokes equations can be equally well carried out if the unknown λ is sought in the quotient space $L^2(\Omega)/P_0(\Omega)$, where $P_0(\Omega)$ denotes the space of constant functions over Ω, instead of the space $L_0^2(\Omega)$ (it is easily seen that there exists a linear isometry between $L_0^2(\Omega)$ and $L^2(\Omega)/P_0(\Omega)$). This observation reflects in particular that the *unknown pressure* λ *is determined only up to an additive constant* (an evident property if the point of departure is the above boundary value problem). If the chosen space is $L_0^2(\Omega)$, this indeterminacy of course disappears since the unknown λ is then subjected to the condition $\int_\Omega \lambda \, dx = 0$.

The boundary condition $\boldsymbol{u} = \boldsymbol{0}$ imposed over the entire boundary Γ on the unknown velocity $\boldsymbol{u}$, or even a more general boundary condition $\boldsymbol{u} = \boldsymbol{u}_0$ again over the entire boundary Γ, is admittedly far from realistic. However, taking into account more physically plausible boundary conditions (such as a free surface boundary condition, for instance) poses considerable mathematical challenges. This explains why more realistic boundary conditions are seldom considered.[59]

Problems

6.14-1[60] Given a domain Ω in $\mathbb{R}^N$, define the spaces

$$\boldsymbol{V}(\Omega) := \{\boldsymbol{v} \in \boldsymbol{H}_0^1(\Omega); \operatorname{div} \boldsymbol{v} = 0 \text{ in } \Omega\} \quad \text{and} \quad \mathcal{V}(\Omega) := \{\boldsymbol{\varphi} \in \mathcal{D}(\Omega; \mathbb{R}^N); \operatorname{div} \boldsymbol{\varphi} = 0 \text{ in } \Omega\}.$$

(1) Let a vector field $\boldsymbol{h} \in \boldsymbol{H}^{-1}(\Omega)$ be such that ${}_{\boldsymbol{H}^{-1}(\Omega)}\langle \boldsymbol{h}, \boldsymbol{v} \rangle_{\boldsymbol{H}_0^1(\Omega)} = 0$ for all $\boldsymbol{\varphi} \in \mathcal{V}(\Omega)$. Show that there exists a function $p \in L^2(\Omega)$ such that $\boldsymbol{h} = \mathbf{grad}\, p$.

Hint: Since, as a domain, Ω is connected, there exists a sequence $(\Omega_k)_{k=1}^\infty$ of connected domains in $\mathbb{R}^N$ with the following properties:

$$\Omega_k \subset \Omega_{k+1} \quad \text{and} \quad \overline{\Omega}_k \subset \Omega \text{ for all } k \geq 1, \quad \text{and} \quad \Omega = \bigcup_{k=1}^\infty \Omega_k.$$

Given any $\boldsymbol{v} = (v_i)_{i=1}^N \in \boldsymbol{H}_0^1(\Omega)$ and any $\varepsilon > 0$, let $\boldsymbol{v}_\varepsilon := (v_{i,\varepsilon})_{i=1}^N$, where each family $(v_{i,\varepsilon})_{\varepsilon>0}$ is a *regularizing family* (Section 2.6) of the function $v_i \in H_0^1(\Omega)$, $1 \leq i \leq N$. Show that, for each integer $k \geq 1$, there exists $\varepsilon(k) > 0$ such that, given any $\boldsymbol{v} \in \boldsymbol{V}(\Omega)$ with $\boldsymbol{v} = \boldsymbol{0}$ on $\Omega - \Omega_k$, then $\boldsymbol{v}_\varepsilon \in \mathcal{V}(\Omega)$ for all $0 < \varepsilon \leq \varepsilon(k)$ and $|\boldsymbol{v}_\varepsilon - \boldsymbol{v}|_{1,\Omega} \to 0$ as $\varepsilon \to 0$. Infer from this result that ${}_{\boldsymbol{H}^{-1}(\Omega)}\langle \boldsymbol{h}, \boldsymbol{v} \rangle_{\boldsymbol{H}_0^1(\Omega)} = 0$; then use Theorem 6.14-2.

Remark The result proved in (1) is a special case of *de Rham's theorem*,[61] a deep result asserting more generally that, given *any* open subset of $\mathbb{R}^N$, *any vector-valued distribution on* Ω *that vanishes on the space* $\mathcal{V}(\Omega)$ *is the gradient of a distribution on* Ω.

[59] A notable exception is found in:

V.A. SOLONNIKOV [1982]: On the Stokes equations in domains with non-smooth boundaries and on viscous incompressible flow with a free surface, in *Nonlinear Partial Differential Equations and Their Applications* (H. BREZIS & J.L. LIONS, editors), pp. 340–423, Pitman, Boston.

[60] Questions (1) and (2) of this problem respectively constitute Theorem 2.3 and Corollary 2.5 of GIRAULT & RAVIART [1986, Chapter 1].

[61] G. de RHAM [1955]: *Variétés Différentiables*, Hermann, Paris.

(2) Using (1) and Theorem 4.3-2, show that the subspace $\mathcal{V}(\Omega)$ of $\boldsymbol{V}(\Omega)$ is dense in the space $(\boldsymbol{V}(\Omega), |\cdot|_{1,\Omega})$. □

6.14-2 Let Ω be a domain in $\mathbb{R}^N$. This problem lists two properties of the operator **grad**, *considered as acting from* $H_0^1(\Omega)$ *into* $\boldsymbol{L}^2(\Omega)$ (instead of from $L_0^2(\Omega)$ into $\boldsymbol{H}^{-1}(\Omega)$ as in Theorem 6.14-1).

(1) Show that $-\operatorname{div} : \boldsymbol{L}^2(\Omega) \to H^{-1}(\Omega)$ is the adjoint operator of $\mathbf{grad} : H_0^1(\Omega) \to \boldsymbol{L}^2(\Omega)$.

(2) Show that the image of the space $H_0^1(\Omega)$ under the operator **grad** is closed in $\boldsymbol{L}^2(\Omega)$ (the proof is much simpler than that of part (b) in Theorem 6.14-1, as it no longer rests on J.L. Lions lemma).

6.14-3 Given any $\boldsymbol{\ell} \in \boldsymbol{H}^{-1}(\Omega)$, let $\boldsymbol{u} := \boldsymbol{\mathcal{A}}(\boldsymbol{\ell}) \in \boldsymbol{H}_0^1(\Omega)$ denote the unique solution to the variational equations $(\boldsymbol{u}, \boldsymbol{v})_{1,\Omega} = \boldsymbol{\ell}(\boldsymbol{v})$ for all $\boldsymbol{v} \in \boldsymbol{H}_0^1(\Omega)$. Show that

$$(\mathbf{Ker}\operatorname{div})^\perp = \{\boldsymbol{\mathcal{A}}(\mathbf{grad}\,\mu) \in \boldsymbol{H}_0^1(\Omega);\ \mu \in L^2(\Omega)\},$$

where the mapping div is considered as acting from $\boldsymbol{H}_0^1(\Omega)$ into $L^2(\Omega)$.

Hint: Use question (1) in Problem 6.14-1.

6.14-4 Let Ω be a domain in $\mathbb{R}^N$. Show that the closure of the space $\boldsymbol{V}(\Omega) := \{\boldsymbol{v} \in \boldsymbol{H}_0^1(\Omega); \operatorname{div}\boldsymbol{v} = 0 \text{ in } \Omega\}$ *with respect to the norm* $\|\cdot\|_{0,\Omega}$ is a *strict* subspace of the space $\{\boldsymbol{v} \in \boldsymbol{L}^2(\Omega);\ \operatorname{div}\boldsymbol{v} = 0 \text{ in } H^{-1}(\Omega)\}$ (naturally, the same property holds *a fortiori* for the closure of the space $\mathcal{V}(\Omega)$ already encountered, like the space $\boldsymbol{V}(\Omega)$, in Problem 6.14-1).

6.15 A second application of J.L. Lions lemma: Korn's inequality

Our second application of *J.L. Lions lemma* will be to prove a basic inequality, which plays a crucial role in linearized elasticity.

Korn's inequality[62] asserts that, given a domain Ω in $\mathbb{R}^N$, there exists a constant C depending solely on Ω such that

$$\Big(\sum_{i=1}^N \|v_i\|_{0,\Omega}^2 + \sum_{i,j=1}^N \|\partial_j v_i\|_{0,\Omega}^2\Big)^{1/2} \le C\Big(\sum_{i=1}^N \|v_i\|_{0,\Omega}^2 + \sum_{i,j=1}^N \|e_{ij}(\boldsymbol{v})\|_{0,\Omega}^2\Big)^{1/2}$$

[62]This inequality appeared for the first time, with a proof under the assumption that the vector fields $\boldsymbol{v}$ vanish on the boundary of Ω, in:

A. KORN [1906]: Die Eigenschwingungen eines elastischen Körpers mit ruhender Oberfläche, *Sitzungsberichte der Mathematisch-physikalischen Klasse der Königlich bayerischen Akademie der Wissenschaften zu München* **36**, 351–402.

A. KORN [1908]: Solution générale du problème d'équilibre dans la théorie de l'élasticité, dans le cas où les efforts sont donnés à la surface, *Annales de la Faculté des Sciences de Toulouse* **10**, 165–269.

A. KORN [1909]: Über einige Ungleichungen, welche in der Theorie der elastischen und elektrischen Schwingungen eine Rolle spielen, *Bulletin International de l'Académie des Sciences de Cracovie* **9**, 705–724.

A second proof, this time under the assumption that the vector fields $\boldsymbol{v}$ satisfy $\int_\Omega \mathbf{curl}\,\boldsymbol{v}\,dx = \mathbf{0}$, was then given in:

K.O. FRIEDRICHS [1947]: On the boundary-value problems of the theory of elasticity and Korn's inequality, *Annals of Mathematics* **48**, 441–471.

The first proof in full generality (based on the Calderón–Zygmund theory of singular integrals) is due to:

J. GOBERT [1962]: Une inégalité fondamentale de la théorie de l'élasticité, *Bulletin de la Société Royale des Sciences de Liège* **31**, 182–191.

for all vector fields $\boldsymbol{v} = (v_i)_{i=1}^N \in H^1(\Omega; \mathbb{R}^N)$, where

$$e_{ij}(\boldsymbol{v}) := \frac{1}{2}(\partial_j v_i + \partial_i v_j) \in L^2(\Omega), \quad 1 \le i,\, j \le N.$$

As we will see in the next section (Theorem 6.16-1), its special case $N = 3$ is crucial to establishing the existence and uniqueness of the solution to the weak formulation of the boundary value problem of three-dimensional linearized elasticity (as the key to proving the coerciveness of the associated bilinear form).

Korn's inequality thus provides an upper bound for the $L^2(\Omega)$-norms of *all* the N^2 partial derivatives $\partial_j v_i$ of a vector field $\boldsymbol{v} = (v_i) \in H^1(\Omega; \mathbb{R}^N)$ in terms of the $L^2(\Omega)$-norms of *only* $\dfrac{N(N+1)}{2}$ particular linear combinations of these partial derivatives, namely the functions $e_{ij}(\boldsymbol{v}) = e_{ji}(\boldsymbol{v})$. This truly remarkable feature suggests that none of its various available proofs[63] should be simple. For instance, the proof given below (Theorem 6.15-1) is short and illuminating, but it depends on the deep, and difficult to prove, *lemma of J.L. Lions* (Theorem 6.11-4); otherwise, there exist more direct proofs, which do not depend on J.L. Lions lemma, but rely instead on delicate computations and estimates[64] (one such proof is proposed in Problem 6.15-4).

In what follows, spaces of vector-valued, *resp. symmetric* matrix-valued, fields are denoted by boldface, *resp.* blackboard bold roman, capitals, while the norms are denoted as in the scalar case. Thus, for instance,

$$\|\boldsymbol{v}\|_{1,\Omega} = \left(\sum_{i=1}^N \|v_i\|_{1,\Omega}^2\right)^{1/2} \qquad \text{for each } \boldsymbol{v} = (v_i) \in \boldsymbol{H}^1(\Omega) := H^1(\Omega; \mathbb{R}^N),$$

$$\|\boldsymbol{e}\|_{0,\Omega} = \left(\sum_{i,j=1}^N \|e_{ij}\|_{0,\Omega}^2\right)^{1/2} \qquad \text{for each } \boldsymbol{e} = (e_{ij}) \in \mathbb{L}^2(\Omega) := L^2(\Omega; \mathbb{S}^N),$$

where $\mathbb{S}^N$ denotes the space of all real $N \times N$ symmetric matrices.

Theorem 6.15-1 (Korn's inequality, *alias* Korn's inequality in $\boldsymbol{H}^1(\Omega)$) *Let Ω be a domain*[65] *in $\mathbb{R}^N$. Then there exists a constant $C = C(\Omega)$ such that*

$$\|\boldsymbol{v}\|_{1,\Omega} \le C\left(\|\boldsymbol{v}\|_{0,\Omega}^2 + \|\boldsymbol{e}(\boldsymbol{v})\|_{0,\Omega}^2\right)^{1/2} \quad \text{for all } \boldsymbol{v} \in \boldsymbol{H}^1(\Omega),$$

where

$$\boldsymbol{e}(\boldsymbol{v}) := (e_{ij}(v)) \quad \text{with } e_{ij}(\boldsymbol{v}) := \frac{1}{2}(\partial_j v_i + \partial_i v_j),\ 1 \le i,\, j \le N.$$

---

[63] See, e.g., the list of references provided in:

C.O. HORGAN [1995]: Korn's inequalities and their applications in continuum mechanics, SIAM Review **37**, 491–511.

[64] See for instance FICHERA [1972a] or:

J.A. NITSCHE [1981]: On Korn's second inequality, *RAIRO Analyse Numérique* **15**, 237–248.

An illuminating account of J.A. Nitsche's approach for domains with a boundary of class $\mathcal{C}^1$ is found in CHIPOT [2002, Section 6.1].

[65] A counterexample showing that the Korn inequality does not necessarily hold if Ω is not a domain is found in:

G. GEYMONAT; G. GILARDI [1998]: Contre-exemple à l'inégalité de Korn et au lemme de Lions dans des domaines irréguliers, in *Equations aux Dérivées Partielles et Applications. Articles Dédiés à Jacques-Louis Lions*, pp. 541–548, Gauthier-Villars, Paris.

Proof [66] (i) *Define the space*

$$\boldsymbol{E}(\Omega) := \{\boldsymbol{v} \in \boldsymbol{L}^2(\Omega);\ \boldsymbol{e}(\boldsymbol{v}) \in \mathbb{L}^2(\Omega)\}.$$

Then, equipped with the norm defined by

$$\|\boldsymbol{v}\| := (\|\boldsymbol{v}\|_{0,\Omega}^2 + \|\boldsymbol{e}(\boldsymbol{v})\|_{0,\Omega}^2)^{1/2} \quad \text{for each } \boldsymbol{v} \in \boldsymbol{E}(\Omega),$$

the space $\boldsymbol{E}(\Omega)$ is a Hilbert space.

The relation $\boldsymbol{e}(\boldsymbol{v}) \in \mathbb{L}^2(\Omega)$ appearing in the definition of the space $\boldsymbol{E}(\Omega)$ is to be understood in the sense of distributions, i.e., it means that there exist functions in the space $L^2(\Omega)$, *denoted* $e_{ij}(\boldsymbol{v}) = e_{ji}(\boldsymbol{v})$, such that

$$\int_\Omega e_{ij}(\boldsymbol{v})\varphi \, \mathrm{d}x = -\frac{1}{2}\int_\Omega (v_i\partial_j\varphi + v_j\partial_i\varphi)\,\mathrm{d}x \quad \text{for all } \varphi \in \mathcal{D}(\Omega).$$

Consider a Cauchy sequence $(\boldsymbol{v}^k)_{k=1}^\infty$ of elements $\boldsymbol{v}^k = (v_i^k)_{i=1}^N \in \boldsymbol{E}(\Omega)$. The definition of the norm $\|\cdot\|$ shows that there exist functions $v_i \in L^2(\Omega)$ and $e_{ij} \in L^2(\Omega)$ such that

$$v_i^k \to v_i \text{ in } L^2(\Omega) \quad \text{and} \quad e_{ij}(\boldsymbol{v}^k) \to e_{ij} \text{ in } L^2(\Omega) \text{ as } k \to \infty,$$

since the space $L^2(\Omega)$ is complete. Given a function $\varphi \in \mathcal{D}(\Omega)$, letting $k \to \infty$ in the relations

$$\int_\Omega e_{ij}(\boldsymbol{v}^k)\varphi \, \mathrm{d}x = -\frac{1}{2}\int_\Omega (v_i^k\partial_j\varphi + v_j^k\partial_i\varphi)\,\mathrm{d}x, \quad k \geq 1,$$

shows that $e_{ij} = e_{ij}(\boldsymbol{v})$.

(ii) *The two spaces $\boldsymbol{E}(\Omega)$ and $\boldsymbol{H}^1(\Omega)$ coincide.*

Clearly, $\boldsymbol{H}^1(\Omega) \subset \boldsymbol{E}(\Omega)$. To prove the other inclusion, let $\boldsymbol{v} = (v_i)_{i=1}^N \in \boldsymbol{E}(\Omega)$. Then for $1 \leq i,\, j, k \leq N$,

$$\partial_k v_i \in H^{-1}(\Omega),$$
$$\partial_j(\partial_k v_i) = \{\partial_j e_{ik}(\boldsymbol{v}) + \partial_k e_{ij}(\boldsymbol{v}) - \partial_i e_{jk}(\boldsymbol{v})\} \in H^{-1}(\Omega),$$

since $w \in L^2(\Omega)$ implies $\partial_\ell w \in H^{-1}(\Omega)$, $1 \leq \ell \leq N$. Hence $\partial_k v_i \in L^2(\Omega)$ by the *lemma of J.L. Lions* (Theorem 6.11-4), and thus $\boldsymbol{v} \in \boldsymbol{H}^1(\Omega)$.

(iii) *Korn's inequality.*

The identity mapping ι from $\boldsymbol{H}^1(\Omega)$ equipped with $\|\cdot\|_{1,\Omega}$ into $\boldsymbol{E}(\Omega)$ equipped with $\|\cdot\|$ is injective, continuous (there clearly exists a constant c such that $\|\boldsymbol{v}\| \leq c\,\|\boldsymbol{v}\|_{1,\Omega}$ for all $\boldsymbol{v} \in \boldsymbol{H}^1(\Omega)$), and surjective by (ii).

Therefore the *corollary to the Banach open mapping theorem* (Theorem 5.6-2) shows that the inverse mapping ι^{-1} is also continuous, which is exactly what is expressed by *Korn's inequality*. □

[66] The proof given here follows that of Theorem 3.3 in Duvaut & Lions [1976, Chapter 3, Section 3].

Similar inequalities can be established on a domain Ω in $\mathbb{R}^N$, such as a *Korn inequality in* $\boldsymbol{W}^{1,p}(\Omega)$, which asserts that for each $1 < p < \infty$, there exists a constant C_p such that[67]

$$\|\boldsymbol{v}\|_{0,\Omega} \le C_p\big(\|\boldsymbol{v}\|^p_{-1,\Omega} + \|\boldsymbol{e}(\boldsymbol{v})\|^p_{-1,\Omega}\big)^{1/p} \quad \text{for all } \boldsymbol{v} \in \boldsymbol{W}^{1,p}(\Omega),$$

or a *Korn inequality in* $\boldsymbol{L}^2(\Omega)$ (this inequality will be established in the course of the proof of Theorem 6.19-2), which asserts that there exists a constant C such that

$$\|\boldsymbol{v}\|_{0,\Omega} \le C\big(\|\boldsymbol{v}\|^2_{-1,\Omega} + \|\boldsymbol{e}(\boldsymbol{v})\|^2_{-1,\Omega}\big)^{1/2} \quad \text{for all } \boldsymbol{v} \in \boldsymbol{L}^2(\Omega).$$

Our next goal is to establish an equivalent form of the Korn inequality in $\boldsymbol{H}^1(\Omega)$, this time in a *quotient space* (Theorem 6.15-3). For this purpose, we first need to identify those vector fields $\boldsymbol{v} \in \boldsymbol{H}^1(\Omega)$ that satisfy $\boldsymbol{e}(\boldsymbol{v}) = \boldsymbol{0}$ in $\mathbb{L}^2(\Omega)$ (Theorem 6.15-2). The notation $\mathbb{A}^N$ designates the space of all real $N \times N$ antisymmetric matrices.

Theorem 6.15-2 *Let* Ω *be a connected open subset of* $\mathbb{R}^N$. *Then*

$$\{\boldsymbol{v} \in \boldsymbol{H}^1(\Omega);\ \boldsymbol{e}(\boldsymbol{v}) = \boldsymbol{0} \textit{ in } \Omega\} = \{\boldsymbol{v} \in \boldsymbol{H}^1(\Omega);\ \textit{there exist } \boldsymbol{B} \in \mathbb{A}^N \textit{ and } \boldsymbol{c} \in \mathbb{R}^N \textit{ such that } \boldsymbol{v}(x) = \boldsymbol{B}x + \boldsymbol{c} \textit{ for almost all } x \in \Omega\}.$$

Proof For each $1 \le i, j, k \le N$, any vector field $\boldsymbol{v} = (v_i) \in \boldsymbol{H}^1(\Omega)$ satisfies

$$\int_\Omega (\partial_j v_i)\partial_k\varphi\,\mathrm{d}x = \int_\Omega \{e_{ij}(\boldsymbol{v})\partial_k\varphi + e_{ik}(\boldsymbol{v})\partial_j\varphi - e_{jk}(\boldsymbol{v})\partial_i\varphi\}\,\mathrm{d}x \quad \text{for all } \varphi \in \mathcal{D}(\Omega),$$

since the two sides of this relation are equal to $-\int_\Omega v_i\partial_{kj}\varphi\,\mathrm{d}x$ (to see this, simply use the definition of the functions $e_{ij}(\boldsymbol{v})$, the definition of a weak derivative, and the observation that each function $\partial_k\varphi$ belongs to $\mathcal{D}(\Omega)$ if $\varphi \in \mathcal{D}(\Omega)$). Consequently,

$$\boldsymbol{e}(\boldsymbol{v}) = \boldsymbol{0} \text{ in } \Omega \quad \text{implies} \quad \int_\Omega (\partial_j v_i)\partial_k\varphi\,\mathrm{d}x = 0 \text{ for all } \varphi \in \mathcal{D}(\Omega).$$

Since $\partial_j v_i \in L^2(\Omega) \subset L^1_{\mathrm{loc}}(\Omega)$, there exist constants b_{ij}, $1 \le i,\, j \le N$, such that

$$\partial_j v_i(x) = b_{ij} \quad \text{for almost all } x \in \Omega$$

(by Theorem 6.3-4; recall that a domain is connected). In addition, $e_{ij}(\boldsymbol{v}) = 0$ implies that

$$b_{ij} = -b_{ji}.$$

Let

$$w_i(x) := \sum_{j=1}^N b_{ij}x_j \quad \text{for each } x \in \Omega,\ 1 \le i \le N.$$

Then

$$\int_\Omega v_i\partial_j\varphi\,\mathrm{d}x = -\int_\Omega (\partial_j v_i)\varphi\,\mathrm{d}x = -b_{ij}\int_\Omega \varphi\,\mathrm{d}x = -\int_\Omega (\partial_j w_i)\varphi\,\mathrm{d}x = \int_\Omega w_i\partial_j\varphi\,\mathrm{d}x$$

[67] G. Geymonat; P. Suquet [1986]: Functional spaces for Norton-Hoff materials, *Mathematical Methods in the Applied Sciences* **8**, 206–222.

for all $\varphi \in \mathcal{D}(\Omega)$ (by definition of the weak derivatives $\partial_j v_i$). There thus exist constants c_i such that $(v_i - w_i)(x) = c_i$, $1 \leq i \leq N$, for almost all $x \in \Omega$ (again by Theorem 6.3-4).

We have therefore shown that, if a vector field $\boldsymbol{v} \in \boldsymbol{H}^1(\Omega)$ satisfies $\boldsymbol{e}(\boldsymbol{v}) = \boldsymbol{0}$ in Ω, there exist an $N \times N$ antisymmetric matrix $\boldsymbol{B} = (b_{ij})$ and a vector $\boldsymbol{c} \in \mathbb{R}^N$ such that

$$\boldsymbol{v}(x) = \boldsymbol{B}x + \boldsymbol{c} \quad \text{for almost all } x \in \Omega. \qquad \square$$

Remark Expressed in terms of *weak derivatives*, the first relation in the above proof asserts that, for each $1 \leq i, j, k \leq N$,

$$\partial_{jk} v_i = \partial_k e_{ij}(\boldsymbol{v}) + \partial_j e_{ik}(\boldsymbol{v}) - \partial_i e_{jk}(v) \quad \text{in } H^{-1}(\Omega),$$

hence also *in the sense of distributions.* $\square$

Theorem 6.15-2 implies that, *when* $N = 3$, a vector field $\boldsymbol{v} \in \boldsymbol{H}^1(\Omega)$ satisfies $\boldsymbol{e}(\boldsymbol{v}) = \boldsymbol{0}$ in $\mathbb{L}^2_s(\Omega)$ if and only if there exist two vectors $\boldsymbol{b} \in \mathbb{R}^3$ and $\boldsymbol{c} \in \mathbb{R}^3$ such that

$$\boldsymbol{v}(x) = \boldsymbol{b} \wedge \boldsymbol{ox} + \boldsymbol{c} \quad \text{for almost all } x \in \Omega$$

(the "if" part is immediately verified). When thought of as a displacement field of the set Ω, such a vector field is called an **infinitesimal rigid displacement.**[68]

Let $\mathbb{M}^N$ denote the space of all $N \times N$ real matrices. Given an open subset Ω of $\mathbb{R}^N$ and a smooth enough vector field $\boldsymbol{v} = (v_i) : \Omega \to \mathbb{R}^N$, the *gradient matrix* field *of* $\boldsymbol{v}$ is the matrix field $\boldsymbol{\nabla v} : \Omega \to \mathbb{M}^N$ defined by $(\boldsymbol{\nabla v})_{ij} := \partial_j v_i$. Hence the matrix field $\boldsymbol{e}(\boldsymbol{v}) : \Omega \to \mathbb{S}^N$ introduced in this section can be also defined by

$$\boldsymbol{e}(\boldsymbol{v}) := \frac{1}{2}(\boldsymbol{\nabla v}^T + \boldsymbol{\nabla v}).$$

For this reason, $\boldsymbol{e}(\boldsymbol{v})$ is also called the **symmetrized gradient** field of $\boldsymbol{v}$ and will be also denoted (as in the next theorem) by the more "operator-like" notation

$$\boldsymbol{\nabla}_s \boldsymbol{v} := \frac{1}{2}(\boldsymbol{\nabla v}^T + \boldsymbol{\nabla v}).$$

Theorem 6.15-3 (Korn's inequality in the quotient space $\boldsymbol{H}^1(\Omega)/\,\mathbf{Ker}\,\boldsymbol{\nabla}_s$) *Let Ω be a domain in $\mathbb{R}^N$. Define the quotient space*

$$\dot{\boldsymbol{H}}^1(\Omega) := \boldsymbol{H}^1(\Omega)/\,\mathbf{Ker}\,\boldsymbol{\nabla}_s,$$

where

$$\mathbf{Ker}\,\boldsymbol{\nabla}_s := \{\boldsymbol{v} \in \boldsymbol{H}^1(\Omega);\ \boldsymbol{\nabla}_s \boldsymbol{v} = \boldsymbol{0} \text{ in } \Omega\}.$$

Equipped with the quotient norm $\|\cdot\|_{1,\Omega}$ defined by

$$\|\dot{\boldsymbol{v}}\|_{1,\Omega} := \inf_{\boldsymbol{r} \in \mathrm{Ker}\,\boldsymbol{\nabla}_s} \|\boldsymbol{v} + \boldsymbol{r}\|_{1,\Omega} \quad \textit{for each } \dot{\boldsymbol{v}} \in \dot{\boldsymbol{H}}^1(\Omega),$$

[68] "Infinitesimal" reflects that the space of such vector fields is the tangent space at the origin of the manifold of rigid deformations of the set Ω; see Theorem 4.1 in:

P.G. Ciarlet; C. Mardare [2003]: On rigid and infinitesimal rigid displacements in three-dimensional elasticity, *Mathematical Models and Methods in Applied Sciences* **13**, 1589–1598.

the space $\dot{\boldsymbol{H}}^1(\Omega)$ *is thus a Hilbert space (Problem 4.1-5). Then:*

(a) *The Korn inequality in* $\boldsymbol{H}^1(\Omega)$ *(Theorem 6.15-1) implies that there exists a constant* $\dot{C} = \dot{C}(\Omega)$ *such that the* **Korn's inequality in** $\dot{\boldsymbol{H}}^1(\Omega)$ *holds, viz.,*

$$\|\dot{\boldsymbol{v}}\|_{1,\Omega} \leq \dot{C}\,\|\boldsymbol{e}(\dot{\boldsymbol{v}})\|_{0,\Omega} \quad \textit{for all } \dot{\boldsymbol{v}} \in \dot{\boldsymbol{H}}^1(\Omega),$$

where $\boldsymbol{e}(\dot{\boldsymbol{v}}) := \boldsymbol{e}(\boldsymbol{w})$ *for any* $\boldsymbol{w} \in \dot{\boldsymbol{v}}$.

(b) *Conversely, the Korn inequality in* $\dot{\boldsymbol{H}}^1(\Omega)$ *implies the Korn inequality in* $\boldsymbol{H}^1(\Omega)$.

Proof By Theorem 6.15-2, the space $\mathbf{Ker}\,\boldsymbol{\nabla}_s$ is finite-dimensional and its dimension is $M := \dfrac{N(N+1)}{2}$.

By the Hahn–Banach theorem in a normed vector space (Theorem 5.9-1), there exist M continuous linear forms ℓ_α on $\boldsymbol{H}^1(\Omega)$, $1 \leq \alpha \leq M$, with the following property: An element $\boldsymbol{r} \in \mathbf{Ker}\,\boldsymbol{\nabla}_s$ is equal to $\boldsymbol{0}$ if and only if $\ell_\alpha(\boldsymbol{r}) = 0$, $1 \leq \alpha \leq M$. We then claim that there exists a constant D such that

$$\|\boldsymbol{v}\|_{1,\Omega} \leq D\Big(\|\boldsymbol{e}(\boldsymbol{v})\|_{0,\Omega} + \sum_{\alpha=1}^{M} |\ell_\alpha(\boldsymbol{v})| \Big) \quad \text{for all } \boldsymbol{v} \in \boldsymbol{H}^1(\Omega).$$

This inequality in turn immediately implies Korn's inequality in $\dot{\boldsymbol{H}}^1(\Omega)$: Given any $\boldsymbol{v} \in \boldsymbol{H}^1(\Omega)$, let $\boldsymbol{r}(\boldsymbol{v}) \in \mathbf{Ker}\,\boldsymbol{\nabla}_s$ be such that $\ell_\alpha(\boldsymbol{v} + \boldsymbol{r}(\boldsymbol{v})) = 0$, $1 \leq \alpha \leq M$; then

$$\|\dot{\boldsymbol{v}}\|_{1,\Omega} = \inf_{\boldsymbol{r} \in \mathbf{Ker}\,\boldsymbol{\nabla}_s} \|\boldsymbol{v} + \boldsymbol{r}\|_{1,\Omega} \leq \|\boldsymbol{v} + \boldsymbol{r}(\boldsymbol{v})\|_{1,\Omega} \leq D\,\|\boldsymbol{e}(\boldsymbol{v})\|_{0,\Omega} = D\,\|\boldsymbol{e}(\dot{\boldsymbol{v}})\|_{0,\Omega}\,.$$

To establish the existence of such a constant D, assume the contrary. Then there exist $\boldsymbol{v}^k \in \boldsymbol{H}^1(\Omega)$, $k \geq 1$, such that

$$\|\boldsymbol{v}^k\|_{1,\Omega} = 1 \ \text{ for all } k \geq 1 \quad \text{and} \quad \Big(\|\boldsymbol{e}(\boldsymbol{v}^k)\|_{0,\Omega} + \sum_{\alpha=1}^{M} |\ell_\alpha(\boldsymbol{v}^k)| \Big) \underset{k\to\infty}{\longrightarrow} 0.$$

By the Rellich–Kondrachov theorem (Theorem 6.6-3), there exists a subsequence $(\boldsymbol{v}^\ell)_{\ell=1}^\infty$ that converges in $\boldsymbol{L}^2(\Omega)$. Since the sequence $(\boldsymbol{e}(\boldsymbol{v}^\ell))_{\ell=1}^\infty$ also converges in $\mathbb{L}^2(\Omega)$, the subsequence $(\boldsymbol{v}^\ell)_{\ell=1}^\infty$ is a Cauchy sequence with respect to the norm $\boldsymbol{v} \to \big(\|\boldsymbol{v}\|_{0,\Omega}^2 + \|\boldsymbol{e}(\boldsymbol{v})\|_{0,\Omega}^2\big)^{1/2}$, and hence also with respect to the norm $\|\cdot\|_{1,\Omega}$ by the *Korn inequality in* $\boldsymbol{H}^1(\Omega)$ (Theorem 6.15-1). Consequently, there exists $\boldsymbol{v} \in \boldsymbol{H}^1(\Omega)$ such that

$$\|\boldsymbol{v}^\ell - \boldsymbol{v}\|_{1,\Omega} \underset{\ell\to\infty}{\longrightarrow} 0.$$

But then $\boldsymbol{v} = \boldsymbol{0}$ since $\boldsymbol{e}(\boldsymbol{v}) = 0$ and $\ell_\alpha(\boldsymbol{v}) = 0$, $1 \leq \alpha \leq M$, in contradiction with the relations $\|\boldsymbol{v}^\ell\|_{1,\Omega} = 1$ for all $\ell \geq 1$. This proves (a).[69]

We next show that, conversely, *Korn's inequality in the quotient space* $\dot{\boldsymbol{H}}^1(\Omega)$ *implies Korn's inequality in the space* $\boldsymbol{H}^1(\Omega)$.

[69] Another proof of (a) is found in Duvaut & Lions [1976, Chapter 3, Theorem 3.4].

Assume the contrary. Then there exist $\boldsymbol{v}^k \in \boldsymbol{H}^1(\Omega)$, $k \geq 1$, such that

$$\|\boldsymbol{v}^k\|_{1,\Omega} = 1 \text{ for all } k \geq 1 \quad \text{and} \quad \left(\|\boldsymbol{v}^k\|_{0,\Omega}^2 + \|\boldsymbol{e}(\boldsymbol{v}^k)\|_{0,\Omega}^2\right)^{1/2} \underset{k\to\infty}{\longrightarrow} 0.$$

Let $\boldsymbol{r}^k \in \mathbf{Ker}\,\boldsymbol{\nabla}_s$ denote for each $k \geq 1$ the projection of $\boldsymbol{v}^k$ on $\mathbf{Ker}\,\boldsymbol{\nabla}_s$ with respect to the inner-product of $\boldsymbol{H}^1(\Omega)$, which thus satisfies

$$\|\boldsymbol{v}^k - \boldsymbol{r}^k\|_{1,\Omega} = \inf_{\boldsymbol{r}\in\mathbf{Ker}\,\boldsymbol{\nabla}_s} \|\boldsymbol{v}^k - \boldsymbol{r}\|_{1,\Omega} \quad \text{and} \quad \|\boldsymbol{v}^k\|_{1,\Omega}^2 = \|\boldsymbol{v}^k - \boldsymbol{r}^k\|_{1,\Omega}^2 + \|\boldsymbol{r}^k\|_{1,\Omega}^2.$$

The space $\mathbf{Ker}\,\boldsymbol{\nabla}_s$ being finite-dimensional, the inequalities $\|\boldsymbol{r}^k\|_{1,\Omega} \leq 1$ for all $k \geq 1$ imply the existence of a subsequence $(\boldsymbol{r}^\ell)_{\ell=1}^\infty$ that converges in $\boldsymbol{H}^1(\Omega)$ to an element $\boldsymbol{r} \in \mathbf{Ker}\,\boldsymbol{\nabla}_s$. Besides, *Korn's inequality in* $\boldsymbol{H}^1(\Omega)$ implies that $\|\boldsymbol{v}^\ell - \boldsymbol{r}^\ell\|_{1,\Omega} \underset{\ell\to\infty}{\longrightarrow} 0$, so that

$$\|\boldsymbol{v}^\ell - \boldsymbol{r}\|_{1,\Omega} \underset{\ell\to\infty}{\longrightarrow} 0.$$

Hence $\|\boldsymbol{v}^\ell - \boldsymbol{r}\|_{0,\Omega} \underset{\ell\to\infty}{\longrightarrow} 0$, which forces $\boldsymbol{r}$ to be $\mathbf{0}$, since $\|\boldsymbol{v}^\ell\|_{0,\Omega} \to 0$ on the other hand. We thus reach the conclusion that $\|\boldsymbol{v}^\ell\|_{1,\Omega} \to 0$, a contradiction. □

Finally, we examine the effect of (homogeneous) *boundary conditions*.

Recall that the seminorm $|\cdot|_{1,\Omega}$ becomes a norm equivalent to $\|\cdot\|_{1,\Omega}$ on the closed subspace $\{v \in H^1(\Omega);\ v = 0 \text{ on } \Gamma_0\}$ of $H^1(\Omega)$ if dΓ-meas $\Gamma_0 > 0$ (Theorem 6.6-6). As shown in the next theorem, the seminorm $\boldsymbol{v} \in \boldsymbol{H}^1(\Omega) \to \|\boldsymbol{e}(\boldsymbol{v})\|_{0,\Omega}$ similarly becomes a norm equivalent to $\|\cdot\|_{1,\Omega}$ over the closed subspace $\{\boldsymbol{v} \in \boldsymbol{H}^1(\Omega);\ \boldsymbol{v} = \mathbf{0} \text{ on } \Gamma_0\}$ of $\boldsymbol{H}^1(\Omega)$ if dΓ-meas $\Gamma_0 > 0$.

Notice that, while the proof for an arbitrary subset $\Gamma_0 \subset \Gamma$ with dΓ-meas $\Gamma_0 > 0$ rests on Korn's inequality (Theorem 6.15-1), the proof of which itself rests on J.L. Lions lemma, the proof in the special case where $\Gamma_0 = \Gamma$ becomes deceptively easy, as an immediate corollary of a simple identity (Problem 6.15-1).

Theorem 6.15-4 (Korn's inequality with boundary conditions) *Let Ω be a domain in $\mathbb{R}^N$, and let Γ_0 be a dΓ-measurable subset of the boundary Γ of Ω such that dΓ-meas $\Gamma_0 > 0$. Then the space*

$$\boldsymbol{V} := \{\boldsymbol{v} \in \boldsymbol{H}^1(\Omega);\ \boldsymbol{v} = \mathbf{0} \text{ on } \Gamma_0\}$$

is a closed subspace of $\boldsymbol{H}^1(\Omega)$ and there exists a constant $C = C(\Omega, \Gamma_0)$ such that

$$\|\boldsymbol{v}\|_{1,\Omega} \leq C\,\|\boldsymbol{e}(\boldsymbol{v})\|_{0,\Omega} \quad \textit{for all } \boldsymbol{v} \in \boldsymbol{V}.$$

Proof (i) *The space $\boldsymbol{V}$ is closed in $\boldsymbol{H}^1(\Omega)$ and the seminorm $|\cdot| : \boldsymbol{H}^1(\Omega) \to \mathbb{R}$ defined by*

$$|\boldsymbol{v}| := \|\boldsymbol{e}(\boldsymbol{v})\|_{0,\Omega} \quad \text{for each } \boldsymbol{v} \in \boldsymbol{H}^1(\Omega)$$

becomes a norm over the space $\boldsymbol{V}$.

That $\boldsymbol{V}$ is closed in $\boldsymbol{H}^1(\Omega)$ is established as in the proof of Theorem 6.7-5. We saw in Theorem 6.15-2 that, if a vector field $\boldsymbol{v} \in \boldsymbol{H}^1(\Omega)$ satisfies $\boldsymbol{e}(\boldsymbol{v}) = \mathbf{0}$ in Ω, then there exist an $N \times N$ antisymmetric matrix $\boldsymbol{B}$ and a vector $\boldsymbol{c} \in \mathbb{R}^N$ such that

$$\boldsymbol{v}(x) = \boldsymbol{B}x + \boldsymbol{c} \quad \text{for almost all } x \in \Omega.$$

Since the subset of $\mathbb{R}^N$ where such a vector field $\boldsymbol{v}$ vanishes is always of zero $\mathbb{R}^{N-1}$-measure unless $\boldsymbol{B} = \boldsymbol{0}$ and $\boldsymbol{c} = \boldsymbol{0}$ (Problem 6.15-2), it follows that $\boldsymbol{v} = \boldsymbol{0}$ when $d\Gamma$-meas $\Gamma_0 > 0$. Hence the seminorm $|\cdot|$ becomes a norm over the space $\boldsymbol{V}$.

(ii) *Korn's inequality with boundary conditions.*

If this inequality is false, there exists a sequence $(\boldsymbol{v}^k)_{k=1}^\infty$ of elements $\boldsymbol{v}^k \in \boldsymbol{V}$ such that

$$\|\boldsymbol{v}^k\|_{1,\Omega} = 1 \text{ for all } k \quad \text{and} \quad \lim_{k\to\infty} \|\boldsymbol{e}(\boldsymbol{v}^k)\|_{0,\Omega} = 0.$$

The sequence $(\boldsymbol{v}^k)_{k=1}^\infty$ being then bounded in $\boldsymbol{H}^1(\Omega)$, there exists a subsequence $(\boldsymbol{v}^\ell)_{\ell=1}^\infty$ that converges in $\boldsymbol{L}^2(\Omega)$ by the *Rellich–Kondrachov theorem* (Theorem 6.6-3); furthermore, the sequence $(\boldsymbol{e}(\boldsymbol{v}^\ell))_{\ell=1}^\infty$ also converges in $\boldsymbol{L}^2(\Omega)$ (to $\boldsymbol{0}$, but this fact is not used at this stage). The subsequence $(\boldsymbol{v}^\ell)_{\ell=1}^\infty$ is thus a Cauchy sequence with respect to the norm $\|\cdot\|$ defined by

$$\|\boldsymbol{v}\| := \left(\|\boldsymbol{v}\|_{0,\Omega}^2 + \|\boldsymbol{e}(\boldsymbol{v})\|_{0,\Omega}^2\right)^{1/2} \quad \text{for each } \boldsymbol{v} \in \boldsymbol{H}^1(\Omega),$$

hence with respect to the norm $\|\cdot\|_{1,\Omega}$, by the *Korn inequality in* $\boldsymbol{H}^1(\Omega)$ (Theorem 6.15-1).

The space $\boldsymbol{V}$ being complete (as a closed subspace of $\boldsymbol{H}^1(\Omega)$), there exists $\boldsymbol{v} \in \boldsymbol{V}$ such that

$$\boldsymbol{v}^\ell \to \boldsymbol{v} \quad \text{in } \boldsymbol{H}^1(\Omega) \text{ as } \ell \to \infty,$$

and the limit $\boldsymbol{v}$ satisfies $\|\boldsymbol{e}(\boldsymbol{v})\|_{0,\Omega} = \lim_{\ell\to\infty} \|\boldsymbol{e}(\boldsymbol{v}^\ell)\|_{0,\Omega} = 0$; hence $\boldsymbol{v} = \boldsymbol{0}$ by (i). But this contradicts the relations $\|\boldsymbol{v}^\ell\|_{1,\Omega} = 1$ for all $\ell \geq 1$. This completes the proof. □

As shown in the next section, the Korn inequalities established above are (with $N = 3$) essential for proving existence theorems in *three-dimensional linearized elasticity*. Other Korn inequalities can be likewise established that are this time essential for proving existence theorems in *linearized shell theory*. They include: a general *Korn inequality on a surface*[70] (a surface being defined as in Section 8.8); a *Korn inequality on an elliptic surface*[71] (i.e., a surface in which all the points are "elliptic"; cf. Section 8.12); a *Korn inequality on a surface without boundary*;[72] a *Korn inequality on an elliptic surface without boundary*;[73] or a *Korn*

[70]Due to:

M. Bernadou; P.G. Ciarlet [1976]: Sur l'ellipticité du modèle linéaire de coques de W.T. Koiter, in *Computing Methods in Applied Sciences and Engineering* (R. Glowinski & J.L. Lions, editors), pp. 89–136, Lecture Notes in Economics and Mathematical Systems, **134**, Springer, Heidelberg.

Other proofs or generalizations have been then given by:

M. Bernadou; P.G. Ciarlet; B. Miara [1994]: Existence theorems for two-dimensional linear shell theories, *Journal of Elasticity* **34**, 111–138.

A. Blouza; H. Le Dret: [1999]: Existence and uniqueness for the linear Koiter model for shells with little regularity, *Quarterly of Applied Mathematics* **57**, 317–337.

P.G. Ciarlet; S. Mardare [2001]: On Korn's inequalities in curvilinear coordinates, *Mathematical Models and Methods in Applied Sciences* **11**, 1379–1391.

J.L. Akian [2003]: A simple proof of the ellipticity of Koiter's model, *Analysis and Applications* **1**, 1–16.

[71]P.G. Ciarlet; V. Lods [1996]: On the ellipticity of linear membrane shell equations, *Journal de Mathématiques Pures et Appliquées* **75**, 107–124.

P.G. Ciarlet; E. Sanchez-Palencia [1996]: An existence and uniqueness theorem for the two-dimensional linear membrane shell equations, *Journal de Mathématiques Pures et Appliquées* **75**, 51–67.

[72]S. Mardare [2003]: Inequality of Korn's type on compact surfaces without boundary, *Chinese Annals of Mathematics, Series B*, **24**, 191–204.

[73]S. Slicaru [1998]: On the ellipticity of the middle surface of a shell and its application to the asymptotic analysis of "membrane shells," *Journal of Elasticity* **46**, 33–42.

inequality on a Riemannian manifold.[74]

Problems

6.15-1 Let Ω be an open subset of $\mathbb{R}^N$.
(1) Given a vector field $\boldsymbol{v} = (v_i)_{i=1}^N : \Omega \to \mathbb{R}^N$ with components $v_i \in \mathcal{C}^\infty(\Omega)$, show that

$$2\sum_{i,j=1}^N |e_{ij}(\boldsymbol{v})|^2 - \sum_{i,j=1}^N |\partial_i v_j|^2 = \Big|\sum_{i=1}^N \partial_i v_i\Big|^2 + \sum_{i,j=1}^N \partial_i(v_j\partial_j v_i - v_i \partial_j v_j) \quad \text{in } \Omega.$$

(2) Deduce from (1) that

$$|\boldsymbol{v}|_{1,\Omega} \leq \sqrt{2}\, \|\boldsymbol{e}(\boldsymbol{v})\|_{0,\Omega} \quad \text{for all } \boldsymbol{v} \in \boldsymbol{H}_0^1(\Omega).$$

(3) Show that, if Ω is of finite width (Section 6.5), the seminorm $\boldsymbol{v} \in \boldsymbol{H}_0^1(\Omega) \to \|\boldsymbol{e}(\boldsymbol{v})\|_{0,\Omega}$ becomes a norm on the space $\boldsymbol{H}_0^1(\Omega)$, equivalent to the norm $\|\cdot\|_{1,\Omega}$ (this result constitutes the special case $\Gamma_0 = \Gamma$ of Theorem 6.15-4, but with a much weaker assumption on Ω since Ω was assumed to be a domain in *ibid.*).

6.15-2 (1) Show that, given two vectors $\boldsymbol{b}, \boldsymbol{c} \in \mathbb{R}^3$, the set $\boldsymbol{E} := \{x \in \mathbb{R}^3;\ \boldsymbol{b} \wedge \boldsymbol{ox} + \boldsymbol{c} = \boldsymbol{0}\}$ is of zero area, unless $\boldsymbol{b} = \boldsymbol{c} = \boldsymbol{0}$.
Hint: Show that $\boldsymbol{E} = \varnothing$ if either $\boldsymbol{b} = \boldsymbol{0}$ and $\boldsymbol{c} \neq \boldsymbol{0}$, or $\boldsymbol{b} \neq \boldsymbol{0}$ and $\boldsymbol{b}\cdot\boldsymbol{c} \neq 0$. Then show that $\boldsymbol{E} = \Big\{\Big(\dfrac{\boldsymbol{b}\wedge\boldsymbol{c}}{\boldsymbol{b}\cdot\boldsymbol{b}} + t\boldsymbol{b}\Big) \in \mathbb{R}^3;\ t \in \mathbb{R}\Big\}$ if $\boldsymbol{b} \neq \boldsymbol{0}$ and $\boldsymbol{b}\cdot\boldsymbol{c} = 0$.
(2) More generally, show that, given an $N \times N$ antisymmetric matrix $\boldsymbol{B}$ and a vector $\boldsymbol{c} \in \mathbb{R}^N$, the set $\boldsymbol{E} := \{x \in \mathbb{R}^N;\ \boldsymbol{B}x + \boldsymbol{c} = \boldsymbol{0}\}$ is of zero $\mathbb{R}^{N-1}$-Lebesgue measure, unless $\boldsymbol{B} = \boldsymbol{0}$ and $\boldsymbol{c} = \boldsymbol{0}$ (this result is used in part (i) of the proof of Theorem 6.15-4).

6.15-3 Let ω be a domain in $\mathbb{R}^2$. Theorem 6.15-1 shows that there exists a constant $c = c(\omega) > 0$ such that

$$\|\boldsymbol{\eta}\|_{1,\omega} \leq c\Big(\|\boldsymbol{\eta}\|_{0,\omega}^2 + \|\boldsymbol{e}(\boldsymbol{\eta})\|_{0,\omega}^2\Big)^{1/2} \quad \text{for all } \boldsymbol{\eta} = (\eta_\alpha) \in H^1(\omega) \times H^1(\omega),$$

where $\boldsymbol{e}(\boldsymbol{\eta}) = (e_{\alpha\beta}(\boldsymbol{\eta}))$, with $e_{\alpha\beta}(\boldsymbol{\eta}) := (\frac{1}{2}\partial_\alpha\eta_\beta + \partial_\beta\eta_\alpha)$. Show that this *two-dimensional* Korn inequality in $\boldsymbol{H}^1(\omega)$ can be also derived from the *three-dimensional* Korn inequality in $\boldsymbol{H}^1(\omega \times]-1,1[)$.

6.15-4 In 1988, Vladimir Aleksandrovich Kondrat'ev and Olga Oleinik published a remarkably self-contained, and to a large extent elementary, proof of Korn's inequality.[75] Indeed, their proof, which is the object of the present problem, does not rely on advanced functional analytic results, such as the lemma of J.L. Lions (as in Theorem 6.15-1) or the Calderón–Zygmund theory of singular integrals as in the original proof of J. Gobert in 1982 (quoted earlier in this section).[76] It relies instead on *two crucial inequalities*, which constitute questions (1) and (2) below (the proof of these inequalities, especially the second one, is somewhat delicate, however), and on the *hypoellipticity of the Laplace operator* Δ (Theorem 6.4-2), which is used in question (2).

[74] W. Chen; J. Jost [2002]: A Riemannian version of Korn's inequality, *Calculus of Variations* **14**, 517–530.

[75] V.A. Kondrat'ev; O.A. Oleinik [1988]: Boundary-value problems for the system of elasticity theory in unbounded domains. Korn's inequalities, *Uspehi Mathematiceskii Nauk* **43**, 55–98 (in Russian) [English translation: *Russian Mathematical Surveys* **43** (1988), 65–119].

[76] Another proof of Korn's inequality that also relies on the Calderón–Zygmund theory of singular integrals (and on the Cesàro-Volterra path integral formula; cf. Theorem 6.18-2) is due to:

P.P. Mosolov; V.P. Mjasnikov [1971]: A proof of Korn's inequality, *Soviet Mathematics Doklady* **12**, 1618–1622.

In what follows, Ω is a domain in $\mathbb{R}^N$, the notations C_1, C_2, etc. designate various constants that only depend on Ω, the function $\rho : \overline{\Omega} \to \mathbb{R}$ is defined by $\rho(x) := \text{dist}(x, \partial\Omega)$ for each $x \in \overline{\Omega}$, and $\boldsymbol{u} = (u_i)_{i=1}^N \in \boldsymbol{\mathcal{C}}^\infty(\overline{\Omega})$ denotes a given vector field. The other notations are the same as elsewhere in the text.

(1) Show that

$$\int_\Omega \rho^2 \sum_{k=1}^N |\partial_k v|^2 \, dx \leq C_1 \left(\|v\|_{0,\Omega}^2 + \|\Delta v\|_{0,\Omega}^2 \right)$$

for all functions $v \in L^2(\Omega) \cap \mathcal{C}^\infty(\Omega)$ such that $\Delta v \in L^2(\Omega)$ (the right-hand side of this inequality is thus finite for all such functions v).

(2) Show that

$$|v|_{1,\Omega}^2 \leq C_2 \bigg(\int_\Omega \rho^2 \sum_{i,j=1}^N |\partial_{ij} v|^2 \, dx + \|v\|_{0,\Omega}^2 \bigg)$$

for all functions $v \in H^1(\Omega) \cap \mathcal{C}^\infty(\Omega)$ that satisfy $\int_\Omega \rho^2 \sum_{i,j=1}^N |\partial_{ij} v|^2 \, dx < \infty$.

(3) Construct a vector field $\boldsymbol{v} \in \boldsymbol{H}^1(\Omega) \cap \boldsymbol{\mathcal{C}}^\infty(\Omega)$ that satisfies

$$\Delta \boldsymbol{v} = \Delta \boldsymbol{u} \text{ in } \Omega \quad \text{and} \quad \|\boldsymbol{v}\|_{1,\Omega} \leq C_3 \, \|\boldsymbol{e}(\boldsymbol{u})\|_{0,\Omega} .$$

Hint: Use the relations $\Delta u_i = \sum_{j=1}^N (2\partial_j e_{ij}(\boldsymbol{u}) - \partial_i e_{jj}(\boldsymbol{u}))$, $1 \leq i \leq N$.

(4) Let $\boldsymbol{w} := \boldsymbol{u} - \boldsymbol{v}$. Using question (1), show that, for all $1 \leq i, j \leq N$,

$$\int_\Omega \rho^2 \sum_{k=1}^N |\partial_k e_{ij}(\boldsymbol{w})|^2 \, dx \leq C_4 \, \|\boldsymbol{e}(\boldsymbol{v})\|_{0,\Omega}^2 \leq C_5 \, \|\boldsymbol{e}(\boldsymbol{u})\|_{0,\Omega}^2 .$$

(5) Using question (4) and the identity

$$\partial_{jk} w_i = \partial_j e_{ik}(\boldsymbol{w}) + \partial_k e_{ij}(\boldsymbol{w}) - \partial_i e_{ik}(\boldsymbol{w}) \quad \text{for all } 1 \leq i, j, k \leq N$$

(it is not a coincidence that the same identity was used in the proof of Theorem 6.15-1), show that, for all $1 \leq k \leq N$,

$$\int_\Omega \rho^2 \sum_{i,j=1}^N |\partial_{ij} w_k|^2 \, dx \leq C_6 \, \|\boldsymbol{e}(\boldsymbol{u})\|_{0,\Omega}^2 .$$

(6) Using questions (2) and (5) and the relation $\boldsymbol{w} = \boldsymbol{u} - \boldsymbol{v}$, conclude that

$$|\boldsymbol{u}|_{1,\Omega}^2 \leq C_7 \left(\|\boldsymbol{u}\|_{0,\Omega}^2 + \|\boldsymbol{e}(\boldsymbol{u})\|_{0,\Omega}^2 \right) .$$

Korn's inequality then follows from this inequality, since the space $\mathcal{C}^\infty(\overline{\Omega})$ is dense in the space $H^1(\Omega)$ (Theorem 6.6-4).

6.16 Application of Korn's inequality: The equations of three-dimensional linearized elasticity

The objective of this section is to establish an *existence theorem for the weak formulation of the equations of three-dimensional linearized elasticity*, which are described in the next theorem in the usual manner (i.e., by prescribing a function space, a bilinear form, and a linear form). To this end, the crucial step consists as usual in verifying that the bilinear form found in this formulation is indeed *coercive*, a property that will follow from the *Korn inequality*.

In this section, Latin indices range in the set $\{1, 2, 3\}$, save when they are used for indexing sequences, and the summation convention with respect to repeated indices is used in conjunction with this rule.

Given a smooth enough 3×3 *matrix field* $\boldsymbol{\sigma} = (\sigma_{ij})$ defined over $\overline{\Omega}$, its *divergence* $\mathbf{div}\,\boldsymbol{\sigma} : \overline{\Omega} \to \mathbb{R}^3$ is the *vector field* defined by

$$\mathbf{div}\,\boldsymbol{\sigma} := \begin{pmatrix} \partial_1\sigma_{11} + \partial_2\sigma_{12} + \partial_3\sigma_{13} \\ \partial_1\sigma_{21} + \partial_2\sigma_{22} + \partial_3\sigma_{23} \\ \partial_1\sigma_{31} + \partial_2\sigma_{32} + \partial_3\sigma_{33} \end{pmatrix}.$$

The notations used for spaces of vector and matrix fields are the same as those used in Section 6.15. The matrix inner product is denoted : (see Section 4.2).

Theorem 6.16-1 (existence of a solution to the equations of three-dimensional linearized elasticity) *Let Ω be a domain in $\mathbb{R}^3$, let Γ_1 be a relatively open subset of $\Gamma := \partial\Omega$ such that*

$$\mathrm{d}\Gamma\text{-meas}\,\Gamma_0 > 0, \quad \textit{where } \Gamma_0 := \Gamma - \Gamma_1,$$

let λ and μ be two constants that satisfy

$$\lambda \geq 0 \quad \textit{and} \quad \mu > 0,$$

let

$$\boldsymbol{f} = (f_i) \in \boldsymbol{L}^2(\Omega) \quad \textit{and} \quad \boldsymbol{g} = (g_i) \in \boldsymbol{L}^2(\Gamma_1)$$

be two given vector fields, and finally, let

$$\boldsymbol{V} := \{\boldsymbol{v} = (v_i) \in \boldsymbol{H}^1(\Omega);\ \boldsymbol{v} = \boldsymbol{0} \text{ on } \Gamma_0\},$$

$$a(\boldsymbol{u}, \boldsymbol{v}) := \int_\Omega \{\lambda \operatorname{tr} \boldsymbol{e}(\boldsymbol{u}) \operatorname{tr} \boldsymbol{e}(\boldsymbol{v}) + 2\mu \boldsymbol{e}(\boldsymbol{u}) : \boldsymbol{e}(\boldsymbol{v})\}\,\mathrm{d}x \quad \textit{for each } \boldsymbol{u}, \boldsymbol{v} \in \boldsymbol{V},$$

where

$$\boldsymbol{e}(\boldsymbol{v}) := (e_{ij}(\boldsymbol{v})) \in \mathbb{L}^2(\Omega) \quad \textit{with } e_{ij}(\boldsymbol{v}) := \frac{1}{2}(\partial_j v_i + \partial_i v_j) \textit{ for each } \boldsymbol{v} = (v_i) \in \boldsymbol{H}^1(\Omega),$$

$$\ell(\boldsymbol{v}) := \int_\Omega \boldsymbol{f} \cdot \boldsymbol{v}\,\mathrm{d}x + \int_{\Gamma_1} \boldsymbol{g} \cdot \boldsymbol{v}\,\mathrm{d}\Gamma \quad \textit{for all } \boldsymbol{v} \in \boldsymbol{V}.$$

Then there exists a unique vector field $\boldsymbol{u} = (u_i) \in \boldsymbol{V}$ that minimizes the functional $J : \boldsymbol{V} \to \mathbb{R}$ defined by

$$J(\boldsymbol{v}) := \frac{1}{2}a(\boldsymbol{v}, \boldsymbol{v}) - \ell(\boldsymbol{v}) = \frac{1}{2}\int_\Omega \{\lambda(\operatorname{tr} \boldsymbol{e}(\boldsymbol{v}))^2 + 2\mu \boldsymbol{e}(\boldsymbol{v}) : \boldsymbol{e}(\boldsymbol{v})\}\,\mathrm{d}x - \left(\int_\Omega \boldsymbol{f} \cdot \boldsymbol{v}\,\mathrm{d}x + \int_{\Gamma_1} \boldsymbol{g} \cdot \boldsymbol{v}\,\mathrm{d}\Gamma\right)$$

for all $\boldsymbol{v} \in \boldsymbol{V}$, or equivalently, that satisfies the variational equations

$$\int_\Omega \{\lambda \operatorname{tr} \boldsymbol{e}(\boldsymbol{u}) \operatorname{tr} \boldsymbol{e}(\boldsymbol{v}) + 2\mu \boldsymbol{e}(\boldsymbol{u}) : \boldsymbol{e}(\boldsymbol{v})\}\,\mathrm{d}x = \int_\Omega \boldsymbol{f} \cdot \boldsymbol{v}\,\mathrm{d}x + \int_{\Gamma_1} \boldsymbol{g} \cdot \boldsymbol{v}\,\mathrm{d}\Gamma \quad \textit{for all } \boldsymbol{v} \in \boldsymbol{V}.$$

Assume in addition that $\boldsymbol{u} \in \boldsymbol{H}^2(\Omega)$. *Then* $\boldsymbol{u}$ *satisfies the boundary value problem*

$$\begin{aligned} -\,\mathbf{div}\{\lambda(\operatorname{tr}\boldsymbol{e}(\boldsymbol{u}))\boldsymbol{I} + 2\mu\boldsymbol{e}(\boldsymbol{u})\} &= \boldsymbol{f} \quad \text{in } \Omega, \\ \boldsymbol{u} &= \boldsymbol{0} \quad \text{on } \Gamma_0, \\ \{\lambda(\operatorname{tr}\boldsymbol{e}(\boldsymbol{u}))\boldsymbol{I} + 2\mu\boldsymbol{e}(\boldsymbol{u})\}\boldsymbol{\nu} &= \boldsymbol{g} \quad \text{on } \Gamma_1, \end{aligned}$$

where $\boldsymbol{\nu} = (\nu_i) : \Gamma \to \mathbb{R}^3$ *denotes the unit outer normal vector field along* Γ.

Proof As a closed subspace of $\boldsymbol{H}^1(\Omega)$ (Theorem 6.15-4), the space $\boldsymbol{V}$ is a Hilbert space. By the Cauchy–Schwarz inequality, the symmetric bilinear form $a(\cdot,\cdot)$ and the linear form ℓ are continuous over the space $\boldsymbol{H}^1(\Omega)$. The bilinear form is $\boldsymbol{V}$-coercive, since

$$\begin{aligned} a(\boldsymbol{v},\boldsymbol{v}) &= \int_\Omega \{\lambda(\operatorname{tr}\boldsymbol{e}(\boldsymbol{v}))^2 + 2\mu\boldsymbol{e}(\boldsymbol{v}) : \boldsymbol{e}(\boldsymbol{v})\}\,\mathrm{d}x \\ &\geq 2\mu \int_\Omega \boldsymbol{e}(\boldsymbol{v}) : \boldsymbol{e}(\boldsymbol{v})\,\mathrm{d}x = 2\mu\,\|\boldsymbol{e}(\boldsymbol{v})\|_{0,\Omega}^2 \quad \text{for all } \boldsymbol{v} \in \boldsymbol{V}, \end{aligned}$$

and, by the *Korn inequality with boundary conditions* (Theorem 6.15-4), there exists a constant $C > 0$ such that $\|\boldsymbol{e}(\boldsymbol{v})\|_{0,\Omega} \geq C^{-1}\|\boldsymbol{v}\|_{1,\Omega}$ for all $\boldsymbol{v} \in \boldsymbol{V}$.

Therefore, by Theorems 6.1-1 and 6.1-2, there exists a unique vector field $\boldsymbol{u}$ that minimizes the announced functional J over the space $\boldsymbol{V}$, or equivalently, that satisfies the announced variational equations.

In view of finding the corresponding boundary value problems, we first rewrite $a(\boldsymbol{u},\boldsymbol{v})$ for any $\boldsymbol{u} = (u_i)$, $\boldsymbol{v} = (v_i) \in \boldsymbol{H}^1(\Omega)$ as

$$a(\boldsymbol{u},\boldsymbol{v}) = \int_\Omega \sigma_{ij}(\boldsymbol{u})e_{ij}(\boldsymbol{v})\,\mathrm{d}x = \int_\Omega \sigma_{ij}(\boldsymbol{u})\partial_j v_i\,\mathrm{d}x,$$

where

$$\sigma_{ij}(\boldsymbol{u}) := \lambda \operatorname{tr}\boldsymbol{e}(\boldsymbol{u})\delta_{ij} + 2\mu e_{ij}(\boldsymbol{u}) = \sigma_{ji}(\boldsymbol{u}).$$

Thanks to the fundamental Green's formula (Theorem 6.6-7), the following *Green's formula* holds:

$$\int_\Omega \sigma_{ij}(\boldsymbol{u})\partial_j v_i\,\mathrm{d}x = -\int_\Omega (\partial_j\sigma_{ij}(\boldsymbol{u}))v_i\,\mathrm{d}x + \int_\Gamma \sigma_{ij}(\boldsymbol{u})\nu_j v_i\,\mathrm{d}\Gamma \quad \text{for all } \boldsymbol{u} \in \boldsymbol{H}^2(\Omega),\ \boldsymbol{v} \in \boldsymbol{H}^1(\Omega).$$

If $\boldsymbol{u} \in \boldsymbol{H}^2(\Omega)\cap\boldsymbol{V}$, the variational equations $a(\boldsymbol{u},\boldsymbol{v}) = \ell(\boldsymbol{v})$ for all $\boldsymbol{v} \in \boldsymbol{V}$ therefore become

$$\int_\Omega (-\partial_j\sigma_{ij}(\boldsymbol{u}) - f_i)v_i\,\mathrm{d}x = \int_{\Gamma_1} (g_i - \sigma_{ij}(\boldsymbol{u})\nu_j)v_i\,\mathrm{d}\Gamma \quad \text{for all } (v_i) \in \boldsymbol{V}.$$

In particular then, for each $1 \leq i \leq 3$,

$$\int_\Omega (-\partial_j\sigma_{ij}(\boldsymbol{u}) - f_i)v_i\,\mathrm{d}x = 0 \quad \text{for all } v_i \in \mathcal{D}(\Omega),$$

which implies that $-\partial_j\sigma_{ij}(\boldsymbol{u}) - f_i = 0$ in $L^2(\Omega)$ (Theorem 6.3-2), or equivalently, in vector form:

$$-\,\mathbf{div}\,\boldsymbol{\sigma}(\boldsymbol{u}) = \boldsymbol{f} \text{ in } \boldsymbol{L}^2(\Omega) \quad \text{with } \boldsymbol{\sigma}(\boldsymbol{u}) := (\sigma_{ij}(\boldsymbol{u})) = \lambda(\operatorname{tr}\boldsymbol{e}(\boldsymbol{u}))\boldsymbol{I} + 2\mu\boldsymbol{e}(\boldsymbol{u}).$$

Taking these equations into account, we are thus left for each $1 \leq i \leq 3$ with

$$\int_{\Gamma_1} (g_i - \sigma_{ij}(\boldsymbol{u})\nu_j) v_i \,\mathrm{d}\Gamma = 0 \quad \text{for all } v_i \in \{w \in H^1(\Omega);\ w = 0 \text{ on } \Gamma_0\},$$

which implies that $g_i - \sigma_{ij}(\boldsymbol{u})\nu_j = 0$ in $L^2(\Gamma_1)$ (Theorem 6.7-3), or equivalently, in vector form:

$$\boldsymbol{\sigma}(\boldsymbol{u})\boldsymbol{\nu} = \boldsymbol{g} \quad \text{in } \boldsymbol{L}^2(\Gamma_1).$$

Finally, $\boldsymbol{u} = \boldsymbol{0}$ on Γ_0 since $\boldsymbol{u} \in \boldsymbol{V}$. □

Remarks (1) The bilinear form $a(\cdot,\cdot)$ remains $\boldsymbol{V}$-coercive if the Lamé constants satisfy the weaker assumptions $3\lambda + 2\mu > 0$ and $\mu > 0$; cf. Problem 6.16-1(1).

(2) The special case $N = 3$ of the Sobolev imbedding theorem (Theorem 6.6-1) combined with the continuity of the trace operator (Theorem 6.6-5) show that the linear form ℓ remains continuous on the space $\boldsymbol{H}^1(\Omega)$ under the weaker assumptions that $\boldsymbol{f} \in \boldsymbol{L}^{6/5}(\Omega)$ and $\boldsymbol{g} \in L^{4/3}(\Gamma_1)$.

(3) The vector equation $-\mathbf{div}\{\lambda(\operatorname{tr}\boldsymbol{e}(\boldsymbol{u}))\boldsymbol{I} + 2\mu\boldsymbol{e}(\boldsymbol{u})\} = \boldsymbol{f}$ in Ω may be equivalently written in the form of the **Navier equations,**[77] viz.,

$$-\mu\Delta\boldsymbol{u} - (\lambda+\mu)\,\mathbf{grad}\operatorname{div}\boldsymbol{u} = \boldsymbol{f} \quad \text{in } \Omega$$

or

$$\mu\,\mathbf{curl}\,\mathbf{curl}\,\boldsymbol{u} - (\lambda+2\mu)\,\mathbf{grad}\operatorname{div}\boldsymbol{u} = \boldsymbol{f} \quad \text{in } \Omega.$$ □

The boundary value problem found in Theorem 6.16-1, viz.,

$$\begin{aligned} -\mathbf{div}\,\boldsymbol{\sigma}(\boldsymbol{u}) &= \boldsymbol{f} \quad \text{in } \Omega, \\ \boldsymbol{u} &= \boldsymbol{0} \quad \text{on } \Gamma_0, \\ \boldsymbol{\sigma}(\boldsymbol{u})\boldsymbol{\nu} &= \boldsymbol{g} \quad \text{on } \Gamma_1, \end{aligned}$$

where

$$\boldsymbol{\sigma}(\boldsymbol{u}) := \lambda(\operatorname{tr}\boldsymbol{e}(\boldsymbol{u}))\boldsymbol{I} + 2\mu\boldsymbol{e}(\boldsymbol{u}),$$

is called the **boundary value problem of three-dimensional linearized elasticity**. Like the Stokes equations in $\mathbb{R}^3$ (Section 6.14), it thus provides an example of a *system of three partial differential equations with three unknowns.*

More specifically, it is a mathematical model for the following physical situation: The set $\overline{\Omega}$ is the *reference configuration*[78] of an *elastic body*, subjected to *applied body forces* acting in its interior, of density $\boldsymbol{f} = (f_i) : \Omega \to \mathbb{R}^3$ *per* unit volume, and to *applied surface forces* acting on a portion Γ_1 of its boundary Γ of the set Ω, of density $\boldsymbol{g} = (g_i) : \Gamma_1 \to \mathbb{R}^3$ *per* unit area.

The *unknown* of the problem is the **displacement vector field** $\boldsymbol{u} = (u_i) : \overline{\Omega} \to \mathbb{R}^3$, i.e., the vector $\boldsymbol{u}(x) = (u_i(x))$ represents the displacement that each point x of the reference configuration $\overline{\Omega}$ undergoes under the action of the applied forces (Figure 9.7-1). The elastic body is assumed to be subjected to a *homogeneous boundary condition of place* on the portion

[77] So named after Claude Louis Navier (1785–1836).

[78] A detailed treatment of all the notions from elasticity theory used here (reference configuration, elastic body, applied forces, dead loads, etc.) is found in, e.g., Ciarlet [1988].

$\Gamma_0 := \Gamma - \Gamma_1$ of its boundary. This means that the boundary condition $\boldsymbol{u} = \mathbf{0}$ on Γ_0 is imposed on the unknown displacement vector field.

Finally, it is assumed that the *elastic material* constituting the body is *homogeneous, isotropic*, and that the reference configuration $\overline{\Omega}$ is a *natural state*. These assumptions imply that the behavior of the material is, "to within the first order," governed by only two constants, λ and μ, called the *Lamé constants*[79] of the material. Experimental evidence shows that the Lamé constants of actual elastic materials satisfy the inequalities $\lambda \geq 0$ and $\mu > 0$, which accordingly have been assumed to hold in Theorem 6.16-1 (the Lamé constants measure the "rigidity" of the constituting material: the larger they are, the more rigid the material is).

The symmetric matrix field $\boldsymbol{e}(\boldsymbol{u}) = (e_{ij}(\boldsymbol{u})) : \overline{\Omega} \to \mathbb{S}^3$ is called the *linearized strain tensor field*, the symmetric matrix field $\boldsymbol{\sigma}(\boldsymbol{u}) = (\sigma_{ij}(\boldsymbol{u})) : \overline{\Omega} \to \mathbb{S}^3$ is called the *linearized stress tensor field*, and their components $e_{ij}(\boldsymbol{u})$ and $\sigma_{ij}(\boldsymbol{u})$ are respectively called *linearized strains* and *linearized stresses*. The linear relation $\boldsymbol{\sigma}(\boldsymbol{u}) = \lambda(\operatorname{tr}\boldsymbol{e}(\boldsymbol{u}))\boldsymbol{I} + 2\mu\boldsymbol{e}(\boldsymbol{u})$ between these two linearized tensors, which is known in elasticity as *Hooke's law*,[80] characterizes a *homogeneous* and *isotropic linearly elastic body*.

The functional $J : \boldsymbol{v} \in \boldsymbol{V} \to \mathbb{R}$ found in Theorem 6.16-1 represents the **energy** of a *homogeneous*, and *isotropic, linearly elastic body*, and the variational equations $a(\boldsymbol{u}, \boldsymbol{v}) = \ell(\boldsymbol{v})$ for all $\boldsymbol{v} \in \boldsymbol{V}$ found in Theorem 6.16-1 constitute the *linearized principle of virtual work*, which thus holds for all *kinematically admissible* displacements $\boldsymbol{v} \in \boldsymbol{V}$, i.e., those vector fields $\boldsymbol{v} \in \boldsymbol{V}$ that satisfy the boundary condition $\boldsymbol{v} = \mathbf{0}$ on Γ_0.

Remark The energy of a *nonhomogeneous* and *anisotropic* linearly elastic body takes the more general form

$$J(\boldsymbol{v}) = \frac{1}{2}\int_\Omega \mathsf{A}\boldsymbol{e}(\boldsymbol{v}) : \boldsymbol{e}(\boldsymbol{v})\,\mathrm{d}x - \left(\int_\Omega \boldsymbol{f}\cdot\boldsymbol{v}\,\mathrm{d}x + \int_{\Gamma_1} \boldsymbol{g}\cdot\boldsymbol{v}\,\mathrm{d}\Gamma\right) \quad \text{for all } \boldsymbol{v} \in \boldsymbol{V},$$

where the *elasticity tensor* $\mathsf{A} = (A_{ijk\ell})$ possesses the following properties: Its components $A_{ijk\ell}$ are in $L^\infty(\Omega)$, they satisfy the symmetries $A_{ijk\ell} = A_{jik\ell} = A_{k\ell ij}$, and there exists a constant $\alpha > 0$ such that

$$\mathsf{A}(x)\boldsymbol{t} : \boldsymbol{t} \geq \alpha\boldsymbol{t} : \boldsymbol{t} \quad \text{for almost all } x \in \Omega \text{ and for all matrices } \boldsymbol{t} = (t_{ij}) \in \mathbb{S}^3,$$

where $(\mathsf{A}(x)\boldsymbol{t})_{ij} := A_{ijk\ell}(x)t_{k\ell}$. In this case, the relation $\boldsymbol{\sigma}(u) = \lambda(\operatorname{tr}\boldsymbol{e}(\boldsymbol{u}))\boldsymbol{I} + 2\mu\boldsymbol{e}(\boldsymbol{u})$ is replaced by the more general linear relation

$$\boldsymbol{\sigma}(\boldsymbol{u}) = \mathsf{A}\boldsymbol{e}(\boldsymbol{u}).$$

The special case of a homogeneous and isotropic linearly elastic body (which corresponds to Hooke's law) corresponds to

$$A_{ijk\ell} = \lambda\delta_{ij}\delta_{k\ell} + \mu(\delta_{ik}\delta_{j\ell} + \delta_{i\ell}\delta_{jk}). \qquad \square$$

The boundary value problem of linearized elasticity is called a **displacement-traction problem** if $\mathrm{d}\Gamma$-meas $\Gamma_0 > 0$ and $\mathrm{d}\Gamma$-meas $\Gamma_1 > 0$, or a **pure displacement problem** if $\Gamma_0 = \Gamma$, or a **pure traction problem** if $\Gamma_1 = \Gamma$ (the analysis of the latter problem, which is not covered by Theorem 6.16-1, is the object of Problem 6.16-2).

[79] So named after Gabriel Lamé (1795–1870).

[80] So named after Robert Hooke (1635–1703).

A *mixed* and a *dual formulation* of the *pure displacement problem*, in the spirit of Section 6.13, are also possible; cf. Problem 6.16-3.

One can show that, if $\Gamma = \Gamma_0$, the weak solution found in Theorem 6.16-1, which is thus in the space $\boldsymbol{V} = \boldsymbol{H}_0^1(\Omega)$ in this case, possesses *additional regularity* if the data (the boundary of Ω and the right-hand side $\boldsymbol{f}$) also possess additional regularity:

$^\flat$**Theorem 6.16-2 (regularity of the weak solution to the pure displacement problem of linearized elasticity[81])** *Let Ω be a domain in $\mathbb{R}^3$ with a boundary Γ of class $\mathcal{C}^2$, let $\boldsymbol{f} \in \boldsymbol{L}^p(\Omega)$ for some $p \geq 6/5$, and let $\Gamma_0 := \Gamma$ in* Theorem 6.16-1. *Then in this case the weak solution $\boldsymbol{u} \in \boldsymbol{H}_0^1(\Omega)$ is in the space $\boldsymbol{W}^{2,p}(\Omega)$ and it satisfies*

$$-\,\mathbf{div}\{\lambda(\operatorname{tr}\boldsymbol{e}(\boldsymbol{u}))\boldsymbol{I} + 2\mu\boldsymbol{e}(\boldsymbol{u})\} = \boldsymbol{f} \quad \text{in } L^p(\Omega). \qquad \square$$

Problems

6.16-1 In what follows, N is an integer ≥ 2.

(1) Let λ and μ be two constants that satisfy $N\lambda + 2\mu > 0$ and $\mu > 0$. Show that there exists a constant $\alpha = \alpha(N, \lambda, \mu) > 0$ such that

$$\lambda(\operatorname{tr}\boldsymbol{B})^2 + 2\mu\operatorname{tr}(\boldsymbol{B}^T\boldsymbol{B}) \geq \alpha\operatorname{tr}(\boldsymbol{B}^T\boldsymbol{B}) \quad \text{for all } \boldsymbol{B} \in \mathbb{M}^N.$$

The special case $N = 3$ of this inequality thus implies that the bilinear form $a(\cdot,\cdot)$ found in Theorem 6.16-1 remains $\boldsymbol{V}$-coercive if $3\lambda + 2\mu > 0$ and $\mu > 0$.

(2) Conversely, let λ and μ be two constants with the property that the inequality of (1) is satisfied for some constant $\alpha > 0$. Show that, necessarily, $N\lambda + 2\mu > 0$ and $\mu > 0$.

6.16-2 This problem extends the existence and uniqueness results of Theorem 6.16-1 to the *pure traction problem of three-dimensional linearized elasticity*.

Let Ω be a domain in $\mathbb{R}^3$ with boundary Γ and let constants $\lambda \geq 0$ and $\mu > 0$ and vector fields $\boldsymbol{f} \in \boldsymbol{L}^2(\Omega)$ and $\boldsymbol{g} \in \boldsymbol{L}^2(\Gamma)$ be given. Show that the following minimization problem: Find $\boldsymbol{u} \in \boldsymbol{H}^1(\Omega)$ such that $J(\boldsymbol{u}) = \inf_{\boldsymbol{v}\in\boldsymbol{H}^1(\Omega)} J(\boldsymbol{v})$, where

$$J(\boldsymbol{v}) := \frac{1}{2}\int_\Omega \{\lambda(\operatorname{tr}\boldsymbol{e}(\boldsymbol{v}))^2 + 2\mu\boldsymbol{e}(\boldsymbol{v}) : \boldsymbol{e}(\boldsymbol{v})\}\,\mathrm{d}x - \left(\int_\Omega \boldsymbol{f}\cdot\boldsymbol{v}\,\mathrm{d}x + \int_\Gamma \boldsymbol{g}\cdot\boldsymbol{v}\,\mathrm{d}\Gamma\right) \quad \text{for all } \boldsymbol{v} \in \boldsymbol{H}^1(\Omega),$$

has a solution if and only if

$$\int_\Omega \boldsymbol{f}\cdot\boldsymbol{v}\,\mathrm{d}x + \int_\Gamma \boldsymbol{g}\cdot\boldsymbol{v}\,\mathrm{d}\Gamma = 0 \quad \text{for all } \boldsymbol{v} \in \mathbf{Ker}\,\boldsymbol{\nabla}_s,$$

i.e., for all infinitesimal rigid displacements (Section 6.15), and that this solution is unique up to the addition of an infinitesimal rigid displacement.

Hint: Use Theorem 6.15-3.

6.16-3 Questions (3) and (5) in this problem respectively provide a *mixed formulation*[82] and a *dual formulation* of the *pure displacement problem of linearized elasticity*, viz., the special case $\Gamma_0 = \Gamma$ of Theorem 6.16-1. The assumptions and notations are the same as in this theorem.

---

[81] A sketch of the proof, which is long and delicate, is provided in CIARLET [1988, Section 6.3].

[82] Other mixed formulations are possible; see, e.g., Section 11 in:

D.N. ARNOLD; R.S. FALK; R. WINTHER [2006]: Finite element exterior calculus, homological techniques, and applications, in *Acta Numerica, Volume* 15 (A. ISERLES, editor), pp. 1–155, Cambridge University Press, Cambridge, UK.

(1) Define the space

$$\mathbb{H}(\mathbf{div};\Omega) := \{\boldsymbol{\tau} \in \mathbb{L}^2(\Omega);\ \mathbf{div}\,\boldsymbol{\tau} \in \boldsymbol{L}^2(\Omega)\}.$$

Show that, equipped with the norm defined by

$$\|\boldsymbol{\tau}\|_{\mathbb{H}(\mathbf{div};\Omega)} := \left(\|\boldsymbol{\tau}\|_{0,\Omega}^2 + \|\,\mathbf{div}\,\boldsymbol{\tau}\|_{0,\Omega}\right)^{1/2} \quad \text{for each } \boldsymbol{\tau} \in \mathbb{H}(\mathbf{div},\Omega),$$

the space $\mathbb{H}(\mathbf{div};\Omega)$ is a Hilbert space.

(2) Show that the mapping $\mathsf{B} : \mathbb{S}^3 \to \mathbb{S}^3$ defined by

$$\mathsf{B}\boldsymbol{\sigma} := \frac{1}{2\mu}\left(\boldsymbol{\sigma} - \frac{\lambda}{3\lambda + 2\mu}(\operatorname{tr}\boldsymbol{\sigma})\boldsymbol{I}\right) \quad \text{for each } \boldsymbol{\sigma} \in \mathbb{S}^3$$

is the inverse of the mapping $\mathsf{A} : \mathbb{S}^3 \to \mathbb{S}^3$ defined by

$$\mathsf{A}\boldsymbol{e} := \lambda(\operatorname{tr}\boldsymbol{e})\boldsymbol{I} + 2\mu\boldsymbol{e} \quad \text{for each } \boldsymbol{e} \in \mathbb{S}^3.$$

(3) Show that there exists a unique pair $(\boldsymbol{\sigma},\boldsymbol{\lambda}) \in \mathbb{H}(\mathbf{div};\Omega) \times \boldsymbol{L}^2(\Omega)$ that satisfies

$$\int_\Omega \mathsf{B}\boldsymbol{\sigma} : \boldsymbol{\tau}\,\mathrm{d}x + \int_\Omega \mathbf{div}\,\boldsymbol{\tau}\cdot\boldsymbol{\lambda}\,\mathrm{d}x = 0 \quad \text{for all } \boldsymbol{\tau} \in \mathbb{H}(\mathbf{div};\Omega),$$

$$\int_\Omega \mathbf{div}\,\boldsymbol{\sigma}\cdot\boldsymbol{\mu}\,\mathrm{d}x = -\int_\Omega \boldsymbol{f}\cdot\boldsymbol{\mu}\,\mathrm{d}x \quad \text{for all } \boldsymbol{\mu} \in \boldsymbol{L}^2(\Omega).$$

(4) Let $\boldsymbol{u} \in \boldsymbol{H}_0^1(\Omega)$ be the unique solution (Theorem 6.16-1) to the following quadratic minimization problem: Find $\boldsymbol{u} \in \boldsymbol{H}_0^1(\Omega)$ such that $J(\boldsymbol{u}) = \inf_{\boldsymbol{v}\in\boldsymbol{H}_0^1(\Omega)} J(\boldsymbol{v})$, where

$$J(\boldsymbol{v}) := \frac{1}{2}\int_\Omega \{\lambda(\operatorname{tr}\boldsymbol{e}(\boldsymbol{v}))^2 + 2\mu\boldsymbol{e}(\boldsymbol{v}) : \boldsymbol{e}(\boldsymbol{v})\}\,\mathrm{d}x - \int_\Omega \boldsymbol{f}\cdot\boldsymbol{v}\,\mathrm{d}x \quad \text{for each } \boldsymbol{v} \in \boldsymbol{H}_0^1(\Omega).$$

Show that

$$\boldsymbol{\sigma} = \mathsf{A}\boldsymbol{e}(\boldsymbol{u}) \quad \text{and} \quad \boldsymbol{\lambda} = \boldsymbol{u}.$$

(5) Show that the matrix field $\boldsymbol{\sigma} = \mathsf{A}\boldsymbol{e}(\boldsymbol{u})$ is the unique solution to the constrained quadratic minimization problem

$$\boldsymbol{\sigma} \in \mathbb{U}_{\boldsymbol{f}} := \{\boldsymbol{\tau} \in \mathbb{H}(\mathbf{div};\Omega);\ \mathbf{div}\,\boldsymbol{\tau} + \boldsymbol{f} = \boldsymbol{0} \text{ in } \boldsymbol{L}^2(\Omega)\},$$

$$I(\boldsymbol{\sigma}) = \inf_{\boldsymbol{\tau}\in\mathbb{U}_{\boldsymbol{f}}} I(\boldsymbol{\tau}), \quad \text{where } I(\boldsymbol{\tau}) := \frac{1}{2}\int_\Omega \mathsf{B}\boldsymbol{\tau} : \boldsymbol{\tau}\,\mathrm{d}x \text{ for each } \boldsymbol{\tau} \in \mathbb{L}_s^2(\Omega).$$

Hint: Mimic the proof of Theorem 6.13-2.

Remark In elasticity theory, $\mathbb{U}_{\boldsymbol{f}}$ is called the *set of admissible stresses*, and the functional $I : \mathbb{L}^2(\Omega) \to \mathbb{R}$ is called the *complementary energy*. □

6.16-4 Greek and Latin indices vary in the sets $\{1,2\}$ and $\{1,2,3\}$ respectively, and the summation convention with respect to repeated indices is used. Let Ω be a domain in $\mathbb{R}^2$, let Γ_1 be a relatively open subset of $\Gamma := \partial\Omega$ such that $\mathrm{d}\Gamma\text{-meas}\,\Gamma_0 > 0$ where $\Gamma_0 := \Gamma - \Gamma_1$, and let $\boldsymbol{f} = (f_i) \in L^2(\Omega)$ be a given vector field. Define the space

$$\boldsymbol{V} := \{\boldsymbol{v} = (v_i) \in H^1(\Omega) \times H^1(\Omega) \times H^2(\Omega);\ v_i = \partial_\nu v_3 = 0 \text{ on } \Gamma_0\},$$

and the functional $J : \boldsymbol{V} \to \mathbb{R}$ by

$$J(\boldsymbol{v}) := \frac{1}{2}\int_\Omega \left\{\frac{\varepsilon^3}{3} a_{\alpha\beta\sigma\tau}\partial_{\sigma\tau}v_3\partial_{\alpha\beta}v_3 + \varepsilon a_{\alpha\beta\sigma\tau}e_{\sigma\tau}(\boldsymbol{v})e_{\alpha\beta}(\boldsymbol{v})\right\}\mathrm{d}x - \int_\Omega f_i v_i\,\mathrm{d}x, \quad \boldsymbol{v} = (v_i) \in \boldsymbol{V},$$

where $\varepsilon > 0$ is a constant, $a_{\alpha\beta\sigma\tau} = a_{\beta\alpha\sigma\tau} = a_{\sigma\tau\alpha\beta}$ are constants with the property that there exists a constant $C > 0$ such that

$$a_{\alpha\beta\sigma\tau} t_{\sigma\tau} t_{\alpha\beta} \geq C t_{\alpha\beta} t_{\alpha\beta} \quad \text{for all } (t_{\alpha\beta}) \in \mathbb{S}^2,$$

and

$$e_{\alpha\beta}(\boldsymbol{v}) := \frac{1}{2}(\partial_\alpha v_\beta + \partial_\beta v_\alpha).$$

(1) Show that there exists a unique vector field $\boldsymbol{u} \in \boldsymbol{V}$ such that $J(\boldsymbol{u}) = \inf_{\boldsymbol{v} \in \boldsymbol{V}} J(\boldsymbol{v})$.

(2) Assume that $\boldsymbol{u} = (u_i) \in H^2(\Omega) \times H^2(\Omega) \times H^4(\Omega)$. Show that $\boldsymbol{u}$ satisfies the following boundary value problem:

$$\partial_{\alpha\beta} m_{\alpha\beta}(\boldsymbol{u}) = f_3 \quad \text{and} \quad -\partial_\beta n_{\alpha\beta}(\boldsymbol{u}) = f_\alpha \text{ in } \Omega,$$
$$u_i = \partial_\nu u_3 = 0 \text{ on } \Gamma_0,$$
$$m_{\alpha\beta}(\boldsymbol{u}) \nu_\alpha \nu_\beta = 0, \quad n_{\alpha\beta}(\boldsymbol{u}) \nu_\beta = 0, \quad \text{and} \quad (\partial_\alpha m_{\alpha\beta}(\boldsymbol{u})) \nu_\beta + \partial_\tau (m_{\alpha\beta}(\boldsymbol{u}) \nu_\alpha \tau_\beta) = 0 \text{ on } \Gamma_1,$$

where

$$m_{\alpha\beta}(\boldsymbol{u}) := \frac{\varepsilon^3}{3} a_{\alpha\beta\sigma\tau} \partial_{\sigma\tau} u_3 \quad \text{and} \quad n_{\alpha\beta}(\boldsymbol{u}) := \varepsilon a_{\alpha\beta\sigma\tau} e_{\sigma\tau}(\boldsymbol{u}).$$

The above boundary value problem constitutes the equations of the **Kirchhoff–Love theory of a linearly elastic plate**[83] of thickness 2ε, *clamped* along a portion Γ_0 of its boundary. Note that this problem consists in fact of two *decoupled* boundary value problems, one (for the unknown u_3) constituting the *flexural equations* (already encountered, but with different notations, in Theorem 6.8-7) and the other (for the unknowns u_1 and u_2) constituting the *membrane equations*.

Remark The constants $a_{\alpha\beta\sigma\tau}$ are the components of the *elasticity tensor of the plate*. They are given by

$$a_{\alpha\beta\sigma\tau} = \frac{4\lambda\mu}{(\lambda + 2\mu)} \delta_{\alpha\beta} \delta_{\sigma\tau} + 2\mu(\delta_{\alpha\sigma}\delta_{\beta\tau} + \delta_{\alpha\tau}\delta_{\beta\sigma}),$$

in terms of the *Lamé constants* $\lambda \geq 0$ and $\mu > 0$ of the elastic material constituting the plate. □

6.17 The classical Poincaré lemma and its weak version as an application of J.L. Lions lemma and of the hypoellipticity of Δ

The summation convention with respect to repeated indices is used throughout this section. Given an open subset Ω of $\mathbb{R}^N$, consider the linear operator $\mathbf{grad} : \mathcal{C}^2(\Omega) \to \mathcal{C}^1(\Omega; \mathbb{R}^N)$ defined by

$$p \in \mathcal{C}^2(\Omega) \to \mathbf{grad}\, p := (\partial_i p) \in \boldsymbol{\mathcal{C}}^1(\Omega) := \mathcal{C}^1(\Omega; \mathbb{R}^N).$$

A natural question then arises as to whether this linear operator is *invertible*, i.e., whether, given a vector field $\boldsymbol{h} = (h_i) \in \mathcal{C}^1(\Omega; \mathbb{R}^N)$, there exists a function $p \in \mathcal{C}^2(\Omega)$ such that

$$\mathbf{grad}\, p = \boldsymbol{h} \quad \text{in } \Omega,$$

or equivalently, such that

$$\partial_i p = h_i \quad \text{in } \Omega, \ 1 \leq i \leq N.$$

[83] These equations are studied at length in CIARLET [1997, Chapter 1].

Since then $\partial_{ij}p = \partial_{ji}p$ if this is the case, it is clear that the functions h_i must *necessarily* satisfy the *compatibility conditions*

$$\partial_i h_j - \partial_j h_i = 0 \quad \text{in } \mathcal{C}(\Omega), \ \ 1 \le i, j \le N,$$

or equivalently,

$$\mathbf{curl}\, \boldsymbol{h} = \mathbf{0} \quad \text{in } \boldsymbol{\mathcal{C}}(\Omega) := \mathcal{C}(\Omega; \mathbb{R}^N),$$

where the *curl operator* $\mathbf{curl} : \mathcal{C}^1(\Omega; \mathbb{R}^N) \to \mathcal{C}(\Omega; \mathbb{R}^{\frac{N(N-1)}{2}})$ is defined for any integer $N \ge 2$ by[84]

$$(\mathbf{curl}\, \boldsymbol{h})_{ij} := (\partial_i h_j - \partial_j h_i), \ \ 1 \le i < j \le N, \quad \text{for each } \boldsymbol{h} \in \mathcal{C}^1(\Omega; \mathbb{R}^N).$$

It is remarkable that these necessary conditions become *sufficient* if the open set Ω is *simply connected*: this is the essence of the classical *Poincaré lemma*, established in Theorem 6.17-2 below ("classical," as opposed to the "weak" form of this lemma, established in Theorem 6.17-4).

Before proving this lemma, we establish a technical, interesting *per se*, result: While the *paths*, resp. *homotopies*, that come into the definition of a *general* simply connected topological space X are only assumed to be *continuous* mappings from $[0,1]$, *resp.* $[0,1] \times [0,1]$, into X (Section 1.9), they may be assumed to be of class $\mathcal{C}^\infty$ when X is an *open subset of* $\mathbb{R}^N$ (that these mappings be of class $\mathcal{C}^2$ would in fact suffice for our subsequent purposes, but proving that they are of class $\mathcal{C}^\infty$ involves no extra cost).

Theorem 6.17-1 *Let Ω be an open and simply connected open subset of $\mathbb{R}^N$. Then:*

(a) *Given any two distinct points $x \in \Omega$ and $y \in \Omega$, there exists a path $\boldsymbol{\gamma} \in \mathcal{C}^\infty([0,1]; \mathbb{R}^N)$ joining x to y in Ω.*

(b) *Given any two distinct points $x \in \Omega$ and $y \in \Omega$, let $\boldsymbol{\gamma}^0 \in \mathcal{C}^\infty([0,1]; \mathbb{R}^N)$ and $\boldsymbol{\gamma}^1 \in \mathcal{C}^\infty([0,1]; \mathbb{R}^N)$ be any two distinct paths joining x to y in Ω. Then there exists a homotopy $\boldsymbol{H} \in \mathcal{C}^\infty([0,1] \times [0,1]; \mathbb{R}^N)$ joining $\boldsymbol{\gamma}^0$ to $\boldsymbol{\gamma}^1$ in Ω.*

Proof (i) Since a simply connected space is arcwise connected, there exists a path $\boldsymbol{\pi} \in \mathcal{C}([0,1]; \mathbb{R})$ joining x to y in Ω. Since $\operatorname{Im} \boldsymbol{\pi}$ is a compact subset of Ω, there exists $\delta > 0$ such that

$$\bigcup_{x \in \operatorname{Im} \boldsymbol{\pi}} \overline{B(x;\delta)} \subset \Omega.$$

Let $\widetilde{\boldsymbol{\pi}} = (\widetilde{\pi}_i)_{i=1}^N \in \mathcal{C}(\mathbb{R}; \mathbb{R}^N)$ be an extension of $\boldsymbol{\pi}$ (such an extension exists by the Tietze–Urysohn theorem; cf. Theorem 1.7-7), i.e., such that $\widetilde{\boldsymbol{\pi}}|_{[0,1]} = \boldsymbol{\pi}$, let $(\widetilde{\pi}_{i,\varepsilon})_{\varepsilon > 0}$ be a regularizing family (Section 2.6) of each component $\widetilde{\pi}_i$, $1 \le i \le N$, and let $\widetilde{\boldsymbol{\pi}}_\varepsilon := (\widetilde{\pi}_{i,\varepsilon})_{i=1}^N$, $\varepsilon > 0$. Then $\widetilde{\boldsymbol{\pi}}_\varepsilon \in \mathcal{C}^\infty(\mathbb{R}; \mathbb{R}^N)$ and

$$\sup_{0 \le t \le 1} |\widetilde{\boldsymbol{\pi}}_\varepsilon(t) - \boldsymbol{\pi}(t)| \le \frac{\delta}{2} \quad \text{for } \varepsilon > 0 \text{ small enough}$$

(Theorem 2.6-1(b)). Fix such an $\varepsilon > 0$ and let $\boldsymbol{\gamma} : [0,1] \to \mathbb{R}$ be defined by

$$\boldsymbol{\gamma}(t) := \widetilde{\boldsymbol{\pi}}_\varepsilon(t) + (1-t)(x - \widetilde{\boldsymbol{\pi}}_\varepsilon(0)) + t(y - \widetilde{\boldsymbol{\pi}}_\varepsilon(1)), \quad 0 \le t \le 1.$$

[84] A justification of this definition is found in, e.g.:
G. Csato; B. Dacorogna; O. Kneuss [2011]: *The Pullback Equation*, Birkhäuser, Basel.

Then $\gamma \in \mathcal{C}^\infty([0,1];\mathbb{R})$, $\gamma(0) = x$, $\gamma(1) = y$, and

$$\begin{aligned}|\gamma(t) - \pi(t)| &\leq |\widetilde{\pi}_\varepsilon(t) - \pi(t)| + (1-t)\,|\pi(0) - \widetilde{\pi}_\varepsilon(0)| + t\,|\pi(1) - \widetilde{\pi}_\varepsilon(1)| \\ &\leq \frac{\delta}{2} + (1-t)\frac{\delta}{2} + t\frac{\delta}{2} = \delta, \quad 0 \leq t \leq 1,\end{aligned}$$

since $\pi(0) = x$ and $\pi(1) = y$, so that $\gamma(t) \in \bigcup_{x \in \operatorname{Im}\pi} \overline{B(x;\delta)} \subset \Omega$ for all $t \in [0,1]$.

(ii) By definition of a simply connected set, there exists a homotopy $\boldsymbol{G} \in \mathcal{C}([0,1] \times [0,1];\mathbb{R}^N)$ joining $\boldsymbol{\gamma}^0$ and $\boldsymbol{\gamma}^1$ in Ω. Since $\operatorname{Im}\boldsymbol{G}$ is a compact subset of Ω, there exists $\delta > 0$ such that

$$\bigcup_{x \in \operatorname{Im}\boldsymbol{G}} \overline{B(x;\delta)} \subset \Omega.$$

Let $\widetilde{\boldsymbol{G}} = (\widetilde{G}_i)_{i=1}^N \in \mathcal{C}(\mathbb{R} \times \mathbb{R};\mathbb{R}^N)$ be an extension of $\boldsymbol{G}$ (which again exists by the Tietze–Urysohn theorem), i.e., such that $\widetilde{\boldsymbol{G}}|_{[0,1]\times[0,1]} = \boldsymbol{G}$, let $(\widetilde{G}_{i,\varepsilon})_{\varepsilon>0}$ be a regularizing family of each component $\widetilde{G}_i$, $1 \leq i \leq N$, and let $\widetilde{\boldsymbol{G}}_\varepsilon := (\widetilde{G}_{i,\varepsilon})_{i=1}^N$, $\varepsilon > 0$. Then $\widetilde{\boldsymbol{G}}_\varepsilon \in \mathcal{C}^\infty(\mathbb{R} \times \mathbb{R};\mathbb{R}^N)$ and (again by Theorem 2.6-1(b)),

$$\sup_{0\leq t\leq 1,\, 0\leq\lambda\leq 1} |\widetilde{\boldsymbol{G}}_\varepsilon(t,\lambda) - \boldsymbol{G}(t,\lambda)| \leq \frac{\delta}{2} \quad \text{for } \varepsilon > 0 \text{ small enough.}$$

Fix such an $\varepsilon > 0$ and let $\boldsymbol{H} : [0,1] \times [0,1] \to \mathbb{R}$ be defined by

$$\boldsymbol{H}(t,\lambda) = \widetilde{\boldsymbol{G}}_\varepsilon(t,\lambda) + (1-\lambda)(\boldsymbol{\gamma}^0(t) - \widetilde{\boldsymbol{G}}_\varepsilon(t,0)) + \lambda(\boldsymbol{\gamma}^1(t) - \widetilde{\boldsymbol{G}}_\varepsilon(t,1)).$$

Then $\boldsymbol{H} \in \mathcal{C}^\infty([0,1] \times [0,1];\mathbb{R}^N)$, $\boldsymbol{H}(t,0) = \boldsymbol{\gamma}^0(t) = \boldsymbol{G}(t,0)$ and $\boldsymbol{H}(t,1) = \boldsymbol{\gamma}^1(t) = \boldsymbol{G}(t,1)$ for $0 \leq t \leq 1$, and

$$\begin{aligned}|\boldsymbol{H}(t,\lambda) - \boldsymbol{G}(t,\lambda)| &\leq |\widetilde{\boldsymbol{G}}_\varepsilon(t,\lambda) - \boldsymbol{G}(t,\lambda)| + (1-\lambda)\,|\boldsymbol{G}(t,0) - \widetilde{\boldsymbol{G}}_\varepsilon(t,0)| \\ &\quad + \lambda\,|\boldsymbol{G}(t,1) - \widetilde{\boldsymbol{G}}_\varepsilon(t,1)| \\ &\leq \delta/2 + (1-\lambda)\delta/2 + \lambda\delta/2 = \delta, \quad 0 \leq t \leq 1,\ 0 \leq \lambda \leq 1.\end{aligned}$$

Hence $\boldsymbol{H}(t,\lambda) \in \bigcup_{x \in \operatorname{Im}\boldsymbol{G}} \overline{B(x;\delta)} \subset \Omega$ for all $0 \leq t \leq 1$, $0 \leq \lambda \leq 1$. □

We now prove the "classical" Poincaré lemma. In the proofs below, Latin indices range in the set $\{1, 2, \ldots, N\}$ and the summation convention with respect to repeated indices is used in conjunction with this rule.

Theorem 6.17-2 (classical Poincaré lemma;[85] *alias* Poincaré lemma in $\mathcal{C}^2(\Omega)$) *Let Ω be a simply connected open subset of $\mathbb{R}^N$, and let there be given a vector field $\boldsymbol{h} \in \boldsymbol{\mathcal{C}}^1(\Omega)$ that satisfies*

$$\mathbf{curl}\,\boldsymbol{h} = \mathbf{0} \quad \textit{in } \Omega.$$

[85] So named after Henri Poincaré, who indeed mentioned in 1886 a generalization of this result (to differential forms of arbitrary degree), a proof of which was then given in 1889 by Vito Volterra. But the "Poincaré lemma" as stated here (i.e., for differential forms of degree one) goes back in effect (for $N = 2$) to Alexis Claude de Clairaut, Leonhard Euler, and Alexis Fontaine des Bertins, who independently proved it around 1740. More details about the genesis of this lemma and its generalizations are found in:

H. Samelson [2001]: Differential forms, the early days; or the stories of Deahna's theorem and of Volterra's theorem, *American Mathematical Monthly* **108**, 552–530.

A masterly account of Henri Poincaré's outstanding achievements is given in Gray [2012].

Then there exists a function $p \in \mathcal{C}^2(\Omega)$ *such that*

$$\mathbf{grad}\, p = \boldsymbol{h} \quad in\ \Omega$$

and any other solution $\widetilde{p} \in \mathcal{C}^2(\Omega)$ *to the equations* $\mathbf{grad}\,\widetilde{p} = \boldsymbol{h}$ *in* Ω *is of the form* $\widetilde{p} = p + C$ *for some constant* C.

Proof (i) Let a point $x^0 \in \Omega$ be given. Since Ω is in particular arcwise-connected, given any point $x^1 \in \Omega$ distinct from x^0, there exists a *path* $\boldsymbol{\gamma} = (\gamma_i) \in \mathcal{C}^\infty([0,1];\mathbb{R}^N)$ *joining* x^0 *to* x^1 *in* Ω (Theorem 6.17-1), i.e., such that $\boldsymbol{\gamma}(0) = x^0$, $\boldsymbol{\gamma}(1) = x^1$, and $\boldsymbol{\gamma}(t) \in \Omega$ for all $0 \le t \le 1$.

Let a vector field $\boldsymbol{h} = (h_i) \in \boldsymbol{\mathcal{C}}^1(\Omega)$ be given. If a function $p \in \mathcal{C}^2(\Omega)$ exists that satisfies $\partial_i p = h_i$ in Ω, $1 \le i \le N$, then the function $P \in \mathcal{C}^1[0,1]$ defined by $P(t) := p(\boldsymbol{\gamma}(t))$, $0 \le t \le 1$, which thus *depends* a priori *on the path* $\boldsymbol{\gamma}$, satisfies

$$\frac{\mathrm{d}P}{\mathrm{d}t}(t) = \partial_i p(\boldsymbol{\gamma}(t))\frac{\mathrm{d}\gamma_i}{\mathrm{d}t}(t) = h_i(\boldsymbol{\gamma}(t))\frac{\mathrm{d}\gamma_i}{\mathrm{d}t}(t), \quad 0 \le t \le 1.$$

Motivated by this observation, we first note that, for any $P^0 \in \mathbb{R}$, there exists a unique solution $P \in \mathcal{C}^1[0,1]$, again *a priori dependent on the path* $\boldsymbol{\gamma}$, to the *linear Cauchy problem*:

$$\frac{\mathrm{d}P}{\mathrm{d}t}(t) = h_i(\boldsymbol{\gamma}(t))\frac{\mathrm{d}\gamma_i}{\mathrm{d}t}(t), \ 0 \le t \le 1, \quad \text{and} \quad P(0) = P^0$$

(Theorem 3.8-2). Incidentally, this result already shows that, if the system

$$\partial_i p(x) = h_i(x) \ \text{ in } \Omega,\ 1 \le i \le N, \quad \text{and} \quad p(x^0) = P^0$$

has a solution, then this solution is *unique*.

(ii) In order that the value $P(1)$ found by solving the Cauchy problem of (i) be an acceptable candidate for the unknown value $p(x^1)$, the number $P(1)$ must be of course *independent of the path chosen for joining* x^0 *to* x^1. As we now show, this property crucially hinges on the *compatibility relations* $\partial_j h_i = \partial_i h_j$ satisfied by the functions h_i, together with the assumption that Ω is *simply connected.*

Let $\boldsymbol{\gamma}_0 \in \mathcal{C}^\infty([0,1];\mathbb{R}^N)$ and $\boldsymbol{\gamma}_1 \in \mathcal{C}^\infty([0,1];\mathbb{R}^N)$ be two paths joining x^0 to x^1 in Ω. Since Ω is simply connected, there exists a homotopy $\boldsymbol{G} = (G_i)_{i=1}^N \in \mathcal{C}^\infty([0,1]\times[0,1];\mathbb{R}^N)$ joining $\boldsymbol{\gamma}^0$ to $\boldsymbol{\gamma}^1$ in Ω (Theorem 6.17-1), i.e., such that

$$\begin{gathered}\boldsymbol{G}(t,\lambda) \in \Omega \quad \text{for all } 0 \le t \le 1,\ 0 \le \lambda \le 1,\\ \boldsymbol{G}(\cdot,0) = \boldsymbol{\gamma}_0 \quad \text{and} \quad \boldsymbol{G}(\cdot,1) = \boldsymbol{\gamma}_1,\\ \boldsymbol{G}(0,\lambda) = x^0 \quad \text{and} \quad \boldsymbol{G}(1,\lambda) = x^1 \quad \text{for all } 0 \le \lambda \le 1.\end{gathered}$$

Let $P(\cdot,\lambda) \in \mathcal{C}^1[0,1]$ denote for each $\lambda \in [0,1]$ the unique solution to the Cauchy problem that corresponds to the particular path $\boldsymbol{G}(\cdot,\lambda) = (G_i(\cdot,\lambda))$ joining x^0 to x^1 in Ω. We thus have

$$\frac{\partial P}{\partial t}(t,\lambda) = h_i(\boldsymbol{G}(t,\lambda))\frac{\partial G_i}{\partial t}(t,\lambda), \ 0 \le t \le 1, \quad \text{and} \quad P(0,\lambda) = P^0, \ 0 \le \lambda \le 1.$$

Our objective then consists in showing that

$$\frac{\partial P}{\partial \lambda}(1, \lambda) = 0, \quad 0 \leq \lambda \leq 1$$

(it is easily seen that $P \in \mathcal{C}^1([0,1] \times [0,1])$), as this relation will imply that $P(1,0) = P(1,1)$, as desired.

For each $0 \leq t \leq 1$, $0 \leq \lambda \leq 1$, let

$$\sigma(t, \lambda) := \frac{\partial P}{\partial \lambda}(t, \lambda) - h_j(\boldsymbol{G}(t, \lambda)) \frac{\partial G_j}{\partial \lambda}(t, \lambda).$$

Then the assumptions $\partial_i h_j = \partial_j h_i$ in Ω, $1 \leq i, j \leq N$, and the relations $\frac{\partial}{\partial t}\left(\frac{\partial G_j}{\partial \lambda}\right) = \frac{\partial}{\partial \lambda}\left(\frac{\partial G_j}{\partial t}\right)$, $1 \leq j \leq N$, together imply that, for each $0 \leq t \leq 1$, $0 \leq \lambda \leq 1$,

$$\begin{aligned}
\frac{\partial \sigma}{\partial t}(t, \lambda) &= \frac{\partial}{\partial t}\left(\frac{\partial P}{\partial \lambda}\right)(t, \lambda) - \partial_i h_j(\boldsymbol{G}(t, \lambda)) \frac{\partial G_i}{\partial t}(t, \lambda) \frac{\partial G_j}{\partial \lambda}(t, \lambda) - h_j(\boldsymbol{G}(t, \lambda)) \frac{\partial}{\partial t}\left(\frac{\partial G_j}{\partial \lambda}\right)(t, \lambda) \\
&= \frac{\partial}{\partial \lambda}\left(\frac{\partial P}{\partial t}\right)(t, \lambda) - \partial_j h_i(\boldsymbol{G}(t, \lambda)) \frac{\partial G_j}{\partial \lambda}(t, \lambda) \frac{\partial G_i}{\partial t}(t, \lambda) - h_j(\boldsymbol{G}(t, \lambda)) \frac{\partial}{\partial \lambda}\left(\frac{\partial G_j}{\partial t}\right)(t, \lambda) \\
&= \frac{\partial}{\partial \lambda}\left(\frac{\partial P}{\partial t}\right)(t, \lambda) - \frac{\partial}{\partial \lambda}\left(h_i(\boldsymbol{G}(t, \lambda)) \frac{\partial G_i}{\partial t}(t, \lambda)\right) \\
&= \frac{\partial}{\partial \lambda}\left(\frac{\partial P}{\partial t}(t, \lambda) - h_i(\boldsymbol{G}(t, \lambda)) \frac{\partial G_i}{\partial t}(t, \lambda)\right) = 0,
\end{aligned}$$

since

$$\frac{\partial P}{\partial t}(t, \lambda) = h_i(\boldsymbol{G}(t, \lambda)) \frac{\partial G_i}{\partial t}(t, \lambda), \quad 0 \leq t \leq 1, \ 0 \leq \lambda \leq 1.$$

Noting that

$$\sigma(0, \lambda) = \frac{\partial P}{\partial \lambda}(0, \lambda) - h_j(\boldsymbol{G}(0, \lambda)) \frac{\partial G_j}{\partial \lambda}(0, \lambda) = 0, \quad 0 \leq \lambda \leq 1$$

(because $\frac{\partial P}{\partial \lambda}(0, \lambda) = 0$ and $\frac{\partial G_j}{\partial \lambda}(0, \lambda) = 0$ for all $0 \leq \lambda \leq 1$, since $P(0, \lambda) = P^0$ and $G_j(0, \lambda) = x_j^0$, $1 \leq j \leq N$, for all $0 \leq \lambda \leq 1$), we thus infer that

$$0 = \sigma(1, \lambda) = \frac{\partial P}{\partial \lambda}(1, \lambda) - h_j(\boldsymbol{G}(1, \lambda)) \frac{\partial G_j}{\partial \lambda}(1, \lambda) = \frac{\partial P}{\partial \lambda}(1, \lambda), \quad 0 \leq \lambda \leq 1$$

(because $G_j(0, \lambda) = x_j^1$ for all $0 \leq \lambda \leq 1$).

(iii) We can now unambiguously define a function $p : \Omega \to \mathbb{R}$ by letting

$$p(x^1) := P(1) \quad \text{for each } x^1 \in \Omega,$$

where $P \in \mathcal{C}^1[0,1]$ is the solution to the Cauchy problem of (i) where $\boldsymbol{\gamma} \in \mathcal{C}^\infty([0,1]; \mathbb{R}^N)$ is *any* path joining x^0 to x^1 in Ω. It thus remains to show that $p \in \mathcal{C}^1(\Omega)$ and that $\partial_i p = h_i$ in Ω, $1 \leq i \leq N$.

Let a point $x \in \Omega$ and an integer $1 \le i \le N$ be given. Then there clearly exist a point $x^1 \in \Omega$, a path $\boldsymbol{\gamma} = (\gamma_i) \in \mathcal{C}^\infty([0,1];\mathbb{R}^N)$ joining x^0 to x^1 in Ω, a number $\tau \in]0,1[$, and an open interval $I \subset [0,1]$ containing τ, such that

$$\boldsymbol{\gamma}(t) = x + (t-\tau)\boldsymbol{e}_i \quad \text{for } t \in I,$$

where $\boldsymbol{e}_i$ is the ith basis vector in $\mathbb{R}^N$.

Let $P \in \mathcal{C}^1[0,1]$ denote the solution of the Cauchy problem of (i) corresponding to this path $\boldsymbol{\gamma}$, so that $p(\boldsymbol{\gamma}(t)) = P(t)$, $0 \le t \le 1$. Then

$$\begin{aligned} P(t) &= P(\tau) + (t-\tau)\frac{\mathrm{d}P}{\mathrm{d}t}(\tau) + o(t-\tau) \\ &= P(\tau) + (t-\tau)h_j(\boldsymbol{\gamma}(\tau))\frac{\mathrm{d}\gamma_j}{\mathrm{d}t}(\tau) + o(t-\tau) \\ &= P(\tau) + (t-\tau)h_i(x) + o(t-\tau) \quad \text{for } |t-\tau| \text{ small enough} \end{aligned}$$

(we use here that $\dfrac{\mathrm{d}\gamma_j}{\mathrm{d}t}(\tau) = \delta_{ij}$). Consequently,

$$p(x + (t-\tau)\boldsymbol{e}_i) = p(x) + (t-\tau)h_i(x) + o(t-\tau) \quad \text{for } |t-\tau| \text{ small enough},$$

which shows that the function p possesses an ith partial derivative $\partial_i p$ at x, given by $\partial_i p(x) = h_i(x)$.

Since the point $x \in \Omega$ and the index $1 \le i \le N$ are arbitrary and the functions h_i are of class $\mathcal{C}^1$ in Ω, the function p is of class $\mathcal{C}^2$ in Ω and satisfies $\partial_i p = h_i$ in $\mathcal{C}^1(\Omega)$, $1 \le i \le N$.

(iv) If a function $\pi \in \mathcal{C}^1(\Omega)$ satisfies $\partial_i \pi = 0$ in a *connected* open subset of $\mathbb{R}^N$ for all $1 \le i \le N$, then π is a constant (a proof of this classical result will be given in greater generality in Theorem 7.2-4). Hence the function p found in (iii) is unique *modulo* the addition of a constant. □

Remark The assumption of simple-connectedness is essential; cf. Problem 6.17-2. □

As a useful complement to Theorem 6.17-2, we now show that any function $p \in \mathcal{C}^2(\Omega)$ can be expressed in terms of its gradient $\mathbf{grad}\, p = (\partial_i p) \in \mathcal{C}^1(\Omega;\mathbb{R}^N)$ by means of a *path integral* in Ω and that, under the assumption of Theorem 6.17-2, the same path integral provides a particular solution p to the equations $\mathbf{grad}\, p = \boldsymbol{h}$ in Ω[86] (hence all the other solutions are of the form $p + C$, with C a constant).

Theorem 6.17-3 *Let Ω be a connected open subset of $\mathbb{R}^N$ and let x^0 be a point in Ω.*

(a) *Given any function $p \in \mathcal{C}^2(\Omega)$, let the vector field $\boldsymbol{h} = (h_i)_{i=1}^N \in \boldsymbol{\mathcal{C}}^1(\Omega)$ be defined by*

$$\boldsymbol{h} := \mathbf{grad}\, p.$$

Then, given any point $x \in \Omega$ and given any path $\boldsymbol{\gamma}_x \in \mathcal{C}^\infty([0,1];\mathbb{R}^N)$ joining x^0 to x in Ω,

$$p(x) = p(x^0) + \int_{\boldsymbol{\gamma}_x} \boldsymbol{h}(y)\cdot \mathrm{d}\boldsymbol{y}, \quad \textit{where } \int_{\boldsymbol{\gamma}_x} \boldsymbol{h}(y)\cdot \mathrm{d}\boldsymbol{y} := \int_{\boldsymbol{\gamma}_x} h_i(y)\,\mathrm{d}y_i.$$

[86] This result is due to Augustin-Louis Cauchy (1789–1857).

(b) *Assume that Ω is simply connected. Then, given any vector field $\boldsymbol{h} \in \boldsymbol{\mathcal{C}}^1(\Omega)$ that satisfies*

$$\mathbf{curl}\, \boldsymbol{h} = \mathbf{0} \quad \text{in } \Omega,$$

and given any point $x \in \Omega$, the path integral $\int_{\boldsymbol{\gamma}_x} \boldsymbol{h}(y) \cdot \mathrm{d}\boldsymbol{y}$ is independent of the path $\boldsymbol{\gamma}_x \in \mathcal{C}^\infty([0,1]; \mathbb{R}^N)$ chosen for joining x^0 to x. Besides, the function $p : \Omega \to \mathbb{R}$ defined for any such path by

$$p(x) := \int_{\boldsymbol{\gamma}_x} \boldsymbol{h}(y) \cdot \mathrm{d}\boldsymbol{y}, \quad x \in \Omega,$$

is of class $\mathcal{C}^2$ in Ω and is a particular solution to the equations $\mathbf{grad}\, p = \boldsymbol{h}$ in Ω.

Proof Given any point $x \in \Omega$ and any path $\boldsymbol{\gamma}_x = (\gamma_x^i)_{i=1}^N \in \mathcal{C}^\infty([0,1]; \mathbb{R}^N)$ joining x^0 to x, the equations $P(t) = p(\boldsymbol{\gamma}_x(t))$, $0 \le t \le 1$, and

$$\frac{\mathrm{d}P}{\mathrm{d}t}(t) = h_i(\boldsymbol{\gamma}_x(t)) \frac{\mathrm{d}\gamma_x^i}{\mathrm{d}t}(t), \quad 0 \le t \le 1,$$

found in part (i) of the proof of Theorem 6.17-2, together imply that

$$P(1) = P(0) + \int_0^1 h_i(\boldsymbol{\gamma}_x(t)) \frac{\mathrm{d}\gamma_x^i}{\mathrm{d}t}(t)\, \mathrm{d}t,$$

i.e., that

$$p(x) = p(x^0) + \int_{\boldsymbol{\gamma}_x} h_i(y)\, \mathrm{d}y_i \quad \text{for any } x \in \Omega.$$

This proves (a).

We next show that, given any vector field $\boldsymbol{h} = (h_i)_{i=1}^N \in \boldsymbol{\mathcal{C}}^1(\Omega)$ that satisfies $\mathbf{curl}\, \boldsymbol{h} = \mathbf{0}$ in Ω, the integral $\int_{\boldsymbol{\gamma}_x} \boldsymbol{h}(y) \cdot \mathrm{d}\boldsymbol{y}$ is independent of the path $\boldsymbol{\gamma}_x \in \mathcal{C}^\infty([0,1]\,; \mathbb{R}^N)$ joining x^0 to x if Ω is *simply connected*. So, let $\boldsymbol{\gamma}_x^0 \in \mathcal{C}^\infty([0,1]\,; \mathbb{R}^N)$ and $\boldsymbol{\gamma}_x^1 \in \mathcal{C}^\infty([0,1]\,; \mathbb{R}^N)$ be two such paths, and let $\boldsymbol{G} = (G_i)_{i=1}^N \in \mathcal{C}^\infty([0,1]\,; \mathbb{R}^N)$ be a homotopy joining $\boldsymbol{\gamma}_x^0$ to $\boldsymbol{\gamma}_x^1$ in Ω.

Then, as shown in part (ii) of the proof of Theorem 6.17-2, for each $0 \le \lambda \le 1$,

$$\int_{\boldsymbol{G}(\cdot,\lambda)} \boldsymbol{h}(y) \cdot \mathrm{d}\boldsymbol{y} = \int_0^1 \frac{\partial P}{\partial t}(t,\lambda)\, \mathrm{d}t = P(1,\lambda) - P^0,$$

where $P(\cdot,\lambda) \in \boldsymbol{\mathcal{C}}^1\,[0,1]$ denotes the unique solution to the Cauchy problem

$$\frac{\partial P}{\partial t}(t,\lambda) = h_i(\boldsymbol{G}(t,\lambda)) \frac{\partial G_i}{\partial t}(t,\lambda), \quad 0 \le t \le 1, \quad \text{and} \quad P(0,\lambda) = P^0.$$

It was also shown there that the relation $\mathbf{curl}\, \boldsymbol{h} = \mathbf{0}$ in Ω implies that $\dfrac{\partial P}{\partial \lambda}(1,\lambda) = 0$, $0 \le \lambda \le 1$, so that

$$\int_{\boldsymbol{\gamma}_x^0} \boldsymbol{h}(y) \cdot \mathrm{d}\boldsymbol{y} = P(1,0) - P^0 = P(1,1) - P^0 = \int_{\boldsymbol{\gamma}_x^1} \boldsymbol{h}(y) \cdot \mathrm{d}\boldsymbol{y},$$

as announced.

The same argument as in part (iii) of the proof of Theorem 6.17-2 then shows that the function $p : \Omega \to \mathbb{R}$ defined by

$$p : x \in \Omega \to p(x) := \int_{\gamma_x} \boldsymbol{h}(y) \cdot \mathrm{d}\boldsymbol{y}$$

(which is unambiguously defined as shown above), is differentiable in Ω, with partial derivatives given by

$$\partial_i p(x) = h_i(x), \quad x \in \Omega, \ \ 1 \le i \le N.$$

These relations also imply that the function p is of class $\mathcal{C}^2$ in Ω. This proves (b). □

Our *third application of J.L. Lions lemma* will now consist in showing that Poincaré's lemma still holds under a substantially *weaker regularity assumption*, viz., that the components h_i, $1 \le i \le N$, of the vector field $\boldsymbol{h}$ be only *distributions in* $H^{-1}(\Omega)$. Note that the *classical* Poincaré lemma and the *hypoellipticity of* Δ also play a key role in the proof.

Also, recall that a *totally different* characterization of vector fields in $\boldsymbol{H}^{-1}(\Omega)$ as gradients of scalar functions in $L^2(\Omega)$ has already been established in Theorem 6.14-2, as an application of *J.L. Lions lemma* (again) and of the *Banach closed range theorem.*

Theorem 6.17-4 (weak Poincaré lemma; *alias* Poincaré lemma in $L^2(\Omega)$[87]) *Let Ω be a simply connected domain in $\mathbb{R}^N$ and let there be given a vector field $\boldsymbol{h} \in \boldsymbol{H}^{-1}(\Omega) := H^{-1}(\Omega; \mathbb{R}^N)$ that satisfies*

$$\mathbf{curl}\, \boldsymbol{h} = \mathbf{0} \quad \textit{in } \boldsymbol{H}^{-2}(\Omega).$$

Then there exists a function $p \in L^2(\Omega)$ such that

$$\mathbf{grad}\, p = \boldsymbol{h} \quad \textit{in } \boldsymbol{H}^{-1}(\Omega).$$

Besides, any other solution $\widetilde{p} \in L^2(\Omega)$ to the equations $\partial_i \widetilde{p} = h_i$ in $H^{-1}(\Omega)$, $1 \le i \le N$, is of the form $\widetilde{p} = p + C$, where C is a constant.

Proof Recall that the *gradient operator* $\mathbf{grad} : \mathcal{D}'(\Omega) \to \boldsymbol{\mathcal{D}}'(\Omega)$ is defined by

$$(\mathbf{grad}\, v)_i := \partial_i v, \ \ 1 \le i \le N, \quad \text{for each } v \in \mathcal{D}'(\Omega),$$

the *divergence operator* $\operatorname{div} : \boldsymbol{\mathcal{D}}'(\Omega) \to \mathcal{D}'(\Omega)$ is defined by

$$\operatorname{div} \boldsymbol{v} := \sum_{i=1}^{N} \partial_i v_i \quad \text{for each } \boldsymbol{v} = (v_i)_{i=1}^N \in \boldsymbol{\mathcal{D}}'(\Omega),$$

[87]This result is due to:

P.G. Ciarlet; P. Ciarlet, Jr. [2005]: Another approach to linearized elasticity and a new proof of Korn's inequality, *Mathematical Models and Methods in Applied Sciences* **15**, 259–271.

The simpler proof given here is due to:

S. Kesavan [2005]: On Poincaré's and J.L. Lions' lemmas, *Comptes Rendus de l'Académie des Sciences de Paris, Série* I, **340**, 27–30.

Poincaré's lemma was later shown to hold in the even weaker sense of distributions in:

S. Mardare [2008]: On Poincaré and De Rham's theorems, *Revue Roumaine de Mathématiques Pures et Appliquées* **53**, 523–541.

the *vector Laplacian operator* $\Delta : \mathcal{D}'(\Omega) \to \mathcal{D}'(\Omega)$ is defined by

$$(\Delta \boldsymbol{v})_i := \Delta v_i, \quad 1 \le i \le N, \quad \text{for each } \boldsymbol{v} = (v_i)_{i=1}^N \in \mathcal{D}'(\Omega),$$

and the *curl operator* $\mathbf{curl} : \mathcal{D}'(\Omega) \to \mathcal{D}'\big(\Omega; \mathbb{R}^{\frac{N(N-1)}{2}}\big)$ is defined for any integer $N \ge 2$ by

$$(\mathbf{curl}\, \boldsymbol{v})_{ij} := (\partial_i v_j - \partial_j v_i), \quad 1 \le i < j \le N, \quad \text{for each } \boldsymbol{v} = (v_i) \in \mathcal{D}'(\Omega).$$

We thus have to show that, if $\boldsymbol{h} \in \boldsymbol{H}^{-1}(\Omega)$ satisfies $\mathbf{curl}\, \boldsymbol{h} = \mathbf{0}$ in $\boldsymbol{H}^{-2}(\Omega)$, then there exists $p \in L^2(\Omega)$ such that $\boldsymbol{h} = \mathbf{grad}\, p$ in $\boldsymbol{H}^{-1}(\Omega)$. To this end, we proceed in two stages.

(i) Since Theorem 6.14-3 (the proof of which relies on J.L. Lions lemma, by way of Theorem 6.14-1) clearly applies as well if the right-hand side $\boldsymbol{h}$ belongs to $\boldsymbol{H}^{-1}(\Omega)$ instead of $\boldsymbol{L}^2(\Omega)$, there exist a vector field $\boldsymbol{u} \in \boldsymbol{H}_0^1(\Omega)$ and a function $\lambda \in L^2(\Omega)$ such that

$$\begin{aligned} -\Delta \boldsymbol{u} + \mathbf{grad}\, \lambda &= \boldsymbol{h} \quad \text{in } \boldsymbol{H}^{-1}(\Omega), \\ \operatorname{div} \boldsymbol{u} &= 0 \quad \text{in } L^2(\Omega). \end{aligned}$$

Note that the assumptions that Ω is simply connected and that $\mathbf{curl}\, \boldsymbol{h} = \mathbf{0}$ in $\boldsymbol{H}^{-2}(\Omega)$ are not needed at this stage.

(ii) The assumption that $\mathbf{curl}\, \boldsymbol{h} = \mathbf{0}$ in $\boldsymbol{H}^{-2}(\Omega)$, together with the relation

$$\mathbf{curl}\, \mathbf{grad}\, \pi = \mathbf{0} \quad \text{in } \mathcal{D}'(\Omega) \text{ for any } \pi \in \mathcal{D}'(\Omega),$$

implies that

$$\Delta(\mathbf{curl}\, \boldsymbol{u}) = \mathbf{curl}(\Delta \boldsymbol{u}) = \mathbf{curl}\, \mathbf{grad}\, \pi - \mathbf{curl}\, \boldsymbol{h} = \mathbf{0}.$$

Since $\mathbf{curl}\, \boldsymbol{u} \in \boldsymbol{L}^2(\Omega) \subset \boldsymbol{L}^1_{\text{loc}}(\Omega)$, the *hypoellipticity of* Δ (Theorem 6.4-2) shows that

$$\mathbf{curl}\, \boldsymbol{u} \in \boldsymbol{C}^\infty(\Omega),$$

so that $(\partial_j u_i - \partial_i u_j) \in \mathcal{C}^\infty(\Omega)$ for all $1 \le i, j \le N$. Therefore

$$\sum_{j=1}^N \partial_j(\partial_j u_i - \partial_i u_j) = \Delta u_i - \partial_i(\operatorname{div} \boldsymbol{u}) = \Delta u_i \in \mathcal{C}^\infty(\Omega), \quad 1 \le i \le N,$$

since $\operatorname{div} \boldsymbol{u} = 0$.

Since $\Delta \boldsymbol{u} \in \boldsymbol{C}^\infty(\Omega)$ and $\mathbf{curl}\, \Delta \boldsymbol{u} = \mathbf{0}$ in Ω, and since Ω is *simply connected*, the *classical Poincaré lemma* (Theorem 6.17-2) can be applied, showing that there exists a function $\widetilde{p} \in \mathcal{C}^\infty(\Omega) \subset L^1_{\text{loc}}(\Omega) \subset \mathcal{D}'(\Omega)$ such that

$$\mathbf{grad}\, \widetilde{p} = \Delta \boldsymbol{u} = \mathbf{grad}\, \lambda - \boldsymbol{h} \quad \text{in } \boldsymbol{H}^{-1}(\Omega).$$

Since the distribution

$$p := \lambda - \widetilde{p} \in L^1_{\text{loc}}(\Omega)$$

is such that

$$\mathbf{grad}\, p = \mathbf{grad}\, \lambda - \mathbf{grad}\, \widetilde{p} = \boldsymbol{h} \in \boldsymbol{H}^{-1}(\Omega),$$

J.L. Lions lemma (Theorem 6.11-4) shows that p is in effect a *function in* $L^2(\Omega)$.

The uniqueness up to the addition of a constant of the solution $p \in L^2(\Omega)$ to the equation $\mathbf{grad}\, p = \boldsymbol{h}$ in $\boldsymbol{H}^{-1}(\Omega)$ is established as in the proof of Theorem 6.14-2. □

Remark Another application of the *hypoellipticity of* Δ shows that the vector field $\boldsymbol{u} \in \boldsymbol{H}_0^1(\Omega)$ is also in the space $\boldsymbol{\mathcal{C}}^\infty(\Omega)$, since $\Delta\boldsymbol{u} \in \boldsymbol{\mathcal{C}}^\infty(\Omega)$. This property is not used in the above proof, however. □

Together with the hypoellipticity of Δ, the J.L. Lions lemma thus plays a key role for proving the weak Poincaré lemma. Remarkably, the *weak Poincaré lemma conversely provides a very simple proof of J.L. Lions lemma* (Problem 6.17-3).

Problems

6.17-1 Let Ω be an open subset of $\mathbb{R}^N$.

(1) Let there be given a vector field $\boldsymbol{h} \in \boldsymbol{\mathcal{C}}^1(\Omega)$ and a point $x_0 \in \Omega$, such that, for any point $x \in \Omega$, the curvilinear integral $\int_{\gamma(x)} \boldsymbol{h}(y) \cdot \mathrm{d}\boldsymbol{y}$ is independent of the path $\gamma(x)$ joining x_0 to x in Ω. Show that $\mathbf{curl}\, \boldsymbol{h} = \mathbf{0}$ in Ω.

(2) Let there be given a vector field $\boldsymbol{h} \in \boldsymbol{\mathcal{C}}^1(\Omega)$. Show that there exists $p \in \mathcal{C}^2(\Omega)$ such that $\mathbf{grad}\, p = \boldsymbol{h}$ in Ω if and only if the curvilinear integral $\int_\gamma \boldsymbol{h}(y) \cdot \mathrm{d}\boldsymbol{y}$ is independent of the path γ joining any two distinct points in Ω.

6.17-2 This exercise provides a *counterexample to both the classical and the weak Poincaré lemmas* (Theorems 6.17-2 and 6.17-4) when Ω is not simply connected.

(1) Let

$$\Omega := \{(x_1, x_2) \in \mathbb{R}^2;\ 1 < x_1^2 + x_2^2 < 2\},$$
$$h_1(x_1, x_2) := \frac{x_2}{x_1^2 + x_2^2}, \quad \text{and} \quad h_2(x_1, x_2) := -\frac{x_1}{x_1^2 + x_2^2} \text{ for } (x_1, x_2) \in \Omega,$$

so that $h_1, h_2 \in \mathcal{C}^\infty(\Omega)$ and $\partial_1 h_2 - \partial_2 h_1 = 0$ in Ω.

Show that there does *not* exist any function $p \in \mathcal{C}^2(\Omega)$ that satisfies $\partial_i p = h_i$ in Ω, $i = 1, 2$.

Hint: Let $\widetilde{\Omega} := \Omega - \gamma$, where $\gamma := \{(x_1, 0) \in \mathbb{R}^2; -2 < x_1 < -1\}$. Then compute explicitly the general solution p of the equations $\partial_i p = h_i$ in $\widetilde{\Omega}$, $i = 1, 2$, and show that $\lim_{x_2 \to 0^+} p(x_1, x_2) \neq \lim_{x_2 \to 0^-} p(x_1, x_2)$ for all $-2 < x_1 < -1$.

(2) Construct a similar counterexample in any dimension $N \geq 3$.

6.17-3 (1) Let Ω be a simply connected domain in $\mathbb{R}^N$, and let $v \in \mathcal{D}'(\Omega)$ be a distribution such that $\mathbf{grad}\, v \in \boldsymbol{H}^{-1}(\Omega)$. Assuming that the weak Poincaré lemma (Theorem 6.17-4) holds, give a two-line proof of *J.L. Lions lemma* (Theorem 6.11-4) by showing that $v \in L^2(\Omega)$.

(2) Assuming that J.L. Lions lemma holds for simply connected domains in $\mathbb{R}^N$, show that it holds for any domain in $\mathbb{R}^N$.

6.18 Application of Poincaré's lemma: The classical and weak Saint-Venant lemmas; the Cesàro–Volterra path integral formula

This section is the "matrix analogue" of Section 6.17, with the *matrix symmetrized gradient operator*

$$\boldsymbol{v} : \mathcal{D}'(\Omega; \mathbb{R}^N) \to \boldsymbol{\nabla}_s(\boldsymbol{v}) := \frac{1}{2}(\boldsymbol{\nabla}\boldsymbol{v}^T + \boldsymbol{\nabla}\boldsymbol{v}) \in \mathcal{D}'(\Omega; \mathbb{S}^N)$$

playing the role of the *vector gradient operator*

$$\mathbf{grad} : p \in \mathcal{D}'(\Omega) \to \mathbf{grad}\, p \in \mathcal{D}'(\Omega; \mathbb{R}^N).$$

This explains why the discourse follows along the same lines as in Section 6.17. Note also that Theorems 6.18-1 and 6.18-3 below both crucially depend on *Poincaré's lemma*, in its classical and weak versions respectively.

The summation convention with respect to repeated indices is used throughout this section. Given an open subset of $\mathbb{R}^N$, consider the linear operator from the space $\boldsymbol{\mathcal{C}}^3(\Omega) := \mathcal{C}^3(\Omega; \mathbb{R}^N)$ into the space $\mathcal{C}^2(\Omega; \mathbb{S}^N)$ defined by

$$\boldsymbol{v} = (v_i) \in \boldsymbol{\mathcal{C}}^3(\Omega) \to \boldsymbol{\nabla}_s \boldsymbol{v} = \left(\frac{1}{2}(\partial_j v_i + \partial_i v_i)\right) \in \mathcal{C}^2(\Omega; \mathbb{S}^N).$$

Recall that the matrix field $\boldsymbol{\nabla}_s \boldsymbol{v}$ appears in the fundamental *Korn inequality* of Theorem 6.15-1 (where it was denoted $\boldsymbol{e}(\boldsymbol{v}) = (e_{ij}(\boldsymbol{v}))$.

A natural question therefore arises as to whether this linear operator is *invertible*, i.e., whether, given a matrix field $\boldsymbol{e} = (e_{ij}) \in \mathcal{C}^2(\Omega; \mathbb{S}^N)$, there exists a vector field $\boldsymbol{v} = (v_i) \in \mathcal{C}^3(\Omega; \mathbb{R}^N)$ such that

$$\frac{1}{2}(\partial_j v_i + \partial_i v_j) = e_{ij} \quad \text{in } \Omega, \ \ 1 \le i, j \le N,$$

or equivalently, such that

$$\boldsymbol{\nabla}_s \boldsymbol{v} = \boldsymbol{e} \quad \text{in } \Omega.$$

If this is the case, it is then immediately verified that the functions $e_{ij} = e_{ji} \in \mathcal{C}^2(\Omega)$ must *necessarily* satisfy the **Saint-Venant compatibility relations**:[88]

$$\partial_{\ell j} e_{ik} + \partial_{ki} e_{j\ell} - \partial_{\ell i} e_{jk} - \partial_{kj} e_{i\ell} = 0 \quad \text{in } \mathcal{C}(\Omega), \ \ 1 \le i, j, k, \ell \le N.$$

Remark When $N = 3$, the Saint-Venant compatibility relations can be conveniently condensed as a single *matrix equation*; cf. Problem 6.18-4. □

It is remarkable that these necessary conditions become *sufficient* if the open set Ω is *simply connected.*

[88]So named after Adhémar-Jean-Claude Barré de Saint-Venant (1797–1886), who published these relations in 1864.

Theorem 6.18-1 (classical Saint-Venant lemma; *alias* Saint-Venant lemma in $\mathcal{C}^3(\Omega)$) *Let Ω be a simply connected open subset of $\mathbb{R}^N$, and let there be given functions $e_{ij} = e_{ji} \in \mathcal{C}^2(\Omega)$, $1 \leq i,j \leq N$, that satisfy the Saint-Venant compatibility relations*

$$\partial_{\ell j}e_{ik} + \partial_{ki}e_{j\ell} - \partial_{\ell i}e_{jk} - \partial_{kj}e_{i\ell} = 0 \quad \text{in } \Omega, \ \ 1 \leq i,j,k,\ell \leq N.$$

Then there exists a vector field $\boldsymbol{v} = (v_i) \in \mathcal{C}^3(\Omega; \mathbb{R}^N)$ such that

$$\frac{1}{2}(\partial_j v_i + \partial_i v_j) = e_{ij} \quad \text{in } \Omega, \ \ 1 \leq i,j \leq N.$$

Besides, any other solution $\widetilde{\boldsymbol{v}} = (\widetilde{v}_i) \in \mathcal{C}^3(\Omega; \mathbb{R}^N)$ to the equations

$$\frac{1}{2}(\partial_j \widetilde{v}_i + \partial_i \widetilde{v}_j) = e_{ij} \quad \text{in } \Omega, \ \ 1 \leq i,j \leq N,$$

is of the form

$$\widetilde{\boldsymbol{v}}(x) = \boldsymbol{v}(x) + \boldsymbol{B}x + \boldsymbol{c}, \quad x \in \Omega,$$

for some $N \times N$ antisymmetric matrix $\boldsymbol{B}$ and some vector $\boldsymbol{c} \in \mathbb{R}^N$.

Proof It is implicitly understood that the various relations found in this proof hold for all the values $1, 2, \ldots, N$ of the Latin indices appearing in them. The Saint-Venant compatibility relations may be equivalently rewritten as

$$\partial_\ell h_{ijk} = \partial_k h_{ij\ell} \ \text{ in } \mathcal{C}(\Omega) \quad \text{with } h_{ijk} := \partial_j e_{ik} - \partial_i e_{jk} \in \mathcal{C}^1(\Omega).$$

Hence the *classical Poincaré lemma* (Theorem 6.17-2) shows that there exist functions $p_{ij} \in \mathcal{C}^2(\Omega)$, unique up to additive constants, such that

$$\partial_k p_{ij} = h_{ijk} = \partial_j e_{ik} - \partial_i e_{jk} \quad \text{in } \mathcal{C}^1(\Omega).$$

Besides, since $\partial_k p_{ij} = -\partial_k p_{ji}$ in $\mathcal{C}^1(\Omega)$, we have the freedom of choosing the functions p_{ij} in such a way that $p_{ij} + p_{ji} = 0$ in $\mathcal{C}^2(\Omega)$.

Noting that the functions $q_{ij} := (e_{ij} + p_{ij}) \in \mathcal{C}^2(\Omega)$ satisfy

$$\begin{aligned}\partial_k q_{ij} &= \partial_k e_{ij} + \partial_k p_{ij} = \partial_k e_{ij} + \partial_j e_{ik} - \partial_i e_{jk} \\ &= \partial_j e_{ik} + \partial_j p_{ik} = \partial_j q_{ik} \quad \text{in } \mathcal{C}^1(\Omega),\end{aligned}$$

we again resort to the *classical Poincaré lemma* to assert the existence of functions $v_i \in \mathcal{C}^3(\Omega)$, unique up to additive constants, such that

$$\partial_j v_i = q_{ij} = e_{ij} + p_{ij} \quad \text{in } \Omega.$$

Consequently,

$$\frac{1}{2}(\partial_j v_i + \partial_i v_j) = e_{ij} + \frac{1}{2}(p_{ij} + p_{ji}) = e_{ij} \quad \text{in } \Omega,$$

as required. That all other solutions are of the indicated form is established as in the proof of Theorem 6.15-2. □

Remark The assumption of simple-connectedness is essential; cf. Problem 6.18-2. □

As a useful complement to Theorem 6.18-1, we now show that each component of any vector field $\boldsymbol{v} \in \mathcal{C}^3(\Omega; \mathbb{R}^N)$ can be expressed in terms of the components of its symmetrized gradient field $\boldsymbol{\nabla}_s \boldsymbol{v} = (\frac{1}{2}(\partial_j v_i + \partial_i v_j)) \in \mathcal{C}^2(\Omega; \mathbb{S}^N)$ by means of a *path integral in* Ω and that, under the assumptions of Theorem 6.18-1, the same path integral provides a particular solution $\boldsymbol{v}$ to the equations $\boldsymbol{\nabla}_s \boldsymbol{v} = \boldsymbol{e}$ in Ω (hence all the other solutions are obtained by adding to this particular solution vector fields of the form $x \in \Omega \to \boldsymbol{B}x + \boldsymbol{c}$, with $\boldsymbol{B}$ an $N \times N$ antisymmetric matrix and $\boldsymbol{c} \in \mathbb{R}^N$).

Theorem 6.18-2 (Cesàro–Volterra path integral formula[89]) *Let Ω be a connected open subset of $\mathbb{R}^N$ and let x^0 be a point in Ω.*

(a) *Given any vector field $(v_i) \in \mathcal{C}^3(\Omega; \mathbb{R}^N)$, define the symmetric tensor field $(e_{ij}) \in \mathcal{C}^2(\Omega; \mathbb{S}^N)$ by*

$$e_{ij} := \frac{1}{2}(\partial_j v_i + \partial_i v_j), \quad 1 \le i, j \le N.$$

Then, given any point $x = (x_k) \in \Omega$ and given any path $\gamma_x \in \mathcal{C}^\infty([0,1]; \mathbb{R}^N)$ joining x^0 to x in Ω, the components of the vector field are given by the **Cesàro–Volterra path integral formula**, *viz.,*

$$\begin{aligned} v_i(x) &= v_i^0 + p_{ik}^0(x_k - x_k^0) \\ &\quad + \int_{\gamma_x} \{e_{ij}(y) + (\partial_k e_{ij}(y) - \partial_i e_{kj}(y))\} (x_k - y_k) \, \mathrm{d}y_j, \quad 1 \le i \le N, \end{aligned}$$

where

$$v_i^0 := v_i(x^0) \quad \text{and} \quad p_{ik}^0 := \frac{1}{2} \left(\partial_k v_i(x^0) - \partial_i v_k(x^0) \right).$$

(b) *Assume that Ω is simply connected. Then, given any symmetric tensor field $(e_{ij}) \in \mathcal{C}^2(\Omega; \mathbb{S}^N)$ whose components satisfy the Saint-Venant compatibility relations*

$$\partial_{\ell j} e_{ik} + \partial_{ki} e_{j\ell} - \partial_{\ell i} e_{jk} - \partial_{kj} e_{i\ell} = 0 \quad \text{in } \Omega, \quad 1 \le i, j, k, \ell \le N,$$

and given any point $x \in \Omega$, each path integral $\int_{\gamma_x} \{e_{ij}(y) + (\partial_k e_{ij}(y) - \partial_i e_{kj}(y))\} (x_k - y_k) \mathrm{d}y_j$, $1 \le i \le N$, is independent of the path $\gamma_x \in \mathcal{C}^\infty([0,1]; \mathbb{R}^N)$ chosen for joining x^0 to x. Besides, the vector field $(v_i) : \Omega \to \mathbb{R}^N$ defined for each $x = (x_k) \in \Omega$ by

$$v_i(x) := \int_{\gamma_x} \{e_{ij}(y) + (\partial_k e_{ij}(y) - \partial_i e_{kj}(y))\} (x_k - y_k) \, \mathrm{d}y_j, \quad 1 \le i \le N,$$

is of class $\mathcal{C}^2$ in Ω and is a particular solution to the equations

$$\frac{1}{2}(\partial_j v_i + \partial_i v_j) = e_{ij} \quad \text{in } \Omega, \quad 1 \le i, j \le N.$$

---

[89]Due to:

E. Cesàro [1906]: Sulle formole del Volterra, fondamentali nella teoria delle distorsioni elastiche, *Rendiconti Napoli* **12**, 311–321.

V. Volterra [1907]: Sur l'équilibre des corps élastiques multiplement connexes, *Annales de l'Ecole Normale* **24**, 401–517.

Proof Given any $x \in \Omega$ and any path $\boldsymbol{\gamma}_x = (\gamma_x^i)_{i=1}^N \in \mathcal{C}^\infty([0,1];\mathbb{R}^N)$ joining x^0 to x, we have

$$\begin{aligned} v_i(x) &= v_i(x^0) + \int_0^1 \frac{\mathrm{d}}{\mathrm{d}t}\,[v_i(\boldsymbol{\gamma}_x(t))]\,\mathrm{d}t = v_i(x^0) + \int_0^1 \partial_j v_i(\boldsymbol{\gamma}_x(t))\frac{\mathrm{d}\gamma_x^j}{\mathrm{d}t}(t)\,\mathrm{d}t \\ &= v_i(x^0) + \int_{\boldsymbol{\gamma}_x} \partial_j v_i(y)\,\mathrm{d}y_j = v_i(x^0) + \int_{\boldsymbol{\gamma}_x} e_{ij}(y)\,\mathrm{d}y_j + \int_{\boldsymbol{\gamma}_x} p_{ij}(y)\,\mathrm{d}y_j, \end{aligned}$$

where the functions $p_{ij} \in \mathcal{C}^2(\Omega)$ are defined by

$$p_{ij} := \frac{1}{2}(\partial_j v_i - \partial_i v_j) \quad \text{in } \Omega.$$

Noting that

$$\begin{aligned} \int_{\boldsymbol{\gamma}_x} p_{ij}(y)\,\mathrm{d}y_j &= \int_0^1 p_{ij}(\boldsymbol{\gamma}(t))\frac{\mathrm{d}\gamma_x^j}{\mathrm{d}t}(t)\,\mathrm{d}t \\ &= -\int_0^1 \left(\frac{\mathrm{d}}{\mathrm{d}t}\,[p_{ij}(\boldsymbol{\gamma}_x(t))]\right)\gamma_x^j(t)\,\mathrm{d}t + p_{ij}(x)\gamma_x^j(1) - p_{ij}(x^0)\gamma_x^j(0) \\ &= -\int_{\boldsymbol{\gamma}_x} \partial_j p_{ik}(y) y_k\,\mathrm{d}y_j + p_{ik}(x)x_k - p_{ik}^0 x_k^0, \end{aligned}$$

and that

$$\int_{\boldsymbol{\gamma}_x} x_k \partial_j p_{ik}(y)\,\mathrm{d}y_j = x_k \int_0^1 \frac{\mathrm{d}}{\mathrm{d}t}\,[p_{ik}(\boldsymbol{\gamma}_x(t))]\,\mathrm{d}t = p_{ik}(x)x_k - p_{ik}^0 x_k,$$

we conclude that

$$\int_{\boldsymbol{\gamma}_x} p_{ij}(y)\,\mathrm{d}y_j = p_{ik}^0(x_k - x_k^0) + \int_{\boldsymbol{\gamma}_x} \partial_j p_{ik}(y)(x_k - y_k)\,\mathrm{d}y_j.$$

The Cesàro–Volterra formula then follows from the relations

$$\partial_j p_{ik} = \partial_k e_{ij} - \partial_i e_{kj}.$$

This proves (a).

The proof of (b) is similar to that of part (b) of Theorem 6.17-3; for this reason, it is left as a problem (Problem 6.18-5). □

Remark Combined with the (delicate) theory of Calderón–Zygmund singular integrals, the above explicit representation of a vector field in terms of its symmetrized gradient by means of the Cesàro–Volterra path integral formula also provides a direct proof of *Korn's inequality*.[90] □

Remark When $N = 3$, the three relations that constitute the Cesàro–Volterra path integral formula can be conveniently condensed into a single *vector equation*; cf. Problem 6.18-4. □

---

[90] This approach is due to:
P.P. Mosolov; V.P. Mjasnikov [1971]: A proof of Korn's inequality, *Soviet Mathematics Doklady* **12**, 1618–1622.

Using the *weak version of Poincaré's lemma*, hence *in fine* again *J.L. Lions lemma*, we now show that the Saint-Venant lemma still holds under a *substantially weaker regularity assumption*, viz., that e_{ij}, $1 \le i,j \le N$, be only *functions in* $L^2(\Omega)$.

Interestingly, this "weak version" of the Saint-Venant lemma also provides a *new proof of Korn's inequality* (Theorem 6.18-5).

Theorem 6.18-3 (weak Saint-Venant lemma; *alias* Saint-Venant lemma in $\boldsymbol{H}^1(\Omega)$[91]) *Let Ω be a simply connected domain in $\mathbb{R}^N$. Let $\boldsymbol{e} = (e_{ij}) \in \mathbb{L}^2(\Omega)$ be a symmetric matrix field that satisfies the Saint-Venant compatibility relations*

$$\partial_{\ell j} e_{ik} + \partial_{ki} e_{j\ell} - \partial_{\ell i} e_{jk} - \partial_{kj} e_{i\ell} = 0 \quad \text{in } H^{-2}(\Omega), \quad 1 \le i,j,k,\ell \le N.$$

Then there exists a vector field $\boldsymbol{v} = (v_i)_{i=1}^N \in \boldsymbol{H}^1(\Omega)$ such that

$$\frac{1}{2}(\partial_j v_i + \partial_i v_j) = e_{ij} \quad \text{in } L^2(\Omega), \quad 1 \le i,j \le N.$$

Besides, all other solutions $\widetilde{\boldsymbol{v}} = (\widetilde{v}_i)_{i=1}^N \in \boldsymbol{H}^1(\Omega)$ to the equations $e_{ij} = \frac{1}{2}(\partial_j \widetilde{v}_i + \partial_i \widetilde{v}_j)$, $1 \le i,j \le N$, are of the form

$$\widetilde{\boldsymbol{v}}(x) = \boldsymbol{v}(x) + \boldsymbol{B}x + \boldsymbol{c} \quad \text{for almost all } x \in \Omega,$$

for some $N \times N$ antisymmetric matrix $\boldsymbol{B}$ and some vector $\boldsymbol{c} \in \mathbb{R}^N$.

Proof The proof is analogous to that of Theorem 6.18-1, save that it is now the *weak* version of Poincaré's lemma (Theorem 6.17-4) that is used; first, to show that there exist functions $p_{ij} \in L^2(\Omega)$, unique up to additive constants, that satisfy

$$\partial_k p_{ij} = h_{ijk} = \partial_j e_{ik} - \partial_i e_{jk} \quad \text{in } H^{-1}(\Omega),$$

and, second, to show that there exist functions $v_i \in H^1(\Omega)$, again unique up to additive constants, that satisfy $\partial_j v_i = q_{ij} = e_{ij} + p_{ij}$ in $L^2(\Omega)$.

Consequently,

$$\frac{1}{2}(\partial_j v_i + \partial_i v_j) = e_{ij} + \frac{1}{2}(p_{ij} + p_{ji}) = e_{ij} \quad \text{in } L^2(\Omega),$$

as desired. That all other solutions are of the indicated form follows from Theorem 6.15-2. □

[91]This result is due to:

P.G. Ciarlet; P. Ciarlet, Jr. [2005]: Another approach to linearized elasticity and a new proof of Korn's inequality, *Mathematical Models and Methods in Applied Sciences* **15**, 259–271.

Various extensions are found in:

G. Geymonat; F. Krasucki [2005]: Some remarks on the compatibility conditions in elasticity, *Accademia Nazionale delle Scienze detta dei* XL. *Rendiconti. Serie* V. *Memorie di Matematica e Applicazioni. Parte* I, **29**, 175–181.

G. Geymonat; F. Krasucki [2006]: Beltrami's solutions of general equilibrium equations in continuum mechanics, *Comptes Rendus de l'Académie des Sciences de Paris, Série* 1, **342**, 359–363.

C. Amrouche; P.G. Ciarlet; L. Gratie; S. Kesavan [2006]: On the characterization of matrix fields as linearized strain tensor fields, *Journal de Mathématiques Pures et Appliquées* **86**, 116–132.

Remark A *different* necessary and sufficient condition for a tensor $\boldsymbol{e} \in \mathbb{L}^2(\Omega)$ to be of the form $\boldsymbol{e} = \frac{1}{2}(\boldsymbol{\nabla v}^T + \boldsymbol{\nabla v})$ for some $\boldsymbol{v} \in \boldsymbol{H}^1(\Omega)$ will be given in the next section (Theorem 6.19-6). It asserts that the tensor $\boldsymbol{e}$ should lie in the orthogonal complement in $\mathbb{H}_0^1(\Omega)$ of the space spanned by all symmetric tensors $\boldsymbol{\sigma} \in \mathbb{H}_0^1(\Omega)$ that satisfy $\mathbf{div}\,\boldsymbol{\sigma} = \mathbf{0}$ in Ω; besides, the open set Ω need not be simply connected. □

Let a symmetric matrix field $\boldsymbol{e} = (e_{ij}) \in \mathbb{L}^2(\Omega)$ satisfy

$$\partial_{\ell j} e_{ik} + \partial_{ki} e_{j\ell} - \partial_{\ell i} e_{jk} - \partial_{kj} e_{i\ell} = 0 \text{ in } H^{-2}(\Omega), \quad 1 \le i, j, k, \ell \le N,$$

i.e., the weak form of Saint-Venant's compatibility relations. By Theorem 6.18-3, there then exists a unique equivalence class $\dot{\boldsymbol{v}} \in \dot{\boldsymbol{H}}^1(\Omega) = \boldsymbol{H}^1(\Omega)/\mathbf{Ker}\,\boldsymbol{\nabla}_s$ such that $\boldsymbol{e} = \boldsymbol{\nabla}_s \dot{\boldsymbol{v}}$ in $\mathbb{L}^2(\Omega)$, where (Theorem 6.15-2)

$$\mathbf{Ker}\,\boldsymbol{\nabla}_s = \{\boldsymbol{v} \in \boldsymbol{H}^1(\Omega);\ \text{there exist } \boldsymbol{B} \in \mathbb{A}^N \text{ and } \boldsymbol{c} \in \mathbb{R}^N \text{ such that } \boldsymbol{v}(x) = \boldsymbol{B}x + \boldsymbol{c} \text{ for almost all } x \in \Omega\}.$$

We now show that the mapping $\boldsymbol{\Xi} : \boldsymbol{e} \in \mathbb{L}^2(\Omega) \to \dot{\boldsymbol{v}} \in \dot{\boldsymbol{H}}^1(\Omega)$ defined in this fashion possesses a remarkable property.

Theorem 6.18-4 *Let Ω be a simply connected domain in $\mathbb{R}^N$. Define the space*

$$\mathbb{E}(\Omega) := \{\boldsymbol{e} = (e_{ij}) \in \mathbb{L}^2(\Omega);\ \partial_{\ell j} e_{ik} + \partial_{ki} e_{j\ell} - \partial_{\ell i} e_{jk} - \partial_{kj} e_{i\ell} = 0 \textit{ in } H^{-2}(\Omega),\ 1 \le i, j, k, \ell \le N\},$$

and let

$$\boldsymbol{\Xi} : \mathbb{E}(\Omega) \to \dot{\boldsymbol{H}}^1(\Omega)$$

be the linear mapping defined for each $\boldsymbol{e} \in \mathbb{E}(\Omega)$ by $\boldsymbol{\Xi}(\boldsymbol{e}) := \dot{\boldsymbol{v}}$, where $\dot{\boldsymbol{v}}$ is the unique element in the quotient space $\dot{\boldsymbol{H}}^1(\Omega)$ that satisfies

$$\boldsymbol{\nabla}_s \dot{\boldsymbol{v}} = \boldsymbol{e} \quad \textit{in } \mathbb{L}^2(\Omega)$$

(Theorem 6.18-3*). Then*

$$\boldsymbol{\Xi} \in \mathcal{L}(\mathbb{E}(\Omega);\ \dot{\boldsymbol{H}}^1(\Omega)), \quad \boldsymbol{\Xi} \textit{ is bijective, and } \boldsymbol{\Xi}^{-1} \in \mathcal{L}(\dot{\boldsymbol{H}}^1(\Omega); \mathbb{E}(\Omega)).$$

Proof Clearly, $\mathbb{E}(\Omega)$ is a Hilbert space as a closed subspace of $\mathbb{L}^2(\Omega)$. The mapping $\boldsymbol{\Xi}$ is injective since $\boldsymbol{\Xi}(\boldsymbol{e}) = \dot{\mathbf{0}}$ means that $\boldsymbol{e} = \boldsymbol{\nabla}_s \dot{\mathbf{0}} = \mathbf{0}$ and surjective since, given any $\dot{\boldsymbol{v}} \in \dot{\boldsymbol{H}}^1(\Omega)$, the matrix field $(e_{ij}) := \boldsymbol{\nabla}_s \dot{\boldsymbol{v}} \in \mathbb{L}^2(\Omega)$ necessarily satisfies $\partial_{\ell j} e_{ik} + \partial_{ki} e_{j\ell} - \partial_{\ell i} e_{jk} - \partial_{kj} e_{i\ell} = 0$ in $H^{-2}(\Omega)$.

Finally, the inverse mapping

$$\boldsymbol{\Xi}^{-1} : \dot{\boldsymbol{v}} \in \dot{\boldsymbol{H}}^1(\Omega) \to \boldsymbol{\nabla}_s \dot{\boldsymbol{v}} \in \mathbb{E}(\Omega)$$

is continuous, since there clearly exists a constant c such that

$$\|\boldsymbol{\nabla}_s \dot{\boldsymbol{v}}\|_{0,\Omega} = \|\boldsymbol{\nabla}_s(\boldsymbol{v} + \boldsymbol{r})\|_{0,\Omega} \le c\,\|\boldsymbol{v} + \boldsymbol{r}\|_{1,\Omega}$$

for any $\boldsymbol{v} \in \boldsymbol{H}^1(\Omega)$ and any $\boldsymbol{r} \in \mathbf{Ker}\, \nabla_s$, so that

$$\|\boldsymbol{\nabla}_s \dot{\boldsymbol{v}}\|_{0,\Omega} \le c \inf_{\boldsymbol{r}\in \mathbf{Ker}\,\boldsymbol{\nabla}_s} \|\boldsymbol{v} + \boldsymbol{r}\|_{1,\Omega} = c\, \|\dot{\boldsymbol{v}}\|_{1,\Omega} \cdot$$

The conclusion thus follows from the *corollary to the Banach open mapping theorem* (Theorem 5.6-2). □

Remarkably, Korn's inequalities of Section 6.15 can now be very simply recovered from Theorem 6.18-4:

Theorem 6.18-5 *That the mapping* $\Xi : \mathbb{E}(\Omega) \to \dot{\boldsymbol{H}}^1(\Omega)$ *is an isomorphism implies Korn's inequalities in both spaces* $\boldsymbol{H}^1(\Omega)$ *and* $\dot{\boldsymbol{H}}^1(\Omega)$ (*Theorems* 6.15-1 *and* 6.15-3).

Proof Since Ξ is an isomorphism, there exists a constant $\dot{C}$ such that

$$\|\Xi(\boldsymbol{e})\|_{1,\Omega} \le \dot{C}\, \|\boldsymbol{e}\|_{0,\Omega} \quad \text{for all } \boldsymbol{e} \in \mathbb{E}(\Omega),$$

or equivalently, such that

$$\|\dot{\boldsymbol{v}}\|_{1,\Omega} \le \dot{C}\, \|\boldsymbol{\nabla}_s \dot{\boldsymbol{v}}\|_{0,\Omega} \quad \text{for all } \dot{\boldsymbol{v}} \in \dot{\boldsymbol{H}}^1(\Omega).$$

But this is exactly the *Korn inequality in the quotient space* $\dot{\boldsymbol{H}}^1(\Omega)$, which is itself equivalent to the *Korn inequality in the space* $\boldsymbol{H}^1(\Omega)$ (Theorem 6.15-3). □

Problems

6.18-1 Show directly that, for $N = 3$, the 81 Saint-Venant compatibility relations reduce in fact to only six independent ones (which are not uniquely defined).

6.18-2 This problem provides a *counterexample to both the classical and weak Saint-Venant lemmas* (Theorems 6.18-1 and 6.18-3) when $N = 3$ and Ω is not simply connected.

Let $\Omega := \{x = (x_1, x_2, x_3) \in \mathbb{R}^3;\ 1 < x_1^2 + x_2^2 < 2 \text{ and } 0 < x_3 < 1\} \subset \mathbb{R}^3$, and let

$$e_{11}(x) := -\frac{x_2}{x_1^2 + x_2^2}, \quad e_{12}(x) = e_{21}(x) := \frac{x_1}{2(x_1^2 + x_2^2)}, \quad e_{ij}(x) = 0 \ \text{ if } i + j \ge 4, \ \text{ for } x \in \Omega,$$

so that $\partial_{\ell j} e_{ik} + \partial_{ki} e_{j\ell} - \partial_{\ell i} e_{jk} - \partial_{kj} e_{i\ell} = 0$ in Ω. Show that there does *not* exist any vector field $\boldsymbol{v} = (v_i) \in \mathcal{C}^3(\Omega)$ that satisfies $\frac{1}{2}(\partial_j v_i + \partial_i v_j) = e_{ij}$ in Ω.

Hint: Let $\widetilde{\Omega} := \Omega - \gamma$, where $\gamma := \{(x_1, 0, x_3) \in \mathbb{R}^3;\ -2 < x_1 < -1 \text{ and } 0 < x_3 < 1\}$. Then compute explicitly the general solution $\boldsymbol{v}$ of the equations $e_{ij}(\boldsymbol{v}) = e_{ij}$ in $\widetilde{\Omega}$, and show that $\lim_{x_2 \to 0^+} \boldsymbol{v}(x_1, x_2, x_3) \neq \lim_{x_2 \to 0^-} \boldsymbol{v}(x_1, x_2, x_3)$ for all $-2 < x_1 < -1$ and $0 < x_3 < 1$.

6.18-3 Let Ω be a domain in $\mathbb{R}^3$ with boundary Γ, and let the space $\mathbb{E}(\Omega)$ and the mapping $\Xi : \mathbb{E}(\Omega) \to \dot{\boldsymbol{H}}^1(\Omega)$ be defined as in Theorem 6.18-4. Given constants $\lambda \ge 0$ and $\mu > 0$ and vector fields $\boldsymbol{f} \in \boldsymbol{L}^2(\Omega)$ and $\boldsymbol{g} \in \boldsymbol{L}^2(\Gamma)$, define the functional

$$j : \boldsymbol{e} \in \mathbb{E}(\Omega) \to j(\boldsymbol{e}) := \frac{1}{2} \int_\Omega \{\lambda (\operatorname{tr} \boldsymbol{e})^2 + 2\mu \boldsymbol{e} : \boldsymbol{e}\}\, \mathrm{d}x - \ell(\Xi(\boldsymbol{e})),$$

where the functional $\ell : \boldsymbol{H}^1(\Omega) \to \mathbb{R}$ is defined by $\ell(\boldsymbol{v}) := \int_\Omega \boldsymbol{f} \cdot \boldsymbol{v}\, \mathrm{d}x + \int_\Gamma \boldsymbol{g} \cdot \boldsymbol{v}\, \mathrm{d}\Gamma$ for each $\boldsymbol{v} \in \boldsymbol{H}^1(\Omega)$ and is assumed to satisfy $\ell(\boldsymbol{r}) = 0$ for all $\boldsymbol{r} = \mathbf{Ker}\, \boldsymbol{\nabla}_s$.

(1) Show that the following quadratic minimization problem: Find $\boldsymbol{\varepsilon} \in \mathbb{E}(\Omega)$ such that $j(\boldsymbol{\varepsilon}) = \inf_{\boldsymbol{e}\in\mathbb{E}(\Omega)} j(\boldsymbol{e})$, has one and only one solution $\boldsymbol{\varepsilon}$.

(2) Show that $\boldsymbol{\varepsilon} = \boldsymbol{\nabla}_s \dot{\boldsymbol{u}}$, where $\boldsymbol{u} \in \boldsymbol{H}^1(\Omega)$ is any solution to the *pure traction problem of three-dimensional linearized elasticity* (Problem 6.16-2).

Remark While the minimization problem over the space $\dot{\boldsymbol{H}}^1(\Omega)$ found in Problem 6.16-2 is an *unconstrained one* with three unknowns, that of (1) over the space $\mathbb{E}(\Omega)$ is in effect a *constrained quadratic minimization problem* over the space $\mathbb{L}^2_s(\Omega)$ with six unknowns, the *constraints* being the compatibility relations $\partial_{\ell j} e_{ik} + \partial_{ki} e_{j\ell} - \partial_{\ell i} e_{jk} - \partial_{kj} e_{i\ell} = 0$ in $H^{-2}(\Omega)$ that the matrix fields $\boldsymbol{e} = (e_{ij}) \in \mathbb{E}(\Omega)$ satisfy (these compatibility relations reduce in fact to six independent ones; cf. Problem 6.18-1). □

6.18-4 Let $\varepsilon_{ijk} := 1$, *resp.* $\varepsilon_{ijk} := -1$, if $\{i, j, k\}$ is an even, *resp.* odd, permutation of $\{1, 2, 3\}$, and $\varepsilon_{ijk} := 0$ if at least two indices are equal. The *matrix curl operator* $\mathbf{CURL} : \mathcal{D}'(\Omega; \mathbb{M}^3) \to \mathcal{D}'(\Omega; \mathbb{M}^3)$ and the *matrix curl-curl operator* $\mathbf{CURL\,CURL} : \mathcal{D}'(\Omega; \mathbb{M}^3) \to \mathcal{D}'(\Omega; \mathbb{S}^3)$ are respectively defined by

$$(\mathbf{CURL}\,\boldsymbol{e})_{ij} := \varepsilon_{i\ell k} \partial_\ell e_{jk} \quad \text{for any matrix field } \boldsymbol{e} = (e_{ij}) \in \mathcal{D}'(\Omega; \mathbb{M}^3),$$
$$(\mathbf{CURL\,CURL}\,\boldsymbol{e})_{ij} := \varepsilon_{ik\ell} \varepsilon_{jmn} \partial_{\ell n} e_{km} \quad \text{for any matrix field } \boldsymbol{e} = (e_{ij}) \in \mathcal{D}'(\Omega; \mathbb{M}^3).$$

(1) Show that, if $N = 3$, the Saint-Venant compatibility relations

$$\partial_{\ell j} e_{ik} + \partial_{ki} e_{j\ell} - \partial_{\ell i} e_{jk} - \partial_{kj} e_{i\ell} = 0 \quad \text{in } \mathcal{C}(\Omega), \quad \text{for } 1 \le i, j, k, \ell \le 3,$$

are equivalent to the matrix equation

$$\mathbf{CURL\,CURL}\,\boldsymbol{e} = \mathbf{0} \quad \text{in } \mathcal{C}(\Omega; \mathbb{S}^3),$$

which also shows that the Saint-Venant compatibility relations reduce in fact to only six independent ones in this case.

(2) Show that, again if $N = 3$, the Cesàro–Volterra path integral formulas of Theorem 6.18-2 are equivalent to the vector equation

$$\boldsymbol{v}(x) = \boldsymbol{v}(x^0) + \boldsymbol{d}^0 \wedge (\boldsymbol{x} - \boldsymbol{x}^0) + \int_{\gamma(x)} \boldsymbol{\nabla}_s \boldsymbol{v}(y)\,\mathrm{d}y + \int_{\gamma(x)} (\boldsymbol{x} - \boldsymbol{y}) \wedge (\mathbf{CURL}\,\boldsymbol{\nabla}_s \boldsymbol{v}(y)\,\mathrm{d}y),$$

where the vector $\boldsymbol{d}^0 \in \mathbb{R}^3$ is defined by

$$\boldsymbol{d}^0 := \frac{1}{2} \begin{pmatrix} \partial_2 v_3 - \partial_3 v_2 \\ \partial_3 v_1 - \partial_1 v_3 \\ \partial_1 v_2 - \partial_2 v_1 \end{pmatrix} (x^0).$$

6.18-5 Let Ω be a simply connected open subset of $\mathbb{R}^N$ and let $(e_{ij}) \in \mathcal{C}^2(\Omega; \mathbb{S}^N)$ be a tensor field that satisfies the Saint-Venant compatibility relations in Ω.

(1) Show that, given any point $x \in \Omega$, each path integral $\int_{\gamma_x} \{e_{ij}(y) + (\partial_k e_{ij}(y) - \partial_i e_{kj}(y))\} (x_k - y_k)\mathrm{d}y_j$, $1 \le i \le N$, is independent of the path $\gamma_x \in \mathcal{C}^\infty([0,1]; \mathbb{R}^N)$ chosen for joining x^0 to x.

Hint: As in the proof of Theorem 6.18-1, rewrite the Saint-Venant relations as

$$\partial_\ell h_{ijk} = \partial_k h_{ij\ell} \text{ in } \mathcal{C}(\Omega) \quad \text{with } h_{ijk} := \partial_j e_{ik} - \partial_i e_{jk} \in \mathcal{C}^1(\Omega).$$

Then mimic the proof of Theorem 6.17-3(b).

(2) Using an argument similar to that used in the proof of *ibid.*, show that the vector field $(v_i) : \Omega \to \mathbb{R}^N$ defined for each $x = (x_k) \in \Omega$ by

$$v_i(x) = \int_{\gamma_x} \{e_{ij}(y) + (\partial_k e_{ij}(y) - \partial_i e_{kj}(y))\} (x_k - y_k)\mathrm{d}y_j, \quad 1 \le i \le N,$$

is differentiable in Ω, with partial derivatives given by

$$\partial_j v_i(x) = e_{ij}(x) + \int_{\gamma_x} \{\partial_j e_{ik}(y) - \partial_i e_{kj}(y)\} \mathrm{d}y_k,$$

$$\partial_i v_j(x) = e_{ji}(x) + \int_{\gamma_x} \{\partial_i e_{jk}(y) - \partial_j e_{ki}(y)\} \mathrm{d}y_k,$$

which shows that $\frac{1}{2}(\partial_j v_i + \partial_i v_j) = e_{ij}$ in Ω, $1 \le i,j \le N$.

(3) Show that the field (v_i) so defined is of class $\mathcal{C}^3$ in Ω.

6.19 Another application of J.L. Lions lemma: The Donati lemmas

The summation convention with respect to repeated indices is used throughout this section. Recall that, given an open subset Ω of $\mathbb{R}^N$, the *vector divergence operator* $\mathbf{div} : \mathcal{D}'(\Omega; \mathbb{M}^N) \to \mathcal{D}'(\Omega; \mathbb{R}^N)$ is defined by

$$(\mathbf{div}\, \boldsymbol{e})_i := \partial_j e_{ij} \quad \text{for any } \boldsymbol{e} = (e_{ij}) \in \mathcal{D}'(\Omega; \mathbb{M}^N).$$

The Saint-Venant compatibility relations (Section 6.18) provide a characterization of matrix fields as symmetrized gradient fields, but *another* characterization is possible.[92] More specifically, Luigi Donati[93] had already noticed in 1890 that, *if a smooth enough symmetric matrix field* $\boldsymbol{e} = (e_{ij})$ *defined on an open subset* $\Omega \subset \mathbb{R}^3$ *satisfies* (with the present notation)

$$\int_\Omega \boldsymbol{e} : \boldsymbol{s}\,\mathrm{d}x = 0 \quad \text{for all } \boldsymbol{s} \in \mathcal{D}(\Omega; \mathbb{S}^3) \text{ that satisfy } \mathbf{div}\, \boldsymbol{s} = \mathbf{0} \text{ in } \Omega,$$

then the components e_{ij} *necessarily satisfy the Saint-Venant compatibility relations in* Ω. Combined with the classical Saint-Venant lemma (Theorem 6.18-1), this observation thus implies that, if these relations are satisfied, then there exists a smooth vector field $\boldsymbol{v}$ such that $\boldsymbol{\nabla}_s \boldsymbol{v} = \boldsymbol{e}$ in Ω (at least if Ω is simply connected).

The objective of this section[94] is to provide several extensions, referred to in the sequel as *Donati's lemmas*, of this classical result to matrix fields $\boldsymbol{e}$ with less regularity. The first

[92] A history of the genesis of the classical characterizations of matrix fields as symmetrized gradient fields is found in:

M.E. GURTIN [1972]: The linear theory of elasticity, in *Handbuch der Physik, Volume* VIa/2 (S. FLÜGGE & C. TRUESDELL, editors), pp. 1–295, Springer, Berlin.

[93] L. DONATI [1890]: Illustrazione al teorema del Menabrea, *Memorie della Accademia delle Scienze dell'Istituto di Bologna* **10**, 267–274.

L. DONATI [1894]: Ulteriori osservazioni intorno al teorema del Menabrea, *Memorie della Accademia delle Scienze dell'Istituto di Bologna* **4**, 449–474.

[94] The content of this section is adapted from:

C. AMROUCHE; P.G. CIARLET; L. GRATIE; S. KESAVAN [2006]: On the characterization of matrix fields as linearized strain tensor fields, *Journal de Mathématiques Pures et Appliquées* **86**, 116–132.

extension is to symmetric matrix fields $\boldsymbol{e} = (e_{ij})$ whose components are only in $H^{-1}(\Omega)$, so that the resulting vector field $\boldsymbol{v}$ (i.e., that satisfies $\boldsymbol{\nabla}_s \boldsymbol{v} = \boldsymbol{e}$ in $\mathbb{H}^{-1}(\Omega)$) is found as expected in $\boldsymbol{L}^2(\Omega)$ (Theorem 6.19-4); the second and third extensions both hold if the components e_{ij} are in $L^2(\Omega)$, but they differ in that the resulting vector field $\boldsymbol{v}$ (i.e., the field that satisfies $\boldsymbol{e} = \boldsymbol{\nabla}_s \boldsymbol{v}$ in $\mathbb{L}^2(\Omega)$) is found either in $\boldsymbol{H}_0^1(\Omega)$ (Theorem 6.19-6) or in $\boldsymbol{H}^1(\Omega)$ (Theorem 6.19-6). Interestingly, these results hold for domains that need *not* be simply connected.

The property of the operator $\boldsymbol{\nabla}_s$ established in the next theorem extends *J.L. Lions lemma* in $H^m(\Omega)$ (i.e., for distributions v in $H^m(\Omega)$ whose gradient $\mathbf{grad}\, v$ is in $\boldsymbol{H}^m(\Omega)$; cf. Theorem 6.11-5) to *vector fields* $\boldsymbol{v} \in \boldsymbol{H}^m(\Omega)$ with *symmetrized gradients* in $\mathbb{H}^m(\Omega)$. Like other results in this section, this is one more illustration that the matrix operator $\boldsymbol{\nabla}_s : \mathcal{D}'(\Omega; \mathbb{R}^N) \to \mathcal{D}'(\Omega; \mathbb{S}^N)$ is indeed the "matrix analogue" of the vector operator $\mathbf{grad} : \mathcal{D}'(\Omega) \to \mathcal{D}'(\Omega; \mathbb{R}^N)$.

Theorem 6.19-1 (J.L. Lions lemma in $\boldsymbol{H}^m(\Omega)$: Vector version) *Let Ω be a domain in $\mathbb{R}^N$ and let $m \in \mathbb{Z}$. Then*

$$\boldsymbol{v} \in \boldsymbol{H}^m(\Omega) \quad \textit{and} \quad \boldsymbol{\nabla}_s \boldsymbol{v} \in \mathbb{H}^m(\Omega) \ \textit{ implies } \boldsymbol{v} \in \boldsymbol{H}^{m+1}(\Omega).$$

Proof Recall that J.L. Lions lemma in $H^m(\Omega)$ (Theorem 6.11-5) asserts that for any $m \in \mathbb{Z}$, $v \in H^m(\Omega)$ and $\mathbf{grad}\, v \in H^m(\Omega)$ implies that $v \in H^{m+1}(\Omega)$.

Let $\boldsymbol{v} = (v_i) \in \boldsymbol{H}^m(\Omega)$ be such that $\boldsymbol{\nabla}_s \boldsymbol{v} \in \mathbb{H}^m(\Omega)$ for some integer $m \in \mathbb{Z}$. Then the identity

$$(\mathbf{grad}(\partial_k v_i))_j = \partial_j\left((\boldsymbol{\nabla}_s \boldsymbol{v})_{ik}\right) + \partial_k\left((\boldsymbol{\nabla}_s \boldsymbol{v})_{ij}\right) - \partial_i\left((\boldsymbol{\nabla}_s \boldsymbol{v})_{jk}\right)$$

shows that each component $\partial_k v_i \in H^{m-1}(\Omega)$ of the matrix $\boldsymbol{\nabla} \boldsymbol{v}$ is such that $\mathbf{grad}(\partial_k v_i) \in \boldsymbol{H}^{m-1}(\Omega)$. Therefore, J.L. Lions lemma in $H^{m-1}(\Omega)$ shows that $\partial_k v_i \in H^m(\Omega)$. Since $v_i \in H^m(\Omega)$ by assumption, another application of the same J.L. Lions lemma, this time in $H^m(\Omega)$, shows that $v_i \in H^{m+1}(\Omega)$. □

Remark The above vector version of J.L. Lions lemma is no longer a triviality for $m = 0$, by contrast with the original J.L. Lions lemma. □

The next theorem lists two properties of the operator $\boldsymbol{\nabla}_s$ *considered as acting from the space $\boldsymbol{L}^2(\Omega)$ into the space $\mathbb{H}^{-1}(\Omega)$*. Note that these are nothing but the *natural matrix analogs* of the properties established for vector fields in parts (a) and (b) of Theorem 6.14-1.

Theorem 6.19-2 *Let Ω be a domain in $\mathbb{R}^N$. Then:*

(a) *The dual of the continuous operator*

$$\boldsymbol{\nabla}_s : \boldsymbol{L}^2(\Omega) \to \mathbb{H}^{-1}(\Omega)$$

is the continuous operator

$$-\,\mathbf{div} : \mathbb{H}_0^1(\Omega) \to \boldsymbol{L}^2(\Omega).$$

(b) *The image of the space $\boldsymbol{L}^2(\Omega)$ under the operator $\boldsymbol{\nabla}_s$ is closed in $\mathbb{H}^{-1}(\Omega)$.*

Proof (i) For any $\boldsymbol{v} = (v_i) \in \boldsymbol{L}^2(\Omega)$ and any $\boldsymbol{e} = (e_{ij}) \in \mathbb{H}_0^1(\Omega)$,

$$\begin{aligned} {}_{\mathbb{H}^{-1}(\Omega)}\langle \boldsymbol{\nabla}_s \boldsymbol{v}, \boldsymbol{e} \rangle_{\mathbb{H}_0^1(\Omega)} &= {}_{H^{-1}(\Omega)}\langle \partial_j v_i, e_{ij} \rangle_{H_0^1(\Omega)} \\ &= {}_{L^2(\Omega)}\langle v_i, \partial_j e_{ij} \rangle_{L^2(\Omega)} = {}_{\boldsymbol{L}^2(\Omega)}\langle \boldsymbol{v}, -\,\mathbf{div}\, \boldsymbol{e} \rangle_{\boldsymbol{L}^2(\Omega)} \end{aligned}$$

(the symmetry of $\boldsymbol{e}$ is used in the first equality). Hence the dual operator of $\boldsymbol{\nabla}_s : \boldsymbol{L}^2(\Omega) \to \mathbb{H}^{-1}(\Omega)$ is $-\,\mathbf{div} : \mathbb{H}^1_0(\Omega) \to \boldsymbol{L}^2(\Omega)$ and the dual operator of $\boldsymbol{\nabla}_s : \dot{\boldsymbol{L}}^2(\Omega) \to \mathbb{H}^{-1}(\Omega)$ (defined for each $\dot{\boldsymbol{v}} \in \dot{\boldsymbol{L}}^2(\Omega)$ by $\boldsymbol{\nabla}_s\dot{\boldsymbol{v}} := \nabla_s \boldsymbol{w}$ for any $\boldsymbol{w} \in \dot{\boldsymbol{v}}$) is $-\,\mathbf{div} : \mathbb{H}^1_0(\Omega) \to \dot{\boldsymbol{L}}^2(\Omega)$. This proves (a).

(ii) *There exists a constant C such that*

$$\|\boldsymbol{v}\|_{0,\Omega} \leq C\left(\|\boldsymbol{v}\|_{-1,\Omega} + \|\boldsymbol{\nabla}_s\boldsymbol{v}\|_{-1,\Omega}\right) \quad \text{for all } \boldsymbol{v} \in \boldsymbol{L}^2(\Omega).$$

It is easily seen that the space

$$\boldsymbol{K}(\Omega) := \{\boldsymbol{v} \in \boldsymbol{H}^{-1}(\Omega);\ \boldsymbol{\nabla}_s\boldsymbol{v} \in \mathbb{H}^{-1}(\Omega)\},$$

equipped with the norm

$$\boldsymbol{v} \in \boldsymbol{K}(\Omega) \to \|\boldsymbol{v}\|_{\boldsymbol{K}(\Omega)} := (\|\boldsymbol{v}\|^2_{-1,\Omega} + \|\boldsymbol{\nabla}_s\boldsymbol{v}\|^2_{-1,\Omega})^{1/2},$$

is *complete* (mimic part (iii) of the proof of Theorem 6.14-1). The identity mapping $\boldsymbol{\iota} : (\boldsymbol{L}^2(\Omega), \|\cdot\|_{0,\Omega}) \to (\boldsymbol{K}(\Omega), \|\cdot\|_{\boldsymbol{K}(\Omega)})$ being injective, continuous (the corresponding inequality clearly holds), and *surjective* by *J.L. Lions lemma in $\boldsymbol{H}^{-1}(\Omega)$* (Theorem 6.19-1), the *corollary to the Banach open mapping theorem* shows that the inverse mapping $\boldsymbol{\iota}$ is also continuous. This is exactly what the inequality announced in (ii) expresses.

(iii) *There exists a constant $\dot{C}$ such that*

$$\|\dot{\boldsymbol{v}}\|_{0,\Omega} \leq \dot{C}\,\|\boldsymbol{\nabla}_s\dot{\boldsymbol{v}}\|_{-1,\Omega} \quad \text{for all } \dot{\boldsymbol{v}} \in \dot{\boldsymbol{L}}^2(\Omega).$$

Assume that such an inequality does not hold. Then there exist $\dot{\boldsymbol{v}}^k \in \dot{\boldsymbol{L}}^2(\Omega)$, $k \geq 1$, such that

$$\|\dot{\boldsymbol{v}}^k\|_{0,\Omega} = 1 \text{ for all } k \geq 1 \quad \text{and} \quad \|\boldsymbol{\nabla}_s\dot{\boldsymbol{v}}^k\|_{-1,\Omega} \to 0 \text{ as } k \to \infty.$$

Since the space $\mathbf{Ker}\,\boldsymbol{\nabla}_s$ is finite-dimensional (Theorem 6.15-2), there exists for each $\dot{\boldsymbol{v}} \in \dot{\boldsymbol{L}}^2(\Omega)$ an element $\boldsymbol{w} \in \dot{\boldsymbol{v}}$ such that $\|\boldsymbol{w}\|_{0,\Omega} = \|\dot{\boldsymbol{v}}\|_{0,\Omega} := \inf_{\boldsymbol{r}\in\mathbf{Ker}\,\boldsymbol{\nabla}_s} \|\boldsymbol{v} + \boldsymbol{r}\|_{0,\Omega}$. Hence, for each integer $k \geq 1$, there exist $\boldsymbol{w}^k \in \dot{\boldsymbol{v}} \subset \boldsymbol{L}^2(\Omega)$ such that

$$\|\boldsymbol{w}^k\|_{0,\Omega} = 1 \text{ for all } k \geq 1 \quad \text{and} \quad \|\boldsymbol{\nabla}_s\boldsymbol{w}^k\|_{-1,\Omega} = \|\boldsymbol{\nabla}_s\dot{\boldsymbol{v}}^k\|_{-1,\Omega} \to 0 \text{ as } k \to \infty.$$

By the *Rellich–Kondrachov compact imbedding theorem in the space $L^2(\Omega)$* (Theorem 6.11-3), there thus exists a subsequence $(\boldsymbol{w}^{\sigma(k)})_{k=1}^\infty$ that converges in $\boldsymbol{H}^{-1}(\Omega)$. Since the subsequence $(\boldsymbol{\nabla}_s\boldsymbol{w}^{\sigma(k)})_{k=1}^\infty$ converges in $\mathbb{H}^{-1}(\Omega)$ (to $\boldsymbol{0}$, but this fact is not used at this stage), the subsequence $(\boldsymbol{w}^{\sigma(k)})_{k=1}^\infty$ is thus a Cauchy sequence in the space $(\boldsymbol{K}(\Omega), \|\cdot\|_{\boldsymbol{K}(\Omega)})$, *hence also in the space $\boldsymbol{L}^2(\Omega)$*, by the inequality established in (ii).

Consequently, there exists $\boldsymbol{w} \in \boldsymbol{L}^2(\Omega)$ such that

$$\boldsymbol{w}^{\sigma(k)} \underset{k\to\infty}{\longrightarrow} \boldsymbol{w} \quad \text{in } \boldsymbol{L}^2(\Omega).$$

Besides,

$$\boldsymbol{\nabla}_s\boldsymbol{w}^{\sigma(k)} \underset{k\to\infty}{\longrightarrow} \boldsymbol{0} = \boldsymbol{\nabla}_s\boldsymbol{w} \quad \text{in } \boldsymbol{H}^{-1}(\Omega),$$

which means that $\boldsymbol{w} \in \mathbf{Ker}\,\boldsymbol{\nabla}_s$. Therefore,

$$\dot{\boldsymbol{w}}^{\sigma(k)} \underset{k\to\infty}{\longrightarrow} \dot{\boldsymbol{w}} = \dot{\mathbf{0}} \quad \text{in } \dot{\boldsymbol{L}}^2(\Omega),$$

which contradicts $\|\dot{\boldsymbol{w}}^{\sigma(k)}\|_{0,\Omega} = \|\boldsymbol{w}^{\sigma(k)}\|_{0,\Omega} = 1$ for all $k \geq 1$. Hence the announced inequality holds.

(iv) Clearly, the images in the space $\mathbb{H}^{-1}(\Omega)$ of the spaces $\boldsymbol{L}^2(\Omega)$ and $\dot{\boldsymbol{L}}^2(\Omega)$ under the operator $\boldsymbol{\nabla}_s$ are identical. The linear operator $\boldsymbol{\nabla}_s : \dot{\boldsymbol{L}}^2(\Omega) \to \mathbb{H}^{-1}(\Omega)$ is injective (since $\dot{\boldsymbol{L}}^2(\Omega) = \boldsymbol{L}^2(\Omega)/\,\mathbf{Ker}\,\boldsymbol{\nabla}_s$), clearly continuous, and has an inverse from $\mathbf{Im}\,\boldsymbol{\nabla}_s \subset \mathbb{H}^{-1}(\Omega)$ onto $\dot{\boldsymbol{L}}^2(\Omega)$ that is also continuous by (iii). Hence the space $\mathbf{Im}\,\boldsymbol{\nabla}_s$ is a complete subspace of $\mathbb{H}^{-1}(\Omega)$ and as such is a *closed subspace* of $\mathbb{H}^{-1}(\Omega)$. This proves (b). □

Note in passing that the inequality established in (ii) constitutes a **Korn inequality in $\boldsymbol{L}^2(\Omega)$**.

The next theorem lists two properties of the operator $\boldsymbol{\nabla}_s$, now *considered as acting from the space $\boldsymbol{H}_0^1(\Omega)$ into the space $\mathbb{L}^2(\Omega)$; note that $\boldsymbol{\nabla}_s$ now becomes injective* since $\mathbf{Ker}\,\boldsymbol{\nabla}_s = \mathbf{0}$ in this case.

Theorem 6.19-3 *Let Ω be a domain in $\mathbb{R}^N$. Then:*

(a) *The dual of the injective continuous operator*

$$\boldsymbol{\nabla}_s : \boldsymbol{H}_0^1(\Omega) \to \mathbb{L}^2(\Omega)$$

is the continuous operator

$$-\,\mathbf{div} : \mathbb{L}^2(\Omega) \to \boldsymbol{H}^{-1}(\Omega).$$

(b) *The image of the space $\boldsymbol{H}_0^1(\Omega)$ under the operator $\boldsymbol{\nabla}_s$ is closed in $\mathbb{L}^2(\Omega)$.*

Proof The proof is similar to that of Theorem 6.19-2; it is even simpler, since elementary computations show that (Problem 6.15-1)

$$|\boldsymbol{v}|_{1,\Omega} := \Big(\sum_{i,j} \|\partial_j v_i\|_{0,\Omega}^2\Big)^{1/2} \leq \sqrt{2}\,\|\boldsymbol{\nabla}_s \boldsymbol{v}\|_{0,\Omega} \quad \text{for all } \boldsymbol{v} = (v_i) \in \boldsymbol{H}_0^1(\Omega).$$

This relation implies that there exists a constant C such that (Theorem 6.5-2)

$$\|\boldsymbol{v}\|_{1,\Omega} \leq C\,\|\boldsymbol{\nabla}_s \boldsymbol{v}\| \quad \text{for all } \boldsymbol{v} \in \boldsymbol{H}_0^1(\Omega),$$

which constitutes the analogue of part (iii) in the proof of Theorem 6.19-2. □

We are now in a position to prove our first *Donati lemma*, which constitutes the "matrix analogue" of Theorem 6.14-2.

Theorem 6.19-4 (Donati lemma in $\boldsymbol{L}^2(\Omega)$) *Let Ω be a domain in $\mathbb{R}^N$. Given a matrix field $\mathbf{e} \in \mathbb{H}^{-1}(\Omega)$, there exists a vector field $\boldsymbol{v}$ such that*

$$\boldsymbol{v} \in \boldsymbol{L}^2(\Omega) \quad \textit{and} \quad \boldsymbol{\nabla}_s \boldsymbol{v} = \mathbf{e} \ \textit{ in } \mathbb{H}^{-1}(\Omega)$$

if and only if

$$ {}_{\mathbb{H}^{-1}(\Omega)}\langle \boldsymbol{e}, \boldsymbol{s}\rangle_{\mathbb{H}_0^1(\Omega)} = 0 \quad \textit{for all } \boldsymbol{s} \in \mathbb{H}_0^1(\Omega) \textit{ that satisfy } \mathbf{div}\, \boldsymbol{s} = \mathbf{0} \text{ in } \boldsymbol{L}^2(\Omega). $$

All other solutions $\widetilde{\boldsymbol{v}} \in \boldsymbol{L}^2(\Omega)$ *of the equation* $\boldsymbol{\nabla}_s \widetilde{\boldsymbol{v}} = \boldsymbol{e}$ *are of the form*

$$ \widetilde{\boldsymbol{v}}(x) = \boldsymbol{v}(x) + \boldsymbol{B}x + \boldsymbol{c} \quad \textit{for almost all } x \in \Omega, $$

for some $N \times N$ *antisymmetric matrix* $\boldsymbol{B}$ *and some vector* $\boldsymbol{c} \in \mathbb{R}^N$.

Proof It was shown in Theorem 6.19-2 that the dual of $\boldsymbol{\nabla}_s : \boldsymbol{L}^2(\Omega) \to \mathbb{H}^{-1}(\Omega)$ is $-\,\mathbf{div} : \mathbb{H}_0^1(\Omega) \to \boldsymbol{L}^2(\Omega)$ and that the image $\operatorname{Im} \boldsymbol{\nabla}_s$ of $\boldsymbol{L}^2(\Omega)$ under $\boldsymbol{\nabla}_s$ is closed in $\mathbb{H}^{-1}(\Omega)$. Therefore, the *Banach closed range theorem* (first part; cf. Theorem 5.11-5) implies that

$$ \operatorname{Im} \boldsymbol{\nabla}_s = \{\boldsymbol{e} \in \mathbb{H}^{-1}(\Omega);\ {}_{\mathbb{H}^{-1}(\Omega)}\langle \boldsymbol{e}, \boldsymbol{s}\rangle_{\mathbb{H}_0^1(\Omega)} = 0 \quad \text{for all } \boldsymbol{s} \in \operatorname{Ker}(-\,\mathbf{div})\}, $$

which is exactly what the theorem asserts. That all other solutions $\widetilde{\boldsymbol{v}} \in \boldsymbol{L}^2(\Omega)$ of the equation $\boldsymbol{\nabla}_s \widetilde{\boldsymbol{v}} = \boldsymbol{e}$ are of the form indicated in the theorem follows from the characterization of the space $\mathbf{Ker}\, \boldsymbol{\nabla}_s$ established in Theorem 6.15-2. □

Remark Theorem 6.19-4 can be extended[95] to matrix fields $\boldsymbol{e} \in \mathbb{W}^{-1,p}(\Omega)$, $1 < p < \infty$, that satisfy ${}_{\mathbb{W}^{-1,p}(\Omega)}\langle \boldsymbol{e}, \boldsymbol{s}\rangle_{\mathbb{W}_0^{1,q}(\Omega)} = 0$ for all $\boldsymbol{s} \in \mathbb{W}_0^{1,q}(\Omega)$ that satisfy $\mathbf{div}\, \boldsymbol{s} = \mathbf{0}$ in $\boldsymbol{L}^q(\Omega)$, where q is the conjugate exponent of p. □

While the Donati lemma above is a corollary of Theorem 6.19-2, another Donati lemma can be similarly obtained, this time as a corollary to Theorem 6.19-3.

Theorem 6.19-5 (Donati Lemma in $\boldsymbol{H}_0^1(\Omega)$) *Let* Ω *be a domain in* $\mathbb{R}^N$. *Given a matrix field* $\boldsymbol{e} \in \mathbb{L}^2(\Omega)$, *there exists a vector field* $\boldsymbol{v}$ *such that*

$$ \boldsymbol{v} \in \boldsymbol{H}_0^1(\Omega) \quad \textit{and} \quad \boldsymbol{\nabla}_s \boldsymbol{v} = \boldsymbol{e} \ \textit{ in } \mathbb{L}^2(\Omega) $$

if and only if

$$ \int_\Omega \boldsymbol{e} : \boldsymbol{s}\, dx = 0 \quad \textit{for all } \boldsymbol{s} \in \mathbb{L}^2(\Omega) \textit{ that satisfy } \mathbf{div}\, \boldsymbol{s} = \mathbf{0} \textit{ in } \boldsymbol{H}^{-1}(\Omega). $$

In this case, the vector field $\boldsymbol{v}$ *is unique.*

Proof Since the dual operator of $\boldsymbol{\nabla}_s : \boldsymbol{H}_0^1(\Omega) \to \mathbb{L}^2(\Omega)$ is $-\,\mathbf{div} : \mathbb{L}^2(\Omega) \to \boldsymbol{H}^{-1}(\Omega)$ and the image of $\boldsymbol{H}_0^1(\Omega)$ under $\boldsymbol{\nabla}_s$ is closed in $\mathbb{L}^2(\Omega)$ (Theorem 6.19-3), the existence of the vector field $\boldsymbol{v}$ follows from the Banach closed range theorem (as in the proof of Theorem 6.19-4, but this time applied to the operator $\boldsymbol{\nabla}_s$ considered as acting from $\boldsymbol{H}_0^1(\Omega)$ into $\mathbb{L}^2(\Omega)$). That $\mathbf{Ker}\, \boldsymbol{\nabla}_s = \{\mathbf{0}\}$ in this case implies that such a vector field $\boldsymbol{v}$ is unique. □

[95] G. Geymonat; F. Krasucki [2005]: Some remarks on the compatibility conditions in elasticity, *Accademia Nazionale delle Scienze detta dei* XL. *Rendiconti. Serie* V. *Memorie di Matematica e Applicazioni. Parte* I, **29**, 175–181.

Remark A similar result holds for more general boundary conditions,[96] of the form $\boldsymbol{v} = \boldsymbol{0}$ on a relatively open subset Γ_0 of $\Gamma := \partial\Omega$ with $\mathrm{d}\Gamma\text{-meas}\,\Gamma_0 > 0$. □

Finally, a third Donati lemma can be derived as a consequence of the vector version of J.L. Lions lemma (Theorem 6.19-1), and of the first Donati lemma (Theorem 6.19-4). Notice that the tensor fields $\boldsymbol{s}$ that satisfy $\mathbf{div}\,\boldsymbol{s} = \boldsymbol{0}$ now range in the space $\mathbb{H}_0^1(\Omega)$, instead of the space $\mathbb{L}^2(\Omega)$ as in Theorem 6.19-5; as a result, the sought vector fields $\boldsymbol{v}$ now lie in the space $\boldsymbol{H}^1(\Omega)$ instead of the space $\boldsymbol{H}_0^1(\Omega)$.

Theorem 6.19-6 (Donati lemma in $\boldsymbol{H}^1(\Omega)$) *Let Ω be a domain in $\mathbb{R}^N$. Given a matrix field $\boldsymbol{e} \in \mathbb{L}^2(\Omega)$, there exists a vector field $\boldsymbol{v}$ such that*

$$\boldsymbol{v} \in \boldsymbol{H}^1(\Omega) \quad \textit{and} \quad \boldsymbol{\nabla}_s \boldsymbol{v} = \boldsymbol{e} \ \textit{in}\ \mathbb{L}^2(\Omega)$$

if and only if

$$\int_\Omega \boldsymbol{e} : \boldsymbol{s}\,\mathrm{d}x = 0 \quad \textit{for all } \boldsymbol{s} \in \mathbb{H}_0^1(\Omega) \textit{ that satisfy } \mathbf{div}\,\boldsymbol{s} = \boldsymbol{0} \textit{ in } \boldsymbol{L}^2(\Omega).$$

All other solutions $\widetilde{\boldsymbol{v}} \in \boldsymbol{L}^2(\Omega)$ of the equation $\boldsymbol{\nabla}_s \widetilde{\boldsymbol{v}} = \boldsymbol{e}$ are of the form

$$\widetilde{\boldsymbol{v}}(x) = \boldsymbol{v}(x) + \boldsymbol{B}x + \boldsymbol{c} \quad \textit{for almost all } x \in \Omega,$$

for some $N \times N$ antisymmetric matrix $\boldsymbol{B}$ and some vector $\boldsymbol{c} \in \mathbb{R}^N$.

Proof Let $\boldsymbol{e} \in \mathbb{L}^2(\Omega)$ be such that $\int_\Omega \boldsymbol{e} : \boldsymbol{s}\,\mathrm{d}x = 0$ for all $\boldsymbol{s} \in \mathbb{H}_0^1(\Omega)$ that satisfy $\mathbf{div}\,\boldsymbol{s} = \boldsymbol{0}$ in $\boldsymbol{L}^2(\Omega)$.

Since $\mathbb{L}^2(\Omega) \subset \mathbb{H}^{-1}(\Omega)$, Theorem 6.19-4 shows that there exists $\boldsymbol{v} \in \boldsymbol{L}^2(\Omega)$ such that $\boldsymbol{\nabla}_s \boldsymbol{v} = \boldsymbol{e}$ in $\mathbb{H}^{-1}(\Omega)$; hence $\boldsymbol{\nabla}_s \boldsymbol{v} \in \mathbb{L}^2(\Omega)$ since $\boldsymbol{e} \in \mathbb{L}^2(\Omega)$ by assumption. Theorem 6.19-1 with $m = 0$ then asserts that $\boldsymbol{v} \in \boldsymbol{H}^1(\Omega)$. The announced relations are therefore sufficient.

Conversely, assume that $\boldsymbol{e} = (e_{ij}) = \boldsymbol{\nabla}_s \boldsymbol{v}$ for some $\boldsymbol{v} = (v_i) \in \boldsymbol{H}^1(\Omega)$. Then the symmetry of $\boldsymbol{e}$ and Green's formula (Theorem 6.6-7) together imply that

$$\begin{aligned} \int_\Omega \boldsymbol{e} : \boldsymbol{s}\,\mathrm{d}x &= \int_\Omega e_{ij} s_{ij}\,\mathrm{d}x = \int_\Omega (\partial_j v_i) s_{ij}\,\mathrm{d}x \\ &= -\int_\Omega v_i \partial_j s_{ij}\,\mathrm{d}x = -\int_\Omega \boldsymbol{v} \cdot \mathbf{div}\,\boldsymbol{s}\,\mathrm{d}x \quad \text{for all } \boldsymbol{s} = (s_{ij}) \in \mathbb{H}_0^1(\Omega). \end{aligned}$$

Consequently, $\int_\Omega \boldsymbol{e} \cdot \boldsymbol{s}\,\mathrm{d}x = 0$ if $\boldsymbol{s} \in \mathbb{H}_0^1(\Omega)$ satisfies $\mathbf{div}\,\boldsymbol{s} = 0$ in $\boldsymbol{L}^2(\Omega)$; the announced relations are therefore necessary.

The nonuniqueness result again follows from Theorem 6.15-2. □

Interesting complements to both Theorems 6.19-4 and 6.19-6 are proposed in Problem 6.19-1.

[96] G. Geymonat; F. Krasucki [2005]: Some remarks on the compatibility conditions in elasticity, *Accademia Nazionale delle Scienze detta dei* XL. *Rendiconti. Serie* V. *Memorie di Matematica e Applicazioni. Parte* I, **29**, 175–181.

Problems

6.19-1 Given a domain Ω in $\mathbb{R}^N$, define the spaces

$$\mathbb{V}(\Omega) := \{\boldsymbol{s} \in \mathbb{H}_0^1(\Omega);\ \mathbf{div}\, \boldsymbol{s} = \mathbf{0} \text{ in } \Omega\} \quad \text{and} \quad \mathbb{W}(\Omega) := \{\boldsymbol{\sigma} \in \mathbb{D}(\Omega);\ \mathbf{div}\, \boldsymbol{\sigma} = \mathbf{0} \text{ in } \Omega\}.$$

(1) Let a matrix field $\boldsymbol{e} \in \mathbb{H}^{-1}(\Omega)$ be such that

$${}_{\mathbb{H}^{-1}(\Omega)}\langle \boldsymbol{e}, \boldsymbol{s}\rangle_{\mathbb{H}_0^1(\Omega)} = 0 \quad \text{for all } \boldsymbol{s} \in \mathbb{W}(\Omega).$$

Show that there exists a vector field $\boldsymbol{v} \in \boldsymbol{L}^2(\Omega)$ such that $\boldsymbol{\nabla}_s \boldsymbol{v} = \boldsymbol{e}$ in $\mathbb{H}^{-1}(\Omega)$.

Hint: See the hint provided in Problem 6.14-1.

(2) Using (1) and Theorem 4.3-2, show that the subspace $\mathbb{W}(\Omega)$ of $\mathbb{V}(\Omega)$ is dense in $(\mathbb{V}(\Omega), |\cdot|_{1,\Omega})$.

(3) Let a matrix field $\boldsymbol{e} \in \mathbb{L}^2(\Omega)$ be such that

$$\int_\Omega \boldsymbol{e} : \boldsymbol{\sigma}\, \mathrm{d}x = 0 \quad \text{for all } \boldsymbol{\sigma} \in \mathbb{W}(\Omega).$$

Show that there exists a vector field $\boldsymbol{v} \in \boldsymbol{H}^1(\Omega)$ such that $\boldsymbol{\nabla}_s \boldsymbol{v} = \boldsymbol{e}$ in $\mathbb{L}^2(\Omega)$.[97]

Remark. One can show that, in fact, the following result[98] holds more generally *in the sense of distributions*: Let Ω be any open subset of $\mathbb{R}^N$. If a matrix field $\boldsymbol{e} = (e_{ij}) \in \mathbb{D}'(\Omega)$ satisfies ${}_{\mathbb{D}'(\Omega)}\langle \boldsymbol{e}, \boldsymbol{\sigma}\rangle_{\mathbb{D}(\Omega)} := {}_{\mathcal{D}'(\Omega)}\langle e_{ij}, \sigma_{ij}\rangle_{\mathcal{D}(\Omega)} = 0$ for all matrix fields $\boldsymbol{\sigma} = (\sigma_{ij}) \in \mathbb{D}(\Omega)$ that satisfy $\mathbf{div}\, \boldsymbol{\sigma} = \mathbf{0}$ in Ω, then there exists a vector field $\boldsymbol{v} \in \boldsymbol{\mathcal{D}}'(\Omega)$ that satisfies $\boldsymbol{\nabla}_s \boldsymbol{v} = \boldsymbol{e}$ in $\mathbb{D}'(\Omega)$. □

6.19-2 Show that the closure of the space $\mathbb{V}(\Omega) := \{\boldsymbol{s} \in \mathbb{H}_0^1(\Omega);\ \mathbf{div}\, \boldsymbol{s} = \mathbf{0} \text{ in } \Omega\}$ (the same as in Problem 6.19-1) with respect to the norm $\|\cdot\|_{0,\Omega}$ is a *strict* subspace of the space $\{\boldsymbol{s} \in \mathbb{L}^2(\Omega);\ \mathbf{div}\, \boldsymbol{s} = \mathbf{0} \text{ in } \boldsymbol{H}^{-1}(\Omega)\}$ (naturally, the same is *a fortiori* true of the closure of the space $\{\boldsymbol{\sigma} \in \mathbb{D}(\Omega);\ \mathbf{div}\, \boldsymbol{\sigma} = \mathbf{0} \text{ in } \Omega\}$).

6.19-3 Let Ω be a domain in $\mathbb{R}^3$ with boundary Γ. Define the Hilbert space

$$\widetilde{\mathbb{E}}(\Omega) := \left\{\boldsymbol{e} \in \mathbb{L}^2(\Omega);\ \int_\omega \boldsymbol{e} : \boldsymbol{s}\, \mathrm{d}x = 0 \text{ for all } \boldsymbol{s} \in \mathbb{L}^2(\Omega) \text{ that satisfy } \mathbf{div}\, \boldsymbol{s} = \mathbf{0} \text{ in } \boldsymbol{H}^{-1}(\Omega)\right\},$$

and, for each $\boldsymbol{e} \in \widetilde{\mathbb{E}}(\Omega)$, let $\widetilde{\boldsymbol{\Xi}}(\boldsymbol{e})$ denote the unique element in the space $\boldsymbol{H}_0^1(\Omega)$ that satisfies $\boldsymbol{\nabla}_s \widetilde{\boldsymbol{\Xi}}(\boldsymbol{e}) = \boldsymbol{e}$ (Theorem 6.19-5).

(1) Show that the linear operator $\widetilde{\boldsymbol{\Xi}} : \widetilde{\mathbb{E}}(\Omega) \to \boldsymbol{H}_0^1(\Omega)$ defined in this fashion is bijective, continuous, and has a continuous inverse.

(2) Given constants $\lambda \geq 0$ and $\mu > 0$ and a vector field $\boldsymbol{f} \in \boldsymbol{L}^2(\Omega)$, define the functional

$$\widetilde{j} : \boldsymbol{e} \in \widetilde{\mathbb{E}}(\Omega) \to \widetilde{j}(\boldsymbol{e}) := \frac{1}{2}\int_\Omega \{\lambda(\operatorname{tr} \boldsymbol{e})^2 + 2\mu \boldsymbol{e} : \boldsymbol{e}\}\, \mathrm{d}x - \int_\Omega \boldsymbol{f} \cdot \widetilde{\boldsymbol{\Xi}}(\boldsymbol{e})\, \mathrm{d}x$$

and show that the following quadratic minimization problem: Find $\widetilde{\boldsymbol{\varepsilon}} \in \widetilde{\mathbb{E}}(\Omega)$ such that $\widetilde{j}(\widetilde{\boldsymbol{\varepsilon}}) = \inf_{\boldsymbol{e} \in \widetilde{\mathbb{E}}(\Omega)} \widetilde{j}(\boldsymbol{e})$, has one and only one solution.

(3) Show that $\widetilde{\boldsymbol{\varepsilon}} = \boldsymbol{\nabla}_s \boldsymbol{u}$, where $\boldsymbol{u} \in \boldsymbol{H}_0^1(\Omega)$ is the solution to the *pure displacement problem of linearized elasticity* (Section 6.16).

Remark A comparison with Problem 6.18-3 is instructive. □

---

[97] This result is due to:
T.W. TING [1974]: St. Venant's compatibility conditions, *Tensors, N.S.* **28**, 5–12.

[98] J.J. MOREAU [1979]: Duality characterization of strain tensor distributions in an arbitrary open set, *Journal of Mathematical Analysis and Applications* **72**, 760–770.

6.20 Pfaff systems

We conclude this chapter by studying a specific class of *systems of linear partial differential equations of the first order*, which (together with the classical Poincaré lemma; cf. Theorem 6.17-2) play in particular a crucial role in the proof of the *fundamental theorem of Riemannian geometry for an open subset of* $\mathbb{R}^n$ (Theorem 8.6-1) and in the proof of the *fundamental theorem of surface theory* (Theorem 8.16-1).

Let Ω be an open subset of $\mathbb{R}^N$ and let $n \geq 1$ be an integer. A **Pfaff system**[99] is a set of N equations of the form

$$\partial_i \boldsymbol{F}(x) = \boldsymbol{F}(x)\boldsymbol{\Gamma}_i(x), \quad x \in \Omega,\ 1 \leq i \leq N,$$

where the matrix fields $\boldsymbol{\Gamma}_i : \Omega \to \mathbb{M}^n$, $1 \leq i \leq N$, are *given* and the *unknown* is the matrix field $\boldsymbol{F} : \Omega \to \mathbb{M}^n$. The unknown is often required *in addition* to satisfy a condition of the form

$$\boldsymbol{F}(x^0) = \boldsymbol{F}^0,$$

where the point $x^0 \in \Omega$ and the matrix $\boldsymbol{F}^0 \in \mathbb{M}^n$ are given (if $n = 1$, this condition is nothing but an initial condition for an ordinary differential equation), which then implies the *uniqueness* of the solution.

Remark Pfaff systems can take more general forms; cf. Problem 6.20-1. □

A *necessary condition* for the existence of a solution to such a Pfaff system immediately emerges, which simply expresses the *commutativity of the second partial derivatives of the solutions* (just like the necessary condition for the system $\partial_i p(x) = h_i(x)$, $x \in \Omega$, $1 \leq i \leq N$, in Section 6.17): Assume that $\boldsymbol{\Gamma}_i \in \mathcal{C}^1(\Omega; \mathbb{M}^n)$, $1 \leq i \leq N$; then a solution $\boldsymbol{F} \in \mathcal{C}^2(\Omega; \mathbb{M}^n)$ necessarily satisfies $\partial_{ki}\boldsymbol{F}(x) = \partial_{ik}\boldsymbol{F}(x)$, $x \in \Omega$, $1 \leq i, k \leq N$, viz.,

$$\boldsymbol{F}(x)(\boldsymbol{\Gamma}_i(x)\boldsymbol{\Gamma}_k(x) + \partial_i\boldsymbol{\Gamma}_k(x)) = \boldsymbol{F}(x)(\boldsymbol{\Gamma}_k(x)\boldsymbol{\Gamma}_i(x) + \partial_k\boldsymbol{\Gamma}_i(x)), \quad x \in \Omega,\ 1 \leq i, k \leq N.$$

So, under the *assumption* that the matrix $\boldsymbol{F}(x) \in \mathbb{M}^n$ is invertible at each $x \in \Omega$,

$$\partial_i\boldsymbol{\Gamma}_k(x) - \partial_k\boldsymbol{\Gamma}_i(x) + \boldsymbol{\Gamma}_i(x)\boldsymbol{\Gamma}_k(x) - \boldsymbol{\Gamma}_k(x)\boldsymbol{\Gamma}_i(x) = \boldsymbol{0}, \quad x \in \Omega,\ 1 \leq i, k \leq N.$$

Remarkably, this necessary condition becomes also *sufficient* for the *existence* of a solution if the open set Ω is *simply connected*, as we now show; note the resemblance of the next proof with that of the *classical Poincaré lemma* (Theorem 6.17-2).

Theorem 6.20-1 (existence and uniqueness of the solution to a Pfaff system[100]**)** *Let* Ω *be a simply connected open subset of* $\mathbb{R}^N$ *and let there be given matrix fields* $\boldsymbol{\Gamma}_i \in$

---

[99] So named after Johann Friedrich Pfaff (1765–1825), who had the honor of counting Carl Friedrich Gauß among his doctoral students.

[100] This result goes back to:

E. Cartan [1927]: Sur la possibilité de plonger un espace riemannien donné dans un espace euclidien, *Annales de la Société Polonaise de Mathématiques* **6**, 1–7.

It was extended later to *nonlinear* Pfaff systems by:

T.Y. Thomas [1934]: Systems of total differential equations defined over simply connected domains, *Annals of Mathematics* **35**, 730–734.

$\mathcal{C}^1(\Omega; \mathbb{M}^n)$, $1 \le i \le N$, *that satisfy*

$$\partial_i \boldsymbol{\Gamma}_k(x) - \partial_k \boldsymbol{\Gamma}_i(x) + \boldsymbol{\Gamma}_i(x)\boldsymbol{\Gamma}_k(x) - \boldsymbol{\Gamma}_k(x)\boldsymbol{\Gamma}_i(x) = \mathbf{0}, \quad x \in \Omega,\ 1 \le i, k \le N.$$

Let a point $x^0 \in \Omega$ *and a matrix* $\boldsymbol{F}^0 \in \mathbb{M}^n$ *be given. Then there exists one, and only one, matrix field* $\boldsymbol{F} \in \mathcal{C}^2(\Omega; \mathbb{M}^n)$ *that satisfies*

$$\begin{aligned} &\partial_i \boldsymbol{F}(x) = \boldsymbol{F}(x)\boldsymbol{\Gamma}_i(x), \quad x \in \Omega,\ 1 \le i \le N, \\ &\boldsymbol{F}(x^0) = \boldsymbol{F}^0. \end{aligned}$$

Proof The notation A_{ij} designates the element at the ith row and jth column of an arbitrary matrix $\boldsymbol{A} \in \mathbb{M}^n$. The *specific* notation Γ_{ij}^p designates the element at the pth row and jth column of a matrix $\boldsymbol{\Gamma}_i \in \mathbb{M}^n$, where i is an integer that ranges in $\{1, \dots, N\}$. The summation convention is used with respect to repeated indices or exponents that range in $\{1, \dots, N\}$ or in $\{1, \dots, n\}$.

(i) Let x^1 be an arbitrary point in the set Ω, *distinct from* x^0. Since Ω is in particular *arcwise-connected*, there exists a *path* $\boldsymbol{\gamma} = (\gamma^i) \in \mathcal{C}^1([0,1]; \mathbb{R}^N)$ *joining* x^0 *to* x^1 *in* Ω; this means that

$$\boldsymbol{\gamma}(0) = x^0,\ \boldsymbol{\gamma}(1) = x^1, \quad \text{and} \quad \boldsymbol{\gamma}(t) \in \Omega \text{ for all } 0 \le t \le 1.$$

Assume that a matrix field $\boldsymbol{F} = (F_{\ell j}) \in \mathcal{C}^1(\Omega; \mathbb{M}^n)$ satisfies

$$\partial_i \boldsymbol{F}(x) = \boldsymbol{F}(x)\boldsymbol{\Gamma}_i(x), \quad \text{or equivalently,} \quad \partial_i F_{\ell j}(x) = \Gamma_{ij}^p(x) F_{\ell p}(x), \quad \text{at each } x \in \Omega.$$

Then, *for each integer* $1 \le \ell \le n$, the n functions $\zeta_j \in \mathcal{C}^1([0,1])$ defined by (for simplicity, the dependence on ℓ is dropped)

$$\zeta_j(t) := F_{\ell j}(\boldsymbol{\gamma}(t)), \quad 0 \le t \le 1, \quad 1 \le j \le n,$$

satisfy the following *Cauchy problem for a linear system of* n *ordinary differential equations with respect to* n *unknowns*:

$$\begin{aligned} &\frac{\mathrm{d}\zeta_j}{\mathrm{d}t}(t) = \Gamma_{ij}^p(\boldsymbol{\gamma}(t)) \frac{\mathrm{d}\gamma^i}{\mathrm{d}t}(t)\zeta_p(t), \quad 0 \le t \le 1, \\ &\zeta_j(0) = \zeta_j^0, \end{aligned}$$

where the *initial values* ζ_j^0 are given by

$$\zeta_j^0 := F_{\ell j}^0$$

(note in passing that these Cauchy problems only differ by their initial values ζ_j^0).

Since a Cauchy problem of the form (with self-explanatory notations)

$$\begin{aligned} &\frac{\mathrm{d}\boldsymbol{\zeta}}{\mathrm{d}t}(t) = \boldsymbol{A}(t)\boldsymbol{\zeta}(t), \quad 0 \le t \le 1, \\ &\boldsymbol{\zeta}(0) = \boldsymbol{\zeta}^0, \end{aligned}$$

has *one* and *only one* solution $\boldsymbol{\zeta} \in \mathcal{C}^1([0,1]; \mathbb{R}^n)$ if $\boldsymbol{A} \in \mathcal{C}([0,1]; \mathbb{M}^n)$ (Theorem 3.8-2), each one of these Cauchy problems has one and only one solution.

Incidentally, this result already shows that, *if it exists, the unknown field* $\boldsymbol{F} = (F_{\ell j})$ *is unique*.

(ii) In order that the values $\zeta_j(1)$ found by solving the above Cauchy problem for a given integer $\ell \in \{1, \dots, n\}$ be acceptable candidates for the unknown values $F_{\ell j}(x^1)$, they must be of course *independent of the path chosen for joining* x^0 *to* x^1.

So, let $\boldsymbol{\gamma}_0 \in \mathcal{C}^1([0,1];\mathbb{R}^N)$ and $\boldsymbol{\gamma}_1 \in \mathcal{C}^1([0,1];\mathbb{R}^N)$ be two paths joining x^0 to x^1 in Ω. The open set Ω being *simply connected*, there exists a *homotopy* $\boldsymbol{G} = (G^i) : [0,1] \times [0,1] \to \mathbb{R}^N$ *joining* $\boldsymbol{\gamma}_0$ *to* $\boldsymbol{\gamma}_1$ *in* Ω (Theorem 6.17-1), i.e., such that

$$\begin{gathered}\boldsymbol{G}(\cdot,0) = \boldsymbol{\gamma}_0, \quad \boldsymbol{G}(\cdot,1) = \boldsymbol{\gamma}_1, \quad \boldsymbol{G}(t,\lambda) \in \Omega \text{ for all } 0 \le t \le 1,\ 0 \le \lambda \le 1, \\ \boldsymbol{G}(0,\lambda) = x^0 \quad \text{and} \quad \boldsymbol{G}(1,\lambda) = x^1 \text{ for all } 0 \le \lambda \le 1,\end{gathered}$$

and smooth enough in the sense that

$$\boldsymbol{G} \in \mathcal{C}^1([0,1]\times[0,1];\mathbb{R}^N) \quad \text{and} \quad \frac{\partial}{\partial t}\left(\frac{\partial \boldsymbol{G}}{\partial \lambda}\right) = \frac{\partial}{\partial \lambda}\left(\frac{\partial \boldsymbol{G}}{\partial t}\right) \in \mathcal{C}([0,1]\times[0,1];\mathbb{R}^N).$$

Let $\boldsymbol{\zeta}(\cdot,\lambda) = (\zeta_j(\cdot,\lambda)) \in \mathcal{C}^1([0,1];\mathbb{R}^N)$ denote for each $0 \le \lambda \le 1$ the solution of the Cauchy problem corresponding to the path $\boldsymbol{G}(\cdot,\lambda)$ joining x^0 to x^1. We thus have

$$\begin{aligned}&\frac{\partial \zeta_j}{\partial t}(t,\lambda) = \Gamma^p_{ij}(\boldsymbol{G}(t,\lambda))\frac{\partial G^i}{\partial t}(t,\lambda)\zeta_p(t,\lambda) \quad \text{for all } 0 \le t \le 1,\ 0 \le \lambda \le 1, \\ &\zeta_j(0,\lambda) = \zeta_j^0 \quad \text{for all } 0 \le \lambda \le 1.\end{aligned}$$

Our objective is to show that

$$\frac{\partial \zeta_j}{\partial \lambda}(1,\lambda) = 0 \quad \text{for all } 0 \le \lambda \le 1,$$

since this relation will imply that $\zeta_j(1,0) = \zeta_j(1,1)$, as desired. For this purpose, a direct differentiation shows that, for all $0 \le t \le 1$, $0 \le \lambda \le 1$,

$$\frac{\partial}{\partial \lambda}\left(\frac{\partial \zeta_j}{\partial t}\right) = \{\Gamma^q_{ij}\Gamma^p_{kq} + \partial_k\Gamma^p_{ij}\}\zeta_p\frac{\partial G^k}{\partial \lambda}\frac{\partial G^i}{\partial t} + \Gamma^p_{ij}\zeta_p\frac{\partial}{\partial \lambda}\left(\frac{\partial G^i}{\partial t}\right) + \sigma_q\Gamma^q_{ij}\frac{\partial G^i}{\partial t},$$

where

$$\sigma_j := \frac{\partial \zeta_j}{\partial \lambda} - \Gamma^p_{kj}\zeta_p\frac{\partial G^k}{\partial \lambda},$$

on the one hand (in the relations above and below, $\Gamma^q_{ij}, \partial_k\Gamma^p_{ij}$, etc. stand for $\Gamma^q_{ij}(\boldsymbol{G}(\cdot,\cdot))$, $\partial_k\Gamma^p_{ij}(\boldsymbol{G}(\cdot,\cdot))$, etc.).

On the other hand, a direct differentiation of the equation defining the functions σ_j shows that, for all $0 \le t \le 1, 0 \le \lambda \le 1$,

$$\frac{\partial}{\partial t}\left(\frac{\partial \zeta_j}{\partial \lambda}\right) = \frac{\partial \sigma_j}{\partial t} + \left\{\partial_i\Gamma^p_{kj}\frac{\partial G^i}{\partial t}\zeta_p + \Gamma^q_{kj}\frac{\partial \zeta_q}{\partial t}\right\}\frac{\partial G^k}{\partial \lambda} + \Gamma^p_{ij}\zeta_p\frac{\partial}{\partial t}\left(\frac{\partial G^i}{\partial \lambda}\right).$$

But $\dfrac{\partial \zeta_j}{\partial t} = \Gamma^p_{ij}\dfrac{\partial G^i}{\partial t}\zeta_p$, so that we also have

$$\frac{\partial}{\partial t}\left(\frac{\partial \zeta_j}{\partial \lambda}\right) = \frac{\partial \sigma_j}{\partial t} + \{\partial_i\Gamma^p_{kj} + \Gamma^q_{kj}\Gamma^p_{iq}\}\zeta_p\frac{\partial G^i}{\partial t}\frac{\partial G^k}{\partial \lambda} + \Gamma^p_{ij}\zeta_p\frac{\partial}{\partial t}\left(\frac{\partial G^i}{\partial \lambda}\right).$$

Hence, subtracting the above relations and noting that $\dfrac{\partial}{\partial\lambda}\Big(\dfrac{\partial\zeta_j}{\partial t}\Big) = \dfrac{\partial}{\partial t}\Big(\dfrac{\partial\zeta_j}{\partial\lambda}\Big)$ and $\dfrac{\partial}{\partial\lambda}\Big(\dfrac{\partial G^i}{\partial t}\Big) = \dfrac{\partial}{\partial t}\Big(\dfrac{\partial G^i}{\partial\lambda}\Big)$ by assumption, we infer that

$$\frac{\partial\sigma_j}{\partial t} + \{\partial_i\Gamma^p_{kj} - \partial_k\Gamma^p_{ij} + \Gamma^q_{kj}\Gamma^p_{iq} - \Gamma^q_{ij}\Gamma^p_{kq}\}\zeta_p\frac{\partial G^k}{\partial\lambda}\frac{\partial G^i}{\partial t} - \Gamma^q_{ij}\frac{\partial G^i}{\partial t}\sigma_q = 0.$$

But $\{\partial_i\Gamma^p_{kj} - \partial_k\Gamma^p_{ij} + \Gamma^q_{kj}\Gamma^p_{iq} - \Gamma^q_{ij}\Gamma^p_{kq}\}$ is nothing but the element at the pth row and jth column of the matrix field $\partial_i\boldsymbol{\Gamma}_k - \partial_k\boldsymbol{\Gamma}_i + \boldsymbol{\Gamma}_i\boldsymbol{\Gamma}_k - \boldsymbol{\Gamma}_k\boldsymbol{\Gamma}_i$, *which vanishes in* Ω *by assumption.* Hence

$$\partial_i\Gamma^p_{kj} - \partial_k\Gamma^p_{ij} + \Gamma^q_{kj}\Gamma^p_{iq} - \Gamma^q_{ij}\Gamma^p_{kq} = 0,$$

on the one hand. On the other hand,

$$\sigma_j(0,\lambda) = \frac{\partial\zeta_j}{\partial\lambda}(0,\lambda) - \Gamma^p_{kj}(\boldsymbol{G}(0,\lambda))\zeta_p(0,\lambda)\frac{\partial G^k}{\partial\lambda}(0,\lambda) = 0,$$

since $\zeta^0_j(0,\lambda) = \zeta^0_j$ and $\boldsymbol{G}(0,\lambda) = x^0$ for all $0 \le \lambda \le 1$. Therefore, for any *fixed* value of the parameter $\lambda \in [0,1]$, each function $\sigma_j(\cdot,\lambda)$ satisfies a *Cauchy problem for an ordinary differential equation*, viz.,

$$\begin{aligned} &\frac{\mathrm{d}\sigma_j}{\mathrm{d}t}(t,\lambda) = \Gamma^q_{ij}(\boldsymbol{G}(t,\lambda))\frac{\partial G^i}{\partial t}(t,\lambda)\sigma_q(t,\lambda), \quad 0 \le t \le 1, \\ &\sigma_j(0,\lambda) = 0. \end{aligned}$$

But the solution of such a Cauchy problem is unique (Theorem 3.8-2); hence $\sigma_j(t,\lambda) = 0$ for all $0 \le t \le 1$. In particular then,

$$\begin{aligned} \sigma_j(1,\lambda) &= \frac{\partial\zeta_j}{\partial\lambda}(1,\lambda) - \Gamma^p_{kj}(\boldsymbol{G}(1,\lambda))\zeta_p(1,\lambda)\frac{\partial G^k}{\partial\lambda}(1,\lambda) \\ &= 0 \quad \text{for all } 0 \le \lambda \le 1, \end{aligned}$$

and thus $\dfrac{\partial\zeta_j}{\partial\lambda}(1,\lambda) = 0$ for all $0 \le \lambda \le 1$, since $\boldsymbol{G}(1,\lambda) = x^1$ for all $0 \le \lambda \le 1$.

For each integer $\ell \in \{1,\ldots,n\}$, we may thus unambiguously define a vector field $(F_{\ell j}) : \Omega \to \mathbb{R}^n$ by letting

$$F_{\ell j}(x^1) := \zeta_j(1) \quad \text{for any } x^1 \in \Omega,$$

where $\boldsymbol{\gamma} \in \mathcal{C}^1([0,1];\mathbb{R}^N)$ is *any* path joining x^0 to x^1 in Ω and the vector field $(\zeta_j) \in \mathcal{C}^1([0,1])$ is the solution to the Cauchy problem

$$\begin{aligned} &\frac{\mathrm{d}\zeta_j}{\mathrm{d}t}(t) = \Gamma^p_{ij}(\boldsymbol{\gamma}(t))\frac{\mathrm{d}\gamma^i}{\mathrm{d}t}(t)\zeta_p(t), \quad 0 \le t \le 1, \\ &\zeta_j(0) = \zeta^0_j, \end{aligned}$$

corresponding to such a path.

(iii) To establish that such a vector field is indeed the ℓth row-vector field of the unknown matrix field that we are seeking, we need to show that $(F_{\ell j})^n_{j=1} \in \mathcal{C}^1(\Omega;\mathbb{R}^n)$ and that this field satisfies the partial differential equations $\partial_i F_{\ell j} = \Gamma^p_{ij} F_{\ell p}$ in Ω corresponding to the fixed integer ℓ used in the above Cauchy problem.

Let x be an arbitrary point in Ω and let the integer $i \in \{1, \dots, n\}$ be *fixed* in what follows. Then there exist $x^1 \in \Omega$, a path $\gamma \in \mathcal{C}^1([0,1]; \mathbb{R}^n)$ joining x^0 to x^1, $\tau \in]0,1[$, and an open interval $I \subset [0,1]$ containing τ such that

$$\gamma(t) = x + (t - \tau)\boldsymbol{e}_i \quad \text{for } t \in I,$$

where $\boldsymbol{e}_i$ is the ith basis vector in $\mathbb{R}^N$. Since each function ζ_j is continuously differentiable in $[0,1]$ and satisfies $\frac{\mathrm{d}\zeta_j}{\mathrm{d}t}(t) = \Gamma^p_{kj}(\gamma(t))\frac{\mathrm{d}\gamma^k}{\mathrm{d}t}(t)\zeta_p(t)$ for all $0 \le t \le 1$, and since $\frac{\mathrm{d}\gamma^k}{\mathrm{d}t}(\tau) = \delta^k_i$, we have

$$\begin{aligned}\zeta_j(t) &= \zeta_j(\tau) + (t-\tau)\frac{\mathrm{d}\zeta_j}{\mathrm{d}t}(\tau) + o(t-\tau)\\ &= \zeta_j(\tau) + (t-\tau)\Gamma^p_{ij}(\gamma(\tau))\zeta_p(\tau) + o(t-\tau)\end{aligned}$$

for all $t \in I$. Equivalently,

$$F_{\ell j}(x + (t-\tau)\boldsymbol{e}_i) = F_{\ell j}(x) + (t-\tau)\Gamma^p_{ij}(x)F_{\ell p}(x) + o(t-x).$$

This relation shows that each function $F_{\ell j}$ possesses partial derivatives in the set Ω, given at each $x \in \Omega$ by

$$\partial_i F_{\ell n}(x) = \Gamma^p_{ij}(x)F_{\ell p}(x),$$

or, in matrix form, $\partial_i \boldsymbol{F}(x) = \boldsymbol{F}(x)\boldsymbol{\Gamma}_i(x)$.

(iv) We know from (iii) that the matrix field $(F_{\ell j})$ is of class $\mathcal{C}^1$ in Ω (its partial derivatives are continuous in Ω) and that it satisfies the partial differential equations $\partial_i F_{\ell j} = \Gamma^p_{ij} F_{\ell p}$ in Ω. Differentiating these equations then shows that the matrix field $(F_{\ell j})$ is in fact of class $\mathcal{C}^2$ in Ω. This completes the proof. □

The *regularity assumptions* on the matrix fields $\boldsymbol{\Gamma}_i : \Omega \to \mathbb{M}^n$, $1 \le i \le N$, can be significantly weakened. More specifically, the existence of a solution to the Pfaff system considered in Theorem 6.20-1 still holds if $\boldsymbol{\Gamma}_i \in \mathcal{C}(\Omega; \mathbb{M}^n)$, $1 \le i \le N$, with a solution $\boldsymbol{F}$ in the space $\mathcal{C}^1(\Omega; \mathbb{M}^n)$ in this case,[101] or if Ω is a domain in $\mathbb{R}^N$ and $\boldsymbol{\Gamma}_i \in L^p(\Omega; \mathbb{M}^n)$ for some $p > n$, with a solution $\boldsymbol{F}$ in the space $W^{1,p}(\Omega; \mathbb{M}^n)$ in this case.[102] Naturally, the compatibility conditions on the matrix fields $\boldsymbol{\Gamma}_i$ are to be understood in such cases in the sense of distributions.

[101] P. Hartman; A. Wintner [1950]: On the fundamental equations of differential geometry, *American Journal of Mathematics* **72**, 757–774.

[102] S. Mardare [2005]: On Pfaff systems with L^p coefficients and their applications in differential geometry, *Journal de Mathématiques Pures et Appliquées* **84**, 1659–1692.

S. Mardare [2007]: On systems of first order linear partial differential equations with L^p coefficients, *Advances in Differential Equations* **12**, 301–306.

Problem

6.20-1 Let Ω be a simply connected open subset of $\mathbb{R}^N$, let there be given matrix fields $\boldsymbol{A}_i \in \mathcal{C}^1(\Omega; \mathbb{M}^n)$, $1 \le i \le k$, $\boldsymbol{B}_j \in \mathcal{C}^1(\Omega; \mathbb{M}^m)$, $1 \le j \le N$, and $\boldsymbol{C}_k \in \mathcal{C}^1(\Omega; \mathbb{M}^{m \times n})$, $1 \le k \le N$, that satisfy

$$\begin{aligned} \partial_i \boldsymbol{A}_j - \partial_j \boldsymbol{A}_i + \boldsymbol{A}_i \boldsymbol{A}_j - \boldsymbol{A}_j \boldsymbol{A}_i &= \boldsymbol{0} \quad \text{in } \Omega, \\ \partial_i \boldsymbol{B}_j - \partial_j \boldsymbol{B}_i + \boldsymbol{B}_j \boldsymbol{B}_i - \boldsymbol{B}_i \boldsymbol{B}_j &= \boldsymbol{0} \quad \text{in } \Omega, \\ \partial_i \boldsymbol{C}_j - \partial_j \boldsymbol{C}_i + \boldsymbol{C}_i \boldsymbol{A}_j - \boldsymbol{C}_j \boldsymbol{A}_i + \boldsymbol{B}_j \boldsymbol{C}_i - \boldsymbol{B}_i \boldsymbol{C}_j &= \boldsymbol{0} \quad \text{in } \Omega, \end{aligned}$$

and let a point $x^0 \in \Omega$ and a matrix $\boldsymbol{F}^0 \in \mathbb{M}^{m \times n}$ be given. Show that there exists one, and only one, matrix field $\boldsymbol{F} \in \mathcal{C}^2(\Omega; \mathbb{M}^{m \times n})$ that solves the following *Pfaff system*:

$$\begin{aligned} \partial_i \boldsymbol{F} &= \boldsymbol{F} \boldsymbol{A}_i + \boldsymbol{B}_i \boldsymbol{F} + \boldsymbol{C}_i \quad \text{in } \Omega, \\ \boldsymbol{F}(x^0) &= \boldsymbol{F}^0. \end{aligned}$$

Remark This problem thus provides a generalization of the Pfaff system considered in Theorem 6.20-1, which corresponds to the special case where $\boldsymbol{B}_j = \boldsymbol{0}$ and $\boldsymbol{C}_k = \boldsymbol{0}$, as well as a generalization of the *classical Poincaré lemma* (Theorem 6.17-2), which corresponds to the special case where $m = n = 1$ and $\boldsymbol{A}_i = \boldsymbol{0}$ and $\boldsymbol{B}_j = \boldsymbol{0}$. □

CHAPTER 7

DIFFERENTIAL CALCULUS IN NORMED VECTOR SPACES

Introduction

Nonlinear functional analysis begins in earnest with this chapter, which is centered on the notion of *derivability* of mappings between *arbitrary normed vector spaces.*

More specifically, given a mapping $f : X \to Y$ between two normed vector spaces X and Y, the *Fréchet derivative* of f at a point $a \in X$ is defined (when it exists) as the unique element $f'(a) \in \mathcal{L}(X;Y)$ that satisfies

$$f(a+h) = f(a) + f'(a)h + \|h\|_X \, \delta(h),$$

with $\delta(h) \to 0$ in Y as $h \to 0$ in X (Section 7.1). From this natural definition follows a wealth of consequences that generalize well-known properties of real-valued functions of a real variable, such as the *chain rule* (Theorem 7.1-3); the all-important *mean value theorem* (in its various forms; cf. Theorems 7.2-1, 7.2-2, and 7.6-1); *Sard's lemma* (Theorem 7.5-1), which will play a key role in the definition of the *Brouwer topological degree* (Chapter 9); the *differentiability of the limit of a sequence of differentiable functions* (Theorem 7.3-1); the *differentiability of a function defined by an integral* (Theorem 7.4-1); and, for functions that possess *higher order derivatives* (defined in Section 7.8), the *Schwarz lemma* (Theorem 7.8-1) and *Taylor formulas* (Theorem 7.9-1).

As an application of the chain rule, we give a proof of the *Piola identity* (Theorem 7.1-4), a fundamental identity that will play a key role in Chapter 9 in the two proofs given there of the *Brouwer fixed point theorem* and in the proof of *Ball's existence theorem.*

The emphasis is also on *applications*, which include an analysis of necessary and sufficient conditions for *extrema of real-valued functions*, in relation to their properties of *differentiability* (Section 7.9) or *convexity* (Section 7.12); a detailed proof of the *Newton–Kantorovich theorem* (Theorem 7.7-3), which provides sufficient conditions for the convergence of *Newton's method in a Banach space*, the *maximum principle for second-order linear elliptic operators* (Theorem 7.10-2), or general *Lagrange interpolation in* $\mathbb{R}^n$ and *multipoint Taylor formulas* (Section 7.11).

One of this chapter's highlights is the *implicit function theorem* (Theorem 7.13-1), *one of the most fundamental theorems of nonlinear functional analysis*, and its special case known as the *local inversion theorem* (Theorem 7.14-1).

As first applications of the implicit function theorem, we show how it provides remarkably simple proofs of the *differentiability of mappings such as* $A \to A^{-1}$ *or* $A \to A^{1/2}$ (Sections 7.13 and 7.14). We also show that it lies at the heart of the proof of the *invariance of domain* *theorem for mappings of class C^1 in Banach spaces* (Theorem 7.14-2); note that, in the finite-dimensional case, this theorem will be later extended, but with a substantially more delicate analysis, to mappings that are only *continuous* (Section 9.17).

This chapter concludes with a proof of existence of *Lagrange multipliers* for general *constrained optimization problems* (Theorems 7.15-1 and 7.15-2) and a brief introduction to *saddle-points* and *Lagrangians* (Section 7.16).

All functions, matrices, and vector spaces considered in this chapter are real, save when otherwise indicated.

7.1 The Fréchet derivative; the chain rule; the Piola identity; application to extrema of real-valued functions

Recall that, given two normed vector spaces X and Y, the notation $\mathcal{L}(X;Y)$, or simply $\mathcal{L}(X)$ if $X = Y$, denotes the vector space formed by all continuous linear mappings from X into Y. Equipped with the norm defined by

$$\|A\|_{\mathcal{L}(X;Y)} := \sup_{\begin{cases} x\in X \\ x\neq 0\end{cases}} \frac{\|Ax\|_Y}{\|x\|_X} \quad \text{for each } A \in \mathcal{L}(X;Y),$$

the space $\mathcal{L}(X;Y)$ becomes itself a normed vector space, which is complete if the space Y is complete. When $Y = \mathbb{R}$, the space $X' := \mathcal{L}(X;\mathbb{R})$ is the *dual space* of the space X (Sections 2.9 and 3.5).

Let X and Y be *normed vector spaces*, and let Ω be an *open* subset of the space X. A mapping $f : \Omega \subset X \to Y$ is **differentiable at a point** $a \in \Omega$ if there exists an element A in the space $\mathcal{L}(X;Y)$ such that

$$f(a+h) = f(a) + Ah + \|h\|_X\, \delta(h) \quad \text{with } \lim_{h\to 0} \delta(h) = 0 \text{ in } Y.$$

Note that it is tacitly understood, here and elsewhere, that only points $(a+h)$ that belong to the set Ω should be considered in the above relation. Two simple, yet crucial, observations are then in order.

First, if $f : \Omega \subset X \to Y$ is differentiable at $a \in \Omega$, the mapping $A \in \mathcal{L}(X;Y)$ is unique. To see this, let $r_0 > 0$ be such that $B(a;r_0) \subset \Omega$ (the set Ω is open by assumption), and (with self-explanatory notations) assume that

$$f(a+h) = f(a) + A_1 h + \|h\|\, \delta_1(h) = f(a) + A_2 h + \|h\|\, \delta_2(h) \quad \text{for all } \|h\| < r_0.$$

Then

$$\|(A_1 - A_2)h\| \le \|h\| \left(\|\delta_1(h) - \delta_2(h)\|\right) \quad \text{for all } \|h\| < r \le r_0,$$

and thus $A_1 = A_2$ since

$$\|A_1 - A_2\| = \sup_{\begin{cases} h\neq 0 \\ \|h\|<r\end{cases}} \frac{\|(A_1 - A_2)h\|}{\|h\|} \le \sup_{\|h\|<r} \|\delta_1(h) - \delta_2(h)\|$$

can be made arbitrarily small by letting $r \to 0$.

Second, a mapping $f : \Omega \subset X \to Y$ differentiable at $a \in \Omega$ is continuous at a, since

$$\|f(a+h) - f(a)\| \leq (\|A\| + \|\delta(h)\|)\,\|h\| \quad \text{for all } \|h\| < r.$$

The linear mapping $A \in \mathcal{L}(X;Y)$ defined in this fashion is denoted $f'(a)$, and

$$f'(a) \in \mathcal{L}(X;Y)$$

is called the **Fréchet derivative**,[1] or simply the **derivative**, of the mapping f **at the point** a. If $X = \mathbb{R}$ and x denotes the generic point of $\mathbb{R}$, the derivative $f'(a)$ at a point $a \in \Omega \subset \mathbb{R}$ is also denoted

$$\frac{\mathrm{d}f}{\mathrm{d}x}(a) := f'(a).$$

Remark In the special case where $X = \mathbb{R}$, the derivative of a function $f : \Omega \subset \mathbb{R} \to Y$ at $a \in \Omega$ is classically defined as $f'(a) = \lim_{\left\{\substack{h \to 0 \\ h \neq 0}\right.} \dfrac{f(a+h) - f(a)}{h}$ (if this limit exists), so that $\lim_{\left\{\substack{h \to 0 \\ h \neq 0}\right.} \delta(h) = 0$ in Y, where $\delta(h) := \dfrac{f(a+h) - f(a) - f'(a)h}{h}$. Therefore the two definitions coincide in this special case, because the spaces Y and $\mathcal{L}(\mathbb{R}, Y)$ can be identified. But, except in this special case, the derivative $f'(a) \in \mathcal{L}(X;Y)$ *cannot* be identified with an element of Y. □

If a mapping $f : \Omega \subset X \to Y$ is differentiable at all points of the open set Ω, it is said to be **differentiable in** Ω. If the mapping

$$f' : x \in \Omega \subset X \to f'(x) \in \mathcal{L}(X;Y),$$

which is well defined in this case, is continuous, the mapping f is said to be **continuously differentiable in** Ω, or simply **of class** $\mathcal{C}^1$ **in** Ω. The space of all continuously differentiable mappings from Ω into Y is denoted

$$\mathcal{C}^1(\Omega; Y), \quad \text{or simply} \quad \mathcal{C}^1(\Omega) \text{ if } Y = \mathbb{R}.$$

It is immediately verified that, if $f : \Omega \subset X \to Y$ and $g : \Omega \subset X \to Y$ are differentiable at $a \in \Omega$, then $(f+g)$ and αf for any $\alpha \in \mathbb{R}$ are also differentiable at $a \in \Omega$, with $(f+g)'(a) = f'(a) + g'(a)$ and $(\alpha f)'(a) = \alpha f'(a)$. *The space $\mathcal{C}^1(\Omega; Y)$ is thus a vector space.*

Remark When $X = \mathbb{R}^n$ and $Y = \mathbb{R}$, the space $\mathcal{C}^1(\Omega; Y) = \mathcal{C}^1(\Omega)$ can be equipped with a metrizable topology, called the *Fréchet topology* (Problem 7.8-3). □

If $f \in \mathcal{C}^1(\Omega; Y)$ and if, *in addition*, $f : \Omega \to Y$ is injective, the direct image $f(\Omega)$ is open in Y, and $f^{-1} \in \mathcal{C}^1(f(\Omega); X)$, the mapping f is said to be a $\mathcal{C}^1$**-diffeomorphism** of Ω onto $f(\Omega)$.

Remark If $X = Y = \mathbb{R}^n$, the direct image $f(\Omega)$ of an open subset Ω of $\mathbb{R}^n$ under an injective mapping $f \in \mathcal{C}(\Omega; \mathbb{R}^n)$ is automatically open in $\mathbb{R}^n$ (remarkably, there is no need to assume in this

---

[1] So named after Maurice Fréchet (1878–1973).

case that f is differentiable): this is the content of the deep *Brouwer invariance of domain theorem* (Theorem 9.17-3). □

We now give various *examples* of Fréchet derivatives of mappings $f : \Omega \subset X \to Y$, where X and Y are normed vector spaces and Ω is open in X. To begin with, consider a *continuous affine mapping*

$$f : x \in X \to f(x) = Ax + b \quad \text{with } A \in \mathcal{L}(X;Y) \text{ and } b \in Y.$$

Since $f(a+h) = f(a) + Ah$ for all $a \in X$ and all $h \in X$, such a mapping is continuously differentiable in X, with

$$f'(x) = A \quad \text{for all } x \in X,$$

and hence the mapping f' is constant in this case.

Remark Using the *mean value theorem* (Theorem 7.2-1 below), we will see later (Theorem 7.2-4) that *conversely*, if $f'(x) = A \in \mathcal{L}(X;Y)$ for all $x \in \Omega$ and Ω is *connected*, then there exists a vector $b \in Y$ such that $f(x) = Ax + b$ for all $x \in \Omega$. □

We now examine the case where one of the two spaces X and Y is a *product*, equipped with any norm that induces the product topology (Section 2.2).

Theorem 7.1-1 *If the space Y is a product $Y = Y_1 \times Y_2 \times \cdots \times Y_m$ of normed vector spaces Y_i, a mapping $f : \Omega \subset X \to Y$ defined by m component mappings $f_i : \Omega \subset X \to Y_i$, $1 \le i \le m$, is differentiable at a point $a \in \Omega$ if and only if each mapping f_i, $1 \le i \le m$, is differentiable at the same point a.*

If this is the case, the derivative $f'(a) \in \mathcal{L}(X;Y)$ can be identified with the element $(f_1'(a), f_2'(a), \dots, f_m'(a))$ of the product space $\mathcal{L}(X;Y_1) \times \mathcal{L}(X;Y_2) \times \cdots \times \mathcal{L}(X;Y_m)$.

Proof To fix ideas, assume that Y is equipped with the norm defined by $y = (y_i)_{i=1}^m \in Y \to \|y\| = \max_{1\le i\le m} \|y_i\|_{Y_i}$.

If $f = (f_i)_{i=1}^m : \Omega \subset X \to Y$ is differentiable at $a \in \Omega$, the relation $f(a+h) = f(a) + f'(a)h + \|h\|\,\delta(h)$ is equivalent to the m relations

$$f_i(a+h) = f_i(a) + A_i h + \|h\|\,\delta_i(h), \quad 1 \le i \le m,$$

where $A_i \in \mathcal{L}(X;Y_i)$ is the ith component of $f'(a) \in \mathcal{L}(X;Y)$. Since $\|\delta_i(h)\|_{Y_i} \le \|\delta(h)\|$, each mapping $f_i : \Omega \subset X \to Y_i$ is differentiable at a, with $f_i'(a) = A_i$.

If each component mapping f_i, $1 \le i \le m$, is differentiable at $a \in \Omega$, then

$$\begin{aligned} f(a+h) - f(a) &= (f_i(a+h) - f_i(a))_{i=1}^m = \big(f_i'(a)h + \|h\|\,\delta_i(h)\big)_{i=1}^m \\ &= \big(f_i'(a)h\big)_{i=1}^m + \|h\|\,(\delta_i(h))_{i=1}^m. \end{aligned}$$

The linear operator $h \in X \to (f_i'(a)h)_{i=1}^m \in Y$ is continuous since

$$\max_{1\le i\le m} \|f_i'(a)h\|_{Y_i} \le \Big(\max_{1\le i\le m} \|f_i'(a)\|\Big)\,\|h\| \quad \text{for all } h \in X,$$

and $\lim_{h\to 0} \delta(h) := (\delta_i(h))_{i=0}^m = 0$ in Y since $\|\delta(h)\| = \max_{1\le i\le m} \|\delta_i(h)\|_{Y_i}$. Hence f is differentiable at a, with $f'(a) = (f_i'(a))_{i=1}^m$. □

Consider next the case where the space X is a *product* $X = X_1 \times X_2 \times \cdots \times X_n$ of normed vector spaces X_j. Given a point $a = (a_1, a_2, \dots, a_n) \in \Omega$, there exists for each index j an open subset Ω_j of the space X_j containing the point a_j such that the open set $\Omega_1 \times \Omega_2 \times \cdots \times \Omega_n$ is contained in Ω. If, for some $1 \leq j \leq n$, the *partial mapping*

$$f(a_1, \dots, a_{j-1}, \cdot, a_{j+1}, \dots, a_n) : \Omega_j \subset X_j \to Y$$

is differentiable at the point $a_j \in \Omega_j$, its derivative is denoted

$$\partial_j f(a) \in \mathcal{L}(X_j; Y)$$

and is called the j**th partial derivative** of the mapping f at the point a. If x_j denotes a generic point in the space X_j, such a partial derivative is also denoted

$$\frac{\partial f}{\partial x_j}(a) := \partial_j f(a).$$

Theorem 7.1-2 *If a mapping $f : \Omega \subset X = X_1 \times X_2 \times \cdots \times X_n$ is differentiable at a point $a \in \Omega$, the n partial derivatives $\partial_j f(a)$, $1 \leq j \leq n$, exist and*

$$f'(a)h = \sum_{j=1}^{n} \partial_j f(a) h_j \quad \textit{for all } h = (h_1, h_2, \dots, h_n) \in X_1 \times X_2 \times \cdots \times X_n.$$

Proof To avoid cumbersome notations, assume that $n = 2$ (the extension to any $n \geq 3$ is clear). Also assume, to fix ideas, that X is equipped with the norm defined by $x = (x_1, x_2) \in X \to \|x\| = \max\{\|x_1\|_{X_1}, \|x_2\|_{X_2}\}$.

It is then immediately verified that the derivative $f'(a) \in \mathcal{L}(X; Y)$ defines continuous linear operators $A_1 \in \mathcal{L}(X_1; Y)$ and $A_2 \in \mathcal{L}(X_2; Y)$ by means of the relations

$$A_1 h_1 = f'(a)(h_1, 0) \text{ for all } h_1 \in X_1 \quad \text{and} \quad A_2 h_2 = f'(a)(0, h_2) \text{ for all } h_2 \in X_2,$$

so that

$$f'(a)(h_1, h_2) = A_1 h_1 + A_2 h_2 \quad \text{for all } (h_1, h_2) \in (X_1 \times X_2).$$

Therefore,

$$f(a_1 + h_1, a_2) = f(a_1, a_2) + f'(a)(h_1, 0) + \|(h_1, 0)\|\, \delta(h_1, 0),$$

which shows that $A_1 = \partial_1 f(a)$, since $\|(h_1, 0)\|\, \delta(h_1, 0) = \|h_1\|_{X_1} \delta_1(h_1)$ with $\lim_{h_1 \to 0} \delta_1(h_1) = 0$. A similar argument shows that $A_2 = \partial_2 f(a)$. $\square$

When Y is a product, a mapping is thus differentiable at a point if and only if all its component mappings are differentiable at the same point (Theorem 7.1-1). *By contrast, when X is a product,* a mapping may no longer be differentiable at a point if all the partial mappings are differentiable at the same point (Problem 7.1-3). What *can* be proved when X is a product is that $f \in \mathcal{C}^1(\Omega; Y)$ if and only if $\partial_j f \in \mathcal{C}(\Omega; \mathcal{L}(X_j, Y))$, $1 \leq j \leq n$ (Theorem 7.2-3).

Finally, suppose that

$$X = X_1 \times X_2 \times \cdots \times X_n \quad \text{and} \quad Y = Y_1 \times Y_2 \times \cdots \times Y_m,$$

so that in this case a function $f : \Omega \subset X \to Y$ is determined by means of m functions $f_i : \Omega \subset X \to Y_i$ of n variables. Then the relation

$$k = f'(a)h \text{ with } h = (h_1, h_2, \ldots, h_n) \in X \quad \text{and} \quad k = (k_1, k_2, \ldots, k_m) \in Y,$$

is equivalent to the relations

$$k_i = \sum_{j=1}^{n} \partial_j f_i(a) h_j, \quad 1 \le i \le m.$$

In the important special case where $X = \mathbb{R}^n$ and $Y = \mathbb{R}^m$, the relation $k = f'(a)h$ may be written in matrix form as

$$\begin{pmatrix} k_1 \\ k_2 \\ \vdots \\ k_m \end{pmatrix} = \begin{pmatrix} \partial_1 f_1(a) & \partial_2 f_1(a) & \ldots & \partial_n f_1(a) \\ \partial_1 f_2(a) & \partial_2 f_2(a) & \ldots & \partial_n f_2(a) \\ \vdots & \vdots & & \vdots \\ \partial_1 f_m(a) & \partial_2 f_m(a) & \ldots & \partial_n f_m(a) \end{pmatrix} \begin{pmatrix} h_1 \\ h_2 \\ \vdots \\ h_n \end{pmatrix},$$

the numbers $\partial_j f_i(a) = (\partial f_i / \partial x_j)(a)$ being the usual partial derivatives of the functions f_i. The Fréchet derivative $f'(a) \in \mathcal{L}(\mathbb{R}^n; \mathbb{R}^m)$ is thus identified in this case with the matrix $(\partial_j f_i(a))$, also called the **gradient matrix** of f at a, and often denoted

$$\nabla f(a) := (\partial_j f_i(a)).$$

If $m = n$, the determinant of the matrix $(\partial_j f_i(a))$ is called the **Jacobian**[2] of the function f at the point a.

Let X_1, X_2, Y be normed vector spaces. Recall (Section 2.11) that a bilinear mapping $B : X_1 \times X_2 \to Y$ is continuous if and only if

$$\|B\| := \sup_{\begin{cases} x_1 \in X_1, \ x_2 \in X_2 \\ x_1 \neq 0, \ x_2 \neq 0 \end{cases}} \frac{\|B(x_1, x_2)\|_Y}{\|x_1\|_{X_1} \|x_2\|_{X_2}} < \infty.$$

A continuous bilinear mapping $B : X_1 \times X_2 \to Y$ is differentiable at all points in the space $X_1 \times X_2$ since

$$B(a_1 + h_1, a_2 + h_2) = B(a_1, a_2) + B(h_1, a_2) + B(a_1, h_2) + B(h_1, h_2)$$

for all $(a_1, a_2) \in X_1 \times X_2$ and all $h = (h_1, h_2) \in X_1 \times X_2$, and

$$\|B(h_1, h_2)\| \le \|B\| \, \|h_1\|_{X_1} \|h_2\|_{X_2}.$$

[2]So named after Carl Gustav Jacob Jacobi (1804–1851).

Therefore $B(h_1, h_2) = \|h\| \, \delta(h)$ with $\lim_{h \to 0} \delta(h) = 0$ in Y (to see this, equip the space $X_1 \times X_2$ with $\|x\| = \max(\|x_1\|_{X_1}, \|x_2\|_{X_2})$ for all $x = (x_1, x_2) \in X_1 \times X_2$). The derivative and partial derivatives are thus respectively defined by the formulas

$$B'(a_1, a_2)(h_1, h_2) = B(h_1, a_2) + B(a_1, h_2),$$
$$\partial_1 B(a_1, a_2)h_1 = B(h_1, a_2) \quad \text{and} \quad \partial_2 B(a_1, a_2)h_2 = B(a_1, h_2).$$

If $X_1 = X_2 = X$, a similar computation shows that the mapping $f : x \in X \to f(x) := B(x, x) \in Y$ is also differentiable, with $f'(a)h = B(a, h) + B(h, a)$ for all $a, h \in X$. If in addition the bilinear mapping $B : X \times X \to Y$ is *symmetric*, i.e., if $B(x, \widetilde{x}) = B(\widetilde{x}, x)$ for all $x, \widetilde{x} \in X$, the above formula reduces to $f'(a)h = 2B(a, h)$.

As exemplified above, *the derivative* $f'(a) \in \mathcal{L}(X; Y)$ *is often computed in terms of its action on vectors of* X, i.e., it is the expression of the *vectors*

$$f'(a)h = \lim_{\begin{cases} \theta \to 0 \\ \theta \neq 0 \end{cases}} \frac{f(a + \theta h) - f(a)}{\theta} \in Y$$

that is computed for arbitrary vectors h of the space X rather than the mapping $f'(a)$ itself.

Note also that $f'(a)h$ is nothing but the derivative at $\theta = 0$ of the function

$$\theta \in I(h) \subset \mathbb{R} \to f(a + \theta h) \in Y,$$

which is defined on an *ad hoc* open interval $I(h)$ of $\mathbb{R}$ containing the origin. This observation motivates the following definition. Given a mapping $f : \Omega \subset X \to Y$, a point $a \in \Omega$, and a nonzero vector $h \in X$, assume that the function $\theta \in I(h) \subset \mathbb{R} \to f(a + \theta h) \in Y$ is differentiable at $\theta = 0$. Then f is said to possess at $a \in \Omega$ a **Gâteaux derivative**[3] **in the direction** h, also called a **directional derivative**, defined by

$$\partial_h f(a) := \lim_{\begin{cases} \theta \to 0 \\ \theta \neq 0 \end{cases}} \frac{f(a + \theta h) - f(a)}{\theta} \in Y.$$

Clearly, *if* f *is differentiable at* $a \in \Omega$, *it possesses Gâteaux derivatives in all directions* $h \in X$. The converse is not necessarily true, however (Problem 7.1-3).

Examples of Gâteaux derivatives include the *usual partial derivatives* when $X = \mathbb{R}^n$ (in which case the vectors h are simply the basis vectors of $\mathbb{R}^n$) and the *outer normal derivative operator* ∂_ν defined by $\partial_\nu f(a) := \sum_{i=1}^n \nu_i \partial_i f(a)$, at a point a of the boundary of an open subset of $\mathbb{R}^N$ where the unit outer normal vector $\nu = (\nu_i)_{i=1}^n$ exists.

To illustrate how derivatives are computed by means of Gâteaux derivatives, let us compute the derivatives of the mappings

$$\iota_1 : \boldsymbol{A} \in \mathbb{M}^n \to \iota_1(\boldsymbol{A}) := \operatorname{tr} \boldsymbol{A} \quad \text{and} \quad \iota_n : \boldsymbol{A} \in \mathbb{M}^n \to \iota_n(\boldsymbol{A}) := \det \boldsymbol{A},$$

[3] So named after Section 25 in:
R. Gâteaux [1919]: Fonctions d'une infinité de variables indépendantes, *Bulletin de la Société Mathématique de France* **47**, 70–96.

where $\mathbb{M}^n$ denotes the space of all square matrices of order n. Since the mapping ι_1 is linear and continuous, it is differentiable at any $\boldsymbol{A} \in \mathbb{M}^n$ (as shown above), with

$$\iota_1'(\boldsymbol{A})\boldsymbol{H} = \iota_1(\boldsymbol{H}) = \mathrm{tr}(\boldsymbol{H}) = \boldsymbol{I} : \boldsymbol{H} \quad \text{for all } \boldsymbol{H} \in \mathbb{M}^n,$$

where : denotes the matrix inner product (Section 4.2).

As a polynomial of degree n with respect to the n^2 elements of the matrix $\boldsymbol{A}$, the mapping ι_n is clearly continuously differentiable over the space $\mathbb{M}^n$. If the matrix $\boldsymbol{A}$ is invertible, we can write

$$\begin{aligned}\iota_n(\boldsymbol{A}+\boldsymbol{H}) &= \det(\boldsymbol{A}+\boldsymbol{H}) = \det \boldsymbol{A} \det(\boldsymbol{I}+\boldsymbol{A}^{-1}\boldsymbol{H}) \\ &= (\det \boldsymbol{A})(1 + \mathrm{tr}(\boldsymbol{A}^{-1}\boldsymbol{H}) + o(\boldsymbol{H})) \quad \text{for all } \boldsymbol{H} \in \mathbb{M}^n,\end{aligned}$$

since (by definition of the determinant)

$$\det(\boldsymbol{I}+\boldsymbol{E}) = 1 + \mathrm{tr}\,\boldsymbol{E} + \{\text{monomials of degree } \geq 2\}.$$

We have thus proved that, when the matrix $\boldsymbol{A}$ is invertible,

$$\iota_n'(\boldsymbol{A})(\boldsymbol{H}) = \det \boldsymbol{A}\,\mathrm{tr}(\boldsymbol{A}^{-1}\boldsymbol{H}) = \mathrm{tr}\{(\mathbf{Cof}\,\boldsymbol{A})^T\boldsymbol{H}\} = \mathbf{Cof}\,\boldsymbol{A} : \boldsymbol{H},$$

where $\mathbf{Cof}\,\boldsymbol{A} \in \mathbb{M}^n$ designates the cofactor matrix of $\boldsymbol{A}$ (recall that $\mathbf{Cof}\,\boldsymbol{A} = (\det \boldsymbol{A})\boldsymbol{A}^{-T}$ if $\boldsymbol{A}$ is invertible). Noting that the mapping $\boldsymbol{A} \in \mathbb{M}^n \to \mathbf{Cof}\,\boldsymbol{A} \in \mathbb{M}^n$ is continuous, we conclude that the relation

$$\iota_n'(\boldsymbol{A})\boldsymbol{H} = \mathbf{Cof}\,\boldsymbol{A} : \mathbf{H} \quad \text{for all } \boldsymbol{H} \in \mathbb{M}^n$$

holds in fact for all $\boldsymbol{A} \in \mathbb{M}^n$.

As another instance, let us compute the derivative of the mapping

$$f : \boldsymbol{A} \in \mathbb{U}^n \subset \mathbb{M}^n \to \boldsymbol{A}^{-1} \in \mathbb{M}^n,$$

where $\mathbb{U}^n$ denotes the set of all invertible matrices of order n. Then, by Theorem 3.6-3, the set $\mathbb{U}^n$ is open in $\mathbb{M}^n$, and $\boldsymbol{A}+\boldsymbol{H} = \boldsymbol{A}(\boldsymbol{I}+\boldsymbol{A}^{-1}\boldsymbol{H})$ is invertible if $\|\boldsymbol{H}\| < \|\boldsymbol{A}^{-1}\|^{-1}$, where $\|\cdot\|$ is any subordinate matrix norm on $\mathbb{M}^n$. Therefore, for such $\boldsymbol{H} \in \mathbb{M}^n$,

$$\begin{aligned}f(\boldsymbol{A}+\boldsymbol{H}) = (\boldsymbol{A}+\boldsymbol{H})^{-1} &= (\boldsymbol{I}+\boldsymbol{A}^{-1}\boldsymbol{H})^{-1}\boldsymbol{A}^{-1} = \left(\boldsymbol{I} - \boldsymbol{A}^{-1}\boldsymbol{H} + o(\boldsymbol{H})\right)\boldsymbol{A}^{-1} \\ &= \boldsymbol{A}^{-1} - \boldsymbol{A}^{-1}\boldsymbol{H}\boldsymbol{A}^{-1} + o(\boldsymbol{H}),\end{aligned}$$

again by Theorem 3.6-3. Consequently, the mapping f is differentiable at any $\boldsymbol{A} \in \mathbb{U}^n$, with

$$f'(\boldsymbol{A})\boldsymbol{H} = -\boldsymbol{A}^{-1}\boldsymbol{H}\boldsymbol{A}^{-1} \quad \text{for all } \boldsymbol{H} \in \mathbb{M}^n.$$

Remark It will be shown later that the above mapping f is even *infinitely differentiable*, and that the above space $\mathbb{M}^n$ can be replaced by the space $\mathcal{L}(X;Y)$ where both X and Y are *infinite-dimensional Banach spaces* (Theorem 7.13-2). □

In various instances, the mapping to be differentiated is itself composed of simpler mappings whose derivatives are known. In this case, the following result is particularly useful:

Theorem 7.1-3 (chain rule) *Let X, Y, Z be normed vector spaces, let U and V be open subsets of the space X and Y respectively, let $f : U \subset X \to Y$ be a mapping differentiable at a point $a \in U$ and such that $f(U) \subset V$, and let $g : V \subset Y \to Z$ be a mapping differentiable at the point $f(a)$. Then the mapping $g \circ f : U \subset X \to Z$ is differentiable at the point $a \in U$,*

$$(g \circ f)'(a) = g'(f(a))f'(a).$$

Besides, $g \circ f \in \mathcal{C}^1(U; Z)$ if $f \in \mathcal{C}^1(U; Y)$ and $g \in \mathcal{C}^1(V; Z)$.

Proof Given any point $(a + h) \in U$, let

$$b := f(a) \quad \text{and} \quad k(h) := f(a + h) - b,$$

so that $\lim_{h \to 0} k(h) = 0$ (the mapping f is continuous at a since it is differentiable at a). By assumption,

$$\begin{aligned} f(a+h) &= f(a) + f'(a) + \|h\|\, \delta(h) \quad &&\text{with } \lim_{h \to 0} \delta(h) = 0, \\ g(b+k) &= g(b) + g'(b) + \|k\|\, \eta(k) \quad &&\text{with } \lim_{k \to 0} \eta(k) = 0, \end{aligned}$$

so that

$$\begin{aligned} (g \circ f)(a+h) - (g \circ f)(a) &= g(f(a+h)) - g(b) = g'(b)\big(f(a+h) - f(a)\big) + \|k(h)\|\, \eta(k(h)) \\ &= g'(b)\big(f'(a)h + \|h\|\, \delta(h)\big) + \|k(h)\|\, \eta(k(h)\big). \end{aligned}$$

The relations

$$\begin{aligned} \big\|g'(b)(\|h\|\, \delta(h))\big\| &\leq \|h\|\, \|g'(b)\|\, \|\delta(h)\|\,, \\ \|k(h)\| = \|f(a+h) - f(a)\| &= \big\|f'(a)h + \|h\|\, \delta(h)\big\| \leq \|h\|\, (\|f'(a)\| + \|\delta(h)\|) \end{aligned}$$

then imply that

$$g'(b)\big(\,\|h\|\, \delta(h)\big) + \|k(h)\|\, \eta\big(k(h)\big) = \|h\|\, \rho(h) \quad \text{with } \lim_{h \to 0} \rho(h) = 0,$$

which shows that $g \circ f : U \subset X \to Z$ is differentiable at $a \in U$, with $(g \circ f)'(a) = g'(f(a))f'(a)$.

Assume next that $f \in \mathcal{C}^1(U; Y)$ and $g \in \mathcal{C}^1(V; Z)$. Since both mappings $f' : U \to \mathcal{L}(X; Y)$ and $g' \circ f : U \to \mathcal{L}(Y; Z)$ are thus continuous (the second one by Theorem 1.7-2, since both mappings $g' : V \to \mathcal{L}(Y; Z)$ and $f : U \to V$ are continuous), the mapping $x \in U \to (f'(x), g'(f(x))) \in \mathcal{L}(X; Y) \times \mathcal{L}(Y; Z)$ is also continuous.

Noting that the bilinear mapping $(A, B) \in \mathcal{L}(X; Y) \times \mathcal{L}(Y; Z) \to B \circ A \in \mathcal{L}(X; Z)$ is continuous (since $\|B \circ A\| \leq \|B\|\, \|A\|$ for all $(A, B) \in \mathcal{L}(X; Y) \times \mathcal{L}(Y; Z)$), we conclude (again by Theorem 1.7-2) that the composite mapping

$$(g \circ f)' = (g' \circ f) \circ f' : U \to \mathcal{L}(X; Z)$$

is also continuous. Hence $g \circ f \in \mathcal{C}^1(U; Z)$. $\square$

In the special case where $X = \mathbb{R}^n, Y = \mathbb{R}^m$, and $Z = \mathbb{R}^\ell$, let $h := g \circ f$ and $b := f(a)$. Then the chain rule shows that, in this case,

$$\begin{pmatrix} \partial_1 h_1(a) & \dots & \partial_n h_1(a) \\ \vdots & & \vdots \\ \partial_1 h_\ell(a) & \dots & \partial_n h_\ell(a) \end{pmatrix} = \begin{pmatrix} \partial_1 g_1(b) & \dots & \partial_m g_1(b) \\ \vdots & & \vdots \\ \partial_1 g_\ell(b) & \dots & \partial_m g_\ell(b) \end{pmatrix} \begin{pmatrix} \partial_1 f_1(a) & \dots & \partial_n f_1(a) \\ \vdots & & \vdots \\ \partial_1 f_m(a) & \dots & \partial_n f_m(a) \end{pmatrix},$$

which is nothing but the matrix form of the well-known formulas

$$\partial_j h_i(a) = \sum_{k=1}^m \partial_k g_i(b) \partial_j f_k(a), \quad 1 \le i \le \ell, \ 1 \le j \le n.$$

When X is a Hilbert space, the derivative of a real-valued function $f : \Omega \subset X \to \mathbb{R}$ at a point $a \in \Omega$ can be identified with an element of the space X: Since the derivative $f'(a)$ is by definition an element of the dual space $X' = \mathcal{L}(X;\mathbb{R})$, and X is a Hilbert space with inner product $(\cdot,\cdot)$, there exists by the *F. Riesz representation theorem* (Theorem 4.6-1) a unique element, denoted $\operatorname{grad} f(a)$, in the space X that satisfies

$$f'(a)h = (\operatorname{grad} f(a), h) \quad \text{for all } h \in X.$$

Note, however that, *when $X = \mathbb{R}^n$, the vector* $\operatorname{grad} f(a)$ *is also often denoted* $\nabla f(a)$ (as in Chapter 6), a possible source of confusion since, according to the definition given earlier in this section, the same notation $\nabla f(a)$ also denotes a $1 \times n$ *row* matrix.

For instance, if the space $\mathbb{M}^n$ is equipped with the matrix inner product, the Fréchet derivatives $\iota_1'(\boldsymbol{A})$ and $\iota_n'(\boldsymbol{A})$ at a matrix $\boldsymbol{A} \in \mathbb{M}^n$ of the mappings $\iota_1 : \boldsymbol{A} \in \mathbb{M}^n \to \iota_1(\boldsymbol{A}) := \operatorname{tr} \boldsymbol{A} \in \mathbb{R}$ and $\iota_n : \boldsymbol{A} \in \mathbb{M}^n \to \iota_n(\boldsymbol{A}) := \det \boldsymbol{A}$ can be respectively identified with the matrices $\boldsymbol{I}$ and $\mathbf{Cof}\,\boldsymbol{A}$.

As a first application of the chain rule, we show that, if two matrix fields are related through the *Piola transform* (defined in Theorem 7.1-4(b) below), their divergences are in turn related through a remarkably simple relation.[4] This relation is itself a consequence of the *fundamental Piola identity* (Theorem 7.1-4(a)), which *inter alia* plays a key role in the derivation of a compensated compactness result used in John Ball's theorem (Section 9.7) and in the two proofs of Brouwer's fixed point theorem given in this book (Sections 9.9 and 9.16).

In what follows, Latin indices vary in $\{1, \dots, n\}$. Given a differentiable matrix field $\boldsymbol{T} = (T_{ij}) : \Omega \to \mathbb{M}^n$, *resp.* $\widehat{\boldsymbol{T}} = (\widehat{T}_{ij}) : \widehat{\Omega} \to \mathbb{M}^n$, where Ω, *resp.* $\widehat{\Omega}$, is an open subset of $\mathbb{R}^n$, its *divergence* is the vector field $\mathbf{div}\,\boldsymbol{T} : \Omega \to \mathbb{R}^n$, *resp.* $\widehat{\mathbf{div}}\widehat{\boldsymbol{T}} : \widehat{\Omega} \to \mathbb{R}^n$, defined by

$$(\mathbf{div}\,\boldsymbol{T}(x))_i := \sum_j \partial_j T_{ij}(x), \quad \textit{resp.} \quad (\widehat{\mathbf{div}}\widehat{\boldsymbol{T}}(\widehat{x}))_i = \sum_j \widehat{\partial}_j \widehat{T}_{ij}(\widehat{x}),$$

where $x = (x_i)$, *resp.* $\widehat{x} = (\widehat{x}_i)$, denotes a generic point in Ω, *resp.* in $\widehat{\Omega}$.

[4]Which plays in particular a key role in the derivation of the equilibrium equations of a three-dimensional continuum; see, e.g., CIARLET [1988, Chapter 2].

Theorem 7.1-4 (Piola identity and Piola transform[5]**)** *Let Ω and $\widehat{\Omega}$ be two open subsets in $\mathbb{R}^n$, and let $\boldsymbol{\varphi} : \Omega \to \widehat{\Omega}$ be a mapping that is twice differentiable in Ω.*

(a) *Then the* **Piola identity** *holds, viz.,*

$$\boldsymbol{div\, Cof\, \nabla\varphi} = \mathbf{0} \quad \text{in } \Omega.$$

(b) *Given a matrix field $\widehat{\boldsymbol{T}} : \widehat{\Omega} \to \mathbb{M}^n$, let the matrix field $\boldsymbol{T} : \Omega \to \mathbb{M}^n$ be defined by means of the* **Piola transform**, *viz.,*

$$\boldsymbol{T}(x) := \widehat{\boldsymbol{T}}(\widehat{x})\, \boldsymbol{Cof\, \nabla\varphi}(x) \quad \text{for all } \widehat{x} = \boldsymbol{\varphi}(x) \in \widehat{\Omega}.$$

Assume that the field $\widehat{T}$ is differentiable in $\widehat{\Omega}$ and that the gradient matrix $\boldsymbol{\nabla\varphi}(x) \in \mathbb{M}^n$ is invertible at all points $x \in \Omega$. Then the matrix field $\boldsymbol{T}$ is also differentiable in Ω, and

$$\mathbf{div}\, \boldsymbol{T}(x) = (\det \boldsymbol{\nabla\varphi}(x))\widehat{\mathbf{div}}\, \widehat{\boldsymbol{T}}(\widehat{x}) \quad \text{for all } \widehat{x} = \boldsymbol{\varphi}(x) \in \widehat{\Omega}.$$

Proof All the indices appearing in this proof under the summation sign range in $\{1, \ldots, n\}$. The notation $\mathbb{M}^p$ designates the set of all $p \times p$ matrices.

(i) To establish the Piola identity, we need to show that, for each $1 \le i \le n$,

$$\sum_j \partial_j (\mathbf{Cof}\, \boldsymbol{\nabla\varphi})_{ij} = 0.$$

So, let an index $i \in \{1, \ldots, n\}$ be fixed. By definition of the cofactor matrix,

$$(\mathbf{Cof}\, \boldsymbol{\nabla\varphi})_{ij} = (-1)^{i+j} \det \boldsymbol{A}_{ij},$$

where $\boldsymbol{A}_{ij} : \Omega \to \mathbb{M}^{n-1}$ denotes the matrix field obtained by deleting the ith column and the jth row of the matrix field $\boldsymbol{\nabla\varphi}^T : \Omega \to \mathbb{M}^n$. Then

$$\sum_j \partial_j (\mathbf{Cof}\, \boldsymbol{\nabla\varphi})_{ij} = \sum_j (-1)^{i+j} \sum_{k \ne j} \det \boldsymbol{A}_{ij}^k,$$

where, for each $k \ne j$, $\boldsymbol{A}_{ij}^k : \Omega \to \mathbb{M}^{n-1}$ denotes the matrix field obtained by replacing the row $(\partial_k \varphi_1 \cdots \partial_k \varphi_{i-1} \partial_k \varphi_{i+1} \cdots \partial_k \varphi_n)$ in $\boldsymbol{A}_{ij}$ by the row $(\partial_{jk} \varphi_1 \cdots \partial_{jk} \varphi_{i-1} \partial_{jk} \varphi_{i+1} \cdots \partial_{jk} \varphi_n)$. That $\partial_{jk}\varphi_i = \partial_{kj}\varphi_i$ then implies that

$$\det \boldsymbol{A}_{ij}^k = (-1)^{k-j-1} \det \boldsymbol{A}_{ik}^j.$$

Consequently,

$$\begin{aligned}
\sum_j (-1)^{i+j} \sum_{k \ne j} \det \boldsymbol{A}_{ij}^k &= \sum_j (-1)^{i+j} \Big(\sum_{k \le j-1} \det \boldsymbol{A}_{ij}^k + \sum_{k \ge j+1} \det \boldsymbol{A}_{ij}^k \Big) \\
&= \sum_{j=1}^n (-1)^{i+j} \Big(\sum_{k \le j-1} \det \boldsymbol{A}_{ij}^k + \sum_{k \ge j+1} (-1)^{k-j-1} \det \boldsymbol{A}_{ik}^j \Big) = 0,
\end{aligned}$$

[5]So named after Gabrio Piola (1794–1850).

and thus the Piola identity holds.

(ii) If the matrix $\boldsymbol{\nabla\varphi}(x) \in \mathbb{M}^n$ is invertible at all points $x \in \Omega$, the Piola identity can be rewritten in this case as

$$\partial_j(\mathbf{Cof}\,\boldsymbol{\nabla\varphi})_{ij} = \partial_j\left((\det\boldsymbol{\nabla\varphi})\boldsymbol{\nabla\varphi}^{-T}\right)_{ij} = 0,$$

since $\mathbf{Cof}\,\boldsymbol{A} = (\det\boldsymbol{A})\boldsymbol{A}^{-T}$ if $\boldsymbol{A}$ is invertible. The relations

$$T_{ij}(x) = (\det\boldsymbol{\nabla\varphi}(x))\sum_k \widehat{T}_{ik}(\widehat{x})(\boldsymbol{\nabla\varphi}(x)^{-T})_{kj}, \quad 1 \le i,\, j \le n,$$

imply that, for each $1 \le i \le n$,

$$\sum_j \partial_j T_{ij}(x) = (\det\boldsymbol{\nabla\varphi}(x))\sum_{j,k}\partial_j\widehat{T}_{ik}(\widehat{x})(\boldsymbol{\nabla\varphi}(x)^{-T})_{kj},$$

since the other terms vanish as a consequence of the *Piola identity*. Next, by the *chain rule* (Theorem 7.1-3),

$$\partial_j\widehat{T}_{ik}(\widehat{x}) = \sum_\ell \widehat{\partial}_\ell\widehat{T}_{ik}(\boldsymbol{\varphi}(x))\partial_j\varphi_\ell(x) = \sum_\ell \widehat{\partial}_\ell\widehat{T}_{ik}(\widehat{x})(\boldsymbol{\nabla\varphi}(x))_{\ell j},$$

and the announced relation between $\mathbf{div}\,\boldsymbol{T}(x)$ and $\widehat{\mathbf{div}}\widehat{\boldsymbol{T}}(\widehat{x})$ follows by noting that

$$\sum_j(\boldsymbol{\nabla\varphi}(x))_{\ell j}\left(\boldsymbol{\nabla\varphi}(x)^{-T}\right)_{kj} = \delta_{\ell k}. \qquad \square$$

To conclude this section, we establish elementary, but basic, *necessary conditions* satisfied by *the Fréchet derivative* at an *extremum of a real-valued function*; other less immediate, but likewise basic, necessary conditions involving the Fréchet derivative, viz., the existence of *Lagrange multipliers*, will be established later (Section 7.15).

Since the real-valued functions that we have in mind include in particular the quadratic functionals encountered in the weak formulations of elliptic boundary value problems (Chapter 6), or more generally the integrals found in the calculus of variations (Chapter 9), we shall momentarily revert to notations such as $v \in V$, $J : V \to \mathbb{R}$, etc., instead of $x \in X$, $f : X \to Y$, etc.

Let Ω be an open subset of a normed vector space V. A function $J : \Omega \subset V \to \mathbb{R}$ is said to have a **local minimum**, *resp.* a **local maximum**, at a point $u \in \Omega$ if there exists a neighborhood $W \subset \Omega$ of u such that

$$J(u) \le J(v), \quad \textit{resp.} \quad J(u) \ge J(v), \quad \text{for all } v \in W.$$

If there is no need to distinguish between maximum and minimum, the function J is said to have a **local extremum** at the point u. If

$$J(u) < J(v), \quad \textit{resp.} \quad J(u) > J(v), \quad \text{for every } v \in W,\, v \ne u,$$

the local minimum, *resp.* local maximum, is said to be **strict**.

Following a common abuse of language, we shall often say that the point u *itself* is a (possibly strict) local minimum, maximum, or extremum.

We begin with the natural extension of a well-known result for real-valued functions of one real variable.

Theorem 7.1-5 (necessary condition for a local extremum over an open set) *Let Ω be an open subset of a normed vector space V and let $J : \Omega \subset V \to \mathbb{R}$ be a function differentiable at a point $u \in \Omega$. If $J : \Omega \to \mathbb{R}$ has a local extremum at u, then*

$$J'(u) = 0.$$

Proof Let v be any vector of V. The set Ω being open, there exists an open interval I containing 0 such that the function

$$\varphi : t \in I \to \varphi(t) := J(u + tv)$$

is well defined. By the chain rule (Theorem 7.1-3), the function φ is differentiable at $t = 0$, with

$$\varphi'(0) = J'(u)v.$$

To fix ideas, suppose that the point u is a local minimum. Then

$$0 \leq \lim_{t \to 0^+} \frac{\varphi(t) - \varphi(0)}{t} = \varphi'(0) = \lim_{t \to 0^-} \frac{\varphi(t) - \varphi(0)}{t} \leq 0,$$

which shows that

$$J'(u)v = 0.$$

Therefore $J'(u) = 0$ since the vector $v \in V$ is arbitrary. □

A point $u \in \Omega$ where $J'(u) = 0$ is called a **stationary point**, or a **critical point**, of the function J, and the equation $J'(u) = 0$ is called the **Euler equation**.[6]

Remark If $V = V_1 \times V_2 \times \cdots \times V_n$, solving the Euler equations $J'(u) = 0$ thus amounts to solving the system of n equations $\partial_j J(u_1, \ldots, u_n) = 0$, $1 \leq j \leq n$. □

If $J'(u) = 0$, *additional* assumptions are evidently needed to insure that u is indeed a local extremum of J (consider for instance the function $J : v \in \mathbb{R} \to J(v) := v^3$ at $u = 0$). Such sufficient conditions, which usually involve the second derivative of J or the convexity of J, will be studied later (Sections 7.9 and 7.12).

The assumption in Theorem 7.1-5 that Ω is *open* is essential (consider for instance the function $J : v \in [0,1] \to J(v) := v$ at $u = 0$).

Let again $J : \Omega \to \mathbb{R}$ be a function defined over an open subset Ω of a normed vector space V, and let U be a *subset of* Ω. Then J is said to have a **constrained local extremum relative to** U at a point $u \in \Omega$ if the *restriction* $J|_U$ of J to the set U has a local extremum at u. In other words, J has a constrained local minimum, *resp.* maximum, if there exists a neighborhood $W \subset \Omega$ of u such that

$$J(u) \leq J(v), \quad \textit{resp.} \quad J(u) \geq J(v), \quad \text{for all } v \in W \cap U.$$

Our first result concerning *constrained* local extrema is an easy extension of the necessary condition $J'(u) = 0$ of Theorem 7.1-5, in the special case where the set U is *convex*. For definiteness, we consider a local minimum.

[6]So named after Leonhard Euler (1707–1783).

Theorem 7.1-6 (necessary condition for a constrained local minimum relative to a convex set) *Let Ω be an open subset of a normed vector space V, let U be a convex subset of Ω, and let $J : \Omega \to \mathbb{R}$ be a function differentiable at a point $u \in U$. If the function J has a constrained local minimum at u relative to the set U, then*

$$J'(u)(v-u) \geq 0 \quad \textit{for all } v \in U.$$

Proof Let v be any point in the set U. Since U is convex, $(u + t(v-u)) \in U$ for all $0 \leq t \leq 1$. The differentiability of J at u then implies that, for all $0 \leq t \leq 1$,

$$0 \leq J(u + t(v-u)) - J(u) = t\left(J'(u)(v-u) + \eta(t)\right) \quad \text{with } \lim_{t \to 0^+} \eta(t) = 0.$$

Hence $J'(u)(v-u) \geq 0$ (otherwise $J(u+tw) - J(u)$ would be < 0 for $t > 0$ sufficiently small). □

The relations $J'(u)(v-u) \geq 0$ for all $v \in U$ constitute the **Euler inequalities**. If U is a *subspace* of V, they clearly imply that $J'(u)w = 0$ for all $w \in U$; in particular then, they reduce to the Euler equation $J'(u) = 0$ if $U = V$ (Theorem 7.1-5).

Naturally it is no coincidence that the same Euler equation and Euler inequalities were found earlier in the special case where the function J is a *quadratic functional* (Theorem 6.1-2).

Problems

7.1-1 (1) Let $(X, \|\cdot\|)$ be a normed vector space. Show that the mapping $x \in X \to \|x\| \in \mathbb{R}$ is not differentiable at $x = 0$.

(2) Let Ω be an open subset of $\mathbb{R}^n$. Show that the mapping $v \in L^2(\Omega) \to \|v\|_{L^2(\Omega)} \in \mathbb{R}$ is differentiable at any nonzero $v \in L^2(\Omega)$.

(3) Let the space $c_0 := \{x = (x_i)_{i=1}^\infty \in \ell^\infty;\ \lim_{i\to\infty} x_i = 0\}$ be equipped with the norm $\|\cdot\|_\infty$ (Section 2.4). Show that the mapping $x \in c_0 \to \|x\|_\infty$ is differentiable at $a = (a_i)_{i=1}^\infty \in c_0$ if and only if there exists i_0 such that $|a_{i_0}| > |a_i|$ for all $i \neq i_0$.

7.1-2 Let X and Y be two normed vector spaces and let U and V be open subsets of X and Y, respectively.

(1) Assume that there exists a bijection $f : U \to V$ and a point $a \in U$ such that f is differentiable at a and $f^{-1} : V \to U$ is differentiable at $f(a)$. Show that $f'(a) : X \to Y$ is a bijection.

(2) Show that, if in addition both spaces X and Y are finite-dimensional, their dimensions are equal.

7.1-3 Let X and Y be two normed vector spaces, let Ω be an open subset of X, and let $a \in \Omega$.

(1) Let $f : \Omega \subset X \to Y$ be a mapping such that, for *any* vector $h \in X$, the function $\theta \in I(h) \subset \mathbb{R} \to f(a + \theta h) \in Y$, which is defined on an open interval $I(h)$ of $\mathbb{R}$ containing the origin, is differentiable at $\theta = 0$; in other words, the Gâteaux derivative $\partial_h f(a)$ exists for *all* vectors $h \in X$. By means of a counterexample, show that f is not necessarily differentiable at a (while the other implication holds, as shown in the text).

(2) Let Ω be an open subset of $\mathbb{R}^2$ containing the origin $(0,0)$, and let $f : \Omega \subset \mathbb{R}^2 \to \mathbb{R}$ be a function such that $f(0,0) = 0$ and the Gâteaux derivatives $\partial_h f(0,0)$ exist for *all* vectors $h \in \mathbb{R}^2$. Let the function $g : [0, 2\pi] \to \mathbb{R}$ be defined by $g(\theta) := \partial_{h(\theta)} f(0,0)$, where $h(\theta) = (\cos\theta, \sin\theta) \in \mathbb{R}^2$.

Show that f is differentiable at $(0,0)$ if and only if the point of coordinates $(\cos\theta, \sin\theta, g(\theta))$ describes an ellipse in $\mathbb{R}^3$ when θ varies in the interval $[0, 2\pi]$.

7.1-4 Let a function $f \in \mathcal{C}(\mathbb{R})$ be such that every point of $\mathbb{R}$ is a local extremum of f. Is f a constant function?

7.1-5 Let Ω be an open subset of $\mathbb{R}^n$ and let $f : \Omega \times \mathbb{R} \to \mathbb{R}$ be a *Carathéodory function*, i.e., a function with the following properties (Carathéodory functions will be introduced in greater generality in Section 9.5): For each $s \in \mathbb{R}$, the function $f(\cdot, s) : \Omega \to \mathbb{R}$ is measurable and, for almost all $x \in \Omega$, the function $f(x, \cdot) : \mathbb{R} \to \mathbb{R}$ is continuous. Given any measurable function $v : \Omega \to \mathbb{R}$, define the measurable function $Av : \Omega \to \mathbb{R}$ by

$$Av(x) := f(x, v(x)), \quad x \in \Omega.$$

The object of this problem is to study the differentiability properties of the operator A defined in this fashion, which is called a **Nemytskii,**[7] or a **substitution, operator**.

(1) Assume that there exists a function $a \in L^2(\Omega)$ and a constant $b \geq 0$ such that

$$|f(x, s)| \leq a(x) + b\,|s| \quad \text{for almost all } x \in \Omega \text{ and all } s \in \mathbb{R}.$$

Show that the corresponding Nemytskii operator A maps $L^2(\Omega)$ into $L^2(\Omega)$ and that $A \in \mathcal{C}(L^2(\Omega); L^2(\Omega))$.

(2) Let $f(x, s) := \sin s$. Show that the corresponding Nemytskii operator $A : L^2(\Omega) \to L^2(\Omega)$, which is continuous by (1), is not Fréchet-differentiable.

(3) Show that $A \in \mathcal{C}^1(L^2(\Omega); L^2(\Omega))$ if and only if[8] there exist functions $a \in L^2(\Omega)$ and $b \in L^\infty(\Omega)$ such that

$$f(x, s) = a(x) + b(x)s \quad \text{for almost all } x \in \Omega \text{ and all } s \in \mathbb{R}.$$

(4) Assume that Ω is a domain and that $f : \Omega \times \mathbb{R}$ is as smooth as necessary. Show that the corresponding Nemytskii operator $A : H^m(\Omega) \to L^2(\Omega)$ is Fréchet-differentiable if the integer m is such that $H^m(\Omega) \hookrightarrow \mathcal{C}(\overline{\Omega})$.

7.2 The mean value theorem in a normed vector space; first applications

A *basic result* from differential calculus in normed vector spaces is a generalization of the *mean value theorem for real-valued functions.* This theorem asserts that, given a real-valued function f that is continuous on a compact interval $[a, b] \subset \mathbb{R}$ and differentiable on the open interval $]a, b[$, there exists a point $c \in\,]a, b[$ such that

$$f(b) - f(a) = f'(c)(b - a).$$

This formula *cannot* be generalized for *vector-valued* functions: for instance, the mapping $f : t \in [0, 2\pi] \to f(t) = (\cos t, \sin t) \in \mathbb{R}^2$ satisfies $f(2\pi) - f(0) = 0$, yet its derivative $f'(t) = (-\sin t, \cos t)$ never vanishes when t varies in $[0, 2\pi]$. As we shall see in the next theorem, what *can* be generalized, however, is the *inequality*

$$|f(b) - f(a)| \leq \sup_{t \in]a,b[} |f'(t)|\,|b - a|,$$

[7] So named after Viktor Vladimirovich Nemytskii (1900–1967).

[8] The spectacular "only if" part is due to:

M.M. Vainberg [1952]: Some questions of differential calculus in linear spaces, *Uspehi Matematicheskii Nauk (New Series)* **7**, 55–102 (in Russian).

which evidently follows from the relation $f(b) - f(a) = f'(c)(b - a)$.

Given two points a and b in a vector space,

$$[a, b] = \{x = ta + (1 - t)b;\ t \in [0, 1]\},$$
$$]a, b[= \{x = ta + (1 - t)b;\ t \in]0, 1[\},$$

denote respectively the *closed segment*, and the *open segment*, with *end-points* a and b.

Theorem 7.2-1 (mean value theorem in a normed vector space) *Let there be given two normed vector spaces X and Y, an open subset Ω of X containing a closed segment $[a, b]$, and a mapping $f : \Omega \subset X \to Y$ continuous on the closed segment $[a, b]$ and differentiable on the open segment $]a, b[$. Then*

$$\|f(b) - f(a)\|_Y \leq \left(\sup_{x \in]a,b[} \|f'(x)\|_{\mathcal{L}(X;Y)} \right) \|b - a\|_X .$$

Proof Since the inequality surely holds if $\sup_{x \in]a,b[} \|f'(x)\| = \infty$, it remains to consider the case where

$$M := \sup_{x \in]a,b[} \|f'(x)\| < \infty.$$

The mapping $\varphi : [0, 1] \to Y$ defined by

$$\varphi(t) := f(a + t(b - a)), \quad 0 \leq t \leq 1,$$

is continuous (as composed of two continuous functions) and, by the chain rule (Theorem 7.1-3), differentiable on $]0, 1[$ with

$$\varphi'(t) = f'(a + t(b - a))(b - a), \quad 0 < t < 1.$$

Consequently,

$$\sup_{0<t<1} \|\varphi'(t)\| \leq M \|b - a\| .$$

For each $\varepsilon > 0$, the set

$$I(\varepsilon) := \{t \in [0, 1]\,;\ \|\varphi(t) - \varphi(0)\| \leq (M \|b - a\| + \varepsilon)\, t + \varepsilon\}$$

is nonempty since $0 \in I(\varepsilon)$, and closed as the inverse image of the closed interval $]-\infty, 0]$ by the continuous function

$$\chi : t \in [0, 1] \to \|\varphi(t) - \varphi(0)\| - (M \|b - a\| + \varepsilon)t - \varepsilon.$$

Let

$$t_0 := \sup\{t \in [0, 1];\ t \in I(\varepsilon)\}.$$

Then $t_0 \in I(\varepsilon)$ because $I(\varepsilon)$ is closed, and $t_0 > 0$ because $\chi(0) = -\varepsilon < 0$. We now show that $t_0 = 1$.

Assume otherwise that $t_0 < 1$. Then, by definition of the derivative $\varphi'(t_0)$, which exists since $0 < t_0 < 1$,

$$\varphi(t_0 + \delta) - \varphi(t_0) = \varphi'(t_0)\delta + |\delta|\, \eta(\delta) \quad \text{with } \lim_{\delta \to 0} \eta(\delta) = 0.$$

Let δ_0 be so chosen that

$$t_0 < t_0 + \delta_0 < 1 \quad \text{and} \quad \|\eta(\delta_0)\| \le \varepsilon.$$

Then

$$\begin{aligned}\|\varphi(t_0+\delta_0)-\varphi(0)\| &\le \|\varphi(t_0+\delta_0)-\varphi(t_0)\| + \|\varphi(t_0)-\varphi(0)\| \\ &\le M\,\|b-a\|\,\delta_0 + \delta_0\varepsilon + (M\,\|b-a\|+\varepsilon)\,t_0 + \varepsilon \\ &= (M\,\|b-a\|+\varepsilon)\,(t_0+\delta_0)+\varepsilon,\end{aligned}$$

which implies that $(t_0+\delta_0) \in I(\varepsilon)$, in contradiction with the definition of t_0. Hence $t_0 = 1$.

That $1 \in I(\varepsilon)$ then implies that

$$\|f(b)-f(a)\| = \|\varphi(1)-\varphi(0)\| \le M\,\|b-a\| + 2\varepsilon,$$

and thus $\|f(b)-f(a)\| \le M\,\|b-a\|$ since $\varepsilon > 0$ is arbitrary. □

The mean value theorem is often used by means of the following immediate, yet very convenient, consequence (see, e.g., the proofs of the next two theorems in this section), referred to in the sequel as *"the" corollary to the mean value theorem.*

Theorem 7.2-2 (corollary to the mean value theorem) *Let there be given two normed vector spaces X and Y, an open subset Ω of X containing a closed segment $[a,b]$, and a mapping $f : \Omega \subset X \to Y$ continuous on the closed segment $[a,b]$ and differentiable on the open segment $]a,b[$. Finally, let there be given a mapping $A \in \mathcal{L}(X;Y)$. Then*

$$\|f(b)-f(a)-A(b-a)\|_Y \le \left(\sup_{x\in]a,b[} \|f'(x)-A\|_{\mathcal{L}(X,Y)}\right) \|b-a\|_X\,.$$

Proof It suffices to apply the mean value theorem to the mapping $x \in \Omega \subset X \to (f(x) - Ax) \in Y$, whose derivative at any $x \in \Omega$ is $f'(x) - A$. □

Our first application of the above corollary is an important relation between *differentiability* and *partial differentiability.*

Theorem 7.2-3 *Let X_j, $1 \le j \le n$, and Y be normed vector spaces, let Ω be an open subset of the product $X_1 \times X_2 \times \cdots \times X_n$, and let $f : \Omega \to Y$ be a mapping. Then $f \in \mathcal{C}^1(\Omega;Y)$ if and only if $\partial_j f \in \mathcal{C}(\Omega;\mathcal{L}(X_j;Y))$ for all $1 \le j \le n$.*

Proof To fix ideas, let

$$\|h\|_X := \max_{1\le j\le n} \|h_j\| \quad \text{for each } h = (h_1,h_2,\ldots,h_n) \in X := X_1 \times X_2 \times \cdots \times X_n.$$

Assume that $f \in \mathcal{C}^1(\Omega;Y)$. In particular then,

$$f'(x)h = \sum_{j=1}^n \partial_j f(x)h_j \quad \text{for all } x \in \Omega \text{ and all } h = (h_1,h_2,\ldots,h_n) \in X_1 \times X_2 \times \cdots \times X_n$$

(Theorem 7.1-2). Noting that $\partial_j f(x)h_j = f'(x)h^j$ for each $1 \le j \le n$, where the vector $h^j \in X_1 \times X_2 \times \cdots \times X_n$ is defined by $h_i^j := h_j \delta_{ij}$, $1 \le i \le n$, we infer from this relation that

$$\|\partial_j f(x)\|_{\mathcal{L}(X_j;Y)} \le \|f'(x)\|_{\mathcal{L}(X;Y)}, \quad 1 \le j \le n.$$

Consequently,

$$\|\partial_j f(a) - \partial_j f(b)\|_{\mathcal{L}(X_j;Y)} \le \|f'(a) - f'(b)\|_{\mathcal{L}(X;Y)} \quad \text{for all } a, b \in \Omega, \ 1 \le j \le n.$$

Hence $\partial_j f \in \mathcal{C}(\Omega; \mathcal{L}(X_j; Y))$ for all $1 \le j \le n$.

To establish the converse property, we assume that $n = 2$ (simply to avoid cumbersome notations; otherwise the extension to any $n \ge 3$ is clear). So, let $f : \Omega \subset X_1 \times X_2 \to Y$ be such that $\partial_j f \in \mathcal{C}(\Omega; \mathcal{L}(X_j, Y))$, $1 \le j \le 2$.

Given $a \in \Omega$, let $r > 0$ be such that $B(a; r) \subset \Omega$ and let $h = (h_1, h_2) \in X_1 \times X_2$ be such that $a+h = (a_1+h_1, a_2+h_2) \in B(a; r)$, so that $[(a_1 + h_1, a_2), (a_1 + h_1, a_2 + h_2)] \subset B(a; r)$. Finally, let Ω_2 be an open subset of X_2 such that $(a_1 + h_1, x_2) \in B(a; r)$ for all $x_2 \in \Omega_2$. Then, on the one hand, Theorem 7.2-2 applied to the function $g : x_2 \in \Omega_2 \subset X_2 \to g(x_2) := f(a_1 + h_1, x_2)$ (which is differentiable for all $x_2 \in \Omega_2$ by assumption) with $A := \partial_2 f(a) \in \mathcal{L}(X_2; Y)$ gives

$$\begin{aligned}\|f(a_1 + h_1, a_2 + h_2) - f(a_1 + h_1, a_2) - \partial_2 f(a)h_2\| &= \|g(a_2 + h_2) - g(a_2) - \partial_2 f(a)h_2\| \\ \le \|h_2\| \sup_{0<\theta<1} \|\partial_2 g(a_2 + \theta h_2) - \partial_2 f(a)\| &= \|h_2\|\, \eta_2(h) \quad \text{with } \lim_{h\to 0} \eta_2(h) = 0,\end{aligned}$$

since $\eta_2(h) := \sup_{0<\theta<1} \|\partial_2 f(a_1 + h_1, a_2 + \theta h_2) - \partial_2 f(a_1, a_2)\|$ and $\partial_2 f \in \mathcal{C}(\Omega; \mathcal{L}(X_2; Y))$ by assumption. On the other hand, the definition of $\partial_1 f(a)$ gives

$$\|f(a_1 + h_1, a_2) - f(a_1, a_2) - \partial_1 f(a)h_1\| = \|h_1\|\, \eta_1(h) \quad \text{with } \lim_{h\to 0} \eta_1(h) = 0.$$

The last two relations together imply that

$$\begin{aligned}\|f(a + h) - f(a) - (\partial_1 f(a)h_1 + \partial_2 f(a)h_2)\| &\le \|h_1\|\, \eta_1(h) + \|h_2\|\, \eta_2(h) \\ &= \|h\|\, \eta(h) \quad \text{with } \lim_{h\to 0} \eta(h) = 0.\end{aligned}$$

The mapping f is thus differentiable at $a \in \Omega$, with

$$f'(a)h := \partial_1 f(a)h_1 + \partial_2 f(a)h_2 \quad \text{for all } h = (h_1, h_2) \in X = X_1 \times X_2.$$

This relation also shows that

$$\|f'(a) - f'(b)\|_{\mathcal{L}(X;Y)} \le \|\partial_1 f(a) - \partial_2 f(b)\|_{\mathcal{L}(X_1;Y)} + \|\partial_2 f(a) - \partial_2 f(b)\|_{\mathcal{L}(X_2;Y)} \quad \text{for all } a, b \in \Omega.$$

Hence $f \in \mathcal{C}^1(\Omega; Y)$. □

We noted in Section 7.1 that a continuous affine mapping $f : x \in \Omega \subset X \to f(x) := Ax + b \in Y$, with $A \in \mathcal{L}(X; Y)$ and $b \in Y$, is differentiable in Ω, with $f'(x) = A$ for all $x \in \Omega$. Thanks to the mean value theorem, we can now show that this necessary condition becomes sufficient if the open set Ω is *connected.*

Theorem 7.2-4 *Let X and Y be normed vector spaces, let Ω be a connected open subset of X, and let $f : \Omega \subset X \to Y$ be a mapping differentiable in Ω. Assume that there exists $A \in \mathcal{L}(X;Y)$ such that*

$$f'(x) = A \in \mathcal{L}(X;Y) \quad \textit{for all } x \in \Omega.$$

Then there exists $b \in Y$ such that

$$f(x) = Ax + b \quad \textit{for all } x \in \Omega.$$

Proof Given any $x \in \Omega$, there exists $r = r(x) > 0$ such that $B(x;r) \subset \Omega$, and for each $y \in B(x;r)$ the segment $[x,y]$ belongs to $B(x;r)$. An application of Theorem 7.2-2 then shows that

$$\|f(y) - f(x) - A(y-x)\| \leq \sup_{z \in]x,y[} \|f'(z) - A\| \, \|y - x\| = 0 \quad \text{for all } y \in B(x;r).$$

Hence the mapping $g : x \in \Omega \to g(x) := (f(x) - Ax) \in Y$ satisfies $g(y) = g(x)$ for all $y \in B(x;r)$.

Fix a point $x_0 \in \Omega$. Then the set

$$U := \{x \in \Omega;\ g(x) = g(x_0)\}$$

is nonempty ($x_0 \in U$), relatively closed in Ω since $g : \Omega \to Y$ is continuous, and open since, given any point $x \in U$, there exists $r > 0$ such that $B(x;r) \subset U$ (as shown above). Therefore $U = \Omega$ since Ω is connected by assumption. □

Other important applications of the mean value theorem will be treated in the next two sections.

Problems

7.2-1 Let X and Y be normed vector spaces, let Ω be an open subset of X, let a be a point in Ω, and let $f : \Omega \subset X \to Y$ be a mapping that is differentiable in $\Omega - \{a\}$ and continuous at a. Show that, if $A := \lim_{x \to a} f'(x)$ exists, then f is differentiable at a and $f'(a) = A$.

7.2-2 Let Ω be a domain in $\mathbb{R}^n$, let $u \in \mathcal{C}^1(\overline{\Omega}; \mathbb{R}^n)$, and let $\|\cdot\|$ denote any subordinate matrix norm over $\mathbb{M}^n$. Show that there exists a constant $c(\Omega) > 0$ such that the mapping $f : x \in \overline{\Omega} \to f(x) := x + u(x) \in \mathbb{R}^n$ is injective in $\overline{\Omega}$ if $\sup_{x \in \overline{\Omega}} \|\nabla u(x)\| < c(\Omega)$.

7.3 Application of the mean value theorem: Differentiability of the limit of a sequence of differentiable functions

Let X and Y be two normed vector spaces, let Ω be an open subset of X, and let $(f_n)_{n=1}^{\infty}$ be a sequence of differentiable functions $f_n : \Omega \subset X \to Y$ that converges *locally uniformly* (Section 2.3) to a differentiable function $f : \Omega \subset X \to Y$ as $n \to \infty$. It should be clear that, without any additional assumption, no conclusion can be drawn in general about the convergence in the space $\mathcal{L}(X;Y)$ of the sequence $(f'_n)_{n=1}^{\infty}$ formed by the derivatives $f'_n \in \mathcal{L}(X;Y)$, let alone about its convergence to f'.

Consider for example the functions $f_n : \mathbb{R} \to \mathbb{R}^2$ defined by

$$f_n : x \in \mathbb{R} \to f_n(x) := \left(\frac{1}{n}\cos(n^2 x), \frac{1}{n}\sin(n^2 x)\right) \in \mathbb{R}^2 \quad \text{for each integer } n \geq 1,$$

which are of class $\mathcal{C}^\infty$ on $\mathbb{R}$. Then the sequence $(f_n)_{n=1}^\infty$ converges uniformly on $\mathbb{R}$ to the mapping $f : x \in \mathbb{R} \to f(x) := (0,0) \in \mathbb{R}^2$, also of class $\mathcal{C}^\infty$. Yet,

$$\text{at each } x \in \mathbb{R}, \quad \|f_n'(x)\| = n \to \infty \ \text{ as } n \to \infty.$$

As shown in the next theorem, whose proof relies in a crucial way on the *mean value theorem in a normed vector space* and on its *corollary* (Theorems 7.2-1 and 7.2-2), it turns out that the proper assumption on the sequence $(f_n')_{n=1}^\infty$ is that it *converges locally uniformly.* Observe that, by contrast, the assumption made below on the sequence $(f_n)_{n=1}^\infty$, viz., that of *simple convergence*, is very mild.

Theorem 7.3-1 (differentiability of the limit of a sequence of differentiable functions) *Let X and Y be normed vector spaces, let Ω be an open subset of X, and let $(f_n)_{n=1}^\infty$ be a sequence of functions $f_n : \Omega \to Y$ with the following properties:*

Each function f_n, $n \geq 1$, is differentiable in Ω, resp. *of class $\mathcal{C}^1$ in Ω, the sequence $(f_n)_{n=1}^\infty$ converges pointwise to a function $f : \Omega \to Y$, and the sequence $(f_n')_{n=1}^\infty$ converges locally uniformly to a function $g : \Omega \to \mathcal{L}(X;Y)$.*

Then the sequence $(f_n)_{n=1}^\infty$ converges locally uniformly to the function f, the function f is differentiable in Ω, resp. *of class $\mathcal{C}^1$ in Ω, and $f' = g$.*

Proof (i) *The sequence $(f_n)_{n=1}^\infty$ converges locally uniformly to f.*

Let $x_0 \in X$ and $\varepsilon > 0$ be given. By assumption, there exists an open ball $B := B(x_0; r)$ such that

$$\lim_{n\to\infty} \sup_{x \in B} \|f_n'(x) - g(x)\|_{\mathcal{L}(X;Y)} = 0.$$

Since $\|f_m'(x) - f_n'(x)\| \leq \|f_m'(x) - g(x)\| + \|f_n'(x) - g(x)\|$ for all $m, n \geq 1$ and all $x \in B$, there exists $n_0 \geq 1$ such that

$$\sup_{x \in B} \|f_m'(x) - f_n'(x)\| \leq \frac{\varepsilon}{2r} \quad \text{for all } m, n \geq n_0,$$

so that by the *mean value theorem in a normed vector space* (which can be applied since B is convex),

$$\begin{aligned}\|f_m(x) - f_n(x) - f_m(x_0) - f_n(x_0)\| &\leq r \sup_{x \in B} \|f_m'(x) - f_n'(x)\| \\ &\leq \frac{\varepsilon}{2} \quad \text{for all } m, n \geq n_0 \text{ and all } x \in B.\end{aligned}$$

By assumption, the sequence $(f_n)_{n=1}^\infty$ converges pointwise to f. So, there exists $n_1 \geq n_0$ such that

$$\|f_m(x_0) - f_n(x_0)\| \leq \|f_m(x_0) - f(x_0)\| + \|f_n(x_0) - f(x_0)\| \leq \frac{\varepsilon}{2} \quad \text{for all } m, n \geq n_1,$$

and thus

$$\|f_m(x) - f_n(x)\| \leq \varepsilon \quad \text{for all } m, n \geq n_1 \text{ and all } x \in B.$$

Fix a point x in the ball B and let $m \to \infty$ in the above inequality; this gives

$$\|f(x) - f_n(x)\| \leq \varepsilon \quad \text{for all } n \geq n_1,$$

again by the assumed pointwise convergence of the sequence $(f_n)_{n=1}^{\infty}$. But the integer n_1 does not depend on x; hence

$$\sup_{x \in B} \|f(x) - f_n(x)\| \leq \varepsilon \quad \text{for all } n \geq n_1.$$

(ii) *The function f is differentiable in Ω,* resp. *of class $\mathcal{C}^1$ in Ω, and $f' = g$.*

Given any point $x_0 \in \Omega$, let $B := B(x_0; r)$ be the ball defined as in (i), and let the auxiliary functions $k_n : B \to Y$, $n \geq 1$, be defined at each $x \in B$ by

$$k_n(x) := \frac{1}{\|x - x_0\|}\left(f_n(x) - f_n(x_0) - f_n'(x_0)(x - x_0)\right) \quad \text{if } x \neq x_0, \text{ and } k_n(x_0) := 0.$$

First, the assumptions made on the sequences $(f_n)_{n=1}^{\infty}$ and $(f_n')_{n=1}^{\infty}$ imply that *the sequence $(k_n)_{n=1}^{\infty}$ converges pointwise in B*, to the function $k : B \to Y$ defined at each $x \in B$ by

$$k(x) := \frac{1}{\|x - x_0\|}\left(f(x) - f(x_0) - g(x_0)(x - x_0)\right) \quad \text{if } x \neq x_0, \text{ and } k(x_0) := 0.$$

Second, the *corollary to the mean value theorem* shows that, at each $x \in B$,

$$\begin{aligned}
\|k_m(x) - k_n(x)\| &= \frac{1}{\|x - x_0\|}\|f_m(x) - f_n(x) - \left(f_m'(x_0) - f_n'(x_0)\right)(x - x_0)\| \\
&\leq \sup_{\xi \in B} \|\left(f_m'(\xi) - f_n'(\xi)\right) - \left(f_m'(x_0) - f_n'(x_0)\right)\| \quad \text{if } x \neq x_0;
\end{aligned}$$

besides,

$$\|k_m(x) - k_n(x)\| = 0 \quad \text{for all } m, n \geq 1 \quad \text{if } x = x_0.$$

Hence the assumption of local uniform convergence made on the sequence $(f_n')_{n=1}^{\infty}$ shows that, given any $\varepsilon > 0$, there exists $n_2 \geq 1$ such that

$$\sup_{x \in B} \|k_m(x) - k_n(x)\| \leq \varepsilon \quad \text{for all } m, n \geq n_2.$$

Therefore the argument made in (i) about the sequence $(f_n)_{n=1}^{\infty}$ can be repeated *verbatim* for the sequence $(k_n)_{n=1}^{\infty}$, thus showing that

$$\lim_{n \to \infty} \sup_{x \in B} \|k_n(x) - k(x)\| = 0.$$

Each function k_n, $n \geq 1$, is continuous at x_0 (by definition of the differentiability of f_n at x_0). Hence, as a limit of a locally uniformly convergent sequence of continuous functions

(Theorem 2.3-3), the function k is also continuous at x_0. But the continuity of k at x_0 means that the function f is differentiable at x_0, with

$$f'(x_0) = g(x_0).$$

If the functions f_n, $n \geq 1$, are of class $\mathcal{C}^1$ in Ω, their derivatives $f'_n : \Omega \to \mathcal{L}(X;Y)$ are continuous in Ω. Hence the function $g : \Omega \to \mathcal{L}(X;Y)$ is also continuous in Ω, again as a limit of a locally uniformly convergent sequence of continuous functions. □

Surprisingly, under the additional assumptions of connectedness of the open set Ω and of completeness of the space Y, the conclusions of Theorem 7.3-1 remain unaltered if the sequence $(f_n)_{n=1}^\infty$ is assumed to pointwise converge at *only one point* of Ω; cf. Problem 7.3-1.

Since *series in normed vector spaces* are defined as *limits* (Section 3.6), Theorem 7.3-1 applies as well to functions defined as limits of *convergent series whose partial sums are differentiable*; cf. Problem 7.3-2 for an example.

Problems

7.3-1 (complement to Theorem 7.3-1) Let X be a normed vector space, let Ω be a connected open subset of X, let Y be a Banach space, and let $(f_n)_{n=1}^\infty$ be a sequence of functions $f_n : \Omega \to Y$ with the following properties: Each function f_n, $n \geq 1$, is differentiable in Ω, *resp.* of class $\mathcal{C}^1$ in Ω, there exists a point $x_0 \in \Omega$ such that the sequence $(f_n(x_0))_{n=1}^\infty$ converges in Y, and the sequence $(f'_n)_{n=1}^\infty$ converges locally uniformly to a function $g : \Omega \to \mathcal{L}(X;Y)$.

Show that the sequence $(f_n)_{n=1}^\infty$ converges locally uniformly to a function $f : \Omega \to Y$, and that f is differentiable in Ω, *resp.* of class $\mathcal{C}^1$ in Ω, with $f' = g$.

7.3-2 Let a function $g \in L^2(0, 2\pi)$ be such that the coefficients appearing in its Fourier series (Theorem 4.9-2) satisfy $|a_k| \leq \dfrac{C}{k^{2+\sigma}}$, $k \geq 0$, and $|b_k| \leq \dfrac{C}{k^{2+\sigma}}$, $k \geq 1$, for some constants $C > 0$ and $\sigma > 0$. Using Theorem 7.3-1, show that $g \in \mathcal{C}^1\,[0,1]$.

7.4 Application of the mean value theorem: Differentiability of a function defined by an integral

The mean value theorem, together with Lebesgue's dominated convergence theorem, provides a very useful criterion of differentiability of a *function defined by a Lebesgue integral*, viz., a function of the form

$$g : y \in U \to g(y) := \int_\Omega f(x,y)\,\mathrm{d}x,$$

where Ω and U are open subsets of $\mathbb{R}^n$ and $\mathbb{R}^m$.

Theorem 7.4-1 *Let Ω and U be open subsets of $\mathbb{R}^n$ and $\mathbb{R}^m$ respectively, and let $f : \Omega \times U \to \mathbb{R}$ be a function with the following properties:*

$$f(\cdot, y) \in \mathcal{L}^1(\Omega) \quad \textit{for each } y \in U,$$

the function $f(x,\cdot) : U \to \mathbb{R}$ is of class $\mathcal{C}^1$ in U for almost all $x \in \Omega$,

$$\partial_j f(\cdot, y) := \frac{\partial f}{\partial y_j}(\cdot, y) \in \mathcal{L}^1(\Omega) \quad \textit{for each } y \in U,\ \ 1 \leq j \leq m,$$

and finally, there exists a function $h \in \mathcal{L}^1(\Omega)$ *with the following property: Given any point* $y \in U$*, there exists a neighborhood* V_y *of* y *in* U *such that*

$$|\partial_j f(x,z)| \leq h(x) \quad \textit{for almost all } x \in \Omega \textit{ and all } z \in V_y.$$

Then the function $g : U \to \mathbb{R}$ *defined by*

$$g(y) := \int_\Omega f(x,y)\,\mathrm{d}x \quad \textit{at each } y \in U,$$

is of class $\mathcal{C}^1$ *in* U *and*

$$\partial_j g(y) = \int_\Omega \partial_j f(x,y)\,\mathrm{d}x \quad \textit{at each } y \in U,\ 1 \leq j \leq m.$$

Proof Throughout the proof, e_j designates one of the vectors of the canonical basis of $\mathbb{R}^m$ and y designates a given point in U. Given any sequence $(h_k)_{k=1}^\infty$ of real numbers such that

$$h_k \neq 0 \quad \text{and} \quad (y + h_k e_j) \in V_y \quad \text{for all } k > 1 \quad \text{and} \quad \lim_{k\to\infty} h_k = 0,$$

define the functions $\delta_k : \Omega \to [-\infty, \infty]$, $k \geq 1$, by

$$\delta_k(x) := \frac{1}{h_k}\big(f(x, y + h_k e_j) - f(x,y) - \partial_j f(x,y) h_k\big), \quad x \in \Omega.$$

Then, by the *corollary to the mean value theorem* (Theorem 7.2-2),

$$|\delta_k(x)| \leq \sup_{\xi \in [y, y+h_k e_j]} |\partial_j f(x,\xi) - \partial_j f(x,y)| \leq 2h(x)$$

for each $k \geq 1$ and almost all $x \in \Omega$. Besides, the assumed differentiability of the function $f(x,\cdot) : U \to \mathbb{R}$ implies that

$$\lim_{k\to\infty} \delta_k(x) = 0 \quad \text{for almost all } x \in \Omega.$$

Therefore, by the *Lebesgue dominated convergence theorem* (Theorem 1.15-3),

$$\lim_{k\to\infty} \int_\Omega \delta_k(x)\,\mathrm{d}x = 0,$$

which implies that the function $g : U \to \mathbb{R}$ has partial derivatives given at each point $y \in U$ by

$$\partial_j g(y) = \int_\Omega \partial_j f(x,y)\,\mathrm{d}x, \quad 1 \leq j \leq m.$$

Given any point $y \in U$, let $y_k \in U$, $k \geq 1$, be such that $y_k \in V_y$ for all $k \geq 1$ and $\lim_{k\to\infty} y_k = y$. Then

$$|\partial_j g(y_k) - \partial_j g(y)| \leq \int_\Omega |\partial_j f(x,y_k) - \partial_j f(x,y)|\,\mathrm{d}x,$$

and

$$|\partial_j f(x,y_k) - \partial_j f(x,y)| \leq 2h(x),$$

for each $k \geq 1$ and almost all $x \in \Omega$. Besides, the assumption that the function $f(\cdot,x) : U \to \mathbb{R}$ is of class $\mathcal{C}^1$ implies that

$$\lim_{k\to\infty} |\partial_j f(x,y_k) - \partial_j f(x,y)| = 0 \quad \text{for almost all } x \in \Omega.$$

Hence $\lim_{k\to\infty} \partial_j g(y_k) = \partial_j g(y)$, again by *Lebesgue's dominated convergence theorem*. That $g \in \mathcal{C}^1(U)$ then follows from Theorem 7.2-3 (another consequence of the corollary to the mean value theorem). □

Remark In fact, a similar theorem holds in the more general situation where $\mathbb{R}^m$ is replaced by an arbitrary normed vector space X and the function f takes its value in an arbitrary Banach space Y.[9] But then such an extension rests on the notion of Lebesgue-integrability of functions with values in a Banach space, viz., Y and $\mathcal{L}(X;Y)$ in this case (in this book we only consider the special case where Ω is an interval in $\mathbb{R}$ and the function to be integrated is continuous; cf. Section 3.3). □

Problem

7.4-1 For each $y \in \mathbb{R}$, let $g(y) := \int_{-\infty}^{\infty} e^{-2i\pi xy} e^{-\pi x^2}\, dx$.
(1) Show that $g(0) = 1$.
(2) Show that the function $g : \mathbb{R} \to \mathbb{R}$ is infinitely differentiable.
(3) Show that $g'(y) + 2\pi y g(y) = 0$, $y \in \mathbb{R}$, and deduce from this observation that $g(y) = e^{-\pi y^2}$, $y \in \mathbb{R}$.

7.5 Application of the mean value theorem: Sard's theorem

The following basic result, which plays in particular a key role in the definition of *Brouwer's topological degree in* $\mathbb{R}^n$ (Section 9.15), constitutes a beautiful application of the mean value theorem.

Theorem 7.5-1 (Sard's theorem[10]) *Given an open subset Ω of $\mathbb{R}^n$ and a function $f \in \mathcal{C}^1(\Omega;\mathbb{R}^n)$, let*

$$S_f := \{x \in \Omega;\ \det \nabla f(x) = 0\}.$$

Then

$$dx - \text{meas}\, f(S_f) = 0.$$

Proof As usual, $|\cdot|$ denotes both the Euclidean norm in $\mathbb{R}^n$ and the associated operator norm; $\text{diam}\, K := \sup\{|x-y|\,;\, x,y \in K\}$ and $B(x;r) := \{y \in \mathbb{R}^n;\ |y-x| < r\}$; a *cube in* $\mathbb{R}^n$ is any set of the form $\{y \in \mathbb{R}^n;\ \|y-x\|_\infty \leq r\}$ with $x \in \mathbb{R}^n$ and $r > 0$; and, for notational brevity, we let $S := S_f$.

[9] See SCHWARTZ [1993b, Theorem 6.3.5].

[10] A. SARD [1942]: The measure of the critical values of differential maps, *Bulletin of the American Mathematical Society* **48**, 883–890.

(i) *Let K be any closed cube contained in Ω. Then*

$$dx - \text{meas}\, f(S \cap K) = 0.$$

By the *mean value theorem in a normed vector space* (*Theorem* 7.2-1),

$$|f(y) - f(x)| \leq \gamma\, |y - x| \quad \textit{for all } x, y \in K, \ \textit{where } \gamma = \gamma(f, K) := \sup_{\xi \in K} |f'(\xi)|.$$

Let $\varepsilon > 0$ be given. Since $f' \in \mathcal{C}(\Omega; \mathbb{M}^n)$ by assumption and K is a compact subset of Ω, there exists $\delta = \delta(\varepsilon, f, K) > 0$ such that

$$|f'(y) - f'(x)| \leq \varepsilon \quad \text{for all } x, y \in K \text{ such that } |x - y| \leq \delta.$$

Let σ denote the length of the sides of K and let $\ell = \ell(\delta, K) = \ell(\varepsilon, f, K)$ be any integer that satisfies $\ell \geq \sqrt{n}\, \sigma \delta^{-1}$. Then the cube K can be written as a union of ℓ^n cubes K_i of side $\sigma \ell^{-1}$; hence

$$K = \bigcup_{i=1}^{\ell^n} K_i \quad \text{with } \operatorname{diam} K_i \leq \sqrt{n}\, \frac{\sigma}{\ell}, \quad 1 \leq i \leq \ell^n.$$

Given any $x \in S \cap K$ (if $S \cap K = \varnothing$, there is nothing to prove), there exists an integer $1 \leq j = j(x) \leq \ell^n$ such that $x \in K_j$. Then

$$|f(y) - f(x)| \leq \gamma\, |y - x| \leq \gamma \operatorname{diam} K_j = \gamma \sqrt{n}\, \frac{\sigma}{\ell} \quad \text{for all } y \in K_j,$$

which shows that

$$f(K_j) \subset \overline{B(f(x);\ \gamma\sqrt{n}\, \frac{\sigma}{\ell})},$$

on the one hand (Figure 7.5-1). On the other hand, the *corollary to the mean value theorem* (Theorem 7.2-2) shows that

$$|f(y) - f(x) - f'(x)(y - x)| \leq \Big(\sup_{\xi \in K_j} |f'(\xi) - f'(x)| \Big)\, |y - x| \leq \varepsilon \sqrt{n}\, \frac{\sigma}{\ell} \quad \text{for all } y \in K_j.$$

Since $\det f'(x) = 0$ by assumption, there exists a subspace H of $\mathbb{R}^n$ with $\dim H < n - 1$ such that *the points* $f(x) + f'(x)(y - x)$, $y \in K_j$, *lie in the hyperplane* $f(x) + H$ (Figure 7.5-1). Therefore,

$$dx - \text{meas}\, f(K_j) \leq \Big(2\gamma\sqrt{n}\, \frac{\sigma}{\ell}\Big)^{n-1} \times \Big(2\varepsilon\sqrt{n}\, \frac{\sigma}{\ell}\Big) = 2^n \gamma^{n-1} n^{\frac{n}{2}} \frac{\sigma^n}{\ell^n},$$

which in turn implies that

$$dx - \text{meas}\, f(S \cap K) \leq \sum_{\substack{i=1,\ldots,\ell^n \\ S \cap K_i \neq \varnothing}} dx - \text{meas}\, f(S \cap K_i) \leq C\varepsilon,$$

where $C = C(\varepsilon, f, K) := 2^n \gamma^{n-1} n^{\frac{n}{2}} \sigma^n$. Since $\varepsilon > 0$ is arbitrary, this shows that $dx -$ meas $f(S \cap K) = 0$.

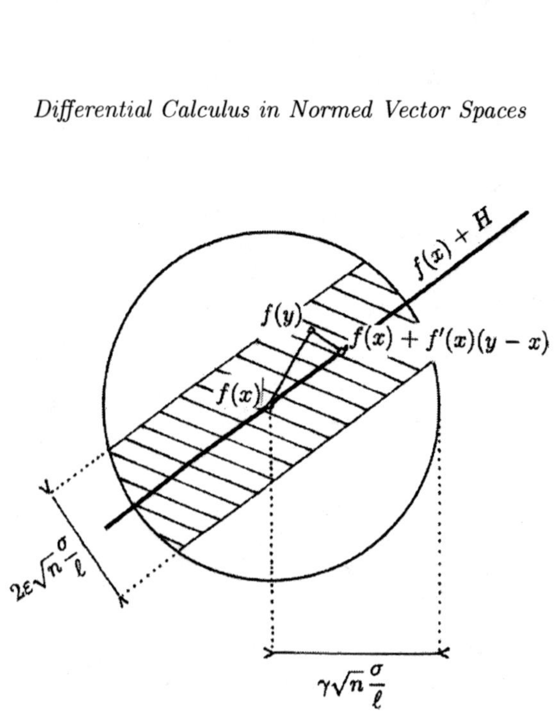

Figure 7.5-1 The direct image $\{f(y) \in \mathbb{R}^n;\ y \in K_j\}$ of the cube K_j under f lies in the hatched region.

(ii) *There exists a countably infinite family* $(K_i)_{i=1}^{\infty}$ *of closed cubes* $K_i \subset \Omega$ *such that* $S \subset \bigcup_{i=1}^{\infty} K_i$.

Let $(C_m)_{m=1}^{\infty}$ be a countably infinite family of compact subsets C_m such that $\Omega = \bigcup_{m=1}^{\infty} C_m$, so that

$$S = \bigcup_{m=1}^{\infty} (C_m \cap S).$$

For each $m \geq 1$, the set

$$C_m \cap S = \{x \in C_m;\ f'(x) = 0\}$$

is compact, as a closed subset of C_m (the function f is assumed to be of class $\mathcal{C}^1$ in Ω). Then the assertion follows by noting that each set $C_m \cap S$ can be covered by a finite number of closed cubes (each point in $C_m \cap S$ is the center of an open ball contained in a closed cube, itself contained in an open ball contained in Ω), by the finite subcovering property of compact sets.

(iii) *Conclusion.*

By (ii), $S = \bigcup_{i=1}^{\infty}(S \cap K_i)$. Consequently,

$$\begin{aligned} \mathrm{d}x - \operatorname{meas} f(S) &= \mathrm{d}x - \operatorname{meas} f\left(\bigcup_{i=1}^{\infty}(S \cap K_i)\right) \\ &\leq \sum_{i=1}^{\infty} (\mathrm{d}x - \operatorname{meas} f(S \cap K_i)) = 0. \end{aligned}$$

□

Problems

7.5-1 Give an example of a function $f \in \mathcal{C}^1(\mathbb{R})$ such that the closure of the image of the set $\{x \in \mathbb{R};\ f'(x) = 0\}$ under f is $\mathbb{R}$.

7.5-2 Let Ω be an open subset of $\mathbb{R}^n$, let $m > n$, and let $f \in \mathcal{C}^1(\Omega; \mathbb{R}^m)$. Show that the image $f(\Omega)$ of Ω under f is a set of zero Lebesgue measure in $\mathbb{R}^m$.

7.5-3 Let $S = \{x \in \mathbb{R}^n;\ |x| = 1\}$ denote the unit sphere (as usual, $|\cdot|$ denotes the Euclidean norm) in $\mathbb{R}^n$, let Ω be an open subset of $\mathbb{R}^n$ that contains S, and let $f \in \mathcal{C}^1(\Omega; \mathbb{R}^n)$. Show that dx-meas $f(S) = 0$.

7.6 A mean value theorem for functions of class $\mathcal{C}^1$ with values in a Banach space

The mean value theorem in a normed vector space (Theorem 7.2-1) admits an interesting complement when the mapping $f : \Omega \subset X \to Y$ is *of class* $\mathcal{C}^1$ and the space Y is a *Banach space*. This complement plays a key role in the proof of the *Newton–Kantorovich theorem* (Theorem 7.7-3) and for establishing the *Taylor formula with integral remainder* (Theorem 7.9-1(d)).

Note that the integral $\int_0^1 f'((1-\theta)a + \theta b)(b-a)\,\mathrm{d}\theta$ found in the next theorem makes sense since the function $\theta \in [0,1] \to f'((1-\theta)a + \theta b)(b-a) \in Y$ is *continuous* and Y is a *Banach space* (Section 3.3).

Theorem 7.6-1 (mean value theorem for functions of class $\mathcal{C}^1$ with values in a Banach space) *Let Ω be an open subset in a normed vector space X, let Y be a Banach space, and let $f \in \mathcal{C}^1(\Omega; Y)$. Then, given any closed segment $[a,b] \subset \Omega$,*

$$f(b) - f(a) = \int_0^1 f'\left((1-\theta)a + \theta b\right)(b-a)\,\mathrm{d}\theta.$$

Proof Let I be an open interval of $\mathbb{R}$ containing the interval $[0,1]$. Given any function $g \in \mathcal{C}(\overline{I}; Y)$, define the function

$$G : \theta \in I \to G(\theta) := \int_0^\theta g(\xi)\,\mathrm{d}\xi \in Y,$$

so that, given any point $\theta \in [0,1]$ and any $h > 0$ such that $(\theta + h) \in I$,

$$G(\theta + h) - G(\theta) - hg(\theta) = \int_\theta^{\theta+h} (g(\xi) - g(\theta))\,\mathrm{d}\xi.$$

Consequently, by Theorem 3.2-1,

$$\|G(\theta+h) - G(\theta) - hg(\theta)\|_Y \le \int_\theta^{\theta+h} \|g(\xi) - g(\theta)\|_Y\,\mathrm{d}\xi \le h \sup_{\theta \le \xi \le \theta+h} \|g(\xi) - g(\theta)\|_Y,$$

which in turn implies that

$$G(\theta+h) = G(\theta) + hg(\theta) + h\delta(h) \quad \text{with} \lim_{h \to 0^+} \delta(h) = 0 \text{ in } Y,$$

since the function g is continuous by assumption. A similar argument shows that the last relation also holds if $h < 0$, this time with $\lim_{h \to 0^-} \delta(h) = 0$. This shows that *the function* $G : I \to Y$ *is differentiable at each point of* $[0, 1]$, with a derivative given by

$$G'(\theta) = g(\theta) \quad \text{in } Y \text{ at each } \theta \in [0, 1]$$

(by definition of the Fréchet derivative, $G'(\theta) \in \mathcal{L}(\mathbb{R}; Y)$; but this relation makes sense as an equality in the space Y, since the space $\mathcal{L}(\mathbb{R}; Y)$ can be identified with Y).

Given a function $f \in \mathcal{C}^1(\Omega; Y)$ and a closed segment $[a, b] \subset \Omega$, there exists an open interval $I \subset \mathbb{R}$ containing $[0, 1]$ such that $\{(1 - \theta)a + \theta b; \theta \in \overline{I}\} \subset \Omega$ since Ω is open. Then the function

$$g : \theta \in \overline{I} \to g(\theta) := f'((1 - \theta)a + \theta b)(b - a) \in Y$$

belongs to the space $\mathcal{C}(\overline{I}; Y)$. Hence, by the above argument,

$$g(\theta) = G'(\theta) \quad \text{in } Y \text{ at each } \theta \in [0, 1],$$

where $G(\theta) := \int_0^\theta g(\xi)\,d\xi$, $0 \leq \theta \leq 1$, on the one hand.

On the other hand, it is easily seen that the same function $g \in \mathcal{C}(\overline{I}; Y)$ satisfies

$$g(\theta) = \widetilde{G}'(\theta) \quad \text{in } Y \text{ at each } \theta \in [0, 1],$$

where $\widetilde{G}(\theta) := f((1 - \theta)a + \theta b) \in Y$, $0 \leq \theta \leq 1$. Since the two functions G and $\widetilde{G}$ therefore share the same derivative at each point of the *connected* open interval $]0, 1[$, they are equal on $]0, 1[$, up to a constant vector in Y (Theorem 7.2-4); hence also on $[0, 1]$ by continuity. There thus exists a vector $c \in Y$ such that $G(\theta) = \widetilde{G}(\theta) + c$ for all $0 \leq \theta \leq 1$. In particular then, $G(1) - G(0) = \widetilde{G}(1) - \widetilde{G}(0)$, or equivalently,

$$\int_0^1 f'((1 - \theta)a + \theta b)(b - a)\,d\theta = f(b) - f(a),$$

as was to be proved. □

Problem

7.6-1 The assumptions are those of Theorem 7.6-1. Applying this theorem to the function $g : x \in \Omega \to g(x) := (f(x) - Ax) \in Y$ shows that, given any continuous linear operator $A \in \mathcal{L}(X; Y)$, the following inequality holds:

$$\|f(b) - f(a) - A(b - a)\|_Y \leq \sup_{x \in [a,b]} \|f'(x) - A\|_{\mathcal{L}(X;Y)} \|b - a\|_X .$$

Remark This provides another way to recover in this case the corollary to the mean value theorem (Theorem 7.2-2). □

7.7 Newton's method for solving nonlinear equations; the Newton–Kantorovich theorem in a Banach space

The Banach fixed point theorem (Theorem 3.7-1) provides in a sense the simplest way to show that a nonlinear equation in a Banach space (written as $f(x) = x$) has a solution and to solve this equation by means of an iterative method. The existence theorems proved in this section provide other, but not as simple, ways to likewise establish *the existence of a solution to a nonlinear equation in a Banach space* (now written as $f(x) = 0$) together with an *iterative method for approximating such a solution.* Like that of the Banach fixed point theorem, their proofs require only a *modicum* of linear and nonlinear functional analysis, viz., the notion of *complete space* and (in this section) the *mean value theorem.*

Remark Other powerful existence theorems for nonlinear equations in $\mathbb{R}^n$ or in an infinite-dimensional Banach space, whose proofs are, however, substantially more delicate, such as *Brouwer's fixed point theorem, Schauder's fixed point theorem,* or the *Minty–Browder theorem for monotone operators,* will be established in Chapter 9. □

Under specific assumptions, these objectives will be achieved in Theorems 7.7-1–7.7-3, by means of generalizations of the well-known *Newton's method*[11] for differentiable functions $f : I \subset \mathbb{R} \to \mathbb{R}$, where I is an open interval. This method, defined in this case by the sequence

$$x_{k+1} = x_k - \frac{f(x_k)}{f'(x_k)}, \quad k \geq 0,$$

where the point $x_0 \in I$ is arbitrarily chosen, has an immediate geometric interpretation (Figure 7.7-1): the point x_{k+1} is the intersection of the axis with the tangent to the curve $y = f(x)$, $x \in \Omega$, at the point x_k. Naturally, this method is well defined only if $f'(x_k) \neq 0$ for all $k \geq 0$.

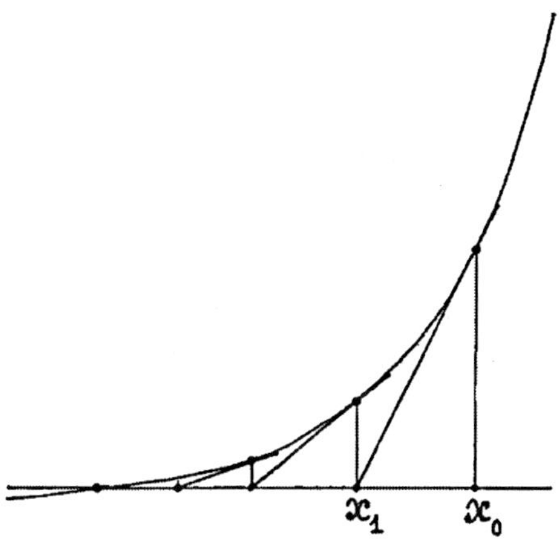

Figure 7.7-1 *Newton's method for a function* $f : I \subset \mathbb{R} \to \mathbb{R}$. Given an arbitrary point $x_0 \in \Omega$, each Newton iterate $x_{k+1} = x_k - (f'(x_k))^{-1}f(x_k)$, $k \geq 0$, is the intersection of the x-axis with the tangent to the curve $y = f(x)$, $x \in \Omega$, at the point x_k. This figure originally appeared in P.G. CIARLET [2007]: *Introduction a l'Analyse Numerique Matricelle et a l'Optimisation*, Dunod, Paris.

[11]This method is due to Sir Isaac Newton (1642–1727), who used it in 1669 for computing zeros of polynomials.

Remark Surprisingly, even in the simplest case where f is a quadratic polynomial, it is not completely obvious to accurately analyze the behavior of the points x_k as $k \to \infty$; see part (i) of the proof of Theorem 7.7-3, where such an analysis is carried out in details on a specific example. □

This simple case suggests the following definition of **Newton's method** for finding the zeros of a differentiable mapping $f : \Omega \subset X \to Y$, where X and Y are now arbitrary *normed vector spaces* and Ω is *open* in X: Given an arbitrary point $x_0 \in \Omega$, the sequence $(x_k)_{k=0}^{\infty}$ is defined by

$$x_{k+1} = x_k - f'(x_k)^{-1} f(x_k), \quad k \geq 0.$$

Of course, this makes sense only if *all the points* x_k, which are called the **Newton iterates** for the mapping f, *remain in* Ω *and the derivatives* $f'(x_k) \in \mathcal{L}(X;Y)$ *are invertible for all* $k \geq 0$.

Remark If the function f is *affine*, i.e., $f(x) := Ax - b$, $x \in X$, for some invertible linear operator $A \in \mathcal{L}(X;Y)$ and some vector $b \in Y$, the iteration described above reduces to the solution of the linear equation $Ax_1 = b$; in other words, Newton's method converges in a *single* iteration in this case. □

Newton's method is thus applicable in particular to the solution of *systems of n nonlinear equations in n unknowns*, which correspond to mappings $\boldsymbol{f} = (f_i) : \Omega \subset \mathbb{R}^n \to \mathbb{R}^n$. In this case, one iteration of Newton's method consists in solving the *linear system*

$$\boldsymbol{f}'(\boldsymbol{x}_k)\boldsymbol{\delta x}_k = -\boldsymbol{f}(\boldsymbol{x}_k), \quad \text{where } \boldsymbol{f}'(\boldsymbol{x}_k) = (\partial_j f_i(\boldsymbol{x}_k)),$$

and then in letting

$$\boldsymbol{x}_{k+1} := \boldsymbol{x}_k + \boldsymbol{\delta x}_k.$$

In practice, it can be costly to calculate at *each* iteration the elements of the new matrix $(\partial_j f_i(\boldsymbol{x}_k))$, and then to solve the corresponding linear system. This observation leads naturally to a *variant* of Newton's method, which consists in *keeping the matrix to be inverted fixed during p consecutive iterations* (where p is some fixed integer ≥ 2), which leads to iterations of the form

$$\begin{aligned}
\boldsymbol{x}_{k+1} &= \boldsymbol{x}_k - \boldsymbol{f}'(\boldsymbol{x}_0)^{-1}\boldsymbol{f}(\boldsymbol{x}_k), \qquad 0 \leq k \leq p-1, \\
\boldsymbol{x}_{k+1} &= \boldsymbol{x}_k - \boldsymbol{f}'(\boldsymbol{x}_p)^{-1}\boldsymbol{f}(\boldsymbol{x}_k), \qquad p \leq k \leq 2p-1, \\
&\vdots \\
\boldsymbol{x}_{k+1} &= \boldsymbol{x}_k - \boldsymbol{f}'(\boldsymbol{x}_{rp})^{-1}\boldsymbol{f}(\boldsymbol{x}_k), \qquad rp \leq k \leq (r+1)p-1.
\end{aligned}$$

One may even never update the matrix, which leads to iterations of the form

$$\boldsymbol{x}_{k+1} = \boldsymbol{x}_k - \boldsymbol{f}'(\boldsymbol{x}_0)^{-1}\boldsymbol{f}(\boldsymbol{x}_k), \quad k \geq 0,$$

or even replace the matrix $\boldsymbol{f}'(\boldsymbol{x}_0)$ by a particular matrix $\boldsymbol{A}_0$ which is "easily invertible," which leads to iterations of the form

$$\boldsymbol{x}_{k+1} = \boldsymbol{x}_k - \boldsymbol{A}_0^{-1}\boldsymbol{f}(\boldsymbol{x}_k), \quad k \geq 0.$$

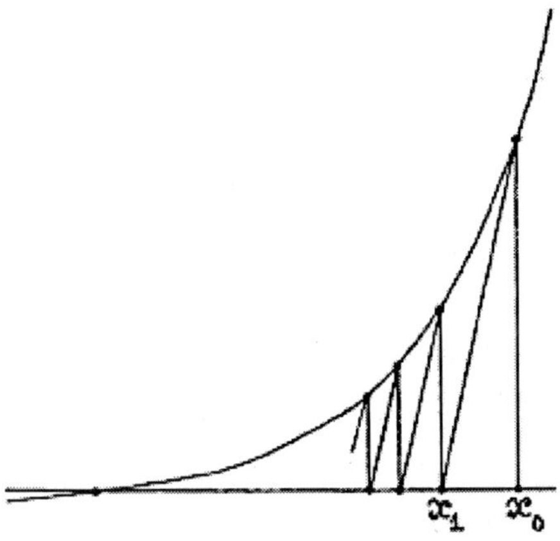

Figure 7.7-2 *A variant of Newton's method for a function $f : I \subset \mathbb{R} \to \mathbb{R}$.* If the initial slope a_0 is sufficiently close to $f'(x_0)$, the sequence $(x_k)_{k=0}^{\infty}$ defined by $x_{k+1} = x_k - a_0^{-1} f(x_k)$, $k \geq 0$, may still converge to a zero of the function f. This figure originally appeared in P.G. CIARLET [2007]: *Introduction a l'Analyse Numerique Matricelle et a l'Optimisation*, Dunod, Paris.

Indeed, in the case of functions $f : I \subset \mathbb{R} \to \mathbb{R}$, convergence may be achieved as long as the initial slope is sufficiently close to $f'(x_0)$ (Figure 7.7-2).

With such variants of Newton's method in mind, we are naturally led to give the following definition of a **generalized Newton's method** for finding the zeros of a function $f : \Omega \subset X \to Y$ from an open subset Ω of a normed vector space X into a normed vector space Y. Given an arbitrary point $x_0 \in \Omega$, and a sequence $(A_k)_{k=0}^{\infty}$ of invertible operators $A_k \in \mathcal{L}(X;Y)$ such that $A_k^{-1} \in \mathcal{L}(Y;X)$ for all $k \geq 0$, the sequence $(x_k)_{k=0}^{\infty}$ is defined by

$$x_{k+1} = x_k - A_k^{-1} f(x_k), \quad k \geq 0.$$

As illustrated by the above examples, the linear operators A_k may, or may not, depend on the function f.

The following theorems provide *sufficient* conditions on the *data* (the function f and its derivative in a neighborhood of the point $x_0 \in \Omega$ and the sequence $(A_k)_{k=0}^{\infty}$) that guarantee the *existence of a zero* of f in a neighborhood of x_0, together with the *convergence of the corresponding generalized Newton's methods* to this zero.

Theorem 7.7-1 (convergence of the generalized Newton's method) *Let there be given two Banach spaces X and Y, an open subset Ω of X, a mapping $f : \Omega \to Y$ differentiable in Ω, and a sequence $(A_k)_{k=0}^{\infty}$ of bijective operators $A_k \in \mathcal{L}(X;Y)$, so that $A_k^{-1} \in \mathcal{L}(Y;X)$.*

Assume that there exist a point $x_0 \in \Omega$ and three constants r, M, β such that

$$r > 0 \quad \textit{and} \quad \overline{B(x_0; r)} \subset \Omega,$$

$$\|A_k^{-1}\|_{\mathcal{L}(Y;X)} \le M \quad \textit{for all } k \ge 0,$$

$$\beta < 1 \quad \textit{and} \quad \|f'(x) - A_k\|_{\mathcal{L}(X;Y)} \le \frac{\beta}{M} \quad \textit{for all } x \in \overline{B(x_0; r)} \textit{ and all } k \ge 0,$$

$$\|f(x_0)\|_Y \le \frac{r}{M}(1 - \beta).$$

Then the sequence $(x_k)_{k=0}^{\infty}$ defined by

$$x_{k+1} := x_k - A_k^{-1} f(x_k), \quad k \ge 0,$$

is contained in the closed ball $\overline{B(x_0; r)}$ and converges as $k \to \infty$ to a zero a of f, which is the only zero of f in $\overline{B(x_0; r)}$. Finally,

$$\|x_k - a\| \le \frac{\beta^k}{1 - \beta} \|x_1 - x_0\|, \quad k \ge 1,$$

and thus the convergence is geometric.

Proof (i) To begin with, we show that, for every integer $k \ge 0$,

$$\|x_{k+1} - x_k\| \le M \|f(x_k)\|,$$

$$\|x_{k+1} - x_0\| \le r,$$

$$\|f(x_{k+1})\| \le \frac{\beta}{M} \|x_{k+1} - x_k\|.$$

In particular then, $x_k \in \overline{B(x_0; r)}$ for all $k \ge 0$, which shows that the *sequence* $(x_k)_{k=0}^{\infty}$ *is well defined.*

The relation

$$x_1 - x_0 = -A_0^{-1} f(x_0)$$

implies that

$$\|x_1 - x_0\| \le M \|f(x_0)\| \le r(1 - \beta) \le r.$$

Since $f(x_1)$ may be also written as

$$f(x_1) = f(x_1) - f(x_0) - A_0(x_1 - x_0),$$

an application of the corollary to the mean value theorem (Theorem 7.2-2) gives

$$\|f(x_1)\| \le \sup_{x \in \overline{B(x_0; r)}} \|f'(x) - A_0\| \|x_1 - x_0\| \le \frac{\beta}{M} \|x_1 - x_0\|.$$

Hence the three announced inequalities hold for $k = 0$. Assume that they hold for $k = 0, \ldots n$, for some integer $n \ge 0$. Since

$$x_{n+1} - x_n = -A_n^{-1} f(x_n),$$

it follows that

$$\|x_{n+1} - x_n\| \leq M \|f(x_n)\|,$$

which shows that the first inequality holds for $k = n + 1$. Since $\|f(x_n)\| \leq \frac{\beta}{M} \|x_n - x_{n-1}\|$ by the induction hypothesis, it further follows that

$$\|x_{n+1} - x_n\| \leq \beta \|x_n - x_{n-1}\| \leq \cdots \leq \beta^n \|x_1 - x_0\|,$$

so that

$$\begin{aligned}\|x_{n+1} - x_0\| &\leq \sum_{\ell=1}^{n+1} \|x_\ell - x_{\ell-1}\| \leq \left(\sum_{\ell=1}^{n+1} \beta^{\ell-1}\right) \|x_1 - x_0\| \\ &\leq \frac{1}{1-\beta} \|x_1 - x_0\| \leq \frac{M}{1-\beta} \|f(x_0)\| \leq r.\end{aligned}$$

Hence the second inequality holds for $k = n + 1$, thus showing that $x_{n+1} \in \overline{B(x_0; r)}$. Since $f(x_{n+1})$ may be also written as

$$f(x_{n+1}) = f(x_{n+1}) - f(x_n) - A_n(x_{n+1} - x_n),$$

another application of the corollary to the mean value theorem gives

$$\|f(x_{n+1})\| \leq \sup_{x \in \overline{B(x_0;r)}} \|f'(x) - A_n\| \|x_{n+1} - x_n\| \leq \frac{\beta}{M} \|x_{n+1} - x_n\|.$$

Hence the three announced inequalities hold for $k = n + 1$.

(ii) We next show that *the mapping f has a zero in the closed ball $\overline{B(x_0; r)}$*. Since

$$\begin{aligned}\|x_{k+\ell} - x_k\| &\leq \sum_{\nu=0}^{\ell-1} \|x_{k+\nu+1} - x_{k+\nu}\| \leq \beta^k \sum_{\nu=0}^{\ell-1} \beta^\nu \|x_1 - x_0\| \\ &\leq \frac{\beta^k}{1-\beta} \|x_1 - x_0\| \quad \text{for all } k, \ell \geq 0,\end{aligned}$$

the sequence $(x_k)_{k=0}^\infty$ is a *Cauchy sequence* in the ball $\overline{B(x_0; r)}$, which is a *complete* metric space (as a closed subset of the complete space X). Therefore there exists a point $a \in \overline{B(x_0; r)}$ such that $\lim_{k\to\infty} x_k = a$. The mapping f being continuous in Ω (since f is differentiable in Ω by assumption),

$$\|f(a)\| = \lim_{k\to\infty} \|f(x_k)\| \leq \frac{\beta}{M} \lim_{k\to\infty} \|x_k - x_{k-1}\| = 0.$$

Hence $f(a) = 0$. Letting ℓ tend to ∞ further shows that

$$\|x_k - a\| \leq \frac{\beta^k}{1-\beta} \|x_1 - x_0\| \quad \text{for each } k \geq 1,$$

as announced.

(iii) Finally, we show that a is the only zero of f in the closed ball $\overline{B(x_0;r)}$.

Let $b \in \overline{B(x_0;r)}$ be a zero of f. Since $f(a) = f(b) = 0$, the difference $(b-a)$ may be also written as

$$b - a = -A_0^{-1}(f(b) - f(a) - A_0(b-a)),$$

so that, by yet another application of the corollary to the mean value theorem,

$$\|b-a\| \leq \|A_0^{-1}\| \sup_{x \in \overline{B(x_0;r)}} \|f'(x) - A_0\| \, \|b-a\| \leq \beta \, \|b-a\| \, ,$$

which implies that $a = b$, since $\beta < 1$. □

The particular choice $A_k := A_0$ for all $k \geq 0$ in Theorem 7.7-1 is simply tantamount to regarding a zero of the mapping f in $\overline{B(x_0;r)}$ as a *fixed point* of the particular mapping (see Problem 7.7-2)

$$g : x \in \overline{B(x_0;r)} \to g(x) := x - A_0^{-1} f(x) \in Y.$$

The particular choice $A_k := f'(x_k)$ for each $k \geq 0$ in Theorem 7.7-1, which thus corresponds to the original *Newton's method*, is more illuminating. It yields the following important corollary to this theorem, where all the assumptions are now made on $f(x_0)$ and on the mappings f' and $(f')^{-1}$ in a neighborhood of the point x_0.

Theorem 7.7-2 (convergence of Newton's method) *Let there be given two Banach spaces X and Y, an open subset Ω of X, a point $x_0 \in \Omega$, and a mapping $f : \Omega \subset X \to Y$ differentiable in Ω. Assume that there exist three constants r, M, and β such that*

$$r > 0 \text{ and } \overline{B(x_0;r)} \subset \Omega,$$

$$f'(x) \in \mathcal{L}(X;Y) \text{ is a bijection, so that } (f'(x))^{-1} \in \mathcal{L}(Y;X) \text{ at each } x \in \overline{B(x_0;r)},$$

$$\|(f'(x))^{-1}\|_{\mathcal{L}(Y;X)} \leq M \quad \text{for all } x \in \overline{B(x_0;r)},$$

$$\beta < 1 \quad \text{and} \quad \|f'(\widetilde{x}) - f'(x)\|_{\mathcal{L}(X;Y)} \leq \frac{\beta}{M} \quad \text{for all } \widetilde{x}, x \in \overline{B(x_0;r)},$$

$$\|f(x_0)\|_Y \leq \frac{r}{M}(1-\beta).$$

Then the sequence $(x_k)_{k=0}^{\infty}$ defined by

$$x_{k+1} := x_k - f'(x_k)^{-1} f(x_k), \quad k \geq 0,$$

is contained in the closed ball $\overline{B(x_0;r)}$ and converges as $k \to \infty$ to a zero a of f, which is the only zero of f in $\overline{B(x_0;r)}$. Finally,

$$\|x_k - a\| \leq \frac{\beta^k}{1-\beta} \|x_1 - x_0\| \quad \text{for each } k \geq 1,$$

and thus the convergence is geometric. □

If the mapping f' is Lipschitz-continuous in a neighborhood of x_0 with a sufficiently small Lipschitz constant, the assumption in Theorem 7.7-2 that $(f'(x))^{-1}$ exists and satisfies $\|(f'(x))^{-1}\|_{\mathcal{L}(Y;X)} \leq M$ for all $x \in \overline{B(x_0;r)}$ can be replaced by the single assumption that

$(f'(x_0))^{-1} \in \mathcal{L}(Y;X)$ exists (in which case $(f'(x))^{-1} \in \mathcal{L}(Y;X)$ also exists for all x in a sufficiently small neighborhood of x_0; cf. Theorem 3.6-3), according to the following result, whose proof is more delicate than those of Theorems 7.7-1 and 7.7-2, however.

The following result is a *basic theorem of nonlinear functional analysis*, as well as a *basic theorem of numerical analysis.*

Theorem 7.7-3 (Newton–Kantorovich theorem in a Banach space[12]**)** *Let there be given two Banach spaces X and Y, an open subset Ω of X, a point $x_0 \in \Omega$, and a mapping $f \in \mathcal{C}^1(\Omega;Y)$ such that*

$$f'(x_0) \in \mathcal{L}(X;Y) \text{ is a bijection, so that } (f'(x_0))^{-1} \in \mathcal{L}(Y;X).$$

Assume that there exist three constants λ, μ, ν such that

$$\begin{aligned}
&0 < \lambda\mu\nu \leq \frac{1}{2} \quad \text{and} \quad B(x_0;r) \subset \Omega, \quad \text{where } r := \frac{1}{\mu\nu}, \\
&\|f'(x_0)^{-1} f(x_0)\|_X \leq \lambda, \\
&\|f'(x_0)^{-1}\|_{\mathcal{L}(Y;X)} \leq \mu, \\
&\|f'(\widetilde{x}) - f'(x)\|_{\mathcal{L}(X;Y)} \leq \nu \|\widetilde{x} - x\|_X \quad \text{for all } \widetilde{x}, x \in B(x_0;r).
\end{aligned}$$

Then $f'(x) \in \mathcal{L}(X;Y)$ is a bijection and thus $(f'(x))^{-1} \in \mathcal{L}(Y;X)$ at each $x \in B(x_0;r)$, and the sequence $(x_k)_{k=0}^{\infty}$ defined by

$$x_{k+1} = x_k - (f'(x_k))^{-1} f(x_k), \quad k \geq 0,$$

is contained in the ball $B(x_0;r_-)$, where

$$r_- := \frac{1 - \sqrt{1 - 2\lambda\mu\nu}}{\mu\nu} \leq r,$$

and converges to a zero $a \in \overline{B(x_0;r_-)}$ of f. Besides, for each $k \geq 0$,

$$\|x_k - a\|_X \leq \frac{r}{2^k}\left(\frac{r_-}{r}\right)^{2^k} \text{ if } \lambda\mu\nu < \frac{1}{2}, \quad \text{or} \quad \|x_k - a\|_X \leq \frac{r}{2^k} \text{ if } \lambda\mu\nu = \frac{1}{2}.$$

[12] L.V. Kantorovich [1948]: Functional analysis and applied mathematics, *Uspehi Matematiceskii Nauk (New Series)* **3**, 89–185 (in Russian).

A different proof was later given in Kantorovich & Akilov [1964]. The proof given here, which follows the latter but is simpler, is adapted from:

J.M. Ortega [1968]: The Newton–Kantorovich theorem, *The American Mathematical Monthly* **75**, 658–660.

Interesting complements and more in-depth treatments are found in:

W.C. Rheinboldt [1968]: A unified convergence theory for a class of iterative processes, *SIAM Journal on Numerical Analysis* **5**, 42–63.

W.B. Gragg; R.A. Tapia [1974]: Optimal error bounds for the Newton–Kantorovich theorem, *SIAM Journal on Numerical Analysis* **11**, 10–13.

P. Deuflhard [2004]: *Newton Methods for Nonlinear Problems – Affine Invariance and Adaptive Algorithms*, Springer, Berlin.

J.P. Dedieu [2006]: *Points Fixes, Zéros et la Méthode de Newton*, Springer, Berlin.

If $\lambda\mu\nu < \frac{1}{2}$, *assume in addition that*

$$B(x_0; r_+) \subset \Omega, \quad \textit{where } r_+ := \frac{1+\sqrt{1-2\lambda\mu\nu}}{\mu\nu},$$
$$\|f'(\widetilde{x}) - f'(x)\|_{\mathcal{L}(X;Y)} \le \nu \, \|\widetilde{x} - x\|_X \quad \textit{for all } \widetilde{x}, x \in B(x_0; r_+).$$

Then the point $a \in \overline{B(x_0; r_-)}$ *is the only zero of* f *in* $B(x_0; r_+)$.

If $\lambda\mu\nu = \frac{1}{2}$ *(in which case* $r_- = r = r_+$*), assume in addition that*

$$\overline{B(x_0; r)} \subset \Omega.$$

Then the point $a \in \overline{B(x_0; r)}$ *is the only zero of* f *in* $\overline{B(x_0; r)}$.

Proof For notational brevity, all norms are denoted by the same symbol $\|\cdot\|$ throughout the proof.

Let the numbers t_k, $k \ge 0$, *with* $t_0 := 0$, be the *Newton iterates for the quadratic polynomial*

$$p : t \in \mathbb{R} \to p(t) := \frac{\mu\nu}{2} t^2 - t + \lambda.$$

The key idea of the proof is based on the so-called *majorant method*, which then consists in showing that *the sequence* $(t_k)_{k=0}^\infty$ *majorizes the sequence* $(x_k)_{k=0}^\infty$ *formed by the Newton iterates* $x_{k+1} = x_k - (f'(x_k))^{-1} f(x_k)$, $k \ge 0$, in the sense that

$$\|x_{k+1} - x_k\| \le t_{k+1} - t_k \quad \text{for all } k \ge 0.$$

This property will in turn imply that the sequence $(x_k)_{k=0}^\infty$ converges to a zero a of f, and that

$$\|x_k - a\| \le r_- - t_k \quad \text{for all } k \ge 0,$$

where $r_- = \lim_{k\to\infty} t_k$ is the smallest root of p. This explains why the proof begins with a careful analysis of the behavior of the Newton iterates t_k, $k \ge 0$, for the polynomial p.

(i) *The Newton iterates*

$$t_0 := 0 \quad \text{and} \quad t_{k+1} := t_k - \frac{p(t_k)}{p'(t_k)} = t_k + \frac{\frac{\mu\nu}{2} t_k^2 - t_k + \lambda}{1 - \mu\nu t_k}, \quad k \ge 0,$$

for the polynomial p *satisfy the relations*

$$t_{k+1} - t_k = \frac{\mu\nu (t_k - t_{k-1})^2}{2(1 - \mu\nu t_k)}, \quad k \ge 1,$$
$$r_- - t_{k+1} = \frac{\mu\nu (r_- - t_k)^2}{2(1-\mu\nu t_k)} \quad \text{and} \quad t_{k+1} - t_k \le \frac{\lambda}{2^k}, \quad k \ge 0,$$
$$1 - \mu\nu t_k \ge \frac{1}{2^k} \quad \text{and} \quad r_- - t_k \le \frac{1}{\mu\nu 2^k} (\mu\nu r_-)^{2^k}, \quad k \ge 0,$$

where $r_- := \dfrac{1-\sqrt{1-2\lambda\mu\nu}}{\mu\nu}$ *is the smallest root of* p *if* $\lambda\mu\nu < \dfrac{1}{2}$ (in which case $\mu\nu r_- < 1$) *and* $r_- = r := \dfrac{1}{\mu\nu}$ *is the double root of* p *if* $\lambda\mu\nu = \dfrac{1}{2}$ (in which case $\mu\nu r_- = 1$).

First, it should be clear (e.g., from a figure) that the sequence $(t_k)_{k=0}^{\infty}$ is well defined, strictly increasing, with $t_k < r_- \le \dfrac{1}{\mu\nu}$, so that $1 - \mu\nu t_k > 0$ for all $k \ge 0$ (these properties hold in fact for any $t_0 < \dfrac{1}{\mu\nu}$).

It is immediately seen that the relation that defines t_k in terms of t_{k-1} can be also written as $\dfrac{\mu\nu}{2}t_k^2 - t_k + \lambda = \dfrac{\mu\nu}{2}(t_k - t_{k-1})^2$, $k \ge 1$. Hence

$$t_{k+1} - t_k = \frac{\frac{\mu\nu}{2}t_k^2 - t_k + \lambda}{1 - \mu\nu t_k} = \frac{\mu\nu(t_k - t_{k-1})^2}{2(1-\mu\nu t_k)}, \quad k \ge 1.$$

By definition of t_{k+1}, proving that

$$r_- - t_{k+1} = \frac{\mu\nu(r_- - t_k)^2}{2(1-\mu\nu t_k)}, \quad k \ge 0,$$

is the same as proving that

$$(1-\mu\nu t_k)(t_k - r_-) + \frac{\mu\nu}{2}t_k^2 - t_k + \lambda = -\frac{\mu\nu}{2}t_k^2 + \mu\nu t_k r_- - \frac{\mu\nu}{2}r_-^2, \quad k \ge 0,$$

a relation that certainly holds since it reduces to $\dfrac{\mu\nu}{2}r_-^2 - r_- + \lambda = 0$.

The inequality $t_{k+1} - t_k \le \dfrac{\lambda}{2^k}$ holds for $k = 0$ (it then reduces to $t_1 - t_0 = t_1 = \lambda$). So, assume it holds for $k = 0, \dots, n-1$ for some integer $n \ge 1$. Then

$$t_n = \sum_{j=1}^{n}(t_j - t_{j-1}) \le \lambda \sum_{j=1}^{n} \frac{1}{2^{j-1}} = 2\lambda\left(1 - \frac{1}{2^n}\right),$$

so that

$$1 - \mu\nu t_n \ge 1 - 2\mu\nu\lambda\left(1 - \frac{1}{2^n}\right) \ge 1 - \left(1 - \frac{1}{2^n}\right) = \frac{1}{2^n}.$$

Combining this inequality with the above expression for the difference $(t_{n+1} - t_n)$ and the induction hypothesis therefore gives

$$t_{n+1} - t_n = \frac{\mu\nu(t_n - t_{n-1})^2}{2(1-\mu\nu t_n)} \le \frac{\mu\nu}{2}\left(\frac{\lambda}{2^{n-1}}\right)^2 2^n \le \frac{\lambda}{2^n},$$

since $\lambda\mu\nu \le \dfrac{1}{2}$ by assumption. Hence the inequality $t_{k+1} - t_k \le \dfrac{\lambda}{2^k}$ holds as well for $k = n$.

The inequality $r_- - t_k \le \dfrac{1}{\mu\nu 2^k}(\mu\nu r_-)^{2^k}$ holds for $k = 0$ (it then reduces to $r_- - t_0 = r_- \le r_-$). So, assume it holds for $k = 0, \dots, n$ for some integer $n \ge 0$. Combining the above expression for the difference $(r_- - t_{n+1})$ with the inequality $1 - \mu\nu t_n \geq \frac{1}{2^n}$ (just established above) and the induction hypothesis therefore gives

$$r_- - t_{n+1} = \frac{\mu\nu(r_- - t_n)^2}{2(1-\mu\nu t_n)} \leq \left(\frac{\mu\nu}{2}\right) 2^n \frac{1}{(\mu\nu)^2 2^{2n}} \left((\mu\nu r_-)^{2^n}\right)^2 = \frac{1}{\mu\nu 2^{n+1}} (\mu\nu r_-)^{2^{n+1}}.$$

Hence the inequality $r_- - t_k \leq \frac{1}{\mu\nu 2^k}(\mu\nu r_-)^{2^k}$ holds as well for $k = n+1$.

(ii) *A first functional analytic preliminary: The mapping* $f : \Omega \subset Y$ *satisfies*

$$\|f(\widetilde{x}) - f(x) - f'(x)(\widetilde{x} - x)\| \leq \frac{\nu}{2} \|\widetilde{x} - x\|^2 \quad \text{for all } \widetilde{x}, x \in B(x_0; r).$$

The proof of this inequality rests on the *mean value theorem for functions of class* $\mathcal{C}^1$ *with values in a Banach space* (Theorem 7.6-1), applied to the function $f \in \mathcal{C}^1(\Omega; Y)$ between any two points x and $\widetilde{x}$ in the open subset $B(x_0; r)$ of Ω (as a convex set, the ball $B(x_0; r)$ contains the closed segment $[x, \widetilde{x}]$). This gives

$$f(\widetilde{x}) - f(x) = \int_0^1 f'\left((1-\theta)x + \theta\widetilde{x}\right)(\widetilde{x} - x)\,\mathrm{d}\theta \quad \text{for all } \widetilde{x}, x \in B(x_0; r).$$

Noting that the expression $f(\widetilde{x}) - f(x) - f'(x)(\widetilde{x} - x)$ can be also written as

$$f(\widetilde{x}) - f(x) - f'(x)(\widetilde{x} - x) = \int_0^1 \left(f'((1-\theta)x + \theta\widetilde{x}) - f'(x)\right)(\widetilde{x} - x)\,\mathrm{d}\theta,$$

we conclude that

$$\begin{aligned}\|f(\widetilde{x}) - f(x) - f'(x)(\widetilde{x} - x)\| &\leq \left(\int_0^1 \|f'\left((1-\theta)x + \theta\widetilde{x}\right) - f'(x)\| \, \|\widetilde{x} - x\|\,\mathrm{d}\theta\right) \\ &\leq \int_0^1 \nu\theta \, \|\widetilde{x} - x\|^2\,\mathrm{d}\theta = \frac{\nu}{2}\|\widetilde{x} - x\|^2.\end{aligned}$$

(iii) *A second functional analytic preliminary: Given any* $x \in B(x_0; r)$*, the derivative* $f'(x) \in \mathcal{L}(X;Y)$ *is a bijection from* X *onto* Y*, so that* $(f'(x))^{-1} \in \mathcal{L}(Y;X)$*. Besides,*

$$\|(f'(x))^{-1}\| \leq \frac{\mu}{1 - \mu\nu\,\|x - x_0\|} \quad \text{for all } x \in B(x_0; r).$$

Noting that

$$\|x - x_0\| < \frac{1}{\mu\nu} \quad \text{implies } \|f'(x_0)^{-1}(f'(x) - f'(x_0))\| \leq \mu\nu\,\|x - x_0\| < 1,$$

we infer from Theorem 3.6-3 (which can be applied since X is a Banach space by assumption) that, if $x \in B(x_0; r)$, then $f'(x) \in \mathcal{L}(X;Y)$ is a bijection from X onto Y and $(f'(x))^{-1} \in \mathcal{L}(Y;X)$ with

$$\|(f'(x))^{-1}\| \leq \frac{\|(f'(x_0))^{-1}\|}{1 - \|f'(x_0)^{-1}(f'(x) - f'(x_0))\|} \leq \frac{\mu}{1 - \mu\nu\,\|x - x_0\|}.$$

(iv) *A third—and last—functional analytic preliminary: Define the auxiliary function*

$$g : x \in B(x_0; r) \to g(x) := x - \left(f'(x)\right)^{-1} f(x) \in X$$

(which is unambiguously defined by (iii)). *Then, given any $x \in B(x_0; r)$ such that $g(x) \in B(x_0; r)$, the following estimate holds:*

$$\|g(g(x)) - g(x)\| \leq \frac{\mu\nu \|g(x) - x\|^2}{2(1 - \mu\nu \|g(x) - x_0\|)}.$$

The estimate of (iii) shows that, given any $x \in B(x_0; r)$ such that $g(x) \in B(x_0; r)$,

$$\|g(g(x)) - g(x)\| = \| \left(f'(g(x))\right)^{-1} f(g(x))\| \leq \frac{\mu \|f(g(x))\|}{1 - \mu\nu \|g(x) - x_0\|}.$$

Noting that $f(x) + f'(x)(g(x) - x) = 0$ for all $x \in B(x_0; r)$ by definition of the function g, we infer from (ii) that

$$\|f(g(x))\| = \left\|f(g(x)) - f(x) - f'(x)(g(x) - x)\right\| \leq \frac{\nu}{2} \|g(x) - x\|^2 \quad \text{for all } x \in B(x_0; r).$$

Hence the announced estimate holds.

(v) *The Newton iterates $x_{k+1} := x_k - (f'(x_k))^{-1} f(x_k)$, $k \geq 0$, for the mapping f belong to the ball $B(x_0; r_-)$* (hence they are well defined) *and they satisfy the estimate*

$$\|x_{k+1} - x_k\| \leq t_{k+1} - t_k \quad \text{for all } k \geq 0,$$

where the numbers t_k, $k \geq 0$, are the Newton iterates for the polynomial $p : t \in \mathbb{R} \to \frac{\mu\nu}{2} t^2 - t + \lambda$ when $t_0 = 0$ (see (i)).

The announced properties hold for $k = 0$ since

$$\|x_1 - x_0\| = \|(f'(x_0))^{-1} f(x_0)\| \leq \lambda = t_1 - t_0 < r_-.$$

So, assume that they hold for $k = 0, \ldots, n - 1$ for some integer $n \geq 1$, so that

$$\|x_n - x_0\| \leq \sum_{\ell=0}^{n-1} \|x_{\ell+1} - x_\ell\| \leq \sum_{\ell=0}^{n-1} (t_{\ell+1} - t_\ell) = t_n - t_0 = t_n.$$

Then

$$x_{n+1} := x_n - (f'(x_n))^{-1} f(x_n) = g(x_n)$$

is well defined (since $x_n \in B(x_0; r_-)$ by the induction hypothesis and thus $(f'(x_n))^{-1} \in \mathcal{L}(Y; X)$ is well defined; cf. (iii)). We thus have

$$x_{n+1} - x_n = g(x_n) - g(x_{n-1}) = g(g(x_{n-1})) - g(x_{n-1}),$$

so that, by (iv) (which can be applied since both x_{n-1} and $g(x_{n-1}) = x_n$ belong to $B(x_0; r_-)$ by the induction hypothesis) and (i),

$$\begin{aligned} \|x_{n+1} - x_n\| = \|g(g(x_{n-1})) - g(x_{n-1})\| &\leq \frac{\mu\nu \|g(x_{n-1}) - x_{n-1}\|^2}{2(1 - \mu\nu \|g(x_{n-1}) - x_0\|)} \\ = \frac{\mu\nu \|x_n - x_{n-1}\|^2}{2(1 - \mu\nu \|x_n - x_0\|)} &\leq \frac{\mu\nu (t_n - t_{n-1})^2}{2(1 - \mu\nu t_n)} = t_{n+1} - t_n. \end{aligned}$$

Finally,

$$\|x_{n+1} - x_0\| \le \sum_{\ell=0}^{n} \|x_{\ell+1} - x_\ell\| \le t_{n+1} < r_-,$$

which shows that the announced properties hold for $k = n$.

(vi) *The Newton iterates $x_k \in B(x_0; r_-)$, $k \ge 0$, converge to a zero $a \in \overline{B(x_0; r_-)}$ of f, and*

$$\|a - x_k\| \le \frac{1}{\mu\nu 2^k}(\mu\nu r_-)^{2^k}, \quad k \ge 0.$$

Since

$$\|x_m - x_n\| \le \sum_{k=n}^{m-1} \|x_{k+1} - x_k\| \le t_m - t_n \quad \text{for all } m > n \ge 0,$$

and the sequence $(t_k)_{k=0}^{\infty}$ converges as $k \to \infty$ (to r_-), the sequence $(x_k)_{k=0}^{\infty}$ is a Cauchy sequence in the complete metric space $\overline{B(x_0; r_-)}$. Hence *the sequence $(x_k)_{k=0}^{\infty}$ converges to a point $a \in \overline{B(x_0; r_-)}$*. Besides,

$$\begin{aligned}\|f(x_k)\| &= \|f'(x_k)(x_{k+1} - x_k)\| \le \left(\|f'(x_0)\| + \|f'(x_k) - f'(x_0)\|\right) \|x_{k+1} - x_k\| \\ &\le \left(\|f'(x_0)\| + \nu \|x_k - x_0\|\right) \|x_{k+1} - x_k\| \le \left(\|f'(x_0)\| + \nu r_-\right)(t_{k+1} - t_k), \quad k \ge 0.\end{aligned}$$

Consequently, $f(a) = \lim_{k\to\infty} f(x_k) = 0$ (the function f is continuous in $\overline{B(x_0; r_-)}$, since it is differentiable there by assumption). Hence *a is a zero of f*.

Letting $\ell \to \infty$ in the inequality $\|x_\ell - x_k\| \le t_\ell - t_k$ further shows that

$$\|a - x_k\| \le r_- - t_k \quad \text{for each } k \ge 0.$$

Hence the announced estimate for $\|a - x_k\|$ follows from (i).

(vii) *Uniqueness of a zero of f in $B(x_0; r_+)$ when $\lambda\mu\nu < \frac{1}{2}$ under the additional assumptions that $B(x_0; r_+) \subset \Omega$ and*

$$\|f'(\widetilde{x}) - f'(x)\| \le \nu \|\widetilde{x} - x\| \quad \text{for all } \widetilde{x}, x \in B(x_0; r_+).$$

Define the auxiliary function

$$h : x \in \Omega \to h(x) := \left(f'(x_0)\right)^{-1} f(x) \in X,$$

whose zeros are thus the same as those of the function f. Clearly then, $h \in \mathcal{C}^1(\Omega; X)$, and the derivative of h at each $x \in \Omega$ is given by $h'(x) = (f'(x_0))^{-1} f'(x)$, so that in particular,

$$h'(x_0) = \mathrm{id}_X,$$

and

$$\|h'(\widetilde{x}) - h'(x)\| \le \|(f'(x_0))^{-1}\| \, \|f'(\widetilde{x}) - f'(x)\| \le \mu\nu \|\widetilde{x} - x\| \quad \text{for all } \widetilde{x}, x \in B(x_0; r_+).$$

First, we show that, if $\lambda\mu\nu \le \frac{1}{2}$, *the function f has at most one zero in the open ball $B(x_0; r)$*.

To this end, assume that $a, b \in B(x_0; r)$ are such that $f(a) = f(b) = 0$. Then, by the corollary to the mean value theorem (Theorem 7.2-2),

$$\|b-a\| = \|h(b) - h(a) - (b-a)\| \le \Big(\sup_{x \in]a,b[} \|h'(x) - \mathrm{id}_X\| \Big) \|b-a\| .$$

Besides,

$$\sup_{x\in]a,b[} \|h'(x) - \mathrm{id}_X\| = \sup_{x \in]a,b[} \|h'(x) - h'(x_0)\| \le \mu\nu \sup_{x\in]a,b[} \|x - x_0\| < \mu\nu r,$$

since

$$\sup_{x\in]a,b[} \|x - x_0\| = \sup_{t \in]0,1[} \|(1-t)(a-x_0) + t(b-x_0)\| \le \max\{\|a-x_0\|, \|b-x_0\|\} < r.$$

But $\mu\nu r = 1$; hence $a = b$.

Second, we show that, *if* $\lambda\mu\nu < \frac{1}{2}$, *the function* f *does not have any zero in the set* $B(x_0; r_+) - \overline{B(x_0; r_-)}$. To this end, we infer from (ii) that

$$\|h(x) - h(x_0) - h'(x_0)(x-x_0)\| \le \frac{\mu\nu}{2} \|x - x_0\|^2 \quad \text{for all } x \in B(x_0; r_+).$$

But $h'(x_0) = \mathrm{id}_X$ and $\|h(x_0)\| \le \lambda$; hence

$$\begin{aligned} \|h(x)\| &\ge \|h(x_0) + h'(x_0)(x-x_0)\| - \frac{\mu\nu}{2}\|x-x_0\|^2 \\ &\ge \|x - x_0\| - \|h(x_0)\| - \frac{\mu\nu}{2}\|x-x_0\|^2 \\ &\ge -\left(\frac{\mu\nu}{2}\|x-x_0\|^2 - \|x-x_0\| + \lambda\right) = -p(\|x-x_0\|) \quad \text{for all } x \in B(x_0, r_+). \end{aligned}$$

Since $p(t) < 0$ for all $r_- < t < r_+$ when $\lambda\mu\nu < \frac{1}{2}$, it follows that

$$\|h(x)\| > 0 \quad \text{for all } r_- < \|x - x_0\| < r_+.$$

Consequently, $f(x) \neq 0$ for all $x \in B(x_0; r_+) - \overline{B(x_0; r_-)}$, on the one hand. Since, on the other hand, f has at most one zero in $B(x_0; r)$, *the zero* $a \in \overline{B(x_0; r_-)}$ *found in* (vi) *is the only zero of* f *in* $B(x_0; r_+)$ *if* $\lambda\mu\nu < \frac{1}{2}$.

If $\lambda\mu\nu = \frac{1}{2}$, the preceding analysis only shows that, if it so happens that the zero a found in (vi) belongs to the *open* ball $B(x_0; r)$, then a is the only zero of f in this open ball; but no conclusion about uniqueness can be reached if $a \in \partial B(x_0; r)$. This is why this case is treated separately, in the next—and last—step of this proof.

(viii) *Uniqueness of a zero of* f *when* $\lambda\mu\nu = \frac{1}{2}$, *under the additional assumption that* $\overline{B(x_0, r)} \subset \Omega$.

First, we notice that

$$\|f'(\widetilde{x}) - f'(x)\| \le \nu \|\widetilde{x} - x\| \quad \text{for all } \widetilde{x}, x \in \overline{B(x_0; r)},$$

since this inequality, which holds by assumption for all $\widetilde{x}, x \in B(x_0; r)$, can be extended by continuity to $\overline{B(x_0;r)}$ if $\overline{B(x_0;r)} \subset \Omega$.

Our objective is to show that, when $\lambda\mu\nu = \frac{1}{2}$, the zero $a \in \overline{B(x_0;r)}$ found in (vi) is the only zero of f in $\overline{B(x_0;r)}$. To this end, we establish that, when $\lambda\mu\nu = \frac{1}{2}$, *if any point* $b \in \overline{B(x_0;r)}$ *satisfies* $f(b) = 0$, *then the Newton iterates* $x_{k+1} = x_k - (f'(x_k)^{-1})f(x_k)$, $k \geq 0$, *satisfy*

$$\|b - x_k\| \leq \frac{r}{2^k} \quad \text{for all } k \geq 0.$$

Clearly, this relation holds for $k = 0$; so, assume that it holds for $k = 0, \ldots, n$ for some integer $n \geq 0$. Since $f(b) = 0$ we may write $\|b - x_{n+1}\|$ as

$$\|b - x_{n+1}\| = \|(f'(x_n))^{-1}(f(b) - f(x_n) - f'(x_n)(b - x_n))\|,$$

and thus, from (ii) and the induction hypothesis,

$$\begin{aligned}\|b - x_{n+1}\| &\leq \|(f'(x_n))^{-1}\| \, \|f(b) - f(x_n) - f'(x_n)(b - x_n)\| \\ &\leq \frac{\nu}{2} \|(f'(x_n))^{-1}\| \, \|b - x_n\|^2 \leq \frac{\nu r^2}{2^{2n+1}} \|(f'(x_n))^{-1}\|.\end{aligned}$$

Besides, the inequality established in (iii) shows that, in particular,

$$\|(f'(x_n))^{-1}\| \leq \frac{\mu}{1 - \mu\nu \|x_n - x_0\|}.$$

Recalling that $t_0 = 0$ and $t_{k+1} - t_k \leq \frac{\lambda}{2^k}$, $k \geq 0$, and that $\|x_n - x_0\| \leq t_n$ (see (i) and (v)), we next infer that

$$\|x_n - x_0\| \leq t_n \leq \lambda\left(1 + \frac{1}{2} + \cdots + \frac{1}{2^{n-1}}\right) = 2\lambda\left(1 - \frac{1}{2^n}\right).$$

Therefore,

$$\|b - x_{n+1}\| \leq \left(\frac{\mu\nu r}{(1 - 2\lambda\mu\nu\,(1 - 2^{-n}))\, 2^n}\right)\frac{r}{2^{n+1}} = \frac{r}{2^{n+1}},$$

since $\mu\nu r = 2\lambda\mu\nu = 1$. Hence

$$\|b - x_k\| \leq \frac{r}{2^k} \quad \text{for all } k \geq 0.$$

Consequently,

$$\|b - a\| = \lim_{k\to\infty} \|b - x_k\| = 0,$$

which shows that $b = a$. This completes the proof. □

Remarks (1) The equalities established for the differences $(t_{k+1} - t_k)$ and $(r_- - t_{k+1})$ at the beginning of part (i) hold in effect *for any* $t_0 < r$ (i.e., not only for $t_0 = 0$), while the estimates established for the differences $(t_{k+1} - t_k), (1 - \mu\nu t_k)$, and $(r_- - t_k)$ in the same part (i) crucially depend on the assumption that $t_0 = 0$.

(2) The assumption that Y is complete is essential for establishing the estimate of (ii). If Y is not complete but f is *twice* differentiable in Ω with $\sup_{x\in\overline{B(x_0;r)}}\|f''(x)\| \le \nu$, the estimate of (ii) then follows from the *generalized mean value theorem* (which will be proved later in this chapter; cf. Theorem 7.9-1(b)).

(3) The inequality $\|f'(x_0)^{-1}f(x)\| \ge -p(\|x-x_0\|)$ for all $x \in B(x_0;r_+)$ established in part (vii) of the above proof provides a motivation for the explicit form of the polynomial p. □

It is worth emphasizing that the Newton–Kantorovich theorem thus provides not only an iterative procedure for approximating solutions of nonlinear equations, but also an *existence theory* for such equations. See Problem 7.7-4, where this observation is illustrated by means of a nonlinear two-point boundary value problem.

We now show[13] how the number of constants appearing in the assumptions of the classical Newton–Kantorovich theorem can be reduced from three to two, then from two to one, thanks to a very simple change in the formulation of the assumptions.

Theorem 7.7-4 (Newton–Kantorovich theorem "with only two constants") *Let there be given two Banach spaces X and Y, an open subset Ω of X, a point $x_0 \in \Omega$, and a mapping $f \in \mathcal{C}^1(\Omega;Y)$ such that*

$$f'(x_0) \in \mathcal{L}(X;Y) \quad \textit{is a bijection, so that } f'(x_0)^{-1} \in \mathcal{L}(Y;X).$$

Assume that there exist two constants λ and r such that

$$0 < \lambda \le \frac{r}{2} \quad \textit{and} \quad B(x_0;r) \subset \Omega,$$
$$\|f'(x_0)^{-1}f(x_0)\|_X \le \lambda,$$
$$\|f'(x_0)^{-1}(f(\widetilde{x}) - f'(x))\|_{\mathcal{L}(X)} \le \frac{1}{r}\|\widetilde{x} - x\|_X \quad \textit{for all } \widetilde{x}, x \in B(x_0;r).$$

Then $f'(x) \in \mathcal{L}(X;Y)$ is a bijection and thus $f'(x)^{-1} \in \mathcal{L}(Y;X)$ at each $x \in B(x_0;r)$, and the sequence $(x_k)_{k=0}^{\infty}$ defined by

$$x_{k+1} = x_k - f'(x_k)^{-1}f(x_k), \quad k \ge 0,$$

is such that

$$x_k \in B(x_0;r_-) \quad \textit{for all } k \ge 0, \textit{ where } r_- := r\left(1 - \sqrt{1 - \frac{2\lambda}{r}}\right) \le r,$$

and converges to a zero $a \in \overline{B(x_0;r_-)}$ of f. Besides, for each $k \ge 0$,

$$\|x_k - a\| \le \frac{r}{2^k}\left(\frac{r_-}{r}\right)^{2^k} \textit{ if } 0 < \lambda < \frac{r}{2}, \quad \textit{or} \quad \|x_k - a\| \le \frac{r}{2^k} \textit{ if } \lambda = \frac{r}{2}.$$

[13]The rest of this section is based on:
P.G. CIARLET; C. MARDARE [2012]: The Newton–Kantorovich theorem, *Analysis and Applications* **10**, 249–269.

If $0 < \lambda < \dfrac{r}{2}$, *assume in addition that*

$$B(x_0; r_+) \subset \Omega, \quad \textit{where } r_+ := r\left(1 + \sqrt{1 - \frac{2\lambda}{r}}\right),$$
$$\|f'(x_0)^{-1}(f(\widetilde{x}) - f'(x))\|_{\mathcal{L}(X)} \leq \frac{1}{r}\|\widetilde{x} - x\|_X \quad \textit{for all } \widetilde{x}, x \in B(x_0; r_+).$$

Then the point $a \in \overline{B(x_0; r_-)}$ *is the only zero of* f *in* $B(x_0; r_+)$.

If $\lambda = \dfrac{r}{2}$ *(in which case* $r_- = r = r_+$*), assume in addition that* $\overline{B(x_0; r)} \subset \Omega$. *Then the point* $a \in \overline{B(x_0; r)}$ *is the only zero of* f *in* $\overline{B(x_0; r)}$.

Proof Rather than adapting step by step the proof of Theorem 7.7-3 under these new assumptions, it is much quicker to use the following simple observation: With the same notations and assumptions as in Theorem 7.7-3, define (as in part (vii) of its proof) the auxiliary function $h \in \mathcal{C}^1(\Omega; X)$ by

$$h(x) := f'(x_0)^{-1} f(x), \quad x \in \Omega,$$

so that $h'(x) = f'(x_0)^{-1} f'(x)$, $x \in \Omega$. *Then the Newton iterates for the mapping* h *coincide with those for the mapping* f since

$$x_{k+1} - x_k = -h'(x_k)^{-1} h(x_k) = -f'(x_k)^{-1} f(x_k), \quad k \geq 0.$$

It thus suffices to check that the assumptions of Theorem 7.7-3 hold for the function h (instead of the function f). Since in this case we can choose

$$\mu := \|h'(x_0)^{-1}\| = \|\mathrm{id}_X\| = 1,$$

these assumptions are therefore satisfied if there exist two constants λ and ν such that

$$0 < \lambda\nu \leq \frac{1}{2} \text{ and } B(x_0; r) \subset \Omega, \text{ where } r := \frac{1}{\nu},$$
$$\|h(x_0)\| = \|f'(x_0)^{-1} f(x_0)\| \leq \lambda,$$
$$\|h'(\widetilde{x}) - h'(x)\| = \|f'(x_0)^{-1}(f'(\widetilde{x}) - f'(x))\| \leq \nu \|\widetilde{x} - x\| \quad \text{for all } \widetilde{x}, x \in B(x_0; r),$$

which are precisely the assumptions made in Theorem 7.7-4. □

To conclude this analysis, we now give a substantially simpler statement (in that only one constant is needed in its assumptions) and a substantially simpler proof of the Newton–Kantorovich theorem when $\lambda = \frac{r}{2}$. *The advantage of this new proof over the traditional proof is that it altogether avoids the Newton iterates* t_k, $k \geq 0$, *for the quadratic polynomial* p.

Its only drawback is that it does not yield the improved error estimates $\|x_k - a\| \leq \frac{r}{2^k}\left(\frac{r_-}{r}\right)^{2^k}$ that hold when $\lambda < \frac{r}{2}$ (indeed, the Newton iterates t_k, $k \geq 0$, used in the majorant method seem unavoidable in order to obtain such improved error estimates when $\lambda < \frac{r}{2}$). But this shortcoming is more than compensated for by the simplicity of the proof.

Note that, like that of the "classical" Newton–Kantorovich theorem (Theorem 7.7-3), the proof of Theorem 7.7-5 is *self-contained*.

Theorem 7.7-5 (Newton–Kantorovich theorem "with only one constant") *Let there be given two Banach spaces X and Y, an open subset Ω of X, a point $x_0 \in \Omega$, and a mapping $f \in \mathcal{C}^1(\Omega; Y)$ such that*

$$f'(x_0) \in \mathcal{L}(X;Y) \quad \text{is a bijection, so that } f'(x_0)^{-1} \in \mathcal{L}(Y;X).$$

Assume that there exists a constant r such that

$$r > 0 \quad \text{and} \quad \overline{B(x_0;r)} \subset \Omega,$$
$$\|f'(x_0)^{-1}f(x_0)\|_X \le \frac{r}{2},$$
$$\|f'(x_0)^{-1}(f'(\widetilde{x}) - f'(x))\|_{\mathcal{L}(X)} \le \frac{1}{r}\,\|\widetilde{x} - x\|_X \quad \text{for all } \widetilde{x}, x \in B(x_0;r).$$

Then $f'(x) \in \mathcal{L}(X;Y)$ is a bijection and thus $f'(x)^{-1} \in \mathcal{L}(Y;X)$ at each $x \in B(x_0;r)$, and the sequence $(x_k)_{k=0}^{\infty}$ defined by

$$x_{k+1} = x_k - (f'(x_k))^{-1}f(x_k), \quad k \ge 0,$$

is such that $x_k \in B(x_0;r)$ for all $k \ge 0$ and converges to a zero $a \in \overline{B(x_0;r)}$ of f. Besides, for each $k \ge 0$,

$$\|x_k - a\| \le \frac{r}{2^k},$$

and the point $a \in \overline{B(x_0;r)}$ is the only zero of f in $\overline{B(x_0;r)}$.

Proof As in the proofs of Theorems 7.7-3 and 7.7-4, we introduce the auxiliary function $h \in \mathcal{C}^1(\Omega;X)$ defined by $h(x) := f'(x_0)^{-1}f(x)$, $x \in \Omega$, so that $h'(x) = f'(x_0)^{-1}f'(x) \in \mathcal{L}(X)$, $x \in \Omega$, and $h'(x_0) = \mathrm{id}_X$. In terms of the function h, the assumptions of Theorem 7.7-5 therefore become

$$\|h(x_0)\| \le \frac{r}{2} \quad \text{and} \quad \|h'(\widetilde{x}) - h'(x)\| \le \frac{1}{r}\,\|\widetilde{x} - x\| \text{ for all } \widetilde{x}, x \in B(x_0;r).$$

(i) *The following estimates hold:*

$$\|h'(x)^{-1}\| \le \frac{1}{1 - \|x - x_0\|/r} \quad \text{for all } x \in B(x_0;r),$$
$$\|h(\widetilde{x}) - h(x) - h'(x)(\widetilde{x} - x)\| \le \frac{1}{2r}\,\|\widetilde{x} - x\|^2 \quad \text{for all } \widetilde{x}, x \in \overline{B(x_0;r)}.$$

By assumption,

$$\|h'(x) - h'(x_0)\| = \|h'(x) - \mathrm{id}_X\|_{\mathcal{L}(X)} \le \frac{1}{r}\,\|x - x_0\| < 1 \quad \text{at each } x \in B(x_0;r).$$

Therefore, at each $x \in B(x_0;r)$, the derivative $h'(x) \in \mathcal{L}(X)$ is a bijection, and by Theorem 3.6-3,

$$\|h'(x)^{-1}\| \le \frac{\|h'(x_0)^{-1}\|}{1 - \|h'(x_0)^{-1}(h'(x) - h'(x_0))\|}$$
$$= \frac{1}{1 - \|h'(x) - h'(x_0)\|} \le \frac{1}{1 - \|x - x_0\|/r}.$$

Hence the first estimate holds.

Using the mean value theorem for functions of class $\mathcal{C}^1$ with values in a Banach space (Theorem 7.6-1), we next have

$$\|h(\widetilde{x}) - h(x) - h'(x)(\widetilde{x} - x)\| = \left\| \int_0^1 \left(h'((1-t)x + t\widetilde{x}) - h'(x)\right)(\widetilde{x} - x)\,\mathrm{d}t \right\|$$

$$\leq \left(\int_0^1 \|h'((1-t)x + t\widetilde{x}) - h'(x)\|\,\mathrm{d}t \right) \|\widetilde{x} - x\|$$

$$\leq \frac{1}{r}\left(\int_0^1 t\,\mathrm{d}t \right) \|\widetilde{x} - x\|^2 = \frac{1}{2r}\|\widetilde{x} - x\|^2 \quad \text{for all } \widetilde{x}, x \in B(x_0; r).$$

But the above inequality holds as well for all $\widetilde{x}, x \in \overline{B(x_0; r)}$ since the functions appearing on each side are continuous. Hence the second estimate holds.

(ii) *The Newton iterates* x_k, $k \geq 0$, *for the function* h, *which are the same as those for the function* f, *belong to the open ball* $B(x_0; r)$ (hence they are well defined) *and they satisfy the following estimates for all* $k \geq 1$:

$$\|x_k - x_{k-1}\| \leq \frac{r}{2^k}, \qquad \|x_k - x_0\| \leq r\left(1 - \frac{1}{2^k}\right),$$

$$\|h'(x_k)^{-1}\| \leq 2^k, \qquad \|h(x_k)\| \leq \frac{r}{2^{2k+1}}.$$

First, let us check that the above estimates hold for $k = 1$. Clearly, the point $x_1 = x_0 - h'(x_0)^{-1}h(x_0) = x_0 - h(x_0)$ is well defined since $h'(x_0)$ is invertible. Besides,

$$\|x_1 - x_0\| = \|h(x_0)\| \leq \frac{r}{2},$$

and, by (i),

$$\|(h'(x_1))^{-1}\| \leq \frac{1}{1 - \|x_1 - x_0\|/r} \leq 2.$$

By definition of x_1, and by (i) again,

$$\|h(x_1)\| = \|h(x_1) - h(x_0) - h'(x_0)(x_1 - x_0)\| \leq \frac{1}{2r}\|x_1 - x_0\|^2 \leq \frac{r}{2^3}.$$

So, assume that the estimates hold for $k = 1, \ldots, n$ for some integer $n \geq 1$. The point $x_{n+1} = x_n - h'(x_n)^{-1}h(x_n)$ is thus well defined since $h'(x_n)$ is invertible. Moreover, by the induction hypothesis and by the estimates of (i) (for the third and fourth estimates),

$$\|x_{n+1} - x_n\| \leq \|h'(x_n)^{-1}\|\,\|h(x_n)\| \leq \frac{r}{2^{n+1}},$$

$$\|x_{n+1} - x_0\| \leq \|x_n - x_0\| + \|x_{n+1} - x_n\| \leq r\left(1 - \frac{1}{2^n}\right) + \frac{r}{2^{n+1}} = r\left(1 - \frac{1}{2^{n+1}}\right),$$

$$\|h'(x_{n+1})^{-1}\| \leq \frac{1}{1 - \|x_{n+1} - x_0\|/r} \leq 2^{n+1},$$

$$\|h(x_{n+1})\| = \|h(x_{n+1}) - h(x_n) - h'(x_n)(x_{n+1} - x_n)\|$$

$$\leq \frac{1}{2r}\|x_{n+1} - x_n\|^2 \leq \frac{r}{2^{2(n+1)+1}}.$$

Hence the estimates also hold for $k = n + 1$.

(iii) *The Newton iterates x_k, $k \geq 0$, converge to a zero a of h, hence of f, which belongs to the closed ball $\overline{B(x_0; r)}$. Besides,*

$$\|x_k - a\| \leq \frac{r}{2^k} \quad \text{for all } k \geq 0.$$

The estimates $\|x_k - x_{k-1}\| \leq r/2^k$, $k \geq 1$, established in (ii) clearly imply that $(x_k)_{k=1}^{\infty}$ is a Cauchy sequence. Since $x_k \in B(x_0; r) \subset \overline{B(x_0; r)}$, and $\overline{B(x_0; r)}$ is a complete metric space (as a closed subset of the Banach space X), there exists $a \in \overline{B(x_0; r)}$ such that

$$a = \lim_{k \to \infty} x_k.$$

Since $\|h(x_k)\| \leq r/2^{2k+1}$, $k \geq 1$, by (ii), and h is a continuous function,

$$h(a) = \lim_{k \to \infty} h(x_k) = 0.$$

Hence the point a is a zero of f.

Given integers $k \geq 1$ and $\ell \geq 1$, we have, again by (ii),

$$\|x_k - x_{k+\ell}\| \leq \sum_{j=k}^{\ell+p-1} \|x_{j+1} - x_j\| \leq \sum_{j=k}^{k+p-1} \frac{r}{2^{j+1}} < \sum_{j=k}^{\infty} \frac{r}{2^{j+1}} = \frac{r}{2^k},$$

so that, for each $k \geq 1$,

$$\|x_k - a\| = \lim_{\ell \to \infty} \|x_k - x_{k+\ell}\| \leq \frac{r}{2^k}.$$

(iv) *Uniqueness of a zero of h, hence of f, in the closed ball $\overline{B(x_0; r)}$.*

We first show that, if $b \in \overline{B(x_0; r)}$ is such that $h(b) = 0$, then

$$\|x_k - b\| \leq \frac{r}{2^k} \quad \text{for all } k \geq 0.$$

Clearly, this is true if $k = 0$; so, assume that this inequality holds for $k = 1, \ldots, n$, for some integer $n \geq 0$. Noting that we can write

$$x_{n+1} - b = x_n - h'(x_n)^{-1} h(x_n) - b = h'(x_n)^{-1}(h(b) - h(x_n) - h'(x_n)(b - x_n)),$$

we infer from (i) and (ii) and from the induction hypothesis that

$$\|x_{n+1} - b\| \leq \|h'(x_n)^{-1}\| \frac{1}{2r} \|b - x_n\|^2 \leq \frac{r}{2^{n+1}}.$$

Hence the inequality $\|x_k - b\| \leq r/2^k$ holds for all $k \geq 1$. Consequently,

$$\lim_{n \to \infty} \|x_k - b\| = \|a - b\| = 0,$$

which shows that $b = a$. This completes the proof. □

Problems

7.7-1 (1) The computation of the *square root* of a number $\alpha > 0$ can be carried out by applying Newton's method to the function $f : x \in \mathbb{R} \to f(x) := x^2 - \alpha$, which in this case consists in defining a sequence $(x_k)_{k=0}^{\infty}$ by

$$x_{k+1} = \frac{1}{2}\left(x_k + \frac{\alpha}{x_k}\right), \quad k \geq 0.$$

Examine how the convergence of this sequence depends on the initial value x_0.

(2) Given a real number $\alpha \neq 0$, let the sequence $(x_k)_{k=0}^{\infty}$ be defined by

$$x_{k+1} = x_k(2 - \alpha x_k), \quad k \geq 0,$$

where $x_0 \in \mathbb{R}$. Show that this is again Newton's method applied to a particular function, for computing the *inverse* of α. Examine how the convergence of this sequence depends on x_0.

(3) Let $\alpha > 0$. Analyze in the same manner the iterative method

$$x_{k+1} = \frac{1}{3}\left(2x_k + \frac{\alpha}{x_k^2}\right), \quad k \geq 0, \quad \text{with } x_0 \neq 0 \text{ given},$$

which provides a somewhat surprising example, where the iterates x_k, $k \geq 0$, are well defined and converge to $\alpha^{1/3}$, *except for a countably infinite number of initial guesses* x_0.

7.7-2 Assume that $A_k = A_0$ for all $k \geq 0$ in Theorem 7.7-1, whose assumptions thus reduce in this case to

$$\|A_0^{-1}\| \leq M, \quad \sup_{x \in \overline{B(x_0;r)}} \|f'(x) - A_0\| \leq \frac{\beta}{M} \text{ with } \beta < 1, \quad \text{and} \quad \|f(x_0)\| \leq \frac{r}{M}(1-\beta).$$

Show that, under these assumptions, the mapping

$$g : x \in \overline{B(x_0;r)} \to g(x) := x - A_0^{-1}f(x) \in Y$$

maps the set $\overline{B(x_0;r)}$ into itself and is a *contraction* in this set. Hence the associated generalized Newton's method is nothing but the *method of successive approximations* (Section 3.7) applied to the contraction g.

This observation thus provides a direct proof of the convergence of Newton's method in this case.

7.7-3 This problem establishes the convergence of a generalized Newton's method to the zero of a function, *when the existence of this zero is already known.* Let there be given two Banach spaces X and Y, an open subset Ω of X, a mapping $f \in \mathcal{C}^1(\Omega; Y)$, a point $a \in \Omega$ such that

$$f(a) = 0 \quad \text{and} \quad f'(a) \in \mathcal{L}(X;Y) \text{ is a bijection, so that } (f'(a))^{-1} \in \mathcal{L}(Y;X),$$

and a sequence $(A_k)_{k=0}^{\infty}$ of bijections $A_k \in \mathcal{L}(X;Y)$ with the property that

$$\sup_{k \geq 0} \|A_k - f'(a)\|_{\mathcal{L}(X;Y)} \leq \frac{\lambda}{\|(f'(a))^{-1}\|_{\mathcal{L}(Y;X)}} \quad \text{for some } \lambda < \frac{1}{2}$$

(the special case $A_k := f'(x_k)$ for each $k \geq 0$ thus corresponds to Newton's method).

(1) Show that there exists a closed ball $B \subset \Omega$ centered at a such that, given any point $x_0 \in B$, the sequence $(x_k)_{k=0}^{\infty}$ defined by

$$x_{k+1} = x_k - A_k^{-1}f(x_k), \quad k \geq 0,$$

is contained in B and there exists β such that

$$\beta < 1 \quad \text{and} \quad \|x_k - a\| \leq \beta^k \|x_0 - a\| \quad \text{for each } k \geq 0,$$

so that $x_k \to a$ as $k \to \infty$.

(2) Show that a is the only zero of f in B.

7.7-4 Consider the *nonlinear two-point boundary value problem*

$$-u''(t) + u(t)^p = \varphi(t), \quad 0 \leq t \leq 1,$$
$$u(0) = u(1) = 0,$$

where $p \geq 2$ is an integer and $\varphi \in \mathcal{C}[0,1]$ is a given function. Note that the results of this problem also apply to the problem $-u''(t) - u(t)^p = \varphi(t)$, $0 \leq t \leq 1$, and $u(0) = u(1) = 0$.

As shown in the proof of Theorem 3.9-1, finding a solution $u \in \mathcal{C}^2[0,1]$ to such a boundary value problem is the same as finding a solution $u \in \mathcal{C}[0,1]$ to a *nonlinear integral equation*, which in this case takes the form

$$u(t) = \int_0^1 G(t,\xi)(\varphi(\xi) - u(\xi)^p)\,\mathrm{d}\xi, \quad 0 \leq t \leq 1,$$

the function G being defined by $G(t,\xi) := \xi(1-t)$ if $0 \leq \xi \leq t \leq 1$ and $G(t,\xi) := t(1-\xi)$ if $0 \leq t < \xi \leq 1$. Solving this integral equation in turn amounts to *finding a zero of the nonlinear mapping* $f : u \in \mathcal{C}[0,1] \to f(u) \in \mathcal{C}[0,1]$ *defined by*

$$(f(u))(t) = u(t) + \int_0^1 G(t,\xi)(u(\xi)^p - \varphi(\xi))\,\mathrm{d}\xi, \quad 0 \leq t \leq 1.$$

In what follows, the space $X := \mathcal{C}[0,1]$ is equipped with the sup-norm, denoted $\|\cdot\|_X$, which thus makes it a Banach space.

(1) Show that the mapping f is of class $\mathcal{C}^1$, with a Fréchet derivative $f'(u) \in \mathcal{L}(X)$ given by

$$f'(u)v = v + p\int_0^1 G(\cdot,\xi)u(\xi)^{p-1}v(\xi)\,\mathrm{d}\xi \quad \text{for all } v \in X.$$

(2) Let u_0 denote the function equal to zero on $[0,1]$. Show that

$$\|f'(u_0)^{-1}f(u_0)\|_X = \frac{1}{8}\|\varphi\|_X, \quad \|f'(u_0)^{-1}\|_{\mathcal{L}(X)} = 1,$$
$$\|f'(\widetilde{u}) - f'(u)\|_{\mathcal{L}(X)} \leq \frac{1}{8}p(p-1)r^{p-2}\|\widetilde{u} - u\|_X \quad \text{for all } \widetilde{u}, u \in B(u_0;r) \text{ and any } r > 0.$$

(3) Let $r_p = \left(\dfrac{8}{p(p-1)}\right)^{\frac{1}{p-1}}$. Show that, if $\|\varphi\|_X \leq 4r_p$, the assumptions of the Newton–Kantorovich theorem are satisfied. This shows that, in this case, the above nonlinear two-point boundary value problem has a solution and that this solution can be approximated by Newton's method.

(4) Show that, given the kth Newton iterate u_k, finding the $(k+1)$st iterate u_{k+1} amounts to solving the *linear* boundary value problem

$$-u''(t) + p(u_k(t))^{p-1}u(t) = (p-1)(u_k(t))^p - \varphi(t), \quad 0 \leq t \leq 1,$$
$$u(0) = u(1) = 0.$$

Remark As we shall see later (Problem 9.14-3), a powerful existence theorem (based on the theory of monotone operators) for a nonlinear boundary value problem of the form $-u''(t) + f(t,u(t)) = 0$,

$0 \le t \le 1$, and $u(0) = u(1) = 0$ asserts that it has a solution if there exists a constant c such that $\frac{\partial f}{\partial u}(t, v) \ge c > -\pi^2$ for all $0 \le t \le 1$ and $v \in \mathbb{R}$, a condition that is *not* satisfied here if the exponent p is even. The above example thus illustrates the power of the Newton–Kantorovich theorem, seen here as an efficient alternative for proving existence theorems when other approaches fail. □

7.8 Higher order derivatives; Schwarz lemma

Let there be given two normed vector spaces X and Y, an open subset Ω of X, and a mapping $f : \Omega \subset X \to Y$ differentiable in Ω. If the mapping

$$f' : x \in \Omega \subset X \to f'(x) \in \mathcal{L}(X; Y),$$

which is thus well defined in this case, is differentiable at a point $a \in \Omega$, its derivative

$$f''(a) := (f')'(a) \in \mathcal{L}(X; \mathcal{L}(X; Y))$$

is called the **second derivative of f at** a, and f is said to be **twice differentiable at** a. If a mapping $f : \Omega \subset X \to Y$ is twice differentiable at all points of Ω, and if the mapping

$$f'' : x \in \Omega \to f'(x) \in \mathcal{L}(X; \mathcal{L}(X; Y)),$$

which is thus well defined in this case, is continuous, the mapping f is said to be **twice continuously differentiable in** Ω, or simply **of class $\mathcal{C}^2$ in** Ω. The notation

$$\mathcal{C}^2(\Omega; Y), \quad \text{or simply} \quad \mathcal{C}^2(\Omega) \text{ if } Y = \mathbb{R},$$

designates the space of all twice continuously differentiable mappings from Ω into Y.

Since the space $\mathcal{L}(X; \mathcal{L}(X; Y))$ can be identified with the space $\mathcal{L}_2(X; Y)$ of all continuous bilinear mappings from $X \times X$ into Y (Theorem 2.11-5), *the second derivative of f at a can be identified with a continuous bilinear mapping from X into Y*, simply by letting

$$(f''(a)h)k = f''(a)(h, k) \quad \text{for all } h, k \in X.$$

Thanks to yet another application of the *mean value theorem in a normed vector space*, the following generalization of the well-known *Schwarz lemma* for real-valued functions of two real variables can be established.

Theorem 7.8-1 (Schwarz lemma[14]) *Let X and Y be two normed vector spaces, let Ω be an open subset of X, and let $f : \Omega \subset X \to Y$ be a mapping twice differentiable at a point $a \in \Omega$. Then the second derivative $f''(a)$ at a point a is a symmetric bilinear mapping, i.e.,*

$$f''(a)(h, k) = f''(a)(k, h) \quad \textit{for all } h, k \in X.$$

Proof Clearly, the above relation holds if $h = 0$ or if $k = 0$. So, let there be given two vectors $h \ne 0$ and $k \ne 0$. Since Ω is open, there exist $r > 0$ and $t_0 > 0$ such that $B(a; r) \subset \Omega$,

[14]So named after Karl Hermann Amandus Schwarz (1843–1921).

and all the points of the form $a + t(\xi + k)$ and $a + t\xi$ belong to $B(a; r)$ for all $|t| \leq t_0$ and all $\xi \in \overline{B(0; s)}$, where $s := \|h\|$.

For each $|t| \leq t_0$, define a function $g_t : \xi \in \overline{B(0; s)} \to Y$ by

$$g_t(\xi) := f(a + t(\xi + k)) - f(a + t\xi) \quad \text{for all } \xi \in \overline{B(0; s)}.$$

By the chain rule (Theorem 7.1-3), each function g_t, $|t| \leq t_0$, is differentiable in $\overline{B(0; s)}$, with

$$g_t'(\xi) = tf'(a + t(\xi + k)) - tf'(a + t\xi) \quad \text{at each } \xi \in \overline{B(0; s)}.$$

An application of the corollary to the mean value theorem (Theorem 7.2-2) with

$$A := t^2 f''(a)k \in \mathcal{L}(X; Y)$$

gives

$$\|g_t(h) - g_t(0) - Ah\| \leq \Big(\sup_{\xi \in B(0;s)} \|g_t'(\xi) - A\| \Big) \|h\| \quad \text{for each } |t| \leq t_0,$$

where, for each $|t| \leq t_0$ and each $\xi \in B(0; s)$,

$$g_t'(\xi) - A = t(f'(a + t(\xi + k)) - f'(a + t\xi) - tf''(a)k).$$

By definition of the second derivative, $f''(a) = (f')'(a)$. Hence, for each $|t| \leq t_0$ and each $\xi \in B(0; h)$,

$$\begin{aligned} f'(a + t(\xi + k)) &= f'(a) + tf''(a)(\xi + k) + |t| \, \|\zeta + k\| \, \alpha(t, \xi), \\ f'(a + t\xi) &= f'(a) + tf''(a)\xi + |t| \, \beta(t, \xi), \end{aligned}$$

with $\lim_{t\to 0} \big(\sup_{\|\xi\| \leq s} |\alpha(t, \xi)| \big) = \lim_{t \to 0} \big(\sup_{\|\xi\| \leq s} |\beta(t, \xi)| \big) = 0$. Consequently,

$$\sup_{\xi \in B(0;s)} \|g_t'(\xi) - A\| = t^2 \varepsilon(t) \quad \text{with } \lim_{t \to 0} \varepsilon(t) = 0,$$

which in turn implies that, for each $|t| \leq t_0$,

$$\left\| \frac{g_t(h) - g_t(0)}{t^2} - \big(f''(a)k\big)\, h \right\| \leq \varepsilon(t) \, \|h\| \quad \text{with } \lim_{t \to 0} \varepsilon(t) = 0.$$

Since then

$$f''(a)(k, h) = (f''(a)k)h = \lim_{t \to 0} \frac{g_t(h) - g_t(0)}{t^2},$$

and since the difference $g_t(h) - g_t(0) = f(a + t(h + k)) - f(a + th) - f(a + tk) + f(a)$ is an expression that is *symmetric with respect to h and k*, it follows that

$$f''(a)(k, h) = f''(a)(h, k),$$

as was to be proved. □

The actual *computation of second derivatives* is often based on the following observation, which in effect reduces it to *two* successive computations of *first* derivatives:

Theorem 7.8-2 *Let X and Y be two normed vector spaces, let Ω be an open subset of X, and let $f : \Omega \subset X \to Y$ be a mapping that is differentiable in Ω and twice differentiable at a point $a \in \Omega$. Then*

$$f''(a)(h,k) = g_k'(a)h \quad \textit{for all } h,k \in X,$$

where, for each vector $k \in X$, the mapping $g_k : \Omega \to Y$ is defined by

$$g_k(x) := f'(x)k \quad \textit{at each } x \in \Omega.$$

Proof Let h and k be two vectors in X. The mapping $g_k : \Omega \to Y$ is in effect a composition mapping $g_k = \psi_k \circ \varphi$, with

$$\varphi : x \in \Omega \subset X \to \varphi(x) := f'(x) \in \mathcal{L}(X;Y) \quad \text{and} \quad \psi_k : A \in \mathcal{L}(X;Y) \to \psi_k(A) := Ak \in Y.$$

Since φ is differentiable at a (by assumption, f is twice differentiable at a), with

$$\varphi'(a)h = f''(a)h \in \mathcal{L}(X;Y) \quad \text{for all } h \in X,$$

and ψ_k is differentiable in $\mathcal{L}(X;Y)$ (as a continuous linear mapping), with

$$\psi_k'(A)B = Bk \quad \text{for all } A, B \in \mathcal{L}(X;Y),$$

the *chain rule* shows that the mapping $g_k = \psi_k \circ \varphi$ is differentiable at a, with

$$g_k'(a)h = \psi_k'(f'(a))\varphi'(a)h = (\varphi'(a)h)k = (f''(a)h)k = f''(a)(h,k).$$

Hence the assertion is established. □

The practical rule for computing $f''(a)(h,k)$ therefore consists in *first* computing the derivative of the function $x \in \Omega \subset X \to f'(x)k \in Y$ at the point $x = a$, *then* in applying this derivative to the vector k.

To illustrate this rule, we compute the second derivative of a mapping of the form $f : x \in X \to f(x) := B(x,x)$, where X is a normed vector space and $B : X \times X \to Y$ is a continuous bilinear mapping. As shown in Section 7.1, the mapping $x \in X \to f'(x)k \in Y$ is given in this case by

$$x \in X \to f'(x)k = B(x,k) + B(k,x).$$

Noting that, for a fixed vector $k \in X$, the above mapping is linear and continuous, we thus conclude that

$$f''(a)(h,k) = B(h,k) + B(k,h) \quad \text{for all } h,k \in X.$$

Note that $f''(a)(h,k) = 2B(h,k)$ if the mapping B is in addition *symmetric*.

If $(X, (\cdot,\cdot))$ is a *Hilbert space*, the second derivative of a *real-valued* function $f : \Omega \subset X \to \mathbb{R}$ at a point $a \in \Omega$ can be identified with an element of $\mathcal{L}(X)$, called the **Hessian of f at a** and denoted $\operatorname{Hess} f(a)$: To see this, note that, since $f''(a) \in \mathcal{L}(X;\mathcal{L}(X;\mathbb{R}))$ and since the dual space $X' = \mathcal{L}(X;\mathbb{R})$ can be identified with X (by the F. Riesz representation theorem), it follows in this case that

$$f''(a)(h,k) = (\operatorname{Hess} f(a)h, k) \quad \text{for all } h,k \in X.$$

In the important special case where $X = \mathbb{R}^n$ and $Y = \mathbb{R}$, the real number $f''(a)(h,k)$ can be written as

$$f''(a)(h,k) = \sum_{i,j=1}^{n} \partial_{ij} f(a) h_i k_j \quad \text{for all } h = (h_i)_{i=1}^n \in \mathbb{R}^n \text{ and } k = (k_i)_{i=1}^n \in \mathbb{R}^n,$$

where

$$\partial_{ij} f(a) := f''(a)(e_i, e_j), \quad 1 \le i,j \le n,$$

and $(e_i)_{i=1}^n$ denotes the canonical basis of $\mathbb{R}^n$. The real numbers $\partial_{ij} f(a)$ denote the usual *partial derivatives of the second order* of the function f at a, since by Theorems 7.8-1 and 7.8-2,

$$\partial_{ij} f(a) = \partial_i(\partial_j f)(a) = \partial_{ji} f(a) = \partial_j(\partial_i f)(a), \quad 1 \le i,j \le n,$$

where $\partial_i f(a)$, $1 \le i \le n$, denote the usual partial derivatives of the function f (Section 7.1). For this reason the numbers $\partial_{ij} f(a)$ are also denoted

$$\frac{\partial^2 f}{\partial x_i \partial x_j}(a) := \partial_{ij} f(a) \text{ if } i \ne j \quad \text{and} \quad \frac{\partial^2 f}{\partial x_i^2}(a) := \partial_{ii} f(a) \text{ if } i = j,$$

it being implicitly understood that $x = (x_1, x_2, \dots, x_n)$ denotes a generic point in $\mathbb{R}^n$. In matrix form, we thus have

$$f''(a)(h,k) = (h_1 \cdots h_n) \begin{pmatrix} \partial_{11} f(a) & \cdots & \partial_{1n} f(a) \\ \vdots & & \vdots \\ \partial_{n1} f(a) & \cdots & \partial_{nn} f(a) \end{pmatrix} \begin{pmatrix} k_1 \\ \vdots \\ k_n \end{pmatrix},$$

where the $n \times n$ matrix $(\partial_{ij} f(a))$, which is *symmetric* as a consequence of the Schwarz lemma, is nothing but the Hessian of f at a expressed in the basis $(e_i)_{i=1}^n$ of $\mathbb{R}^n$ equipped with the Euclidean inner product. For this reason, the matrix $(\partial_{ij} f(a))$ is called the **Hessian matrix of f at a.**

Higher order derivatives are similarly defined. Let again X and Y be two normed vector spaces and let Ω be an open subset of X. Recall that, for each integer $k \ge 2$, the space $\mathcal{L}_k(X;Y)$ of all continuous k-linear mappings from X into Y can be identified with the space $\mathcal{L}(X; \mathcal{L}_{k-1}(X;Y))$, where $\mathcal{L}_1(X;Y) = \mathcal{L}(X;Y)$ (Theorem 2.11-5).

Let

$$f^{(0)} := f, \quad f^{(1)} := f', \quad f^{(2)} := f''.$$

The **mth derivative**

$$f^{(m)}(a) \in \mathcal{L}(X; \mathcal{L}_{m-1}(X;Y)) = \mathcal{L}_m(X;Y)$$

at a point $a \in \Omega$ of a mapping $f : \Omega \subset X \to Y$ is then defined by induction for any integer $m \ge 3$ as the derivative at the point a of the mapping

$$f^{(m-1)} : x \in \Omega \to f^{(m-1)}(x) \in \mathcal{L}_{m-1}(X;Y).$$

If the mth derivative $f^{(m)}(a)$ exists, the mapping f is said to be m **times differentiable at the point a.**

The mapping f is said to be m **times differentiable in** Ω if it is m times differentiable at all points in Ω. If the mth *derivative mapping* $f^{(m)} : \Omega \to \mathcal{L}_m(X;Y)$ is continuous, the mapping f is said to be m **times continuously differentiable** in Ω, or **of class** $\mathcal{C}^m$ in Ω. The notation

$$\mathcal{C}^m(\Omega;Y), \quad \text{or simply} \quad \mathcal{C}^m(\Omega) \text{ if } Y = \mathbb{R},$$

designates the space of all m times continuously differentiable mappings from Ω into Y. Note that, like the space $\mathcal{C}(\Omega)$ (Problem 2.3-2), the space $\mathcal{C}^m(\Omega)$ can be equipped with a *metrizable topology* (Problem 7.8-3).

Finally,

$$\mathcal{C}^\infty(\Omega;Y) := \bigcap_{m=0}^{\infty} \mathcal{C}^m(\Omega;Y), \quad \text{or simply} \quad \mathcal{C}^\infty(\Omega) \text{ if } Y = \mathbb{R},$$

designates the space of all **infinitely differentiable mappings** from Ω into Y.

If $f \in \mathcal{C}^m(\Omega;Y)$ for some $1 \leq m \leq \infty$ and if, *in addition*, $f : \Omega \to Y$ is injective, the direct image $f(\Omega)$ is open in Y, and $f^{-1} \in \mathcal{C}^m(f(\Omega);X)$, the mapping f is said to be a $\mathcal{C}^m$**-diffeomorphism** of Ω onto $f(\Omega)$.

Remark An interesting example of a polynomial $\mathcal{C}^\infty$-diffeomorphism of the plane, with an inverse that is also a polynomial mapping, is given in Problem 7.8-4. □

The following theorem gathers properties of higher order derivatives that generalize analogous properties of second-order derivatives. Since their proofs are similar to those of Theorems 7.8-1 and 7.8-2, they are omitted.

Theorem 7.8-3 *Let X and Y be two normed vector spaces, let Ω be an open subset of X, and let $f : \Omega \subset X \to Y$ be a mapping that is m times differentiable at a point $a \in \Omega$ for some integer $m \geq 2$. Then*

$$(f^{(p)})^{(m-p)}(a) = f^{(m)}(a) \quad \textit{for all } 0 \leq p \leq m,$$

$$f^{(m)}(a)(h_1, h_2, \ldots, h_m) = \left(\left(\cdots\left(\left(f'(a)h_m\right)h_{m-1}\right)\cdots\right)h_2\right)h_1 \quad \textit{for all } h_1, h_2, \ldots, h_m \in X.$$

Besides, the mapping $f^{(m)}(a) \in \mathcal{L}_m(X;Y)$ is symmetric, in the sense that (Section 2.11)

$$f^{(m)}(a)(h_1, h_2, \ldots, h_m) = f^{(m)}(h_{\sigma(1)}, h_{\sigma(2)}, \ldots, h_{\sigma(m)})$$

for all $h_1, h_2, \ldots, h_m \in X$ and all permutation $\sigma \in \mathfrak{S}_n$. □

In the special case where $X = \mathbb{R}^n$ and $Y = \mathbb{R}$, the *usual partial derivatives of order m* at a point $a \in \Omega \subset \mathbb{R}^n$ are thus recovered as

$$\frac{\partial^m f}{\partial x_1^{\alpha_1} \partial x_2^{\alpha_2} \cdots \partial x_n^{\alpha_n}}(a) = f^{(m)}(a)(e_1, \ldots, e_1, e_2, \ldots, e_2, \ldots, e_n, \ldots, e_n)$$

where each basis vector e_i of $\mathbb{R}^n$ occurs α_i times, with $0 \leq \alpha_i$, $1 \leq i \leq n$, and $\sum_{i=1}^m \alpha_i = m$. Such a partial derivative can be also written using the *multi-index notation* (Section 1.18), viz.,

$$\frac{\partial^m f}{\partial x_1^{\alpha_1} \partial x_2^{\alpha_2} \cdots \partial x_n^{\alpha_n}}(a) = \partial^{\boldsymbol{\alpha}} f(a) \quad \text{where } \boldsymbol{\alpha} := (\alpha_1, \alpha_2, \ldots, \alpha_m).$$

Note also that in this case there exist constants $C(m,n)$ such that

$$\max_{|\alpha|=m} |\partial^\alpha f(x)| \leq \|f^{(m)}\|_{\mathcal{L}_m(\mathbb{R}^n;\mathbb{R})} \leq C(m,n) \max_{|\alpha|=m} |\partial^\alpha f(x)| \quad \text{for all } x \in \Omega.$$

The following result constitutes a natural complement to the *chain rule* (Theorem 7.1-3).

Theorem 7.8-4 *Let X, Y, Z be normed vector spaces, let U and V be open subsets of the spaces X and Y respectively, let m be an integer ≥ 1, and let $f : U \subset X \to Y$ and $g : V \subset Y \to Z$ be two mappings of class $\mathcal{C}^m$ in U and V respectively, with the property that $f(U) \subset V$. Then the composition mapping $g \circ f : U \subset X \to Z$ is of class $\mathcal{C}^m$ in U.*

Proof The assertion holds if $m = 1$ (Theorem 7.1-3); so, assume that it holds for $m = 1, \ldots, k-1$, for some integer $k \geq 2$.

Let then $f \in \mathcal{C}^k(U;Y)$ and $g \in \mathcal{C}^k(V;Z)$ be two given mappings. Arguing as in the proof of Theorem 7.1-3, we conclude that both mappings $f' : U \to \mathcal{L}(X;Y)$ and $g' \circ f : U \to \mathcal{L}(Y;Z)$ are of class $\mathcal{C}^{k-1}$ (the second one by the induction hypothesis).

Since the bilinear mapping $(A, B) \in \mathcal{L}(X;Y) \times \mathcal{L}(Y;Z) \to B \circ A \in \mathcal{L}(X;Z)$ is of class $\mathcal{C}^{k-1}$ (it is in effect of class $\mathcal{C}^\infty$), we conclude (again from the induction hypothesis) that the composite mapping

$$(g \circ f)' = (g' \circ f) \circ f' : U \to \mathcal{L}(X;Z)$$

is also of class $\mathcal{C}^{k-1}$; hence $(g \circ f) : U \to Z$ is of class $\mathcal{C}^k$. So, the assertion also holds for $m = k$. □

Remark Under the weaker assumptions that f and g are of class $\mathcal{C}^{m-1}$ in U and V respectively, and m times differentiable at a point $a \in U$ and at the point $f(a) \in V$ respectively, a similar argument shows that the composite mapping $g \circ f : U \subset X \to Z$ is m times differentiable at a. □

Examples of mappings of class $\mathcal{C}^\infty$ include *continuous affine mappings*, i.e., of the form

$$f : x \in \Omega \subset X \to f(x) := (Ax + b) \in Y, \quad \text{with } A \in \mathcal{L}(X;Y) \text{ and } b \in Y,$$

in which case $f'(x) = A$ for all $x \in \Omega$ (Section 7.1), and thus $f''(x) = 0 \in \mathcal{L}_2(X;Y)$ for all $x \in \Omega$.

They also include *continuous multilinear mappings* (Section 2.11): Consider for instance a continuous bilinear mapping $B : X_1 \times X_2 \to Y$, in which case (Section 7.1),

$$B'(x_1, x_2)(h_1, h_2) = B(h_1, x_2) + B(x_1, h_2) \quad \text{for all } (x_1, x_2) \in X_1 \times X_2 \text{ and } (h_1, h_2) \in X_1 \times X_2.$$

This relation shows that the mapping

$$(x_1, x_2) \in X_1 \times X_2 \to B'(x_1, x_2) \in \mathcal{L}(X_1 \times X_2; Y)$$

is *linear* (by the assumed bilinearity of B), and *continuous* since

$$\begin{aligned} \|B'(x_1; x_2)\|_{\mathcal{L}(X_1 \times X_2; Y)} &= \sup_{(h_1,h_2) \neq (0,0)} \frac{\|B(h_1, x_2) + B(x_1, h_2)\|}{\|h_1\| + \|h_2\|} \\ &\leq \|B\| \left(\|x_1\| + \|x_2\|\right) \quad \text{for all } (x_1, x_2) \in X_1 \times X_2 \end{aligned}$$

(by the assumed continuity of B; without loss of generality, we assume here that the product space $X_1 \times X_2$ is equipped with the norm $(x_1, x_2) \in X_1 \times X_2 \to \|x_1\|_{X_1} + \|x_2\|_{X_2}$). Hence

$$B'''(x_1, x_2) = 0 \in \mathcal{L}_3(X_1 \times X_2; Y) \quad \text{for all } (x_1, x_2) \in X_1 \times X_2.$$

A similar argument shows that, for any integer $k \geq 3$, any continuous k-linear mapping is of class $\mathcal{C}^\infty$ and that its $(k+1)$st derivative vanishes.

Further important examples of mappings of class $\mathcal{C}^\infty$ will be provided later in this chapter (see in particular Theorems 7.13-2 and 7.14-3).

Problems

7.8-1 Let Ω be a connected open subset of $\mathbb{R}^n$, and let $f : x = (x_1, x_2, \ldots, x_n) \in \Omega \to f(x) \in \mathbb{R}$ be a function that is m times differentiable in Ω for some integer $m \geq 2$. Show that, if $f^{(m)}(x) = 0$ for all $x \in \Omega$, then f is the restriction to Ω of a polynomial of degree $\leq m-1$ of the variables x_i, $1 \leq i \leq n$.

7.8-2 Let $I \subset \mathbb{R}$ be an open interval. Given a mapping $\boldsymbol{F} : t \in I \to \boldsymbol{F}(t) \in \mathbb{M}^n$ such that $\boldsymbol{F}(t)$ is invertible for all $t \in I$, compute the second derivative at a point $t \in I$ of the mapping $t \in I \to (\boldsymbol{F}(t))^{-1} \in \mathbb{M}^n$ in terms of $\boldsymbol{F}(t)^{-1}$, $\boldsymbol{F}'(t)$, and $\boldsymbol{F}''(t)$.

7.8-3 Let Ω be an open subset of $\mathbb{R}^n$ and let $m \geq 1$ be an integer. Given any function $f \in \mathcal{C}^m(\Omega)$ and any compact subset K of Ω, let

$$|f|_{m,K} := \sup_{\begin{cases} x \in K \\ |\alpha| \leq m \end{cases}} |\partial^\alpha f(x)|.$$

Note that each mapping $|\cdot|_{m,K} : \mathcal{C}^m(\Omega) \to \mathbb{R}$ thus defined is a *seminorm*, but *not* a norm, on the space $\mathcal{C}^m(\Omega)$.

Let $(K_i)_{i=1}^\infty$ be a sequence of compact subsets of Ω such that $K_i \subset \operatorname{int} K_{i+1}$ for all $i \geq 1$ and $\Omega = \bigcup_{i=1}^\infty K_i$ (as in Problem 2.3-2), and let $\sum_{i=1}^\infty \alpha_i$ with $\alpha_i > 0$ for all $i \geq 1$ be a convergent series.

(1) Show that the mapping $d_m : \mathcal{C}^m(\Omega) \times \mathcal{C}^m(\Omega) \to \mathbb{R}$ defined by

$$d_m(f, g) = \sum_{i=1}^\infty \alpha_i \frac{|f - g|_{m,K_i}}{1 + |f - g|_{m,K_i}} \quad \text{for each } f, g \in \mathcal{C}^m(\Omega)$$

is a distance on the space $\mathcal{C}^m(\Omega)$.

(2) Show that a sequence $(f_k)_{k=1}^\infty$ of functions $f_k \in \mathcal{C}^m(\Omega)$ converges to a function $f \in \mathcal{C}^m(\Omega)$ in the metric space $(\mathcal{C}^m(\Omega), d_m)$ if and only if

$$\lim_{k \to \infty} |f_k - f|_{m,K} = 0 \quad \text{for each compact subset } K \subset \Omega.$$

This problem thus shows that the space $\mathcal{C}^m(\Omega)$ can be equipped with a *metrizable topology*, called the *Fréchet topology associated with the family of seminorms* $(|\cdot|_{m,K})_{K \in \mathcal{K}}$, where $\mathcal{K}$ denotes the family of all compact subsets of Ω.

7.8-4 The **Hénon map**[15] is defined by

$$f : (x, y) \in \mathbb{R}^2 \to (y + x^2 + a, -bx) \in \mathbb{R}^2$$

[15] M. Hénon [1976]: A two-dimensional mapping with a strange attractor, *Communications in Mathematics and Physics* **50**, 69–77.

where a and $b \neq 0$ are real constants.

(1) Show that the Hénon map is a $\mathcal{C}^\infty$-diffeomorphism of the plane.

(2) Show that any composition $g := f \circ f \circ \cdots \circ f : \mathbb{R}^2 \to \mathbb{R}^2$ of Hénon maps has an inverse $g^{-1} : \mathbb{R}^2 \to \mathbb{R}^2$ whose components are polynomials of the same degree as those of g.

Remark The Hénon map provides the simplest example of a nontrivial $\mathcal{C}^\infty$-diffeomorphism of the plane; it has been extensively studied because, in spite of its simplicity, it displays some essential features of general *dynamical systems*. □

7.9 Taylor formulas; application to extrema of real-valued functions

We now state and prove several *Taylor formulas*[16] *in normed vector spaces*. The first one generalizes the definition of the derivative; the second one generalizes the mean value theorem; the third and fourth ones give explicit forms of the remainder; more specifically, the third one is a generalization of the classical mean value theorem, viz., $f(a+h) - f(a) = hf'(a+\theta h)$ for some $0 < \theta < 1$, and the fourth one generalizes the formula $f(a+h) - f(a) = \int_a^{a+h} f'(\eta)\,\mathrm{d}\eta = \int_0^1 f'(a+th)h\,\mathrm{d}t$, which both apply to real-valued functions of one real variable.

These Taylor formulas in normed vector spaces play in particular a key role in the local analysis of real-valued functions (as shown at the end of this section), in the derivation of the maximum principle for elliptic operators (as shown in the next section), or in the interpolation theory for multivariate functions (as shown in Section 7.11).

Given normed vector spaces X and Y, an open subset Ω of X, and a mapping $f : \Omega \subset X \to Y$ that is m times differentiable at a point $a \in \Omega$, the shorter notation

$$f^{(m)}(a)h^m := f^{(m)}(a)(h, h, \dots, h) \in Y$$

will be used whenever the m vectors in the space X found in the m-uple, to which the m-linear mapping $f^{(m)}(a) \in \mathcal{L}_m(X; Y)$ is applied, are all equal to the same vector $h \in X$.

Theorem 7.9-1 (Taylor formulas in normed vector spaces) *Let X and Y be normed vector spaces, let Ω be an open subset of X, let $[a, a+h]$ be a closed segment contained in Ω, let $f : \Omega \subset X \to Y$ be a given mapping, and let m be an integer ≥ 1.*

(a) **(Taylor–Young formula)**[17] *If f is $(m-1)$ times differentiable in Ω and m times differentiable at the point a, then*

$$f(a+h) = f(a) + f'(a)h + \cdots + \frac{1}{m!}f^{(m)}(a)h^m + \|h\|^m\,\delta(h) \quad \textit{with} \lim_{h\to 0}\delta(h) = 0.$$

(b) **(generalized mean value theorem)** *If f is $(m-1)$ times continuously differentiable*

[16]So named after Brook Taylor (1685–1731), who introduced such formulas around 1715 for real-valued functions of a real variable.

[17]W.H. Young [1910]: *The Fundamental Theorems of the Differentiable Calculus*, Cambridge University Press, Cambridge, UK.

in Ω *and* m *times differentiable on the open segment* $]a, a+h[$, *then*

$$\left\| f(a+h) - \left(f(a) + f'(a)h + \cdots + \frac{1}{(m-1)!} f^{(m-1)}(a)h^{m-1} \right) \right\| \leq \frac{1}{m!} \sup_{x \in]a,a+h[} \| f^{(m)}(x) \| \, \|h\|^m .$$

(c) (**Taylor–MacLaurin formula**)[18] *If* $Y = \mathbb{R}$ *and* f *is* $(m-1)$ *times continuously differentiable in* Ω *and* m *times differentiable on the open segment* $]a, a+h[$, *there exists* $0 < \theta < 1$ *such that*

$$f(a+h) = f(a) + f'(a)h + \cdots + \frac{1}{(m-1)!} f^{(m-1)}(a)h^{m-1} + \frac{1}{m!} f^m(a+\theta h)h^m.$$

(d) (**Taylor formula with integral remainder**) *If* Y *is a Banach space and* f *is* m *times continuously differentiable in* Ω, *then*

$$f(a+h) = f(a) + f'(a)h + \cdots + \frac{1}{(m-1)!} f^{(m-1)}(a)h^{m-1} + \frac{1}{(m-1)!} \int_0^1 (1-t)^{m-1} \big(f^{(m)}(a+th)h^m \big) \, dt.$$

Proof (i) *Proof of* (a): Property (a) holds for $m = 1$ by definition of the derivative (Section 7.1); so, assume that, for some integer $k \geq 2$, property (a) holds for $m = 1, \ldots, k-1$.

Let $f : \Omega \subset X \to Y$ be a function that is $(k-1)$ times differentiable in Ω and k times differentiable at $a \in \Omega$, and let $r > 0$ be such that $B(a, r) \subset \Omega$. Then the auxiliary function

$$g : \xi \in B(a;r) \to g(\xi) := f(a+\xi) - \left(f(a) + f'(a)\xi + \cdots + \frac{1}{k!} f^{(k)}(a)\xi^k \right) \in Y$$

is differentiable in $B(a;r)$, with

$$g'(\xi) = f'(a+\xi) - \left(f'(a) + \cdots + \frac{1}{(k-1)!} f^{(k)}(a)\xi^{k-1} \right) \quad \text{for all } \xi \in B(a;r).$$

Noting that $f' : \Omega \subset X \to \mathcal{L}(X;Y)$ is $(k-2)$ times differentiable in Ω and $(k-1)$ times differentiable at a, we infer from the induction hypothesis that

$$f'(a+\xi) = f'(a) + \cdots + \frac{1}{(k-1)!} f^{(k)}(a)\xi^{k-1} + \|\xi\|^{k-1} \delta(\xi) \quad \text{for all } \xi \in B(a;r),$$

with $\lim_{\xi \to 0} \delta(\xi) = 0$ in Y, which in turn implies that

$$\|g'(\xi)\| \leq \|\xi\|^{k-1} \widetilde{\delta}(\xi) \quad \text{for all } \xi \in B(a;r), \text{ with } \lim_{\xi \to 0^+} \widetilde{\delta}(\xi) = 0.$$

Let now $a+h$ be any point in the ball $B(a;r)$. Then, by the *mean value theorem in a normed vector space* (Theorem 7.2-1),

$$\|g(h) - g(0)\| \leq \left(\sup_{\xi \in]0,h[} \|g'(\xi)\| \right) \|h\| ,$$

[18] So named after Colin MacLaurin (1698–1746).

and thus property (a) holds for $m = k$, since

$$\begin{aligned}\|f(a+h) - \Big(f(a) + f'(a)h + \cdots + \frac{1}{(m-1)!}f^{(k-1)}h^{k-1}\Big)\| \\ = \|g(h) - g(0)\| \leq \|h\|^k \eta(h), \quad \text{with } \lim \eta(h) = 0.\end{aligned}$$

(ii) *Proof of* (b): Property (b) holds for $m = 1$ by the mean value theorem; so, assume that, for some integer $k \geq 2$, (b) holds for $m = 1, \ldots, k-1$.

Let $f : \Omega \subset X \to Y$ be a function that is $(k-1)$ times continuously differentiable in Ω and k times differentiable on $]a, a+h[\subset \Omega$. The auxiliary function

$$\widetilde{g} : t \in [0,1] \to \widetilde{g}(t) := f(a+th) - \Big(f(a) + f'(a)(th) + \cdots + \frac{1}{(k-1)!}f^{k-1}(a)(th)^{k-1}\Big) \in Y$$

is differentiable on $[0,1]$ (clearly, $\widetilde{g}$ is differentiable on an open interval containing $[0,1]$), with

$$\widetilde{g}'(t) = f'(a+th)h - \Big(f'(a) + \cdots + \frac{1}{(k-2)!}f^{(k-1)}(a)(th)^{k-2}\Big)h, \quad 0 \leq t \leq 1.$$

Noting that $f' : \Omega \subset X \to \mathcal{L}(X;Y)$ is $(k-2)$ times continuously differentiable in Ω and $(k-1)$ times differentiable on $]a, a+h[$, we infer from the induction hypothesis that

$$\begin{aligned}\|f'(a+th) - \Big(f'(a) + \cdots + \frac{1}{(k-2)!}f^{(k-1)}(a)(th)^{k-2}\Big)\| \\ \leq \frac{1}{(k-1)!}\Big(\sup_{x\in]a,a+h[} \|f^{(k)}(x)\|\Big)t^{k-1}\|h\|^{k-1}, \quad 0 \leq t \leq 1,\end{aligned}$$

or equivalently, in terms of the function $\widetilde{g}$,

$$\|\widetilde{g}'(t)\| \leq \chi'(t), \quad 0 \leq t \leq 1,$$

where the monomial $\chi : [0,1] \to \mathbb{R}$ is defined by

$$\chi : t \in [0,1] \to \chi(t) := \frac{1}{k!}\Big(\sup_{x\in]a,a+h[} \|f^{(k)}(x)\|\Big)t^k \|h\|^k.$$

Given any integer $\ell \geq 1$, an application of the mean value theorem gives

$$\begin{aligned}\|\widetilde{g}(1) - \widetilde{g}(0)\| &\leq \sum_{j=0}^{\ell-1} \Big\|\widetilde{g}\Big(\frac{j+1}{\ell}\Big) - \widetilde{g}\Big(\frac{j}{\ell}\Big)\Big\| \leq \frac{1}{\ell}\sum_{j=0}^{\ell-1}\Big(\sup\Big\{\|\widetilde{g}'(t)\|;\ \frac{j}{\ell} < t < \frac{j+1}{\ell}\Big\}\Big) \\ &\leq \frac{1}{\ell}\sum_{j=0}^{\ell-1}\Big(\sup\Big\{\chi'(t);\ \frac{j}{\ell} < t < \frac{j+1}{\ell}\Big\}\Big).\end{aligned}$$

Hence

$$\begin{aligned}\|\widetilde{g}(1) - \widetilde{g}(0)\| &\leq \lim_{\ell\to\infty} \frac{1}{\ell}\sum_{j=0}^{\ell-1}\Big(\sup\Big\{\chi'(t);\ \frac{j}{\ell} < t < \frac{j+1}{\ell}\Big\}\Big) = \int_0^1 \chi'(t)\,\mathrm{d}t \\ &= \chi(1) - \chi(0) = \frac{1}{k!}\Big(\sup_{x\in]a,a+h[} \|f^{(k)}(x)\|\Big)\|h\|^k.\end{aligned}$$

Property (b) therefore holds for $m = k$ since

$$\widetilde{g}(1) - \widetilde{g}(0) = f(a+h) - \Big(f(a) + f'(a)h + \cdots + \frac{1}{(k-1)!} f^{(k-1)}(a)h^{k-1}\Big).$$

(iii) *Proof of* (c): Recall that $Y = \mathbb{R}$ now. The auxiliary function

$$\varphi : t \in [0,1] \to \varphi(t) := f(a+th) \in \mathbb{R}$$

is $(m-1)$ times continuously differentiable in an open interval of $\mathbb{R}$ containing $[0,1]$ and m times differentiable on $]0,1[$, with

$$\varphi^{(\ell)}(t) = f^{(\ell)}(a+th)h^{\ell}, \quad 0 \le \ell \le m,\ 0 \le t \le 1.$$

Then, by the Taylor–MacLaurin formula for real-valued functions of one real variable (assumed to be known),

$$\varphi(1) = \varphi(0) + \varphi'(0) + \cdots + \frac{1}{(m-1)!}\varphi^{(m-1)}(0) + \frac{1}{m!}\varphi^{(m)}(\theta) \quad \text{for some } 0 < \theta < 1,$$

which, expressed in terms of the function f, is exactly the announced Taylor–MacLaurin formula for the function $f : \Omega \subset X \to \mathbb{R}$.

(iv) *Proof of* (d): Recall that Y is now a Banach space. The same auxiliary function as in (iii), viz.,

$$\varphi : t \in [0,1] \to \varphi(t) := f(a+th) \in Y,$$

is now m times continuously differentiable in an open interval of $\mathbb{R}$ containing $[0,1]$. The auxiliary function

$$\psi : t \in [0,1] \to \psi(t) := \Big(\varphi(t) + (1-t)\varphi'(t) + \cdots + \frac{1}{(m-1)!}(1-t)^{m-1}\varphi^{(m-1)}(t)\Big) \in Y$$

is thus differentiable in an open interval containing $[0,1]$, with

$$\psi'(t) = \frac{1}{(m-1)!}(1-t)^{m-1}\varphi^{(m)}(t), \quad 0 \le t \le 1,$$

(as is immediately verified), so that, by the *mean value theorem for functions of class C^1 with values in a Banach space* (Theorem 7.6-1),

$$\begin{aligned}
\psi(1) - \psi(0) &= \varphi(1) - \Big(\varphi(0) + \varphi'(0) + \cdots + \frac{1}{(m-1)!}\varphi^{(m-1)}(0)\Big) \\
&= f(a+h) - \Big(f(a) + f'(a)h + \cdots + \frac{1}{(m-1)!} f^{(m-1)}(a)h^{m-1}\Big) \\
&= \int_0^1 \psi'(t)\,\mathrm{d}t = \frac{1}{(m-1)!}\int_0^1 (1-t)^{m-1}\Big(f^{(m)}(a+th)h^m\Big)\,\mathrm{d}t.
\end{aligned}$$

□

Remark Under the stronger assumptions of (d), the Taylor–MacLaurin formula of (c) becomes a consequence of (d) with $Y = \mathbb{R}$, since

$$\int_0^1 \frac{(1-t)^{m-1}}{(m-1)!}\left(f^{(m)}(a+th)h^m\right)\mathrm{d}t = f^{(m)}(a+\theta h)h^m \int_0^1 \frac{(1-t)^{m-1}}{(m-1)!}\mathrm{d}t = \frac{1}{m!}f^{(m)}(a+\theta h)h^m. \ \square$$

Let Ω be an open subset of a normed vector space V. Thanks to the Taylor formulas established in Theorem 7.7-1, the necessary condition $J'(u) = 0$ that a real-valued function $J : \Omega \subset V \to \mathbb{R}$ must satisfy at a local extremum $a \in \Omega$ (Theorem 7.1-5) can now be provided with worthwhile complements when J is *twice differentiable* either at u, or in Ω. Such complements take either the form of *sufficient conditions* (Theorem 7.9-2) or the form of further *necessary conditions* (Theorem 7.9-3). Note in this respect that there is *no* converse to either assertion (a) or assertion (b) of Theorem 7.9-2 (Problem 7.9-1).

For definiteness, we treat the case of a *minimum*.

Theorem 7.9-2 (sufficient conditions for a local minimum) *Let Ω be an open subset of a normed vector space V, let $J : \Omega \subset V \to \mathbb{R}$ be a function differentiable in Ω, and let $u \in \Omega$ be such that $J'(u) = 0$.*

(a) *If the function J is twice differentiable at u and if there exists a number α such that*

$$\alpha > 0 \quad \text{and} \quad J''(u)(v,v) \geq \alpha \|v\|^2 \quad \text{for all } v \in V,$$

then u is a strict local minimum of the function $J : \Omega \subset V \to \mathbb{R}$.

(b) *If the function J is twice differentiable in Ω and if there exists a neighborhood $W \subset \Omega$ of the point u such that*

$$J''(w)(v,v) \geq 0 \quad \text{for all } w \in W \text{ and all } v \in V,$$

then u is a local minimum of the function $J : \Omega \subset V \to \mathbb{R}$. If, in addition,

$$J''(w)(v,v) > 0 \quad \text{for all } w \in W \text{ and all } v \in V,\ v \neq 0,$$

the local minimum u is strict.

Proof If $J'(u) = 0$ and J is twice differentiable at $u \in \Omega$, then

$$J(u+v) - J(u) = \frac{1}{2}J''(u)(v,v) + \|v\|^2\,\delta(v) \quad \text{with } \lim_{v\to 0}\delta(v) = 0,$$

by the Taylor–Young formula. Let $r > 0$ be such that $B(u;r) \subset \Omega$ and $|\delta(v)| < \frac{\alpha}{2}$ for all $\|v\| < r$. Then the assumption that $J''(u)(v,v) \geq \alpha\|v\|^2$ for all $v \in V$ implies that

$$J(u+v) - J(u) \geq \left(\frac{\alpha}{2} + \delta(v)\right)\|v\|^2 > 0 \quad \text{for all } \|v\| < r,\ v \neq 0.$$

Hence u is a strict local minimum of the function J. This proves (a).

Assume now that J is twice differentiable in Ω and that there exists $r > 0$ such that $J''(w)(v,v) \geq 0$ for all $w \in B(u;r)$ and all $v \in V$, *resp.* $J''(w)(v,v) > 0$ for all $w \in B(u;r)$ and all $v \in V$, $v \neq 0$. Then, for each $\|v\| < r$, there exists $\theta = \theta(v)$ such that

$$0 < \theta < 1 \quad \text{and} \quad J(u+v) - J(u) = \frac{1}{2}J''(u+\theta v)(v,v),$$

by the Taylor–MacLaurin formula. Hence u is a local minimum, *resp.* a strict local minimum, of the function J. This proves (b). $\square$

Theorem 7.9-3 (necessary conditions for a local minimum) *Let Ω be an open subset of a normed vector space V and let $J : \Omega \subset V \to \mathbb{R}$ be a function differentiable in Ω and twice differentiable at a point $u \in \Omega$. If u is a local minimum of the function J, then*

$$J'(u) = 0 \quad \textit{and} \quad J''(u)(v,v) \geq 0 \quad \textit{for all } v \in V.$$

Proof Given a nonzero vector $v \in V$ (if $v = 0$, there is nothing to prove), there exists an open interval I of $\mathbb{R}$ containing 0 such that the function

$$\varphi : t \in I \to \varphi(t) := J(u + tv)$$

is differentiable in I and twice differentiable at $t = 0$, with

$$\varphi'(0) = J'(u)v = 0, \quad \varphi''(0) = J''(u)(v,v), \quad \text{and} \quad \varphi(t) \geq \varphi(0) \quad \text{for all } t \in I, \ t \neq 0.$$

Consequently

$$\varphi(t) - \varphi(0) = \frac{t^2}{2} J''(u)(v,v) + t^2 \delta(t), \quad \text{with } \lim_{t \to 0} \delta(t) = 0,$$

by the Taylor–Young formula. If $J''(u)(v,v) < 0$, let $t_0 \in I$ be such that $t_0 \neq 0$ and $|\delta(t_0)| < \frac{1}{2} |J''(u)(v,v)|$; then $\varphi(t_0) - \varphi(0) < 0$, a contradiction. □

Note that, should the second derivative of J at u vanish, conclusions similar to those of Theorems 7.9-2 and 7.9-3 can still be derived, but instead in terms of derivatives of J at u of order higher than two (Problem 7.9-2).

Problems

7.9-1 (1) Give an example of a twice differentiable function J having a strict local minimum at a point u, but such that $J''(u)(v,v) = 0$ for at least one vector $v \neq 0$ (hence assertion (a) of Theorem 7.9-2 has no converse).

(2) Give an example of a twice differentiable function J having a strict minimum at a point u, but such that, in every ball B centered at u, there exist $v \in B$ and $w \in V$ satisfying $J''(v)(w,w) < 0$ (hence assertion (b) of Theorem 7.9-2 has no converse).

7.9-2 This problem generalizes Theorems 7.9-2 and 7.9-3. Let Ω be an open subset of a normed vector space V, let $m \geq 2$ be an integer, and let $J : \Omega \subset X \to \mathbb{R}$ be a function that is $(m-1)$ times differentiable in Ω and m times differentiable at a point $u \in \Omega$, with

$$J'(u) = 0, \ldots, J^{(m-1)}(u) = 0, \quad \text{and} \quad J^{(m)}(u) \neq 0.$$

(1) Show that, if J has a local minimum at u, then m is even and $J^{(m)}(u)v^m \geq 0$ for all $v \in V$.

(2) Show that, if there exists $\alpha > 0$ such that $J^{(m)}(u)v^m \geq \alpha \|v\|^m$ for all $v \in V$, then J has a strict local minimum at u, and thus m is even by (1).

(3) Give an example of an infinitely differentiable function that has a strict local minimum at a point u, but such that $J^{(m)}(u) = 0$ for all integers $m \geq 1$.

7.9-3 Let $m \geq 1$ be an integer. Let $f : \mathbb{R} \to \mathbb{R}$ be a function that is m times differentiable in $\mathbb{R}$, and let $a \in \mathbb{R}$ and $b_k \in \mathbb{R}$, $1 \leq k \leq m$, be such that

$$f(a+h) = b_0 + b_1 h + \cdots + b_m h^m + |h|^m \varepsilon(h) \quad \text{with } \lim_{h \to 0} \varepsilon(h) = 0.$$

(1) Show that $b_k = \frac{1}{k!} f^{(k)}(a)$, $0 \leq k \leq m$.

(2) By means of a counterexample, show that (1) does not necessarily hold if f is not assumed to be m times differentiable in $\mathbb{R}$.

7.10 Application: Maximum principle for second-order linear elliptic operators

Let Ω be a bounded and connected open subset of $\mathbb{R}^N$, and let $\Gamma := \partial\Omega$. One aim of this section is to derive crucial properties, such as *uniqueness* or *continuous dependence on the data* (Theorems 7.10-3 and 7.10-4), of *classical solutions* $u \in \mathcal{C}(\overline{\Omega}) \cap \mathcal{C}^2(\Omega)$ of *linear second-order elliptic boundary value problems* of the form

$$\mathcal{L}u = f \text{ in } \Omega \quad \text{and} \quad u = u_0 \text{ on } \Gamma.$$

These properties will be derived from a basic property, called the *maximum principle for linear elliptic operators* (Theorem 7.10-2), itself a simple corollary of the fundamental *Hopf lemma* (Theorem 7.10-1). Such results are established here, simply because the necessary condition satisfied by the second derivative of a real-valued function at a maximum established in the preceding section (Theorem 7.9-3) plays a key role in the next proof.

It is worth emphasizing that these properties hold under *very weak assumptions on the set* Ω *and on the coefficient functions* a_{ij}, b_i, and c found either in the operator $\mathcal{M}$ of Theorem 7.10-1 or in the operator $\mathcal{L}$ of Theorems 7.10-2–7.10-4: Apart from the basic assumption that the operators $\mathcal{M}$ and $\mathcal{L}$ are *uniformly elliptic on compact subsets of* Ω (an assumption that is intermediary between those of ellipticity and uniform ellipticity given in Section 6.7), the other assumptions are indeed very mild: they simply express that the coefficients a_{ii} and b_i are *uniformly bounded on compact subsets of* Ω and that the function c is either ≥ 0 (Theorems 7.10-2 and 7.10-3) or bounded below by an ad hoc constant c_0 that may be < 0 (Theorem 7.10-4).

It is in particular striking that *no regularity assumption is needed on the coefficients* a_{ij}, b_i, and c, which may thus be in particular *discontinuous functions* and *unbounded* in Ω.

It is likewise striking that, apart from the assumptions of boundedness and connectedness on the open set Ω, *no regularity assumption is made on its boundary* Γ.

Theorem 7.10-1 (Hopf's lemma[19]) *Let* Ω *be a bounded and connected open subset of* $\mathbb{R}^N$ *and let* $\Gamma := \partial\Omega$. *Let a linear partial differential operator* $\mathcal{M}$ *be defined for functions* $v \in \mathcal{C}^2(\Omega)$ *by*

$$\mathcal{M}v(x) := -\sum_{i,j=1}^{N} a_{ij}(x)\partial_{ij}v(x) + \sum_{i=1}^{N} b_i(x)\partial_i v(x) \quad \textit{for all } x \in \Omega,$$

[19] E. Hopf [1927]: Elementare Bemerkungen über die Lösungen partieller Differentialgleichungen zweiter Ordnung vom elliptischen Typus, in *Sitzungsberichte der Preußischen Akademie der Wissenschaften, Berlin*, 147–152.

where the functions $a_{ij} = a_{ji} : \Omega \to \mathbb{R}$ *and* $b_i : \Omega \to \mathbb{R}$ *satisfy the following properties: Given any compact subset* K *of* Ω*, there exist constants* $\mu(K)$ *and* $C(K)$ *such that*

$$\mu(K) > 0 \quad \textit{and} \quad \sum_{i,j=1}^{N} a_{ij}(x)\xi_i\xi_j \geq \mu(K)\sum_{i=1}^{N}|\xi_i|^2 \;\; \textit{for all } x \in K \textit{ and all } (\xi_i)_{i=1}^N \in \mathbb{R}^N,$$

$$|a_{ii}(x)| \leq C(K) \quad \textit{and} \quad |b_i(x)| \leq C(K) \;\; \textit{for all } x \in K,\; 1 \leq i \leq N.$$

Assume that a function $v \in \mathcal{C}(\overline{\Omega}) \cap \mathcal{C}^2(\Omega)$ *satisfies*

$$\mathcal{M}v(x) \leq 0 \quad \textit{for all } x \in \Omega.$$

Then either v *is a constant function, or*

$$v(x) < \sup_{y\in\Gamma} v(y) \quad \textit{for all } x \in \Omega.$$

Proof (i) *Idea of the proof.* First, notice that $\sup_{y\in\Gamma} v(y) < \infty$ and $\sup_{y\in\overline{\Omega}} v(y) < \infty$ if $v \in \mathcal{C}(\overline{\Omega})$ since the sets Γ and $\overline{\Omega}$ are compact (Ω is bounded by assumption).

The proof amounts to showing that, if

$$\widetilde{\Omega} := \left\{ x \in \Omega;\; v(x) = \sup_{y\in\overline{\Omega}} v(y) \right\}$$

is a nonempty subset of Ω, then $\widetilde{\Omega} = \Omega$. Since $\widetilde{\Omega}$ is closed for the induced topology of Ω (by the continuity of v) and Ω is connected (by assumption), it thus suffices to show that, if $\widetilde{\Omega} \neq \varnothing$, then $\widetilde{\Omega}$ is open. So, assume that $\widetilde{\Omega}$ contains a point x_0, in which case there exists $\delta > 0$ such that $B(x_0; 2\delta) \subset \Omega$ since Ω is open (by assumption). *The objective is to establish that* $B(x_0;\delta) \subset \widetilde{\Omega}$*, which will imply that* $\widetilde{\Omega}$ *is open.*

(ii) Assume the contrary, i.e., that there exists $x_1 = (x_i^1)_{i=1}^N \in B(x_0;\delta)$ such that

$$v(x_1) < v(x_0) = \sup_{y\in\overline{\Omega}} v(y),$$

and let

$$2R := \sup\{\rho > 0;\; v(x) < v(x_0) \text{ for all } x \in B(x_1;\rho)\}.$$

The definition of $2R$ then implies that

$$0 < 2R \leq \delta \quad \text{and} \quad \overline{B(x_1; 2R)} \subset B(x_0; 2\delta) \subset \Omega,$$
$$v(x) < v(x_0) \quad \text{for all } x \in B(x_1; 2R).$$

Besides, there exists a point x_2 such that

$$x_2 \in \partial B(x_1; 2R) \quad \text{and} \quad v(x_2) = v(x_0)$$

(otherwise the compactness of $\partial B(x_1; 2R)$ and the continuity of the function v would together contradict the definition of $2R$).

(iii) For any $\alpha > 0$, consider the *auxiliary function*

$$w_\alpha : x = (x_i)_{i=1}^N \in \mathbb{R}^N \to w_\alpha(x) := e^{-\alpha|x-x_1|^2} - e^{-4\alpha R^2},$$

where $|\cdot|$ denotes as usual the Euclidean norm in $\mathbb{R}^N$. Then

$$\mathcal{M}w_\alpha(x) = e^{-\alpha|x-x_1|^2}\Big(- 4\alpha^2 \sum_{i,j=1}^N a_{ij}(x)(x_i - x_i^1)(x_j - x_j^1) + 2\alpha \sum_{i=1}^N a_{ii}(x) - 2\alpha \sum_{i=1}^N b_i(x)(x_i - x_i^1)\Big) \quad \text{for all } x \in \Omega,$$

and

$$w_\alpha(x) = 0 \quad \text{for all } x \in \partial B(x_1; 2R).$$

Since $\overline{B(x_1; 2R)} \subset \Omega$, the set

$$K := \{x \in \mathbb{R}^N;\ R \le |x - x_1| \le 2R\}$$

is a compact subset of Ω. The assumptions made on the functions a_{ij} and b_i thus imply that

$$\sup_{x\in K} \mathcal{M}w_\alpha(x) \le e^{-\alpha|x-x_1|^2}\big(- 4\alpha^2\mu(K)R^2 + 2\alpha NC(K)(1 + 2R)\big).$$

So, *we henceforth choose* $\alpha > 0$ *so that*

$$\mathcal{M}w_\alpha(x) < 0 \quad \text{for all } x \in K.$$

(iv) *A simple result about matrices* (needed in part (v)): *Let* $\boldsymbol{A} = (a_{ij})$ *and* $\boldsymbol{B} = (b_{ij})$ *be two* $N \times N$ *real symmetric nonnegative-definite matrices. Then* $\sum_{i,j=1}^N a_{ij}b_{ij} \ge 0$.

To see this, let $\boldsymbol{Q}$ be an orthogonal matrix such that $\boldsymbol{A} = \boldsymbol{QDQ}^T$, with $\boldsymbol{D} = \mathbf{Diag}(\lambda_i(\boldsymbol{A}))$. Let $(\widetilde{b}_{ij}) := \boldsymbol{Q}^T\boldsymbol{BQ}$; then

$$\sum_{i,j=1}^N a_{ij}b_{ij} = \text{tr}(\boldsymbol{AB}) = \text{tr}(\boldsymbol{QDQ}^T\boldsymbol{B}) = \text{tr}(\boldsymbol{DQ}^T\boldsymbol{BQ}) = \sum_{i=1}^N \lambda_i(\boldsymbol{A})\widetilde{b}_{ii} \ge 0$$

since $\lambda_i(\boldsymbol{A}) \ge 0$ and $\widetilde{b}_{ii} \ge 0$, $1 \le i \le N$ (the symmetric matrix $(\widetilde{b}_{ij})$ is also nonnegative-definite).

(v) Noting that $v(x) < v(x_0)$ for all $x \in \partial B(x_1; R) \subset B(x_1; 2R)$, *we henceforth choose* $\varepsilon > 0$ *so that the auxiliary function*

$$v_\varepsilon : x \in \Omega \to v_\varepsilon(x) := v(x) + \varepsilon w_\alpha(x)$$

satisfies

$$v_\varepsilon(x) < v(x_0) \quad \text{for all } x \in \partial B(x_1; R)$$

(that there exists such an $\varepsilon > 0$ follows from the compactness of $\partial B(x_1; R)$ and the continuity of the functions v and w_α).

Consider the function v_ε *on the compact set* K. On the one hand, *its maximum on* K *cannot be attained on* $\partial B(x_1; R)$, since $x_2 \in K$ and

$$v_\varepsilon(x_2) = v(x_2) + \varepsilon w_\alpha(x_2) = v(x_0) + \varepsilon w_\alpha(x_2) > v_\varepsilon(x) \quad \text{for all } x \in \partial B(x_1; R)$$

(recall that $v(x_2) = v(x_0)$, $w_\alpha(x_2) = 0$ since $x_2 \in \partial B(x_1; 2R)$, and $v(x_0) > v_\varepsilon(x)$ for all $x \in \partial B(x_1; R)$).

On the other hand, *its maximum on* K *cannot be attained in*

$$\text{int}\, K = \{x \in \mathbb{R}^N;\ R < |x - x_1| < 2R\}.$$

To see this, assume on the contrary that there exists a point $\widetilde{x} \in \text{int}\, K$ such that

$$v_\varepsilon(\widetilde{x}) = \sup_{x \in K} v_\varepsilon(x) = \sup_{x \in \text{int}\, K} v_\varepsilon(x).$$

Since the set $\text{int}\, K$ is open and v_ε is twice differentiable at $\widetilde{x}$, Theorem 7.9-3 shows that, first,

$$\partial_i v_\varepsilon(\widetilde{x}) = 0, \quad 1 \le i \le N,$$

and, second, *the* $N \times N$ *symmetric matrix* $(-\partial_{ij} v_\varepsilon(\widetilde{x}))$ *is nonnegative-definite. Since the* $N \times N$ *symmetric matrix* $(a_{ij}(\widetilde{x}))$ *is positive-definite by assumption*, we would then infer from part (iv) that

$$-\sum_{i,j=1}^{N} a_{ij}(\widetilde{x}) \partial_{ij} v_\varepsilon(\widetilde{x}) \ge 0,$$

but this is impossible, since

$$\mathcal{M} v_\varepsilon(\widetilde{x}) = -\sum_{i,j=1}^{N} a_{ij}(\widetilde{x}) \partial_{ij} v_\varepsilon(\widetilde{x}) = \mathcal{M} v(\widetilde{x}) + \varepsilon \mathcal{M} w_\alpha(\widetilde{x}) < 0$$

(recall that $\mathcal{M} v(x) \le 0$ for all $x \in \Omega$ by assumption and that $\alpha > 0$ has been so chosen that $\mathcal{M} w_\alpha(x) < 0$ for all $x \in K$; cf. (iii)).

Hence *the function* v_ε *attains its maximum on* K *on* $\partial B(x_1; 2R)$.

(vi) Since the auxiliary function w_α constructed in part (iii) satisfies $w_\alpha(x) = 0$ for all $x \in \partial B(x_1; 2R)$, the auxiliary function $v_\varepsilon = v + \varepsilon w_\alpha$ constructed in part (v) satisfies $v_\varepsilon(x) = v(x)$ for all $x \in \partial B(x_1; 2R)$. Since

$$x_2 \in \partial B(x_1; 2R) \quad \text{and} \quad v(x_2) = \sup_{y \in \overline{\Omega}} v(y)$$

(by part (ii)), *the function* v_ε *attains its maximum on* K *on* $\partial B(x_1; 2R)$ (by part (v)), and thus *in particular at the point* x_2.

Denoting as usual by ∂_ν the outer normal derivative operator along $\partial B(x_1; 2R)$, we must therefore have

$$\partial_\nu v_\varepsilon(x_2) \ge 0,$$

on the one hand. But, since $\partial_\nu v(x_2) = 0$ (recall that $x_2 \in \Omega$ and $v(x_2) = \sup_{y\in\Omega} v(y)$) and $\partial_\nu w_\alpha(x_2) = -4\alpha Re^{-4\alpha R^2} < 0$, we must also have

$$\partial_\nu v_\varepsilon(x_2) = \partial_\nu v(x_2) + \varepsilon \partial_\nu w_\alpha(x_2) < 0,$$

on the other hand.

We have therefore reached a contradiction. This completes the proof. □

Hopf's lemma will now be put to use for establishing the following fundamental property of linear second-order elliptic operators.

Theorem 7.10-2 (maximum principle for second-order linear elliptic operators) *Let Ω be a bounded and connected open subset of $\mathbb{R}^N$, and let $\Gamma := \partial\Omega$. Let a linear partial differential operator $\mathcal{L}$ be defined for functions $v \in \mathcal{C}^2(\Omega)$ by*

$$\mathcal{L}v(x) := -\sum_{i,j=1}^N a_{ij}(x)\partial_{ij}v(x) + \sum_{i=1}^N b_i(x)\partial_i v(x) + c(x)v(x) \quad \textit{for all } x \in \Omega,$$

where the functions $a_{ij} = a_{ji} : \Omega \to \mathbb{R}$, $b_i : \Omega \to \mathbb{R}$, and $c : \Omega \to \mathbb{R}$ satisfy the following properties: Given any compact subset K of Ω, there exist constants $\mu(K)$ and $C(K)$ such that

$$\mu(K) > 0 \quad \textit{and} \quad \sum_{i,j=1}^N a_{ij}(x)\xi_i\xi_j \geq \mu(K)\sum_{i=1}^N |\xi_i|^2 \quad \textit{for all } x \in K \textit{ and all } (\xi_i)_{i=1}^N \in \mathbb{R}^N,$$

$$|a_{ii}(x)| \leq C(K) \quad \textit{and} \quad |b_i(x)| \leq C(K) \quad \textit{for all } x \in K,\ 1 \leq i \leq N,$$

$$c(x) \geq 0 \quad \textit{for all } x \in \Omega.$$

Assume that a function $v \in \mathcal{C}(\overline{\Omega}) \cap \mathcal{C}^2(\Omega)$ satisfies

$$\mathcal{L}v(x) \leq 0 \quad \textit{for all } x \in \Omega.$$

Then

$$v(x) \leq \max\{0, \sup_{y\in\Gamma} v(y)\} \quad \textit{for all } x \in \Omega.$$

Proof Assume that the property is false, i.e., that there exists a point $\widetilde{x} \in \Omega$ such that

$$v(\widetilde{x}) = \sup_{x\in\overline{\Omega}} v(x) > \max\left\{0, \sup_{y\in\Gamma} v(y)\right\}.$$

The set

$$\widetilde{\Omega} := \{x \in \Omega;\ v(x) = v(\widetilde{x})\}$$

is thus nonempty since $\widetilde{x} \in \widetilde{\Omega}$, and closed for the induced topology of Ω since $v \in \mathcal{C}(\Omega)$. Since $v(\widetilde{x}) > 0$, there exists $r > 0$ such that $v(x) \geq 0$ for all $x \in B(\widetilde{x}; r)$, again since $v \in \mathcal{C}(\Omega)$. Consequently,

$$\mathcal{M}v(x) := \mathcal{L}v(x) - c(x)v(x) \leq 0 \quad \text{for all } x \in B(\widetilde{x}; r),$$

since $c(x) \geq 0$ for all $x \in \Omega$ by assumption.

Hopf's lemma can thus be applied on the open set $B(\widetilde{x}; r)$: Since $v(\widetilde{x}) = \sup_{x\in\overline{\Omega}} v(x) \geq \sup_{y\in\partial B(\widetilde{x};r)} v(y)$, the function v is necessarily a constant function on $B(\widetilde{x}; r)$. Hence the set $\widetilde{\Omega}$ is also open.

The assumed connectedness of the open set Ω thus implies that $\widetilde{\Omega} = \Omega$, i.e., that $v(x) = v(\widetilde{x})$ for all $x \in \Omega$, and hence also for all $x \in \overline{\Omega}$ since $v \in \mathcal{C}(\overline{\Omega})$. But this contradicts the assumed inequality $v(\widetilde{x}) > \sup_{y\in\Gamma} v(y)$. This completes the proof. □

Remark In the special case where $\mathcal{L} = -\Delta$, the maximum principle can be proved directly, in a simpler way; cf. Problem 6.7-3. □

The next theorem gathers two useful properties of *classical solutions to linear elliptic boundary value problems of the second order*, which are immediate corollaries of the maximum principle.

Theorem 7.10-3 (uniqueness and continuous dependence on the boundary values) *Let there be given an open subset Ω of $\mathbb{R}^N$ and a linear partial differential operator $\mathcal{L}$ that satisfy all the assumptions of Theorem* 7.10-2.

(a) *If a function $v \in \mathcal{C}(\overline{\Omega}) \cap \mathcal{C}^2(\Omega)$ satisfies $\mathcal{L}v(x) = 0$ for all $x \in \Omega$, then*

$$\sup_{x\in\overline{\Omega}} |v(x)| \leq \sup_{y\in\Gamma} |v(y)| .$$

(b) *Given functions $f : \Omega \to \mathbb{R}$ and $u_0 \in \mathcal{C}(\Gamma)$, the boundary value problem*

$$\mathcal{L}u = f \text{ in } \Omega, \quad u = u_0 \text{ on } \Gamma,$$

has at most one solution $u \in \mathcal{C}(\overline{\Omega}) \cap \mathcal{C}^2(\Omega)$.

Proof Given a function $v \in \mathcal{C}(\overline{\Omega}) \cap \mathcal{C}^2(\Omega)$ that satisfies $\mathcal{L}v(x) = 0$ for all $x \in \Omega$, define two auxiliary functions $w^+, w^- \in \mathcal{C}(\overline{\Omega}) \cap \mathcal{C}^2(\Omega)$ by

$$w^\pm : x \in \overline{\Omega} \to w^\pm(x) := \pm v(x) - \|v\|_\Gamma , \quad \text{where } \|v\|_\Gamma := \sup_{y\in\Gamma} |v(y)| .$$

Then

$$\mathcal{L}w^\pm(x) = -\|v\|_\Gamma \, c(x) \leq 0 \text{ for all } x \in \Omega \quad \text{and} \quad w^\pm(x) \leq 0 \text{ for all } x \in \Gamma,$$

so that

$$w^\pm(x) \leq 0 \quad \text{for all } x \in \Omega$$

by the *maximum principle* applied to the operator $\mathcal{L}$. This proves (a), which in turn clearly implies (b). □

The maximum principle on $\mathcal{L}$ thus implies the *uniqueness of classical solutions $u \in \mathcal{C}(\overline{\Omega}) \cap \mathcal{C}^2(\Omega)$* to the boundary value problem $\mathcal{L}u = f$ in Ω and $u = u_0$ on Γ, as well as their *continuous dependence on the function $u_0 : \Gamma \to \mathbb{R}$* with respect to sup-norms, in the following sense: *If two functions $u \in \mathcal{C}(\overline{\Omega}) \cap \mathcal{C}^2(\Omega)$ and $\widetilde{u} \in \mathcal{C}(\overline{\Omega}) \cap \mathcal{C}^2(\Omega)$ satisfy*

$$\mathcal{L}u = f \text{ in } \Omega \quad \text{and} \quad u = u_0 \text{ on } \Gamma, \quad \text{and} \quad \mathcal{L}\widetilde{u} = f \text{ in } \Omega \quad \text{and} \quad \widetilde{u} = \widetilde{u}_0 \text{ on } \Gamma,$$

then

$$\sup_{x\in\overline{\Omega}} |u(x) - \widetilde{u}(x)| \le \sup_{y\in\Gamma} |u_0(y) - \widetilde{u}_0(y)| .$$

Remark Uniqueness may fail if $u \in \mathcal{C}^2(\Omega)$ satisfies $\mathcal{L}u = f$ in Ω, but the boundary condition $u = u_0$ does not hold on the whole boundary Γ. For example, the boundary value problem

$$\begin{aligned} -\Delta u &= 0 \quad \text{in } \Omega := \{(x_1, x_2) \in \mathbb{R}^2;\ x_1^2 + x_2^2 < 1 \text{ and } x_2 > 0\},\\ u &= u_0 \quad \text{on } \Gamma - \{(0,0)\},\ \text{where } u_0(x_1,x_2) := x_1x_2 \text{ for } (x_1,x_2) \in \Gamma - \{(0,0)\} \end{aligned}$$

possesses two distinct solutions $u, \widetilde{u} : \overline{\Omega} - \{(0,0)\} \to \mathbb{R}$, respectively given by

$$u(x_1,x_2) := x_1x_2 \quad\text{and}\quad \widetilde{u}(x_1,x_2) := \frac{x_1x_2}{(x_1^2+x_2^2)^2} \quad \text{for } (x_1,x_2) \in \overline{\Omega} - \{(0,0)\}. \qquad \square$$

Under a mild additional assumption, the upper bound of Theorem 7.10-3(a) can be considerably improved, in that it now covers the case where the function $\mathcal{L}v$ no longer necessarily vanishes in Ω and the coefficient function c is allowed to be slightly negative on Ω.

Theorem 7.10-4 (continuous dependence on the right-hand side and on the boundary values) *Let Ω be a bounded and connected open subset of $\mathbb{R}^N$ and let $\Gamma := \partial\Omega$. Let a linear partial differential operator $\mathcal{L}$ be defined for functions $v \in \mathcal{C}^2(\Omega)$ by*

$$\mathcal{L}v(x) := -\sum_{i,j=1}^N a_{ij}(x)\partial_{ij}v(x) + \sum_{i=1}^N b_i(x)\partial_i v(x) + c(x)v(x) \quad \textit{for all } x \in \Omega,$$

where the functions $a_{ij} = a_{ji} : \Omega \to \mathbb{R}$, $b_i : \Omega \to \mathbb{R}$, and $c : \Omega \to \mathbb{R}$ satisfy the following properties: Given any compact subset K of Ω, there exist constants $\mu(K)$ and $C(K)$ such that

$$\mu(K) > 0 \quad \textit{and} \quad \sum_{i,j=1}^N a_{ij}(x)\xi_i\xi_j \ge \mu(K)\sum_{i=1}^N |\xi_i|^2 \quad \textit{for all } x \in K \textit{ and all } (\xi_i)_{i=1}^N \in \mathbb{R}^N,$$

$$|a_{ii}(x)| \le C(K) \quad \textit{and} \quad |b_i(x)| \le C(K) \quad \textit{for all } x \in K,\ 1 \le i \le N.$$

(a) *Assume that there exists a function $w \in \mathcal{C}(\overline{\Omega}) \cap \mathcal{C}^2(\Omega)$ that satisfies*

$$\mathcal{M}w(x) := -\sum_{i,j=1}^N a_{ij}(x)\partial_{ij}w(x) + \sum_{i=1}^N b_i(x)\partial_i w(x) \ge 1 \quad \textit{for all } x \in \Omega,$$

$$w(x) \ge 0 \quad \textit{for all } x \in \overline{\Omega},$$

and that there exists a constant $c_0 \le 0$ such that

$$c(x) \ge c_0 > -\frac{1}{\sup_{y\in\overline{\Omega}} |w(y)|} \quad \textit{for all } x \in \Omega.$$

Then there exists a constant $C = C(w, c_0)$ such that

$$\sup_{x\in\overline{\Omega}} |v(x)| \le C\Big(\sup_{y\in\Gamma} |v(y)| + \sup_{y\in\Omega} |\mathcal{L}v(y)|\Big) \quad \textit{for all } v \in \mathcal{C}(\overline{\Omega}) \cap \mathcal{C}^2(\Omega)$$

(*note that* $\sup_{y\in\Omega} |\mathcal{L}v(y)| = \infty$ *is not excluded in this inequality*).

(b) *Assume that, for some index* $1 \le i_0 \le N$, *there exist constants* α *and* β *such that*

$$0 < \alpha \le a_{i_0 i_0}(x) \quad \text{and} \quad b_{i_0}(x) \le \beta \text{ for all } x \in \Omega,$$

an assumption satisfied in particular by any uniformly elliptic operator (*Section* 6.7) *whose coefficients are continuous functions on* $\overline{\Omega}$. *Then there exists a function* $w \in \mathcal{C}^\infty(\mathbb{R}^N)$ *that satisfies*

$$\mathcal{M}w(x) \ge 1 \text{ for all } x \in \Omega \quad \text{and} \quad w(x) \ge 0 \text{ for all } x \in \overline{\Omega}.$$

Proof (i) *Proof of* (a) *when* $c_0 = 0$, *i.e., when* $c(x) \ge 0$ *for all* $x \in \Omega$. Given a function $v \in \mathcal{C}(\overline{\Omega}) \cap \mathcal{C}^2(\Omega)$ that satisfies $\|\mathcal{L}v\|_\Omega := \sup_{y\in\Omega} |\mathcal{L}v(y)| < \infty$ (if $\|\mathcal{L}v\|_\Omega = \infty$, the announced inequality surely holds), let $\|v\|_\Gamma := \sup_{y\in\Gamma} |v(y)|$ and define two auxiliary functions $w^+, w^- \in \mathcal{C}(\overline{\Omega}) \cap \mathcal{C}^2(\Omega)$ by

$$w^\pm : x \in \overline{\Omega} \to w^\pm(x) := \pm v(x) - \|v\|_\Gamma - \|\mathcal{L}v\|_\Omega \, w(x).$$

Then

$$\mathcal{L}w^\pm(x) = \pm\mathcal{L}v(x) - \|v\|_\Gamma \, c(x) - \|\mathcal{L}v\|_\Omega \, \mathcal{L}w(x) \le 0 \quad \text{for all } x \in \Omega,$$

since $\mathcal{L}w(x) = \mathcal{M}w(x) + c(x)w(x) \ge 1$ for all $x \in \Omega$ by assumption. Besides, by definition of the functions w^+ and w^-,

$$w^\pm(x) \le 0 \quad \text{for all } x \in \Gamma,$$

so that

$$w^\pm(x) \le 0 \quad \text{for all } x \in \Omega,$$

by the *maximum principle* applied to the operator $\mathcal{L}$. Hence in this case,

$$\sup_{x\in\overline{\Omega}} |v(x)| \le \|v\|_\Gamma + c_1 \|\mathcal{L}v\|_\Omega \quad \text{with } c_1 := \sup_{y\in\overline{\Omega}} |w(y)|.$$

(ii) *Proof of* (a) *when* $c(x) \ge c_0 > -\dfrac{1}{c_1} = -\dfrac{1}{\sup_{y\in\overline{\Omega}} |w(y)|}$ *for all* $x \in \Omega$. Let

$$c^+(x) := \max\{0, c(x)\} \quad \text{and} \quad c^-(x) := -\min\{0, c(x)\} \quad \text{for all } x \in \Omega,$$

so that, given any function $v \in \mathcal{C}(\overline{\Omega}) \cap \mathcal{C}^2(\Omega)$,

$$\mathcal{L}^+v(x) := \mathcal{M}v(x) + c^+(x)v(x) = \mathcal{L}v(x) + c^-(x)v(x) \quad \text{for all } x \in \Omega.$$

The inequality established in (i) can be applied to the operator $\mathcal{L}^+$, thus showing that

$$\sup_{x\in\overline{\Omega}} |v(x)| \le \|v\|_\Gamma + c_1 \|\mathcal{L}^+v\|_\Omega \le \|v\|_\Gamma + c_1 \left(\|\mathcal{L}v\|_\Omega - c_0 \sup_{x\in\overline{\Omega}} |v(x)| \right),$$

since $0 \le c^-(x) \le -c_0$ for all $x \in \Omega$. Hence in this case (note that $1 + c_0c_1 > 0$),

$$\sup_{x\in\overline{\Omega}} |v(x)| \le \frac{1}{(1 + c_0c_1)} \left(\|v\|_\Gamma + c_1 \|\mathcal{L}v\|_\Omega \right).$$

(iii) *Proof of* (b). Let $\delta > 0$ be so chosen that $\alpha\delta^2 - \beta\delta \geq 1$. Since Ω is bounded by assumption, there exists γ such that $|x_{i_0}| \leq \gamma$ if $x \in \overline{\Omega}$. Then the function

$$w : x = (x_i) \in \overline{\Omega} \to w(x) := e^{2\gamma\delta} - e^{\delta(x_{i_0}+\gamma)}$$

satisfies

$$\mathcal{M}w(x) = \left(\delta^2 a_{i_0 i_0}(x) - \delta b_{i_0}(x)\right) e^{\delta(x_{i_0}+\gamma)} \geq (\alpha\delta^2 - \beta\delta) \geq 1 \quad \text{for all } x \in \Omega,$$

and $w(x) \geq 0$ for all $x \in \overline{\Omega}$. □

Under the stronger assumptions of Theorem 7.10-4, a consequence of the maximum principle is thus the continuous dependence with respect to sup-norms of classical solutions to the boundary value problem $\mathcal{L}u = f$ in Ω and $u = u_0$ on Γ on *both* functions $f : \Omega \to \mathbb{R}$ and $u_0 : \Gamma \to \mathbb{R}$, in the following sense: *If two functions* $u \in \mathcal{C}(\overline{\Omega}) \cap \mathcal{C}^2(\Omega)$ *and* $\widetilde{u} \in \mathcal{C}(\overline{\Omega}) \cap \mathcal{C}^2(\Omega)$ *satisfy*

$$\mathcal{L}u = f \text{ in } \Omega \quad \text{and} \quad u = u_0 \text{ on } \Gamma, \quad \text{and} \quad \mathcal{L}\widetilde{u} = \widetilde{f} \text{ in } \Omega \quad \text{and} \quad u = \widetilde{u}_0 \text{ on } \Gamma,$$

and if $\sup_{y\in\Omega} |f(y)| < \infty$ *and* $\sup_{y\in\Omega} |\widetilde{f}| < \infty$, *then*

$$\sup_{x\in\overline{\Omega}} |u(x) - \widetilde{u}(x)| \leq C\left(\sup_{y\in\Gamma} |u_0(y) - \widetilde{u}_0(y)| + \sup_{y\in\Omega} |f(y) - \widetilde{f}(y)|\right).$$

Recall that uniqueness and continuous dependence on the data, *albeit* with respect to *different norms* (viz., those of the spaces $H^1(\Omega)$ for the solutions u or of the space $L^2(\Omega)$ for the right-hand sides f), were also obtained in Section 6.7 for the *weak solutions* of second-order elliptic boundary problems of the form $\mathcal{L}u = f$ in Ω and $u = 0$ on Γ (under the assumption that the function $c \in L^\infty(\Omega)$ be ≥ 0 almost everywhere in Ω).

In fact, a *weak maximum principle*[20] analogous to that of Theorem 7.10-2 can be established for functions u that are *only in* $H^1(\Omega)$ and that satisfy $\mathcal{L}u = f$ *only in the sense of distributions.* The next theorem, which for the sake of comparison applies to the operator $\mathcal{L}$ of Theorem 6.7-6, viz., that defined for functions $v \in \mathcal{C}^2(\Omega)$ by

$$\mathcal{L}v(x) = -\sum_{i,j=1}^{N} \partial_j\left(a_{ij}(x)\partial_i v(x)\right) + c(x)v(x) \quad \text{for all } x \in \Omega,$$

gives a flavor of the type of result that can be proved.[21]

♭Theorem 7.10-5 (weak maximum principle for a second-order elliptic operator) *Let* Ω *be a domain in* $\mathbb{R}^N$, *and let functions* $a_{ij} = a_{ji} \in L^\infty(\Omega)$, $c \in L^\infty(\Omega)$, *and* $f \in L^2(\Omega)$

[20]Due to:
G. Stampacchia [1965]: Le problème de Dirichlet pour les équations elliptiques du second ordre à coefficients discontinus, *Annales de l'Institut Fourier (Grenoble)* **15**, 189–258.

[21]A proof of Theorem 7.10.5 is found in Brezis [2011, Theorem 9.27]. A proof of the weak maximum principle for more general second-order elliptic operators is found in Gilbarg & Trudinger [1998, Theorem 8.1].

be given that satisfy the following properties: There exists a constant μ *such that*

$$\mu > 0 \quad \textit{and} \quad \sum_{i,j=1}^{N} a_{ij}(x)\xi_i\xi_j \geq \mu \sum_{i=1}^{N} |\xi_i|^2 \quad \textit{for almost all } x \in \Omega \textit{ and all } (\xi_i)_{i=1}^N \in \mathbb{R}^N,$$

$$c(x) \geq 0 \quad \textit{for almost all } x \in \Omega,$$

$$f(x) \leq 0 \quad \textit{for almost all } x \in \Omega.$$

Finally, let there be given a function $v \in H^1(\Omega)$ *that satisfies*

$$\int_\Omega \Big(\sum_{i,j=1}^{N} a_{ij}\partial_i v \partial_j \varphi + cv\varphi \Big) \mathrm{d}x = \int_\Omega f\varphi \mathrm{d}x \quad \textit{for all } \varphi \in \mathcal{D}(\Omega).$$

Then

$$v(x) \leq \max\{0, \operatorname*{ess\,sup}_{y\in\Gamma} v(y)\} \quad \textit{for almost all } x \in \Omega,$$

where

$$\operatorname*{ess\,sup}_{y\in\Gamma} v(y) := \inf\{\tau \geq 0;\ \operatorname{tr} v(y) \leq \tau \textit{ for } \mathrm{d}\Gamma\textit{-almost all } y \in \Gamma\}.$$ □

Problems

7.10-1 Let $c :]0,1[\to \mathbb{R}$ be a function that satisfies $c(x) \geq 0$ for all $0 < x < 1$. Show directly that, if a function $v \in \mathcal{C}\,[0,1] \cap \mathcal{C}^2\,]0,1[$ satisfies $-v''(x) + c(x)v(x) \leq 0$ for all $0 < x < 1$ and $v(0) \leq 0$ and $v(1) \leq 0$, then $v(x) \leq 0$ for all $0 \leq x \leq 1$.
Hint: Show that, for each $\varepsilon > 0$, $v(x) - \frac{\varepsilon}{2}x(1-x) \leq 0$ for all $0 \leq x \leq 1$.

7.10-2 Let Ω be a bounded and connected open subset of $\mathbb{R}^N$, and let $\Gamma := \partial\Omega$. Let a uniformly elliptic linear partial differential operator $\mathcal{L}$ be defined for functions $v \in \mathcal{C}^2(\Omega)$ by

$$\mathcal{L}v(x) := -\sum_{i,j=1}^{N} a_{ij}(x)\partial_{ij}v(x) + \sum_{i=1}^{N} b_i(x)\partial_i v(x) + c(x)v(x) \quad \text{for all } x \in \Omega,$$

where the functions a_{ij}, b_i, and c are continuous over $\overline{\Omega}$; no further assumption such as $c(x) \geq c_0$ for all $x \in \overline{\Omega}$ (as in the text) is made on the function c.

Show that, if there exist a function $w \in \mathcal{C}(\overline{\Omega}) \cap \mathcal{C}^2(\Omega)$ such that $\mathcal{L}w(x) = 0$ for all $x \in \Omega$ and $w(x) > 0$ for all $x \in \overline{\Omega}$, then the boundary value problem $\mathcal{L}u = f$ in Ω and $u = g$ on Γ has at most one classical solution $u \in \mathcal{C}(\overline{\Omega}) \cap \mathcal{C}^2(\Omega)$.

7.11 Application: Lagrange interpolation in $\mathbb{R}^n$ and multipoint Taylor formulas

Lagrange interpolation in $\mathbb{R}^n$ consists in prescribing a *finite* set A in $\mathbb{R}^n$ and then in interpolating the *values* of a given function v at the points of A by a *polynomial* Πv in n variables. *Hermite interpolation in* $\mathbb{R}^n$ consists in interpolating the values of a given function v at some points of A and in interpolating *in addition* the values of some *derivatives* of v at some points of A (sometimes also at points not in A), again by a polynomial Πv in n variables.

Our basic objective is to establish the following general *interpolation error estimate:*[22] *Under the assumptions that the Lagrange interpolation polynomial Πv is uniquely defined, that $\Pi p = p$ whenever p is a polynomial of degree $\leq k$, and that $v \in \mathcal{C}^{k+1}(T)$, then*

$$\max_{|\boldsymbol{\alpha}|=m} \sup_{x \in T} |\partial^{\boldsymbol{\alpha}} \Pi v(x) - \partial^{\boldsymbol{\alpha}} v(x)| \leq C \frac{h_T^{k+1}}{\rho_T^m} \max_{|\boldsymbol{\alpha}|=k+1} \sup_{\xi \in T} |\partial^{\boldsymbol{\alpha}} v(\xi)| \quad \text{for each } 0 \leq m \leq k.$$

In this estimate (where the multi-index notation for denoting partial derivatives is used; cf. Section 1.18), T is the convex hull of A, h_T is the diameter of T, ρ_T is the supremum of the diameters of the spheres inscribed in T, and C is a numerical constant that is "independent of A" in the sense that C is the same for all *affine-equivalent Lagrange interpolation schemes* (this key notion will be defined below).

To begin with, we give some general definitions. For each integer $k \geq 0$, the notation P_k designates the space of all *polynomials p of degree $\leq k$ in the variables $x_1, x_2, \dots, x_n$*, thus of the form

$$p : x = (x_i)_{i=1}^n \in \mathbb{R}^n \to p(x) := \sum_{\alpha_1+\alpha_2+\cdots+\alpha_n \leq k} c_{\alpha_1\alpha_2\cdots\alpha_n} x_1^{\alpha_1} x_2^{\alpha_2} \cdots x_n^{\alpha_n},$$

where $\alpha_i \in \mathbb{N}$, $1 \leq i \leq n$, and the coefficients $c_{\alpha_1\alpha_2\cdots\alpha_n}$ are real numbers; or equivalently, of the form

$$p : x \in \mathbb{R}^n = \sum_{|\boldsymbol{\alpha}| \leq k} c_{\boldsymbol{\alpha}} x^{\boldsymbol{\alpha}}$$

if the multi-index notation is used, with the convention that $x^{\mathbf{0}} := 1$. The dimension of the space P_k is given by

$$\dim P_k = \binom{n+k}{k} = \frac{(n+k)!}{k!n!}.$$

If S is any subset of $\mathbb{R}^n$, we let

$$P_k(S) := \{p_{|S};\, p \in P_k\}.$$

Clearly, the dimension of the space $P_k(S)$ is the same as that of the space $P_k = P_k(\mathbb{R}^n)$ if the interior of the set S is nonempty.

Recall that an n-**simplex** in $\mathbb{R}^n$ is the convex hull T of $(n+1)$ points $a_j = (a_{ij})_{i=1}^n \in \mathbb{R}^n$, $1 \leq j \leq n+1$, which are called the *vertices* of the n-simplex, and which are such that the $(n+1) \times (n+1)$ matrix

$$A = \begin{pmatrix} a_{11} & a_{12} & \cdots & a_{1,n+1} \\ a_{21} & a_{22} & \cdots & a_{2,n+1} \\ \vdots & \vdots & & \vdots \\ a_{n1} & a_{n2} & \cdots & a_{n,n+1} \\ 1 & 1 & \cdots & 1 \end{pmatrix}$$

[22] The first results of this kind, for both Lagrange and Hermite interpolation over triangles, are due to: M. Zlámal [1968]: On the finite element method, *Numerische Mathematik* **12**, 394–409.

is *invertible*, or equivalently, such that the $(n+1)$ points a_j are not contained in a hyperplane; cf. Section 2.16. The set T is thus of the form

$$T = \left\{ \sum_{j=1}^{n+1} \lambda_j a_j;\ 0 \le \lambda_j \le 1,\ 1 \le j \le n+1,\ \sum_{j=1}^{n+1} \lambda_j = 1 \right\}.$$

Notice that *a 2-simplex is a triangle* and that *a 3-simplex is a tetrahedron.*

For any integer m with $0 \le m \le n$, an *m-face* of an n-simplex T is any m-simplex whose $(m+1)$ vertices are also vertices on T. In particular, an $(n-1)$-face is called a *face* and a 1-face is called an *edge*, or a *side*.

Given any point $x = (x_i)_{i=1}^n \in \mathbb{R}^n$, its **barycentric coordinates** $\lambda_j(x)$, $1 \le j \le n+1$, *with respect to the* $(n+1)$ *vertices* a_j are the unique solutions of the linear system

$$\sum_{j=1}^{n+1} a_{ij}\lambda_j(x) = x_i, \quad 1 \le i \le n, \quad \sum_{j=1}^{n+1} \lambda_j(x) = 1,$$

whose matrix is precisely the above matrix A, and the *functions* $\lambda_j : \mathbb{R}^n \to \mathbb{R}$ defined in this fashion are called the **barycentric coordinates** with respect to the $(n+1)$ vertices a_j. It thus follows from their definition that the *barycentric coordinates of* $x \in \mathbb{R}^n$ *are affine functions of the coordinates* $x_1, x_2, \ldots, x_n$ of x (equivalently they belong to the space P_1), since

$$\lambda_i = \sum_{j=1}^{n} b_{ij}x_j + b_{i,n+1}, \quad 1 \le i \le n+1,$$

where the $(n+1) \times (n+1)$ matrix $B = (b_{ij})$ is the inverse of the matrix A.

The *barycenter, or center of gravity*, of an n-simplex T is the point of T all of whose barycentric coordinates are equal (to $1/(n+1)$).

We now describe a few *basic examples of Lagrange interpolation in* $\mathbb{R}^n$. To begin with, we show that *a polynomial* $p : x \in \mathbb{R}^n \to \sum_{|\boldsymbol{\alpha}|\le 1} c_{\boldsymbol{\alpha}} x^{\boldsymbol{\alpha}}$ *of degree* 1 *is uniquely determined by its values* $p(a_j)$ *at the* $(n+1)$ *vertices* a_j *of an* n*-simplex in* $\mathbb{R}^n$, $1 \le j \le n+1$. To this end, it suffices to show that the linear system $\sum_{|\boldsymbol{\alpha}|\le 1} c_{\boldsymbol{\alpha}} a_i^{\boldsymbol{\alpha}} = \mu_i$, $1 \le i \le n+1$, has one and only one solution $c_{\boldsymbol{\alpha}}, |\boldsymbol{\alpha}| \le 1$, for each right-hand side μ_i, $1 \le i \le n+1$. Since

$$\dim P_1 = \operatorname{card} A_1 = n+1, \quad \text{where } A_1 := \bigcup_{j=1}^{n+1} \{a_j\}$$

(Figure 7.11-1), the matrix of this linear system is *square*, and therefore it suffices to prove either *uniqueness* or *existence*. In this case, existence is clear: The barycentric coordinates $\lambda_i \in P_1$ verify $\lambda_i(a_j) = \delta_{ij}$, $1 \le i, j \le n+1$, and thus the polynomial $x \in \mathbb{R}^n \to \sum_{i=1}^{n+1} \mu_i \lambda_i(x)$ has the desired interpolation property. The resulting *identity*

$$p = \sum_{i=1}^{n+1} p(a_i)\lambda_i \quad \text{for all } p \in P_1,$$

then shows that, given a function v defined over a domain containing the set A, the unique polynomial of degree ≤ 1 interpolating the values $v(a_i)$, $1 \leq i \leq n+1$, is given by

$$\Pi_1 v = \sum_{i=1}^{n+1} v(a_i)\lambda_i.$$

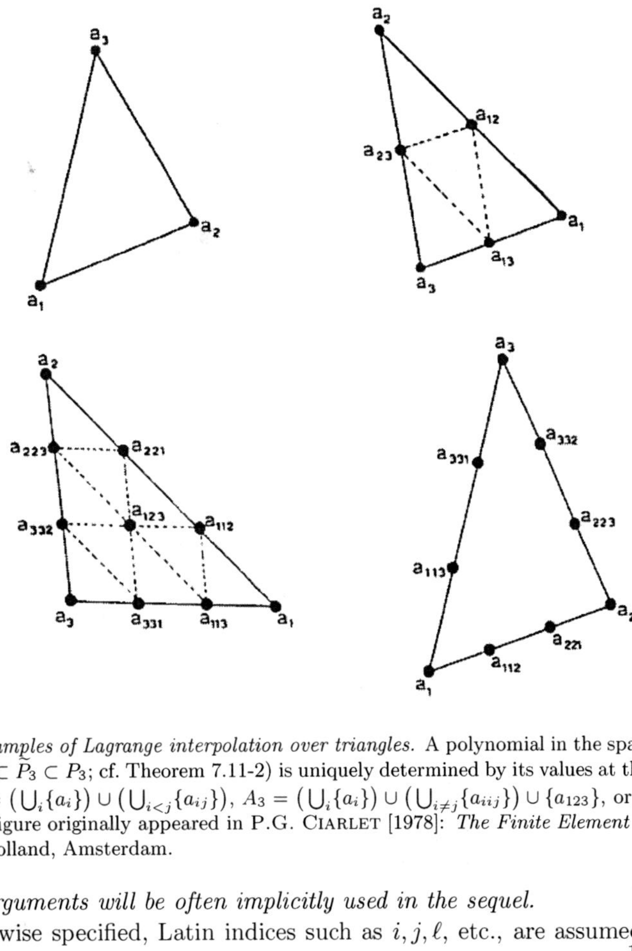

Figure 7.11-1 *Examples of Lagrange interpolation over triangles.* A polynomial in the space P_1, P_2, P_3, or $\widetilde{P}_3$ (which satisfies $P_2 \subset \widetilde{P}_3 \subset P_3$; cf. Theorem 7.11-2) is uniquely determined by its values at the points of the sets $A_1 = \bigcup_i \{a_i\}$, $A_2 = \left(\bigcup_i \{a_i\}\right) \cup \left(\bigcup_{i<j} \{a_{ij}\}\right)$, $A_3 = \left(\bigcup_i \{a_i\}\right) \cup \left(\bigcup_{i\neq j} \{a_{iij}\}\right) \cup \{a_{123}\}$, or $\widetilde{A}_3 = A_3 - \{a_{123}\}$, respectively. This figure originally appeared in P.G. Ciarlet [1978]: *The Finite Element Method for Elliptic Problems*, North-Holland, Amsterdam.

The above arguments will be often implicitly used in the sequel.

Unless otherwise specified, Latin indices such as i, j, ℓ, etc., are assumed until Theorem 7.11-2 (included) to take their values in the set $\{1, 2, \ldots, n+1\}$. Let $a_{ij} = \dfrac{1}{2}(a_i + a_j)$, $i < j$, denote the midpoints of the edges of an n-simplex T. Observing that $\lambda_\ell(a_{ij}) = \dfrac{1}{2}(\delta_{\ell i} + \delta_{\ell j})$ for $i < j$ and that

$$\dim P_2 = \operatorname{card} A_2, \quad \text{where } A_2 := \left(\bigcup_i \{a_i\}\right) \cup \left(\bigcup_{i<j} \{a_{ij}\}\right),$$

we obtain the *identity*

$$p = \sum_i \lambda_i (2\lambda_i - 1) p(a_i) + \sum_{i<j} 4\lambda_i\lambda_j p(a_{ij}) \quad \text{for all } p \in P_2.$$

Consequently, given a function defined over a domain containing the set A_2 (Figure 7.11-1), the unique polynomial of degree ≤ 2 interpolating the values $v(a_i)$ and $v(a_{ij})$, $i < j$, is given by

$$\Pi_2 v = \sum_i \lambda_i (2\lambda_i - 1) v(a_i) + \sum_{i<j} 4\lambda_i\lambda_j v(a_{ij}).$$

Let $a_{iij} := \frac{1}{3}(2a_i + a_j)$ for $i \neq j$, and $a_{ij\ell} = \frac{1}{3}(a_i + a_j + a_\ell)$ for $i < j < \ell$. From the identity

$$p = \sum_i \frac{1}{2}\lambda_i (3\lambda_i - 1)(3\lambda_i - 2) p(a_i) + \sum_{i \neq j} \frac{9}{2}\lambda_i\lambda_j (3\lambda_i - 1) p(a_{iij})$$
$$+ \sum_{i<j<\ell} 27\lambda_i\lambda_j\lambda_\ell p(a_{ij\ell}) \quad \text{for all } p \in P_3$$

(established in the same manner as above), we likewise infer that, given a function v defined over a domain containing the set

$$A_3 := \left(\bigcup_i \{a_i\}\right) \cup \left(\bigcup_{i \neq j} \{a_{iij}\}\right) \cup \left(\bigcup_{i<j<k} \{a_{ijk}\}\right)$$

(Figure 7.11-1), the unique polynomial of degree ≤ 3 interpolating the values $v(a_i)$, $v(a_{iij})$, $i \neq j$, and $v(a_{ijk})$, $i < j < k$, is given by

$$\Pi_3 v = \sum_i \frac{1}{2}\lambda_i (3\lambda_i - 1)(3\lambda_i - 2) v(a_i) + \sum_{i \neq j} \frac{9}{2}\lambda_i\lambda_j (3\lambda_i - 1) v(a_{iij}) + \sum_{i<j<\ell} 27\lambda_i\lambda_j\lambda_\ell v(a_{ij\ell}).$$

More generally, Lagrange interpolating polynomials of *arbitrary* degree $k \geq 1$ can be similarly defined, according to the following result (which contains the above three examples as special cases):

Theorem 7.11-1 *Let T be an n-simplex with vertices a_j, $1 \leq j \leq n+1$. Then for a given integer $k \geq 1$, any polynomial $p \in P_k$ is uniquely determined by its values on the set*

$$A_k := \left\{ \sum_{j=1}^{n+1} \mu_j a_j \in \mathbb{R}^n;\ \sum_{j=1}^{n+1} \mu_j = 1, \mu_j \in \left\{0, \frac{1}{k}, \ldots, \frac{k-1}{k}, 1\right\},\ 1 \leq j \leq n+1 \right\}.$$

Proof [23] Let $N_k := \dim P_k = \operatorname{card} A_k = \binom{n+k}{k}$ and let $A_k = \bigcup_{\ell=1}^{N_k} \{b_\ell\}$. Any point b_ℓ, $1 \leq \ell \leq N_k$, of the set A_k is of the form

$$b_\ell = \frac{1}{k}\sum_{i=1}^{n+1} m_i^\ell a_i, \quad \text{with } m_i^\ell \in \{0, 1, \ldots, k\},\ 1 \leq i \leq n+1, \text{ and } \sum_{i=0}^{k} m_i^\ell = k.$$

[23]This proof is found in:
R.A. NICOLAIDES [1972]: On a class of finite elements generated by Lagrange interpolation, *SIAM Journal on Numerical Analysis* **9**, 435–445.

It is then easily verified that each function

$$p_\ell : x \in \mathbb{R}^n \to p_\ell(x) := \frac{1}{m_1^\ell! m_2^\ell! \cdots m_{n+1}^\ell!} \prod_{\substack{i=1 \\ m_i^\ell \geq 1}}^{n+1} \prod_{j=0}^{m_i^\ell - 1} (k\lambda_i(x) - j), \quad 1 \leq \ell \leq N_k,$$

where the functions λ_i, $1 \leq i \leq n+1$, are as before the barycentric coordinates with respect to the vertices a_i, $1 \leq i \leq n+1$, has the following properties:

$$p_\ell \in P_k \quad \text{and} \quad p_\ell(b_m) = \delta_{\ell m}, \quad 1 \leq m \leq N_k.$$

Hence the following identity holds:

$$p = \sum_{\ell=1}^{N_k} p(b_\ell) p_\ell \quad \text{for all } p \in P_k.$$

This proves the assertion. □

Remarks (1) An identity such as

$$p = \sum_i \frac{1}{2}\lambda_i(3\lambda_i - 1)(3\lambda_i - 2)p(a_i) + \sum_{i \neq j} \frac{9}{2}\lambda_i\lambda_j(3\lambda_i - 1)p(a_{iij}) + \sum_{i<j<\ell} 27\lambda_i\lambda_j\lambda_k p(a_{ij\ell}) \quad \text{for all } p \in P_3$$

is thus a special case of the above identity.

(2) The set A_k as defined in Theorem 7.11-1 is called the *principal lattice of order k* of the n-simplex T. □

In each one of the above examples, the interpolating polynomial is assumed to belong to a space P that coincides for some $k \geq 1$ with the space P_k, i.e., the space of *all* polynomials of degree $\leq k$. In order to achieve greater generality (besides at no extra cost, as it will turn out), it is, however, desirable to relax this assumption, by assuming instead that the interpolating function belongs to a space P that may only *strictly contain* a space P_k for some $k \geq 1$; otherwise, the space P may be itself a space of polynomials (as in the next examples), or may even contain functions that are not polynomials.[24]

Theorem 7.11-2 *For each triple (i, j, ℓ) with $i < j < \ell$, let*

$$\varphi_{ij\ell}(p) := 12p(a_{ij\ell}) + 2 \sum_{m=i,j,\ell} p(a_m) - 3 \sum_{\substack{r,s=i,j,\ell \\ r \neq s}} p(a_{rrs}).$$

Then any polynomial in the space

$$\widetilde{P}_3 := \{p \in P_3;\ \varphi_{ij\ell}(p) = 0,\ i < j < \ell\}$$

[24]Such spaces P are usually associated with *Hermite interpolation*; see for instance the interpolation scheme analyzed in:

P.G. Ciarlet [1978]: Interpolation error estimates for the reduced Hsieh-Clough-Tocher triangle, *Mathematics of Computation* **32**, 335–344.

is uniquely determined by its values on the set

$$\widetilde{A}_3 := \Big(\bigcup_i \{a_i\}\Big) \cup \Big(\bigcup_{i\neq j} \{a_{iij}\}\Big).$$

In addition, the strict inclusion

$$P_2 \subsetneqq \widetilde{P}_3$$

holds.

Proof The (straightforward) proof is left as a problem (Problem 7.11-1). □

Remark Examples of *Hermite interpolation over n-simplices* are provided in Problems 7.11-6 and 7.11-7. □

We now describe another kind of Lagrange interpolation, also corresponding to a strict inclusion $P_k \subsetneqq P$. To this end, we need again a few definitions. For each integer $k \geq 0$, the notation Q_k designates the space of all *polynomials p that are of degree* $\leq k$ *with respect to each one of the n variables* $x_1, x_2, \dots, x_n$, thus of the form

$$p : x = (x_1, x_2, \dots, x_n) \in \mathbb{R}^n \to p(x) = \sum_{\alpha_i \leq k, 1\leq i\leq n} c_{\alpha_1\alpha_2\cdots\alpha_n} x_1^{\alpha_1} x_2^{\alpha_2} \cdots x_n^{\alpha_n},$$

where $\alpha_i \in \mathbb{N}$, $0 \leq i \leq n$, and the coefficients $c_{\alpha_1\alpha_2\cdots\alpha_n}$ are real numbers. The dimension of the space Q_k is given by

$$\dim Q_k = (k+1)^n,$$

and the *strict* inclusion

$$P_k \subsetneqq Q_k$$

holds for each integer $k \geq 1$ (clearly, $Q_0 = P_0$). If S is a subset of $\mathbb{R}^n$ with a nonempty interior, the dimension of the space

$$Q_k(S) := \{p_{|S};\ p \in Q_k\}$$

is clearly the same as that of the space $Q_k = Q_k(\mathbb{R}^n)$.

An n-**rectangle** in $\mathbb{R}^n$, or simply a *rectangle* if $n = 2$, is a set of the form

$$T = \prod_{i=1}^{n} [a_i, b_i] = \{x = (x_1, x_2, \dots, x_n);\ a_i \leq x_i \leq b_i,\ 1 \leq i \leq n\},$$

with $-\infty < a_i < b_i < \infty$ for each i; in particular, the **unit hypercube** $[0,1]^n$ is an n-rectangle. A *face* of an n-rectangle T is any one of the sets

$$\{a_j\} \times \prod_{\left\{\substack{i=1\\ i\neq j}\right.}^{n} [a_i, b_i] \quad \text{or} \quad \{b_j\} \times \prod_{\left\{\substack{i=1\\ i\neq j}\right.}^{n} [a_i, b_i], \quad 1 \leq j \leq n,$$

while an *edge* of T, also called a *side*, is any one of the sets

$$[a_j, b_j] \times \prod_{\begin{cases} i=1 \\ i \neq j \end{cases}}^{n} \{c_i\},$$

with $c_i = a_i$ or b_i, $1 \leq i \leq n$, $i \neq j$, $1 \leq j \leq n$. A *vertex* of T is any point $x = (x_1, x_2, \ldots, x_n)$ of T with $x_i = a_i$ or b_i, $1 \leq i \leq n$. Clearly, *an n-rectangle is the convex hull of its vertices.*

Note that, *according to the above definition, any side of an n-rectangle is parallel to one of the coordinate axes of* $\mathbb{R}^n$.

We now show that, given any integer $k \geq 1$ and any n-rectangle T, a polynomial $p \in Q_k$ is uniquely determined by its values at $(k+1)^n$ judiciously chosen points of T. See Figure 7.11-2 for the special cases $k = 1, 2, 3$ and $n = 2$; see also Problems 7.11-2 and 7.11-3 for similar examples of interpolation over rectangles, but where the values at interior points are no longer used for defining the interpolating polynomial.

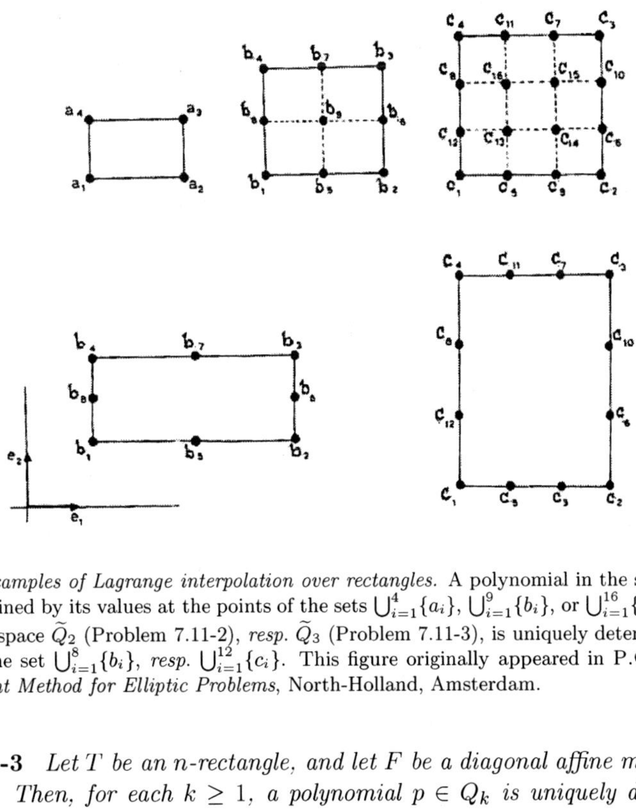

Figure 7.11-2 *Examples of Lagrange interpolation over rectangles.* A polynomial in the space Q_1, Q_2, or Q_3 is uniquely determined by its values at the points of the sets $\bigcup_{i=1}^{4}\{a_i\}$, $\bigcup_{i=1}^{9}\{b_i\}$, or $\bigcup_{i=1}^{16}\{c_i\}$, respectively. A polynomial in the space $\widetilde{Q}_2$ (Problem 7.11-2), *resp.* $\widetilde{Q}_3$ (Problem 7.11-3), is uniquely determined by its values at the points of the set $\bigcup_{i=1}^{8}\{b_i\}$, *resp.* $\bigcup_{i=1}^{12}\{c_i\}$. This figure originally appeared in P.G. CIARLET [1978]: *The Finite Element Method for Elliptic Problems*, North-Holland, Amsterdam.

Theorem 7.11-3 *Let T be an n-rectangle, and let F be a diagonal affine mapping such that $T = F([0,1]^n)$. Then, for each $k \geq 1$, a polynomial $p \in Q_k$ is uniquely determined by its values on the set $F(B_k)$, where*

$$B_k := \left\{ \left(\frac{i_1}{k}, \frac{i_2}{k}, \ldots, \frac{i_n}{k} \right) \in \mathbb{R}^n;\ i_j \in \{0, 1, \ldots, k\},\ 1 \leq j \leq n \right\}.$$

Proof Given an n-rectangle T, there exists an *invertible diagonal affine mapping*, i.e., of the form $x \in \mathbb{R}^n \to F(x) = Bx + b$, where B is an $n \times n$ *invertible* diagonal matrix and b is a vector in $\mathbb{R}^n$, such that

$$T = F([0,1]^n).$$

Hence, it suffices to consider the case where $T = [0,1]^n$, in which case the result follows from the identity

$$p = \sum_{\substack{0 \le i_j \le k \\ 1 \le j \le n}} \left(\prod_{j=1}^{n} \left(\prod_{\substack{i'_j = 0 \\ i'_j \ne i_j}}^{k} \frac{kx_j - i'_j}{i_j - i'_j} \right) \right) p\left(\frac{i_1}{k}, \frac{i_2}{k}, \ldots, \frac{i_n}{k} \right) \quad \text{for all } p \in Q_k. \qquad \square$$

We now describe a *general framework*,[25] which encompasses all the above examples of *Lagrange interpolation in* $\mathbb{R}^n$: In each case, we are given a set

$$A = \bigcup_{i=1}^{N} \{a_i\}$$

of N *distinct points* of $\mathbb{R}^n$, with the property that their convex hull

$$T := \operatorname{co} A$$

has a *nonempty* interior, and we are also given an *N-dimensional space P of real-valued functions defined over T*, together with a set of N *linear forms* $\varphi_i : P \to \mathbb{R}$, $1 \le i \le N$, *of the particular form*

$$\varphi_i : p \in P \to p(a_i), \quad 1 \le i \le N,$$

with the property that, *given any real numbers μ_i, $1 \le i \le N$, there exists one and only one function p such that*

$$\varphi_i(p) = \mu_i, \quad \text{or equivalently,} \quad p(a_i) = \mu_i, \quad 1 \le i \le N.$$

Remark The reason for introducing such linear forms (which may seem a bit artificial in the present case, since they are all of the same form, i.e., point values) is that their consideration provides a unified framework that works as well for *Hermite interpolation* (not treated here). $\square$

If all the above conditions are satisfied, we say that (A, P) constitutes a **Lagrange interpolation scheme** in $\mathbb{R}^n$.

Several general remarks are in order about this definition. First, *the set T is closed* (hence compact since it is clearly bounded): By Theorem 2.16-1, the convex hull T of the set A is of the form

$$T = \left\{ x \in \mathbb{R}^n;\ x = \sum_{i=1}^{N} \mu_i a_i,\ \sum_{i=1}^{N} \mu_i = 1 \text{ and } \mu_i \ge 0,\ 1 \le i \le N \right\}.$$

[25] Which is in effect the definition of a *Lagrange finite element* proposed in:
P.G. Ciarlet [1975]: *Lectures on the Finite Element Method*, Tata Institute of Fundamental Research, Bombay.

So let $x^k = \sum_{i=1}^N \mu_i^k a_i \in T$ converge to $x \in \mathbb{R}^n$ as $k \to \infty$. Since each sequence $(\mu_i^k)_{k=1}^\infty$, $1 \le i \le N$, is bounded, there exists a subsequence $(x^{\sigma(k)})_{k=1}^\infty$ such that $\mu_i^{\sigma(k)} \to \mu_i$ as $k \to \infty$ for each $1 \le i \le N$. Clearly then, $x^k \to x = \sum_{i=1}^N \mu_i a_i \in T$ as $k \to \infty$.

Second, *the linear forms φ_i, $1 \le i \le N$, are linearly independent and they form a basis of the dual space* of P.

Third, *there exist uniquely defined functions $p_i \in P$, $1 \le i \le N$,* such that $\varphi_j(p_i) = \delta_{ij}$, $1 \le j \le N$, *and the following identity holds*:

$$p = \sum_{i=1}^N \varphi_i(p)p_i, \quad \text{or equivalently,} \quad p = \sum_{i=1}^N p(a_i)p_i \quad \text{for all } p \in P.$$

Hence *the functions p_i, $1 \le i \le N$, form a basis of the space P.*

Fourth, *given a function $v : T \to \mathbb{R}$, there exists one and only one function $\Pi v \in P$ that satisfies*

$$\varphi_i(\Pi v) = \varphi_i(v), \quad \text{or equivalently,} \quad \Pi v(a_i) = v(a_i), \quad 1 \le i \le N.$$

This function Πv, which is thus given by

$$\Pi v = \sum_{i=1}^N \varphi_i(v)p_i = \sum_{i=1}^N v(a_i)p_i,$$

is called the **Lagrange interpolant** *of* v (it being implicitly understood that it corresponds to a given Lagrange interpolation scheme (A, P)).

Given a normed vector space $V(T)$ of functions $v : T \to \mathbb{R}$, typically such as $\mathcal{C}^m(T)$ or $W^{m,r}(\mathring{T})$, the *basic problem of Lagrange interpolation in* $\mathbb{R}^n$ then consists in seeking sufficient conditions guaranteeing that *the* **interpolation error** $\|\Pi v - v\|_{V(T)}$ *can be made as small as one pleases if the diameter of T is small enough.*

When $V(T) = \mathcal{C}^m(T)$, the estimate of the interpolation error rests in particular on the following result, which in essence asserts that, if $P_k(T) \subset P$, then *any* mth *derivative,* $0 \le m \le k$, *of the difference $\Pi v - v$ depends only on the* $(k+1)$st *derivative of the function* v.

Recall that, for each integer $m \ge 1$, $v^{(m)}(x) \in \mathcal{L}_m(\mathbb{R}^n; \mathbb{R})$ denotes the mth order derivative of a function v at x, and that $v^{(0)}(x) := v(x)$ (Section 7.8).

Theorem 7.11-4[26] *Let (A, P) be a Lagrange interpolation scheme, where $A := \bigcup_{i=1}^N \{a_i\}$ and the space P satisfies in inclusions*

$$P_k(T) \subset P \subset \mathcal{C}^k(T) \quad \textit{for some integer } k \ge 0, \textit{ where } T := \operatorname{co} A.$$

Then, given a function $v \in \mathcal{C}^{k+1}(T)$, its Lagrange interpolant $\Pi v \in P$, given by

$$\Pi v = \sum_{i=1}^N v(a_i)p_i,$$

[26]Due to:
P.G. Ciarlet; P.A. Raviart [1972]: General Lagrange and Hermite interpolation in $\mathbb{R}^n$ with applications to finite element methods, *Archive for Rational Mechanics and Analysis* **46**, 177–199.
The simpler proof given here rests on an observation due to Rémi Arcangéli (private communication).

satisfies

$$\Pi v^{(m)}(x) - v^{(m)}(x) = \frac{1}{k!}\sum_{i=1}^{N}\left(\int_0^1 (1-t)^k \left(v^{(k+1)}(ta_i + (1-t)x)(a_i - x)^{k+1}\right) \mathrm{d}t\right) p_i^{(m)}(x)$$

at each $x \in T$ *and for each integer* $0 \le m \le k$.

Proof (i) Since the boundary of T is Lipschitz-continuous (as the boundary of the convex hull of a finite subset of $\mathbb{R}^N$), each space $\mathcal{C}^m(T)$, $0 \le m \le k+1$, is defined as

$$\mathcal{C}^m(T) := \{v_{|T};\ v \in \mathcal{C}^m(\mathbb{R}^n)\}$$

(Theorem 1.18-1). Given a function $v \in \mathcal{C}^{k+1}(T)$, the *Taylor formula with integral remainder* (Theorem 7.9-1(d)) therefore holds for all points a, x in the convex set T:

$$v(a) = v(x) + v'(x)(a-x) + \cdots + \frac{1}{k!}v^{(k)}(x)(a-x)^k + \mathcal{R}\big(v^{(k+1)}; a, x\big),$$

where

$$\mathcal{R}(v^{(k+1)}; a, x) := \frac{1}{k!}\int_0^1 (1-t)^k \left(v^{(k+1)}(ta + (1-t)x)(a-x)^{k+1}\right) \mathrm{d}t.$$

Consequently, at each point $x \in T$, the mth derivative of the Lagrange interpolant is given for each integer $0 \le m \le k$ by

$$\begin{aligned}(\Pi v)^{(m)}(x) &= \sum_{i=1}^{N} v(a_i) p_i^{(m)}(x) \\ &= \sum_{\ell=0}^{k}\left(\frac{1}{\ell!}\sum_{i=1}^{N}\left(v^{(\ell)}(x)(a_i - x)^{\ell}\right) p_i^{(m)}(x)\right) + \sum_{i=1}^{N}\mathcal{R}(v^{(k+1)}; a_i, x) p_i^{(m)}(x).\end{aligned}$$

(ii) *Let there be given a symmetric* ℓ*-linear continuous mapping* $A_\ell \in \mathcal{L}_\ell(\mathbb{R}^n;\mathbb{R})$, *for some integer* ℓ *satisfying* $0 \le \ell \le k$ (with $\mathcal{L}_0(\mathbb{R}^n;\mathbb{R})$ identified with $\mathbb{R}$). *Then*

$$\frac{1}{\ell!}\sum_{i=1}^{N}\left(A_\ell (a_i - x)^{\ell}\right) p_i^{(m)}(x) = A_\ell \delta_{\ell m}$$

at each $x \in T$ *and for each integer* $0 \le m \le k$.

By assumption, $p(x) = \sum_{i=1}^{N} p(a_i)p_i(x)$ for all $p \in P$ and all $x \in T$, and $P_k(T) \subset P$. Fix a point $y \in \mathbb{R}^n$. That the function

$$p_y : x \in \mathbb{R}^n \to p_y(x) := A_\ell (x-y)^\ell$$

is a polynomial of degree $\ell \le k$ then implies that

$$p_y(x) = \sum_{i=1}^{N}\left(A_\ell (a_i - y)^\ell\right) p_i(x) \quad \text{for all } x \in T,$$

which in turn implies that

$$p_y^{(m)}(x) = \sum_{i=1}^N \left(A_\ell(a_i - y)^\ell\right) p_i^{(m)}(x) \quad \text{for all } x \in T \text{ and all } 0 \le m \le k.$$

If $m \le \ell - 1$, the mth derivative $p_y^{(m)}(x)$ at any $x \in T$ is a sum of terms containing A_ℓ applied to an m-uple of vectors of $\mathbb{R}^n$ that contains at least once the vector $(x - y)$. The continuity of A_ℓ therefore implies that

$$0 = \lim_{y \to x} p_y^{(m)}(x) = \lim_{y \to x} \sum_{i=1}^N \left(A_\ell(a_i - y)^\ell\right) p_i^{(m)}(x) = \sum_{i=1}^N \left(A_\ell(a_i - x)^\ell\right) p_i^{(m)}(x) \quad \text{for all } x \in T.$$

If $m \ge \ell$,

$$p_y^{(m)}(x) = \ell! A_\ell \delta_{\ell m} = \sum_{i=1}^N \left(A_\ell(a_i - y)^\ell\right) p_i^{(m)}(x) \quad \text{for all } x \in T$$

(the assumed symmetry of A_ℓ is used here), so that, thanks again to the continuity of A_ℓ,

$$\ell! A_\ell \delta_{\ell m} = \lim_{y \to x} \sum_{i=1}^N \left(A_\ell(a_i - y)^\ell\right) p_i^{(m)}(x) = \sum_{i=1}^N \left(A_\ell(a_i - x)^\ell\right) p_i^{(m)}(x) \quad \text{for all } x \in T.$$

Hence the relation announced in (ii) holds for each integer $0 \le m \le k$.

(iii) The particular choices $A_\ell := v^{(\ell)}(x)$, $0 \le \ell \le k$, in (ii) show that

$$\sum_{\ell=0}^k \frac{1}{\ell!} \left(\sum_{i=1}^N \left(v^{(\ell)}(x)(a_i - x)^\ell\right) p_i^{(m)}(x) \right) = v^{(m)}(x) \quad \text{for all } x \in T,$$

which completes the proof. □

If the function v is only assumed to be in the space $\mathcal{C}^k(T)$ and $(k+1)$ times differentiable in the open set $\mathring{T}$, the Taylor formula with integral remainders has to be replaced by the *Taylor–MacLaurin formula* (Theorem 7.9-1(c)). As a result, the remainder $\sum_{i=1}^N \mathcal{R}(v^{(k+1)}; a_i, x) p_i^{(m)}(x)$ has to be replaced in this case by

$$\frac{1}{(k+1)!} \sum_{i=1}^N \left(v^{(k+1)}(\eta_i(x))(a_i - x)^{k+1}\right) p_i^{(m)}(x) \quad \text{for some points } \eta_i(x) \in \,]x, a_i[\,, \ 1 \le i \le N.$$

The notations and assumptions being those of Theorem 7.11-4, its special case $m = 0$ shows that any function $v \in \mathcal{C}^{k+1}(T)$, *resp.* any function $v \in \mathcal{C}^k(T)$ that is k times differentiable in $\mathring{T}$, can be expanded at each $x \in T$ as

$$v(x) = \sum_{i=1}^N v(a_i) p_i(x) + \mathsf{R}(v^{(k+1)}; x),$$

where

$$\mathsf{R}(v^{(k+1)};x) := -\frac{1}{k!}\sum_{i=1}^{N}\left(\int_0^1 (1-t)^k\left(v^{(k+1)}(ta_i+(1-t)x)(a_i-x)^{k+1}\right)\mathrm{d}t\right)p_i(x),$$

resp.,

$$\mathsf{R}(v^{(k+1)};x) := -\frac{1}{(k+1)!}\sum_{i=1}^{N}\left(v^{(k+1)}(\eta_i(x))(a_i-x)^{k+1}\right)p_i(x)$$

for some points $\eta_i(x) \in \left]x, a_i\right[$, $1 \le i \le N$.

Since the factors of the point values $v(a_i)$ are functions *independent of the function* v, and since the function v appears *only by means of its* $(k+1)$*st derivative* in either *remainder* $\mathsf{R}(v^{(k+1)};x)$, such an expansion thus provides an example of a **multipoint Taylor formula.**[27]

As an illustration, let us return for instance to our first and second examples and apply Theorem 7.11-4 with $m=0$. This shows that, given an n-simplex T with vertices a_i, $1 \le i \le n+1$, any function $v \in \mathcal{C}^1(T)$ that is two times differentiable in $\mathring{T}$ can be expanded as the following *multipoint Taylor formula* (recall that the functions λ_i, $1 \le i \le n+1$, denote the barycentric coordinates with respect to the vertices of T):

$$v(x) = \sum_{i=1}^{n+1} v(a_i)\lambda_i(x) - \frac{1}{2}\sum_{i=1}^{n+1}\left(v''(\eta_i(x))(a_i-x)^2\right)\lambda_i(x) \quad \text{for all } x \in \mathbb{R}^n,$$

where $\eta_i(x) \in \left]x, a_i\right[$, $1 \le i \le n+1$. Likewise, given an n-simplex T with vertices a_i, $1 \le i \le n+1$, and midpoints of the edges $a_{ij} = \frac{1}{2}(a_i + a_j)$, $1 \le i \le j \le n+1$, any function $v \in \mathcal{C}^2(T)$ that is three times differentiable in $\mathring{T}$ can be expanded as the following *multipoint Taylor formula*:

$$\begin{aligned} v(x) = {} & \sum_{i=1}^{n+1} v(a_i)\lambda_i(x)(2\lambda_i(x)-1) + \sum_{1\le i<j\le n+1} v(a_{ij})4\lambda_i(x)\lambda_j(x) \\ & - \frac{1}{6}\sum_{i=1}^{n+1}\left(v^{(3)}(\eta_i(x))(a_i-x)^3\right)\lambda_i(x)(2\lambda_i(x)-1) \\ & - \frac{2}{3}\sum_{1\le i<j\le n+1}\left(v^{(3)}(\eta_{ij}(x))(a_{ij}-x)^3\right)\lambda_i(x)\lambda_j(x) \quad \text{for all } x \in \mathbb{R}^n, \end{aligned}$$

where $\eta_i(x) \in \left]x, a_i\right[$, $1 \le i \le n+1$, and $\eta_{ij}(x) \in \left]x, a_{ij}\right[$, $1 \le i, j \le n+1$.

The estimate of the interpolation error also crucially rests on the notion (defined below) of *affine-equivalent Lagrange interpolation schemes,* which itself rests on the following result.

[27]The first examples of such multipoint Taylor formulas were given in:

C. Coatmélec [1966]: Approximation et interpolation des fonctions différentiables de plusieurs variables, *Annales Scientifiques de l'Ecole Normale Supérieure* **83**, 271–341.

P.G. Ciarlet; C. Wagschal [1971]: Multipoint Taylor formulas and applications to the finite element method, *Numerische Mathematik* **17**, 84–100.

Theorem 7.11-5 *Let $(\widehat{A}, \widehat{P})$ be a Lagrange interpolation scheme, where*

$$\widehat{A} = \bigcup_{i=1}^{N} \{\widehat{a}_i\},$$

and let

$$F : x \in \mathbb{R}^n \to F(x) := Bx + b \in \mathbb{R}^n$$

be an invertible affine mapping (i.e., B is an invertible $n \times n$ matrix and $b \in \mathbb{R}^n$).

(a) *Define the set*

$$A := \bigcup_{i=1}^{N} \{F(\widehat{a}_i)\}$$

and the space

$$P := \{p : T \to \mathbb{R};\ p = \widehat{p} \circ F^{-1},\ \widehat{p} \in \widehat{P}\}, \quad \textit{where } T := \operatorname{co} A.$$

Then (A, P) is also a Lagrange interpolation scheme.

(b) *Let*

$$\begin{aligned} \widehat{h} &:= \operatorname{diam} \widehat{T}, \quad & \widehat{\rho} &:= \sup\{\operatorname{diam} \widehat{U};\ \widehat{U} \textit{ is a ball contained in } \widehat{T}\}, \quad \textit{where } \widehat{T} := \operatorname{co} \widehat{A}, \\ h_T &:= \operatorname{diam} T, \quad & \rho_T &:= \sup\{\operatorname{diam} U;\ U \textit{ is a ball contained in } T\}. \end{aligned}$$

Then

$$|B| \le \frac{h_T}{\widehat{\rho}} \quad \textit{and} \quad |B^{-1}| \le \frac{\widehat{h}}{\rho_T}.$$

Proof It is clear that $\operatorname{int} F(\widehat{T}) \neq \varnothing$ and $T = F(\widehat{T})$ and that functions $\widehat{p}_i \in \widehat{P}$ uniquely defined by the relations $\widehat{p}_i(\widehat{a}_j) = \delta_{ij}$, $1 \le i, j \le N$, form a basis in the space $\widehat{P}$. Let

$$a_i := F(\widehat{a}_i), \quad 1 \le i \le N.$$

It is then immediately verified that the functions

$$p_i := \widehat{p}_i \circ F^{-1}, \quad 1 \le i \le N,$$

which belong to the space P, satisfy $p_i(a_i) = \delta_{ij}$, $1 \le i, j \le N$.

Hence the functions p_i, $1 \le i \le N$, form a basis of the space P, and (A, P) is thus also a Lagrange interpolation scheme. This proves (a).

Since $\widehat{\rho} > 0$ (the interior of $\widehat{T}$ is nonempty by assumption),

$$|B| = \frac{1}{\widehat{\rho}} \sup_{\begin{cases} \widehat{\xi} \in \mathbb{R}^n \\ |\widehat{\xi}| = \widehat{\rho} \end{cases}} |B\widehat{\xi}|.$$

Each vector $\widehat{\xi} \in \mathbb{R}^n$ with $|\widehat{\xi}| = \widehat{\rho}$ can be written as $\widehat{\xi} = \widehat{y} - \widehat{z}$ with $\widehat{y}, \widehat{z} \in \widehat{T}$ (by definition of $\widehat{\rho}$); hence

$$B\widehat{\xi} = (B\widehat{y} + b) - (B\widehat{z} + b) = y - z \quad \text{with } y, z \in T.$$

Consequently, $|B\widehat{\xi}| \leq h_T$ for such a vector $\widehat{\xi}$ (by definition of h_T), which shows that $|B| \leq \dfrac{h_T}{\widehat{\rho}}$.

The proof of the inequality $|B| \leq \dfrac{\widehat{h}}{\rho_T}$ is similar. This proves (b). $\square$

Motivated by the theorem above, we say that two Lagrange interpolation schemes (A, P) and $(\widehat{A}, \widehat{P})$, where $\widehat{A} = \bigcup_{i=1}^N \{\widehat{a}_i\}$ and $A = \bigcup_{i=1}^N \{a_i\}$, are **affine-equivalent** if *there exists an invertible affine* mapping $F : \mathbb{R}^n \to \mathbb{R}^n$ such that

$$a_i = F(\widehat{a}_i), \quad 1 \leq i \leq N, \quad \text{and} \quad P = \{p : \operatorname{co} A \to \mathbb{R},\, p = \widehat{p} \circ F^{-1},\, \widehat{p} \in \widehat{P}\}.$$

We are now in a position to prove the main result of this section (similar error estimates, but this time in terms of Sobolev norms and seminorms, can be also derived; cf. Problem 7.11-5). In this respect, recall that there exist constants $C(m, n)$ such that, for any function $w \in \mathcal{C}^{k+1}(T)$ and any integer $1 \leq m \leq k+1$,

$$\max_{|\boldsymbol{\alpha}|=m} |\partial^{\boldsymbol{\alpha}} w(x)| \leq \|w^{(m)}(x)\| \leq C(m,n) \max_{|\boldsymbol{\alpha}|=m} |\partial^{\boldsymbol{\alpha}} w(x)| \quad \text{for all } x \in T,$$

where $\|\cdot\|$ denotes here the norm in the space $\mathcal{L}_m(\mathbb{R}^n; \mathbb{R})$ (Section 7.8).

Theorem 7.11-6 (Lagrange interpolation error estimates)[28] *Let $(\widehat{A}, \widehat{P})$ be a Lagrange interpolation scheme such that*

$$P_k(\widehat{T}) \subset \widehat{P} \subset \mathcal{C}^k(\widehat{T}) \quad \textit{for some integer } k \geq 0, \textit{ where } \widehat{T} := \operatorname{co} \widehat{A}.$$

Then there exist constants $C_m = C_m(\widehat{A}, \widehat{P})$, $0 \leq m \leq k$, which are the same for all Lagrange interpolation schemes (A, P) that are affine-equivalent to $(\widehat{A}, \widehat{P})$, such that the Lagrange interpolant $\Pi v \in P$ of any function $v \in \mathcal{C}^{k+1}(T)$, where $T = \operatorname{co} A$, satisfies

$$\sup_{x \in T} |\Pi v(x) - v(x)| \leq C_0 h_T^{k+1} \sup_{\xi \in T} \|v^{(k+1)}(\xi)\|,$$

$$\sup_{x \in T} \|\Pi v^{(m)}(x) - v^{(m)}(x)\| \leq C_m \frac{h_T^{k+1}}{\rho_T^m} \sup_{\xi \in T} \|v^{(k+1)}(\xi)\|, \quad 1 \leq m \leq k,$$

where

$$h_T := \operatorname{diam} T \quad \textit{and} \quad \rho_T = \sup\{\operatorname{diam} U;\ U \textit{ is a ball contained in } T\}.$$

Proof By Theorem 7.11-4 and with the notations of this theorem,

$$\Pi v^{(m)}(x) - v^{(m)}(x) = \frac{1}{k!} \sum_{i=1}^N \left(\int_0^1 (1-t)^k \left(v^{(k+1)}(ta_i + (1-t)x)(a_i - x)^{k+1} \right) \mathrm{d}t \right) p_i^{(m)}(x)$$

at each $x \in T$ and for each integer $0 \leq m \leq k$.

---

[28]Like the notion of affine-equivalence, this theorem is due to:
P.G. Ciarlet, P.A. Raviart [1972]: General Lagrange and Hermite interpolation in $\mathbb{R}^n$ with applications to finite element methods, *Archive for Rational Mechanics and Analysis* **46**, 177–199.
It is also shown in *ibid.* that similar error estimates can be derived for *Hermite interpolation schemes.*

By definition of h_T,

$$|v^{(k+1)}(ta_i + (1-t)x)(a_i - x)^{k+1}| \le \sup_{\xi \in T} \|v^{(k+1)}(\xi)\| \, h_T^{k+1} \quad \text{at each } x \in T,$$

since $[a_i, x] \subset T$, $1 \le i \le N$.

Given any Lagrange interpolation scheme (A, P) that is affine-equivalent to $(\widehat{A}, \widehat{P})$, let $F : x \in \mathbb{R}^n \to F(x) = Bx + b \in \mathbb{R}^n$ denote the associated invertible affine mapping. For each $1 \le i \le N$, the functions $p_i : T \to \mathbb{R}$ and $\widehat{p}_i := p_i \circ F : \widehat{T} \to \mathbb{R}$ are related at each $x \in T$ by

$$p_i(x) = \widehat{p}_i(F^{-1}(x)),$$
$$p_i^{(m)}(x)(\xi_1, \xi_2, \dots, \xi_m) = \widehat{p}_i^{(m)}(F^{-1}(x))(B^{-1}\xi_1, B^{-1}\xi_2, \dots, B^{-1}\xi_m)$$

for each integer $1 \le m \le k$ and for all vectors $\xi_\mu \in \mathbb{R}^n$, $1 \le \mu \le m$ (to see this, use the chain rule and that F is affine). Therefore,

$$\|p_i^{(m)}(x)\| = \sup_{\left\{\substack{\|\xi_\mu\|=1 \\ 1 \le \mu \le m}\right.} |p_i^{(m)}(x)(\xi_1, \xi_2, \dots, \xi_m)| \le \|\widehat{p}_i^{(m)}(F^{-1}(x))\| \, |B^{-1}|^m \quad \text{at each } x \in T,$$

so that, by Theorem 7.11-5,

$$\sup_{x \in T} \|p_i^{(m)}(x)\| \le \sup_{\widehat{x} \in \widehat{T}} \|\widehat{p}_i^{(m)}(\widehat{x})\| \, \frac{\widehat{h}^m}{\rho_T^m}, \quad 1 \le m \le k.$$

The announced error estimates therefore hold with

$$C_0 := \frac{1}{(k+1)!} \sum_{i=1}^{N} \sup_{\widehat{x} \in \widehat{T}} |\widehat{p}_i(\widehat{x})| \quad \text{and} \quad C_m := \frac{\widehat{h}^m}{(k+1)!} \sum_{i=1}^{N} \sup_{\widehat{x} \in \widehat{T}} \|\widehat{p}_i^{(m)}(\widehat{x})\|, \quad 1 \le m \le k. \ \square$$

Naturally, *estimates for the usual partial derivatives* immediately follow from the above estimates, since for each integer $1 \le m \le k$,

$$\max_{|\boldsymbol{\alpha}|=m} |\partial^{\boldsymbol{\alpha}} \Pi v(x) - \partial^{\boldsymbol{\alpha}} v(x)| \le \|\Pi v^{(m)}(x) - v^{(m)}(x)\| \quad \text{at each } x \in T.$$

This observation will be also put to use in Theorem 7.11-7.

Also note that the constant $C_0 = \frac{1}{(k+1)!} \sum_{i=1}^N \sup_{x \in \widehat{T}} |\widehat{p}_i(\widehat{x})|$ found above is nothing but the n-dimensional analogue of the *Lebesgue constants* found in the analysis of *Lagrange interpolation in dimension one* (Section 5.4).

Theorem 7.11-6 applies to *all* the examples given in this section. Consider for instance our second and third examples.

Let $\widehat{T}$ denote an n-simplex with vertices $\widehat{a}_i$, $1 \le i \le n+1$, considered as *fixed* once and for all. Then, given any n-simplex with vertices a_i, $1 \le i \le n+1$, *there exists a unique invertible affine mapping* $F : \mathbb{R}^n \to \mathbb{R}^n$ *such that* $F(\widehat{a}_i) = a_i$, $1 \le i \le n+1$. Then it *automatically follows* that (with self-explanatory notations)

$$F(\widehat{a}_{ij}) = a_{ij}, \ \ i < j, \quad F(\widehat{a}_{iij}) = a_{iij}, \ \ i \ne j, \quad F(\widehat{a}_{ij\ell}) = a_{ij\ell}, \ \ i < j < \ell.$$

Besides, it is clear that

$$P_k(T) = \{\widehat{p} \circ F^{-1};\ \widehat{p} \in P_K(\widehat{T})\} \quad \text{for any integer } k \geq 0.$$

Therefore there exist constants, all denoted by the same letter C for convenience, such that the Lagrange interpolant $\Pi_2 v \in P_2(T)$ of any function $v \in \mathcal{C}^3(T)$ satisfies

$$\max_{|\alpha|=m} \sup_{x \in T} |\partial^\alpha \Pi_2 v(x) - \partial^\alpha v(x)| \leq C \frac{h_T^3}{\rho_T^m} \sup_{\xi \in T} \|v^{(3)}(\xi)\|, \quad 0 \leq m \leq 2,$$

while the Lagrange interpolant $\Pi_3 v \in P_3(T)$ of any function $v \in \mathcal{C}^4(T)$ satisfies

$$\max_{|\alpha|=m} \sup_{x \in T} |\partial^\alpha \Pi_3 v(x) - \partial^\alpha v(x)| \leq C \frac{h_T^4}{\rho_T^m} \sup_{\xi \in T} \|v^{(4)}(\xi)\|, \quad 0 \leq m \leq 3.$$

To conclude this analysis, we show how to dispose of the parameter ρ_T in the interpolation error estimates of Theorem 7.11-6 for the derivatives, simply by considering interpolation schemes where the sets T are not "too flat" in the following sense.[29]

We say that $(A_T, P_T)_{T \in \mathcal{T}}$ is a **regular family** *of Lagrange interpolation schemes* if there exists a constant σ such that

$$\frac{h_T}{\rho_T} \leq \sigma \quad \text{for all } T \in \mathcal{T}$$

(here, $T = \text{co}\, A_T$ is in effect viewed as the parameter that defines the family). Thanks to this definition, the error estimates of Theorem 7.11-6 can be immediately converted into estimates that involve only the diameter h_T.

Theorem 7.11-7 (Lagrange interpolation error estimates for a regular family) *Let there be given a regular family $(A_T, P_T)_{T \in \mathcal{T}}$ of Lagrange interpolation schemes that are all affine-equivalent to a Lagrange interpolation scheme $(\widehat{A}, \widehat{P})$ that satisfies*

$$P_k(\widehat{T}) \subset \widehat{P} \subset \mathcal{C}^k(\widehat{T}) \quad \textit{for some integer } k \geq 0, \ \textit{where } \widehat{T} := \text{co}\, \widehat{A}.$$

Then there exists a constant C such that, for any $T \in \mathcal{T}$, the Lagrange interpolant $\Pi_T v \in P_T$ of any function $v \in \mathcal{C}^{k+1}(T)$ satisfies

$$\max_{|\alpha|=m} \sup_{x \in T} |\partial^\alpha \Pi_T(x) - \partial^\alpha v(x)| \leq C h_T^{k+1-m} \sup_{\xi \in T} \|v^{(k+1)}(\xi)\|, \quad 0 \leq m \leq k. \qquad \square$$

In our analysis of Lagrange interpolation in *dimension one* (Section 5.4), *the set $T = [a, b]$ was fixed, while the degree of the interpolating polynomials was increasing.* By contrast, the

[29] This notion can be further refined, as first noted by:

P. Jamet [1976]: Estimation d'erreur pour des éléments finis droits presque dégénérés, *Revue Française d'Automatique, Informatique, Recherche Opérationnelle, Série Rouge: Analyse Numérique* **10**, 43–61.

I. Babuška; A.K. Aziz [1976]: On the angle condition in the finite element method, *SIAM Journal on Numerical Analysis* **13**, 214–226.

Recent developments and references about this notion are found in:

J. Brandts; S. Korotov; M. Křížek [2011]: Generalization of the Zlámal condition for simplicial finite elements in $\mathbb{R}^d$, *Applied Mathematics* **56**, 417–424.

present analysis applies to a family of affine-equivalent Lagrange interpolation schemes where *the degree k is fixed, while the diameter h of T approaches zero.*

Problems

7.11-1 The notations are those of Theorem 7.11-2.

(1) Show that $\dim \widetilde{P}_3 = \operatorname{card} \widetilde{A}_3$; then infer from this relation and Theorem 7.11-1 (with $k = 3$) that any polynomial in the space $\widetilde{P}_3$ is uniquely determined by its values on the set $\widetilde{A}_3$.

(2) Given a polynomial $p \in P_2$ (in which case $p \in P_2 \to p'' \in \mathcal{L}_2(\mathbb{R}^n; \mathbb{R})$ is a constant mapping), deduce from the Taylor formulas $p(a_m) = p(a_{ij\ell}) + \cdots$ and $p(a_{rrs}) = p(a_{ij\ell}) + \cdots$ that $\varphi_{ij\ell}(p) = 0$, $i < j < \ell$, thus showing that $P_2 \subset \widetilde{P}_3$.

7.11-2 (1) Let the points b_i, $1 \le i \le 9$, be as in Figure 7.11-2. Show that any polynomial p in the space

$$\widetilde{Q}_2 := \left\{ p \in Q_2;\ 4p(b_9) + \sum_{i=1}^{4} p(b_i) - 2\sum_{i=5}^{8} p(b_i) = 0 \right\}$$

is uniquely defined by its values on the set $\bigcup_{i=1}^{8} \{b_i\}$.

(2) Show that the inclusion $P_2 \subset \widetilde{Q}_2$ holds.

7.11-3 (1) Let the points c_i, $1 \le i \le 16$, be as in Figure 7.11-2. Show that any polynomial p in the space

$$\widetilde{Q}_3 := \{p \in Q_3;\ \psi_i(p) = 0,\ 0 \le i \le 3\},$$

where

$$\begin{aligned} \psi_i(p) := {} & 4p(c_{1+i}) + 2p(c_{2+i}) + p(c_{3+i}) + 2p(c_{4+i}) - 6p(c_{5+i}) \\ & - 3p(c_{6+i}) - 3p(c_{11+i}) - 6p(c_{12+i}) + 9p(c_{13+i}), \quad 0 \le i \le 3, \end{aligned}$$

is uniquely defined by its values on the set $\bigcup_{i=1}^{12} \{c_i\}$.

(2) Show that the inclusion $P_3 \subset \widetilde{Q}_3$ holds.

7.11-4 The notations and assumptions are those of Theorem 7.11-4. Can the expressions of the difference $(\Pi v^{(m)}(x) - v^{(m)}(x))$, $x \in T$, $1 \le m \le k$, found in this theorem be obtained by differentiating the expression found in this theorem for $m = 0$?

7.11-5 The object of this problem is to derive interpolation error estimates similar to those of Theorem 7.11-6, but instead expressed in terms of *Sobolev seminorms*.[30]

(1) Let $\widehat{\Omega}$ and Ω be two domains in $\mathbb{R}^n$ with the following property: There exist an $n \times n$ invertible matrix B and a vector $b \in \mathbb{R}^n$ such that $\Omega = F(\widehat{\Omega})$, where $F(x) := Bx + b$ for all $x \in \mathbb{R}^n$. Show that, if a function v belongs to the Sobolev space $W^{m,q}(\Omega)$ for some integer $m \ge 0$ and some extended real number $1 \le q \le \infty$, the function $\widehat{v} := v \circ F$ belongs to the space $W^{m,q}(\widehat{\Omega})$ and there exists a constant $C = C(m, n)$ such that (Sobolev seminorms such as $|\cdot|_{m,q,\Omega}$ are defined in Section 6.5)

$$\begin{aligned} |\widehat{v}|_{m,q,\widehat{\Omega}} &\le C\,|B|^m\,|\det B|^{-1/q}\,|v|_{m,q,\Omega} && \text{for all } v \in W^{m,q}(\Omega), \\ |v|_{m,q,\Omega} &\le C\,|B^{-1}|^m\,|\det B|^{1/q}\,|\widehat{v}|_{m,q,\widehat{\Omega}} && \text{for all } \widehat{v} \in W^{m,q}(\widehat{\Omega}). \end{aligned}$$

[30]Such error estimates for affine-equivalent Lagrange interpolation schemes are due to:

P.G. Ciarlet; P.A. Raviart [1972]: General Lagrange and Hermite interpolation in $\mathbb{R}^n$ with applications to finite element methods, *Archive for Rational Mechanics and Analysis* **46**, 177–199.

(2) Let there be given a Lagrange interpolation scheme $(\widehat{A}, \widehat{P})$ such that the following inclusions hold for some integers $k \geq 0$ and $m \geq 0$ and extended real numbers $1 \leq q \leq \infty$ and $1 \leq r \leq \infty$:

$$\begin{aligned} W^{k+1,q}(\operatorname{int} \widehat{T}) &\hookrightarrow \mathcal{C}^0(\widehat{T}) \quad \text{where } \widehat{T} := \operatorname{co} \widehat{A}, \\ W^{k+1,q}(\operatorname{int} \widehat{T}) &\hookrightarrow W^{m,r}(\operatorname{int} \widehat{T}), \\ P_k(\widehat{T}) \subset \widehat{P} &\subset W^{m,r}(\operatorname{int} \widehat{T}). \end{aligned}$$

Show that there exists a constant $C = C(\widehat{A}, \widehat{P})$, which is the same for all Lagrange interpolation schemes (A, P) that are affine-equivalent to $(\widehat{A}, \widehat{P})$, such that the Lagrange interpolant $\Pi v \in P$ of any function $v \in W^{k+1,q}(\operatorname{int} T)$, where $T := \operatorname{co} A$ (the first inclusion above insures that Πv is well defined) satisfies

$$|v - \Pi v|_{m,r,\operatorname{int} T} \leq C(\operatorname{meas} T)^{1/r-1/q} \frac{h_T^{k+1}}{\rho_T^m} |v|_{k+1,q,\operatorname{int} T},$$

where $h_T := \operatorname{diam} T$ and $\rho_T := \sup\{\operatorname{diam} B;\ B \text{ is a ball contained in } T\}$.

Hint: Use Problem 6.6-5 on the set $\widehat{T}$, combined with question (1).

7.11-6 Let T be an n-simplex with vertices a_i, $1 \leq i \leq n+1$, and let $a_{ij\ell} := \frac{1}{3}(a_i + a_j + a_\ell)$, $1 \leq i < j < \ell \leq n+1$. Show that any polynomial in the space P_3 is uniquely determined by its values $(p(a_i))$ and by those of its Fréchet derivatives $p'(a_i) \in \mathcal{L}(\mathbb{R}^n)$ at the vertices a_i, $1 \leq i \leq n+1$, and by its values $p(a_{ij\ell})$ at the points $a_{ij\ell}$, $1 \leq i < j < \ell \leq n+1$.

7.11-7 Let T be a triangle with vertices a_i, $1 \leq i \leq 3$, and for each $1 \leq i \leq 3$, let b_i denote the midpoint of the side of T opposite to a_i. Show that any polynomial in the space P_5 is uniquely determined by its values $p(a_i)$ and those of its first and second derivatives $p'(a_i) \in \mathcal{L}(\mathbb{R}^2)$ and $p''(a_i) \in \mathcal{L}_2(\mathbb{R}^2)$ at the vertices a_i, $1 \leq i \leq 3$, and by the values of the Gâteaux derivatives $p'(b_i)\,(a_i - b_i)$, $1 \leq i \leq 3$.

7.12 Convex functions and differentiability; application to extrema of real-valued functions

Our first objective is to characterize *convex* and *strictly convex* functions (Section 2.17) in terms of the *first derivative* (Theorem 7.12-1), or in terms of the second derivative (Theorem 7.12-2).

Theorem 7.12-1 (convexity and the first derivative) *Let Ω be an open subset of a normed vector space V, let $J : \Omega \subset V \to \mathbb{R}$ be a function differentiable in Ω, and let U be a convex subset of Ω. Then:*

(a) *The function J is convex over U if and only if*

$$J(v) \geq J(u) + J'(u)(v-u) \quad \textit{for all } u, v \in U.$$

(b) *The function J is strictly convex over U if and only if*

$$J(v) > J(u) + J'(u)(v-u) \quad \textit{for all } u, v \in U, u \neq v.$$

Proof Let u and v be two distinct points of U and let $0 < \theta < 1$ be given. If the function J is convex, then

$$J(u + \theta(v-u)) \leq (1-\theta)J(u) + \theta J(v),$$

which can also be written as

$$\frac{J(u+\theta(v-u))-J(u)}{\theta} \le J(v)-J(u).$$

Consequently,

$$J'(u)(v-u) = \lim_{\theta\to 0} \frac{J(u+\theta(v-u))-J(u)}{\theta} \le J(v)-J(u).$$

If the function J is strictly convex, the preceding argument needs to be refined, since it does not produce a strict inequality when θ approaches zero. So, let $0<\omega<1$ be a fixed number. Since

$$u+\theta(v-u) = \frac{\omega-\theta}{\omega}u + \frac{\theta}{\omega}(u+\omega(v-u)) \quad \text{for all } 0\le\theta\le\omega,$$

the convexity of J implies that

$$J(u+\theta(v-u)) \le \frac{\omega-\theta}{\omega}J(u) + \frac{\theta}{\omega}J(u+\omega(v-u)) \quad \text{for all } 0\le\theta\le\omega.$$

Hence, if the function J is strictly convex,

$$\frac{J(u+\theta(v-u))-J(u)}{\theta} \le \frac{J(u+\omega(v-u))-J(u)}{\omega} < J(v)-J(u) \quad \text{for all } 0<\theta\le\omega,$$

since $\omega<1$ by assumption. Consequently,

$$J'(u)(v-u) = \lim_{\theta\to 0}\frac{J(u+\theta(v-u))-J(u)}{\theta} \le \frac{J(u+\omega(v-u))-J(u)}{\omega} < J(v)-J(u)$$

in this case.

Conversely, assume that

$$J(v) \ge J(u)+J'(u)(v-u) \quad \text{for all } u,v\in U.$$

Let u and v be two distinct points of U and let $0<\theta<1$; hence in particular,

$$J(v) \ge J(v+\theta(u-v)) - \theta J'(v+\theta(u-v))(u-v),$$
$$J(u) \ge J(v+\theta(u-v)) + (1-\theta)J'(v+\theta(u-v))(u-v).$$

Adding the two above inequalities multiplied respectively by $(1-\theta)$ and θ then gives

$$J(\theta u+(1-\theta)v) \le \theta J(u)+(1-\theta)J(v),$$

which establishes the convexity of the function J, or its strict convexity if the inequalities are strict. □

Note that the geometric interpretation of the inequalities of (a) is clear if $V=\mathbb{R}$ or if $V=\mathbb{R}^2$ (Figure 7.12-1).

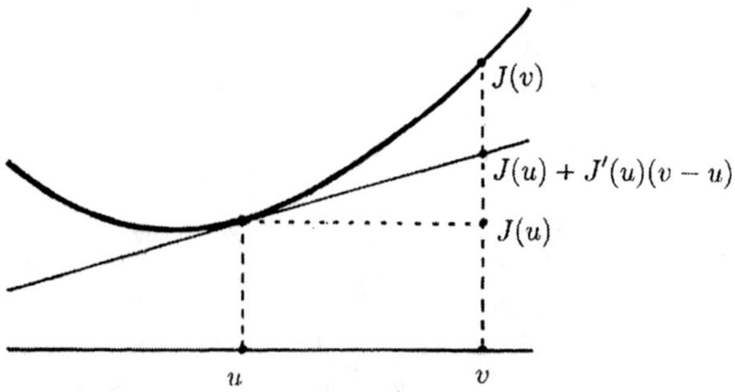

Figure 7.12-1 The inequalities $J(v) \geq J(u) + J'(u)(v-u)$ for all $u, v \in U$ (Theorem 7.12-1(a)) mean that the function is always "above" its tangents if $V = \mathbb{R}$, or "above" its tangent planes if $V = \mathbb{R}^2$. This figure originally appeared in P.G. CIARLET [2007]: *Introduction a l'Analyse Numerique Matricelle et a l'Optimisation*, Dunod, Paris.

Theorem 7.12-2 (convexity and the second derivative) *Let Ω be an open subset of a normed vector space V, let $J : \Omega \subset V \to \mathbb{R}$ be a function twice differentiable in Ω, and let U be a convex subset of Ω: Then:*

(a) *The function J is convex over U if and only if*

$$J''(u)(v-u, v-u) \geq 0 \quad \textit{for all } u, v \in U.$$

(b) *If*

$$J''(u)(v-u, v-u) > 0 \quad \textit{for all } u, v \in U, \quad u \neq v.$$

the function J is strictly convex over U.

Proof Assume that either the inequalities of (a), or those of (b), are satisfied. Let u and v be two distinct points of U. By the Taylor–MacLaurin formula (Theorem 7.9-1(c)), there exists a point $w = u + \theta(v-u)$ with $0 < \theta < 1$ such that

$$\begin{aligned} J(v) - J(u) - J'(u)(v-u) &= \frac{1}{2} J''(w)(v-u, v-u) \\ &= \frac{1}{2\theta^2} J''(w)(u-w, u-w). \end{aligned}$$

The convexity, or the strict convexity, of the function J then follows from Theorem 7.12-1.

Assume that J is convex over U. Given any point $u \in U$, define the auxiliary function $G : \Omega \to \mathbb{R}$ by

$$G : v \in \Omega \to G(v) := J(v) - J'(u)v.$$

Then the function G has a minimum at u *relative to the set* U, since

$$G(v) - G(u) = J(v) - J(u) - J'(u)(v-u) \geq 0 \quad \text{for all } v \in U,$$

by Theorem 7.12-1(a). The function G being twice differentiable in Ω, with $G'' = J''$, the Taylor–Young formula (Theorem 7.9-1(a)) can be applied, showing that, given any $v \in U$,

$$0 \leq G(u+t(v-u)) - G(u) = \frac{t^2}{2}(J''(u)(v-u, v-u) + \delta(t)) \quad \text{for all } 0 \leq t \leq 1, \text{ with } \lim_{t \to 0} \delta(t) = 0,$$

since $G'(u) = 0$. Letting $t \to 0$ then implies that $J''(u)(v-u, v-u) \geq 0$. □

The strictly convex function $J : v \in \mathbb{R} \to J(v) := v^4$ shows that there does not exist in general a converse to (b).

The converse does hold, however, in the particular case of a *quadratic functional over* $\mathbb{R}^n$. Since in this case,

$$J(\boldsymbol{v}) = \frac{1}{2}\boldsymbol{v}^T\boldsymbol{A}\boldsymbol{v} - \boldsymbol{b}^T\boldsymbol{v} \quad \text{for all } \boldsymbol{v} \in \mathbb{R}^n, \text{ with } \boldsymbol{A} = \boldsymbol{A}^T,$$

it follows that

$$J(\boldsymbol{v}) - J(\boldsymbol{u}) - J'(\boldsymbol{u})(\boldsymbol{v} - \boldsymbol{u}) = \frac{1}{2}(\boldsymbol{v} - \boldsymbol{u})^T\boldsymbol{A}(\boldsymbol{v} - \boldsymbol{u}) \quad \text{for all } \boldsymbol{u}, \boldsymbol{v} \in \mathbb{R}^n.$$

Then Theorem 7.12-1 shows that *a quadratic functional over* $\mathbb{R}^n$ *is convex if and only if the symmetric matrix* $\boldsymbol{A}$ *is nonnegative-definite, and strictly convex if and only if the matrix* $\boldsymbol{A}$ *is positive-definite.* Naturally, similar conclusions hold for the more general *quadratic functionals over an arbitrary normed vector space* considered in Section 6.1.

We now focus our attention on *extrema of convex functions*. As shown in Theorem 7.12-3(a) below, an important consequence of the assumption of convexity is that any *local* minimum (as defined in Section 7.1) is in fact a "*global*" one, according to the following definition.

Let $J : U \to \mathbb{R}$ be a function defined over a set U. The function J is said to have a **minimum**, or a **maximum**, at a point $u \in U$ if

$$J(u) \leq J(v), \quad \text{or} \quad J(u) \geq J(v), \quad \text{for all } v \in U,$$

and a **strict minimum**, or a **strict maximum**, if

$$J(u) < J(v), \quad \text{or} \quad J(u) > J(v), \quad \text{for all } v \in U,\ v \neq u.$$

Similar definitions hold for a *constrained* minimum, or maximum, *relative to a subset of the set* U.

The following theorem gathers a number of constantly used properties of *minima of convex functions*. Note that property (c) considerably improves upon Theorem 7.1-6 where, without the assumption of convexity, the Euler inequalities could only be shown to constitute a *necessary* condition for a constrained minimum. Likewise, property (d) considerably improves upon Theorem 7.1-5.

Theorem 7.12-3 (minima of convex functions) *Let U be a convex subset of a normed vector space V.*

(a) *If a convex function $J : U \subset V \to \mathbb{R}$ has a local minimum at a point $u \in U$, then J has a minimum at u.*

(b) *A strictly convex function $J : U \subset V \to \mathbb{R}$ has at most one minimum, and this minimum is strict.*

(c) *Let Ω be an open subset of V that contains U and let $J : \Omega \subset V \to \mathbb{R}$ be a function convex on U and differentiable at a point $u \in U$. Then J has a constrained minimum at u relative to the set U if and only if the* **Euler inequalities** *hold, viz.,*

$$J'(u)(v - u) \geq 0 \quad \textit{for every } v \in U.$$

(d) *Assume in addition that the convex set U is open. Let $J : U \subset V \to \mathbb{R}$ be a convex function, differentiable at a point $u \in U$. Then J has a minimum at u if and only if the* **Euler equation** *holds, viz.,*

$$J'(u) = 0.$$

Proof Let $v = u + w$ be any point of the convex set U distinct from u. By the convexity of the function $J : U \to \mathbb{R}$,

$$J(u + \theta w) \leq (1 - \theta)J(u) + \theta J(v) \quad \text{for all } 0 \leq \theta \leq 1,$$

which can also be written as

$$J(u + \theta w) - J(u) \leq \theta(J(v) - J(u)) \quad \text{for all } 0 \leq \theta \leq 1.$$

Since the point u is a local minimum, there exists a number θ_0 such that

$$\theta_0 > 0 \quad \text{and} \quad 0 \leq J(u + \theta_0 w) - J(u),$$

which implies that $J(v) \geq J(u)$ for all $v \in U$; hence u is a minimum of J. This proves (a).

If the function $J : U \to \mathbb{R}$ is strictly convex and J has a minimum at $u \in U$, the same argument leads to the existence of θ_0 such that

$$\theta_0 > 0 \quad \text{and} \quad 0 \leq J(u + \theta_0 w) - J(u) < \theta_0(J(v) - J(u)),$$

which show that the minimum is strict, and therefore unique. This proves (b).

In Theorem 7.1-6, the necessity of the condition $J'(u)(v - u) \geq 0$ for all $v \in U$ was established under the sole assumption that J is differentiable at u. That this condition becomes also sufficient if $J : U \to \mathbb{R}$ is convex follows from the inequalities

$$J(v) - J(u) \geq J'(u)(v - u) \quad \text{for every } v \in U,$$

established in Theorem 7.12-1(a). This proves (c).

Property (d) clearly follows from property (c). □

As an application of the above results, consider the *least-squares solution of a linear system* (Section 4.4): Given a real $m \times n$ matrix $\boldsymbol{A}$ and a vector $\boldsymbol{c} \in \mathbb{R}^m$, one seeks a vector $\boldsymbol{u} \in \mathbb{R}^n$ such that

$$\|\boldsymbol{A}\boldsymbol{u} - \boldsymbol{c}\|_m = \inf_{\boldsymbol{v} \in \mathbb{R}^n} \|\boldsymbol{A}\boldsymbol{v} - \boldsymbol{c}\|_m,$$

where $\|\cdot\|_m$ denotes the Euclidean norm in $\mathbb{R}^m$. Define the quadratic functional

$$\begin{aligned} J : \boldsymbol{v} \in \mathbb{R}^n \to J(\boldsymbol{v}) &:= \frac{1}{2}\|\boldsymbol{A}\boldsymbol{v} - \boldsymbol{c}\|_m^2 - \frac{1}{2}\|\boldsymbol{c}\|_m^2 \\ &= \frac{1}{2}(\boldsymbol{A}\boldsymbol{v}, \boldsymbol{A}\boldsymbol{v})_m - (\boldsymbol{c}, \boldsymbol{A}\boldsymbol{v})_m \\ &= \frac{1}{2}(\boldsymbol{A}^T\boldsymbol{A}\boldsymbol{v}, \boldsymbol{v})_n - (\boldsymbol{A}^T\boldsymbol{c}, \boldsymbol{v})_n, \quad \boldsymbol{v} \in \mathbb{R}^n, \end{aligned}$$

where $(\cdot,\cdot)_m$ and $(\cdot,\cdot)_n$ denote the Euclidean inner products in the spaces $\mathbb{R}^m$ and $\mathbb{R}^n$, respectively.

The symmetric matrix $\boldsymbol{A}^T\boldsymbol{A}$ being nonnegative-definite, the function J is convex (Theorem 7.12-2). Since the above least-squares problem is equivalent to finding a vector $\boldsymbol{u} \in \mathbb{R}^n$ such that

$$J(\boldsymbol{u}) = \inf_{\boldsymbol{v}\in\mathbb{R}^n} J(\boldsymbol{v}),$$

Theorem 7.12-3 therefore shows that *the set of solutions coincides with the set of solutions to the equation*

$$J'(\boldsymbol{u}) = \boldsymbol{A}^T\boldsymbol{A}\boldsymbol{u} - \boldsymbol{A}^T\boldsymbol{c} = \boldsymbol{0},$$

which are precisely the *normal equations* found earlier in Section 4.4, by an application of the *projection theorem*.

Note in passing that the same conclusions could be also drawn from the identity

$$\|\boldsymbol{A}(\boldsymbol{u}+\boldsymbol{w}) - \boldsymbol{c}\|_m^2 = \|\boldsymbol{A}\boldsymbol{u} - \boldsymbol{c}\|_m^2 + 2(\boldsymbol{A}^T\boldsymbol{A}\boldsymbol{u} - \boldsymbol{A}^T\boldsymbol{c}, \boldsymbol{w})_n + \|\boldsymbol{A}\boldsymbol{w}\|_m^2 \quad \text{for all } \boldsymbol{u}, \boldsymbol{w} \in \mathbb{R}^n,$$

which is nothing but *the Taylor formula*

$$J(\boldsymbol{u}+\boldsymbol{w}) = J(\boldsymbol{u}) + J'(\boldsymbol{u})\boldsymbol{w} + \frac{1}{2}(\boldsymbol{A}^T\boldsymbol{A}\boldsymbol{w}, \boldsymbol{w})_n$$

applied to the quadratic function J, whose Hessian is the constant matrix $\boldsymbol{A}^T\boldsymbol{A}$ (constant in that it does not depend on $\boldsymbol{u} \in \mathbb{R}^n$).

Problems

7.12-1 Let $(V, (\cdot,\cdot))$ be a real Hilbert space, and let $J \in \mathcal{C}^1(V)$ be an *α-coercive functional*, in the sense that there exists a constant α such that

$$\alpha > 0 \quad \text{and} \quad (\operatorname{grad} J(v) - \operatorname{grad} J(u), v-u) \geq \alpha \|v-u\|^2 \quad \text{for all } u, v \in V.$$

Clearly, *α-coercive functionals generalize the coercive quadratic functionals* $v \in V \to \frac{1}{2}a(v,v) - \ell(v)$ introduced in Section 6.1.

(1) Show that

$$J(v) - J(u) \geq (\operatorname{grad} J(u), v-u) + \frac{\alpha}{2}\|v-u\|^2 \quad \text{for all } u, v \in V.$$

(2) Show that $J : V \to \mathbb{R}$ is strictly convex.

(3) Let U be a nonempty, closed, convex subset of V. Show that the following minimization problem: Find $u \in U$ such that $J(u) = \inf_{v\in U} J(v)$, has one and only one solution.

Hint: Using (1), show that $\lim_{\|v\|\to\infty} J(v) = \infty$. Then, using the *Banach–Eberlein–Šmulian theorem* (Theorem 5.14-4), show that any *infimizing sequence* $(u_k)_{k=1}^\infty$ of the functional J on U, i.e., such that $u_k \in U$, $k \geq 1$, and $\lim_{k\to\infty} J(u_k) = \inf_{v\in U} J(v)$, contains a subsequence that weakly converges in U. Finally, show that the limit of this subsequence is a solution to the minimization problem.

(4) Show that $u \in U$ is a solution to the minimization problem of (3) if and only if $(\operatorname{grad} J(u), v-u) \geq 0$ for all $v \in U$, or if and only if $\operatorname{grad} J(u) = 0$ if $U = V$.

(5) Show that, if J is twice differentiable in V, then J is α-coercive if and only if

$$(\operatorname{Hess} J(u)w, w) \geq \alpha \|w\|^2 \quad \text{for all } w \in V.$$

7.12-2 This problem analyzes one instance of a *gradient method*,[31] which approximates by means of an *iterative method* the solution u of the minimization problem considered in question (3) of Problem 7.12-1.

In what follows, $(V, (\cdot, \cdot))$ is a real Hilbert space, and $J \in \mathcal{C}^1(V)$ is an *α-coercive functional*, according to the definition given in Problem 7.12-1, with the *additional* property that there exists a constant M such that

$$\|\text{grad}\, J(v) - \text{grad}\, J(u)\| \leq M \|v - u\| \quad \text{for all } u, v \in V.$$

(1) Assume first that $U = V$. Given any point $u_0 \in V$, and a sequence $(\rho_k)_{k=0}^{\infty}$ of real numbers, define the sequence $(u_k)_{k=1}^{\infty}$ by

$$u_{k+1} = u_k - \rho_k \,\text{grad}\, J(u_k), \quad k \geq 0.$$

Show that, if there exist two numbers a and b such that

$$0 < a \leq \rho_k \leq b < \frac{2\alpha}{M^2} \quad \text{for all } k \geq 0,$$

then there exists a constant $\beta < 1$ such that

$$\|u_k - u\| \leq \beta^k \|u_0 - u\| \quad \text{for all } k \geq 1,$$

where u is the unique solution of the following *unconstrained minimization problem*: Find $u \in U$ such that $J(u) = \inf_{v \in V} J(v)$.

(2) Assume next that U is a nonempty, closed, convex subset of V and let $P : V \to U$ denote the *projection operator* of V onto U (Section 4.3). Given any point $u_0 \in U$ and a sequence $(\rho_k)_{k=0}^{\infty}$ of real numbers, define the sequence $(u_k)_{k=1}^{\infty}$ by

$$u_{k+1} = P(u_k - \rho_k \,\text{grad}\, J(u_k)), \quad k \geq 0.$$

Show that, if there exist two numbers a and b such that

$$0 < a \leq \rho_k \leq b < \frac{2\alpha}{M^2} \quad \text{for all } k \geq 0,$$

then there exists a constant $\beta < 1$ such that

$$\|u_k - u\| \leq \beta^k \|u_0 - u\| \quad \text{for all } k \geq 1,$$

where u is the unique solution of the *constrained minimization problem*: Find $u \in U$ such that $J(u) = \inf_{v \in U} J(v)$.

7.12-3 This problem analyzes a *penalty method*, i.e., one that approximates the solution of a *constrained* minimization problem of a specific form by means of solutions of *unconstrained* minimization problems.

Let $J : \mathbb{R}^n \to \mathbb{R}$ be a strictly convex functional (hence in particular continuous; cf. Theorem 2.17-1) such that $\lim_{\|v\| \to \infty} J(v) = \infty$, and let U be a nonempty convex subset of $\mathbb{R}^n$ of the form $U := \{v \in \mathbb{R}^n;\ \psi(v) = 0\}$, where the function $\psi : \mathbb{R}^n \to \mathbb{R}$ is convex (hence continuous) and satisfies $\psi(v) \geq 0$ for all $v \in \mathbb{R}^n$.

(1) Show that the following *constrained* minimization problem: Find $u \in U$ such that $J(u) = \inf_{v \in U} J(v)$, has a unique solution.

[31] Gradient methods are analyzed at length in, e.g., CIARLET [1987, Chapter 8].

(2) Show that, for each $\varepsilon > 0$, the following *unconstrained* minimization problem: Find $u_\varepsilon \in \mathbb{R}^n$ such that

$$J_\varepsilon(u_\varepsilon) = \inf_{v\in\mathbb{R}^n} J_\varepsilon(v), \quad \text{where } J_\varepsilon(v) := J(v) + \frac{1}{\varepsilon}\psi(v) \text{ for all } v \in \mathbb{R}^n,$$

has a unique solution.

(3) Let $\varepsilon(k) > 0$, $k \geq 0$, be such that $\lim_{k\to\infty} \varepsilon(k) = 0$. Show that $\lim_{k\to\infty} u_{\varepsilon(k)} = u$.

7.12-4 Let $J : \boldsymbol{v} \in \mathbb{R}^n \to J(\boldsymbol{v}) := \frac{1}{2}\boldsymbol{v}^T\boldsymbol{A}v - \boldsymbol{b}^T\boldsymbol{v}$, where $\boldsymbol{A}$ is an $n \times n$ real symmetric matrix and $\boldsymbol{b} \in \mathbb{R}^n$, be a quadratic functional. Prove the following assertions:

(1) There exists a vector $\boldsymbol{u} \in \mathbb{R}^n$ such that

$$J(\boldsymbol{u}) < J(\boldsymbol{v}) \quad \text{for every } \boldsymbol{v} \in \mathbb{R}^n, \ \ \boldsymbol{v} \neq \boldsymbol{u},$$

if and only if the matrix $\boldsymbol{A}$ is positive-definite (J is then strictly convex).

(2) There exists a vector $\boldsymbol{u} \in \mathbb{R}^n$ such that

$$J(\boldsymbol{u}) \leq J(\boldsymbol{v}) \quad \text{for every } \boldsymbol{v} \in \mathbb{R}^n$$

if and only if the matrix $\boldsymbol{A}$ is nonnegative-definite (J is then convex) and the set $\{\boldsymbol{w} \in \mathbb{R}^n,\ \boldsymbol{A}\boldsymbol{w} = \boldsymbol{b}\}$ is nonempty.

(3) If the matrix $\boldsymbol{A}$ is nonnegative-definite and the set $\{\boldsymbol{w} \in \mathbb{R}^n,\ \boldsymbol{A}\boldsymbol{w} = \boldsymbol{b}\}$ is empty, then $\inf_{\boldsymbol{v}\in\mathbb{R}^n} J(\boldsymbol{v}) = -\infty$.

(4) If $\inf_{\boldsymbol{v}\in\mathbb{R}^n} J(\boldsymbol{v}) > -\infty$, then the matrix $\boldsymbol{A}$ is nonnegative-definite and the set $\{\boldsymbol{w} \in \mathbb{R}^n,\ \boldsymbol{A}\boldsymbol{w} = \boldsymbol{b}\}$ is nonempty.

7.12-5 (1) Let $\boldsymbol{E}$ be the square matrix of order n, all of whose components are equal to one. Calculate the eigenvalues of $\boldsymbol{E}$ and determine the corresponding eigenspaces.

(2) Let the (open and convex) set Ω be defined by

$$\Omega = \{\boldsymbol{v} = (v_i) \in \mathbb{R}^n : v_i > 0,\ 1 \leq i \leq n\}$$

and define the function

$$J : \boldsymbol{v} \in \Omega \subset \mathbb{R}^n \to J(\boldsymbol{v}) := -\left(\prod_{i=1}^n v_i\right)^{1/n} \in \mathbb{R}.$$

Compute the numbers

$$J'(\boldsymbol{u})\boldsymbol{v} \quad \text{and} \quad J''(\boldsymbol{u})(\boldsymbol{v}, \boldsymbol{w}) \ \text{ for } \boldsymbol{u} \in \Omega,\ \boldsymbol{v} \in \mathbb{R}^n,\ \boldsymbol{w} \in \mathbb{R}^n.$$

(3) Show that the function $J : \Omega \subset \mathbb{R}^n \to \mathbb{R}$ is convex, but not strictly convex.

(4) Denote by j the restriction of the function J to the convex subset

$$U = \left\{\boldsymbol{v} = (v_i) \in \Omega : \sum_{i=1}^n v_i = n\right\}$$

of the open set Ω. Show that the function $j : U \subset \mathbb{R}^n \to \mathbb{R}$ is strictly convex.

(5) Denote by $\boldsymbol{e}$ the vector of Ω, all of whose components are equal to one. Show that

$$J'(\boldsymbol{e})(\boldsymbol{v} - \boldsymbol{e}) = 0 \quad \text{for every } \boldsymbol{v} \in U.$$

Conclude that there exists a unique vector $\boldsymbol{u}$ such that

$$\boldsymbol{u} \in U \quad \text{and} \quad J(\boldsymbol{u}) = \inf_{\boldsymbol{v}\in U} J(\boldsymbol{v}).$$

(6) Show that

$$\left(\prod_{i=1}^{n} v_i\right)^{1/n} \leq \frac{1}{n}\sum_{i=1}^{n} v_i \quad \text{for every } \boldsymbol{v} = (v_i) \in \Omega,$$

and describe the subset of Ω for which the inequality becomes an equality.

Remark The inequality of (6) constitutes the *arithmetic mean-geometric inequality*, already encountered in Problem 2.17-10. □

7.12-6 Given the vertices $a_i \in \mathbb{R}^n$, $1 \leq i \leq 3$, of a nondegenerate triangle in $\mathbb{R}^n$, let the function $J : \mathbb{R}^n \to \mathbb{R}$ defined by $J(v) := \sum_{i=1}^{3} |v - a_i|$, where $|\cdot|$ denotes the Euclidean norm in $\mathbb{R}^n$.

(1) Show that there exists one and only one point $u \in \mathbb{R}^n$ such that $J(u) = \inf_{v\in\mathbb{R}^n} J(v)$.

(2) Give a geometric characterization of u, by means of the angles between the vectors $(a_i - u)$, and $(a_{i+1} - u)$, $1 \leq i \leq 3$ (modulo 3).

7.13 The implicit function theorem; first application: Class $\mathcal{C}^\infty$ of the mapping $A \to A^{-1}$

Using the mean value theorem and Banach fixed point theorem, we now prove the *implicit function theorem*,[32] a *basic result*, not only in differential calculus *per se* but in *nonlinear functional analysis* in general. This result provides sufficient conditions under which an equation of the form $\varphi(x, y) = 0$ is *locally* equivalent to an equation of the form $y = f(x)$ ("locally" means in a neighborhood of a particular solution of the equation $\varphi(x, y) = 0$). Such a function f is called an **implicit function** (Figure 7.13-1).

In what follows, $\dfrac{\partial\varphi}{\partial x}(a, b)$ and $\dfrac{\partial\varphi}{\partial y}(a, b)$ denote the partial derivatives of the mapping φ with respect to the generic variables x and y in the spaces X and Y, respectively, at a point $(a, b) \in X \times Y$. Note also that frequent use is made in the statements of the theorems of this section and the next on the *corollary to the Banach open mapping theorem* (Theorem 5.6-2); for instance, to insure that $\left(\dfrac{\partial\varphi}{\partial y}(a, b)\right)^{-1} \in \mathcal{L}(Z; Y)$ in Theorem 7.13-1.

Theorem 7.13-1 (implicit function theorem) *Let there be given a normed vector space X and two Banach spaces Y and Z, an open subset Ω of the space $X \times Y$ containing a point (a, b), and a mapping $\varphi \in \mathcal{C}(\Omega; Z)$ with the following properties:*

$$\varphi(a, b) = 0,$$

$$\frac{\partial\varphi}{\partial y}(x, y) \in \mathcal{L}(Y; Z) \text{ exists at all points } (x, y) \in \Omega \text{ and } \frac{\partial\varphi}{\partial y} \in \mathcal{C}(\Omega; \mathcal{L}(Y; Z)),$$

$$\frac{\partial\varphi}{\partial y}(a, b) \in \mathcal{L}(Y; Z) \text{ is a bijection, so that } \left(\frac{\partial\varphi}{\partial y}(a, b)\right)^{-1} \in \mathcal{L}(Z; Y).$$

[32] The first implicit function theorem (where the function denoted φ in Theorem 7.13-1 is a real-valued function of two real variables) is due to:

U. DINI [1878]: *Analisi Infinitesimale. Lezioni dettate nella Reale Università di Pisa, Anno Accademico* 1877–1878.

A nice historical perspective is given in:

G.M. SCARPELLO; D. RITELLI [2002]: A historical outline of the theorem of implicit functions, *Divulgaciones Matemáticas* **10**, 171–180.

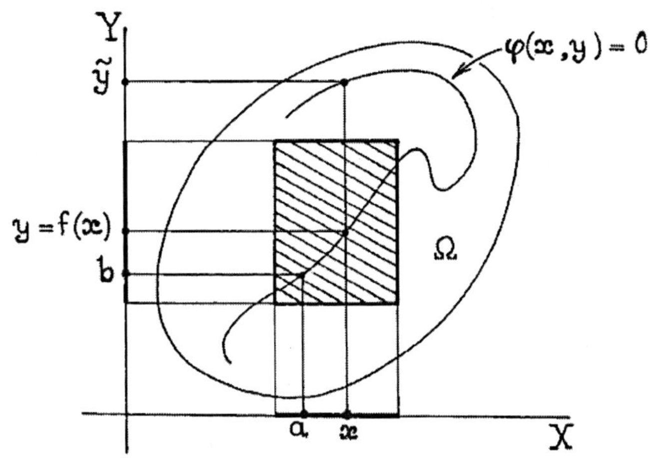

Figure 7.13-1 Under the assumptions of Theorem 7.13-1, there exists a neighborhood $V \times W \subset \Omega$ of a point (a, b) such that $\varphi(a, b) = 0$, where all the solutions (x, y) to the equation $\varphi(x, y) = 0$ are of the form $(x, f(x))$, $x \in V$, where the mapping $f : V \to W$ is an *implicit function.* This result is essentially *local*: it may happen that there exist points $x \in V$ and $\widetilde{y} \in Y - W$ such that $\varphi(x, \widetilde{y}) = 0$. This figure originally appeared in P.G. CIARLET [1988]: *Mathematical Elasticity, Volume* I: *Three-Dimensional Elasticity,* North-Holland, Amsterdam.

(a) *Then there exist an open neighborhood V of a in X, a neighborhood W of b in Y, and an* **implicit function** $f \in \mathcal{C}(V; W)$ *such that*

$$V \times W \subset \Omega \quad \textit{and} \quad \{(x, y) \in V \times W;\ \varphi(x, y) = 0\} = \{(x, y) \in V \times W;\ y = f(x)\}.$$

(b) *Assume in addition that φ is differentiable at $(a, b) \in \Omega$. Then f is differentiable at a and*

$$f'(a) = -\Big(\frac{\partial \varphi}{\partial y}(a, b)\Big)^{-1} \frac{\partial \varphi}{\partial x}(a, b) \in \mathcal{L}(X; Y).$$

(c) *Assume in addition that $\varphi \in \mathcal{C}^m(\Omega; Z)$ for some integer $m \geq 1$, resp. $\varphi \in \mathcal{C}^\infty(\Omega; Z)$. Then there exists an open neighborhood $\widetilde{V} \subset V$ of a in X and a neighborhood $\widetilde{W} \subset W$ of b in Y such that*

$$\frac{\partial \varphi}{\partial y}(x, y) \in \mathcal{L}(Y; Z) \ \textit{ is a bijection, so that } \Big(\frac{\partial \varphi}{\partial y}(x, y)\Big)^{-1} \in \mathcal{L}(Z; Y) \ \textit{ at each } (x, y) \in \widetilde{V} \times \widetilde{W},$$

$$f \in \mathcal{C}^m(\widetilde{V}; Y), \ \textit{ resp. } f \in \mathcal{C}^\infty(\widetilde{V}; Y),$$

$$f'(x) = -\Big(\frac{\partial \varphi}{\partial y}(x, f(x))\Big)^{-1} \frac{\partial \varphi}{\partial y}(x, f(x)) \in \mathcal{L}(X; Y) \ \textit{ at each } x \in \widetilde{V}.$$

Proof For clarity, the proof is broken into seven parts.

(i) *Establishing the existence of the implicit function $x \in V \to f(x) \in W$ amounts to finding each $f(x)$, $x \in V$, as the unique fixed point of an* ad hoc *mapping that depends on x.*

Define a mapping $\psi \in \mathcal{C}(\Omega; Y)$ by

$$\psi(x, y) := y - \Big(\frac{\partial \varphi}{\partial y}(a, b)\Big)^{-1} \varphi(x, y) \in Y \quad \text{at each } (x, y) \in \Omega.$$

Then $\dfrac{\partial \psi}{\partial y}(x, y) = I - \Big(\dfrac{\partial \varphi}{\partial y}(a, b)\Big)^{-1} \dfrac{\partial \varphi}{\partial y}(x, y) \in \mathcal{L}(Y)$ exists at all points $(x, y) \in \Omega$ and $\dfrac{\partial \psi}{\partial y} \in \mathcal{C}(\Omega; \mathcal{L}(Y))$. Besides,

$$\psi(a, b) = b \in Y, \quad \frac{\partial \psi}{\partial y}(a, b) = 0 \in \mathcal{L}(Y),$$

and a point $(x, y) \in \Omega$ satisfies $\varphi(x, y) = 0$ if and only if $\psi(x, y) = y$, i.e., if and only if y is a *fixed point* of the mapping $\psi(x, \cdot)$, which *depends on* x. We are thus naturally led to seek whether such mappings can become *contractions* in an appropriate *complete* metric space, so as to apply the *Banach fixed point theorem*. We now show that this is indeed the case if the points $(x, y) \in \Omega$ are restricted to lie in a sufficiently small neighborhood of (a, b).

(ii) *Existence of the implicit function.*

Since $\dfrac{\partial \psi}{\partial y} \in \mathcal{C}(\Omega; \mathcal{L}(Y))$ and $\dfrac{\partial \psi}{\partial y}(a, b) = 0$, there exists a neighborhood V' of a in X and a neighborhood W of b in Y such that

$$V' \times W \subset \Omega \quad \text{and} \quad \Big\|\frac{\partial \psi}{\partial y}(x, y)\Big\| \le \frac{1}{2} \quad \text{for all } (x, y) \in V' \times W.$$

Besides, there is no loss of generality in assuming that $W = \overline{B(b; r)}$ for some $r > 0$. For each $x \in V'$, we can therefore apply the *mean value theorem* (Theorem 7.2-1) in W (as a closure of a ball, W is a convex subset of the Banach space Y) to the mapping $T_x : W \to Y$ defined by

$$T_x(y) := \psi(x, y) \in Y \quad \text{at each } y \in W.$$

This gives, for each $x \in V'$,

$$\|T_x(\widetilde{y}) - T_x(y)\| \le \frac{1}{2} \|\widetilde{y} - y\| \quad \text{for all } \widetilde{y} \in W \text{ and all } y \in W,$$

which shows that $T_x : W \to Y$ is a contraction for each $x \in V'$.

However, nothing guarantees at this stage that T_x maps W *into itself* for each $x \in V'$. But such a property holds for those points $x \in V'$ that lie in a neighborhood of a smaller than V'. More specifically, let V be a neighborhood of a with the following properties:

$$V \text{ is open}, \quad V \subset V', \quad \text{and} \quad \|\psi(x, b) - \psi(a, b)\| \le \frac{r}{2} \quad \text{for all } x \in V$$

(this is possible since $\psi \in \mathcal{C}(\Omega; Y)$). Then

$$\begin{aligned} \|\psi(x, y) - b\| = \|\psi(x, y) - \psi(a, b)\| &\le \|\psi(x, y) - \psi(x, b)\| + \|\psi(x, b) - \psi(a, b)\| \\ &\le \frac{1}{2}\|y - b\| + \frac{r}{2} \le r \quad \text{for all } (x, y) \in V \times W, \end{aligned}$$

so that $T_x(y) = \psi(x,y) \in W = \overline{B(b;r)}$ for all $(x,y) \in V \times W$.

For each $x \in V$, the mapping $T_x : W \to W$ is thus a *contraction* in the *complete metric space* W. By the *Banach fixed point theorem* (Theorem 3.7-1), this contradiction has a *unique fixed point* $f(x) \in W$, which thus satisfies $\psi(x, f(x)) = f(x)$, or equivalently $\varphi(x, f(x)) = 0$.

Besides, the uniqueness of the fixed point shows that, for each $x \in V$, there is no other point $\widetilde{y}$ *in* W such that $(x, \widetilde{y}) \in \Omega$ and $\varphi(x, \widetilde{y}) = 0$ (of course there might be such a point $\widetilde{y}$ in $Y - W$; cf. Figure 7.13-1).

The *existence of the implicit function* $x \in V \to f(x) \in W$ is thus established.

(iii) *Continuity of the implicit function.*

Given any two points $x_0 \in V$ and $x \in V$,

$$\begin{aligned}\|f(x) - f(x_0)\| &= \|T_x(f(x)) - T_{x_0}(f(x_0))\| \\ &\le \|T_x(f(x)) - T_x(f(x_0))\| + \|T_x(f(x_0)) - T_{x_0}(f(x_0))\| \\ &\le \frac{1}{2}\|f(x) - f(x_0)\| + \|T_x(f(x_0)) - T_{x_0}(f(x_0))\|,\end{aligned}$$

so that

$$\|f(x) - f(x_0)\| \le 2\,\|T_x(f(x_0)) - T_{x_0}(f(x_0))\| = 2\,\|\psi(x, f(x_0)) - \psi(x_0, f(x_0))\|.$$

That $\psi \in \mathcal{C}(\Omega; Y)$ then implies that $\lim_{x\to x_0} \psi(x, f(x_0)) = \psi(x_0, f(x_0))$, hence that $\lim_{x\to x_0} f(x) = f(x_0)$, which shows that $f \in \mathcal{C}(V, W)$. This completes the proof of (a).

(iv) *Differentiability of the implicit function at* $a \in V$, *under the additional assumption that the mapping* φ *is differentiable at* $(a, b) \in V \times W \subset \Omega$.

Given any point $(a + h) \in V$, let $k(h) := f(a + h) - f(a)$. Then

$$\begin{aligned}0 &= \varphi(a + h, f(a + h)) - \varphi(a, f(a)) \\ &= \frac{\partial \varphi}{\partial x}(a, b)h + \frac{\partial \varphi}{\partial y}(a, b)k(h) + (\|h\| + \|k(h)\|)\delta(h, k(h))\end{aligned}$$

with

$$\lim_{(h,k)\to(0,0)} \delta(h, k) = 0 \quad \text{in } X \times Y,$$

so that

$$k(h) = -\Big(\frac{\partial \varphi}{\partial y}(a, b)\Big)^{-1} \frac{\partial \varphi}{\partial x}(a, b)h - (\|h\| + \|k(h)\|)\Big(\frac{\partial \varphi}{\partial y}(a, b)\Big)^{-1} \delta(h, k(h)).$$

Consequently, there exist constants

$$\alpha := \Big\|\Big(\frac{\partial \varphi}{\partial y}(a, b)\Big)^{-1} \frac{\partial \varphi}{\partial x}(a, b)\Big\|_{\mathcal{L}(X;Y)} \quad \text{and} \quad \beta := \Big\|\Big(\frac{\partial \varphi}{\partial y}(a, b)\Big)^{-1}\Big\|_{\mathcal{L}(Z;Y)}$$

such that

$$\|k(h)\| \le \alpha\,\|h\| + \beta\,(\|h\| + \|k(h)\|)\,\|\delta(h, k(h))\|.$$

Besides,

$$\lim_{h\to 0} k(h) = 0 \quad \text{in } Y,$$

since the implicit function is continuous (part (iii)). Therefore there exists r_0 such that $\beta\,\|\delta(h,k(h))\| \le \frac{1}{2}$ if $\|h\| \le r_0$, which in turn implies that

$$\|k(h)\| \le (2\alpha+1)\,\|h\| \quad \text{if } \|h\| \le r_0.$$

It therefore follows that

$$k(h) = f(a+h) - f(a) = -\Big(\frac{\partial\varphi}{\partial y}(a,b)\Big)^{-1}\frac{\partial\varphi}{\partial x}(a,b)h + \|h\|\,\varepsilon(h) \quad \text{with } \lim_{h\to 0}\varepsilon(h) = 0 \text{ in } Y,$$

thus showing that f is differentiable at $a \in V$, with

$$f'(a) = -\Big(\frac{\partial\varphi}{\partial y}(a,b)\Big)^{-1}\frac{\partial\varphi}{\partial x}(a,b).$$

This proves (b).

(v) *Class $\mathcal{C}^1$ of the implicit function, under the additional assumption that the mapping φ is of class $\mathcal{C}^1$ in Ω.*

Since $\dfrac{\partial\varphi}{\partial y} \in \mathcal{C}(\Omega;\mathcal{L}(Y;Z))$ in this case (Theorem 7.2-3), Theorem 3.6-3 shows that there exists an open set $\widetilde{\Omega} \subset \Omega$ containing (a,b) such that $\dfrac{\partial\varphi}{\partial y}(x,y) \in \mathcal{L}(Y;Z)$ is a bijection and $\Big(\dfrac{\partial\varphi}{\partial y}(x,y)\Big)^{-1} \in \mathcal{L}(Z;Y)$ at each $(x,y) \in \widetilde{\Omega}$.

The arguments from part (i) to part (iii) can then be reproduced *verbatim* with $\widetilde{\Omega}$ in lieu of Ω, leading to the existence of an open neighborhood $\widetilde{V} \subset V$ of a in X, of a neighborhood $\widetilde{W} \subset W$ of b in Y, and of an implicit function $f \in \mathcal{C}(\widetilde{V};\widetilde{W})$ that is differentiable at each point $x \in \widetilde{V}$ (since the set $\widetilde{V}$ is open, the argument of part (iv) establishing the differentiability of the implicit function at the point a also applies to any point $x \in \widetilde{V}$).

It remains to show that $f' \in \mathcal{C}(\widetilde{V};\mathcal{L}(X;Y))$. Given any $x \in \widetilde{V}$, let

$$A(x) := -\Big(\frac{\partial\varphi}{\partial y}(x,f(x))\Big)^{-1} \quad \text{and} \quad B(x) := \frac{\partial\varphi}{\partial x}(x,f(x)),$$

so that, given any two points $x \in \widetilde{V}$ and $\widetilde{x} \in \widetilde{V}$,

$$\begin{aligned} f'(x) - f'(\widetilde{x}) &= A(x)B(x) - A(\widetilde{x})B(\widetilde{x}) \\ &= A(x)(B(x) - B(\widetilde{x})) + (A(x) - A(\widetilde{x}))B(\widetilde{x}). \end{aligned}$$

Let (x_n) be a sequence of points $x_n \in \widetilde{V}$ such that $x_n \to \widetilde{x}$ as $n \to \infty$. Then, again by Theorem 3.6-3, $A(x_n) \to A(\widetilde{x})$ in $\mathcal{L}(Z;Y)$ since $\dfrac{\partial\varphi}{\partial y} \in \mathcal{C}(\Omega;\mathcal{L}(Y;Z))$ and $B(x_n) \to B(\widetilde{x})$ in $\mathcal{L}(X;Z)$ since $\dfrac{\partial\varphi}{\partial x} \in \mathcal{C}(\Omega;\mathcal{L}(X;Z))$. Hence $f'(x_n) \to f'(\widetilde{x})$ in $\mathcal{L}(X;Y)$ as $n \to \infty$. This proves (c).

(vi) *An application of parts* (i)–(v) *to a special case* (this application will be needed in part (vii)).

We also showed in Theorem 3.6-3 that, given a Banach space X and a normed vector space Y, the set

$$\mathcal{U} := \{A \in \mathcal{L}(\mathsf{X};\mathsf{Y});\ A : \mathsf{X} \to \mathsf{Y} \text{ is a bijection and } A^{-1} \in \mathcal{L}(\mathsf{Y};\mathsf{X})\}$$

is open in $\mathcal{L}(\mathsf{X};\mathsf{Y})$ and the mapping $A \in \mathcal{U} \to A^{-1} \in \mathcal{L}(\mathsf{Y};\mathsf{X})$ is *continuous*. We now establish that this mapping *is of class $\mathcal{C}^1$ in $\mathcal{U}$*, a result needed in the last part of the proof (where we will show that, in fact, this mapping is of class $\mathcal{C}^\infty$ in $\mathcal{U}$).

To this end, the idea is to apply parts (i)–(v) above to the *particular mapping*

$$\Phi : (A, B) \in \mathcal{L}(\mathsf{X};\mathsf{Y}) \times \mathcal{L}(\mathsf{Y};\mathsf{X}) \to \Phi(A, B) := (AB - I_{\mathsf{Y}}) \in \mathcal{L}(\mathsf{Y}),$$

which is of class $\mathcal{C}^\infty$ in $\mathcal{L}(\mathsf{X};\mathsf{Y}) \times \mathcal{L}(\mathsf{Y};\mathsf{X})$ (a continuous bilinear mapping is of class $\mathcal{C}^\infty$), and to the *particular open subset*

$$\mathcal{O} := \mathcal{U} \times \mathcal{L}(\mathsf{Y};\mathsf{X})$$

of the space $\mathcal{L}(\mathsf{X};\mathsf{Y}) \times \mathcal{L}(\mathsf{Y};\mathsf{X})$. Since (Section 7.1)

$$\partial_2\Phi(A, B)K = AK \quad \text{for all } K \in \mathcal{L}(\mathsf{Y};\mathsf{X}) \text{ at each } (A, B) \in \mathcal{O},$$

it follows that $\partial_2\Phi(A, B) \in \mathcal{L}(\mathcal{L}(\mathsf{Y};\mathsf{X}); \mathcal{L}(\mathsf{Y}))$ is a bijection and $(\partial_2\Phi(A, B))^{-1} \in \mathcal{L}(\mathcal{L}(\mathsf{Y}); \mathcal{L}(\mathsf{Y};\mathsf{X}))$ at each $(A, B) \in \mathcal{O}$, since

$$(\partial_2\Phi(A, B))^{-1}H = A^{-1}H \quad \text{for all } H \in \mathcal{L}(\mathsf{Y}) \text{ at each } (A, B) \in \mathcal{O}.$$

Given any pair $(A_0, A_0^{-1}) \in \mathcal{O}$, which therefore satisfies $\Phi(A_0, A_0^{-1}) = 0 \in \mathcal{L}(\mathsf{Y})$, there thus exist by parts (i)–(v) an open neighborhood $\mathcal{V}$ of A_0 in $\mathcal{U}$, an open neighborhood $\mathcal{W}$ of A_0^{-1} in $\mathcal{L}(\mathsf{Y};\mathsf{X})$, and an implicit function $F \in \mathcal{C}^1(\mathcal{V}; \mathcal{L}(\mathsf{Y};\mathsf{X}))$, such that

$$\{(A, B) \in \mathcal{V} \times \mathcal{W};\ AB = I_{\mathsf{Y}}\} = \{(A, B) \in \mathcal{V} \times \mathcal{W};\ B = F(A)\}.$$

But in this particular case, the implicit function is simply given by

$$F(A) = A^{-1} \quad \text{for all } A \in \mathcal{V}.$$

Noting that the mapping $A \in \mathcal{V} \to A^{-1} \in \mathcal{L}(\mathsf{Y};\mathsf{X})$ is thus of class $\mathcal{C}^1$, we conclude that *the mapping $A \in \mathcal{U} \to A^{-1} \in \mathcal{L}(\mathsf{Y};\mathsf{X})$ is of class $\mathcal{C}^1$*.

(vii) *Class $\mathcal{C}^m$ of the implicit function, under the additional assumption that the mapping φ is of class $\mathcal{C}^m$ in Ω*, for $m \geq 1$ or $m = \infty$.

By (v), the assertion holds for $m = 1$; so, assume that it holds for $m = 1, \ldots, k-1$, for some integer $k \geq 2$.

Under the same assumptions as in part (vi), the induction hypothesis applied to the particular mapping Φ of part (vi) implies that the mapping $A \in \mathcal{U} \to A^{-1} \in \mathcal{L}(\mathsf{Y};\mathsf{X})$ is of class $\mathcal{C}^{k-1}$.

Since the mapping $\varphi : \Omega \to Z$ is by assumption of class $\mathcal{C}^k$ in Ω, both mappings $\dfrac{\partial\varphi}{\partial x} : \Omega \to \mathcal{L}(X; Z)$ and $\dfrac{\partial\varphi}{\partial y} : \Omega \to \mathcal{L}(Y; Z)$ are of class $\mathcal{C}^{k-1}$ in Ω. Besides, the above observation shows that the mapping $\left(\dfrac{\partial\varphi}{\partial y}\right)^{-1} : \widetilde{\Omega} \to \mathcal{L}(Z;Y)$ is of class $\mathcal{C}^{k-1}$ in $\widetilde{\Omega}$ (the open set $\widetilde{\Omega} \subset \Omega$ has been defined in part (v)).

Since the implicit function $f : \widetilde{V} \to W \subset Y$ is of class $\mathcal{C}^{k-1}$ in $\widetilde{V}$ by the induction hypothesis, both mappings

$$x \in \widetilde{V} \to \left(\frac{\partial\varphi}{\partial y}(x, f(x))\right)^{-1} \in \mathcal{L}(Z;Y) \quad \text{and} \quad x \in \widetilde{V} \to \left(\frac{\partial\varphi}{\partial x}(x, f(x))\right) \in \mathcal{L}(X;Z)$$

are of class $\mathcal{C}^{k-1}$ in $\widetilde{V}$, by Theorem 7.8-4. Hence the mapping

$$f' : x \in V \to f'(x) = -\left(\frac{\partial\varphi}{\partial y}(x, f(x))\right)^{-1} \frac{\partial\varphi}{\partial x}(x, f(x)) \in \mathcal{L}(X;Y)$$

is also of class $\mathcal{C}^{k-1}$ in $\widetilde{V}$, again by Theorem 7.8-4. Consequently, $f : \widetilde{V} \to Y$ is of class $\mathcal{C}^k$ in $\widetilde{V}$. Therefore the assertion holds as well for $m = k$. □

An important property *per se* has been established in parts (vi) and (vii) of the above proof, which as such deserves to be recorded.

Theorem 7.13-2 (class $\mathcal{C}^\infty$ of the mapping $A \to A^{-1}$) *Let X and Y be two Banach spaces and let*

$$\mathcal{U} := \{A \in \mathcal{L}(X;Y);\ A : X \to Y \text{ is a bijection, so that } A^{-1} \in \mathcal{L}(Y;X)\}$$

(which is an open subset of $\mathcal{L}(X;Y)$; cf. Theorem 3.6-3). Then the mapping $F : A \in \mathcal{U} \to A^{-1} \in \mathcal{L}(Y;X)$ is of class $\mathcal{C}^\infty$ in $\mathcal{U}$. □

Problems

7.13-1 Let the assumptions of Theorem 7.13-1 be satisfied with $X = \mathbb{R}^2$, $Y = Z = \mathbb{R}$, and $m = 2$. Compute the partial derivatives of the first and second order of the implicit function f at a point a (the function f, *resp.* φ, appearing in this theorem is thus a real-valued function of two, *resp.* three, real variables in this case) in terms of the partial derivatives of the first and second order of the function φ at the point (a, b).

7.13-2 Let Ω be a domain in $\mathbb{R}^n$, let $m \geq 1$, and let the space $\mathcal{C}^m(\overline{\Omega})$ be equipped with the norm defined by (Problem 3.2-1)

$$v \in \mathcal{C}^m(\overline{\Omega}) \to \max_{|\alpha| \leq m} \sup_{x \in \Omega} |\partial^\alpha v(x)|.$$

(1) Show that $U := \{v \in \mathcal{C}^m(\overline{\Omega});\ v(x) > 0 \text{ for all } x \in \overline{\Omega}\}$ is an open subset of the space $\mathcal{C}^m(\overline{\Omega})$.

(2) Show that the mapping $f : v \in U \to f(v) := \dfrac{1}{v} \in U \subset \mathcal{C}^m(\overline{\Omega})$ is of class $\mathcal{C}^\infty$ in U.

7.14 The local inversion theorem; the invariance of domain theorem for mappings of class $\mathcal{C}^1$ in Banach spaces; class $\mathcal{C}^\infty$ of the mapping $A \to A^{1/2}$

The rest of this chapter is devoted to various applications of the implicit function theorem. While the first two applications (the local inversion theorem and the invariance of domain theorem for mappings of class $\mathcal{C}^1$ in Banach spaces; cf. Theorems 7.14-1 and 7.14-2) are of a general nature, those treated later apply to specific situations.

In the special case where $Z = X$ and the mapping φ is of the form $\varphi(x;y) := x - g(y)$ in the implicit function theorem, applying this theorem amounts to "locally inverting the relation $x = g(y)$ by means of a relation of the form $y = f(x)$." For brevity, we only state this *corollary to the implicit function theorem* (Theorem 7.13-1) under regularity assumptions that correspond to its part (c).

Theorem 7.14-1 (local inversion theorem) *Let there be given two Banach spaces X and Y, an open subset O of the space Y containing a point b, and a mapping $g \in \mathcal{C}^m(O;X)$ for some integer $m \geq 1$, resp. $g \in \mathcal{C}^\infty(O;X)$, with the following property:*

$$g'(b) \in \mathcal{L}(Y;X) \text{ is a bijection, so that } (g'(b))^{-1} \in \mathcal{L}(X;Y).$$

Then there exist an open neighborhood V of $a := g(b)$ in X, an open neighborhood $W \subset O$ of b in Y, and an **implicit function** *$f \in \mathcal{C}^m(V;Y)$, resp. $f \in \mathcal{C}^\infty(V;Y)$, such that $f(V) \subset W$ and*

$$\{(x,y) \in V \times W;\ x = g(y)\} = \{(x,y) \in V \times W;\ y = f(x)\}.$$

Besides,

$$g'(y) \in \mathcal{L}(Y;X) \text{ is a bijection, so that } \big(g'(y)\big)^{-1} \in \mathcal{L}(X;Y), \text{ at each } y \in W,$$
$$f'(x) = \big(g'(f(x))\big)^{-1} \text{ at each } x \in V.$$

Proof All the conclusions follow from Theorem 7.13-1(c) applied to the mapping $\varphi : \Omega \subset X \times Y \to X$ defined by

$$\varphi(x,y) := x - g(y) \quad \text{for all } (x,y) \in \Omega := X \times O.$$

More specifically, let $\widetilde{V}$ and $\widetilde{W}$ be respectively the open neighborhood of a and the neighborhood of b found in Theorem 7.13-1(c). If $\widetilde{W}$ is open, let $V := \widetilde{V}$ and $W := \widetilde{W}$. If $\widetilde{W}$ is not open, let W be any open neighborhood of b contained in $\widetilde{W}$; then let $V := f^{-1}(W)$. □

Recall that a mapping $f : X \to Y$ from a topological space X into a topological space Y is *open* if the direct image $f(U)$ of any open subset of X under f is an open subset of Y.

The *Banach open mapping theorem* (Theorem 5.6-1) provides sufficient conditions for a *linear* mapping between infinite-dimensional Banach spaces to be *open*; for this reason, it constitutes one of the basic theorems of *linear functional analysis* (as was abundantly illustrated at various places in Chapter 5). The next theorem provides sufficient conditions for a *nonlinear* mapping between infinite-dimensional Banach spaces to be open (of course it *a fortiori* applies to a linear mapping, but then the result becomes a triviality); as such, it constitutes one of the basic theorems of *nonlinear functional analysis.*

Notice that its proof essentially hinges on the local inversion theorem, and hence *in fine* on the *implicit function theorem*.

Theorem 7.14-2 (invariance domain theorem for mappings of class $\mathcal{C}^1$ in Banach spaces) *Let there be given two Banach spaces X and Y, an open subset Ω of X, and a* *mapping* $f \in \mathcal{C}^1(\Omega; Y)$ *with the following property:*

$$f'(x) \in \mathcal{L}(X;Y) \text{ is invertible, so that } \big(f'(x)\big)^{-1} \in \mathcal{L}(Y;X) \text{ at each } x \in \Omega.$$

(a) *Then* $f : \Omega \to Y$ *is an open mapping. In particular,* $f(\Omega)$ *is open in* Y.

(b) *If, in addition, the mapping* $f : \Omega \to Y$ *is injective, then* f *is a* $\mathcal{C}^1$*-diffeomorphism of* Ω *onto its image* $f(\Omega)$.

Proof (i) *Given any point* $a \in \Omega$ *and any neighborhood* V *of* a *in* Ω, *the direct image* $f(V)$ *of* V *under* f *is a neighborhood of* $f(a)$ *in* Y.

The key idea is to use the *local inversion theorem* (Theorem 7.14-1) *with* X *exchanged with* Y *and* f *exchanged with* g.

More specifically, there exist by this theorem an open neighborhood $\widehat{V} \subset \Omega$ of a in X, an open neighborhood W of $b := f(a)$ in Y, and a mapping $g \in \mathcal{C}^1(W; X)$ such that $g(W) \subset \widehat{V}$, and the equation $y = f(x)$ has one and only one solution $x = g(y) \in \widehat{V}$ for each $y \in W$.

Since $g(W)$ is the reciprocal image of W under the continuous mapping $f : \widehat{V} \subset \Omega \to Y$ and W is open in Y, $g(W)$ is thus open in X. Therefore the set $\widetilde{V} := g(W)$ is an open neighborhood of a in Ω and the mapping $f|_{\widetilde{V}} : \widetilde{V} \to W$ is a *homeomorphism*, with $g : W \to \widetilde{V}$ as its inverse homeomorphism.

Let now V be any neighborhood of a in Ω. Then $V \cap \widetilde{V}$ is also a neighborhood of a in Ω, and the direct image $f(V \cap \widetilde{V})$ is therefore a neighborhood of b in W since $f|_{\widetilde{V}} : \widetilde{V} \to W$ is a homeomorphism. Consequently, $f(V)$ is *a fortiori* a neighborhood of b in W.

(ii) Let now U be an open subset of Ω. Given any point $y \in f(U)$, there exists at least one point $x \in U$ such that $y = f(x)$. As an open subset of Ω containing x, the set U is a neighborhood of x in Ω. Consequently, its direct image $f(U)$ is a neighborhood of y in Y by (i).

As a neighborhood of each one of its points, the set $f(U)$ is thus open. This proves (a).

(iii) If the mapping $f : \Omega \to Y$ is in addition injective, then $f : \Omega \to f(\Omega)$ is a homeomorphism since the direct image under f of any open subset of Ω is open in $f(\Omega)$ by (ii). Since $f \in \mathcal{C}^1(\Omega; Y)$ by assumption and $f^{-1} \in \mathcal{C}^1(f(\Omega); X)$ by the local inversion theorem (differentiability is a local property), the mapping $f : \Omega \to f(\Omega)$ is a $\mathcal{C}^1$-diffeomorphism. This proves (b). □

An interesting complement to Theorem 7.14-2, proposed in Problem 7.14-3, asserts that it suffices that $f'(x)$ be *surjective* (in other words, $f(x)$ no longer needs to be bijective) if Y is *finite-dimensional*; besides, X needs no longer to be complete in this case.

Remarkably, *if* $X = Y = \mathbb{R}^n$, a conclusion similar to that of Theorem 7.14-2(a) holds for an *injective* mapping $f : \Omega \subset \mathbb{R}^n \to \mathbb{R}^n$ that is *only continuous*: this result constitutes the *Brouwer invariance of domain theorem in* $\mathbb{R}^n$ (from which Theorem 7.14-2 borrows its name). As we shall see later (Section 9.17), the proof of this theorem is, however, substantially harder than that of Theorem 7.14-2, as it rests on the *Brouwer topological degree in* $\mathbb{R}^n$.

To further illustrate the efficiency of the local inversion theorem, this time by means of a specific application, we now establish that the mapping that associates with any symmetric positive-definite matrix $\boldsymbol{C}$ its *square root* $\boldsymbol{C}^{1/2}$ is of class $\mathcal{C}^\infty$. Note that, remarkably, this property is established *without* computing explicitly the successive derivatives of this mapping.[33]

It what follows, $\mathbb{S}^n$ denotes the set of all symmetric matrices of order n and $\mathbb{S}^n_>$ denotes the set of all matrices in $\mathbb{S}^n$ that are positive-definite. Note that $\mathbb{S}^n_>$ is open in $\mathbb{S}^n$ (Problem 2.2-1).

Theorem 7.14-3 (class $\mathcal{C}^\infty$ of the mapping $\boldsymbol{A} \to \boldsymbol{A}^{1/2}$) *Given any matrix $\boldsymbol{A} \in \mathbb{S}^n_>$, there exists a unique matrix $\boldsymbol{A}^{1/2} \in \mathbb{S}^n_>$ such that $(\boldsymbol{A}^{1/2})^2 = \boldsymbol{A}$, and the mapping*

$$\boldsymbol{\Phi} : \boldsymbol{A} \in \mathbb{S}^n_> \to \boldsymbol{\Phi}(\boldsymbol{A}) = \boldsymbol{A}^{1/2} \in \mathbb{S}^n_>$$

defined in this fashion is of class $\mathcal{C}^\infty$.

Proof For completeness, we also provide a proof of the existence and uniqueness of the square root. Surprisingly, while the *existence* of the square root is immediate (part (i)), its *uniqueness* is not so obvious (part (ii)).

(i) Let $\boldsymbol{A}$ be a symmetric positive-definite matrix. Then the *existence* of a symmetric positive-definite matrix $\boldsymbol{B}$ satisfying $\boldsymbol{B}^2 = \boldsymbol{A}$ is clear: Let $\boldsymbol{P}$ be an orthogonal matrix that diagonalizes the matrix $\boldsymbol{A}$, i.e., $\boldsymbol{A} = \boldsymbol{P}^T\boldsymbol{D}\boldsymbol{P}$ with $\boldsymbol{D} = \mathbf{Diag}\,\mu_i$ and $\mu_i > 0$, $1 \le i \le n$. Then the matrix

$$\boldsymbol{B} := \boldsymbol{P}^T\,(\mathbf{Diag}\,\sqrt{\mu_i})\,\boldsymbol{P}$$

is symmetric positive-definite and satisfies $\boldsymbol{B}^2 = \boldsymbol{A}$.

(ii) In view of establishing the *uniqueness* of the square root, we first establish a preliminary result: Let $\boldsymbol{B}$ be a symmetric positive-definite matrix; then any eigenvector of the matrix $\boldsymbol{B}^2$, associated with an eigenvalue μ, is also an eigenvector of the matrix $\boldsymbol{B}$, associated with the eigenvalue $\sqrt{\mu}$ (the eigenvalue μ is necessarily > 0 since the matrix $\boldsymbol{B}^2$ is also symmetric and positive-definite). In other words,

$$\boldsymbol{B}^2\boldsymbol{v} = \mu\boldsymbol{v} \text{ and } \boldsymbol{v} \neq 0 \text{ implies that } \boldsymbol{B}\boldsymbol{v} = \sqrt{\mu}\boldsymbol{v}.$$

To see this, observe that the relation $\boldsymbol{B}^2\boldsymbol{v} = \mu\boldsymbol{v}$ can be rewritten as

$$(\boldsymbol{B} + \sqrt{\mu}\boldsymbol{I})(\boldsymbol{B} - \sqrt{\mu}\boldsymbol{I})\boldsymbol{v} = \mathbf{0}.$$

Then, necessarily, $\boldsymbol{w} := (\boldsymbol{B} - \sqrt{\mu}\boldsymbol{I})\boldsymbol{v} = \mathbf{0}$, for otherwise $\boldsymbol{w}$ would be an eigenvector of the matrix $\boldsymbol{B}$ corresponding to the eigenvalue $-\sqrt{\mu} < 0$.

Let then $\boldsymbol{B}_1$ and $\boldsymbol{B}_2$ be two symmetric positive-definite matrices that satisfy

$$\boldsymbol{A} = \boldsymbol{B}_1^2 = \boldsymbol{B}_2^2.$$

Then $\boldsymbol{A}\boldsymbol{v} = \mu\boldsymbol{v}$ and $\boldsymbol{v} \neq \mathbf{0}$ implies that $\boldsymbol{B}_1^2\boldsymbol{v} = \boldsymbol{B}_2^2\boldsymbol{v} = \mu\boldsymbol{v}$, and thus that $\boldsymbol{B}_1\boldsymbol{v} = \boldsymbol{B}_2\boldsymbol{v} = \sqrt{\mu}\boldsymbol{v}$. The matrices $\boldsymbol{B}_1$ and $\boldsymbol{B}_2$, which have the same eigenvectors and the same eigenvalues, are therefore equal. Hence the *uniqueness* of the square root is established.[34]

[33] For explicit formulas, see, e.g.:
C. Padovani [2000]: On the derivative of some tensor-valued functions, *Journal of Elasticity* **58**, 257–268.

[34] This short proof is due to:
R.A. Stephenson [1980]: On the uniqueness of the square-root of a symmetric, positive-definite tensor, *Journal of Elasticity* **10**, 213–214.

(iii) Let $\psi : \mathbb{S}^n_> \to \mathbb{S}^n_>$ denote the *inverse mapping* of $\boldsymbol{\Phi}$, thus defined by $\psi(\boldsymbol{B}) = \boldsymbol{B}^2$ for all $\boldsymbol{B} \in \mathbb{S}^n_>$. Then the Fréchet derivative $\psi'(\boldsymbol{B}) \in \mathcal{L}(\mathbb{S}^n)$ of the mapping ψ at each $\boldsymbol{B} \in \mathbb{S}^n_>$, which is given by

$$\psi'(\boldsymbol{B})\boldsymbol{H} = \boldsymbol{BH} + \boldsymbol{HB} \quad \text{for any } \boldsymbol{H} \in \mathbb{S}^n,$$

has an inverse, which is also in $\mathcal{L}(\mathbb{S}^n)$. To see this, let $\boldsymbol{H} \in \mathbb{S}^n$ be such that $\psi'(\boldsymbol{B})\boldsymbol{H} = \boldsymbol{0}$, let $(\boldsymbol{p}_i)_{i=1}^n$ be a basis of $\mathbb{R}^n$ consisting of eigenvectors of $\boldsymbol{B}$, and let $\lambda_i > 0$, $1 \le i \le n$, be the corresponding eigenvalues of $\boldsymbol{B}$. Then

$$\psi'(\boldsymbol{B})\boldsymbol{H}\boldsymbol{p}_i = \boldsymbol{B}\boldsymbol{H}\boldsymbol{p}_i + \lambda_i \boldsymbol{H}\boldsymbol{p}_i = \boldsymbol{0}, \quad 1 \le i \le n,$$

so that $\boldsymbol{H}\boldsymbol{p}_i = \boldsymbol{0}$, $1 \le i \le n$; for otherwise $\boldsymbol{H}\boldsymbol{p}_i$ would be an eigenvector of $\boldsymbol{B}$ corresponding to the eigenvalue $-\lambda_i < 0$. Hence $\boldsymbol{H} = \boldsymbol{0}$, which shows that $\psi'(\boldsymbol{B}) \in \mathcal{L}(\mathbb{S}^n)$ has an inverse, which is thus also in $\mathcal{L}(\mathbb{S}^n)$ (the space $\mathbb{S}^n$ is finite-dimensional). Consequently, all the assumptions of the local inversion theorem (Theorem 7.14-1) are satisfied.

Since the mapping $\psi : \mathbb{S}^n_> \to \mathbb{S}^n_>$ is of class $\mathcal{C}^\infty$, its inverse mapping $\boldsymbol{\Phi} : \mathbb{S}^n_> \to \mathbb{S}^n_>$ is thus also of class $\mathcal{C}^\infty$. □

Problems

7.14-1 Let Ω be a domain in $\mathbb{R}^n$, and let $f \in \mathcal{C}^1(\overline{\Omega}; \mathbb{R}^n)$ be a mapping that satisfies

$$\det \nabla f(x) > 0 \text{ for all } x \in \Omega \quad \text{and} \quad \int_\Omega \det \nabla f(x) \mathrm{d}x \le \int_{f(\Omega)} \mathrm{d}x.$$

Show that the restriction of f to Ω is injective.[35]

Hint: Use the local inversion theorem to show that, if f is not injective on Ω, there exists an open subset W of $f(\Omega)$ such that $\operatorname{card} f^{-1}(x) \ge 2$ for all $x \in W$.

7.14-2 Let there be given a normed vector space X, a *finite-dimensional* vector space Y, an open subset Ω of X, and a mapping $f \in \mathcal{C}^1(\Omega; Y)$ with the following property:

$$\text{at each } x \in \Omega,\ f'(x) \in \mathcal{L}(X;Y) \text{ is a surjection of } X \text{ onto } Y.$$

Show that $f : \Omega \to Y$ is an *open mapping*.

7.14-3 Let there be given two Banach spaces X and Y and a mapping $f \in \mathcal{C}^1(X;Y)$ with the following properties:

$$\text{at each } x \in X,\ f'(x) \in \mathcal{L}(X;Y) \text{ is a bijection and } \sup_{x \in X} \|f'(x)^{-1}\|_{\mathcal{L}(Y;X)} < \infty.$$

Show that f is a surjection[36] of X onto Y.

[35] This result is due to:

P.G. Ciarlet; J. Nečas [1987]: Injectivity and self-contact in nonlinear elasticity, *Archive for Rational Mechanics and Analysis* **97**, 171–188.

[36] Various sufficient conditions for a mapping between two Banach spaces to be either injective or surjective (as here) are found in:

G. Zampieri [1992]: Diffeomorphisms with Banach space domains, *Nonlinear Analysis, Theory, Methods & Applications* **19**, 923–932.

7.14-4 For each matrix $\boldsymbol{F} \in \mathbb{U}^n$, where $\mathbb{U}^n$ denotes the set of all invertible real matrices of order n (which is an open subset of $\mathbb{M}^n$; cf. Theorem 3.6-3), let $\boldsymbol{F} = \boldsymbol{RU}$ denote its unique *polar factorization* (Problem 4.3-5). Show that the mappings $\boldsymbol{F} \in \mathbb{U}^n \to \boldsymbol{R} \in \mathbb{M}^n$ and $\boldsymbol{F} \in \mathbb{U}^n \to \boldsymbol{U} \in \mathbb{M}^n$ defined in this fashion are of class $\mathcal{C}^\infty$.

7.14-5 Greek and Latin indices vary in the sets $\{1,2\}$ and $\{1,2,3\}$ respectively, and the summation convention with respect to repeated indices is used. Let Ω be a domain in $\mathbb{R}^2$. Given a smooth enough vector field $\boldsymbol{v} = (v_i) : \overline{\Omega} \to \mathbb{R}^3$, let

$$\boldsymbol{A}(\boldsymbol{v}) := (-\partial_\beta N_{1\beta}(\boldsymbol{v}), -\partial_\beta N_{2\beta}(\boldsymbol{v}), \partial_{\alpha\beta} m_{\alpha\beta}(\boldsymbol{v}) - \partial_\beta (N_{\alpha\beta}(\boldsymbol{v})\partial_\alpha v_3)),$$

where

$$m_{\alpha\beta}(\boldsymbol{v}) := \frac{\varepsilon^3}{3} a_{\alpha\beta\sigma\tau}\partial_{\sigma\tau} v_3, \quad N_{\alpha\beta}(\boldsymbol{v}) := \varepsilon a_{\alpha\beta\sigma\tau} E_{\sigma\tau}(\boldsymbol{v}), \quad E_{\alpha\beta}(\boldsymbol{v}) := \frac{1}{2}(\partial_\alpha v_\beta + \partial_\beta v_\alpha + \partial_\alpha v_3 \partial_\beta v_3),$$

where $\varepsilon > 0$ is a constant and $a_{\alpha\beta\sigma\tau} = a_{\beta\alpha\sigma\tau} = a_{\sigma\tau\alpha\beta}$ are constants with the property that there exists a constant C such that

$$a_{\alpha\beta\sigma\tau} t_{\sigma\tau} t_{\alpha\beta} \geq C t_{\alpha\beta} t_{\alpha\beta} \quad \text{for all } (t_{\alpha\beta}) \in \mathbb{S}^2.$$

(1) Given any $p > 2$, show that the nonlinear operator $\boldsymbol{A}$ defined in this fashion maps the space $W^{3,p}(\Omega) \times W^{3,p}(\Omega) \times W^{4,p}(\Omega)$ into the space $W^{1,p} \times W^{1,p}(\Omega) \times L^p(\Omega)$ and that $\boldsymbol{A}$ is infinitely differentiable between these spaces.

(2) Show that, if the boundary Γ of Ω is smooth enough, the derivative of $\boldsymbol{A}$ at the origin is for any $p > 2$ a continuous bijection from the space

$$\boldsymbol{V}^p(\Omega) := \{\boldsymbol{v} = (v_i) \in W^{3,p}(\Omega) \times W^{3,p}(\Omega) \times W^{4,p}(\Omega);\ v_i = \partial_\nu v_3 = 0 \text{ on } \Gamma\}$$

onto the space

$$\boldsymbol{W}^p(\Omega) := W^{1,p}(\Omega) \times W^{1,p}(\Omega) \times L^p(\Omega).$$

Hint: Use the following regularity result:[37] If the boundary Γ is smooth enough, the solution $\boldsymbol{u} \in H_0^1(\Omega) \times H_0^1(\Omega) \times H_0^2(\Omega)$ of the minimization problem of Problem 6.16-4 in the special case where $\Gamma_0 = \Gamma$ is in the space $\boldsymbol{V}^p(\Omega)$ for any vector field $\boldsymbol{f}$ in the space $\boldsymbol{W}^p(\Omega)$.

(3) Using the *local inversion theorem*, show that, if the boundary Γ is smooth enough, there exist for each $p > 2$ a neighborhood $\boldsymbol{F}^p$ of the origin in $\boldsymbol{W}^p(\Omega)$ and a neighborhood $\boldsymbol{U}^p$ of the origin in $\boldsymbol{V}^p(\Omega)$ with the following property: For each $\boldsymbol{f} = (f_i) \in \boldsymbol{F}^p$, the following nonlinear boundary value problem has a unique solution $\boldsymbol{u}$ in $\boldsymbol{U}^p$:

$$\begin{aligned}
\partial_{\alpha\beta} m_{\alpha\beta}(\boldsymbol{u}) - \partial_\beta(N_{\alpha\beta}(\boldsymbol{u})\partial_\alpha u_3) &= f_3 \quad \text{in } \Omega,\\
-\partial_\beta N_{\alpha\beta}(\boldsymbol{u}) &= f_\alpha \quad \text{in } \Omega,\\
u_i = \partial_\nu u_3 &= 0 \quad \text{on } \Gamma.
\end{aligned}$$

This boundary value problem constitutes the equations of the **Kirchhoff–Love theory of nonlinearly elastic plates.**[38]

Remark The existence of a *weak solution* to this boundary value problem can be also established, in effect in greater generality, by using the methods of the *calculus of variations* (Problem 9.3-3). □

---

[37] For a proof, see:
P.G. Ciarlet; P. Destuynder [1979]: A justification of a nonlinear model in plate theory, *Computer Methods in Applied Mechanics and Engineering* **17/18**, 227–258.

[38] These equations are studied at length in Ciarlet [1997, Chapter 4].

7.15 Constrained extrema of real-valued functions; Lagrange multipliers

As another application of the *implicit function theorem*, we now give a *necessary* condition for a point u to be a *constrained local extremum* of a real-valued function $J : \Omega \to \mathbb{R}$ relative to a subset U of Ω, in the following *special case*: The set Ω is an open subset of a product $V_1 \times V_2$ of two normed vector spaces, and the subset U of Ω is of the form

$$U = \{(v_1, v_2) \in \Omega : \varphi(v_1, v_2) = 0\},$$

for some given mapping

$$\varphi : \Omega \subset V_1 \times V_2 \to V_2.$$

Observe that a set U defined in this fashion is *not* open in general (think of a curve in $\mathbb{R}^2$ when $V_1 = V_2 = \mathbb{R}$ and the function φ is continuous). This is why the necessary condition established in Theorem 7.1-5 is of no use in this situation.

Theorem 7.15-1 (necessary condition for a constrained local extremum) *Let Ω be an open subset of a product $V_1 \times V_2$, where V_1 is a normed vector space and V_2 is a Banach space. Given a mapping $\varphi \in \mathcal{C}^1(\Omega; V_2)$, let*

$$u = (u_1, u_2) \in U := \{(v_1, v_2) \in \Omega;\ \varphi(v_1, v_2) = 0\}$$

be such that

$$\partial_2\varphi(u_1, u_2) \in \mathcal{L}(V_2) \textit{ is a bijection, so that } (\partial_2\varphi(u_1, u_2))^{-1} \in \mathcal{L}(V_2).$$

Let $J : \Omega \to \mathbb{R}$ be a function differentiable at u. If J has a constrained local extremum at u relative to U, then there exists an element $\Lambda(u) \in \mathcal{L}(V_2; \mathbb{R})$ such that

$$J'(u) + \Lambda(u)\varphi'(u) = 0.$$

Proof The assumptions made on the spaces V_1 and V_2, the set Ω, and the mapping φ allow us to apply the implicit function theorem (Theorem 7.13-1) in a neighborhood of the point u. This theorem shows that there exist an open neighborhood O_1 of u_1 in V_1, a neighborhood W_2 of u_2 in V_2, and an *implicit function* $f \in \mathcal{C}(O_1; W_2)$ such that

$$O_1 \times W_2 \subset \Omega \quad \text{and} \quad (O_1 \times W_2) \cap U = \{(v_1, v_2) \in O_1 \times W_2 : v_2 = f(v_1)\}.$$

Moreover, the implicit function f is differentiable at the point $u_1 \in O_1$ and its derivative is given by

$$f'(u_1) = -(\partial_2\varphi(u))^{-1}\partial_1\varphi(u).$$

Thanks to the implicit function theorem, the restriction of the function J to the set $(O_1 \times W_2) \cap U$ thus becomes a function of a single variable in the *open* set O_1, defined by

$$G : v_1 \in O_1 \to G(v_1) := J(v_1, f(v_1)) \in \mathbb{R},$$

and this function G has a *local extremum at the point* $u_1 \in O_1$. Besides, the function G is differentiable at the point u_1 by the chain rule (Theorem 7.1-3), and its derivative is given by

$$G'(u_1) = \partial_1 J(u) + \partial_2 J(u) f'(u_1) = \partial_1 J(u) - \partial_2 J(u)(\partial_2 \varphi(u))^{-1} \partial_1 \varphi(u).$$

Therefore, we can apply the necessary condition of Theorem 7.1-5 (because the set O_1 is open), which gives

$$G'(u_1) = 0 \in \mathcal{L}(V_1; \mathbb{R}).$$

Hence we have

$$\partial_1 J(u) = \partial_2 J(u)(\partial_2 \varphi(u))^{-1} \partial_1 \varphi(u),$$

on the one hand. Since we evidently have

$$\partial_2 J(u) = \partial_2 J(u)(\partial_2 \varphi(u))^{-1} \partial_2 \varphi(u),$$

on the other hand, the announced result follows by setting

$$\Lambda(u) := -\partial_2 J(u)(\partial_2 \varphi(u))^{-1}. \qquad \square$$

The mapping $\Lambda(u) \in \mathcal{L}(V_2; \mathbb{R})$ found in Theorem 7.15-1 is called the **generalized Lagrange multiplier**[39] *associated with the constrained local extremum* $u \in U$.

The preceding result is frequently used in the following often encountered situation. Given two integers m and n satisfying $1 \leq m \leq n-1$ and functions

$$J : \Omega \subset \mathbb{R}^n \to \mathbb{R} \quad \text{and} \quad \varphi_i : \Omega \subset \mathbb{R}^n \to \mathbb{R}, \ \ 1 \leq i \leq m,$$

all defined over the same open subset Ω of $\mathbb{R}^n$, one seeks a *necessary condition* satisfied by a *constrained local extremum of the function* J *relative to the set*

$$U := \{v \in \Omega : \varphi_i(v) = 0, \ 1 \leq i \leq m\}.$$

It is clear that this problem is a particular case of the preceding one (with V_1 and V_2 respectively identified with the spaces $\mathbb{R}^{n-m}$ and $\mathbb{R}^m$), so that Theorem 7.15-1 leads to the following result.

Theorem 7.15-2 (necessary condition for a constrained local extremum) *Let* Ω *be an open subset of* $\mathbb{R}^n$, *let* $\varphi_i : \Omega \to \mathbb{R}$, $1 \leq i \leq m \leq n-1$, *be functions of class* $\mathcal{C}^1$ *over* Ω, *and let* u *be a point of the set*

$$U := \{v \in \Omega : \varphi_i(v) = 0, \ 1 \leq i \leq m\},$$

such that the derivatives $\varphi_i'(u) \in \mathcal{L}(\mathbb{R}^n; \mathbb{R})$, $1 \leq i \leq m$, *are linearly independent.*

Let $J : \Omega \to \mathbb{R}$ *be a function differentiable at* u. *If* J *has a constrained local extremum at* u *relative to the set* U, *then there exist* m *numbers* $\lambda_i = \lambda_i(u)$, $1 \leq i \leq m$, *such that*

$$J'(u) + \sum_{i=1}^{m} \lambda_i \varphi_i'(u) = 0,$$

and these numbers λ_i, $1 \leq i \leq m$, *are uniquely defined.*

[39] So named after Joseph-Louis Lagrange (1736–1813), who is at the origin of this notion, as well as of several other basic notions that pervade the *calculus of variations* (such as those considered in Sections 7.16 and 9.1).

Proof The linear independence of the derivatives $\varphi_i'(u)$ implies that the matrix with elements $\partial_j\varphi_i(u)$, $1 \le i \le m$, $1 \le j \le n$, has rank m. Suppose (simply to fix ideas) that the submatrix with elements $\partial_j\varphi_i(u)$, $1 \le i, j \le m$, is invertible. It then suffices to apply Theorem 7.15-1 with

$$V_1 := \{(v_j)_{j=m+1}^n \in \mathbb{R}^{n-m}\} \quad \text{and} \quad V_2 := \{(v_i)_{i=1}^m \in \mathbb{R}^m\},$$
$$\varphi : v \in \Omega \subset V_1 \times V_2 \to \varphi(v) := (\varphi_i(v))_{i=1}^m \in V_2.$$

This theorem shows that there exists an element $\Lambda(u)$ of the space $\mathcal{L}(\mathbb{R}^m; \mathbb{R})$ such that $J'(u) + \Lambda(u)\varphi'(u) = 0$; equivalently, there exist m real numbers $\lambda_i = \lambda_i(u)$, $1 \le i \le m$, such that

$$J'(u) + \sum_{i=1}^m \lambda_i\varphi_i'(u) = 0.$$

The uniqueness of the numbers λ_i is a consequence of the linear independence of the derivatives $\varphi_i'(u)$. □

The numbers $\lambda_i = \lambda_i(u)$, $1 \le i \le m$, found in the above theorem are called the **Lagrange multipliers**, and the vector $\boldsymbol{\lambda} = (\lambda_i)_{i=1}^m \in \mathbb{R}$ is called the **Lagrange multiplier**, *associated with the constrained local extremum* $u \in U$.

The vectors $u = (u_i)_{i=1}^n \in \mathbb{R}^n$ and $\lambda = (\lambda_i)_{i=1}^m \in \mathbb{R}^m$ that satisfy the necessary condition of Theorem 7.15-2 are thus obtained by solving the following *system of* $(m+n)$ *equations*:

$$\begin{aligned}
\partial_1 J(u) + \lambda_1\partial_1\varphi_1(u) + \cdots + \lambda_m\partial_1\varphi_m(u) &= 0,\\
&\vdots\\
\partial_n J(u) + \lambda_1\partial_n\varphi_1(u) + \cdots + \lambda_m\partial_n\varphi_m(u) &= 0,\\
\varphi_1(u) &= 0,\\
&\vdots\\
\varphi_m(u) &= 0.
\end{aligned}$$

Note that the first n equations may be also conveniently written in vector form as

$$\begin{pmatrix} \partial_1 J(u) \\ \vdots \\ \partial_n J(u) \end{pmatrix} + \begin{pmatrix} \partial_1\varphi_1(u) & \dots & \partial_1\varphi_m(u) \\ \vdots & & \vdots \\ \partial_n\varphi_1(u) & \dots & \partial_n\varphi_m(u) \end{pmatrix} \begin{pmatrix} \lambda_1 \\ \vdots \\ \lambda_m \end{pmatrix} = \operatorname{grad} J(u) + (\nabla\varphi(u))^T\lambda = 0,$$

To conclude, consider the example of a *quadratic functional* over the space $\mathbb{R}^n$, thus a function of the form

$$J : \boldsymbol{v} \in \mathbb{R}^n \to J(\boldsymbol{v}) := \frac{1}{2}\boldsymbol{v}^T\boldsymbol{A}\boldsymbol{v} - \boldsymbol{c}^T\boldsymbol{v},$$

where $\boldsymbol{A}$ is a real *symmetric* matrix of order n and $\boldsymbol{c} \in \mathbb{R}^n$. Such a function J is differentiable in $\mathbb{R}^n$ and its derivative $J'(\boldsymbol{u}) \in \mathcal{L}(\mathbb{R}^n; \mathbb{R})$ can be identified (by means of the Euclidean inner product) at each $\boldsymbol{u} \in \mathbb{R}^n$ with the vector $(\boldsymbol{A}\boldsymbol{u} - \boldsymbol{c}) \in \mathbb{R}^n$.

Assume then that we seek the constrained local *extrema of the functional* J *relative to a set of the particular form*

$$\boldsymbol{U} := \{\boldsymbol{v} \in \mathbb{R}^n;\ \boldsymbol{B}\boldsymbol{v} = \boldsymbol{d}\},$$

where $\boldsymbol{B}$ is a real $m \times n$ matrix and $\boldsymbol{d} \in \mathbb{R}^m$, with $m \le n-1$. The derivative $\boldsymbol{\varphi}' : \mathbb{R}^n \to \mathcal{L}(\mathbb{R}^n; \mathbb{R}^m)$ of the function

$$\boldsymbol{\varphi} : \boldsymbol{v} \in \mathbb{R}^n \to \boldsymbol{\varphi}(\boldsymbol{v}) := \boldsymbol{B}\boldsymbol{v} - \boldsymbol{d} \in \mathbb{R}^m$$

being the constant function equal to the matrix $\boldsymbol{B}$, it follows from Theorem 7.15-2 that, *if the matrix $\boldsymbol{B}$ has rank m* (with the notations of Theorem 7.15-2, this assumption means that the derivatives $\varphi_i'(u)$, $1 \le i \le m$, are linearly independent), then a *necessary* condition for the functional J to have a constrained local extremum at $u \in U$ relative to the set U is the existence of a solution $(\boldsymbol{u}, \boldsymbol{\lambda}) \in \mathbb{R}^n \times \mathbb{R}^m$ to the *linear system*

$$\begin{aligned} \boldsymbol{A}\boldsymbol{u} + \boldsymbol{B}^T\boldsymbol{\lambda} &= \boldsymbol{c} \\ \boldsymbol{B}\boldsymbol{u} \qquad &= \boldsymbol{d}. \end{aligned}$$

Observe that the *same* linear system can be also obtained (Problem 6.12-2) as a consequence of the *Babuška–Brezzi inf-sup theorem* (Theorem 6.12-1), and hence by completely different means.

Taking into consideration the *constraint* $\boldsymbol{B}\boldsymbol{v} = \boldsymbol{d}$ thus results in having to solve a *larger* linear system than that when there is no constraint. Note in this respect that it is not possible to avoid the computation of the vector $\boldsymbol{\lambda} \in \mathbb{R}^m$ even if, as is often the case, one is only interested in finding the *constrained local extrema* $\boldsymbol{u} \in \boldsymbol{U}$. In other words, the unknown *Lagrange multiplier* $\boldsymbol{\lambda}$ appears simply as a necessary "intermediary."

The extension to sets of the form $\boldsymbol{U} := \{\boldsymbol{v} \in \mathbb{R}^n;\ \boldsymbol{B}\boldsymbol{v} \le \boldsymbol{d}\}$ is the object of Problem 7.15-3.

Problems

7.15-1 Find the constrained local extrema and the associated Lagrange multipliers of the function $J : v = (v_1, v_2) \in \mathbb{R}^2 \to J(v) := -v_2$ relative to the set $U := \{(v_1, v_2) \in \mathbb{R}^2;\ v_1^2 + v_2^2 = 1\}$.

Hint: Use a local analysis to show that the points are indeed constrained local extrema of J relative to U.

7.15-2 Let $U := \{v \in \mathbb{R}^n;\ \varphi(v) = 0\}$ where $\varphi \in \mathcal{C}^1(\mathbb{R}^n)$, let a and b be two distinct points in $\mathbb{R}^n$ that do not belong to the set U, let the function $J : \mathbb{R}^n \to \mathbb{R}$ be defined by $J(v) := |v-a| + |v-b|$, $v \in \mathbb{R}^n$, where $|\cdot|$ denotes the Euclidean norm in $\mathbb{R}^n$, and assume that $u \in U$ is a local extremum of J relative to U.

Show that, if $\varphi'(u) \ne 0$ and $\dfrac{u-a}{|u-a|} + \dfrac{u-b}{|u-b|} \ne 0$, the normal at u to the hypersurface U lies on the bissectrix at the vertex u in the triangle with vertices a, b, u.

Remark When $n = 2$ or $n = 3$, the geometric interpretation of this result is nothing but the celebrated **Fermat principle**[40] of geometrical optics. □

7.15-3 The object of this problem is to establish (in question (2)) the analogue of Theorem 7.15-2 when the "equality constraints" $\varphi_i(v) = 0$, $1 \le i \le m$, are replaced by "*inequality constraints*" of the form $\varphi_i(v) \le 0$, $1 \le i \le m$. Here we shall confine ourselves for simplicity to mappings φ_i that are all *affine.*[41]

[40] So named after Pierre de Fermat (1601–1665).

[41] More general mappings φ_i can be considered; see, e.g., CIARLET [1987, Section 9.2].

Let $(V, (\cdot,\cdot))$ be a real Hilbert space, let c_i, $1 \le i \le m$, be vectors in V, and let d_i, $1 \le i \le m$, be real numbers.

(1) Let

$$U := \{v \in V;\ (c_i, v) = d_i,\ 1 \le i \le m\},$$

let Ω be an open subset of V containing U, and let there be given a function $J : \Omega \to \mathbb{R}$. Show that, if the vectors c_i are linearly independent and J has a constrained local extremum relative to U at a point $u \in U$ and is differentiable at u, then there exist uniquely defined *Lagrange multipliers* $\lambda_i = \lambda_i(u)$, $1 \le i \le m$, such that

$$\operatorname{grad} J(u) + \sum_{i=1}^{m} \lambda_i c_i = 0.$$

(2) Let

$$\widetilde{U} := \{v \in V;\ (c_i, v) \le d_i,\ 1 \le i \le m\},$$

let Ω be an open subset of V containing $\widetilde{U}$, and let there be given a function $J : \Omega \to \mathbb{R}$. Assume that J has a constrained local minimum relative to $\widetilde{U}$ at a point $\widetilde{u} \in \widetilde{U}$ and is differentiable at $\widetilde{u}$ (note that the vectors c_i, $1 \le i \le m$, are no longer assumed to be linearly independent), and let

$$I(\widetilde{u}) := \{i \in \{1, 2, \ldots, m\};\ (c_i, \widetilde{u}) = d_i\}.$$

Then show that there exist numbers $\widetilde{\lambda}_i = \widetilde{\lambda}_i(\widetilde{u})$, $i \in I(\widetilde{u})$, such that

$$\widetilde{\lambda}_i \ge 0,\ \ i \in I(\widetilde{u}), \quad \text{and} \quad \operatorname{grad} J(\widetilde{u}) + \sum_{i \in I(u)} \widetilde{\lambda}_i c_i = 0.$$

This result constitutes the **Kuhn–Tucker**[42] **theorem**, which plays a key role in *nonlinear programming*; the numbers $\widetilde{\lambda}_i = \widetilde{\lambda}_i(\widetilde{u})$, $i \in I(\widetilde{u})$, are called the **Kuhn–Tucker multipliers** *associated with the constrained local extremum* $\widetilde{u} \in \widetilde{U}$.

Hint: Show that

$$(\operatorname{grad} J(\widetilde{u}), w) \ge 0 \quad \text{for all } w \in C(\widetilde{u}) := \{v \in V;\ (c_i, v) \le 0,\ i \in I(\widetilde{u})\}.$$

Then show that the existence of the Kuhn–Tucker multipliers follows from the *Farkaš lemma* (Problem 4.3-11).

Remark Kuhn–Tucker multipliers $\widetilde{\lambda}_i = \widetilde{\lambda}_i(\widetilde{u})$ can be in fact defined for *all* $1 \le i \le m$, simply by letting $\widetilde{\lambda}_i := 0$ for those indices i for which $(c_i, \widetilde{u}) < d_i$. In this case the last relation of (2) can be recast in a form more reminiscent of that of (1), viz.,

$$\widetilde{\lambda}_i \ge 0,\ \ 1 \le i \le m, \quad \sum_{i=1}^{m} \widetilde{\lambda}_i \left((c_i, \widetilde{u}) - d_i\right) = 0, \quad \text{and} \quad \operatorname{grad} J(\widetilde{u}) + \sum_{i=1}^{m} \widetilde{\lambda}_i c_i = 0.$$

For instance, let there be given a *quadratic functional*

$$J : \boldsymbol{v} \in \mathbb{R}^n \to J(\boldsymbol{v}) := \frac{1}{2} \boldsymbol{v}^T \boldsymbol{A} \boldsymbol{v} - \boldsymbol{c}^T \boldsymbol{v},$$

where $\boldsymbol{A}$ is a *positive-definite symmetric* matrix of order n and $\boldsymbol{b} \in \mathbb{R}^n$, and a set $\widetilde{\boldsymbol{U}}$ of the form

$$\widetilde{\boldsymbol{U}} := \{\boldsymbol{v} \in \mathbb{R}^n;\ \boldsymbol{B}\boldsymbol{v} \le \boldsymbol{d}\},$$

[42] H.W. Kuhn; A.W. Tucker [1951]: Nonlinear programming, in *Proceedings of the Second Berkeley Symposium on Mathematical Statistics and Probability* (J. Neyman, editor), pp. 481–492, University of California Press, Berkeley.

where $\boldsymbol{B}$ is a real $m \times n$ matrix and $\boldsymbol{d} \in \mathbb{R}^m$, and $\boldsymbol{Bv} \leq \boldsymbol{d}$ means that $(\boldsymbol{Bv})_i \leq d_i$, $1 \leq i \leq m$. Then a *necessary* condition for the functional J to have a constrained local minimum at $\widetilde{\boldsymbol{u}} \in \widetilde{\boldsymbol{U}}$ relative to $\widetilde{U}$ is the existence of a solution $(\widetilde{\boldsymbol{u}}, \widetilde{\boldsymbol{\lambda}}) \in \mathbb{R}^{n+m}$ to the *nonlinear system of equations* (again with self-explanatory notation)

$$\boldsymbol{A}\widetilde{\boldsymbol{u}} + \boldsymbol{B}^T\widetilde{\boldsymbol{\lambda}} = \boldsymbol{c}, \quad \boldsymbol{B}\widetilde{\boldsymbol{u}} \leq \boldsymbol{d}, \quad \widetilde{\boldsymbol{\lambda}} \geq \boldsymbol{0}, \quad \text{and} \quad (\widetilde{\boldsymbol{\lambda}}, \boldsymbol{B}\widetilde{\boldsymbol{u}} - \boldsymbol{d}) = 0.$$

This system should be compared with the *linear* system of equations found in the text when $\boldsymbol{U} := \{\boldsymbol{v} \in \mathbb{R}^n;\ \boldsymbol{Bv} = \boldsymbol{d}\}$ and the matrix $\boldsymbol{B}$ is of rank m. □

7.15-4 Let Ω be a domain in $\mathbb{R}^2$, let x_i, $1 \leq i \leq m$, be m distinct points in Ω, and let $f \in L^2(\Omega)$.
(1) Show that the following minimization problem: Find

$$u \in U := \{v \in H_0^2(\Omega);\ v(x_i) = 0,\ 1 \leq i \leq m\}$$

such that

$$J(u) = \inf_{v \in U} J(v), \quad \text{where } J(v) := \frac{1}{2}\int_\Omega |\Delta v|^2 \, dx - \int_\Omega f v dx \text{ for each } v \in H_0^2(\Omega),$$

has a unique solution u (use results from Sections 6.1, 6.2, and 6.8).

(2) Show that there exist real numbers $\lambda_i = \lambda_i(u)$, $1 \leq i \leq m$, such that u satisfies the partial differential equation

$$\Delta^2 u = f + \sum_{i=1}^m \lambda_i \delta_{x_i} \quad \text{in } \mathcal{D}'(\Omega),$$

i.e., *in the sense of distributions*, where, for each $1 \leq i \leq m$, δ_{x_i} denotes the Dirac distribution at x_i.

(3) Let now $\widetilde{U} := \{v \in H_0^2(\Omega);\ v(x_i) \geq 0,\ 1 \leq i \leq m\}$. Show that the following minimization problem: Find $\widetilde{u} \in \widetilde{U}$ such that $J(\widetilde{u}) = \inf_{u \in \widetilde{U}} J(v)$, has a unique solution $\widetilde{u}$; then, using Problem 7.15-3, show that there exist numbers $\widetilde{\lambda}_i = \widetilde{\lambda}_i(\widetilde{u})$, $1 \leq i \leq m$, such that

$$\Delta^2 \widetilde{u} = f + \sum_{i=1}^m \widetilde{\lambda}_i \delta_{x_i} \quad \text{in } \mathcal{D}'(\Omega),$$

$$\widetilde{\lambda}_i \geq 0, \quad 1 \leq i \leq m, \quad \text{and} \quad \sum_{i=1}^m \widetilde{\lambda}_i \widetilde{u}(x_i) = 0.$$

Remarks (1) This result thus significantly improves upon Problem 6.9-3(2).

(2) The minimization problem of (2) models a linearly elastic plate (Section 6.8) attached to every point x_i, while that of (3) models a plate subjected to *unilateral contact* at every point x_i (only an "upward" displacement is allowed at such a point). Then the *Lagrange multipliers* λ_i, $1 \leq i \leq m$, found in (2), or the *Kuhn–Tucker multipliers* $\widetilde{\lambda}_i$, $1 \leq i \leq m$, found in (3), have a remarkable *mechanical interpretation*: Each λ_i represents the magnitude of the *reaction force* concentrated at the point x_i that is needed to keep the plate from moving at that point, while $\widetilde{\lambda}_i$ either represents such a reaction force if $u(x_i) = 0$ (i.e., if contact occurs at x_i) while $\widetilde{\lambda}_i = 0$ if $u(x_i) > 0$ (i.e., if there is no contact at x_i). □

7.16 Lagrangians and saddle-points; primal and dual problems

The aim of this section is to show how a variety of *constrained optimization problems* can be cast in a single framework. Doing so will explain in particular the appearance of an auxiliary unknown in such problems, such as the pressure $\lambda \in L_0^2(\Omega)$ in the formulation of the Stokes equations as a minimization problem (Section 6.14), or the vector $\boldsymbol{\lambda} \in \mathbb{R}^m$ in the constrained quadratic minimization problem in $\mathbb{R}^n$ described at the end of the preceding section.

Let V and M be any two sets, and let

$$\mathcal{L} : V \times M \to \mathbb{R}$$

be a function. A point $(u, \lambda) \in V \times M$ is said to be a **saddle-point** (Figure 7.16-1) of the function $\mathcal{L}$ if the point u is a minimum of the function $\mathcal{L}(\cdot, \lambda) : V \to \mathbb{R}$ and if the point λ is a maximum of the function $\mathcal{L}(u, \cdot) : M \to \mathbb{R}$, i.e., if

$$\sup_{\mu \in M} \mathcal{L}(u, \mu) = \mathcal{L}(u, \lambda) = \inf_{v \in V} \mathcal{L}(v, \lambda).$$

A function $\mathcal{L} : V \times M \to \mathbb{R}$ that has a saddle-point $(u, \lambda) \in V \times M$ is called a **Lagrangian**.

Remark "*Lagrangian*" also refers to a different notion, which will be introduced in Section 9.1. □

An important property of saddle-points is that they are *de facto* solutions of sup-inf and inf-sup problems:

Theorem 7.16-1 *If (u, λ) is a saddle-point of a function $\mathcal{L} : V \times M \to \mathbb{R}$, then*

$$\inf_{v \in V} \sup_{\mu \in M} \mathcal{L}(v, \mu) = \sup_{\mu \in M} \mathcal{L}(u, \mu) = \mathcal{L}(u, \lambda) = \inf_{v \in V} \mathcal{L}(v, \lambda) = \sup_{\mu \in M} \inf_{v \in V} \mathcal{L}(v, \mu).$$

Proof First, we show that *the inequality*

$$\sup_{\mu \in M} \inf_{v \in V} \mathcal{L}(v, \mu) \leq \inf_{v \in V} \sup_{\mu \in M} \mathcal{L}(v, \mu)$$

always hold, i.e., irrespectively of the existence of a saddle-point.

Given any elements $\widetilde{v} \in V$ and $\widetilde{\mu} \in M$, we clearly have

$$\inf_{v \in V} \mathcal{L}(v, \widetilde{\mu}) \leq \mathcal{L}(\widetilde{v}, \widetilde{\mu}) \leq \sup_{\mu \in M} \mathcal{L}(\widetilde{v}, \mu)$$

(not excluding the values $-\infty$ and ∞ for the left- and right-hand sides of these inequalities). Since $\inf_{v \in V} L(v, \widetilde{\mu})$ is a function of $\widetilde{\mu} \in M$ and $\sup_{\mu \in M} L(\widetilde{v}, \mu)$ is a function of $\widetilde{v} \in V$, the desired inequality follows.

In order to establish the converse inequality, we simply note that, if (u, λ) is a saddle-point of $\mathcal{L} : V \times M \to \mathbb{R}$, then

$$\inf_{v \in V} \sup_{\mu \in M} \mathcal{L}(v, \mu) \leq \sup_{\mu \in M} \mathcal{L}(u, \mu) = \mathcal{L}(u, \lambda) = \inf_{v \in V} \mathcal{L}(v, \lambda) \leq \sup_{\mu \in M} \inf_{v \in V} \mathcal{L}(v, \mu). \qquad \square$$

We next show that the solution (u, λ) of any variational problem that is amenable to the *Babuška–Brezzi inf-sup theorem* (Theorem 6.12-1) is a *saddle-point of an* ad hoc *Lagrangian* (under additional, but mild, assumptions bearing on the bilinear form $a(\cdot, \cdot)$). The next result thus provides sufficient conditions guaranteeing the *existence of a saddle-point*, at least for a specific class of functions.

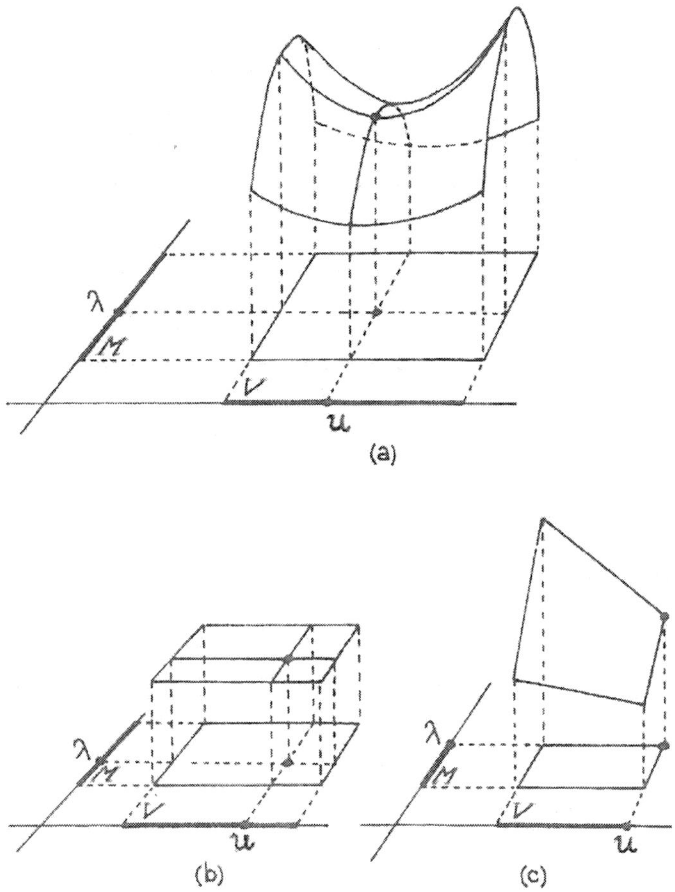

Figure 7.16-1 *Various kinds of saddle-points of a function* $\mathcal{L} : V \times M \to \mathbb{R}$; here, both sets V and M are assumed to be compact intervals of $\mathbb{R}$. Usually, a saddle-point is thought of as having the shape of a saddle, which explains the terminology; cf. (a). Other shapes are possible, however; cf. (b) and (c). This figure originally appeared in P.G. CIARLET [2007]: *Introduction a l'Analyse Numerique Matricelle et a l'Optimisation*, Dunod, Paris.

Theorem 7.16-2 (existence of saddle-points) *Let V and M be two Hilbert spaces, and let $a(\cdot,\cdot) : V \times V \to \mathbb{R}$ and $b : V \times M \to \mathbb{R}$ be two continuous bilinear forms with the following properties: The bilinear form $a(\cdot,\cdot)$ is symmetric and satisfies*

$$a(v,v) \geq 0 \quad \textit{for all } v \in V,$$

there exists a constant α such that

$$\alpha > 0 \quad \textit{and} \quad a(v,v) \geq \alpha \, \|v\|_V^2 \ \textit{ for all } v \in U_0 := \{v \in V;\ b(v,\mu) = 0 \textit{ for all } \mu \in M\},$$

i.e., $a(\cdot,\cdot)$ is U_0-coercive, and there exists a constant β such that

$$\beta > 0 \quad and \quad \inf_{\begin{cases}\mu\in M\\ \mu\neq 0\end{cases}} \sup_{\begin{cases}v\in V\\ v\neq 0\end{cases}} \frac{|b(v,\mu)|}{\|v\|_V\,\|\mu\|_M} \geq \beta.$$

Finally, let $\ell : V \to \mathbb{R}$ and $\chi : M \to \mathbb{R}$ be two continuous linear forms.

Then the unique solution $(u,\lambda) \in V \times M$ of the variational problem (Theorem 6.12-1)

$$\begin{aligned} a(u,v) + b(v,\lambda) &= \ell(v) \quad \text{for all } v \in V,\\ b(u,\mu) &= \chi(\mu) \quad \text{for all } \mu \in M,\end{aligned}$$

is the unique saddle-point of the Lagrangian $\mathcal{L} : V \times M \to \mathbb{R}$ defined by

$$\mathcal{L}(v,\mu) = \frac{1}{2}a(v,v) - \ell(v) + b(v,\mu) - \chi(\mu) \quad \text{for each } (v,\mu) \in V \times M.$$

Conversely, if this function $\mathcal{L} : V \times M \to \mathbb{R}$ is a Lagrangian, i.e., $\mathcal{L}$ has a saddle-point $(u,\lambda) \in V \times M$, then (u,λ) is the unique solution of the above variational problem.

Proof First, the relation

$$\mathcal{L}(u,\mu) \leq \mathcal{L}(u,\lambda) \quad \text{for all } \mu \in M$$

is satisfied if and only if $b(u,\mu-\lambda) \leq \chi(\mu-\lambda)$ for all $\mu \in M$, which is in turn equivalent to

$$b(u,\mu) = \chi(\mu) \quad \text{for all } \mu \in M,$$

since M is a vector space. Second, for a *fixed* $\lambda \in M$, the function $v \in V \to \mathcal{L}(v,\lambda)$ is a quadratic functional, whose second derivative with respect to the variable $v \in V$ satisfies

$$\frac{\partial^2\mathcal{L}}{\partial v^2}(w,\lambda)(v,v) = a(v,v) \geq 0 \quad \text{for all } v, w \in V.$$

Hence, by Theorem 7.9-2,

$$\mathcal{L}(u,\lambda) = \inf_{v\in V} \mathcal{L}(v,\lambda)$$

if and only if $\frac{\partial \mathcal{L}}{\partial v}(u,\lambda) = 0$, which is the same as

$$a(u,v) - \ell(v) + b(v,\lambda) = 0 \quad \text{for all } v \in V.$$

This shows that $(u,\lambda) \in V \times M$ is a solution of the variational problem if and only if

$$\sup_{\mu\in M} \mathcal{L}(u,\mu) = \mathcal{L}(u,\lambda) = \inf_{v\in V} \mathcal{L}(v,\lambda),$$

i.e., if and only if (u,λ) is a saddle-point of the function $\mathcal{L} : V \times M \to \mathbb{R}$. □

We showed in Theorem 6.12-2 that the *first argument* $u \in V$ of the saddle-point (u,λ) of the Lagrangian $\mathcal{L} : V \times M \to \mathbb{R}$ defined in Theorem 7.16-2 is the unique solution of a *constrained quadratic minimization problem* (reproduced in the next theorem), called in the present context the *primal problem*.

We now show that (under the stronger assumption that the bilinear form $a(\cdot,\cdot)$ is coercive over the whole space V) the *second argument* $\lambda \in M$ of the saddle-point (u, λ) is also the unique solution of an optimization problem. This problem takes the form of an *unconstrained maximization problem*, called in the present context the *dual problem* (of the above primal problem).

Remark The "*primal problem*" and "*dual problem*" as defined now are to be carefully distinguished from the "*primal formulation*" and "*dual formulation*" defined in Section 6.13. □

Theorem 7.16-3 (primal and dual problems) *Let the assumptions on the spaces V and M, on the bilinear forms $a(\cdot,\cdot) : V \times V \to \mathbb{R}$ and $b : V \times M \to \mathbb{R}$, and on the linear forms $\ell : V \to \mathbb{R}$ and $\chi : M \to \mathbb{R}$ be as in* Theorem 7.16-2, *the symmetric bilinear form $a(\cdot,\cdot)$ being in addition assumed to be V-coercive, i.e., there exists a constant α such that*

$$\alpha > 0 \quad \textit{and} \quad a(v,v) \geq \alpha \|v\|^2 \textit{ for all } v \in V.$$

(a) *Let the subset U_χ of the space V and the functional $J : V \to \mathbb{R}$ be respectively defined by*

$$U_\chi := \{v \in V;\ b(v,\mu) = \chi(\mu) \textit{ for all } \mu \in M\},$$
$$J(v) := \frac{1}{2}a(v,v) - \ell(v) \textit{ for each } v \in V,$$

and let u be the unique solution to the **primal problem**:

$$u \in U_\chi \quad \textit{and} \quad J(u) = \inf_{u \in U_\chi} J(v).$$

Then $u \in U_\chi \subset V$ is the first argument of the unique saddle-point (u, λ) of the Lagrangian $\mathcal{L} : V \times M \to \mathbb{R}$ defined by

$$\mathcal{L}(v,\mu) := \frac{1}{2}a(v,v) - \ell(v) + b(v,\mu) - \chi(\mu) \quad \textit{for each } (v,\mu) \in V \times M.$$

(b) *Let the functional $K : M \to \mathbb{R}$ be defined by*

$$\mu \in M \to K(\mu) := -\frac{1}{2}a(u_\mu, u_\mu) - \chi(\mu),$$

where, for each $\mu \in M$, the element u_μ is the unique solution to the unconstrained quadratic minimization problem:

$$u_\mu \in V \quad \textit{and} \quad \mathcal{L}(u_\mu, \mu) = \inf_{v \in V} \mathcal{L}(v,\mu).$$

Then the second argument $\lambda \in M$ of the saddle-point (u, λ) of the Lagrangian $\mathcal{L} : V \times M \to \mathbb{R}$ is the unique solution to the **dual problem**:

$$\lambda \in M \quad \textit{and} \quad K(\lambda) = \sup_{\mu \in M} K(\mu).$$

Proof Part (a) follows from Theorems 6.12-2 and 7.16-2.

Since the bilinear form $a(\cdot,\cdot)$ is now assumed to be V-coercive, there exists for a fixed $\mu \in M$ a unique element $u_\mu \in V$ such that

$$\mathcal{L}(u_\mu,\mu) = \inf_{v\in V} \mathcal{L}(v,\mu) := -\frac{1}{2}a(u_\mu,u_\mu) - \chi(\mu) = K(\mu).$$

Noting that $u_\lambda = u$, we then infer from Theorem 7.16-2 that

$$\mathcal{L}(u,\lambda) = \inf_{v\in V} \mathcal{L}(v,\lambda) = K(\lambda) = \sup_{\mu\in M}\inf_{v\in V} \mathcal{L}(v,\mu) = \sup_{\mu\in M} K(\mu),$$

which proves (b). □

The solution $\lambda \in M$ to the *dual problem* of Theorem 7.16-3 is called the **Lagrange multiplier** *associated with the constraint* $u \in U_\chi$ (or equivalently $b(u,\mu) = \chi(\mu)$ for all $\mu \in M$) that the solution to the *primal problem* must satisfy.

A first application of Theorem 7.16-3 is provided by the *Stokes equations* introduced and analyzed in Section 6.14. It was shown there (Theorem 6.14-3) that *the unknown velocity* $\boldsymbol{u} \in \boldsymbol{H}_0^1(\Omega)$ *is the unique solution to the constrained quadratic minimization problem*

$$\boldsymbol{u} \in \boldsymbol{U}_0 := \{\boldsymbol{v} \in \boldsymbol{H}_0^1(\Omega);\ \operatorname{div}\boldsymbol{v} = 0 \text{ in } \Omega\},$$

$$I(\boldsymbol{u}) = \inf_{\boldsymbol{v}\in\boldsymbol{U}_0} I(\boldsymbol{v}),\ \textit{where } I(\boldsymbol{v}) := \frac{\nu}{2}\int_\Omega \boldsymbol{\nabla v} : \boldsymbol{\nabla v}\,\mathrm{d}x - \int_\Omega \boldsymbol{f}\cdot\boldsymbol{v}\,\mathrm{d}x.$$

All the assumptions of Theorem 7.16-3 being satisfied in this case, it follows that $\boldsymbol{u}$ *is the first argument of the unique saddle-point* $(\boldsymbol{u},\lambda) \in \boldsymbol{H}_0^1(\Omega) \times L_0^2(\Omega)$ *of the Lagrangian* $\mathcal{L} : \boldsymbol{H}_0^1(\Omega) \times L_0^2(\Omega) \to \mathbb{R}$ defined for each $(\boldsymbol{v},\mu) \in \boldsymbol{H}_0^1(\Omega) \times L_0^2(\Omega)$ by

$$\mathcal{L}(\boldsymbol{v},\mu) := \frac{\nu}{2}\int_\Omega \boldsymbol{\nabla v} : \boldsymbol{\nabla v}\,\mathrm{d}x - \int_\Omega \boldsymbol{f}\cdot\boldsymbol{v}\,\mathrm{d}x - \int_\Omega (\operatorname{div}\boldsymbol{v})\mu\,\mathrm{d}x.$$

Noting that, by Theorem 7.16-2, the saddle-point $(\boldsymbol{u},\lambda)$ also satisfies the variational equations of Theorem 6.14-3(a), we thus conclude that *the unknown pressure* $\lambda \in L_0^2(\Omega)$ *is the Lagrange multiplier associated with the incompressible constraint* $\operatorname{div}\boldsymbol{u} = 0$ *in* Ω.

Remark Similar examples of dual problems are proposed in Problem 7.16-1. □

As another application of Theorem 7.16-3, consider the problem (already encountered in Section 7.15), now viewed as a *primal problem*, of minimizing a quadratic functional

$$J : \boldsymbol{v} \in \mathbb{R}^n \to J(v) := \frac{1}{2}\boldsymbol{v}^T\boldsymbol{A}\boldsymbol{v} - \boldsymbol{c}^T\boldsymbol{v},$$

where $\boldsymbol{A}$ is a positive-definite symmetric matrix of order n and $\boldsymbol{c} \in \mathbb{R}^n$, over a subset $\boldsymbol{U}$ of $\mathbb{R}^n$ of the form

$$\boldsymbol{U} := \{\boldsymbol{v} \in \mathbb{R}^n;\ \boldsymbol{B}\boldsymbol{v} = \boldsymbol{d}\},$$

where $\boldsymbol{B}$ is an $m \times n$ matrix of rank m (hence $m \le n$) and $\boldsymbol{c} \in \mathbb{R}^m$. Then Theorem 7.16-3 (all the assumptions of which are satisfied) asserts that the unique solution $\boldsymbol{u}$ to this primal problem is the first argument of the unique saddle-point $(\boldsymbol{u}, \boldsymbol{\lambda}) \in \mathbb{R}^n \times \mathbb{R}^m$ of the *Lagrangian* $\mathcal{L} : \mathbb{R}^n \times \mathbb{R}^m \to \mathbb{R}$ defined by

$$\mathcal{L}(\boldsymbol{v}, \boldsymbol{\mu}) := \frac{1}{2}\boldsymbol{v}^T \boldsymbol{A}\boldsymbol{v} - \boldsymbol{c}^T \boldsymbol{v} + \boldsymbol{v}^T \boldsymbol{B}^T \boldsymbol{\mu} - \boldsymbol{d}^T \boldsymbol{\mu} \quad \text{for each } (\boldsymbol{v}, \boldsymbol{\mu}) \in \mathbb{R}^n \times \mathbb{R}^m,$$

and that the second argument $\boldsymbol{\lambda} \in \mathbb{R}^m$ of the saddle-point $(\boldsymbol{u}, \boldsymbol{\lambda})$ is the *Lagrange multiplier associated with* the constraint $\boldsymbol{B}\boldsymbol{u} = \boldsymbol{d}$.

Incidentally, this shows that the definition given here of a Lagrange multiplier is indeed a special case of that given in the previous section.

Note that dual problems and Lagrange multipliers associated with *inequality constraints* of the form $\boldsymbol{B}\boldsymbol{u} \leq \boldsymbol{d}$, instead of *equality constraints* of the form $\boldsymbol{B}\boldsymbol{u} = \boldsymbol{d}$ as above, can be as well defined; cf. Problem 7.16-2.

Problems

7.16-1 (1) What is the dual problem of the constrained quadratic minimization problem of Theorem 6.13-1(c), now viewed as a primal problem?

(2) What is the dual problem of the constrained quadratic minimization problem of Theorem 6.13-2(c), now viewed as a primal problem?

Remark As already noted, the adjectives "primal" and "dual" are used in the present section according to the usual practice in *optimization theory*, while the same adjectives were used with a *different* meaning in Section 6.13, then according to the usual practice in *finite element approximation theory*. □

7.16-2 Given a positive-definite symmetric matrix $\boldsymbol{A}$ of order n, an $m \times n$ matrix $\boldsymbol{B}$, and vectors $\boldsymbol{c} \in \mathbb{R}^n$ and $\boldsymbol{d} \in \mathbb{R}^m$, define the functional $J : \mathbb{R}^n \to \mathbb{R}$ and the subset $\boldsymbol{U}$ of $\mathbb{R}^n$ by

$$J(\boldsymbol{v}) := \frac{1}{2}\boldsymbol{v}^T \boldsymbol{A}\boldsymbol{v} - \boldsymbol{c}^T \boldsymbol{v}, \ \ \boldsymbol{v} \in \mathbb{R}^n, \quad \text{and} \quad \boldsymbol{U} := \{\boldsymbol{v} \in \mathbb{R}^n;\ \boldsymbol{B}\boldsymbol{v} \leq \boldsymbol{d}\},$$

where $\boldsymbol{B}\boldsymbol{v} \leq \boldsymbol{d}$ means $(\boldsymbol{B}\boldsymbol{v})_i \leq d_i$, $1 \leq i \leq m$. Assume that $\boldsymbol{U} \neq \varnothing$.

(1) Show that there is one and only one solution to the *primal problem* defined here as

$$\boldsymbol{u} \in \boldsymbol{U} \quad \text{and} \quad J(\boldsymbol{u}) = \inf_{\boldsymbol{v} \in \boldsymbol{U}} J(\boldsymbol{v}).$$

(2) Let $\mathbb{R}^m_+ := \{\boldsymbol{\mu} = (\mu_i)_{i=1}^m \in \mathbb{R}^m;\ \mu_i \geq 0,\ 1 \leq i \leq m\}$ and define the *function* $\mathcal{L} : \mathbb{R}^n \times \mathbb{R}^m_+$ by

$$\mathcal{L}(\boldsymbol{v}, \boldsymbol{\mu}) := \frac{1}{2}\boldsymbol{v}^T \boldsymbol{A}\boldsymbol{v} - \boldsymbol{c}^T \boldsymbol{v} + \boldsymbol{v}^T \boldsymbol{B}^T \boldsymbol{\mu} - \boldsymbol{d}^T \boldsymbol{\mu}, \quad \text{for each } (\boldsymbol{v}, \boldsymbol{\mu}) \in \mathbb{R}^n \times \mathbb{R}^m_+.$$

Show that the function $\mathcal{L}$ is a *Lagrangian*, i.e., that $\mathcal{L}$ has at least one saddle-point over the set $\mathbb{R}^n \times \mathbb{R}^m_+$ and that the first argument of this saddle-point is uniquely defined.

(3) Assume in addition that rank $\boldsymbol{B} = m$. Show that $\mathcal{L}$ has a unique saddle-point $(\boldsymbol{u}, \boldsymbol{\lambda})$ over $\mathbb{R}^n \times \mathbb{R}^m_+$ and that $\boldsymbol{\lambda}$ is the unique solution of the *dual problem*, defined here as

$$K(\boldsymbol{\lambda}) = \sum_{\boldsymbol{\mu} \in \mathbb{R}^m_+} K(\boldsymbol{\mu}),$$

where $K(\boldsymbol{\mu}) := \inf_{\boldsymbol{v} \in \mathbb{R}^n} \mathcal{L}(\boldsymbol{v}, \boldsymbol{\mu})$ for each $\boldsymbol{\mu} \in \mathbb{R}^m_+$.

7.16-3 The assumptions are the same as in Problem 7.16-2. The objective of this problem is to analyze **Uzawa's method**,[43] an iterative method that approximates the solution of a *constrained* optimization problem (the primal problem of Problem 7.16-2(1)) by means of a sequence of solutions of *unconstrained* optimization problems, as follows.

Given any $\boldsymbol{\lambda}_0 \in \mathbb{R}^m_+$, define iteratively $(\boldsymbol{u}_k, \boldsymbol{\lambda}_k) \in \mathbb{R}^n \times \mathbb{R}^m_+$ by

$$J(\boldsymbol{u}_k) + (\boldsymbol{B}\boldsymbol{u}_k - \boldsymbol{d})^T\boldsymbol{\lambda}_k = \inf_{\boldsymbol{v}\in\mathbb{R}^n} \{J(\boldsymbol{v}) + (\boldsymbol{B}\boldsymbol{v} - \boldsymbol{d})^T\boldsymbol{\lambda}_k\}, \quad k \geq 0,$$

$$\boldsymbol{\lambda}_{k+1} = \boldsymbol{P}(\boldsymbol{\lambda}_k + \rho(\boldsymbol{B}\boldsymbol{u}_k - \boldsymbol{d})), \quad k \geq 0,$$

where $\boldsymbol{P} : \mathbb{R}^m \to \mathbb{R}^m_+$ denotes the projection operator from $\mathbb{R}^m$ onto $\mathbb{R}^m_+$ (Section 4.3) and ρ is a real parameter.

(1) Let $\alpha > 0$ denote the smallest eigenvalue of $\boldsymbol{A}$. Show that, if $0 < \rho < \dfrac{2\alpha}{|\boldsymbol{B}|^2}$, the sequence $(\boldsymbol{u}_k)_{k=0}^{\infty}$ converges to the unique solution of the primal problem of Problem 7.16-2(1).

(2) Assume in addition that rank $\boldsymbol{B} = m$. Show that the sequence $(\boldsymbol{\lambda}_k)_{k=0}^{\infty}$ converges to the unique solution of the dual problem found in Problem 7.16-2(3).

(3) Assuming again that rank $\boldsymbol{B} = m$, show that Uzawa's method is a *gradient method* as defined in Problem 7.12-2, but now applied to the *dual problem* found in Problem 7.16-2(3).

7.16-4 The objective of this problem is to give sufficient conditions for the *existence* of a saddle-point, the result of question (5) constituting a particular case of the **Ky Fan–Sion theorem.**[44]

Let V and M be nonempty convex and compact subsets of finite-dimensional vector spaces, and let $\mathcal{L} : V \times M \to \mathbb{R}$ be a continuous function with the following properties:

$$\begin{aligned} &\mathcal{L}(v, \cdot) : M \to \mathbb{R} \quad &&\text{is concave for every } v \in V,\\ &\mathcal{L}(\cdot, \mu) : V \to \mathbb{R} \quad &&\text{is convex for every } \mu \in M. \end{aligned}$$

(1) Show that the function

$$K : \mu \in M \to K(\mu) := \inf_{v\in V} \mathcal{L}(v, \mu)$$

is concave and continuous.

(2) Assume until question (4) that the function $\mathcal{L}(\cdot, \mu) : V \to \mathbb{R}$ is strictly convex for every $\mu \in M$, so that

$$K(\mu) = \mathcal{L}(u(\mu), \mu),$$

where the element $u(\mu) \in V$ is uniquely defined. Show that the function $\mu \in M \to u(\mu) \in V$ defined in this fashion is continuous.

(3) Let $\lambda \in M$ be a point satisfying (by (1), at least one such point exists)

$$K(\lambda) = \sup_{\mu\in M} K(\mu).$$

Show that, for any $\mu \in M$,

$$K(\lambda) \geq \mathcal{L}(u(\theta\mu + (1 - \theta)\lambda), \mu) \quad \text{for all } 0 \leq \theta \leq 1.$$

(4) Show that

$$\sup_{\mu\in M} \inf_{v\in V} \mathcal{L}(v, \mu) \geq \inf_{v\in V} \sup_{\mu\in M} \mathcal{L}(v, \mu),$$

[43] H. Uzawa [1958]: Iterative methods for concave programming, in *Studies in Linear and Nonlinear Programming* (K.J. Arrow, L. Hurwicz, & H. Uzawa, editors), pp. 154–165, Stanford University Press, Stanford, CA.

[44] Ky Fan [1953]: Minimax theorems, *Proceedings of the National Academy of Sciences* **39**, 42–47.
M. Sion [1958]: On general mini-max theorems, *Pacific Journal of Mathematics* **8**, 171–176.

and conclude that the point $(u(\lambda), \lambda)$ is a saddle-point of the function $\mathcal{L}$.

(5) If the function $\mathcal{L}(\cdot, \mu) : V \to \mathbb{R}$ is convex but not necessarily strictly convex for every $\mu \in M$, introduce the auxiliary functions

$$\mathcal{L}_\varepsilon : (v, \mu) \in V \times M \to \mathcal{L}_\varepsilon(v, \mu) := \mathcal{L}(v, \mu) + \varepsilon \|v\|^2, \quad \varepsilon > 0,$$

and, by letting $\varepsilon \to 0$, show that the function $\mathcal{L}$ has at least one saddle-point.

CHAPTER 8

DIFFERENTIAL GEOMETRY IN $\mathbb{R}^n$

Introduction

Why such a chapter? Simply because, even though *differential geometry in* $\mathbb{R}^n$ may be correctly viewed as only a brief introduction to differential geometry in general, its modest scope already provides beautiful *existence and uniqueness theorems for two highly nonlinear systems of partial differential equations*, a topic at the core of *nonlinear functional analysis.* Besides, its frequent usage of notions from differential calculus makes it a natural sequel to the previous chapter.

This chapter first reviews (Sections 8.1–8.5) basic notions, such as the *metric tensor*, *covariant derivatives*, and the fundamental *Riemann curvature tensor*, that naturally arise when an open subset of the n-dimensional Euclidean space $\mathbb{E}^n$ is equipped with curvilinear coordinates; it also provides a brief *introduction to tensor analysis*, whose aim is simply to put notions such as "covariant indices" versus "contravariant exponents," or "tensors," in their proper perspective.

A detailed proof is then given of the *fundamental theorem of Riemannian geometry* (Theorem 8.6-1) for an open subset of $\mathbb{R}^n$. This theorem answers the following question:

Given an open subset Ω of $\mathbb{R}^n$ and a smooth enough symmetric and positive-definite $n \times n$ matrix field (g_{ij}) defined on Ω, when can the *Riemannian manifold* $(\Omega; (g_{ij}))$ be isometrically immersed in the Euclidean space $\mathbb{E}^n$ of the same dimension? Or equivalently, when does there exist an immersion $\boldsymbol{\Theta} : \Omega \to \mathbb{E}^n$ that satisfies the following *nonlinear system of* $\dfrac{n(n+1)}{2}$ *partial differential equations*:

$$\partial_i \boldsymbol{\Theta} \cdot \partial_j \boldsymbol{\Theta} = g_{ij} \quad \text{in } \Omega, \;\; 1 \le i \le j \le n?$$

As shown in Theorem 8.6-1, the answer to this question turns out to be remarkably simple to state (but not so simple to prove): Under the assumption that Ω is *simply connected, the necessary condition that the Riemann curvature tensor associated with* (g_{ij}) *vanish in* Ω *is also sufficient for the existence of such an immersion* $\boldsymbol{\Theta}$.

Besides, *if* Ω *is connected, this immersion is unique up to compositions with isometries of* $\mathbb{E}^n$. This means that, if $\widetilde{\boldsymbol{\Theta}} : \Omega \to \mathbb{E}^n$ is any other smooth immersion that satisfies the above nonlinear system of $\dfrac{n(n+1)}{2}$ partial differential equations in Ω, then there exist a vector $\boldsymbol{c} \in \mathbb{E}^n$ and an orthogonal matrix $\boldsymbol{Q}$ of order n such that

$$\widetilde{\boldsymbol{\Theta}}(x) = \boldsymbol{c} + \boldsymbol{Q}\boldsymbol{\Theta}(x) \quad \text{for all } x \in \Omega.$$

This uniqueness result is the content of the aptly called *rigidity theorem* (Theorem 8.7-1).

This chapter then reviews (Sections 8.8–8.14) basic notions, such as the *two fundamental forms*, the *Gauß and Codazzi–Mainardi equations*, the *Gaussian curvature*, and *covariant derivatives*, that are naturally associated with a *surface* in $\mathbb{E}^3$ defined by means of two *curvilinear coordinates*, that is, components of points that vary in an open subset ω of $\mathbb{R}^2$. A spectacular application of the beautiful *Gauß Theorem Egregium* (Theorem 8.15-1) to *cartography* is given in passing (Theorem 8.15-2), according to which it is not possible to draw a flat map of a portion of the surface of the earth that would preserve distances (up to a scale).

This chapter concludes with a detailed proof of the *fundamental theorem of surface theory* (Theorem 8.16-1). This theorem answers the following question:

Given an open subset ω of $\mathbb{R}^2$ and a smooth enough symmetric and positive-definite matrix field $(a_{\alpha\beta})$ together with a smooth enough symmetric matrix field $(b_{\alpha\beta})$ defined over ω, when are they the first and second fundamental forms of a surface $\boldsymbol{\theta}(\omega) \subset \mathbb{E}^3$, i.e., when does there exist an immersion $\boldsymbol{\theta} : \omega \to \mathbb{E}^3$ that satisfies the following *nonlinear system of six partial differential equations*:

$$\partial_\alpha \boldsymbol{\theta} \cdot \partial_\beta \boldsymbol{\theta} = a_{\alpha\beta} \quad \text{and} \quad \partial_{\alpha\beta} \boldsymbol{\theta} \cdot \left\{ \frac{\partial_1 \boldsymbol{\theta} \wedge \partial_2 \boldsymbol{\theta}}{|\partial_1 \boldsymbol{\theta} \wedge \partial_2 \boldsymbol{\theta}|} \right\} = b_{\alpha\beta} \text{ in } \omega, \quad 1 \le \alpha \le \beta \le 2?$$

As shown in Theorem 8.16-1, the answer to this question turns out to be again remarkably simple to state (but its proof is by no means easy): Under the assumption that ω is *simply connected, the necessary conditions expressed by the Gauß and Codazzi–Mainardi equations are also sufficient for the existence of such an immersion* $\boldsymbol{\theta}$.

Besides, *if ω is connected, this immersion is unique up to composition with proper isometries of* $\mathbb{E}^3$. This means that, if $\widetilde{\boldsymbol{\theta}} : \omega \to \mathbb{E}^3$ is any other smooth immersion that satisfies the above nonlinear system of six partial differential equations in ω, then there exist a vector $\boldsymbol{c} \in \mathbb{E}^3$ and a proper orthogonal matrix $\boldsymbol{Q}$ of order three such that

$$\widetilde{\boldsymbol{\theta}}(y) = \boldsymbol{c} + \boldsymbol{Q}\boldsymbol{\theta}(y) \quad \text{for all } y \in \omega.$$

This uniqueness result constitutes another *rigidity theorem* (Theorem 8.17-1).

Note that the proofs of both the fundamental theorem of Riemannian geometry and the fundamental theorem of surface theory crucially hinge on the *classical Poincaré lemma* and on the *existence theorem for Pfaff systems* established in Chapter 6.

8.1 Curvilinear coordinates in an open subset of $\mathbb{R}^n$

To begin with, we list some notations and conventions that will be consistently used throughout this chapter.

Save when otherwise indicated, e.g., when they are used for indexing sequences, *Latin* indices and exponents range in the set $\{1, \dots, n\}$, and the *summation convention* with respect to repeated indices or exponents is systematically used in conjunction with this rule. For instance,

$$\boldsymbol{g}_i(x) = g_{ij}(x)\boldsymbol{g}^j(x) \quad \text{means} \quad \boldsymbol{g}_i(x) = \sum_{j=1}^n g_{ij}(x)\boldsymbol{g}^j(x) \quad \text{for } i = 1, \dots, n.$$

Kronecker's symbols are designated by δ_i^j, δ_{ij}, or δ^{ij} according to the context.

Let $\mathbb{E}^n$ denote the *n-dimensional Euclidean space*, with $\boldsymbol{a} \cdot \boldsymbol{b}$ denoting the Euclidean inner product of $\boldsymbol{a}, \boldsymbol{b} \in \mathbb{E}^n$, and $|\boldsymbol{a}| = \sqrt{\boldsymbol{a} \cdot \boldsymbol{a}}$ denoting the Euclidean norm of $\boldsymbol{a} \in \mathbb{E}^n$. The vectors of the *canonical orthonormal basis* of $\mathbb{E}^n$ are denoted $\widehat{\boldsymbol{e}}^i = \widehat{\boldsymbol{e}}_i$. The *Cartesian coordinates* of a point $\widehat{x} \in \mathbb{E}^n$ are denoted $\widehat{x}_i$; finally, we let $\widehat{\partial}_i := \partial/\partial\widehat{x}_i$.

In addition, let there be given an *n-dimensional vector space* in which n vectors, denoted $\boldsymbol{e}^i = \boldsymbol{e}_i$, form a basis. *This space will be identified with* $\mathbb{R}^n$. Let x_i denote the coordinates of a point $x \in \mathbb{R}^n$ and let $\partial_i := \partial/\partial x_i$, $\partial_{ij} := \partial^2/\partial x_i \partial x_j$, and $\partial_{ijk} := \partial^3/\partial x_i \partial x_j \partial x_k$.

Let there be given an *open* subset $\widehat{\Omega}$ of $\mathbb{E}^n$ and assume that there exist an *open* subset Ω of $\mathbb{R}^n$ and an *injective* mapping $\boldsymbol{\Theta} : \Omega \to \mathbb{E}^n$ such that $\boldsymbol{\Theta}(\Omega) = \widehat{\Omega}$. Then each point $\widehat{x} \in \widehat{\Omega}$ can be unambiguously written as

$$\widehat{x} = \boldsymbol{\Theta}(x), \quad x \in \Omega,$$

and the n coordinates x_i of x are called the **curvilinear coordinates** of $\widehat{x}$ (cf. Figure 8.1-1 when $n = 3$). Naturally, *there are infinitely many ways of defining curvilinear coordinates in a given open set* $\widehat{\Omega}$, depending on how the open set Ω and the mapping $\boldsymbol{\Theta}$ are chosen.

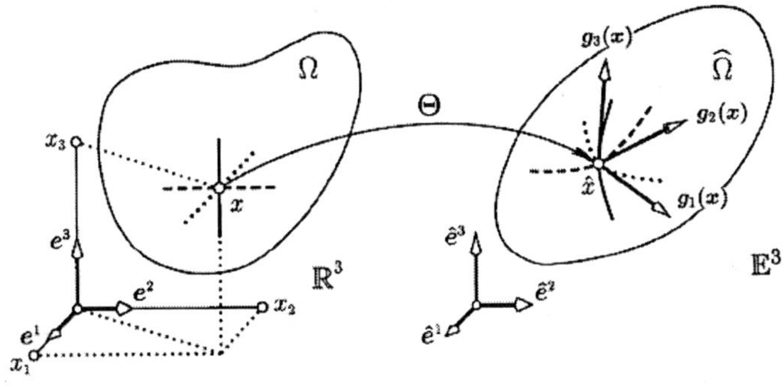

Figure 8.1-1 *Curvilinear coordinates and covariant bases in an open set* $\widehat{\Omega} \subset \mathbb{E}^3$. The three coordinates x_1, x_2, x_3 of $x \in \Omega$ are the *curvilinear coordinates* of $\widehat{x} = \boldsymbol{\Theta}(x) \in \widehat{\Omega}$. If the three vectors $\boldsymbol{g}_i(x) = \partial_i \boldsymbol{\Theta}(x)$ are linearly independent, they form the *covariant basis* at $\widehat{x} = \boldsymbol{\Theta}(x)$ and they are tangent to the coordinate lines passing through $\widehat{x}$ (Section 8.2). This figure originally appeared in P.G. CIARLET [2005]: *An Introduction to Differential Geometry with Applications to Elasticity*, Springer, Dordrecht.

Examples of curvilinear coordinates when $n = 3$ include the well-known *cylindrical* and *spherical coordinates* (Figure 8.1-2).

Alternatively, an open subset Ω of $\mathbb{R}^n$ together with a mapping $\boldsymbol{\Theta} : \Omega \to \mathbb{E}^n$ are instead *a priori* given. If $\boldsymbol{\Theta} \in \mathcal{C}(\Omega; \mathbb{E}^n)$ and $\boldsymbol{\Theta}$ is *injective*, the set $\widehat{\Omega} := \boldsymbol{\Theta}(\Omega)$ is *open* by *Brouwer's invariance of domain theorem* (which will be established later; cf. Theorem 9.17-3), and curvilinear coordinates inside $\widehat{\Omega}$ are unambiguously defined in this case.

If $\boldsymbol{\Theta} \in \mathcal{C}^1(\Omega; \mathbb{E}^n)$ and the n vectors $\partial_i \boldsymbol{\Theta}(x)$ are linearly independent at each point $x \in \Omega$, the set $\widehat{\Omega}$ is again *open*, this time by the *invariance of domain theorem for mappings of class* $\mathcal{C}^1$ *in Banach spaces* (Theorem 7.14-2), but curvilinear coordinates can be unambiguously defined *only locally* in this case: Given $x \in \Omega$, all that can be asserted (by the local inversion

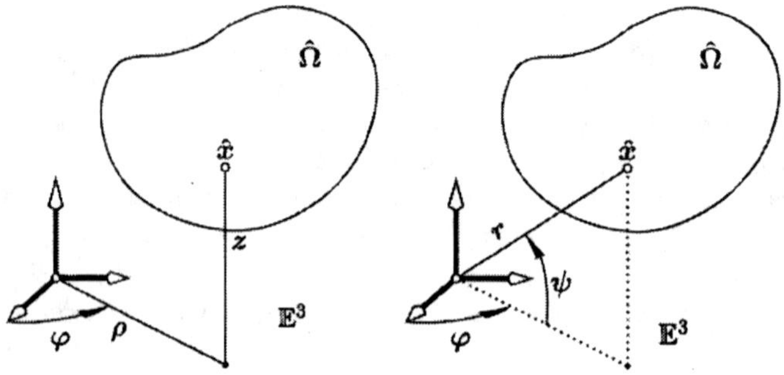

Figure 8.1-2 *Two familiar examples of curvilinear coordinates.* Let the mapping $\boldsymbol{\Theta}$ be defined by

$$\boldsymbol{\Theta} : (\varphi, \rho, z) \in \Omega \to (\rho \cos \varphi, \rho \sin \varphi, z) \in \mathbb{E}^3.$$

Then (φ, ρ, z) are the *cylindrical coordinates* of $\widehat{x} = \boldsymbol{\Theta}(\varphi, \rho, z)$. Note that $(\varphi + 2k\pi, \rho, z)$ or $(\varphi + \pi + 2k\pi, -\rho, z)$, $k \in \mathbb{Z}$, are also cylindrical coordinates of the same point $\widehat{x}$ and that φ is not defined if $\widehat{x}$ is the origin of $\mathbb{E}^3$.

Let the mapping $\boldsymbol{\Theta}$ be defined by

$$\boldsymbol{\Theta} : (\varphi, \psi, r) \in \Omega \to (r \cos \psi \cos \varphi, r \cos \psi \sin \varphi, r \sin \psi) \in \mathbb{E}^3.$$

Then (φ, ψ, r) are the *spherical coordinates* of $\widehat{x} = \boldsymbol{\Theta}(\varphi, \psi, r)$. Note that $(\varphi + 2k\pi, \psi + 2\ell\pi, r)$ or $(\varphi + 2k\pi, \psi + \pi + 2\ell\pi, -r)$, $k, \ell \in \mathbb{Z}$, are also spherical coordinates of the same point $\widehat{x}$ and that φ and ψ are not defined if $\widehat{x}$ is the origin of $\mathbb{E}^3$. This figure originally appeared in P.G. Ciarlet [2005]: *An Introduction to Differential Geometry with Applications to Elasticity*, Springer, Dordrecht.

theorem; cf. Theorem 7.14-1) is the existence of an open neighborhood V of x in Ω such that the restriction of $\boldsymbol{\Theta}$ to V is a $\mathcal{C}^1$-diffeomorphism, hence an injection, of V onto $\boldsymbol{\Theta}(V)$.

8.2 Metric tensor; volumes and lengths in curvilinear coordinates

Let Ω be an open subset of $\mathbb{R}^n$ and let

$$\boldsymbol{\Theta} = \Theta_i \widehat{\boldsymbol{e}}^i : \Omega \to \mathbb{E}^n$$

be a mapping that is *differentiable at a point* $x \in \Omega$. If $\boldsymbol{\delta x} = \delta x_i \boldsymbol{e}^i$ is such that $(x + \boldsymbol{\delta x}) \in \Omega$, then

$$\boldsymbol{\Theta}(x + \boldsymbol{\delta x}) = \boldsymbol{\Theta}(x) + \boldsymbol{\nabla\Theta}(x)\boldsymbol{\delta x} + |\boldsymbol{\delta x}|\, \boldsymbol{\varepsilon}(\boldsymbol{\delta x}) \quad \text{with} \lim_{\boldsymbol{\delta x} \to \boldsymbol{0}} \boldsymbol{\varepsilon}(\boldsymbol{\delta x}) = \boldsymbol{0},$$

where the $n \times n$ matrix

$$\boldsymbol{\nabla\Theta}(x) = (\partial_j \Theta_i(x))$$

(the row index is i) is the *gradient matrix of* $\boldsymbol{\Theta}$ at x (Section 7.1) and $\boldsymbol{\delta x} \in \mathbb{R}^n$ designates the column vector with components δx_i.

Let the n column vectors $\boldsymbol{g}_i(x) \in \mathbb{E}^n$ be defined by

$$\boldsymbol{g}_i(x) := \partial_i \boldsymbol{\Theta}(x),$$

i.e., $\boldsymbol{g}_i(x)$ *is the* ith *column vector of the matrix* $\boldsymbol{\nabla\Theta}(x)$. Then $\boldsymbol{\Theta}(x+\boldsymbol{\delta x})$ may be also written as

$$\boldsymbol{\Theta}(x+\boldsymbol{\delta x}) = \boldsymbol{\Theta}(x) + \delta x^i \boldsymbol{g}_i(x) + |\boldsymbol{\delta x}|\,\boldsymbol{\varepsilon}(\boldsymbol{\delta x}) \quad \text{with} \lim_{\boldsymbol{\delta x}\to\boldsymbol{0}} \boldsymbol{\varepsilon}(\boldsymbol{\delta x}) = \boldsymbol{0}.$$

If in particular $\boldsymbol{\delta x}$ is of the form $\boldsymbol{\delta x} = \delta t \boldsymbol{e}_i$, where $\delta t \in \mathbb{R}$ and $\boldsymbol{e}_i$ is one of the basis vectors in $\mathbb{R}^n$, this relation reduces to

$$\boldsymbol{\Theta}(x+\delta t\boldsymbol{e}_i) = \boldsymbol{\Theta}(x) + \delta t \boldsymbol{g}_i(x) + |\delta t|\,\boldsymbol{\chi}(\delta t) \quad \text{with} \lim_{\delta t\to 0} \boldsymbol{\chi}(\delta t) = 0.$$

A mapping $\boldsymbol{\Theta}: \Omega \to \mathbb{E}^n$ is an **immersion at** $x \in \Omega$ if it is differentiable at x and the matrix $\boldsymbol{\nabla\Theta}(x)$ is invertible, or equivalently, if the n vectors $\boldsymbol{g}_i(x) = \partial_i\boldsymbol{\Theta}(x)$ are linearly independent.

Assume from now on in this section that *the mapping* $\boldsymbol{\Theta}$ *is an immersion at* x, in which case *the* n *vectors* $\boldsymbol{g}_i(x)$ are said to constitute the **covariant basis** at the point $\widehat{x} = \boldsymbol{\Theta}(x)$. Then the last relation shows that *each vector* $\boldsymbol{g}_i(x)$ *is tangent to the* ith **coordinate line** *passing through* $\widehat{x} = \boldsymbol{\Theta}(x)$, which is defined as the image by $\boldsymbol{\Theta}$ of the points of Ω that lie on the line parallel to $\boldsymbol{e}_i$ passing through x (there exist t_0 and t_1 with $t_0 < 0 < t_1$ such that the ith coordinate line is given by $t \in]t_0, t_1[\to \boldsymbol{f}_i(t) := \boldsymbol{\Theta}(x+t\boldsymbol{e}_i)$ in a neighborhood of $\widehat{x}$; hence $\boldsymbol{f}_i'(0) = \partial_i\boldsymbol{\Theta}(x) = \boldsymbol{g}_i(x)$). Examples of coordinate lines are shown in Figures 8.1-1 and 8.1-2.

Returning to a general increment $\boldsymbol{\delta x} = \delta x^i \boldsymbol{e}_i$, we also infer from the expression of $\boldsymbol{\Theta}(x+\boldsymbol{\delta x})$ that

$$\begin{aligned}|\boldsymbol{\Theta}(x+\boldsymbol{\delta x}) - \boldsymbol{\Theta}(x)| &= \sqrt{\boldsymbol{\delta x}^T \boldsymbol{\nabla\Theta}(x)^T \boldsymbol{\nabla\Theta}(x)\boldsymbol{\delta x}} + |\boldsymbol{\delta x}|\,\eta(\boldsymbol{\delta x}) \\ &= \sqrt{\delta x^i \boldsymbol{g}_i(x)\cdot\boldsymbol{g}_j(x)\delta x^j} + |\boldsymbol{\delta x}|\,\eta(\boldsymbol{\delta x}) \quad \text{with} \lim_{\boldsymbol{\delta x}\to 0}\eta(\boldsymbol{\delta x}) = 0.\end{aligned}$$

In other words, the principal part with respect to $\boldsymbol{\delta x}$ of the distance between the points $\boldsymbol{\Theta}(x+\boldsymbol{\delta x})$ and $\boldsymbol{\Theta}(x)$ is $\sqrt{\delta x^i \boldsymbol{g}_i(x)\cdot\boldsymbol{g}_j(x)\delta x^j}$. This observation suggests defining an $n\times n$ *matrix* $(g_{ij}(x))$ by letting (the row index is i)

$$g_{ij}(x) := \boldsymbol{g}_i(x)\cdot\boldsymbol{g}_j(x) = \left(\boldsymbol{\nabla\Theta}(x)^T\boldsymbol{\nabla\Theta}(x)\right)_{ij}.$$

The elements $g_{ij}(x)$ of this *symmetric* matrix are called the **covariant components of the metric tensor** at $\widehat{x} = \boldsymbol{\Theta}(x)$. Note that *the matrix* $\boldsymbol{\nabla\Theta}(x)$ *is invertible* and that *the symmetric matrix* $(g_{ij}(x))$ *is positive-definite*, since the vectors $\boldsymbol{g}_i(x)$ are assumed to be linearly independent.

The n vectors $\boldsymbol{g}_i(x)$ being linearly independent, *the* n^2 *relations*

$$\boldsymbol{g}^i(x)\cdot\boldsymbol{g}_j(x) = \delta^i_j$$

unambiguously define n *linearly independent vectors* $\boldsymbol{g}^i(x)$. To see this, let *a priori* $\boldsymbol{g}^i(x) = X^{ik}(x)\boldsymbol{g}_k(x)$ in the relations $\boldsymbol{g}^i(x)\cdot\boldsymbol{g}_j(x) = \delta^i_j$. This gives $X^{ik}(x)g_{kj}(x) = \delta^i_j$; consequently, $X^{ik}(x) = g^{ik}(x)$, where

$$(g^{ij}(x)) := (g_{ij}(x))^{-1}.$$

Hence $\boldsymbol{g}^i(x) = g^{ik}(x)\boldsymbol{g}_k(x)$. These relations in turn imply that

$$\begin{aligned}\boldsymbol{g}^i(x)\cdot\boldsymbol{g}^j(x) &= \big(g^{ik}(x)\boldsymbol{g}_k(x)\big)\cdot\big(g^{j\ell}(x)\boldsymbol{g}_\ell(x)\big)\\ &= g^{ik}(x)g^{j\ell}(x)g_{k\ell}(x) \;=\; g^{ik}(x)\delta_k^j \;=\; g^{ij}(x),\end{aligned}$$

and thus the vectors $\boldsymbol{g}^i(x)$ are *linearly independent* since the matrix $(g^{ij}(x))$ is positive-definite. We would likewise establish that $\boldsymbol{g}_i(x) = g_{ij}(x)\boldsymbol{g}^j(x)$.

The n vectors $\boldsymbol{g}^i(x)$ form the **contravariant basis** at the point $\widehat{x} = \boldsymbol{\Theta}(x)$, and the elements $g^{ij}(x)$ of the symmetric positive-definite matrix $(g^{ij}(x))$ are the **contravariant components of the metric tensor** at $\widehat{x} = \boldsymbol{\Theta}(x)$.

Let us record for convenience the fundamental relations that are satisfied by the vectors of the covariant and contravariant bases and the covariant and contravariant components of the metric tensor at a point $x \in \Omega$ where the mapping $\boldsymbol{\Theta}$ is an immersion:

$$\begin{gathered}\boldsymbol{g}_i(x) = \partial_i\boldsymbol{\Theta}(x) \quad\text{and}\quad \boldsymbol{g}^i(x)\cdot\boldsymbol{g}_i(x) = \delta_i^j,\\ g_{ij}(x) = \boldsymbol{g}_i(x)\cdot\boldsymbol{g}_j(x),\quad g^{ij}(x) = \boldsymbol{g}^i(x)\cdot\boldsymbol{g}^j(x),\quad\text{and}\quad (g^{ij}(x)) = (g_{ji}(x))^{-1},\\ \boldsymbol{g}_i(x) = g_{ij}(x)\boldsymbol{g}^j(x) \quad\text{and}\quad \boldsymbol{g}^i(x) = g^{ij}(x)\boldsymbol{g}_j(x).\end{gathered}$$

A mapping $\boldsymbol{\Theta} : \Omega \subset \mathbb{R}^n \to \mathbb{E}^n$ is an **immersion** if it is an immersion at each point in Ω, i.e., if $\boldsymbol{\Theta}$ is differentiable in Ω and the n vectors $\boldsymbol{g}_i(x) = \partial_i\boldsymbol{\Theta}(x)$ are linearly independent at each $x \in \Omega$. In this case, the vector fields $\boldsymbol{g}_i : \Omega \to \mathbb{E}^n$ and $\boldsymbol{g}^i : \Omega \to \mathbb{E}^n$ respectively form the **covariant** and **contravariant bases**.

Such an immersion $\boldsymbol{\Theta} : \Omega \subset \mathbb{R}^n \to \mathbb{E}^n$ is called an n**-dimensional parametrized manifold**. If in addition $\boldsymbol{\Theta}$ is *injective*, the *image* $\boldsymbol{\Theta}(\Omega) \subset \mathbb{E}^n$ is called[1] an n**-dimensional manifold** in $\mathbb{E}^n$.

Remark What is exactly the "tensor" hidden behind its covariant components $g_{ij}(x)$ or its contravariant exponents $g^{ij}(x)$ will be explained in Section 8.4; the meaning of the adjectives "covariant" or "contravariant" attached to the components $g_{ij}(x)$ or $g^{ij}(x)$ will be also explained there. □

We now review fundamental formulas showing how *volume* and *lengths inside* $\widehat{\Omega} = \boldsymbol{\Theta}(\Omega)$ are computed by means of integrals *inside* Ω, whose integrands are functions of the covariant or contravariant components of the metric tensor, which are themselves functions of the *curvilinear coordinates* used in the open set $\widehat{\Omega}$ (see Figure 8.2-1 when $n = 3$); see also Problem 8.2-1, where *areas* inside $\widehat{\Omega}$ are likewise computed in the special case $n = 3$.

These formulas highlight the crucial role played by the metric tensor field $(g_{ij}) : \Omega \to \mathbb{S}^n_>$ for computing such "metric" notions inside $\widehat{\Omega}$, thus justifying its name.

Theorem 8.2-1 (volumes and lengths in curvilinear coordinates) *Let* Ω *be an open subset of* $\mathbb{R}^n$, *let* $\boldsymbol{\Theta} : \Omega \to \mathbb{E}^n$ *be a* $\mathcal{C}^1$*-diffeomorphism* (*Section* 7.1), *and let* $\widehat{\Omega} := \boldsymbol{\Theta}(\Omega)$.

(a) *Let* V *be an open subset of* Ω, *let* $\widehat{V} := \boldsymbol{\Theta}(V)$, *and let a function* $\widehat{f} \in L^1(\widehat{V})$ *be given. Then*

$$\int_{\widehat{V}} \widehat{f}(\widehat{x})\,\mathrm{d}\widehat{x} = \int_V (\widehat{f}\circ\boldsymbol{\Theta})(x)\,|\det\boldsymbol{\nabla}\boldsymbol{\Theta}(x)|\,\mathrm{d}x = \int_V (\widehat{f}\circ\boldsymbol{\Theta})(x)\sqrt{g(x)}\,\mathrm{d}x,$$

[1]The definitions given here are in effect special cases of more general definitions; cf. SCHLICHTKRULL [2012].

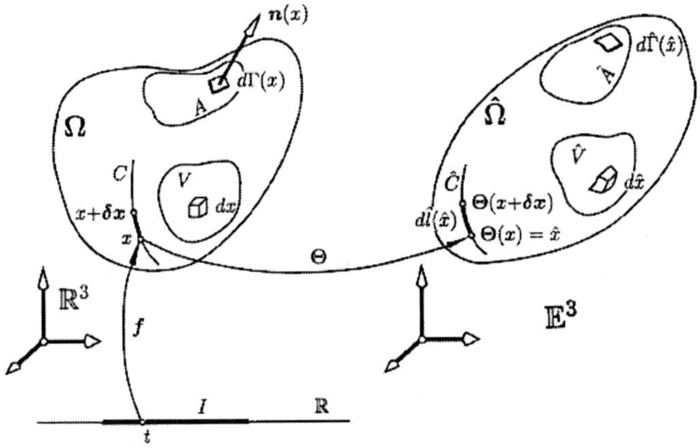

Figure 8.2-1 *Volumes, areas, and lengths in curvilinear coordinates.* Let V be an open subset of Ω, let A be a $\mathrm{d}\Gamma$-measurable subset of a domain D such that $\overline{D} \subset \Omega$, and let I be a compact interval of $\mathbb{R}$. Then the volume of $\widehat{V} := \boldsymbol{\Theta}(V) \subset \widehat{\Omega}$, the area of $\widehat{A} := \boldsymbol{\Theta}(A) \subset \Omega$ (when $n = 3$), and the length of a curve $\widehat{C} = \boldsymbol{\Theta}(C) \subset \widehat{\Omega}$ are computed by means of the covariant and contravariant components of the metric tensor; cf. Theorem 8.2-1 and Problem 8.2-1. This figure originally appeared in P.G. CIARLET [2005]: *An Introduction to Differential Geometry with Applications to Elasticity*, Springer, Dordrecht.

where

$$g(x) := \det(g_{ij}(x)) \quad \textit{at each } x \in \Omega.$$

In particular, the **volume** *of* $\widehat{V}$ *is given by*

$$\mathrm{vol}\,\widehat{V} := \int_{\widehat{V}} \mathrm{d}\widehat{x} = \int_V \sqrt{g(x)}\,\mathrm{d}x.$$

(b) *Let* $C = \boldsymbol{f}(I)$ *be a curve in* Ω, *where* I *is a compact interval of* $\mathbb{R}$ *and* $\boldsymbol{f} = f^i \boldsymbol{e}_i \in \mathcal{C}^1(I;\mathbb{R}^n)$ *is an injective mapping such that* $\boldsymbol{f}(I) \subset \Omega$ *and* $\dfrac{\mathrm{d}f^i}{\mathrm{d}t}(t)\boldsymbol{e}_i \neq \boldsymbol{0}$ *for all* $t \in I$. *Then the* **length** *of the curve* $\widehat{C} := \boldsymbol{\Theta}(C) \subset \widehat{\Omega}$ *is given by*

$$\mathrm{length}\,\widehat{C} = \int_I \sqrt{g_{ij}(\boldsymbol{f}(t))\frac{\mathrm{d}f^i}{\mathrm{d}t}(t)\frac{\mathrm{d}f^j}{\mathrm{d}t}(t)}\,\mathrm{d}t.$$

Proof By the formula for changes of variables in Lebesgue integrals (Theorem 1.16-1), the function $(\widehat{f} \circ \boldsymbol{\Theta})\sqrt{g}$ belongs to the space $L^1(\Omega)$ and

$$\int_{\widehat{V}} \widehat{f}(\widehat{x})\,\mathrm{d}\widehat{x} = \int_V (\widehat{f} \circ \boldsymbol{\Theta})(x)\,|\det \boldsymbol{\nabla\Theta}(x)|\,\mathrm{d}x.$$

The relation $(g_{ij}(x)) = \boldsymbol{\nabla\Theta}(x)^T \boldsymbol{\nabla\Theta}(x)$ then shows that

$$g(x) := \det(g_{ij}(x)) = |\det \boldsymbol{\nabla\Theta}(x)|^2 \quad \text{at each } x \in \Omega,$$

which proves (a).

By the formula giving the length of a curve (Section 1.17),

$$\text{length}\, \widehat{C} := \int_I \left| \frac{\mathrm{d}\widehat{\boldsymbol{f}}}{\mathrm{d}t}(t) \right| \mathrm{d}t, \quad \text{where } \widehat{\boldsymbol{f}} := \boldsymbol{\Theta} \circ \boldsymbol{f}.$$

Then, at each $t \in I$, the relation

$$\frac{\mathrm{d}\widehat{\boldsymbol{f}}}{\mathrm{d}t}(t) = \frac{\mathrm{d}}{\mathrm{d}t}\boldsymbol{\Theta}(\boldsymbol{f}(t)) = \frac{\mathrm{d}f^i}{\mathrm{d}t}(t)\partial_i\boldsymbol{\Theta}(\boldsymbol{f}(t)) = \frac{\mathrm{d}f^i}{\mathrm{d}t}(t)\boldsymbol{g}_i(\boldsymbol{f}(t))$$

shows that

$$\left|\frac{\mathrm{d}\widehat{\boldsymbol{f}}}{\mathrm{d}t}(t)\right|^2 = \left(\frac{\mathrm{d}f^i}{\mathrm{d}t}(t)\boldsymbol{g}_i(\boldsymbol{f}(t))\right) \cdot \left(\frac{\mathrm{d}f^j}{\mathrm{d}t}(t)\boldsymbol{g}_j(\boldsymbol{f}(t))\right) = g_{ij}(\boldsymbol{f}(t))\frac{\mathrm{d}f^i}{\mathrm{d}t}(t)\frac{\mathrm{d}f^j}{\mathrm{d}t}(t),$$

which proves (b). □

Remark The result of (b) shows that the length element $\mathrm{d}\widehat{\ell}(\widehat{x})$ at $\widehat{x} = \boldsymbol{\Theta}(x) \in \widehat{\Omega}$ is given by

$$\mathrm{d}\widehat{\ell}(\widehat{x}) = \sqrt{\boldsymbol{\delta x}^T \boldsymbol{\nabla\Theta}(x)^T \boldsymbol{\nabla\Theta}(x)\boldsymbol{\delta x}} = \sqrt{\delta x^i g_{ij}(x)\delta x^j},$$

where $\boldsymbol{\delta x} = \delta x^i \boldsymbol{e}_i$. Either expression recalls that $\mathrm{d}\widehat{\ell}(\widehat{x})$ is by definition the principal part with respect to $\boldsymbol{\delta x} = \delta x^i \boldsymbol{e}_i$ of $|\boldsymbol{\Theta}(x + \boldsymbol{\delta x}) - \boldsymbol{\Theta}(x)|$, whose expression precisely led to the introduction of the matrix $(g_{ij}(x))$. □

The relation established in (b) expresses that *the lengths of curves inside the n-dimensional manifold* $\boldsymbol{\Theta}(\Omega) \subset \mathbb{E}^n$ *are precisely those induced by the Euclidean metric of the space* $\mathbb{E}^n$.

Problem

8.2-1 Let Ω be an open subset of $\mathbb{R}^3$, let $\boldsymbol{\Theta} : \Omega \to \mathbb{E}^3$ be a $\mathcal{C}^1$-diffeomorphism, and let $\widehat{\Omega} := \boldsymbol{\Theta}(\Omega)$. Given a *domain* D (Section 1.18) such that $\overline{D} \subset \Omega$, let $\boldsymbol{n} = n_i \boldsymbol{e}^i$ denote the unit outer normal vector along ∂D, and let A be a $\mathrm{d}\Gamma$-measurable subset of $\Gamma := \partial D$.

(1) Let $\widehat{D} := \boldsymbol{\Theta}(D)$ and $\widehat{A} := \boldsymbol{\Theta}(A)$. Show that $\overline{\boldsymbol{\Theta}(D)} = \boldsymbol{\Theta}(\overline{D})$, and $\partial\widehat{D} = \boldsymbol{\Theta}(\partial D)$.

(2) Let $\widehat{A}$ be any $\mathrm{d}\widehat{\Gamma}$-measurable subset of $\widehat{\Gamma} := \partial\widehat{D}$. Show that, given any function $\widehat{h} \in L^1(\widehat{A})$,

$$\begin{aligned}\int_{\widehat{A}} \widehat{h}(\widehat{x})\mathrm{d}\widehat{\Gamma} &= \int_A (\widehat{h} \circ \boldsymbol{\Theta})(x)\, |\mathbf{Cof}\,\boldsymbol{\nabla\Theta}(x)\boldsymbol{n}(x)|\, \mathrm{d}\Gamma \\ &= \int_A (\widehat{h} \circ \boldsymbol{\Theta})(x)\sqrt{g(x)}\sqrt{n_i(x)g^{ij}(x)n_j(x)}\,\mathrm{d}\Gamma.\end{aligned}$$

In particular, the *area* of $\widehat{A}$ is given by

$$\text{area}\,\widehat{A} := \int_{\widehat{A}} \mathrm{d}\widehat{\Gamma} = \int_A \sqrt{g(x)}\sqrt{n_i(x)g^{ij}(x)n_j(x)}\,\mathrm{d}\Gamma,$$

where the functions $g^{ij} : \Omega \to \mathbb{R}$ are the contravariant components of the metric tensor.

8.3 Covariant derivative of a vector field

Let there be given a *vector field* defined in an open subset $\widehat{\Omega}$ of $\mathbb{E}^n$ by means of its *Cartesian components* $\widehat{v}_i : \widehat{\Omega} \to \mathbb{R}$, i.e., this field is defined by its values $\widehat{v}_i(\widehat{x})\widehat{\boldsymbol{e}}^i \in \mathbb{E}^n$ at each $\widehat{x} \in \widehat{\Omega}$, where the vectors $\widehat{\boldsymbol{e}}^i$ constitute the orthonormal basis of $\mathbb{E}^n$ (Figure 8.3-1). Assume that the open set $\widehat{\Omega}$ is equipped with *curvilinear coordinates* from an open subset Ω of $\mathbb{R}^n$, by means of an injective mapping $\boldsymbol{\Theta} : \Omega \to \mathbb{E}^n$ satisfying $\boldsymbol{\Theta}(\Omega) = \widehat{\Omega}$ (Section 8.1).

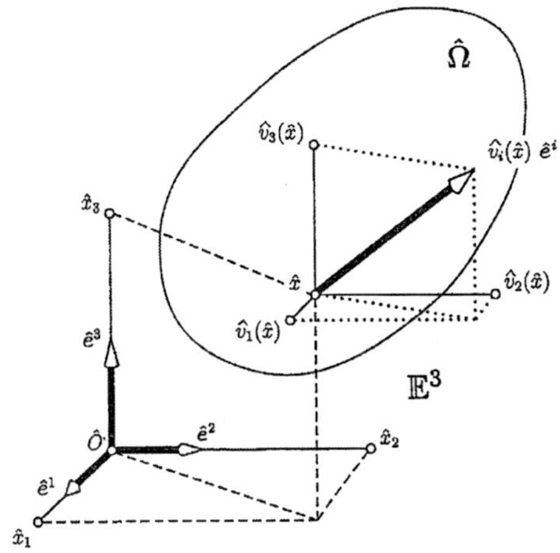

Figure 8.3-1 *A vector field in Cartesian coordinates.* At each point $\widehat{x} \in \widehat{\Omega}$, the vector $\widehat{v}_i(\widehat{x})\widehat{\boldsymbol{e}}^i$ is defined by its Cartesian components $\widehat{v}_i(\widehat{x})$ over an orthonormal basis of $\mathbb{E}^3$ formed by the vectors $\widehat{\boldsymbol{e}}^i$. This figure originally appeared in P.G. CIARLET [2005]: *An Introduction to Differential Geometry with Applications to Elasticity*, Springer, Dordrecht.

How do we define appropriate components of this vector field, but this time in terms of these curvilinear coordinates? It turns out that one proper way to do so consists in defining n functions $v_i : \Omega \to \mathbb{R}$ by the requirement that (Figure 8.3-2)

$$v_i(x)\boldsymbol{g}^i(x) = \widehat{v}_i(\widehat{x})\widehat{\boldsymbol{e}}^i \quad \text{for all } \widehat{x} = \boldsymbol{\Theta}(x), \ \ x \in \Omega,$$

where the vectors $\boldsymbol{g}^i(x)$ form the *contravariant basis* at $\widehat{x} = \boldsymbol{\Theta}(x)$ (Section 8.2) and the components $v_i(x)$ are called the **covariant components** *of the vector* $v_i(x)\boldsymbol{g}^i(x)$ *at* $\widehat{x}$. Using the relations $\boldsymbol{g}^i(x) \cdot \boldsymbol{g}_j(x) = \delta^i_j$ and $\widehat{\boldsymbol{e}}^i \cdot \widehat{\boldsymbol{e}}_j = \delta^i_j$, one immediately finds how the Cartesian and covariant components are related, viz.,

$$\begin{aligned} v_j(x) &= v_i(x)\boldsymbol{g}^i(x) \cdot \boldsymbol{g}_j(x) = \widehat{v}_i(\widehat{x})\widehat{\boldsymbol{e}}^i \cdot \boldsymbol{g}_j(x), \\ \widehat{v}_i(\widehat{x}) &= \widehat{v}_j(\widehat{x})\widehat{\boldsymbol{e}}^j \cdot \widehat{\boldsymbol{e}}_i = v_j(x)\boldsymbol{g}^j(x) \cdot \widehat{\boldsymbol{e}}_i. \end{aligned}$$

Another proper way consists in defining n functions $v^i : \Omega \to \mathbb{R}$ by the requirement that

$$v^i(x)\boldsymbol{g}_i(x) := \widehat{v}_i(\widehat{x})\widehat{\boldsymbol{e}}^i \quad \text{at each } \widehat{x} = \boldsymbol{\Theta}(x), \ \ x \in \Omega,$$

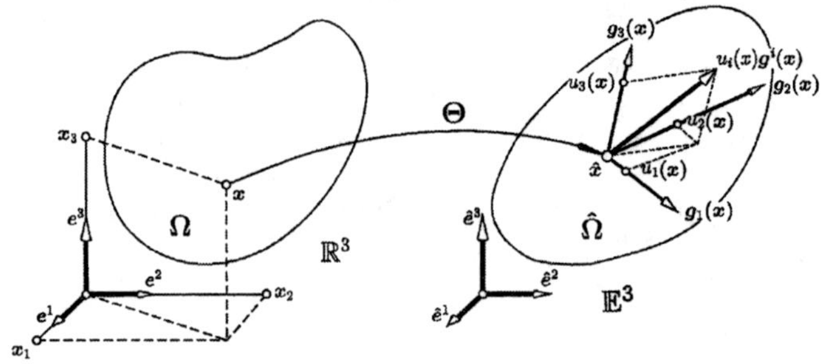

Figure 8.3-2 *A vector field in curvilinear coordinates.* Let a vector field be defined at each $\widehat{x} \in \widehat{\Omega}$ by its Cartesian components $\widehat{v}_i(\widehat{x})$ over the vectors $\widehat{\boldsymbol{e}}^i$ (Figure 8.3-1). The same vector field in curvilinear coordinates is defined at each $x \in \Omega$ by its covariant components $v_i(x)$ over the contravariant basis vectors $\boldsymbol{g}^i(x)$, by the requirement that $v_i(x)\boldsymbol{g}^i(x) = \widehat{v}_i(\widehat{x})\widehat{\boldsymbol{e}}^i$, $\widehat{x} = \boldsymbol{\Theta}(x)$. This figure originally appeared in P.G. Ciarlet [2005]: *An Introduction to Differential Geometry with Applications to Elasticity*, Springer, Dordrecht.

where the vectors $\boldsymbol{g}_i(x)$ form the *covariant basis* at $\widehat{x} = \boldsymbol{\Theta}(x)$. The components $v^i(x)$ are called the **contravariant components of the vector** $v^i(x)\boldsymbol{g}_i(x)$ at $\widehat{x}$. It is then immediately verified that the covariant and contravariant components are related by the relations

$$v_i(x) = g_{ij}(x)v^j(x) \quad \text{and} \quad v^j(x) = g^{ij}(x)v_i(x)$$

(since, e.g., $v_i(x) = v_j(x)\boldsymbol{g}^j(x) \cdot \boldsymbol{g}_i(x) = v^j(x)\boldsymbol{g}_j(x) \cdot \boldsymbol{g}_i(x)$, etc.).

Remark In Section 8.4, we will explain in what sense the components $v_i(x)$ are "covariant," while the components $v^i(x)$ are "contravariant." □

Suppose next that we wish to *compute a partial derivative* $\widehat{\partial}_j\widehat{v}_i(\widehat{x})$ *at a point* $\widehat{x} = \boldsymbol{\Theta}(x) \in \widehat{\Omega}$ *in terms of the partial derivatives* $\partial_\ell v_k(x)$ *and of the values* $v_q(x)$ (which are also expected to appear by virtue of the chain rule). Such a computation (perhaps not as easy as it seems at first sight; see the proof of the next theorem) is required for example in order to write a system of partial differential equations whose unknown is a vector field in terms of curvilinear coordinates.

As we now show, carrying out such a transformation naturally leads to the fundamental notion of *covariant derivative of a vector field*.

Theorem 8.3-1 *Let Ω be an open subset of $\mathbb{R}^n$ and let $\boldsymbol{\Theta} : \Omega \to \mathbb{E}^n$ be a $\mathcal{C}^2$-diffeomorphism (Section 7.8) of Ω onto $\widehat{\Omega} := \boldsymbol{\Theta}(\Omega)$. Given a vector field $\widehat{v}_i\widehat{\boldsymbol{e}}^i : \widehat{\Omega} \to \mathbb{E}^n$ in Cartesian coordinates with components $\widehat{v}_i \in \mathcal{C}^1(\widehat{\Omega})$, let $v_i\boldsymbol{g}^i : \Omega \to \mathbb{E}^n$ be the same vector field in curvilinear coordinates, i.e., that defined by*

$$\widehat{v}_i(\widehat{x})\widehat{\boldsymbol{e}}^i = v_i(x)\boldsymbol{g}^i(x) \quad \textit{at each } \widehat{x} = \boldsymbol{\Theta}(x), \;\; x \in \Omega.$$

Then the functions $v_i : \Omega \to \mathbb{R}$ defined in this fashion are of class $\mathcal{C}^1$ in Ω and

$$\widehat{\partial}_j\widehat{v}_i(\widehat{x}) = \left(v_{k\|\ell}[\boldsymbol{g}^k]_i[\boldsymbol{g}^\ell]_j\right)(x), \quad \textit{at each } \widehat{x} = \boldsymbol{\Theta}(x),$$

where

$$v_{i\|j} := \partial_j v_i - \Gamma_{ij}^p v_p \quad \textit{with } \Gamma_{ij}^p := \boldsymbol{g}^p \cdot \partial_i \boldsymbol{g}_j,$$

and

$$[\boldsymbol{g}^i(x)]_k := \boldsymbol{g}^i(x) \cdot \widehat{\boldsymbol{e}}_k$$

denotes at each $x \in \Omega$ *the* k*th component of* $\boldsymbol{g}^i(x)$ *over the basis* $\{\widehat{\boldsymbol{e}}_1, \ldots, \widehat{\boldsymbol{e}}_n\}$.

Proof The following convention holds throughout this proof: The simultaneous appearance of $\widehat{x}$ and x in an equality means that they are related by $\widehat{x} = \boldsymbol{\Theta}(x)$ and that the equality in question holds at each $x \in \Omega$.

(i) *Another expression of* $[\boldsymbol{g}^i(x)]_k := \boldsymbol{g}^i(x) \cdot \widehat{\boldsymbol{e}}_k$.

Let $\boldsymbol{\Theta}(x) = \Theta^k(x)\widehat{\boldsymbol{e}}_k$ and $\widehat{\boldsymbol{\Theta}}(\widehat{x}) = \widehat{\Theta}^i(\widehat{x})\boldsymbol{e}_i$, where $\widehat{\boldsymbol{\Theta}} : \widehat{\Omega} \to \Omega$ denotes the inverse mapping of $\boldsymbol{\Theta} : \Omega \to \widehat{\Omega}$. Since $\widehat{\boldsymbol{\Theta}}(\boldsymbol{\Theta}(x)) = x$ at each $x \in \Omega$, the chain rule (Theorem 7.1-3) shows that the matrices $\boldsymbol{\nabla\Theta}(x) := (\partial_j \Theta^k(x))$ (the row index is k) and $\widehat{\boldsymbol{\nabla}}\widehat{\boldsymbol{\Theta}}(\widehat{x}) := (\widehat{\partial}_k \widehat{\Theta}^i(\widehat{x}))$ (the row index is i) satisfy

$$\widehat{\boldsymbol{\nabla}}\widehat{\boldsymbol{\Theta}}(\widehat{x})\boldsymbol{\nabla\Theta}(x) = \boldsymbol{I},$$

or equivalently, for each i and each j,

$$\widehat{\partial}_k \widehat{\Theta}^i(\widehat{x})\partial_j \Theta^k(x) = \left(\widehat{\partial}_1 \widehat{\Theta}^i(\widehat{x}) \cdots \partial_n \widehat{\Theta}^i(\widehat{x})\right) \begin{pmatrix} \partial_j \Theta^1(x) \\ \vdots \\ \partial_j \Theta^n(x) \end{pmatrix} = \delta_j^i.$$

The components $\partial_j \Theta^k(x)$, $1 \leq k \leq n$, of the above column vector being precisely those of the vector $\boldsymbol{g}_j(x)$, the components $\widehat{\partial}_k \widehat{\Theta}^i(\widehat{x})$, $1 \leq k \leq n$, of the row vector above must be those of the vector $\boldsymbol{g}^i(x)$ since $\boldsymbol{g}^i(x)$ is uniquely defined for each exponent i by the n relations $\boldsymbol{g}^i(x) \cdot \boldsymbol{g}_j(x) = \delta_j^i, 1 \leq j \leq n$. Hence the kth component of $\boldsymbol{g}^i(x)$ over the basis $\{\widehat{\boldsymbol{e}}_1, \ldots, \widehat{\boldsymbol{e}}_n\}$ can be also expressed in terms of the inverse mapping $\widehat{\boldsymbol{\Theta}}$ as

$$[\boldsymbol{g}^i(x)]_k = \widehat{\partial}_k \widehat{\Theta}^i(\widehat{x}).$$

(ii) *Introduction of the functions* $\Gamma_{\ell k}^q := \boldsymbol{g}^q \cdot \partial_\ell \boldsymbol{g}_k \in \mathcal{C}(\Omega)$.

We next compute the derivatives $\partial_\ell \boldsymbol{g}^q(x)$ (the fields $\boldsymbol{g}^q = g^{qr}\boldsymbol{g}_r$ are of class $\mathcal{C}^1$ in Ω since $\boldsymbol{\Theta}$ is assumed to be of class $\mathcal{C}^2$ in Ω), which will be needed in (iii) for expressing the derivatives $\widehat{\partial}_j \widehat{v}_i(\widehat{x})$ as functions of x (recall that $\widehat{v}_i(\widehat{x}) = v_k(x)[\boldsymbol{g}^k(x)]_i$). Since the n vectors $\boldsymbol{g}^k(x)$ form a basis, we may write *a priori*

$$\partial_\ell \boldsymbol{g}^q(x) = -\Gamma_{\ell k}^q(x)\boldsymbol{g}^k(x),$$

thereby unambiguously defining functions $\Gamma_{\ell k}^q : \Omega \to \mathbb{R}$. To find the expressions of these functions in terms of the mappings $\boldsymbol{\Theta}$ and $\widehat{\boldsymbol{\Theta}}$, we observe that

$$\Gamma_{\ell k}^q(x) = \Gamma_{\ell m}^q(x)\delta_k^m = \Gamma_{\ell m}^q(x)\boldsymbol{g}^m(x) \cdot \boldsymbol{g}_k(x) = -\partial_\ell \boldsymbol{g}^q(x) \cdot \boldsymbol{g}_k(x).$$

Noting that $\partial_\ell(\boldsymbol{g}^q(x) \cdot \boldsymbol{g}_k(x)) = 0$ and $[\boldsymbol{g}^q(x)]_p = \widehat{\partial}_p \widehat{\Theta}^q(\widehat{x})$, we thus obtain

$$\Gamma_{\ell k}^q(x) = \boldsymbol{g}^q(x) \cdot \partial_\ell \boldsymbol{g}_k(x) = \widehat{\partial}_p \widehat{\Theta}^q(\widehat{x})\partial_{\ell k}\Theta^p(x) = \Gamma_{k\ell}^q(x).$$

Since $\boldsymbol{\Theta} : \Omega \to \mathbb{E}^n$ is a $\mathcal{C}^2$-diffeomorphism by assumption, the last relations show that $\Gamma^q_{\ell k} \in \mathcal{C}(\Omega)$.

(iii) *The partial derivatives $\widehat{\partial}_j \widehat{v}_i(\widehat{x})$ of the Cartesian components of the vector field $\widehat{v}_i \widehat{\boldsymbol{e}}^i \in \mathcal{C}^1(\widehat{\Omega}; \mathbb{E}^n)$ are given at each $\widehat{x} = \boldsymbol{\Theta}(x) \in \widehat{\Omega}$ by*

$$\widehat{\partial}_j \widehat{v}_i(\widehat{x}) = v_{k\|\ell}(x)[\boldsymbol{g}^k(x)]_i [\boldsymbol{g}^\ell(x)]_j,$$

where

$$v_{k\|\ell}(x) := \partial_\ell v_k(x) - \Gamma^q_{\ell k}(x) v_q(x),$$

and $[\boldsymbol{g}^k(x)]_i$ and $\Gamma^q_{\ell k}(x)$ are defined as in (i) *and* (ii).

To compute the partial derivatives $\widehat{\partial}_j \widehat{v}_i(\widehat{x})$ as functions of x, we simply use the relation $\widehat{v}_i(\widehat{x}) = v_k(x)[\boldsymbol{g}^k(x)]_i$. Noting that, by the chain rule and by (i), a differentiable function $w : \Omega \to \mathbb{R}$ satisfies

$$\widehat{\partial}_j w(\widehat{\boldsymbol{\Theta}}(\widehat{x})) = \partial_\ell w(x) \widehat{\partial}_j \widehat{\Theta}^\ell(\widehat{x}) = \partial_\ell w(x)[\boldsymbol{g}^\ell(x)]_j,$$

we conclude that

$$\begin{aligned}
\widehat{\partial}_j \widehat{v}_i(\widehat{x}) &= \widehat{\partial}_j v_k(\widehat{\boldsymbol{\Theta}}(\widehat{x}))[\boldsymbol{g}^k(x)]_i + v_q(x) \widehat{\partial}_j [\boldsymbol{g}^q(\widehat{\boldsymbol{\Theta}}(\widehat{x}))]_i \\
&= \partial_\ell v_k(x)[\boldsymbol{g}^\ell(x)]_j [\boldsymbol{g}^k(x)]_i + v_q(x)\big(\partial_\ell [\boldsymbol{g}^q(x)]_i\big)[\boldsymbol{g}^\ell(x)]_j \\
&= \big(\partial_\ell v_k(x) - \Gamma^q_{\ell k}(x) v_q(x)\big) [\boldsymbol{g}^k(x)]_i [\boldsymbol{g}^\ell(x)]_j,
\end{aligned}$$

since $\partial_\ell \boldsymbol{g}^q(x) = -\Gamma^q_{\ell k}(x) \boldsymbol{g}^k(x)$ by (ii). This completes the proof. □

The functions

$$v_{i\|j} = \partial_j v_i - \Gamma^p_{ij} v_p$$

that appeared in Theorem 8.3-1 are called the **covariant components of the covariant derivative of the vector field $v_i \boldsymbol{g}^i : \Omega \to \mathbb{E}^n$**.

Remark We will see in Section 8.4 that, like the functions g_{ij} (Section 8.2), the functions $v_{i\|j}$ are the *covariant components* of a *second-order tensor*, but of a *different* nature than that of the metric tensor. □

The functions

$$\Gamma^p_{ij} = \boldsymbol{g}^p \cdot \partial_i \boldsymbol{g}_j : \Omega \to \mathbb{R}$$

are called the **Christoffel symbols[2] of the second kind**; Christoffel symbols of the *first* kind will be introduced in Section 8.5.

The following result summarizes properties of covariant derivatives and Christoffel symbols that are constantly used.

---

[2]So named after:

E.B. CHRISTOFFEL [1869]: Über die Transformation der homogenen Differentialausdrücke zweiten Grades, *Journal für die Reine und Angewandte Mathematik* **70**, 46–70.

Theorem 8.3-2 *Let the assumptions on the mapping* $\boldsymbol{\Theta} : \Omega \to \mathbb{E}^n$ *be as in* Theorem 8.3-1, *and let there be given a vector field* $v_i \boldsymbol{g}^i : \Omega \to \mathbb{E}^n$ *with covariant components* $v_i \in \mathcal{C}^1(\Omega)$.

(a) *The covariant components* $v_{i\|j} \in \mathcal{C}(\Omega)$ *of the covariant derivative of the vector field* $v_i \boldsymbol{g}^i : \Omega \to \mathbb{E}^n$, *which are defined by*

$$v_{i\|j} := \partial_j v_i - \Gamma^p_{ij} v_p, \quad \textit{where } \Gamma^p_{ij} := \boldsymbol{g}^p \cdot \partial_i \boldsymbol{g}_j,$$

can be also defined by the relations

$$v_{i\|j} \boldsymbol{g}^i = \partial_j (v_i \boldsymbol{g}^i), \quad \textit{or equivalently,} \quad v_{i\|j} = \{\partial_j (v_k \boldsymbol{g}^k)\} \cdot \boldsymbol{g}_i.$$

(b) *The Christoffel symbols* $\Gamma^p_{ij} := \boldsymbol{g}^p \cdot \partial_i \boldsymbol{g}_j = \Gamma^p_{ji} \in \mathcal{C}(\Omega)$ *satisfy the relations*

$$\partial_i \boldsymbol{g}^p = -\Gamma^p_{ij} \boldsymbol{g}^j \quad \textit{and} \quad \partial_j \boldsymbol{g}_q = \Gamma^i_{jq} \boldsymbol{g}_i.$$

Proof It remains to verify that the covariant components $v_{i\|j}$, defined in Theorem 8.3-1 by

$$v_{i\|j} = \partial_j v_i - \Gamma^p_{ij} v_p,$$

may be equivalently defined by the relations

$$\partial_j (v_i \boldsymbol{g}^i) = v_{i\|j} \boldsymbol{g}^i,$$

which unambiguously define the functions $v_{i\|j} := \{\partial_j (v_k \boldsymbol{g}^k)\} \cdot \boldsymbol{g}_i$ since the vectors $\boldsymbol{g}^i$ are linearly independent at all points of Ω by assumption.

To this end, we simply note that, by definition, the Christoffel symbols satisfy $\partial_i \boldsymbol{g}^p = -\Gamma^p_{ij} \boldsymbol{g}^j$ (cf. part (ii) of the proof of Theorem 8.3-1); hence

$$\partial_j (v_i \boldsymbol{g}^i) = (\partial_j v_i) \boldsymbol{g}^i + v_i \partial_j \boldsymbol{g}^i = (\partial_j v_i) \boldsymbol{g}^i - v_i \Gamma^i_{jk} \boldsymbol{g}^k = v_{i\|j} \boldsymbol{g}^i.$$

To establish the other relations $\partial_j \boldsymbol{g}_q = \Gamma^i_{jq} \boldsymbol{g}_i$, we note that, since $\boldsymbol{g}^p \cdot \boldsymbol{g}_q = \delta^p_q$,

$$0 = \partial_j (\boldsymbol{g}^p \cdot \boldsymbol{g}_q) = -\Gamma^p_{ji} \boldsymbol{g}^i \cdot \boldsymbol{g}_q + \boldsymbol{g}^p \cdot \partial_j \boldsymbol{g}_q = -\Gamma^p_{qj} + \boldsymbol{g}^p \cdot \partial_j \boldsymbol{g}_q.$$

Hence

$$\partial_j \boldsymbol{g}_q = (\partial_j \boldsymbol{g}_q \cdot \boldsymbol{g}^p) \boldsymbol{g}_p = \Gamma^p_{qj} \boldsymbol{g}_p. \qquad \square$$

Remark A crucial property of the Christoffel symbols Γ^p_{ij} is that they can be also defined *solely in terms of the components* g_{ij} *of the metric tensor and their derivatives* $\partial_k g_{ij}$ (Theorem 8.5-1). $\square$

If the space $\mathbb{E}^n$ is identified with $\mathbb{R}^n$ and $\boldsymbol{\Theta}(x) = x$ for all $x \in \Omega$, the relation $\partial_j (v_i \boldsymbol{g}^i)(x) = (v_{i\|j} \boldsymbol{g}^i)(x)$ reduces to $\widehat{\partial}_j (\widehat{v}_i(\widehat{x}) \widehat{\boldsymbol{e}}^i) = (\widehat{\partial}_j \widehat{v}_i(\widehat{x})) \widehat{\boldsymbol{e}}^i$. In this sense, *a covariant component of the covariant derivative of a vector field constitutes a generalization of a partial derivative in Cartesian coordinates.*

The classical *gradient, Laplacian, divergence,* and *curl* operators in Cartesian coordinates can be likewise expressed in terms of *curvilinear coordinates*; cf. Problem 8.3-3.

Problems

8.3-1 Compute the vectors of the covariant and contravariant bases, the covariant and contravariant components of the metric tensor, and the Christoffel symbols corresponding to cylindrical and spherical coordinates (Figure 8.1-2).

8.3-2 In part (iii) of the proof of Theorem 8.3-1, it is shown that

$$\widehat{\partial}_j \widehat{v}_i(\widehat{x}) = \left(v_{k\|\ell}[\boldsymbol{g}^k]_i[\boldsymbol{g}^\ell]_j\right)(x) \quad \text{for all } \widehat{x} = \boldsymbol{\Theta}(x) \in \widehat{\Omega}.$$

Show that, conversely, each covariant derivative $v_{i\|j}(x)$ can be expressed as a linear combination of the partial derivatives $\widehat{\partial}_\ell \widehat{v}_k(\widehat{x})$.

8.3-3 This problem provides the expression of the *gradient, Laplacian, divergence*, and *curl operators* in *curvilinear coordinates*. The notations and assumptions are those of Theorem 8.3-1.

(1) Given a smooth enough function $\widehat{v} : \widehat{\Omega} \to \mathbb{R}$, let $\widehat{\mathbf{grad}}\, \widehat{v} : \Omega \to \mathbb{E}^n$ be the vector field with components $\widehat{\partial}_i \widehat{v}$, let $\widehat{\Delta} \widehat{v} := \sum_i \dfrac{\partial^2 \widehat{v}}{\partial \widehat{x}_i^2} : \Omega \to \mathbb{R}$, and let the function $v : \Omega \to \mathbb{R}$ be defined by $v(x) = \widehat{v}(\boldsymbol{\Theta}(x))$ at each $x \in \Omega$. Show that

$$\left(\widehat{\mathbf{grad}}\, \widehat{v}\right)(\widehat{x}) = \left((\partial_i v)\boldsymbol{g}^i\right)(x),$$

$$\widehat{\Delta} \widehat{v}(\widehat{x}) = \left(\frac{1}{\sqrt{g}} \partial_i \left(\sqrt{g} g^{ij} \partial_j v\right)\right)(x),$$

where $g := \det(g_{ij})$.

(2) Given a smooth enough vector field $\widehat{\boldsymbol{v}} = \widehat{v}_i \widehat{\boldsymbol{e}}^i : \widehat{\Omega} \to \mathbb{E}^n$, let $\widehat{\operatorname{div}}\, \widehat{\boldsymbol{v}} := \widehat{\partial}_i \widehat{v}_i : \widehat{\Omega} \to \mathbb{R}$. Show that

$$\widehat{\operatorname{div}}\, \widehat{\boldsymbol{v}}(\widehat{x}) = \left(v_{k\|\ell} g^{k\ell}\right)(x) \quad \text{at each } \widehat{x} = \boldsymbol{\Theta}(x) \in \widehat{\Omega}.$$

(3) Given a smooth enough vector field $\widehat{\boldsymbol{v}} = \widehat{v}_i \widehat{\boldsymbol{e}}^i : \widehat{\Omega} \to \mathbb{E}^3$, let $\widehat{\mathbf{curl}}\, \widehat{\boldsymbol{v}} : \widehat{\Omega} \to \mathbb{E}^3$ be the vector field with components $\widehat{\varepsilon}^{ijk} \widehat{\partial}_i \widehat{v}_j$, where $\widehat{\varepsilon}^{ijk} = 1$ if $\{i, j, k\}$ is an even permutation of $\{1, 2, 3\}$, $\widehat{\varepsilon}^{ijk} = -1$ if $\{i, j, k\}$ is an odd permutation of $\{1, 2, 3\}$, and $\widehat{\varepsilon}^{ijk} = 0$ otherwise. Show that

$$\widehat{\mathbf{curl}}\, \widehat{\boldsymbol{v}}(\widehat{x}) = \left(\varepsilon^{ijk} v_{j\|i} \boldsymbol{g}_k\right)(x),$$

where $\varepsilon^{ijk}(x) := \dfrac{1}{\sqrt{g(x)}} \widehat{\varepsilon}^{ijk}$.

8.4 Tensors — A brief introduction

Tensor analysis is a vast subject; so, the aims of this short section are necessarily modest.

Its first aim is to explain the meaning of the adjective *"covariant," resp. "contravariant,"* attached to the indices or exponents in the components $v_i(x)$, $g_{ij}(x)$, or $v_{i\|j}(x)$, *resp.* $v^i(x)$ or $g^{ij}(x)$, encountered in the previous sections. Note that the adjective "covariant" has *another* meaning when it is attached to "derivative."

Its second aim is to give the definition of a *tensor field* in the special case thus far considered, viz., that of a tensor field defined on an open subset $\widehat{\Omega}$ of $\mathbb{E}^n$ of the form $\widehat{\Omega} = \boldsymbol{\Theta}(\Omega)$, where Ω is an open subset Ω of $\mathbb{R}^n$ and $\boldsymbol{\Theta} : \Omega \to \widehat{\Omega}$ is a $\mathcal{C}^1$-diffeomorphism.

Remark This special case is the simplest one. For instance, defining tensors on a *two*-dimensional surface in the *three*-dimensional Euclidean space $\mathbb{E}^3$ already requires substantially more care, as we shall briefly indicate at the end of Section 8.13. □

For brevity, we shall essentially focus our attention on first-order and second-order tensors, leaving the definitions and properties of higher order tensors as problems.

A *first viewpoint* consists in looking at *how the above components vary under a change of curvilinear coordinates* inside a *given* open subset of $\mathbb{E}^n$.

More specifically, let there be given a $\mathcal{C}^1$-diffeomorphism $\boldsymbol{\Theta}$ from an open subset of $\mathbb{R}^n$ onto an open subset $\widehat{\Omega}$ of $\mathbb{E}^n$ and a point $\widehat{x} = \boldsymbol{\Theta}(x) \in \widehat{\Omega}$. Then we *define* the vectors $\boldsymbol{g}_i(x) := \partial_i \boldsymbol{\Theta}_i(x) \in \mathbb{E}^n$ as forming the **covariant basis** at the point $\widehat{x} \in \widehat{\Omega}$ (associated with $\boldsymbol{\Theta}$), and we *define* the (unique) vectors $\boldsymbol{g}^j(x)$ that satisfy $\boldsymbol{g}^j(x) \cdot \boldsymbol{g}_i(x) = \delta_i^j$ as forming the **contravariant basis** at $\widehat{x} \in \widehat{\Omega}$ (again associated with $\boldsymbol{\Theta}$). At this stage then, the adjectives "covariant" and "contravariant" are to be understood simply as a means to distinguish between the two kinds of bases (associated with $\boldsymbol{\Theta}$) defined at the same point $\widehat{x} \in \widehat{\Omega}$.

Let now Ω and $\widetilde{\Omega}$ be two open subsets of $\mathbb{R}^n$ and let $\boldsymbol{\Theta} : \Omega \to \mathbb{E}^n$ and $\widetilde{\boldsymbol{\Theta}} : \widetilde{\Omega} \to \mathbb{E}^n$ be two $\mathcal{C}^1$-diffeomorphisms *such that* $\boldsymbol{\Theta}(\Omega) = \widetilde{\boldsymbol{\Theta}}(\widetilde{\Omega})$. Let then $\boldsymbol{g}_i(x) := \partial_i \boldsymbol{\Theta}(x)$ and $\widetilde{\boldsymbol{g}}_i(\widetilde{x}) = \widetilde{\partial}_i \widetilde{\boldsymbol{\Theta}}(\widetilde{x})$ denote the associated vectors of the covariant bases at the *same* point $\boldsymbol{\Theta}(x) = \widetilde{\boldsymbol{\Theta}}(x) \in \mathbb{E}^n$, and let $\boldsymbol{g}^i(x)$ and $\widetilde{\boldsymbol{g}}^i(\widetilde{x})$ be the vectors of the corresponding contravariant bases at the same point $\widehat{x}$. A simple computation then shows that

$$\boldsymbol{g}_i(x) = \frac{\partial \chi^k}{\partial x_i}(x) \widetilde{\boldsymbol{g}}_k(\widetilde{x}) \quad \text{and} \quad \boldsymbol{g}^i(x) = \frac{\partial \widetilde{\chi}^i}{\partial \widetilde{x}_k}(x) \widetilde{\boldsymbol{g}}^k(\widetilde{x}),$$

where

$$\boldsymbol{\chi} = (\chi^j) := \widetilde{\boldsymbol{\Theta}}^{-1} \circ \boldsymbol{\Theta} \in \mathcal{C}^1(\Omega; \widetilde{\Omega}) \quad \text{and} \quad \widetilde{\boldsymbol{\chi}} = (\widetilde{\chi}^i) := \boldsymbol{\chi}^{-1} \in \mathcal{C}^1(\widetilde{\Omega}; \Omega).$$

Then the "covariant" components $v_i(x)$ and $\widetilde{v}_i(\widetilde{x})$, and the "contravariant" components $v^i(x)$ and $\widetilde{v}^i(\widetilde{x})$ (both with self-explanatory notations), of a *vector field* at the *same* point $\boldsymbol{\Theta}(x) = \widetilde{\boldsymbol{\Theta}}(\widetilde{x}) \in \widehat{\Omega} \subset \mathbb{E}^n$ satisfy by definition (Section 8.3)

$$v_i(x) \boldsymbol{g}^i(x) = \widetilde{v}_i(\widetilde{x}) \widetilde{\boldsymbol{g}}^i(\widetilde{x}) = v^i(x) \widetilde{\boldsymbol{g}}_i(x) = \widetilde{v}^i(\widetilde{x}) \widetilde{\boldsymbol{g}}_i(\widetilde{x}).$$

It is then easily verified that, since $\widetilde{x} = \boldsymbol{\chi}(x)$,

$$v_i(x) = \frac{\partial \chi^k}{\partial x_i}(x) \widetilde{v}_k(\widetilde{x}) \quad \text{and} \quad v^i(x) = \frac{\partial \widetilde{\chi}^i}{\partial \widetilde{x}_k}(x) \widetilde{v}^k(\widetilde{x}).$$

In other words, *under a change of curvilinear coordinates*, the components $v_i(x)$ "vary like" the vectors $\boldsymbol{g}_i(x)$ of the *covariant* basis while the components $v^i(x)$ of a vector "vary like" the vectors $\boldsymbol{g}^i(x)$ of the *contravariant* basis. This is why they are respectively called *"covariant"* and *"contravariant."*

Likewise, let $g_{ij}(x)$ and $\widetilde{g}_{ij}(\widetilde{x})$ denote the "covariant" components, and let $g^{ij}(x)$ and $\widetilde{g}^{ij}(\widetilde{x})$ denote the "contravariant" components, of the *metric tensor field* at the *same* point $\boldsymbol{\Theta}(x) = \widetilde{\boldsymbol{\Theta}}(\widetilde{x}) \in \widehat{\Omega} \subset \mathbb{E}^n$. Then a simple computation shows that

$$g_{ij}(x) = \frac{\partial \chi^k}{\partial x_i}(x) \frac{\partial \chi^\ell}{\partial x_j}(x) \widetilde{g}_{k\ell}(\widetilde{x}) \quad \text{and} \quad g^{ij}(x) = \frac{\partial \widetilde{\chi}^i}{\partial \widetilde{x}_k}(x) \frac{\partial \widetilde{\chi}^j}{\partial \widetilde{x}_\ell}(x) \widetilde{g}^{k\ell}(\widetilde{x}).$$

These formulas again explain why the components $g_{ij}(x)$ and $g^{ij}(x)$ are respectively called "covariant" and "contravariant": *under a change of curvilinear coordinates, each* index in

$g_{ij}(x)$ "varies like" that of the corresponding vector of the *covariant* basis, while *each* exponent in $g^{ij}(x)$ "varies like" that of the corresponding vector of the *contravariant* basis.

An analogous analysis shows that the components $v_{i\|j}(x)$ of the *covariant derivative of a vector field* (Section 8.3) are likewise "covariant"; cf. Problem 8.4-1.

A *second viewpoint*, and certainly a more illuminating one, consists in viewing the *same* scalars $v_i(x)$ or $v^i(x)$ as the components of a *vector*, and $g_{ij}(x)$ or $g^{ij}(x)$ as the components of a *linear operator*, over appropriate *bases* constructed by means of the vectors $\boldsymbol{g}_i(x)$ and $\boldsymbol{g}^j(x)$, which are thus considered as *given* in this approach, i.e., by means of a given $\mathcal{C}^1$-diffeomorphism $\boldsymbol{\Theta} : \Omega \to \widehat{\Omega}$.

Thus, for instance, a vector

$$v_j(x)\boldsymbol{g}^j(x) = v^i(x)\boldsymbol{g}_i(x) \in \mathbb{E}^n$$

may be defined either by means of its *covariant components* $v_j(x)$ over the vectors $\boldsymbol{g}^j(x)$ of the *contravariant basis*, or by means of its *contravariant components* $v^i(x)$ over the vectors $\boldsymbol{g}_i(x)$ of the *covariant basis*, its *intrinsic character* as a vector in $\mathbb{E}^n$ being reflected by the relation (Section 8.3)

$$v_i(x)\boldsymbol{g}^i(x) = \widehat{v}_i(\widehat{x})\widehat{\boldsymbol{e}}^i.$$

In this fashion, a vector provides an instance of a **first-order tensor**, "first-order" simply reflecting that its components are defined by means of either *one* index or *one* exponent.

The vector space spanned by the vectors $\boldsymbol{g}_i(x)$ is in effect the **tangent space**[3]

$$\mathbb{T}_{\widehat{x}}\widehat{\Omega}$$

to the n-dimensional manifold $\widehat{\Omega} = \boldsymbol{\Theta}(\Omega)$ *at the point* $\widehat{x} = \boldsymbol{\Theta}(x)$, while the vector space spanned by the vectors $\boldsymbol{g}^j(x)$ is the *dual* space of $\mathbb{T}_{\widehat{x}}\widehat{\Omega}$. *In the present situation*, the space $\mathbb{T}_{\widehat{x}}\widehat{\Omega}$ and its dual are *identified* by means of the *Euclidean inner product*, an identification already used in the defining relations $\boldsymbol{g}^j(x) \cdot \boldsymbol{g}_i(x) = \delta_i^j$.

Since the vectors $\boldsymbol{g}_i(x)$ are linearly independent, the space $\mathbb{T}_{\widehat{x}}\widehat{\Omega}$ *may be identified with* $\mathbb{E}^n$ *at each* $\widehat{x} = \boldsymbol{\Theta}(x) \in \widehat{\Omega}$. This explains why it was thus far considered that $\boldsymbol{g}_i(x)$ and $\boldsymbol{g}^j(x)$ were vectors in $\mathbb{E}^n$. But this identification is specific to the present situation, where the manifold $\widehat{\Omega}$ is an open subset of $\mathbb{E}^n$; for instance, the situation is *different* if the manifold is a *surface in* $\mathbb{E}^3$ (it is intuitively clear that, by contrast, the tangent spaces "vary" in general along a surface; cf. Section 8.9).

Remark The relations $\boldsymbol{g}^j(x)\cdot\boldsymbol{g}_i(x) = \delta_i^j$ are *equivalent* to the relations $\boldsymbol{g}_i(x) = g_{ij}(x)\boldsymbol{g}^j(x)$, or to the relations $\boldsymbol{g}^j(x) = g^{ij}(x)\boldsymbol{g}_j(x)$, either of which may be thus also used for *defining* the components $g_{ij}(x)$ or $g^{ij}(x)$ (instead of $g_{ij}(x) := \boldsymbol{g}_i(x) \cdot \boldsymbol{g}_j(x)$, or $g^{ij}(x) := \boldsymbol{g}^i(x) \cdot \boldsymbol{g}^j(x)$, as in Section 8.2). $\square$

A vector field defined by its values $v_j(x)\boldsymbol{g}^i(x)$, or $v^i(x)\boldsymbol{g}_i(x)$, at each point $x \in \Omega$, thus maps Ω into the *set*

$$\mathbb{T}\widehat{\Omega} := \bigsqcup_{\widehat{x}\in\widehat{\Omega}} \mathbb{T}_{\widehat{x}}\widehat{\Omega},$$

where $\bigsqcup$ denotes the *disjoint union* sign (Section 1.3). The set $\mathbb{T}\widehat{\Omega}$ is called the **tangent bundle** of $\widehat{\Omega}$.

[3]For a detailed presentation of the notions of tangent space, see, e.g., SCHLICHTKRULL [2012, Chapter 3].

Note in passing a useful (and immediately verified) property that will be used repeatedly in the sequel: Any vector $\boldsymbol{w} \in \mathbb{T}_{\widehat{x}}\widehat{\Omega}$ can be expanded over either basis as

$$\boldsymbol{w} = (\boldsymbol{w} \cdot \boldsymbol{g}^i(x))\boldsymbol{g}_i(x) = (\boldsymbol{w} \cdot \boldsymbol{g}_j(x))\boldsymbol{g}^j(x).$$

Next, let linear operators $\boldsymbol{g}_i(x) \otimes \boldsymbol{g}_j(x) \in \mathcal{L}(\mathbb{T}_{\widehat{x}}\widehat{\Omega}; \mathbb{T}_{\widehat{x}}\widehat{\Omega})$ be *defined* at each $x \in \Omega$ by

$$(\boldsymbol{g}_i(x) \otimes \boldsymbol{g}_j(x))\boldsymbol{w} := (\boldsymbol{g}_j(x) \cdot \boldsymbol{w})\boldsymbol{g}_i(x) \quad \text{for each } \boldsymbol{w} \in \mathbb{T}_{\widehat{x}}\widehat{\Omega}.$$

Then *these n^2 linear operators form a basis in the space $\mathcal{L}(\mathbb{T}_{\widehat{x}}\widehat{\Omega}; \mathbb{T}_{\widehat{x}}\widehat{\Omega})$*, since any linear operator $\boldsymbol{T}(x) \in \mathcal{L}(\mathbb{T}_{\widehat{x}}\widehat{\Omega}; \mathbb{T}_{\widehat{x}}\widehat{\Omega})$ can be expanded as

$$\boldsymbol{T}(x) = T^{ij}(x)(\boldsymbol{g}_i(x) \otimes \boldsymbol{g}_j(x)) \quad \text{with } T^{ij}(x) := \boldsymbol{g}^i(x) \cdot (\boldsymbol{T}(x)\boldsymbol{g}^j(x)).$$

To see this, it suffices to verify that, when they are applied to the vectors $\boldsymbol{g}^k(x)$ (which form a basis of $\mathbb{T}_{\widehat{x}}\widehat{\Omega}$), both sides of the resulting vector are equal to the same vector in $\mathbb{T}_{\widehat{x}}\widehat{\Omega}$; indeed,

$$\begin{aligned} T^{ij}(x)(\boldsymbol{g}_i(x) \otimes \boldsymbol{g}_j(x))\boldsymbol{g}^k(x) &= \left(\boldsymbol{g}^i(x) \cdot \boldsymbol{T}(x)\boldsymbol{g}^j(x)\right)\left(\boldsymbol{g}_j(x) \cdot \boldsymbol{g}^k(x)\right)\boldsymbol{g}_i(x) \\ &= \left(\boldsymbol{g}^i(x) \cdot \boldsymbol{T}(x)\boldsymbol{g}^k(x)\right)\boldsymbol{g}_i(x) = \boldsymbol{T}(x)\boldsymbol{g}^k(x). \end{aligned}$$

The real numbers $T^{ij}(x)$ are called the *contravariant components* of the linear operator $\boldsymbol{T}(x)$ (that they are contravariant immediately follows from the definition $T^{ij}(x) := \boldsymbol{g}^i(x) \cdot (\boldsymbol{T}(x)\boldsymbol{g}^j(x))$ since the vectors $\boldsymbol{g}^i(x)$ are themselves contravariant).

A linear operator $\boldsymbol{T}(x) \in \mathcal{L}(\mathbb{T}_{\widehat{x}}\widehat{\Omega}; \mathbb{T}_{\widehat{x}}\widehat{\Omega})$ can thus be defined by means of its *contravariant components* $T^{ij}(x)$ over the *covariant basis* $(\boldsymbol{g}_i(x) \otimes \boldsymbol{g}_j(x))$. But the *same* operator $\boldsymbol{T}(x)$ can be also defined by means of its *covariant components*

$$T_{ij}(x) := \boldsymbol{g}_i(x) \cdot \left(\boldsymbol{T}(x)\boldsymbol{g}_j(x)\right)$$

over the *contravariant basis* $(\boldsymbol{g}^i(x) \otimes \boldsymbol{g}^j(x))$ of the space $\mathcal{L}(\mathbb{T}_{\widehat{x}}\widehat{\Omega}; \mathbb{T}_{\widehat{x}}\widehat{\Omega})$ defined by

$$\left(\boldsymbol{g}^i(x) \otimes \boldsymbol{g}^j(x)\right)\boldsymbol{w} := \left(\boldsymbol{g}^j(x) \cdot \boldsymbol{w}\right)\boldsymbol{g}^i(x) \quad \text{for each } \boldsymbol{w} \in \mathbb{T}_{\widehat{x}}\widehat{\Omega},$$

or by means of its *mixed components*

$$T^{i\cdot}_{\cdot j}(x) := \boldsymbol{g}^i(x) \cdot (\boldsymbol{T}(x)\boldsymbol{g}_j(x)), \quad \textit{resp.} \quad T^{\cdot j}_{i\cdot}(x) := \boldsymbol{g}_i(x) \cdot \left(\boldsymbol{T}(x)\boldsymbol{g}^j(x)\right),$$

over the *mixed basis* $(\boldsymbol{g}_i(x) \otimes \boldsymbol{g}^j(x))$, *resp.* $(\boldsymbol{g}^i(x) \otimes \boldsymbol{g}_j(x))$, defined for each $\boldsymbol{w} \in \mathbb{T}_{\widehat{x}}\widehat{\Omega}$ by

$$\left(\boldsymbol{g}_i(x) \otimes \boldsymbol{g}^j(x)\right)\boldsymbol{w} := (\boldsymbol{g}^j(x) \cdot \boldsymbol{w})\boldsymbol{g}_i(x), \quad \textit{resp.} \quad \left(\boldsymbol{g}^i(x) \otimes \boldsymbol{g}_j(x)\right)\boldsymbol{w} := (\boldsymbol{g}_j(x) \cdot \boldsymbol{w})\boldsymbol{g}^i(x).$$

To sum up, any linear operator $\boldsymbol{T}(x) \in \mathcal{L}(\mathbb{T}_{\widehat{x}}\widehat{\Omega}; \mathbb{T}_{\widehat{x}}\widehat{\Omega})$ can be written in four different ways, viz.,

$$\begin{aligned} \boldsymbol{T}(x) &= T^{ij}(x)(\boldsymbol{g}_i(x) \otimes \boldsymbol{g}_j(x)) = T_{ij}(x)\left(\boldsymbol{g}^i(x) \otimes \boldsymbol{g}^j(x)\right) \\ &= T^{i\cdot}_{\cdot j}\left(\boldsymbol{g}_i(x) \otimes \boldsymbol{g}^j(x)\right) = T^{\cdot j}_{i\cdot}(x)\left(\boldsymbol{g}^i(x) \otimes \boldsymbol{g}_j(x)\right), \end{aligned}$$

according to which basis is chosen in the space $\mathcal{L}(\mathbb{T}_{\widehat{x}}\widehat{\Omega};\mathbb{T}_{\widehat{x}}\widehat{\Omega})$. Note that *the mixed components $T^{i\cdot}_{\cdot j}(x)$ and $T^{\cdot j}_{i\cdot}(x)$ of $\boldsymbol{T}(x)$ are in general different.*

If, however, the tensor $\boldsymbol{T}(x)$ is *symmetric with respect to the Euclidean inner product*, i.e., if $(\boldsymbol{T}(x)\boldsymbol{v}\cdot\boldsymbol{w}) = (\boldsymbol{v}\cdot\boldsymbol{T}(x)\boldsymbol{w})$ for all $\boldsymbol{v},\boldsymbol{w}\in\mathbb{T}_{\widehat{x}}\widehat{\Omega}$, then

$$T^{\cdot j}_{i\cdot}(x) = \boldsymbol{g}_i(x)\cdot\big(\boldsymbol{T}(x)\boldsymbol{g}^j(x)\big) = \big(\boldsymbol{T}(x)\boldsymbol{g}_i(x)\big)\cdot\boldsymbol{g}^j(x) = T^{j\cdot}_{\cdot i}(x).$$

The shorter notation

$$T^j_i(x) := T^{\cdot j}_{i\cdot}(x) = T^{j\cdot}_{\cdot i}(x)$$

is then preferred in this case.

A linear operator $\boldsymbol{T}(x)\in\mathcal{L}(\mathbb{T}_{\widehat{x}}\widehat{\Omega};\mathbb{T}_{\widehat{x}}\widehat{\Omega})$ provides an instance of a **second-order tensor**, "second-order" simply reflecting that its components are defined by either *two* exponents, or *two* indices, or *one* exponent and *one* index, or *one* index and *one* exponent.

Note also that the definition of the four types of components of such a tensor $\boldsymbol{T}(x)$ immediately shows that they are related by

$$\begin{aligned} T^{i\cdot}_{\cdot j}(x) &= g^{ik}(x)T_{kj(x)}, \qquad & T^{ij}(x) &= g^{ik}(x)T^{\cdot j}_{k\cdot}(x),\\ T^{\cdot j}_{i\cdot}(x) &= g_{ik}(x)T^{kj}(x), \qquad & T_{ij}(x) &= g_{ik}(x)T^{\cdot k}_{i\cdot}(x). \end{aligned}$$

As a first instance of a second-order tensor, consider the *identity mapping* $\boldsymbol{I}$ *of the space* $\mathcal{L}(\mathbb{T}_{\widehat{x}}\widehat{\Omega};\mathbb{T}_{\widehat{x}}\widehat{\Omega})$, the *covariant*, resp. *contravariant*, *components* of which (i.e., with respect to the basis $(\boldsymbol{g}^i(x)\otimes\boldsymbol{g}^j(x))$, *resp.* $(\boldsymbol{g}_i(x)\otimes\boldsymbol{g}_j(x))$) are simply $g_{ij}(x)$, *resp.* $g^{ij}(x)$, since, for each vector $\boldsymbol{w}\in\mathbb{T}_{\widehat{x}}\widehat{\Omega}$,

$$\begin{aligned} g_{ij}(x)\left(\boldsymbol{g}^i(x)\otimes\boldsymbol{g}^j(x)\right)\boldsymbol{w} &= g_{ij}(x)\left(\boldsymbol{g}^j(x)\cdot\boldsymbol{w}\right)\boldsymbol{g}^i(x) = \left(\boldsymbol{g}^j(x)\cdot\boldsymbol{w}\right)\boldsymbol{g}_j(x) = \boldsymbol{w},\\ g^{ij}(x)(\boldsymbol{g}_i(x)\otimes\boldsymbol{g}_j(x))\boldsymbol{w} &= g^{ij}(x)(\boldsymbol{g}_j(x)\cdot\boldsymbol{w})\boldsymbol{g}_i(x) = (\boldsymbol{g}_j(x)\cdot\boldsymbol{w})\boldsymbol{g}^j(x) = \boldsymbol{w}. \end{aligned}$$

In other words,

$$\boldsymbol{I} = g_{ij}(x)\left(\boldsymbol{g}^i(x)\otimes\boldsymbol{g}^j(x)\right) = g^{ij}(x)(\boldsymbol{g}_i(x)\otimes\boldsymbol{g}_j(x)).$$

Like any element in the space $\mathcal{L}(\mathbb{T}_{\widehat{x}}\widehat{\Omega};\mathbb{T}_{\widehat{x}}\widehat{\Omega})$, the identity mapping, which is clearly *symmetric*, has also mixed components $g^j_i(x) = g^{\cdot j}_{i\cdot}(x) = g^{j\cdot}_{\cdot i}(x)$, which are simply equal to the Kronecker symbol δ^j_i, since

$$\boldsymbol{I} = \delta^j_i\,\left(\boldsymbol{g}^i(x)\otimes\boldsymbol{g}_j(x)\right)$$

as is immediately verified, by applying both sides to an arbitrary vector $\boldsymbol{w}\in\mathbb{T}_{\widehat{x}}\widehat{\Omega}$.

The identity mapping thus provides an example of a *symmetric second-order tensor in the space* $\mathcal{L}(\mathbb{T}_{\widehat{x}}\widehat{\Omega};\mathbb{T}_{\widehat{x}}\widehat{\Omega})$.

Note that the *same* components $g_{ij}(x)$, *resp.* $g^{ij}(x)$, can be also viewed as the covariant, *resp.* contravariant, components of a *different* second-order tensor, this time viewed as an element of the *space* $\mathcal{L}^s_2(\mathbb{T}_{\widehat{x}}\widehat{\Omega}\times\mathbb{T}_{\widehat{x}}\widehat{\Omega};\mathbb{R})$ *formed by all symmetric bilinear forms on* $\mathbb{T}_{\widehat{x}}\widehat{\Omega}\times\mathbb{T}_{\widehat{x}}\widehat{\Omega}$ and defined by

$$(v^i(x)\boldsymbol{g}_i(x), w^j(x)\boldsymbol{g}_j(x))\in\mathbb{T}_{\widehat{x}}\widehat{\Omega}\times\mathbb{T}_{\widehat{x}}\widehat{\Omega}\to g_{ij}(x)v^i(x)w^j(x)\in\mathbb{R}.$$

Note that $g_{ij}(x)v^i(x)w^j(x) \in \mathbb{R}$ is nothing but the *Euclidean inner product* of the vectors $v^i(x)\boldsymbol{g}_i(x) \in \mathbb{T}_{\widehat{x}}\widehat{\Omega}$ and $w^j\boldsymbol{g}_j(x) \in \mathbb{T}_{\widehat{x}}\widehat{\Omega}$. The corresponding basis of the space $\mathcal{L}_2^s(\mathbb{T}_{\widehat{x}}\widehat{\Omega}\times\mathbb{T}_{\widehat{x}}\widehat{\Omega};\mathbb{R})$ is then given by the $\dfrac{n(n+1)}{2}$ bilinear forms

$$(v^i(x)\boldsymbol{g}_i(x), w^j\boldsymbol{g}_j(x)) \in \mathbb{T}_{\widehat{x}}\widehat{\Omega} \times \mathbb{T}_{\widehat{x}}\widehat{\Omega} \to v^k(x)w^\ell(x) \in \mathbb{R}, \quad 1 \leq k \leq \ell \leq n.$$

In fact, these two seemingly different definitions of a second-order tensor can be easily reconciled into a single one, since $\mathcal{L}_2^s(\mathbb{T}_{\widehat{x}}\widehat{\Omega} \times \mathbb{T}_{\widehat{x}}\widehat{\Omega};\mathbb{R})$ is a subspace of the space $\mathcal{L}_2(\mathbb{T}_{\widehat{x}}\widehat{\Omega} \times \mathbb{T}_{\widehat{x}}\widehat{\Omega};\mathbb{R})$, which can be identified with the space $\mathcal{L}(\mathbb{T}_{\widehat{x}}\widehat{\Omega};\mathbb{T}_{\widehat{x}}\widehat{\Omega})$ (Theorem 2.11-5).

Another instance of a second-order tensor in the space $\mathcal{L}(\mathbb{T}_{\widehat{x}}\widehat{\Omega};\mathbb{T}_{\widehat{x}}\widehat{\Omega})$ is provided by the **covariant derivative at** $x \in \Omega$ of a vector field $v_i\boldsymbol{g}^i : \Omega \to \mathbb{T}\widehat{\Omega}$, which is *by definition* the tensor $v_{i\|j}(x)(\boldsymbol{g}^i(x) \otimes \boldsymbol{g}^j(x)) \in \mathcal{L}(\mathbb{T}_{\widehat{x}}\widehat{\Omega};\mathbb{T}_{\widehat{x}}\widehat{\Omega})$, whose *covariant components* are those introduced in Theorem 8.3-1, viz.,

$$v_{i\|j}(x) := \partial_j v_i(x) - \Gamma_{ij}^p(x)v_p(x)$$

(that these components $v_{i\|j}(x)$ are indeed covariant can be easily checked directly; cf. Problem 8.4-1).

To decipher the nature of this tensor, let $\widehat{\boldsymbol{v}} : \widehat{\Omega} \to \mathbb{E}^n$ denote the vector field whose covariant components are the functions $v_i : \Omega \to \mathbb{R}$; i.e., such that $v_i(x)\boldsymbol{g}^i(x) = \widehat{v}_i(\widehat{x})\widehat{\boldsymbol{e}}^i$ at each $\widehat{x} = \boldsymbol{\Theta}(x)$, $x \in \Omega$ (Section 8.3). Then *the covariant derivative at* $x \in \Omega$ *is simply the Fréchet derivative* $\widehat{\boldsymbol{v}}'(\widehat{x})$ (Section 7.1) *of the vector field* $\widehat{\boldsymbol{v}} : \widehat{\Omega} \to \mathbb{R}^n$ (whose matrix is $(\widehat{\partial}_j\widehat{v}_i(\widehat{x})) \in \mathbb{M}^n$ in Cartesian coordinates) *expressed in curvilinear coordinates.* To see this, let $\widehat{\boldsymbol{w}} = \widehat{w}^j\widehat{\boldsymbol{e}}_j = w^m(x)\boldsymbol{g}_m(x)$ be an arbitrary vector in $\mathbb{T}_{\widehat{x}}\widehat{\Omega}$, so that $\widehat{w}^j = (w^m(x)\boldsymbol{g}_m(x)) \cdot \widehat{\boldsymbol{e}}^j = [\boldsymbol{g}_m(x)]^j$. Then

$$\begin{aligned}\widehat{\boldsymbol{v}}'(\widehat{x})\widehat{\boldsymbol{w}} &= \widehat{\partial}_j\widehat{v}_i(\widehat{x})\widehat{w}^j\widehat{\boldsymbol{e}}^i = v_{k\|\ell}(x)\,[\boldsymbol{g}^k(x)]_i\,[\boldsymbol{g}^\ell(x)]_j w^m(x)\,[\boldsymbol{g}_m(x)]^j\widehat{\boldsymbol{e}}^i \\ &= v_{k\|\ell}(x)w^\ell(x)\boldsymbol{g}^k(x),\end{aligned}$$

since $[\boldsymbol{g}^\ell(x)]_j\,[\boldsymbol{g}_m(x)]^j = \boldsymbol{g}^\ell(x)\cdot\boldsymbol{g}_m(x) = \delta_m^\ell$ and $[\boldsymbol{g}^k(x)]_i\widehat{\boldsymbol{e}}^i = \boldsymbol{g}^k(x)$, on the one hand. On the other hand,

$$\begin{aligned}v_{k\|\ell}(x)\big(\boldsymbol{g}^k(x)\otimes\boldsymbol{g}^\ell(x)\big)w^i(x)\boldsymbol{g}_i(x) &= v_{k\|\ell}(x)w^i(x)\big(\boldsymbol{g}^\ell(x)\cdot\boldsymbol{g}_i(x)\big)\boldsymbol{g}^k(x) \\ &= v_{k\|\ell}(x)w^\ell(x)\boldsymbol{g}^k(x),\end{aligned}$$

since $\boldsymbol{g}^\ell(x)\cdot\boldsymbol{g}_i(x) = \delta_i^\ell$. Hence the assertion is established.

The above examples lead to the following *definitions*: After possible identifications between the spaces $\mathcal{L}_2(\mathbb{T}_{\widehat{x}}\widehat{\Omega}\times\mathbb{T}_{\widehat{x}}\widehat{\Omega};\mathbb{R})$ and $\mathcal{L}(\mathbb{T}_{\widehat{x}}\widehat{\Omega};\mathbb{T}_{\widehat{x}}\widehat{\Omega})$ (Theorem 2.11-5), a **second-order tensor at** $x \in \Omega$ is an element in the space $\mathcal{L}(\mathbb{T}_{\widehat{x}}\widehat{\Omega};\mathbb{T}_{\widehat{x}}\widehat{\Omega})$; after possible identifications between the spaces $\mathcal{L}_2(\mathbb{T}_{\widehat{x}}\widehat{\Omega}\times\mathbb{T}_{\widehat{x}}\widehat{\Omega};\mathbb{R})$ and $\mathcal{L}(\mathbb{T}_{\widehat{x}}\widehat{\Omega};\mathbb{T}_{\widehat{x}}\widehat{\Omega})$ and between the spaces $\mathcal{L}(\mathbb{T}_{\widehat{x}}\widehat{\Omega}\times\mathbb{T}_{\widehat{x}}\widehat{\Omega}\times\mathbb{T}_{\widehat{x}}\widehat{\Omega};\mathbb{R})$ and $\mathcal{L}(\mathbb{T}_{\widehat{x}}\widehat{\Omega};\mathcal{L}(\mathbb{T}_{\widehat{x}}\widehat{\Omega};\mathbb{T}_{\widehat{x}}\widehat{\Omega}))$, a **third-order tensor at** $x \in \Omega$ is an element of the space $\mathcal{L}(\mathbb{T}_{\widehat{x}}\widehat{\Omega};\mathcal{L}(\mathbb{T}_{\widehat{x}}\widehat{\Omega};\mathbb{T}_{\widehat{x}}\widehat{\Omega}))$ or of the space $\mathcal{L}(\mathcal{L}(\mathbb{T}_{\widehat{x}}\widehat{\Omega};\mathbb{T}_{\widehat{x}}\widehat{\Omega});\mathbb{T}_{\widehat{x}}\widehat{\Omega})$; and so forth.

Examples of *third-order tensors* are given in Problems 8.4-3–8.4-5; examples of fourth-order tensors are given in Problems 8.4-4 and 8.4-5.

Problems

8.4-1 Given two $\mathcal{C}^1$-diffeomorphisms $\boldsymbol{\Theta} : \Omega \subset \mathbb{R}^n \to \mathbb{E}^n$ and $\widetilde{\boldsymbol{\Theta}} : \widetilde{\Omega} \subset \mathbb{R}^n \to \mathbb{E}^n$ such that $\boldsymbol{\Theta}(\Omega) = \widetilde{\boldsymbol{\Theta}}(\widetilde{\Omega})$, let $v_{i\|j}(x)$ and $\widetilde{v}_{i\|j}(\widetilde{x})$ denote the covariant components at $x \in \Omega$ of the covariant derivative of a given vector field at the same point $\boldsymbol{\Theta}(x) = \widetilde{\boldsymbol{\Theta}}(\widetilde{x})$. Using the notations used in the text for the changes of coordinates, show directly that

$$v_{i\|j}(x) = \frac{\partial \chi^k}{\partial x_i}(x)\frac{\partial \chi^\ell}{\partial x_j}(x)\widetilde{v}_{k\|\ell}(\widetilde{x}).$$

8.4-2 (1) Show that the mixed components

$$v^i\|_j(x) := g^{ik}(x)v_{k\|j}(x), \quad x \in \Omega,$$

of the covariant derivative of a vector field $v^i\boldsymbol{g}_i : \Omega \to \mathrm{T}\widehat{\Omega}$ are also given by

$$v^i\|_j(x) = \partial_j v^i(x) + \Gamma^i_{jq}(x)v^q(x).$$

(2) Show that the same mixed components $v^i\|_j : \Omega \to \mathbb{R}$ can be also defined by means of the relations

$$\partial_j(v^i\boldsymbol{g}_i) = v^i\|_j\boldsymbol{g}_i.$$

8.4-3 Given $x \in \Omega \subset \mathbb{R}^3$, let $\varepsilon^{ijk}(x) := \dfrac{1}{\sqrt{g(x)}}$ if $\{i, j, k\}$ is an even permutation of $\{1, 2, 3\}$, $\varepsilon^{ijk}(x) := -\dfrac{1}{\sqrt{g(x)}}$ if $\{i, j, k\}$ is an odd permutation of $\{1, 2, 3\}$, and $\varepsilon^{ijk}(x) := 0$ otherwise. Show that, at each $x \in \Omega$, the mapping

$$(v_i(x)\boldsymbol{g}^i(x), w_j(x)\boldsymbol{g}^j(x)) \in \mathrm{T}_{\widehat{x}}\widehat{\Omega} \times \mathrm{T}_{\widehat{x}}\widehat{\Omega} \to \varepsilon^{ijk}(x)v_i(x)w_j(x)\boldsymbol{g}_k(x) \in \mathrm{T}_{\widehat{x}}\widehat{\Omega}$$

defines a *third-order tensor*, called the **orientation tensor**, as an element of the *space of all antisymmetric bilinear mappings from* $\mathrm{T}_{\widehat{x}}\widehat{\Omega} \times \mathrm{T}_{\widehat{x}}\widehat{\Omega}$ *into* $\mathrm{T}_{\widehat{x}}\widehat{\Omega}$. Note that $\varepsilon^{ijk}(x)v_i(x)w_j(x)\boldsymbol{g}_k(x)$ is nothing but the *vector product* of the two vectors $v_i\boldsymbol{g}^i(x)$ and $w_j\boldsymbol{g}^j(x)$, expressed in curvilinear coordinates.

8.4-4 Let Ω be an open subset of $\mathbb{R}^n$ and let $\boldsymbol{\Theta} : \Omega \to \mathbb{E}^n$ be a $\mathcal{C}^3$-diffeomorphism of Ω onto $\widehat{\Omega} := \boldsymbol{\Theta}(\Omega)$.

(1) Show that

$$\partial_k\left(\boldsymbol{g}^i \otimes \boldsymbol{g}^j\right) = -\Gamma^i_{k\ell}\boldsymbol{g}^\ell \otimes \boldsymbol{g}^j - \Gamma^j_{k\ell}\boldsymbol{g}^i \otimes \boldsymbol{g}^\ell.$$

(2) Let there be given tensors $T_{ij}(x)\boldsymbol{g}^i(x) \otimes \boldsymbol{g}^j(x), x \in \Omega$, with covariant components $T_{ij} \in \mathcal{C}^1(\Omega)$, and let the functions $T_{ij\|k} : \Omega \to \mathbb{R}$ be defined by the relations

$$T_{ij\|k}\boldsymbol{g}^i \otimes \boldsymbol{g}^j = \partial_k(T_{ij}\boldsymbol{g}^i \otimes \boldsymbol{g}^j).$$

Show that

$$T_{ij\|k} = \partial_k T_{ij} - \Gamma^\ell_{ki}T_{\ell j} - \Gamma^\ell_{kj}T_{i\ell}.$$

(3) Show that, at each $x \in \Omega$, the numbers $T_{ij\|k}(x)$ are the covariant components of a *third-order tensor*, as an element in the space $\mathcal{L}(\mathcal{L}(\mathrm{T}_{\widehat{x}}\widehat{\Omega}; \mathrm{T}_{\widehat{x}}\widehat{\Omega}); \mathrm{T}_{\widehat{x}}\widehat{\Omega})$.

(4) Assume that $T_{ij} \in \mathcal{C}^2(\Omega)$, and let the functions $T_{ij\|k\ell} : \Omega \to \mathbb{R}$ be defined by the relations

$$T_{ij\|k\ell}\boldsymbol{g}^i \otimes \boldsymbol{g}^j = (\partial_{\ell k} - \Gamma^p_{\ell k}\partial_p)(T_{ij}\boldsymbol{g}^i \otimes \boldsymbol{g}^j).$$

Show that

$$T_{ij\|k\ell} = \partial_\ell T_{ij\|k} - \Gamma^p_{\ell i} T_{pj\|k} - \Gamma^p_{\ell j} T_{ip\|k} - \Gamma^p_{\ell k} T_{ij\|p},$$

which shows in particular that

$$T_{ij\|k\ell} = T_{ij\|\ell k}.$$

(5) Show that, at each $x \in \Omega$, the numbers $T_{ij\|k\ell}(x)$ are the covariant components of a *fourth-order tensor*, as an element in the space $\mathcal{L}(\mathbb{T}_{\widehat{x}}\widehat{\Omega}; \mathcal{L}(\mathbb{T}_{\widehat{x}}\widehat{\Omega}; \mathcal{L}(\mathbb{T}_{\widehat{x}}\widehat{\Omega}; \mathbb{T}_{\widehat{x}}\widehat{\Omega})))$.

8.4-5 Consider the *pure displacement problem of three-dimensional linearized elasticity*, which (with self-explanatory notations) takes the following form in *Cartesian coordinates* (Section 6.16):

$$-\widehat{\partial}_j \widehat{\sigma}^{ij} = \widehat{f}^i \text{ in } \widehat{\Omega} \quad \text{and} \quad \widehat{u}^i = 0 \text{ on } \widehat{\Gamma},$$

where

$$\widehat{\sigma}^{ij} := \widehat{A}^{ijkl} \widehat{e}_{k\ell}(\widehat{\boldsymbol{u}}), \quad \widehat{A}^{ijk\ell} := \lambda \delta^{ij} \delta^{k\ell} + \mu(\delta^{ik}\delta^{j\ell} + \delta^{i\ell}\delta^{jk}), \quad \widehat{e}_{ij}(\widehat{\boldsymbol{u}}) = \frac{1}{2}(\widehat{\partial}_j \widehat{u}_i + \widehat{\partial}_i \widehat{u}_j).$$

(1) Show directly that this boundary value problem expressed in terms of curvilinear coordinates becomes

$$-\sigma^{ij}\|_j = f^i \text{ in } \Omega \quad \text{and} \quad u^i = 0 \text{ on } \Gamma,$$

where

$$\begin{aligned}
\sigma^{ij} &= A^{ijk\ell} e_{k\ell}(\boldsymbol{u}),\\
A^{ijk\ell} &:= \lambda g^{ij} g^{k\ell} + \mu(g^{ik} g^{j\ell} + g^{i\ell} g^{jk}), \quad e_{ij}(\boldsymbol{u}) := \frac{1}{2}(u_{i\|j} + u_{j\|i}),\\
\sigma^{ij}\|_k &:= \partial_k \sigma^{ij} + \Gamma^i_{pk} \sigma^{pj} + \Gamma^j_{kq} \sigma^{iq}.
\end{aligned}$$

(2) Show that the numbers $A^{ijk\ell}(x)$ are at each $x \in \Omega$ the contravariant components of a *fourth-order tensor*, as an element in the space

$$\mathcal{L}^s_2(\mathcal{L}^s_2(\mathbb{T}_{\widehat{x}}\widehat{\Omega} \times \mathbb{T}_{\widehat{x}}\widehat{\Omega}; \mathbb{R}) \times \mathcal{L}^s_2(\mathbb{T}_{\widehat{x}}\widehat{\Omega} \times \mathbb{T}_{\widehat{x}}\widehat{\Omega}; \mathbb{R}); \mathbb{R}).$$

(3) Show that the numbers $\sigma^{ij}\|_k(x)$ defined in (1) are at each $x \in \Omega$ the mixed components of a *third-order tensor*, as an element in the space $\mathcal{L}(\mathcal{L}(\mathbb{T}_{\widehat{x}}\widehat{\Omega}; \mathbb{T}_{\widehat{x}}\widehat{\Omega}); \mathbb{T}_{\widehat{x}}\widehat{\Omega})$.

(4) Show that $\sigma^{ij}\|_k = g^{i\ell} g^{mj} \sigma_{\ell m\|k}$ where the covariant components $\sigma_{\ell m\|k}$ of the same third-order tensor field are defined as in Problem 8.4-4(2).

8.4-6 Given a vector field $\boldsymbol{u} = u_i \boldsymbol{g}^i : \Omega \to \mathbb{T}\widehat{\Omega}$ with components $u_i \in \mathcal{C}^3(\Omega)$, show that the *Saint-Venant compatibility relations* (Section 6.18) take the following form in *curvilinear coordinates* (with self-explanatory notations):

$$e_{ki\|j\ell}(\boldsymbol{u}) + e_{\ell j\|ik}(\boldsymbol{u}) - e_{kj\|i\ell}(\boldsymbol{u}) - e_{\ell i\|jk}(\boldsymbol{u}) = 0 \quad \text{in } \Omega,$$

where $e_{ij}(\boldsymbol{u}) := \frac{1}{2}(u_{i\|j} + u_{j\|i})$ and the functions $e_{ij\|k\ell} \in \mathcal{C}(\Omega)$ are defined as in Problem 8.4-4(4).

8.5 Necessary conditions satisfied by the metric tensor; the Riemann curvature tensor

As expected, *the components* $g_{ij} = g_{ji} = (\boldsymbol{\nabla}\boldsymbol{\Theta}^T \boldsymbol{\nabla}\boldsymbol{\Theta})_{ij} : \Omega \to \mathbb{R}$ *of the metric tensor* (Section 8.2) *defined by a smooth immersion* $\boldsymbol{\Theta} : \Omega \to \mathbb{E}^n$ *cannot be arbitrary functions.*

As shown in the next theorem, they must satisfy relations that take the form

$$\partial_j \Gamma_{ikq} - \partial_k \Gamma_{ijq} + \Gamma^p_{ij}\Gamma_{kqp} - \Gamma^p_{ik}\Gamma_{jqp} = 0 \quad \text{in } \Omega,$$

where the functions Γ_{ijq} and Γ^p_{ij} have simple expressions in terms of the functions g_{ij} and of some of their partial derivatives (although here they are given a different definition, the functions Γ^p_{ij} are nothing but the Christoffel symbols of the second kind introduced in Section 8.3). Recall that, according to the rule governing Latin indices and exponents, these relations are meant to hold for all $i, j, k, q \in \{1, \ldots, n\}$.

Theorem 8.5-1 (necessary conditions satisfied by the metric tensor) *Let Ω be an open set in $\mathbb{R}^n$, let $\boldsymbol{\Theta} \in \mathcal{C}^3(\Omega; \mathbb{E}^n)$ be an immersion, and let*

$$g_{ij} := \partial_i \boldsymbol{\Theta} \cdot \partial_j \boldsymbol{\Theta}$$

denote the covariant components of the metric tensor. Let the functions $\Gamma_{ijq} \in \mathcal{C}^1(\Omega)$ and $\Gamma^p_{ij} \in \mathcal{C}^1(\Omega)$ be defined by

$$\Gamma_{ijq} := \frac{1}{2}(\partial_j g_{iq} + \partial_i g_{jq} - \partial_q g_{ij}) \quad \textit{and} \quad \Gamma^p_{ij} := g^{pq}\Gamma_{ijq} \quad \textit{where } (g^{pq}) := (g_{ij})^{-1}.$$

Then, necessarily,

$$\partial_j \Gamma_{ikq} - \partial_k \Gamma_{ijq} + \Gamma^p_{ij}\Gamma_{kqp} - \Gamma^p_{ik}\Gamma_{jqp} = 0 \quad \textit{in } \Omega, \quad 1 \le i, j, k, q \le n.$$

Proof Let the covariant and contravariant bases be defined as in Section 8.2, viz., by $\boldsymbol{g}_i = \partial_i \boldsymbol{\Theta}$ and $\boldsymbol{g}^j \cdot \boldsymbol{g}_i = \delta^j_i$. It is then immediately verified that the functions $\Gamma_{ijq} := \frac{1}{2}(\partial_j g_{iq} + \partial_i g_{jq} - \partial_q g_{ij})$ are also given by

$$\Gamma_{ijq} = \partial_i \boldsymbol{g}_j \cdot \boldsymbol{g}_q.$$

Since $\boldsymbol{g}^j = g^{ij}\boldsymbol{g}_i$, the last relations imply that the functions $\Gamma^p_{ij} := g^{pq}\Gamma_{ijq}$ are also given by

$$\Gamma^p_{ij} = \partial_i \boldsymbol{g}_j \cdot \boldsymbol{g}^p.$$

Therefore,

$$\partial_i \boldsymbol{g}_j = \Gamma^p_{ij}\boldsymbol{g}_p,$$

since $\partial_i \boldsymbol{g}_j = (\partial_i \boldsymbol{g}_j \cdot \boldsymbol{g}^p)\boldsymbol{g}_p$. Together, the above relations give

$$\partial_k \Gamma_{ijq} = \partial_{ik}\boldsymbol{g}_j \cdot \boldsymbol{g}_q + \partial_i \boldsymbol{g}_j \cdot \partial_k \boldsymbol{g}_q \quad \text{and} \quad \partial_i \boldsymbol{g}_j \cdot \partial_k \boldsymbol{g}_q = \Gamma^p_{ij}\boldsymbol{g}_p \cdot \partial_k \boldsymbol{g}_q = \Gamma^p_{ij}\Gamma_{kqp}.$$

Consequently,

$$\partial_{ik}\boldsymbol{g}_j \cdot \boldsymbol{g}_q = \partial_k \Gamma_{ijq} - \Gamma^p_{ij}\Gamma_{kqp}.$$

Since $\partial_{ik}\boldsymbol{g}_j = \partial_{ij}\boldsymbol{g}_k$, we also have

$$\partial_{ik}\boldsymbol{g}_j \cdot \boldsymbol{g}_q = \partial_j \Gamma_{ikq} - \Gamma^p_{ik}\Gamma_{jqp},$$

and thus the required necessary conditions immediately follow. □

As shown in the above proof, *the necessary conditions*

$$\partial_j \Gamma_{ikq} - \partial_k \Gamma_{ijq} + \Gamma^p_{ij}\Gamma_{kqp} - \Gamma^p_{ik}\Gamma_{jqp} = 0$$

simply constitute a rewriting of the relations $\partial_{ik}\boldsymbol{g}_j = \partial_{ki}\boldsymbol{g}_j$, *in the form of the equivalent relations*

$$\partial_{ik}\boldsymbol{g}_j \cdot \boldsymbol{g}_q = \partial_{ki}\boldsymbol{g}_j \cdot \boldsymbol{g}_q.$$

Hence, the key to these necessary conditions is simply the *Schwarz lemma* (Theorem 7.8-1).

The functions

$$\Gamma_{ijq} = \frac{1}{2}(\partial_j g_{iq} + \partial_i g_{jq} - \partial_q g_{ij}) = \partial_i \boldsymbol{g}_j \cdot \boldsymbol{g}_q = \Gamma_{jiq}$$

and

$$\Gamma^p_{ij} = g^{pq}\Gamma_{ijq} = \partial_i \boldsymbol{g}_j \cdot \boldsymbol{g}^p = \Gamma^p_{ji}$$

are the **Christoffel symbols of the first**, and **second, kinds**, *associated with the metric tensor field* (g_{ij}). We saw in Section 8.3 that the same Christoffel symbols of the second kind also naturally appear in the definition of *covariant derivatives*.

The functions

$$R_{qijk} := \partial_j \Gamma_{ikq} - \partial_k \Gamma_{ijq} + \Gamma^p_{ij}\Gamma_{kqp} - \Gamma^p_{ik}\Gamma_{jqp}$$

are the **covariant components** of a *fourth-order tensor* (Problem 8.5-1), called the **Riemann curvature tensor**[4] *associated with the metric tensor field* (g_{ij}). The relations $R_{qijk} = 0$ found in Theorem 8.5-1 thus express that the *Riemann curvature tensor associated with the metric tensor field of a smooth enough immersion* $\boldsymbol{\Theta} : \Omega \subset \mathbb{R}^n \to \mathbb{E}^n$ *vanishes.*[5]

Note that a *different* set of necessary conditions can also be found, this time expressed in terms of the *square root* of the matrix field (g_{ij}); cf. Problem 8.5-2.

Problems

8.5-1 Let Ω be an open subset of $\mathbb{R}^n$ and let $\boldsymbol{\Theta} \in \mathcal{C}^2(\Omega; \mathbb{E}^n)$ be an immersion.

(1) Show that the Christoffel symbols $\Gamma^p_{ij} \in \mathcal{C}(\Omega)$ or $\Gamma_{ijq} \in \mathcal{C}(\Omega)$ are not components of a third-order tensor.

(2) Assuming now that $\boldsymbol{\Theta} \in \mathcal{C}^3(\Omega; \mathbb{E}^n)$, show that the covariant components $R_{qijk} \in \mathcal{C}(\Omega)$ of the Riemann curvature tensor are indeed the covariant components of a *fourth-order tensor*, according to the definition given in Section 8.4.

8.5-2 Let Ω be an open subset of $\mathbb{R}^n$ and let $\boldsymbol{\Theta} \in \mathcal{C}^3(\Omega; \mathbb{R}^n)$ be a *given* immersion. At each point $x \in \Omega$, let

$$\boldsymbol{\nabla\Theta}(x) = \boldsymbol{R}(x)\boldsymbol{U}(x),$$

where $\boldsymbol{U}(x) := (\boldsymbol{\nabla\Theta}(x)^T\boldsymbol{\nabla\Theta}(x))^{1/2} \in \mathbb{S}^n_>$ and $\boldsymbol{R}(x) := \boldsymbol{\nabla\Theta}(x)\boldsymbol{U}(x)^{-1} \in \mathbb{O}^n$ denote the unique *polar factorization* of the matrix $\boldsymbol{\nabla\Theta}(x) \in \mathbb{M}^n$, so that $\boldsymbol{U} \in \mathcal{C}^2(\Omega, \mathbb{S}^n_>)$ and $\boldsymbol{R} \in \mathcal{C}^2(\Omega; \mathbb{O}^n)$ (Problem 4.3-5).

[4]This tensor was introduced in the landmark lecture *Über die Hypothesen, welche der Geometrie zu Grunde liegen*, that Bernhard Riemann (1826–1866) delivered on 10 June 1854, as the complement to his "Habilitation" (where, among other things, he introduced the Riemann integral).

[5]This result is due to:

E.B. Christoffel [1869]: Über die Transformation der homogenen Differentialausdrücke zweiten Grades, *Journal für die Reine und Angewandte Mathematik* **70**, 46–70.

Let then the antisymmetric matrix fields $\boldsymbol{A}_j \in \mathcal{C}^1(\Omega; \mathbb{A}^n)$, $1 \le j \le n$, be defined in terms of the matrix field $\boldsymbol{U}$ by

$$\boldsymbol{A}_j := \frac{1}{2}\{\boldsymbol{U}^{-1}(\boldsymbol{\nabla c}_j - (\boldsymbol{\nabla c}_j)^T)\boldsymbol{U}^{-1} + \boldsymbol{U}^{-1}\partial_j\boldsymbol{U} - (\partial_j\boldsymbol{U})\boldsymbol{U}^{-1}\},$$

where $\boldsymbol{c}_j \in \mathcal{C}^2(\Omega; \mathbb{R}^n)$ denotes the jth column vector field of the matrix field $\boldsymbol{U}^2 \in \mathcal{C}^2(\Omega; \mathbb{S}^n_>)$.

Show that the matrix field $\boldsymbol{U}$ *necessarily* satisfies the compatibility conditions[6]

$$\partial_i \boldsymbol{A}_j - \partial_j \boldsymbol{A}_i + \boldsymbol{A}_i\boldsymbol{A}_j - \boldsymbol{A}_j\boldsymbol{A}_i = \boldsymbol{0} \quad \text{in } \Omega.$$

8.6 Existence of an immersion on an open subset of $\mathbb{R}^n$ with a prescribed metric tensor; the fundamental theorem of Riemannian geometry

Recall that $\mathbb{M}^n, \mathbb{S}^n$, and $\mathbb{S}^n_>$ denote the sets of all square matrices of order n, of all symmetric matrices of order n, and of all symmetric positive-definite matrices of order n.

So far, we have considered that we are *given* an open set $\Omega \subset \mathbb{R}^n$ and a smooth enough immersion $\boldsymbol{\Theta} : \Omega \to \mathbb{E}^n$, thus allowing us to define a matrix field

$$\boldsymbol{C} = (g_{ij}) = \boldsymbol{\nabla\Theta}^T\boldsymbol{\nabla\Theta} : \Omega \to \mathbb{S}^n_>,$$

where $g_{ij} : \Omega \to \mathbb{R}$ are the covariant components of the associated *metric tensor*.

We now turn to the *reciprocal questions*:

Given an open subset Ω of $\mathbb{R}^n$ and a smooth enough matrix field $\boldsymbol{C} = (g_{ij}) : \Omega \to \mathbb{S}^n_>$, *when does there exist an immersion* $\boldsymbol{\Theta} : \Omega \to \mathbb{E}^n$ *such that*

$$\boldsymbol{\nabla\Theta}^T\boldsymbol{\nabla\Theta} = \boldsymbol{C} \quad \text{in } \Omega,$$

or equivalently, such that

$$\partial_i\boldsymbol{\Theta} \cdot \partial_j\boldsymbol{\Theta} = g_{ij} \quad \text{in } \Omega?$$

If such an immersion exists, *to what extent is it unique*?

The answers are remarkably simple: *If* Ω *is simply connected, the necessary conditions*

$$\partial_j\Gamma_{ikq} - \partial_k\Gamma_{ijq} + \Gamma^p_{ij}\Gamma_{kqp} - \Gamma^p_{ik}\Gamma_{jqp} = 0 \quad \text{in } \Omega$$

found in Theorem 8.5-1 *are also sufficient for the existence of such an immersion* and *this immersion is unique up to isometries in* $\mathbb{E}^n$. Accordingly, these results comprise two essentially distinct parts, a *global existence result* (Theorem 8.6-1) and a *uniqueness result* (Theorem 8.7-1). Note that these two results are established under *different* assumptions on the set Ω and on the smoothness of the field (g_{ij}).

Remark Whether an immersion $\boldsymbol{\Theta} : \Omega \to \mathbb{E}^n$ found in this fashion is *injective* is a different issue, which accordingly should be resolved by different means. □

[6]These compatibility conditions are due to:
R.T. Shield [1973]: The rotation associated with large strains, *SIAM Journal on Applied Mathematics* **25**, 483–491.

In order to put these results in their proper perspective, let us make a brief incursion into *Riemannian geometry*.

Considered as an *n-dimensional manifold*, an open set $\Omega \subset \mathbb{R}^n$ equipped with an immersion $\boldsymbol{\Theta} : \Omega \to \mathbb{E}^n$ provides an example of a **Riemannian manifold** $(\Omega; (g_{ij}))$, i.e., a *manifold*, in this case the set Ω, equipped with a *Riemannian metric*,[7] in this case the symmetric positive-definite matrix field $(g_{ij}) : \Omega \to \mathbb{S}^n_>$ defined by $g_{ij} := \partial_i \boldsymbol{\Theta} \cdot \partial_j \boldsymbol{\Theta}$ in Ω. More generally, a **Riemannian metric on a manifold** is a twice covariant (Section 8.4), symmetric, positive-definite tensor field acting on vectors in the tangent spaces to the manifold (these tangent spaces coincide with $\mathbb{R}^n$ in the present instance; cf. again Section 8.4).

This particular Riemannian manifold $(\Omega; (g_{ij}))$ is **isometrically immersed** in the Euclidean space $\mathbb{E}^n$, in the sense that there exists an immersion $\boldsymbol{\Theta} : \Omega \to \mathbb{E}^n$ that satisfies the relations $g_{ij} = \partial_i \boldsymbol{\Theta} \cdot \partial_j \boldsymbol{\Theta}$ in Ω; equivalently, the length of any curve in the Riemannian manifold $(\Omega; (g_{ij}))$ is the same as the length of its image by $\boldsymbol{\Theta}$ in the Euclidean space $\mathbb{E}^n$ (Theorem 8.2-1(b)).

The first question above can thus be rephrased as follows: *Given an open subset Ω of $\mathbb{R}^n$ and a positive-definite symmetric matrix field $(g_{ij}) : \Omega \to \mathbb{S}^n_>$, when is the Riemannian manifold $(\Omega; (g_{ij}))$* **flat**, in the sense that it can be isometrically immersed in a Euclidean space of the *same* dimension n?

The answer to this question can then be rephrased as follows (compare with the statement of Theorem 8.6-1 below): *Let Ω be a simply connected open subset of $\mathbb{R}^n$. Then a Riemannian manifold $(\Omega; (g_{ij}))$ with a Riemannian metric (g_{ij}) of class $\mathcal{C}^2$ in Ω is flat if and only if its Riemannian curvature tensor vanishes in Ω.* Recast as such, this existence result becomes a special case of the **fundamental theorem on flat Riemannian manifolds**, which applies to general finite-dimensional Riemannian manifolds.

The answer to the second question, viz., the issue of uniqueness, can be rephrased as follows (compare with the statement of Theorem 8.7-1 in the next section): *Let Ω be a connected open subset of $\mathbb{R}^n$. Then the isometric immersions of a flat Riemannian manifold $(\Omega; (g_{ij}))$ into a Euclidean space $\mathbb{E}^n$ are unique up to isometries of $\mathbb{E}^n$.* Recast as such, this result is called the **rigidity theorem**.

Recast as such, these two theorems together constitute a special case (where the dimensions of the manifold and of the Euclidean space are equal) of the **fundamental theorem of Riemannian geometry**. This theorem addresses the same *existence* and *uniqueness* issues in the more general setting where Ω is replaced by a p-dimensional manifold and $\mathbb{E}^n$ is replaced by a $(p+q)$-dimensional Euclidean space[8] (the fundamental theorem of surface theory, together with the rigidity theorem for surfaces, established in Sections 8.16 and 8.17, constitutes another important special case).

Another fascinating question (which will not be addressed here) is the following: Given again an open subset Ω of $\mathbb{R}^n$ equipped with a symmetric, positive-definite matrix field $(g_{ij}) : \Omega \to \mathbb{S}^n$, assume this time that *the Riemannian manifold $(\Omega; (g_{ij}))$ is no longer flat*, i.e., its Riemannian curvature tensor no longer vanishes in Ω. *Can such a Riemannian*

[7]The notion of a *Riemannian manifold* was introduced by Bernhard Riemann on 10 June 1854, in the same landmark lecture (*op. cit.*) where he also introduced the *Riemann curvature tensor* (Section 8.5).

[8]When the p-dimensional manifold is an open subset of $\mathbb{R}^p$, a proof of this theorem is given in:

M. Szopos [2005]: On the recovery and continuity of a submanifold with boundary, *Analysis and Applications* **3**, 119–143.

manifold still be isometrically immersed, but this time in a higher-dimensional Euclidean space? Equivalently, does there exist a Euclidean space $\mathbb{E}^m$ with $m > n$, and does there exist an immersion $\boldsymbol{\Theta} : \Omega \to \mathbb{E}^m$ such that $g_{ij} = \partial_i \boldsymbol{\Theta} \cdot \partial_j \boldsymbol{\Theta}$ in Ω?

The answer is yes, according to the following beautiful **Nash theorem**:[9] *Any p-dimensional Riemannian manifold equipped with a continuous metric can be isometrically immersed in a Euclidean space of dimension 2p with an immersion of class* $\mathcal{C}^1$*; it can also be isometrically immersed in a Euclidean space of dimension* $(2p+1)$ *with an injective immersion of class* $\mathcal{C}^1$.

Let us now humbly return to the question of existence[10] raised at the beginning of this section, i.e., when the manifold is an open set in $\mathbb{R}^n$. In what follows, we let

$$\mathcal{C}^2(\Omega; \mathbb{S}^n_>) := \{\boldsymbol{C} \in \mathcal{C}^2(\Omega; \mathbb{S}^n);\ \boldsymbol{C}(x) \in \mathbb{S}^n_> \text{ for all } x \in \Omega\}.$$

Theorem 8.6-1 (existence of an immersion on an open subset of $\mathbb{R}^n$ with a prescribed metric tensor) *Let* Ω *be a simply connected open set in* $\mathbb{R}^n$ *and let* $\boldsymbol{C} = (g_{ij}) \in \mathcal{C}^2(\Omega; \mathbb{S}^n_>)$ *be a matrix field that satisfies*

$$R_{qijk} := \partial_j \Gamma_{ikq} - \partial_k \Gamma_{ijq} + \Gamma^p_{ij} \Gamma_{kqp} - \Gamma^p_{ik} \Gamma_{jqp} = 0 \quad \text{in } \Omega,$$

where

$$\Gamma_{ijq} := \frac{1}{2}(\partial_j g_{iq} + \partial_i g_{jq} - \partial_q g_{ij}) \quad \text{and} \quad \Gamma^p_{ij} := g^{pq}\Gamma_{ijq} \quad \text{with } (g^{pq}) := (g_{ij})^{-1}.$$

Then there exists an immersion $\boldsymbol{\Theta} \in \mathcal{C}^3(\Omega; \mathbb{E}^n)$ *such that*

$$\boldsymbol{\nabla\Theta}^T \boldsymbol{\nabla\Theta} = \boldsymbol{C} \quad \text{in } \Omega.$$

Proof The proof relies on a simple, yet crucial, observation. When a smooth enough immersion $\boldsymbol{\Theta} = (\Theta_\ell) : \Omega \to \mathbb{E}^n$ is *a priori* given (as it was so far), its components Θ_ℓ, $1 \le \ell \le n$, satisfy the relations $\partial_{ij}\Theta_\ell = \Gamma^p_{ij}\partial_p \Theta_\ell$, which are nothing but another way of writing the relations $\partial_i \boldsymbol{g}_j = \Gamma^p_{ij} \boldsymbol{g}_p$ that were found in the proof of Theorem 8.5-1. This observation thus suggests to begin by solving (see part (ii)) the Pfaff system of partial differential equations (Section 6.20)

$$\partial_i F_{\ell j} = \Gamma^p_{ij} F_{\ell p} \quad \text{in } \Omega,$$

whose solutions $F_{\ell j} : \Omega \to \mathbb{R}$ then constitute natural candidates for the partial derivatives $\partial_j \Theta_\ell$ of the unknown immersion $\boldsymbol{\Theta} = (\Theta_\ell) : \Omega \to \mathbb{E}^n$ (see part (iii)).

To begin with, we establish in (i) relations that will in turn allow us to rewrite the assumed relations

$$\partial_j \Gamma_{ikq} - \partial_k \Gamma_{ijq} + \Gamma^p_{ij} \Gamma_{kqp} - \Gamma^p_{ik} \Gamma_{jqp} = 0 \quad \text{in } \Omega$$

in the *equivalent* form

$$\partial_i \Gamma^p_{kj} - \partial_k \Gamma^p_{ij} + \Gamma^q_{kj} \Gamma^p_{iq} - \Gamma^q_{ij} \Gamma^p_{kq} = 0 \quad \text{in } \Omega,$$

[9] J. NASH [1954]: $\mathcal{C}^1$ isometric imbeddings, *Annals of Mathematics* **60**, 383–396.
John Forbes Nash was awarded the Nobel Prize in Economic Sciences in 1994.

[10] The first proofs of a local version of this theorem are due to:
M. JANET [1926]: Sur la possibilité de plonger un espace riemannien donné dans un espace euclidien, *Annales de la Société Polonaise de Mathématiques* **5**, 38–43.
E. CARTAN [1927]: Sur la possibilité de plonger un espace riemannien donné dans un espace euclidien, *Annales de la Société Polonaise de Mathématiques* **6**, 1–7.

which is more appropriate for the existence result of part (ii). Note that the positive-definiteness of the symmetric matrices (g_{ij}) is not needed for this purpose; only their invertibility is used in (i).

(i) *Let Ω be an open subset of $\mathbb{R}^n$ and let there be given a field $(g_{ij}) \in \mathcal{C}^2(\Omega; \mathbb{S}^n)$ of symmetric invertible matrices. The functions $\Gamma_{ijq}, \Gamma^p_{ij}$, and g^{pq} being defined by*

$$\Gamma_{ijq} := \frac{1}{2}(\partial_j g_{iq} + \partial_i g_{jq} - \partial_q g_{ij}), \quad \Gamma^p_{ij} := g^{pq}\Gamma_{ijq}, \quad (g^{pq}) := (g_{ij})^{-1},$$

define the functions

$$R_{qijk} := \partial_j \Gamma_{ikq} - \partial_k \Gamma_{ijq} + \Gamma^p_{ij}\Gamma_{kqp} - \Gamma^p_{ik}\Gamma_{jqp},$$
$$R^p_{\cdot ijk} := \partial_j \Gamma^p_{ik} - \partial_k \Gamma^p_{ij} + \Gamma^\ell_{ik}\Gamma^p_{j\ell} - \Gamma^\ell_{ij}\Gamma^p_{k\ell}.$$

Then

$$R^p_{\cdot ijk} = g^{pq}R_{qijk} \quad \text{and} \quad R_{qijk} = g_{pq}R^p_{\cdot ijk}.$$

Using the relations

$$\Gamma_{jq\ell} + \Gamma_{\ell jq} = \partial_j g_{q\ell} \quad \text{and} \quad \Gamma_{ikq} = g_{q\ell}\Gamma^\ell_{ik},$$

which themselves follow from the definitions of the functions Γ_{ijq} and Γ^p_{ij}, and noting that

$$(g^{pq}\partial_j g_{q\ell} + g_{q\ell}\partial_j g^{pq}) = \partial_j(g^{pq}g_{q\ell}) = \partial_j(\delta^p_q) = 0,$$

we obtain

$$\begin{aligned} g^{pq}(\partial_j \Gamma_{ikq} - \Gamma^r_{ik}\Gamma_{jqr}) &= \partial_j \Gamma^p_{ik} - \Gamma_{ikq}\partial_j g^{pq} - \Gamma^\ell_{ik}g^{pq}(\partial_j g_{q\ell} - \Gamma_{\ell jq}) \\ &= \partial_j \Gamma^p_{ik} + \Gamma^\ell_{ik}\Gamma^p_{j\ell} - \Gamma^\ell_{ik}(g^{pq}\partial_j g_{q\ell} + g_{q\ell}\partial_j g^{pq}) \\ &= \partial_j \Gamma^p_{ik} + \Gamma^\ell_{ik}\Gamma^p_{j\ell}. \end{aligned}$$

Likewise,

$$g^{pq}(\partial_k \Gamma_{ijq} - \Gamma^r_{ij}\Gamma_{kqr}) = \partial_k \Gamma^p_{ij} + \Gamma^\ell_{ij}\Gamma^p_{k\ell}.$$

Hence the relations $R^p_{\cdot ijk} = g^{pq}R_{qijk}$ hold, and so do the relations $R_{qijk} = g_{pq}R^p_{\cdot ijk}$ (which are clearly equivalent).

(ii) *Let Ω be a simply connected open subset of $\mathbb{R}^n$ and let there be given functions $\Gamma^p_{ij} = \Gamma^p_{ji} \in \mathcal{C}^1(\Omega)$ that satisfy the relations*

$$\partial_j \Gamma_{ikq} - \partial_k \Gamma_{ijq} + \Gamma^p_{ij}\Gamma_{kqp} - \Gamma^p_{ik}\Gamma_{jqp} = 0 \quad \text{in } \Omega,$$

which (in view of part (i) and of the symmetry relations $\Gamma^p_{ij} = \Gamma^p_{ji}$) *are equivalent to the relations*

$$\partial_i \Gamma^p_{kj} - \partial_k \Gamma^p_{ij} + \Gamma^q_{kj}\Gamma^p_{iq} - \Gamma^q_{ij}\Gamma^p_{kq} = 0 \quad \text{in } \Omega.$$

Let a point $x^0 \in \Omega$ and a matrix $(F^0_{\ell j}) \in \mathbb{M}^n$ be given. Then there exists one, and only one, matrix field $(F_{\ell j}) \in \mathcal{C}^2(\Omega; \mathbb{M}^n)$ that satisfies the Pfaff system

$$\partial_i F_{\ell j}(x) = \Gamma^p_{ij}(x)F_{\ell p}(x), \quad x \in \Omega, \quad \text{and} \quad F_{\ell j}(x^0) = F^0_{\ell j}.$$

The existence and uniqueness of the field $(F_{\ell j}) \in \mathcal{C}^2(\Omega; \mathbb{M}^n)$ follow from the existence and uniqueness theorem for *Pfaff systems* (Theorem 6.20-1), which can be applied since the matrix fields $(\Gamma^p_{ij}) \in \mathcal{C}^1(\Omega; \mathbb{M}^n)$ (the row index is p and the column index is j), $1 \le i \le n$, satisfy the *compatibility conditions* $R^p_{\cdot ijk} = 0$ in the open set Ω, and Ω is *simply connected* by assumption.

(iii) *Let* Ω *be a simply connected open subset of* $\mathbb{R}^n$ *and let* $(g_{ij}) \in \mathcal{C}^2(\Omega; \mathbb{S}^n_>)$ *be a matrix field that satisfies*

$$\partial_j \Gamma_{ikq} - \partial_k \Gamma_{ijq} + \Gamma^p_{ij}\Gamma_{kqp} - \Gamma^p_{ik}\Gamma_{jqp} = 0 \quad \text{in } \Omega,$$

the functions Γ_{ijq}, Γ^p_{ij}, *and* g^{pq} *being defined by*

$$\Gamma_{ijq} := \frac{1}{2}(\partial_j g_{iq} + \partial_i g_{jq} - \partial_q g_{ij}), \quad \Gamma^p_{ij} := g^{pq}\Gamma_{ijq}, \quad (g^{pq}) := (g_{ij})^{-1}.$$

Given an arbitrary point $x^0 \in \Omega$, *let* $(F^0_{ij}) \in \mathbb{M}^n$ *be any invertible matrix that satisfies*

$$F^0_{ki}F^0_{kj} = g^0_{ij}, \quad \text{where } (g^0_{ij}) := (g_{ij}(x^0))$$

(for instance, $(F^0_{ij}) := (g^0_{ij})^{1/2}$), *let* $\boldsymbol{\Theta}^0 \in \mathbb{E}^n$ *be a given vector, and let* $(F_{\ell j}) \in \mathcal{C}^2(\Omega; \mathbb{M}^n)$ *denote the solution to the Pfaff system*

$$\partial_i F_{\ell j}(x) = \Gamma^p_{ij}(x) F_{\ell p}(x), \ x \in \Omega, \quad \text{and} \quad F_{\ell j}(x^0) = F^0_{\ell j},$$

which exists and is unique by parts (i) *and* (ii). *Then there exists one, and only one, immersion* $\boldsymbol{\Theta} = (\Theta_\ell) \in \mathcal{C}^3(\Omega; \mathbb{E}^n)$ *such that*

$$\partial_j \Theta_\ell = F_{\ell j} \quad \text{and} \quad \partial_i \boldsymbol{\Theta} \cdot \partial_j \boldsymbol{\Theta} = g_{ij} \ \text{in } \Omega, \quad \text{and} \quad \boldsymbol{\Theta}(x^0) = \boldsymbol{\Theta}^0.$$

To begin with, we show that *the* n *vector fields defined by*

$$\boldsymbol{g}_j := (F_{\ell j})_{\ell=1}^n \in \mathcal{C}^2(\Omega; \mathbb{R}^n)$$

satisfy

$$\boldsymbol{g}_i \cdot \boldsymbol{g}_j = g_{ij} \quad \text{in } \Omega.$$

To this end, we note that, by construction, these fields satisfy

$$\partial_i \boldsymbol{g}_j = \Gamma^p_{ij} \boldsymbol{g}_p \ \text{ in } \Omega \quad \text{and} \quad \boldsymbol{g}_j(x^0) = \boldsymbol{g}^0_j,$$

where $\boldsymbol{g}^0_j$ is the jth column vector of the matrix $(F^0_{\ell j}) \in \mathbb{M}^n$. Hence the matrix field $(\boldsymbol{g}_i \cdot \boldsymbol{g}_j) \in \mathcal{C}^2(\Omega; \mathbb{M}^n)$ satisfies

$$\partial_k(\boldsymbol{g}_i \cdot \boldsymbol{g}_j) = \Gamma^m_{kj}(\boldsymbol{g}_m \cdot \boldsymbol{g}_i) + \Gamma^m_{ki}(\boldsymbol{g}_m \cdot \boldsymbol{g}_j) \ \text{ in } \Omega, \quad \text{and} \quad (\boldsymbol{g}_i \cdot \boldsymbol{g}_j)(x^0) = g^0_{ij}.$$

The definitions of the functions Γ_{ijq} and Γ^p_{ij} imply that

$$\partial_k g_{ij} = \Gamma_{ikj} + \Gamma_{jki} \quad \text{and} \quad \Gamma_{ijq} = g_{pq}\Gamma^p_{ij}.$$

Hence the matrix field $(g_{ij}) \in \mathcal{C}^2(\Omega; \mathbb{S}^n_>)$ satisfies

$$\partial_k g_{ij} = \Gamma^m_{kj} g_{mi} + \Gamma^m_{ki} g_{mj} \ \text{ in } \Omega, \quad \text{and} \quad g_{ij}(x^0) = g^0_{ij}.$$

Viewed as a system of partial differential equations, together with given values at x^0, with respect to the matrix field $(g_{ij}) : \Omega \to \mathbb{M}^n$, the above system can have *at most one solution* in the space $\mathcal{C}^2(\Omega; \mathbb{M}^n)$.

To see this, let $x^1 \in \Omega$ be distinct from x^0 and let $\gamma \in \mathcal{C}^1([0,1]; \mathbb{R}^n)$ be any path joining x^0 to x^1 in Ω. Then the n^2 functions $g_{ij}(\gamma(t))$, $0 \le t \le 1$, satisfy a Cauchy problem for a linear system of n^2 ordinary differential equations, which as such has *at most one* solution (Theorem 3.8-2).

An inspection of the two above systems therefore shows that their solutions are identical, i.e., that $\boldsymbol{g}_i \cdot \boldsymbol{g}_j = g_{ij}$ in Ω.

It remains to show that *there exists one, and only one, immersion* $\boldsymbol{\Theta} \in \mathcal{C}^3(\Omega; \mathbb{E}^n)$ *such that*

$$\partial_i \boldsymbol{\Theta} = \boldsymbol{g}_i \text{ in } \Omega \quad \text{and} \quad \boldsymbol{\Theta}(x^0) = \boldsymbol{\Theta}^0,$$

where $\boldsymbol{g}_i := (F_{\ell i})_{\ell=1}^n$.

Since the functions Γ_{ij}^p satisfy $\Gamma_{ij}^p = \Gamma_{ji}^p$, any solution $(F_{\ell j}) \in \mathcal{C}^2(\Omega; \mathbb{M}^n)$ of the system

$$\partial_i F_{\ell j}(x) = \Gamma_{ij}^p(x) F_{\ell p}(x), \quad x \in \Omega, \quad \text{and} \quad F_{\ell j}(x^0) = F_{\ell j}^0,$$

satisfies

$$\partial_i F_{\ell j} = \partial_j F_{\ell i} \quad \text{in } \Omega.$$

The open set Ω being *simply connected*, the *classical Poincaré lemma* (Theorem 6.17-2) shows that, *for each integer* $\ell \in \{1, \ldots, n\}$, there exists a function $\Theta_\ell \in \mathcal{C}^3(\Omega)$, unique up to the addition of a constant, such that

$$\partial_i \Theta_\ell = F_{\ell i} \quad \text{in } \Omega,$$

or equivalently, there exists one, and only one, mapping $\boldsymbol{\Theta} := (\Theta_\ell) \in \mathcal{C}^3(\Omega; \mathbb{E}^n)$ that satisfies

$$\partial_i \boldsymbol{\Theta} = \boldsymbol{g}_i \text{ in } \Omega \quad \text{and} \quad \boldsymbol{\Theta}(x^0) = \boldsymbol{\Theta}^0.$$

That $\boldsymbol{\Theta}$ is an immersion follows from the assumed invertibility of the matrices (g_{ij}). The proof is thus complete. $\square$

Incidentally, it is remarkable that the solution $\boldsymbol{\Theta}$ of the *nonlinear* equation $\boldsymbol{\nabla\Theta}^T \boldsymbol{\nabla\Theta} = \boldsymbol{C}$ in Ω is obtained by successively solving a *linear* Pfaff system in Ω (part (ii) of the above proof) and *linear* equations (viz., $\partial_i \boldsymbol{\Theta} = \boldsymbol{g}_i$ in Ω; cf. part (iii)).

Since the solution $(F_{\ell j})$ of the Pfaff system found in part (ii) is unique, and since each function $\boldsymbol{\Theta}_\ell$ found in part (iii) is likewise uniquely determined, Theorem 8.6-1 can be conveniently rephrased as the following *existence and uniqueness result.*

Theorem 8.6-2 *Let Ω be a simply connected open set in $\mathbb{R}^n$ and let $\boldsymbol{C} = (g_{ij}) \in \mathcal{C}^2(\Omega; \mathbb{S}_>^n)$ be a matrix field that satisfies*

$$R_{qijk} := \partial_j \Gamma_{ikq} - \partial_k \Gamma_{ijq} + \Gamma_{ij}^p \Gamma_{kqp} - \Gamma_{ik}^p \Gamma_{jqp} = 0 \quad \textit{in } \Omega,$$

where

$$\Gamma_{ijq} := \frac{1}{2}(\partial_j g_{iq} + \partial_i g_{jq} - \partial_q g_{ij}) \quad \textit{and} \quad \Gamma_{ij}^p := g^{pq} \Gamma_{ijq} \quad \textit{with } (g^{pq}) := (g_{ij})^{-1}.$$

Finally, let there be given a point $x^0 \in \Omega$*, a vector* $\boldsymbol{\Theta}^0 \in \mathbb{E}^n$*, and a matrix* $\boldsymbol{F}^0 \in \mathbb{M}^n$ *such that* $(\boldsymbol{F}^0)^T \boldsymbol{F}^0 = \boldsymbol{C}(x^0)$.

Then there exists one, and only one, immersion $\boldsymbol{\Theta} \in \mathcal{C}^3(\Omega; \mathbb{E}^n)$ *such that*

$$\boldsymbol{\nabla}\boldsymbol{\Theta}^T \boldsymbol{\nabla}\boldsymbol{\Theta} = \boldsymbol{C} \quad \textit{in } \Omega,$$
$$\boldsymbol{\Theta}(x^0) = \boldsymbol{\Theta}^0 \quad \textit{and} \quad \boldsymbol{\nabla}\boldsymbol{\Theta}(x^0) = \boldsymbol{F}^0. \qquad \square$$

Otherwise, the uniqueness issue *in general*, i.e., when no conditions such as $\boldsymbol{\Theta}(x^0) = \boldsymbol{\Theta}^0$ and $\boldsymbol{\nabla}\boldsymbol{\Theta}(x^0) = \boldsymbol{F}^0$ are imposed as in Theorem 8.6-2, is addressed in the next section, in effect under *weaker regularity assumptions* than in Theorem 8.6-2 and *without the assumption of simple-connectedness* of Ω.

Let Ω be a simply connected open subset of $\mathbb{R}^n$ and let a point $x_0 \in \Omega$, a vector $\boldsymbol{\Theta}_0 \in \mathbb{E}^n$, and an $n \times n$ invertible matrix $\boldsymbol{F}_0$ be given. Theorem 8.6-2 thus establishes the existence of a (clearly nonlinear) mapping $\boldsymbol{\mathcal{F}}$ that associates with any matrix field $\boldsymbol{C} = (g_{ij}) \in \mathcal{C}^2(\Omega; \mathbb{S}^n_>)$ satisfying

$$\boldsymbol{C}(x_0) = \boldsymbol{F}_0^T \boldsymbol{F}_0 \quad \text{and} \quad R_{qijk} := \partial_j \Gamma_{ikq} - \partial_k \Gamma_{ijq} + \Gamma^p_{ij} \Gamma_{kqp} - \Gamma^p_{ik} \Gamma_{jqp} = 0 \ \text{ in } \Omega$$

(where the functions Γ_{ijq} and Γ^p_{ij} are defined in terms of the functions g_{ij} as in Theorem 8.6-1) a well-defined immersion $\boldsymbol{\Theta} = \boldsymbol{\mathcal{F}}(\boldsymbol{C}) \in \mathcal{C}^3(\Omega; \mathbb{E}^n)$ that satisfies

$$\boldsymbol{\nabla}\boldsymbol{\Theta}^T \boldsymbol{\nabla}\boldsymbol{\Theta} = \boldsymbol{C} \ \text{in } \Omega \quad \text{and} \quad \boldsymbol{\Theta}(x_0) = \boldsymbol{\Theta}_0 \quad \text{and} \quad \boldsymbol{\nabla}\boldsymbol{\Theta}(x_0) = \boldsymbol{F}_0.$$

Then there exist natural topologies on the spaces $\mathcal{C}^2(\Omega; \mathbb{S}^3)$ and $\mathcal{C}^3(\Omega; \mathbb{E}^n)$ such that the mapping $\boldsymbol{\mathcal{F}}$ defined in this fashion is *continuous.* In other words, *an immersion is a continuous function of its metric tensor*, between such spaces of continuously differentiable functions; cf. Problem 8.6-3.

Remark A similar conclusion holds, but this time in terms of *Sobolev norms*, as a consequence of a *nonlinear Korn inequality*[11] (so called because it generalizes to the nonlinear case the linear Korn inequality established in Theorem 6.15-3). $\square$

While the approach in the proof of Theorem 8.6-1 consists in first determining a matrix field $\boldsymbol{F} : \Omega \to \mathbb{M}^n$, then in determining an immersion $\boldsymbol{\Theta} : \Omega \to \mathbb{E}^n$ such that $\boldsymbol{\nabla}\boldsymbol{\Theta} = \boldsymbol{F}$ in Ω, *another* approach consists in first determining an *orthogonal matrix field* $\boldsymbol{R} : \Omega \to \mathbb{O}^n$, then in determining an immersion $\boldsymbol{\Theta} : \Omega \to \mathbb{E}^n$ such that $\boldsymbol{\nabla}\boldsymbol{\Theta} = \boldsymbol{R}\boldsymbol{C}^{1/2}$ in Ω; in this case, the compatibility conditions are expressed in terms of the *matrix field* $\boldsymbol{C}^{1/2} : \Omega \to \mathbb{S}^n$; cf. Problems 8.6-1 and 8.6-2.

Remarks (1) The assumptions

$$\partial_j \Gamma^p_{ik} - \partial_k \Gamma^p_{ij} + \Gamma^\ell_{ik} \Gamma^p_{j\ell} - \Gamma^\ell_{ij} \Gamma^p_{k\ell} = 0 \quad \text{in } \Omega,$$

made in part (ii) on the functions $\Gamma^p_{ij} = \Gamma^p_{ji}$, are thus *sufficient* conditions for the equations $\partial_i F_{\ell j} = \Gamma^p_{ij} F_{\ell p}$ in Ω to have solutions. Conversely, a simple computation, in effect quite similar to that carried

[11] P.G. CIARLET; C. MARDARE [2004]: Continuity of a deformation in H^1 as a function of its Cauchy-Green tensor in L^1, *Journal of Nonlinear Science* **14**, 415–427.

out in the proof of Theorem 8.5-1, shows that they are also *necessary* conditions, simply expressing that, if these equations have a solution invertible everywhere in Ω, then necessarily $\partial_{ik}F_{\ell j} = \partial_{ki}F_{\ell j}$ in Ω. It is no surprise that these necessary conditions are of the same nature as those of Theorem 8.5-1, in that they again hinge on the *Schwarz lemma*.

(2) The assumed *positive-definiteness* of the matrices (g_{ij}) is used only in part (iii), for defining *ad hoc* initial vectors $\boldsymbol{g}_i^0$.

(3) The definitions of the functions Γ_{ij}^p and Γ_{ijq} imply that the functions

$$R_{qijk} := \partial_j \Gamma_{ikq} - \partial_k \Gamma_{ijq} + \Gamma_{ij}^p \Gamma_{kqp} - \Gamma_{ik}^p \Gamma_{jqp}$$

satisfy, for all i, j, k, p,

$$R_{qijk} = R_{jkqi} = -R_{qikj}, \quad \text{and} \quad R_{qijk} = 0 \quad \text{if } j = k \text{ or } q = i.$$

These relations in turn imply that, *when* $n = 3$, *the* 81 *sufficient conditions*

$$R_{qijk} = 0 \text{ in } \Omega \quad \text{for all } 1 \leq i, j, k, q \leq 3,$$

are satisfied if and only if the six relations

$$R_{1212} = R_{1213} = R_{1223} = R_{1313} = R_{1323} = R_{2323} = 0 \quad \text{in } \Omega$$

are satisfied (it is easily seen that there are other sets of six relations that will suffice as well when $n = 3$). □

The existence result of Theorem 8.6-1 also holds[12] "*up to the boundary of the set* Ω" in the following sense: Assume that the set Ω has a Lipschitz-continuous boundary (Section 1.18) and that the functions g_{ij} and their partial derivatives of order ≤ 2 can be extended by continuity to the closure $\overline{\Omega}$, the symmetric matrix field extended in this fashion remaining positive-definite over the set $\overline{\Omega}$. Then the immersion $\boldsymbol{\Theta}$ and its partial derivatives of order ≤ 3 can be also extended by continuity to $\overline{\Omega}$.

The *regularity assumptions* on the components g_{ij} of the symmetric positive-definite matrix field $\boldsymbol{C} = (g_{ij})$ made in Theorem 8.6-1 (viz., that $g_{ij} \in \mathcal{C}^2(\Omega)$) can be significantly *weakened.*

More specifically, the existence theorem still holds[13] if $g_{ij} \in \mathcal{C}^1(\Omega)$, with a resulting mapping $\boldsymbol{\Theta}$ in the space $\mathcal{C}^2(\Omega; \mathbb{E}^n)$; it also holds[14] if Ω is a domain in $\mathbb{R}^n$ and $g_{ij} \in W^{1,p}(\Omega)$ for some $p > n$, with a resulting mapping $\boldsymbol{\Theta}$ in the space $W^{2,p}(\Omega; \mathbb{E}^n)$. As expected, the sufficient conditions $R_{qijk} = 0$ in Ω of Theorem 8.6-1 are then assumed to hold only *in the sense of distributions*, viz., as

$$\int_\Omega \{-\Gamma_{ikq}\partial_j \varphi + \Gamma_{ijq}\partial_k \varphi + \Gamma_{ij}^p \Gamma_{kqp}\varphi - \Gamma_{ik}^p \Gamma_{jqp}\varphi\} \, dx = 0 \quad \text{for all } \varphi \in \mathcal{D}(\Omega).$$

---

[12]P.G. Ciarlet; C. Mardare [2004]: Recovery of a manifold with boundary and its continuity as a function of its metric tensor, *Journal de Mathématiques Pures et Appliquées* **83**, 811–843.

[13]C. Mardare [2003]: On the recovery of a manifold with prescribed metric tensor, *Analysis and Applications* **1**, 433–453.

[14]S. Mardare [2007]: On systems of first order linear partial differential equations with L^p-coefficients, *Advances in Differential Equations* **12**, 301–360.

Problems

8.6-1 The objective of this problem is to show that the necessary conditions of Problem 8.5-2 become also *sufficient* for the *existence*[15] of an immersion $\boldsymbol{\Theta} \in \mathcal{C}^3(\Omega; \mathbb{E}^n)$ if the open set $\Omega \subset \mathbb{R}^n$ is *simply connected*, an assumption that accordingly holds throughout this problem.

(1) Let there be given antisymmetric matrix fields $\boldsymbol{A}_j \in \mathcal{C}^1(\Omega; \mathbb{A}^n)$, $1 \le j \le n$, that satisfy

$$\partial_i \boldsymbol{A}_j - \partial_j \boldsymbol{A}_i + \boldsymbol{A}_i \boldsymbol{A}_j - \boldsymbol{A}_j \boldsymbol{A}_i = \boldsymbol{0} \quad \text{in } \Omega,$$

a point $x^0 \in \Omega$, and an orthogonal matrix $\boldsymbol{R}^0 \in \mathbb{O}^n$. Then the *Pfaff system*

$$\partial_j \boldsymbol{R}(x) = \boldsymbol{R}(x)\boldsymbol{A}_j(x), \quad x \in \Omega, \quad \text{and} \quad \boldsymbol{R}(x^0) = \boldsymbol{R}^0,$$

has a unique solution $\boldsymbol{R} \in \mathcal{C}^2(\Omega; \mathbb{M}^n)$ (Theorem 6.20-1). Show that $\boldsymbol{R}(x) \in \mathbb{O}^n$ at each $x \in \Omega$.

(2) In the remainder of this problem, a matrix field $\boldsymbol{U} \in \mathcal{C}^2(\Omega; \mathbb{S}^n_>)$ is given that satisfies the compatibility conditions

$$\partial_i \boldsymbol{A}_j - \partial_j \boldsymbol{A}_i + \boldsymbol{A}_i \boldsymbol{A}_j - \boldsymbol{A}_j \boldsymbol{A}_i = \boldsymbol{0} \quad \text{in } \Omega,$$

where the matrix fields $\boldsymbol{A}_j \in \mathcal{C}^1(\Omega; \mathbb{M}^n)$, $1 \le j \le n$, are defined in terms of the matrix field $\boldsymbol{U}$ by

$$\boldsymbol{A}_j := \frac{1}{2}\{\boldsymbol{U}^{-1}(\boldsymbol{\nabla} \boldsymbol{c}_j - (\boldsymbol{\nabla} \boldsymbol{c}_j)^T)\boldsymbol{U}^{-1} + \boldsymbol{U}^{-1}\partial_j \boldsymbol{U} - (\partial_j \boldsymbol{U})\boldsymbol{U}^{-1}\},$$

where $\boldsymbol{c}_j \in \mathcal{C}^2(\Omega; \mathbb{R}^n)$ denotes the jth column vector field of the matrix field $\boldsymbol{U}^2 \in \mathcal{C}^2(\Omega; \mathbb{S}^n_>)$.

Show that $\boldsymbol{A}_j(x) \in \mathbb{A}^n$ at each $x \in \Omega$ and that each matrix field $\boldsymbol{A}_j$ may be also written as

$$\boldsymbol{A}_j = \boldsymbol{U}\boldsymbol{\Gamma}_j \boldsymbol{U}^{-1} - (\partial_j \boldsymbol{U})\boldsymbol{U}^{-1}, \quad \text{where } \boldsymbol{\Gamma}_j := \frac{1}{2}\boldsymbol{U}^{-2}(\partial_j(\boldsymbol{U}^2) + \boldsymbol{\nabla} \boldsymbol{c}_j - (\boldsymbol{\nabla} \boldsymbol{c}_j)^T) \in \mathcal{C}^1(\Omega; \mathbb{M}^n).$$

(3) Let there be given a vector $\boldsymbol{\Theta}^0 \in \mathbb{E}^n$. The matrix field $\boldsymbol{R} \in \mathcal{C}^1(\Omega; \mathbb{O}^n)$ being determined as in (1), show that there exists one, and only one, vector field $\boldsymbol{\Theta} \in \mathcal{C}^3(\Omega; \mathbb{E}^n)$ that satisfies

$$\boldsymbol{\nabla}\boldsymbol{\Theta}(x) = \boldsymbol{R}(x)\boldsymbol{U}(x), \quad x \in \Omega, \quad \text{and} \quad \boldsymbol{\Theta}(x^0) = \boldsymbol{\Theta}^0.$$

Hint: Note that solving $\boldsymbol{\nabla}\boldsymbol{\Theta} = \boldsymbol{R}\boldsymbol{U}$ in Ω is the same as solving the equations $\partial_j \boldsymbol{\Theta} = \boldsymbol{R}\boldsymbol{u}_j$ in Ω, $1 \le j \le n$, where $\boldsymbol{u}_j \in \mathcal{C}^2(\Omega; \mathbb{R}^n)$ denotes the jth column vector field of the matrix field $\boldsymbol{U}$. Then show that these equations can be solved by means of the *classical Poincaré lemma* (Theorem 6.17-2).

8.6-2 Let Ω be an open subset of $\mathbb{R}^n$. Show that a matrix field $\boldsymbol{C} = (g_{ij}) \in \mathcal{C}^2(\Omega; \mathbb{S}^2_>)$ satisfies the relations $R_{qijk} = 0$ in Ω of Theorem 8.6-1 if and only if the matrix field $\boldsymbol{U} := \boldsymbol{C}^{1/2} \in \mathcal{C}^2(\Omega; \mathbb{S}^2_>)$ satisfies the compatibility conditions

$$\partial_i \boldsymbol{A}_j - \partial_j \boldsymbol{A}_i + \boldsymbol{A}_i \boldsymbol{A}_j - \boldsymbol{A}_j \boldsymbol{A}_i = \boldsymbol{0} \quad \text{in } \Omega.$$

where the antisymmetric matrix fields $\boldsymbol{A}_j \in \mathcal{C}^1(\Omega; \mathbb{A}^n)$, $1 \le j \le n$, are defined in terms of the matrix field $\boldsymbol{U}$ by

$$\boldsymbol{A}_j := \frac{1}{2}\{\boldsymbol{U}^{-1}(\boldsymbol{\nabla} \boldsymbol{c}_j - (\boldsymbol{\nabla} \boldsymbol{c}_j)^T)\boldsymbol{U}^{-1} + \boldsymbol{U}^{-1}\partial_j \boldsymbol{U} - (\partial_j \boldsymbol{U})\boldsymbol{U}^{-1}\},$$

where $\boldsymbol{c}_j \in \mathcal{C}^2(\Omega; \mathbb{R}^n)$ denotes the jth column vector field of the matrix field $\boldsymbol{U}^2 \in \mathcal{C}^2(\Omega; \mathbb{S}^n_>)$.

Hint: Use Theorems 8.5-1 and 8.6-1 and Problems 8.5-2 and 8.6-1.

---

[15]This existence result is due to:

P.G. CIARLET; L. GRATIE; O. IOSIFESCU; C. MARDARE; C. VALLÉE [2007]: Another approach to the fundamental theorem of Riemannian geometry in $\mathbb{R}^3$, by way of rotation fields, *Journal de Mathématiques Pures et Appliquées* **87**, 237–252.

A detailed analysis of the special case $n = 3$ is also carried out in this paper.

8.6-3 Given an open subset Ω of $\mathbb{R}^n$, the notation $K \Subset \Omega$ means that K is a compact subset of Ω. Given any integer $m \geq 0$ and any $K \Subset \Omega$, define the *seminorms* $|\cdot|_{m,K}$ and $\|\cdot\|_{m,K}$ by

$$|f|_{m,K} = \sup_{\begin{cases} x \in K \\ |\boldsymbol{\alpha}| = m \end{cases}} |\partial^{\boldsymbol{\alpha}} f(x)| \quad \text{and} \quad \|f\|_{m,K} = \sup_{\begin{cases} x \in K \\ |\boldsymbol{\alpha}| \leq m \end{cases}} |\partial^{\boldsymbol{\alpha}} f(x)| \quad \text{for each } f \in \mathcal{C}^m(\Omega),$$

and define analogous seminorms for vector-valued and matrix-valued functions, $|\cdot|$ now designating either the Euclidean vector norm or its subordinate matrix norm.

In what follows, Ω is a simply connected open subset of $\mathbb{R}^n$, and a point $x_0 \in \Omega$ is given.

(1) Let $\boldsymbol{C}^\ell = (g_{ij}^\ell) \in \mathcal{C}^2(\Omega; \mathbb{S}^n_>)$, $\ell \geq 0$, be matrix fields satisfying $R^\ell_{qijk} = 0$ in Ω, $\ell \geq 0$, with the property that

$$\lim_{\ell \to \infty} \|\boldsymbol{C}^\ell - \boldsymbol{I}\|_{2,K} = 0 \quad \text{for each } K \Subset \Omega.$$

By Theorem 8.6-2, there thus exist uniquely determined immersions $\boldsymbol{\Theta}^\ell \in \mathcal{C}^3(\Omega; \mathbb{E}^3)$ that satisfy

$$(\boldsymbol{\nabla}\boldsymbol{\Theta}^\ell)^T \boldsymbol{\nabla}\boldsymbol{\Theta}^\ell = \boldsymbol{C}^\ell \quad \text{in } \Omega, \ \ell \geq 0,$$

and

$$\boldsymbol{\Theta}^\ell(x_0) = x_0 \quad \text{and} \quad \boldsymbol{\nabla}\boldsymbol{\Theta}^\ell(x_0) = \boldsymbol{I}, \ \ell \geq 0.$$

Show that, for each $K \Subset \Omega$,

$$\lim_{\ell \to \infty} |\boldsymbol{\Theta}^\ell - \mathbf{id}|_{m,K} = \lim_{\ell \to \infty} |\boldsymbol{\Theta}^\ell|_{m,K} = 0 \quad \text{for } m = 2, \text{ then for } m = 3.$$

Hint: Show that (the notations should be self-explanatory) the seminorms $|(\boldsymbol{g}^q)^\ell|_{0,K}$ are bounded independently of $\ell \geq 0$, and use the relations

$$\partial_{ij}\boldsymbol{\Theta}^\ell = \frac{1}{2}(\partial_j g^\ell_{iq} + \partial_i g^\ell_{jq} - \partial_q g^\ell_{ij})(\boldsymbol{g}^q)^\ell, \quad \ell \geq 0.$$

(2) Show that, for each $K \Subset \Omega$,

$$\lim_{\ell \to \infty} |\boldsymbol{\Theta}^\ell - \mathbf{id}|_{m,K} = 0 \quad \text{for } m = 1, \text{ then for } m = 0.$$

Hint: Use the differentiability of the limit of a sequence of continuously differentiable mappings (Theorem 7.3-1).

(3) Let a vector $\boldsymbol{\Theta}^0 \in \mathbb{E}^n$ and an $n \times n$ invertible matrix $\boldsymbol{F}_0$ be given, and let $\boldsymbol{C}^\ell = (g^\ell_{ij}) \in \mathcal{C}^2(\Omega; \mathbb{S}^n_>)$, $\ell \geq 0$, *resp.* $\boldsymbol{C} = (g_{ij}) \in \mathcal{C}^2(\Omega; \mathbb{S}^n_>)$, be matrix fields satisfying $R^\ell_{qijk} = 0$ in Ω, $\ell > 0$, resp. $R_{qijk} = 0$ in Ω, with the property that

$$\lim_{\ell \to \infty} \|\boldsymbol{C}^\ell - \boldsymbol{C}\|_{2,K} = 0 \quad \text{for each } K \Subset \Omega.$$

By Theorem 8.6-2, there thus exist uniquely determined immersions $\boldsymbol{\Theta}^\ell \in \mathcal{C}^3(\Omega; \mathbb{E}^n)$, $\ell \geq 0$, *resp.* $\boldsymbol{\Theta} \in \mathcal{C}^3(\Omega; \mathbb{E}^n)$, that satisfy

$$(\boldsymbol{\nabla}\boldsymbol{\Theta}^\ell)^T \boldsymbol{\nabla}\boldsymbol{\Theta}^\ell = \boldsymbol{C}^\ell \text{ in } \Omega, \ell \geq 0, \quad \text{resp.} \quad \boldsymbol{\nabla}\boldsymbol{\Theta}^T \boldsymbol{\nabla}\boldsymbol{\Theta} = \boldsymbol{C} \text{ in } \Omega,$$

$$\boldsymbol{\Theta}^\ell(x_0) = \boldsymbol{\Theta}_0 \text{ and } \boldsymbol{\nabla}\boldsymbol{\Theta}^\ell(x_0) = \boldsymbol{F}_0, \ell \geq 0, \quad \text{resp.} \quad \boldsymbol{\Theta}(x_0) = \boldsymbol{\Theta}_0 \text{ and } \boldsymbol{\nabla}\boldsymbol{\Theta}(x_0) = \boldsymbol{F}_0.$$

Assuming that the immersion $\boldsymbol{\Theta}$ is injective in Ω, show that

$$\lim_{\ell \to \infty} \|\boldsymbol{\Theta}^\ell - \boldsymbol{\Theta}\|_{3,K} = 0 \quad \text{for each } K \Subset \Omega.$$

Hint: Introduce the matrix fields $\boldsymbol{\nabla}\boldsymbol{\Theta}^{-T}\boldsymbol{C}^\ell\boldsymbol{\nabla}\boldsymbol{\Theta}^{-1}$, $\ell \geq 0$, and use questions (1) and (2).

(4) Show that (3) holds even if the immersion $\boldsymbol{\Theta}$ is not injective in Ω.

Remark Define the sets

$$\boldsymbol{\mathcal{X}} := \{\boldsymbol{C} = (g_{ij}) \in \mathcal{C}^2(\Omega; \mathbb{S}^n_>);\ \boldsymbol{C}(x_0) = \boldsymbol{F}_0^T \boldsymbol{F}_0 \text{ and } R_{qijk} = 0 \text{ in } \Omega\},$$
$$\boldsymbol{\mathcal{Y}} := \{\boldsymbol{\Theta} = \mathcal{C}^3(\Omega; \mathbb{E}^n);\ \boldsymbol{\Theta}(x_0) = \boldsymbol{\Theta}_0 \text{ and } \boldsymbol{\nabla\Theta}(x_0) = \boldsymbol{F}_0\},$$

and let the distances d_2 and d_3 be defined as in Problem 7.8-3. Then the sequential continuity established in question (3) shows that *the mapping defined by*

$$\boldsymbol{\mathcal{F}} : \boldsymbol{C} \in (\boldsymbol{\mathcal{X}}; d_2) \to \boldsymbol{\mathcal{F}}(\boldsymbol{C}) := \boldsymbol{\Theta} \in (\boldsymbol{\mathcal{Y}}; d_3),$$

where $\boldsymbol{\Theta}$ *designates for each* $\boldsymbol{C} \in \boldsymbol{\mathcal{X}}$ *the unique element in the set* $\boldsymbol{\mathcal{Y}}$ *that satisfies* $\boldsymbol{\nabla\Theta}^T\boldsymbol{\nabla\Theta} = \boldsymbol{C}$ *in* Ω, *is continuous.*[16] □

8.7 Uniqueness up to isometries of immersions with the same metric tensor; the rigidity theorem for an open subset of $\mathbb{R}^n$

A mapping $\boldsymbol{\Phi} : \mathbb{E}^n \to \mathbb{E}^n$ is called an **isometry of** $\mathbb{E}^n$ if it *preserves the Euclidean distance*, i.e., if

$$|\boldsymbol{\Phi}(x) - \boldsymbol{\Phi}(y)| = |x - y| \quad \text{for all } x, y \in \mathbb{E}^n,$$

or equivalently (Problem 8.7-1), *if and only if there exist a vector* $\mathbf{c} \in \mathbb{R}^n$ *and an orthogonal matrix* $\boldsymbol{Q} \in \mathbb{O}^n$ *such that*

$$\boldsymbol{\Phi}(x) = \mathbf{c} + \boldsymbol{Q}x \quad \text{for all } x \in \mathbb{E}^n.$$

If $\boldsymbol{Q}$ is a *proper* orthogonal matrix, the mapping $\boldsymbol{\Phi}$ is said to be a **proper isometry of** $\mathbb{E}^n$.

In Section 8.6, we have established the *existence* of an immersion $\boldsymbol{\Theta} : \Omega \subset \mathbb{R}^n \to \mathbb{E}^n$ with a prescribed metric tensor, under the assumptions that the Riemann curvature tensor associated with this tensor vanishes in Ω and that the open set Ω is simply connected. We now turn to the question of *uniqueness* of such immersions.

This uniqueness result is the object of the next theorem,[17] aptly called a *rigidity theorem* in view of its geometrical interpretation: It asserts that, if two immersions $\boldsymbol{\Theta} \in \mathcal{C}^1(\Omega; \mathbb{E}^n)$ and $\widetilde{\boldsymbol{\Theta}} \in \mathcal{C}^1(\Omega; \mathbb{E}^n)$ share the same metric tensor field, then the set $\boldsymbol{\Theta}(\Omega)$ is obtained by subjecting the set $\widetilde{\boldsymbol{\Theta}}(\Omega)$ either to a *rotation* (represented by a proper orthogonal matrix $\boldsymbol{Q} \in \mathbb{O}^n_+$), or to a *symmetry with respect to a hyperplane followed by a rotation*, then by subjecting the resulting set to a *translation* (represented by a vector $\mathbf{c}$). The composition of two such mappings is called a **rigid deformation** of the set $\boldsymbol{\Theta}(\Omega)$.

Note that the assumption of simple-connectedness of Ω is no longer needed here.

[16]This result is due to:

P.G. Ciarlet; F. Laurent [2003]: Continuity of a deformation as a function of its Cauchy-Green tensor. *Archive for Rational Mechanics and Analysis* **167**, 255–269.

[17]Stated without proof in:

E. Cosserat; F. Cosserat [1896]: Sur la théorie de l'élasticité. Premier mémoire, *Annales de la Faculté des Sciences de l'Université de Toulouse* **10**, 1–116.

Then proved in Section 30 of:

E. Cartan [1928]: *Leçons sur la Géométrie des Espaces de Riemann*, Gauthier-Villars, Paris.

Theorem 8.7-1 (rigidity theorem for an open subset of $\mathbb{R}^n$) *Let Ω be a connected open subset of $\mathbb{R}^n$ and let $\widetilde{\boldsymbol{\Theta}} \in \mathcal{C}^1(\Omega; \mathbb{E}^n)$ and $\boldsymbol{\Theta} \in \mathcal{C}^1(\Omega; \mathbb{E}^n)$ be two immersions whose associated metric tensors are the same, i.e.,*

$$\boldsymbol{\nabla}\widetilde{\boldsymbol{\Theta}}^T \boldsymbol{\nabla}\widetilde{\boldsymbol{\Theta}} = \boldsymbol{\nabla}\boldsymbol{\Theta}^T \boldsymbol{\nabla}\boldsymbol{\Theta} \quad \textit{in } \Omega.$$

Then there exist a vector $\boldsymbol{c} \in \mathbb{E}^n$ and an orthogonal matrix $\boldsymbol{Q} \in \mathbb{O}^n$ such that

$$\widetilde{\boldsymbol{\Theta}}(x) = \boldsymbol{c} + \boldsymbol{Q}\boldsymbol{\Theta}(x) \quad \textit{for all } x \in \Omega.$$

Proof *The space $\mathbb{R}^n$ is identified throughout this proof with the Euclidean space $\mathbb{E}^n$.* In particular then, $\mathbb{R}^n$ inherits the inner product and norm of $\mathbb{E}^n$. Recall that

$$|\boldsymbol{A}| := \sup\{|\boldsymbol{A}\boldsymbol{b}|\,;\ \boldsymbol{b} \in \mathbb{R}^n,\ |\boldsymbol{b}| = 1\}$$

denotes the *spectral norm* of a matrix $\boldsymbol{A} \in \mathbb{M}^n$.

To begin with, we consider the *special case* where $\widetilde{\boldsymbol{\Theta}} : \Omega \to \mathbb{E}^n = \mathbb{R}^n$ is the *identity mapping of* $\mathbb{E}^n$. The issue of uniqueness reduces in this case to identifying all the mappings $\boldsymbol{\Theta} \in \mathcal{C}^1(\Omega; \mathbb{E}^n)$ that satisfy

$$\boldsymbol{\nabla}\boldsymbol{\Theta}(x)^T \boldsymbol{\nabla}\boldsymbol{\Theta}(x) = \boldsymbol{I} \quad \text{at each } x \in \Omega.$$

Parts (i) to (iii) are devoted to finding an *explicit solution* to this nonlinear system of partial differential equations.

(i) We first establish that *a mapping $\boldsymbol{\Theta} \in \mathcal{C}^1(\Omega; \mathbb{E}^n)$ that satisfies*

$$\boldsymbol{\nabla}\boldsymbol{\Theta}(x)^T \boldsymbol{\nabla}\boldsymbol{\Theta}(x) = \boldsymbol{I} \quad \text{at each } x \in \Omega$$

is locally an isometry. This means that, given any point $x^0 \in \Omega$, there exists an open neighborhood V of x^0 contained in Ω such that

$$|\boldsymbol{\Theta}(y) - \boldsymbol{\Theta}(x)| = |y - x| \quad \text{for all } x, y \in V.$$

Let B be an open ball centered at x^0 and contained in Ω. Since the set B is convex, the *mean value theorem in a normed vector space* (Theorem 7.2-1) can be applied, showing that

$$|\boldsymbol{\Theta}(y) - \boldsymbol{\Theta}(x)| \le \sup_{z \in]x,y[} |\boldsymbol{\nabla}\boldsymbol{\Theta}(z)|\,|y - x| \quad \text{for all } x, y \in B.$$

Since the spectral norm of an orthogonal matrix is one, we thus have

$$|\boldsymbol{\Theta}(y) - \boldsymbol{\Theta}(x)| \le |y - x| \quad \text{for all } x, y \in B.$$

Since the matrix $\boldsymbol{\nabla}\boldsymbol{\Theta}(x^0)$ is invertible, the *local inversion theorem* (Theorem 7.14-1) shows that there exist an open neighborhood V of x^0 contained in Ω and an open neighborhood $\widehat{V}$ of $\boldsymbol{\Theta}(x^0)$ in $\mathbb{E}^n$ such that the restriction of $\boldsymbol{\Theta}$ to V is a $\mathcal{C}^1$-diffeomorphism from V onto $\widehat{V}$. Besides, there is no loss of generality in assuming that V is contained in B and that $\widehat{V}$ is convex (to see this, apply the local inversion theorem first to the restriction of $\boldsymbol{\Theta}$ to B, thus producing a first neighborhood V' of x^0 contained in B, then to the restriction of the inverse mapping obtained in this fashion to an open ball $\widehat{V}$ centered at $\boldsymbol{\Theta}(x^0)$ and contained in $\boldsymbol{\Theta}(V')$).

Let $\boldsymbol{\Theta}^{-1} : \widehat{V} \to V$ denote the inverse mapping of $\boldsymbol{\Theta} : V \to \widehat{V}$. The chain rule (Theorem 7.1-3) applied to the relation $\boldsymbol{\Theta}^{-1}(\boldsymbol{\Theta}(x)) = x$ for all $x \in V$ then shows that

$$\widehat{\boldsymbol{\nabla}}\boldsymbol{\Theta}^{-1}(\widehat{x}) = \boldsymbol{\nabla}\boldsymbol{\Theta}(x)^{-1} \quad \text{for all } \widehat{x} = \boldsymbol{\Theta}(x), \ \ x \in V.$$

The matrix $\widehat{\boldsymbol{\nabla}}\boldsymbol{\Theta}^{-1}(\widehat{x})$ being thus orthogonal at each $\widehat{x} \in \widehat{V}$, the mean value theorem applied in the convex set $\widehat{V}$ shows that

$$|\boldsymbol{\Theta}^{-1}(\widehat{y}) - \boldsymbol{\Theta}^{-1}(\widehat{x})| \le |\widehat{y} - \widehat{x}| \quad \text{for all } \widehat{x}, \widehat{y} \in \widehat{V},$$

or equivalently, that

$$|y - x| \le |\boldsymbol{\Theta}(y) - \boldsymbol{\Theta}(x)| \quad \text{for all } x, y \in V.$$

The restriction of the mapping $\boldsymbol{\Theta}$ to the open neighborhood V of x^0 is thus an isometry.

(ii) We next establish that, *if a mapping $\boldsymbol{\Theta} \in \mathcal{C}^1(\Omega; \mathbb{E}^n)$ is locally an isometry, then its derivative is locally constant.* This means that, given any $x^0 \in \Omega$, there exists an open neighborhood V of x^0 contained in Ω such that

$$\boldsymbol{\nabla}\boldsymbol{\Theta}(x) = \boldsymbol{\nabla}\boldsymbol{\Theta}(x^0) \quad \text{for all } x \in V.$$

Given $x^0 \in \Omega$, let $V \subset \Omega$ denote the open neighborhood of x^0 found as in (i), and let the differentiable function $F : V \times V \to \mathbb{R}$ be defined at each $x = (x_p) \in V$ and $y = (y_p) \in V$ by

$$F(x, y) := (\Theta_\ell(y) - \Theta_\ell(x))(\Theta_\ell(y) - \Theta_\ell(x)) - (y_\ell - x_\ell)(y_\ell - x_\ell).$$

Then $F(x, y) = 0$ for all $x, y \in V$ by (i). Hence

$$G_i(x, y) := \frac{1}{2}\frac{\partial F}{\partial y_i}(x, y) = \frac{\partial \Theta_\ell}{\partial y_i}(y)(\Theta_\ell(y) - \Theta_\ell(x)) - \delta_{i\ell}(y_\ell - x_\ell) = 0$$

for all $x, y \in V$. For a fixed $y \in V$, each function $G_i(\cdot, y) : V \to \mathbb{R}$ is differentiable and its derivative vanishes. Consequently,

$$\frac{\partial G_i}{\partial x_j}(x, y) = -\frac{\partial \Theta_\ell}{\partial y_i}(y)\frac{\partial \Theta_\ell}{\partial x_j}(x) + \delta_{ij} = 0 \quad \text{for all } x, y \in V,$$

or equivalently, in matrix form,

$$\boldsymbol{\nabla}\boldsymbol{\Theta}(y)^T \boldsymbol{\nabla}\boldsymbol{\Theta}(x) = \boldsymbol{I} \quad \text{for all } x, y \in V.$$

Letting $y = x^0$ in this relation shows that

$$\boldsymbol{\nabla}\boldsymbol{\Theta}(x) = \boldsymbol{\nabla}\boldsymbol{\Theta}(x^0) \quad \text{for all } x \in V.$$

(iii) By (ii), the mapping $\boldsymbol{\nabla}\boldsymbol{\Theta} : \Omega \to \mathbb{M}^n$ is differentiable and its derivative vanishes in Ω. Therefore, by Theorem 7.2-4, the mapping $\boldsymbol{\nabla}\boldsymbol{\Theta}$ *is a constant* since the set Ω is connected. This means that there exists a matrix $\boldsymbol{Q} \in \mathbb{M}^n$ such that

$$\boldsymbol{\nabla}\boldsymbol{\Theta}(x) = \boldsymbol{Q} \quad \text{for all } x \in \Omega.$$

Another application of the same theorem then shows that the mapping $\boldsymbol{\Theta}$ is *affine* in Ω, i.e., that there exists a vector $\boldsymbol{c} \in \mathbb{E}^n$ and a matrix $\boldsymbol{Q} \in \mathbb{M}^n$ such that

$$\boldsymbol{\Theta}(x) = \boldsymbol{c} + \boldsymbol{Q}x \quad \text{for all } x \in \Omega.$$

Since $\boldsymbol{Q} = \boldsymbol{\nabla\Theta}(x^0)$ and $\boldsymbol{\nabla\Theta}(x^0)^T\boldsymbol{\nabla\Theta}(x^0) = \boldsymbol{I}$ by assumption, the matrix $\boldsymbol{Q}$ is *orthogonal.*

(iv) We now consider the general case, where

$$\boldsymbol{\nabla}\widetilde{\boldsymbol{\Theta}}(x)^T\boldsymbol{\nabla}\widetilde{\boldsymbol{\Theta}}(x) = \boldsymbol{\nabla\Theta}(x)^T\boldsymbol{\nabla\Theta}(x) \quad \text{at each } x \in \Omega.$$

Given any point $x^0 \in \Omega$, let the neighborhoods V of x^0 and $\widehat{V}$ of $\boldsymbol{\Theta}(x^0)$ and the mapping $\boldsymbol{\Theta}^{-1} : \widehat{V} \to V$ be defined as in part (i) (by assumption, the mapping $\boldsymbol{\Theta}$ is an immersion; hence the matrix $\boldsymbol{\nabla\Theta}(x^0)$ is invertible).

Consider the composite mapping

$$\widehat{\boldsymbol{\Phi}} := \widetilde{\boldsymbol{\Theta}} \circ \boldsymbol{\Theta}^{-1} : \widehat{V} \to \mathbb{E}^n.$$

Clearly, $\widehat{\boldsymbol{\Phi}} \in \mathcal{C}^1(\widehat{V}; \mathbb{E}^n)$ and

$$\widehat{\boldsymbol{\nabla}}\widehat{\boldsymbol{\Phi}}(\widehat{x}) = \boldsymbol{\nabla}\widetilde{\boldsymbol{\Theta}}(x)\widehat{\boldsymbol{\nabla}}\boldsymbol{\Theta}^{-1}(\widehat{x}) = \boldsymbol{\nabla}\widetilde{\boldsymbol{\Theta}}(x)\boldsymbol{\nabla\Theta}(x)^{-1} \quad \text{at each } \widehat{x} = \boldsymbol{\Theta}(x),\ \ x \in V.$$

Hence the assumed relations $\boldsymbol{\nabla\Theta}(x)^T\boldsymbol{\nabla\Theta}(x) = \boldsymbol{\nabla}\widetilde{\boldsymbol{\Theta}}(x)^T\boldsymbol{\nabla}\widetilde{\boldsymbol{\Theta}}(x)$ at each $x \in \Omega$ imply that

$$\widehat{\boldsymbol{\nabla}}\widehat{\boldsymbol{\Phi}}(\widehat{x})^T\widehat{\boldsymbol{\nabla}}\widehat{\boldsymbol{\Phi}}(\widehat{x}) = \boldsymbol{I} \quad \text{at each } x \in V.$$

By parts (i)–(iii), there thus exist a vector $\boldsymbol{c} \in \mathbb{R}^n$ and a matrix $\boldsymbol{Q} \in \mathbb{O}^n$ such that

$$\widehat{\boldsymbol{\Phi}}(\widehat{x}) = \widetilde{\boldsymbol{\Theta}}(x) = \boldsymbol{c} + \boldsymbol{Q}\boldsymbol{\Theta}(x) \quad \text{for all } \widehat{x} = \boldsymbol{\Theta}(x),\ \ x \in V,$$

and hence such that

$$\boldsymbol{\Xi}(x) := \boldsymbol{\nabla}\widetilde{\boldsymbol{\Theta}}(x)\boldsymbol{\nabla\Theta}(x)^{-1} = \boldsymbol{Q} \quad \text{for all } x \in V.$$

The *continuous* mapping $\boldsymbol{\Xi} : V \to \mathbb{M}^n$ defined in this fashion is thus *locally constant* in Ω. As in part (iii), we conclude from the assumed *connectedness* of Ω that the mapping $\boldsymbol{\Xi}$ is *constant in* Ω. Thus the proof is complete. □

The special case in Theorem 8.7-1, where $\boldsymbol{\Theta}$ is the identity mapping of $\mathbb{R}^n$ identified with $\mathbb{E}^n$, constitutes the classical **Liouville theorem**;[18] this theorem asserts that *if a mapping* $\boldsymbol{\Theta} \in \mathcal{C}^1(\Omega; \mathbb{E}^n)$ *is such that* $\boldsymbol{\nabla\Theta}(x) \in \mathbb{O}^n$ *for all* $x \in \Omega$ *and* Ω *is an open connected subset of* $\mathbb{R}^n$, *then there exist a vector* $\boldsymbol{c} \in \mathbb{R}^n$ *and an orthogonal matrix* $\boldsymbol{Q} \in \mathbb{O}^n$ *such that*

$$\boldsymbol{\Theta}(x) = \boldsymbol{c} + \boldsymbol{Q}x \quad \text{for all } x \in \Omega.$$

Remarks (1) Liouville's theorem still applies to mappings $\boldsymbol{\Theta} \in H^1(\Omega; \mathbb{E}^n)$ that satisfy $\boldsymbol{\nabla\Theta}(x) \in \mathbb{O}^3_+$ for almost all $x \in \Omega$ (note the restriction on the sign of det $\boldsymbol{\nabla\Theta}(x)$); cf. Problem 8.7-3.

[18] Actually, this result is a *corollary* to a more general one, which applies to *conformal mappings* in $\mathbb{R}^n$ (i.e., mappings that preserve angles), due to:

J. Liouville [1850]: Extension au cas des trois dimensions de la question du tracé géographique, Note VI in the Appendix to G. Monge: *Application de l'Analyse à la Géométrie, Cinquième Edition*, Bachelier, Paris.

(2) More generally, Theorem 8.7-1 still applies[19] if $\boldsymbol{\Theta} \in \mathcal{C}^1(\Omega; \mathbb{E}^n)$ and $\widetilde{\boldsymbol{\Theta}} \in H^1(\Omega; \mathbb{E}^n)$ under the additional assumptions that $\det \boldsymbol{\nabla}\boldsymbol{\Theta} > 0$ in Ω and $\det \boldsymbol{\nabla}\widetilde{\boldsymbol{\Theta}} > 0$ almost everywhere in Ω. □

While the immersions $\boldsymbol{\Theta} \in \mathcal{C}^3(\Omega; \mathbb{E}^n)$ found in Theorem 8.6-1 are by Theorem 8.7-1 only defined up to isometries of $\mathbb{E}^n$, Theorem 8.6-2 shows that they become *uniquely determined* if they are required to satisfy *ad hoc* additional conditions. We now show that *the same uniqueness result* (i.e., with the same additional conditions) *already holds under weaker regularity assumptions on the immersions* $\boldsymbol{\Theta}$.

Theorem 8.7-2 *Let Ω be a connected open subset of $\mathbb{R}^n$, and let there be given an immersion $\boldsymbol{\Phi} \in \mathcal{C}^1(\Omega; \mathbb{E}^n)$, a point $x_0 \in \Omega$, a vector $\boldsymbol{\Theta}_0 \in \mathbb{E}^n$, and a matrix $\boldsymbol{F}_0 \in \mathbb{M}^n$ that satisfy*

$$\boldsymbol{F}_0^T \boldsymbol{F}_0 = \boldsymbol{\nabla}\boldsymbol{\Phi}(x_0)^T \boldsymbol{\nabla}\boldsymbol{\Phi}(x_0).$$

Then there exists one and only one immersion $\boldsymbol{\Theta} \in \mathcal{C}^1(\Omega; \mathbb{E}^n)$ that satisfies

$$\boldsymbol{\nabla}\boldsymbol{\Theta}(x)^T \boldsymbol{\nabla}\boldsymbol{\Theta}(x) = \boldsymbol{\nabla}\boldsymbol{\Phi}(x)^T \boldsymbol{\nabla}\boldsymbol{\Phi}(x) \quad \textit{for all } x \in \Omega,$$
$$\boldsymbol{\Theta}(x_0) = \boldsymbol{\Theta}_0 \quad \textit{and} \quad \boldsymbol{\nabla}\boldsymbol{\Theta}(x_0) = \boldsymbol{F}_0.$$

Proof Given an immersion $\boldsymbol{\Phi} \in \mathcal{C}^1(\Omega; \mathbb{E}^n)$, the mapping $\boldsymbol{\Theta} : \Omega \to \mathbb{E}^n$ defined by

$$\boldsymbol{\Theta}(x) := \boldsymbol{\Theta}_0 + \boldsymbol{F}_0 \boldsymbol{\nabla}\boldsymbol{\Phi}(x_0)^{-1}(\boldsymbol{\Phi}(x) - \boldsymbol{\Phi}(x_0)) \quad \text{at each } x \in \Omega,$$

satisfies the announced properties.

Besides, it is uniquely determined. To see this, let $\boldsymbol{\Theta} \in \mathcal{C}^1(\Omega; \mathbb{E}^n)$ and $\boldsymbol{\psi} \in \mathcal{C}^1(\Omega; \mathbb{E}^n)$ be two immersions that satisfy

$$\boldsymbol{\nabla}\boldsymbol{\Theta}(x)^T \boldsymbol{\nabla}\boldsymbol{\Theta}(x) = \boldsymbol{\nabla}\boldsymbol{\psi}(x)^T \boldsymbol{\nabla}\boldsymbol{\psi}(x) \quad \text{for all } x \in \Omega.$$

Hence there exist (by Theorem 8.7-1) a vector $\boldsymbol{c} \in \mathbb{R}^n$ and an orthogonal matrix $\boldsymbol{Q} \in \mathbb{O}^n$ such that

$$\boldsymbol{\psi}(x) = \boldsymbol{c} + \boldsymbol{Q}\boldsymbol{\Theta}(x) \quad \text{for all } x \in \Omega,$$

so that $\boldsymbol{\nabla}\boldsymbol{\psi}(x) = \boldsymbol{Q}\boldsymbol{\nabla}\boldsymbol{\Theta}(x)$ for all $x \in \Omega$. The relation $\boldsymbol{\nabla}\boldsymbol{\Theta}(x_0) = \boldsymbol{\nabla}\boldsymbol{\psi}(x_0)$ then implies that $\boldsymbol{Q} = \boldsymbol{I}$ and the relation $\boldsymbol{\Theta}(x_0) = \boldsymbol{\psi}(x_0)$ in turn implies that $\boldsymbol{c} = \boldsymbol{0}$. □

Remark One possible choice for the matrix $\boldsymbol{F}_0$ is the *square root* of the symmetric positive-definite matrix $\boldsymbol{\nabla}\boldsymbol{\Phi}(x_0)^T \boldsymbol{\nabla}\boldsymbol{\Phi}(x_0)$ (Theorem 7.14-3). □

Problems

8.7-1 Let Ω be a connected open subset of $\mathbb{R}^n$ and let $\boldsymbol{\Phi} : \Omega \to \mathbb{R}^n$ be a mapping that preserves the Euclidean distance, i.e., that satisfies

$$|\boldsymbol{\Phi}(x) - \boldsymbol{\Phi}(y)| = |x - y| \quad \text{for all } x, y \in \Omega.$$

[19] P.G. Ciarlet; C. Mardare [2003]: On rigid and infinitesimal rigid displacements in three-dimensional elasticity, *Mathematical Models and Methods in Applied Sciences* **13**, 1589–1598.

Show that there exist a vector $\boldsymbol{c} \in \mathbb{R}^n$ and an orthogonal matrix $\boldsymbol{Q} \in \mathbb{O}^n$ such that

$$\boldsymbol{\Phi}(x) = \boldsymbol{x} + \boldsymbol{Q}x \quad \text{for all } x \in \Omega.$$

Remark This result constitutes the classical **Mazur–Ulam theorem**[20] (a proof of which is already provided in parts (ii) and (iii) of Theorem 8.7-1, but under the additional assumption that the mapping $\boldsymbol{\Phi}$ is of class $\mathcal{C}^1$ in Ω). An infinite-dimensional version also holds; cf. Problem 4.1-4. □

8.7-2 Let $\boldsymbol{\Phi} : \mathbb{E}^n \to \mathbb{E}^n$ be a mapping that preserves *one* nonzero Euclidean distance; in other words, there exists $\delta > 0$ such that

$$|x - y| = \delta \quad \text{implies} \quad |\boldsymbol{\Phi}(x) - \boldsymbol{\Phi}(y)| = \delta.$$

Show that $\boldsymbol{\Phi}$ preserves in fact *any* Euclidean distance, i.e., that $\boldsymbol{\Phi}$ *is an isometry of* $\mathbb{E}^n$; note that it is *not* assumed *a priori* that $\boldsymbol{\Phi}$ is continuous.

8.7-3 (1) Let Ω be a connected open subset of $\mathbb{R}^n$ and let $\boldsymbol{\Theta} \in H^1(\Omega; \mathbb{E}^n)$ be a mapping that satisfies

$$\det \boldsymbol{\nabla\Theta} > 0 \text{ a.e. in } \Omega \quad \text{and} \quad \boldsymbol{\nabla\Theta}^T\boldsymbol{\nabla\Theta} = \boldsymbol{I} \text{ a.e. in } \Omega.$$

Show that there exists a vector $\boldsymbol{c} \in \mathbb{E}^n$ and a proper orthogonal matrix $\boldsymbol{Q} \in \mathbb{O}^n_+$ such that[21]

$$\boldsymbol{\Theta}(x) = \boldsymbol{c} + \boldsymbol{Q}x \quad \text{for almost all } x \in \Omega.$$

Hint: Use the *Piola identity* (Theorem 7.1-4) to conclude that $\boldsymbol{\Delta\Theta} = \mathbf{div}\,\mathbf{Cof}\,\boldsymbol{\nabla\Theta} = 0$ in $\mathcal{D}'(\Omega; \mathbb{E}^n)$, hence that $\boldsymbol{\Theta} \in \mathcal{C}^\infty(\Omega; \mathbb{E}^n)$ by the *hypoellipticity of* Δ (Theorem 6.4-2); then use the identity

$$\Delta(\partial_i\Theta_j\partial_i\Theta_j) = 2\partial_i\Theta_j\partial_i(\Delta\Theta_j) + 2\partial_{ik}\Theta_j\partial_{ik}\Theta_j$$

to infer that $\partial_{ik}\Theta_j = 0$ in $\mathcal{D}'(\Omega)$; then use an argument analogous to that used in the proof of Theorem 6.3-4.[22]

(2) Let $\Omega = \{x \in \mathbb{R}^3;\ |x| < 1\}$ and let the mapping $\boldsymbol{\Theta} : \Omega \to \mathbb{E}^3$ be defined by $\boldsymbol{\Theta}(x) = x$ if $x_1 > 0$ and $\boldsymbol{\Theta}(x) = (-x_1, x_2, x_3)$ if $x_1 < 0$. Verify that $\boldsymbol{\Theta} \in H^1(\Omega; \mathbb{E}^3)$ and that $\boldsymbol{\nabla\Theta}^T\boldsymbol{\nabla\Theta} = \boldsymbol{I}$ almost everywhere in Ω, yet that there does *not* exist any orthogonal matrix $\boldsymbol{Q} \in \mathbb{O}_3$ such that $\boldsymbol{\Theta}(x) = \boldsymbol{Q}x$ for almost all $x \in \Omega$. This is why an assumption about the sign of $\det \boldsymbol{\nabla\Theta}$ is needed in this case.

8.8 Curvilinear coordinates on a surface in $\mathbb{R}^3$

In the rest of this chapter the integer n is equal to three; hence *Latin* indices and exponents vary in the set $\{1, 2, 3\}$. In addition, *Greek* indices and exponents vary in the set $\{1, 2\}$, and the *summation convention* is systematically used in conjunction with these rules. For instance, the relation

$$\partial_\alpha(\eta_i\boldsymbol{a}^i) = (\eta_{\beta|\alpha} - b_{\alpha\beta}\eta_3)\boldsymbol{a}^\beta + (\eta_{3|\alpha} + b^\beta_\alpha\eta_\beta)\boldsymbol{a}^3$$

[20] S. Mazur; S. Ulam [1932]: Sur les transformations isométriques d'espaces vectoriels normés, *Comptes Rendus de l'Académie des Sciences de Paris* **194**, 946–948.

[21] This extension of Liouville's theorem is due to:

Y.G. Reshetnyak [1967]: Liouville's theory on conformal mappings under minimal regularity assumptions, *Siberian Mathematical Journal* **8**, 69–85.

[22] This elegant proof is found in:

G. Friesecke; R.D. James; S. Müller [2002]: A theorem on geometric rigidity and the derivation of nonlinear plate theory from three dimensional elasticity, *Communications on Pure and Applied Mathematics* **55**, 1461–1506.

means that

$$\partial_\alpha\Big(\sum_{i=1}^3 \eta_i \boldsymbol{a}^i\Big) = \sum_{\beta=1}^2 (\eta_{\beta|\alpha} - b_{\alpha\beta}\eta_3)\boldsymbol{a}^\beta + \Big(\eta_{3|\alpha} + \sum_{\beta=1}^2 b_\alpha^\beta \eta_\beta\Big)\boldsymbol{a}^3 \quad \text{for } \alpha = 1, 2.$$

Kronecker's symbols are designated by $\delta_\alpha^\beta, \delta_{\alpha\beta}$, or $\delta^{\alpha\beta}$ according to the context.

Let there be given a *three-dimensional Euclidean space* $\mathbb{E}^3$, equipped with an orthonormal basis consisting of three vectors $\widehat{\boldsymbol{e}}^i = \widehat{\boldsymbol{e}}_i$, and let $\boldsymbol{a} \cdot \boldsymbol{b}$, $|\boldsymbol{a}|$, and $\boldsymbol{a} \wedge \boldsymbol{b}$ denote the Euclidean inner product, the Euclidean norm, and the vector product of vectors $\boldsymbol{a}, \boldsymbol{b}$ in the space $\mathbb{E}^3$.

In addition, let there be given a *two-dimensional vector space*, in which two vectors $\boldsymbol{e}^\alpha = \boldsymbol{e}_\alpha$ form a basis. *This space will be identified with* $\mathbb{R}^2$. Let y_α denote the coordinates of a point $y \in \mathbb{R}^2$ and let $\partial_\alpha := \partial/\partial y_\alpha$ and $\partial_{\alpha\beta} := \partial^2/\partial y_\alpha \partial y_\beta$.

Finally, let there be given an *open* subset ω of $\mathbb{R}^2$ and a smooth enough *injective* mapping $\boldsymbol{\theta} : \omega \to \mathbb{E}^3$ (specific smoothness assumptions on $\boldsymbol{\theta}$ will be made later, according to each context). Then the set

$$\widehat{\omega} := \boldsymbol{\theta}(\omega)$$

is called a **surface** in $\mathbb{E}^3$. Since the mapping $\boldsymbol{\theta} : \omega \to \mathbb{E}^3$ is injective, each point $\widehat{y} \in \widehat{\omega}$ can be unambiguously written as

$$\widehat{y} = \boldsymbol{\theta}(y), \quad y \in \omega,$$

and the two coordinates y_α of y are then called the **curvilinear coordinates** of $\widehat{y}$ (Figure 8.8-1). Well-known *examples* of surfaces and of curvilinear coordinates and their corresponding coordinate lines (which will be defined in the next section) are given in Figures 8.8-2 and 8.8-3.

Naturally, once a surface is defined as $\widehat{\omega} = \boldsymbol{\theta}(\omega)$, *there are infinitely many other ways of defining curvilinear coordinates on* $\widehat{\omega}$, depending on how the domain ω and the mapping $\boldsymbol{\theta}$ are chosen. For instance, a portion $\widehat{\omega}$ of a sphere may be represented by means of *Cartesian coordinates, spherical coordinates*, or *stereographic coordinates* (Figure 8.8-3). Incidentally, this example illustrates the variety of restrictions that have to be imposed on $\widehat{\omega}$ according to which kind of curvilinear coordinates it is equipped with.

8.9 First fundamental form of a surface; areas, lengths, and angles on a surface

Let ω be an open subset of $\mathbb{R}^2$ and let

$$\boldsymbol{\theta} = \theta_i \widehat{\boldsymbol{e}}^i : \omega \subset \mathbb{R}^2 \to \widehat{\omega} := \boldsymbol{\theta}(\omega) \quad \text{in } \mathbb{E}^3$$

be an injective mapping that is *differentiable at a point* $y \in \omega$. If $\boldsymbol{\delta y} = \delta y_\alpha \boldsymbol{e}^\alpha$ is such that $(y + \boldsymbol{\delta y}) \in \omega$, then (Section 7.1)

$$\boldsymbol{\theta}(y + \boldsymbol{\delta y}) = \boldsymbol{\theta}(y) + \boldsymbol{\nabla\theta}(y)\boldsymbol{\delta y} + |\boldsymbol{\delta y}|\, \boldsymbol{\varepsilon}(\boldsymbol{\delta y}) \quad \text{with } \lim_{\boldsymbol{\delta y} \to \boldsymbol{0}} \boldsymbol{\varepsilon}(\boldsymbol{\delta y}) = \boldsymbol{0},$$

where the 3×2 matrix $\boldsymbol{\nabla\theta}(y)$ and the column vector $\boldsymbol{\delta y}$ are given by

$$\boldsymbol{\nabla\theta}(y) = \begin{pmatrix} \partial_1\theta_1 & \partial_2\theta_1 \\ \partial_1\theta_2 & \partial_2\theta_2 \\ \partial_1\theta_3 & \partial_2\theta_3 \end{pmatrix}(y) \quad \text{and} \quad \boldsymbol{\delta y} = \begin{pmatrix} \delta y_1 \\ \delta y_2 \end{pmatrix}.$$

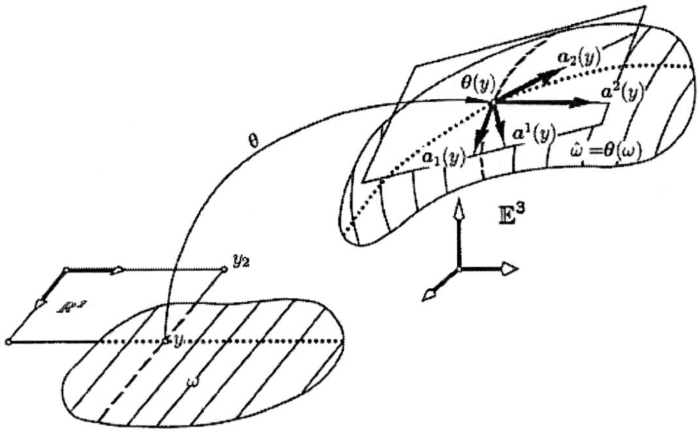

Figure 8.8-1 *Curvilinear coordinates on a surface and covariant and contravariant bases of the tangent plane.* Let $\widehat{\omega} = \boldsymbol{\theta}(\omega)$ be a surface in $\mathbb{E}^3$. The two coordinates y_1, y_2 of $y \in \omega$ are the curvilinear coordinates of $\widehat{y} = \boldsymbol{\theta}(y) \in \widehat{\omega}$. If the two vectors $\boldsymbol{a}_\alpha(y) = \partial_\alpha \boldsymbol{\theta}(y)$ are linearly independent, they are tangent to the coordinate lines passing through $\widehat{y}$ and they form the covariant basis of the tangent plane to $\widehat{\omega}$ at $\widehat{y} = \boldsymbol{\theta}(y)$ (Section 8.9). The two vectors $\boldsymbol{a}^\alpha(y)$ from this tangent plane defined by $\boldsymbol{a}^\alpha(y) \cdot \boldsymbol{a}_\beta(y) = \delta^\alpha_\beta$ form its contravariant basis (Section 8.9). This figure originally appeared in P.G. Ciarlet [2005]: *An Introduction to Differential Geometry with Applications to Elasticity*, Springer, Dordrecht.

Let the two column vectors $\boldsymbol{a}_\alpha(y)$ be defined by

$$\boldsymbol{a}_\alpha(y) := \partial_\alpha \boldsymbol{\theta}(y) = \begin{pmatrix} \partial_\alpha \theta_1 \\ \partial_\alpha \theta_2 \\ \partial_\alpha \theta_3 \end{pmatrix} (y),$$

i.e., $\boldsymbol{a}_\alpha(y)$ *is the* αth *column vector of the matrix* $\boldsymbol{\nabla\theta}(y)$. Then $\boldsymbol{\theta}(y + \boldsymbol{\delta y})$ may be also written as

$$\boldsymbol{\theta}(y + \boldsymbol{\delta y}) = \boldsymbol{\theta}(y) + \delta y^\alpha \boldsymbol{a}_\alpha(y) + |\boldsymbol{\delta y}|\, \boldsymbol{\varepsilon}(\boldsymbol{\delta y}) \quad \text{with} \lim_{\boldsymbol{\delta y} \to \boldsymbol{0}} \boldsymbol{\varepsilon}(\boldsymbol{\delta y}) = \boldsymbol{0}.$$

If in particular $\boldsymbol{\delta y}$ is of the form $\boldsymbol{\delta y} = \delta t \boldsymbol{e}_\alpha$, where $\delta t \in \mathbb{R}$ and $\boldsymbol{e}_\alpha$ is one of the basis vectors in $\mathbb{R}^2$, this relation reduces to

$$\boldsymbol{\theta}(y + \delta t \boldsymbol{e}_\alpha) = \boldsymbol{\theta}(y) + \delta t \boldsymbol{a}_\alpha(y) + |\delta t|\, \boldsymbol{\chi}(\delta t) \quad \text{with} \lim_{\delta t \to 0} \boldsymbol{\chi}(\delta t) = \boldsymbol{0}.$$

A mapping $\boldsymbol{\theta} : \omega \to \mathbb{E}^3$ is an **immersion at** $y \in \omega$ if it is differentiable at y and the 3×2 matrix $\boldsymbol{\nabla\theta}(y)$ is of rank two, or equivalently if the two vectors $\boldsymbol{a}_\alpha(y) = \partial_\alpha \boldsymbol{\theta}(y)$ are linearly independent.

Assume that *the mapping* $\boldsymbol{\theta}$ *is an immersion at* y. In this case, the last relation shows that *each vector* $\boldsymbol{a}_\alpha(y)$ *is tangent to the* αth **coordinate line** *passing through* $\widehat{y} = \boldsymbol{\theta}(y)$, defined as the image by $\boldsymbol{\theta}$ of the points of ω that lie on a line parallel to $\boldsymbol{e}_\alpha$ passing through y (there exist t_0 and t_1 with $t_0 < 0 < t_1$ such that the *αth* coordinate line is given by $t \in]t_0, t_1[\to \boldsymbol{f}_\alpha(t) := \boldsymbol{\theta}(y + t\boldsymbol{e}_\alpha)$ in a neighborhood of $\widehat{y}$; hence $\boldsymbol{f}'_\alpha(0) = \partial_\alpha \boldsymbol{\theta}(y) = \boldsymbol{a}_\alpha(y)$). Examples of coordinate lines are shown in Figures 8.8-2 and 8.8-3.

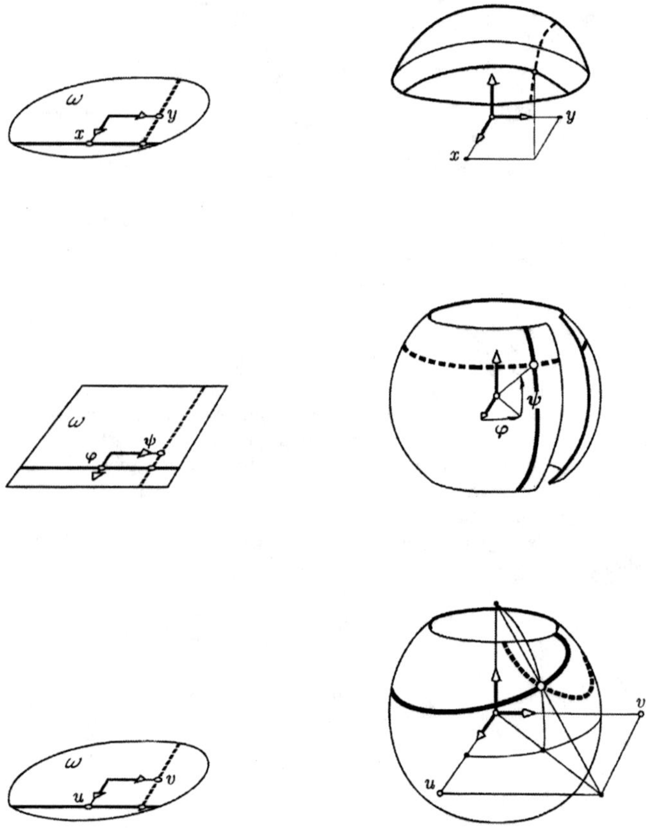

Figure 8.8-2 *Three examples of curvilinear coordinates on a portion of a sphere.* Let Σ be a sphere of radius R. A portion of Σ "contained in the northern hemisphere" can be represented by means of *Cartesian coordinates*, with a mapping $\boldsymbol{\theta}$ of the form

$$\boldsymbol{\theta} : (x, y) \in \omega \to (x, y, \{R^2 - (x^2 + y^2)\}^{1/2}) \in \mathbb{E}^3.$$

A portion of Σ that excludes both "poles" and a "meridian" (to fix ideas) can be represented by means of *spherical coordinates*, with a mapping $\boldsymbol{\theta}$ of the form

$$\boldsymbol{\theta} : (\varphi, \psi) \in \omega \to (R \cos\psi \cos\varphi, R\cos\psi \sin\varphi, R \sin\psi) \in \mathbb{E}^3.$$

A portion of Σ that excludes the "North pole" can be represented by means of *stereographic coordinates*, with a mapping $\boldsymbol{\theta}$ of the form

$$\boldsymbol{\theta} : (u, v) \in \omega \to \Big(\frac{2R^2 u}{u^2 + v^2 + R^2}, \frac{2R^2 v}{u^2 + v^2 + R^2}, R\frac{u^2 + v^2 - R^2}{u^2 + v^2 + R^2}\Big) \in \mathbb{E}^3.$$

The corresponding coordinate lines (Section 8.9) are represented in each case, with self-explanatory graphical conventions. This figure originally appeared in P.G. Ciarlet [2005]: *An Introduction to Differential Geometry with Applications to Elasticity*, Springer, Dordrecht.

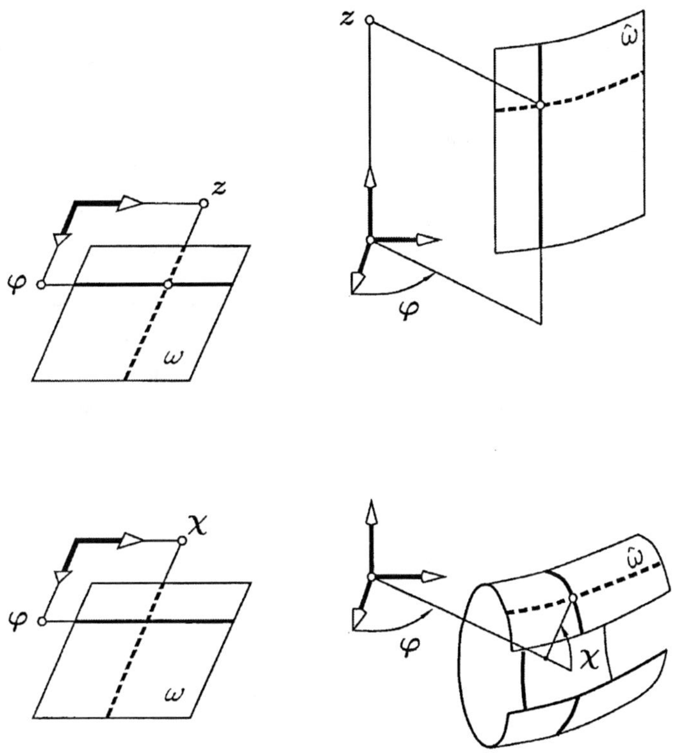

Figure 8.8-3 *Two familiar examples of surfaces and curvilinear coordinates.* A portion $\widehat{\omega}$ of a *circular cylinder* of radius R can be represented by a mapping $\boldsymbol{\theta}$ of the form

$$\boldsymbol{\theta} : (\varphi, z) \in \omega \to (R\cos\varphi, R\sin\varphi, z) \in \mathbb{E}^3.$$

A portion $\widehat{\omega}$ of a *torus* can be represented by a mapping $\boldsymbol{\theta}$ of the form

$$\boldsymbol{\theta} : (\varphi, \chi) \in \omega \to ((R + r\cos\chi)\cos\varphi, (R + r\cos\chi)\sin\varphi, r\sin\chi) \in \mathbb{E}^3,$$

with $R > r$.

The corresponding coordinate lines are represented in each case, with self-explanatory graphical conventions. This figure originally appeared in P.G. CIARLET [2005]: *An Introduction to Differential Geometry with Applications to Elasticity*, Springer, Dordrecht.

More generally, let a curve in ω be defined by an injective mapping $\boldsymbol{g} = g^\alpha \boldsymbol{e}_\alpha \in \mathcal{C}^1(I; \mathbb{R}^2)$, where I is an interval of $\mathbb{R}$ containing 0, $\boldsymbol{g}(0) = y$, $\boldsymbol{g}(I) \subset \omega$, and $\boldsymbol{g}'(0) \neq \boldsymbol{0}$. Then the tangent vector to the curve $(\boldsymbol{\theta} \circ \boldsymbol{g})(I) \subset \widehat{\omega}$ at the point $\boldsymbol{\theta}(y)$ is given by

$$(\boldsymbol{\theta} \circ \boldsymbol{g})'(0) = \frac{\mathrm{d}g^\alpha}{\mathrm{d}t}(0)\partial_\alpha \boldsymbol{\theta}(\boldsymbol{g}(0)) = \frac{\mathrm{d}g^\alpha}{\mathrm{d}t}(0)\boldsymbol{a}_\alpha(y).$$

This shows that the *vectors* $\boldsymbol{a}_\alpha(y)$ *span the* **tangent plane** *to the surface* $\widehat{\omega}$ *at* $\widehat{y} = \boldsymbol{\theta}(y)$. The vectors $\boldsymbol{a}_\alpha(y)$ are said to form the **covariant basis of the tangent plane** to $\widehat{\omega}$ at $\widehat{y}$; see Figure 8.8-1.

Returning to a general increment $\boldsymbol{\delta y} = \delta y^\alpha \boldsymbol{e}_\alpha$, we also infer from the expression of $\boldsymbol{\theta}(y + \boldsymbol{\delta y})$ that

$$\begin{aligned}|\boldsymbol{\theta}(y + \boldsymbol{\delta y}) - \boldsymbol{\theta}(y)| &= \sqrt{\boldsymbol{\delta y}^T \boldsymbol{\nabla\theta}(y)^T \boldsymbol{\nabla\theta}(y) \boldsymbol{\delta y}} + |\boldsymbol{\delta y}|\, \eta(\boldsymbol{\delta y}) \\ &= \sqrt{\delta y^\alpha \boldsymbol{a}_\alpha(y) \cdot \boldsymbol{a}_\beta(y) \delta y^\beta} + |\boldsymbol{\delta y}|\, \eta(\boldsymbol{\delta y}) \quad \text{with} \lim_{\boldsymbol{\delta y} \to \boldsymbol{0}} \boldsymbol{\eta}(\boldsymbol{\delta y}) = \boldsymbol{0}.\end{aligned}$$

In other words, the principal part with respect to $\boldsymbol{\delta y}$ of the length between the points $\boldsymbol{\theta}(y + \boldsymbol{\delta y})$ and $\boldsymbol{\theta}(y)$ is $\sqrt{\delta y^\alpha \boldsymbol{a}_\alpha(y) \cdot \boldsymbol{a}_\beta(y) \delta y^\beta}$. This observation suggests defining a matrix $(a_{\alpha\beta}(y))$ of order two by letting

$$a_{\alpha\beta}(y) := \boldsymbol{a}_\alpha(y) \cdot \boldsymbol{a}_\beta(y) = \left(\boldsymbol{\nabla\theta}(y)^T \boldsymbol{\nabla\theta}(y)\right)_{\alpha\beta}.$$

The elements $a_{\alpha\beta}(y)$ of this matrix, which is clearly *symmetric*, are called the **covariant components of the first fundamental form**, also called the **metric tensor**, *of the surface* $\widehat{\omega}$ at $\widehat{y} = \boldsymbol{\theta}(y)$. Note that the symmetric matrix $(a_{\alpha\beta}(y))$ is *positive-definite* since the vectors $\boldsymbol{a}_\alpha(y)$ are assumed to be linearly independent (as is immediately verified).

The two vectors $\boldsymbol{a}_\alpha(y)$ being thus defined, *the four relations*

$$\boldsymbol{a}^\alpha(y) \cdot \boldsymbol{a}_\beta(y) = \delta^\alpha_\beta$$

unambiguously define two linearly independent vectors $\boldsymbol{a}^\alpha(y)$ *in the tangent plane.* To see this, let *a priori* $\boldsymbol{a}^\alpha(y) = Y^{\alpha\sigma}(y)\boldsymbol{a}_\sigma(y)$ in the relations $\boldsymbol{a}^\alpha(y) \cdot \boldsymbol{a}_\beta(y) = \delta^\alpha_\beta$. This gives $Y^{\alpha\sigma}(y) a_{\sigma\beta}(y) = \delta^\alpha_\beta$, which means that $Y^{\alpha\sigma}(y) = a^{\alpha\sigma}(y)$, where

$$(a^{\alpha\beta}(y)) := (a_{\alpha\beta}(y))^{-1}.$$

Hence $\boldsymbol{a}^\alpha(y) = a^{\alpha\sigma}(y)\boldsymbol{a}_\sigma(y)$. These relations in turn imply that

$$\begin{aligned}\boldsymbol{a}^\alpha(y) \cdot \boldsymbol{a}^\beta(y) &= a^{\alpha\sigma}(y) a^{\beta\tau}(y) \boldsymbol{a}_\sigma(y) \cdot \boldsymbol{a}_\tau(y) \\ &= a^{\alpha\sigma}(y) a^{\beta\tau}(y) a_{\sigma\tau}(y) \;=\; a^{\alpha\sigma}(y)\delta^\beta_\sigma \;=\; a^{\alpha\beta}(y),\end{aligned}$$

and thus the vectors $\boldsymbol{a}^\alpha(y)$ are linearly independent since the matrix $(a^{\alpha\beta}(y))$ is positive-definite. One would likewise establish that $\boldsymbol{a}_\alpha(y) = a_{\alpha\beta}(y)\boldsymbol{a}^\beta(y)$.

The two vectors $\boldsymbol{a}^\alpha(y)$ form the **contravariant basis of the tangent plane** to the surface $\widehat{\omega}$ at $\widehat{y} = \boldsymbol{\theta}(y)$ (Figure 8.8-1), and the elements $a^{\alpha\beta}(y)$ of the symmetric matrix $(a^{\alpha\beta}(y))$ are called the **contravariant components of the first fundamental form**, or **metric tensor**, of the surface $\widehat{\omega}$ at $\widehat{y} = \boldsymbol{\theta}(y)$.

Let us record for convenience the fundamental relations satisfied by the vectors of the covariant and contravariant bases of the tangent plane and the covariant and contravariant components of the first fundamental tensor at a point $y \in \omega$ where the mapping $\boldsymbol{\theta}$ is an immersion:

$$\boldsymbol{a}_\alpha(y) = \partial_\alpha \boldsymbol{\theta}(y) \quad \text{and} \quad \boldsymbol{a}^\beta(y) \cdot \boldsymbol{a}_\alpha(y) = \delta_\alpha^\beta,$$

$$a_{\alpha\beta}(y) = \boldsymbol{a}_\alpha(y) \cdot \boldsymbol{a}_\beta(y), \quad a^{\alpha\beta}(y) = \boldsymbol{a}^\alpha(y) \cdot \boldsymbol{a}^\beta(y), \quad \text{and} \quad (a^{\alpha\beta}(y)) = (a_{\alpha\beta}(y))^{-1},$$

$$\boldsymbol{a}_\alpha(y) = a_{\alpha\beta}(y)\boldsymbol{a}^\beta(y) \quad \text{and} \quad \boldsymbol{a}^\alpha(y) = a^{\alpha\beta}(y)\boldsymbol{a}_\beta(y).$$

A mapping $\boldsymbol{\theta} : \omega \to \mathbb{E}^3$ is an **immersion** if it is an immersion at each point in ω, i.e., if $\boldsymbol{\theta}$ is differentiable in ω and the two vectors $\partial_\alpha \boldsymbol{\theta}(y)$ are linearly independent at each $y \in \omega$. In this case, the vector fields $\boldsymbol{a}_\alpha : \omega \to \mathbb{R}^3$ and $\boldsymbol{a}^\alpha : \omega \to \mathbb{R}^3$ respectively form the **covariant**, and **contravariant, bases of the tangent planes**.

Remark The presentation in this section closely follows that of Section 8.2, the mapping $\boldsymbol{\theta} : \omega \subset \mathbb{R}^2 \to \mathbb{E}^3$ "replacing" the mapping $\boldsymbol{\Theta} : \Omega \subset \mathbb{R}^n \to \mathbb{E}^n$. There are indeed strong *similarities* between the two presentations, such as the way the metric tensor is defined in both cases, but there are also sharp *differences*. In particular, $\boldsymbol{\nabla\theta}(y)$ is *not* a square matrix, while $\boldsymbol{\nabla\Theta}(x)$ is a square matrix. □

We now review fundamental formulas showing how *areas and lengths on the surface* $\widehat{\omega} = \boldsymbol{\theta}(\omega)$ (Figure 8.9-1) are computed by means of integrals *inside* ω, whose integrands are functions of the *covariant components of the first fundamental form of the surface*, thus *in fine* in terms of the curvilinear coordinates used in $\widehat{\omega}$.

These formulas highlight the crucial role played by the matrix field $(a_{\alpha\beta}) : \omega \to \mathbb{S}^2_>$ for computing "metric" notions on the surface $\boldsymbol{\theta}(\omega)$.

Theorem 8.9-1 (areas and lengths on a surface) *Let ω be an open subset of $\mathbb{R}^2$, let $\boldsymbol{\theta} \in \mathcal{C}^1(\omega; \mathbb{E}^3)$ be an injective immersion of class $\mathcal{C}^1$, and let $\widehat{\omega} = \boldsymbol{\theta}(\omega)$.*

(a) *Let A be an open subset of ω, let $\widehat{A} := \boldsymbol{\theta}(A)$, and let a function $\widehat{f} \in L^1(\widehat{A})$ be given, and let $\mathrm{d}\widehat{a}(\widehat{y})$ denote the area element along the surface $\widehat{\omega}$ at the point $\widehat{y} \in \widehat{\omega}$. Then*

$$\int_{\widehat{A}} \widehat{f}(\widehat{y})\,\mathrm{d}\widehat{a}(\widehat{y}) = \int_A (\widehat{f} \circ \boldsymbol{\theta})(y)\sqrt{a(y)}\,\mathrm{d}y, \quad \textit{where } a(y) := \det(a_{\alpha\beta}(y)) \textit{ at each } y \in \omega.$$

In particular, the **area** *of $\widehat{A}$ is given by*

$$\operatorname{area} \widehat{A} = \int_A \sqrt{a(y)}\,\mathrm{d}y.$$

(b) *Let $C = \boldsymbol{f}(I)$ be a curve in ω, where I is a compact interval of $\mathbb{R}$ and $\boldsymbol{f} = f^\alpha \boldsymbol{e}_\alpha \in \mathcal{C}^1(I; \mathbb{R}^2)$ is an injective mapping such that $\boldsymbol{f}(I) \subset \omega$ and $\dfrac{\mathrm{d}f^\alpha}{\mathrm{d}t}(t)\boldsymbol{e}_\alpha \neq \boldsymbol{0}$ for all $t \in I$. Then*

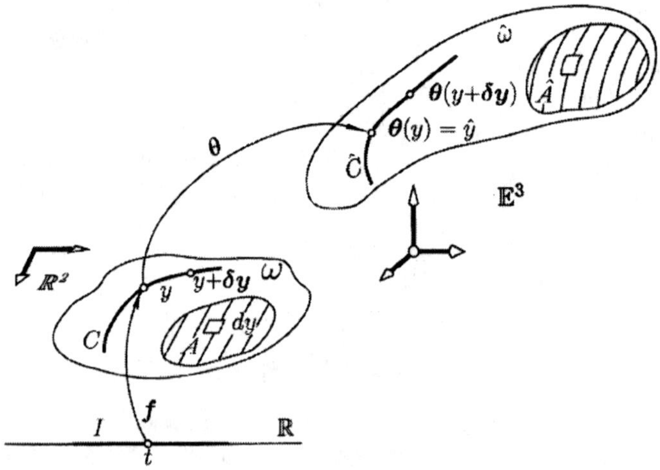

Figure 8.9-1 *Areas and lengths on a surface.* Let A be an open subset of ω and let $C = \boldsymbol{f}(I)$ be a curve in ω, where I is a compact interval of $\mathbb{R}$. Then the area of $\widehat{A} := \boldsymbol{\theta}(A) \subset \widehat{\omega}$ and the length of the curve $\widehat{C} = \boldsymbol{\theta}(C) \subset \widehat{\omega}$ are computed by means of the covariant components of the first fundamental form of the surface $\widehat{\omega}$; cf. Theorem 8.9-1. This figure originally appeared in P.G. CIARLET [2005]: *An Introduction to Differential Geometry with Applications to Elasticity*, Springer, Dordrecht.

the **length** *of the curve* $\widehat{C} := \boldsymbol{\theta}(C) \subset \widehat{\omega}$ *is given by*

$$\operatorname{length} \widehat{C} = \int_I \sqrt{a_{\alpha\beta}(\boldsymbol{f}(t)) \frac{\mathrm{d}f^\alpha}{\mathrm{d}t}(t) \frac{\mathrm{d}f^\beta}{\mathrm{d}t}(t)}\, \mathrm{d}t.$$

Proof The formula given in (a) is the special case $n = 2$ of the formula defining the n-dimensional area (Section 1.17).

By the formula giving the length of a curve (Section 1.17),

$$\operatorname{length} \widehat{C} := \int_I \left| \frac{\mathrm{d}\widehat{\boldsymbol{f}}}{\mathrm{d}t}(t) \right| \mathrm{d}t \quad \text{where } \widehat{\boldsymbol{f}} := \boldsymbol{\theta} \circ \boldsymbol{f}.$$

At each $t \in I$, the relation

$$\frac{\mathrm{d}\widehat{\boldsymbol{f}}}{\mathrm{d}t}(t) = \frac{\mathrm{d}}{\mathrm{d}t}(\boldsymbol{\theta}(\boldsymbol{f}(t))) = \frac{\mathrm{d}f^\alpha}{\mathrm{d}t}(t)\partial_\alpha \boldsymbol{\theta}(\boldsymbol{f}(t)) = \frac{\mathrm{d}f^\alpha}{\mathrm{d}t}(t)\boldsymbol{a}_\alpha(\boldsymbol{f}(t))$$

then shows that

$$\left| \frac{\mathrm{d}\widehat{\boldsymbol{f}}}{\mathrm{d}t}(t) \right|^2 = \left(\frac{\mathrm{d}f^\alpha}{\mathrm{d}t}(t)\boldsymbol{a}_\alpha(f(t)) \right) \cdot \left(\frac{\mathrm{d}f^\beta}{\mathrm{d}t}(t)\boldsymbol{a}_\beta(\boldsymbol{f}(t)) \right) = a_{\alpha\beta}(\boldsymbol{f}(t)) \frac{\mathrm{d}f^\alpha}{\mathrm{d}t}(t) \frac{\mathrm{d}f^\beta}{\mathrm{d}t}(t),$$

which proves (b). □

Remark The result of (b) shows that the length element $\mathrm{d}\widehat{\ell}(\widehat{y})$ at $\widehat{y} = \boldsymbol{\theta}(y) \in \widehat{\omega}$ is given by

$$\mathrm{d}\widehat{\ell}(\widehat{y}) = \sqrt{\delta y^\alpha a_{\alpha\beta}(y) \delta y^\beta}.$$

This expression recalls that $\mathrm{d}\widehat{\ell}(\widehat{y})$ is by definition the principal part with respect to $\boldsymbol{\delta y} = \delta y^\alpha \boldsymbol{e}_\alpha$ of the length $|\boldsymbol{\theta}(y+\boldsymbol{\delta y}) - \boldsymbol{\theta}(y)|$, whose expression precisely led to the introduction of the matrix $(a_{\alpha\beta}(y))$. □

The relation established in (b) expresses that *the lengths of curves inside the surface $\boldsymbol{\theta}(\omega)$ are precisely those induced by the Euclidean metric of the space $\mathbb{E}^3$.*

Finally, we indicate how *angles between intersecting curves* drawn on a surface can be computed.

Theorem 8.9-2 (angles on a surface) *Let ω be an open subset of $\mathbb{R}^2$ and let $\boldsymbol{\theta} \in \mathcal{C}^1(\omega; \mathbb{E}^3)$ be an injective immersion. Let $\boldsymbol{\theta}(\boldsymbol{f}(I))$ and $\boldsymbol{\theta}(\boldsymbol{g}(J))$ be two curves drawn on the surface $\widehat{\omega} := \boldsymbol{\theta}(\omega)$, where $\boldsymbol{f} = f^\alpha \boldsymbol{e}_\alpha \in \mathcal{C}^1(I; \mathbb{R}^2)$ and $\boldsymbol{g} = g^\alpha \boldsymbol{e}_\alpha \in \mathcal{C}^1(J; \mathbb{R}^2)$ are injective mappings such that $\boldsymbol{f}(I) \subset \omega$ and $\boldsymbol{g}(J) \subset \omega$. Assume that these two curves intersect at a point $\widehat{y} := \boldsymbol{\theta}(\boldsymbol{f}(t)) = \boldsymbol{\theta}(\boldsymbol{g}(\tau))$ and that $\frac{\mathrm{d}f^\alpha}{\mathrm{d}t}(t)\boldsymbol{e}_\alpha \neq \mathbf{0}$ and $\frac{\mathrm{d}g^\alpha}{\mathrm{d}t}(\tau)\boldsymbol{e}_\alpha \neq \mathbf{0}$. Then the cosine of the angle χ between the tangents to these two curves at $\widehat{y}$ is given by*

$$\cos\chi = \frac{a_{\alpha\beta}(\boldsymbol{f}(t))\frac{\mathrm{d}f^\alpha}{\mathrm{d}t}(t)\frac{\mathrm{d}g^\beta}{\mathrm{d}t}(\tau)}{\sqrt{a_{\alpha\beta}(\boldsymbol{f}(t))\frac{\mathrm{d}f^\alpha}{\mathrm{d}t}(t)\frac{\mathrm{d}f^\beta}{\mathrm{d}t}(t)}\sqrt{a_{\alpha\beta}(\boldsymbol{g}(\tau))\frac{\mathrm{d}g^\alpha}{\mathrm{d}t}(\tau)\frac{\mathrm{d}g^\beta}{\mathrm{d}t}(\tau)}}.$$

Proof The tangent vectors at the point $\widehat{y}$ to the curves $\boldsymbol{\theta}(\boldsymbol{f}(I))$ and $\boldsymbol{\theta}(\boldsymbol{g}(J))$ are respectively given by

$$\frac{\mathrm{d}(\boldsymbol{\theta}\circ\boldsymbol{f})}{\mathrm{d}t}(t) = \partial_\alpha\boldsymbol{\theta}(\boldsymbol{f}(t))\,\frac{\mathrm{d}f^\alpha}{\mathrm{d}t}(t) = \frac{\mathrm{d}f^\alpha}{\mathrm{d}t}(t)\boldsymbol{a}_\alpha(\boldsymbol{f}(t)),$$
$$\frac{\mathrm{d}(\boldsymbol{\theta}\circ\boldsymbol{g})}{\mathrm{d}\tau}(\tau) = \partial_\beta\boldsymbol{\theta}(\boldsymbol{g}(\tau))\frac{\mathrm{d}g^\beta}{\mathrm{d}\tau}(\tau) = \frac{\mathrm{d}g^\beta}{\mathrm{d}\tau}(\tau)\boldsymbol{a}_\beta(\boldsymbol{g}(\tau)).$$

The cosine of the angle χ between these two vectors therefore satisfies

$$\left(\frac{\mathrm{d}f^\alpha}{\mathrm{d}t}(t)\boldsymbol{a}_\alpha(\boldsymbol{f}(t))\right)\cdot\left(\frac{\mathrm{d}g^\beta}{\mathrm{d}\tau}(\tau)\boldsymbol{a}_\beta(\boldsymbol{g}(\tau))\right) = \left|\frac{\mathrm{d}f^\alpha}{\mathrm{d}t}(t)\boldsymbol{a}_\alpha(\boldsymbol{f}(t))\right|\,\left|\frac{\mathrm{d}g^\beta}{\mathrm{d}\tau}(\tau)\boldsymbol{a}_\beta(\boldsymbol{g}(\tau))\right|\cos\chi,$$

which shows that $\cos\chi$ is given by the announced formula. □

Remark In particular, the cosine of the angle $\varphi(y)$ between the two coordinate lines passing through the point $\widehat{y} = \boldsymbol{\theta}(y)$ is given by $\cos\varphi(y) = \dfrac{a_{12}(y)}{\sqrt{a_{11}(y)a_{22}(y)}}$. □

Problems

8.9-1 (1) Let a portion of a sphere in $\mathbb{E}^3$ be equipped with one of the curvilinear coordinates shown in Figure 8.8-2. Show that, in each instance, the coordinate lines are portions of circles.

(2) Let a portion of a torus in $\mathbb{E}^3$ be equipped with the curvilinear coordinates shown in Figure 8.8-3. Show that the coordinate lines are portions of circles.

(3) Are there other types of surfaces in $\mathbb{E}^3$ whose coordinate lines are portions of circles?

8.9-2 Compute the area of a torus, parametrized as in Figure 8.8-3.

8.10 Isometric, equiareal, and conformal surfaces

Let ω be an open subset of $\mathbb{R}^2$ and let $\boldsymbol{\theta} : \omega \to \mathbb{E}^3$ and $\widetilde{\boldsymbol{\theta}} : \omega \to \mathbb{E}^3$ be two injective immersions of class $\mathcal{C}^1$. The two surfaces $\boldsymbol{\theta}(\omega)$ and $\widetilde{\boldsymbol{\theta}}(\omega)$ are said to be **isometric** if *lengths are preserved*, i.e., if given *any* curve $C = \boldsymbol{f}(I)$ in ω, where I is a compact interval of $\mathbb{R}$ and $\boldsymbol{f} = f^\alpha \boldsymbol{e}_\alpha \in \mathcal{C}^1(I;\mathbb{R}^2)$ is an injective mapping such that $\boldsymbol{f}(I) \subset \omega$ and $\frac{\mathrm{d}f^\alpha}{\mathrm{d}t}(t)\boldsymbol{e}_\alpha \neq \boldsymbol{0}$ for all $t \in I$, the lengths of the curves $\boldsymbol{\theta}(C)$ and $\widetilde{\boldsymbol{\theta}}(C)$ are equal.

The two surfaces $\boldsymbol{\theta}(\omega)$ and $\widetilde{\boldsymbol{\theta}}(\omega)$ are said to be **equiareal** if *areas are preserved*, i.e., if, given any open set $A \subset \omega$ such that $\overline{A}$ is compact, the areas of the surfaces $\boldsymbol{\theta}(A)$ and $\widetilde{\boldsymbol{\theta}}(A)$ are equal.

The two surfaces $\boldsymbol{\theta}(\omega)$ and $\widetilde{\boldsymbol{\theta}}(\omega)$ are said to be **conformal** if *angles between tangents to intersecting curves are preserved.*

It thus follows from the formulas giving lengths, areas, and angles on a surface (Theorems 8.9-1 and 8.9-2) that, *if the two surfaces $\boldsymbol{\theta}(\omega)$ and $\widetilde{\boldsymbol{\theta}}(\omega)$ share the same first fundamental form, they are isometric, equiareal,* and *conformal.* We now examine to what extend the converse properties hold.

Remark Such results will be put to a crucial use in our brief incursion into cartography (Section 8.15). □

To begin with, we show that *two isometric surfaces necessarily share the same first fundamental form.*

Theorem 8.10-1 *Let ω be an open subset of $\mathbb{R}^2$ and let $\boldsymbol{\theta} \in \mathcal{C}^1(\omega;\mathbb{E}^3)$ and $\widetilde{\boldsymbol{\theta}} \in \mathcal{C}^1(\omega;\mathbb{E}^3)$ be two injective immersions such that the two surfaces $\boldsymbol{\theta}(\omega)$ and $\widetilde{\boldsymbol{\theta}}(\omega)$ are isometric.*

Then the first fundamental forms $(a_{\alpha\beta}) : \omega \to \mathbb{S}^2_>$ and $(\widetilde{a}_{\alpha\beta}) : \widetilde{\omega} \to \mathbb{S}^2_>$ of the surfaces $\boldsymbol{\theta}(\omega)$ and $\widetilde{\boldsymbol{\theta}}(\widetilde{\omega})$ are equal, i.e.,

$$a_{\alpha\beta}(y) = \widetilde{a}_{\alpha\beta}(y) \quad \text{at each } y \in \omega.$$

Proof Without loss of generality, assume that $I = [0,1]$, and let $I(t) := [0,t]$ for each $t \in I$. Assume that, for all curves $C = \boldsymbol{f}(I)$ in ω, where $\boldsymbol{f} = f^\alpha \boldsymbol{e}_\alpha \in \mathcal{C}^1(I;\mathbb{R}^2)$ is any injective mapping such that $\boldsymbol{f}(I) \subset \omega$ and $\frac{\mathrm{d}f^\alpha}{\mathrm{d}t}(t)\boldsymbol{e}_\alpha \neq \boldsymbol{0}$ at each $t \in I$, the lengths of the curves $\boldsymbol{\theta}(\boldsymbol{f}(I))$ and $\widetilde{\boldsymbol{\theta}}(\boldsymbol{f}(I))$ are equal. Then, by assumption,

$$\int_{I(t)} \left|\frac{\mathrm{d}(\boldsymbol{\theta} \circ \boldsymbol{f})}{\mathrm{d}\tau}(\tau)\right| \mathrm{d}\tau = \int_{I(t)} \left|\frac{\mathrm{d}(\widetilde{\boldsymbol{\theta}} \circ \boldsymbol{f})}{\mathrm{d}\tau}(\tau)\right| \mathrm{d}\tau \quad \text{at each } t \in I,$$

for any injective mapping $\boldsymbol{f} = f^\alpha \boldsymbol{e}_\alpha \in \mathcal{C}^1(I;\mathbb{R}^2)$ such that $\boldsymbol{f}(I) \subset \omega$ and $\frac{\mathrm{d}f^\alpha}{\mathrm{d}\tau}(\tau)\boldsymbol{e}_\alpha \neq \boldsymbol{0}$ at each $\tau \in I$. Differentiating this equality with respect to $t \in I$ then shows that

$$\left|\frac{\mathrm{d}(\boldsymbol{\theta} \circ \boldsymbol{f})}{\mathrm{d}t}(t)\right| = \left|\frac{\mathrm{d}(\widetilde{\boldsymbol{\theta}} \circ \boldsymbol{f})}{\mathrm{d}t}(t)\right| \quad \text{at each } t \in I,$$

or equivalently, that

$$a_{\alpha\beta}(\boldsymbol{f}(t))\frac{\mathrm{d}f^\alpha}{\mathrm{d}t}(t)\frac{\mathrm{d}f^\beta}{\mathrm{d}t}(t) = \widetilde{a}_{\sigma\tau}(\boldsymbol{f}(t))\frac{\mathrm{d}f^\sigma}{\mathrm{d}t}(t)\frac{\mathrm{d}f^\tau}{\mathrm{d}t}(t) \quad \text{at each } t \in I.$$

Given any $t \in I$ and any nonzero vector $(\xi^\alpha) \in \mathbb{R}^2$, there exists a mapping $\boldsymbol{f}$ of the above type such that $\dfrac{\mathrm{d}f^\alpha}{\mathrm{d}t}(t) = \xi^\alpha$. Consequently

$$a_{\alpha\beta}(\boldsymbol{f}(t)) = \widetilde{a}_{\alpha\beta}(\boldsymbol{f}(t)) \quad \text{at each } t \in I,$$

which in turn implies that

$$a_{\alpha\beta}(y) = \widetilde{a}_{\alpha\beta}(y) \quad \text{at each } y \in \omega. \qquad \square$$

Remark Theorem 8.10-1 can be easily extended to the more general case where the two surfaces are defined by means of injective immersions $\boldsymbol{\theta} \in \mathcal{C}^1(\omega; \mathbb{E}^3)$ and $\widetilde{\boldsymbol{\theta}} \in \mathcal{C}^1(\widetilde{\omega}; \mathbb{E}^3)$ defined on *different* open subsets $\omega \in \mathbb{R}^2$ and $\widetilde{\omega} \in \mathbb{R}^2$ and there exists a $\mathcal{C}^1$-diffeomorphism $\boldsymbol{\chi} = \chi^\alpha \boldsymbol{e}_\alpha$ from ω onto $\widetilde{\omega}$. Then the first fundamental forms $(a_{\alpha\beta}) : \omega \to \mathbb{S}^2_>$ and $(\widetilde{a}_{\alpha\beta}) : \widetilde{\omega} \to \mathbb{S}^2_>$ are necessarily related in this case by

$$a_{\alpha\beta}(y) = \widetilde{a}_{\sigma\tau}(\widetilde{y})\partial_\alpha \chi^\sigma(y)\partial_\beta \chi^\tau(y) \quad \text{at each } \widetilde{y} = \chi(y) \in \widetilde{\omega},$$

or equivalently, by

$$\partial_\alpha \boldsymbol{\theta} \cdot \partial_\beta \boldsymbol{\theta} = \partial_\alpha(\widetilde{\boldsymbol{\theta}} \circ \chi) \cdot \partial_\beta(\widetilde{\boldsymbol{\theta}} \circ \chi) \quad \text{in } \omega. \qquad \square$$

Examples of isometric surfaces include of course surfaces $\boldsymbol{\theta}(\omega)$ and $\widetilde{\boldsymbol{\theta}}(\omega)$ that are equal *up to an isometry of* $\mathbb{R}^3$, i.e., such that $\widetilde{\boldsymbol{\theta}}(y) = \boldsymbol{c} + \boldsymbol{Q}\boldsymbol{\theta}(y)$, $y \in \omega$, for some vector $\boldsymbol{c} \in \mathbb{E}^3$ and some orthogonal matrix $\boldsymbol{Q} \in \mathbb{O}^3$, since

$$\widetilde{a}_{\alpha\beta}(y) = \partial_\alpha \widetilde{\boldsymbol{\theta}}(y) \cdot \partial_\beta \widetilde{\boldsymbol{\theta}}(y) = \partial_\alpha \boldsymbol{\theta}(y) \cdot \partial_\beta \boldsymbol{\theta}(y) = a_{\alpha\beta}(y) \quad \text{at each } y \in \omega$$

in this case.

Remark Such surfaces share in addition the same *second* fundamental form, which will be introduced in the next section. $\square$

Examples that are not as simple include *developable surfaces*; these surfaces, which are at least locally *isometric to a portion of a plane*, will be briefly introduced in Section 8.12. Even more difficult examples include *nonspherical surfaces that are isometric to a portion of a sphere*[23]). Note, however, that *any "closed" surface that is isometric to a sphere is a sphere.*[24]

We next characterize equiareal and conformal surfaces and their relation to isometric surfaces.

Theorem 8.10-2 *Let the notations and assumptions be as in* Theorem 8.10-1.

(a) *The two surfaces* $\boldsymbol{\theta}(\omega)$ *and* $\widetilde{\boldsymbol{\theta}}(\omega)$ *are equiareal if and only if*

$$\det(a_{\alpha\beta}(y)) = \det(\widetilde{a}_{\alpha\beta}(y)) \quad \textit{at each } y \in \omega.$$

---

[23]Such examples were already known as far back as 1888 to Augustus Edward Hough Love (1863–1940), the author of the famous two-volume *Treatise on the Mathematical Theory of Elasticity*, published in 1893.

[24]This result is due to:

H. LIEBMANN [1899]: Eine neue Eigenschaft der Kugel, *Nachrichten von der Gesellschaft der Wissenschaften zu Göttingen, Mathematisch-Physikalische Klasse*, 45–55.

For a "modern" proof, see DO CARMO [1976, Section 5.2, Theorem 1].

(b) *The two surfaces $\boldsymbol{\theta}(\omega)$ and $\widetilde{\boldsymbol{\theta}}(\widetilde{\omega})$ are conformal if and only if there exists at each $y \in \omega$ a constant $C(y) > 0$ such that*

$$a_{\alpha\beta}(y) = C(y)\widetilde{a}_{\alpha\beta}(y).$$

(c) *The two surfaces $\boldsymbol{\theta}(\omega)$ and $\widetilde{\boldsymbol{\theta}}(\widetilde{\omega})$ are both equiareal and conformal if and only if they are isometric.*

Proof The "if" parts for (a), (b), and (c) follow from Theorems 8.9-1, 8.9-2, and 8.10-1. If the two surfaces are equiareal,

$$\int_A \sqrt{\det(a_{\alpha\beta}(y))}\,dy = \int_A \sqrt{\det(\widetilde{a}_{\alpha\beta}(y))}\,dy$$

for each open subset A of ω such that $\overline{A}$ is compact by Theorem 8.9-1. Therefore, $\det(a_{\alpha\beta}(y)) = \det(\widetilde{a}_{\alpha\beta}(y))$ at each $y \in \omega$, since both functions $y \in \omega \to \sqrt{\det(a_{\alpha\beta}(y))}$ and $y \in \omega \to \sqrt{\det(\widetilde{a}_{\alpha\beta}(y))}$ are continuous by assumptions.

The "only if" part of (a) is thus proved. The "only if" part of (b) reduces to a simple exercise about matrices and for this reason is left as a problem (Problem 8.10-1). The "only if" part of (c) is an immediate consequence of the "only if" parts of (a) and (b) combined with Theorem 8.9-1. □

Problem

8.10-1 Let the notations and assumptions be as in Theorem 8.9-2. Show that, if the two surfaces $\boldsymbol{\theta}(\omega)$ and $\widetilde{\boldsymbol{\theta}}(\omega)$ are conformal, there exists at each $y \in \omega$ a constant $C(y) > 0$ such that $a_{\alpha\beta}(y) = C(y)\widetilde{a}_{\alpha\beta}(y)$.

Hint: Given $y \in \omega$, let $\boldsymbol{A}(y) := (a_{\alpha\beta}(y))$ and $\widetilde{\boldsymbol{A}}(y) := (\widetilde{a}_{\alpha\beta}(y))$. Show that, for all nonzero vectors $\boldsymbol{\xi} \in \mathbb{R}^2$ and $\boldsymbol{\eta} \in \mathbb{R}^2$,

$$\frac{(\boldsymbol{A}(y)\boldsymbol{\xi}) \cdot \boldsymbol{\eta}}{\sqrt{(\boldsymbol{A}(y)\boldsymbol{\xi}) \cdot \boldsymbol{\xi}}\sqrt{(\boldsymbol{A}(y)\boldsymbol{\eta}) \cdot \boldsymbol{\eta}}} = \frac{(\widetilde{\boldsymbol{A}}(y)\boldsymbol{\xi}) \cdot \boldsymbol{\eta}}{\sqrt{(\widetilde{\boldsymbol{A}}(y)\boldsymbol{\xi}) \cdot \boldsymbol{\xi}}\sqrt{(\widetilde{\boldsymbol{A}}(y)\boldsymbol{\eta}) \cdot \boldsymbol{\eta}}}.$$

Therefore the assertion reduces to establishing a property of positive-definite symmetric matrices.

8.11 Second fundamental form of a surface; curvature on a surface

Letting $n = 3$ in Theorems 8.6-1 and 8.7-1 shows that the image $\boldsymbol{\Theta}(\Omega) \subset \mathbb{E}^3$ of a *three-dimensional* open set $\Omega \subset \mathbb{R}^3$ by a smooth enough immersion $\boldsymbol{\Theta} : \Omega \subset \mathbb{R}^n \to \mathbb{E}^3$ is well defined by its metric (uniquely up to isometries in $\mathbb{E}^3$), provided that the compatibility conditions $R_{qijk} = 0$ in Ω are satisfied by the covariant components $g_{ij} : \Omega \to \mathbb{R}$ of its *metric tensor*. By contrast, a *surface* given as the image $\boldsymbol{\theta}(\omega) \subset \mathbb{E}^3$ of a *two-dimensional* open set $\omega \subset \mathbb{R}^2$ by a smooth enough immersion $\boldsymbol{\theta} : \omega \subset \mathbb{R}^2 \to \mathbb{E}^3$ *cannot* be defined by its metric alone.

As intuitively suggested by Figure 8.11-1, the missing information is provided by the "curvature" of a surface. A proper way to give substance to this otherwise vague notion

$$\left|\frac{\mathrm{d}\rho}{\mathrm{d}s}(s)\right|^{-1} = \left|\frac{\mathrm{d}(\boldsymbol{\theta}\circ\boldsymbol{f})}{\mathrm{d}t}(t)\right| = \sqrt{a_{\alpha\beta}(\boldsymbol{f}(t))\frac{\mathrm{d}f^\alpha}{\mathrm{d}t}(t)\frac{\mathrm{d}f^\beta}{\mathrm{d}t}(t)}. \qquad \square$$

The elements $b_{\alpha\beta}(y)$ of the symmetric matrix $(b_{\alpha\beta}(y))$ defined in Theorem 8.11-1 are called the **covariant components of the second fundamental form** of the surface $\widehat{\omega} = \boldsymbol{\theta}(\omega)$ at $\widehat{y} = \boldsymbol{\theta}(y)$.

Problems

8.11-1 Let ω be an open subset of $\mathbb{R}^2$ and let $\boldsymbol{\theta} : \omega \to \mathbb{R}^3$ be a mapping that is differentiable at a point $y \in \omega$. Show that $|\boldsymbol{a}_1(y) \wedge \boldsymbol{a}_2(y)| = \sqrt{a(y)}$, where $\boldsymbol{a}_\alpha(y) := \partial_\alpha \boldsymbol{\theta}(y)$ and $a(y) := \det(\boldsymbol{a}_\alpha(y) \cdot \boldsymbol{a}_\beta(y))$.

Remark This relation can be also derived from **Lagrange's identity**,[25] which asserts that

$$|x|^2\,|y|^2 - |(x,y)|^2 = \sum_{1\le i<j\le n} |x_i y_j - x_j y_i|^2 \quad \text{for any vectors } x = (x_i)_{i=1}^n \in \mathbb{C}^n,\ y = (y_i)_{i=1}^n \in \mathbb{C}^n,$$

where $(\cdot,\cdot)$ and $|\cdot|$ respectively denote the Hermitian inner product and the associated norm over $\mathbb{C}^n$ (incidentally, note that this identity implies the Cauchy–Schwarz inequality in $\mathbb{C}^n$). Hence in particular,

$$|x|^2\,|y|^2 - (x,y)^2 = |x \wedge y|^2 \quad \text{for any } x, y \in \mathbb{R}^3. \qquad \square$$

8.11-2 (1) Compute the vectors of the covariant and contravariant bases and the covariant and contravariant components of the first and second fundamental form of a portion of a sphere equipped with the curvilinear coordinates shown in Figure 8.8-2.

(2) In each case, verify that the inverse of the radius of the sphere satisfies the relation established in Theorem 8.11-1.

(3) Carry out the same computations as in (1) for a portion of a torus equipped with the curvilinear coordinates shown in Figure 8.8-3.

(4) Carry out the same computation as in (1) for a portion of a hyperbolic paraboloid represented by a mapping of the form $\boldsymbol{\theta} : (x,y) \in \omega \subset \mathbb{R}^2 \to \left(x, y, \dfrac{c}{ab}xy\right) \in \mathbb{E}^3$, where a, b, and c are > 0 constants.

8.12 Principal curvatures; Gaussian curvature

The analysis of the previous section suggests that precise information about the shape of a surface $\widehat{\omega} = \boldsymbol{\theta}(\omega)$ in a neighborhood of one of its points $\widehat{y} = \boldsymbol{\theta}(y)$ can be gathered by letting the plane P turn around the normal vector $\boldsymbol{a}_3(y)$ and by following in this process the variations of the curvatures at $\widehat{y}$ of the corresponding planar curves $P \cap \widehat{\omega}$, as given in Theorem 8.11-1.

As a first step in this direction, we show that these curvatures span a *compact interval of* $\mathbb{R}$. In particular then, they "stay away from infinity."

Note that this compact interval contains 0 if, and only if, the radius of curvature of the curve $P \cap \widehat{\omega}$ is infinite for at least one such plane P.

[25] J.L. Lagrange [1773]: Solutions analytiques de quelques problèmes sur les pyramides triangulaires, *Mémoire de l'Académie Royale de Berlin.*

Theorem 8.12-1 *Consider the set $\mathcal{P}$ of all planes P normal to the surface $\widehat{\omega} = \boldsymbol{\theta}(\omega)$ at a point $\widehat{y} = \boldsymbol{\theta}(y)$, and assume that the assumptions of Theorem 8.11-1 hold for each $P \in \mathcal{P}$.*

(a) *When P varies in $\mathcal{P}$, the set of curvatures of the associated planar curves $P \cap \widehat{\omega}$ spans a compact interval of $\mathbb{R}$, denoted $\left[\frac{1}{R_1(y)}, \frac{1}{R_2(y)}\right]$.*

(b) *Let the matrix $(b_\alpha^\beta(y))$, α being the row index, be defined by*

$$b_\alpha^\beta(y) := a^{\beta\sigma}(y) b_{\alpha\sigma}(y),$$

where $(a^{\alpha\beta}(y)) = (a_{\alpha\beta}(y))^{-1}$ (Section 8.9) and the matrix $(b_{\alpha\beta}(y))$ is defined as in Theorem 8.11-1. Then

$$\frac{1}{R_1(y)} + \frac{1}{R_2(y)} = \operatorname{tr}(b_\alpha^\beta(y)) = b_1^1(y) + b_2^2(y),$$

$$\frac{1}{R_1(y)R_2(y)} = \det(b_\alpha^\beta(y)) = b_1^1(y)b_2^2(y) - b_1^2(y)b_2^1(y) = \frac{\det(b_{\alpha\beta}(y))}{\det(a_{\alpha\beta}(y))}.$$

(c) *If $\frac{1}{R_1(y)} \neq \frac{1}{R_2(y)}$, there is a unique pair of orthogonal planes $P_1 \in \mathcal{P}$ and $P_2 \in \mathcal{P}$ such that the curvatures of the associated planar curves $P_1 \cap \widehat{\omega}$ and $P_2 \cap \widehat{\omega}$ are precisely $\frac{1}{R_1(y)}$ and $\frac{1}{R_2(y)}$.*

Proof (i) Let Δ_P denote the intersection of $P \in \mathcal{P}$ with the tangent plane T to the surface $\widehat{\omega}$ at $\widehat{y}$, and let $\widehat{C}_P$ denote the intersection of P with $\widehat{\omega}$. Hence Δ_P is tangent to $\widehat{C}_P$ at $\widehat{y} \in \widehat{\omega}$.

In a sufficiently small neighborhood of $\widehat{y}$ the restriction of the curve $\widehat{C}_P$ to this neighborhood is given by $\widehat{C}_P = (\boldsymbol{\theta} \circ \boldsymbol{f}_P)(I_P)$, where $I_P \subset \mathbb{R}$ is an open interval and $\boldsymbol{f}_P = f_P^\alpha \boldsymbol{e}_\alpha : I_P \to \mathbb{R}^2$ is a smooth enough injective mapping that satisfies $\frac{\mathrm{d} f_P^\alpha}{\mathrm{d}t}(t)\boldsymbol{e}_\alpha \neq \mathbf{0}$, where $t \in I_P$ is such that $y = \boldsymbol{f}_P(t)$. Hence the line Δ_P is given by

$$\Delta_P = \left\{\widehat{y} + \lambda \frac{\mathrm{d}(\boldsymbol{\theta} \circ \boldsymbol{f}_P)}{\mathrm{d}t}(t);\ \lambda \in \mathbb{R}\right\} = \{\widehat{y} + \lambda \xi_P^\alpha \boldsymbol{a}_\alpha(y);\ \lambda \in \mathbb{R}\},$$

where $\xi_P^\alpha := \frac{\mathrm{d} f_P^\alpha}{\mathrm{d}t}(t)$ and $\xi_P^\alpha \boldsymbol{e}_\alpha \neq \mathbf{0}$ by assumption.

Since the line $\{y + \mu \xi_P^\alpha \boldsymbol{e}_\alpha;\ \mu \in \mathbb{R}\}$ is tangent to the curve $C_P := \boldsymbol{\theta}^{-1}(\widehat{C}_P)$ at $y \in \omega$ (the mapping $\boldsymbol{\theta} : \omega \to \mathbb{E}^3$ is injective by assumption) for each such parametrizing function $\boldsymbol{f}_P : I_P \to \mathbb{R}^2$ and since the vectors $\boldsymbol{a}_\alpha(y)$ are linearly independent, there exists a bijection between the set of all lines $\Delta_P \subset T$, $P \in \mathcal{P}$, and the set of all lines supporting the nonzero tangent vectors to the curves C_P.

Hence Theorem 8.11-1 shows that when P varies in $\mathcal{P}$, the curvature of the corresponding curves $\widehat{C}_P$ at $\widehat{y}$ takes the same values as does the ratio $\frac{b_{\alpha\beta}(y)\xi^\alpha\xi^\beta}{a_{\alpha\beta}(y)\xi^\alpha\xi^\beta}$ when $\boldsymbol{\xi} := \begin{pmatrix}\xi_1\\ \xi_2\end{pmatrix}$ varies in $\mathbb{R}^2 - \{\mathbf{0}\}$.

(ii) Let the symmetric matrices $\boldsymbol{A}$ and $\boldsymbol{B}$ of order two be defined by

$$\boldsymbol{A} := (a_{\alpha\beta}(y)) \quad \text{and} \quad \boldsymbol{B} := (b_{\alpha\beta}(y)).$$

Since $\boldsymbol{A}$ is positive-definite, it has a (unique) *square root* $\boldsymbol{C}$, i.e., a symmetric positive-definite matrix $\boldsymbol{C}$ such that $\boldsymbol{A} = \boldsymbol{C}^2$ (Theorem 7.14-3). Hence the ratio

$$\frac{b_{\alpha\beta}(y)\xi^\alpha\xi^\beta}{a_{\alpha\beta}(y)\xi^\alpha\xi^\beta} = \frac{\boldsymbol{\xi}^T\boldsymbol{B}\boldsymbol{\xi}}{\boldsymbol{\xi}^T\boldsymbol{A}\boldsymbol{\xi}} = \frac{\boldsymbol{\eta}^T\boldsymbol{C}^{-1}\boldsymbol{B}\boldsymbol{C}^{-1}\boldsymbol{\eta}}{\boldsymbol{\eta}^T\boldsymbol{\eta}}, \quad \text{where } \boldsymbol{\eta} := \boldsymbol{C}\boldsymbol{\xi},$$

is nothing but the *Rayleigh quotient* associated with the symmetric matrix $\boldsymbol{C}^{-1}\boldsymbol{B}\boldsymbol{C}^{-1}$. When $\boldsymbol{\eta}$ varies in $\mathbb{R}^2 - \{\boldsymbol{0}\}$, this Rayleigh quotient thus spans the compact interval of $\mathbb{R}$ whose end-points are the smallest and largest eigenvalue, respectively denoted $\frac{1}{R_1(y)}$ and $\frac{1}{R_2(y)}$, of the matrix $\boldsymbol{C}^{-1}\boldsymbol{B}\boldsymbol{C}^{-1}$[26]. This proves (a).

Furthermore, the relation

$$\boldsymbol{C}(\boldsymbol{C}^{-1}\boldsymbol{B}\boldsymbol{C}^{-1})\boldsymbol{C}^{-1} = \boldsymbol{B}\boldsymbol{C}^{-2} = \boldsymbol{B}\boldsymbol{A}^{-1}$$

shows that the eigenvalues of the symmetric matrix $\boldsymbol{C}^{-1}\boldsymbol{B}\boldsymbol{C}^{-1}$ coincide with those of the matrix $\boldsymbol{B}\boldsymbol{A}^{-1}$. Note that

$$\boldsymbol{B}\boldsymbol{A}^{-1} = \left(b_\alpha^\beta(y)\right) \quad \text{with } b_\alpha^\beta(y) := a^{\beta\sigma}(y)b_{\alpha\sigma}(y),$$

α being the row index, since $\boldsymbol{A}^{-1} = (a^{\alpha\beta}(y))$.

Hence the relations in (b) simply express that the sum and the product of the eigenvalues of the matrix $\boldsymbol{B}\boldsymbol{A}^{-1}$ are respectively equal to its trace and to its determinant, which may be also written as $\frac{\det(b_{\alpha\beta}(y))}{\det(a_{\alpha\beta}(y))}$ since $\boldsymbol{B}\boldsymbol{A}^{-1} = (b_\alpha^\beta(y))$. This proves (b).

(iii) Let $\boldsymbol{\eta}_1 = \begin{pmatrix}\eta_1^1\\ \eta_1^2\end{pmatrix} = \boldsymbol{C}\boldsymbol{\xi}_1$ and $\boldsymbol{\eta}_2 = \begin{pmatrix}\eta_2^1\\ \eta_2^2\end{pmatrix} = \boldsymbol{C}\boldsymbol{\xi}_2$, with $\boldsymbol{\xi}_1 = \begin{pmatrix}\xi_1^1\\ \xi_1^2\end{pmatrix}$ and $\boldsymbol{\xi}_2 = \begin{pmatrix}\xi_2^1\\ \xi_2^2\end{pmatrix}$, be two orthogonal eigenvectors of the symmetric matrix $\boldsymbol{C}^{-1}\boldsymbol{B}\boldsymbol{C}^{-1}$ corresponding to the eigenvalues $\frac{1}{R_1(y)}$ and $\frac{1}{R_2(y)}$, respectively. Hence

$$0 = \boldsymbol{\eta}_1^T\boldsymbol{\eta}_2 = \boldsymbol{\xi}_1^T\boldsymbol{C}^T\boldsymbol{C}\boldsymbol{\xi}_2 = \boldsymbol{\xi}_1^T\boldsymbol{A}\boldsymbol{\xi}_2,$$

since $\boldsymbol{C}^T = \boldsymbol{C}$. By (i), the corresponding lines Δ_{P_1} and Δ_{P_2} of the tangent plane are parallel to the vectors $\xi_1^\alpha\boldsymbol{a}_\alpha(y)$ and $\xi_2^\beta\boldsymbol{a}_\beta(y)$, which are orthogonal since

$$(\xi_1^\alpha\boldsymbol{a}_\alpha(y)) \cdot (\xi_2^\beta\boldsymbol{a}_\beta(y)) = a_{\alpha\beta}(y)\xi_1^\alpha\xi_2^\beta = \boldsymbol{\xi}_1^T\boldsymbol{A}\boldsymbol{\xi}_2.$$

If $\frac{1}{R_1(y)} \neq \frac{1}{R_2(y)}$, the directions of the vectors $\boldsymbol{\eta}_1$ and $\boldsymbol{\eta}_2$ are uniquely determined and the lines Δ_{P_1} and Δ_{P_2} are likewise uniquely determined and orthogonal. This proves (c). □

[26]For a proof, see, e.g., CIARLET [1987, Theorem 1.3-1].

We are now in a position to state several fundamental *definitions*:

The elements $b_\alpha^\beta(y)$ of the (in general nonsymmetric) matrix $(b_\alpha^\beta(y))$ defined in Theorem 2.6-1 are called the **mixed components of the second fundamental form** of the surface $\widehat{\omega} = \boldsymbol{\theta}(\omega)$ at $\widehat{y} = \boldsymbol{\theta}(y)$.

The real numbers $\dfrac{1}{R_1(y)}$ and $\dfrac{1}{R_2(y)}$ (one or both being possibly equal to 0) found in Theorem 8.12-1 are called the **principal curvatures** of $\widehat{\omega}$ at $\widehat{y}$.

If $\dfrac{1}{R_1(y)} = \dfrac{1}{R_2(y)}$, the curvatures of the planar curves $P \cap \widehat{\omega}$ are the same in all directions, i.e., for all $P \in \mathcal{P}$. If $\dfrac{1}{R_1(y)} = \dfrac{1}{R_2(y)} = 0$, the point $\widehat{y} = \boldsymbol{\theta}(y)$ is called a **planar point**. If $\dfrac{1}{R_1(y)} = \dfrac{1}{R_2(y)} \neq 0$, the point $\widehat{y}$ is called an **umbilical point**.

If $\dfrac{1}{R_1(y)} \neq 0$ *and* $\dfrac{1}{R_2(y)} \neq 0$, the real numbers $R_1(y)$ and $R_2(y)$ are called the **principal radii of curvature** of $\widehat{\omega}$ at $\widehat{y}$. Recall that if (for instance) $\dfrac{1}{R_1(y)} = 0$, the corresponding radius of curvature $R_1(y)$ is said to be *infinite*, according to the convention made in Section 8.11. While the principal radii of curvature may simultaneously change their signs in another system of curvilinear coordinates, the associated *centers of curvature* are intrinsically defined.

The numbers $\dfrac{1}{2}\Big(\dfrac{1}{R_1(y)} + \dfrac{1}{R_2(y)}\Big)$ and $\dfrac{1}{R_1(y)R_2(y)}$, which are the principal invariants of the matrix $(b_\alpha^\beta(y))$ (Theorem 8.12-1), are respectively called the **mean curvature** and the **Gaussian**, or **total, curvature** of the surface $\widehat{\omega}$ at $\widehat{y}$.

A point on a surface is an **elliptic, parabolic**, or **hyperbolic, point** according to whether its *Gaussian curvature* is > 0, $= 0$ but the point is not planar, or < 0; see Figure 8.12-1.

As already noted in Section 8.11, a surface in $\mathbb{E}^3$ cannot be defined by its *metric* alone, i.e., through its first fundamental form alone, since its *curvature* must be in addition specified through its second fundamental form. But quite surprisingly, the *Gaussian curvature* at a point can be expressed solely in terms of the functions $a_{\alpha\beta}$ and their derivatives! This is the celebrated *Theorema Egregium* ("astonishing theorem") of Gauß (which will be proved later; cf. Theorem 8.15-1).

Another striking result involving the *Gaussian curvature* is the equally celebrated **Gauß–Bonnet theorem**:[27] *Let S be a smooth enough, "closed," "orientable,"[28] and compact surface in $\mathbb{R}^3$ and let $K : S \to \mathbb{R}$ denote its Gaussian curvature. Then*

$$\int_S K(\widehat{y})\,\mathrm{d}\widehat{a}(\widehat{y}) = 2\pi(2 - 2g(S)),$$

where the **genus** $g(S)$ *is the number of "holes" of S* (for instance, a sphere has genus zero,

[27] A first proof (in a special case) appeared in:

C.F. GAUß [1827]: Disquisitiones generales circa superficies curvas, *Commentationes Societatis Regiae Scientiarum Gottingensis Recentiores* **6**, 99–146.

The first proof in the general case is due to:

O. BONNET [1848]: Mémoire sur la théorie générale des surfaces, *Journal de l'Ecole Polytechnique* **19**, 1–146.

For a "modern" proof, see, e.g., KLINGENBERG [1973, Theorem 6.3-5].

[28] A "closed" compact surface is one "without boundary," such as a sphere or a torus; "orientable" surfaces, which exclude for instance Klein bottles, are defined in, e.g., KLINGENBERG [1973, Section 5.5].

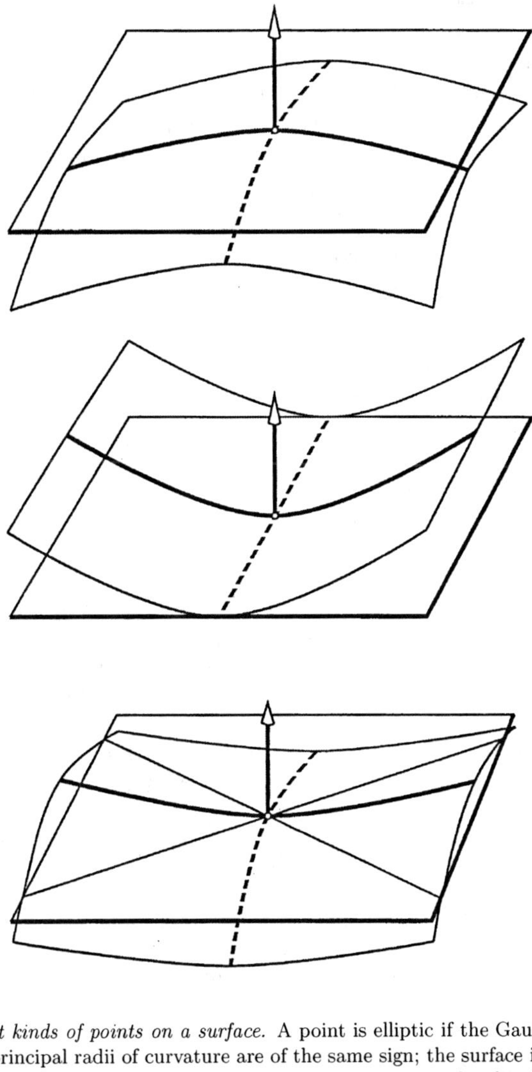

Figure 8.12-1 *Different kinds of points on a surface.* A point is elliptic if the Gaussian curvature is > 0, or equivalently, if the two principal radii of curvature are of the same sign; the surface is then locally on one side of its tangent plane. A point is parabolic if exactly one of the two principal radii of curvature is infinite; the surface is in general locally on one side of its tangent plane. A point is hyperbolic if the Gaussian curvature is < 0, or equivalently, if the two principal radii of curvature are of different signs; the surface then intersects its tangent plane along two curves.

Note that the surfaces on this figure are assumed to be portions of *quadrics*; this explains why on these particular surfaces some curves are in effect segments. This figure originally appeared in P.G. Ciarlet [2005]: *An Introduction to Differential Geometry with Applications to Elasticity*, Springer, Dordrecht.

while a torus has genus one); cf. Figure 8.12-2. The *integer* $\chi(S) \in \mathbb{Z}$ defined by $\chi(S) := (2 - 2g(S))$ is the **Euler characteristic** of S.

A **developable surface** is one whose *Gaussian curvature* vanishes everywhere.[29] A portion of a plane provides a first example, the only one of a developable surface in which *all* points are planar. Any developable surface in which *all* points are *parabolic* can be likewise fully described: It is *either* a portion of a cylinder, *or* a portion of a cone (Figure 8.11-1), *or* a portion of a surface spanned by the tangents to a skewed curve. The description of a developable surface comprising both planar and parabolic points is more subtle.[30]

The interest of developable surfaces is that they can be, at least locally, continuously "rolled out," or "developed" (hence their name), onto a *plane without changing the metric* of the intermediary surfaces in the process.

Problems

8.12-1 Let ω be an open subset of $\mathbb{R}^2$, let $\boldsymbol{\theta} \in \mathcal{C}^2(\omega; \mathbb{E}^3)$ be an injective immersion, and assume that the assumptions of Theorem 8.11-1 are satisfied at each point $y \in \omega$.

(1) Show that, if all the points of the surface $\widehat{\omega} := \boldsymbol{\theta}(\omega)$ are planar ($\frac{1}{R_1(y)} = \frac{1}{R_2(y)} = 0$ at each $y \in \omega$), then $\widehat{\omega}$ is a portion of a plane.[31]

(2) Show that, if all the points of the surface $\widehat{\omega} := \theta(\omega)$ are umbilical ($\frac{1}{R_1(y)} = \frac{1}{R_2(y)} \neq 0$ at each $y \in \omega$), then $\widehat{\omega}$ is a portion of a sphere.[32]

8.12-2 The notations and assumptions being as in Theorem 8.12-1, assume that $\widehat{y}$ is neither planar nor umbilical; in other words, the principal curvatures at $\widehat{y}$ are not equal. Then the two orthogonal lines tangent to the planar curves $P_1 \cap \widehat{\omega}$ and $P_2 \cap \widehat{\omega}$ (Theorem 8.12-1(c)) are called the *principal directions* at $\widehat{y}$. A *line of curvature* is a curve on $\widehat{\omega}$ that is tangent to a principal direction at each one of its points.

Show that a point that is neither planar nor umbilical possesses a neighborhood where two orthogonal families of lines of curvature can be chosen as coordinate lines.[33]

8.12-3 Let ω be an open subset of $\mathbb{R}^2$, let $\boldsymbol{\theta} \in \mathcal{C}^2(\omega; \mathbb{E}^3)$ be an injective immersion, and assume that the assumptions of Theorem 8.11-1 are satisfied at each point $y \in \omega$.

An *asymptotic line* is a curve on a surface that is everywhere tangent to a direction along which the radius of curvature is infinite; any point along an asymptotic line is thus either parabolic or hyperbolic. Show that if all the points of the surface $\boldsymbol{\theta}(\omega)$ are hyperbolic, any point possesses a neighborhood where two intersecting families of asymptotic lines can be chosen as coordinate lines.[34]

8.12-4 Let $K : S \to \mathbb{R}$ denote the Gaussian curvature along a torus S. Show by a direct computation that $\int_S \widehat{K}(\widehat{y})\,d\widehat{a}(\widehat{y}) = 0$.

---

[29] According to the definition in STOKER [1969, Chapter 5, Section 2]. A slightly different definition is given in KLINGENBERG [1973, Section 3.7].

[30] Although the above examples are in a sense the only ones possible, at least locally; see STOKER [1969, Chapter 5, Sections 2–6].

[31] See, e.g., STOKER [1969, Chapter 4, Section 11].

[32] See, e.g., STOKER [1969, Chapter 4, Section 18].

[33] See, e.g., KLINGENBERG [1973, Lemma 3.6.6].

[34] See, e.g., KLINGENBERG [1973, Lemma 3.6.12].

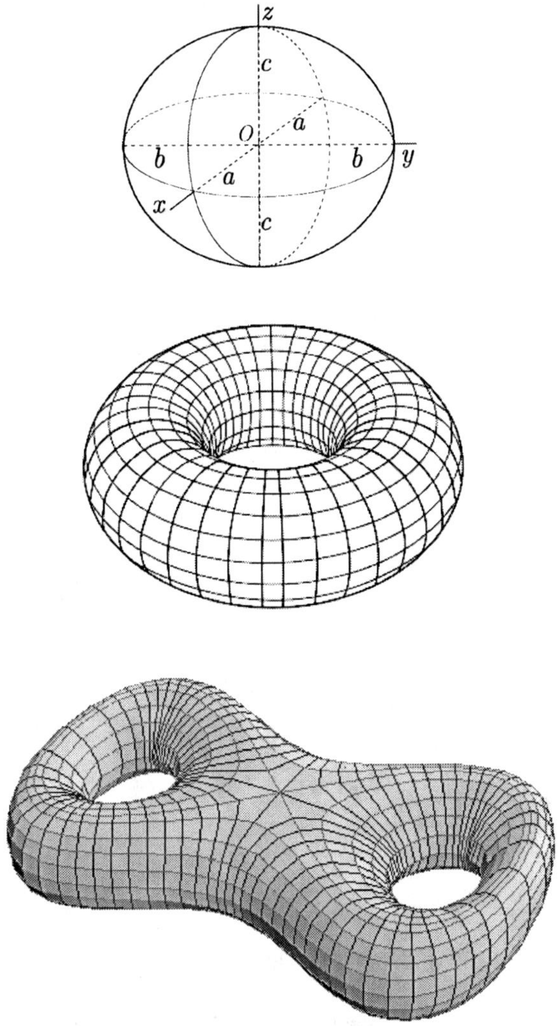

Figure 8.12-2 *Compact, orientable, and closed surfaces in* $\mathbb{E}^3$, *and their genus* (the coordinates in $\mathbb{E}^3$ are denoted x, y, z). An *ellipsoid*, defined for instance by the equation

$$\frac{x^2}{a^2} + \frac{y^2}{b^2} + \frac{z^2}{c^2} = 1,$$

where a, b, and c are > 0 constants, has genus zero. A *torus*, defined for instance by the equation

$$(x^2 + y^2 + z^2 + R^2 - r^2)^2 - 4R^2(x^2 + y^2) = 0,$$

where $0 < r < R$, has genus one. A *"double torus,"* defined for instance by the equation

$$x^8 + 2x^4(y^2 - x^2) + (y^2 - x^2)^2 + z^2 - 1/25 = 0,$$

has genus two. Top image reprinted courtesy of Wikipedia and Peter Mercator. Middle image reprinted courtesy of Wikipedia and YassineMrabet. Bottom image reprinted courtesy of Stan Wagon and with kind permission of Springer Science+Business Media.

8.13 Covariant derivatives of a vector field defined on a surface; the Gauß and Weingarten formulas

As in the previous sections, consider a surface $\widehat{\omega} = \boldsymbol{\theta}(\omega)$ in $\mathbb{E}^3$, where $\boldsymbol{\theta} : \omega \subset \mathbb{R}^2 \to \mathbb{E}^3$ is a smooth enough injective immersion, and let

$$\boldsymbol{a}_3(y) = \boldsymbol{a}^3(y) := \frac{\boldsymbol{a}_1(y) \wedge \boldsymbol{a}_2(y)}{|\boldsymbol{a}_1(y) \wedge \boldsymbol{a}_2(y)|}, \quad y \in \omega.$$

Then the two vectors $\boldsymbol{a}_\alpha(y) = \partial_\alpha \boldsymbol{\theta}(y)$ (which form the covariant basis of the tangent plane to $\widehat{\omega}$ at $\widehat{y} = \boldsymbol{\theta}(y)$; cf. Section 8.9) together with the vector $\boldsymbol{a}_3(y)$ (which is normal to $\widehat{\omega}$ and has Euclidean norm one) form the **covariant basis** at each point $\widehat{y} = \boldsymbol{\theta}(y)$, $y \in \omega$.

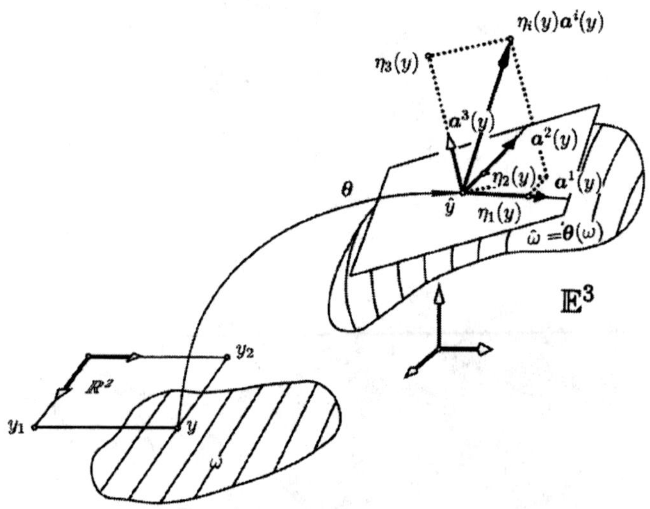

Figure 8.13-1 *Contravariant bases and vector fields along a surface.* At each point $\widehat{y} = \boldsymbol{\theta}(y)$, $y \in \omega$, the three vectors $\boldsymbol{a}^i(y)$, where $\boldsymbol{a}^\alpha(y)$ form the contravariant basis of the tangent plane to $\widehat{\omega} = \boldsymbol{\theta}(\omega)$ at $\widehat{y}$ (Figure 8.8-1) and $\boldsymbol{a}^3(y) = \dfrac{\boldsymbol{a}_1(y) \wedge \boldsymbol{a}_2(y)}{|\boldsymbol{a}_1(y) \wedge \boldsymbol{a}_2(y)|}$, form the contravariant basis at $\widehat{y}$. An arbitrary vector field defined on $\widehat{\omega}$ may then be defined by its covariant components $\eta_i : \omega \to \mathbb{R}$ over the vector fields $\boldsymbol{a}^i$. This means that $\eta_i(y)\boldsymbol{a}^i(y)$ is the vector at the point $\widehat{y}$.

Recall that the vectors $\boldsymbol{a}^\alpha(y)$ of the tangent plane to $\widehat{\omega}$ at $\widehat{y}$ are defined by the relations $\boldsymbol{a}^\alpha(y) \cdot \boldsymbol{a}_\beta(y) = \delta^\alpha_\beta$ (Section 8.9). Then the vectors $\boldsymbol{a}^\alpha(y)$ (which form the contravariant basis of the tangent plane at $\widehat{y}$; cf. Section 8.9) together with the vector $\boldsymbol{a}^3(y)$ form the **contravariant basis** *at* $\widehat{y}$; see Figure 8.13-1. Note that the vectors of the covariant and contravariant bases at $\widehat{y}$ satisfy

$$\boldsymbol{a}^i(y) \cdot \boldsymbol{a}_j(y) = \delta^i_j.$$

How do we define a *vector field given on the surface* $\widehat{\omega}$? One way to do so in terms of the *curvilinear coordinates* used for defining the surface $\widehat{\omega}$ consists in writing it as $\eta_i \boldsymbol{a}^i : \omega \to \mathbb{E}^3$, i.e., in specifying its **covariant components** $\eta_i : \omega \to \mathbb{R}$ over the vector fields $\boldsymbol{a}^i$ formed by the *contravariant bases.* This means that $\eta_i(y)\boldsymbol{a}^i(y)$ is the value of the vector field at each point $\widehat{y} = \boldsymbol{\theta}(y) \in \widehat{\omega}$ (Figure 8.13-1).

Our objective in this section is to compute the *partial derivatives* $\partial_\alpha(\eta_i \boldsymbol{a}^i)$ of such a vector field. These are found in the next theorem, as immediate consequences of two basic formulas, those of *Gauß* and *Weingarten.* The *Christoffel symbols "on a surface"* and the *covariant derivatives of a vector field defined on a surface* are also naturally introduced in this process.

Note that the Christoffel symbols "on a surface" $\Gamma^\sigma_{\alpha\beta}$ and $\Gamma_{\alpha\beta\tau}$ introduced in this section and the next are denoted by the *same* symbols as the "n-dimensional" Christoffel symbols introduced in Sections 8.3 and 8.5, viz., Γ^p_{ij} and Γ_{ijq}. No confusion should arise, however.

Theorem 8.13-1 *Let ω be an open subset of $\mathbb{R}^2$ and let $\boldsymbol{\theta} \in \mathcal{C}^2(\omega; \mathbb{E}^3)$ be an immersion.*

(a) *The derivatives of the vectors of the covariant and contravariant bases are given by*

$$\partial_\alpha \boldsymbol{a}_\beta = \Gamma^\sigma_{\alpha\beta}\boldsymbol{a}_\sigma + b_{\alpha\beta}\boldsymbol{a}_3 \quad \textit{and} \quad \partial_\alpha \boldsymbol{a}^\beta = -\Gamma^\beta_{\alpha\sigma}\boldsymbol{a}^\sigma + b^\beta_\alpha \boldsymbol{a}^3,$$
$$\partial_\alpha \boldsymbol{a}_3 = \partial_\alpha \boldsymbol{a}^3 = -b_{\alpha\beta}\boldsymbol{a}^\beta = -b^\sigma_\alpha \boldsymbol{a}_\sigma,$$

where

$$\Gamma^\sigma_{\alpha\beta} := \boldsymbol{a}^\sigma \cdot \partial_\alpha \boldsymbol{a}_\beta = \Gamma^\sigma_{\beta\alpha}, \quad b_{\alpha\beta} = \boldsymbol{a}_3 \cdot \partial_\alpha \boldsymbol{a}_\beta, \quad \textit{and} \quad b^\beta_\alpha = a^{\beta\sigma} b_{\alpha\sigma}$$

(*the functions $b_{\alpha\beta}$ and b^β_α are the covariant and mixed components of the second fundamental form of $\widehat{\omega}$, introduced in Theorems* 8.11-1 *and* 8.12-1).

(b) *Let there be given a vector field $\eta_i \boldsymbol{a}^i : \omega \to \mathbb{E}^3$ with covariant components $\eta_i \in \mathcal{C}^1(\omega)$. Then $\eta_i \boldsymbol{a}^i \in \mathcal{C}^1(\omega; \mathbb{E}^3)$ and the partial derivatives $\partial_\alpha(\eta_i \boldsymbol{a}^i) \in \mathcal{C}^1(\omega; \mathbb{E}^3)$ are given by*

$$\partial_\alpha(\eta_i \boldsymbol{a}^i) = (\partial_\alpha \eta_\beta - \Gamma^\sigma_{\alpha\beta}\eta_\sigma - b_{\alpha\beta}\eta_3)\boldsymbol{a}^\beta + (\partial_\alpha \eta_3 + b^\beta_\alpha \eta_\beta)\boldsymbol{a}^3$$
$$= (\eta_{\beta|\alpha} - b_{\alpha\beta}\eta_3)\boldsymbol{a}^\beta + (\eta_{3|\alpha} + b^\beta_\alpha \eta_\beta)\boldsymbol{a}^3,$$

where

$$\eta_{\beta|\alpha} := \partial_\alpha \eta_\beta - \Gamma^\sigma_{\alpha\beta}\eta_\sigma \quad \textit{and} \quad \eta_{3|\alpha} := \partial_\alpha \eta_3.$$

Proof Since any vector $\boldsymbol{c}$ in the tangent plane can be expanded as $\boldsymbol{c} = (\boldsymbol{c} \cdot \boldsymbol{a}_\beta)\boldsymbol{a}^\beta = (\boldsymbol{c} \cdot \boldsymbol{a}^\sigma)\boldsymbol{a}_\sigma$, since $\partial_\alpha \boldsymbol{a}^3$ is in the tangent plane ($\partial_\alpha \boldsymbol{a}^3 \cdot \boldsymbol{a}^3 = \frac{1}{2}\partial_\alpha(\boldsymbol{a}^3 \cdot \boldsymbol{a}^3) = 0$), and since $\partial_\alpha \boldsymbol{a}^3 \cdot \boldsymbol{a}_\beta = -b_{\alpha\beta}$ (Theorem 8.11-1), it follows that

$$\partial_\alpha \boldsymbol{a}^3 = (\partial_\alpha \boldsymbol{a}^3 \cdot \boldsymbol{a}_\beta)\boldsymbol{a}^\beta = -b_{\alpha\beta}\boldsymbol{a}^\beta.$$

This formula, together with the definition of the functions b^β_α (Theorem 8.12-1), implies in turn that

$$\partial_\alpha \boldsymbol{a}_3 = (\partial_\alpha \boldsymbol{a}_3 \cdot \boldsymbol{a}^\sigma)\boldsymbol{a}_\sigma = -b_{\alpha\beta}(\boldsymbol{a}^\beta \cdot \boldsymbol{a}^\sigma)\boldsymbol{a}_\sigma = -b_{\alpha\beta}a^{\beta\sigma}\boldsymbol{a}_\sigma = -b^\sigma_\alpha \boldsymbol{a}_\sigma.$$

Any vector $\boldsymbol{c}$ can be expanded as $\boldsymbol{c} = (\boldsymbol{c} \cdot \boldsymbol{a}^i)\boldsymbol{a}_i = (\boldsymbol{c} \cdot \boldsymbol{a}_j)\boldsymbol{a}^j$. In particular,

$$\partial_\alpha \boldsymbol{a}_\beta = (\partial_\alpha \boldsymbol{a}_\beta \cdot \boldsymbol{a}^\sigma)\boldsymbol{a}_\sigma + (\partial_\alpha \boldsymbol{a}_\beta \cdot \boldsymbol{a}^3)\boldsymbol{a}_3 = \Gamma^\sigma_{\alpha\beta}\boldsymbol{a}_\sigma + b_{\alpha\beta}\boldsymbol{a}_3,$$

by definition of $\Gamma^\sigma_{\alpha\beta}$ and $b_{\alpha\beta}$. Finally,

$$\partial_\alpha \boldsymbol{a}^\beta = (\partial_\alpha \boldsymbol{a}^\beta \cdot \boldsymbol{a}_\sigma)\boldsymbol{a}^\sigma + (\partial_\alpha \boldsymbol{a}^\beta \cdot \boldsymbol{a}_3)\boldsymbol{a}^3 = -\Gamma^\beta_{\alpha\sigma}\boldsymbol{a}^\sigma + b^\beta_\alpha \boldsymbol{a}^3,$$

since

$$\partial_\alpha \boldsymbol{a}^\beta \cdot \boldsymbol{a}_\sigma = -\boldsymbol{a}^\beta \cdot \partial_\alpha \boldsymbol{a}_\sigma = -\Gamma^\beta_{\alpha\sigma} \quad \text{and} \quad \partial_\alpha \boldsymbol{a}^\beta \cdot \boldsymbol{a}_3 = -\boldsymbol{a}^\beta \cdot \partial_\alpha \boldsymbol{a}_3 = b^\sigma_\alpha \boldsymbol{a}_\sigma \cdot \boldsymbol{a}^\beta = b^\beta_\alpha.$$

That $\eta_i \boldsymbol{a}^i \in \mathcal{C}^1(\omega; \mathbb{E}^3)$ if $\eta_i \in \mathcal{C}^1(\omega)$ is clear since $\boldsymbol{a}^i \in \mathcal{C}^1(\omega; \mathbb{E}^3)$ if $\boldsymbol{\theta} \in \mathcal{C}^2(\omega; \mathbb{E}^3)$. The formulas established *supra* immediately lead to the announced expression of $\partial_\alpha(\eta_i \boldsymbol{a}^i)$. □

The relations established in Theorem 8.13-1, viz.,

$$\partial_\alpha \boldsymbol{a}_\beta = \Gamma^\sigma_{\alpha\beta} \boldsymbol{a}_\sigma + b_{\alpha\beta} \boldsymbol{a}_3 \quad \text{and} \quad \partial_\alpha \boldsymbol{a}^\beta = -\Gamma^\beta_{\alpha\sigma} \boldsymbol{a}^\sigma + b^\beta_\alpha \boldsymbol{a}^3$$

and

$$\partial_\alpha \boldsymbol{a}_3 = \partial_\alpha \boldsymbol{a}^3 = -b_{\alpha\beta} \boldsymbol{a}^\beta = -b^\sigma_\alpha \boldsymbol{a}_\sigma,$$

respectively constitute the **formulas of Gauß**[35] and **Weingarten**.[36]

If the vector field is tangent to the surface $\widehat{\omega}$ (i.e., if $\eta_3 = 0$), the functions (appearing in Theorem 8.13-1)

$$\eta_{\beta|\alpha} = \partial_\alpha \eta_\beta - \Gamma^\sigma_{\alpha\beta} \eta_\sigma$$

are called the **covariant components** of the **covariant derivative** of the *tangent vector field* $\eta_\beta \boldsymbol{a}^\beta : \omega \to \mathbb{E}^3$, and the functions

$$\Gamma^\sigma_{\alpha\beta} := \boldsymbol{a}^\sigma \cdot \partial_\alpha \boldsymbol{a}_\beta = -\partial_\alpha \boldsymbol{a}^\sigma \cdot \boldsymbol{a}_\beta$$

are the **Christoffel symbols of the second kind** (the Christoffel symbols of the first kind will be introduced in the next section).

Remark The Christoffel symbols $\Gamma^\sigma_{\alpha\beta}$ can be also defined solely in terms of the covariant components of the first fundamental form; see the proof of Theorem 8.14-1 in the next section. □

The definition of the covariant components $\eta_{\alpha|\beta} = \partial_\beta \eta_\alpha - \Gamma^\sigma_{\alpha\beta} \eta_\sigma$ of the covariant derivative of a vector field tangent to the surface $\boldsymbol{\theta}(\omega)$ given in Theorem 8.13-1 is reminiscent of the definition of the covariant components $v_{i\|j} = \partial_j v_i - \Gamma^p_{ij} v_p$ of the covariant derivative of a vector field defined on an open set $\boldsymbol{\Theta}(\Omega)$ (Theorem 8.13-1). However, the former are more subtle to apprehend than the latter.[37] To see this, recall that the covariant components $v_{i\|j} = \partial_j v_i - \Gamma^p_{ij} v_p$ may be also defined by the relations (Theorem 8.3-2)

$$v_{i\|j} \boldsymbol{g}^i = \partial_j (v_i \boldsymbol{g}^i).$$

By contrast, *the covariant components* $\eta_{\alpha|\beta} = \partial_\beta \eta_\alpha - \Gamma^\sigma_{\alpha\beta} \eta_\sigma$ *satisfy only the relations*

$$\eta_{\alpha|\beta} \boldsymbol{a}^\alpha = \boldsymbol{P} \left(\partial_\beta (\eta_\alpha \boldsymbol{a}^\alpha) \right),$$

---

[35] C.F. Gauß [1827]: Disquisitiones generales circa superficies curvas, *Commentationes Societatis Regiae Scientiarum Gottingensis Recentiores* **6**, 99–146.

[36] J. Weingarten [1861]: Über eine Klasse auf einander abwickelbarer Flächen, *Journal für Reine und Angewandte Mathematik* **59**, 382–393.

[37] For a more detailed analysis of the notion of covariant derivative on a surface, see, e.g., Kühnel [2002, Chapter 4].

where $\boldsymbol{P}$ denotes the projection operator on the tangent plane in the direction of the normal vector (i.e., $\boldsymbol{P}(c_i\boldsymbol{a}^i) := c_\alpha \boldsymbol{a}^\alpha$), since

$$\partial_\beta(\eta_\alpha \boldsymbol{a}^\alpha) = \eta_{\alpha|\beta}\boldsymbol{a}^\alpha + b_\beta^\alpha \eta_\alpha \boldsymbol{a}^3$$

for such tangential fields by Theorem 8.13-1. This is so because a surface has in general a *nonzero curvature*, manifesting itself here by the extra term $b_\beta^\alpha \eta_\alpha \boldsymbol{a}^3$. This term vanishes in ω if $\widehat{\omega}$ is a *portion of a plane*, since in this case $b_\beta^\alpha = b_{\alpha\beta} = 0$. Note that, *again in this case*, the formula giving the partial derivatives in Theorem 8.13-1(b) reduces to

$$\partial_\alpha(\eta_i \boldsymbol{a}^i) = (\eta_{i|\alpha})\boldsymbol{a}^i.$$

Problems

8.13-1 Given an open subset ω of $\mathbb{R}^2$ and an immersion $\boldsymbol{\theta} \in \mathcal{C}^2(\omega, \mathbb{E}^3)$, define at each point $y \in \omega$ the matrices

$$\boldsymbol{a}^i(y) \otimes \boldsymbol{a}^j(y) := \boldsymbol{a}^i(y)(\boldsymbol{a}^j(y))^T \in \mathbb{M}^3,$$

where the vectors $\boldsymbol{a}^i(y)$ of the contravariant basis at $\widehat{y} = \boldsymbol{\theta}(y)$ are viewed here as column vectors.

(1) Show that the nine matrices $\boldsymbol{a}^i(y) \otimes \boldsymbol{a}^j(y)$ form a basis of the space $\mathbb{M}^3$ at each $y \in \omega$.

(2) Show that

$$\begin{aligned}
\partial_\sigma(\boldsymbol{a}^\alpha \otimes \boldsymbol{a}^\beta) &= -\Gamma_{\sigma\tau}^\alpha \boldsymbol{a}^\tau \otimes \boldsymbol{a}^\beta - \Gamma_{\sigma\tau}^\beta \boldsymbol{a}^\alpha \otimes \boldsymbol{a}^\tau + b_\sigma^\beta \boldsymbol{a}^\alpha \otimes \boldsymbol{a}^3 + b_\sigma^\alpha \boldsymbol{a}^3 \otimes \boldsymbol{a}^\beta, \\
\partial_\sigma(\boldsymbol{a}^\alpha \otimes \boldsymbol{a}^3) &= -b_{\sigma\tau}\boldsymbol{a}^\alpha \otimes \boldsymbol{a}^\tau - \Gamma_{\sigma\tau}^\alpha \boldsymbol{a}^\tau \otimes \boldsymbol{a}^3 + b_\sigma^\alpha \boldsymbol{a}^3 \otimes \boldsymbol{a}^3, \\
\partial_\sigma(\boldsymbol{a}^3 \otimes \boldsymbol{a}^\beta) &= -b_{\sigma\tau}\boldsymbol{a}^\tau \otimes \boldsymbol{a}^\beta - \Gamma_{\sigma\tau}^\beta \boldsymbol{a}^3 \otimes \boldsymbol{a}^\tau + b_\sigma^\beta \boldsymbol{a}^3 \otimes \boldsymbol{a}^3, \\
\partial_\sigma(\boldsymbol{a}^3 \otimes \boldsymbol{a}^3) &= -b_{\sigma\tau}\boldsymbol{a}^\tau \otimes \boldsymbol{a}^3 - b_{\sigma\tau}\boldsymbol{a}^3 \otimes \boldsymbol{a}^\tau.
\end{aligned}$$

(3) Let $(T_{\alpha\beta}) : \omega \to \mathbb{M}^2$ be a matrix field with components $T_{\alpha\beta} \in \mathcal{C}^1(\omega)$. The *covariant components* $T_{\alpha\beta|\sigma}$ *of the covariant derivative* of this matrix field are defined by

$$T_{\alpha\beta|\sigma} := \partial_\sigma T_{\alpha\beta} - \Gamma_{\sigma\alpha}^\nu T_{\nu\beta} - \Gamma_{\sigma\beta}^\nu T_{\alpha\nu}.$$

Using (2), show that the same components $T_{\alpha\beta|\sigma}$ can be also defined by means of the relations

$$\partial_\sigma(T_{\alpha\beta}\boldsymbol{a}^\alpha \otimes \boldsymbol{a}^\beta) = T_{\alpha\beta|\sigma}\boldsymbol{a}^\alpha \otimes \boldsymbol{a}^\beta + b_\sigma^\alpha T_{\alpha\beta}\boldsymbol{a}^3 \otimes \boldsymbol{a}^\beta + b_\sigma^\beta T_{\alpha\beta}\boldsymbol{a}^\alpha \otimes \boldsymbol{a}^3.$$

Remark If the surface $\widehat{\omega}$ is a *portion of a plane*, the last formula becomes analogous to that found in question (2) of Problem 8.4-4. □

8.13-2 Let ω be an open subset of $\mathbb{R}^2$ and let $\boldsymbol{\theta} \in \mathcal{C}^3(\omega; \mathbb{E}^3)$ be an immersion. The *covariant components* $\eta_{\alpha|\sigma\tau}$ *of the second covariant derivative* of a *tangent* vector field $\eta_\alpha \boldsymbol{a}^\alpha$ are defined by

$$\eta_{\alpha|\sigma\tau} := \partial_\tau \eta_{\alpha|\sigma} - \Gamma_{\tau\alpha}^\nu \eta_{\nu|\sigma} - \Gamma_{\tau\sigma}^\nu \eta_{\alpha|\nu}.$$

Show that the components $\eta_{\alpha|\sigma\tau}$ can be also defined by means of the relations

$$\partial_{\tau\sigma}(\eta_\alpha \boldsymbol{a}^\alpha) = (\eta_\alpha|_{\sigma\tau} + \Gamma_{\tau\sigma}^\mu \eta_\alpha|_\mu - b_{\alpha\tau} b_\sigma^\nu \eta_\nu)\boldsymbol{a}^\alpha + (b_\tau^\alpha \eta_{\alpha|\sigma} + b_\sigma^\alpha \eta_\alpha|_\tau + (b_\sigma^\alpha|_\tau + \Gamma_{\tau\sigma}^\mu b_\mu^\alpha)\eta_\alpha)\boldsymbol{a}^3,$$

where $b_\sigma^\alpha|_\tau := \partial_\tau b_\sigma^\alpha - \Gamma_{\tau\sigma}^\mu b_\mu^\alpha + \Gamma_{\tau\mu}^\alpha b_\sigma^\mu$.

8.14 Necessary conditions satisfied by the first and second fundamental forms: The Gauß and Codazzi–Mainardi equations

As expected, *the components* $a_{\alpha\beta} = a_{\beta\alpha} : \omega \to \mathbb{R}$ *and* $b_{\alpha\beta} = b_{\beta\alpha} : \omega \to \mathbb{R}$ *of the first and second fundamental forms of a surface* $\boldsymbol{\theta}(\omega)$ *defined by a smooth immersion* $\boldsymbol{\theta} : \omega \to \mathbb{E}^3$ *cannot be arbitrary functions.*

As shown in the next theorem, they must satisfy relations that take the form

$$\partial_\beta \Gamma_{\alpha\sigma\tau} - \partial_\sigma \Gamma_{\alpha\beta\tau} + \Gamma^\mu_{\alpha\beta}\Gamma_{\sigma\tau\mu} - \Gamma^\mu_{\alpha\sigma}\Gamma_{\beta\tau\mu} = b_{\alpha\sigma}b_{\beta\tau} - b_{\alpha\beta}b_{\sigma\tau} \quad \text{in } \omega,$$
$$\partial_\beta b_{\alpha\sigma} - \partial_\sigma b_{\alpha\beta} + \Gamma^\mu_{\alpha\sigma} b_{\beta\mu} - \Gamma^\mu_{\alpha\beta} b_{\sigma\mu} = 0 \quad \text{in } \omega,$$

where the functions $\Gamma_{\alpha\beta\tau}$ and $\Gamma^\sigma_{\alpha\beta}$ have simple expressions in terms of the functions $a_{\alpha\beta}$ and of some of their partial derivatives (although they will be *a priori* differently defined, the functions $\Gamma^\sigma_{\alpha\beta}$ are nothing but the Christoffel symbols of the second kind introduced in the previous section). Recall that, according to the rule governing Greek indices and exponents, these relations are meant to hold for all $\alpha, \beta, \sigma, \tau \in \{1, 2\}$, but they reduce in fact to only *three* independent relations, as we shall see later.

Remark A *different* set of necessary conditions can be found that are instead expressed in terms of the *square root of the matrix field* $(a_{\alpha\beta})$ and of the matrix field $(b_{\alpha\beta})$; cf. Problem 8.14-4. □

Theorem 8.14-1 (necessary conditions satisfied by the first and second fundamental forms) *Let* ω *be an open subset of* $\mathbb{R}^2$, *let* $\boldsymbol{\theta} \in \mathcal{C}^3(\omega; \mathbb{E}^3)$ *be an immersion, and let*

$$a_{\alpha\beta} := \partial_\alpha \boldsymbol{\theta} \cdot \partial_\beta \boldsymbol{\theta} \quad \text{and} \quad b_{\alpha\beta} := \partial_{\alpha\beta}\boldsymbol{\theta} \cdot \left\{ \frac{\partial_1 \boldsymbol{\theta} \wedge \partial_2 \boldsymbol{\theta}}{|\partial_1 \boldsymbol{\theta} \wedge \partial_2 \boldsymbol{\theta}|} \right\}$$

denote the covariant components of the first and second fundamental forms of the surface $\boldsymbol{\theta}(\omega)$. *Let the functions* $\Gamma_{\alpha\beta\tau} \in \mathcal{C}^1(\omega)$ *and* $\Gamma^\sigma_{\alpha\beta} \in \mathcal{C}^1(\omega)$ *be defined by*

$$\Gamma_{\alpha\beta\tau} := \frac{1}{2}(\partial_\beta a_{\alpha\tau} + \partial_\alpha a_{\beta\tau} - \partial_\tau a_{\alpha\beta}) \quad \text{and} \quad \Gamma^\sigma_{\alpha\beta} := a^{\sigma\tau}\Gamma_{\alpha\beta\tau}, \quad \text{where } (a^{\sigma\tau}) := (a_{\alpha\beta})^{-1}.$$

Then the functions $a_{\alpha\beta}$ *and* $b_{\alpha\beta}$ *necessarily satisfy the* **Gauß equations**[38]

$$\partial_\beta \Gamma_{\alpha\sigma\tau} - \partial_\sigma \Gamma_{\alpha\beta\tau} + \Gamma^\mu_{\alpha\beta}\Gamma_{\sigma\tau\mu} - \Gamma^\mu_{\alpha\sigma}\Gamma_{\beta\tau\mu} = b_{\alpha\sigma}b_{\beta\tau} - b_{\alpha\beta}b_{\sigma\tau} \quad \text{in } \omega,$$

and the **Codazzi–Mainardi equations**[39]

$$\partial_\beta b_{\alpha\sigma} - \partial_\sigma b_{\alpha\beta} + \Gamma^\mu_{\alpha\sigma} b_{\beta\mu} - \Gamma^\mu_{\alpha\beta} b_{\sigma\mu} = 0 \quad \text{in } \omega.$$

---

[38] So named after:

C.F. GAUß [1827]: Disquisitiones generales circa superficies curvas, *Commentationes Societatis Regiae Scientiarum Gottingensis Recentiores* **6**, 99–146.

[39] So named after:

D. CODAZZI [1868–1869]: Sulle coordinate curvilinee d'una superficie dello spazio, *Annali di Mathematica Pura e Applicata* **2**, 101–119.

G. MAINARDI [1856]: Su la teoria generale delle superficie, *Giornale dell' Istituto Lombardo* **9**, 385–404.

Proof Let $\boldsymbol{a}_i$ and $\boldsymbol{a}^j$ denote as before the vectors of the covariant and contravariant bases. It is then immediately verified that the functions

$$\Gamma_{\alpha\beta\tau} := \frac{1}{2}(\partial_\beta a_{\alpha\tau} + \partial_\alpha a_{\beta\tau} - \partial_\tau a_{\alpha\beta})$$

are also given by

$$\Gamma_{\alpha\beta\tau} = \partial_\alpha \boldsymbol{a}_\beta \cdot \boldsymbol{a}_\tau.$$

Since $\boldsymbol{a}^\sigma = a^{\sigma\tau}\boldsymbol{a}_\tau$, the functions $\Gamma^\sigma_{\alpha\beta} := a^{\sigma\tau}\Gamma_{\alpha\beta\tau}$ are also given by

$$\Gamma^\sigma_{\alpha\beta} = \partial_\alpha \boldsymbol{a}_\beta \cdot \boldsymbol{a}^\sigma.$$

Differentiating and using the *formula of Gauß* (Theorem 8.13-1), we thus obtain

$$\partial_\sigma \Gamma_{\alpha\beta\tau} = \partial_{\alpha\sigma}\boldsymbol{a}_\beta \cdot \boldsymbol{a}_\tau + \partial_\alpha \boldsymbol{a}_\beta \cdot \partial_\sigma \boldsymbol{a}_\tau = \partial_{\alpha\sigma}\boldsymbol{a}_\beta \cdot \boldsymbol{a}_\tau + \Gamma^\mu_{\alpha\beta}\Gamma_{\sigma\tau\mu} + b_{\alpha\beta}b_{\sigma\tau}.$$

Consequently,

$$\partial_{\alpha\sigma}\boldsymbol{a}_\beta \cdot \boldsymbol{a}_\tau = \partial_\sigma \Gamma_{\alpha\beta\tau} - \Gamma^\mu_{\alpha\beta}\Gamma_{\sigma\tau\mu} - b_{\alpha\beta}b_{\sigma\tau}.$$

Since $\partial_{\alpha\sigma}\boldsymbol{a}_\beta = \partial_{\alpha\beta}\boldsymbol{a}_\sigma$, we also have

$$\partial_{\alpha\sigma}\boldsymbol{a}_\beta \cdot \boldsymbol{a}_\tau = \partial_\beta \Gamma_{\alpha\sigma\tau} - \Gamma^\mu_{\alpha\sigma}\Gamma_{\beta\tau\mu} - b_{\alpha\sigma}b_{\beta\tau}.$$

Hence the Gauß equations immediately follow.

Differentiating the relations $b_{\alpha\beta} = \partial_\alpha \boldsymbol{a}_\beta \cdot \boldsymbol{a}_3$ and using the *formula of Weingarten* (Theorem 8.13-1), we obtain

$$\partial_\sigma b_{\alpha\beta} = \partial_{\alpha\sigma}\boldsymbol{a}_\beta \cdot \boldsymbol{a}_3 + \partial_\alpha \boldsymbol{a}_\beta \cdot \partial_\sigma \boldsymbol{a}_3 = \partial_{\alpha\sigma}\boldsymbol{a}_\beta \cdot \boldsymbol{a}_3 - \Gamma^\mu_{\alpha\beta}b_{\sigma\mu}.$$

Consequently,

$$\partial_{\alpha\sigma}\boldsymbol{a}_\beta \cdot \boldsymbol{a}_3 = \partial_\sigma b_{\alpha\beta} + \Gamma^\mu_{\alpha\beta}b_{\sigma\mu}.$$

Since $\partial_{\alpha\sigma}\boldsymbol{a}_\beta = \partial_{\alpha\beta}\boldsymbol{a}_\sigma$, we also have

$$\partial_{\alpha\sigma}\boldsymbol{a}_\beta \cdot \boldsymbol{a}_3 = \partial_\beta b_{\alpha\sigma} + \Gamma^\mu_{\alpha\sigma}b_{\beta\mu},$$

from which the Codazzi–Mainardi equations immediately follow. □

As shown in the above proof, the Gauß and Codazzi–Mainardi equations thus constitute a simple, but clever, rewriting of the relations $\partial_{\alpha\sigma}\boldsymbol{a}_\beta = \partial_{\alpha\beta}\boldsymbol{a}_\sigma$ in the form of the equivalent relations $\partial_{\alpha\sigma}\boldsymbol{a}_\beta \cdot \boldsymbol{a}_\tau = \partial_{\alpha\beta}\boldsymbol{a}_\sigma \cdot \boldsymbol{a}_\tau$ and $\partial_{\alpha\sigma}\boldsymbol{a}_\beta \cdot \boldsymbol{a}_3 = \partial_{\alpha\beta}\boldsymbol{a}_\sigma \cdot \boldsymbol{a}_3$. Hence, as in Theorem 8.5-1, the key to these necessary conditions is simply the *Schwarz lemma* (Theorem 7.8-1).

The functions

$$\Gamma_{\alpha\beta\tau} = \frac{1}{2}(\partial_\beta a_{\alpha\tau} + \partial_\alpha a_{\beta\tau} - \partial_\tau a_{\alpha\beta}) = \partial_\alpha \boldsymbol{a}_\beta \cdot \boldsymbol{a}_\tau = \Gamma_{\beta\alpha\tau}$$

and

$$\Gamma^\sigma_{\alpha\beta} = a^{\sigma\tau}\Gamma_{\alpha\beta\tau} = \partial_\alpha \boldsymbol{a}_\beta \cdot \boldsymbol{a}^\sigma = \Gamma^\sigma_{\beta\alpha}$$

are the **Christoffel symbols of the first**, and **second, kind**. Recall that the Christoffel symbols of the second kind also naturally appeared in a different context (that of covariant differentiation; cf. Section 8.13).

Finally, the functions

$$R_{\tau\alpha\beta\sigma} := \partial_\beta \Gamma_{\alpha\sigma\tau} - \partial_\sigma \Gamma_{\alpha\beta\tau} + \Gamma^\mu_{\alpha\beta}\Gamma_{\sigma\tau\mu} - \Gamma^\mu_{\alpha\sigma}\Gamma_{\beta\tau\mu}$$

are the **covariant components of the Riemann curvature tensor of the surface** $\boldsymbol{\theta}(\omega)$. The notations $R_{\tau\alpha\beta\sigma}$ used for these components are thus similar to those used for the covariant components R_{qijk} of the Riemann curvature tensor introduced in Section 8.5; no confusion should arise, however.

The definitions of the functions $\Gamma^\sigma_{\alpha\beta}$ and $\Gamma_{\alpha\beta\tau}$ imply that *the* 16 *Gauß equations are satisfied if and only if they are satisfied for* $\alpha = 1$, $\beta = 2$, $\sigma = 1$, $\tau = 2$ and that *the eight Codazzi–Mainardi equations are satisfied if and only if they are satisfied* for $\alpha = 1$, $\beta = 2$, $\sigma = 1$ and $\alpha = 1$, $\beta = 2$, $\sigma = 2$ (other choices of indices with the same properties are clearly possible).

In other words, the Gauß equations and the Codazzi–Mainardi equations in fact respectively reduce to *one* and *two* equations.

Problems

8.14-1 Given an open subset ω of $\mathbb{R}^3$ and an immersion $\boldsymbol{\theta} \in \mathcal{C}^3(\omega; \mathbb{E}^3)$, let the *mixed components* of the *Riemann curvature tensor of the surface* $\boldsymbol{\theta}(\omega)$ be defined by

$$R^{\mu}_{\cdot\alpha\beta\sigma} := \partial_\beta \Gamma^\mu_{\sigma\alpha} - \partial_\sigma \Gamma^\mu_{\beta\alpha} + \Gamma^\tau_{\sigma\alpha}\Gamma^\mu_{\beta\tau} - \Gamma^\tau_{\beta\alpha}\Gamma^\mu_{\sigma\tau}.$$

(1) Show that the *Gauß equations* (Theorem 8.14-1) are equivalent to the equations

$$R^{\mu}_{\cdot\alpha\beta\sigma} = b_{\sigma\alpha} b^\mu_\beta - b_{\beta\alpha} b^\mu_\sigma.$$

Hint: Imitate part (i) of the proof of Theorem 8.6-1 to show that $R^{\mu}_{\cdot\alpha\beta\sigma} = a^{\mu\tau} R_{\tau\alpha\beta\sigma}$.

(2) Show that the *Codazzi–Mainardi equations* (Theorem 8.14-1) are equivalent to the equations

$$b_{\alpha\beta|\sigma} = b_{\alpha\sigma|\beta},$$

where the functions $b_{\alpha\beta|\sigma}$ are the covariant components of the covariant derivative of the second fundamental form $(b_{\alpha\beta}) : \omega \to \mathbb{M}^2$ (cf. question (3) of Problem 8.13-1).

8.14-2 Show that, when they are expressed in terms of the *mixed* components of the second fundamental form (instead of its covariant components as in Theorem 8.13-1), the *Codazzi–Mainardi equations* take the form

$$\partial_\alpha b^\sigma_\beta - \partial_\beta b^\sigma_\alpha + \Gamma^\sigma_{\alpha\tau} b^\tau_\beta - \Gamma^\sigma_{\beta\tau} b^\tau_\alpha = 0 \quad \text{in } \omega.$$

Hint: Use (and prove first) the relations $\partial_\alpha a_{\beta\sigma} = \Gamma^\tau_{\alpha\beta} a_{\sigma\tau} + \Gamma^\tau_{\alpha\sigma} a_{\beta\tau}$.

8.14-3 Let ω be an open subset of $\mathbb{R}^2$ and let $\boldsymbol{\theta} \in \mathcal{C}^3(\omega; \mathbb{E}^3)$ be an immersion. Show that the covariant components of the second covariant derivative of a vector field $\eta_\alpha \boldsymbol{a}^\alpha$ tangent to the surface $\boldsymbol{\theta}(\omega)$ satisfy the *Ricci identities*, viz.,

$$\eta_{\alpha|\sigma\tau} - \eta_{\alpha|\tau\sigma} = R^{\nu}_{\cdot\alpha\sigma\tau} \eta_\nu$$

(the covariant components $\eta_{\alpha|\sigma\tau}$ and the mixed components $R^{\nu}_{\cdot\alpha\sigma\tau}$ are defined in Problems 8.13-2 and 8.14-1, respectively).

8.14-4 Let ω be an open subset of $\mathbb{R}^2$ and let $\boldsymbol{\theta} \in \mathcal{C}^3(\omega; \mathbb{E}^3)$ be a *given* immersion. Then let, as usual,

$$\boldsymbol{a}_\alpha := \partial_\alpha \boldsymbol{\theta}, \quad \boldsymbol{a}_3 := \frac{\boldsymbol{a}_1 \wedge \boldsymbol{a}_2}{|\boldsymbol{a}_1 \wedge \boldsymbol{a}_2|}, \quad a_{\alpha\beta} := \boldsymbol{a}_\alpha \cdot \boldsymbol{a}_\beta, \quad (a^{\sigma\tau}) := (a_{\alpha\beta})^{-1},$$

$$\Gamma_{\alpha\beta\tau} := \frac{1}{2}(\partial_\beta a_{\alpha\tau} + \partial_\alpha a_{\beta\tau} - \partial_\tau a_{\alpha\beta}), \quad \Gamma^\sigma_{\alpha\beta} := a^{\sigma\tau}\Gamma_{\alpha\beta\tau},$$

$$b_{\alpha\beta} := \partial_\alpha \boldsymbol{a}_\beta \cdot \boldsymbol{a}_3, \quad b^\sigma_\alpha := a^{\beta\sigma} b_{\alpha\beta},$$

and let, *in addition,*

$$\boldsymbol{\Gamma}_\alpha := \begin{pmatrix} \Gamma^1_{\alpha 1} & \Gamma^1_{\alpha 2} & -b^1_\alpha \\ \Gamma^2_{\alpha 1} & \Gamma^2_{\alpha 2} & -b^2_\alpha \\ b_{\alpha 1} & b_{\alpha 2} & 0 \end{pmatrix}, \quad \boldsymbol{C} := \begin{pmatrix} a_{11} & a_{12} & 0 \\ a_{21} & a_{22} & 0 \\ 0 & 0 & 1 \end{pmatrix}, \quad \boldsymbol{U} := \boldsymbol{C}^{1/2}, \quad \boldsymbol{A}_\alpha := (\boldsymbol{U}\boldsymbol{\Gamma}_\alpha - \partial_\alpha \boldsymbol{U})\boldsymbol{U}^{-1}.$$

Show that the matrix fields $\boldsymbol{A}_\alpha \in \mathcal{C}^1(\omega; \mathbb{M}^3)$ are antisymmetric and that they *necessarily* satisfy the *compatibility conditions*

$$\partial_1 \boldsymbol{A}_2 - \partial_2 \boldsymbol{A}_1 + \boldsymbol{A}_1\boldsymbol{A}_2 - \boldsymbol{A}_2\boldsymbol{A}_1 = \boldsymbol{0} \quad \text{in } \omega.$$

8.15 Gauß Theorema Egregium; application to cartography

Letting $\alpha = 1$, $\beta = 2$, $\sigma = 1$, $\tau = 2$ in the Gauß equations (Theorem 8.14-1) gives in particular

$$R_{2121} = b_{11}b_{22} - b_{12}b_{12} = \det(b_{\alpha\beta}).$$

Consequently, the *Gaussian curvature* (Section 8.12) at each point $\boldsymbol{\theta}(y)$ of the surface $\boldsymbol{\theta}(\omega)$ can be written as

$$\frac{1}{R_1(y)R_2(y)} = \frac{R_{2121}(y)}{\det(a_{\alpha\beta}(y))}, \quad y \in \omega,$$

since $\dfrac{1}{R_1(y)R_2(y)} = \dfrac{\det(b_{\alpha\beta}(y))}{\det(a_{\alpha\beta}(y))}$ (Theorem 8.12-1). An inspection of the function R_{2121} thus leads to the astonishing conclusion that, *at each point of the surface, a notion involving the "curvature" of the surface, viz., the Gaussian curvature, is entirely determined by the knowledge of the "metric" of the surface in a neighborhood of the same point*, viz., the components of the first fundamental forms and their partial derivatives of order ≤ 2 at the same point! This startling observation constitutes one of the most beautiful theorems of mathematics:

Theorem 8.15-1 (Gauß Theorema Egregium[40]) *Let ω be an open subset of $\mathbb{R}^2$, let $\boldsymbol{\theta} \in \mathcal{C}^3(\omega; \mathbb{E}^3)$ be an immersion, let $a_{\alpha\beta} = \partial_\alpha \boldsymbol{\theta} \cdot \partial_\beta \boldsymbol{\theta}$ denote the covariant components of the first fundamental form of the surface $\boldsymbol{\theta}(\omega)$, and let the functions $\Gamma_{\alpha\beta\tau}$ and R_{2121} be defined by*

$$\Gamma_{\alpha\beta\tau} := \frac{1}{2}(\partial_\beta a_{\alpha\tau} + \partial_\alpha a_{\beta\tau} - \partial_\tau a_{\alpha\beta}),$$

$$R_{2121} := \frac{1}{2}(2\partial_{12}a_{12} - \partial_{11}a_{22} - \partial_{22}a_{11}) + a^{\alpha\beta}(\Gamma_{12\alpha}\Gamma_{12\beta} - \Gamma_{11\alpha}\Gamma_{22\beta}).$$

[40] C. F. Gauß [1828]: Disquisitiones generales circas superficies curvas, *Commentationes societatis regiae scientiarum Gottingensis recentiores* **6**, Göttingen.

Then, at each point $\boldsymbol{\theta}(y)$, $y \in \omega$, *of the surface* $\boldsymbol{\theta}(\omega)$, *the Gaussian curvature is given by*

$$\frac{1}{R_1(y)R_2(y)} = \frac{R_{2121}(y)}{\det(a_{\alpha\beta}(y))}. \qquad \square$$

We now briefly enter the fascinating field of *mathematical cartography*,[41] i.e., the *science of maps that represent a portion of the surface of the earth*, which will be for simplicity assumed here to be a sphere (which is of course only an approximation).

A **map** is a pair $(\omega, \boldsymbol{\theta})$ where ω is a bounded open subset of $\mathbb{R}^2$ and $\boldsymbol{\theta} \in \mathcal{C}^1(\omega; \mathbb{E}^3)$ is an injective immersion such that $\boldsymbol{\theta}(\omega) \subset S_R = \{\widehat{x} \in \mathbb{E}^3;\ |\widehat{x}| = R\}$, where $R > 0$ is the radius of the earth. A common example of map is shown in Figure 8.15-1.

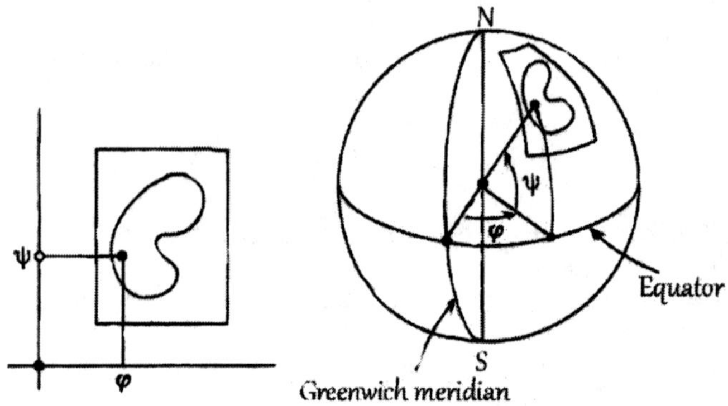

Figure 8.15-1 *An example of map.* The set ω is an open rectangle contained in $]-\pi, \pi[\times]-\frac{\pi}{2}, \frac{\pi}{2}[\subset \mathbb{R}^2$ and the mapping $\boldsymbol{\theta} : \omega \to \mathbb{E}^3$ is defined by

$$\boldsymbol{\theta} : (\varphi, \psi) := R\cos\varphi\cos\psi\widehat{\boldsymbol{e}}_1 + R\sin\varphi\cos\psi\widehat{\boldsymbol{e}}_2 + R\sin\psi\widehat{\boldsymbol{e}}_3 \quad \text{at each } (\varphi, \psi) \in \omega.$$

The curvilinear coordinates φ and ψ are thus none other than the *spherical coordinates* (Figure 8.8-2), renamed here *longitude* and *latitude*.

What are the ideal properties of a map?

First and foremost, a map should *preserve distances* (up to a scaling factor, ignored here), i.e., the planar set $\omega \subset \mathbb{R}^2$ (identified here with a surface in $\mathbb{E}^3$ corresponding to the mapping $y \in \omega \to (y, 0) \in \mathbb{E}^3$) and the surface $\boldsymbol{\theta}(\omega)$ should be *isometric*, according to the definition given in Section 8.10.

Second, the map should be *equiareal*, in the sense that it *preserve areas* (up to a scaling factor, again ignored here); cf. Section 8.10.

[41] Detailed accounts are found in, e.g.:

D.H. Maling [1992]: *Coordinate Systems and Map Projections*, Second Edition, Pergamon Press, Oxford.

J.P. Snyder [1993]: *Flattening the Earth: Two Thousand Years of Map Projection*, University of Chicago Press, Chicago.

Q. Yang; J.P. Snyder; W.R. Tobler [2000]: *Map Projection Transformation—Principle and Applications*, Taylor and Francis, London.

T.G. Freeman [2002]: *Portraits of the Earth. A Mathematician Looks at Maps*, American Mathematical Society, Providence.

Third, it should be *conformal*, in the sense that it *preserve angles* between intersecting curves; cf. again Section 8.10.

Alas, this beautiful program must be considerably scaled down; an actual map can only posses *either* the *second* property, *or* the *third* one, but *never* the *first* one:

Theorem 8.15-2 (a) *There is no map that preserves distances.*[42]

(b) *There is no map that preserves both areas and angles.*

Proof Let $(\omega, \boldsymbol{\theta})$ be a map that preserves distances, which means that the surfaces $\boldsymbol{\theta}(\omega) \subset \mathbb{E}^3$ and $\boldsymbol{\iota}(\omega)$, where $\boldsymbol{\iota}(y) := y^\alpha \widehat{\boldsymbol{e}}_\alpha$ for all $y = (y^\alpha) \in \omega$, are isometric. Therefore they share the same first fundamental form (Theorem 8.10-1). Consequently, their *Gaussian curvature* is the same since it depends only on the first fundamental form by *Gauß Theorema Egregium* (Theorem 8.15-1). But this is impossible since the Gaussian curvature of a surface contained in a plane (here $\boldsymbol{\iota}(\omega)$) vanishes everywhere, while the Gaussian curvature of a portion of a sphere with radius R is everywhere equal to $1/R^2$. This proves (a).

It is likewise impossible that a map preserves both areas and angles, because the surfaces $\omega \subset \mathbb{R}^2$ and $\boldsymbol{\theta}(\omega) \subset \mathbb{E}^3$ would then be isometric by Theorem 8.10-2(c). This proves (b). □

Examples of maps that preserve areas or angles are provided in Figures 8.15-2[43] and 8.15-3; see also Problems 8.15-1–8.15-3.

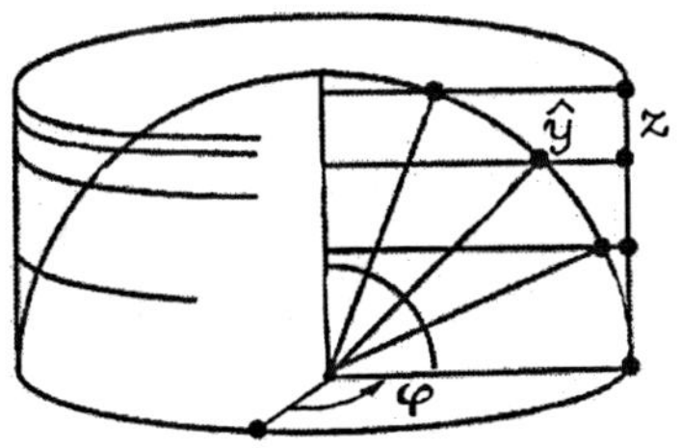

Figure 8.15-2 *An example of a map that preserves areas.* The curvilinear coordinates of a point $\widehat{y}$ different from a pole are its latitude φ and the coordinate z of its projection onto the *"cylindrical wrapping"* of the earth, as indicated in the figure; cf. Problem 8.15-3.

Problems

8.15-1 Let $\omega =]-\pi R, \pi R[\times]0, R[\subset \mathbb{R}^2$. Give the expression of the mapping $\boldsymbol{\theta} : \omega \to \mathbb{E}^3$ that corresponds to the *cylindrical wrapping of the earth* (Figure 8.15-2) and verify that the map $(\omega, \boldsymbol{\theta})$ preserves areas.

8.15-2 Show that a map that uses *stereographical coordinates* (Figure 8.8-2) preserves angles.

8.15-3 Let ω be an open rectangle contained in $]-\pi, \pi[\times \mathbb{R}$ and let the mapping $\boldsymbol{\theta} : \omega \to \mathbb{E}^3$ be

---

[42]This impossibility was first established, by means of a direct proof, in:

L. Euler [1775]: On representations of a spherical surface on the plane, *Proceedings of the Saint Petersburg Academy of Sciences.*

[43]This example was known to Archimedes (287–212 B.C.), who used it for computing the area of a sphere.

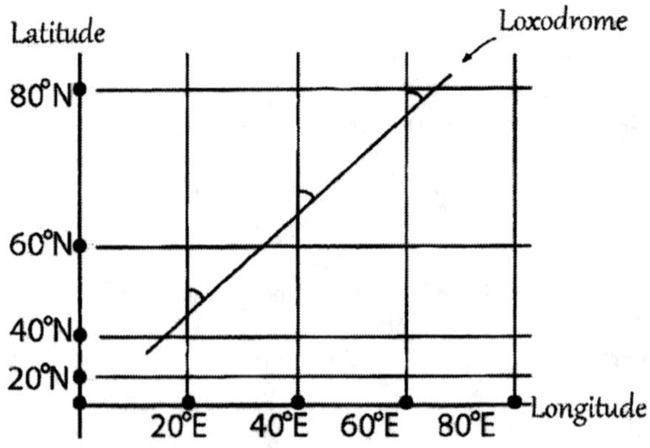

Figure 8.15-3 *An example of a map that preserves angles.* The *Mercator map* $(\omega, \boldsymbol{\theta})$ is a map where parallels and meridians are still orthogonal lines as in Figure 8.15-1, but where the latitude is distorted in such a way that the map *preserves angles.* As a result, the image by $\boldsymbol{\theta}$ of a *loxodrome*, i.e., a straight segment inside the set ω, intersects all the meridians at a constant angle on the earth itself; cf. Problem 8.15-3.

defined at each $(\varphi, \chi) \in \omega$ by

$$\boldsymbol{\theta}(\varphi, \chi) := R\cos\varphi\cos F(\chi)\widehat{\boldsymbol{e}}_1 + R\sin\varphi\cos F(\chi)\widehat{\boldsymbol{e}}_2 + R\sin F(\chi)\widehat{\boldsymbol{e}}_3, \quad \text{where } F(\chi) := \log\tan\frac{\chi}{2}.$$

Show that the map $(\omega, \boldsymbol{\theta})$, which is a *Mercator map*[44] (Figure 8.15-3), preserves angles.

8.16 Existence of a surface with prescribed first and second fundamental forms; the fundamental theorem of surface theory

Let $\mathbb{M}^2$, $\mathbb{S}^2$, and $\mathbb{S}^2_>$ denote the sets of all square matrices of order two, of all symmetric matrices of order two, and of all symmetric, positive-definite matrices of order two.

So far, we have considered that we are *given* an open set $\omega \subset \mathbb{R}^2$ and a smooth enough immersion $\boldsymbol{\theta} : \omega \to \mathbb{E}^3$, thus allowing us to define the fields $(a_{\alpha\beta}) : \omega \to \mathbb{S}^2_>$ and $(b_{\alpha\beta}) : \omega \to \mathbb{S}^2$, where $a_{\alpha\beta} : \omega \to \mathbb{R}$ and $b_{\alpha\beta} : \omega \to \mathbb{R}$ are the covariant components of the *first* and *second fundamental forms* of the surface $\boldsymbol{\theta}(\omega) \subset \mathbb{E}^3$.

Note that *the immersion $\boldsymbol{\theta}$ need not be injective* in order that these matrix fields be well defined.

We now turn to the reciprocal questions:

Given an open subset ω of $\mathbb{R}^2$ and two smooth enough matrix fields $(a_{\alpha\beta}) : \omega \to \mathbb{S}^2_>$ and $(b_{\alpha\beta}) : \omega \to \mathbb{S}^2$, *when are they the first and second fundamental forms of a surface $\boldsymbol{\theta}(\omega) \subset \mathbb{E}^3$; or equivalently, when does there exist an immersion $\boldsymbol{\theta} : \omega \to \mathbb{E}^3$ such that*

$$\partial_\alpha\boldsymbol{\theta} \cdot \partial_\beta\boldsymbol{\theta} = a_{\alpha\beta} \quad \text{and} \quad \partial_{\alpha\beta}\boldsymbol{\theta} \cdot \left\{ \frac{\partial_1\boldsymbol{\theta} \wedge \partial_2\boldsymbol{\theta}}{|\partial_1\boldsymbol{\theta} \wedge \partial_2\boldsymbol{\theta}|} \right\} = b_{\alpha\beta} \text{ in } \omega?$$

[44] So named after Gerardus Mercator, who first drew in 1569 such a map of the earth. The ensuing combined use of loxodromes and compass revolutionized marine navigation.

If such an immersion exists, *to what extent is it unique?*

The answers to these questions turn out to be remarkably simple to state (but not to prove): *If ω is simply connected, the necessary conditions found in* Theorem 8.14-1, viz., *the Gauß and Codazzi–Mainardi equations, are also sufficient for the existence of such an immersion. If ω is connected, this immersion is unique up to isometries in* $\mathbb{E}^3$.

Whether an immersion found in this fashion is *injective* is a different issue, which accordingly should be resolved by different means.

This result is another special case of the *fundamental theorem of Riemannian geometry* alluded to in Section 8.6. This theorem asserts that a simply connected Riemannian manifold of dimension p can be isometrically immersed into a Euclidean space of dimension $(p+q)$ if and only if there exist tensors satisfying together generalized *Gauß*, and *Codazzi–Mainardi, equations* and that the corresponding isometric immersions are unique up to isometries in the Euclidean space.[45]

Like the fundamental theorem of Riemannian geometry for an open subset of $\mathbb{R}^n$ (Theorems 8.6-1 and 8.7-1), this theorem comprises two essentially distinct parts, a *global existence result* (Theorem 8.16-1) called the **fundamental theorem of surface theory**, or **Bonnet's theorem,**[46] and a *uniqueness result* (Theorem 8.17-1), called the **rigidity theorem for surfaces**. Note that these two results are established under *different assumptions* on the set ω and on the smoothness of the fields $(a_{\alpha\beta})$ and $(b_{\alpha\beta})$.

Not surprisingly, the proof of existence relies essentially on the *existence theorem for a Pfaff system* (Theorem 6.20-1) and on the (classical) *Poincaré lemma* (Theorem 6.17-2), exactly like the proof of Theorem 8.6-1. In what follows, we let

$$\mathcal{C}^2(\omega;\mathbb{S}^2_>) = \{\boldsymbol{A} \in \mathcal{C}^2(\omega;\mathbb{S}^2);\ \boldsymbol{A}(y) \in \mathbb{S}^2_> \text{ for all } y \in \omega\}.$$

Theorem 8.16-1 (fundamental theorem of surface theory) *Let ω be a simply connected open subset of $\mathbb{R}^2$ and let $(a_{\alpha\beta}) \in \mathcal{C}^2(\omega;\mathbb{S}^2_>)$ and $(b_{\alpha\beta}) \in \mathcal{C}^1(\omega;\mathbb{S}^2)$ be two matrix fields that satisfy the Gauß and Codazzi–Mainardi equations, viz.,*

$$R_{\tau\alpha\beta\sigma} := \partial_\beta \Gamma_{\alpha\sigma\tau} - \partial_\sigma \Gamma_{\alpha\beta\tau} + \Gamma^\mu_{\alpha\beta}\Gamma_{\sigma\tau\mu} - \Gamma^\mu_{\alpha\sigma}\Gamma_{\beta\tau\mu} = b_{\alpha\sigma}b_{\beta\tau} - b_{\alpha\beta}b_{\sigma\tau} \quad \textit{in } \omega,$$
$$\partial_\beta b_{\alpha\sigma} - \partial_\sigma b_{\alpha\beta} + \Gamma^\mu_{\alpha\sigma} b_{\beta\mu} - \Gamma^\mu_{\alpha\beta} b_{\sigma\mu} = 0 \quad \textit{in } \omega,$$

where

$$\Gamma_{\alpha\beta\tau} := \frac{1}{2}(\partial_\beta a_{\alpha\tau} + \partial_\alpha a_{\beta\tau} - \partial_\tau a_{\alpha\beta}) \quad \textit{and} \quad \Gamma^\sigma_{\alpha\beta} := a^{\sigma\tau}\Gamma_{\alpha\beta\tau} \quad \textit{where } (a^{\sigma\tau}) := (a_{\alpha\beta})^{-1}.$$

[45] A substantial literature has been devoted to this theorem and its various proofs; see in particular:

R. H. Szczarba [1970]: On isometric immersions of Riemannian manifolds in Euclidean space, *Boletim da Sociedade Brasileira de Matemática* **1**, 31–45.

K. Tenenblat [1971]: On isometric immersions of Riemannian manifolds, *Boletim da Sociedade Brasileira de Matemática* **2**, 23–36.

H. Jacobowith [1982]: The Gauß-Codazzi equations, *Tensor (N.S.)* **39**, 15–22.

M. Szopos [2005]: On the recovery and continuity of a submanifold with boundary, *Analysis and Applications* **3**, 119–143.

[46] The first proof of a local form of this theorem appeared in:

P.O. Bonnet [1867]: Mémoire sur la théorie des surfaces applicables sur une surface donnée, *Journal de l'Ecole Polytechnique* **42**, 1–151.

Then there exists an immersion $\boldsymbol{\theta} \in \mathcal{C}^3(\omega; \boldsymbol{E}^3)$ *such that*

$$\partial_\alpha \boldsymbol{\theta} \cdot \partial_\beta \boldsymbol{\theta} = a_{\alpha\beta} \quad \textit{and} \quad \partial_{\alpha\beta}\boldsymbol{\theta} \cdot \left\{ \frac{\partial_1 \boldsymbol{\theta} \wedge \partial_2 \boldsymbol{\theta}}{|\partial_1 \boldsymbol{\theta} \wedge \partial_2 \boldsymbol{\theta}|} \right\} = b_{\alpha\beta} \ \textit{in } \omega.$$

Proof [47] (i) *Define matrix fields* $\boldsymbol{\Gamma}_\alpha \in \mathcal{C}^1(\omega; \mathbb{M}^3)$, $\alpha = 1, 2$, *by*

$$\boldsymbol{\Gamma}_\alpha := \begin{pmatrix} \Gamma^1_{\alpha 1} & \Gamma^1_{\alpha 2} & -b^1_\alpha \\ \Gamma^2_{\alpha 1} & \Gamma^2_{\alpha 2} & -b^2_\alpha \\ b_{\alpha 1} & b_{\alpha 2} & 0 \end{pmatrix} \quad \textit{where } b^\beta_\alpha := a^{\beta\sigma} b_{\sigma\alpha}.$$

Then the Gauß and Codazzi–Mainardi equations are satisfied by the matrix fields $(a_{\alpha\beta}) \in \mathcal{C}^2(\omega; \mathbb{S}^2_>)$ *and* $(b_{\alpha\beta}) \in \mathcal{C}^1(\omega; \mathbb{S}^2)$ *if and only if the matrix fields* $\boldsymbol{\Gamma}_\alpha$ *satisfy the relations*

$$\partial_\alpha \boldsymbol{\Gamma}_\beta - \partial_\beta \boldsymbol{\Gamma}_\alpha + \boldsymbol{\Gamma}_\alpha \boldsymbol{\Gamma}_\beta - \boldsymbol{\Gamma}_\beta \boldsymbol{\Gamma}_\alpha = \boldsymbol{0} \quad \textit{in } \omega.$$

Rewritten componentwise, the above relations read

$$\begin{aligned}
\partial_\alpha \Gamma^\mu_{\beta\sigma} - \partial_\beta \Gamma^\mu_{\alpha\sigma} + \Gamma^\tau_{\beta\sigma}\Gamma^\mu_{\alpha\tau} - \Gamma^\tau_{\alpha\sigma}\Gamma^\mu_{\beta\tau} - b_{\beta\sigma}b^\mu_\alpha + b_{\alpha\sigma}b^\mu_\beta &= 0, \\
\partial_\alpha b_{\beta\sigma} - \partial_\beta b_{\alpha\sigma} + \Gamma^\tau_{\beta\sigma} b_{\alpha\tau} - \Gamma^\tau_{\alpha\sigma} b_{\beta\tau} &= 0, \\
\partial_\alpha b^\mu_\beta - \partial_\beta b^\mu_\alpha + \Gamma^\mu_{\alpha\tau} b^\tau_\beta - \Gamma^\mu_{\beta\tau} b^\tau_\alpha &= 0, \\
b_{\alpha\tau} b^\tau_\beta - b_{\beta\tau} b^\tau_\alpha &= 0.
\end{aligned}$$

It is easily seen (Problem 8.14-1) that *the Gauß equations* $R_{\tau\alpha\beta\sigma} = b_{\alpha\sigma}b_{\beta\tau} - b_{\alpha\beta}b_{\sigma\tau}$ *in* ω *are equivalent to the equations*

$$R^{\mu}_{\cdot\alpha\beta\sigma} := \partial_\beta \Gamma^\mu_{\alpha\sigma} - \partial_\sigma \Gamma^\mu_{\alpha\beta} + \Gamma^\tau_{\alpha\sigma}\Gamma^\mu_{\beta\tau} - \Gamma^\tau_{\alpha\beta}\Gamma^\mu_{\sigma\tau} = b_{\alpha\sigma}b^\mu_\beta - b_{\alpha\beta}b^\mu_\sigma \quad \text{in } \omega.$$

Hence the first relations are equivalent to the Gauß equations; the second equations are nothing but the Codazzi–Mainardi equations; the third equations are equivalent to the Codazzi–Mainardi equations, as is easily seen (Problem 8.14-2); the fourth equations are always satisfied since

$$b_{\alpha\tau} b^\tau_\beta = b_{\alpha\tau} b_{\sigma\beta} a^{\tau\sigma} = b_{\sigma\beta} b^\sigma_\alpha.$$

(ii) *Given a point* $y^0 \in \omega$, let $\boldsymbol{a}^0_\alpha \in \mathbb{E}^3$, $\alpha = 1, 2$, denote *two vectors that satisfy*

$$\boldsymbol{a}^0_\alpha \cdot \boldsymbol{a}^0_\beta = a_{\alpha\beta}(y^0)$$

(for instance, let $(\boldsymbol{a}^0_\alpha)_\beta$ be the component at the αth row and βth column of the square root of the matrix $(a_{\alpha\beta}(y^0))$ and let $(\boldsymbol{a}^0_\alpha)_3 := 0$), and let $\boldsymbol{F}^0 \in \mathbb{M}^3$ denote the *matrix whose* i*th column* is $\boldsymbol{a}^0_i$, *where*

$$\boldsymbol{a}^0_3 := \frac{\boldsymbol{a}^0_1 \wedge \boldsymbol{a}^0_2}{|\boldsymbol{a}^0_1 \wedge \boldsymbol{a}^0_2|}.$$

[47]The elegant proof given here is adapted from:
S. Mardare [2005]: On Pfaff systems with L^p coefficients and their applications in differential geometry, *Journal de Mathématiques Pures et Appliquées* **84**, 1659–1692.

Then there exists one, and only one, matrix field $\boldsymbol{F} \in \mathcal{C}^2(\omega; \mathbb{M}^3)$ *that satisfies*

$$\partial_\alpha \boldsymbol{F}(y) = \boldsymbol{F}(y)\boldsymbol{\Gamma}_\alpha(y),\ y \in \omega, \quad \text{and} \quad \boldsymbol{F}(y^0) = \boldsymbol{F}^0.$$

That such a field $\boldsymbol{F} \in \mathcal{C}^2(\omega; \mathbb{M}^3)$ exists and is unique follows from the *existence and uniqueness theorem for Pfaff systems* (Theorem 6.20-1), which can be applied since the open set is *simply connected* by assumption and the matrix fields $\boldsymbol{\Gamma}_\alpha \in \mathcal{C}^1(\omega; \mathbb{M}^3)$ verify the *compatibility condition*

$$\partial_\alpha \boldsymbol{\Gamma}_\beta - \partial_\beta \boldsymbol{\Gamma}_\alpha + \boldsymbol{\Gamma}_\alpha \boldsymbol{\Gamma}_\beta - \boldsymbol{\Gamma}_\beta \boldsymbol{\Gamma}_\alpha = \mathbf{0} \quad \text{in } \omega.$$

(iii) *Let* $\boldsymbol{a}_i \in \mathcal{C}^2(\omega; \mathbb{E}^3)$, $1 \leq i \leq 3$, *denote the* ith *column vector field of the matrix field* $\boldsymbol{F} \in \mathcal{C}^2(\omega; \mathbb{M}^3)$ *found in* (ii), *and let a vector* $\boldsymbol{\theta}^0 \in \mathbb{E}^3$ *be given. Then there exists one, and only one, vector field* $\boldsymbol{\theta} \in \mathcal{C}^3(\omega; \mathbb{E}^3)$ *that satisfies*

$$\partial_\alpha \boldsymbol{\theta}(y) = \boldsymbol{a}_\alpha(y),\ y \in \omega, \quad \text{and} \quad \boldsymbol{\theta}(y^0) = \boldsymbol{\theta}^0.$$

Taking the first and second columns of the matrix equation $\partial_\alpha \boldsymbol{F} = \boldsymbol{F}\boldsymbol{\Gamma}_\alpha$ in ω solved in (ii) then gives

$$\partial_\alpha \boldsymbol{a}_\beta = \Gamma^\sigma_{\alpha\beta} \boldsymbol{a}_\sigma + b_{\alpha\beta} \boldsymbol{a}_3 \quad \text{in } \omega,$$

which, combined with the symmetry relations $\Gamma^\sigma_{\alpha\beta} = \Gamma^\sigma_{\beta\alpha}$ and $b_{\alpha\beta} = b_{\beta\alpha}$, shows that

$$\partial_\alpha \boldsymbol{a}_\beta = \partial_\beta \boldsymbol{a}_\alpha \quad \text{in } \omega.$$

Hence the existence and uniqueness of $\boldsymbol{\theta} \in \mathcal{C}^3(\omega; \mathbb{E}^3)$ follows from the *classical Poincaré lemma* (Theorem 6.17-2) applied to each component of the vector equation $\partial_\alpha \boldsymbol{\theta} = \boldsymbol{a}_\alpha$ in ω; the assumption that ω is *simply connected* is again essential here.

(iv) *The mapping* $\boldsymbol{\theta} \in \mathcal{C}^3(\omega; \mathbb{E}^3)$ *found in* (iii) *satisfies*

$$\partial_\alpha \boldsymbol{\theta} \cdot \partial_\beta \boldsymbol{\theta} = a_{\alpha\beta} \quad \text{and} \quad \partial_{\alpha\beta} \boldsymbol{\theta} \cdot \frac{\partial_1 \boldsymbol{\theta} \wedge \partial_2 \boldsymbol{\theta}}{|\partial_1 \boldsymbol{\theta} \wedge \partial_2 \boldsymbol{\theta}|} = b_{\alpha\beta} \quad \text{in } \omega.$$

Let

$$a_{i3}(y) = a_{3i}(y) := \delta_{i3} \quad \text{and} \quad \Gamma^i_{\alpha j}(y) := (\boldsymbol{\Gamma}_\alpha)_{ij},\ y \in \omega.$$

Then the definitions of the functions $\Gamma_{\alpha\beta\tau}$ and $\Gamma^\sigma_{\alpha\beta}$ in terms of the functions $a_{\alpha\beta}$ and $a^{\alpha\beta}$ imply that the functions $a_{ij} \in \mathcal{C}^2(\omega)$ satisfy

$$\partial_\alpha a_{ij} = \Gamma^m_{\alpha i} a_{mj} + \Gamma^m_{\alpha j} a_{mi} \quad \text{in } \omega \quad \text{and} \quad a_{ij}(y^0) = \boldsymbol{a}^0_i \cdot \boldsymbol{a}^0_j.$$

The equations $\partial_\alpha \boldsymbol{F} = \boldsymbol{F}\boldsymbol{\Gamma}_\alpha$ in ω and $\boldsymbol{F}(y^0) = \boldsymbol{F}^0$ satisfied by the matrix field $\boldsymbol{F}$ (part (iii)) imply that the functions $\boldsymbol{a}_i \cdot \boldsymbol{a}_j \in \mathcal{C}^2(\omega)$ satisfy

$$\begin{gathered}\partial_\alpha(\boldsymbol{a}_i \cdot \boldsymbol{a}_j) = \partial_\alpha \boldsymbol{a}_i \cdot \boldsymbol{a}_j + \boldsymbol{a}_i \cdot \partial_\alpha \boldsymbol{a}_j = \Gamma^m_{\alpha i}(\boldsymbol{a}_m \cdot \boldsymbol{a}_j) + \Gamma^m_{\alpha j}(\boldsymbol{a}_m \cdot \boldsymbol{a}_i) \quad \text{in } \omega, \\ (\boldsymbol{a}_i \cdot \boldsymbol{a}_j)(y^0) = \boldsymbol{a}^0_i \cdot \boldsymbol{a}^0_j.\end{gathered}$$

But either one of these systems of partial differential equations, together with given values at y^0, can have *at most one solution*. To see this, let $\boldsymbol{\gamma} \in \mathcal{C}^1([0,1]; \mathbb{R}^2)$ be a path joining y^0 to any given point $y \in \omega$; then the matrix fields $(a_{ij} \circ \gamma) \in \mathcal{C}^1([0,1]\,;\mathbb{M}^3)$ and $((\boldsymbol{a}_i \cdot \boldsymbol{a}_j) \circ \gamma) \in \mathcal{C}^1([0,1]\,;\mathbb{M}^3)$ satisfy the same *linear Cauchy problem*, which can have at most one solution; cf. Theorem 3.8-2.

Consequently, *the solutions to these two systems coincide*, i.e.,

$$\boldsymbol{a}_\alpha(y) \cdot \boldsymbol{a}_\beta(y) = a_{\alpha\beta}(y) \quad \text{and} \quad \boldsymbol{a}_i(y) \cdot \boldsymbol{a}_3(y) = \delta_{i3}, \;\; y \in \omega.$$

These relations in turn imply that

$$\partial_\alpha \boldsymbol{\theta}(y) \cdot \partial_\beta \boldsymbol{\theta}(y) = a_{\alpha\beta}(y), \quad y \in \omega,$$

$$\boldsymbol{F}^T(y)\boldsymbol{F}(y) = \begin{pmatrix} a_{11} & a_{12} & 0 \\ a_{21} & a_{22} & 0 \\ 0 & 0 & 1 \end{pmatrix}(y), \quad y \in \omega,$$

$$\boldsymbol{a}_3(y) = \varepsilon \frac{\boldsymbol{a}_1(y) \wedge \boldsymbol{a}_2(y)}{|\boldsymbol{a}_1(y) \wedge \boldsymbol{a}_2(y)|}, \quad y \in \omega, \text{ with either } \varepsilon = 1 \text{ or } \varepsilon = -1$$

(the number ε does not depend on $y \in \omega$ since the field $\boldsymbol{a}_3 : \omega \to \mathbb{E}^3$ is continuous on the connected set ω). The symmetric matrices $(a_{\alpha\beta})(y)$ being positive-definite at each $y \in \omega$ by assumption, it also follows that $(\det \boldsymbol{F}(y))^2 > 0$, hence that $\det \boldsymbol{F}(y) \neq 0$, at each $y \in \omega$. Since

$$\det \boldsymbol{F}(y) = (\boldsymbol{a}_1(y) \wedge \boldsymbol{a}_2(y)) \cdot \boldsymbol{a}_3(y) = \varepsilon\, |\boldsymbol{a}_1(y) \wedge \boldsymbol{a}_2(y)|\,, \quad y \in \omega,$$
$$\det \boldsymbol{F}(y^0) = ((\boldsymbol{a}_1^0 \wedge \boldsymbol{a}_2^0) \cdot \boldsymbol{a}_3^0) = |\boldsymbol{a}_1^0 \wedge \boldsymbol{a}_2^0| > 0,$$

and $\det \boldsymbol{F} : \omega \to \mathbb{R}$ is a continuous function that does not vanish on the connected set ω, the only possibility is $\varepsilon = 1$, i.e.,

$$\boldsymbol{a}_3(y) = \frac{\boldsymbol{a}_1(y) \wedge \boldsymbol{a}_2(y)}{|\boldsymbol{a}_1(y) \wedge \boldsymbol{a}_2(y)|}, \quad y \in \omega.$$

The relation $\partial_\alpha \boldsymbol{a}_\beta = \Gamma^\sigma_{\alpha\beta} \boldsymbol{a}_\sigma + b_{\alpha\beta} \boldsymbol{a}_3$ then implies that $\partial_\alpha \boldsymbol{a}_\beta \cdot \boldsymbol{a}_3 = b_{\alpha\beta}$, i.e., that

$$\partial_{\alpha\beta} \boldsymbol{\theta}(y) \cdot \frac{\partial_1 \boldsymbol{\theta}(y) \wedge \partial_2 \boldsymbol{\theta}(y)}{|\partial_1 \boldsymbol{\theta}(y) \wedge \partial_2 \boldsymbol{\theta}(y)|} = b_{\alpha\beta}(y), \quad y \in \omega,$$

which completes the proof. □

Incidentally, it is remarkable that the solution $\boldsymbol{\theta}$ of the *nonlinear* equations

$$\partial_\alpha \boldsymbol{\theta} \cdot \partial_\beta \boldsymbol{\theta} = a_{\alpha\beta} \quad \text{and} \quad \partial_{\alpha\beta} \boldsymbol{\theta} \cdot \left\{ \frac{\partial_1 \boldsymbol{\theta} \wedge \partial_2 \boldsymbol{\theta}}{|\partial_1 \boldsymbol{\theta} \wedge \partial_2 \boldsymbol{\theta}|} \right\} = b_{\alpha\beta} \;\text{ in } \omega,$$

is obtained by successively solving a *linear* Pfaff system (part (ii) of the above proof) and *linear* equations (viz., $\partial_\alpha \boldsymbol{\theta} = \boldsymbol{a}_\alpha$ in ω; cf. part (iii)).

Since the solution $\boldsymbol{F}$ of the Pfaff system found in part (ii) is unique, and since the mapping $\boldsymbol{\theta}$ found in part (iv) is uniquely determined, Theorem 8.16-1 can also be rephrased as the following *existence and uniqueness theorem*.

Theorem 8.16-2 *Let the assumptions on the set* ω *and on the matrix fields* $(a_{\alpha\beta})$ *and* $(b_{\alpha\beta})$ *be as in Theorem* 8.16-1, *let a point* $y_0 \in \omega$ *and a vector* $\boldsymbol{\theta}_0 \in \mathbb{E}^3$ *be given, and let* $\boldsymbol{a}_\alpha^0 \in \mathbb{R}^3$ *be two vectors that satisfy*

$$\boldsymbol{a}_\alpha^0 \cdot \boldsymbol{a}_\beta^0 = (a_{\alpha\beta}(y_0)).$$

Then there exists one and only one immersion $\boldsymbol{\theta} \in \mathcal{C}^3(\omega; \mathbb{E}^3)$ *that satisfies*

$$\partial_\alpha \boldsymbol{\theta} \cdot \partial_\beta \boldsymbol{\theta} = a_{\alpha\beta} \quad \textit{and} \quad \partial_{\alpha\beta} \boldsymbol{\theta} \cdot \frac{\partial_1 \boldsymbol{\theta} \wedge \partial_2 \boldsymbol{\theta}}{|\partial_1 \boldsymbol{\theta} \wedge \partial_2 \boldsymbol{\theta}|} = b_{\alpha\beta} \ \textit{in } \omega,$$

$$\boldsymbol{\theta}(y_0) = \boldsymbol{\theta}_0 \quad \textit{and} \quad \partial_\alpha \boldsymbol{\theta}(y_0) = \boldsymbol{a}_\alpha^0. \qquad \square$$

Otherwise the uniqueness issue *in general*, i.e., when no conditions such as $\boldsymbol{\theta}(y^0) = \boldsymbol{\theta}^0$ and $\partial_\alpha \boldsymbol{\theta}(y^0) = \boldsymbol{a}_\alpha^0$ are imposed as in Theorem 8.16-2, is addressed in the next section, in effect under *weaker regularity assumptions* than in Theorem 8.16-2.

Let ω be a simply connected open subset of $\mathbb{R}^2$, and let a point $y_0 \in \omega$, a vector $\boldsymbol{\theta}_0 \in \mathbb{E}^3$, and two linearly independent vectors $\boldsymbol{a}_\alpha^0 \in \mathbb{R}^3$, be given. Theorem 8.16-2 thus establishes the existence of a (clearly nonlinear) mapping that associates with any matrix fields $(a_{\alpha\beta}) \in \mathcal{C}^2(\omega; \mathbb{S}^2_>)$ and $(b_{\alpha\beta}) \in \mathcal{C}^1(\omega; \mathbb{S}^2)$ satisfying the Gauß and Codazzi–Mainardi relations in ω and $a_{\alpha\beta}(y^0) = \boldsymbol{a}_\alpha^0 \cdot \boldsymbol{a}_\beta^0$, a well-defined immersion $\boldsymbol{\theta} \in \mathcal{C}^3(\omega; \mathbb{E}^3)$ that satisfies $\boldsymbol{\theta}(y_0) = \boldsymbol{\theta}_0$ and $\partial_\alpha \boldsymbol{\theta}(y_0) = \boldsymbol{a}_\alpha^0$ and such that $(a_{\alpha\beta})$ and $(b_{\alpha\beta})$ are the two fundamental forms of the surface $\boldsymbol{\theta}(\omega)$.

Then there exist natural topologies such that this mapping is *continuous*. In other words, *a surface is a continuous function of its two fundamental forms*, between such spaces of continuously differentiable functions; cf. Problem 8.16-1.

Remark A similar conclusion holds, but this time in terms of *Sobolev norms*, as a consequence of a *nonlinear Korn inequality on a surface*.[48] $\square$

The fundamental theorem of surface theory (Theorem 8.16-1) *can be also proved as a corollary to the fundamental theorem of Riemannian geometry for an open subset of* $\mathbb{E}^3$ (Theorem 8.6-1), under the stronger assumption that $(b_{\alpha\beta}) \in \mathcal{C}^2(\omega; \mathbb{S}^2)$, however. This different proof[49], relies on the following elementary observation: Given a smooth enough immersion $\boldsymbol{\theta} : \omega \to \mathbb{E}^3$ and $\varepsilon > 0$, let the mapping $\boldsymbol{\Theta} : \omega \times \left]-\varepsilon, \varepsilon\right[\to \mathbb{E}^3$ be defined by

$$\boldsymbol{\Theta}(y, x_3) := \boldsymbol{\theta}(y) + x_3 \boldsymbol{a}_3(y) \quad \text{for all } (y, x_3) \in \omega \times \left]-\varepsilon, \varepsilon\right[,$$

where $\boldsymbol{a}_3 := \dfrac{\partial_1 \boldsymbol{\theta} \wedge \partial_2 \boldsymbol{\theta}}{|\partial_1 \boldsymbol{\theta} \wedge \partial_2 \boldsymbol{\theta}|}$, and let

$$g_{ij} := \partial_i \boldsymbol{\Theta} \cdot \partial_j \boldsymbol{\Theta}.$$

Then an immediate computation shows that

$$g_{\alpha\beta} = a_{\alpha\beta} - 2x_3 b_{\alpha\beta} + x_3^2 c_{\alpha\beta} \quad \text{and} \quad g_{i3} = \delta_{i3} \ \text{in } \omega \times \left]-\varepsilon, \varepsilon\right[,$$

[48] P.G. Ciarlet; L. Gratie; C. Mardare [2005]: A nonlinear Korn inequality on a surface, *Journal de Mathématiques Pures et Appliquées* **85**, 2–16.

[49] Due to:

P.G. Ciarlet; F. Larsonneur [2002]: On the recovery of a surface with prescribed first and second fundamental forms, *Journal de Mathématiques Pures et Appliquées* **81**, 167–185.

where $a_{\alpha\beta}$ and $b_{\alpha\beta}$ are the covariant components of the first and second fundamental forms of the surface $\boldsymbol{\theta}(\omega)$ and $c_{\alpha\beta} := a^{\sigma\tau} b_{\alpha\sigma} b_{\beta\tau}$.

Assume that the matrices (g_{ij}) constructed in this fashion are *invertible*, hence positive-definite, over the set $\omega \times]-\varepsilon, \varepsilon[$ (they may not be, of course; but the resulting difficulty is easily circumvented). Then the field $(g_{ij}) : \omega \times]-\varepsilon, \varepsilon[\to \mathbb{S}^3_>$ becomes a natural candidate for applying the "three-dimensional" existence result of Theorem 8.6-1, provided of course that *the "three-dimensional" sufficient conditions* of this theorem, viz.,

$$\partial_j \Gamma_{ikq} - \partial_k \Gamma_{ijq} + \Gamma^p_{ij} \Gamma_{kqp} - \Gamma^p_{ik} \Gamma_{jqp} = 0 \quad \text{in } \Omega,$$

can be shown to hold, as a consequence of the assumed "two-dimensional" Gauß and Codazzi–Mainardi equations. That this is indeed the case is the essence of this proof, but proving this implication rests on exceedingly delicate computations, however.

By Theorem 8.6-1, there then exists an immersion $\boldsymbol{\Theta} : \omega \times]-\varepsilon, \varepsilon[\to \mathbb{E}^3$ that satisfies $g_{ij} = \partial_i \boldsymbol{\Theta} \cdot \partial_j \boldsymbol{\Theta}$ in $\omega \times]-\varepsilon, \varepsilon[$, and it is then easy to check that $\boldsymbol{\theta} := \boldsymbol{\Theta}(\cdot, 0)$ indeed satisfies $\partial_\alpha \boldsymbol{\theta} \cdot \partial_\beta \boldsymbol{\theta} = a_{\alpha\beta}$ and $\partial_{\alpha\beta} \boldsymbol{\theta} \cdot \left\{ \dfrac{\partial_1 \boldsymbol{\theta} \wedge \partial_2 \boldsymbol{\theta}}{|\partial_1 \boldsymbol{\theta} \wedge \partial_2 \boldsymbol{\theta}|} \right\} = b_{\alpha\beta}$ in ω.

A *different* set of compatibility conditions, expressed again in terms of the matrix fields $(a_{\alpha\beta})$ and $(b_{\alpha\beta})$ but where the field $(a_{\alpha\beta})$ now appears through its *square root*, can be identified that likewise lead to a similar existence and uniqueness theorem; cf. Problem 8.16-2.

The *regularity assumptions* made in Theorem 8.16-1 on the matrix fields $(a_{\alpha\beta})$ and $(b_{\alpha\beta})$ can be significantly weakened in several ways (with self-explanatory notation, such as $W^{1,p}(\omega; \mathbb{S}^2_>)$). For instance, an existence theorem still holds[50] if $(a_{\alpha\beta}) \in \mathcal{C}^1(\omega; \mathbb{S}^2_>)$ and $(b_{\alpha\beta}) \in \mathcal{C}(\omega; \mathbb{S}^2)$, with a resulting mapping $\boldsymbol{\theta}$ in the space $\mathcal{C}^2(\omega; \mathbb{E}^3)$.

The existence result of Theorem 8.16-1 also holds *"up to the boundary of the set ω"* in the following sense:[51] Assume that the functions $a_{\alpha\beta}$, *resp.* $b_{\alpha\beta}$, and their partial derivatives of order ≤ 2, *resp.* ≤ 1, can be extended by continuity to the closure $\overline{\omega}$, the symmetric field $(a_{\alpha\beta})$ extended in this fashion remaining positive-definite over the set $\overline{\omega}$. Then the immersion $\boldsymbol{\theta}$ and its partial derivatives of order ≤ 3 can be also extended by continuity to $\overline{\omega}$.

Theorem 8.16-1 can be also extended to *Sobolev spaces*: If for some $p > 2$, $(a_{\alpha\beta}) \in W^{1,p}(\omega; \mathbb{S}^2_>)$ and $(b_{\alpha\beta}) \in L^p(\omega; \mathbb{S}^2)$ are two matrix fields that satisfy the Gauß and Codazzi–Mainardi equations *in the sense of distributions*, and ω is a simply connected domain in $\mathbb{R}^2$, then[52] there exists a mapping $\boldsymbol{\theta} \in W^{2,p}(\omega; \mathbb{E}^3)$ such that $(a_{\alpha\beta})$ and $(b_{\alpha\beta})$ are the fundamental forms of the surface $\boldsymbol{\theta}(\omega)$.

Problems

8.16-1 Given an open subset $\omega \in \mathbb{R}^2$, the notation $\mathcal{K} \Subset \omega$ means that $\mathcal{K}$ is a compact subset of ω. Given any integer $m \geq 0$ and any $\mathcal{K} \Subset \omega$, the *seminorm* $|\cdot|_{m,\mathcal{K}}$ is defined over the space $\mathcal{C}^m(\omega)$ by

$$|g|_{m,\mathcal{K}} := \sup_{\left\{ \substack{y \in \mathcal{K} \\ |\alpha| \leq m} \right.} |\partial^\alpha g(y)| \quad \text{for each } g \in \mathcal{C}^m(\omega).$$

[50]P. Hartman; A. Wintner [1950]: On the embedding problem in differential geometry, *American Journal of Mathematics* **72**, 553–564.

[51]P.G. Ciarlet; C. Mardare [2005]: Recovery of a surface with boundary and its continuity as a function of its two fundamental forms, *Analysis and Applications* **3**, 99–117.

[52]S. Mardare [2007]: On systems of first order linear partial differential equations with L^p coefficients, *Advances in Differential Equations* **73**, 301–360.

Analogous seminorms are also defined for vector-valued and matrix-valued functions, $|\cdot|$ now designating the Euclidean vector norm or its subordinate matrix norm.

Let ω be a simply connected open subset of $\mathbb{R}^2$, let a point $y_0 \in \omega$, a vector $\boldsymbol{\theta}_0 \in \mathbb{E}^3$, and two linearly independent vectors $\boldsymbol{a}_\alpha^0 \in \mathbb{R}^3$ be given. Let $(a_{\alpha\beta}^\ell) \in \mathcal{C}^2(\omega; \mathbb{S}^2_>)$ and $(b_{\alpha\beta}^\ell) \in \mathcal{C}^2(\omega; \mathbb{S}^2)$, $\ell \geq 1$, and $(a_{\alpha\beta}) \in \mathcal{C}^2(\omega; \mathbb{S}^2_>)$ and $(b_{\alpha\beta}) \in \mathcal{C}^2(\omega; \mathbb{S}^2)$, be matrix fields satisfying the *Gauß and Codazzi–Mainardi relations* in ω, and

$$a_{\alpha\beta}^\ell(y_0) = \boldsymbol{a}_\alpha^0 \cdot \boldsymbol{a}_\beta^0, \ \ \ell \geq 1, \quad \text{and} \quad a_{\alpha\beta}(y_0) = \boldsymbol{a}_\alpha^0 \cdot \boldsymbol{a}_\beta^0,$$

$$\lim_{\ell\to\infty} |(a_{\alpha\beta}^\ell - (a_{\alpha\beta})|_{2,\mathcal{K}} = 0 \quad \text{and} \quad \lim_{\ell\to\infty} |(b_{\alpha\beta}^\ell) - (b_{\alpha\beta})|_{2,\mathcal{K}} = 0 \ \text{ for each } \mathcal{K} \Subset \omega.$$

By Theorem 8.16-2, there thus exist uniquely determined immersions $\boldsymbol{\theta}^\ell \in \mathcal{C}^3(\omega; \mathbb{E}^3)$, $\ell \geq 1$, *resp.* $\boldsymbol{\theta} \in \mathcal{C}^3(\omega; \mathbb{E}^3)$, such that $(a_{\alpha\beta}^\ell)$ and $(b_{\alpha\beta}^\ell)$, $\ell \geq 1$, *resp.* $(a_{\alpha\beta})$ and $(b_{\alpha\beta})$, are the first and second fundamental forms of the surface $\boldsymbol{\theta}^\ell(\omega)$, *resp.* $\boldsymbol{\theta}(\omega)$, and such that

$$\boldsymbol{\theta}^\ell(y_0) = \boldsymbol{\theta}_0 \text{ and } \partial_\alpha \boldsymbol{\theta}^\ell(y_0) = \boldsymbol{a}_\alpha^0, \ \ \ell \geq 1, \quad \text{resp.} \quad \boldsymbol{\theta}(y_0) = \boldsymbol{\theta}_0 \text{ and } \partial_\alpha \boldsymbol{\theta}(y_0) = \boldsymbol{a}_\alpha^0.$$

(1) Define the matrix fields $(g_{ij}^\ell) \in \mathcal{C}^2(\omega \times \mathbb{R}; \mathbb{S}^3)$, $\ell \geq 1$, and $(g_{ij}) \in \mathcal{C}^2(\omega \times \mathbb{R}; \mathbb{S}^3)$ by (for brevity, the dependence on $y \in \omega$ is omitted)

$$g_{\alpha\beta}^\ell := a_{\alpha\beta}^\ell - 2x_3 b_{\alpha\beta}^\ell + x_3^2 a^{\sigma\tau,\ell} b_{\alpha\sigma}^\ell b_{\beta\tau}^\ell \quad \text{and} \quad g_{i3}^\ell := \delta_{i3}, \ \ \ell \geq 1,$$
$$g_{\alpha\beta} := a_{\alpha\beta} - 2x_3 b_{\alpha\beta} + x_3^2 a^{\sigma\tau} b_{\alpha\sigma} b_{\beta\tau} \quad \text{and} \quad g_{i3} := \delta_{i3}.$$

Then show that the fields $(g_{i\ell}^\ell)$, $\ell \geq 1$, and (g_{ij}) are positive-definite over an open set $\Omega \subset \mathbb{R}^3$ of the form $\Omega := \bigcup_{k=0}^\infty \omega_k \times]-\varepsilon_k, \varepsilon_k[$ where $\omega_k \Subset \omega$ and $\varepsilon_k > 0$ for each $k \geq 0$.

(2) Show that

$$\lim_{\ell\to\infty} |\boldsymbol{\theta}^\ell - \boldsymbol{\theta}|_{3,\mathcal{K}} = 0 \quad \text{for each } \mathcal{K} \Subset \omega.$$

Hint: Use (1) combined with Problem 8.6-3.

Remark Define the sets

$$\boldsymbol{X} := \{((a_{\alpha\beta}), (b_{\alpha\beta})) \in \mathcal{C}^2(\omega; \mathbb{S}^3_>) \times \mathcal{C}^2(\omega; \mathbb{S}^2); (a_{\alpha\beta}) \text{ and } (b_{\alpha\beta}) \text{ satisfy the Gauß and Codazzi–Mainardi relations in } \omega \text{ and } a_{\alpha\beta}(y_0) = \boldsymbol{a}_\alpha^0 \cdot \boldsymbol{a}_\beta^0\},$$

$$\boldsymbol{Y} := \{\boldsymbol{\theta} \in \mathcal{C}^3(\omega; \mathbb{E}^3);\ \boldsymbol{\theta}(y_0) = \boldsymbol{\theta}_0 \text{ and } \partial_\alpha \boldsymbol{\theta}(y_0) = \boldsymbol{a}_\alpha^0\}.$$

Then question (2) shows that the mapping defined by

$$((a_{\alpha\beta}), (b_{\alpha\beta})) \in (\boldsymbol{X}; d_2) \to \boldsymbol{\theta} \in (\boldsymbol{Y}; d_3),$$

where $\boldsymbol{\theta}$ is the immersion found in Theorem 8.16-2, is continuous[53] (the distances d_2 and d_3 are defined as in Problem 7.8-3). □

8.16-2 The objective of this problem is to show that the necessary conditions of Problem 8.14-4 become also *sufficient* for the *existence*[54] of an immersion $\boldsymbol{\theta} \in \mathcal{C}^3(\omega; \mathbb{E}^3)$ if the open set $\omega \subset \mathbb{R}^2$ is *simply connected*, an assumption that accordingly holds throughout this problem.

---

[53] This result is due to:

P.G. CIARLET [2003]: The continuity of a surface as a function of its two fundamental forms, *Journal de Mathématiques Pures et Appliquées* **82**, 253–274.

[54] The compatibility conditions of Problem 8.16-2 and this existence result are due to:

P.G. CIARLET; L. GRATIE; C. MARDARE [2008]: A new approach to the fundamental theorem of surface theory, *Archive for Rational Mechanics and Analysis* **188**, 457–473.

Yet *another* set of related necessary and sufficient (if ω is simply connected) compatibility conditions is

In what follows, $(a_{\alpha\beta}) \in \mathcal{C}^2(\omega; \mathbb{S}^2_>)$ and $(b_{\alpha\beta}) \in \mathcal{C}^1(\omega; \mathbb{S}^2)$ are two given matrix fields that satisfy

$$\partial_1 \boldsymbol{A}_2 - \partial_2 \boldsymbol{A}_1 + \boldsymbol{A}_1 \boldsymbol{A}_2 - \boldsymbol{A}_2 \boldsymbol{A}_1 = \boldsymbol{0} \quad \text{in } \omega,$$

where the matrix fields $\boldsymbol{A}_\alpha \in \mathcal{C}^1(\omega; \mathbb{M}^3)$ are constructed from the matrix fields $(a_{\alpha\beta})$ and $(b_{\alpha\beta})$ through the following series of definitions:

$$\Gamma_{\alpha\beta\tau} := \frac{1}{2}(\partial_\beta a_{\alpha\tau} + \partial_\alpha a_{\beta\tau} - \partial_\tau a_{\alpha\beta}), \quad (a^{\sigma\tau}) := (a_{\alpha\beta})^{-1}, \quad \Gamma^\sigma_{\alpha\beta} := a^{\sigma\tau}\Gamma_{\alpha\beta\tau}, \quad b^\sigma_\alpha := a^{\beta\sigma} b_{\alpha\beta},$$

$$\boldsymbol{\Gamma}_\alpha := \begin{pmatrix} \Gamma^1_{\alpha 1} & \Gamma^1_{\alpha 2} & -b^1_\alpha \\ \Gamma^2_{\alpha 1} & \Gamma^2_{\alpha 2} & -b^2_\alpha \\ b_{\alpha 1} & b_{\alpha 2} & 0 \end{pmatrix}, \quad \boldsymbol{C} := \begin{pmatrix} a_{11} & a_{12} & 0 \\ a_{21} & a_{22} & 0 \\ 0 & 0 & 1 \end{pmatrix}, \quad \boldsymbol{U} := \boldsymbol{C}^{1/2}, \quad \boldsymbol{A}_\alpha := (\boldsymbol{U}\boldsymbol{\Gamma}_\alpha - \partial_\alpha \boldsymbol{U})\boldsymbol{U}^{-1}.$$

(1) Show that the matrix fields $\boldsymbol{A}_\alpha \in \mathcal{C}^1(\omega; \mathbb{M}^3)$ are antisymmetric.

(2) Let a point $y^0 \in \omega$ and a proper orthogonal matrix $\boldsymbol{R}^0 \in \mathbb{O}^3_+$ be given. Show that there exists one, and only one, proper orthogonal matrix field $\boldsymbol{R} \in \mathcal{C}^2(\omega; \mathbb{O}^3_+)$ that satisfies

$$\partial_\alpha \boldsymbol{R} = \boldsymbol{R}\boldsymbol{A}_\alpha \quad \text{in } \omega \quad \text{and} \quad \boldsymbol{R}(y^0) = \boldsymbol{R}^0,$$

where $\mathcal{C}^2(\omega; \mathbb{O}^3_+) := \{\boldsymbol{R} \in \mathcal{C}^2(\omega; \mathbb{M}^3);\ \boldsymbol{R}(y) \in \mathbb{O}^3_+ \text{ for all } y \in \omega\}$.

(3) Let $\boldsymbol{u}_\alpha \in \mathcal{C}^2(\omega; \mathbb{E}^3)$ denote the αth column vector field of the matrix field $\boldsymbol{R} \in \mathcal{C}^2(\omega; \mathbb{O}^3_+)$ found in (2). Show that there exists an immersion $\boldsymbol{\theta} \in \mathcal{C}^3(\omega; \mathbb{E}^3)$ that satisfies

$$\partial_\alpha \boldsymbol{\theta} = \boldsymbol{R}\boldsymbol{u}_\alpha \quad \text{in } \omega.$$

(4) Show that the immersion found in (3) satisfies

$$\partial_\alpha \boldsymbol{\theta} \cdot \partial_\beta \boldsymbol{\theta} = a_{\alpha\beta} \text{ in } \omega \quad \text{and} \quad \partial_{\alpha\beta}\boldsymbol{\theta} \cdot \frac{\partial_1 \boldsymbol{\theta} \wedge \partial_2 \boldsymbol{\theta}}{|\partial_1 \boldsymbol{\theta} \wedge \partial_2 \boldsymbol{\theta}|} = b_{\alpha\beta} \text{ in } \omega.$$

8.16-3 Let ω be an open subset of $\mathbb{R}^2$. Show that two matrix fields $(a_{\alpha\beta}) \in \mathcal{C}^2(\omega; \mathbb{S}^2_>)$ and $(b_{\alpha\beta}) \in \mathcal{C}^1(\omega; \mathbb{S}^2)$ satisfy the Gauß and Codazzi–Mainardi equations in ω if and only if they satisfy

$$\partial_1 \boldsymbol{A}_2 - \partial_2 \boldsymbol{A}_1 + \boldsymbol{A}_1 \boldsymbol{A}_2 - \boldsymbol{A}_2 \boldsymbol{A}_1 = \boldsymbol{0} \quad \text{in } \omega,$$

where the matrix fields $\boldsymbol{A}_\alpha \in \mathcal{C}^1(\omega; \mathbb{A}^3)$ are constructed from the matrix fields $(a_{\alpha\beta})$ and $(b_{\alpha\beta})$ as in Problem 8.16-2.

8.17 Uniqueness of surfaces with the same fundamental forms; the rigidity theorem for surfaces

In Section 8.16, we have established the *existence* of an immersion $\boldsymbol{\theta} : \omega \subset \mathbb{R}^2 \to \mathbb{E}^3$ giving rise to a surface $\boldsymbol{\theta}(\omega)$ with prescribed first and second fundamental forms under the assumptions that these forms satisfy the Gauß and Codazzi–Mainardi conditions in ω and that the open set ω is simply connected. We now turn to the question of *uniqueness* of such immersions.

---

possible, this time in *vector form*, see:

C. Vallée; D. Fortuné [1976]: Compatibility equations in shell theory, *International Journal of Engineering Science* **34**, 495–499.

P.G. Ciarlet; O. Iosifescu [2009]: A new approach to the fundamental theorem of surface theory, by means of the Darboux-Vallée-Fortunée compatibility relation, *Journal de Mathématiques Pures et Appliquées* **91**, 384–401.

This is the object of the next theorem, which, like Theorem 8.7-1, constitutes another *rigidity theorem*. It asserts that, if two immersions $\boldsymbol{\theta} \in \mathcal{C}^2(\omega; \mathbb{E}^3)$ and $\widetilde{\boldsymbol{\theta}} \in \mathcal{C}^2(\omega; \mathbb{E}^3)$ share the same fundamental forms, then the surface $\widetilde{\boldsymbol{\theta}}(\omega)$ is obtained by subjecting the surface $\boldsymbol{\theta}(\omega)$ to a *rotation* (represented by a proper orthogonal matrix $\boldsymbol{Q}$), then by subjecting the rotated surface to a *translation* (represented by a vector $\mathbf{c}$). In other words, *the immersion found in* Theorem 8.16-1 *is unique up to proper isometries of* $\mathbb{E}^3$ (Section 8.7).

As shown in the next proof, the issue of uniqueness can be resolved as a corollary to the rigidity theorem for an open subset of $\mathbb{R}^3$ (Theorem 8.7-1); this is why weaker smoothness assumptions than in the existence theorem (Theorem 8.16-1) suffice. Recall that $\mathbb{O}^3$ denotes the set of all orthogonal matrices of order three and that $\mathbb{O}^3_+ = \{\boldsymbol{Q} \in \mathbb{O}^3; \det \boldsymbol{Q} = 1\}$ denotes the set of all proper orthogonal matrices of order three.

Note that the assumption that ω be simply connected is no longer needed here.

Theorem 8.17-1 (rigidity theorem for surfaces) *Let ω be a connected open subset of $\mathbb{R}^2$ and let $\widetilde{\boldsymbol{\theta}} \in \mathcal{C}^2(\omega; \mathbb{E}^3)$ and $\boldsymbol{\theta} \in \mathcal{C}^2(\omega; \mathbb{E}^3)$ be two immersions whose associated first and second fundamental forms satisfy (with self-explanatory notations)*

$$\widetilde{a}_{\alpha\beta} = a_{\alpha\beta} \quad \text{and} \quad \widetilde{b}_{\alpha\beta} = b_{\alpha\beta} \ \text{ in } \omega.$$

Then there exist a vector $\mathbf{c} \in \mathbb{E}^3$ and a matrix $\boldsymbol{Q} \in \mathbb{O}^3_+$ such that

$$\widetilde{\boldsymbol{\theta}}(y) = \mathbf{c} + \boldsymbol{Q}\boldsymbol{\theta}(y) \quad \text{for all } y \in \omega.$$

Proof Let the matrix field $(g_{ij}) \in \mathcal{C}(\omega \times \mathbb{R}; \mathbb{S}^3)$ be defined in $\omega \times \mathbb{R}$ by

$$g_{\alpha\beta}(y, x_3) := a_{\alpha\beta}(y) - 2x_3 b_{\alpha\beta}(y) + x_3^2 a^{\sigma\tau}(y) b_{\alpha\sigma}(y) b_{\beta\tau}(y) \quad \text{at each } (y, x_3) \in \omega \times \mathbb{R},$$
$$g_{i3}(y, x_3) := \delta_{i3} \quad \text{at each } (y, x_3) \in \omega \times \mathbb{R}.$$

There exist open subsets ω_ℓ, $\ell \geq 0$, of ω such that $\overline{\omega}_\ell$ is a compact subset of ω for each $\ell \geq 0$ and such that

$$\omega = \bigcup_{\ell=0}^{\infty} \omega_\ell.$$

Then, for each $\ell \geq 0$, there exists $\varepsilon_\ell = \varepsilon_\ell(\omega_\ell) > 0$ such that the symmetric matrices $(g_{ij}(y, x_3))$ are positive-definite at all $(y, x_3) \in \overline{\omega}_\ell \times [-\varepsilon_\ell, \varepsilon_\ell]$ (since the functions $g_{ij} := \omega \times \mathbb{R} \to \mathbb{R}$ are continuous and the symmetric matrices $(a_{\alpha\beta}(y)) \in \mathbb{S}^2$ are positive-definite at each $y \in \overline{\omega}_\ell$).

Define the open set

$$\Omega := \bigcup_{\ell=0}^{\infty} (\omega_\ell \times]-\varepsilon_\ell, \varepsilon_\ell[) \subset \omega \times \mathbb{R},$$

which is connected by Theorem 1.9-9 (clearly, Ω is arcwise-connected since the set ω is open and connected).

The two mappings $\widetilde{\boldsymbol{\Theta}} \in \mathcal{C}^1(\Omega; \mathbb{E}^3)$ and $\boldsymbol{\Theta} \in \mathcal{C}^1(\Omega; \mathbb{E}^3)$ defined by (with self-explanatory notations)

$$\widetilde{\boldsymbol{\Theta}}(y, x_3) := \widetilde{\boldsymbol{\theta}}(y) + x_3 \widetilde{\boldsymbol{a}}_3(y) \quad \text{and} \quad \boldsymbol{\Theta}(y, x_3) := \boldsymbol{\theta}(y) + x_3 \boldsymbol{a}_3(y) \ \text{ at each } (y, x_3) \in \Omega,$$

therefore satisfy

$$\nabla\widetilde{\Theta}^T\nabla\widetilde{\Theta} = \nabla\Theta^T\nabla\Theta = (g_{ij}) \quad \text{in } \Omega,$$

which shows in particular that they are both *immersions* since the symmetric matrix field (g_{ij}) is positive-definite in Ω.

Therefore, by Theorem 8.7-1 there exist a vector $\boldsymbol{c} \in \mathbb{E}^3$ and an orthogonal matrix $\boldsymbol{Q} \in \mathbb{O}^3$ such that

$$\widetilde{\Theta}(y, x_3) = \boldsymbol{c} + \boldsymbol{Q}\Theta(y, x_3) \quad \text{for all } (y, x_3) \in \Omega.$$

Hence, on the one hand,

$$\det \nabla\widetilde{\Theta}(y, x_3) = \det \boldsymbol{Q} \det \nabla\Theta(y, x_3) \quad \text{for all } (y, x_3) \in \Omega.$$

On the other hand, a simple computation shows that

$$\det \nabla\Theta(y, x_3) = \sqrt{\det(a_{\alpha\beta}(y))}\,\big(1 - x_3(b_1^1 + b_2^2)(y) + x_3^2(b_1^1 b_2^2 - b_1^2 b_2^1)(y)\big)$$

for all $(y, x_3) \in \Omega$, where

$$b_\alpha^\beta(y) := a^{\beta\sigma}(y) b_{\alpha\sigma}(y), \quad y \in \omega,$$

so that

$$\det \nabla\widetilde{\Theta}(y, x_3) = \det \nabla\Theta(y, x_3) \quad \text{for all } (y, x_3) \in \Omega.$$

Therefore $\det \boldsymbol{Q} = 1$, which shows that $\boldsymbol{Q} \in \mathbb{O}^3$ is in fact a proper orthogonal matrix. The conclusion then follows by letting $x_3 = 0$ in the relation

$$\widetilde{\Theta}(y, x_3) = \boldsymbol{c} + \boldsymbol{Q}\Theta(y, x_3) \quad \text{for all } (y, x_3) \in \Omega. \qquad \square$$

Remarks (1) By contrast, the rigidity theorem for an open subset of $\mathbb{R}^n$ (Theorem 8.7-1) involves isometries of $\mathbb{E}^n$ that are not necessarily proper.

(2) The rigidity theorem for a surface can be extended to mappings Θ with components in *Sobolev spaces*.[55] □

[55] P.G. CIARLET; C. MARDARE [2003]: On rigid and infinitesimal rigid displacements in shell theory, *Journal de Mathématiques Pures et Appliquées* **83**, 1–15.

CHAPTER 9

THE "GREAT THEOREMS" OF NONLINEAR FUNCTIONAL ANALYSIS

Introduction

The title of this chapter is slightly misleading, for two reasons. First, such basic results as the *Banach fixed point theorem*, *Sard's lemma*, the *Newton–Kantorovich theorem*, or the *implicit function theorem* also count among the "great theorems" of nonlinear functional analysis; yet they do not appear here (since they were treated in Chapters 3 and 7, respectively).

Second, while the treatment of the *basic* notions of *linear functional analysis* given in this book can be considered as reasonably complete, that of *nonlinear functional analysis* is by necessity not as thorough, in view of the vastness of the subject. Our more modest objective in this chapter is simply to give a reasonably complete treatment only of those notions that are the *most basic*, thus leaving aside more advanced or specialized topics (such as, e.g., gamma-convergence, concentration-compactness, compensated compactness, the mountain pass lemma, or the Leray–Schauder degree in infinite-dimensional Banach spaces), which are only briefly introduced here (specific references are then provided in each instance).

The first part of the present chapter constitutes an introduction to the *calculus of variations*, in the sense that it considers *minimization problems for nonquadratic functionals*, typically defined over the Sobolev space $W^{1,p}(\Omega)$, where $1 < p < \infty$ and Ω is a domain in $\mathbb{R}^n$. As expected, the solutions of such minimization problems satisfy, at least in the sense of distributions, *nonlinear partial differential equations* posed over Ω, which constitute *Euler–Lagrange equations* in the language of the calculus of variations (these equations are introduced on an ideal model problem in Section 9.1); recall in this respect that minimizers of *quadratic* functionals satisfy by contrast *linear* partial differential equations in Ω (Chapter 6).

General existence theorems for such minimization problems are then established (Theorems 9.3-1 and 9.5-2) for functionals that are *sequentially weakly lower semicontinuous*, a property that plays a fundamental role in the calculus of variations. This property is usually derived by assuming the *coerciveness* of the functional, the *convexity* of its integrand, and the *reflexivity* of the space over which the functionals are to be minimized. These assumptions explain why a crucial use is made in the proofs of these theorems of the fundamental notions developed at the end of Chapter 5: weak convergence, the Banach–Saks–Mazur theorem, or the Banach–Eberlein–Šmulian theorem.

Applications of these general theorems include the *von Kármán equations* (Theorem 9.4-3), the *Dirichlet problem for the p-Laplacian* (Theorem 9.6-1), and especially, the remarkable *existence theorem of John Ball in three-dimensional nonlinear elasticity* (Theorem 9.7-4), which itself rests on the introduction of two fundamental notions, *polyconvexity* and *compensated compactness* (Section 9.7).

It is also shown that, thanks to *Ekeland's variational principle* (Theorems 9.8-1 and 9.8-2), the existence of minimizers can still be obtained over *nonreflexive Banach spaces* when the functionals are of class $\mathcal{C}^1$, bounded from below, and satisfy the *Palais–Smale condition* (Theorem 9.8-3).

The second part of this chapter, in effect often closely intertwined with the first one, is centered on *one of the most basic theorems of nonlinear functional analysis: Brouwer's fixed point theorem.* This theorem simply asserts that any continuous mapping from a compact and convex subset of $\mathbb{R}^n$ into itself has at least one fixed point.

A first, and to a large extent elementary, proof of Brouwer's theorem is given in Theorem 9.9-2, which is based on the observation that if two smooth functions v and $\widetilde{v}$ coincide on the boundary of a domain Ω in $\mathbb{R}^n$, then $\int_\Omega \det \nabla v(x)\,dx = \int_\Omega \det \nabla \widetilde{v}(x)\,dx$, a relation that itself immediately follows from the fundamental *Piola's identity* (Section 7.1).

We then begin to describe some of the numerous far-reaching *applications* of Brouwer's theorem, which include a brief incursion into the *Perron–Frobenius theory of nonnegative matrices* (Theorem 9.9-4), or the effectiveness of the *Galerkin method* for establishing the existence of solutions to the von Kármán equations (cf. Theorem 9.10-1; then without recourse to a functional as in Section 9.4), and to the *Navier–Stokes equations* (Theorem 9.11-1).

It is also shown how Brouwer's fixed point theorem can be extended to *infinite-dimensional* normed vector spaces, in the form of *Schauder's fixed point theorem* (Theorem 9.12-1) or of *Schäfer's fixed point theorem* (Theorem 9.12-2), itself a special case of another basic theorem of nonlinear functional analysis, the *Leray–Schauder fixed point theorem* (Theorem 9.12-3).

Another approach for establishing the existence of solutions to nonlinear partial differential equations is based on the fundamental *Minty–Browder theorem* (Theorem 9.14-1), which applies to a large class of nonlinear operators called *monotone operators* (Section 9.14). Its proof again essentially relies on Brouwer's fixed point theorem used in conjunction with the Galerkin method. For instance, this approach provides another way of establishing the existence of solutions to the *Dirichlet problem for the p-Laplacian* (cf. Theorem 9.14-2; then without recourse to a functional as in Section 9.6).

In the last three sections of this chapter, we provide a detailed construction of the *Brouwer topological degree in* $\mathbb{R}^n$, another fundamental notion of nonlinear functional analysis (Section 9.15). We then show how the Brouwer degree provides a second, and strikingly short, proof of *Brouwer's fixed point theorem* (Theorem 9.16-1), as well as the key to the proofs of some of the most spectacular results of *nonlinear functional analysis in* $\mathbb{R}^n$, the *hairy ball theorem* (Theorem 9.16-2), *Borsuk's* and the *Borsuk–Ulam theorems* (Theorems 9.17-1 and 9.17-2), and, finally, the deep *Brouwer invariance of domain theorem in* $\mathbb{R}^n$ (Theorem 9.17-3).

9.1 Nonlinear partial differential equations as the Euler–Lagrange equations associated with the minimization of a functional

Minimizers of *quadratic* functionals over Sobolev spaces such as $H_0^1(\Omega)$ or $\boldsymbol{H}_0^1(\Omega) := H_0^1(\Omega; \mathbb{R}^n)$ also solve *linear* second-order boundary value problems posed over Ω, at least if they are smooth enough (otherwise the partial differential equations are at least satisfied in the sense of distributions, i.e., in the space $\mathcal{D}'(\Omega)$). For instance, under the assumptions of Theorem 6.7-2, if the minimizer $u \in H_0^1(\Omega)$ of the functional

$$J : v \in H_0^1(\Omega) \to \int_\Omega \left(|\nabla v|^2 + cv^2 \right) \mathrm{d}x - \int_\Omega fv \,\mathrm{d}x$$

over the space $H_0^1(\Omega)$ is in the space $H^2(\Omega)$, then u also solves the boundary value problem

$$-\Delta u + cu = f \text{ in } \Omega \quad \text{and} \quad u = 0 \text{ on } \Gamma := \partial\Omega.$$

For instance, under the assumptions of Theorem 6.16-1, if the minimizer $\boldsymbol{u} \in \boldsymbol{H}_0^1(\Omega)$ of the functional

$$J : \boldsymbol{v} \in \boldsymbol{H}_0^1(\Omega) \to \int_\Omega \{\lambda(\operatorname{tr} \boldsymbol{e}(\boldsymbol{v}))^2 + 2\mu \boldsymbol{e}(v) : \boldsymbol{e}(\boldsymbol{v})\} \mathrm{d}x - \int_\Omega \boldsymbol{f} \cdot \boldsymbol{v} \,\mathrm{d}x$$

over the space $\boldsymbol{H}_0^1(\Omega)$ is in the space $\boldsymbol{H}^2(\Omega)$, then $\boldsymbol{u}$ also solves the boundary value problem

$$-\,\mathbf{div}\{\lambda(\operatorname{tr} \boldsymbol{e}(\boldsymbol{u}))\boldsymbol{I} + 2\mu \boldsymbol{e}(\boldsymbol{u})\} = \boldsymbol{f} \text{ in } \Omega \quad \text{and} \quad \boldsymbol{u} = \boldsymbol{0} \text{ on } \Gamma.$$

In such examples, the partial differential equations in Ω are derived as follows: Let $u \in V$ be such that $J(u) = \inf_{v \in V} J(v)$, where the space V and the functional $J : v \in V \to J(v) = \frac{1}{2}a(v,v) - \ell(v)$ verify the assumptions of Theorem 6.1-1. Since in this case $a(u,v) = \ell(v)$ for all $v \in V$ (Theorem 6.1-2), the partial differential equations are obtained, first, by applying *Green's formula* to these variational equations (this is licit since u is assumed for this purpose to possess extra regularity), and, second, by using that, if $w \in L^2(\Omega)$ satisfies $\int_\Omega w\varphi \,\mathrm{d}x = 0$ for all $\varphi \in \mathcal{D}(\Omega)$, then $w = 0$.

Note that the variational equations $a(u,v) = \ell(v)$ for all $v \in V$ simply express that *the Gâteaux derivatives $a(u,v) - \ell(v)$ of J at u vanish in all the directions $v \in V$*, or equivalently, that the *Fréchet derivative $J'(u)$ vanishes at u* (recall that the functional J is Fréchet differentiable in this case).

Such considerations are in fact of a much wider applicability, because they likewise apply to functionals that are *no longer quadratic*, thereby yielding a powerful means of relating a wide class of *nonlinear* boundary value problems to the minimization of functionals, as shown in the next theorem.

Note that, in this theorem, each partial derivative $\frac{\partial \mathcal{L}}{\partial \boldsymbol{a}}(x, \boldsymbol{a}, \boldsymbol{F})$ is identified with the column vector $\left(\frac{\partial \mathcal{L}}{\partial a_i}(x, \boldsymbol{a}, \boldsymbol{F})\right)_{i=1}^m \in \mathbb{R}^m$, each partial derivative $\frac{\partial \mathcal{L}}{\partial \boldsymbol{F}}(x, \boldsymbol{a}, \boldsymbol{F})$ is identified with the matrix $\left(\frac{\partial \mathcal{L}}{\partial F_{ij}}(x, \boldsymbol{a}, \boldsymbol{F})\right) \in \mathbb{M}^{m \times n}$ (the row index is i), and $\nabla \boldsymbol{v}(x) = (\partial_j v_i(x)) \in \mathbb{M}^{m \times n}$, $x \in \Omega$ (the row index is i) (in conformity with the notations defined in Section 7.1).

Note also that *all the assumptions made in the statement of parts* (a) *and* (b), resp. *of part* (c), *in* Theorem 9.1-1 *about the function $\mathcal{L}$ are satisfied if $\mathcal{L} \in \mathcal{C}^1(\overline{\Omega} \times \mathbb{R}^m \times \mathbb{M}^{m \times n})$*, resp. *if $\mathcal{L} \in \mathcal{C}^2(\overline{\Omega} \times \mathbb{R}^m \times \mathbb{R}^{m \times n})$*.

Theorem 9.1-1 *Let $m \geq 1$ and $n \geq 1$ be two integers, let Ω be a domain in $\mathbb{R}^n$ with boundary Γ, and let there be given a function $\mathcal{L} : \overline{\Omega} \times \mathbb{R}^m \times \mathbb{M}^{m \times n} \to \mathbb{R}$ with the following* *properties: at each* $x \in \overline{\Omega}$, *the function* $\mathcal{L}(x,\cdot,\cdot) : \mathbb{R}^m \times \mathbb{M}^{m\times n} \to \mathbb{R}$ *is Fréchet-differentiable; for any* $r > 0$, *there exists a constant* $k(r)$ *such that its partial derivatives satisfy*

$$\left|\frac{\partial \mathcal{L}}{\partial \boldsymbol{a}}(x,\boldsymbol{b},\boldsymbol{G}) - \frac{\partial \mathcal{L}}{\partial \boldsymbol{a}}(x,\boldsymbol{a},\boldsymbol{F})\right| + \left|\frac{\partial \mathcal{L}}{\partial \boldsymbol{F}}(x,\boldsymbol{b},\boldsymbol{G}) - \frac{\partial \mathcal{L}}{\partial \boldsymbol{F}}(x,\boldsymbol{a},\boldsymbol{F})\right| \le k(r)\left(|\boldsymbol{b}-\boldsymbol{a}| + |\boldsymbol{G}-\boldsymbol{F}|\right)$$

for all $x \in \overline{\Omega}$ *and for all* $|\boldsymbol{a}| + |\boldsymbol{F}| \le r$ *and* $|\boldsymbol{b}| + |\boldsymbol{G}| \le r$; *and the functions* $x \in \Omega \to \mathcal{L}(x,\boldsymbol{v}(x),\boldsymbol{\nabla}\boldsymbol{v}(x))$, $x \in \Omega \to \frac{\partial \mathcal{L}}{\partial \boldsymbol{a}}(x,\boldsymbol{v}(x),\boldsymbol{\nabla}\boldsymbol{v}(x))$, *and* $x \in \Omega \to \frac{\partial \mathcal{L}}{\partial \boldsymbol{F}}(x,\boldsymbol{v}(x),\boldsymbol{\nabla}\boldsymbol{v}(x))$ *are Lebesgue-integrable in* Ω *for each vector field* $\boldsymbol{v} \in \mathcal{C}^1(\overline{\Omega};\mathbb{R}^m)$. *Finally, given a vector field* $\boldsymbol{u}_0 \in \mathcal{C}(\Gamma;\mathbb{R}^m)$ *and a vector field* $\boldsymbol{f} \in \mathcal{C}(\overline{\Omega};\mathbb{R}^m)$, *define the space*

$$\boldsymbol{V} := \{\boldsymbol{v} \in \mathcal{C}^1(\overline{\Omega};\mathbb{R}^m);\ \boldsymbol{v} = \boldsymbol{u}_0 \text{ on } \Gamma\},$$

and the functional

$$J : \boldsymbol{v} \in \mathcal{C}^1(\overline{\Omega};\mathbb{R}^m) \to J(\boldsymbol{v}) := \int_\Omega \mathcal{L}(x,\boldsymbol{v}(x),\boldsymbol{\nabla}\boldsymbol{v}(x))\,\mathrm{d}x - \int_\Omega \boldsymbol{f}(x)\cdot\boldsymbol{v}(x)\,\mathrm{d}x.$$

(a) *The functional* J *is Fréchet-differentiable* (Section 7.1) *over the space* $\mathcal{C}^1(\overline{\Omega};\mathbb{R}^m)$, *equipped with the norm defined by*

$$\boldsymbol{v} \in \mathcal{C}^1(\overline{\Omega};\mathbb{R}^m) \to \|\boldsymbol{v}\| := \sup_{x\in\overline{\Omega}}|\boldsymbol{v}(x)| + \sup_{x\in\overline{\Omega}}|\boldsymbol{\nabla}\boldsymbol{v}(x)|,$$

with Gâteaux derivatives given for each $\boldsymbol{u},\boldsymbol{w} \in \mathcal{C}^1(\overline{\Omega};\mathbb{R}^m)$ *by*

$$\begin{aligned} J'(\boldsymbol{u})\boldsymbol{w} = \int_\Omega &\left\{\frac{\partial \mathcal{L}}{\partial \boldsymbol{a}}(x,\boldsymbol{u}(x),\boldsymbol{\nabla}\boldsymbol{u}(x))\cdot\boldsymbol{w}(x) + \frac{\partial \mathcal{L}}{\partial \boldsymbol{F}}(x,\boldsymbol{u}(x),\boldsymbol{\nabla}\boldsymbol{u}(x)) : \boldsymbol{\nabla}\boldsymbol{w}(x)\right\}\mathrm{d}x \\ &- \int_\Omega \boldsymbol{f}(x)\cdot\boldsymbol{w}(x)\,\mathrm{d}x. \end{aligned}$$

(b) *Assume that* $\boldsymbol{u}$ *is a minimizer of* J *over* $\boldsymbol{V}$, *i.e., that*

$$\boldsymbol{u} \in \boldsymbol{V} \quad \text{and} \quad J(\boldsymbol{u}) = \inf_{\boldsymbol{v}\in\boldsymbol{V}} J(\boldsymbol{v}),$$

and let the space $\boldsymbol{W}$ *be defined by*

$$\boldsymbol{W} := \{\boldsymbol{w} \in \mathcal{C}^1(\overline{\Omega};\mathbb{R}^m);\ \boldsymbol{w} = \boldsymbol{0} \text{ on } \Gamma\}.$$

Then the minimizer $\boldsymbol{u}$ *satisfies the variational equations*

$$\begin{aligned} \int_\Omega &\left\{\frac{\partial \mathcal{L}}{\partial \boldsymbol{a}}(x,\boldsymbol{u}(x),\boldsymbol{\nabla}\boldsymbol{u}(x))\cdot\boldsymbol{w}(x) + \frac{\partial \mathcal{L}}{\partial \boldsymbol{F}}(x,\boldsymbol{u}(x),\boldsymbol{\nabla}\boldsymbol{u}(x)) : \boldsymbol{\nabla}\boldsymbol{w}(x)\right\}\mathrm{d}x \\ &= \int_\Omega \boldsymbol{f}(x)\cdot\boldsymbol{w}(x)\,\mathrm{d}x \quad \text{for all } \boldsymbol{w} \in \boldsymbol{W}. \end{aligned}$$

(c) *If in addition the matrix field* $x \in \overline{\Omega} \to \frac{\partial \mathcal{L}}{\partial \boldsymbol{F}}(x,\boldsymbol{u}(x),\boldsymbol{\nabla}\boldsymbol{u}(x)) \in \mathbb{M}^{m\times n}$ *is in the space* $\mathcal{C}^1(\overline{\Omega};\mathbb{M}^{m\times n})$, *then* $\boldsymbol{u}$ *satisfies the boundary value problem*

$$\begin{aligned} -\mathbf{div}\,\frac{\partial \mathcal{L}}{\partial \boldsymbol{F}}(x,\boldsymbol{u}(x),\boldsymbol{\nabla}\boldsymbol{u}(x)) + \frac{\partial \mathcal{L}}{\partial \boldsymbol{a}}(x,\boldsymbol{u}(x),\boldsymbol{\nabla}\boldsymbol{u}(x)) &= \boldsymbol{f}(x), \quad x \in \Omega, \\ \boldsymbol{u}(x) &= \boldsymbol{u}_0(x), \quad x \in \Gamma. \end{aligned}$$

Proof (i) *The functional $J : \mathcal{C}^1(\overline{\Omega}; \mathbb{R}^n) \to \mathbb{R}$ is Fréchet-differentiable.*

Let there be given any vector fields $\boldsymbol{u}, \boldsymbol{w} \in \mathcal{C}^1(\overline{\Omega}; \mathbb{R}^n)$. The Taylor–MacLaurin formula (Theorem 7.9-1(c)) shows that, at each $x \in \overline{\Omega}$, there exists $0 < \theta(x) < 1$ such that

$$\begin{aligned} J(\boldsymbol{u}+\boldsymbol{w}) - J(\boldsymbol{u}) &= \int_\Omega \frac{\partial \mathcal{L}}{\partial \boldsymbol{a}}(x, \boldsymbol{u}(x) + \theta(x)\boldsymbol{w}(x), \boldsymbol{\nabla u}(x) + \theta(x)\boldsymbol{\nabla w}(x)) \cdot \boldsymbol{w}(x)\,\mathrm{d}x \\ &+ \int_\Omega \frac{\partial \mathcal{L}}{\partial \boldsymbol{F}}(x, \boldsymbol{u}(x) + \theta(x)\boldsymbol{w}(x), \boldsymbol{\nabla u}(x) + \theta(x)\boldsymbol{\nabla w}(x)) : \boldsymbol{\nabla w}(x)\,\mathrm{d}x \\ &- \int_\Omega \boldsymbol{f}(x) \cdot \boldsymbol{w}(x)\,\mathrm{d}x. \end{aligned}$$

Then, by assumption, for any $r > 0$,

$$\begin{aligned} &\Big|\frac{\partial \mathcal{L}}{\partial \boldsymbol{a}}(x, \boldsymbol{u}(x) + \theta(x)\boldsymbol{w}(x), \boldsymbol{\nabla u}(x) + \theta(x)\boldsymbol{\nabla w}(x)) - \frac{\partial \mathcal{L}}{\partial \boldsymbol{a}}(x, \boldsymbol{u}(x), \boldsymbol{\nabla u}(x))\Big| \\ &\quad + \Big|\frac{\partial \mathcal{L}}{\partial \boldsymbol{F}}(x, \boldsymbol{u}(x) + \theta(x)\boldsymbol{w}(x), \boldsymbol{\nabla u}(x) + \theta(x)\boldsymbol{\nabla w}(x)) - \frac{\partial \mathcal{L}}{\partial \boldsymbol{F}}(x, \boldsymbol{u}(x), \boldsymbol{\nabla} u(x))\Big| \le k(r)\, \|\boldsymbol{w}\| \end{aligned}$$

for all $x \in \overline{\Omega}$ and for all $\boldsymbol{u}, \boldsymbol{w} \in \mathcal{C}^1(\overline{\Omega}; \mathbb{R}^n)$ that satisfy $\|\boldsymbol{u}\| \le r$ and $\|\boldsymbol{u}+\boldsymbol{w}\| \le r$. Hence, for such vector fields,

$$\begin{aligned} J(\boldsymbol{u}+\boldsymbol{w}) - J(\boldsymbol{u}) &= \int_\Omega \left\{ \frac{\partial \mathcal{L}}{\partial \boldsymbol{a}}(x, \boldsymbol{u}(x), \boldsymbol{\nabla u}(x)) \cdot \boldsymbol{w}(x) + \frac{\partial \mathcal{L}}{\partial \boldsymbol{F}}(x, \boldsymbol{u}(x), \boldsymbol{\nabla u}(x)) : \boldsymbol{\nabla w}(x) \right\} \mathrm{d}x \\ &- \int_\Omega \boldsymbol{f}(x) \cdot \boldsymbol{w}(x)\,\mathrm{d}x + \|\boldsymbol{w}\|\,\delta(\boldsymbol{w}) \quad \text{with } \lim \delta(\boldsymbol{w}) = 0 \text{ as } \|\boldsymbol{w}\| \to 0. \end{aligned}$$

The functional $J : \mathcal{C}^1(\overline{\Omega}; \mathbb{R}^n) \to \mathbb{R}$ is thus Fréchet-differentiable at each $\boldsymbol{u} \in \mathcal{C}^1(\overline{\Omega}; \mathbb{R}^n)$, and its derivative $J'(\boldsymbol{u}) \in \mathcal{L}(\mathcal{C}^1(\overline{\Omega}; \mathbb{R}^m); \mathbb{R})$ is given by

$$\begin{aligned} J'(\boldsymbol{u})\boldsymbol{w} &= \int_\Omega \left\{ \frac{\partial \mathcal{L}}{\partial \boldsymbol{a}}(x, \boldsymbol{u}(x), \boldsymbol{\nabla u}(x)) \cdot \boldsymbol{w}(x) + \frac{\partial \mathcal{L}}{\partial \boldsymbol{F}}(x, \boldsymbol{u}(x), \boldsymbol{\nabla u}(x)) : \boldsymbol{\nabla w}(x) \right\} \mathrm{d}x \\ &- \int_\Omega \boldsymbol{f}(x) \cdot \boldsymbol{w}(x)\,\mathrm{d}x \quad \text{for all } \boldsymbol{w} \in \boldsymbol{W}. \end{aligned}$$

(ii) Let $\boldsymbol{u} \in \boldsymbol{V}$ be such that $J(\boldsymbol{u}) = \inf_{\boldsymbol{v} \in \boldsymbol{V}} J(\boldsymbol{v})$. Since $\boldsymbol{V}$ is a *convex* subset of the space $\mathcal{C}^1(\overline{\Omega}; \mathbb{R}^m)$, then necessarily (Theorem 7.1-6) the *Euler inequalities* are satisfied, viz.,

$$J'(\boldsymbol{u})(\boldsymbol{v} - \boldsymbol{u}) \ge 0 \text{ for all } \boldsymbol{v} \in \boldsymbol{V}, \quad \text{or equivalently,} \quad J'(\boldsymbol{u})\boldsymbol{w} \ge 0 \text{ for all } \boldsymbol{w} \in \boldsymbol{W}.$$

Hence

$$J'(\boldsymbol{u})\boldsymbol{w} = 0 \quad \text{for all } \boldsymbol{w} \in \boldsymbol{W},$$

since $\boldsymbol{W}$ is a vector space. The variational equations announced in (b) are thus satisfied.

(iii) Given any vector field $\boldsymbol{w} = (w_i) \in \mathcal{C}^1(\overline{\Omega}; \mathbb{R}^m)$ and any matrix field $\boldsymbol{T} = (T_{ij}) \in \mathcal{C}^1(\overline{\Omega}; \mathbb{M}^{m \times n})$, the following *Green's formula* holds (as a consequence of the fundamental Green's formula, which can be applied since Ω is a domain; cf. Theorem 1.18-2):

$$\int_\Omega \sum_{i=1}^m \sum_{j=1}^n T_{ij} \partial_j w_i \,\mathrm{d}x = -\int_\Omega \sum_{i=1}^m \Big(\sum_{j=1}^n \partial_j T_{ij} \Big) w_i \,\mathrm{d}x + \int_\Gamma \sum_{i=1}^m \Big(\sum_{j=1}^n T_{ij} \nu_j \Big) w_i \,\mathrm{d}\Gamma,$$

where $\boldsymbol{\nu} = (\nu_j)_{j=1}^n$ denotes the unit outer normal vector field along Γ; equivalently,

$$\int_\Omega \boldsymbol{T} : \boldsymbol{\nabla w}\,\mathrm{d}x = -\int_\Omega \mathbf{div}\,\boldsymbol{T}\cdot\boldsymbol{w}\,\mathrm{d}x + \int_\Gamma \boldsymbol{T\nu}\cdot\boldsymbol{w}\,\mathrm{d}\Gamma.$$

If $\boldsymbol{u} \in \boldsymbol{V}$ is such that $\dfrac{\partial\mathcal{L}}{\partial\boldsymbol{F}}(\cdot,\boldsymbol{u}(\cdot),\boldsymbol{\nabla u}(\cdot)) \in \mathcal{C}^1(\overline{\Omega};\mathbb{M}^{m\times n})$, the variational equations found in (b) can therefore be rewritten as

$$\int_\Omega \Big\{-\mathbf{div}\,\frac{\partial\mathcal{L}}{\partial\boldsymbol{F}}(x,\boldsymbol{u}(x),\boldsymbol{\nabla u}(x)) + \frac{\partial\mathcal{L}}{\partial\boldsymbol{a}}(x,\boldsymbol{u}(x),\boldsymbol{\nabla u}(x)) - \boldsymbol{f}(x)\Big\}\cdot\boldsymbol{w}(x)\,\mathrm{d}x = 0 \quad \text{for all } \boldsymbol{w}\in\boldsymbol{W}$$

(the boundary integral vanishes in the Green's formula since $\boldsymbol{w} = \boldsymbol{0}$ on Γ).

The partial differential equations in Ω announced in (c) then follow from Theorem 6.3-2, which can be applied since the inclusion $\mathcal{D}(\Omega;\mathbb{R}^m) \subset \boldsymbol{W}$ holds. □

In the language of the *calculus of variations*, the function $\mathcal{L} : \overline{\Omega}\times\mathbb{R}^m\times\mathbb{M}^{m\times n}\to\mathbb{R}$ that appears in the functional J is called a **Lagrangian**, and the partial differential equations appearing in the boundary value problem found in this theorem constitutes the associated **Euler–Lagrange equations.**[1]

Remark The same terminology *Lagrangian* was also introduced, but with a completely different meaning, in Section 7.16. □

Several comments are in order about Theorem 9.1-1:

The crucial use made of Theorem 6.3-2 at the end of the above proof explains why this theorem is often referred to as the *fundamental lemma of the calculus of variations*.

For simplicity, the function space where the unknown is sought was chosen in Theorem 9.1-1 to be $\mathcal{C}^1(\overline{\Omega};\mathbb{R}^m)$. But, as illustrated by the various examples considered later in this chapter, the unknown is typically sought in a Sobolev space $W^{1,p}(\Omega;\mathbb{R}^m)$ for some $1 < p < \infty$ (this was already the case, then with $p = 2$, of the examples treated in Chapter 6).

As expected, the Fréchet differentiability of a functional J over such a space is usually not as easy to establish as in Theorem 9.1-1 (except for quadratic functionals); in some instance, it may even fail to hold.

Be that as it may, Theorem 9.1-1 shows what kind of partial differential equations can be expected to be solved by minimizing a functional. Since the computations that need to be carried out for finding these equations are *formally* the same (even when they make sense only in the sense of distributions), the *expression* of these partial differential equations is independent of the normed vector spaces over which the functional is differentiable.

Theorem 9.1-1 provides the basis for stating the *two basic problems of the calculus of variations* (which will be studied in the next sections): *first*, given a subset U of a function space V and a functional $J : V \to \mathbb{R}$ of the form considered in Theorem 9.1-1, find sufficient conditions guaranteeing the *existence* of a minimizer u of J over U; *second*, identify the associated Euler–Lagrange equations, either in the sense of distributions, or in the classical sense under appropriate regularity assumptions.

[1]So named after Leonhard Euler (1707–1783) and Joseph-Louis Lagrange (1736–1813), who discovered how to solve the *isochrone curve problem* by means of such equations; the same isochrone problem had been already solved by means of a geometrical approach by Christiaan Huyghens (1629–1695).

It turns out that a key property for establishing the existence of a minimizer is the *sequential weak lower semicontinuity* of the functional. This is why we begin by studying this notion in the next section, together with its link with the notion of *convexity*.

Problem

9.1-1 Let Ω be a domain in $\mathbb{R}^2$ with boundary Γ and let $u_0 : \Gamma \to \mathbb{R}$ be a given function. The *minimal surface problem in nonparametric form*[2] consists in seeking a function $u : \overline{\Omega} \to \mathbb{R}$ that minimizes the functional J defined by $J(v) := \int_\Omega \sqrt{1 + |\boldsymbol{\nabla} v|^2}\, dx$ over an appropriate space V of functions $v : \overline{\Omega} \to \mathbb{R}$ that are equal to u_0 on Γ.

(1) Show that the functional J is well defined and Fréchet-differentiable over the Sobolev space $W^{1,1}(\Omega)$, with a derivative $J'(u)$ given at each $u \in W^{1,1}(\Omega)$ by

$$J'(u)v = \int_\Omega \frac{\boldsymbol{\nabla} u \cdot \boldsymbol{\nabla} v}{\sqrt{1 + |\boldsymbol{\nabla} u|^2}} dx \quad \text{for all } v \in W^{1,1}(\Omega).$$

(2) Show that a smooth enough solution u to the minimal surface problem satisfies the nonlinear boundary value problem

$$\operatorname{div}\left(\frac{\boldsymbol{\nabla} u}{\sqrt{1 + |\boldsymbol{\nabla} u|^2}}\right) = 0 \quad \text{in } \Omega.$$

(3) Show that the partial differential equation in Ω can be equivalently rewritten as[3]

$$\left(1 + (\partial_2 u)^2\right) \partial_{11} u - 2\partial_1 u \partial_2 u \partial_{12} u + \left(1 + (\partial_1 u)^2\right) \partial_{22} u = 0 \text{ in } \Omega \quad \text{and} \quad u = u_0 \text{ on } \Gamma.$$

(4) Let $\Omega := \{x \in \mathbb{R}^2; 1 < |x| < 2\}$, let $u_0(x) := \gamma > 0$ if $|x| = 1$ and $u_0(x) := 0$ if $|x| = 2$, and assume that a solution to the corresponding minimal surface problem is a function of $|x|$ only, in which case the minimization problem reduces to one for functions of only one variable. Show that there exists a constant γ^* such that there exists a unique such solution if $\gamma < \gamma^*$, while there is no solution if $\gamma \geq \gamma^*$.

---

[2]There always exists a *classical solution* to this problem if Ω is convex (which is not the case in question (4)) and $u_0 \in \mathcal{C}(\Gamma)$; see:

T. RADO [1930]: The problem of the least area and the problem of Plateau, *Mathematische Zeitschrift* **32**, 763–796.

Otherwise, *generalized solutions* (defined in a specific sense) always exist as long as Ω is bounded (as in question (4)); see:

R. TEMAM [1971]: Solutions généralisées de certaines équations du type hypersurfaces minima, *Archive for Rational Mechanics and Analysis* **44**, 121–156.

The *minimal surface problem in parametric form* consists in seeking a minimal surface defined by means of *curvilinear coordinates* (cf. Section 8.8; the unknown is then a vector field $\boldsymbol{u} : \overline{\Omega} \subset \mathbb{R}^2 \to \mathbb{R}^3$); see:

B. DACOROGNA [1982]: Minimal hypersurfaces in parametric form with nonconvex integrands, *Indiana University Mathematics Journal* **31**, 531–552.

For a thorough historical perspective and an in-depth survey of the minimal surface problem, see:

W.H. MEEKS III; J. PÉREZ [2011]: The classical theory of minimal surfaces, *Bulletin of the American Mathematical Society* **48**, 325–407.

[3]This equation was discovered by:

J.L. LAGRANGE [1760]: Essai d'une nouvelle méthode pour déterminer les maxima et les minima des formules intégrales indéfinies, *Miscellanea Taurinensia* **325**, 173–199.

9.2 Convex functions and sequentially lower semicontinuous functions with values in $\mathbb{R} \cup \{\infty\}$

In what follows, we consider functions with values in the subset $\mathbb{R} \cup \{\infty\}$ of the set $\{-\infty\} \cup \mathbb{R} \cup \{\infty\}$ of *extended real numbers*, equipped with the natural operations and ordering that it inherits from the set $\mathbb{R}$, with specific rules concerning the symbol $-\infty$ and ∞.[4]

Given a set X, a function $f : X \to \mathbb{R} \cup \{\infty\}$ is said to be **proper** if the set $\{x \in X;\ f(x) < \infty\}$ is nonempty. The **epigraph** $\operatorname{epi} f$ of a proper function $f : X \to \mathbb{R} \cup \{\infty\}$ is defined as the nonempty subset

$$\operatorname{epi} f := \{(x, \alpha) \in X \times \mathbb{R};\ f(x) \leq \alpha\}$$

of the set $X \times \mathbb{R}$. Note that $f(x) < \infty$ if and only if there exists $\alpha \in \mathbb{R}$ such that $(x, \alpha) \in \operatorname{epi} f$, by definition of the set $\operatorname{epi} f$; note also that $\operatorname{epi} f$ cannot be the whole product space $X \times \mathbb{R}$ because $\operatorname{epi} f = X \times \mathbb{R}$ would mean that $f(x) = -\infty$ for all $x \in X$, which is precisely excluded.

It will be tacitly assumed in the sequel that all functions that are considered are proper.

The notion of *convexity* for real-valued functions can be extended to *functions with values in the set* $\mathbb{R} \cup \{\infty\}$ as follows: Let U be a *convex* subset of a vector space. A function $J : U \to \mathbb{R} \cup \{\infty\}$ is said to be **convex** if

$$J(\lambda u + (1 - \lambda)v) \leq \lambda J(u) + (1 - \lambda)J(v) \quad \text{for all } u, v \in U \text{ and all } 0 \leq \lambda \leq 1,$$

or **strictly convex** if

$$J(\lambda u + (1 - \lambda)v) < \lambda J(u) + (1 - \lambda)J(v) \quad \text{for all } u, v \in U,\ u \neq v, \text{ and all } 0 < \lambda < 1.$$

Notice that, since the value $-\infty$ is excluded, the right-hand side of the above inequalities is always a well-defined number in the set $\mathbb{R} \cup \{\infty\}$.

Remarks (1) One interest of allowing the value ∞ lies in the observation that a *real-valued convex* function defined over a *convex* set can be identified with a convex function with values in the set $\mathbb{R} \cup \{\infty\}$, now defined over the *whole* space; cf. Problem 9.2-1.

(2) Allowing the value ∞ is also needed in the definition of the *Legendre–Fenchel transform*, which plays a key role in *duality theory*; in this direction, see Problems 9.2-6 and 9.2-7. □

The next theorem characterizes a convex function with values in the set $\mathbb{R} \cup \{\infty\}$ and defined over a whole vector space in terms of its *epigraph*.

Theorem 9.2-1 *Let V be a vector space. A function $J : V \to \mathbb{R} \cup \{\infty\}$ is convex if and only if its epigraph* $\operatorname{epi} J$ *is a convex subset of the space $V \times \mathbb{R}$.*

Proof Assume that $J : V \to \mathbb{R} \cup \{\infty\}$ is convex. Then, given any two points (u, α) and (v, β) in $\operatorname{epi} J$, the inequalities $J(u) \leq \alpha$ and $J(v) \leq \beta$, together with the assumed convexity of J, imply that, for any $0 \leq \lambda \leq 1$,

$$J(\lambda u + (1 - \lambda)v) \leq \lambda J(u) + (1 - \lambda)J(v) \leq \lambda\alpha + (1 - \lambda)\beta,$$

[4]For details, see, e.g., BOURBAKI [1966a, Chapter 4, Section 4] or TAYLOR [1965, Sections 1.7 and 4.1]. The value $-\infty$ is excluded in order to avoid pathological situations; see for instance the discussion in EKELAND & TEMAM [1976, Chapter 1, Section 2.1]).

which means that

$$\lambda(u,\alpha) + (1-\lambda)(v,\beta) = (\lambda u + (1-\lambda)v,\ \lambda\alpha + (1-\lambda)\beta) \in \operatorname{epi} J.$$

Hence epi J is convex.

Conversely, assume that the epigraph of a function $J : V \to \mathbb{R} \cup \{\infty\}$ is convex. Given any $u \in V$ and $v \in V$ such that $J(u) < \infty$ and $J(v) < \infty$, both points $(u, J(u))$ and $(v, J(v))$ belong to epi J. Hence the assumption that epi J is convex implies that, for any $0 \leq \lambda \leq 1$,

$$\lambda(u, J(u)) + (1-\lambda)(v, J(v)) = (\lambda u + (1-\lambda)v, \lambda J(u) + (1-\lambda)J(v)) \in \operatorname{epi} J,$$

which means that

$$J(\lambda u + (1-\lambda)v) \leq \lambda J(u) + (1-\lambda)J(v).$$

The function J is thus convex (if $J(u) = \infty$, or $J(v) = \infty$, or $J(u) = J(v) = \infty$, the last inequality is surely satisfied). □

We next study the relation between *convexity* and the important notion of *sequential weakly lower semicontinuity*, which plays a key role in establishing the *existence of minimizers* for such functionals, as will be shown in the next section; see Theorem 9.3-1.

Let V be a topological space. A function $J : V \to \mathbb{R} \cup \{\infty\}$ is said to be **lower semicontinuous** if, for each $\alpha \in \mathbb{R}$, the inverse image

$$J^{-1}\left(]-\infty, \alpha]\right) = \{v \in V;\ J(v) \leq \alpha\}$$

is a *closed* subset of V, or equivalently, if, for each $\alpha \in \mathbb{R}$, the inverse image

$$J^{-1}\left(]\alpha, \infty]\right) = \{v \in V;\ \alpha < J(v) \leq \infty\}$$

is an *open* subset of V. Clearly, *a continuous function $J : V \to \mathbb{R}$ is lower semicontinuous* and, conversely, a lower semicontinuous function $J : V \to \mathbb{R}$ is continuous if and only if the function $-J : V \to \mathbb{R}$ is also lower semicontinuous.

The next theorem characterizes a lower semicontinuous function in terms of its *epigraph* (Figure 9.2-1), and of *sequences*.

Recall that the *limit inferior* of a sequence $(\alpha_k)_{k=0}^{\infty}$ of extended real numbers is the extended real number

$$\liminf_{k\to\infty} \alpha_k := \lim_{k\to\infty}\left(\inf_{\ell \geq k} \alpha_\ell\right),$$

which is well defined since a monotone sequence is always convergent in the set $\{-\infty\} \cup \mathbb{R} \cup \{\infty\}$; equivalently, $\liminf_{k\to\infty} \alpha_k$ can be defined as the smallest limit of convergent subsequences that can be extracted from the sequence $(\alpha_k)_{k=0}^{\infty}$.

Theorem 9.2-2 (a) *Let V be a topological space. A function $J : V \to \mathbb{R} \cup \{\infty\}$ is lower semicontinuous if and only if its epigraph* epi $J = \{(v, \alpha) \in V \times \mathbb{R};\ J(v) \leq \alpha\}$ *is a closed subset of the space $V \times \mathbb{R}$.*

(b) *Let V be a topological space. If a function $J : V \to \mathbb{R} \cup \{\infty\}$ is lower semicontinuous, then J is* **sequentially lower semicontinuous**, *in the sense that*

$$\lim_{k\to\infty} u_k = u \text{ in } V \quad \textit{implies} \quad J(u) \leq \liminf_{k\to\infty} J(u_k).$$

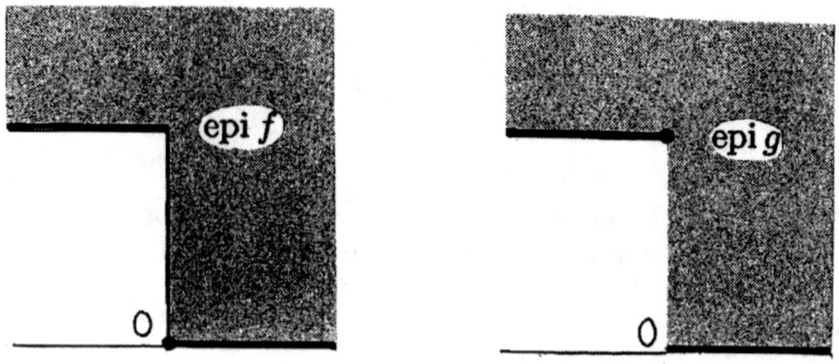

Figure 9.2-1 The function $f : x \in \mathbb{R} \to f(x) := 0$ if $x \geq 0$ and $f(x) := 1$ if $x < 0$ is lower semicontinuous, while the function $g : x \in \mathbb{R} \to g(x) := 0$ if $x > 0$ and $g(x) := 1$ if $x \leq 0$ is not lower semicontinuous: the set epi f is a closed subset of $\mathbb{R}^2$, while epi g is not.

(c) *Let V be a topological space whose topology is metrizable and let $J : V \to \mathbb{R} \cup \{\infty\}$ be a function with the following property:*

$$\lim_{k\to\infty} u_k = u \text{ in } V \quad \text{implies} \quad J(u) \leq \liminf_{k\to\infty} J(u_k).$$

Then the function J is lower semicontinuous.

Proof (i) *Proof of* (a): Assume that, for each $\alpha \in \mathbb{R}$, the set $\{v \in V;\ \alpha < J(v) \leq \infty\}$ is open in V. Given any point

$$(v_0, \alpha_0) \in (V \times \mathbb{R} - \text{epi}\, J) = \{(v, \alpha) \in V \times \mathbb{R};\ \alpha < J(v)\},$$

let $\beta_0 \in \mathbb{R}$ be such that $\alpha_0 < \beta_0 < J(v_0)$. Then the set $\{v \in V;\ \beta_0 < J(v)\} \times]-\infty, \beta_0[$ is open in $V \times \mathbb{R}$, contains the point (v_0, α_0), and is contained in the set $(V \times \mathbb{R} - \text{epi}\, J)$, which is thus open in $V \times \mathbb{R}$; hence epi J is closed in $V \times \mathbb{R}$.

Conversely, assume that the set $\{(v, \alpha) \in V \times \mathbb{R};\ \alpha < J(v)\}$ is open in $V \times \mathbb{R}$. Then, for each $\alpha \in \mathbb{R}$, the set $\{v \in V;\ \alpha < J(v)\}$ is open in V by definition of the product topology of $V \times \mathbb{R}$ (Section 1.6).

(ii) *Proof of* (b): Let $u_k \to u$ in V as $k \to \infty$. Assume first that $J(u) < \infty$. Given any $\varepsilon > 0$, the lower semicontinuity of the function $J : V \to \mathbb{R} \cup \{\infty\}$ implies that the set $V(\varepsilon) := \{v \in V;\ J(u) - \varepsilon < J(v)\}$ is an open neighborhood of u. Therefore there exists $k_0 = k_0(\varepsilon)$ such that $u_k \in V(\varepsilon)$ for all $k \geq k_0$, which means that

$$J(u) - \varepsilon < J(u_k) \quad \text{for all } k \geq k_0,$$

which in turn implies that

$$J(u) - \varepsilon \leq \liminf_{k\to\infty} J(u_k).$$

Hence $J(u) \leq \liminf_{k\to\infty} J(u_k)$ since $\varepsilon > 0$ is arbitrary.

Assume next that $J(u) = \infty$. Given any $\alpha > 0$, the lower semicontinuity of J implies that the set $\widetilde{V}(\alpha) := \{v \in V;\ \alpha < J(v)\}$ is an open neighborhood of u. Therefore there exists $\widetilde{k}_0 = \widetilde{k}_0(\alpha)$ such that $u_k \in \widetilde{V}(\alpha)$ for all $k \geq \widetilde{k}_0$, which means that

$$\alpha < J(u_k) \quad \text{for all } k \geq \widetilde{k}_0,$$

which in turn implies that

$$\alpha \le \liminf_{k\to\infty} J(u_k).$$

Hence $\liminf_{k\to\infty} J(u_k) = \infty = J(u)$ since $\alpha > 0$ is arbitrary.

(iii) *Proof of* (c): Showing that J is lower semicontinuous is by (i) equivalent to showing that $\operatorname{epi} J$ is closed in $V \times \mathbb{R}$, i.e., to showing that

$$(u_k, \alpha_k) \in \operatorname{epi} J, \;\; k \ge 1, \quad \text{and} \quad \lim_{k\to\infty} (u_k, \alpha_k) = (u, \alpha) \text{ in } V \times \mathbb{R} \quad \text{implies } (u, \alpha) \in \operatorname{epi} J,$$

since the topology of V is now assumed to be metrizable (Theorem 1.10-2). Since such a sequence satisfies $\lim_{k\to\infty} u_k = u$ in V and $J(u_k) \le \alpha_k$ for all $k \ge 1$, it follows that

$$J(u) \le \liminf_{k\to\infty} J(u_k) \le \lim_{k\to\infty} \alpha_k = \alpha,$$

i.e., that $(u, \alpha) \in \operatorname{epi} J$ as desired. □

Let V be a *normed vector space* and let U be a nonempty subset of V. A function $J : U \to \mathbb{R} \cup \{\infty\}$ is said to be **strongly lower semicontinuous** if it is lower semicontinuous when U is endowed with the strong topology of V, i.e., the topology induced by the norm of V. A function $J : U \to \mathbb{R} \cup \{\infty\}$ is said to be **sequentially weakly lower semicontinuous** if

$$u_k \in U \rightharpoonup u \in U \text{ as } k \to \infty \quad \text{implies} \quad J(u) \le \liminf_{k\to\infty} J(u_k),$$

where $\rightharpoonup$ denotes the *weak convergence* in V (Section 5.12).

The following sufficient condition for a function to be sequentially weakly lower semicontinuous is fundamental. Notice that the next proof rests on no less than the *geometric form of the Hahn–Banach theorem* (part (i)), the *Banach–Steinhaus theorem* (part (ii)), and the *Banach–Saks–Mazur theorem* (part (iii)).

Theorem 9.2-3 (sufficient condition for sequential weak lower semicontinuity) *Let V be a normed vector space. Then a convex and strongly lower semicontinuous function $J : V \to \mathbb{R} \cup \{\infty\}$ is sequentially weakly lower semicontinuous on V.*

Proof (i) *There exist a continuous linear functional $\ell \in V'$ and $c \in \mathbb{R}$ such that*

$$J(v) > \ell(v) + c \quad \text{for all } v \in V.$$

Let $(v_0, \alpha_0) \notin \operatorname{epi} J$ (recall that $\operatorname{epi} J$ is a strict subset of $V \times \mathbb{R}$), so that $\alpha_0 < J(v_0)$. Since $\operatorname{epi} J$ is a convex and closed subset of $V \times \mathbb{R}$ by Theorems 9.2-1 and 9.2-2, the *geometric form of the Hahn–Banach theorem* (Theorem 5.10-2) shows that the sets $\{(v_0, \alpha_0)\}$ and $\operatorname{epi} J$ are *strictly separated by a hyperplane*; this means that there exist a continuous linear functional $\widetilde{\ell} \in (V \times \mathbb{R})' = V' \times \mathbb{R}$ and $\gamma \in \mathbb{R}$ such that

$$\widetilde{\ell}(v_0, \alpha_0) < \gamma < \widetilde{\ell}(v, \alpha) \quad \text{for all } (v, \alpha) \in \operatorname{epi} J.$$

Since $\widetilde{\ell} \in V' \times \mathbb{R}$, there exist $\widehat{\ell} \in V'$ and $a \in \mathbb{R}$ such that

$$\widetilde{\ell}(v, \alpha) = \widehat{\ell}(v) + a\alpha \quad \text{for all } (v, \alpha) \in V \times \mathbb{R}.$$

Since $(v, J(v)) \in \operatorname{epi} J$ for each $v \in V$,

$$\widehat{\ell}(v_0) + a\alpha_0 < \gamma < \widehat{\ell}(v) + aJ(v) \quad \text{for all } v \in V.$$

Letting $v = v_0$ in this relation gives $a(\alpha_0 - J(v_0)) < 0$, which implies that $a > 0$ since $\alpha_0 < J(v_0)$. Finally then,

$$J(v) > a^{-1}(-\widehat{\ell}(v) + \gamma) \quad \text{for all } v \in V.$$

Hence the assertion follows with $\ell := -a^{-1}\widehat{\ell}$ and $c := a^{-1}\gamma$.

(ii) *Let $u_k \in V$, $k \geq 0$, and $u \in V$. Then*

$$u_k \rightharpoonup u \text{ as } k \to \infty \quad \text{implies} \quad \Lambda := \liminf_{k\to\infty} J(u_k) > -\infty.$$

By definition of the limit inferior, there exists a subsequence $(u_m)_{m=0}^{\infty}$ of the sequence $(u_k)_{k=0}^{\infty}$ such that

$$\Lambda = \lim_{m\to\infty} J(u_m).$$

Since $u_m \rightharpoonup u$ as $m \to \infty$, the sequence $(u_m)_{m=0}^{\infty}$ is *bounded* in V, as a consequence of the *Banach–Steinhaus theorem*; cf. Theorems 5.3-2 and 5.12-2. Let $M := \sup_{m\geq 0} \|u_m\| < \infty$; then, by (i),

$$J(u_m) > -M \|\ell\|_{V'} + c \quad \text{for all } m \geq 0,$$

which proves the assertion.

(iii) *The functional J is sequentially weakly lower semicontinuous on V.*

Let $u_k \rightharpoonup u$ in V as $k \to \infty$, and let again $(u_m)_{m=0}^{\infty}$ denote a subsequence such that

$$\Lambda = \lim_{m\to\infty} J(u_m).$$

If $\Lambda = \infty$, the assertion surely holds. So, the only remaining case is $\Lambda \in \mathbb{R}$ (by (ii), $\Lambda = -\infty$ is excluded), so that $(u, \Lambda) \in V \times \mathbb{R}$. We thus have

$$(u_m, J(u_m)) \in \operatorname{epi} J, \ m \geq 0, \quad \text{and} \quad (u_m, J(u_m)) \rightharpoonup (u, \Lambda) \text{ in } V \times \mathbb{R}.$$

As a convex and closed subset of $V \times \mathbb{R}$, the set epi J is *sequentially weakly closed* by the *Banach–Saks–Mazur theorem* (Theorem 5.13-1). Hence

$$(u, \Lambda) \in \operatorname{epi} J,$$

which means that

$$J(u) \leq \Lambda = \lim_{m\to\infty} J(u_m) = \liminf_{k\to\infty} J(u_k),$$

as was to be proved. □

Note that, if the convex function J is *real-valued* and *differentiable*, the proof is much easier: Given a sequence $(u_k)_{k=1}^{\infty}$ that weakly converges to an element $u \in V$, the characterization of convexity for differentiable functions (Theorem 7.12-1) implies that

$$J(u) \leq J(u_k) - J'(u)(u_k - u) \quad \text{for all } k,$$

and, by definition of weak convergence, $\lim_{k\to\infty} J'(u)(u_k - u) = 0$ since $J'(u) \in V'$. Hence

$$J(u) \leq \liminf_{k\to\infty} J(u_k),$$

and thus the function J is sequentially weakly lower semicontinuous.

One can more generally define a *weakly lower semicontinuous function* $J : V \to \mathbb{R} \cup \{\infty\}$ as one that is lower semicontinuous when the normed vector space V is equipped with its *weak topology* (Section 5.12), or equivalently, as one whose epigraph is closed with respect to the weak topology of V (recall that, in an infinite-dimensional normed vector space, the strong and the weak topologies are always distinct; cf. Theorem 5.12-5(b)). But, since the weak topology is *not* metrizable when V is infinite-dimensional (Theorem 5.12-5(b)), the sequential weak lower semicontinuity does not necessarily imply the weak lower semicontinuity in this case (by contrast, it does if V is a topological space whose topology is metrizable; cf. Theorem 9.2-2(c)). Be that as it may, we shall see that *the weaker notion of sequential weak lower semicontinuity is sufficient for our purposes.*

As a convex and strongly continuous, hence *a fortiori* strongly lower semicontinuous, function, the *norm* in a normed vector space provides an *example* of a sequentially weakly lower semicontinuous function. Therefore, by Theorem 9.2-3,

$$u_k \rightharpoonup u \text{ as } k \to \infty \quad \text{implies} \quad \|u\| \leq \liminf_{k\to\infty} \|u_k\|,$$

a property already established in Theorem 5.12-2, by means of the *Banach–Steinhaus theorem.*

Problems

9.2-1 Let U be a subset of a vector space V, and let $J : U \to \mathbb{R}$ be a real-valued function. Show that the function $\widetilde{J} : V \to \mathbb{R} \cup \{\infty\}$ defined by

$$\widetilde{J} : v \in V \to \widetilde{J}(v) := J(v) \text{ if } v \in U \quad \text{and} \quad \widetilde{J}(v) := \infty \text{ if } v \notin U$$

is convex if and only if the set U is convex and the function $J : U \to \mathbb{R}$ is convex.

9.2-2 Let V be a vector space.

(1) Let $f, g : V \to \mathbb{R} \cup \{\infty\}$ be convex functions. Show that the functions $f + g$ and αf, $\alpha > 0$, are convex.

(2) Let $(J_i)_{i\in I}$ be a family of convex functions $J_i : V \to \mathbb{R}\cup\{\infty\}$. Show that the function $\sup_{i\in I} J_i$ is convex.

9.2-3 Let V be a topological space.

(1) Show that a function $J : V \to \mathbb{R}\cup\{\infty\}$ is lower semicontinuous if and only if, given any $u \in V$ and any $\varepsilon > 0$, there exists a neighborhood $W = W(u, \varepsilon)$ of u such that $J(v) \geq J(u) - \varepsilon$ for all $v \in W$.

(2) Let $f, g : V \to \mathbb{R} \cup \{\infty\}$ be lower semicontinuous functions. Show that the functions $f + g$ and αf, $\alpha > 0$, are lower semicontinuous.

(3) Let $(J_i)_{i\in I}$ be a family of lower semicontinuous functions $J_i : V \to \mathbb{R} \cup \{\infty\}$. Show that the function $\sup_{i\in I} J_i$ is lower semicontinuous.

9.2-4 Let V be a set and A be a subset of V. The *indicator function* $I_A : V \to \mathbb{R} \cup \{\infty\}$ of A is defined by $I_A(x) := 0$ if $x \in A$ and $I_A(x) := \infty$ if $x \notin A$.

(1) Let V be a vector space. Show that A is a convex subset of V if and only if I_A is convex.

(2) Let V be a topological space. Show that A is a closed subset of V if and only if I_A is lower semicontinuous.

9.2-5 Let V be a normed vector space. Show that a convex function $J : V \to \mathbb{R} \cup \{\infty\}$ is continuous on the interior of the set $\{v \in V;\ J(v) < \infty\}$[5] (if V is finite-dimensional, this property follows from Theorem 2.17-1).

9.2-6 Let Σ be a reflexive Banach space and let Σ' and Σ'' denote its dual and bidual space. The **Legendre–Fenchel transform** of a function $g : \Sigma \to \mathbb{R}\cup\{\infty\}$ is the function $g^* : \Sigma' \to \mathbb{R}\cup\{\infty\}$ defined by

$$g^* : \sigma' \in \Sigma' \to g^*(\sigma') := \sup_{\sigma\in\Sigma}\{{}_{\Sigma'}\langle \sigma', \sigma\rangle_\Sigma - g(\sigma)\}.$$

(1) Show that, if g is a proper, convex, and lower semicontinuous function, then g^* is also proper, convex, and lower semicontinuous.

(2) Show that, if g is a proper, convex, and lower semicontinuous function, then the Legendre–Fenchel transform $g^{**} : \Sigma \to \mathbb{R} \cup \{\infty\}$ of g^* (the space X'' is here identified with X by means of the canonical isometry; cf. Section 5.14) satisfies $g^{**} = g$; this result constitutes the **Fenchel–Moreau theorem.**[6,7]

9.2-7 This problem shows how the Fenchel–Moreau theorem (Problem 9.2-6) can be put to use for defining *dual problems* of minimization problems of a specific form, by means of ad hoc *Lagrangians* (dual problems and Lagrangians have been defined in Section 7.16).

Let Σ and V be two reflexive Banach spaces; let $g : \Sigma \to \mathbb{R} \cup \{\infty\}$ and $h : V' \to \mathbb{R} \cup \{\infty\}$ be two proper, convex, and lower semicontinuous functions; let $\Lambda : \Sigma \to V'$ be a linear and continuous mapping; let the function $G : \Sigma \to \mathbb{R} \cup \{\infty\}$ be defined by

$$G : \sigma \in \Sigma \to G(\sigma) := g(\sigma) + h(\Lambda\sigma);$$

and let the two functions

$$\mathcal{L} : \Sigma \times \Sigma' \to \{-\infty\} \cup \mathbb{R} \cup \{\infty\} \quad \text{and} \quad \widetilde{\mathcal{L}} : \Sigma \times V \to \{-\infty\} \cup \mathbb{R} \cup \{\infty\}$$

be defined by

$$\mathcal{L} : (\sigma, e) \in \Sigma \times \Sigma' \to \mathcal{L}(\sigma, e) := {}_{\Sigma'}\langle e, \sigma\rangle_\Sigma + h(\Lambda\sigma) - g^*(e),$$
$$\widetilde{\mathcal{L}} : (\sigma, v) \in \Sigma \times V \to \widetilde{\mathcal{L}}(\sigma, v) := {}_{V'}\langle \Lambda\sigma, v\rangle_V + g(\sigma) - h^*(v).$$

Using the *Fenchel–Moreau theorem* (Problem 9.2-6), show that

$$\inf_{\sigma\in\Sigma} G(\sigma) = \inf_{\sigma\in\Sigma} \sup_{e\in\Sigma'} \mathcal{L}(\sigma, e) = \inf_{\sigma\in\Sigma} \sup_{v\in V} \widetilde{\mathcal{L}}(\sigma, v).$$

Remark The replacement of the minimization problem $\inf_{\sigma\in\Sigma} G(\sigma)$ by an *inf-sup problem*, such as either one found above, is the basis for defining a *dual problem* of the minimization problem $\inf_{\sigma\in\Sigma} G(\sigma)$, as the corresponding *sup-inf problem*. This means that the dual problem corresponding to the first inf-sup problem is defined as

$$\sup_{e\in\Sigma'} G^*(e), \quad \text{where } G^*(e) := \inf_{\sigma\in\Sigma} \mathcal{L}(\sigma, e) \text{ for each } e \in \Sigma^*,$$

[5] See EKELAND & TEMAM [1976, Chapter 1, Corollary 2.3].

[6] W. FENCHEL [1949]: On conjugate convex functions, *Canadian Journal of Mathematics* **1**, 73–77.
J.J. MOREAU [1970]: Inf-convolution, sous-additivité, convexité des fonctions numériques, *Journal de Mathématiques Pures et Appliquées* **49**, 109–154.

[7] For proofs, see, e.g., EKELAND & TEMAM [1976, Chapter 1, Section 3] or BREZIS [2011, Section 1.4].

while the dual problem corresponding to the second inf-sup problem is defined as

$$\sup_{v \in V} \widetilde{G}^*(v), \quad \text{where } \widetilde{G}^*(v) := \inf_{\sigma \in \Sigma} \widetilde{\mathcal{L}}(\sigma, v) \text{ for each } v \in V.$$

A key issue then consists in deciding whether the infimum $\inf_{\sigma \in \Sigma} G(\sigma)$ is equal to the supremum found in either one of its dual problems, i.e., for instance in the case of the first dual problem (to fix ideas), whether $\inf_{\sigma \in \Sigma} G(\sigma) = \sup_{e \in \Sigma'} G^*(e)$, or equivalently, whether

$$\inf_{\sigma \in \Sigma} \sup_{e \in \Sigma'} \mathcal{L}(\sigma, e) = \sup_{e \in \Sigma'} \inf_{\sigma \in \Sigma} \mathcal{L}(\sigma, e).$$

If this is the case, the next issue consists in deciding whether the Lagrangian $\mathcal{L}$ possesses a *saddle-point* $(\overline{\sigma}, \overline{e}) \in \Sigma \times \Sigma'$ (Section 7.16), i.e., that satisfies[8]

$$\inf_{\sigma \in \Sigma} \sup_{e \in \Sigma'} \mathcal{L}(\sigma, e) = \inf_{\sigma \in \Sigma} \mathcal{L}(\sigma, \overline{e}) = \mathcal{L}(\overline{\sigma}, \overline{e}) = \sup_{e \in \Sigma'} \mathcal{L}(\overline{\sigma}, e) = \sup_{e \in \Sigma'} \inf_{\sigma \in \Sigma} \mathcal{L}(\sigma, e). \qquad \square$$

9.3 Existence of minimizers for coercive and sequentially weakly lower semicontinuous functionals

As shown in the next theorem and its subsequent applications, the notion of sequential weak lower semicontinuity provides a very simple, but highly effective, means of establishing existence of minimizers. Note that the functions to be minimized will henceforth be called *functionals*, to reflect that, in the applications that we have in mind, their arguments will be themselves functions.

A functional $J : U \to \mathbb{R} \cup \{\infty\}$ defined on a nonempty *unbounded* subset U of a *normed vector space* V is said to be **coercive on** U if

$$v \in U \quad \text{and} \quad \|v\| \to \infty \text{ implies } J(v) \to \infty.$$

Recall that a subset U of a normed vector space is *sequentially weakly closed* if the weak limit of any weakly convergent sequence of elements of U also belongs to U, and that this is the case in particular if U is strongly closed and convex (Theorem 5.13-1(b)).

Theorem 9.3-1 (existence of minimizers for coercive and sequentially weakly lower semicontinuous functionals) *Let V be a reflexive Banach space, let U be a nonempty, sequentially weakly closed, subset of V, and let $J : U \to \mathbb{R} \cup \{\infty\}$ be a functional that is sequentially weakly lower semicontinuous, and coercive on U if U is unbounded.*

Then there exists at least one element $u \in U$ such that

$$J(u) = \inf_{v \in U} J(v),$$

and thus $\inf_{v \in U} J(v) > -\infty$.

[8]An instance of application of the Legendre–Fenchel transform (to three-dimensional linearized elasticity; cf. Section 6.16) where this is the case is found in:
P.G. Ciarlet; G. Geymonat; F. Krasucki [2012]: A new duality approach to elasticity, *Mathematical Models and Methods in Applied Sciences* **22**, 1150003.

Proof Assume that $\inf_{v\in U} J(v) < \infty$ (if $J(v) = \infty$ for all $v \in U$, there is nothing to prove). Let $(u_k)_{k=1}^{\infty}$ be an *infimizing sequence* of the functional $J : U \to \mathbb{R} \cup \{\infty\}$, i.e., a sequence $(u_k)_{k=1}^{\infty}$ that satisfies

$$u_k \in U \quad \text{and} \quad \lim_{k\to\infty} J(u_k) = \inf_{v\in U} J(v).$$

Note that $\inf_{v\in U} J(v) = -\infty$ is not excluded at this stage.

If the set U is bounded, so is the sequence $(u_k)_{k=1}^{\infty}$; if U is unbounded, the sequence $(u_k)_{k=1}^{\infty}$ is also bounded since the sequence $(J(u_k))_{k=1}^{\infty}$ is bounded above (otherwise there would exist a subsequence $(u_m)_{m=1}^{\infty}$ such that $J(u_m) \to \infty$ as $m \to \infty$).

Since V is a reflexive Banach space, there exists by the *Banach–Eberlein–Šmulian theorem* (Theorem 5.14-4) a subsequence $(u_m)_{m=1}^{\infty}$ of the sequence $(u_k)_{k=1}^{\infty}$ and an element $u \in V$ such that $u_m \rightharpoonup u$ as $m \to \infty$, where $\rightharpoonup$ denotes weak convergence. Besides, $u \in U$ since U is sequentially weakly closed by assumption. Therefore, $J(u)$ is well defined.

Then, by the assumed sequential weak lower semicontinuity of J on U,

$$-\infty < J(u) \leq \liminf_{m\to\infty} J(u_m) = \inf_{v\in U} J(v),$$

which completes the proof. □

Problems 9.3-1–9.3-3 illustrate the efficiency and applicability of Theorem 9.3-1 for solving *nonlinear boundary value problems* by minimizing ad hoc functionals. First, in Problem 9.3-1, the functional $v \in H_0^1(\Omega) \to \int_\Omega |\nabla v|^2 \, dx$, which, as a convex and continuous function, is sequentially weakly lower semicontinuous over $H_0^1(\Omega)$ by Theorem 9.2-3, is minimized over a sequentially weakly closed subset of $H_0^1(\Omega)$ that is *not* a subspace.

Then Problems 9.3-2 and 9.3-3 provide examples of *nonquadratic* functionals that are coercive and sequentially weakly lower semicontinuous over the spaces $H_0^1(\Omega)$ and $H_0^1(\Omega) \times H_0^1(\Omega) \times H_0^2(\Omega)$, respectively.

In each case, the functional is Fréchet-differentiable, and the associated *nonlinear partial differential equations* can be identified (with a little extra care in Problem 9.3-1). Note that Problem 9.3-3 provides an example of a nonlinear *system* of partial differential equations.

The whole of Section 9.4 will be devoted to another application of Theorem 9.3-1.

Problems

9.3-1 Let Ω be a domain in $\mathbb{R}^N$ with $N \geq 2$, let $1 < p < \infty$ if $N = 2$ or let $1 < p < \dfrac{N+2}{N-2}$ if $N \geq 3$, let the functional $J : H_0^1(\Omega) \to \mathbb{R}$ be defined by

$$v \in H_0^1(\Omega) \to J(v) := \frac{1}{2}\int_\Omega |\nabla v|^2 \, dx,$$

and let the subset U of $H_0^1(\Omega)$ be defined by

$$U = \left\{ v \in H_0^1(\Omega); \int_\Omega |v|^{p+1} \, dx = 1 \right\}.$$

(1) Show that the set U is well defined, and sequentially weakly closed in $H_0^1(\Omega)$.

(2) Show that the function $v \in L^{p+1}(\Omega) \to f(v) := \int_\Omega |v|^{p+1}\, dx$ has the following property:

$$v_k \rightharpoonup v \text{ in } H_0^1(\Omega) \quad \text{or} \quad v_k \to v \text{ in } L^{p+1}(\Omega) \quad \text{implies} \quad f(v_k) \to f(v) \text{ as } k \to \infty.$$

(3) Show that there exists at least one element $u \in U$ such that $J(u) = \inf_{v \in U} J(v)$.

(4) Show that the function $F : v \in H_0^1(\Omega) \to F(v) := \int_\Omega |v|^{p+1}\, dx$ is of class $\mathcal{C}^1$ and that, at each $v \in H_0^1(\Omega)$,

$$F'(v)w = (p+1) \int_\Omega |v|^{p-1}\, vw\, dx \quad \text{for all } w \in H_0^1(\Omega).$$

(5) Let $u \in U$ be a minimizer of J over U. Show that

$$T := \left\{ v \in H_0^1(\Omega); \int_\Omega |u|^{p-1}\, uv\, dx = 0 \right\}$$

is a closed subspace of $H_0^1(\Omega)$.

(6) Using the implicit function theorem (Theorem 7.13-1), show that there exists a mapping $\varphi : T \to H_0^1(\Omega)$ with the following properties: There exists a neighborhood W of 0 in T such that $(u + \varphi(w)) \in U$ for all $w \in W$, $\varphi(0) = 0$, and $\varphi'(0) = \mathrm{id}_T$.

(7) Show that $J'(u)w = 0$ for all $w \in T$.

(8) Show that there exists $\lambda \in \mathbb{R}$ such that

$$J'(u)v = \lambda \int_\Omega (p+1)\, |u|^{p-1}\, uv\, dx \quad \text{for all } v \in H_0^1(\Omega).$$

(9) Show that

$$-\Delta u - \lambda(p+1)\, |u|^{p-1}\, u = 0 \quad \text{in } \mathcal{D}'(\Omega).$$

(10) Show that $\lambda \neq 0$. Conclude that there exists at least one *nonzero* solution (again denoted) $u \in H_0^1(\Omega)$ to the nonlinear boundary value problem

$$-\Delta u - |u|^{p-1}\, u = 0 \text{ in } \mathcal{D}'(\Omega) \quad \text{and} \quad u = 0 \text{ on } \partial\Omega.$$

9.3-2 Let Ω be a domain in $\mathbb{R}^N$, $N \le 3$, let a constant c and a function $f \in L^2(\Omega)$ be given, and let the functional $J : H_0^1(\Omega) \to \mathbb{R}$ be defined by

$$v \in H_0^1(\Omega) \to J(v) := \frac{1}{2} \int_\Omega \left(|\nabla v|^2 + cv^2 + \frac{1}{2} v^4 \right) dx - \int_\Omega fv\, dx.$$

(1) Show that J is Fréchet-differentiable over $H_0^1(\Omega)$ and that, at each $u \in H_0^1(\Omega)$,

$$J'(u)v = \int_\Omega \left(\nabla u \cdot \nabla v + cuv + u^3 v \right) dx - \int_\Omega fv\, dx \quad \text{for all } v \in H_0^1(\Omega).$$

(2) Show that J is coercive on $H_0^1(\Omega)$ and sequentially weakly lower semicontinuous on $H_0^1(\Omega)$. *Hint*: Use the compact injection $H^1(\Omega) \Subset L^4(\Omega)$ (which holds if $N \le 3$).

(3) By (2), there exists at least one function $u \in H_0^1(\Omega)$ such that $J(u) = \inf_{v \in H_0^1(\Omega)} J(v)$. Show that such a minimizer u solves the following nonlinear boundary value problem:

$$-\Delta u + cu + u^3 = f \text{ in } \mathcal{D}'(\Omega) \quad \text{and} \quad u = 0 \text{ on } \partial\Omega.$$

9.3-3 The minimization problem studied in this problem is a mathematical model for the *Kirchhoff–Love theory of nonlinearly elastic plates* (see Problem 7.14-5 for references).

Greek indices and Latin indices vary in the sets $\{1,2\}$ and $\{1,2,3\}$, respectively, and the summation convention with respect to repeated indices is used. Given a domain Ω in $\mathbb{R}^2$ and functions $f_i \in L^2(\Omega)$, define the space $\boldsymbol{V} := H_0^1(\Omega) \times H_0^1(\Omega) \times H_0^2(\Omega)$ and the functional

$$J : \boldsymbol{v} = (v_i) \in \boldsymbol{V} \to J(\boldsymbol{v}) := \frac{1}{2}\int_\Omega \left\{ \frac{\varepsilon^3}{3} a_{\alpha\beta\sigma\tau}\partial_{\sigma\tau}v_3\partial_{\alpha\beta}v_3 + \varepsilon a_{\alpha\beta\sigma\tau}E_{\sigma\tau}(\boldsymbol{v})E_{\alpha\beta}(\boldsymbol{v})\right\} \mathrm{d}x - \int_\Omega f_i v_i \,\mathrm{d}x,$$

where $\varepsilon > 0$ is a constant, the constants $a_{\alpha\beta\sigma\tau} = a_{\beta\alpha\sigma\tau} = a_{\sigma\tau\alpha\beta}$ have the property that there exists a constant $C > 0$ such that

$$a_{\alpha\beta\sigma\tau}t_{\sigma\tau}t_{\alpha\beta} \geq C t_{\alpha\beta}t_{\alpha\beta} \quad \text{for all } (t_{\alpha\beta}) \in \mathbb{S}^2,$$

and

$$E_{\alpha\beta}(\boldsymbol{v}) := \frac{1}{2}(\partial_\alpha v_\beta + \partial_\beta v_\alpha + \partial_\alpha v_3 \partial_\beta v_3).$$

(1) Show that the functional J is sequentially weakly lower semicontinuous over the space $\boldsymbol{V}$.

(2) Show that, if the norms $\|f_\alpha\|_{0,\Omega}$ are small enough, the functional J is coercive on $\boldsymbol{V}$. Hence, by Theorem 9.3-1, there exists in this case[9] at least one vector field $\boldsymbol{u} \in \boldsymbol{V}$ such that $J(\boldsymbol{u}) = \inf_{\boldsymbol{v}\in\boldsymbol{V}} J(\boldsymbol{v})$.

Hint: Use the compact imbedding $H^1(\Omega) \Subset L^4(\Omega)$ and the two-dimensional Korn inequality in the space $H_0^1(\Omega) \times H_0^1(\Omega)$.

(3) Show that the functional J is of class $\mathcal{C}^\infty$ over the space $\boldsymbol{V}$.

(4) Assuming that a minimizer $\boldsymbol{u} \in \boldsymbol{V}$ is smooth enough, show that $\boldsymbol{u}$ satisfies the nonlinear systems of partial differential equations of Problem 7.14-5(3). There, the existence of a solution (when the vector field (f_i) is in a small enough neighborhood of the origin in the space $W^{1,p}(\Omega) \times W^{1,p}(\Omega) \times L^p(\Omega)$ for some $p > 2$) was established by a completely different method, based on the *local inversion theorem* (Theorem 7.14-1).

9.4 Application to the von Kármán equations

The *von Kármán equations*, whose derivation goes back to 1910,[10] constitute one of the most studied nonlinear systems of partial differential equations originating from continuum mechanics. They model nonlinearly elastic plates subjected to specific boundary conditions along their lateral face.

---

[9] This result is due to:

P.G. Ciarlet; P. Destuynder [1979]: A justification of a nonlinear model in plate theory, *Computer Methods in Applied Mechanics and Engineering* **17/18**, 227–258.

The existence of a minimizer holds in fact without any restriction on the magnitude of the norms $\|f_\alpha\|_{0,\Omega}$, but then the proof is more delicate, however; see:

P. Rabier [1979]: Résultats d'existence dans des modèles non linéaires de plaques, *Comptes Rendus de l'Académie des Sciences de Paris, Série A*, **289**, 515–518.

[10] T. von Kármán [1910]: Festigkeitsprobleme im Maschinenbau, in *Encyclopädie der Mathematischen Wissenschaften, Volume* IV/4, pp. 311–385, Leipzig.

A rigorous justification (by means of Gamma-convergence theory) of these equations from nonlinear three-dimensional elasticity is due to:

G. Friesecke; R.D. James; S. Müller [2006]: A hierarchy of plate models derived from nonlinear elasticity by Gamma-convergence, *Archive for Rational Mechanics and Analysis* **180**, 183–236.

A detailed analysis of the von Kármán equations is found in Ciarlet & Rabier [1980].

More specifically, given a domain Ω in $\mathbb{R}^2$ with boundary Γ, one seeks two functions $\xi : \overline{\Omega} \to \mathbb{R}$ and $\psi : \overline{\Omega} \to \mathbb{R}$ that satisfy the **von Kármán equations**

$$\begin{aligned} \Delta^2\xi &= [\psi, \xi] + f \quad \text{in } \Omega, \\ \Delta^2\psi &= -[\xi,\xi] \quad \text{in } \Omega, \\ \xi &= \partial_\nu \xi = 0 \quad \text{on } \Gamma, \\ \psi &= \partial_\nu \psi = 0 \quad \text{on } \Gamma, \end{aligned}$$

where the *Monge–Ampère form* $[\cdot,\cdot]$ is defined by

$$[\eta, \chi] = \partial_{11}\eta\partial_{22}\chi + \partial_{22}\eta\partial_{11}\chi - 2\partial_{12}\eta\partial_{12}\chi,$$

and $f \in L^2(\Omega)$ is a given function, which measures the density of the transverse body force applied to the plate. The unknown ξ is (up to a constant factor) the transverse displacement of the middle surface of the plate, and the function ψ is (again up to a constant factor) the *Airy function*, from which the stress resultants inside the plate can be computed.

Remark The analysis that follows can be extended to the case where the function ψ satisfies nonhomogeneous boundary conditions of the form $\psi = \psi_0$ and $\partial_\nu\psi = \psi_1$ on Γ; cf. Problem 9.4-1. □

The objective of this section[11] is to establish the *existence* of at least one solution $(\xi, \psi) \in H_0^2(\Omega) \times H_0^2(\Omega)$ (Theorem 9.4-3) of these equations by means of Theorem 9.3-1 applied to the *minimization of a* (nonquadratic) *coercive and sequentially weakly lower semicontinuous functional over the space* $H_0^2(\Omega)$. The unknown in this minimization problem is the *first argument* ξ in the pair (ξ, ψ).

Accordingly, we first transform the von Kármán equations into a more condensed form, by reducing their solutions to that of a *single nonlinear equation in the unknown* ξ. Not only is this equation particularly convenient for proving the existence of a solution, but it also shows that *the nonlinearity in the von Kármán equations is "cubic"* (in the sense specified in Theorem 9.4-1).

Remark We shall see in Section 9.10 that the existence of a solution to the von Kármán equations can be also obtained by means of a completely different approach, based on the *Galerkin's method* and on *Brouwer's fixed point theorem*. □

The various results from Sobolev space theory used in the next proofs as well as the notations for the norms and seminorms are found in Sections 6.5, 6.6, and 6.11.

Theorem 9.4-1 *Let Ω be a domain in $\mathbb{R}^2$ and let the bilinear and symmetric operator $B : H^2(\Omega) \times H^2(\Omega) \to H_0^2(\Omega)$ be defined as follows: For each $(\xi, \eta) \in H^2(\Omega) \times H^2(\Omega)$, the function $B(\xi, \eta)$ denotes the unique solution of*

$$B(\xi,\eta) \in H_0^2(\Omega) \quad \textit{and} \quad \Delta^2 B(\xi,\eta) = [\xi,\eta] \quad \textit{in } \mathcal{D}'(\Omega).$$

[11]The content of this section is based on:

M.S. Berger [1967]: On the von Kármán equations and the buckling of a thin elastic plate. I. The clamped plate, *Communications on Pure and Applied Mathematics* **20**, 687–719.

M.S. Berger [1977]: *Nonlinearity and Functional Analysis*, Academic Press, New York.

P.G. Ciarlet; P. Rabier [1980]: *Les Equations de von Kármán*, Lecture Notes in Mathematics, Volume 826, Springer, Berlin.

Let then the operator $C : H_0^2(\Omega) \to H_0^2(\Omega)$ *be defined by*

$$C : \xi \in H_0^2(\Omega) \to C(\xi) := B(B(\xi,\xi),\xi) \in H_0^2(\Omega),$$

so that C *is "cubic" in the sense that* $C(\alpha\xi) = \alpha^3 C(\xi)$ *for all* $\alpha \in \mathbb{R}$ *and all* $\xi \in H_0^2(\Omega)$. *Finally, let* F *be the unique solution of*

$$F \in H_0^2(\Omega) \quad \text{and} \quad \Delta^2 F = f \ \text{ in } \mathcal{D}'(\Omega).$$

Then $(\xi,\psi) \in H_0^2(\Omega) \times H_0^2(\Omega)$ *satisfies the von Kármán equations if and only if* ξ *satisfies the* **reduced von Kármán equation**

$$\xi \in H_0^2(\Omega) \quad \text{and} \quad C(\xi) + \xi - F = 0,$$

and ψ *is given by* $\psi = -B(\xi,\xi)$.

Proof If $(\xi,\eta) \in H^2(\Omega) \times H^2(\Omega)$, the function $[\xi,\eta]$ belongs to $L^1(\Omega)$; hence $B(\xi,\eta)$ is uniquely determined since $L^1(\Omega) \hookrightarrow H^{-2}(\Omega)$, as we now show. Let $g \in L^1(\Omega)$; since $H^2(\Omega) \hookrightarrow \mathcal{C}^0(\overline{\Omega})$, there exists a constant c such that ($\langle \cdot,\cdot \rangle$ denotes the duality between $\mathcal{D}'(\Omega)$ and $\mathcal{D}(\Omega)$)

$$|\langle g,\varphi\rangle| \le \|g\|_{0,1,\Omega} \|\varphi\|_{0,\infty,\Omega} \le c \|g\|_{0,1,\Omega} \|\varphi\|_{2,\Omega}$$

for all $\varphi \in \mathcal{D}(\Omega)$, hence for all $\varphi \in H_0^2(\Omega) = \overline{\mathcal{D}(\Omega)}$ (the closure of $\mathcal{D}(\Omega)$ is meant here with respect to the norm $\|\cdot\|_{2,\Omega}$); this shows that g can be identified with a distribution in $H^{-2}(\Omega)$. By the same inequalities,

$$\|g\|_{-2,\Omega} = \sup_{\substack{\varphi \in \mathcal{D}(\Omega) \\ \varphi \ne 0}} \frac{|\langle g,\varphi\rangle|}{\|\varphi\|_{2,\Omega}} \le c \|g\|_{0,1,\Omega},$$

which shows that $L^1(\Omega) \hookrightarrow H^{-2}(\Omega)$, as announced. Then the pair $(\xi - F, \psi) \in H_0^2(\Omega) \times H_0^2(\Omega)$ satisfies

$$\Delta^2(\xi - F) = [\psi,\xi] \quad \text{and} \quad \Delta^2\psi = -[\xi,\xi]$$

if and only if

$$\xi - F = B(\psi,\xi) \quad \text{and} \quad \psi = -B(\xi,\xi),$$

or equivalently, if and only if ξ satisfies the announced *reduced von Kármán equation*, viz.,

$$\xi - F = B(-B(\xi,\xi),\xi) = -C(\xi),$$

and ψ is given by $\psi = -B(\xi,\xi)$. □

The next theorem gathers useful properties of the Monge–Ampère form $[\cdot,\cdot]$, and of the operators B and C defined in Theorem 9.4-1.

Theorem 9.4-2 *Let* Ω *be a domain in* $\mathbb{R}^2$.

(a) *The following implication holds:*

$$\xi \in H_0^2(\Omega) \quad \text{and} \quad [\xi,\xi] = 0 \ \text{ implies } \xi = 0.$$

(b) *For each* $\xi, \eta \in H^2(\Omega)$, *let*

$$(\xi, \eta)_\Delta := \int_\Omega \Delta\xi \Delta\eta \, \mathrm{d}x.$$

Then

$$(B(\xi, \eta), \chi)_\Delta = (B(\xi, \chi), \eta)_\Delta \quad \textit{for all } (\xi, \eta, \chi) \in H^2(\Omega) \times H_0^2(\Omega) \times H_0^2(\Omega).$$

Consequently, for any $\xi \in H_0^2(\Omega)$,

$$(C(\xi), \xi)_\Delta = (B(\xi, \xi), B(\xi, \xi))_\Delta \geq 0,$$
$$(C(\xi), \xi)_\Delta = 0 \quad \textit{if and only if } \xi = 0.$$

(c) *The nonlinear operators* $B : H^2(\Omega) \times H^2(\Omega) \to H_0^2(\Omega)$ *and* $C : H_0^2(\Omega) \to H_0^2(\Omega)$ *have the following properties (as usual, strong and weak convergences are noted* $\to$ *and* $\rightharpoonup$*):*

$$(\xi^k, \eta^k) \rightharpoonup (\xi, \eta) \textit{ in } H^2(\Omega) \times H^2(\Omega) \quad \textit{implies} \quad B(\xi^k, \eta^k) \to B(\xi, \eta) \textit{ in } H_0^2(\Omega),$$
$$\xi^k \rightharpoonup \xi \textit{ in } H_0^2(\Omega) \quad \textit{implies} \quad C(\xi^k) \to C(\xi) \textit{ in } H_0^2(\Omega),$$

which shows in particular that both B *and* C *are continuous.*

Proof (i) *The trilinear form*

$$T : (\xi, \eta, \chi) \in H^2(\Omega) \times H^2(\Omega) \times H^2(\Omega) \to \int_\Omega [\xi, \eta] \, \chi \, \mathrm{d}x$$

is continuous; moreover, T *becomes a symmetric trilinear form if at least one of its three arguments is in* $H_0^2(\Omega)$, *and in this case there also exists a constant* c *such that, for all such arguments,*

$$\left| \int_\Omega [\xi, \eta] \, \chi \, \mathrm{d}x \right| \leq c \, |\xi|_{2,\Omega} \, |\eta|_{1,4,\Omega} \, |\chi|_{1,4,\Omega} \, .$$

The definition of $[\xi, \eta]$ and the continuous imbedding $H^2(\Omega) \hookrightarrow \mathcal{C}^0(\overline{\Omega})$ show that there exists a constant c_0 such that, for all $(\xi, \eta, \chi) \in H^2(\Omega) \times H^2(\Omega) \times H^2(\Omega)$,

$$\left| \int_\Omega [\xi, \eta] \, \chi \, \mathrm{d}x \right| \leq \| [\xi, \eta] \|_{0,1,\Omega} \, \|\chi\|_{0,\infty,\Omega} \leq c_0 \, |\xi|_{2,\Omega} \, |\eta|_{2,\Omega} \, \|\chi\|_{2,\Omega} \, ,$$

which shows that the trilinear form $T : H^2(\Omega) \times H^2(\Omega) \times H^2(\Omega) \to \mathbb{R}$ is continuous (Theorem 2.11-1). Given three functions $\xi, \eta, \chi \in \mathcal{C}^\infty(\overline{\Omega})$, we have

$$\begin{aligned}
\int_\Omega [\xi, \eta] \, \chi \, \mathrm{d}x &= \int_\Omega (\chi \partial_{11}\xi \partial_{22}\eta - \chi \partial_{12}\xi \partial_{12}\eta) \, \mathrm{d}x + \int_\Omega (\chi \partial_{22}\xi \partial_{11}\eta - \chi \partial_{12}\xi \partial_{12}\eta) \, \mathrm{d}x \\
&= \int_\Omega \partial_2(\chi \partial_{11}\xi \partial_2\eta - \chi \partial_{12}\xi \partial_1\eta) \, \mathrm{d}x - \int_\Omega \partial_2\eta \partial_2(\chi \partial_{11}\xi) \, \mathrm{d}x + \int_\Omega \partial_1\eta \partial_2(\chi \partial_{12}\xi) \, \mathrm{d}x \\
&\quad + \int_\Omega \partial_1(\chi \partial_{22}\xi \partial_1\eta - \chi \partial_{12}\xi \partial_2\eta) \, \mathrm{d}x - \int_\Omega \partial_1\eta \partial_1(\chi \partial_{22}\xi) \, \mathrm{d}x + \int_\Omega \partial_2\eta \partial_1(\chi \partial_{12}\xi) \, \mathrm{d}x.
\end{aligned}$$

Clearly, the integrals $\int_\Omega \partial_1(\cdots)\,dx$ and $\int_\Omega \partial_2(\cdots)\,dx$ vanish if *at least one* of the three functions ξ, η, χ is in $\mathcal{D}(\Omega)$; hence *in this case* we are left with

$$\begin{aligned}\int_\Omega [\xi,\eta]\,\chi\,dx &= \int_\Omega \partial_{12}\xi(\partial_1\eta\partial_2\chi + \partial_2\eta\partial_1\chi)\,dx - \int_\Omega (\partial_{11}\xi\partial_2\eta\partial_2\chi + \partial_{22}\xi\partial_1\eta\partial_1\chi)\,dx \\ &= \int_\Omega [\xi,\chi]\,\eta\,dx.\end{aligned}$$

Since $\overline{\mathcal{C}^\infty(\overline{\Omega})} = H^2(\Omega)$ and $\overline{\mathcal{D}(\Omega)} = H_0^2(\Omega)$, and since both sides are continuous trilinear forms with respect to the norm $\|\cdot\|_{2,\Omega}$ (recall that $H^2(\Omega) \hookrightarrow W^{1,4}(\Omega)$ if Ω is a domain in $\mathbb{R}^2$), the last relation remains valid if the functions ξ, η, and χ belong to $H^2(\Omega)$, one of them being in $H_0^2(\Omega)$. Hence the announced inequality holds, and the trilinear form T becomes symmetric in this case: The left-hand side is unaltered if ξ and η are exchanged and, likewise, the right-hand side is unaltered if η and χ are exchanged.

(ii) Let $\xi \in H_0^2(\Omega)$ be such that $[\xi,\xi] = 0$ and let the function $\chi \in H^2(\Omega)$ be defined by $\chi(x_1,x_2) = \frac{1}{2}(x_1^2 + x_2^2)$. Hence $[\xi,\chi] = \Delta\xi$ and, by the symmetry of T established in (i),

$$0 = \int_\Omega [\xi,\xi]\,\chi\,dx = \int_\Omega [\xi,\chi]\,\xi\,dx = \int_\Omega \xi\Delta\xi\,dx = |\xi|_{1,\Omega}^2.$$

Therefore $\xi = 0$ and (a) is proved.

(iii) Let $(\xi,\eta,\chi) \in H^2(\Omega) \times H_0^2(\Omega) \times H_0^2(\Omega)$. By definition of B and by the symmetry of T,

$$\begin{aligned}(B(\xi,\eta),\chi)_\Delta &= \int_\Omega \Delta B(\xi,\eta)\Delta\chi\,dx = \int_\Omega [\xi,\eta]\,\chi\,dx \\ &= \int_\Omega [\xi,\chi]\,\eta\,dx = \int_\Omega \Delta B(\xi,\chi)\Delta\eta\,dx = (B(\xi,\chi),\eta)_\Delta.\end{aligned}$$

Recall that (Theorem 6.8-1(a))

$$|\xi|_\Delta := \|\Delta\xi\|_{0,\Omega} = |\xi|_{2,\Omega} \quad \text{for all } \xi \in H_0^2(\Omega).$$

Hence $|\cdot|_\Delta$ is a norm over the space $H_0^2(\Omega)$, which corresponds precisely to the inner product $(\cdot,\cdot)_\Delta$. Let $\xi \in H_0^2(\Omega)$; then, by definition of C and by the relation just established,

$$(C(\xi),\xi)_\Delta = (B(B(\xi,\xi),\xi),\xi)_\Delta = (B(\xi,B(\xi,\xi)),\xi)_\Delta = (B(\xi,\xi),B(\xi,\xi))_\Delta \geq 0$$

so that

$$(C(\xi),\xi)_\Delta = 0 \quad \text{implies} \quad [\xi,\xi] = \Delta^2 B(\xi,\xi) = 0$$

(since then $B(\xi,\xi) = 0$), which in turn implies that $\xi = 0$ by (a). Hence all the assertions of (b) are proved.

(iv) By definition of the operator B and by (i),

$$(B(\xi,\eta),\chi)_\Delta = \int_\Omega [\xi,\eta]\,\chi\,dx = \int_\Omega [\chi,\xi]\,\eta\,dx \leq c\,|\chi|_\Delta\,|\xi|_{1,4,\Omega}\,|\eta|_{1,4,\Omega}$$

for all $(\xi, \eta, \chi) \in H^2(\Omega) \times H^2(\Omega) \times H_0^2(\Omega)$. Hence

$$|B(\xi,\eta)|_\Delta = \sup_{\begin{cases}\chi \in H_0^2(\Omega) \\ \chi \neq 0\end{cases}} \frac{(B(\xi,\eta),\chi)_\Delta}{|\chi|_\Delta} \leq c\,|\xi|_{1,4,\Omega}\,|\eta|_{1,4,\Omega}$$

for all $(\xi, \eta) \in H^2(\Omega) \times H^2(\Omega)$. Let $(\xi^k, \eta^k) \rightharpoonup (\xi, \eta)$ in $H^2(\Omega) \times H^2(\Omega)$; by the bilinearity of B,

$$B(\xi^k, \eta^k) - B(\xi, \eta) = B(\xi^k - \xi, \eta) + B(\xi, \eta^k - \eta) + B(\xi^k - \xi, \eta^k - \eta),$$

and thus, by the last inequality,

$$\begin{aligned}&\big|B(\xi^k,\eta^k) - B(\xi,\eta)\big|_\Delta \\ &\leq c\big(|\xi^k - \xi|_{1,4,\Omega}\,|\eta|_{1,4,\Omega} + |\xi|_{1,4,\Omega}\,|\eta^k - \eta|_{1,4,\Omega} + |\xi^k - \xi|_{1,4,\Omega}|\eta^k - \eta|_{1,4,\Omega}\big).\end{aligned}$$

The compact imbedding $H^2(\Omega) \Subset W^{1,4}(\Omega)$ then shows that

$$B(\xi^k, \eta^k) \to B(\xi, \eta) \quad \text{in } H_0^2(\Omega).$$

Let $\xi^k \rightharpoonup \xi$ in $H_0^2(\Omega)$. The above property of the operator B together with the definition of the operator C then shows that

$$C(\xi^k) \to C(\xi) \quad \text{in } H_0^2(\Omega).$$

□

Remark The equation $[\xi, \xi] = 2\det(\partial_{\alpha\beta}\xi) = 0$ solved in (a) is called the *Monge–Ampère equation.* □

We are now in position to establish the announced *existence* result for the von Kármán equations.

Theorem 9.4-3 (existence of solutions to the von Kármán equations) *Let Ω be a domain in $\mathbb{R}^2$ and let the cubic operator $C : H_0^2(\Omega) \to H_0^2(\Omega)$ and the function $F \in H_0^2(\Omega)$ be defined as in Theorem* 9.4-1.

(a) *Define the quartic functional $j : H_0^2(\Omega) \to \mathbb{R}$ by*

$$j : \eta \in H_0^2(\Omega) \to j(\eta) := \frac{1}{4}(C(\eta), \eta)_\Delta + \frac{1}{2}(\eta, \eta)_\Delta - (F, \eta)_\Delta,$$

where $(\xi, \eta)_\Delta = \int_\Omega \Delta\xi\Delta\eta\,\mathrm{d}x$. Then solving the reduced von Kármán equation, i.e., finding ξ such that

$$\xi \in H_0^2(\Omega) \quad \textit{and} \quad C(\xi) + \xi - F = 0,$$

is equivalent to finding the stationary points of the functional j, i.e., those ξ that satisfy

$$\xi \in H_0^2(\Omega) \quad \textit{and} \quad j'(\xi) = 0.$$

(b) *There exists at least one ξ such that*

$$\xi \in H_0^2(\Omega) \quad \textit{and} \quad j(\xi) = \inf_{\eta \in H_0^2(\Omega)} j(\eta).$$

Hence any such minimizer ξ is a solution of the reduced von Kármán equation, to which there corresponds (Theorem 9.4-1) *a solution $(\xi, -B(\xi, \xi)) \in H_0^2(\Omega) \times H_0^2(\Omega)$ of the von Kármán equations.*

Proof (i) *The functional j is differentiable over the space $H_0^2(\Omega)$, and solving the reduced von Kármán equation is equivalent to finding the critical points of this functional.*

Define the functional $j_4 : H_0^2(\Omega) \to \mathbb{R}$ by letting for all $\eta \in H_0^2(\Omega)$:

$$j_4(\eta) := \frac{1}{4}(C(\eta), \eta)_\Delta = \frac{1}{4}(B(B(\eta,\eta),\eta),\eta)_\Delta = \frac{1}{4}(B(\eta,\eta), B(\eta,\eta))_\Delta$$

(note that Theorem 9.4-2(b) is used here). Clearly, $j_4(\eta) \geq 0$ for all $\eta \in H_0^2(\Omega)$ and j_4 is "quartic" in the sense that $j_4(\alpha\eta) = \alpha^4 j_4(\eta)$ for all $\alpha \in \mathbb{R}$ and all $\eta \in H_0^2(\Omega)$. As a continuous bilinear operator (Theorem 9.4-2(c)), B is (infinitely) differentiable (Sections 7.1 and 7.8), and for the same reason, the inner product $(\cdot,\cdot)_\Delta$ is (infinitely) differentiable.

Hence j_4 is also differentiable by the chain rule (Theorem 7.1-3). A simple computation, combined with another application of Theorem 9.4-2(b), then shows that $j_4'(\xi)\eta$, i.e., the linear part with respect to η in the difference $(j_4(\xi+\eta) - j_4(\xi))$ is given by

$$j_4'(\xi)\eta = (B(\xi,\xi), B(\xi,\eta))_\Delta = (B(B(\xi,\xi),\xi),\eta)_\Delta = (C(\xi),\eta)_\Delta.$$

The quadratic functional $j_2(\eta) : H_0^2(\Omega) \to \mathbb{R}$ defined by

$$j_2(\eta) := \frac{1}{2}(\eta,\eta)_\Delta$$

is likewise differentiable, with $j_2'(\xi)\eta = (\xi,\eta)_\Delta$. The continuous linear functional $j_1 : H_0^2(\Omega) \to \mathbb{R}$ defined by

$$j_1(\eta) := (F,\eta)_\Delta$$

is differentiable, with $j_1'(\xi)\eta = (F,\eta)_\Delta$.

To sum up, we have shown that the functional j is differentiable, and that

$$j'(\xi)\eta = (C(\xi) + \xi - F, \eta)_\Delta \quad \text{for all } \xi, \eta \in H_0^2(\Omega).$$

As $(\cdot,\cdot)_\Delta$ is an inner product over $H_0^2(\Omega)$, finding the critical points of the functional j is thus equivalent to solving the reduced von Kármán equation.

(ii) *The functional j is sequentially weakly lower semicontinuous over $H_0^2(\Omega)$.*

Let $\eta^k \rightharpoonup \eta$ in $H_0^2(\Omega)$. Then, by Theorem 9.4-2(c), $B(\eta^k,\eta^k) \to B(\eta,\eta)$ in $H_0^2(\Omega)$, and thus

$$j_4(\eta^k) = \frac{1}{4}(B(\eta^k,\eta^k), B(\eta^k,\eta^k))_\Delta \to j_4(\eta).$$

Since the square of the norm associated with the inner product $(\cdot,\cdot)_\Delta$ is sequentially weakly lower semicontinuous (as a convex and continuous function; cf. Theorem 9.2-3), we have

$$j_2(\eta) \leq \liminf_{k\to\infty} j_2(\eta^k).$$

Finally, $j_1(\eta^k) \to j_1(\eta)$ by definition of weak convergence. We have thus shown that

$$j(\eta) \leq \liminf_{k\to\infty} j(\eta^k).$$

(iii) *The functional j is coercive on $H_0^2(\Omega)$*, i.e.,

$$\eta \in H_0^2(\Omega) \quad \text{and} \quad |\eta|_\Delta := \|\Delta\eta\|_{0,\Omega} \to \infty \text{ implies } j(\eta) \to \infty.$$

Assume the contrary. Then there exists $M \geq 0$ and a sequence $(\eta^k)_{k=1}^{\infty}$ such that

$$\eta^k \in H_0^2(\Omega), \quad |\eta^k|_\Delta \to \infty \text{ as } k \to \infty, \quad \text{and} \quad j(\eta^k) \leq M \text{ for all } k \geq 1.$$

Without loss of generality, we may assume that $\eta^k \neq 0$ for all k. Let

$$\theta^k := \frac{1}{|\eta^k|_\Delta}\eta^k,$$

so that $|\theta^k|_\Delta = 1$. Dividing the inequalities $j(\eta^k) \leq M$ by $|\eta^k|_\Delta^2$ and using that j_4 is quartic, we obtain

$$\frac{1}{2} \leq \frac{1}{2} + |\eta^k|_\Delta^2 j_4(\theta^k) \leq \frac{M}{|\eta^k|_\Delta^2} + \frac{1}{|\eta^k|_\Delta}(F, \theta^k)_\Delta \quad \text{for all } k \geq 1.$$

Passing to the limit in this inequality then leads to a contradiction, since the right-hand side approaches 0 as $k \to \infty$. Hence j is coercive on $H_0^2(\Omega)$.

(iv) *The functional j has at least one minimizer ξ over $H_0^2(\Omega)$. Besides, given any such minimizer $\xi \in H_0^2(\Omega)$, the pair $(\xi, -B(\xi,\xi)) \in H_0^2(\Omega) \times H_0^2(\Omega)$ is a solution to the von Kármán equations.*

The existence of at least one minimizer ξ of j over $H_0^2(\Omega)$ follows from Theorem 9.3-1, since j is sequentially weakly lower semicontinuous over $H_0^2(\Omega)$ (part (ii)) and coercive over the same space (part (iii)).

Hence $\xi \in H_0^2(\Omega)$ satisfies the reduced von Kármán equation (Theorem 9.4-3(a)) and thus the pair $(\xi, -B(\xi,\xi)) \in H_0^2(\Omega) \times H_0^2(\Omega)$ satisfies the von Kármán equations (Theorem 9.4-1). This completes the proof. □

One can further show[12] that, if the boundary Γ is smooth enough, both functions ξ and ψ are in fact in the space $H^4(\Omega) \cap H_0^2(\Omega)$.

Problems

9.4-1 This problem establishes the existence of a solution to the *nonhomogeneous von Kármán equations* posed over a domain $\Omega \subset \mathbb{R}^2$, viz.,

$$\begin{aligned}
\Delta^2\xi &= [\psi,\xi] + f \quad \text{in } \Omega,\\
\Delta^2\psi &= -[\xi,\xi] \quad \text{in } \Omega,\\
\xi &= \partial_\nu\xi = 0 \quad \text{on } \Gamma,\\
\psi = \psi_0 &\quad \text{and} \quad \partial_\nu\psi = \partial_\nu\psi_1 \quad \text{on } \Gamma,
\end{aligned}$$

where $f \in L^2(\Omega)$ and $\psi_0, \psi_1 \in H^2(\Omega)$ are given functions.

(1) Let θ_0 be the unique solution of

$$\theta_0 \in H^2(\Omega), \quad \Delta^2\theta_0 = 0 \text{ in } \mathcal{D}'(\Omega), \quad \theta_0 = \psi_0, \quad \text{and} \quad \partial_\nu\theta_0 = \partial_\nu\psi_1 \text{ on } \Gamma,$$

and define the linear operator

$$\Lambda : \xi \in H_0^2(\Omega) \to \Lambda(\xi) := B(\theta_0, \xi) \in H_0^2(\Omega).$$

[12] See Lions [1969, Chapter 1, Section 4.4].

Let the cubic operator $C : H_0^2(\Omega) \to H_0^2(\Omega)$ and the function $F \in H_0^2(\Omega)$ be defined as in Theorem 9.4-1. Show that $(\xi, \psi) \in H_0^2(\Omega) \times H^2(\Omega)$ satisfies the nonhomogeneous von Kármán equations if and only if ξ satisfies

$$\xi \in H_0^2(\omega) \quad \text{and} \quad C(\xi) + (I - \Lambda)\xi - F = 0,$$

the function ψ being then given by $\psi = \theta_0 - B(\xi, \xi)$.

(2) Show that the linear operator $\Lambda : H_0^2(\omega) \to H_0^2(\omega)$ is compact and symmetric with respect to the inner product $(\cdot, \cdot)_\Delta$.

(3) Define the functional $j : H_0^2(\Omega) \to \mathbb{R}$ by

$$j : \eta \in H_0^2(\Omega) \to j(\eta) := \frac{1}{4}(C(\eta), \eta)_\Delta + \frac{1}{2}((I - \Lambda)\eta, \eta)_\Delta - (F, \eta)_\Delta.$$

Show that finding $\xi \in H_0^2(\Omega)$ such that $C(\xi) + (I - \Lambda)\xi - F = 0$ is equivalent to finding the stationary points of the functional j.

(4) Show that the functional $j : H_0^2(\Omega) \to \mathbb{R}$ is sequentially weakly lower semicontinuous and coercive over the space $H_0^2(\Omega)$. Hence there exists at least one $\xi \in H_0^2(\Omega)$ such that $j(\xi) = \inf_{\eta \in H_0^2(\Omega)} j(\eta)$, so that $(\xi, \theta_0 - B(\xi, \xi)) \in H_0^2(\Omega) \times H^2(\Omega)$ is a solution to the nonhomogeneous von Kármán equations.

9.4-2 Let Ω be a domain in $\mathbb{R}^2$. Solving the **Marguerre–von Kármán equations**[13] consists in finding two functions $\xi : \overline{\Omega} \to \mathbb{R}$ and $\psi : \overline{\Omega} \to \mathbb{R}$ that satisfy

$$\begin{aligned} \Delta^2 \xi &= [\psi, \xi + \theta] + f \quad \text{in } \Omega, \\ \Delta^2 \psi &= -[\xi, \xi + 2\theta] \quad \text{in } \Omega, \\ \xi &= \partial_\nu \xi = 0 \quad \text{on } \Gamma, \\ \psi &= \partial_\nu \psi = 0 \quad \text{on } \Gamma, \end{aligned}$$

where $\theta \in H_0^2(\Omega)$ and $f \in L^2(\Omega)$ are given functions; these equations therefore reduce to the *von Kármán equations* if $\theta = 0$.

The objective of this problem is to show that the existence theory of this section applies as well to these equations. Let the bilinear operator $B : H^2(\Omega) \times H^2(\Omega) \to H_0^2(\Omega)$, the cubic operator $C : H_0^2(\Omega) \to H_0^2(\Omega)$, and the function $F \in H_0^2(\Omega)$ be defined as in Theorem 9.4-1, and let χ denote the unique solution of

$$\chi \in H_0^2(\Omega) \quad \text{and} \quad \Delta^2 \chi = [\theta, \theta] \quad \text{in } \mathcal{D}'(\Omega).$$

(1) Show that $(\xi, \psi) \in H_0^2(\Omega) \times H_0^2(\Omega)$ satisfies the Marguerre–von Kármán equations if and only if $\widetilde{\xi} := \xi + \theta$ satisfies the following *reduced Marguerre–von Kármán equation*:

$$C(\widetilde{\xi}) + \widetilde{\xi} - B(\chi, \widetilde{\xi}) - (\theta + F) = 0 \quad \text{in } H_0^2(\Omega),$$

and ψ is then the unique solution of

$$\psi \in H_0^2(\Omega) \quad \text{and} \quad \Delta^2 \psi = -[\widetilde{\xi} - \theta, \widetilde{\xi} + \theta] \quad \text{in } \mathcal{D}'(\Omega).$$

(2) Show that solving the reduced Marguerre–von Kármán equation is equivalent to finding the stationary point of a quartic functional over the space $H_0^2(\Omega)$ and that this functional has at least one minimizer over this space.

[13]These equations, which constitute a mathematical model of nonlinearly elastic shallow shells, are due to: K. MARGUERRE [1939]: Zur Theorie der gekrümmten Platte großer Formänderung, Jahrbuch der deutschen Luftfahrt-forschung, 413–418.

9.5 Existence of minimizers in $W^{1,p}(\Omega)$

To begin with, we prove a fundamental *sufficient condition for the sequential weak lower semicontinuity of functionals of the specific form*

$$\boldsymbol{\zeta} \in \boldsymbol{L}^1(\Omega) \to \int_\Omega h(x, \boldsymbol{\zeta}(x))\,\mathrm{d}x \in \mathbb{R} \cup \{\infty\},$$

the key assumption being the *convexity of the function $h(x,\cdot)$ for almost all $x \in \Omega$*. This criterion will be in turn the basis for establishing the existence of minimizers for a large class of functionals (Theorem 9.5-2).

First, we need a definition: Let Ω be an open subset of $\mathbb{R}^n$ and let $M \geq 1$ be an integer. Let B be a Borel set in $\mathbb{R}^m$. A function $h : \Omega \times B \to \mathbb{R} \cup \{\infty\}$ is said to be a **Carathéodory function**[14] if $h(x,\cdot) : \boldsymbol{\zeta} \in B \to h(x,\boldsymbol{\zeta}) \in \mathbb{R} \cup \{\infty\}$ is continuous for almost all $x \in \Omega$ and $h(\cdot,\boldsymbol{\zeta}) : x \in \Omega \to h(x,\boldsymbol{\zeta}) \in \mathbb{R} \cup \{\infty\}$ is measurable for all $\boldsymbol{\zeta} \in B$.

Theorem 9.5-1 (sequential weak lower semicontinuity and convexity) *Let Ω be a bounded open subset of $\mathbb{R}^n$, let $M \geq 1$ be an integer, and let $h : \Omega \times \mathbb{R}^M \to \mathbb{R} \cup \{\infty\}$ be a Carathéodory function such that, for almost all $x \in \Omega$, the function $h(x,\cdot) : \boldsymbol{\zeta} \in \mathbb{R}^M \to h(x,\boldsymbol{\zeta}) \in \mathbb{R} \cup \{\infty\}$ is convex, and*

$$\inf_{(x,\boldsymbol{\zeta}) \in \Omega \times \mathbb{R}^M} h(x,\boldsymbol{\zeta}) > -\infty.$$

Then

$$\boldsymbol{\zeta}_k \rightharpoonup \boldsymbol{\zeta} \text{ in } (L^1(\Omega))^M \quad \text{implies} \quad \int_\Omega h(x,\boldsymbol{\zeta}(x))\,\mathrm{d}x \leq \liminf_{k\to\infty} \int_\Omega h(x,\boldsymbol{\zeta}_k(x))\,\mathrm{d}x.$$

Proof (i) Since the set Ω is bounded, constant functions are integrable over Ω, and consequently there is no loss of generality in assuming that $\beta := \inf_{(x,\boldsymbol{\zeta})\in\Omega\times\mathbb{R}^M} h(x,\boldsymbol{\zeta}) = 0$ (if $\beta \neq 0$, replace the function h by the function $h - \beta$).

Since the function h is a *Carathéodory function*, the function $x \in \Omega \to h(x,\boldsymbol{\zeta}(x))$ is measurable whenever the function $\boldsymbol{\zeta} : x \in \Omega \to \boldsymbol{\zeta}(x) \in \mathbb{R}^M$ is itself measurable.[15] Since the function h takes its values in the set $[0,\infty]$, *the integral $\int_\Omega h(x,\boldsymbol{\zeta}(x))\,\mathrm{d}x$ is a well-defined extended real number in the interval $[0,\infty]$ for each function $\boldsymbol{\zeta} \in \boldsymbol{L}^1(\Omega) := (L^1(\Omega))^M$.*

(ii) We next show that *the functional*

$$H : \boldsymbol{\zeta} \in \boldsymbol{L}^1(\Omega) \to H(\boldsymbol{\zeta}) := \int_\Omega h(x,\boldsymbol{\zeta}(x))\,\mathrm{d}x \in [0,\infty]$$

is lower semicontinuous with respect to the strong topology *of the space $\boldsymbol{L}^1(\Omega)$*, i.e., that

$$\boldsymbol{\zeta}_k \underset{k\to\infty}{\longrightarrow} \boldsymbol{\zeta} \text{ in } \boldsymbol{L}^1(\Omega) \quad \text{implies} \quad \int_\Omega h(x,\boldsymbol{\zeta}(x))\,\mathrm{d}x \leq \liminf_{k\to\infty} \int_\Omega h(x,\boldsymbol{\zeta}_k(x))\,\mathrm{d}x$$

[14] So named after:
C. Carathéodory [1965]: *Calculus of Variations and Partial Differential Equations of the First Order*, Holden Day, San Francisco.

[15] See, e.g., Ekeland & Temam [1976, Chapter 8, Section 1].

(if the topology of a normed vector space is metrizable, lower semicontinuity is equivalent to sequential lower semicontinuity; cf. Theorem 9.2-2).

Let then $(\boldsymbol{\zeta}_k)$ be a sequence that strongly converges in the space $\boldsymbol{L}^1(\Omega)$ to a limit $\boldsymbol{\zeta}$, and let $(\boldsymbol{\zeta}_\ell)$ be any subsequence such that the sequence of extended real numbers $(\int_\Omega h(x, \boldsymbol{\zeta}_\ell(x))\,\mathrm{d}x)$ converges in the interval $[0, \infty]$. By definition of the limit inferior, we must show that

$$\int_\Omega h(x, \boldsymbol{\zeta}(x))\,\mathrm{d}x \leq \lim_{\ell\to\infty} \int_\Omega h(x, \boldsymbol{\zeta}_\ell(x))\,\mathrm{d}x.$$

Since the subsequence $(\boldsymbol{\zeta}_\ell)$ strongly converges to $\boldsymbol{\zeta}$ in $\boldsymbol{L}^1(\Omega)$, there exists a subsequence $(\boldsymbol{\zeta}_m)$ of $(\boldsymbol{\zeta}_\ell)$ such that $\boldsymbol{\zeta}_m(x) \to \boldsymbol{\zeta}(x)$ for almost all $x \in \Omega$ (Theorem 3.4-3). Consequently, by the assumed *continuity* of the functions $h(x, \cdot)$ for almost all $x \in \Omega$,

$$\lim_{m\to\infty} h(x, \boldsymbol{\zeta}_m(x)) = h(x, \boldsymbol{\zeta}(x)) \quad \text{in } [0,\infty] \text{ for almost all } x \in \Omega.$$

Therefore, by *Fatou's lemma* (Theorem 1.15-2),

$$\begin{aligned}\int_\Omega h(x, \boldsymbol{\zeta}(x))\,\mathrm{d}x &= \int_\Omega \lim_{m\to\infty} h(x, \boldsymbol{\zeta}_m(x))\,\mathrm{d}x \\ &\leq \liminf_{m\to\infty} \int_\Omega h(x, \boldsymbol{\zeta}_m(x))\,\mathrm{d}x = \lim_{\ell\to\infty} \int_\Omega h(x, \boldsymbol{\zeta}_\ell(x))\,\mathrm{d}x,\end{aligned}$$

which shows that the functional $H : \boldsymbol{L}^1(\Omega) \to [0, \infty]$ is *strongly lower semicontinuous*, on the one hand.

(iii) On the other hand, the functional $H : \boldsymbol{L}^1(\Omega) \to [0, \infty]$ is *convex*, since the assumed convexity of the function h with respect to its second argument implies that, for all $\lambda \in [0, 1]$ and all $\boldsymbol{\zeta}, \boldsymbol{\eta} \in \boldsymbol{L}^1(\Omega)$,

$$\begin{aligned}H(\lambda\boldsymbol{\zeta} + (1-\lambda)\boldsymbol{\eta}) &= \int_\Omega h(x, \lambda\boldsymbol{\zeta}(x) + (1-\lambda)\boldsymbol{\eta}(x))\,\mathrm{d}x \\ &\leq \int_\Omega \big(\lambda h(x, \boldsymbol{\zeta}(x)) + (1-\lambda) h(x, \boldsymbol{\eta}(x))\big)\,\mathrm{d}x \\ &= \lambda H(\boldsymbol{\zeta}) + (1-\lambda) H(\boldsymbol{\eta}).\end{aligned}$$

As a convex and strongly lower semicontinuous functional, H is therefore sequentially weakly lower semicontinuous, by Theorem 9.2-3. □

Remarks (1) The continuity of the functions $h(x, \cdot)$ is not a superfluous assumption since the value ∞ is allowed (convexity implies continuity only in the interior of the set $\{\boldsymbol{\zeta} \in \mathbb{R}^M;\ h(x, \boldsymbol{\zeta}) < \infty\}$; cf. Problem 9.2-5).

(2) If the function h is independent of $x \in \Omega$, the assumption of measurability is automatically satisfied.

(3) If Ω is bounded, weak convergence in any space $\boldsymbol{L}^p(\Omega)$, $1 \leq p < \infty$, implies weak convergence in the space $\boldsymbol{L}^1(\Omega)$. □

As an application of the criterion of sequential weak lower semicontinuity of Theorem 9.5-1, we now establish the existence of minimizers in the Sobolev space $\boldsymbol{W}^{1,p}(\Omega)$, $p > 1$, where Ω is a domain in $\mathbb{R}^n$, for a class of functionals often found in applications.

Theorem 9.5-2 (existence of minimizers in $\boldsymbol{W}^{1,p}(\Omega)$ for functionals with convex integrands) *Let Ω be a domain in $\mathbb{R}^n$ with boundary Γ and let $h : \Omega \times \mathbb{M}^{m\times n} \to \mathbb{R}\cup\{\infty\}$ be a function with the following properties: for almost all $x \in \Omega$, the function $h(x,\cdot) : \boldsymbol{F} \in \mathbb{M}^{m\times n} \to h(x,\boldsymbol{F}) \in \mathbb{R}\cup\{\infty\}$ is convex and continuous; the function $h(\cdot,\boldsymbol{F}) : x \in \Omega \to h(x,\boldsymbol{F}) \in \mathbb{R}\cup\{\infty\}$ is measurable for all $\boldsymbol{F} \in \mathbb{M}^{m\times n}$; and there exist constants α, β, and p such that*

$$\alpha > 0,\ p > 1, \text{ and } h(x,\boldsymbol{F}) \geq \alpha\,|\boldsymbol{F}|^p + \beta \quad \text{for almost all } x \in \Omega \text{ and for all } \boldsymbol{F} \in \mathbb{M}^{m\times n}.$$

Let Γ_0 be a $\mathrm{d}\Gamma$-measurable subset of Γ with $\mathrm{d}\Gamma$-meas $\Gamma_0 > 0$, *let $\boldsymbol{u}_0 : \Gamma_0 \to \mathbb{R}^m$ be a* $\mathrm{d}\Gamma$*-measurable function such that the set*

$$\boldsymbol{U} = \{\boldsymbol{v} \in \boldsymbol{W}^{1,p}(\Omega);\ \boldsymbol{v} = \boldsymbol{u}_0 \text{ on } \Gamma_0\} \quad \text{where } \boldsymbol{W}^{1,p}(\Omega) := (W^{1,p}(\Omega))^m,$$

is nonempty, and let L be a continuous linear functional over the space $\boldsymbol{W}^{1,p}(\Omega)$. Finally, define the functional $J : \boldsymbol{W}^{1,p}(\Omega) \to \mathbb{R}\cup\{\infty\}$ by

$$J(\boldsymbol{v}) := \int_\Omega h(x, \boldsymbol{\nabla v}(x))\,\mathrm{d}x - L(\boldsymbol{v}) \quad \text{for each } \boldsymbol{v} \in \boldsymbol{W}^{1,p}(\Omega),$$

and assume that

$$\inf_{\boldsymbol{v}\in\boldsymbol{U}} J(\boldsymbol{v}) < \infty.$$

Then there exists at least one function $\boldsymbol{u}$ such that

$$\boldsymbol{u} \in \boldsymbol{U} \quad \text{and} \quad J(\boldsymbol{u}) = \inf_{\boldsymbol{v}\in\boldsymbol{U}} J(\boldsymbol{v}).$$

If the function $h(x,\cdot) : \boldsymbol{F} \in \mathbb{M}^{m\times n} \to h(x,\boldsymbol{F})$ is strictly convex for almost all $x \in \Omega$, the minimizer $\boldsymbol{u}$ is unique.

Proof (i) The *Banach space* $\boldsymbol{W}^{1,p}(\Omega)$ is *reflexive* since $1 < p < \infty$ (Theorem 6.5-1) and the set $\boldsymbol{U}$ is *sequentially weakly closed* by the *Banach–Saks–Mazur theorem* (Theorem 5.13-1) since it is strongly closed and convex.

The inequality satisfied by the function h and the continuity of the linear form L imply that

$$J(\boldsymbol{v}) \geq \alpha \int_\Omega |\boldsymbol{\nabla v}|^p\,\mathrm{d}x + \beta \operatorname{vol}\Omega - \|L\|\,\|\boldsymbol{v}\|_{1,p,\Omega} \quad \text{for all } \boldsymbol{v} \in \boldsymbol{W}^{1,p}(\Omega).$$

By the generalized Poincaré inequality (Theorem 6.6-6(c)), there exists a constant $c_1 > 0$ such that

$$\int_\Omega |\psi|^p\,\mathrm{d}x \leq c_1 \Big\{ \int_\Omega |\boldsymbol{\nabla}\psi|^p\,\mathrm{d}x + \Big|\int_{\Gamma_0} \psi\,\mathrm{d}a\Big|^p \Big\} \quad \text{for all } \psi \in W^{1,p}(\Omega).$$

Hence there exist constants $c_2 > 0$ and c_3 such that

$$J(\boldsymbol{v}) \geq c_2\,\|\boldsymbol{v}\|^p_{1,p,\Omega} - \|L\|\,\|\boldsymbol{v}\|_{1,p,\Omega} + c_3 \quad \text{for all } \boldsymbol{v} \in \boldsymbol{U},$$

and since $p > 1$, there exist constants c and d such that

$$c > 0 \quad \text{and} \quad J(\boldsymbol{v}) \geq c\,\|\boldsymbol{v}\|^p_{1,p,\Omega} + d \ \text{ for all } \boldsymbol{v} \in \boldsymbol{U}.$$

Therefore,

$$\boldsymbol{v}^k \in \boldsymbol{U} \quad \text{and} \quad \|\boldsymbol{v}^k\|_{1,p,\Omega} \to \infty \text{ implies } J(\boldsymbol{v}^k) \to \infty,$$

which implies that the functional J is *coercive over the set* $\boldsymbol{U}$.

Since

$$\boldsymbol{u}^\ell \rightharpoonup \boldsymbol{u} \text{ in } \boldsymbol{W}^{1,p}(\Omega) \quad \text{implies} \quad \boldsymbol{\nabla u}^\ell \rightharpoonup \boldsymbol{\nabla u} \text{ in } (L^p(\Omega))^{m\times n}, \quad \text{hence in } (L^1(\Omega))^{m\times n},$$

we conclude from Theorem 9.5-1 (with $M = m \times n$ and $\mathbb{R}^M$ identified with $\mathbb{M}^{m\times n}$) that

$$\boldsymbol{u}^\ell \rightharpoonup \boldsymbol{u} \text{ in } \boldsymbol{W}^{1,p}(\Omega) \quad \text{implies} \quad \int_\Omega h(x, \boldsymbol{\nabla u}(x))\,\mathrm{d}x \le \liminf_{\ell\to\infty} \int_\Omega h(x, \boldsymbol{\nabla u}^\ell(x))\,\mathrm{d}x,$$

on the one hand. On the other hand, since L is a continuous linear form on $\boldsymbol{W}^{1,p}(\Omega)$,

$$\boldsymbol{u}^\ell \rightharpoonup \boldsymbol{u} \text{ in } \boldsymbol{W}^{1,p}(\Omega) \quad \text{implies} \quad L(\boldsymbol{u}) = \lim_{\ell\to\infty} L(\boldsymbol{u}^\ell),$$

by definition of weak convergence. Hence the functional $J : \boldsymbol{W}^{1,p}(\Omega) \to \mathbb{R}$ is *sequentially weakly lower semicontinuous.*

The *existence* of a minimizer of the functional J over the set $\boldsymbol{U}$ then follows from Theorem 9.3-1.

(ii) Assume that the function $h(x,\cdot) : \boldsymbol{F} \in \mathbb{M}^{m\times n} \to h(x, \boldsymbol{F})$ is *strictly* convex for almost all $x \in \Omega$, and let $\boldsymbol{u}_1 \in \boldsymbol{U}$ and $\boldsymbol{u}_2 \in \boldsymbol{U}$ be such that

$$\boldsymbol{u}_1 \neq \boldsymbol{u}_2 \quad \text{and} \quad J(\boldsymbol{u}_1) = J(\boldsymbol{u}_2) = \inf_{\boldsymbol{u}\in\boldsymbol{U}} J(\boldsymbol{v}).$$

Since $\boldsymbol{v} \to (\int_\Omega |\boldsymbol{\nabla v}|^p)^{1/p}\,\mathrm{d}x$ is a norm on the space $\boldsymbol{V} := \{\boldsymbol{v} \in \boldsymbol{W}^{1,p}(\Omega);\ \boldsymbol{v} = \boldsymbol{0} \text{ on } \Gamma_0\}$ (because $\mathrm{d}\Gamma$-meas $\Gamma_0 > 0$; cf. Theorem 6.6-6(b)) and since $(\boldsymbol{u}_1 - \boldsymbol{u}_2) \in \boldsymbol{V}$, the assumption $\boldsymbol{u}_1 \neq \boldsymbol{u}_2$ implies that

$$\mathrm{d}x\text{-meas}\, A > 0 \quad \text{where } A := \{x \in \Omega;\ \boldsymbol{\nabla u}_1(x) \neq \boldsymbol{\nabla u}_2(x)\}.$$

Given any $0 < \lambda < 1$, we then have

$$\begin{aligned} J(\lambda\boldsymbol{u}_1 + (1-\lambda)\boldsymbol{u}_2) &= \int_\Omega h(x, \lambda\boldsymbol{\nabla u}_1(x) + (1-\lambda)\boldsymbol{\nabla u}_2(x))\,\mathrm{d}x - L(\lambda\boldsymbol{u}_1 + (1-\lambda)\boldsymbol{u}_2) \\ &< \lambda \int_A h(x, \boldsymbol{\nabla u}_1(x))\,\mathrm{d}x + (1-\lambda)\int_A h(x, \boldsymbol{\nabla u}_2(x))\,\mathrm{d}x \\ &\quad + \lambda \int_{\Omega - A} h(x, \boldsymbol{\nabla u}_1(x))\,\mathrm{d}x + (1-\lambda)\int_{\Omega - A} h(x, \boldsymbol{\nabla u}_2(x))\,\mathrm{d}x \\ &\quad - \lambda L(\boldsymbol{u}_1) - (1-\lambda)L(\boldsymbol{u}_2), \\ &= \lambda J(\boldsymbol{u}_1) + (1-\lambda)J(\boldsymbol{u}_2) = \inf_{\boldsymbol{v}\in\boldsymbol{U}} J(\boldsymbol{v}), \end{aligned}$$

a contradiction. Hence the minimizer is *unique* in this case. □

Remarks (1) That $\boldsymbol{U}$ is sequentially closed can also be derived by noting that the trace operator $\mathrm{tr} \in \mathcal{L}(W^{1,p}(\Omega); L^p(\Gamma))$ is *compact* (Theorem 6.6-5(b)). Consequently, $\boldsymbol{v}^\ell \rightharpoonup \boldsymbol{v}$ in $\boldsymbol{W}^{1,p}(\Omega)$ implies that $\operatorname{tr} \boldsymbol{v}^\ell \to \operatorname{tr} \boldsymbol{v}$ in $\boldsymbol{L}^p(\Gamma)$ (Theorem 5.12-4(b)). Extracting a subsequence of $(\operatorname{tr} \boldsymbol{v}^\ell)$ that pointwise converges dΓ-almost everywhere on Γ then shows that $\operatorname{tr} \boldsymbol{v}(y) = \boldsymbol{u}_0(y)$ for dΓ-almost all $y \in \Gamma$.

(2) Theorem 9.5-2 can be extended to more general functionals[16] of the form

$$\boldsymbol{v} \in \boldsymbol{W}^{1,p}(\Omega) \to \int_\Omega h(x, \boldsymbol{v}(x), \boldsymbol{\nabla v}(x))\,\mathrm{d}x - L(\boldsymbol{v}),$$

if the function $h : (x, \boldsymbol{a}, \cdot) : \mathbb{M}^{m\times n} \to \mathbb{R} \cup \{\infty\}$ is *convex* for almost all $x \in \Omega$ and all $\boldsymbol{a} \in \mathbb{R}^m$, and there exist constants $\alpha_1 > 0$, $\alpha_2 > 0$, $\beta \in \mathbb{R}$, and $p > q \geq 1$, such that

$$h(x, \boldsymbol{a}, \boldsymbol{F}) \geq \alpha_1 |\boldsymbol{F}|^p + \alpha_2 |\boldsymbol{a}|^q + \beta \quad \text{for almost all } x \in \Omega \text{ and for all } (\boldsymbol{a}, \boldsymbol{F}) \in \mathbb{R}^m \times \mathbb{M}^{m\times n}.$$

(3) The assumption in Theorem 9.5-2 that the integrand $h(x, \cdot)$ is a *convex* function of the variable $\boldsymbol{F} \in \mathbb{M}^{m\times n}$ is essential for establishing the sequential weak lower semicontinuity; cf. Problem 9.5-1 for a counterexample.

(4) The assumption that the integrand is bounded below by a function of the form $\alpha |\boldsymbol{F}|^p + \beta$ for some $\alpha > 0$ and $p > 1$ is likewise essential; cf. Problem 9.5-2 for a counterexample. □

The proof of Theorem 9.5-2 shows that the *convexity* of the integrand with respect to its argument $\boldsymbol{F} \in \mathbb{M}^{m\times n}$ implies the *sequential weak lower semicontinuity* of a functional over $\boldsymbol{W}^{1,p}(\Omega)$. But such a weak lower semicontinuity is in effect related to a notion *more general than convexity*, that of *quasi-convexity*.[17]

A measurable and locally integrable function $h : \mathbb{M}^{m\times n} \to \mathbb{R}$ is **quasi-convex** if, for all bounded open subsets $U \subset \mathbb{R}^n$, all $\boldsymbol{F} \in \mathbb{M}^{m\times n}$, and all $\boldsymbol{\theta} \in W_0^{1,\infty}(\Omega; \mathbb{R}^m)$,

$$h(\boldsymbol{F}) \leq \frac{1}{\mathrm{d}x\text{-meas } U} \int_U h(\boldsymbol{F} + \boldsymbol{\nabla\theta}(x))\,\mathrm{d}x.$$

More specifically, one can establish the following beautiful result: *For any* $1 \leq p \leq \infty$, *a functional of the form*

$$\boldsymbol{v} \in \boldsymbol{W}^{1,p}(\Omega) \to \int_\Omega h(\boldsymbol{\nabla v}(x))\,\mathrm{d}x$$

is sequentially weakly lower semicontinuous if and only if the function h is quasi-convex.[18]

Quasi-convexity also plays a key role in another, remarkably efficient, approach in the calculus of variations, called **Gamma-convergence.**[19] Let V be a normed vector space

[16] See DACOROGNA [2010, Chapter 3, Theorem 3.30].

[17] The notion of quasi-convexity is due to:

C.B. MORREY, JR. [1952]: Quasi-convexity and the lower semicontinuity of multiple integrals, *Pacific Journal of Mathematics* **2**, 25–53.

C.B. MORREY, JR. [1966]: *Multiple Integrals in the Calculus of Variations*, Springer, Berlin.

[18] Various authors contributed to this result. For references and proofs (which apply even to more general functionals, of the form $\boldsymbol{v} \in \boldsymbol{W}^{1,p}(\Omega) \to \int_\Omega h(x, \boldsymbol{v}(x), \boldsymbol{\nabla v}(x)\,\mathrm{d}x)$, see the illuminating account provided in DACOROGNA [2010, Chapters 5 and 9].

[19] This theory originated in two seminal papers:

E. DE GIORGI [1975]: Sulla convergenza di alcune successioni di integrali del tipo dell'area, *Rendiconti Mathematica Roma* **8**, 227–294.

E. DE GIORGI [1977]: Γ-convergenza e G-convergenza, *Bolletina Unione Mathematica Italiana* **5**, 213–220.

An illuminating introduction is given in:

E. DE GIORGI; G. DAL MASO [1983]: *Γ-Convergence and Calculus of Variations*, Lecture Notes in Mathematics, Volume 979, Springer, Berlin.

and let $J(\varepsilon) : V \to \mathbb{R}$ be functionals defined for all $\varepsilon > 0$. The family $(J(\varepsilon))_{\varepsilon>0}$ is said to **Gamma-converge** as $\varepsilon \to 0$ if there exists a functional $J : V \to \mathbb{R} \cup \{\infty\}$, called the **Gamma-limit** of the functionals $J(\varepsilon)$ as $\varepsilon \to 0$, such that

$$v(\varepsilon) \rightharpoonup v \text{ as } \varepsilon \to 0 \quad \text{implies} \quad J(v) \leq \liminf_{\varepsilon \to 0} J(\varepsilon)(v(\varepsilon)),$$

and, given any $v \in V$, there exist $v(\varepsilon) \in V$, $\varepsilon > 0$, such that

$$v(\varepsilon) \rightharpoonup v \text{ as } \varepsilon \to 0 \quad \text{and} \quad J(v) = \lim_{\varepsilon \to 0} J(\varepsilon)(v(\varepsilon)),$$

where $v(\varepsilon) \rightharpoonup v$ as $\varepsilon \to 0$ means that, for each $v' \in V'$, $\lim_{\varepsilon\to 0} {}_{V'}\langle v', v(\varepsilon)\rangle_V = {}_{V'}\langle v', v\rangle_V$). It is then easily seen that the Gamma-limit is unique if it exists. Note also that the Gamma-limit may be equal to ∞ on some subset of V.

Then one can prove the following theorem, which gives the flavor of the type of results that can be established by the Gamma-convergence theory: *Let V be a reflexive Banach space, and let $(J(\varepsilon))_{\varepsilon>0}$ be a family of functionals $J(\varepsilon) : V \to \mathbb{R}$ that Gamma-converges to a functional $J : V \to \mathbb{R} \cup \{\infty\}$ as $\varepsilon \to 0$. Assume in addition that, for each $\varepsilon > 0$, there exists $u(\varepsilon) \in V$ such that $J(\varepsilon)(u(\varepsilon)) = \inf_{v\in V} J(\varepsilon)(v)$ and that all the minimizers $u(\varepsilon)$ are bounded independently of $\varepsilon > 0$.*

Then there exist a subsequence $(u(\varepsilon_k))_{k=1}^{\infty}$ of $(u(\varepsilon))_{\varepsilon>0}$ and $u \in V$ such that

$$u(\varepsilon_k) \rightharpoonup u \text{ as } \varepsilon_k \to 0 \quad \text{and} \quad J(u) = \inf_{v \in V} J(v).$$

In addition,

$$J(\varepsilon_k)(u(\varepsilon_k)) \to J(u) \quad \text{as } \varepsilon_k \to 0.$$

In particular, Gamma-convergence has proved to be extremely efficient for finding, and fully justifying, two-dimensional mathematical models of "thin" nonlinearly elastic structures (such as plates and shells) as limits of three-dimensional nonlinearly elastic models when the thickness, viewed as a small parameter, approaches zero.[20] Computing the Gamma-limit found in this fashion often requires the computation of *quasi-convex envelopes*, according to the following definition: Given any function $h : \mathbb{M}^{m\times n} \to \mathbb{R}$, its **quasi-convex envelope** is the function $Qh : \mathbb{M}^{m\times n} \to \mathbb{R}$ defined by

$$Qh = \sup\{g : \mathbb{M}^{m\times n} \to \mathbb{R};\ g \text{ is quasi-convex and } g \leq h\}.$$

[20] As beautifully shown in the following series of landmark papers:

H. Le Dret; A. Raoult [1995]: The nonlinear membrane model as variational limit of nonlinear three-dimensional elasticity, *Journal de Mathématiques Pures et Appliquées* **74**, 549–578.

H. Le Dret; A. Raoult [1996]: The membrane shell model in nonlinear elasticity: A variational asymptotic derivation, *Journal of Nonlinear Science* **6**, 59–94.

G. Friesecke; R.D. James; S. Müller [2002]: A theorem on geometric rigidity and the derivation of nonlinear plate theory from three-dimensional elasticity, *Communications on Pure and Applied Mathematics* **LV**, 1461–1506.

G. Friesecke; R.D. James; M.G. Mora; S. Müller [2003]: Derivation of nonlinear bending theory for shells from three-dimensional nonlinear elasticity by Gamma-convergence, *Comptes Rendus de l'Académie des Sciences de Paris, Série* 1, **336**, 697–702.

G. Friesecke; R.D. James; S. Müller [2006]: A hierarchy of plate models derived from nonlinear elasticity by Gamma-convergence, *Archive for Rational Mechanics and Analysis* **180**, 183–236.

Finally, it should be emphasized that the applicability of Theorem 9.5-2 is essentially limited to minimization problems posed over open sets Ω that are *domains* in $\mathbb{R}^n$, hence in particular *bounded*. Yet there is a wide array of minimization problems of outstanding physical interest (nonlinear field equations, nonlinear Schrödinger equations, solitary waves, etc.) that are posed over $\Omega = \mathbb{R}^n$. A powerful method, called **concentration-compactness**, has then been devised by Pierre-Louis Lions[21] for successfully solving such problems, by means of ad hoc assumptions on the functional, which somehow allow us to "recover some kind of compactness in infimizing sequences" when the methods that work for domains fail. As this method falls outside the scope of this book (where only boundary value problems posed over domains are considered), we refer the reader to the original publications[22] as well as to more recent references.[23]

Problems

9.5-1 The minimization problem described below constitutes the *Bolza example.*[24] Define the functional $J : W_0^{1,4}(0,1) \to \mathbb{R}$ by

$$v \in W_0^{1,4}(0,1) \to J(v) := \int_0^1 \{(v'(x))^2 - 1)^2 + v(x)^2\} \, dx.$$

(1) Show that the functional J is coercive, but not sequentially weakly lower semicontinuous, over $W_0^{1,4}(0,1)$.

(2) Show that, given any $a \in \mathbb{R}$, the function $y \in \mathbb{R} \to (y^2-1)^2 + a^2$ is not convex and that the function $J : W_0^{1,4}(0,1) \to \mathbb{R}$ is not convex.

(3) Show that $\inf_{v \in W_0^{1,4}(0,1)} J(v) = 0$, but that there is no minimizer of J over $W_0^{1,4}(0,1)$.

9.5-2[25] Define the functional $J : H_0^1(0,1) \to \mathbb{R}$ by

$$v \in H_0^1(0,1) \to J(v) := \int_0^1 x(v'(x) - 1)^2 \, dx.$$

(1) Show that J is not coercive over $H_0^1(0,1)$.

(2) Show that $\inf_{v \in H_0^1(\Omega)} J(v) = 0$, but that there is no minimizer of J over $H_0^1(0,1)$.

9.5-3 Show that the functional $J : W^{1,4}(0,1) \to \mathbb{R}$ defined by

$$v \in W^{1,4}(0,1) \to J(v) := \int_0^1 \left(\frac{1}{2}(v'(x))^2 + v'(x)\right)^2 dx$$

---

[21] Pierre-Louis Lions was awarded the Fields Medal in 1994, notably for his fundamental contributions to the theory of partial differential equations.

[22] P.L. LIONS [1984]: The concentration-compactness principle in the calculus of variations. The locally compact case – Part 1, *Annales de l'Institut Henri Poincaré – Analyse Non Linéaire* **1**, 109–145.

P.L. LIONS [1984]: The concentration-compactness principle in the calculus of variations. The locally compact case – Part 2, *Annales de l'Institut Henri Poincaré – Analyse Non Linéaire* **1**, 223–283.

P.L. LIONS [1985]: The concentration-compactness principle in the calculus of variations. The limit case – Part 1, *Revista Matematica Iberoamericana* **1.1**, 145–201.

P.L. LIONS [1985]: The concentration-compactness principle in the calculus of variations. The limit case – Part 2, *Revista Matematica Iberoamericana* **1.2**, 45–121.

[23] Such as STRUWE [1990, Chapter 1, Section 4], KAVIAN [1993, Chapter 6, Section 8], or TINTAREV & FIESELER [2007].

[24] O. BOLZA [1946]: *Lectures on the Calculus of Variations*, Chelsea Publishing Company, New York.

[25] O. BOLZA [1946] (*op. cit.*).

is not sequentially weakly lower semicontinuous.

9.5-4 Given a function $h \in \mathcal{C}\,[0,\infty[$ that is bounded from below, the functional

$$H : \zeta \in W^{1,p}(0,1) \to H(\zeta) := \int_0^1 h(\zeta'(x))\,\mathrm{d}x \in [0,\infty]$$

is a well-defined number in $\mathbb{R}\cup\{\infty\}$ for each $1 \leq p < \infty$. Show that, if the functional H is sequentially weakly lower semicontinuous, then h is convex (the converse property holds by Theorem 9.5-1).[26]

Hint: For any $0 < \lambda < 1$ and $a, b \in \mathbb{R}$, show that the sequence $(\zeta_k)_{k=1}^{\infty}$ defined by

$$\zeta_k(x) := a \quad \text{if} \quad \frac{j}{k} \leq x < \frac{j+\lambda}{k} \quad \text{and} \quad \xi_k(x) := b \quad \text{if} \quad \frac{j+\lambda}{k} \leq x \leq \frac{j+1}{k}, \quad 0 \leq j \leq k-1,$$

weakly converges in $L^1(0,1)$ to a constant function.

9.5-5 For any $6 \leq p \leq \infty$, define the functional $J : W^{1,p}(0,1) \to \mathbb{R}$ by

$$v \in W^{1,p}(0,1) \to J(v) := \int_0^1 \left(v'(x)\{(v'(x))^2 - 1\}\right)^2 \mathrm{d}x.$$

It is then clear that, for any $6 \leq p \leq \infty$, the function $u := 0$ is a minimizer of the functional J over the space

$$V_p := \{v \in W^{1,p}(0,1);\ v(0) = v(1) = 0\}.$$

(1) Show that there exists a convex neighborhood U of u in V_∞ such that the restriction of J to U is strictly convex; consequently, u is a strict local minimum of J over U (Theorem 7.12-3(b)).

(2) Assuming now that $6 \leq p < \infty$, show that, given any $\varepsilon > 0$, there exists a function u_p such that

$$u_p \in V_p, \quad J(u_p) = J(u) = \inf_{v \in V_p} J(v), \quad u_p \neq u, \quad \text{and} \quad \|u_p - u\|_{W^{1,p}(0,1)} < \varepsilon.$$

This problem[27] thus shows that, if $6 \leq p < \infty$, the minimizer u of J over V_p is no longer isolated, in sharp contrast with the case $p = \infty$ considered in (1).

9.5-6 For any $1 \leq p \leq \infty$, define the functional $J : W^{1,p}(0,1) \to \mathbb{R}$ by

$$v \in W^{1,p}(0,1) \to J(v) := \int_0^1 \left((v(x)+x)^3 - x\right)^2 (v'(x)+1)^6\,\mathrm{d}x,$$

and the space[28]

$$V_p := \{v \in W^{1,p}(0,1);\ v(0) = v(1) = 0\}.$$

(1) Show that the function $u : x \in [0,1] \to u(x) := x^{1/3} - x$ belongs to the space V_1 and satisfies $J(u) = \inf_{v \in V_1} J(v)$.

(2) Show that $\inf_{v \in V_\infty} J(v) > \inf_{v \in V_1} J(v)$.

This example provides an example of the **Lavrentiev phenomenon**,[29] whereby the infimum of a functional to be minimized over a subspace of $W^{1,p}(\Omega)$ may be affected by the value of p.

---

[26] This result constitutes the special case in dimension one of *Tonelli's theorem*, so named after:
L. TONELLI [1920]: La semicontinuità nel calcolo delle variazioni, *Rendiconti del Circolo Matematico di Palermo* **44**, 167–249.

[27] Adapted from:
J.M. BALL; R.J. KNOPS; J.E. MARSDEN [1978]: Two examples in nonlinear elasticity, in *Proceedings – Conference in Nonlinear Analysis, Besançon*, pp. 41–49, Lecture Notes in Mathematics, Volume 466, Springer, Berlin.

[28] This example is due to:
B. MANIÀ [1934]: Sopra un esempio di Lavrentieff, *Bolletone dell Unione Mathematica Italiana* **13**, 147–153.

[29] M. LAVRENTIEV [1926]: Sur quelques problèmes du calcul des variations, *Annales de Mathématiques Pures et Appliquées* **4**, 7–18.

9.6 Application to the p-Laplace operator

As an application of Theorem 9.5-2, we now consider a minimization problem that generalizes the quadratic minimization problem (studied at length in Section 6.7): Find $u \in H_0^1(\Omega)$ such that $J(u) = \inf_{v \in H_0^1(\Omega)} J(v)$, where

$$J(v) = \frac{1}{2}\int_\Omega |\nabla v|^2 \, dx - \int_\Omega fv \, dx,$$

and $f \in L^2(\Omega)$ is a given function. Recall that the unique solution to this minimization problem is also a solution (at least in the sense of distributions) of the homogeneous Dirichlet problem for the Laplace operator Δ.

This minimization problem can be seen as the special case $p = 2$ of the following minimization problem, where p is now any real number satisfying $1 < p < \infty$: Find $u \in W_0^{1,p}(\Omega)$ such that $J_p(u) = \inf_{v \in W_0^{1,p}(\Omega)} J_p(v)$, where

$$J_p(v) := \frac{1}{p}\int_\Omega |\nabla v|^p \, dx - \int_\Omega fv \, dx,$$

and $f \in L^q(\Omega)$, where q denotes the conjugate exponent of p.

We now show that, thanks to Theorem 9.5-2, this minimization problem has a unique solution u (Theorem 9.6-1(a)). We also show (Theorem 9.6-1(b)) that u satisfies a Dirichlet problem for the p-**Laplace operator**, or p-**Laplacian**, defined by

$$\Delta_p : v \to \Delta_p v := \operatorname{div}\left(|\nabla v|^{p-2}\nabla v\right) = \sum_{i=1}^n \partial_i\left(|\nabla v|^{p-2}\partial_i v\right), \quad 1 < p < \infty,$$

The p-Laplacian, which clearly reduces to the Laplacian Δ when $p = 2$, constitutes one of the most commonly studied nonlinear partial differential operators.

Theorem 9.6-1 (application to the Dirichlet problem for the p-Laplacian) *Let there be given a domain $\Omega \subset \mathbb{R}^n$, a $d\Gamma$-measurable subset Γ_0 of $\Gamma := \partial\Omega$ with $d\Gamma$-meas $\Gamma_0 > 0$, a number $1 < p < \infty$, a $d\Gamma$-measurable function $u_0 : \Gamma_0 \to \mathbb{R}$ such that the set*

$$U := \{v \in W^{1,p}(\Omega);\ v = u_0 \text{ on } \Gamma_0\}$$

is nonempty, and a function $f \in L^q(\Omega)$, where q denotes the conjugate exponent of p. Let

$$J_p(v) := \frac{1}{p}\int_\Omega |\nabla v|^p \, dx - \int_\Omega fv \, dx \quad \text{for each } v \in W^{1,p}(\Omega).$$

(a) *There exists a unique function u such that*

$$u \in U \quad \text{and} \quad J_p(u) = \inf_{v \in U} J_p(v).$$

(b) *The minimizer $u \in U$ satisfies the variational equations*

$$\int_\Omega |\nabla u|^{p-2}\nabla u \cdot \nabla v \, dx = \int_\Omega fv \, dx \quad \text{for all } v \in W_0^{1,p}(\Omega),$$

and is a solution to the nonlinear (if $p \neq 2$) partial differential equation

$$\Delta_p u := -\operatorname{div}\left(|\nabla u|^{p-2}\nabla u\right) = f \quad \text{in } \mathcal{D}'(\Omega).$$

Proof (i) The function

$$h : a \in \mathbb{R}^n \to h(a) := |a|^p$$

is strictly convex for $1 < p < \infty$ and evidently satisfies $h(a) \geq |a|^p$ for all $a \in \mathbb{R}^n$; besides, $\inf_{v \in U} J_p(v) < \infty$ since $U \neq \varnothing$. Hence all the assumptions of Theorem 9.5-2 are satisfied; therefore the minimization problem of (a) has a unique solution u.

(ii) Let now a nonzero function $v \in W_0^{1,p}(\Omega)$ be given. Since then $(u + tv) \in U$ for all $t \in \mathbb{R}$, the function

$$f_v : t \in \mathbb{R} \to f_v(t) := J_p(u + tv) \in \mathbb{R}$$

has a minimum at $t = 0$. But f_v is differentiable on $\mathbb{R}$, with a derivative given by (Problem 9.6-1)

$$f_v'(t) = \int_\Omega |\nabla(u + tv)|^{p-2} \left(\nabla(u + tv) \cdot \nabla v\right) \mathrm{d}x - \int_\Omega f v \, \mathrm{d}x \quad \text{at each } t \in \mathbb{R}.$$

Hence

$$\int_\Omega |\nabla u|^{p-2} \nabla u \cdot \nabla v \, \mathrm{d}x - \int_\Omega f v \, \mathrm{d}x = f_v'(0) = 0.$$

Letting v vary in $\mathcal{D}(\Omega) \subset W_0^{1,p}(\Omega)$ then yields the announced partial differential equation in $\mathcal{D}'(\Omega)$. □

Remark Assume that $\Gamma_0 = \Gamma$ and $u_0 = 0$, so that the minimizer u of J_p satisfies in this case

$$-\operatorname{div}\left(|\nabla u|^{p-2} \nabla u\right) = f \quad \text{in } \mathcal{D}'(\Omega),$$
$$u = 0 \quad \text{on } \Gamma,$$

since then $U = W_0^{1,p}(\Omega)$. Using the theory of *monotone operators* (Section 9.13), of which the p-Laplace operator Δ_p provides a basic example, we will show (Theorem 9.14-2) that the solution to this boundary value problem (which exists by Theorem 9.6-1) is also *unique*[30] (like that of the minimization problem, but the uniqueness of a minimizer does not necessarily imply the uniqueness of the solution to the associated boundary value problem). □

Problems

9.6-1 (1) Let u and $v \neq 0$ be two given functions in the space $W^{1,p}(\Omega)$, $1 < p < \infty$, and let $x \in \Omega$ be such that $|\nabla u(x)| < \infty$ and $|\nabla v(x)| < \infty$. Show that the function

$$g : t \in \mathbb{R} \to g(t) := \frac{1}{p} |\nabla(u + tv)(x)|^p \in \mathbb{R}$$

is differentiable on $\mathbb{R}$, with a derivative given at each $t \in \mathbb{R}$ by

$$g'(t) = |\nabla(u + tv)(x)|^{p-2} \left(\nabla(u + tv)(x) \cdot \nabla v(x)\right),$$

with

$$g'(t) = 0 \text{ if } \nabla(u + tv)(x) = 0 \quad \text{and} \quad 1 < p < 2.$$

[30] The uniqueness can be also proved directly, by means of a series of elementary inequalities; see CHIPOT [2009, Proposition 17.5].

(2) Using the Lebesgue dominated convergence theorem, show that, for each $1 < p < \infty$, the function

$$f : t \in \mathbb{R} \to \frac{1}{p} \int_\Omega |\nabla(u + tv)|^p \, dx \in \mathbb{R}$$

is differentiable on $\mathbb{R}$, with a derivative given at each $t \in \mathbb{R}$ by

$$f'(t) = \int_\Omega |\nabla(u + tv)|^{p-2} \left(\nabla u \cdot \nabla v + t \, |\nabla v|^2 \right) dx.$$

(3) Show that, for each $1 < p < \infty$, the functional

$$J : v \in W_0^{1,p}(\Omega) \to \frac{1}{p} \int_\Omega |\nabla v|^p \, dx$$

is *Fréchet-differentiable*, with a derivative $J'(u) \in W^{-1,q}(\Omega)$ given at each $u \in W_0^{1,p}(\Omega)$ by

$$J'(u)v = \int_\Omega |\nabla u|^{p-2} \, \nabla u \cdot \nabla v \, dx \quad \text{for all } v \in W_0^{1,p}(\Omega).$$

9.6-2 Show directly that, when $\Gamma_0 = \Gamma$ and $u_0 = 0$ (in which case $U = W_0^{1,p}(\Omega)$), Theorem 9.6-1 holds under the weaker assumption that Ω is an open subset of $\mathbb{R}^n$ with finite width.

Hint: Using Theorem 9.2-3, show that $J_p : W_0^{1,p}(\Omega) \to \mathbb{R}$ is sequentially weakly lower semicontinuous; then use Theorem 9.3-1.

9.7 Polyconvexity; compensated compactness; John Ball's existence theorem in nonlinear elasticity

In the previous section, we considered minimization problems of the following form: Find $\boldsymbol{u} \in \boldsymbol{U} \subset \boldsymbol{W}^{1,p}(\Omega)$ such that $J(\boldsymbol{u}) = \inf_{\boldsymbol{v} \in \boldsymbol{U}} J(\boldsymbol{v})$, where the functional J is of the form $J(\boldsymbol{v}) = \int_\Omega h(x, \boldsymbol{\nabla v}(x)) \, dx - L(\boldsymbol{v})$ for all $\boldsymbol{v} \in \boldsymbol{W}^{1,p}(\Omega)$. The *convexity* of the functions $\boldsymbol{F} \in \mathbb{M}^{m\times n} \to h(x, \boldsymbol{F})$ for almost all $x \in \Omega$, the *convexity* of the set $\boldsymbol{U}$, and the *coerciveness* of the integrand were then the key assumptions for establishing the existence of a minimizer (Theorem 9.5-2).

In this section, we consider a similar minimization problem that arises in three-dimensional nonlinear elasticity, but with the distinctive feature that *the above convexity assumptions fail*: The integrand is no longer convex with respect to the variable $\boldsymbol{F} \in \mathbb{M}^{m\times n}$, and the set $\boldsymbol{U}$ is no longer convex.

The existence of a minimizer can nevertheless still be established by a proof similar in its *principle* to that of Theorem 9.5-2, thanks to the introduction of *two basic notions: polyconvexity*, a weaker notion of convexity adapted to the problem under consideration, and *compensated compactness*, a property guaranteeing that the limits of weakly converging sequences belong to U even though U is not convex.

Consider an *elastic body*[31] occupying the closure $\overline{\Omega}$ of a domain $\Omega \subset \mathbb{R}^3$ as its *reference configuration*, subjected to a *boundary condition of place* (this condition is defined below) on a portion Γ_0 of the boundary Γ of Ω, and subjected to *body forces* and *surface forces*, of

[31]The notions from elasticity theory, such as elastic body, reference configuration, etc., mentioned in this section are explained in detail in, e.g., GURTIN [1981] or CIARLET [1988].

respective densities $\boldsymbol{f} : \Omega \to \mathbb{R}^3$ and $\boldsymbol{g} : \Gamma_1 \to \mathbb{R}^3$, where $\Gamma_1 := \Gamma - \Gamma_0$ (Figure 9.7-1). Under the influence of these forces and boundary conditions, each point $x \in \overline{\Omega}$ occupies a position denoted $\boldsymbol{\varphi}(x)$, and the vector field $\boldsymbol{\varphi} : \overline{\Omega} \to \mathbb{R}^3$ thus defined is called a **deformation** of the reference configuration $\overline{\Omega}$. In order to be physically admissible, such a deformation must clearly be *injective in* Ω and *orientation-preserving*.

Let

$$\mathbb{M}^3_+ := \{\boldsymbol{F} \in \mathbb{M}^3;\ \det \boldsymbol{F} > 0\}.$$

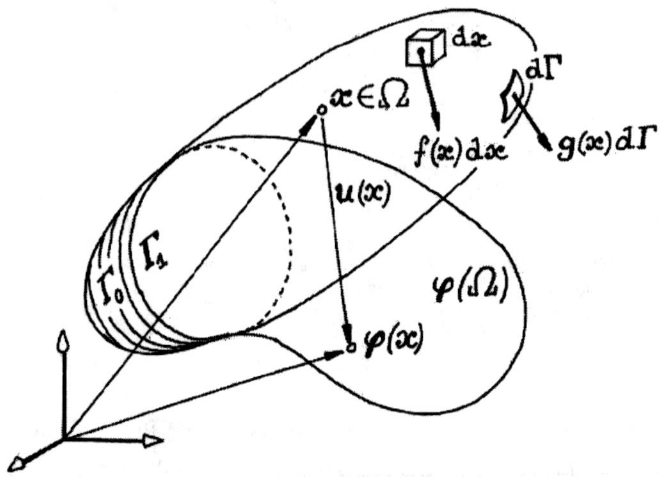

Figure 9.7-1 *Three-dimensional elasticity.* An elastic body with the closure of a domain Ω in $\mathbb{R}^3$ as its reference configuration is subjected to body forces of density $\boldsymbol{f} : \Omega \to \mathbb{R}^3$, to surface forces of density $\boldsymbol{g} : \Gamma_1 \to \mathbb{R}^3$, and to a boundary condition of place $\boldsymbol{\varphi} = \boldsymbol{\varphi}_0$ on Γ_0 (for definiteness, it is assumed in this figure that $\boldsymbol{\varphi}_0 = \mathbf{id}\,|_{\varphi_0}$). The unknown is a vector field $\boldsymbol{\varphi} : \overline{\Omega} \to \mathbb{R}^3$ that is orientation-preserving and injective except possibly on Γ_1, called a *deformation* of the elastic body.

In *linearized elasticity* (Section 6.16), the unknown is instead usually chosen as the displacement vector field $\boldsymbol{u} := \boldsymbol{\varphi} - \mathbf{id}$.

If the material constituting the body is *hyperelastic*, the unknown *deformation* $\boldsymbol{\varphi} : \overline{\Omega} \to \mathbb{R}^3$ undergone by the body is a *stationary point* of the *total energy* I defined by

$$I(\boldsymbol{\psi}) := \int_\Omega W(x, \boldsymbol{\nabla}\boldsymbol{\psi}(x))\,\mathrm{d}x - L(\boldsymbol{\psi}),$$

where

$$W : (x, \boldsymbol{F}) \in \overline{\Omega} \times \mathbb{M}^3_+ \to W(x, \boldsymbol{F}) \in \mathbb{R},$$

denotes the *stored energy function* of the hyperelastic material and

$$L(\boldsymbol{\psi}) := \int_\Omega \boldsymbol{f} \cdot \boldsymbol{\psi}\,\mathrm{d}x + \int_{\Gamma_1} \boldsymbol{g} \cdot \boldsymbol{\psi}\,\mathrm{d}\Gamma,$$

when $\boldsymbol{\psi}$ varies in a set of *admissible deformations* of the form

$$\boldsymbol{\Phi} := \{\boldsymbol{\psi} : \overline{\Omega} \to \mathbb{R}^3;\ \boldsymbol{\psi} \text{ is injective on } \Omega, \det \boldsymbol{\nabla}\boldsymbol{\psi} > 0 \text{ in } \overline{\Omega}, \boldsymbol{\psi} = \boldsymbol{\varphi}_0 \text{ on } \Gamma_0\}.$$

Note, however, that we shall be concerned here with seeking only *particular* stationary points, viz., those that are *minimizers* of the total energy.

In the definition of the set $\boldsymbol{\Phi}$, the condition that $\boldsymbol{\psi}$ be injective on Ω *prevents the interpenetration of matter* ($\boldsymbol{\psi}$ need not be injective on $\overline{\Omega}$ since an admissible deformation loses its injectivity on Γ_1 if self-contact occurs), while the condition $\det \boldsymbol{\nabla}\boldsymbol{\psi} > 0$ in $\overline{\Omega}$, or equivalently, that $\det \boldsymbol{\nabla}\boldsymbol{\psi}(x) \in \mathbb{M}^3_+$ at all points $x \in \overline{\Omega}$, insures that an admissible deformation is *orientation-preserving*. This last condition explains why $W(x, \boldsymbol{F})$ is not defined for $\boldsymbol{F}$ in the whole space $\mathbb{M}^3$, but only for $\boldsymbol{F}$ in the subset $\mathbb{M}^3_+$ of $\mathbb{M}^3$. The condition $\boldsymbol{\psi} = \boldsymbol{\varphi}_0$ on Γ_0, where $\boldsymbol{\varphi}_0 : \Gamma_0 \to \mathbb{R}^3$ is a given vector field, is a *boundary condition of place*.

Remarks (1) It can be shown that the *axiom of material frame-indifference* implies that, as a function of $\boldsymbol{F} \in \mathbb{M}^3_+$, the stored energy function is in fact a function of $\boldsymbol{F}^T\boldsymbol{F} \in \mathbb{S}^3_>$, where $\mathbb{S}^3_>$ denotes the set of all symmetric, positive-definite, symmetric matrices of order three. In other words, there exists a function $\widetilde{W} : \overline{\Omega} \times \mathbb{S}^3_> \to \mathbb{R}$ such that, at each $x \in \overline{\Omega}$, $W(x, \boldsymbol{F}) = \widetilde{W}(x, \boldsymbol{F}^T\boldsymbol{F})$ for all $\boldsymbol{F} \in \mathbb{M}^3_+$. It therefore follows that $W(x, \boldsymbol{\nabla}\boldsymbol{\psi}(x)) = \widetilde{W}\left(x, \boldsymbol{\nabla}\boldsymbol{\psi}(x)^T\boldsymbol{\nabla}\boldsymbol{\psi}(x)\right)$ at each $x \in \overline{\Omega}$, where $\boldsymbol{\nabla}\boldsymbol{\psi}(x)^T\boldsymbol{\nabla}\boldsymbol{\psi}(x) \in \mathbb{S}^3$ is none other than the *metric tensor* at x associated with the deformation $\boldsymbol{\psi}$ (Section 8.2), also called in elasticity theory the *Cauchy–Green strain tensor* at x.

(2) It can be further shown that, if the hyperelastic material is in addition *isotropic* and *homogeneous*, and if the reference configuration is a *natural state*, then the expansion of the function $\widetilde{W}$ (which is then independent of $x \in \overline{\Omega}$) in terms of the matrix $\boldsymbol{E} := \frac{1}{2}(\boldsymbol{C} - \boldsymbol{I})$, where $\boldsymbol{C} := \boldsymbol{F}^T\boldsymbol{F}$ for each $\boldsymbol{F} \in \mathbb{M}^3_+$, must be of the following form for small $|\boldsymbol{E}|$:

$$\widetilde{W}(\boldsymbol{C}) = \frac{\lambda}{2}(\operatorname{tr} \boldsymbol{E})^2 + \mu \operatorname{tr} \boldsymbol{E}^2 + |\boldsymbol{E}|^2\delta(\boldsymbol{E}) \quad \text{with} \quad \lim_{\boldsymbol{E}\to\boldsymbol{0}} \delta(\boldsymbol{E}) = 0,$$

where $\lambda \geq 0$ and $\mu > 0$ are the *Lamé constants* of the material.[32] □

We now list various specific features that the above mathematical model must display in order to be physically acceptable; we also list the difficulties that arise from these specific features.

The *behavior of the stored energy function for large strains*, which mathematically reflects the intuitive idea that "infinite stress must accompany extreme strains,"[33] takes the form of the following *behavior as* $\det \boldsymbol{F} \to 0^+$:

$$\text{For almost all } x \in \Omega, \quad W(x, \boldsymbol{F}) \to \infty \quad \text{as } \det \boldsymbol{F} \to 0^+,$$

a condition that will also insure that any minimizer of the total energy is *orientation-preserving*, and of the following *coerciveness inequality*: there exist sufficiently large constants $p > 0$, $q > 0$, $r > 0$, and constants $\alpha > 0$ and $\beta \in \mathbb{R}$ such that

$$W(x, \boldsymbol{F}) \geq \alpha\{|\boldsymbol{F}|^p + |\,\mathbf{Cof}\,\boldsymbol{F}|^q + (\det \boldsymbol{F})^r\} + \beta \quad \text{for all } \boldsymbol{F} \in \mathbb{M}^3_+ \text{ and for almost all } x \in \Omega.$$

[32] Examples of stored energy functions satisfying such a relation for small $|\boldsymbol{E}|$ for any given constants $\lambda \geq 0$ and $\mu > 0$, as well as all the assumptions of the existence theorem of John Ball (Theorem 9.7-4), have been proposed in:

P.G. Ciarlet; G. Geymonat [1982]: Sur les lois de comportement en élasticité non-linéaire compressible, *Comptes Rendus de l'Académie des Sciences de Paris, Série* II, **295**, 423–426.

[33] How to mathematically express this idea is discussed at length in:

S.S. Antman [1983]: Regular and singular problems for large elastic deformations of tubes, wedges, and cylinders, *Archive for Rational Mechanics and Analysis* **82**, 1–52.

That the *matrix* $\boldsymbol{F}$, the *matrix* $\mathbf{Cof}\,\boldsymbol{F}$, and the *scalar* $\det \boldsymbol{F}$, appear in the right-hand side of the coerciveness inequality reflects the fact that the matrix field $\boldsymbol{\nabla}\boldsymbol{\varphi}$ (through the metric tensor field $\boldsymbol{\nabla}\boldsymbol{\varphi}^T\boldsymbol{\nabla}\boldsymbol{\varphi}$), the matrix field $\mathbf{Cof}\,\boldsymbol{\nabla}\boldsymbol{\varphi}$, and the function $\det \boldsymbol{\nabla}\boldsymbol{\varphi}$, respectively govern the changes of *lengths, surfaces*, and *volumes*, associated with a deformation $\boldsymbol{\varphi}$ (Theorem 8.2-1 and Problem 8.2-1).

A basic fact is that the *stored energy function* $W : (x, \boldsymbol{F}) \in \overline{\Omega} \times \mathbb{M}^3_+ \to \mathbb{R}$ *cannot be convex with respect to the variable* $\boldsymbol{F} \in \mathbb{M}^3_+$ (this means that, given any $x \in \overline{\Omega}$, there is no convex function $\widehat{W}(x, \cdot) : \mathrm{co}\,\mathbb{M}^3_+ = \mathbb{M}^3 \to \mathbb{R}$ such that $\widehat{W}(x, \boldsymbol{F}) = W(x, \boldsymbol{F})$ for all $\boldsymbol{F} \in \mathbb{M}^3_+$; cf. Problem 2.16-3 and Section 2.17). For, it can be shown that such a convexity would contradict both the behavior as $\det \boldsymbol{F} \to 0^+$[34] and the axiom of material frame-indifference.[35] Note that this fact alone already precludes using Theorem 9.5-2.

The lack of convexity of the stored energy function and its behavior for large strains stood for a long while as major difficulties in the mathematical analysis of three-dimensional hyperelasticity, until John Ball was able to overcome them in a landmark paper,[36] notably by means of the weaker requirement of *polyconvexity* (this notion will be defined below).

As in the proof of Theorem 9.3-1, we are naturally led to consider an *infimizing sequence* $(\boldsymbol{\varphi}^k)$ of the total energy

$$I : \boldsymbol{\psi} \to I(\boldsymbol{\psi}) = \int_\Omega W(x, \boldsymbol{\nabla}\boldsymbol{\psi}(x))\,\mathrm{d}x - L(\boldsymbol{\psi})$$

over an appropriate *set* $\boldsymbol{\Phi}$ *of admissible deformations* $\boldsymbol{\psi}$, defined later as an *ad hoc* subset of the space $\boldsymbol{W}^{1,p}(\Omega)$ for some $p > 1$; then to show that this sequence is *bounded* as a consequence of the coerciveness inequality satisfied by the stored energy function; then to extract a *subsequence* $(\boldsymbol{\varphi}^\ell)$ *that weakly converges* to an element $\boldsymbol{\varphi}$; then to show that *the weak limit* $\boldsymbol{\varphi}$ *belongs to the set* $\boldsymbol{\Phi}$; and finally, to show that

$$\int_\Omega W(x, \boldsymbol{\nabla}\boldsymbol{\varphi}(x))\,\mathrm{d}x \le \liminf_{\ell\to\infty} \int_\Omega W(x, \boldsymbol{\nabla}\boldsymbol{\varphi}^\ell(x))\,\mathrm{d}x$$

(as the remaining part $L : \boldsymbol{\psi} \to L(\boldsymbol{\psi})$ of the total energy is a linear functional, it will suffice to assume that L is continuous over the space $\boldsymbol{W}^{1,p}(\Omega)$, as in the proof of Theorem 9.5-2). It will then follow that $\boldsymbol{\varphi} \in \boldsymbol{\Phi}$ is a *minimizer* of the energy, i.e., that $I(\boldsymbol{\varphi}) = \inf_{\boldsymbol{\varphi}\in\boldsymbol{\Phi}} I(\boldsymbol{\varphi})$.

Establishing the *sequential weak lower semicontinuity of the functional*

$$\boldsymbol{\psi} \to \int_\Omega W(x, \boldsymbol{\nabla}\boldsymbol{\psi}(x))\,\mathrm{d}x$$

will be, however, substantially more delicate than in Theorem 9.5-2 since, given any $x \in \Omega$, the function

$$\boldsymbol{F} \to W(x, \boldsymbol{F})$$

[34]S.S. Antman [1970]: Existence of solutions of the equilibrium equations for nonlinearly elastic rings and arches, *Indiana University Mathematics Journal* **20**, 281–302.

[35]B.D. Coleman; W. Noll [1959]: On the thermostatics of continuous media, *Archive for Rational Mechanics and Analysis* **4**, 97–128.

[36]J. Ball [1977]: Convexity conditions and existence theorems in nonlinear elasticity, *Archive for Rational Mechanics and Analysis* **63**, 337–403.

is *not convex* and is *not defined for* $\det \boldsymbol{F} \le 0$.

A closer look at those steps yields various observations and guidelines, which form the basis of John Ball's approach to existence theory in hyperelasticity.

The "impossible convexity" of the stored energy function W with respect to its argument $\boldsymbol{F}$ will be replaced by the weaker assumption of *polyconvexity* of the stored energy function according to the following definition: A function $W : \Omega \times \mathbb{M}^3_+ \to \mathbb{R}$ is **polyconvex** if, for almost all $x \in \Omega$, there exists a *convex* function $\mathbb{W}(x, \cdot) : \mathbb{M}^3 \times \mathbb{M}^3 \times]0, \infty[\to \mathbb{R}$ such that

$$W(x, \boldsymbol{F}) = \mathbb{W}(x, \boldsymbol{F}, \mathbf{Cof}\, \boldsymbol{F}, \det \boldsymbol{F}) \quad \text{for all } \boldsymbol{F} \in \mathbb{M}^3_+.$$

Note that the set $\mathbb{M}^3 \times \mathbb{M}^3 \times]0, \infty[$ naturally appears here, simply because it is the smallest convex subset of $\mathbb{M}^3 \times \mathbb{M}^3 \times \mathbb{R}$ that contains the set $\{(\boldsymbol{F}, \mathbf{Cof}\, \boldsymbol{F}, \det \boldsymbol{F}) \in \mathbb{M}^3 \times \mathbb{M}^3 \times \mathbb{R};\ \boldsymbol{F} \in \mathbb{M}^3_+\}$; cf. Problem 9.7-1.

It was mentioned earlier that the behavior of the stored energy function for large strains is reflected in part by a *coerciveness inequality* of the form

$$W(x, \boldsymbol{F}) \ge \alpha \left\{ |\boldsymbol{F}|^p + |\mathbf{Cof}\, \boldsymbol{F}|^q + (\det \boldsymbol{F})^r \right\} + \beta \quad \text{for all } \boldsymbol{F} \in \mathbb{M}^3_+ \text{ and for almost all } x \in \Omega,$$

with $\alpha > 0$, $\beta \in \mathbb{R}$, and sufficiently large exponents p, q, r. Since this inequality in turn implies that

$$\int_\Omega W(x, \boldsymbol{\nabla}\boldsymbol{\psi}(x))\, \mathrm{d}x \ge \alpha \left\{ \|\boldsymbol{\nabla}\boldsymbol{\psi}\|^p_{0,p,\Omega} + \|\mathbf{Cof}\, \boldsymbol{\nabla}\boldsymbol{\psi}\|^q_{0,q,\Omega} + \|\det \boldsymbol{\nabla}\boldsymbol{\psi}\|^r_{0,r,\Omega} \right\} + \beta \operatorname{vol} \Omega,$$

any function $\boldsymbol{\psi}$ *that satisfies* $\int_\Omega W(x, \boldsymbol{\nabla}\boldsymbol{\psi}(x))\, \mathrm{d}x < \infty$ (such as the functions in an infimizing sequence of the total energy) *must be such that*

$$\boldsymbol{\nabla}\boldsymbol{\psi} \in \boldsymbol{L}^p(\Omega), \quad \mathbf{Cof}\, \boldsymbol{\nabla}\boldsymbol{\psi} \in \boldsymbol{L}^q(\Omega), \quad \det \boldsymbol{\nabla}\boldsymbol{\psi} \in L^r(\Omega).$$

If the remaining part of the total energy is assumed to be a continuous linear form $L : \boldsymbol{W}^{1,p}(\Omega) \to \mathbb{R}$, the following *lower bound for the total energy* therefore holds: There exist constants $a > 0$ and $b \in \mathbb{R}$ such that, for all functions $\boldsymbol{\psi} \in \boldsymbol{W}^{1,p}(\Omega)$ satisfying $\boldsymbol{\psi} = \boldsymbol{\varphi}_0$ on Γ_0,

$$I(\boldsymbol{\psi}) = \int_\Omega W(x, \boldsymbol{\nabla}\boldsymbol{\psi})\, \mathrm{d}x - L(\boldsymbol{\psi}) \ge a \left\{ \|\boldsymbol{\psi}\|^p_{1,p,\Omega} + \|\mathbf{Cof}\, \boldsymbol{\nabla}\boldsymbol{\psi}\|^q_{0,q,\Omega} + \|\det \boldsymbol{\nabla}\boldsymbol{\psi}\|^r_{0,r,\Omega} \right\} + b.$$

How large must then be the exponents p, q, r in the coerciveness inequality? A first observation is that they must all be > 1 in order that the spaces $\boldsymbol{L}^p(\Omega)$, $\boldsymbol{L}^q(\Omega)$, and $L^r(\Omega)$ be *reflexive*, so that we may extract weakly convergent subsequences from bounded sequences. If the functions $\boldsymbol{\psi} \in \boldsymbol{W}^{1,p}(\Omega)$ satisfy a *boundary condition of place* $\boldsymbol{\psi} = \boldsymbol{\varphi}_0$ on $\Gamma_0 \subset \Gamma$ and area $\Gamma_0 > 0$, the generalized Poincaré inequality implies (as in the proof of Theorem 9.5-2) that *the seminorm* $\|\boldsymbol{\nabla}\boldsymbol{\psi}\|_{0,p,\Omega}$ *can be replaced by the norm* $\|\boldsymbol{\psi}\|_{1,p,\Omega}$ *in the lower bound of the integral* $\int_\Omega \boldsymbol{W}(x, \boldsymbol{\nabla}\boldsymbol{\varphi}(x))\, \mathrm{d}x$.

The definition of the *set of admissible deformations* is thus imposing itself in a natural way: We first infer from the above considerations that it should consist of vector fields $\boldsymbol{\psi} \in \boldsymbol{W}^{1,p}(\Omega)$ satisfying the boundary conditions of place $\boldsymbol{\psi} = \boldsymbol{\varphi}_0$ on Γ_0, and such that $\mathbf{Cof}\, \boldsymbol{\nabla}\boldsymbol{\psi} \in \boldsymbol{L}^q(\Omega)$ and $\det \boldsymbol{\nabla}\boldsymbol{\psi} \in L^r(\Omega)$. From the definition of a deformation, we next infer that the functions $\boldsymbol{\psi} \in \boldsymbol{W}^{1,p}(\Omega)$ should also be *orientation-preserving*. If, following John

Ball, we take only these requirements into account, we conclude that *the set of admissible deformations is of the form*

$$\begin{aligned}\boldsymbol{\Phi} = \{\boldsymbol{\psi} \in \boldsymbol{W}^{1,p}(\Omega);\ \mathbf{Cof}\,\boldsymbol{\nabla}\boldsymbol{\psi} \in \boldsymbol{L}^q(\Omega),\ \det \boldsymbol{\nabla}\boldsymbol{\psi} \in L^r(\Omega),\\ \boldsymbol{\psi} = \boldsymbol{\varphi}_0 \quad d\Gamma\text{-a.e. on } \Gamma_0,\ \det \boldsymbol{\nabla}\boldsymbol{\psi} > 0 \text{ a.e. in } \Omega\},\end{aligned}$$

where the exponents p, q, r are those appearing in the coerciveness inequality satisfied by the stored energy function. Notice that the *orientation-preserving condition* $\det \boldsymbol{\nabla}\boldsymbol{\psi} > 0$ *can be only asked to hold almost everywhere in* Ω, since $\det \boldsymbol{\nabla}\boldsymbol{\psi}$ is only in $L^r(\Omega)$.

For ease of exposition, *injectivity* is *not* imposed here on the admissible deformations $\boldsymbol{\psi} \in \boldsymbol{\Phi}$. However, under suitable assumptions, it can be also taken care of by means of more refined arguments.[37]

As in the previous section, the basic idea then consists in considering an *infimizing sequence* $(\boldsymbol{\varphi}^k)$ of the total energy, with $\boldsymbol{\varphi}^k \in \boldsymbol{\Phi}$ for all k. Since the sequences $(\boldsymbol{\varphi}^k)$, $(\mathbf{Cof}\,\boldsymbol{\nabla}\boldsymbol{\varphi}^k)$, and $(\det \boldsymbol{\nabla}\boldsymbol{\varphi}^k)$ are then bounded in the reflexive spaces $\boldsymbol{L}^p(\Omega)$, $\boldsymbol{L}^q(\Omega)$, and $L^r(\Omega)$, respectively (thanks to the coerciveness inequality satisfied by the stored energy function), they contain subsequences $(\boldsymbol{\varphi}^\ell)$, $(\mathbf{Cof}\,\boldsymbol{\nabla}\boldsymbol{\varphi}^\ell)$, and $(\det \boldsymbol{\nabla}\boldsymbol{\varphi}^\ell)$, that weakly converge in these spaces. It is thus expected that their weak limits will provide a minimizer of the total energy over the set $\boldsymbol{\Phi}$ of admissible deformations.

Hence a crucial task will consist in verifying that these weak limits do belong to the set $\boldsymbol{\Phi}$.

Regarding the *orientation-preserving condition*, we shall see that, interestingly, the *behavior of the stored energy function as* $\det \boldsymbol{F} \to 0^+$ implies that the weak limit $\boldsymbol{\varphi}$ of the infimizing sequence also satisfies the orientation-preserving condition $\det \boldsymbol{\nabla}\boldsymbol{\varphi} > 0$ almost everywhere in Ω. In other words, the behavior as $\det \boldsymbol{F} \to 0^+$ compensates the restriction that the stored energy function be only defined for matrices $\boldsymbol{F}$ with $\det \boldsymbol{F} > 0$.

Clearly, *the set $\boldsymbol{\Phi}$ cannot be expected to be convex* (in this direction, see Problems 9.7-2 and 9.7-4); this observation indicates that *difficulties will certainly arise when taking weak limits*, since the Banach–Saks–Mazur theorem cannot be applied in the present situation. Accordingly, following John Ball's approach, we will have to find sufficient conditions insuring that the weak convergences

$$\boldsymbol{\varphi}^\ell \rightharpoonup \boldsymbol{\varphi} \text{ in } \boldsymbol{W}^{1,p}(\Omega), \quad \mathbf{Cof}\,\boldsymbol{\nabla}\boldsymbol{\varphi}^\ell \rightharpoonup \boldsymbol{H} \text{ in } \boldsymbol{L}^q(\Omega), \quad \text{and} \quad \det \boldsymbol{\nabla}\boldsymbol{\varphi}^\ell \rightharpoonup \boldsymbol{\delta} \text{ in } L^r(\Omega)$$

imply that

$$\boldsymbol{H} = \mathbf{Cof}\,\boldsymbol{\nabla}\boldsymbol{\varphi} \quad \text{and} \quad \boldsymbol{\delta} = \det \boldsymbol{\nabla}\boldsymbol{\varphi}.$$

The next two theorems[38] will show that *this is the case* if $p \geq 2$ and $q \geq p/(p-1)$ (hence this imposes further restrictions on the exponents p and q, which, like r, were only required so far to be > 1), by establishing various basic properties of the *nonlinear* mappings $\boldsymbol{\psi} \in \boldsymbol{W}^{1,p}(\Omega) \to \mathbf{Cof}\,\boldsymbol{\nabla}\boldsymbol{\psi}$ and $\boldsymbol{\psi} \in \boldsymbol{W}^{1,p}(\Omega) \to \det \boldsymbol{\nabla}\boldsymbol{\psi}$, notably *with respect to weak convergence* (which is as usual denoted by $\rightharpoonup$).

[37] J. Ball [1981]: Global invertibility of Sobolev functions and the interpenetration of matter, *Proceedings of the Royal Society, Edinburgh* **88A**, 315–328.

P.G. Ciarlet; J. Nečas [1987]: Injectivity and self-contact in nonlinear elasticity, *Archive for Rational Mechanics and Analysis* **19**, 171–188.

[38] The next theorems, as well as the exercises that complement them, are all due to Ball [1977] (*op. cit*).

Theorem 9.7-1 *Let Ω be a domain in $\mathbb{R}^3$. For each $p \geq 2$, the mapping*

$$\boldsymbol{\psi} \in \boldsymbol{W}^{1,p}(\Omega) \to \mathbf{Cof}\,\boldsymbol{\nabla}\boldsymbol{\psi} \in \boldsymbol{L}^{p/2}(\Omega)$$

is well defined and continuous. Furthermore,

$$\boldsymbol{\varphi}^\ell \rightharpoonup \boldsymbol{\varphi} \textit{ in } \boldsymbol{W}^{1,p}(\Omega),\, p \geq 2, \quad \textit{and} \quad \mathbf{Cof}\,\boldsymbol{\nabla}\boldsymbol{\varphi}^\ell \rightharpoonup \boldsymbol{H} \textit{ in } \boldsymbol{L}^q(\Omega),\, q \geq 1,$$

implies that

$$\boldsymbol{H} = \mathbf{Cof}\,\boldsymbol{\nabla}\boldsymbol{\varphi}.$$

Proof (i) By Hölder's inequality, the bilinear mapping

$$(\xi, \eta) \in (L^p(\Omega))^2 \to \xi\eta \in L^{p/2}(\Omega)$$

is well defined and continuous for $p \geq 2$. Consequently, the mapping $\boldsymbol{\psi} \in \boldsymbol{W}^{1,p}(\Omega) \to \mathbf{Cof}\,\boldsymbol{\nabla}\boldsymbol{\psi} \in \boldsymbol{L}^{p/2}(\Omega)$ is well defined and continuous for $p \geq 2$.

(ii) For sufficiently smooth functions $\boldsymbol{\psi}$, for instance in the space $\boldsymbol{\mathcal{C}}^2(\overline{\Omega})$, we can also write, counting the indices *modulo* 3,

$$\begin{aligned}(\mathbf{Cof}\,\boldsymbol{\nabla}\boldsymbol{\psi})_{ij} &= \partial_{i+1}\psi_{j+1}\partial_{i+2}\psi_{j+2} - \partial_{i+2}\psi_{j+1}\partial_{i+1}\psi_{j+2} \\ &= \partial_{i+2}(\psi_{j+2}\partial_{i+1}\psi_{j+1}) - \partial_{i+1}(\psi_{j+2}\partial_{i+2}\psi_{j+1}).\end{aligned}$$

Consequently, an application of the fundamental Green's formula shows that, for all functions $\boldsymbol{\psi} \in \boldsymbol{\mathcal{C}}^2(\overline{\Omega})$ and all functions $\theta \in \mathcal{D}(\Omega)$,

$$\int_\Omega (\mathbf{Cof}\,\boldsymbol{\nabla}\boldsymbol{\psi})_{ij}\theta\,\mathrm{d}x = -\int_\Omega \psi_{j+2}\partial_{i+1}\psi_{j+1}\partial_{i+2}\theta\,\mathrm{d}x + \int_\Omega \psi_{j+2}\partial_{i+2}\psi_{j+1}\partial_{i+1}\theta\,\mathrm{d}x.$$

For a fixed function $\theta \in \mathcal{D}(\Omega)$, the two sides of this relation are continuous if the space $\boldsymbol{\mathcal{C}}^2(\overline{\Omega})$ is equipped with the norm $\|\cdot\|_{1,\Omega}$, since there exist constants $c_1(\theta)$ and $c_2(\theta)$ such that

$$\begin{aligned}\left|\int_\Omega (\mathbf{Cof}\,\boldsymbol{\nabla}\boldsymbol{\psi})_{ij}\theta\,\mathrm{d}x\right| &\leq \|(\mathbf{Cof}\,\boldsymbol{\nabla}\boldsymbol{\psi})_{ij}\|_{0,1,\Omega}\,\|\theta\|_{0,\infty,\Omega} \leq c_1(\theta)\,\|\boldsymbol{\psi}\|^2_{1,\Omega}\,, \\ \left|\int_\Omega \psi_i\partial_j\psi_k\partial_\ell\theta\,\mathrm{d}x\right| &\leq \|\psi_i\|_{0,\Omega}\,|\psi_k|_{1,\Omega}\,|\theta|_{1,\infty,\Omega} \leq c_2(\theta)\,\|\boldsymbol{\psi}\|^2_{1,\Omega}\,.\end{aligned}$$

Therefore this relation remains valid for functions $\boldsymbol{\psi}$ in the space $\boldsymbol{H}^1(\Omega)$, whence in any space $\boldsymbol{W}^{1,p}(\Omega)$, $p \geq 2$, since the space $\mathcal{C}^2(\overline{\Omega})$ is dense in the space $H^1(\Omega)$ when Ω is a domain (Theorem 6.6-4).

(iii) *Let $p \geq 2$. Given an arbitrary function $\theta \in \mathcal{D}(\Omega)$, we next show that*

$$\boldsymbol{\varphi}^\ell \rightharpoonup \boldsymbol{\varphi} \text{ in } \boldsymbol{W}^{1,p}(\Omega) \quad \text{implies} \quad \int_\Omega \varphi_i^\ell\partial_j\varphi_k^\ell\partial_m\theta\,\mathrm{d}x \underset{\ell\to\infty}{\longrightarrow} \int_\Omega \varphi_i\partial_j\varphi_k\partial_m\theta\,\mathrm{d}x,$$

so that (by part (ii)),

$$\boldsymbol{\varphi}^\ell \rightharpoonup \boldsymbol{\varphi} \text{ in } \boldsymbol{W}^{1,p}(\Omega) \quad \text{implies} \quad \int_\Omega (\mathbf{Cof}\,\boldsymbol{\nabla}\boldsymbol{\varphi}^\ell)_{ij}\theta\,\mathrm{d}x \underset{\ell\to\infty}{\longrightarrow} \int_\Omega (\mathbf{Cof}\,\boldsymbol{\nabla}\boldsymbol{\varphi})_{ij}\theta\,\mathrm{d}x.$$

By Hölder's inequality, the bilinear mapping

$$(\xi,\chi) \in L^r(\Omega) \times W^{1,p}(\Omega) \to \int_\Omega \xi \partial_j \chi \partial_m \theta \, dx, \quad \text{with } \frac{1}{p} + \frac{1}{r} \leq 1,$$

is continuous (the function $\theta \in \mathcal{D}(\Omega)$ is held fixed in the argument). Hence, by Theorem 5.12-4(c),

$$\xi^\ell \to \xi \text{ in } L^r(\Omega) \text{ and } \chi^\ell \rightharpoonup \chi \text{ in } W^{1,p}(\Omega) \quad \text{implies} \quad \int_\Omega \xi^\ell \partial_j \chi^\ell \partial_m \theta \, dx \underset{\ell\to\infty}{\longrightarrow} \int_\Omega \xi \partial_j \chi \partial_m \theta \, dx.$$

From the compact imbedding (Theorem 6.6-3)

$$W^{1,p}(\Omega) \Subset L^r(\Omega) \quad \text{for all } 1 \leq r < p^*,$$

where $p^* = \dfrac{3p}{3-p}$ if $p < 3$ and $p^* = \infty$ if $p \geq 3$, we then infer that

$$\boldsymbol{\varphi}^\ell \rightharpoonup \boldsymbol{\varphi} \text{ in } \boldsymbol{W}^{1,p}(\Omega) \quad \text{implies} \quad \boldsymbol{\varphi}^\ell \to \boldsymbol{\varphi} \text{ in } \boldsymbol{L}^r(\Omega) \text{ for all } 1 \leq r < p^*.$$

Hence the assertion is proved since, for any $p \geq 2$ (in fact for any $p > \frac{3}{2}$), there exists a number r that simultaneously satisfies $\frac{1}{p} + \frac{1}{r} \leq 1$ and $r < p^*$.

(iv) Let $p \geq 2$ and $q \geq 1$, and let $(\boldsymbol{\varphi}^\ell)$ be a sequence in the space $\boldsymbol{W}^{1,p}(\Omega)$ such that

$$\boldsymbol{\varphi}^\ell \rightharpoonup \boldsymbol{\varphi} \text{ in } \boldsymbol{W}^{1,p}(\Omega), \quad \mathbf{Cof}\, \boldsymbol{\nabla}\boldsymbol{\varphi}^\ell \in \boldsymbol{L}^q(\Omega), \quad \text{and} \quad \mathbf{Cof}\, \boldsymbol{\nabla}\boldsymbol{\varphi}^\ell \rightharpoonup \boldsymbol{H} \text{ in } \boldsymbol{L}^q(\Omega).$$

Therefore,

$$\int_\Omega (\mathbf{Cof}\, \boldsymbol{\nabla}\boldsymbol{\varphi}^\ell)_{ij} \theta \, dx \longrightarrow \int_\Omega (\mathbf{Cof}\, \boldsymbol{\nabla}\boldsymbol{\varphi})_{ij} \theta \, dx \quad \text{for all } \theta \in \mathcal{D}(\Omega),$$

by part (iii), and

$$\int_\Omega (\mathbf{Cof}\, \boldsymbol{\nabla}\boldsymbol{\varphi}^\ell)_{ij} \theta \, dx \longrightarrow \int_\Omega H_{ij} \theta \, dx,$$

by assumption. We conclude that each function $(\mathbf{Cof}\, \boldsymbol{\nabla}\boldsymbol{\varphi} - \boldsymbol{H})_{ij} \in L^1(\Omega)$ satisfies

$$\int_\Omega (\mathbf{Cof}\, \boldsymbol{\nabla}\boldsymbol{\varphi} - \boldsymbol{H})_{ij} \theta \, dx = 0 \quad \text{for all } \theta \in \mathcal{D}(\Omega).$$

By Theorem 6.3-2, this implies that $(\mathbf{Cof}\, \boldsymbol{\nabla}\boldsymbol{\varphi} - \boldsymbol{H})_{ij} = 0$ almost everywhere in Ω, which completes the proof. □

Remarks (1) Theorem 9.7-1 implies that *the nonconvex set* (Problem 9.7-2)

$$\{(\boldsymbol{\psi}, \boldsymbol{K}) \in \boldsymbol{W}^{1,p}(\Omega) \times \boldsymbol{L}^q(\Omega);\ \boldsymbol{K} = \mathbf{Cof}\, \boldsymbol{\nabla}\boldsymbol{\psi}\}, \quad p \geq 2,\ q \geq 1,$$

is sequentially weakly closed in the space $\boldsymbol{W}^{1,p}(\Omega) \times \boldsymbol{L}^q(\Omega)$. This does *not* mean, however, that the set

$$\{\boldsymbol{\psi} \in \boldsymbol{W}^{1,p}(\Omega);\ \mathbf{Cof}\, \boldsymbol{\nabla}\boldsymbol{\psi} \in \boldsymbol{L}^q(\Omega)\}, \quad p \geq 2,\ q \geq 1,$$

is sequentially weakly closed in the space $\boldsymbol{W}^{1,p}(\Omega)$, and indeed this is not always the case (Problem 9.7-2).

(2) In part (ii), we showed that the functions $\boldsymbol{\psi} \in \boldsymbol{W}^{1,p}(\Omega)$, $p \geq 2$, satisfy

$$\int_\Omega (\mathbf{Cof}\,\boldsymbol{\nabla}\boldsymbol{\psi})_{ij}\theta\,\mathrm{d}x = -\int_\Omega \psi_{j+2}\partial_{i+1}\psi_{j+1}\partial_{i+2}\theta\,\mathrm{d}x + \int_\Omega \psi_{j+2}\partial_{i+2}\psi_{j+1}\partial_{i+1}\theta\,\mathrm{d}x \quad \text{for all } \theta \in \mathcal{D}(\Omega).$$

Hence, for $p \geq 2$, we also have

$$\mathbf{Cof}\,\boldsymbol{\nabla}\boldsymbol{\psi} = \mathbf{Cof}^\sharp\,\boldsymbol{\nabla}\boldsymbol{\psi} \quad \text{in } \mathcal{D}'(\Omega),$$

where

$$(\mathbf{Cof}^\sharp\,\boldsymbol{\nabla}\boldsymbol{\psi})_{ij} := \partial_{i+2}(\psi_{j+2}\partial_{i+1}\psi_{j+1}) - \partial_{i+1}(\psi_{j+2}\partial_{i+2}\psi_{j+1}).$$

The merit of this alternative expression is to allow an extension of the definition of $\mathbf{Cof}\,\boldsymbol{\nabla}\boldsymbol{\psi}$ to functions $\boldsymbol{\psi} \in \boldsymbol{W}^{1,p}(\Omega)$ with $\frac{3}{2} \leq p < 2$, in which case $\mathbf{Cof}\,\boldsymbol{\nabla}\boldsymbol{\psi}$ is then not necessarily an integrable function (Problem 9.7-3). □

From now on in this section, *Latin indices range in the set* $\{1, 2, 3\}$ *and the summation convention with respect to repeated indices is used.* Since, by Hölder's inequality, the trilinear mapping

$$(\xi, \eta, \zeta) \in (L^p(\Omega))^3 \to \xi\eta\zeta \in L^{p/3}(\Omega)$$

is well defined and continuous, and since (ε_{ijk} denote the components of the orientation tensor)

$$\det \boldsymbol{\nabla}\boldsymbol{\psi} = \frac{1}{6}\varepsilon_{ijk}\varepsilon_{\ell mn}\partial_\ell\psi_i\partial_m\psi_j\partial_n\psi_k,$$

it seems that we need $p \geq 3$ in order that the mapping $\boldsymbol{\psi} \in \boldsymbol{W}^{1,p}(\Omega) \to \det \boldsymbol{\nabla}\boldsymbol{\psi} \in L^1(\Omega)$ be well defined and continuous. However, with some specific *additional* information on the function $\mathbf{Cof}\,\boldsymbol{\nabla}\boldsymbol{\psi}$, we can weaken this requirement by taking advantage of the expansion of $\det \boldsymbol{\nabla}\boldsymbol{\psi}$ as

$$\det \boldsymbol{\nabla}\boldsymbol{\psi} = \partial_j\psi_1(\mathbf{Cof}\,\boldsymbol{\nabla}\boldsymbol{\psi})_{1j}$$

(the choice of the first row is arbitrary; we could likewise consider the expansion of $\det \boldsymbol{\nabla}\boldsymbol{\psi}$ along any other row or any one of the columns of the matrix $\boldsymbol{\nabla}\boldsymbol{\psi}$).

Then another application of Hölder's inequality shows that $\det \boldsymbol{\nabla}\boldsymbol{\psi}$ is well determined as an element of the space $L^s(\Omega)$ if $\boldsymbol{\psi} \in \boldsymbol{W}^{1,p}(\Omega)$ with $p \geq 2$ and $\mathbf{Cof}\,\boldsymbol{\nabla}\boldsymbol{\psi} \in \boldsymbol{L}^q(\Omega)$ with $s^{-1} := p^{-1} + q^{-1} \leq 1$. If $p \geq 3$, there is no need to assume that $\mathbf{Cof}\,\boldsymbol{\nabla}\boldsymbol{\psi} \in \boldsymbol{L}^q(\Omega)$ with $p^{-1} + q^{-1} \leq 1$, since then $\mathbf{Cof}\,\boldsymbol{\nabla}\boldsymbol{\psi} \in \boldsymbol{L}^{p/2}(\Omega)$ and $p^{-1} + 2p^{-1} \leq 1$.

We now establish some basic properties of the *nonlinear* mapping

$$(\boldsymbol{\psi}, \mathbf{Cof}\,\boldsymbol{\nabla}\boldsymbol{\psi}) \in \boldsymbol{W}^{1,p}(\Omega) \times \boldsymbol{L}^q(\Omega) \to \det \boldsymbol{\nabla}\boldsymbol{\psi} \in L^s(\Omega)$$

defined in this fashion, notably *with respect to weak convergence.*

Theorem 9.7-2 *Let Ω be a domain in $\mathbb{R}^3$. For each number $p \geq 2$ and each number q such that*

$$\frac{1}{s} := \frac{1}{p} + \frac{1}{q} \leq 1,$$

the mapping

$$(\boldsymbol{\psi}, \mathbf{Cof}\,\boldsymbol{\nabla}\boldsymbol{\psi}) \in \boldsymbol{W}^{1,p}(\Omega) \times \boldsymbol{L}^q(\Omega) \to \det \boldsymbol{\nabla}\boldsymbol{\psi} := \partial_j \psi_1 (\mathbf{Cof}\,\boldsymbol{\nabla}\boldsymbol{\psi})_{1j} \in L^s(\Omega)$$

is well defined and continuous. Furthermore, the weak convergences

$$\begin{aligned} \boldsymbol{\varphi}^\ell &\rightharpoonup \boldsymbol{\varphi} && \text{in } \boldsymbol{W}^{1,p}(\Omega),\ p \geq 2, \\ \mathbf{Cof}\,\boldsymbol{\nabla}\boldsymbol{\varphi}^\ell &\rightharpoonup \boldsymbol{H} && \text{in } \boldsymbol{L}^q(\Omega),\ \frac{1}{p} + \frac{1}{q} \leq 1, \\ \det \boldsymbol{\nabla}\boldsymbol{\varphi}^\ell &\rightharpoonup \delta && \text{in } L^r(\Omega),\ r \geq 1 \end{aligned}$$

imply that

$$\boldsymbol{H} = \mathbf{Cof}\,\boldsymbol{\nabla}\boldsymbol{\varphi} \quad \text{and} \quad \delta = \det \boldsymbol{\nabla}\boldsymbol{\varphi}.$$

Proof (i) The bilinear mapping

$$(\boldsymbol{\psi}, \mathbf{Cof}\,\boldsymbol{\nabla}\boldsymbol{\psi}) \in \boldsymbol{W}^{1,p}(\Omega) \times \boldsymbol{L}^q(\Omega) \to \partial_j \psi_1 (\mathbf{Cof}\,\boldsymbol{\nabla}\boldsymbol{\psi})_{1j} \in L^s(\Omega)$$

is well defined and continuous by Hölder's inequality.

(ii) Any sufficiently smooth vector field $\boldsymbol{\psi}$, for instance in the space $\boldsymbol{\mathcal{C}}^2(\overline{\Omega})$, satisfies

$$\partial_j (\mathbf{Cof}\,\boldsymbol{\nabla}\boldsymbol{\psi})_{1j} = 0,$$

as a consequence of the *Piola identity* $\mathbf{div}\,\mathbf{Cof}\,\boldsymbol{\nabla}\boldsymbol{\psi} = \mathbf{0}$ (Theorem 7.1-4). Therefore,

$$\partial_j \psi_1 (\mathbf{Cof}\,\boldsymbol{\nabla}\boldsymbol{\psi})_{1j} = \partial_j \left\{ \psi_1 (\mathbf{Cof}\,\boldsymbol{\nabla}\boldsymbol{\psi})_{1j} \right\} = \det \boldsymbol{\nabla}\boldsymbol{\psi}$$

for such smooth fields $\boldsymbol{\psi}$. An application of Green's formula then shows that, for all fields $\boldsymbol{\psi} \in \boldsymbol{\mathcal{C}}^2(\overline{\Omega})$ and all functions $\theta \in \mathcal{D}(\Omega)$,

$$\int_\Omega \partial_j \psi_1 (\mathbf{Cof}\,\boldsymbol{\nabla}\boldsymbol{\psi})_{1j} \theta \,\mathrm{d}x = - \int_\Omega \psi_1 (\mathbf{Cof}\,\boldsymbol{\nabla}\boldsymbol{\psi})_{1j} \partial_j \theta \,\mathrm{d}x.$$

Our aim is to show that *this relation still holds for all fields* $\boldsymbol{\psi} \in \boldsymbol{W}^{1,p}(\Omega)$, $p \geq 2$, *such that* $\mathbf{Cof}\,\boldsymbol{\nabla}\boldsymbol{\psi} \in \boldsymbol{L}^{p'}(\Omega)$, with $p^{-1} + (p')^{-1} = 1$, hence *a fortiori* such that $\mathbf{Cof}\,\boldsymbol{\nabla}\boldsymbol{\psi} \in \boldsymbol{L}^q(\Omega)$, with $p^{-1} + q^{-1} \leq 1$. There is, however, a difficulty in applying a straightforward density argument as in part (ii) of the proof of Theorem 9.7-1, since the function

$$\boldsymbol{\psi} \to \int_\Omega \partial_j \psi_1 (\mathbf{Cof}\,\boldsymbol{\nabla}\boldsymbol{\psi})_{1j} \theta \,\mathrm{d}x$$

is *not* continuous with respect to the norm $\|\cdot\|_{1,p,\Omega}$, unless $p \geq 3$. On the other hand, the bilinear form

$$(\boldsymbol{\psi}, \boldsymbol{H}) \in \boldsymbol{W}^{1,p}(\Omega) \times \boldsymbol{L}^{p'}(\Omega) \to \int_\Omega \partial_j \psi_1 H_{1j} \theta \,\mathrm{d}x$$

is clearly continuous if $p^{-1} + (p')^{-1} = 1$; but then the relation

$$\int_\Omega \partial_j \psi_1 H_{1j} \theta \,\mathrm{d}x = - \int_\Omega \psi_1 H_{1j} \partial_j \theta \,\mathrm{d}x$$

does not hold for smooth functions ψ and H_{1j} in general, unless the functions H_{1j} satisfy $\partial_j H_{1j} = 0$, as is the case of the functions $(\mathbf{Cof}\,\boldsymbol{\nabla}\boldsymbol{\psi})_{1j}$ when $\boldsymbol{\psi}$ is smooth. We therefore have to resort to a more refined argument.

The relation $\partial_j(\mathbf{Cof}\,\boldsymbol{\nabla}\boldsymbol{\psi})_{1j} = 0$ for all $\boldsymbol{\psi} \in \boldsymbol{\mathcal{C}}^2(\overline{\Omega})$ implies that

$$\int_\Omega (\mathbf{Cof}\,\boldsymbol{\nabla}\boldsymbol{\psi})_{1j}\partial_j\chi\,\mathrm{d}x = 0 \quad \text{for all } \chi \in \mathcal{D}(\Omega).$$

For each $\chi \in \mathcal{D}(\Omega)$, the mapping

$$\boldsymbol{\psi} \in \boldsymbol{\mathcal{C}}^2(\overline{\Omega}) \to \int_\Omega (\mathbf{Cof}\,\boldsymbol{\nabla}\boldsymbol{\psi})_{1j}\partial_j\chi\,\mathrm{d}x$$

is continuous if the space $\boldsymbol{\mathcal{C}}^2(\overline{\Omega})$ is equipped with the norm $\|\cdot\|_{1,p,\Omega}$, $p \geq 2$, since

$$\left|\int_\Omega (\mathbf{Cof}\,\boldsymbol{\nabla}\boldsymbol{\psi})_{1j}\partial_j\chi\,\mathrm{d}x\right| \leq \|\mathbf{Cof}\,\boldsymbol{\nabla}\boldsymbol{\psi}\|_{0,1,\Omega}\,|\chi|_{1,\infty,\Omega}\,.$$

From the density of $\mathcal{C}^2(\overline{\Omega})$ in $W^{1,p}(\Omega)$, we thus infer that

$$\int_\Omega (\mathbf{Cof}\,\boldsymbol{\nabla}\boldsymbol{\psi})_{1j}\partial_j\chi\,\mathrm{d}x = 0 \quad \text{for all } \boldsymbol{\psi} \in \boldsymbol{W}^{1,p}(\Omega),\ p \geq 2, \text{ and all } \chi \in \mathcal{D}(\Omega).$$

We now show that, given any function $\psi \in W^{1,p}(\Omega)$ and any function $\boldsymbol{w} = (w_j) \in \boldsymbol{L}^{p'}(\Omega)$, with $p^{-1} + (p')^{-1} = 1$, that satisfies $\int_\Omega w_j \partial_j\chi\,\mathrm{d}x = 0$ for all $\chi \in \mathcal{D}(\Omega)$, we have

$$-\int_\Omega \psi w_j \partial_j\theta\,\mathrm{d}x = \int_\Omega (\partial_j\psi)w_j\theta\,\mathrm{d}x \quad \text{for all } \theta \in \mathcal{D}(\Omega).$$

This being the case, the assertion will then follow by letting $\psi = \psi_1$ and $w_j = (\mathbf{Cof}\,\boldsymbol{\nabla}\boldsymbol{\psi})_{1j}$.

When $\boldsymbol{w} \in \boldsymbol{L}^{p'}(\Omega)$ and $\theta \in \mathcal{D}(\Omega)$ are held fixed, both sides of the above relation define continuous linear forms with respect to $\psi \in W^{1,p}(\Omega)$. Hence it suffices to consider the case where $\psi \in \mathcal{C}^\infty(\overline{\Omega})$ since $\{\mathcal{C}^\infty(\overline{\Omega})\}^- = W^{1,p}(\Omega)$. But then $\psi\theta \in \mathcal{D}(\Omega)$ and thus, by assumption,

$$0 = \int_\Omega w_j \partial_j(\psi\theta)\,\mathrm{d}x = \int_\Omega \psi w_j \partial_j\theta\,\mathrm{d}x + \int_\Omega (\partial_j\psi)w_j\theta\,\mathrm{d}x.$$

(iii) We next show that the weak convergences

$$\boldsymbol{\varphi}^\ell \rightharpoonup \boldsymbol{\varphi} \text{ in } \boldsymbol{W}^{1,p}(\Omega),\ p \leq 2, \quad \text{and} \quad \mathbf{Cof}\,\boldsymbol{\nabla}\boldsymbol{\varphi}^\ell \rightharpoonup \mathbf{Cof}\,\boldsymbol{\nabla}\boldsymbol{\varphi} \text{ in } \boldsymbol{L}^{p'}(\Omega),\ \frac{1}{p} + \frac{1}{p'} = 1,$$

together imply that, for any function $\theta \in \mathcal{D}(\Omega)$,

$$\int_\Omega (\det \boldsymbol{\nabla}\boldsymbol{\varphi}^\ell)\theta\,\mathrm{d}x \to \int_\Omega (\det \boldsymbol{\nabla}\boldsymbol{\varphi})\theta\,\mathrm{d}x.$$

By definition of $\det \boldsymbol{\nabla}\boldsymbol{\varphi}$ and by the result of part (ii), it suffices to show that

$$\int_\Omega \varphi_1^\ell(\mathbf{Cof}\,\boldsymbol{\nabla}\boldsymbol{\varphi}^\ell)_{1j}\partial_j\theta\,\mathrm{d}x \longrightarrow \int_\Omega \varphi_1(\mathbf{Cof}\,\boldsymbol{\nabla}\boldsymbol{\varphi})_{1j}\partial_j\theta\,\mathrm{d}x.$$

Arguing as in part (iii) of the proof of Theorem 9.7-1, we infer that this will be the case if

$$\varphi^\ell \rightharpoonup \varphi \text{ in } \boldsymbol{W}^{1,p}(\Omega) \quad \text{implies that} \quad \varphi^\ell \to \varphi \text{ in } \boldsymbol{L}^s(\Omega) \text{ with } \frac{1}{s} + \frac{1}{p'} \leq 1,$$

i.e., if the compact inclusion $W^{1,p}(\Omega) \Subset L^s(\Omega)$ holds. If $2 \leq p < 3$ (the only case that needs to be considered), this inclusion holds provided $s < p^* = 3p/(3-p)$; since

$$\frac{1}{p^*} + \frac{1}{p'} = \left(\frac{1}{p} - \frac{1}{3}\right) + \left(1 - \frac{1}{p}\right) = \frac{2}{3},$$

there do exist numbers $s < p^*$ such that $s^{-1} + (p')^{-1} \leq 1$.

(iv) The implications announced in the theorem are then proved in the same manner as in part (iv) of the proof of Theorem 9.7-1. □

Remarks (1) It follows from Theorem 9.7-2 that *the nonconvex set*

$$\left\{(\boldsymbol{\psi}, \boldsymbol{K}, \delta) \in \boldsymbol{W}^{1,p}(\Omega) \times \boldsymbol{L}^q(\Omega) \times L^r(\Omega);\ \boldsymbol{K} = \mathbf{Cof}\,\boldsymbol{\nabla}\boldsymbol{\psi},\ \delta = \det \boldsymbol{\nabla}\boldsymbol{\psi}\right\}, \quad p \geq 2,\ \frac{1}{p} + \frac{1}{q} \leq 1,\ r \geq 1,$$

is sequentially weakly closed in the space $\boldsymbol{W}^{1,p}(\Omega) \times \boldsymbol{L}^q(\Omega) \times L^r(\Omega)$. This does *not* mean that the set

$$\left\{\boldsymbol{\psi} \in \boldsymbol{W}^{1,p}(\Omega);\ \mathbf{Cof}\,\boldsymbol{\nabla}\boldsymbol{\psi} \in \boldsymbol{L}^q(\Omega),\ \det \boldsymbol{\nabla}\boldsymbol{\psi} \in L^r(\Omega)\right\}, \quad p \geq 2,\ \frac{1}{p} + \frac{1}{q} \leq 1,\ r \geq 1,$$

is sequentially weakly closed in the space $\boldsymbol{W}^{1,p}(\Omega)$, and indeed this is not always the case (Problem 9.7-4).

(2) The results of part (ii) of the above proof can be restated *in the sense of distributions*. First, the relation

$$\int_\Omega (\mathbf{Cof}\,\boldsymbol{\nabla}\boldsymbol{\psi})_{1j} \partial_j \chi \, dx = 0 \quad \text{for all } \chi \in \mathcal{D}(\Omega)$$

means that

$$\partial_j (\mathbf{Cof}\,\boldsymbol{\nabla}\boldsymbol{\psi})_{1j} = 0 \quad \text{in } \mathcal{D}'(\Omega).$$

Hence this relation, which holds for smooth vector fields $\boldsymbol{\psi}$ by *Piola's identity*, also holds in the sense of distributions for fields $\boldsymbol{\psi} \in \boldsymbol{W}^{1,p}(\Omega)$, $p \geq 2$. Likewise, the main result of part (ii) can be equivalently stated as follows:

$$\boldsymbol{\psi} \in \boldsymbol{W}^{1,p}(\Omega),\ p \geq 2, \quad \text{and} \quad \mathbf{Cof}\,\boldsymbol{\nabla}\boldsymbol{\psi} \in \boldsymbol{L}^{p'}(\Omega),\ \frac{1}{p} + \frac{1}{p'} = 1,$$

implies that

$$\partial_j \psi_1 (\mathbf{Cof}\,\boldsymbol{\nabla}\boldsymbol{\psi})_{1j} = \partial_j (\psi_1 (\mathbf{Cof}\,\boldsymbol{\nabla}\boldsymbol{\psi})_{1j}) \quad \text{in } \mathcal{D}'(\Omega).$$

This relation can be used for extending the definition of $\det \boldsymbol{\nabla}\boldsymbol{\psi}$ as a distribution, which is then not necessarily an integrable function (Problem 9.7-5). □

Theorems 9.7-1 and 9.7-2 can be put in a more general perspective: Let (φ^k) be a sequence such that

$$\varphi^k \rightharpoonup \varphi \quad \text{in } \boldsymbol{W}^{1,p}(\Omega),\ \ p \geq 2,$$

and assume in addition that the sequence $(\mathbf{Cof}\,\boldsymbol{\nabla}\boldsymbol{\varphi}^k)$ is *bounded* in the space $\boldsymbol{L}^q(\Omega)$, where $p^{-1}+q^{-1}=1$. Since $q>1$, the space $L^q(\Omega)$ is reflexive and so, by the Banach–Eberlein–Šmulian theorem (Theorem 5.14-4), there exists a subsequence $(\boldsymbol{\varphi}^\ell)$ such that $\mathbf{Cof}\,\boldsymbol{\nabla}\boldsymbol{\varphi}^\ell \rightharpoonup \boldsymbol{H}$ in $\boldsymbol{L}^q(\Omega)$. Besides, $\boldsymbol{H}=\mathbf{Cof}\,\boldsymbol{\nabla}\boldsymbol{\varphi}$ by Theorem 9.5-1, so that the limit $\boldsymbol{H}$ is unique. Therefore the whole sequence weakly converges, i.e.,

$$\mathbf{Cof}\,\boldsymbol{\nabla}\boldsymbol{\varphi}^k \rightharpoonup \mathbf{Cof}\,\boldsymbol{\nabla}\boldsymbol{\varphi} \quad \text{in } L^q(\Omega).$$

Part (iii) of the proof of Theorem 9.7-2 then implies that

$$\int_\Omega (\det \boldsymbol{\nabla}\boldsymbol{\varphi}^k)\theta\,dx \to \int_\Omega (\det \boldsymbol{\nabla}\boldsymbol{\varphi})\theta\,dx \quad \text{for all } \theta\in\mathcal{D}(\Omega),$$

or equivalently, in the sense of distributions,

$$\det \boldsymbol{\nabla}\boldsymbol{\varphi}^k \to \det \boldsymbol{\nabla}\boldsymbol{\varphi} \quad \text{in } \mathcal{D}'(\Omega).$$

In other words, *if some appropriate combinations of partial derivatives* (the components of the matrix $\mathbf{Cof}\,\boldsymbol{\nabla}\boldsymbol{\varphi}^k$) *remain bounded in* $L^q(\Omega)$, *a nonlinear function* (the function $\boldsymbol{\varphi}\in \boldsymbol{W}^{1,p}(\Omega)\to \det\boldsymbol{\nabla}\boldsymbol{\varphi} := \partial_j\varphi_1(\mathbf{Cof}\,\boldsymbol{\nabla}\boldsymbol{\varphi})_{1j}\in\mathcal{D}'(\Omega)$) *becomes continuous with respect to sequential weak convergence*, in the sense that

$$\boldsymbol{\varphi}^k \rightharpoonup \boldsymbol{\varphi} \text{ in } \boldsymbol{W}^{1,p}(\Omega) \quad \text{implies that} \quad \det \boldsymbol{\nabla}\boldsymbol{\varphi}^k \to \det \boldsymbol{\nabla}\boldsymbol{\varphi} \text{ in } \mathcal{D}'(\Omega).$$

This is a special case of the general phenomenon of **compensated compactness**, introduced by François Murat and Luc Tartar.[39] Their first result was the following div-**curl** *lemma*, which plays a crucial role in homogenization theory:

$^\flat$**Theorem 9.7-3 (Murat–Tartar** div-**curl lemma)** *Let Ω be a bounded open subset of $\mathbb{R}^n$, and let there be given two sequences $(\boldsymbol{u}^k)$ and $(\boldsymbol{v}^k)$ such that*

$$\begin{aligned} \boldsymbol{u}^k \rightharpoonup \boldsymbol{u} \text{ in } \boldsymbol{L}^2(\Omega) \quad &\textit{and} \quad \boldsymbol{v}^k \rightharpoonup \boldsymbol{v} \text{ in } \boldsymbol{L}^2(\Omega),\\ (\operatorname{div}\boldsymbol{u}^k) \textit{ is bounded in } L^2(\Omega) \quad &\textit{and} \quad (\mathbf{curl}\,\boldsymbol{v}^k) \textit{ is bounded in } \boldsymbol{L}^2(\Omega),\end{aligned}$$

where $\mathbf{curl}\,\boldsymbol{v} := (\partial_j v_i - \partial_i v_j)_{i<j}$. *Then*

$$\boldsymbol{u}^k\cdot\boldsymbol{v}^k \to \boldsymbol{u}\cdot\boldsymbol{v} \quad \textit{in } \mathcal{D}'(\Omega). \qquad \square$$

The essence of this result is that the Euclidean inner product $(\boldsymbol{u},\boldsymbol{v})\in \boldsymbol{L}^2(\Omega)\times\boldsymbol{L}^2(\Omega)\to \boldsymbol{u}\cdot\boldsymbol{v}\in\mathbb{R}$ remains continuous with respect to weak convergence even though neither sequence $(\boldsymbol{u}^k)$ nor $(\boldsymbol{v}^k)$ is assumed to be relatively compact in $\boldsymbol{L}^2(\Omega)$ (if one of the sequences

[39] F. MURAT [1978]: Compacité par compensation, *Annali Scuola Normale Superiore de Pisa, Serie* IV, **5**, 489–507.

L. TARTAR [1979]: Compensated compactness and partial differential equations, in *Nonlinear Analysis and Mechanics, Heriot-Watt Symposium, Volume* IV (R. J. KNOPS, editor), pp. 136–212, Pitman, Boston.

L. TARTAR [1983]: The compensated compactness method applied to systems of conservation laws, in *Systems of Nonlinear Partial Differential Equations* (J.M. BALL, editor), pp. 263–285, Reidel, Dordrecht.

F. MURAT [1987]: A survey on compensated compactness, in *Contributions to Modern Calculus of Variations* (L. CESARI, editor), pp. 145–183, Longman, Harlow.

A direct proof of Theorem 9.7-3 is proposed as a problem in KAVIAN [1993, Chapter 1, Exercise 34].

were bounded in $\boldsymbol{H}^1(\Omega)$, the conclusion would follow from the Rellich–Kondrachov theorem combined with Theorem 5.12-4(c)). *The lack of compactness is thus compensated by the boundedness in* $L^2(\Omega)$ *of specific linear combinations of partial derivatives* (here $\sum_i \partial_i u_i^k$ and $\partial_j v_i^k - \partial_i v_j^k$), *themselves adapted to the mapping under consideration* (here the mapping $(\boldsymbol{u}, \boldsymbol{v}) \in \boldsymbol{L}^2(\Omega) \times \boldsymbol{L}^2(\Omega) \to \boldsymbol{u} \cdot \boldsymbol{v} \in \mathbb{R}$).

All prerequisite ground has now been laid for establishing the existence of minimizers in hyperelasticity. Notice that, while the statement and proof of this existence result are both reminiscent of the statement and proof of Theorem 9.5-2, the proof is exceedingly more delicate in the present case.

Theorem 9.7-4 (Ball's theorem[40]) *Let Ω be a domain in $\mathbb{R}^3$, and let $W : \Omega \times \mathbb{M}^3_+ \to \mathbb{R}$ be a function with the following properties:*

(a) *Polyconvexity: For almost all $x \in \Omega$, there exists a convex function $\mathbb{W}(x, \cdot) : \mathbb{M}^3 \times \mathbb{M}^3 \times]0, \infty[\to \mathbb{R}$ such that*

$$W(x, \boldsymbol{F}) = \mathbb{W}(x, \boldsymbol{F}, \mathbf{Cof}\,\boldsymbol{F}, \det \boldsymbol{F}) \quad \text{for all } \boldsymbol{F} \in \mathbb{M}^3_+.$$

(b) *Measurability: The function $\mathbb{W}(\cdot, \boldsymbol{F}, \boldsymbol{H}, \delta) : \Omega \to \mathbb{R}$ is measurable for all $(\boldsymbol{F}, \boldsymbol{H}, \delta) \in \mathbb{M}^3 \times \mathbb{M}^3 \times]0, \infty[$.*

(c) *Coerciveness: There exist constants α, p, q, r, and β, such that*

$$\alpha > 0, \quad p \geq 2, \quad q \geq \frac{p}{p-1}, \quad r > 1,$$

$$W(x, \boldsymbol{F}) \geq \alpha \{|\boldsymbol{F}|^p + |\mathbf{Cof}\,\boldsymbol{F}|^q + (\det \boldsymbol{F})^r\} + \beta \text{ for all } \boldsymbol{F} \in \mathbb{M}^3_+ \text{ and almost all } x \in \Omega.$$

(d) *Behavior as $\det \boldsymbol{F} \to 0^+$:*

$$W(x, \boldsymbol{F}) \to \infty \quad \text{as } \det \boldsymbol{F} \to 0^+ \text{ for almost all } x \in \Omega.$$

Let Γ_0 be a $\mathrm{d}\Gamma$-measurable subset of the boundary Γ of Ω with area $\Gamma_0 > 0$, and let $\boldsymbol{\varphi}_0 : \Gamma_0 \to \mathbb{R}^3$ be a $\mathrm{d}\Gamma$-measurable function such that the set

$$\begin{aligned}\boldsymbol{\Phi} := \{\boldsymbol{\psi} \in \boldsymbol{W}^{1,p}(\Omega);\ &\mathbf{Cof}\,\boldsymbol{\nabla}\boldsymbol{\psi} \in \boldsymbol{L}^q(\Omega),\ \det \boldsymbol{\nabla}\boldsymbol{\psi} \in L^r(\Omega),\\ &\boldsymbol{\psi} = \boldsymbol{\varphi}_0 \quad \mathrm{d}\Gamma\text{-a.e. on } \Gamma_0,\ \det \boldsymbol{\nabla}\boldsymbol{\psi} > 0 \text{ a.e. in } \Omega\}\end{aligned}$$

is nonempty. Finally, let L be a continuous linear form over the space $\boldsymbol{W}^{1,p}(\Omega)$, let

$$I(\boldsymbol{\psi}) := \int_\Omega W(x, \boldsymbol{\nabla}\boldsymbol{\psi}(x))\,\mathrm{d}x - L(\boldsymbol{\psi}) \quad \text{for each } \boldsymbol{\psi} \in \boldsymbol{\Phi},$$

and assume that $\inf_{\boldsymbol{\psi} \in \boldsymbol{\Phi}} I(\boldsymbol{\psi}) < \infty$.

Then there exists at least one function $\boldsymbol{\varphi}$ such that

$$\boldsymbol{\varphi} \in \boldsymbol{\Phi} \quad \text{and} \quad I(\boldsymbol{\varphi}) = \inf_{\boldsymbol{\psi} \in \boldsymbol{\Phi}} I(\boldsymbol{\psi}).$$

[40] See Ball [1977, Theorems 7.3 and 7.6] (*op. cit.*).

Proof (i) *The integrals $\int_\Omega W(x, \boldsymbol{\nabla\psi}(x))\,\mathrm{d}x$ are well defined for all $\boldsymbol{\psi} \in \boldsymbol{\Phi}$.* To see this, we first note the following consequences of assumptions (a) and (b): For almost all $x \in \Omega$, the function $\mathbb{W}(x, \cdot) : \mathbb{M}^3 \times \mathbb{M}^3 \times]0, \infty[\to \mathbb{R}$ is continuous (as a convex and real-valued function on a convex open subset of a finite-dimensional space; cf. Theorem 2.17-1); for all $(\boldsymbol{F}, \boldsymbol{H}, \delta) \in \mathbb{M}^3 \times \mathbb{M}^3 \times]0, \infty[$, the function $\mathbb{W}(\cdot, \boldsymbol{F}, \boldsymbol{H}, \delta) : \Omega \to \mathbb{R}$ is measurable, and $\mathbb{M}^3 \times \mathbb{M}^3 \times]0, \infty[$ is a Borel set. Therefore the function $\mathbb{W} : \Omega \times \mathbb{M}^3 \times \mathbb{M}^3 \times]0, \infty[\to \mathbb{R}$ is a *Carathéodory function* (Section 9.5), and consequently the function

$$x \in \Omega \to \mathbb{W}(x, \boldsymbol{\nabla\psi}(x),\ \mathbf{Cof}\,\boldsymbol{\nabla\psi}(x),\ \det \boldsymbol{\nabla\psi}(x)) \in \mathbb{R}$$

is measurable for each $\boldsymbol{\psi} \in \boldsymbol{\Phi}$, since then $\det \boldsymbol{\nabla\psi} > 0$ almost everywhere in Ω. The function W being in addition bounded below (by the coerciveness inequality), the integral

$$\int_\Omega W(x, \boldsymbol{\nabla\psi}(x))\,\mathrm{d}x = \int_\Omega \mathbb{W}(x, \boldsymbol{\nabla\psi}(x), \mathbf{Cof}\,\boldsymbol{\nabla\psi}(x), \det \boldsymbol{\nabla\psi}(x))\,\mathrm{d}x$$

is thus a well-defined extended real number in the interval $[\beta \operatorname{vol}\Omega, \infty]$ for each $\boldsymbol{\psi} \in \boldsymbol{\Phi}$.

(ii) *We next find a lower bound for $I(\boldsymbol{\psi})$ when $\boldsymbol{\psi} \in \boldsymbol{\Phi}$.*

First, we infer from the assumed coerciveness (c) of the function W and the assumed continuity of the linear form L that

$$I(\boldsymbol{\psi}) \geq \alpha \int_\Omega \{ |\boldsymbol{\nabla\psi}|^p + |\mathbf{Cof}\,\boldsymbol{\nabla\psi}|^q + (\det \boldsymbol{\nabla\psi})^r \}\,\mathrm{d}x + \beta \operatorname{vol}\Omega - \|\boldsymbol{L}\|\,\|\boldsymbol{\psi}\|_{1,p,\Omega} \quad \text{for all } \boldsymbol{\psi} \in \boldsymbol{\Phi}.$$

Combining the boundary condition $\boldsymbol{\psi} = \boldsymbol{\varphi}_0$ on Γ_0 with the generalized Poincaré inequality (as in the proof of Theorem 9.5-2), we thus conclude that there exist constants c and d such that

$$c > 0 \quad \text{and} \quad I(\boldsymbol{\psi}) \geq c\{ \|\boldsymbol{\psi}\|^p_{1,p,\Omega} + \|\mathbf{Cof}\,\boldsymbol{\nabla\psi}\|^q_{0,q,\Omega} + \|\det \boldsymbol{\nabla\psi}\|^r_{0,r,\Omega} \} + d \text{ for all } \boldsymbol{\psi} \in \boldsymbol{\Phi}.$$

(iii) *Let $(\boldsymbol{\varphi}^k)$ be an infimizing sequence for the functional I, i.e., a sequence that satisfies*

$$\boldsymbol{\varphi}^k \in \boldsymbol{\Phi} \text{ for all } k, \quad \text{and} \quad \lim_{k\to\infty} I(\boldsymbol{\varphi}^k) = \inf_{\boldsymbol{\psi}\in\boldsymbol{\Phi}} I(\boldsymbol{\psi}).$$

By assumption, $\inf_{\boldsymbol{\psi}\in\boldsymbol{\Phi}} I(\boldsymbol{\psi}) < \infty$, and thus, by part (ii), the sequence $(\boldsymbol{\varphi}^k, \mathbf{Cof}\,\boldsymbol{\nabla\varphi}^k, \det \boldsymbol{\nabla\varphi}^k)$ is bounded in the reflexive Banach space $\boldsymbol{W}^{1,p}(\Omega) \times \boldsymbol{L}^q(\Omega) \times L^r(\Omega)$ (each number p, q, r is > 1). Hence, by the *Banach–Eberlein–Šmulian theorem* (Theorem 5.14-4), there exists a subsequence $(\boldsymbol{\varphi}^\ell, \mathbf{Cof}\,\boldsymbol{\nabla\varphi}^\ell, \det \boldsymbol{\nabla\varphi}^\ell)$ that converges weakly to an element $(\boldsymbol{\varphi}, \boldsymbol{H}, \delta)$ in the space $\boldsymbol{W}^{1,p}(\Omega) \times \boldsymbol{L}^q(\Omega) \times L^r(\Omega)$; thus, by Theorem 9.7-2,

$$\boldsymbol{H} = \mathbf{Cof}\,\boldsymbol{\nabla\varphi} \quad \text{and} \quad \delta = \det \boldsymbol{\nabla\varphi}.$$

To sum up, *there exists a subsequence of the infimizing sequence that satisfies*

$$\begin{aligned} \boldsymbol{\varphi}^\ell &\rightharpoonup \boldsymbol{\varphi} && \text{in } \boldsymbol{W}^{1,p}(\Omega), \\ \mathbf{Cof}\,\boldsymbol{\nabla\varphi}^\ell &\rightharpoonup \mathbf{Cof}\,\boldsymbol{\nabla\varphi} && \text{in } \boldsymbol{L}^q(\Omega), \\ \det \boldsymbol{\nabla\varphi}^\ell &\rightharpoonup \det \boldsymbol{\nabla\varphi} && \text{in } L^r(\Omega). \end{aligned}$$

(iv) *In order to show that* $\boldsymbol{\varphi} \in \boldsymbol{\Phi}$, *it thus remains to establish that* $\det \boldsymbol{\nabla}\boldsymbol{\varphi} > 0$ *almost everywhere in* Ω *and that* $\boldsymbol{\varphi} = \boldsymbol{\varphi}_0$ *on* Γ_0.

Since $\det \boldsymbol{\nabla}\boldsymbol{\varphi}^\ell \rightharpoonup \det \boldsymbol{\nabla}\boldsymbol{\varphi}$ in $L^r(\Omega)$, the *Banach–Saks–Mazur theorem* (Theorem 5.13-1(c)) shows that there exist for each ℓ integers $i(\ell) \geq \ell$ and numbers λ_s^ℓ, $\ell \leq s \leq i(\ell)$, such that

$$\lambda_s^\ell \geq 0, \quad \sum_{s=\ell}^{i(\ell)} \lambda_s^\ell = 1, \quad \mathrm{d}^\ell := \sum_{s=\ell}^{i(\ell)} \lambda_s^\ell \det \boldsymbol{\nabla}\boldsymbol{\varphi}^s \underset{\ell\to\infty}{\longrightarrow} \det \boldsymbol{\nabla}\boldsymbol{\varphi} \quad \text{in } L^r(\Omega).$$

Hence there exists a subsequence (d^m) of (d^ℓ) that converges almost everywhere to $\det \boldsymbol{\nabla}\boldsymbol{\varphi}$. Since the functions d^ℓ are > 0 almost everywhere (≥ 0 would suffice here), we conclude that $\det \boldsymbol{\nabla}\boldsymbol{\varphi} \geq 0$ almost everywhere in Ω.

Assume that $\det \boldsymbol{\nabla}\boldsymbol{\varphi} = 0$ on a subset A of Ω with $\mathrm{d}x$-meas $A > 0$. Since $\det \boldsymbol{\nabla}\boldsymbol{\varphi}^\ell > 0$ almost everywhere on A (again the inequality ≥ 0 would suffice here) and $\det \boldsymbol{\nabla}\boldsymbol{\varphi}^\ell \rightharpoonup \det \boldsymbol{\nabla}\boldsymbol{\varphi}$,

$$\int_A |\det \boldsymbol{\nabla}\boldsymbol{\varphi}^\ell| \,\mathrm{d}x = \int_A \det \boldsymbol{\nabla}\boldsymbol{\varphi}^\ell \,\mathrm{d}x \to \int_A \det \boldsymbol{\nabla}\boldsymbol{\varphi} \,\mathrm{d}x = 0,$$

by definition of weak convergence (the characteristic function of the set A belongs to the dual space of $L^r(\Omega)$), which shows that $\det \boldsymbol{\nabla}\boldsymbol{\varphi}^\ell \to 0$ in $L^1(A)$. Therefore there exists a subsequence $(\boldsymbol{\varphi}^m)$ of $(\boldsymbol{\varphi}^\ell)$ such that

$$\det \boldsymbol{\nabla}\boldsymbol{\varphi}^m(x) \to 0 \quad \text{for almost all } x \in A.$$

Consider next the sequence of measurable functions (f_m) defined by

$$f^m : x \in A \to f^m(x) := W(x, \boldsymbol{\nabla}\boldsymbol{\varphi}^m(x)).$$

Since $f^m \geq \beta$ for all m, we can apply *Fatou's lemma* (Theorem 1.15-2), which shows that

$$\int_A \liminf_{m\to\infty} f^m(x)\,\mathrm{d}x \leq \liminf_{m\to\infty} \int_A f^m(x)\,\mathrm{d}x$$

on the one hand. On the other hand, the *behavior of the function* W *as* $\det \boldsymbol{F} \to 0^+$ (assumption (d)) implies that

$$\liminf_{m\to\infty} f^m(x) = \lim_{m\to\infty} W(x, \boldsymbol{\nabla}\boldsymbol{\varphi}^m(x)) = \lim_{\det \boldsymbol{F}\to 0^+} W(x, \boldsymbol{F}) = \infty \quad \text{for almost all } x \in A,$$

and thus

$$\lim_{m\to\infty} \int_A f^m(x)\,\mathrm{d}x = \lim_{m\to\infty} \int_A W(x, \boldsymbol{\nabla}\boldsymbol{\varphi}^m(x))\,\mathrm{d}x = \infty.$$

But this last relation contradicts the relation $\lim_{m\to\infty} I(\boldsymbol{\varphi}^m) = \inf_{\boldsymbol{\psi}\in\boldsymbol{\Phi}} I(\boldsymbol{\psi}) < \infty$ and the inequalities

$$I(\boldsymbol{\varphi}^m) \geq \int_A W(x, \boldsymbol{\nabla}\boldsymbol{\varphi}^m(x))\,\mathrm{d}x + \beta\,\mathrm{vol}(\Omega - A) - \|\boldsymbol{L}\|\,\|\boldsymbol{\varphi}^m\|_{1,p,\Omega}$$

(a weakly convergent sequence is bounded; cf. Theorem 5.12-2(b)). Hence $\det \boldsymbol{\nabla}\boldsymbol{\varphi} > 0$ *almost everywhere in* Ω.

That $\varphi = \varphi_0$ on Γ_0 is established as in the proof of Theorem 9.5-2.

(v) *Finally, we show that*

$$\int_\Omega W(x, \nabla\varphi(x))\,dx \le \liminf_{\ell\to\infty} \int_\Omega W(x, \nabla\varphi^\ell(x))\,dx.$$

By definition of the limit inferior, we must show that, given any subsequence (φ^m) of (φ^ℓ) such that the sequence $(\int_\Omega W(x, \nabla\varphi^m(x))\,dx)$ converges, then

$$\int_\Omega W(x, \nabla\varphi(x))\,dx \le \lim_{m\to\infty} \int_\Omega W(x, \nabla\varphi^m(x))\,dx.$$

So, let us consider such a subsequence. Using the result of part (iii) and the *Banach–Saks–Mazur theorem* again, we infer that for each m, there exist integers $j(m) \ge m$ and numbers μ_t^m, $m \le t \le j(m)$, such that

$$\mu_t^m \ge 0, \quad \sum_{t=m}^{j(m)} \mu_t^m = 1,$$
$$\boldsymbol{D}^m := \sum_{t=m}^{j(m)} \mu_t^m (\nabla\varphi^t, \mathbf{Cof}\,\nabla\varphi^t, \det\nabla\varphi^t) \underset{m\to\infty}{\longrightarrow} (\nabla\varphi, \mathbf{Cof}\,\nabla\varphi, \det\nabla\varphi)$$

in $\boldsymbol{L}^p(\Omega) \times \boldsymbol{L}^q(\Omega) \times L^r(\Omega)$. Hence there exists a subsequence $(\boldsymbol{D}^n)$ of $(\boldsymbol{D}^m)$ such that, for almost all $x \in \Omega$,

$$\sum_{t=n}^{j(n)} \mu_t^n \left(\nabla\varphi^t(x), \mathbf{Cof}\,\nabla\varphi^t(x), \det\nabla\varphi^t(x)\right) \underset{n\to\infty}{\longrightarrow} (\nabla\varphi(x), \mathbf{Cof}\,\nabla\varphi(x), \det\nabla\varphi(x)).$$

Since the function $\mathbb{W}(x, \cdot)$ is continuous on the set $\mathbb{M}^3 \times \mathbb{M}^3 \times]0, \infty[$ for almost all $x \in \Omega$ (see part (i)), and since $\det\nabla\varphi(x) > 0$ for almost all $x \in \Omega$ by part (iv), it follows that

$$\begin{aligned} W(x, \nabla\varphi(x)) &= \mathbb{W}(x, (\nabla\varphi(x), \mathbf{Cof}\,\nabla\varphi(x), \det\nabla\varphi(x))) \\ &= \lim_{n\to\infty} \mathbb{W}\left(x, \sum_{t=n}^{j(n)} \mu_t^n \left(\nabla\varphi^t(x), \mathbf{Cof}\,\nabla\varphi^t(x), \det\nabla\varphi^t(x)\right)\right) \end{aligned}$$

for almost all $x \in \Omega$. Using this relation, *Fatou's lemma*, and the assumed convexity of the function $\mathbb{W}(x, \cdot)$ for almost all $x \in \Omega$, we next obtain, on the one hand,

$$\begin{aligned} \int_\Omega W(x, \nabla\varphi(x))\,dx &\le \liminf_{n\to\infty} \int_\Omega \mathbb{W}\left(x, \sum_{t=n}^{j(n)} \mu_t^n \left(\nabla\varphi^t(x), \mathbf{Cof}\,\nabla\varphi^t(x), \det\nabla\varphi^t(x)\right)\right) dx \\ &\le \liminf_{n\to\infty} \sum_{t=n}^{j(n)} \mu_t^n \int_\Omega W\left(x, \nabla\varphi^t(x)\right) dx = \lim_{n\to\infty} \int_\Omega W\left(x, \nabla\varphi^n(x)\right) dx \\ &= \lim_{m\to\infty} \int_\Omega W\left(x, \nabla\varphi^m(x)\right) dx. \end{aligned}$$

Note that we have also used here a simple observation: Let (α^n) be a convergent sequence of real numbers, and let

$$\beta^n := \sum_{t=n}^{j(n)} \mu_t^n \alpha^t \text{ with } \mu_t^n \geq 0 \text{ and } \sum_{t=n}^{j(n)} \mu_t^n = 1 \text{ for each } n.$$

Then the sequence (β^n) is also convergent, and $\lim_{n\to\infty} \beta^n = \lim_{n\to\infty} \alpha^n$.

Since, on the other hand, $L(\boldsymbol{\varphi}) = \lim_{\ell\to\infty} L(\boldsymbol{\varphi}^\ell)$ by definition of weak convergence, we have therefore proved that

$$I(\boldsymbol{\varphi}) \leq \liminf_{\ell\to\infty} I(\boldsymbol{\varphi}^\ell).$$

(vi) *The function $\boldsymbol{\varphi}$ is thus a solution of the minimization problem*, since $\boldsymbol{\varphi} \in \boldsymbol{\Phi}$ by parts (iii) and (iv), and since

$$I(\boldsymbol{\varphi}) \leq \liminf_{\ell\to\infty} I(\boldsymbol{\varphi}^\ell) = \inf_{\boldsymbol{\psi}\in\boldsymbol{\Phi}} I(\boldsymbol{\psi}) \quad \text{implies} \quad I(\boldsymbol{\varphi}) = \inf_{\boldsymbol{\psi}\in\boldsymbol{\Phi}} I(\boldsymbol{\psi}). \qquad \square$$

Problems

9.7-1 Recall that co A designates the *convex hull* of a set A (Section 2.16). Show that

$$\text{co}\{(\boldsymbol{F}, \mathbf{Cof}\,\boldsymbol{F}, \det \boldsymbol{F}) \in \mathbb{M}^3 \times \mathbb{M}^3 \times \mathbb{R};\ \boldsymbol{F} \in \mathbb{M}^3_+\} = \mathbb{M}^3 \times \mathbb{M}^3 \times\,]0, \infty[.$$

9.7-2 This problem is a complement to Theorem 9.7-1.

(1) Show that the set

$$\{(\boldsymbol{\psi}, \boldsymbol{K}) \in \boldsymbol{W}^{1,p}(\Omega) \times \boldsymbol{L}^q(\Omega);\ \boldsymbol{K} = \mathbf{Cof}\,\boldsymbol{\nabla\psi}\}, \quad p \geq 2,\ q \geq 1,$$

which is sequentially weakly closed by Theorem 9.7-1, is a nonconvex subset of the space $\boldsymbol{W}^{1,p}(\Omega) \times \boldsymbol{L}^q(\Omega)$.

(2) Let X and Y be normed vector spaces. A (possibly nonlinear) mapping $f : X \to Y$ is said to be *sequentially weakly continuous* if $x^k \rightharpoonup x$ in X implies that $f(x^k) \rightharpoonup f(x)$ in Y. Show that the mapping $\boldsymbol{\psi} \in \boldsymbol{W}^{1,p}(\Omega) \to \mathbf{Cof}\,\boldsymbol{\nabla\psi} \in \boldsymbol{L}^{p/2}(\Omega)$ is sequentially weakly continuous if $p > 2$.

(3) For which values of p and q is the set $\{\boldsymbol{\psi} \in \boldsymbol{W}^{1,p}(\Omega);\ \mathbf{Cof}\,\boldsymbol{\nabla\psi} \in \boldsymbol{L}^q(\Omega)\}$ sequentially weakly closed in the space $\boldsymbol{W}^{1,p}(\Omega)$?

9.7-3 This problem is a complement to Theorem 9.7-1.

(1) Show that the expression

$$(\mathbf{Cof}^\sharp\, \boldsymbol{\nabla\psi})_{ij} := \partial_{i+2}(\psi_{j+2}\partial_{i+1}\psi_{j+1}) - \partial_{i+1}(\psi_{j+2}\partial_{i+2}\psi_{j+1})$$

defines a distribution when $\boldsymbol{\psi} \in \boldsymbol{W}^{1,p}(\Omega)$ for some $p \geq 3/2$. Note that $\mathbf{Cof}^\sharp\,\boldsymbol{\nabla\psi} = \mathbf{Cof}\,\boldsymbol{\nabla\psi}$ if $p \geq 2$.

(2) Show that

$$\boldsymbol{\varphi}^\ell \rightharpoonup \boldsymbol{\varphi} \text{ in } \boldsymbol{W}^{1,p}(\Omega),\ p > \frac{3}{2}, \quad \text{implies} \quad \langle(\mathbf{Cof}^\sharp\,\boldsymbol{\nabla\varphi}^\ell)_{ij}, \theta\rangle \to \langle(\mathbf{Cof}^\sharp\,\boldsymbol{\nabla\varphi})_{ij}, \theta\rangle$$

for all $\theta \in \mathcal{D}(\Omega)$; observe that the inequality $p \geq 3/2$ of (1) has to be replaced by the corresponding strict inequality in (2).

9.7-4 This problem is a complement to Theorem 9.7-2.

(1) Show that the set

$$\{(\boldsymbol{\psi}, \boldsymbol{K}, \delta) \in \boldsymbol{W}^{1,p}(\Omega) \times \boldsymbol{L}^q(\Omega) \times L^r(\Omega);\ \boldsymbol{K} = \mathbf{Cof}\,\boldsymbol{\nabla\psi}, \delta = \det \boldsymbol{\nabla\psi}\}, \quad p \geq 2,\ \frac{1}{p} + \frac{1}{q} \leq 1,\ r \geq 1,$$

which is sequentially weakly closed by Theorem 9.7-2, is a nonconvex subset of the space $\boldsymbol{W}^{1,p}(\Omega) \times \boldsymbol{L}^q(\Omega) \times L^r(\Omega)$.

(2) Show that the mapping $\boldsymbol{\psi} \in \boldsymbol{W}^{1,p}(\Omega) \to \det \boldsymbol{\nabla\psi} \in L^{p/3}(\Omega)$ is sequentially weakly continuous if $p > 3$ (according to the definition given in Problem 9.7-2).

(3) For which values of p, q, r is the set $\{\boldsymbol{\psi} \in \boldsymbol{W}^{1,p}(\Omega);\ \mathbf{Cof}\,\boldsymbol{\nabla\psi} \in \boldsymbol{L}^q(\Omega),\ \det \boldsymbol{\nabla\psi} \in L^r(\Omega)\}$ weakly closed in the space $\boldsymbol{W}^{1,p}(\Omega)$?

(4) Show that the set

$$\{\boldsymbol{\psi} \in \boldsymbol{W}^{1,p}(\Omega);\ \mathbf{Cof}\,\boldsymbol{\nabla\psi} \in \boldsymbol{L}^q(\Omega), \det \boldsymbol{\nabla\psi} \in L^r(\Omega), \det \boldsymbol{\nabla\psi} > 0 \text{ a.e. in } \Omega\}$$

is not convex if $p \geq 2,\ \frac{1}{p} + \frac{1}{q} \leq 1,\ r \geq 1$.

9.7-5 This problem is a complement to Theorem 9.7-2.

(1) Show that the expression

$$\det{}^\sharp\, \boldsymbol{\nabla\psi} := \partial_j(\psi_1(\mathbf{Cof}^\sharp\, \boldsymbol{\nabla\psi})_{ij})$$

defines a distribution when $\boldsymbol{\psi} \in \boldsymbol{W}^{1,p}(\Omega)$ and $\mathbf{Cof}^\sharp\, \boldsymbol{\nabla\psi} \in \boldsymbol{L}^q(\Omega)$, with $p \geq 3/2$ and $p^{-1} + q^{-1} \leq 4/3$, where the distribution $\mathbf{Cof}^\sharp\, \boldsymbol{\nabla\psi}$ is defined as in Problem 9.7-3. Note that $\det{}^\sharp\, \boldsymbol{\nabla\psi} = \det \boldsymbol{\nabla\psi}$ if $p \geq 2$ and $p^{-1} + q^{-1} \leq 1$.

(2) Show that the weak convergences

$$\boldsymbol{\varphi}^\ell \rightharpoonup \boldsymbol{\varphi} \text{ in } \boldsymbol{W}^{1,p}(\Omega),\ p \geq \frac{3}{2}, \quad \text{and} \quad \mathbf{Cof}^\sharp\, \boldsymbol{\nabla\varphi}^\ell \rightharpoonup \mathbf{Cof}^\sharp\, \boldsymbol{\nabla\varphi} \text{ in } \boldsymbol{L}^q(\Omega),\ \frac{1}{p} + \frac{1}{q} < \frac{4}{3},$$

together imply that

$$\langle \det{}^\sharp\, \boldsymbol{\nabla\varphi}^\ell, \theta\rangle \to \langle \det{}^\sharp\, \boldsymbol{\nabla\varphi}, \theta\rangle \quad \text{for all } \theta \in \mathcal{D}(\Omega)$$

(observe that the inequality $p^{-1} + q^{-1} \leq 4/3$ of (1) has to be replaced by the corresponding strict inequality in (2)).

9.8 Ekeland's variational principle; existence of minimizers for functionals that satisfy the Palais–Smale condition

Until now, the existence of minimizers of a functional $J : V \to \mathbb{R} \cup \{\infty\}$ has been established under three basic assumptions, whether in the general case (Theorem 9.3-1) or in specific situations (Theorem 9.5-2): First, the Banach space V is *reflexive*; second, the functional J is *coercive*; third, the functional J is *sequentially weakly lower semicontinuous.* Recall that this last property holds if (as is often the case in practice) J is assumed to be *strongly lower semicontinuous* and *convex* (Theorem 9.2-3).

The objective of this section is to show that the existence of minimizers can still be established when some of these assumptions no longer hold, provided the functional J satisfies instead another set of three basic assumptions: first, J is *of class* $\mathcal{C}^1$ *over* V, hence continuous and *a fortiori* lower semicontinuous over V; second, J is *bounded below on* V, a property that by contrast was a *consequence* of the assumptions of Theorem 9.3-1; third, and most importantly, J satisfies the *Palais–Smale condition*, a key assumption about minimizing sequences of J (the statement of this condition is given in Theorem 9.8-3).

Remark Interestingly, it can be shown that the conjunction of these three assumptions implies that the functional J is necessarily coercive; cf. Problem 9.8-1. □

In order to establish the existence of minimizers under these new assumptions, we first need to establish an important *per se* property of any functional J that is *lower semicontinuous* and *bounded below* over a *closed* subset U of a *Banach space*. This property asserts that, given any $\varepsilon > 0$, one can find an element $u_\varepsilon \in U$ that satisfies

$$J(u_\varepsilon) = \inf_{v\in U} J_\varepsilon(v) \leq \inf_{v\in U} J(v) + \varepsilon, \quad \text{where } J_\varepsilon(v) := J(v) + \varepsilon \|v - u_\varepsilon\|, \quad v \in U;$$

besides, u_ε is the unique solution to this perturbed minimization problem.

In what follows, notions such as lower semicontinuity, closed subsets, convergent sequences, etc., are understood with respect to the *norm topology*.

Theorem 9.8-1 (Ekeland's variational principle for lower semicontinuous functionals[41]) *Let $(V, \|\cdot\|)$ be a Banach space, let U be a nonempty closed subset of V, and let $J : U \to \mathbb{R}$ be a lower semicontinuous functional with the property that*

$$\gamma := \inf_{v\in U} J(v) > -\infty.$$

Then, given any $\varepsilon > 0$, there exists $u_\varepsilon \in U$ such that

$$\gamma \leq J(u_\varepsilon) \leq \gamma + \varepsilon,$$
$$J(u_\varepsilon) < J(v) + \varepsilon \|v - u_\varepsilon\| \quad \textit{for all } v \in U, \quad v \neq u_\varepsilon.$$

Proof Throughout the proof, $\varepsilon > 0$ is given and kept fixed. First, we note that the *epigraph of J*, viz.,

$$\operatorname{epi} J = \{(v, \alpha) \in U \times \mathbb{R},\ J(v) \leq \alpha\}$$

is closed (Theorem 9.2-2), so that, as a subset of $V \times \mathbb{R}$, $\operatorname{epi} J$ *is a complete metric space*. The idea of the proof then consists in constructing, by means of an iterative procedure, a decreasing sequence of closed subsets A_n, $n \geq 1$, of $\operatorname{epi} J$, the intersection of which will be $(u_\varepsilon, J(u_\varepsilon))$.

(i) *The iterative procedure.*

By definition of γ, there exists v_1 such that

$$v_1 \in U \quad \text{and} \quad \gamma \leq J(v_1) \leq \gamma + \varepsilon.$$

Then define the set

$$A_1 := \{(v, \alpha) \in \operatorname{epi} J;\ \alpha \leq J(v_1) - \varepsilon \|v - v_1\|\},$$

[41] I. EKELAND [1974]: On the variational principle, *Journal of Mathematical Analysis and Applications* **47**, 324–353.

I. EKELAND [1979]: Nonconvex minimization problems, *Bulletin of the American Mathematical Society* **1**, 443–473.

which is thus a closed subset of epi J containing $(v_1, J(v_1))$.

Assume first that $A_1 = \{(v_1, J(v_1))\}$, which means that if $(v, \alpha) \in \text{epi}\, J$ but $(v, \alpha) \neq (v_1, J(v_1))$, then $(v, \alpha) \notin A_1$. Since in particular $(v, J(v)) \in \text{epi}\, J$ for all $v \in U$, it thus follows in this case that

$$v \in U \text{ and } v \neq v_1 \quad \text{implies } J(v_1) < J(v) + \varepsilon \|v - v_1\| .$$

Hence it suffices to let $u_\varepsilon = v_1$, and the iterative procedure stops here.

Assume next that $A_1 \supsetneqq \{(v_1, J(v_1))\}$, or equivalently, that

$$U_1 := \{v \in U;\ \text{there exists } \alpha \in \mathbb{R} \text{ such that } (v, \alpha) \in A_1\} \supsetneqq \{v_1\}.$$

Since then $J(v) < J(v_1)$ for all $v \in U_1$, $v \neq v_1$, it follows that

$$\gamma_1 := \inf_{v \in U_1} J(v) < J(v_1).$$

Therefore there exists v_2 such that

$$v_2 \in U_1, \quad (v_2, J(v_2)) \in A_1, \quad \text{and} \quad 0 \leq J(v_2) - \gamma_1 \leq \frac{1}{2}(J(v_1) - \gamma_1).$$

Then define the set

$$A_2 := \{(v, \alpha) \in \text{epi}\, J;\ \alpha \leq J(v_2) - \varepsilon \|v - v_2\|\}$$

which is thus a closed subset of epi J containing $(v_2, J(v_2))$. Besides, given any $(v, \alpha) \in A_2$, the inequality $\alpha + \varepsilon \|v - v_2\| \leq J(v_2)$, combined with the inequality $J(v_2) + \varepsilon \|v_2 - v_1\| \leq J(v_1)$ (which expresses that $(v_2, J(v_2)) \in A_1$) and the triangle inequality, implies that $(v, \alpha) \in A_1$. Hence

$$A_2 \subset A_1.$$

If $A_2 = \{(v_2, J(v_2))\}$, it suffices to let $u_\varepsilon = v_2$, and the iterative procedure stops here. Otherwise, the procedure continues as above, providing points $v_n \in U$, $n \geq 1$, and sets

$$A_n := \{(v, \alpha) \in \text{epi}\, J;\ \alpha \leq J(v_n) - \varepsilon \|v - v_n\|\}, \quad n \geq 1,$$
$$U_n := \{v \in U;\ \text{there exists } \alpha \in \mathbb{R} \text{ such that } (v, \alpha) \in A_n\}, \quad n \geq 1,$$

with the following properties:

$$A_n \text{ is a closed subset of epi}\, J \text{ containing } (v_n, J(v_n)), \text{ and } A_{n+1} \subset A_n,$$
$$U_{n+1} \subset U_n \text{ and thus } \gamma_n := \inf_{v \in U_n} J(v) \leq \gamma_{n+1} = \inf_{v \in U_{n+1}} J(v),$$
$$0 \leq J(v_{n+1}) - \gamma_n \leq \frac{1}{2}(J(v_n) - \gamma_n).$$

If $A_n = \{(v_n, J(v_n))\}$ for some $n \geq 1$, it suffices to let $u_\varepsilon = v_n$, and the iterative procedure stops here; hence the proof is complete in this case. It thus remains to consider the other case, where the *procedure continues ad infinitum*.

(ii) *If the iterative procedure continues* ad infinitum, *the diameter of the sets A_n approaches zero as $n \to \infty$.*

Since $(v_n, J(v_n)) \in A_n$, the definition of the set U_n implies that

$$\gamma_n = \inf_{v \in U_n} J(v) \le J(v_n) \quad \text{for each } n \ge 1.$$

Then the inequalities

$$0 \le J(v_n) - \gamma_n \le J(v_n) - \gamma_{n-1} \le \frac{1}{2}(J(v_{n-1}) - \gamma_{n-1}) \le \cdots \le \frac{1}{2^{n-1}}(J(v_1) - \gamma_1)$$

imply that

$$\lim_{n \to \infty} (J(v_n) - \gamma_n) = 0.$$

Given an arbitrary element (v, α) of the set A_n, the definitions of the sets U_n and A_n show that

$$\gamma_n \le \alpha \le J(v_n) - \varepsilon \|v - v_n\| \le J(v_n).$$

The resulting inequalities

$$\|v - v_n\| \le \frac{1}{\varepsilon}(J(v_n) - \alpha) \le \frac{1}{\varepsilon}(J(v_n) - \gamma_n),$$
$$|\alpha - J(v_n)| = J(v_n) - \alpha \le J(v_n) - \gamma_n$$

then clearly imply that $\operatorname{diam} A_n \to 0$ as $n \to \infty$ (recall that $\varepsilon > 0$ is fixed).

Therefore, by *Cantor's intersection theorem* (Theorem 5.1-1), *there exists a unique element $(u_\varepsilon, \beta_\varepsilon)$ such that*

$$(u_\varepsilon, \beta_\varepsilon) \in \operatorname{epi} J \quad \text{and} \quad (u_\varepsilon, \beta_\varepsilon) \in A_n \text{ for all } n \ge 1.$$

(iii) *The element $u_\varepsilon \in U$ found in* (ii) *satisfies*

$$\gamma \le J(u_\varepsilon) \le \gamma + \varepsilon \quad \text{and} \quad J(u_\varepsilon) < J(v) + \varepsilon \|v - u_\varepsilon\| \text{ for all } v \in U, \;\; v \neq u_\varepsilon.$$

First, the inequalities

$$J(u_\varepsilon) \le \beta_\varepsilon \le J(v_n) - \varepsilon \|u_\varepsilon - v_n\|, \quad n \ge 1,$$

which express that $(u_\varepsilon, \beta_\varepsilon) \in A_n \subset \operatorname{epi} J$, imply that $(u_\varepsilon, J(u_\varepsilon)) \in A_n$ for all $n \ge 1$. Hence, by the uniqueness property,

$$\beta_\varepsilon = J(u_\varepsilon).$$

Next, we claim that

$$J(u_\varepsilon) < J(v) + \varepsilon \|v - u_\varepsilon\| \quad \text{for all } v \in U, \;\; v \neq u_\varepsilon.$$

For, assume otherwise that there exists $v \in U$, $v \neq u_\varepsilon$, such that

$$J(v) - J(u_\varepsilon) + \varepsilon \|v - u_\varepsilon\| \le 0.$$

Since $J(u_\varepsilon)-J(v_n)+\varepsilon\,\|u_\varepsilon - v_n\| \le 0$ for each $n \ge 1$, it would then follow that $(v, J(v)) \in \operatorname{epi} J$ satisfies

$$J(v) - J(v_n) + \varepsilon\,\|v - v_n\| \le 0 \quad \text{for each } n \ge 1,$$

i.e., that $(v, J(v)) \in A_n$ for each $n \ge 1$, a contradiction if $v \neq u_\varepsilon$.

Finally, $(u_\varepsilon, J(u_\varepsilon)) \in A_1$ implies that

$$\gamma \le J(u_\varepsilon) \le J(v_1) - \varepsilon\,\|u_\varepsilon - v_1\| \le J(v_1) \le \gamma + \varepsilon,$$

which completes the proof. □

Remark *For each $\varepsilon > 0$, the set* epi J *is partially ordered* (Section 1.3) *by the relation* $(v,\alpha) \le (w,\beta)$ defined by $\alpha \le \beta - \varepsilon\,\|v - w\|$. Then, the element $(u_\varepsilon, J(u_\varepsilon)) \in \operatorname{epi} J$ found above is *minimal* for this total ordering, in the sense that, if $(v,\alpha) \in \operatorname{epi} J$ satisfies $(v,\alpha) \le (u_\varepsilon, J(u_\varepsilon))$, then necessarily $(v,\alpha) = (u_\varepsilon, J(u_\varepsilon))$. □

Under the additional assumption that the functional J is *of class* $\mathcal{C}^1$ over the whole space, Ekeland's variational principle has the following important consequence: *Even though J may not attain its minimum* (think of the function $J : v \in \mathbb{R} \to e^v$), *there exist minimizing sequences $(u_k)_{k=1}^\infty$ with the property that $J'(u_k) \to 0$ as $k \to \infty$.* This property thus appears as the natural extension to the present situation of the *Euler equation* $J'(u) = 0$ satisfied at a minimum u of a differentiable function (Theorem 7.1-5).

Theorem 9.8-2 (Ekeland's variational principle for functionals of class $\mathcal{C}^1$) *Let V be a Banach space and let $J \in \mathcal{C}^1(V)$ be a functional that is bounded from below. Then there exists a sequence $(u_k)_{k=1}^\infty$ of elements $u_k \in V$ such that*

$$\lim_{k\to\infty} J(u_k) = \inf_{v\in V} J(v) \quad \text{and} \quad \lim_{k\to\infty} J'(u_k) = 0 \text{ in } V'.$$

Proof The functional J being continuous, hence *a fortiori* lower semicontinuous, on V since it is assumed to be of class $\mathcal{C}^1$ on V, Ekeland's variational principle for lower semicontinuous functionals (Theorem 9.8-1) can be applied, showing that there exists a sequence $(u_k)_{k=1}^\infty$ of elements of V such that, for all $k \ge 1$,

$$\inf_{v\in V} J(v) \le J(u_k) \le \inf_{v\in V} J(v) + \frac{1}{k},$$
$$J(u_k) \le J(v) + \frac{1}{k}\,\|v - u_k\| \quad \text{for all } v \in V.$$

But, for each $k \ge 1$, the definition of the derivative $J'(u_k) \in V'$ implies that, for any $v \in V$,

$$J(v) = J(u_k) + J'(u_k)(v - u_k) + \|v - u_k\|\,\delta\,(v - u_k) \quad \text{with } \delta(h) \to 0 \text{ as } h \to 0 \text{ in } V.$$

Hence, for any $w \in V$ with $\|w\| = 1$ and any $t > 0$, letting $v = u_k + tw$ gives

$$-\frac{t}{k} \le J(u_k + tw) - J(u_k) = t(J'(u_k)w + \eta(t)) \quad \text{with } \eta(t) \to 0 \text{ as } t \to 0^+.$$

Dividing by t and letting t approach 0 then shows that

$$-\frac{1}{k} \leq J'(u_k)w \quad \text{for any } w \in V \text{ with } \|w\| = 1,$$

hence also that

$$-\frac{1}{k} \leq -J'(u_k)w = J'(u_k)(-w) \quad \text{for any } w \in V \text{ with } \|w\| = 1,$$

since $-w$ also satisfies $\|-w\| = 1$ in this case. Consequently,

$$\|J'(u_k)\|_{V'} = \sup_{\begin{cases} w \in V \\ \|w\|=1 \end{cases}} |J'(u_k)w| \leq \frac{1}{k}. \qquad \square$$

We now establish the existence of minimizers of functionals $J : V \to \mathbb{R}$ when the Banach space V is not necessarily reflexive and J is not necessarily convex, but J satisfies instead specific additional conditions. Note that *Ekeland's variational principle for functionals of class $\mathcal{C}^1$* (Theorem 9.8-2) plays a crucial role in the next proof.

Theorem 9.8-3 (existence of minimizers for functionals that satisfy the Palais–Smale condition) *Let V be a Banach space and let $J \in \mathcal{C}^1(V)$ be a functional that is bounded below on V and satisfies the* **Palais–Smale condition**:[42] *Any sequence $(u_k)_{k=1}^\infty$ of elements of V such that*

$$(J(u_k))_{k=1}^\infty \text{ converges in } \mathbb{R} \quad \text{and} \quad \lim_{k\to\infty} J'(u_k) = 0 \text{ in } V'$$

contains a convergent subsequence. Then there exists at least one element $u \in V$ such that

$$J(u) = \inf_{v \in V} J(v).$$

Proof By Theorem 9.8-2, there exists a sequence $(u_k)_{k=1}^\infty$ of elements of V such that

$$\lim_{k\to\infty} J(u_k) = \inf_{v\in V} J(v) \quad \text{and} \quad \lim_{k\to\infty} J'(u_k) = 0 \text{ in } V'.$$

By the *Palais–Smale condition*, there then exists a subsequence $(u_{\sigma(k)})_{k=1}^\infty$ that converges to an element $u \in V$, which therefore satisfies

$$J(u) = \lim_{k\to\infty} J(u_{\sigma(k)}) = \inf_{v\in V} J(v). \qquad \square$$

Remark In this proof, the Palais–Smale condition is only used for minimizing sequences. $\square$

Since $J' : V \to V'$ is assumed to be continuous, the above proof also shows that the minimum u found in Theorem 9.8-3 satisfies in this case the Euler equation, since

$$J'(u) = \lim_{k\to\infty} J'(u_{\sigma(k)}) = 0.$$

[42] R.S. Palais; S. Smale [1964]: A generalized Morse theory, *Bulletin of the American Mathematical Society* **70**, 165–171.

Stephen Smale was awarded the Fields Medal in 1966. A fascinating account of his impressive accomplishments until 1999, in mathematics and beyond, is given by Batterson [2000].

In fact, the Palais–Smale condition is especially useful for proving the existence of stationary points (i.e., points that satisfy $J'(u) = 0$; cf. Section 7.1) that are *saddle-points* (Section 7.16), rather than *minima* as in Theorem 9.8-3. In this direction, we will only quote the following beautiful result,[43] which has proved to be a powerful means of establishing existence of solutions to specific classes of nonlinear boundary value problems that are not amenable to the methods described so far.[44]

$^\flat$**Theorem 9.8-4 (mountain pass lemma[45])** *Let V be a Banach space and let $J \in \mathcal{C}^1(V)$ be a functional that satisfies the Palais–Smale condition. Assume in addition that there exist $u_0, u_1 \in V$ and $r > 0$ such that*

$$\|u_1 - u_0\| > r \quad \textit{and} \quad \max\{J(u_0);\ J(u_1)\} < \inf\{J(v);\ \|v - u_0\| = r\}.$$

Then there exists $u \in V$ such that

$$J'(u) = 0 \quad \textit{and} \quad J(u) = \inf_{\pi\in P} \sup_{0\leq t\leq 1} J(\pi(t)),$$

where

$$P = \{\pi \in \mathcal{C}([0,1]\,; V);\ \pi(0) = u_0 \textit{ and } \pi(1) = u_1\}.$$ □

How the "mountain pass lemma" got its name is suggested in Figure 9.8-1.

Problems

9.8-1 Let a Banach space V and a functional $J \in \mathcal{C}^1(V)$ be given.
(1) Using Ekeland's variational principle, show that, if

$$\alpha := \lim_{r\to\infty} (\inf\{J(v);\ \|v\| \geq r\}) < \infty,$$

then there exist $v_k \in V$, $k \geq 1$, such that

$$\|v_k\| \to \infty, \quad J(v_k) \to \alpha, \quad \text{and} \quad J'(v_k) \to 0 \text{ in } V', \text{ as } k \to \infty.$$

(2) Show that if, in addition, J is bounded below on V and satisfies the Palais–Smale condition, then J is coercive on V.

9.8-2 (1) Let Ω be a domain in $\mathbb{R}^n$ with $n \leq 3$ and let $f \in L^2(\Omega)$. Show that the functional $J : H_0^1(\Omega) \to \mathbb{R}$ defined by

$$J(v) := \frac{1}{2}\int_\Omega |\nabla v|^2 \,\mathrm{d}x + \frac{1}{4}\int_\Omega v^4 \,\mathrm{d}x - \int_\Omega fv \mathrm{d}x, \quad v \in H_0^1(\Omega),$$

satisfies the Palais–Smale condition.

---

[43] Due to:
A. Ambrosetti; P.H. Rabinowitz [1973]: Dual variational methods in critical point theory and applications, *Journal of Functional Analysis* **14**, 349–381.
[44] For example, $-\Delta u = u^2 + f$ in $\mathcal{D}'(\Omega)$ and $u \in H_0^1(\Omega)$, where Ω is a domain in $\mathbb{R}^n$, $n \leq 3$; see Kesavan [2004, Section 5.5].
[45] For detailed treatments of the mountain pass lemma and examples, see, e.g., Kavian [1993, Chapter 3, Section 8], Struwe [1990, Chapter 2, Section 6], or Kesavan [2004, Section 5.5].

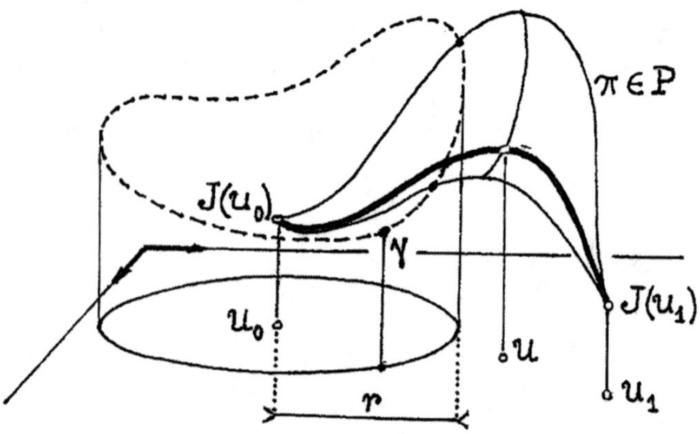

Figure 9.8-1 *The mountain pass lemma.* In a mountainous region, let $(u_0, J(u_0))$ and $(u_1, J(u_1))$ be two distinct points, where $u_0, u_1 \in \mathbb{R}^2$, and $J(u_0), J(u_1) \in \mathbb{R}$ denote their respective altitudes. Assume that these two points are "separated" by a set of the form $\{(v, J(v)) \in \mathbb{R}^2 \times \mathbb{R};\ |v - u_0| = r > 0\}$ (represented by a dashed line on the figure), in the sense that there exists $r > 0$ such that $|u_1 - u_0| > r$ and $\max\{J(u_0), J(u_1)\} < \inf\{J(v);\ v \in \mathbb{R}^2, |v - u_0| = r\}$. Then, if $J \in \mathcal{C}^1(\mathbb{R})$ satisfies the Palais–Smale condition, the mountain pass lemma asserts that, among all the continuous paths $\pi \in P$ joining these two distinct points, there exists at least one path that "climbs the least" (the heavy line on the figure), i.e., a "mountain pass"; the summit $(u, J(u))$ of such a pass is such that $J'(u) = 0$ and $J(u) = \inf_{\pi \in P} \sup_{0 \le t \le 1} J(\pi(t))$.

(2) Using Theorem 9.8-3, show that there exists $u \in H_0^1(\Omega)$ such that $J(u) = \inf_{v \in H_0^1(\Omega)} J(v)$, and that u satisfies the following nonlinear boundary value problem:

$$-\Delta u + u^3 = f \text{ if } \mathcal{D}'(\Omega) \quad \text{and} \quad u = 0 \text{ on } \partial\Omega.$$

Remark Compare with Problem 9.3-2, where the same conclusion was obtained by different means. □

9.8-3 Let Ω be a domain in $\mathbb{R}^n$, let $\lambda \in \mathbb{R}$, let $1 < p < \infty$ if $n = 2$ or $1 < p < \dfrac{n+2}{n-2}$ if $n \ge 3$, and let $f \in L^2(\Omega)$. Show that the functional $J : H_0^1(\Omega) \to \mathbb{R}$ defined by

$$J(v) := \frac{1}{2}\int_\Omega |\nabla v|^2 \, dx + \frac{\lambda}{p+1}\int_\Omega |v|^{p+1} \, dx - \int_\Omega f v \, dx, \quad v \in H_0^1(\Omega),$$

satisfies the Palais–Smale condition.

9.9 Brouwer's fixed point theorem — a first proof

To begin with, we define and briefly study a special class of *Lagrangians* introduced in Section 9.1, one example of which (that of part (b) in the next theorem) will play a key role in the proof given in this section of Brouwer's fixed point theorem.

Theorem 9.9-1 *Let $m \ge 1$ and $n \ge 1$ be two integers, let Ω be a domain in $\mathbb{R}^n$, and let $\mathcal{L} : \overline{\Omega} \times \mathbb{R}^m \times \mathbb{M}^{m \times n} \to \mathbb{R}$ be a Lagrangian that satisfies all the assumptions of Theorem 9.1-1.*

In addition, assume that $\mathcal{L}$ is a **null Lagrangian**, *in the following sense: Any vector field $\boldsymbol{u} \in \mathcal{C}^2(\overline{\Omega}; \mathbb{R}^m)$ such that the matrix field $\frac{\partial \mathcal{L}}{\partial \boldsymbol{F}}(\cdot, \boldsymbol{u}(\cdot), \boldsymbol{\nabla u}(\cdot))$ is in the space $\mathcal{C}^1(\overline{\Omega}; \mathbb{M}^{m\times n})$ satisfies the homogeneous Euler-Lagrange equations associated with the Lagrangian $\mathcal{L}$ (Section 9.1), viz.,*

$$-\,\mathbf{div}\,\frac{\partial \mathcal{L}}{\partial \boldsymbol{F}}(x, \boldsymbol{u}(x), \boldsymbol{\nabla u}(x)) + \frac{\partial \mathcal{L}}{\partial \boldsymbol{a}}(x, \boldsymbol{u}(x), \boldsymbol{\nabla u}(x)) = \mathbf{0} \quad \text{at all } x \in \overline{\Omega}.$$

(a) *Define the functional*

$$J : \boldsymbol{v} \in \mathcal{C}^2(\overline{\Omega}; \mathbb{R}^m) \to J(\boldsymbol{v}) := \int_\Omega \mathcal{L}(x, \boldsymbol{v}(x), \boldsymbol{\nabla v}(x))\,\mathrm{d}x.$$

Then the real number $J(\boldsymbol{v})$ depends only on the trace of $\boldsymbol{v}$ on Γ; in other words,

$$\boldsymbol{v}, \widetilde{\boldsymbol{v}} \in \mathcal{C}^2(\overline{\Omega}; \mathbb{R}^m) \quad \text{and} \quad \boldsymbol{v}|_\Gamma = \widetilde{\boldsymbol{v}}|_\Gamma \text{ implies } J(\boldsymbol{v}) = J(\widetilde{\boldsymbol{v}}).$$

(b) *The function*

$$(x, \boldsymbol{a}, \boldsymbol{F}) \in \overline{\Omega} \times \mathbb{R}^n \times \mathbb{M}^n \to \det \boldsymbol{F} \in \mathbb{R}$$

is a null Lagrangian; as a consequence,

$$\boldsymbol{v}, \widetilde{\boldsymbol{v}} \in \mathcal{C}^2(\overline{\Omega}; \mathbb{R}^m) \quad \text{and} \quad \boldsymbol{v}|_\Gamma = \widetilde{\boldsymbol{v}}|_\Gamma \text{ implies } \int_\Omega \det \boldsymbol{\nabla v}(x)\,\mathrm{d}x = \int_\Omega \det \boldsymbol{\nabla}\widetilde{\boldsymbol{v}}(x)\,\mathrm{d}x.$$

Proof (i) Given two vector fields $\boldsymbol{v}, \widetilde{\boldsymbol{v}} \in \mathcal{C}^2(\overline{\Omega}; \mathbb{R}^m)$ such that $\boldsymbol{v}|_\Gamma = \widetilde{\boldsymbol{v}}|_\Gamma$, let $\boldsymbol{w} := \widetilde{\boldsymbol{v}} - \boldsymbol{v}$, so that $\boldsymbol{w}|_\Gamma = \mathbf{0}$. Define the function

$$f : t \in [0, 1] \to f(t) := J(\boldsymbol{v}_t) \in \mathbb{R} \quad \text{where } \boldsymbol{v}_t := \boldsymbol{v} + t\boldsymbol{w}.$$

Since the function $J : \mathcal{C}^1(\overline{\Omega})$ is Fréchet-differentiable (Theorem 9.1-1) and the affine function $t \in [0, 1] \to \boldsymbol{v}_t \in \mathcal{C}^1(\overline{\Omega}, \mathbb{R}^m)$ is also Fréchet-differentiable (with a derivative equal to $\boldsymbol{w}$ at each $t \in [0, 1]$), the chain rule and the Green's formula (which can be applied since, by assumption, each matrix field $\frac{\partial \mathcal{L}}{\partial \boldsymbol{F}}(\cdot, \boldsymbol{v}_t(\cdot), \boldsymbol{\nabla v}_t(\cdot))$, $0 \le t \le 1$, is in the space $\mathcal{C}^1(\overline{\Omega}; \mathbb{M}^{m\times n})$) together give

$$\begin{aligned} f'(t) = J'(\boldsymbol{v}_t)\boldsymbol{w} &= \int_\Omega \left\{ \frac{\partial \mathcal{L}}{\partial \boldsymbol{a}}(x, \boldsymbol{v}_t(x), \boldsymbol{\nabla v}_t(x)) \cdot \boldsymbol{w}(x) + \frac{\partial \mathcal{L}}{\partial \boldsymbol{F}}(x, \boldsymbol{v}_t(x), \boldsymbol{\nabla v}_t(x)) : \boldsymbol{\nabla w}(x) \right\} \\ &= \int_\Omega \left\{ -\,\mathbf{div}\,\frac{\partial \mathcal{L}}{\partial \boldsymbol{F}}(x, \boldsymbol{v}_t(x), \boldsymbol{\nabla v}_t(x)) + \frac{\partial \mathcal{L}}{\partial \boldsymbol{a}}(x, \boldsymbol{v}_t(x), \boldsymbol{\nabla v}_t(x)) \right\} \cdot \boldsymbol{w}(x)\,\mathrm{d}x = 0 \end{aligned}$$

since $\mathcal{L}$ is by assumption a *null Lagrangian* (the boundary integral vanishes in the Green's formula since $\boldsymbol{w}|_\Gamma = \mathbf{0}$).

Since the function f is thus constant on $[0, 1]$,

$$J(\boldsymbol{v}) = f(0) = f(1) = J(\widetilde{\boldsymbol{v}}),$$

which proves (a).

(ii) Let a vector field $\boldsymbol{v} \in \mathcal{C}^2(\overline{\Omega}; \mathbb{R}^n)$ be given. Since the function

$$\iota_n : \boldsymbol{F} \in \mathbb{M}^n \to \iota_n(\boldsymbol{F}) = \det \boldsymbol{F} \in \mathbb{R}$$

depends neither on $x \in \overline{\Omega}$ nor on $\boldsymbol{a} \in \mathbb{R}^n$, it remains to check that $\mathbf{div}\, \dfrac{\partial \iota_n}{\partial \boldsymbol{F}}(\boldsymbol{\nabla v}(x)) = 0$, $x \in \Omega$, or equivalently, that

$$\sum_{j=1}^n \partial_j \Big(\frac{\partial \iota_n}{\partial F_{ij}}(\boldsymbol{\nabla v}(x)) \Big) = 0, \quad x \in \Omega, \ 1 \le i \le n.$$

The derivative $\iota_n'(\boldsymbol{F}) \in \mathcal{L}(\mathbb{M}^n; \mathbb{R})$ is given at any $\boldsymbol{F} \in \mathbb{M}^n$ by (Section 7.1)

$$\iota_n'(\boldsymbol{F})\boldsymbol{G} = \frac{\partial \iota_n}{\partial F_{ij}}(\boldsymbol{F}) G_{ij} = \mathbf{Cof}\, \boldsymbol{F} : \boldsymbol{G} = \sum_{i,j=1}^n (\mathbf{Cof}\, \boldsymbol{F})_{ij} G_{ij} \quad \text{for all } \boldsymbol{G} = (G_{ij}) \in \mathbb{M}^n.$$

Hence

$$\sum_{j=1}^n \partial_j \Big(\frac{\partial \iota_n}{\partial F_{ij}}(\boldsymbol{\nabla v}(x)) \Big) = \sum_{j=1}^n \partial_j \Big(\mathbf{Cof}\, \boldsymbol{\nabla v}(x) \Big)_{ij} = 0, \quad x \in \Omega, \ 1 \le i \le n,$$

by the *Piola identity* (Theorem 7.1-4). This proves (b). □

Brouwer's fixed point theorem[46] *is one of the most basic theorems of nonlinear functional analysis.* While its classical proof (which will be given in Section 9.16) is substantially more delicate, as it relies on *Brouwer's topological degree*, simpler proofs have been found more recently, such as the one given here.[47]

Theorem 9.9-2 (Brouwer's fixed point theorem — a first proof) *Let K be a compact and convex subset of a finite-dimensional normed vector space, and let $f : K \to K$ be a continuous mapping. Then f has at least one fixed point.*

Proof It clearly suffices to consider the case where the vector space is $\mathbb{R}^n$. For notational brevity, we let

$$B := B(0;1) = \{x \in \mathbb{R}^n;\ |x| < 1\}.$$

(i) *There is no mapping $v \in \mathcal{C}^2(\overline{B}; \mathbb{R}^n)$ that satisfies*

$$v(x) \in \partial B \text{ for all } x \in \overline{B} \quad \text{and} \quad v(x) = x \text{ for all } x \in \partial B.$$

Assume that such a mapping v exists. Let the mapping $\widetilde{v} \in \mathcal{C}^2(\overline{B}; \mathbb{R}^n)$ be defined by $\widetilde{v}(x) := x$ at each $x \in \overline{B}$. Since $v|_{\partial B} = \widetilde{v}|_{\partial B}$ and $F \in \mathbb{M}^n \to \det F \in \mathbb{R}$ is a *null Lagrangian*, it follows from Theorem 9.9-1 (which can be applied since the open ball B is a domain) that

$$\int_B \det \nabla v(x)\, dx = \int_B \det \nabla \widetilde{v}(x)\, dx = \int_B dx > 0.$$

[46] L. BROUWER [1912]: Über Abbildungen von Mannigfaltigkeiten, *Mathematische Annalen* **71**, 97–115.

[47] The clever proof given here is due to:
Y. KANNAI [1981]: An elementary proof of the no-retraction theorem, *American Mathematical Monthly* **88**, 264–268.

Define the function $\varphi : \overline{B} \to \mathbb{R}$ by

$$\varphi(x) := |v(x)|^2 = v(x) \cdot v(x) \quad \text{at each } x \in \overline{B}.$$

Then the function φ is differentiable at each $x \in B$, with (Section 7.1)

$$\varphi'(x)h = 2(\nabla v(x)h) \cdot v(x) \quad \text{for all } h \in \mathbb{R}^n.$$

But φ is a constant function (equal to one) on $\overline{B}$. Hence $\varphi'(x) = 0$ at each $x \in B$, and thus

$$0 = \varphi'(x)h = 2h^T \nabla v(x)^T v(x) = 0 \quad \text{for all } h \in \mathbb{R}^n,$$

which implies that

$$\nabla v(x)^T v(x) = 0 \quad \text{at each } x \in B.$$

Since $v(x) \neq 0$ at each $x \in B$, this means that 0 is an eigenvalue of the matrix $\nabla v(x)^T$ at each $x \in B$. Hence

$$\det \nabla v(x) = 0 \quad \text{for all } x \in B,$$

in contradiction with $\int_B \det \nabla v(x)\,\mathrm{d}x > 0$.

(ii) *There is no mapping* $w \in \mathcal{C}(\overline{B}; \mathbb{R}^n)$ *that satisfies*

$$w(x) \in \partial B \text{ for all } x \in \overline{B} \quad \text{and} \quad w(x) = x \text{ for all } x \in \partial B.$$

Assume that such a mapping w exists, and extend it to a mapping (still denoted) $w : \mathbb{R}^n \to \mathbb{R}^n$ by letting $w(x) := x$ for $|x| > 1$. The extended mapping w thus satisfies

$$w \in \mathcal{C}(\mathbb{R}^n; \mathbb{R}^n), \quad |w(x)| = 1 \text{ if } |x| < 1 \quad \text{and} \quad w(x) = x \text{ if } |x| \geq 1.$$

For each $1 \leq i \leq n$, let $(w_{i,\varepsilon})_{\varepsilon > 0}$ denote a *regularizing family* (Section 2.6) of the ith component $w_i \in \mathcal{C}(\mathbb{R}^n; \mathbb{R})$ of $w \in \mathcal{C}(\mathbb{R}^n; \mathbb{R}^n)$, thus defined by

$$w_{i,\varepsilon}(x) = \int_{\mathbb{R}^n} \omega_\varepsilon(x - y) w_i(y)\,\mathrm{d}y = \int_{B(0;\varepsilon)} \omega_\varepsilon(y) w_i(x - y)\,\mathrm{d}y \quad \text{at each } x \in \mathbb{R}^n,$$

where the function $\omega : x \in \mathbb{R}^n \to [0, \infty[$ used in the definition of the *mollifiers* ω_ε is chosen as a function of $|x|$ only. Then Theorem 2.6-1 shows that $w_{i,\varepsilon} \in \mathcal{C}^\infty(\mathbb{R}^n)$ for all $\varepsilon > 0$ and that there exists $0 < \varepsilon_0 \leq 1$ such that the mapping $w_\varepsilon := (w_{i,\varepsilon})_{i=1}^n \in \mathcal{C}^\infty(\mathbb{R}^n; \mathbb{R}^n)$ satisfies

$$|w_\varepsilon(x)| > 0 \quad \text{for all } \varepsilon \leq \varepsilon_0 \text{ and all } |x| \leq 2,$$

since $|w(x)| \geq 1$ for all $|x| \leq 2$ (the number 2 can be replaced by any real number > 1 in this argument). Besides,

$$w_\varepsilon(x) = x \quad \text{for all } \varepsilon \leq \varepsilon_0 \text{ and all } |x| \geq 2,$$

since, for each $1 \leq i \leq n$, the definition of the functions ω_ε shows that

$$w_{i,\varepsilon}(x) = \int_{B(0;\varepsilon)} \omega_\varepsilon(y) x_i\,\mathrm{d}y - \int_{B(0;\varepsilon)} \omega_\varepsilon(y) y_i\,\mathrm{d}y = x_i \quad \text{for all } |x| \geq 2.$$

The mapping $v \in \mathcal{C}^\infty(\overline{B}; \mathbb{R}^n)$ defined by

$$v(x) := \frac{w_{\varepsilon_0}(2x)}{|w_{\varepsilon_0}(2x)|} \quad \text{at each } |x| \leq 1,$$

thus satisfies

$$|v(x)| = 1 \text{ for all } x \in \overline{B} \quad \text{and} \quad v(x) = x \text{ for all } x \in \partial B,$$

but this contradicts (i).

(iii) *Any continuous mapping* $g : \overline{B} \to \overline{B}$ *has at least one fixed point in* $\overline{B}$.

Assume that such a mapping g has no fixed point in $\overline{B}$. Given any $x \in \overline{B}$, there exist a uniquely defined point $w(x)$ and a uniquely defined real number $\alpha(x) \geq 1$ such that

$$w(x) \in \partial B \quad \text{and} \quad w(x) = g(x) + \alpha(x)(x - g(x)).$$

Note that $\alpha(x) = 1$ if $x \in \partial B$, so that $w(x) = x$ if $x \in \partial B$.

The function $\alpha : \overline{B} \to [1, \infty[$ so defined is continuous, since, at each $x \in \overline{B}$, $\alpha(x)$ is the unique root ≥ 1 of the quadratic polynomial

$$\lambda \in \mathbb{R} \to \lambda^2 |x - g(x)|^2 + 2\lambda(x - g(x)) \cdot g(x) + |g(x)|^2 - 1,$$

whose coefficients are continuous functions of $x \in \overline{B}$.

Consequently, the mapping $w : \overline{B} \to \mathbb{R}^n$ defined in this fashion is also continuous and, by construction, w satisfies

$$w(x) \in \partial B \text{ for all } x \in \overline{B} \quad \text{and} \quad w(x) = x \text{ for all } x \in \partial B.$$

But this is impossible by (ii). Hence the mapping g has at least one fixed point in $\overline{B}$.

(iv) *The result of* (iii) *holds if* $\overline{B}$ *is replaced by any compact and convex subset of* $\mathbb{R}^n$.

First, notice that the result of (iii) clearly holds if the closed unit ball $\overline{B}$ is replaced by any closed ball centered at the origin. Given any compact and convex subset K of $\mathbb{R}^n$, there exist $r > 0$ such that $K \subset \overline{B(0;r)}$. Let $P : \mathbb{R}^n \to K$ denote the projection operator from $\mathbb{R}^n$ onto K, which is thus continuous (Theorem 4.3-1(c)).

Let now $f : K \to K$ be any continuous mapping. The composition $g := f \circ P : \overline{B(0;r)} \to K \subset \overline{B(0;r)}$ is then also continuous. Hence by (iii), g has a fixed point $x_0 \in \overline{B(0;r)}$, which necessarily belongs to K since $g(\overline{B(0;r)}) \subset K$ by construction. Hence

$$x_0 = g(x_0) = f(P(x_0)) = f(x_0),$$

as was to be proved. □

Remarks (1) *Uniqueness* does not hold in general (let for instance $f = \mathrm{id}_K$).

(2) The assumptions that K is *compact* and *convex* are essential (let for instance $f : \mathbb{R}^n \to \mathbb{R}^n$ be defined by $f(x) = x + a$, $x \in \mathbb{R}^n$, for some nonzero vector $a \in \mathbb{R}^n$; let for instance $f : \partial B \to \partial B$ be defined by $f(x) = -x$, $x \in \partial B$).

(3) The assumption of *finite dimensionality* is likewise essential. Otherwise the natural extension of Brouwer's fixed point theorem to an infinite-dimensional normed vector space X applies to a *compact*

mapping that maps a closed, bounded, and convex subset of X into itself; this extension constitutes *Schauder's fixed point theorem* (Theorem 9.12-1). □

If A is a subset of a set B, a mapping $f : B \to A$ such that $f(x) = x$ for all $x \in A$ is called a **retraction** from B onto A. Part (ii) of the above proof thus asserts that *there is no continuous retraction of the closed unit ball of* $\mathbb{R}^n$ *onto its boundary*, and part (iii) shows that *this property implies Brouwer's theorem* (with $K = \overline{B}$).

Remarks (1) The converse implication also holds; cf. Problem 9.9-1.

(2) By contrast, given any point $a \in B$, there exists an obvious continuous retraction from $\overline{B} - \{a\}$ onto ∂B.

(3) Surprisingly, in *any infinite-dimensional* normed vector space, there exists a continuous retraction from the closed unit ball onto its boundary; cf. Problem 9.9-2. □

The following corollary of Brouwer's theorem is often used; see for instance the proofs of Theorems 9.10-1, 9.11-1, and 9.14-1.

Theorem 9.9-3 (corollary to Brouwer's fixed point theorem) *Let* $(V, \|\cdot\|)$ *be a finite-dimensional normed vector space and let* $f : V \to V'$ *be a continuous mapping with the following property: There exists* $r > 0$ *such that*

$$ {}_{V'}\langle f(v), v\rangle_V \geq 0 \quad \textit{for all } v \in V \textit{ such that } \|v\|_V = r. $$

Then there exists $v_0 \in V$ *such that*

$$ \|v_0\|_V \leq r \quad \textit{and} \quad f(v_0) = 0. $$

Proof For brevity, the duality pairing ${}_{V'}\langle \cdot, \cdot\rangle_V$ is denoted $\langle \cdot, \cdot\rangle$.

(i) Let $(e_i)_{i=1}^n$ and $(e_i')_{i=1}^n$ denote dual bases in V and V', i.e., such that $\langle e_i', e_j\rangle = \delta_{ij}$, $1 \leq i, j \leq n$. With any mapping $f : V \to V'$, which can thus be written as

$$ v \in V \to f(v) = \sum_{i=1}^n \langle f(v), e_i\rangle e_i' \in V', $$

we associate a mapping $\widetilde{f} : V \to V$ by letting

$$ v \in V \to \widetilde{f}(v) := \sum_{i=1}^n \langle f(v), e_i\rangle e_i \in V. $$

It is then clear that

$$ \langle e_i', \widetilde{f}(v)\rangle = \langle f(v), e_i\rangle, \quad 1 \leq i \leq n, $$

and that $f(v) = 0$ if and only if $\widetilde{f}(v) = 0$.

(ii) Let then $f : V \to V'$ be a continuous mapping such that, for some $r > 0$,

$$ \langle f(v), v\rangle \geq 0 \quad \text{for all } \|v\| = r. $$

Let $\widetilde{f} : V \to V$ be the associated mapping as in (i), and assume that $f(v) \neq 0$ for all $\|v\| \leq r$. The mapping $h : \overline{B(0;r)} \subset V \to V$ defined by

$$h(v) := r\frac{\widetilde{f}(v)}{\|\widetilde{f}(v)\|_V}, \quad v \in \overline{B(0;r)},$$

is then continuous and maps the closed ball $\overline{B(0;r)}$ into $\partial B(0;r) \subset \overline{B(0;r)}$. Therefore, by *Brouwer's fixed point theorem*, there exists v_0 such that

$$v_0 \in \overline{B(0;r)} \quad \text{and} \quad h(v_0) = v_0,$$

and thus $\|v_0\| = \|h(v_0)\| = r \neq 0$. But, by assumption,

$$\begin{aligned} 0 \leq \langle f(v_0), v_0\rangle &= \sum_{i=1}^{n} \langle f(v_0), e_i\rangle\langle e_i', v_0\rangle \\ &= \sum_{i=1}^{n} \langle e_i', \widetilde{f}(v_0)\rangle\langle e_i', v_0\rangle = -\frac{\|\widetilde{f}(v_0)\|}{r}\sum_{i=1}^{n} |\langle e_i', v_0\rangle|^2, \end{aligned}$$

which implies that $v_0 = 0$, a contradiction. □

Remarks (1) Naturally, the conclusion of Theorem 9.9-3 holds if instead $\langle f(v), v\rangle \leq 0$ for all $v \in V$ such that $\|v\| = r$.

(2) As shown by the above proof, the continuous mapping f need not be defined over the whole space. It suffices that f be defined and continuous over a closed ball centered at the origin and that f satisfy (for instance) $\langle f(v), v\rangle \geq 0$ for all v on the boundary of this ball. □

As a first application of Brouwer's fixed point theorem, we establish an interesting spectral property of *nonnegative square matrices*. Recall that a matrix $(a_{ij}) \in \mathbb{M}^{m\times n}$, *resp.* a vector $(x_i) \in \mathbb{R}^n$, is said to be *nonnegative* if $a_{ij} \geq 0$ for all $1 \leq i \leq m$ and $1 \leq j \leq n$, *resp.* $x_i \geq 0$ for all $1 \leq i \leq n$.

Theorem 9.9-4 *Let A be a nonnegative square matrix of order n. Then there exist $\lambda \in \mathbb{R}$ and a nonzero vector $p \in \mathbb{R}^n$ such that*

$$Ap = \lambda p, \quad \lambda \geq 0, \quad \textit{and} \quad p \geq 0.$$

Proof Define the set

$$K := \left\{x = (x_i) \in \mathbb{R}^n;\ x_i \geq 0,\ 1 \leq i \leq n,\ \sum_{i=1}^{n} x_i = 1\right\}.$$

If there exists $p \in K$ such that $Ap = 0$, the theorem holds with $\lambda = 0$.

Assume otherwise that $Ax \neq 0$ for all $x \in K$, so that $\sum_{i=1}^{n}(Ax)_i > 0$ for all $x \in K$. Then the mapping $f : K \to \mathbb{R}^n$ defined by

$$f(x) := \frac{1}{\sum_{i=1}^{n}(Ax)_i}Ax \quad \text{for each } x \in K,$$

is continuous and maps the compact and convex subset K of $\mathbb{R}^n$ into itself. Therefore, by *Brouwer's fixed point theorem*, there exists $p \in K$ such that $f(p) = p$, i.e., such that

$$Ap = \lambda p \quad \text{with } p \neq 0 \text{ and } p \geq 0, \quad \text{and} \quad \lambda := \sum_{i=1}^{n} (Ap)_i > 0. \qquad \square$$

Theorem 9.9-4 constitutes a small incursion into the **Perron–Frobenius theory of nonnegative matrices.**[48] This theory asserts in particular that *the spectral radius* $\rho(A)$ of an $n \times n$ nonnegative matrix A is an eigenvalue of A, and that the eigensubspace corresponding to $\rho(A)$ contains eigenvectors $p \geq 0$. Further properties hold if the matrix A is *irreducible* (which is in particular the case if all its elements are > 0).[49]

Remark This theory can be extended to nonnegative linear operators acting in infinite-dimensional normed vector spaces, in which an infinite-dimensional "nonnegative hyperoctant" (the analogue of the subset $\{x = (x_i) \in \mathbb{R}^n;\ x_i \geq 0,\ 1 \leq i \leq n\}$ of $\mathbb{R}^n$) can be defined by means of a suitable "order cone": this is the content of the **Krein–Rutman theorem,**[50] another *basic theorem of nonlinear functional analysis.* $\square$

Problems

9.9-1 Show that Brouwer's fixed point theorem implies that, in any finite-dimensional normed vector space, there is no continuous retraction of the closed unit ball onto its boundary.

9.9-2 Let X be any *infinite-dimensional* normed vector space. Show that, *by contrast with the finite-dimensional case*, there exists a *retraction* of X (hence *a fortiori* of the closed unit ball of X) onto the unit sphere of X[51] (i.e., a continuous mapping f from X onto the unit sphere $S := \{x \in X;\ \|x\| = 1\}$ of X that satisfies $f(x) = x$ for all $x \in S$).

---

[48]So named after:

O. Perron [1907]: Grundlagen für eine Theorie des Jacobischen Kettenbruchalgorithmus, *Mathematische Annalen* **64**, 11–76.

G. Frobenius [1912]: Über Matrizen aus nicht negativen Elementen, *Sitzungsberichte Preußische Akademie der Wissenschaft*, Berlin, 456–477.

[49]For more on the Perron–Frobenius theory of irreducible nonnegative matrices, see in particular the illuminating account given in Varga [1962, Chapter 2]; for more details, historical perspectives, and applications, see in particular:

C.R. MacCluer [2000]: The many proofs and applications of Perron's theorem, *SIAM Review* **42**, 487–498.

A. Berman; R.J. Plemmons [1994]: *Nonnegative Matrices in the Mathematical Sciences*, Classics in Applied Mathematics, Vol. 9, SIAM, Philadelphia.

[50]M. Krein; M. Rutman [1948]: Linear operators leaving invariant a cone in a Banach space, *Uspehi Mathematiceskii Nauk* **3**, 3–95 [in Russian; English translation: *American Mathematical Society Translations* 1950, No. 26].

A proof of the Krein–Rutman theorem is found in Deimling [1985, Chapter 6] or in Zeidler [1986, Chapter 7]. In this direction, see also:

S. Karlin [1959]: Positive operators, *Journal of Mathematics and Mechanics* **8**, 907–937.

I. Marek [1970]: Frobenius theory of positive operators: Comparison theorems and applications, *SIAM Journal on Applied Mathematics* **19**, 607–628.

[51]For a proof of this result, which is due to J. Dugundji, see:

H. Steinlein [1979]: Two results of J. Dugundji about extensions of maps and retractions, *Proceedings of the American Mathematical Society* **77**, 298–290.

9.10 Application of Brouwer's theorem to the von Kármán equations, by means of the Galerkin method

The **Galerkin's method**[52] is an often quite effective method for establishing the existence of solutions to *nonlinear problems* (and *a fortiori* to linear ones) posed as a set of *variational equations* (as in Theorems 9.10-1 and 9.11-1), or *variational inequalities* (as in Theorem 9.14-1), over an *infinite-dimensional, separable, reflexive Banach space* V.

Its *principles* are very simple: The space V being *separable*, there exists a countably infinite linearly independent family $(v_i)_{i=1}^{\infty}$ of vectors $v_i \in V$ such that the union of the finite-dimensional subspaces $V_n := \text{Span}(v_i)_{i=1}^{n}$ of V is *dense* in V (Theorem 2.2-7). One then tries to show, *first*, that for each $n \geq 1$ there exists at least one solution $u_n \in V_n$ to analogous variational equations, or variational inequalities, but *now posed over* V_n; *second*, that such "appropriate solutions" u_n are *bounded in* V *independently of* $n \geq 1$ (while these two objectives are achieved in the examples that follow by means of Brouwer's fixed point theorem, different means may be more appropriate in other examples).

If this is the case, the *Banach–Eberlein–Šmulian theorem* (Theorem 5.14-4), which can be applied since V is reflexive, shows that there exist a subsequence $(u_m)_{m=1}^{\infty}$ of $(u_n)_{n=1}^{\infty}$ that *weakly converges* to a vector $u \in V$. By passing to the limit as $n \to \infty$, it is then generally possible to show that the weak limit u is a solution to the original variational equations, or variational inequalities; of course some care has to be exercised at this stage since the sequence $(u_m)_{m=1}^{\infty}$ converges only *weakly*.

We now illustrate how these general principles can be applied to a specific example.

In Section 9.4, we showed that solving the *von Kármán equations*, which are posed over a domain Ω in $\mathbb{R}^2$, amounts to finding a solutions $\xi \in H_0^2(\Omega)$ of the *reduced von Kármán equation*

$$C(\xi) + \xi - F = 0 \quad \text{in } H_0^2(\Omega),$$

where $C : H_0^2(\Omega) \to H_0^2(\Omega)$ is a cubic operator (whose definition is recalled in the next theorem) and $F \in H_0^2(\Omega)$ is a given function. We then introduced a sequentially weakly lower semicontinuous and coercive functional over the space $H_0^2(\Omega)$ whose stationary points coincide with the solution of the reduced von Kármán equation. Hence its minimizers (the existence of which was then guaranteed by Theorem 9.3-1) provide particular solutions to the reduced von Kármán equation.

But, as shown in the next theorem, it is also possible to *directly* establish, i.e., *without* a recourse to a functional, the existence of solutions to this equation, once recast as a set of variational equations, thanks this time to *Brouwer's fixed point theorem.*[53]

Theorem 9.10-1 (existence of solutions to the von Kármán equation) *Let Ω be a domain in $\mathbb{R}^2$. Let the bilinear and symmetric operator $B : H^2(\Omega) \times H^2(\Omega) \to H_0^2(\Omega)$ be defined as follows: For each $(\xi, \eta) \in H^2(\Omega) \times H^2(\Omega)$, let*

$$[\xi, \eta] := \partial_{11}\xi\partial_{22}\eta + \partial_{22}\xi\partial_{11}\eta - 2\partial_{12}\xi\partial_{12}\eta,$$

[52] So named after:
B.G. Galerkin [1915]: *Rods and Plates*, Vestnik Inženerov **19**, 897–908 (in Russian).

[53] This approach is due to Lions [1969, Chapter 1, Section 4.3].

and let the function $B(\xi, \eta)$ denote the unique solution of

$$B(\xi, \eta) \in H_0^2(\Omega) \quad \textit{and} \quad \Delta^2 B(\xi, \eta) = [\xi, \eta] \quad \textit{in } \mathcal{D}'(\Omega).$$

Let then the operator $C : H_0^2(\Omega) \to H_0^2(\Omega)$ be defined by

$$C : \xi \in H_0^2(\Omega) \to C(\xi) := B(B(\xi, \xi), \xi) \in H_0^2(\Omega),$$

so that C is "cubic" in the sense that $C(\alpha\xi) = \alpha^3 C(\xi)$ for all $\alpha \in \mathbb{R}$ and all $\xi \in H_0^2(\Omega)$. Finally, let a function $F \in H_0^2(\Omega)$ be given.

Then the reduced von Kármán equation

$$C(\xi) + \xi - F = 0$$

has at least one solution $\xi \in H_0^2(\Omega)$.

Proof If $F = 0$, there is nothing to prove, since $\xi = 0$ is clearly a solution to the reduced von Kármán equation in this case. So, assume that $F \neq 0$.

Let the space $H_0^2(\Omega)$ be equipped with the inner product $(\cdot, \cdot)_\Delta$ and norm $|\cdot|_\Delta$ defined by

$$(\xi, \eta)_\Delta := \int_\Omega \Delta\xi \Delta\eta \, dx \quad \text{and} \quad |\xi|_\Delta := \sqrt{(\xi, \xi)_\Delta} \quad \text{for each } \xi, \eta \in H_0^2(\Omega).$$

Solving the reduced von Kármán equations is thus the same as solving the *variational equations*

$$(C(\xi) + \xi - F, \eta)_\Delta = 0 \quad \text{for all } \eta \in H_0^2(\Omega).$$

To this end, we use the *Galerkin method*. Since $(H_0^2(\Omega), (\cdot, \cdot)_\Delta)$ is a separable Hilbert space (Section 6.5), it possesses a Hilbert basis $(w_i)_{i=1}^\infty$ (Section 4.9). For each integer $n \geq 1$, define the finite-dimensional inner-product space

$$V^n := \text{Span}(w_i)_{i=1}^n,$$

and define the mapping $f^n : V^n \to V^n$ by

$$f^n(\xi) := P^n(C(\xi) + \xi - F) \in V^n \quad \text{for each } \xi \in V^n,$$

where P^n denotes the projection operator of $H_0^2(\Omega)$ onto V^n, which thus satisfies $(P^n\eta, \xi)_\Delta = (\eta, \xi)_\Delta$ for all $\eta \in H_0^2(\Omega)$ and all $\xi \in V^n$ (Theorem 4.3-1(d)). Therefore, for each $\xi \in V^n$,

$$\begin{aligned}(f^n(\xi), \xi)_\Delta &= (P^n(C(\xi) + \xi - F), \xi)_\Delta = (C(\xi) + \xi - F, \xi)_\Delta \\ &\geq |\xi|_\Delta^2 - |F|_\Delta \, |\xi|_\Delta \,,\end{aligned}$$

since $(C(\xi), \xi)_\Delta \geq 0$ for all $\xi \in H_0^2(\Omega)$ (Theorem 9.4-2(b)). Consequently,

$$(f^n(\xi), \xi)_\Delta \geq 0 \quad \text{for all } \xi \in V^n \text{ such that } |\xi|_\Delta = |F|_\Delta \,.$$

Each mapping $f^n : V^n \to V^n$, $n \geq 1$, is continuous (both mappings $P^n : H_0^2(\Omega) \to V^n$ and $C : H_0^2(\Omega) \to H_0^2(\Omega)$ are continuous; cf. Theorems 4.3-1 and 9.4-2). Hence the *corollary to Brouwer's fixed point theorem* (Theorem 9.9-3) can be applied (recall that $|F|_\Delta > 0$ since we assume that $F \neq 0$), showing that, for each $n \geq 1$, there exists ξ^n such that

$$\xi^n \in V^n, \quad |\xi^n|_\Delta \leq |F|_\Delta, \quad \text{and} \quad f^n(\xi^n) = 0,$$

the last relation being equivalent to

$$(P^n(C(\xi^n) + \xi^n - F), \eta)_\Delta = (C(\xi^n) + \xi^n - F, \eta)_\Delta = 0 \quad \text{for all } \eta \in V^n.$$

Since $(\xi^n)_{n=1}^\infty$ is a bounded sequence in the Hilbert space $H_0^2(\Omega)$, there exist a subsequence $(\xi^m)_{m=1}^\infty$ of $(\xi^n)_{n=1}^\infty$ and $\xi \in H_0^2(\Omega)$ such that ($\rightharpoonup$ denotes weak convergence)

$$\xi^m \rightharpoonup \xi \quad \text{in } H_0^2(\Omega) \text{ as } m \to \infty,$$

by the *Banach–Eberlein–Šmulian theorem* (Theorem 5.14-4).

Given any $\eta \in H_0^2(\Omega)$, let $\eta^m \in V^m$, $m \geq 1$, be such that

$$|\eta^m - \eta|_\Delta \to 0 \quad \text{as } m \to \infty$$

(for instance, choose η^m as the projection of η onto V^m). Hence

$$(C(\xi^m) + \xi^m - F, \eta^m) = 0 \quad \text{for each } m \geq 1.$$

By Theorem 9.4-2(c),

$$\xi^m \rightharpoonup \xi \text{ in } H_0^2(\Omega) \quad \text{implies} \quad C(\xi^m) \to C(\xi) \text{ in } H_0^2(\Omega),$$

and thus $(C(\xi^m), \eta^m)_\Delta \to (C(\xi), \eta)_\Delta$ as $m \to \infty$. By Theorem 5.12-4(c),

$$\xi^m \rightharpoonup \xi \text{ in } H_0^2(\Omega) \text{ and } \eta^m \to \eta \text{ in } H_0^2(\Omega) \quad \text{implies} \quad (\xi^m, \eta^m)_\Delta \to (\xi, \eta)_\Delta \text{ as } m \to \infty.$$

Hence $(C(\xi) + \xi - F, \eta)_\Delta = 0$ for each $\eta \in H_0^2(\Omega)$. This completes the proof. □

Problem

9.10-1 Proceeding as in the proof of Theorem 9.10-1, show that the *reduced Marguerre–von Kármán equation* (Problem 9.4-2) has at least one solution in the space $H_0^2(\Omega)$.

9.11 Application of Brouwer's theorem to the Navier–Stokes equations, by means of the Galerkin method

In this section, Latin indices vary in $\{1, 2, 3\}$ (except for indexing sequences) and the summation convention with respect to such indices is used.

The **Navier–Stokes equations**[54] model the *stationary* (i.e., time-independent) *flow of an incompressible viscous fluid* filling up a domain Ω in $\mathbb{R}^3$. They take the form

$$\begin{aligned} -\nu\Delta \boldsymbol{u} + (\boldsymbol{\nabla u})\boldsymbol{u} + \mathbf{grad}\,\lambda = \boldsymbol{f} &\quad \text{in } \Omega, \\ \operatorname{div} \boldsymbol{u} = 0 &\quad \text{in } \Omega, \\ \boldsymbol{u} = \mathbf{0} &\quad \text{on } \Gamma, \end{aligned}$$

or, componentwise,

$$\begin{aligned} -\nu\Delta u_i + u_j\partial_j u_i + \partial_i\lambda = f_i &\quad \text{in } \Omega, \\ \partial_i u_i = 0 &\quad \text{in } \Omega, \\ u_i = 0 &\quad \text{on } \Gamma. \end{aligned}$$

In these equations, the two *unknowns* are the *vector field* $\boldsymbol{u} = (u_i) : \overline{\Omega} \to \mathbb{R}^3$ and the *function* $\lambda : \overline{\Omega} \to \mathbb{R}$ (clearly, λ is only determined up to an additive constant), which respectively represent the *velocity* of the fluid and the *pressure* inside the fluid; the *data* are a constant $\nu > 0$ and a vector field $\boldsymbol{f} = (f_i) : \Omega \to \mathbb{R}^3$, which respectively represent the *kinematic viscosity* of the fluid and the density per unit mass of the *applied forces.* The relation $\operatorname{div}\boldsymbol{u} = 0$ in Ω means that the fluid is *incompressible.* The boundary condition $\boldsymbol{u} = \mathbf{0}$ on Γ (which expresses that the velocity of the fluid is assumed to vanish along the entire boundary Γ) is chosen for simplicity, as treating a nonhomogeneous boundary condition $\boldsymbol{u} = \boldsymbol{u}_0$ on Γ requires extra care.[55]

Remark In the literature, the notation $(\boldsymbol{u} \cdot \boldsymbol{\nabla})\boldsymbol{u}$ is often preferred to the notation $(\boldsymbol{\nabla u})\boldsymbol{u}$ used here. □

When $N = 3$, the *Stokes equations* (Section 6.14) thus represent a formal linearization of the Navier–Stokes equations, in that the nonlinear term $(\boldsymbol{\nabla u})\boldsymbol{u}$ appearing in the left-hand sides of the partial differential equations above is deemed "negligible" compared with their linear term $-\nu\Delta\boldsymbol{u} + \mathbf{grad}\,\lambda$. In this respect, note that the proof of existence of the unknown $\boldsymbol{u}$ (part (iii) in the next proof) is carried out quite differently from that of the unknown $\boldsymbol{u}$ appearing in the Stokes equations; compare with the proof of Theorem 6.14-3.

As in Section 6.14, the Hilbert space $\boldsymbol{H}_0^1(\Omega)$ is equipped with the inner product $(\cdot,\cdot)_{1,\Omega}$ and norm $|\cdot|_{1,\Omega}$ respectively defined by

$$(\boldsymbol{v}, \boldsymbol{w})_{1,\Omega} = \int_\Omega \boldsymbol{\nabla v} : \boldsymbol{\nabla w}\,\mathrm{d}x \quad \text{and} \quad |\boldsymbol{v}|_{1,\Omega} = \sqrt{(\boldsymbol{v}, \boldsymbol{v})_{1,\Omega}} \text{ for each } \boldsymbol{u}, \boldsymbol{v} \in \boldsymbol{H}_0^1(\Omega),$$

and the Hilbert space $L_0^2(\Omega)$ is defined by

$$L_0^2(\Omega) := \left\{\mu \in L^2(\Omega);\ \int_\Omega \mu\,\mathrm{d}x = 0\right\}.$$

[54] So named after:

C.L.M.H. Navier [1823]: Mémoire sur les lois du mouvement des fluides, *Mémoires de l'Académie Royale des Sciences de Paris* **6**, 389–416.

G.G. Stokes [1845]: On the theories of the internal friction of fluids in motion, *Transactions of the Cambridge Philosophical Society* **8**, 287–305.

[55] See Temam [1977, Chapter 2, Section 1.4].

Theorem 9.11-1 (existence of a solution to the Navier–Stokes equations[56]**)** *Let Ω be a domain in $\mathbb{R}^3$ and let a constant $\nu > 0$ and an element $\boldsymbol{f} \in \boldsymbol{H}^{-1}(\Omega)$ be given. Then there exists $(\boldsymbol{u}, \lambda) \in \boldsymbol{H}_0^1(\Omega) \times L_0^2(\Omega)$ such that*

$$\begin{aligned} -\nu\Delta\boldsymbol{u} + (\boldsymbol{\nabla u})\boldsymbol{u} + \mathbf{grad}\,\lambda = \boldsymbol{f} \quad & \text{in } \boldsymbol{H}^{-1}(\Omega), \\ \operatorname{div}\boldsymbol{u} = 0 \quad & \text{in } \Omega, \\ \boldsymbol{u} = \mathbf{0} \quad & \text{on } \Gamma. \end{aligned}$$

Besides, given any $\boldsymbol{u} \in \boldsymbol{H}_0^1(\Omega)$ such that $(\boldsymbol{u}, \lambda) \in \boldsymbol{H}_0^1(\Omega) \times L_0^2(\Omega)$ is a solution to this boundary value problem, $\lambda \in L_0^2(\Omega)$ is unique.

Proof If $\boldsymbol{f} = \mathbf{0}$, there is nothing to prove since $(\mathbf{0}, 0) \in \boldsymbol{H}_0^1(\Omega) \times L_0^2(\Omega)$ is clearly a solution to the Navier–Stokes equations in this case. Hence we assume that $\boldsymbol{f} \neq \mathbf{0}$.

The idea of the proof consists in writing the Navier–Stokes equations as a set of variational equations in the space $\boldsymbol{H}_0^1(\Omega)$, then in restricting these equations to the subspace $\{\boldsymbol{v} \in \boldsymbol{H}_0^1(\Omega); \operatorname{div}\boldsymbol{v} = 0 \text{ in } \Omega\}$ of $\boldsymbol{H}_0^1(\Omega)$. As a result, only the unknown $\boldsymbol{u}$ appears in these restricted equations, which are then shown to have a solution by making use of the *Galerkin method* and *Brouwer's fixed point theorem* (in a manner reminiscent of that used for the von Kármán equations; cf. the proof of Theorem 9.10-1).[57] Given such a solution $\boldsymbol{u} \in \boldsymbol{H}_0^1(\Omega)$ to the restricted variational equations, the existence of a unique function $\lambda \in L_0^2(\Omega)$ such that the pair $(\boldsymbol{u}, \lambda)$ satisfies the original variational equations is then established as for the Stokes equations.

(i) *Finding a solution $(\boldsymbol{u}, \lambda) \in \boldsymbol{H}_0^1(\Omega) \times L_0^2(\Omega)$ to the above boundary value problem amounts to finding $(\boldsymbol{u}, \lambda) \in \boldsymbol{H}_0^1(\Omega) \times L_0^2(\Omega)$ that satisfies*

$$\begin{aligned} a(\boldsymbol{u}, \boldsymbol{v}) + b(\boldsymbol{u}; \boldsymbol{u}, \boldsymbol{v}) - \int_\Omega \lambda \operatorname{div}\boldsymbol{v}\,dx = \ell(\boldsymbol{v}) \quad & \text{for all } \boldsymbol{v} \in \boldsymbol{H}_0^1(\Omega), \\ \operatorname{div}\boldsymbol{u} = 0 \quad & \text{in } \Omega, \end{aligned}$$

where the bilinear form $a : (\boldsymbol{H}_0^1(\Omega))^2 \to \mathbb{R}$, the trilinear form $b : (\boldsymbol{H}_0^1(\Omega))^3 \to \mathbb{R}$, and the continuous linear form $\ell : \boldsymbol{H}_0^1(\Omega) \to \mathbb{R}$ are defined by

$$a(\boldsymbol{u}, \boldsymbol{v}) := \nu \int_\Omega \boldsymbol{\nabla u} : \boldsymbol{\nabla v}\,dx, \quad b(\boldsymbol{w}; \boldsymbol{u}, \boldsymbol{v}) := \int_\Omega ((\boldsymbol{\nabla u})\boldsymbol{w}) \cdot \boldsymbol{v}\,dx,$$
$$\ell(\boldsymbol{v}) := {}_{\boldsymbol{H}^{-1}(\Omega)}\langle \boldsymbol{f}, \boldsymbol{v}\rangle_{\boldsymbol{H}_0^1(\Omega)}.$$

The bilinear form a is clearly continuous on $(\boldsymbol{H}_0^1(\Omega))^2$. By Hölder's inequality,

$$\begin{aligned} |b(\boldsymbol{w}; \boldsymbol{u}, \boldsymbol{v})| = \left| \int_\Omega w_j(\partial_j u_i)v_i\,dx \right| &\leq \|w_j\|_{0,4,\Omega} \|\partial_j u_i\|_{0,\Omega} \|v_i\|_{0,4,\Omega}, \\ &\leq \sqrt{3}\,\|\boldsymbol{w}\|_{0,4,\Omega} |\boldsymbol{u}|_{1,\Omega} \|\boldsymbol{v}\|_{0,4,\Omega}, \end{aligned}$$

[56]The existence of solutions to the Navier–Stokes equations was established for the first time in two fundamental papers, which together constitute a milestone in the history of mathematical fluid mechanics:

J. Leray [1933]: Essai sur le mouvement plan d'un liquide visqueux que limitent des parois, *Journal de Mathématiques Pures et Appliquées* **13**, 331–418.

J. Leray [1933]: Sur le mouvement d'un liquide visqueux emplissant l'espace, *Acta Mathematica* **63**, 193–248.

[57]The proof given here for the existence of $\boldsymbol{u}$ is based on that of Lions [1969, Chapter 1, Section 7.1].

which implies that the trilinear form b is continuous over $(\boldsymbol{H}_0^1(\Omega))^3$ since $H^1(\Omega) \hookrightarrow L^4(\Omega)$ if Ω is a domain in $\mathbb{R}^n$ and $n \leq 4$ (Theorem 6.6-1).

That $(\boldsymbol{u}, \lambda) \in \boldsymbol{H}_0^1(\Omega) \times L_0^2(\Omega)$ satisfies the variational equations for all $\boldsymbol{v} \in \boldsymbol{H}_0^1(\Omega)$ if and only if $(\boldsymbol{u}, \lambda)$ satisfies $-\nu\Delta\boldsymbol{u} + (\boldsymbol{\nabla u})\boldsymbol{u} + \mathbf{grad}\,\lambda = \boldsymbol{f}$ in $\boldsymbol{H}^{-1}(\Omega)$ immediately follows from the definition of differentiation in the sense of distributions.

(ii) *A technical preliminary: Let* $\boldsymbol{w} \in \boldsymbol{H}_0^1(\Omega)$ *be such that* $\operatorname{div}\boldsymbol{w} = 0$ *in* Ω. *Then*

$$b(\boldsymbol{w}; \boldsymbol{v}, \boldsymbol{v}) = 0 \quad \text{for all } \boldsymbol{v} \in \boldsymbol{H}_0^1(\Omega),$$
$$b(\boldsymbol{w}; \boldsymbol{u}, \boldsymbol{v}) = -b(\boldsymbol{w}; \boldsymbol{v}, \boldsymbol{u}) \quad \text{for all } \boldsymbol{u}, \boldsymbol{v} \in \boldsymbol{H}_0^1(\Omega).$$

By Green's formula,

$$b(\boldsymbol{w}; \boldsymbol{v}, \boldsymbol{v}) = \int_\Omega w_j(\partial_j v_i)v_i\,\mathrm{d}x = \frac{1}{2}\int_\Omega w_j\partial_j(v_iv_i)\,\mathrm{d}x = 0 \quad \text{for all } \boldsymbol{v} \in \boldsymbol{\mathcal{D}}(\Omega),$$

since $\partial_j w_j = 0$ in Ω and $v_i = 0$ on Γ. Consequently, $b(\boldsymbol{w}; \boldsymbol{v}, \boldsymbol{v}) = 0$ for all $\boldsymbol{v} \in \boldsymbol{H}_0^1(\Omega)$ since $\mathcal{D}(\Omega)$ is dense in $H_0^1(\Omega)$ and the trilinear form b is continuous over $(\boldsymbol{H}_0^1(\Omega))^3$. Combined with the bilinearity of $b(\boldsymbol{w}; \cdot, \cdot)$, this result implies that, for any $\boldsymbol{u}, \boldsymbol{v} \in \boldsymbol{H}_0^1(\Omega)$,

$$0 = b(\boldsymbol{w}; \boldsymbol{u} + \boldsymbol{v}, \boldsymbol{u} + \boldsymbol{v}) = b(\boldsymbol{w}; \boldsymbol{u}, \boldsymbol{v}) + b(\boldsymbol{w}; \boldsymbol{v}, \boldsymbol{u}).$$

(iii) *Define the space*

$$\boldsymbol{V}(\Omega) := \{\boldsymbol{v} \in \boldsymbol{H}_0^1(\Omega);\ \operatorname{div}\boldsymbol{v} = 0 \text{ in } \Omega\}.$$

Then there exists $\boldsymbol{u} \in \boldsymbol{V}(\Omega)$ *such that*

$$|\boldsymbol{u}|_{1,\Omega} \leq \frac{1}{\nu}\,\|\boldsymbol{f}\|_{\boldsymbol{H}^{-1}(\Omega)} \quad \text{and} \quad a(\boldsymbol{u}, \boldsymbol{v}) + b(\boldsymbol{u}; \boldsymbol{u}, \boldsymbol{v}) = \ell(\boldsymbol{v}) \ \text{ for all } \boldsymbol{v} \in \boldsymbol{V}(\Omega).$$

Note that the variational equations of (i) reduce to these when the vector fields $\boldsymbol{v}$ are restricted to vary in the subspace $\boldsymbol{V}(\Omega)$ of $\boldsymbol{H}_0^1(\Omega)$.

Since $\boldsymbol{V}(\Omega)$ is a separable Hilbert space (as a closed subspace of $(\boldsymbol{H}_0^1(\Omega), (\cdot,\cdot)_{1,\Omega})$), it possesses a Hilbert basis $(\boldsymbol{w}_i)_{i=1}^\infty$. For each integer $n \geq 1$, define the finite-dimensional inner-product space

$$\boldsymbol{V}^n := \operatorname{span}(\boldsymbol{w}_i)_{i=1}^n.$$

Then, given any element $\boldsymbol{w} \in \boldsymbol{V}^n$, there exists one and only one element $\boldsymbol{F}^n(\boldsymbol{w}) \in \boldsymbol{V}^n$ that satisfies

$$(\boldsymbol{F}^n(\boldsymbol{w}), \boldsymbol{v})_{1,\Omega} = a(\boldsymbol{w}, \boldsymbol{v}) + b(\boldsymbol{w}; \boldsymbol{w}, \boldsymbol{v}) - \ell(\boldsymbol{v}) \quad \text{for all } \boldsymbol{v} \in \boldsymbol{V}^n,$$

since the symmetric bilinear form $(\cdot,\cdot)_{1,\Omega}$ and the linear form

$$\ell_{\boldsymbol{w}} : \boldsymbol{v} \in \boldsymbol{V} \to \ell_{\boldsymbol{w}}(\boldsymbol{v}) := a(\boldsymbol{w}, \boldsymbol{v}) + b(\boldsymbol{w}; \boldsymbol{w}, \boldsymbol{v}) - \ell(\boldsymbol{v})$$

satisfy all the assumptions of Theorem 6.1-2. Besides, the mapping $\boldsymbol{F}^n : \boldsymbol{V}^n \to \boldsymbol{V}^n$ defined in this fashion is continuous since the mapping $\boldsymbol{w} \in \boldsymbol{V}(\Omega) \to \ell_{\boldsymbol{w}} \in \boldsymbol{V}(\Omega)'$ is continuous (the continuity of $b : (\boldsymbol{H}_0^1(\Omega))^3 \to \mathbb{R}$ implies that the bilinear mapping $\boldsymbol{w} \in \boldsymbol{V}(\Omega) \to \boldsymbol{a}(\boldsymbol{w}; \boldsymbol{w}; \cdot) \in \boldsymbol{V}(\Omega)'$ is continuous).

Letting $\boldsymbol{v} = \boldsymbol{w}$ in these equations and noting that $b(\boldsymbol{w}; \boldsymbol{w}, \boldsymbol{w}) = 0$ by (ii) gives

$$(\boldsymbol{F}^n(\boldsymbol{w}), \boldsymbol{w})_{1,\Omega} = a(\boldsymbol{w}, \boldsymbol{w}) - \ell(\boldsymbol{w}) \geq \nu\, |\boldsymbol{w}|_{1,\Omega}^2 - \|\boldsymbol{f}\|_{\boldsymbol{H}^{-1}(\Omega)}\, |\boldsymbol{w}|_{1,\Omega}.$$

Consequently,

$$(\boldsymbol{F}^n(\boldsymbol{w}), \boldsymbol{w})_{1,\Omega} \geq 0 \quad \text{for all } \boldsymbol{w} \in \boldsymbol{V}^n \text{ such that } |\boldsymbol{w}|_{1,\Omega} = \frac{1}{\nu} \|\boldsymbol{f}\|_{\boldsymbol{H}^{-1}(\Omega)}.$$

By the *corollary to Brouwer's fixed point theorem* (Theorem 9.9-3; recall that $\boldsymbol{f} \neq \boldsymbol{0}$ by assumption), there thus exists $\boldsymbol{u}^n \in \boldsymbol{V}^n$ such that

$$|\boldsymbol{u}^n|_{1,\Omega} \leq \frac{1}{\nu} \|\boldsymbol{f}\|_{\boldsymbol{H}^{-1}(\Omega)} \quad \text{and} \quad a(\boldsymbol{u}^n, \boldsymbol{v}) + b(\boldsymbol{u}^n; \boldsymbol{u}^n, \boldsymbol{v}) = \ell(\boldsymbol{v}) \text{ for all } \boldsymbol{v} \in \boldsymbol{V}^n,$$

since these variational equations are equivalent to $\boldsymbol{F}^n(\boldsymbol{u}^n) = \boldsymbol{0}$.

The sequence $(\boldsymbol{u}^n)_{n=1}^{\infty}$ obtained in this fashion being bounded in the space $\boldsymbol{V}(\Omega)$, there exists $\boldsymbol{u} \in \boldsymbol{V}(\Omega)$ such that

$$\boldsymbol{u}^n \underset{n\to\infty}{\rightharpoonup} \boldsymbol{u} \text{ in } \boldsymbol{V}(\Omega) \quad \text{and} \quad |\boldsymbol{u}|_{1,\Omega} \leq \liminf_{n\to\infty} |\boldsymbol{u}_n|_{1,\Omega} \leq \frac{1}{\nu} \|\boldsymbol{f}\|_{\boldsymbol{H}^{-1}(\Omega)},$$

by the Banach–Eberlein–Šmulian theorem (Theorem 5.14-4) and by Theorem 5.12-2. Besides, the compact injection $H^1(\Omega) \Subset L^4(\Omega)$ and Theorem 5.12-4(b) together imply that

$$\boldsymbol{u}^n \to \boldsymbol{u} \quad \text{in } \boldsymbol{L}^4(\Omega).$$

Given any element $\boldsymbol{v} \in \boldsymbol{V}(\Omega)$, let $\boldsymbol{v}^n \in \boldsymbol{V}^n$, $n \geq 1$, be such that

$$\lim_{n\to\infty} |\boldsymbol{v}^n - \boldsymbol{v}|_{1,\Omega} = 0.$$

Then the continuity of $b : \boldsymbol{L}^4(\Omega) \times \boldsymbol{H}^1(\Omega) \times \boldsymbol{L}^4(\Omega)$ (cf. the proof of (i)) combined with two applications of (ii) gives

$$\lim_{n\to\infty} b(\boldsymbol{u}^n; \boldsymbol{u}^n, \boldsymbol{v}^n) = -\lim_{n\to\infty} b(\boldsymbol{u}^n; \boldsymbol{v}^n; \boldsymbol{u}^n) = -b(\boldsymbol{u}; \boldsymbol{v}, \boldsymbol{u}) = b(\boldsymbol{u}; \boldsymbol{u}, \boldsymbol{v}).$$

Hence

$$a(\boldsymbol{u}, \boldsymbol{v}) + b(\boldsymbol{u}; \boldsymbol{u}, \boldsymbol{v}) - \ell(\boldsymbol{v}) = \lim_{n\to\infty} \{a(\boldsymbol{u}^n, \boldsymbol{v}^n) + b(\boldsymbol{u}^n; \boldsymbol{u}^n, \boldsymbol{v}^n) - \ell(\boldsymbol{v}^n)\} = 0$$

(that $\lim_{n\to\infty}\{a(\boldsymbol{u}^n, \boldsymbol{v}^n) - \ell(\boldsymbol{v}^n)\} = a(\boldsymbol{u}, \boldsymbol{v}) - \boldsymbol{\ell}(\boldsymbol{v})$ is clear). Since $\boldsymbol{v} \in \boldsymbol{V}(\Omega)$ is arbitrary, the assertion of (iii) is established.

(iv) *Given any* $\boldsymbol{u} \in \boldsymbol{V}(\Omega) = \{\boldsymbol{v} \in \boldsymbol{H}_0^1(\Omega); \operatorname{div} \boldsymbol{v} = 0 \text{ in } \Omega\}$ *such that*

$$a(\boldsymbol{u}, \boldsymbol{v}) + b(\boldsymbol{u}; \boldsymbol{u}, \boldsymbol{v}) = \ell(\boldsymbol{v}) \quad \text{for all } \boldsymbol{v} \in \boldsymbol{V}(\Omega)$$

(the existence of at least one such $\boldsymbol{u} \in \boldsymbol{V}(\Omega)$ is established in (iii)), *there exists one and only one* $\lambda \in L_0^2(\Omega)$ *such that*

$$a(\boldsymbol{u}, \boldsymbol{v}) + b(\boldsymbol{u}; \boldsymbol{u}, \boldsymbol{v}) - \int_\Omega \lambda \operatorname{div} \boldsymbol{v} \, dx = \ell(\boldsymbol{v}) \quad \text{for all } \boldsymbol{v} \in \boldsymbol{H}_0^1(\Omega).$$

We showed in Theorem 6.14-1 that the operator $\mathbf{grad} \in \mathcal{L}(L_0^2(\Omega);\ \boldsymbol{H}^{-1}(\Omega))$ defined by

$$ {}_{H^{-1}(\Omega)}\langle \mathbf{grad}\,\mu, v\rangle_{H_0^1(\Omega)} = -\int_\Omega \mu \operatorname{div} \boldsymbol{v}\, dx \quad \text{for all } \boldsymbol{v} \in \boldsymbol{H}_0^1(\Omega) $$

is injective with a closed image in $\boldsymbol{H}^{-1}(\Omega)$ and that its *dual* (in the normed vector space sense) is the operator $-\operatorname{div} \in \mathcal{L}(\boldsymbol{H}_0^1(\Omega); L_0^2(\Omega))$. Let $\sigma \in \mathcal{L}(\boldsymbol{H}^{-1}(\Omega); \boldsymbol{H}_0^1(\Omega))$ denote the F. Riesz isometry of the space $\boldsymbol{H}_0^1(\Omega)$ (Section 4.6); then

$$ A^* := -\operatorname{div} \in \mathcal{L}(\boldsymbol{H}_0^1(\Omega), L_0^2(\Omega)) $$

becomes the *adjoint operator* (in the Hilbert space sense; cf. Theorem 4.7-2) of

$$ A := \sigma\,\mathbf{grad} \in \mathcal{L}(L_0^2(\Omega); \boldsymbol{H}_0^1(\Omega)). $$

Let

$$ \ell_{\boldsymbol{u}}(\boldsymbol{v}) := a(\boldsymbol{u}, \boldsymbol{v}) + b(\boldsymbol{u}; \boldsymbol{u}, \boldsymbol{v}) - \ell(\boldsymbol{v}) \quad \text{for each } \boldsymbol{v} \in \boldsymbol{H}_0^1(\Omega), $$

and let $\boldsymbol{u} \in \boldsymbol{V}(\Omega)$ be such that

$$ \ell_{\boldsymbol{u}}(\boldsymbol{v}) = 0 \quad \text{for all } \boldsymbol{v} \in \boldsymbol{V}(\Omega). $$

In other words, the continuous linear form $\ell_{\boldsymbol{u}} \in \boldsymbol{H}^{-1}(\Omega)$ vanishes on $\operatorname{Ker} A^* = \boldsymbol{V}(\Omega)$. Therefore,

$$ (\sigma \ell_{\boldsymbol{u}}, \boldsymbol{v})_{1,\Omega} = 0 \quad \text{for all } \boldsymbol{v} \in \operatorname{Ker} A^*, $$

which means that $\sigma\ell_{\boldsymbol{u}} \in (\operatorname{Ker} A^*)^\perp$. But $(\operatorname{Ker} A^*)^\perp = \overline{\operatorname{Im} A}$ (Theorem 4.7-2), and $\overline{\operatorname{Im} A} = \operatorname{Im} A$ in the present case. Hence

$$ \sigma\ell_{\boldsymbol{u}} \in (\operatorname{Ker} A^*)^\perp = \operatorname{Im} A. $$

This means that there exists $\lambda \in L_0^2(\Omega)$ such that

$$ \sigma\ell_{\boldsymbol{u}} = A\lambda = \sigma\,\mathbf{grad}\,\lambda \in \boldsymbol{H}_0^1(\Omega). $$

Therefore, $\ell_{\boldsymbol{u}} = \mathbf{grad}\,\lambda$ and λ is unique because $\mathbf{grad} : L_0^2(\Omega) \to \boldsymbol{H}^{-1}(\Omega)$ is injective. In other words,

$$ \ell_{\boldsymbol{u}}(\boldsymbol{v}) = {}_{\boldsymbol{H}^{-1}(\Omega)}\langle \mathbf{grad}\,\lambda, \boldsymbol{v}\rangle_{\boldsymbol{H}_0^1(\Omega)} = -\int_\Omega \lambda \operatorname{div} \boldsymbol{v}\, dx \quad \text{for all } \boldsymbol{v} \in \boldsymbol{H}_0^1(\Omega), $$

as was to be proved. □

Remarks (1) While the above proof shows that there is at least one solution $(\boldsymbol{u}, \lambda) \in \boldsymbol{H}_0^1(\Omega) \times L_0^2(\Omega)$ to the Navier–Stokes equations that satisfies $|\boldsymbol{u}|_{1,\Omega} \le \frac{1}{\nu} \|\boldsymbol{f}\|_{\boldsymbol{H}^{-1}(\Omega)}$, it does not imply that *any* solution should satisfy this inequality.

(2) The solution to the Navier–Stokes equations is *unique* if the number $\frac{1}{\nu^2} \|\boldsymbol{f}\|_{\boldsymbol{H}^{-1}(\Omega)}$ is small enough; cf. Problem 9.11-1. □

Problem

9.11-1 Show that the solution of the Navier–Stokes equations is unique if $\frac{1}{\nu^2} \|\boldsymbol{f}\|_{\boldsymbol{H}^{-1}(\Omega)} < \frac{1}{\|b\|}$, where $\|b\|$ denotes the norm (as defined in Theorem 2.11-1) of the trilinear form $b : (\boldsymbol{H}_0^1(\Omega))^3 \to \mathbb{R}$ introduced in Theorem 9.11-1.

9.12 Schauder's fixed point theorem; Schäfer's fixed point theorem; Leray–Schauder fixed point theorem

Another *basic theorem of nonlinear functional analysis* is *Schauder's fixed point theorem* (Theorem 9.12-1), which extends Brouwer's fixed point theorem (Theorem 9.9-2) to *infinite-dimensional* normed vector spaces (part(a)), or to *Banach spaces* (part (b)).

Note that, in each case, *compactness* plays again an essential role in this theorem.

Theorem 9.12-1 (Schauder's fixed point theorem[58]) (a) *Let K be a compact and convex subset of a normed vector space X, and let $f : K \to K$ be a continuous mapping. Then f has at least one fixed point.*

(b) *Let C be a closed and convex subset of a Banach space X and let $f : C \to C$ be a continuous mapping with the property that $\overline{f(C)}$ is compact. Then f has at least one fixed point.*

Proof (i) *Let K be a compact and convex subset of a normed vector space X. Then, for each $\varepsilon > 0$, there exists a finite-dimensional subspace Y^ε of X and a continuous mapping $g^\varepsilon : K \to K \cap Y^\varepsilon$ such that*

$$\|g^\varepsilon(x) - x\| \leq \varepsilon \quad \text{for each } x \in K.$$

Let $\varepsilon > 0$ be given. Since K is compact, there exist an integer $N^\varepsilon \geq 1$ and points $x_i^\varepsilon \in K$, $1 \leq i \leq N_\varepsilon$, such that

$$K \subset \bigcup_{i=1}^{N^\varepsilon} B(x_i^\varepsilon; \varepsilon).$$

The functions $g_i^\varepsilon : X \to \mathbb{R}$, $1 \leq i \leq N^\varepsilon$, defined by

$$g_i^\varepsilon(x) := \varepsilon - \|x - x_i^\varepsilon\| \text{ if } x \in B(x_i^\varepsilon; \varepsilon) \quad \text{and} \quad g_i^\varepsilon(x) := 0 \text{ if } x \notin B(x_i^\varepsilon; \varepsilon)$$

satisfy

$$g_i^\varepsilon \in \mathcal{C}(X) \quad \text{and} \quad g_i^\varepsilon(x) \geq 0 \text{ for all } x \in X.$$

Besides,

$$\sum_{i=1}^{N^\varepsilon} g_i^\varepsilon(x) > 0 \quad \text{for all } x \in K,$$

since each point $x \in K$ belongs to at least one open ball $B(x_i^\varepsilon; \varepsilon)$. For each $x \in K$, let

$$\lambda_i^\varepsilon(x) := \left(\sum_{j=1}^{N^\varepsilon} g_j^\varepsilon(x) \right)^{-1} g_i^\varepsilon(x), \quad 1 \leq i \leq N^\varepsilon,$$

and

$$g^\varepsilon(x) := \sum_{i=1}^{N^\varepsilon} \lambda_i^\varepsilon(x) x_i^\varepsilon \in Y^\varepsilon := \text{Span}(x_i^\varepsilon)_{i=1}^{N^\varepsilon}.$$

---

[58] J. SCHAUDER [1930]: Der Fixpunktsatz in Funktionalräumen, *Studia Mathematica* **2**, 171–180.

Then the function $g^\varepsilon : K \to Y^\varepsilon \subset X$ defined in this fashion is clearly continuous and g^ε maps the set K into itself since K is convex (for each $x \in K$, the vector $g_\varepsilon(x)$ is a linear convex combination of the points $x_i^\varepsilon \in K$). For each $x \in K$, let

$$I^\varepsilon(x) = \{1 \le i \le N^\varepsilon;\ \lambda_i^\varepsilon(x) > 0\} = \{1 \le i \le N^\varepsilon;\ x \in B(x_i^\varepsilon;\varepsilon)\}.$$

Then

$$\|g^\varepsilon(x) - x\| = \Big\| \sum_{i\in I^\varepsilon(x)} \lambda_i^\varepsilon(x)(x_i^\varepsilon - x)\Big\| \le \sum_{i\in I^\varepsilon(x)} \lambda_i^\varepsilon(x)\, \|x_i^\varepsilon - x\| \le \varepsilon.$$

(ii) *Proof of* (a). For each $\varepsilon > 0$, let the points x_i^ε, $1 \le i \le N^\varepsilon$, the finite-dimensional space Y^ε, and the mapping $g^\varepsilon : K \to K \cap Y^\varepsilon$ be defined as in (i). In addition, let K^ε denote the convex hull of the set $\bigcup_{i=1}^{N^\varepsilon}\{x_i^\varepsilon\}$, which is thus a *compact* subset of the finite-dimensional space Y^ε (Theorem 2.16-2). *The mapping*

$$f^\varepsilon := (g^\varepsilon \circ f)|_{K^\varepsilon}$$

maps the set K^ε into itself, since $g^\varepsilon(x) \in K^\varepsilon$ for each $x \in K$ and $K^\varepsilon \subset K$ (the set K is convex). Besides, f^ε is continuous as a composition of two continuous mappings. *Brouwer's fixed point theorem* (Theorem 9.9-2) therefore shows that there exists $x^\varepsilon \in K^\varepsilon$ such that

$$f^\varepsilon(x^\varepsilon) = g^\varepsilon(f(x^\varepsilon)) = x^\varepsilon.$$

Since $K^\varepsilon \subset K$ and K is *compact*, there exist $\varepsilon(n) > 0$, $n \ge 1$, with $\lim_{n\to\infty}\varepsilon(n) = 0$ and a point $x \in K$ such that $\lim_{n\to\infty} x^{\varepsilon(n)} = x$. Let $x^n := x^{\varepsilon(n)}$ and $g^n := g^{\varepsilon(n)}$ for each $n \ge 1$; then

$$\|f(x) - x\| \le \|f(x) - f(x^n)\| + \|f(x^n) - x^n\| + \|x^n - x\| \quad \text{for each } n \ge 1.$$

But $\lim_{n\to\infty} f(x_n) = f(x)$ (the mapping f is continuous) and

$$\|f(x^n) - x^n\| = \|f(x^n) - g^n(f(x^n))\| \le \varepsilon(n), \quad n \ge 1, \text{ with } \lim_{n\to\infty}\varepsilon(n) = 0.$$

Hence $f(x) = x$. This proves (a).

(iii) *Proof of* (b). Let K denote the closed convex hull of $\overline{f(C)}$, which is thus *convex* and *compact*, because X is now assumed to be a *Banach space* (Theorem 3.1-5). Since $f(C) \subset C$ implies that $\overline{f(C)} \subset C$ (the set C is closed), which in turn implies that $K \subset C$ (by definition of the closed convex hull, since the set C is closed and convex; cf. Section 2.16), *the continuous mapping $f|_K$ maps K into itself*, since $f(K) \subset f(C) \subset \overline{f(C)} \subset K$.

By (ii), there thus exists $x \in K \subset C$ such that $f(x) = x$. This proves (b). □

Remark The following example shows why compactness is a crucial assumption in Schauder's theorem. Let $X = \ell^2$ and let the mapping $f : C := \overline{B(0,1)} \subset \ell^2 \to \ell^2$ be defined by

$$x = (x_1, x_2, x_3, \dots) \to f(x) := \left(\sqrt{1 - \|x\|_2^2}, x_1, x_2, \dots\right).$$

Then it is immediately verified that f is continuous and maps the closed and convex subset C of ℓ^2 into itself; yet f does *not* have a fixed point in C. □

The notion of *compact linear operator* (Section 2.10) can be extended as follows to *nonlinear* mappings: Let X and Y be normed vector spaces and let A be a subset of X. A mapping $f : A \subset X \to Y$ is said to be **compact** if f is *continuous* and the image $f(B)$ of any *bounded* subset of A is *relatively compact* (i.e., $\overline{f(B)}$ is a compact subset of Y). Note that, *if X is finite-dimensional, any continuous mapping $f : A \subset X \to Y$ is compact.*

Remark The assumption of *continuity* is essential here: While a *linear* mapping that maps any bounded set into a relatively compact set one is automatically continuous (Theorem 2.9-2(d)), there exist *nonlinear* mappings with the same property but that are not continuous. Consider for example the function $f : \mathbb{R} \to \mathbb{R}$ defined for each $x \in \mathbb{R}$ by $f(x) = n$ if $n \leq x < n+1$, $n \in \mathbb{Z}$. □

Schauder's fixed point theorem (Theorem 9.12-1(b)) can therefore be rephrased as follows in terms of compact mappings: *Let C be a closed, bounded, and convex subset of a Banach space and let $f : C \to C$ be a compact mapping. Then f has at least one fixed point.*

An application of Schauder's theorem to an existence theorem for ordinary differential equations is proposed in Problem 9.12-1.

Remark The **Krasnoselskii fixed point theorem**[59] generalizes the Schauder theorem to normed vector spaces equipped with a *partial ordering* (Section 1.3); as such, it provides existence of *nonnegative solutions* to some specific classes of nonlinear boundary value problems. □

The following corollary to Schauder's theorem provides an efficient means of establishing the existence of solutions to specific *nonlinear boundary value problems.*[60] Using this corollary, one can prove for instance that the semilinear problem

$$-\Delta u = f(u) \text{ in } \Omega \quad \text{and} \quad u = 0 \text{ on } \partial\Omega$$

has a solution $u \in H_0^1(\Omega)$ under the only assumptions that $f : \mathbb{R} \to \mathbb{R}$ is continuous and bounded; cf. Problem 9.12-2.

Theorem 9.12-2 (Schäfer's fixed point theorem[61]) *Let X be a Banach space and let $f : X \to X$ be a compact mapping with the property that there exists $r > 0$ such that*

$$\{x \in X;\ \sigma f(x) = x \text{ for some } 0 \leq \sigma \leq 1\} \subset B(0;r).$$

Then f has at least one fixed point in the closed ball $\overline{B(0;r)}$.

---

[59] M.A. KRASNOSELSKII [1960]: Fixed points of cone-compressive or cone-extending operators, *Soviet Mathematics Doklady* **1**, 1285–1288.

See also the illuminating expository paper:

H. AMANN [1976]: Fixed point equations and nonlinear eigenvalue problems in ordered Banach spaces, *SIAM Review* **18**, 620–709.

[60] See for instance GILBARG & TRUDINGER [1998, Section 11.3], where Schäfer's theorem (Theorem 9.12-2) is used for establishing the existence of solutions in the spaces $\mathcal{C}^{2,\alpha}(\overline{\Omega})$, $0 < \alpha < 1$, to a large class of quasi-linear elliptic boundary value problems; see also EVANS [1998, Section 6.5.2], where Schäfer's theorem is used to prove that any uniformly elliptic operator $\mathcal{L}$ has an eigenvalue $\lambda_1 > 0$ such that any other eigenvalue λ of $\mathcal{L}$ satisfies $\operatorname{Re}\lambda \leq \lambda_1$.

[61] H. SCHÄFER [1955]: Über die Methode der a priori Schranken, *Mathematische Annalen* **129**, 415–416.

Proof Given $x \in X$, let

$$g(x) := f(x) \text{ if } \|f(x)\| \le r \quad \text{and} \quad g(x) := r\frac{f(x)}{\|f(x)\|} \text{ if } \|f(x)\| > r.$$

Then it is easily seen that the mapping $g : X \to X$ defined in this fashion is continuous (consider convergent sequences) and that $g(\overline{B(0;r)})$ is relatively compact; to see this, write

$$g(\overline{B(0;r)}) = \{f(x);\ x \in \overline{B(0;r)} \text{ and } \|f(x)\| < r\}$$
$$\cup \left\{r\frac{f(x)}{\|f(x)\|};\ x \in \overline{B(0;r)} \text{ and } \|f(x)\| \ge r\right\},$$

and observe that $\{f(x); x \in \overline{B(0;r)} \text{ and } \|f(x)\| \le r\}$ is relatively compact (as a subset of the relatively compact set $f(\overline{B(0;r)}))$ and that $\left\{r\frac{f(x)}{\|f(x)\|};\ x \in \overline{B(0;r)} \text{ and } \|f(x)\| > r\right\}$ is likewise relatively compact (consider any sequence of points in this set and use the assumed compactness of f).

Since $g(\overline{B(0;r)}) \subset \overline{B(0;r)}$, *Schauder's fixed point theorem* (Theorem 9.12-1(b)) shows that there exists $x \in X$ such that

$$\|x\| \le r \quad \text{and} \quad g(x) = x.$$

We then claim that, necessarily, $\|f(x)\| \le r$. Otherwise $\|f(x)\| > r$ would imply that

$$g(x) = r\frac{f(x)}{\|f(x)\|} = x,$$

but then $\dfrac{r}{\|f(x)\|} < 1$ and $\|x\| = r$, in contradiction with the assumption. The only possibility is thus $\|f(x)\| \le r$, in which case $f(x) = g(x) = x$. □

Schäfer's fixed point theorem is in fact a special case (in that the dependence on the parameter $\sigma \in [0,1]$ is linear) of another *basic theorem of nonlinear functional analysis*, in effect published much earlier. Its proof, which like that of Theorem 9.12-2 essentially relies on *Schauder's fixed point theorem*, is left as a problem; cf. Problem 9.12-4.

$^\flat$**Theorem 9.12-3 (Leray–Schauder fixed point theorem[62])** *Let X be a Banach space and let $f : X \times [0,1] \to X$ be a compact mapping with the following properties:*

$$f(x,0) = 0 \quad \textit{for all } x \in X,$$

and there exists $r > 0$ such that

$$\{x \in X;\ f(x,\sigma) = x \textit{ for some } 0 \le \sigma \le 1\} \subset B(0;r).$$

Then the mapping $f(\cdot,1) : X \to X$ has at least one fixed point in the closed ball $\overline{B(0;r)}$. □

[62] J. Leray; J. Schauder [1934]: Topologie et équations fonctionnelles, *Annales Scientifiques de l'Ecole Normale Supérieure* **51**, 45–78.

Problems

9.12-1 Let $\|\cdot\|$ denote any norm in $\mathbb{R}^n$. Given $T > 0$, $r > 0$, and $\boldsymbol{u}_0 \in \mathbb{R}^n$, let there be given a function $\boldsymbol{g} : [0,T] \times \overline{B(\boldsymbol{u}_0; r)} \to \mathbb{R}^n$ with the following properties: For each $\boldsymbol{v} \in \overline{B(\boldsymbol{u}_0; r)}$, the function $\boldsymbol{g}(\cdot, \boldsymbol{v}) : [0,T] \to \mathbb{R}^n$ is measurable; for each $t \in [0,T]$, the function $\boldsymbol{g}(t,\cdot) : \overline{B(\boldsymbol{u}_0; r)} \to \mathbb{R}^n$ is continuous; finally, there exists a function $h \in L^1(0,T)$ such that $\|\boldsymbol{g}(t,x)\| \le h(t)$ for all $(t,x) \in [0,T] \times \overline{B(\boldsymbol{u}_0; r)}$.

(1) Show that there exists $0 < \tau \le T$ such that the *integral equation*

$$\boldsymbol{u}(t) = \boldsymbol{u}_0 + \int_0^t \boldsymbol{g}(s, \boldsymbol{u}(s))\,\mathrm{d}s, \quad 0 \le t \le \tau,$$

has at least one solution $\boldsymbol{u} \in \mathcal{C}([0,\tau]; \mathbb{R}^n)$.

Hint: Equip the space $\mathcal{C}([0,\tau]; \mathbb{R}^n)$, $\tau > 0$, with the norm $|||\boldsymbol{v}||| = \sup_{0 \le t \le \tau} \|\boldsymbol{v}(t,\cdot)\|$ and show that, if $\tau > 0$ is small enough, there exists $\rho > 0$ such that the mapping $\boldsymbol{f}$ defined by

$$\boldsymbol{f} : \boldsymbol{v} \in C := \{\boldsymbol{v} \in \mathcal{C}([0,\tau]; \mathbb{R}^n);\ |||\boldsymbol{v} - \boldsymbol{u}_0||| \le \rho\} \to \boldsymbol{f}(\boldsymbol{v}) : t \in [0,\tau] \to \boldsymbol{u}_0 + \int_0^t \boldsymbol{g}(s, \boldsymbol{v}(s))\,\mathrm{d}s,$$

maps the closed ball C into itself. Then, using in particular *Ascoli–Arzelà theorem* (Theorem 3.10-1), show that the mapping $C \to C$ defined in this fashion satisfies all the assumptions of *Schauder's fixed point theorem.*

(2) Show that any solution $\boldsymbol{u} \in \mathcal{C}([0,\tau]; \mathbb{R}^n)$ of the integral equation of (1) is *differentiable almost everywhere on* $[0,\tau]$, and that $\boldsymbol{u}'(t) = \boldsymbol{g}(t, \boldsymbol{u}(t))$ at those points $t \in [0,\tau]$ where it is differentiable. Such a function $\boldsymbol{u} \in \mathcal{C}([0,\tau]; \mathbb{R}^n)$ thus provides a generalization of the notion of solution to the *initial value problem*

$$\boldsymbol{u}'(t) = \boldsymbol{g}(t, \boldsymbol{u}(t)),\ 0 \le t \le \tau, \quad \text{and} \quad \boldsymbol{u}(0) = \boldsymbol{u}_0.$$

The results of (1) and (2) together constitute **Carathéodory's existence theorem** for *systems of ordinary differential equations.*[63] Notice that, since its assumptions are weaker than those of the *Cauchy–Peano theorem* (Theorem 3.11-1), so are accordingly its conclusions.

Remark Such generalized solutions are in effect *absolutely continuous.* Recall that a function $f : [a,b] \to \mathbb{R}$ defined on a compact interval $[a,b]$ of $\mathbb{R}$ is absolutely continuous[64] if, given any $\varepsilon > 0$, there exists $\delta = \delta(\varepsilon) > 0$ such that, given any finite family of subintervals $[a_i, b_i] \subset [a,b]$, $1 \le i \le m$, such that $]a_i, b_i[\cap]a_j, b_j[= \varnothing$ if $i \ne j$ and $\sum_{i=1}^m |b_i - a_i| < \delta$, then $\sum_{i=1}^m |f(b_i) - f(a_i)| < \varepsilon$.

A fundamental theorem, due to Henri Lebesgue, then asserts that a function $f : [a,b] \to \mathbb{R}$ is absolutely continuous if and only if it possesses the following properties: f is differentiable almost everywhere, its derivative f' is in the space $L^1[a,b]$, and $f(x) = f(a) + \int_a^x f'(t)\,\mathrm{d}t$ for all $a \le x \le b$.
□

9.12-2 Let Ω be a domain in $\mathbb{R}^n$ with boundary Γ and let $f : \mathbb{R} \to \mathbb{R}$ be a continuous and bounded function. Show that the nonlinear (unless f is a constant function) boundary value problem

$$-\Delta u = f(u) \ \text{ in } \mathcal{D}'(\Omega) \quad \text{and} \quad u = 0 \ \text{ on } \Gamma$$

has at least one solution $u \in H_0^1(\Omega)$.

Hint: Given any $w \in L^2(\Omega)$, let $G(w) \in H_0^1(\Omega)$ denote the unique solution of $-\Delta G(w) = w$ in $\mathcal{D}'(\Omega)$ and $G(w) = 0$ on Γ, and let the mapping $\widetilde{f} : H_0^1(\Omega) \to L^2(\Omega)$ be defined for each $w \in H_0^1(\Omega)$ by $\widetilde{f}(w)(x) := f(w(x))$ for almost all $x \in \Omega$. Then show that $F = G \circ \widetilde{f} : H_0^1(\Omega) \to H_0^1(\Omega)$ is a well-defined compact mapping, and apply *Schäfer's theorem* to this mapping.

[63] So named after Constantin Carathéodory (1873–1950).

[64] For a detailed analysis of absolutely continuous functions, see, e.g., TAYLOR [1965, Section 9.8].

9.12-3 Let Ω be a domain in $\mathbb{R}^n$ with boundary Γ, let $a_{ij} = a_{ji} \in L^\infty(\Omega)$, $1 \le i,j \le n$, be given functions with the property that there exists $\alpha > 0$ such that, for almost all $x \in \Omega$ and all $(\xi_i)_{i=1}^n \in \mathbb{R}^n$, $\sum_{i,j=1}^n a_{ij}(x)\xi_i\xi_j \ge \alpha \sum_{i=1}^n |\xi_i|^2$, and let $f : \mathbb{R} \to \mathbb{R}$ be a function with the property that there exists a constant k such that

$$|f(s) - f(t)| \le k\,|s-t| \quad \text{for all } s,t \in \mathbb{R}.$$

Using *Schäfer's theorem*, show that the nonlinear boundary value problem

$$-\sum_{i,j=1}^n \partial_j(a_{ij}\partial_i u) = f(u) \ \text{ in } \mathcal{D}'(\Omega) \quad \text{and} \quad u = 0 \ \text{ on } \Gamma$$

has at least one solution $u \in H_0^1(\Omega)$ if the Lipschitz constant k is small enough.

Remark Surprisingly, *sharper* results can be obtained by a *simpler* method in this case; cf. Problem 6.10-5. □

9.12-4 This problem provides a proof[65] of the *Leray-Schauder fixed point theorem* (Theorem 9.12-3). In what follows, X is a Banach space and $f : X \times [0,1] \to X$ is a compact mapping that satisfies the assumptions of Theorem 9.12-3; without loss of generality, it is assumed that $r = 1$.

(1) For each $0 < \varepsilon \le 1$, let

$$g_\varepsilon(x) := f\Big(\frac{x}{1-\varepsilon}, 1\Big) \text{ if } \|x\| < 1-\varepsilon \quad \text{and} \quad g_\varepsilon(x) := f\Big(\frac{x}{\|x\|}, \frac{1-\|x\|}{\varepsilon}\Big) \text{ if } 1-\varepsilon \le \|x\| \le 1.$$

Using *Schauder's fixed point theorem* (Theorem 9.12-1(b)), show that the mapping $g_\varepsilon : \overline{B(0;1)} \to X$ defined in this fashion has a fixed point $x(\varepsilon)$ (note that $g_\varepsilon(\partial B(0;1)) = \{0\}$).

(2) For each integer $k \ge 1$, let

$$x_k := x\Big(\frac{1}{k}\Big) \quad \text{and} \quad \sigma_k := 1 \text{ if } \|x_k\| < 1 - \frac{1}{k} \quad \text{and} \quad \sigma_k := k(1-\|x_k\|) \text{ if } 1 - \frac{1}{k} \le \|x_k\| \le 1.$$

Show that there exists a subsequence of $((x_k, \sigma_k))_{k=1}^\infty$ that converges in $X \times [0,1]$ to a limit (x, σ), where x is a fixed point of the mapping $f(\cdot, 1) : X \to X$ and $\sigma = 1$.

9.13 Monotone operators

Monotone operators have acquired a special status among nonlinear operators, especially because they provide an efficient means for establishing the existence of solutions to specific classes of *nonlinear boundary value problems*.[66] Accordingly, they have been extensively studied.[67]

Our purpose here is simply to establish some of their basic properties (Theorems 9.13-1 and 9.13-2) and, especially, a basic *existence theorem* (Theorem 9.14-1); this theorem will

[65] Adapted from GILBARG & TRUDINGER [1998, Section 11.4].

[66] As evidenced by the seminal contributions of:

F.E. BROWDER [1965]: Existence and uniqueness theorems for solutions of nonlinear boundary value problems, in *Proceedings of Symposia in Applied Mathematics, Volume* XVII: *Applications of Nonlinear Partial Differential Equations in Mathematical Physics*, pp. 24–49, American Mathematical Society, Providence, RI.

J. LERAY; J.L. LIONS [1965]: Quelques résultats de Visik sur les problèmes elliptiques non linéaires par les méthodes de Minty-Browder, *Bulletin de la Société Mathématique de France* **93**, 97–107.

[67] See notably the in-depth treatments of monotone operators found in the books of BREZIS [1973] and ZEIDLER [1990a, 1990b].

then be applied to the *p-Laplace operator* (already encountered in Section 9.6). Recall that $\langle \cdot, \cdot \rangle$ designates the duality pairing between a normed vector space V and its dual V', i.e.,

$$\langle \ell, v \rangle = \ell(v) \quad \text{for all } \ell \in V', \quad v \in V.$$

A mapping $A : V \to V'$ is said to be **monotone** if

$$\langle A(v) - A(u), v - u \rangle \geq 0 \quad \text{for all } u, v \in V,$$

and **strictly monotone** if

$$\langle A(v) - A(u), v - u \rangle > 0 \quad \text{for all } u, v \in V, \quad u \neq v.$$

If the space V is *complete*, this seemingly innocuous definition has already two significant implications. Note, however, that no less than the *Banach–Steinhaus theorem* is needed to establish these.

Theorem 9.13-1 *Let V be a real Banach space and let $A : V \to V'$ be a monotone operator. Then A is locally bounded, in the sense that, given any $u \in V$, there exist $r = r(u) > 0$ and $\rho = \rho(u) > 0$ such that*

$$\|v - u\|_V \leq r \quad \textit{implies} \quad \|A(v) - A(u)\|_{V'} \leq \rho.$$

If in addition A is linear, then $A : V \to V'$ is continuous.

Proof It suffices to consider the case where $u = 0$ and $A(0) = 0$ (otherwise introduce the monotone operator $v \in V \to (A(v + u) - A(u))$.

So, assume that $A(0) = 0$ and that A is not locally bounded at 0, in which case there exist $u_n \in V$, $n \geq 1$, such that

$$\|u_n\| \to 0 \quad \text{and} \quad \|Au_n\| \to \infty \quad \text{as } n \to \infty.$$

For each $n \geq 1$ and each $v \in V$, the monotonicity of A implies that

$$-\langle A(u_n), u_n \rangle + \langle A(-v), u_n + v \rangle \leq \langle A(u_n), v \rangle \leq \langle A(u_n), u_n \rangle - \langle A(v), u_n - v \rangle,$$

and thus

$$|\langle A(u_n), v \rangle| \leq \|A(u_n)\| \, \|u_n\| + \max \{\|A(v)\| \, \|u_n - v\|, \|A(-v)\| \, \|u_n + v\|\}.$$

The continuous linear forms defined by

$$\ell_n := \frac{1}{1 + \|A(u_n)\| \, \|u_n\|} A(u_n) \in V', \quad n \geq 1,$$

therefore satisfy

$$\text{for each } v \in V, \quad \sup_{n \geq 1} |\langle \ell_n, v \rangle| \leq C(v) < \infty,$$

with

$$C(v) := 1 + \sup_{n \geq 1} \Big(\max \{\|A(v)\| \, \|u_n - v\|, \|A(-v)\| \, \|u_n + v\|\} \Big).$$

Consequently, by the *Banach–Steinhaus theorem* (cf. Theorem 5.3-1; the assumption that V is complete is used here), there exists a constant C such that

$$\|\ell_n\| = \frac{1}{1 + \|A(u_n)\|\,\|u_n\|}\|A(u_n)\| \leq C \quad \text{for all } n \geq 1.$$

Since $\|u_n\| \to 0$, there exists $n_0 \geq 1$ such that $1 - \|u_n\|\,C \geq \frac{1}{2}$ for all $n \geq n_0$. Then

$$\|A(u_n)\| \leq \frac{C}{1 - \|u_n\|\,C} \leq 2C \quad \text{for all } n \geq n_0,$$

a contradiction. Hence A *is locally bounded.*

If A is *linear*, the direct image under A of *any* bounded subset of V is thus bounded in V' (by the linearity of A). Consequently, A is continuous (Theorem 2.9-2(d)). □

We now introduce another definition (which admittedly may look odd at first glance; at least, it can be expected to be easy to verify on a specific example), which turns out to be one of the essential assumptions in the basic existence theorem for monotone operators (Theorem 9.14-1).

Let V be a normed vector space. A mapping $A : V \to V'$ is said to be **hemicontinuous** if, given any $u, v, w \in V$, there exists $t_0 = t_0(u, v, w) > 0$ such that the function

$$t \in \,]-t_0, t_0[\, \to \langle A(u + tv), w\rangle \in \mathbb{R}$$

is continuous at $t = 0$.

This definition leads to two further significant consequences when it is combined with that of a monotone operator (as usual $\rightharpoonup$ denotes weak convergence).

Theorem 9.13-2 *Let V be a real normed vector space and let $A : V \to V'$ be a hemicontinuous monotone operator.*

(a) *Let $u_n \in V$, $n \geq 1$, be such that*

$$u_n \rightharpoonup u \text{ in } V, \quad A(u_n) \rightharpoonup b \text{ in } V', \quad \text{and} \quad \langle A(u_n), u_n\rangle \to \langle b, u\rangle \text{ in } \mathbb{R} \quad \text{as } n \to \infty.$$

Then $A(u) = b$.

(b) *If the space V is finite-dimensional, then $A : V \to V'$ is continuous.*

Proof Let $(u_n)_{n=1}^{\infty}$ be a sequence of vectors of V that satisfy the assumptions of (a). Then, for each $v \in V$,

$$\langle b - A(v), u - v\rangle = \lim_{n\to\infty} \langle A(u_n) - A(v), u_n - v\rangle \geq 0,$$

since A is monotone. Given any vector $w \in V$ and $t > 0$, letting $v = u + tw$ in the above inequality shows that

$$t\langle b - A(u + tw), -w\rangle \geq 0 \quad \text{for all } t > 0.$$

Consequently,

$$\langle b - A(u + tw), w\rangle \leq 0 \quad \text{for all } t > 0.$$

The assumed hemicontinuity of A then gives

$$\langle b - A(u), w\rangle = \lim_{t\to 0}\langle b - A(u+tw), w\rangle \le 0 \quad \text{for all } w \in V.$$

Hence $A(u) = b$. This proves (a).

Assume next that V is finite-dimensional and let $u_n \in V$, $n \ge 1$, and $u \in V$ be such that

$$u_n \to u \quad \text{in } V \text{ as } n \to \infty.$$

Since A is then locally bounded by Theorem 9.13-1 (which can be applied since V is then a Banach space; cf. Theorem 3.2-1), there exists $r > 0$ such that the direct image of $B(u;r)$ under A is bounded in V'.

Let $n_0 \ge 1$ be such that $u_n \in B(u;r)$ for all $n \ge n_0$. Since the sequence $(A(u_n))_{n=n_0}^\infty$ is then bounded in V' and V' is also finite-dimensional, there exists $b \in V'$ and a subsequence $(u_m)_{m=1}^\infty$ of $(u_n)_{n=n_0}^\infty$ such that

$$A(u_m) \to b \quad \text{in } V' \text{ as } m \to \infty,$$

and thus

$$\langle A(u_m), u_m\rangle \to \langle b, u\rangle \quad \text{as } m \to \infty.$$

Therefore, $A(u) = b$ by (a), which shows that $A : V \to V'$ is continuous since the limit of the subsequence $(A(u_m))_{m=1}^\infty$ is unique (this limit is equal to $A(u)$). This proves (b). □

Problems

9.13-1 Let V be a real Banach space and let $A : V \to V'$ be a hemicontinuous monotone operator. Show that A is *sequentially demicontinuous*, in the sense that

$$v_n \to v \text{ in } V \quad \text{implies} \quad A(v_n) \rightharpoonup A(v) \text{ in } V'.$$

9.13-2 Let V be a real normed vector space. Show that a differentiable function $A : V \to \mathbb{R}$ is convex if and only if its Fréchet derivative $A' \in \mathcal{L}(V;V')$ is monotone.

9.14 The Minty–Browder theorem for monotone operators; application to the p-Laplace operator

Let V be a real normed vector space. A mapping $A : V \to V'$ is said to be **coercive** if

$$\frac{\langle Av, v\rangle}{\|v\|} \to \infty \quad \text{as } \|v\| \to \infty.$$

Remark The same adjective "coercive" has already been used in two related, but slightly different, contexts, viz., to define "V-coercive bilinear forms" (Section 6.1), or "coercive functionals" (Section 9.3). No confusion should arise, however. □

The next result, which gives sufficient conditions guaranteeing that a hemicontinuous monotone operator (Section 9.13) is *surjective*, constitutes another *basic theorem of nonlinear functional analysis*.

Theorem 9.14-1 (Minty–Browder theorem[68]) *Let V be a real separable reflexive Banach space and let $A : V \to V'$ be a coercive and hemicontinuous monotone operator. Then A is surjective, i.e., given any $f \in V'$, there exists u such that*

$$u \in V \quad \textit{and} \quad A(u) = f.$$

If A is strictly monotone, then A is also injective.

Proof The idea is to use the *Galerkin method* (Section 9.10).

(i) Assume that V is infinite-dimensional (if V is finite-dimensional, the surjectivity of A holds by part (ii) of this proof). Since V is separable, there exists a countably infinite linearly independent family $(v_i)_{i=1}^{\infty}$ of vectors $v_i \in V$ such that $\bigcup_{n=1}^{\infty} V_n$ is *dense* in V, where $V_n := \text{Span}(v_i)_{i=1}^{n}$ (Theorem 2.2-7).

(ii) *For each $n \geq 1$, there exists $u_n \in V_n$ such that*

$$\langle A(u_n), v\rangle = \langle f, v\rangle \quad \text{for all } v \in V_n \text{ and } \|u_n\| \leq C,$$

where the constant C is independent of n.

For each $n \geq 1$, let $A_n := A|_{V_n}$. Then the operator $A_n : V_n \to V_n'$ defined in this fashion is monotone since

$$\langle A_n(v) - A_n(u), v - u\rangle = \langle A(v) - A(u), v - u\rangle \geq 0 \quad \text{for all } u, v \in V_n.$$

Since A_n is also hemicontinuous (like A) and V_n is finite-dimensional, $A_n : V_n \to V_n'$ is *continuous* (Theorem 9.13-2(b)).

Let $f_n := f|_{V_n} \in V_n'$, so that $\|f_n\|_{V_n'} \leq \|f\|_{V'}$. Then

$$\frac{1}{\|v\|}\big(\langle A_n(v), v\rangle - \langle f_n, v\rangle\big) \geq \frac{\langle A(v), v\rangle}{\|v\|} - \|f\|_{V'} \quad \text{for each } v \in V_n,\ v \neq 0.$$

By assumption, $\dfrac{\langle A(v), v\rangle}{\|v\|} \to \infty$ as $\|v\| \to \infty$. Hence there exists a constant C *independent of $n \geq 1$* such that

$$\langle A_n(v) - f_n, v\rangle \geq 0 \quad \text{for all } v \in V_n \text{ with } \|v\| = C.$$

Since $A_n : V_n \to V_n'$ is *continuous*, the *corollary to Brouwer's fixed point theorem* (Theorem 9.9-3) can be applied, showing that the mapping $v \in V_n \to (A_n(v) - f_n) \in V_n'$ has a zero u_n in the ball $\overline{B(0; C)}$. To sum up, we have shown that there exists $u_n \in V_n$ such that

$$\langle A(u_n) - f, v\rangle = \langle A_n(u_n) - f_n, v\rangle = 0 \quad \text{for all } v \in V_n \text{ and } \|u_n\| \leq C.$$

---

[68] So named after:

G.J. Minty [1962]: Monotone (nonlinear) operators in Hilbert space, *Duke Mathematical Journal* **29**, 341–346.

G.J. Minty [1963]: On a monotonicity method for the solution of nonlinear equations in Banach spaces, *Proceedings of the National Academy of Sciences USA* **50**, 1038–1041.

F.E. Browder [1963]: Nonlinear elliptic boundary value problems, *Bulletin of the American Mathematical Society* **69**, 862–874.

(iii) *The sequence $(A(u_n))_{n=1}^\infty$ is bounded in V'.*
Since A is locally bounded (Theorem 9.13-1), there exist $r > 0$ and $\rho > 0$ such that

$$\|v\|_V \le r \quad \text{implies} \quad \|A(v)\| \le \rho.$$

Combining this property with the assumed monotonicity of A and the relation $\langle A(u_n), u_n\rangle = \langle f, u_n\rangle$ (which follows from (ii)) gives

$$\begin{aligned}\langle A(u_n), v\rangle &\le \langle A(u_n), u_n\rangle - \langle A(v), u_n\rangle + \langle A(v), v\rangle \\ &= \langle f, u_n\rangle - \langle A(v), u_n\rangle + \langle A(v), v\rangle \\ &\le \|f\|_{V'} C + \rho C + \rho r \quad \text{for all } n \ge 1 \text{ and all } \|v\| \le r.\end{aligned}$$

The boundedness of $(A(u_n))_{n=1}^\infty$ then follows from the relation

$$\|A(u_n)\|_{V'} = \frac{1}{r} \sup_{\|v\|\le r} \langle A(u_n), v\rangle.$$

(iv) *There exists a subsequence $(u_m)_{m=1}^\infty$ of the sequence $(u_n)_{n=1}^\infty$ with the following properties:*

$$u_m \rightharpoonup u \text{ in } V, \quad A(u_m) \rightharpoonup f \text{ in } V', \quad \text{and} \quad \langle A(u_m), u_m\rangle \to \langle f, u\rangle \text{ as } m \to \infty.$$

Since the sequence $(u_n)_{n=1}^\infty$ is bounded in V and the sequence $(A(u_n))_{n=1}^\infty$ is bounded in V' (cf. (ii) and (iii)) and V is *reflexive*, so that V' is also *reflexive* (Theorem 5.14-2(d)), the *Banach–Eberlein–Šmulian theorem* (Theorem 5.14-4) shows that there exist a subsequence $(u_m)_{m=1}^\infty$ of $(u_n)_{n=1}^\infty$ and $u \in V$ and $g \in V'$ such that

$$u_m \rightharpoonup u \text{ in } V \quad \text{and} \quad A(u_m) \rightharpoonup g \text{ in } V' \quad \text{as } m \to \infty.$$

By definition of the sequence $(u_n)_{n=1}^\infty$ (cf. (ii)), for each integer $k \ge 1$,

$$\langle A(u_m), v_k\rangle = \langle f, v_k\rangle \quad \text{for all } m \ge k.$$

Hence

$$\langle g, v_k\rangle = \lim_{m\to\infty} \langle A(u_m), v_k\rangle = \langle f, v_k\rangle.$$

Since this relation holds for any integer $k \ge 1$, this means that

$$\langle g, v\rangle = \langle f, v\rangle \quad \text{for all } v \in \bigcup_{m=1}^\infty V_m.$$

But, by construction, $\bigcup_{m=1}^\infty V_m = \bigcup_{n=1}^\infty V_n$ is dense in V. Hence $g = f$, and thus

$$A(u_m) \rightharpoonup f \quad \text{as } m \to \infty.$$

Finally,

$$\lim_{m\to\infty} \langle A(u_m), u_m\rangle = \lim_{m\to\infty} \langle f, u_m\rangle = \langle f, u\rangle.$$

(v) By Theorem 9.13-2(a), any sequence $(u_m)_{m=1}^{\infty}$ with the properties of (iv) is such that $A(u) = f$. Hence A is *surjective*. That A is *injective* if in addition A is strictly monotone is clear. $\square$

If the space V appearing in Theorem 9.14-1 is a *Hilbert space*, the duality pairing $\langle \cdot, \cdot \rangle$ can be replaced by the inner product in V, in which case the operator $A : V \to V'$ is replaced by the operator $\sigma A : V \to V$, where $\sigma \in \mathcal{L}(V'; V)$ denotes the F. Riesz isometry of V.

In Theorem 9.6-1, we showed that, given any $1 < p < \infty$ and any function $f \in L^q(\Omega)$, with $q = p/(p-1)$, there exists a unique minimizer $u \in W_0^{1,p}(\Omega)$ to the functional J_p defined by

$$J_p(v) := \frac{1}{p} \int_\Omega |\nabla v|^p \, \mathrm{d}x - \int_\Omega f v \, \mathrm{d}x \quad \text{for each } v \in W_0^{1,p}(\Omega),$$

and that this minimizer is also a solution to the *Dirichlet problem for the p-Laplacian*, viz.,

$$-\Delta_p v := -\operatorname{div}\left(|\nabla u|^{p-2} \nabla u\right) = f \text{ in } \mathcal{D}'(\Omega) \quad \text{and} \quad u = 0 \text{ on } \partial\Omega,$$

where Δ_p is the *p-Laplace operator*, or *p-Laplacian*. We now show that the *Minty–Browder theorem* provides a *direct* way to establish the *existence* of a solution to this boundary value problem and, *in addition*, its *uniqueness*.

Theorem 9.14-2 (application to the Dirichlet problem for the p-Laplacian) *Let Ω be a domain in $\mathbb{R}^n$, let $1 < p < \infty$, and let q denote the conjugate exponent of p.*

(a) *The operator*

$$-\Delta_p : v \in W_0^{1,p}(\Omega) \to -\operatorname{div}\left(|\nabla v|^{p-2} \nabla v\right) \in W^{-1,q}(\Omega) := (W_0^{1,p}(\Omega))'$$

is hemicontinuous, coercive, and strictly monotone.

(b) *For each $f \in W^{-1,q}(\Omega)$, the nonlinear boundary value problem*

$$\Delta_p u = f \text{ in } \mathcal{D}'(\Omega) \quad \text{and} \quad u = 0 \text{ on } \partial\Omega$$

has one and only one solution $u \in W_0^{1,p}(\Omega)$.

Proof The duality is given in this case by

$$\langle \Delta_p v, w \rangle = -\int_\Omega |\nabla v|^{p-2} \nabla v \cdot \nabla w \, \mathrm{d}x \quad \text{for each } v, w \in W_0^{1,p}(\Omega).$$

Note that the right-hand side of this relation is well defined since, by Hölder's inequality,

$$\left| \int_\Omega |\nabla v|^{p-2} \nabla v \cdot \nabla w \, \mathrm{d}x \right| \leq \|\nabla v\|_{0,p,\Omega}^{p-1} \|\nabla w\|_{0,p,\Omega},$$

and $w \to \|\nabla w\|_{0,p,\Omega}$ is a norm on the space $W_0^{1,p}(\Omega)$.

Since it is clear that the mapping

$$t \in \mathbb{R} \to \langle \Delta_p(u + tv, w \rangle \in \mathbb{R}$$

is continuous, the operator $-\Delta_p : W_0^{1,p}(\Omega) \to W^{-1,q}(\Omega)$ is *hemicontinuous*.

The proof of Theorem 9.6-1 showed in particular that the functional $I_p : W_0^{1,p}(\Omega) \to \mathbb{R}$ defined by

$$I_p(v) := \frac{1}{p} \int_\Omega |\nabla v|^p \, dx$$

is strictly convex and Gâteaux-differentiable, with a Gâteaux derivative given at each $u, v \in W_0^{1,p}(\Omega)$ by

$$\lim_{t \to 0} \frac{1}{t} (I_p(u + tv) - I_p(u)) = \langle -\Delta_p u, v \rangle.$$

The strict convexity of the functional I_p then easily implies that

$$\langle \Delta_p u - \Delta_p v, u - v \rangle < 0 \quad \text{for all } u, v \in W_0^{1,p}(\Omega),\ u \neq v.$$

Hence $-\Delta_p : W_0^{1,p}(\Omega) \to W^{-1,q}(\Omega)$ is *strictly monotone*.

Finally, for each nonzero $v \in W_0^{1,p}(\Omega)$,

$$\frac{\langle -\Delta_p v, v \rangle}{\|\nabla v\|_{0,p,\Omega}} = \|\nabla v\|_{0,p,\Omega}^{p-1},$$

and thus $-\Delta_p : W_0^{1,p}(\Omega) \to W^{-1,q}(\Omega)$ is *coercive* (recall that $p > 1$ by assumption).

The space $W_0^{1,p}(\Omega)$ being separable and reflexive if $1 < p < \infty$, the conclusions follow from the Minty–Browder theorem (Theorem 9.14-1), since all its assumptions are satisfied. □

Remarks (1) Further properties of the p-Laplace operator are left as problems; cf. Problems 9.14-1 and 9.14-2.

(2) An application of the Minty–Browder theorem to another nonlinear boundary value problem is proposed in Problem 9.14-3.

(3) Let V be a normed vector space and let $\varphi : [0, \infty[\to [0, \infty[$ be a strictly increasing continuous function such that $\varphi(0) = 0$ and $\varphi(r) \to \infty$ as $r \to \infty$. A mapping $J_\varphi : V \to V'$ is said to be a *duality mapping relative to* φ if, for all $v \in V$,

$$\langle J_\varphi(v), v \rangle = \|J_\varphi(v)\|_{V'} \|v\|_V \quad \text{and} \quad \|J_\varphi(v)\|_{V'} = \varphi(\|v\|_V).$$

Such duality mappings play a key role in studying the *geometry of Banach spaces.*[69] The mapping $v \in W_0^{1,p}(\Omega) \to -\Delta_p v \in W^{-1,q}(\Omega)$ thus provides an example of a duality mapping, relative to the function $\varphi : r \to r^{p-1}$. □

Problems

9.14-1 Let Ω be a domain in $\mathbb{R}^n$, let $1 < p < \infty$, and let q denote the conjugate exponent of p. By Theorem 9.14-2, for each $f \in L^q(\Omega)$, there exists a unique $u \in W_0^{1,p}(\Omega)$ such that $\Delta_p u = f$ in $\mathcal{D}'(\Omega)$. Show that the nonlinear mapping $f \in L^q(\Omega) \to u \in W_0^{1,p}(\Omega)$ defined in this fashion is *compact* (Section 9.12).

9.14-2 Let Ω be a domain in $\mathbb{R}^n$, let $1 < p < \infty$, and let q denote the conjugate exponent of p.

[69] G. Dinca [2004]: Duality mappings on infinite dimensional reflexive and smooth Banach spaces are not compact, *Bulletin de l'Académie Royale de Belgique, Classes des Sciences* **6**, 33–40.

(1) Show that, given any function $f \in L^q(\Omega)$, there exists a unique solution $u \in W^{1,p}(\Omega)$ to the variational equations

$$\int_\Omega \left(|\nabla u|^{p-2}\nabla u \cdot \nabla v \mathrm{d}x + |u|^{p-2}uv\right)\mathrm{d}x = \int_\Omega fv\mathrm{d}x \quad \text{for all } v \in W^{1,p}(\Omega).$$

(2) Show that u satisfies a Neumann problem for the operator $v \in W^{1,p}(\Omega) \to -\Delta_p v + |v|^{p-2}v$.

9.14-3 Let $f \in \mathcal{C}([0,1] \times \mathbb{R})$ be a function differentiable with respect to its second argument, with the property that there exists a constant c_0 such that[70]

$$\frac{\partial f}{\partial v}(x,v) \geq c_0 > -\pi^2 \quad \text{for all } (x,v) \in [0,1] \times \mathbb{R}.$$

(1) Show that the nonlinear boundary value problem

$$-u''(x) + f(x,u(x)) = 0, \quad 0 < x < 1, \quad \text{and} \quad u(0) = u(1) = 0,$$

has one and only one weak solution $u \in H_0^1(0,1)$.

Hint: Show that, given any $u \in H_0^1(0,1)$, there exists a unique distribution $A(u) \in H^{-1}(0,1)$ such that

$$\int_0^1 \{u'(x)v'(x) + f(x,u(x))v(x)\}\,\mathrm{d}x = {}_{H^{-1}(0,1)}\langle A(u), v\rangle_{H_0^1(0,1)} \quad \text{for all } v \in H_0^1(0,1).$$

Then show that the nonlinear operator $A : H_0^1(0,1) \to H^{-1}(0,1)$ defined in this fashion satisfies all the assumptions of the Minty–Browder theorem.

(2) Show that $u \in \mathcal{C}^2[0,1]$; hence u is a classical solution to this boundary value problem.

9.14-4 Let U be a nonempty closed convex subset of a separable reflexive real Banach space and let $A : V \to V'$ be a coercive and hemicontinuous monotone operator. Show that, given any $f \in V'$, there exists u such that

$$u \in U \quad \text{and} \quad \langle A(u), v-u\rangle \geq \langle f, v-u\rangle \text{ for all } v \in U$$

(clearly, u is unique if A is strictly monotone).

Hint: First, show that this problem has a solution if V is finite-dimensional (to begin with, consider the case where U is bounded). Then show that $u \in U$ is a solution to this problem if and only if

$$\langle A(v), v-u\rangle \geq \langle f, v-u\rangle \quad \text{for all } v \in U.$$

Remark This result, which constitutes the **Hartman–Stampacchia theorem,**[71] is thus an extension of *Stampacchia's theorem*, where the corresponding operator $A : V \to V'$ is *linear* and *continuous*; cf. Problem 6.2-1. □

9.14-5 Let $(V, (\cdot,\cdot))$ be a real Hilbert space and let $A : V \to V$ be a Lipschitz-continuous and *strongly monotone* mapping, in the sense that there exists α such that

$$\alpha > 0 \quad \text{and} \quad (A(v) - A(u), v-u) \geq \alpha \|v-u\|^2 \quad \text{for all } u, v \in V.$$

[70]For the extension to the n-dimensional case, see Theorem 4.4 in:
P.G. Ciarlet; M.H. Schultz; R.S. Varga [1969]: Numerical methods of high-order accuracy for nonlinear boundary value problems: V. Monotone operator theory, *Numerische Mathematik* **13**, 51–77.

[71]P. Hartman; G. Stampacchia [1966]: On some nonlinear elliptic differential functional equations, *Acta Mathematica* **115**, 271–310.

Show that, given any $b \in V$, the equation $A(u) = b$ has a unique solution u and that the inverse mapping $A^{-1} : V \to V$ is also Lipschitz-continuous.[72]

Hint: Show that the mapping $v \in V \to v - \theta(A(v) - b) \in V$ is a contraction if $\theta > 0$ is small enough.

9.15 The Brouwer topological degree in $\mathbb{R}^n$: Definition and properties

As will be amply illustrated in the subsequent sections, the *Brouwer topological degree*[73] *in* $\mathbb{R}^n$ is a fundamental notion, which serves as a basis for proving basic properties of *nonlinear* mappings in $\mathbb{R}^n$, such as the *existence*, or *nonexistence*, of solutions to *nonlinear equations in* $\mathbb{R}^n$, or their *multiplicity*. The present section is essentially devoted to carrying out the several stages that eventually lead to the *definition* of the degree in its full generality,[74] and to establishing some of its basic properties.

Throughout this section, $|\cdot|$ denotes as usual the Euclidean norm in $\mathbb{R}^n$; in particular, open balls and distances from a point to a set will be meant with respect to $|\cdot|$, and, given a bounded open subset of $\mathbb{R}^n$, the sup-norm of a function $g \in \mathcal{C}(\overline{\Omega}; \mathbb{R}^n)$ is defined by

$$\|g\| := \sup_{x \in \overline{\Omega}} |g(x)| .$$

Let Ω be a bounded open subset of $\mathbb{R}^n$. To begin with, we give a *first* definition of the *degree* $\deg(f, \Omega, b)$ *of a function* $f : \overline{\Omega} \to \mathbb{R}^n$ *with respect to a point* $b \notin f(\partial\Omega)$, which makes sense only for a specific class of functions f, viz., those that are *continuous over* $\overline{\Omega}$ and *continuously differentiable over* Ω; later on, this assumption of differentiability will be removed.

We then show that there is no ambiguity in this definition, in the sense that the number $\deg(f, \Omega, b)$ defined in the next theorem is indeed independent of the function φ appearing in the integral that defines it. Note that an essential use is made in the proof of this independence of the *Piola identity* (Theorem 7.1-4). Not unexpectedly, the same Piola identity already played a key role in the proof of *Brouwer's fixed point theorem* given in Section 9.9, a second proof of which will be given in Section 9.16, this time by means of the degree.

Theorem 9.15-1 *Let Ω be a bounded open subset of $\mathbb{R}^n$ and let a function $f \in \mathcal{C}(\overline{\Omega}; \mathbb{R}^n) \cap \mathcal{C}^1(\Omega; \mathbb{R}^n)$ and a point $b \notin f(\partial\Omega)$ be given. Let $\varphi : [0, \infty[\to \mathbb{R}$ be any function with the following properties:*

$$\varphi \in \mathcal{C}\,[0, \infty[\,, \ \ \operatorname{supp} \varphi \Subset \,]0, \varepsilon_0[\quad \text{where } \varepsilon_0 := \operatorname{dist}(b, f(\partial\Omega)) > 0, \ \text{ and } \int_{\mathbb{R}^n} \varphi(|y|)\, dy = 1.$$

[72] This result is due to:

F. Zarantonello [1960]: Solving functional equations by contractive averaging, *Mathematics Research Center Report No.* 160, University of Wisconsin-Madison, Madison, WI.

[73] So named after the seminal paper:

L.E.J. Brouwer [1912]: Über Abbildung der Mannigfaltigkeiten, *Mathematische Annalen* **71**, 97–115.

[74] The approach followed in this section, which is probably the simplest one for defining the degree, is essentially based on:

E. Heinz [1959]: An elementary analytic theory of the degree of mapping in n-dimensional space, *Journal of Mathematics and Mechanics* **8**, 231–247.

Then the real number

$$\deg(f,\Omega,b) := \int_\Omega \varphi\left(|f(x)-b|\right)\det\nabla f(x)\,\mathrm{d}x$$

is well defined and independent of the function φ. In particular,

$$\deg\left(f,\Omega,b\right)=0 \quad \textit{if } b\notin f(\overline{\Omega}).$$

Proof (i) First, note that $\mathrm{dist}(b, f(\partial\Omega)) > 0$ if $b \notin f(\partial\Omega)$ since the function $x\in\mathbb{R}^n \to d(b,x)$ is continuous and $f(\partial\Omega)$ is compact. Next, let $\varphi \in \mathcal{C}\,[0,\infty[$ be such that $\operatorname{supp}\varphi \Subset]0,\varepsilon_0[$. If $b\notin f(\overline{\Omega})$, then $|f(x)-b|\geq \varepsilon_0$ for all $x\in\Omega$, and thus *in this case*,

$$\deg(f,\Omega,b) = \int_\Omega \varphi\left(|f(x)-b|\right)\det\nabla f(x)\,\mathrm{d}x = 0$$

is *well defined* and *independent of* φ.

If $b\in (f(\overline{\Omega}) - f(\partial\Omega))$,

$$\int_\Omega \varphi\left(|f(x)-b|\right)\det\nabla f(x)\,\mathrm{d}x = \int_{f^{-1}(B(b;\varepsilon_0))} \varphi\left(|f(x)-b|\right)\det\nabla f(x)\,\mathrm{d}x,$$

and thus $\deg(f,\Omega,b)$ is again *well defined* in this case since $f^{-1}(B(b;\varepsilon_0))\Subset\Omega$ and $f\in\mathcal{C}^1(\Omega;\mathbb{R}^n)$ by assumption.

(ii) *Assume again that $b\in (f(\overline{\Omega}) - f(\partial\Omega))$ and let $\varphi \in \mathcal{C}\,[0,\infty[$ and $\widetilde{\varphi} \in \mathcal{C}\,[0,\infty[$ be any two functions that satisfy*

$$\operatorname{supp}\varphi \Subset]0,\varepsilon_0[\,, \quad \operatorname{supp}\widetilde{\varphi} \Subset]0,\varepsilon_0[\,, \quad \text{and} \quad \int_{\mathbb{R}^n}\varphi\left(|x|\right)\,\mathrm{d}x = \int_{\mathbb{R}^n}\widetilde{\varphi}\left(|x|\right)\,\mathrm{d}x = 1.$$

In order to show that $\deg(f,\Omega,b)$ *is independent of φ, we thus have to prove that*

$$\int_\Omega \psi\left(|f(x)-b|\right)\det\nabla f(x)\,\mathrm{d}x = 0$$

for any function $\psi := (\varphi-\widetilde{\varphi})\in\mathcal{C}\,]0,\infty[$ that satisfies

$$\operatorname{supp}\psi \Subset]0,\varepsilon_0[\quad \text{and} \quad \int_0^\infty r^{n-1}\psi(r)\,\mathrm{d}r = 0$$

(the well-known formula $\int_{\mathbb{R}^n}\varphi(|x|)\,\mathrm{d}x = \sigma_n\int_0^\infty r^{n-1}\varphi(r)\,\mathrm{d}r$, where σ_n denotes the area of the unit sphere in $\mathbb{R}^n$, is used here).

To this end, we will first show that, *under the additional assumption that* $f\in\mathcal{C}(\overline{\Omega};\mathbb{R}^n)\cap\mathcal{C}^2(\Omega;\mathbb{R}^n)$ (instead of $f\in\mathcal{C}(\overline{\Omega};\mathbb{R}^n)\cap\mathcal{C}^1(\Omega;\mathbb{R}^n)$; this additional assumption will be later removed, in part (iii) of the proof), *the integrand* $x\in\Omega\to\psi(|f(x)-b|)\det\nabla f(x)$ *can be rewritten as the divergence of an* ad hoc *vector field* $w\in\mathcal{C}^1(\Omega;\mathbb{R}^n)$ *with* $\operatorname{supp} w\Subset\Omega$.

Given any function ψ with the above properties, define the function $\gamma : [0,\infty[\,\to\mathbb{R}$ by

$$r\in[0,\infty[\,\to\gamma(r) := \frac{1}{r^n}\int_0^r s^{n-1}\psi(s)\,\mathrm{d}s,$$

so that

$$\gamma \in \mathcal{C}^1\,[0,\infty[\,, \quad \operatorname{supp}\gamma \Subset\,]0,\varepsilon_0[\,, \quad \text{and} \quad r\gamma'(r)+n\gamma(r)=\psi(r) \quad \text{for all } r \geq 0.$$

Next, define the function $F : \mathbb{R}^n \to \mathbb{R}^n$ by

$$y \in \mathbb{R}^n \to F(y) = (F_k(y))_{k=1}^n := \gamma\,(|y|)\,y.$$

Noting that the function F vanishes in a neighborhood of $0 \in \mathbb{R}^n$ (since the function γ vanishes in a neighborhood of $0 \in \mathbb{R}$), we conclude that

$$F \in \mathcal{C}^1(\mathbb{R}^n;\mathbb{R}^n)$$

(otherwise this would not be necessarily the case since the Euclidean norm $|\cdot|$, like any norm for that matter, is not differentiable at $0 \in \mathbb{R}^n$; cf. Problem 7.1-1); then a simple computation shows that

$$\operatorname{div}_y F(y) := \sum_{j=1}^n \frac{\partial F_j}{\partial y_j}(y) = \gamma'\,(|y|)\,|y| + n\gamma\,(|y|) = \psi\,(|y|) \quad \text{for each } y \in \mathbb{R}^n.$$

Finally, define the function $w : \Omega \to \mathbb{R}^n$ by

$$x \in \Omega \to w(x) := (\operatorname{Cof}\nabla f(x))^T F(f(x)-b).$$

Then $w \in \mathcal{C}^1(\Omega;\mathbb{R}^n)$ (this is why the assumption $f \in \mathcal{C}^2(\Omega;\mathbb{R}^n)$ is needed in this part of the proof) and

$$\begin{aligned}\operatorname{div} w(x) = \sum_{i=1}^n \partial_i w_i(x) &= \sum_{j=1}^n \left\{\sum_{i=1}^n \partial_i\,(\operatorname{Cof}\nabla f(x))_{ji}\right\} F_j\,(f(x)-b)\\ &+ \sum_{j,k=1}^n \left\{\sum_{i=1}^n (\operatorname{Cof}\nabla f(x))_{ji}\,\partial_i f_k(x)\right\}\partial_k F_j\,(f(x)-b)\,, \quad x \in \Omega.\end{aligned}$$

But

$$\sum_{i=1}^n \partial_i\,(\operatorname{Cof}\nabla f(x))_{ji} = 0, \quad 1 \leq i \leq n,$$

since this relation is nothing but the *Piola identity* (Theorem 7.1-4), and

$$\sum_{i=1}^n (\operatorname{Cof}\nabla f(x))_{ji}\,\partial_i f_k(x) = \delta_{jk}\det\nabla f(x), \quad 1 \leq j,k \leq n,$$

since $A(\operatorname{Cof} A)^T = (\det A)I$ for any matrix $A \in \mathbb{M}^n$.

Combining the above relations, we therefore conclude that, for each $x \in \Omega$,

$$\begin{aligned}\operatorname{div} w(x) &= \sum_{j,k=1}^n \delta_{jk}\,(\partial_k F_j(f(x)-b))\det\nabla f(x)\\ &= \Big((\operatorname{div}_y F)\,(f(x)-b)\Big)\det\nabla f(x) = \psi\,(|f(x)-b|)\det\nabla f(x).\end{aligned}$$

Consequently,

$$\int_\Omega \psi(|f(x)-b|)\det\nabla f(x)\,dx = \int_\Omega \operatorname{div} w(x)\,dx = 0,$$

since $\operatorname{supp}\gamma \Subset\,]0,\varepsilon_0[$ implies that the support of the vector field $w \in \mathcal{C}^1(\Omega;\mathbb{R}^n)$ is a compact subset of Ω (as a result, no regularity assumption is needed on the boundary $\partial\Omega$ to infer that $\int_\Omega \operatorname{div} w(x)\,dx = 0$; to see this, simply extend w by 0 on $\mathbb{R}^n - \Omega$ and integrate over a hypercube containing $\overline{\Omega}$). Hence the assertion of (ii) is established under the additional assumption that $f \in \mathcal{C}(\overline{\Omega};\mathbb{R}^n)\cap\mathcal{C}^2(\Omega;\mathbb{R}^n)$.

(iii) We now show that the *assertion of* (ii) *holds as well under the weaker assumption that* $f \in \mathcal{C}(\overline{\Omega};\mathbb{R}^n)\cap\mathcal{C}^1(\Omega;\mathbb{R}^n)$.

Given a function $f \in \mathcal{C}(\overline{\Omega};\mathbb{R}^n)\cap\mathcal{C}^1(\Omega;\mathbb{R}^n)$, let $\widetilde{f} \in \mathcal{C}(\mathbb{R}^n;\mathbb{R}^n)$ be an extension of f (such an extension exists by the Tietze–Urysohn extension theorem; cf. Theorem 1.7-7), and let $(\widetilde{f}_\eta)_{\eta>0}$ be a *regularizing family* (Section 2.6) of $\widetilde{f}$ (i.e., for each $\eta > 0$, $\widetilde{f}_\eta = (\widetilde{f}^i_\eta)_{i=1}^n$, where, for each $1 \le i \le n$, $(\widetilde{f}^i_\eta)_{\eta>0}$ is a regularizing family of the ith component $\widetilde{f}^i$ of $\widetilde{f}$). Then $\widetilde{f}_\eta \in \mathcal{C}^\infty(\mathbb{R}^n;\mathbb{R}^n)$ for each $\eta > 0$, and (Theorem 2.6-1(b))

$$\lim_{\eta\to 0}\|\widetilde{f}_\eta - f\| = 0 \quad\text{and, for each } K \Subset \Omega, \quad \lim_{\eta\to 0}\sup_{x\in K}|\nabla\widetilde{f}_\eta(x) - \nabla f(x)| = 0.$$

Given a point $b \in (f(\overline{\Omega}) - f(\partial\Omega))$, let $\varepsilon_0 := \operatorname{dist}(b, f(\partial\Omega)) > 0$ as before. Since the functions $\widetilde{f}_\eta$ converge uniformly to f on $\overline{\Omega}$ as $\eta \to 0$, for any $0 < \widetilde{\varepsilon}_0 < \varepsilon_0$, there exists $\eta_0 = \eta_0(\widetilde{\varepsilon}_0) > 0$ such that

$$0 < \widetilde{\varepsilon}_0 \le \operatorname{dist}(b, f_\eta(\partial\Omega)) \quad \text{for all } 0 < \eta \le \eta_0.$$

Let then $\psi \in \mathcal{C}\,[0,\infty[$ be any function that satisfies $\operatorname{supp}\psi \Subset\,]0,\widetilde{\varepsilon}_0[$ and $\int_0^\infty r^{n-1}\psi(r)\,dr = 0$, so that, by (ii),

$$\int_\Omega \psi(|f_\eta(x)-b|)\det\nabla f_\eta(x)\,dx = 0 \quad \text{for all } 0 < \eta \le \eta_0.$$

Consequently,

$$\begin{aligned}\int_\Omega \psi(|f(x)-b|)\det\nabla f(x)\,dx &= \int_{f^{-1}(B(b;\widetilde{\varepsilon}_0))} \psi(|f(x)-b|)\det\nabla f(x)\,dx\\ &= \lim_{\eta\to 0}\int_{f_\eta^{-1}(B(b;\widetilde{\varepsilon}_0))} \psi(|f_\eta(x)-b|)\det\nabla f_\eta(x)\,dx\\ &= \lim_{\eta\to 0}\int_\Omega \psi(|f_\eta(x)-b|)\det\nabla f_\eta(x)\,dx = 0,\end{aligned}$$

since there exists a compact subset K of Ω such that

$$f^{-1}(B(b;\widetilde{\varepsilon}_0)) \subset K \quad\text{and}\quad \bigcup_{0<\eta\le\eta_0} f_\eta^{-1}(B(b;\widetilde{\varepsilon}_0)) \subset K$$

(to establish the second inclusion, consider a sequence $(\eta_k)_{k=1}^\infty$ such that $\eta_k > 0$, $k \ge 1$, and $\eta_k \to 0$ as $k \to \infty$).

The conclusion then follows by noting that we may choose $\widetilde{\varepsilon}_0 < \varepsilon_0$ as close as we please to ε_0. □

We next show that *the degree* $\deg(f, \Omega, b)$ *is stable with respect to small enough variations of* $f \in \mathcal{C}(\overline{\Omega}; \mathbb{R}^n) \cap \mathcal{C}^1(\Omega; \mathbb{R}^n)$ *measured with respect to the sup-norm* $\|\cdot\|$ *over* $\overline{\Omega}$. This result will be later on extended to functions $f \in \mathcal{C}(\overline{\Omega}; \mathbb{R}^n)$ (Theorem 9.15-4(b)).

Remark A particularly short proof of the Tietze–Urysohn extension theorem in the special case considered in Theorem 9.15-1 (viz., that of a continuous extension to $\mathbb{R}^n$ of a function continuous on a compact subset of $\mathbb{R}^n$) is proposed in Problem 9.15-1. □

Theorem 9.15-2 *Let* Ω *be a bounded open subset of* $\mathbb{R}^n$.

(a) *Let a function* $f \in \mathcal{C}(\overline{\Omega}; \mathbb{R}^n) \cap \mathcal{C}^1(\Omega; \mathbb{R}^n)$ *and a point* $b \notin f(\partial\Omega)$ *be given, and let* $r = r(f, b)$ *be any number that satisfies*

$$0 < r < \frac{1}{5} \operatorname{dist}(b, f(\partial\Omega)).$$

Then any function $g \in \mathcal{C}(\overline{\Omega}, \mathbb{R}^n) \cap \mathcal{C}^1(\Omega; \mathbb{R}^n)$ *that satisfies*

$$\|g - f\| < r$$

also satisfies

$$b \notin g(\partial\Omega) \quad \text{and} \quad \|g - f\| < \frac{1}{4} \operatorname{dist}(b, f(\partial\Omega) \cup g(\partial\Omega)).$$

(b) *Let two functions* $f, g \in \mathcal{C}(\overline{\Omega}; \mathbb{R}^n) \cap \mathcal{C}^1(\Omega; \mathbb{R}^n)$ *and a point* $b \notin f(\partial\Omega) \cup g(\partial\Omega)$ *be given with the property that*

$$\|g - f\| < \frac{1}{4} \operatorname{dist}(b, f(\partial\Omega) \cup g(\partial\Omega)).$$

Then

$$\deg(g, \Omega, b) = \deg(f, \Omega, b).$$

Proof (i) Given a function $f \in \mathcal{C}(\overline{\Omega}, \mathbb{R}^n)$ and a point $b \notin f(\partial\Omega)$, let r be any number that satisfies

$$0 < r < \frac{1}{5} \operatorname{dist}(b, f(\partial\Omega)).$$

Since $\operatorname{dist}(b, g(\partial\Omega)) \geq \operatorname{dist}(b, f(\partial\Omega)) - \|g - f\|$ for any function $g \in \mathcal{C}(\overline{\Omega}, \mathbb{R}^n)$,

$$g \in \mathcal{C}(\overline{\Omega}, \mathbb{R}^n) \quad \text{and} \quad \|g - f\| < r \quad \text{implies} \quad \operatorname{dist}(b, g(\partial\Omega)) > 4r.$$

Hence, if $g \in \mathcal{C}(\overline{\Omega}; \mathbb{R}^n)$ satisfies $\|g - f\| < r$, we have

$$\begin{aligned} \|g - f\| < r < \min\left\{\frac{1}{4} \operatorname{dist}(b, g(\partial\Omega)), \frac{1}{5} \operatorname{dist}(b, f(\partial\Omega))\right\} \\ \leq \frac{1}{4} \min\left\{\operatorname{dist}(b, f(\partial\Omega)), \operatorname{dist}(b, g(\partial\Omega))\right\} \leq \frac{1}{4} \operatorname{dist}(b, f(\partial\Omega) \cup g(\partial\Omega)). \end{aligned}$$

This proves (a).

(ii) Let now $f, g \in \mathcal{C}(\overline{\Omega}; \mathbb{R}^n) \cap \mathcal{C}^1(\Omega; \mathbb{R}^n)$ and $b \notin f(\partial\Omega) \cup g(\partial\Omega)$ be such that

$$\|g - f\| < \frac{1}{4} \operatorname{dist}(b, f(\partial\Omega) \cup g(\partial\Omega)).$$

To show that $\deg(f, \Omega, b) = \deg(g, \Omega, b)$, there is no loss of generality in assuming (for notational brevity) that $b = 0$, since it is clear that $\deg(f, \Omega, b) = \deg(f - b, \Omega, 0)$. Let then ε be any number that satisfies

$$\|g - f\| < \varepsilon < \frac{1}{4} \operatorname{dist}(0, f(\partial\Omega) \cup g(\partial\Omega)),$$

let $\chi : [0, \infty[\to [0, 1]$ be any function that satisfies

$$\chi \in \mathcal{C}^1 [0, \infty[, \quad \chi(r) = 1 \text{ if } 0 \leq r \leq 2\varepsilon, \quad \text{and} \quad \chi(r) = 0 \text{ if } r \geq 3\varepsilon,$$

and let the function $h : \overline{\Omega} \to \mathbb{R}^n$ be defined by

$$x \in \overline{\Omega} \to h(x) := (1 - \chi(|f(x)|))\, f(x) + \chi(|f(x)|)\, g(x).$$

Then it is clear that $h \in \mathcal{C}(\overline{\Omega}, \mathbb{R}^n) \cap \mathcal{C}^1(\Omega; \mathbb{R}^n)$, since

$$\begin{aligned}
&h(x) = g(x) \quad \text{if } |f(x)| < 2\varepsilon, \\
&h(x) = g(x) + (1 - \chi(|f(x)|))(f(x) - g(x)) \quad \text{if } 2\varepsilon = |f(x)| \leq 3\varepsilon, \\
&h(x) = f(x) \quad \text{if } 3\varepsilon < |f(x)|,
\end{aligned}$$

and h is of class $\mathcal{C}^1$ on the two open sets $\{x \in \Omega;\ |f(x)| < 2\varepsilon\}$ and $\{x \in \Omega,\ |f(x)| > \varepsilon\}$, the union of which is Ω. The above relations also show that $\|h - f\| \leq \|f - g\|$ and $\|h - g\| \leq \|f - g\|$. Hence

$$\|h - f\| < \varepsilon \quad \text{and} \quad \|h - g\| < \varepsilon.$$

Since $h(x) = f(x)$ if $x \in \partial\Omega$ (because then $|f(x)| \geq \operatorname{dist}(0, f(\partial\Omega)) > 4\varepsilon$), it follows that $\operatorname{dist}(0, h(\partial\Omega)) = \operatorname{dist}(0, f(\partial\Omega)) > 4\varepsilon$.

Let now $\varphi : [0, \infty[\to \mathbb{R}$ and $\psi : [0, \infty[\to \mathbb{R}$ be any two functions with the following properties

$$\varphi \in \mathcal{C}[0, \infty[, \quad \operatorname{supp} \varphi \Subset]3\varepsilon, 4\varepsilon[, \quad \text{and} \quad \int_{\mathbb{R}^n} \varphi(|y|)\, dy = 1,$$

$$\psi \in \mathcal{C}[0, \infty[, \quad \operatorname{supp} \psi \Subset]0, \varepsilon[, \quad \text{and} \quad \int_{\mathbb{R}^n} \psi(|y|)\, dy = 1.$$

Since $4\varepsilon < \min\{\operatorname{dist}(0, f(\partial\Omega)), \operatorname{dist}(0, g(\partial\Omega)), \operatorname{dist}(0, h(\partial\Omega))\}$, we infer from Theorem 9.15-1 that $\deg(f, \Omega, 0)$ and $\deg(h, \Omega, 0)$ may be defined as

$$\deg(f, \Omega, 0) = \int_\Omega \varphi(|f(x)|) \det \nabla f(x)\, dx = \int_{3\varepsilon < |f(x)| < 4\varepsilon} \varphi(|f(x)|) \det \nabla f(x)\, dx,$$

$$\deg(h, \Omega, 0) = \int_\Omega \varphi(|h(x)|) \det \nabla h(x)\, dx = \int_{3\varepsilon < |h(x)| < 4\varepsilon} \varphi(|h(x)|) \det \nabla h(x)\, dx.$$

Hence $\deg(f,\Omega,0) = \deg(h,\Omega,0)$ since $h(x) = f(x)$ if $3\varepsilon < |f(x)|$. Likewise, $\deg(g,\Omega,0)$ and $\deg(h,\Omega,0)$ may be defined as

$$\deg(g,\Omega,0) = \int_\Omega \psi(|g(x)|)\det\nabla g(x)\,dx = \int_{|g(x)|<\varepsilon} \psi(|g(x)|)\det\nabla g(x)\,dx,$$

$$\deg(h,\Omega,0) = \int_\Omega \psi(|h(x)|)\det\nabla h(x)\,dx = \int_{|h(x)|<\varepsilon} \psi(|h(x)|)\det\nabla h(x)\,dx.$$

Hence $\deg(g,\Omega,0) = \deg(h,\Omega,0)$, since $|g(x)| < \varepsilon$ implies $|f(x)| \le |g(x)| + |f(x) - g(x)| < 2\varepsilon$, which in turn implies that $h(x) = g(x)$ if $|g(x)| < \varepsilon$. Consequently,

$$\deg(f,\Omega,0) = \deg(h,\Omega,0) = \deg(g,\Omega,0).$$

This proves (b). □

The next theorem, which makes an essential use of Theorem 9.15-2, will pave the way for extending the definition of $\deg(f,\Omega,b)$ to functions $f : \overline{\Omega} \to \mathbb{R}^n$ that are only continuous over $\overline{\Omega}$.

Theorem 9.15-3 *Let Ω be a bounded open subset of $\mathbb{R}^n$ and let a function $f \in \mathcal{C}(\overline{\Omega};\mathbb{R}^n)$ be given.*

(a) *There exist sequences $(f_k)_{k=1}^\infty$ with the following properties:*

$$f_k \in \mathcal{C}(\overline{\Omega};\mathbb{R}^n) \cap \mathcal{C}^1(\Omega;\mathbb{R}^n) \quad \textit{for all } k \ge 1 \textit{ and } \|f_k - f\| \to 0 \textit{ as } k \to \infty.$$

(b) *Let a point $b \notin f(\partial\Omega)$ be given. Then, given any such sequence $(f_k)_{k=1}^\infty$, there exists an integer $k_0 \ge 1$ such that $\deg(f_k,\Omega,b)$ is well defined for all $k \ge k_0$, and there exists an integer $k_1 \ge k_0$ such that*

$$\deg(f_k,\Omega,b) = \deg(f_{k_1},\Omega,b) \quad \textit{for all } k \ge k_1.$$

(c) *Besides, such a number $\lim_{k\to\infty}\deg(f_k,\Omega,b)$ is independent of the sequence $(f_k)_{k=1}^\infty$.*

Proof (i) *Proof of* (a): Since $f \in \mathcal{C}(\overline{\Omega};\mathbb{R}^n)$ and $\overline{\Omega}$ is a closed subset of $\mathbb{R}^n$, there exists a function $\widetilde{f} \in \mathcal{C}(\mathbb{R}^n;\mathbb{R}^n)$ that extends f by the *Tietze–Urysohn extension theorem* (Theorem 1.7-7). Then any *regularizing family* $(\widetilde{f}_\varepsilon)_{\varepsilon>0}$ of $\widetilde{f}$ is such that $\widetilde{f}_\varepsilon \in \mathcal{C}^\infty(\Omega;\mathbb{R}^n)$ and $(\widetilde{f}_\varepsilon)_{\varepsilon>0}$ converges uniformly to $\widetilde{f}$ over any compact subset of $\mathbb{R}^n$ (Theorem 2.6-1(b)), hence in particular over $\overline{\Omega}$. Therefore the functions $f_k := \widetilde{f}_{\varepsilon_k}|_{\overline{\Omega}}$, $k \ge 1$, where $\lim_{k\to\infty}\varepsilon_k = 0^+$, possess the required properties.

(ii) *Proof of* (b): Given a point $b \notin f(\partial\Omega)$, let ε be any number that satisfies $0 < 4\varepsilon < \operatorname{dist}(b, f(\partial\Omega))$, and let $(f_k)_{k=1}^\infty$ be any sequence with the properties of (a). Since

$$|\operatorname{dist}(b, f_k(\partial\Omega)) - \operatorname{dist}(b, f(\partial\Omega))| \le \|f_k - f\| \quad \text{and} \quad \lim_{k\to\infty}\|f - f_k\| = 0,$$

there exists $k_0 \ge 1$ such that

$$\operatorname{dist}(b, f_k(\partial\Omega)) > 4\varepsilon \quad \text{for all } k \ge k_0,$$

so that $b \notin f_k(\partial\Omega)$ for $k \geq k_0$. Therefore $\deg(f_k, \Omega, b)$ is well defined for all $k \geq k_0$. Since there exists $k_1 \geq k_0$ such that $\|f_k - f_\ell\| < \varepsilon$ for all $k, \ell \geq k_1$ (naturally, the integer k_1 depends on the particular sequence $(f_k)_{k=1}^\infty$ considered), it thus follows that

$$\|f_k - f_\ell\| < \frac{1}{4}\,\mathrm{dist}\,(b, f_k(\partial\Omega) \cup f_\ell(\partial\Omega)) \quad \text{for all } k, \ell \geq k_1.$$

Theorem 9.15-2(b) therefore implies that

$$\deg(f_k, \Omega, b) = \deg(f_\ell, \Omega, b) \quad \text{for all } k, \ell \geq k_1,$$

which shows that the sequence $(\deg(f_k, \Omega, b))_{k \geq k_1}$ is stationary.

(iii) *Proof of* (c): Let now $(\widetilde{f}_k)_{k=1}^\infty$ be another sequence with the properties of (a), so that there exists $\widetilde{k}_1$ such that the sequence $(\deg(\widetilde{f}_k, \Omega, b))_{k \geq \widetilde{k}_1}$ is stationary by (i). Noting that the sequence $(f_1, \widetilde{f}_1, f_2, \widetilde{f}_2, \ldots, f_k, \widetilde{f}_k, \ldots)$ is also a sequence with the properties of (a), we conclude that the limits of the stationary sequences $(\deg(f_k, \Omega, b))_{k \geq k_1}$ and $(\deg(\widetilde{f}_k, \Omega, b))_{k \geq \widetilde{k}_1}$ are necessarily the same since they are both subsequences of the same convergent sequence. Hence $\lim_{k\to\infty} \deg(f_k, \Omega, b)$ is independent of the sequence $(f_k)_{k=1}^\infty$. □

Remark An alternate proof of (a) consists in using the *Weierstraß polynomial approximation theorem in several variables* (Theorem 2.15-2), which asserts that there exists for each $1 \leq i \leq n$ a sequence $(f_k^i)_{k=1}^\infty$ of polynomials $f_k^i : \mathbb{R}^n \to \mathbb{R}$ such that $(f_k^i|_{\overline{\Omega}})_{k=1}^\infty$ uniformly converges to the ith component of f on $\overline{\Omega}$. Hence the functions $f_k := (f_k^i|_{\overline{\Omega}})_{i=1}^n$, $k \geq 1$, possess the required properties. □

We now extend the definition of $\deg(f, \Omega, b)$, heretofore restricted to functions $f \in \mathcal{C}(\overline{\Omega}; \mathbb{R}^n) \cap \mathcal{C}^1(\Omega; \mathbb{R}^n)$, to functions $f \in \mathcal{C}(\overline{\Omega}; \mathbb{R}^n)$: Given a bounded open subset Ω of $\mathbb{R}^n$, a function $f \in \mathcal{C}(\overline{\Omega}; \mathbb{R}^n)$, and a point $b \notin f(\partial\Omega)$, the **Brouwer topological degree** *of f with respect to b* is defined as

$$\deg(f, \Omega, b) := \lim_{k\to\infty} \deg(f_k, \Omega, b),$$

where $(f_k)_{k=1}^\infty$ is *any* sequence of functions that satisfies

$$f_k \in \mathcal{C}(\overline{\Omega}; \mathbb{R}^n) \cap \mathcal{C}^1(\Omega; \mathbb{R}^n) \text{ and } b \notin f_k(\partial\Omega) \quad \text{for all } k \geq 1 \text{ and } \lim_{k\to\infty} \|f_k - f\| = 0,$$

and $\deg(f_k, \Omega, b)$ is defined for each $k \geq 1$ as in Theorem 9.15-1: this definition makes perfect sense because *the sequence $(\deg(f_k, \Omega, b))_{k=1}^\infty$ is stationary for k large enough and its limit is independent of the sequence considered* (Theorem 9.15-3).

Note that, if $f \in \mathcal{C}(\overline{\Omega}; \mathbb{R}^n) \cap \mathcal{C}^1(\Omega; \mathbb{R}^n)$, the consideration of the sequence $(f_k)_{k=1}^\infty$ defined by $f_k := f$ for all $k \geq 1$ shows that the degree as defined above does coincide with the degree as defined in Theorem 9.15-1, and hence there is no ambiguity in using the same notation $\deg(f, \Omega, b)$.

The next theorem establishes various simple properties of the degree. Properties (b) and (c) mean that $\deg(f, \Omega, b)$ *is stable with respect to* (small enough) *variations of f measured with the sup-norm* $\|\cdot\|$ (a property already established in Theorem 9.15-2 for functions $f \in \mathcal{C}(\overline{\Omega}; \mathbb{R}^n) \cap \mathcal{C}^1(\Omega; \mathbb{R}^n)$) *and to variations of b in a connected component of* $\mathbb{R}^n - f(\partial\Omega)$.

Note that *each connected component of* $\mathbb{R}^n - f(\partial\Omega)$ *is open* (Theorem 2.2-6).

Theorem 9.15-4 *Let Ω be a bounded open subset of $\mathbb{R}^n$, and let $f \in \mathcal{C}(\overline{\Omega}; \mathbb{R}^n)$ and $b \notin f(\partial\Omega)$.*

(a) *If $b \notin f(\overline{\Omega})$, then*

$$\deg(f, \Omega, b) = 0.$$

Hence

$$\deg(f, \Omega, b) \neq 0 \quad \textit{implies that } b = f(x) \textit{ for some } x \in \Omega.$$

(b) *Let r be any number that satisfies*

$$0 < r < \frac{1}{5} \operatorname{dist}(b, f(\partial\Omega)).$$

Then

$$g \in \mathcal{C}(\overline{\Omega}; \mathbb{R}^n) \textit{ and } \|g - f\| < r \textit{ implies } b \notin g(\partial\Omega) \quad \textit{and} \quad \deg(g, \Omega, b) = \deg(f, \Omega, b).$$

Besides, given any $0 < \varepsilon \le r$, there exists $g_\varepsilon \in \mathcal{C}(\overline{\Omega}, \mathbb{R}^n) \cap \mathcal{C}^1(\Omega; \mathbb{R}^n)$ such that

$$\|g_\varepsilon - f\| < \varepsilon, \quad b \notin g_\varepsilon(\partial\Omega), \quad \textit{and} \quad \deg(g_\varepsilon, \Omega, b) = \deg(f, \Omega, b).$$

(c) *The following relation holds:*

$$\deg(f, \Omega, b) = \deg(f - b, \Omega, 0).$$

(d) *The function*

$$b \in (\mathbb{R}^n - f(\partial\Omega)) \to \deg(f, \Omega, b)$$

is constant in each connected component of $\mathbb{R}^n - f(\partial\Omega)$.

Proof (i) *Proof of* (a): Let $f \in \mathcal{C}(\overline{\Omega}; \mathbb{R}^n)$ and let $b \notin f(\overline{\Omega})$. Let a sequence $(f_k)_{k=1}^{\infty}$ with $f_k \in \mathcal{C}(\overline{\Omega}, \mathbb{R}^n) \cap \mathcal{C}^1(\Omega; \mathbb{R}^n)$ be such that $\|f_k - f\| \to 0$ as $k \to \infty$, so that there exists $k_0 \ge 1$ such that $b \notin f_k(\overline{\Omega})$ for all $k \ge k_0$. Since then $\deg(f_k, \Omega, b) = 0$ for all $k \ge k_0$ by Theorem 9.15-1, we infer from Theorem 9.15-3 that, in this case,

$$\deg(f, \Omega, b) = \lim_{k \to \infty} \deg(f_k, \Omega, b) = 0$$

(ii) *Proof of* (b): Let r be such that $0 < r < (1/5) \operatorname{dist}(b, f(\partial\Omega))$, and let $g \in \mathcal{C}(\overline{\Omega}; \mathbb{R}^n)$ be any function that satisfies $\|g - f\| < r$.

Let $(f_k)_{k=1}^{\infty}$ and $(g_k)_{k=1}^{\infty}$ with $f_k, g_k \in \mathcal{C}(\overline{\Omega}; \mathbb{R}^n) \cap \mathcal{C}^1(\Omega; \mathbb{R}^n)$ be such that $\|f_k - f\| \to 0$ and $\|g_k - g\| \to 0$ as $k \to \infty$. Then it is clear that there exists $k_0 \ge 1$ such that

$$0 < r < \frac{1}{5} \operatorname{dist}(b, f_k(\partial\Omega)) \quad \text{and} \quad \|g_k - f_k\| < r \text{ for all } k \ge k_0,$$

so that $\deg(f_k, \Omega, b) = \deg(g_k, \Omega, b)$ for all $k \ge k_0$ by Theorem 9.15-2; consequently,

$$\deg(f, \Omega, b) = \lim_{k \to \infty} \deg(f_k, \Omega, b) = \lim_{k \to \infty} \deg(g_k, \Omega, b) = \deg(g, \Omega, b).$$

The second property in (b) then follows by Theorem 9.15-3(a).

(iii) *Proof of* (c): Let $(f_k)_{k=1}^{\infty}$ with $f_k \in \mathcal{C}(\overline{\Omega};\mathbb{R}^n) \cap \mathcal{C}^1(\Omega;\mathbb{R}^n)$ be such that $\|f_k - f\| \to 0$ as $k \to \infty$, so that there exists ε_0 and an integer $k_0 \geq 1$ such that

$$0 < \varepsilon_0 \leq \operatorname{dist}(b, f_k(\partial\Omega)) \quad \text{for all } k \geq k_0.$$

Pick any function $\varphi \in \mathcal{C}\,[0,\infty[$ with $\operatorname{supp}\varphi \Subset\,]0,\varepsilon_0[$. Then, by Theorem 9.15-1,

$$\begin{aligned}\deg(f_k,\Omega,b) &= \int_\Omega \varphi(|f_k(x)-b|)\det\nabla f_k(x)\,\mathrm{d}x \\ &= \int_\Omega \varphi(|(f_k-b)(x)|)\det\nabla(f_k-b)(x)\,\mathrm{d}x = \deg(f_k-b,\Omega,0) \quad \text{for all } k\geq k_0.\end{aligned}$$

Passing to the limit as $k \to \infty$ then shows that $\deg(f,\Omega,b) = \deg(f-b,\Omega,0)$.

(iv) *Proof of* (d): It suffices to show that the function $b \in (\mathbb{R}^n - f(\partial\Omega)) \to \deg(f,\Omega,b)$ is *locally constant.*

Given any $b \notin f(\partial\Omega)$, let the function $g \in \mathcal{C}(\overline{\Omega};\mathbb{R}^n)$ be defined by

$$g(x) := f(x) - b, \quad x \in \overline{\Omega}.$$

By (b), there exists $r = r(f,b) > 0$ such that, if $\widetilde{g} \in \mathcal{C}(\overline{\Omega};\mathbb{R}^n)$ satisfies $\|\widetilde{g} - g\| < r$, then $b \notin \widetilde{g}(\partial\Omega)$ and $\deg(g,\Omega,0) = \deg(\widetilde{g},\Omega,0)$ (note that $b \notin f(\partial\Omega)$ implies $0 \notin g(\partial\Omega)$). Given any point $\widetilde{b} \in (\mathbb{R}^n - f(\partial\Omega))$ such that $|\widetilde{b} - b| < r$, define the function $\widetilde{g} \in \mathcal{C}(\overline{\Omega};\mathbb{R}^n)$ by

$$\widetilde{g}(x) = f(x) - \widetilde{b}, \quad x \in \overline{\Omega},$$

so that $\|\widetilde{g} - g\| = |\widetilde{b} - b| < r$. Then, on the one hand,

$$\deg(\widetilde{g},\Omega,0) = \deg(g,\Omega,0),$$

and, on the other hand,

$$\deg(\widetilde{g},\Omega,0) = \deg(f,\Omega,\widetilde{b}) \quad \text{and} \quad \deg(g,\Omega,0) = \deg(f,\Omega,b),$$

by (c). Hence $\deg(f,\Omega,\widetilde{b}) = \deg(f,\Omega,b)$ if $|\widetilde{b} - b| < r$. □

We now establish two fundamental properties of the Brouwer degree:

First, *if $f \in \mathcal{C}(\overline{\Omega};\mathbb{R}^n) \cap \mathcal{C}^1(\Omega;\mathbb{R}^n)$, then $\deg(f,\Omega,b)$ can be also defined by a remarkably simple formula,* save when the point $b \notin f(\partial\Omega)$ belongs to a set (denoted $f(S_f)$ below) of zero Lebesgue measure.

Second, *if $f \in \mathcal{C}(\overline{\Omega};\mathbb{R}^n)$ and $b \notin f(\partial\Omega)$, then $\deg(f,\Omega,b)$ is an integer in $\mathbb{Z}$* (all that we already know in this respect is that $\deg(f,\Omega,b) = 0$ if $b \notin f(\overline{\Omega})$; cf. Theorem 9.15-4(a)); see Figures 9.15-1 and 9.15-2.

Theorem 9.15-5 *Let Ω be a bounded open subset of $\mathbb{R}^n$.*

(a) *Given $f \in \mathcal{C}(\overline{\Omega};\mathbb{R}^n) \cap \mathcal{C}^1(\Omega;\mathbb{R}^n)$, let*

$$S_f = \{x \in \Omega;\ \det\nabla f(x) = 0\}.$$

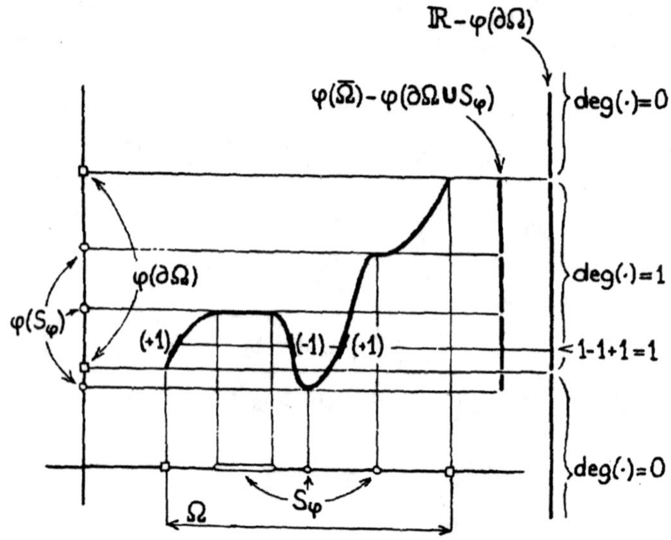

Figure 9.15-1 *The topological degree of a function* $f : \overline{\Omega} \subset \mathbb{R} \to \mathbb{R}$. This figure originally appeared in P.G. CIARLET [1988]: *Mathematical Elasticity, Volume* I: *Three-Dimensional Elasticity*, North-Holland, Amsterdam.

Then, given any point $b \in (f(\overline{\Omega}) - f(\partial\Omega \cup S_f))$ *the inverse image* $f^{-1}(b)$ *of* $\{b\}$ *under* f *is finite, and* $\deg(f, \Omega, b)$ *is an integer in* $\mathbb{Z}$ *given by*

$$\deg(f, \Omega, b) = \sum_{x \in f^{-1}(b)} \operatorname{sgn}(\det \nabla f(x)).$$

In particular then,

$$\deg(\mathrm{id}, \Omega, b) = 1 \textit{ if } b \in \Omega, \quad \textit{and} \quad \deg(-\mathrm{id}, \Omega, b) = (-1)^n \textit{ if } b \in \Omega.$$

(b) *Given* $f \in \mathcal{C}(\overline{\Omega}; \mathbb{R}^n)$, *the function*

$$b \in (\mathbb{R}^n - f(\partial\Omega)) \to \deg(f, \Omega, b),$$

which is constant in each connected component of the open set $\mathbb{R}^n - f(\partial\Omega)$ (*Theorem* 9.15-4(d)), *takes its values in* $\mathbb{Z}$.

Proof (i) *Let* $f \in \mathcal{C}(\overline{\Omega}; \mathbb{R}^n) \cap \mathcal{C}^1(\Omega; \mathbb{R}^n)$ *and let* $b \notin f(\partial\Omega \cup S_f)$ *be such that* $f^{-1}(b) \neq \varnothing$. *Then* $f^{-1}(b)$ *is a finite subset of the open set* Ω.

To see this, note that, by the *local inversion theorem* (Theorem 7.14-1), each point $x \in f^{-1}(b)$ possesses an open neighborhood $V_x \subset \Omega$ such that the restriction $f|_{V_x} \to \mathbb{R}^n$ is a $\mathcal{C}^1$-diffeomorphism onto an open neighborhood W_x of b.

Since then $y \notin f^{-1}(b)$ for all $y \in V_x - \{x\}$, the set $f^{-1}(b)$ is *discrete* (each point $x \in f^{-1}(b)$ possesses a neighborhood V_x such that $(V_x - \{x\}) \cap f^{-1}(b) = \varnothing$) and *compact* (the set $f^{-1}(b)$ is closed since f is continuous on $\overline{\Omega}$, and bounded since Ω is bounded). Hence $f^{-1}(b)$ is finite.

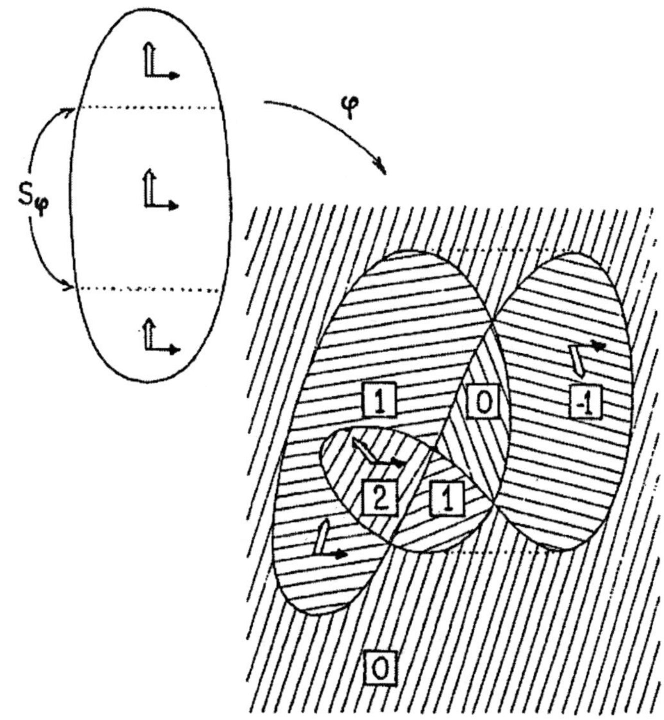

Figure 9.15-2 *The topological degree of a mapping $f : \overline{\Omega} \subset \mathbb{R}^2 \to \mathbb{R}^2$.* Each hatched region is a connected component of $\mathbb{R}^2 - f(\partial\Omega)$, in which the topological degree has a constant value, indicated in a box. This figure originally appeared in P.G. CIARLET [1988]: *Mathematical Elasticity, Volume* I: *Three-Dimensional Elasticity*, North-Holland, Amsterdam.

(ii) *Let $f \in \mathcal{C}(\overline{\Omega}; \mathbb{R}^n) \cap \mathcal{C}^1(\Omega; \mathbb{R}^n)$. Then, if $b \notin f(\partial\Omega \cup S_f)$ is such that $f^{-1}(b) \neq \varnothing$, the degree* $\deg(f, \Omega, b)$ *is also given by the same formula as in* (a).

To see this, we first note that, by (i),

$$f^{-1}(b) = \bigcup_{j \in J} \{x_j\},$$

where $J = J(b)$ is a finite set of indices, the points x_j, $j \in J$, belong to Ω, and, for each $j \in J$, there exist a neighborhood W_j of b and disjoint open connected neighborhoods $\widetilde{V}_j \subset \Omega$ of x_j such that $f|_{\widetilde{V}_j} : \widetilde{V}_j \to W_j$ is a $\mathcal{C}^1$-diffeomorphism. This shows in particular that, for each $j \in J$, $\det \nabla f(x) \neq 0$ for all $x \in \widetilde{V}_j$, which in turn implies that the function $\det \nabla f$ keeps a constant sign in each neighborhood $\widetilde{V}_j$. Besides,

$$\operatorname{dist}(b, f(S_f)) > 0,$$

since $\partial\Omega \cup S_f$ is compact, as a closed subset of $\overline{\Omega}$ (as is immediately verified); hence $f(\partial\Omega \cup S_f)$ is compact and thus $b \notin f(\partial\Omega \cup S_f)$ implies that $\operatorname{dist}(b, f(S_f)) \geq \operatorname{dist}(b, f(\partial\Omega \cup S_f)) > 0$.

Since

$$\inf\left\{|f(x)-b|\,;\ x\in\left(\overline{\Omega}-\bigcup_{j\in J}V_j\right)\right\}>0$$

(the function f is continuous on the compact set $\overline{\Omega}-\bigcup_{j\in J}V_j$), there exist $\varepsilon_0=\varepsilon_0(b)>0$ such that

$$0<\varepsilon_0<\operatorname{dist}(b,f(\partial\Omega\cup S_f))$$

and disjoint open neighborhoods $V_j\subset\widetilde{V}_j$ of x_j, $j\in J$, such that

$$B(b;\varepsilon_0)\subset\bigcap_{j\in J}W_j,\quad V_j=(f|_{\widetilde{V}_j})^{-1}(B(b;\varepsilon_0))\quad\text{for each } j\in J,$$

$$f^{-1}(B(b;\varepsilon_0))=\bigcup_{j\in J}V_j,\quad\text{and}\quad\overline{\bigcup_{j\in J}V_j}\subset\Omega.$$

Let $\varphi\in\mathcal{C}[0,1]$ be such that $\operatorname{supp}\varphi\Subset\,]0,\varepsilon_0[$ and $\int_{\mathbb{R}^n}\varphi(|y|)\,dy=1$. Hence, in particular,

$$\varphi(|f(x)-b|)=0\quad\text{if } x\in\left(\Omega-\bigcup_{j\in J}V_j\right),$$

since $f(x)\notin B(b;\varepsilon_0)$ if $x\notin\bigcup_{j\in J}V_j$. Consequently,

$$\begin{aligned}\deg(f,\Omega,b)&=\int_\Omega\varphi(|f(x)-b|)\det\nabla f(x)\,dx=\int_{\bigcup_{j\in J}V_j}\varphi(|f(x)-b|)\det\nabla f(x)\,dx\\&=\sum_{j\in J}\int_{V_j}\varphi(|f(x)-b|)\det\nabla f(x)\,dx\\&=\sum_{j\in J}\{\operatorname{sgn}\det\nabla f(x);\ x\in V_j\}\int_{V_j}\varphi(|f(x)-b|)\,|\det\nabla f(x)|\,dx.\end{aligned}$$

Note that each integral $\int_{V_j}\varphi(|f(x)-b|)\,|\det\nabla f(x)|\,dx$ is well defined since $\overline{V}_j\subset\Omega$ and $f\in\mathcal{C}^1(\Omega;\mathbb{R}^n)$. Then, for each $i\in I$, the *formula for change of variables in Lebesgue integrals* (Theorem 1.16-1) gives

$$\int_{V_j}\varphi(|f(x)-b|)\,|\det\nabla f(x)|\,dx=\int_{f(V_j)}\varphi(|y-b|)\,dy=\int_{B(b;\varepsilon_0)}\varphi(|y-b|)\,dy=1.$$

We have thus established that, under the assumptions of (ii), $\deg(f,\Omega,b)$ is also given by

$$\deg(f,\Omega,b)=\sum_{j\in J}\{\operatorname{sgn}(\det\nabla f(x));\ x\in V_j\}=\sum_{x\in f^{-1}(b)}\operatorname{sgn}(\det\nabla f(x)).$$

Hence $\deg(f,\Omega,b)\in\mathbb{Z}$ if $b\in(f(\overline{\Omega})-f(\partial\Omega\cup S_f))$. This proves (a).

(iii) *Let now* $f\in\mathcal{C}(\overline{\Omega};\mathbb{R}^n)$ *and* $b\notin f(\partial\Omega)$. *Then* $\deg(f,\Omega,b)\in\mathbb{Z}$.

By Theorem 9.15-4(b), there exists a function $g\in\mathcal{C}(\overline{\Omega};\mathbb{R}^n)\cap\mathcal{C}^1(\Omega;\mathbb{R}^n)$ such that

$$b\notin g(\partial\Omega)\quad\text{and}\quad\deg(f,\Omega,b)=\deg(g,\Omega,b).$$

Since $b \notin g(\partial\Omega)$, there exists $r > 0$ such that $\widetilde{b} \notin g(\partial\Omega)$ and

$$\deg(g, \Omega, b) = \deg(g, \Omega, \widetilde{b}) \quad \text{for all } \widetilde{b} \in B(b, r),$$

by Theorem 9.15-4(d). Let $S_g := \{x \in \Omega;\ \det \nabla g(x) = 0\}$. Since then dx-meas $S_g = 0$ by *Sard's theorem* (Theorem 7.5-1), the intersection $B(b; r) \cap (\mathbb{R}^n - S_g)$ contains at least one point $\widetilde{b}$. Then either

$$\deg(g, \Omega, \widetilde{b}) = 0 \quad \text{if } \widetilde{b} \notin g(\overline{\Omega}),$$

by Theorem 9.15-4(a), or

$$\deg(g, \Omega, \widetilde{b}) \in \mathbb{Z} \quad \text{if } \widetilde{b} \in g(\overline{\Omega}),$$

by (ii). This proves (b). □

Remark The degree is often first defined for functions $f \in \mathcal{C}(\overline{\Omega}; \mathbb{R}^n) \cap \mathcal{C}^1(\Omega; \mathbb{R}^n)$ and points $b \notin (f(\overline{\Omega}) - f(\partial\Omega \cup S_f))$ by the formula found in Theorem 9.15-5(a). But then special care must be taken for extending the definition of the degree to points $b \in f(S_f)$. □

We conclude this section by establishing the *invariance of the degree under homotopy* (Section 1.9), a property with several important consequences; for instance, it implies that *the degree depends only on boundary values* (see part (b) in the next theorem); more importantly, it provides the key to an elegant proof of *Brouwer's fixed point theorem* (see Theorem 9.16-1 in the next section).

Theorem 9.15-6 (homotopic invariance of the degree) *Let Ω be a bounded open subset of $\mathbb{R}^n$.*

(a) *Let there be given two functions $f, g \in \mathcal{C}(\overline{\Omega}; \mathbb{R}^n)$ and a homotopy $H \in \mathcal{C}(\overline{\Omega} \times [0,1]; \mathbb{R}^n)$ joining f to g in the space $\mathcal{C}(\overline{\Omega}; \mathbb{R}^n)$, i.e., such that*

$$H(\cdot, 0) = f \quad \textit{and} \quad H(\cdot, 1) = g.$$

Let a point $b \in \mathbb{R}^n$ be such that

$$b \notin H(\partial\Omega \times [0,1]).$$

Then

$$\deg(H(\cdot, \lambda), \Omega, b) = \deg(f, \Omega, b) \quad \textit{for all } 0 \le \lambda \le 1.$$

Hence in particular, $\deg(g, \Omega, b) = \deg(f, \Omega, b)$.

(b) *Let two functions $f, g \in \mathcal{C}(\overline{\Omega}; \mathbb{R}^n)$ and a point $b \in \mathbb{R}^n$ be such that*

$$f(x) = g(x) \textit{ for all } x \in \partial\Omega \quad \textit{and} \quad b \notin f(\partial\Omega).$$

Then

$$\deg(g, \Omega, b) = \deg(f, \Omega, b).$$

Proof Since $b \notin H(\partial\Omega \times [0,1])$ and the function $H : \partial\Omega \times [0,1] \to \mathbb{R}^n$ is continuous on the compact set $\partial\Omega \times [0,1]$, there exists ε_0 such that

$$0 < \varepsilon_0 \le \operatorname{dist}(b, H(\partial\Omega \times \{\lambda\})) \quad \text{for all } 0 \le \lambda \le 1.$$

Therefore, by Theorem 9.15-4(b), there exists $r > 0$ such that

$$\deg(H(\cdot,\lambda),\Omega,b) = \deg(H(\cdot,\mu),\Omega,b) \quad \text{if } \|H(\cdot,\lambda) - H(\cdot,\mu)\| < r,$$

for some $0 \le \lambda, \mu \le 1$. Since the function $H : \overline{\Omega} \times [0,1] \to \mathbb{R}^n$ is uniformly continuous (the set $\overline{\Omega} \times [0,1]$ is compact), there exists δ such that

$$\|H(\cdot,\lambda) - H(\cdot,\mu)\| < r \quad \text{if } |\lambda - \mu| < \delta.$$

To conclude, it then suffices to write $[0,1]$ as a union of intervals of length $< \delta$. This proves (a).

To prove(b), consider the homotopy $H : \overline{\Omega} \times [0,1] \to \mathbb{R}^n$ defined by

$$H(x,\lambda) := (1-\lambda)f(x) + \lambda g(x), \quad (x,\lambda) \in \overline{\Omega} \times [0,1],$$

which is clearly continuous and such that $b \notin H(\partial\Omega \times [0,1])$; then use (a). □

Other properties of the degree are left as problems (Problems 9.15-2 to 9.15-4).

It is still possible to define a *topological degree* $\deg(f,\Omega,b)$ when Ω is a bounded open subset of an *infinite-dimensional Banach space* X and the mapping $f : \overline{\Omega} \to X$ is of the form $f = I - T$, where the mapping $T : \overline{\Omega} \to X$ is *compact* (i.e., $T \in \mathcal{C}(\overline{\Omega}; X)$ and the image $T(B)$ of any bounded subset B of $\overline{\Omega}$ is relatively compact; cf. Section 9.12), and the point $b \in X$ again satisfies $b \notin f(\partial\Omega)$. This is the fundamental **Leray–Schauder degree**,[75] the definition of which essentially relies on the Brouwer topological degree in $\mathbb{R}^n$ defined in this section, and which possesses properties that are to a large extent similar.

The Leray–Schauder degree provides a powerful means to obtain *existence results for nonlinear partial differential equations.* For instance, the Leray–Schauder degree combined with the *mountain pass lemma* (Theorem 9.8-4), provides existence results[76] for the nonlinear boundary value problem

$$-\Delta_p u + f(x,u) = 0 \text{ in } \Omega \quad \text{and} \quad u = 0 \text{ on } \partial\Omega,$$

where Δ_p denotes the *p-Laplace operator* (Sections 9.6 and 9.14) and $f : \Omega \times \mathbb{R} \to \mathbb{R}$ is a Carathéodory function that satisfies suitable growth conditions.

[75] Introduced and analyzed in one of the most influential papers in nonlinear functional analysis:

J. LERAY; J. SCHAUDER [1934]: Topologie et équations fonctionnelles, *Annales Scientifiques de l'Ecole Normale Supérieure* **51**, 45–78.

An illuminating historical perspective of the Leray–Schauder degree is given in:

J. MAWHIN [1999]: Leray–Schauder degree: A half century of extensions and applications, *Topological Methods in Nonlinear Analysis* **14**, 195–228.

Detailed treatments of the Leray–Schauder degree and examples of its application to nonlinear partial differential equations are found in more specialized texts, such as GILBARG & TRUDINGER [1998], DEIMLING [1985], ZEIDLER [1986], KAVIAN [1993], KESAVAN [2004].

[76] G. DINCA; P. JEBELEAN; J. MAWHIN [2001]: Variational and topological methods for Dirichlet problems with p-Laplacian, *Portugaliae Mathematica* **58**, 339–378.

Problems

9.15-1 This problem provides a simple proof[77] of the version of the Tietze–Urysohn extension theorem used in the proofs of Theorems 9.15-1 and 9.15-3(a).

(1) Let K be a compact subset of $\mathbb{R}^n$. Show that there exists a subset A of K of the form $A = \bigcup_{i=1}^{\infty}\{a_i\}$ such that $\overline{A} = K$.

(2) Given a function $f \in \mathcal{C}(K;\mathbb{R}^n)$, let

$$\widetilde{f}(x) := f(x), \quad x \in K,$$

$$\widetilde{f}(x) := \Big(\sum_{i=1}^{\infty}\frac{1}{2^i}\theta_i(x)\Big)^{-1}\sum_{i=1}^{\infty}\frac{1}{2^i}\theta_i(x)f(a_i), \quad x \in (\mathbb{R}^n - K),$$

where

$$\theta_i(x) = \max\Big\{2 - \frac{|x-a_i|}{\mathrm{dist}(x,K)}, 0\Big\}, \quad x \in (\mathbb{R}^n - K), \quad i \geq 1.$$

Show that the function $\widetilde{f} : \mathbb{R}^n \to \mathbb{R}^n$ defined in this fashion is continuous and extends f.

9.15-2 Let Ω be a bounded open subset of $\mathbb{R}^n$, let $K \subset \overline{\Omega}$ be compact, let $f \in \mathcal{C}(\overline{\Omega};\mathbb{R}^n)$, and let $b \notin f(\partial\Omega \cup K)$. Show that $\deg(f, \Omega - K, b) = \deg(f, \Omega, b)$.

9.15-3 Let Ω be a bounded open subset of $\mathbb{R}^n$ and let Ω_i, $i \in I$, be any family of disjoint open subsets of Ω. Let $f \in \mathcal{C}(\overline{\Omega};\mathbb{R}^n)$ and $b \in \mathbb{R}^n$ be such that $f^{-1}(b) \subset \bigcup_{i\in I}\Omega_i$. Show that there exists a *finite* set $I(b) \subset I$ such that $\deg(f,\Omega_i,b) = 0$ if $i \notin I(b)$ and that $\deg(f,\Omega,b) = \sum_{i\in I(b)}\deg(f,\Omega_i,b)$.

9.15-4 Let Ω be a bounded open subset of $\mathbb{R}^n$. Given $f \in \mathcal{C}(\overline{\Omega};\mathbb{R}^n)$, let U_i, $i \in I$, denote the bounded connected components of the set $\mathbb{R}^n - f(\partial\Omega)$. For each $i \in I$, the integer

$$\deg(f,\Omega,U_i) := \deg(f,\Omega,b) \quad \text{for any } b \in U_i$$

is thus well defined (Theorem 9.15-4(d)). The objective of this problem is to establish **Leray's product formula:**[78] *Let* $g \in \mathcal{C}(\mathbb{R}^n;\mathbb{R}^n)$ *and let* $b \notin (g\circ f)(\partial\Omega)$. *Then*

$$\deg(g\circ f,\Omega,b) = \sum_{i\in I}\deg(f,\Omega,U_i)\deg(g,U_i,b)$$

(since each set U_i is open and bounded, and $b \notin g(\partial U_i)$ as is easily verified, $\deg(g,U_i,b)$ is well defined for each $i \in I$).

(1) Show that the set $\{i \in I;\ \deg(g,U_i,b) \neq 0\}$ is finite, so that the above sum is always well defined.

(2) Show that Leray's product formula holds if $f \in \mathcal{C}(\overline{\Omega},\mathbb{R}^n) \cap \mathcal{C}^1(\Omega;\mathbb{R}^n)$, $g \in \mathcal{C}^1(\mathbb{R}^n;\mathbb{R}^n)$, $b \notin (g\circ f)(\partial\Omega)$, and $\det\nabla(g\circ f)(b) \neq 0$.

(3) Using (2), show that Leray's product formula holds if $f \in \mathcal{C}(\overline{\Omega};\mathbb{R}^n)$, $g \in \mathcal{C}(\mathbb{R}^n;\mathbb{R}^n)$, and $b \notin (g\circ f)(\partial\Omega)$ (this is the difficult part[79]).

9.15-5 Let $B := \{x \in \mathbb{R}^n;\ |x| < 1\}$ and let $f : \partial B \to \mathbb{R}^n$ be a homeomorphism of ∂B onto its image $f(\partial B)$. Show that, if $n \geq 2$, the set $\mathbb{R}^n - f(\partial B)$ has exactly two connected components, one bounded and one unbounded.

---

[77] Due to:
M. Nagumo [1951]: A theory of degree of mapping based on infinitesimal analysis, *American Journal of Mathematics* **73**, 485–496.

[78] J. Leray [1935]: Topologie des espaces abstraits de M. Banach, *Comptes Rendus de l'Académie des Sciences de Paris* **200**, 1082–1084.

[79] For a proof, see, e.g., Deimling [1985, Chapter 1, Theorem 5.1].

Hint: Show that the set $\mathbb{R}^n - f(\partial B)$ has at most a countably infinite number of bounded connected components U_i. Then extend $f : \partial B \to \mathbb{R}^n$ and $f^{-1} : f(\partial B) \to \mathbb{R}^n$ to $\widetilde{f} : \mathbb{R}^n \to \mathbb{R}^n$ and $\widetilde{g} : \mathbb{R}^n \to \mathbb{R}^n$ by the *Tietze–Urysohn extension theorem* and apply the *Leray product formula* (Problem 9.15-4) to $\deg(\widetilde{g} \circ \widetilde{f}, B, b)$ with $b \in B$ and to $\deg(\widetilde{f} \circ \widetilde{g}, U_i, b_i)$ with $b_i \in U_i$.

Remarks (1) This result constitutes the **Jordan–Brouwer separation theorem**, so named after Camille Jordan,[80] who first proved it for $n = 2$, and L.E.J. Brouwer,[81] who then extended it to any $n \geq 2$. Much later, Jean Leray[82] noted that this difficult to prove result (even for $n = 2$) could be easily derived from his product formula (as indicated above).

(2) This result can be further extended as follows:[83] Let f be a homeomorphism from a compact set $K_1 \subset \mathbb{R}^n$ onto a compact set $K_2 \subset \mathbb{R}^n$. Then either the sets $\mathbb{R}^n - K_1$ and $\mathbb{R}^n - K_2$ have the same finite number of connected components, or they both have countably infinitely many connected components.

(3) For $n = 2$, the image $f(\partial B)$ is called a *Jordan curve*. □

9.15-6 This problem provides another proof of the *fundamental theorem of algebra* (Theorem 2.8-1). Let $p : \mathbb{C} \to \mathbb{C}$ be a complex polynomial of degree $n \geq 1$ of the form

$$p(z) := z^n + a_{n-1}z^{n-1} + \cdots + a_1 z + a_0, \quad z \in \mathbb{C}.$$

In what follows, $z = x + iy \in \mathbb{C}$ is identified with $(x, y) \in \mathbb{R}^2$, so that the set $\Omega := \{z \in \mathbb{C};\ |z| < 1\}$ is identified with an open set in $\mathbb{R}^2$ and p is identified with a function $p : \mathbb{R}^2 \to \mathbb{R}^2$.

(1) Show that $\det \nabla p(x, y) = |p'(z)|^2$ at each $(x, y) \in \mathbb{R}^2$ (more generally, given any analytic function on an open subset U of $\mathbb{C}$, this relation holds at each point of U).

(2) Compute $\deg(p, \Omega, 0)$ in the particular case where $p(z) := z^n$, $z \in \mathbb{C}$.

(3) Assuming that $|a_{n-1}| + \cdots + |a_1| + |a_0| < 1$, show that p has at least one root in Ω.

(4) Infer from (3) that any complex polynomial of degree ≥ 1 has at least one root in $\mathbb{C}$.

9.15-7 Let a mapping $f \in \mathcal{C}(\mathbb{R}^n; \mathbb{R}^n)$ be such that $\dfrac{f(x) \cdot x}{|x|} \to \infty$ as $|x| \to \infty$. Show that, given any $b \in \mathbb{R}^n$, $\deg(f, B(0; r), b) = 1$ for r large enough, and hence that f is *surjective* (incidentally, this provides another proof of the corollary to Brouwer's fixed point theorem; cf. Theorem 9.9-3).

Remark The surjectivity of f can be also deduced from the *Minty-Browder theorem* (Theorem 9.14-1), the proof of which uses, not coincidentally, Brouwer's fixed point theorem. □

9.16 Brouwer's fixed point theorem — a second proof — and the hairy ball theorem

The Brouwer degree provides a remarkably short proof of Brouwer's fixed point theorem (compare with the proof of Theorem 9.9-2):

Theorem 9.16-1 (Brouwer's fixed point theorem — a second proof) *Let K be a compact and convex subset of a finite-dimensional normed vector space, and let $f : K \to K$ be a continuous mapping. Then f has at least one fixed point.*

[80] C. Jordan [1887]: *Cours d'Analyse*, Volume 3, Paris.

[81] L.E.J. Brouwer [1911]: Beweis des Jordanschen Satzes für den n-dimensionalen Raum, *Mathematische Annalen* **71**, 314–319 and 598.

[82] J. Leray [1950]: La théorie des points fixes et ses applications en analyse, in *Proceedings—International Congress of Mathematicians, Volume* 2, pp. 202–208, Cambridge.

[83] For a proof, see, e.g., Deimling [1985, Chapter 1, Theorem 5.2].

Proof It suffices to show that *there is no continuous retraction of the closed unit ball of $\mathbb{R}^n$ onto its boundary* (see part (iii) of the proof of Theorem 9.9-2). So, let $B = \{x \in \mathbb{R}^n;\ |x| < 1\}$, and assume that there exists a function $f \in \mathcal{C}(\overline{B})$ such that

$$f(x) = x \text{ for all } x \in \partial\Omega \quad \text{and} \quad f(\overline{B}) = \partial B.$$

Since then $f|_{\partial\Omega} = \mathrm{id}|_{\partial\Omega}$ and $0 \notin f(\partial\Omega)$, we infer from Theorems 9.15-5(a) and 9.15-6(b) that

$$\deg(f, \Omega, 0) = \det(\mathrm{id}, \Omega, 0) = 1,$$

and then from Theorem 9.15-4(a) that there exists $x \in \Omega$ such that $f(x) = 0$; but this contradicts the assumption that $f(\overline{B}) = \partial B$. This completes the proof. □

Another application of the Brouwer degree shows that, when the dimension n is *odd*, any continuously varying vector field that is tangent to the unit sphere ∂B of $\mathbb{R}^n$ (such a field is denoted τ in the next theorem) necessarily vanishes at at least one point of ∂B (by contrast, there exist continuously varying tangent vector fields that never vanish along ∂B when n is even; cf. Figure 9.16-1 and Problem 9.16-1).

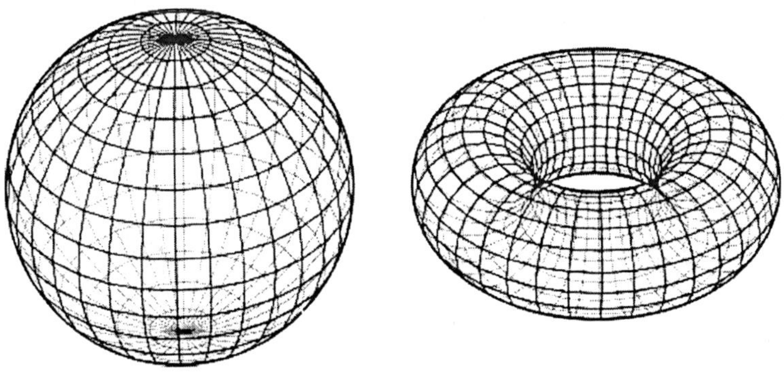

Figure 9.16-1 *How the "hairy ball theorem" (Theorem* 9.16-2) *got its name*: It is impossible to comb a "hairy ball" (the unit sphere in $\mathbb{R}^3$) without leaving a tuft of hair uncombed at at least one point: Either the tangent vector field is discontinuous or it vanishes at that point. By contrast, it is possible to "continuously" comb a *torus* with a continuously varying vector field that never vanishes on it. This image originally appeared in V. V. Isaeva; N. V. Kasyanov; E. V. Presnov [2012]: Topological singularities and symmetry breaking in development, *Biosystems* **109**, 280–298.

Theorem 9.16-2 (hairy ball theorem[84]) *For each integer $n \geq 1$, let $B := \{x \in \mathbb{R}^n;\ \|x\| < 1\}$, and let a mapping $\tau \in \mathcal{C}(\partial B; \mathbb{R}^n)$ be such that*

[84]This theorem is also due to Luitzen Egbertus Jan Brouwer (1881–1966), who proved it in 1912. Many other proofs have appeared since then, among which is a strikingly ingenious, and to a large extent elementary, one in:

J. Milnor [1978]: Analytic proofs of the "hairy ball theorem" and the Brouwer fixed point theorem, *The American Mathematical Monthly* **85**, 521–524.

As its title indicates, this little gem of a paper also provides a proof of Brouwer's fixed point theorem, this time as a *corollary* to the hairy ball theorem.

John Willard Milnor was awarded the Fields Medal in 1962, and the Abel Prize in 2010 for "pioneering discoveries in topology, geometry, and algebra."

$$\tau(x) \cdot x = 0 \quad \textit{for all } x \in \partial B.$$

Then, if n is odd, there exists at least one point $x \in \partial B$ such that

$$\tau(x) = 0.$$

Proof To begin with, we prove a result interesting by itself, asserting that, if n is odd, the unit sphere ∂B of $\mathbb{R}^n$ cannot be continuously transformed into itself in such a way that each point $x \in \partial B$ becomes the symmetric point $-x \in \partial B$ in this process.

(i) *If n is odd, there is no homotopy $H \in \mathcal{C}(\partial B \times [0,1]\,; \partial B)$ such that*

$$H(\cdot, 0) = \mathrm{id}|_{\partial B} \quad \text{and} \quad H(\cdot, 1) = -\mathrm{id}|_{\partial B}.$$

By the *Tietze–Urysohn extension theorem* (Theorem 1.7-7), any such homotopy $H \in \mathcal{C}(\partial B \times [0,1]\,; \partial B)$ can be extended to a mapping $\widetilde{H} \in \mathcal{C}(\overline{B} \times [0,1]\,; \mathbb{R}^n)$. Then, on the one hand, by Theorems 9.15-5(a) and 9.15-6(b),

$$\begin{aligned}\deg(\widetilde{H}(\cdot, 0), B, 0) &= \deg(\mathrm{id}, B, 0) = 1,\\ \deg(\widetilde{H}(\cdot, 1), B, 0) &= \deg(-\mathrm{id}, B, 0) = (-1)^n,\end{aligned}$$

since $0 \notin \partial B$ and $\widetilde{H}(\cdot, 0)|_{\partial B} = \mathrm{id}|_{\partial B}$ and $\widetilde{H}(\cdot, 1)|_{\partial B} = -\mathrm{id}|_{\partial B}$ by assumption. But on the other hand, by the homotopic invariance of the degree (Theorem 9.15-6(a)),

$$\deg(\widetilde{H}(\cdot, 0), B, 0) = \deg(\widetilde{H}(\cdot, 1), B, 0),$$

since $0 \notin \widetilde{H}(\partial B \times \{\lambda\}) = H(\partial B \times \{\lambda\}) \subset \partial B$ for all $0 \le \lambda \le 1$. Hence n is necessarily even if such a homotopy exists.

(ii) Given a mapping $\tau \in \mathcal{C}(\partial B; \mathbb{R}^n)$ such that $\tau(x) \neq 0$ and $\tau(x) \cdot x = 0$ for all $x \in \partial B$, let

$$H(x, \lambda) := (\cos \pi\lambda)x + (\sin \pi\lambda)\frac{\tau(x)}{|\tau(x)|}, \quad (x, \lambda) \in \partial B \times [0,1].$$

Then the mapping $H : \partial B \times [0,1] \to \mathbb{R}^n$ defined in this fashion is continuous, maps $\partial B \times [0,1]$ into ∂B, and is such that $H(\cdot, 0) = \mathrm{id}|_{\partial B}$ and $H(\cdot, 1) = -\mathrm{id}|_{\partial B}$. Hence n is necessarily even by (i). □

Problem

9.16-1 Give an example of a continuous tangent vector field that never vanishes along the unit sphere of $\mathbb{R}^n$ when n is even.

9.17 Borsuk's and Borsuk–Ulam theorems; Brouwer's invariance of domain theorem

Let Ω be a bounded open subset of $\mathbb{R}^n$ and let $f \in \mathcal{C}(\overline{\Omega}; \mathbb{R}^n)$. If $0 \notin f(\partial\Omega)$, one way to prove the *existence of a zero of f in Ω* is to show that $\deg(f, \Omega, 0) \neq 0$ (Theorem 9.15-4(a)). Hence it is crucial to identify additional assumptions implying that this is the case.

The next theorem constitutes another *basic theorem of nonlinear functional analysis*, not only because it identifies such additional assumptions, but also because, among its corollaries it counts two other *basic theorems of nonlinear functional analysis*, viz., the *Borsuk–Ulam theorem* (Theorem 9.17-2) and the *invariance of domain theorem in $\mathbb{R}^n$* (Theorem 9.17-3).

Theorem 9.17-1 (Borsuk's theorem[85]) *Let Ω be a bounded open subset of $\mathbb{R}^n$ that contains 0 and is symmetric with respect to 0, and let $f \in \mathcal{C}(\overline{\Omega}; \mathbb{R}^n)$ be an odd function (i.e., that satisfies $f(x) = -f(-x)$ for all $x \in \overline{\Omega}$) such that $0 \notin f(\partial\Omega)$. Then*

$$\deg(f, \Omega, 0) \quad \textit{is odd.}$$

Consequently, there exists at least one point $x \in \Omega$ such that

$$f(x) = 0.$$

Proof (i) We first show that *there exists a function $\widetilde{g} : \overline{\Omega} \to \mathbb{R}^n$ with the following properties:*

$$\widetilde{g} \in \mathcal{C}(\overline{\Omega}; \mathbb{R}^n) \cap \mathcal{C}^1(\Omega; \mathbb{R}^n), \quad \widetilde{g} \text{ is odd}, \quad \det \nabla\widetilde{g}(0) \neq 0,$$
$$0 \notin \widetilde{g}(\partial\Omega), \quad \deg(\widetilde{g}, \Omega, 0) = \deg(f, \Omega, 0).$$

By Theorem 9.15-4(b), there exists $r = r(f) > 0$ such that

$$\widetilde{g} \in \mathcal{C}(\overline{\Omega}; \mathbb{R}^n) \quad \text{and} \quad \|\widetilde{g} - f\| < r \quad \text{implies } 0 \notin \widetilde{g}(\partial\Omega) \text{ and } \deg(\widetilde{g}, \Omega, 0) = \deg(f, \Omega, 0).$$

By Theorem 9.15-3(a), there exists a function $g_1 \in \mathcal{C}(\overline{\Omega}; \mathbb{R}^n) \cap \mathcal{C}^1(\Omega; \mathbb{R}^n)$ such that $\|g_1 - f\| < \frac{r}{2}$. Let

$$g_2(x) := \frac{1}{2}\left(g_1(x) - g_1(-x)\right), \quad x \in \overline{\Omega},$$

so that the function $g_2 \in \mathcal{C}(\overline{\Omega}; \mathbb{R}^n) \cap \mathcal{C}^1(\Omega; \mathbb{R}^n)$ defined in this fashion is *odd.* The matrix $\nabla g_2(0)$ may not be invertible, but there surely exists $0 < \alpha < (2\sup_{x\in\overline{\Omega}} |x|)^{-1} r$ such that the matrix $(\nabla g_2(0) - \alpha I)$ is invertible. Given such a number α, let

$$\widetilde{g}(x) := g_2(x) - \alpha x, \quad x \in \overline{\Omega}.$$

Then the function $\widetilde{g} \in \mathcal{C}(\overline{\Omega}; \mathbb{R}^n) \cap \mathcal{C}^1(\Omega; \mathbb{R}^n)$ defined in this fashion is odd and has the following properties:

$$\det \nabla\widetilde{g}(0) \neq 0, \quad 0 \notin \widetilde{g}(\partial\Omega), \quad \text{and} \quad \det(\widetilde{g}, \Omega, 0) = \det(f, \Omega, 0),$$

[85]K. Borsuk [1933]: Drei Sätze über die n-dimensionale euklidische Sphäre, *Fundamenta Mathematicae* **21**, 177–190.

the last two properties being consequences of the relation

$$\sup_{x\in\overline{\Omega}} |\widetilde{g}(x) - f(x)| = \sup_{x\in\overline{\Omega}} \left|\frac{1}{2}(g_1(x) - f(x)) - \frac{1}{2}(g_1(-x) - f(-x)) - \alpha x\right| < r.$$

(ii) We next show that *there exists a function* $g : \overline{\Omega} \to \mathbb{R}$ *with the following properties:*

$$g \in \mathcal{C}(\overline{\Omega}; \mathbb{R}^n) \cap \mathcal{C}^1(\Omega; \mathbb{R}^n), \quad g \text{ is odd}, \quad \det \nabla g(0) \neq 0,$$
$$0 \notin g(\partial\Omega), \quad 0 \notin g(S_g), \quad \text{and} \quad \deg(g, \Omega, 0) = \deg(\widetilde{g}, \Omega, 0).$$

To this end, the "hard" part naturally consists in satisfying[86] the relation $0 \notin g(S_g)$.

Again by Theorem 9.15-4(b), there exists $\widetilde{r} = \widetilde{r}(\widetilde{g}) = \widetilde{r}(f) > 0$ such that

$$g \in \mathcal{C}(\overline{\Omega}; \mathbb{R}^n) \quad \text{and} \quad \|g - \widetilde{g}\| < \widetilde{r} \text{ implies } 0 \notin g(\partial\Omega) \quad \text{and} \quad \deg(g, \Omega, 0) = \deg(\widetilde{g}, \Omega, 0).$$

Let $R > 0$ be such that $\overline{\Omega} \subset \overline{B(0; R)}$, let $\varphi : \mathbb{R} \to \mathbb{R}$ be any *odd* function of class $\mathcal{C}^1$ such that

$$\varphi'(0) = 0 \quad \text{and} \quad \varphi(t) \neq 0 \text{ if } t \neq 0,$$

and let

$$\delta := \left(n \sup_{|t|\leq R} |\varphi(t)|\right)^{-1} \widetilde{r}.$$

The basic idea then consists in recursively defining functions

$$h_j : \Omega_j := \{x = (x_i)_{i=1}^n \in \Omega;\ x_j \neq 0\} \to \mathbb{R}^n \quad \text{and} \quad g_j : \overline{\Omega} \to \mathbb{R}^n, \quad j = 1, 2, \ldots, n,$$

as follows: First, let

$$h_1(x) := \frac{\widetilde{g}(x)}{\varphi(x_1)}, \quad x \in \Omega_1, \quad \text{and} \quad g_1(x) := \widetilde{g}(x) - \varphi(x_1)y_1, \quad x \in \overline{\Omega},$$

where the function $\widetilde{g} \in \mathcal{C}(\overline{\Omega}; \mathbb{R}^n)$ is that constructed in (i) and $y_1 \in \mathbb{R}^n$ is any vector that satisfies

$$|y_1| < \delta \quad \text{and} \quad y_1 \notin h_1(S_1), \quad \text{where } S_1 := \{x \in \Omega_1;\ \det \nabla h_1(x) = 0\}.$$

Note that such a vector y_1 surely exists since the set $\mathbb{R}^n - h_1(S_1)$ is dense in $\mathbb{R}^n$, as a consequence of *Sard's theorem* (Theorem 7.5-1). Second, let

$$h_j(x) := \frac{g_{j-1}(x)}{\varphi(x_j)}, \quad x \in \Omega_j, \quad \text{and} \quad g_j(x) := g_{j-1}(x) - \varphi(x_j)y_j, \quad x \in \overline{\Omega}, \quad j = 2, \ldots, n,$$

where $y_j \in \mathbb{R}^n$ is any vector that satisfies

$$|y_j| < \delta \quad \text{and} \quad y_j \notin h_j(S_j), \quad \text{where } S_j := \{x \in \Omega_j;\ \det \nabla h_j(x) = 0\}, \quad j = 2, \ldots, n$$

[86]We follow here the clever construction of:
W. GROMES [1981]: Ein einfacher Beweis des Satzes von Borsuk, *Mathematische Zeitschrift* **178**, 399–400.

(that such a vector y_j exists again follows from Sard's theorem).

It is then clear that the function $g : \overline{\Omega} \to \mathbb{R}^n$ defined by

$$g(x) := g_n(x) = \widetilde{g}(x) - \sum_{j=1}^{n} \varphi(x_j) y_j, \quad x \in \overline{\Omega},$$

has the following properties: First,

$$g \in \mathcal{C}(\overline{\Omega}; \mathbb{R}^n) \cap \mathcal{C}^1(\Omega; \mathbb{R}^n), \quad g \text{ is odd}, \quad \text{and} \quad \det \nabla g(0) \neq 0,$$

since $\widetilde{g} \in \mathcal{C}(\overline{\Omega}; \mathbb{R}^n) \cap \mathcal{C}^1(\Omega; \mathbb{R}^n)$, $\widetilde{g}$ and φ are odd, and $\nabla g(0) = \nabla \widetilde{g}(0)$ (recall that $\varphi'(0) = 0$). Second,

$$0 \notin g(\partial\Omega) \quad \text{and} \quad \deg(g, \Omega, 0) = \deg(\widetilde{g}, \Omega, 0),$$

since

$$\|g - \widetilde{g}\| = \sup_{x \in \overline{\Omega}} \left| \sum_{j=1}^{n} \varphi(x_j) y_j \right| < n\delta \sup_{|t| \le R} |\varphi(t)| = \widetilde{r}.$$

It remains to show that $0 \notin g(S_g)$, i.e., that, if $x^* \in \Omega$ is such that $x^* \neq 0$ and $g(x^*) = 0$, then $\det \nabla g(x^*) \neq 0$ (we already know that $\det \nabla g(0) = \det \nabla \widetilde{g}(0) \neq 0$).

If $x \in \Omega$ and $x \neq 0$, it is readily seen from the recursive definitions of the functions $h_j : \Omega_j \to \mathbb{R}^n$ and $g_j : \overline{\Omega} \to \mathbb{R}^n$, $1 \le j \le n$, that the vector $g(x) \in \mathbb{R}^n$ is given by at least one of the following expressions:

$$g(x) = \varphi(x_n)(h_n(x) - y_n) \quad \text{for any } x \in \Omega_n,$$

$$g(x) = \varphi(x_j)(h_j(x) - y_j) - \sum_{k=j+1}^{n} \varphi(x_k) y_k \quad \text{for any } x \in \Omega_j, \ 1 \le j \le n-1,$$

so that the $n \times n$ matrix $\nabla g(x)$ is given by

$$\nabla g(x) = \varphi(x_n) \nabla h_n(x) + \varphi'(x_n)(H_n(x) - Y_n) \quad \text{for any } x \in \Omega_n,$$

$$\nabla g(x) = \varphi(x_j) \nabla h_j(x) + \varphi'(x_j)(H_j(x) - Y_j) = \sum_{k=j+1}^{n} \varphi'(x_k) Y_k \quad \text{for any } x \in \Omega_j, \ 1 \le j \le n-1,$$

where, for each $1 \le j \le n$, the $n \times n$ matrix $H_j(x)$, *resp.* Y_j, denotes the matrix whose ℓth column is $h_j(x)$, *resp.* y_j, if $\ell = j$ or 0 if $\ell \neq j$.

Given any $x^* = (x_j^*)_{j=1}^n \in \Omega$ such that $x^* \neq 0$ and $g(x^*) = 0$, let $1 \le j = j(x^*) \le n$ be such that $x_j^* \neq 0$ and $x_k^* = 0$ if $j+1 \le k \le n$, this last condition being of course void if $j = n$. Since then $x^* \in \Omega_j$, the relations

$$0 = g(x^*) = \varphi(x_n^*)(h_n(x^*) - y_n) \quad \text{if } j = n,$$

$$0 = g(x^*) = \varphi(x_j^*)(h_j(x^*) - y_j) - \sum_{k=j+1}^{n} \varphi(x_k^*) y_k = \varphi(x_j^*)(h_j(x^*) - y_j) \quad \text{if } j < n$$

(if $j+1 \le k \le n$, $x_k^* = 0$ implies $\varphi(x_k^*) = 0$), show that

$$h_j(x^*) = y_j,$$

since $x_j^* \neq 0$ implies $\varphi(x_j^*) \neq 0$, which in turn implies that

$$\det \nabla h_j(x^*) \neq 0,$$

since $y_j \notin h_j(S_j)$ by construction. Furthermore, the relation $h_j(x^*) = y_j$ also implies that

$$H_j(x^*) = Y_j.$$

Consequently, either

$$\nabla g(x^*) = \varphi(x_n^*)\nabla h_n(x^*) + \varphi'(x_n^*)(H_n(x^*) - Y_n) = \varphi(x_n^*)\nabla h_n(x^*)$$

if $j = n$, or

$$\nabla g(x^*) = \varphi(x_j^*)\nabla h_j(x^*) + \varphi'(x_j^*)(H_j(x^*) - Y_j) - \sum_{k=j+1}^{n} \varphi'(x_k^*)Y_k = \varphi(x_j^*)\nabla h_j(x^*)$$

if $j < n$ ($x_k^* = 0$ for all $j+1 \leq k \leq n$ implies $\varphi'(x_k^*) = \varphi'(0) = 0$). Hence

$$\det \nabla g(x^*) = (\varphi(x_j^*))^n \det \nabla h_j(x^*) \neq 0,$$

since $x_j^* \neq 0$ implies $\varphi(x_j^*) \neq 0$.

(iii) The conclusion is now clear. First, parts (i) and (ii) combined imply that

$$\deg(f, \Omega, 0) = \deg(\widetilde{g}, \Omega, 0) = \deg(g, \Omega, 0).$$

But, since $g \in \mathcal{C}(\overline{\Omega}; \mathbb{R}^n) \cap \mathcal{C}^1(\Omega; \mathbb{R}^n), 0 \notin g(\partial\Omega)$, and $0 \notin g(S_g)$, the degree $\deg(g, \Omega, 0)$ is given by the formula (Theorem 9.15-5(a))

$$\deg(g, \Omega, 0) = \operatorname{sgn}(\det \nabla g(0)) + \sum_{\left\{\substack{x \in g^{-1}(0) \\ x \neq 0}\right.} \operatorname{sgn}(\det \nabla g(x)).$$

Hence $\deg(g, \Omega, 0)$ *is an odd number*, since $\sum_{\left\{\substack{x \in g^{-1}(0) \\ x \neq 0}\right.} \operatorname{sgn}(\det \nabla g(x))$ is an even number (the function g is odd) and $\operatorname{sgn}(\det \nabla g(0))$ is equal to either 1 or -1. □

As a first corollary of Borsuk's theorem, we next prove:

Theorem 9.17-2 (Borsuk–Ulam theorem[87]) *Let Ω be a bounded open subset of $\mathbb{R}^n$ that contains 0 and is symmetric with respect to 0, and let $f \in \mathcal{C}(\partial\Omega; \mathbb{R}^m)$ for some integer $m < n$. Then there exists at least one point $x \in \partial\Omega$ such that $f(x) = f(-x)$.*

[87]Although this theorem appeared in BORSUK [1933] (op. cit.), its name reflects that Stanislaw Ulam was also aware of this result (but he did not publish a proof). The Borsuk–Ulam theorem plays in particular a key role in the study of *critical points of functionals*, as developed at length in the book of KAVIAN [1993].

Proof By the *Tietze–Urysohn extension theorem* (Theorem 1.7-7), the function f can be extended to a continuous function from $\overline{\Omega}$ into $\mathbb{R}^m$, which can be identified with a function $\widetilde{f} \in \mathcal{C}(\overline{\Omega}; \mathbb{R}^n)$ since $\mathbb{R}^m \subset \mathbb{R}^n$.

Assume that the property is false, i.e., that $f(x) \neq f(-x)$ for all $x \in \partial\Omega$, and let

$$g(x) := \widetilde{f}(x) - \widetilde{f}(-x), \quad x \in \overline{\Omega}.$$

Then the function $g : \overline{\Omega} \to \mathbb{R}^n$ defined in this fashion has the following properties:

$$g \in \mathcal{C}(\overline{\Omega}; \mathbb{R}^n), \quad g \text{ is odd}, \quad 0 \notin g(\partial\Omega).$$

Therefore, by *Borsuk's theorem*,

$$\delta := \deg(g, \Omega, 0) \text{ is an odd number}.$$

But then, by Theorem 9.15-4(d), there exists $s > 0$ such that

$$\deg(g, \Omega, b) = \delta \neq 0 \quad \text{for all } b \in B(0; s) \subset \mathbb{R}^n.$$

Therefore, by Theorem 9.15-4(a), given any $b \in B(0; s)$, there exists $x \in \Omega$ such that $g(x) = b$. Consequently,

$$B(0; s) \subset g(\Omega) \subset \mathbb{R}^m,$$

but this is impossible since the dx-measure of $B(0; s)$ in $\mathbb{R}^n$ is > 0, while $\mathbb{R}^m$ has zero dx-measure in $\mathbb{R}^n$. Hence we have reached a contradiction. □

The Borsuk–Ulam theorem has a surprising consequence in meteorology. Assume that, at any given time, the temperature and the air pressure vary continuously along the surface of the earth. Then, at any given time, there is (at least) one pair of diametrically opposite points of the earth where *both* the temperature and the air pressure are the same.

Another, perhaps even more surprising, consequence of the Borsuk–Ulam theorem is suggested in Figure 9.17-1; a proof of this consequence is proposed in Problem 9.17-2.

As a second corollary of Borsuk's theorem, we now prove a deep theorem with many far-reaching consequences. Although this theorem seems intuitively clear (as suggested in Figure 9.17-2), its proof is by no means trivial: like those of the previous theorems, it ultimately relies on the *Brouwer topological degree in* $\mathbb{R}^n$.

Remark By contrast, the proof of this theorem under the additional assumptions that f is of class $\mathcal{C}^1$ in Ω and that $\nabla f(x) \in \mathbb{M}^n$ is invertible at each point $x \in \Omega$ is comparatively much easier; cf. Theorem 7.14-2, which in addition holds in any *infinite-dimensional* Banach space. □

Theorem 9.17-3 (Brouwer's invariance of domain theorem[88] in $\mathbb{R}^n$) *Let Ω be an open subset of $\mathbb{R}^n$ and let $f \in \mathcal{C}(\Omega; \mathbb{R}^n)$ be a locally injective mapping, i.e., each point $x \in \Omega$ possesses a neighborhood $V(x)$ such that $f|_{V(x)} : V(x) \to \mathbb{R}^n$ is injective.*

Then f is an open mapping, i.e., the image $f(U)$ of any open subset U of Ω under f is an open subset of $\mathbb{R}^n$.

In particular, any injective mapping $f \in \mathcal{C}(\Omega; \mathbb{R}^n)$ is a homeomorphism of Ω onto its image $f(\Omega)$.

[88] L.E.J. Brouwer [1912]: Beweis der Invarianz des n-dimensionalen Gebiets, *Mathematische Annalen* **71**, 305–315.

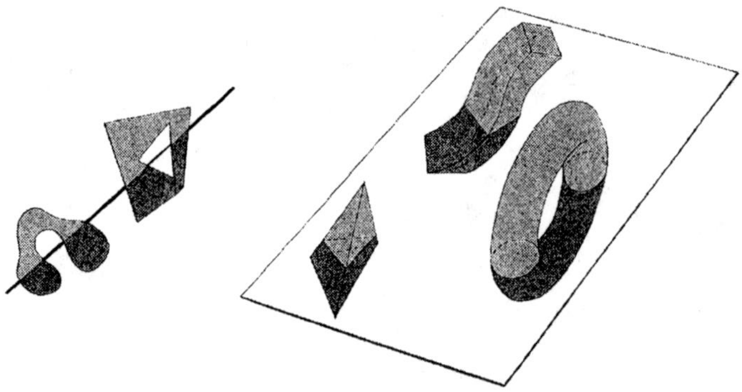

Figure 9.17-1 *A spectacular application of the Borsuk–Ulam theorem.* Consider *any* two bounded measurable sets in $\mathbb{R}^2$; then, whatever their shapes and relative positions, there always exists (at least) one line that separates each set into two subsets of equal area. Consider likewise *any* three bounded measurable sets in $\mathbb{R}^3$; then, again whatever their shapes and relative positions, there always exists (at least) one plane that separates each set into two subsets of equal volume.

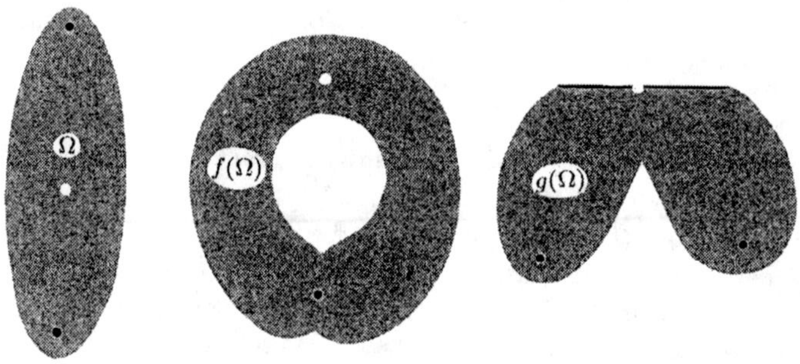

Figure 9.17-2 *The invariance of domain theorem in* $\mathbb{R}^2$. Let Ω be an open subset of $\mathbb{R}^2$ and let a function $f \in \mathcal{C}(\Omega; \mathbb{R}^2)$ be locally injective; then $f(\Omega)$ is open. By contrast, let $g \in \mathcal{C}(\Omega; \mathbb{R}^2)$ be not locally injective; then $g(\Omega)$ is not necessarily open.

Proof (i) It suffices to show that, given any $x_0 \in \Omega$, there exist open balls $B(x_0; r) \subset \Omega$ and $B(f(x_0); s)$ such that

$$B(f(x_0); s) \subset f(B(x_0; r)).$$

Besides, there is no loss of generality in assuming that $x_0 = f(x_0) = 0$ (otherwise replace the mapping f by $x \in \Omega \to (f(x + x_0) - f(x_0)) \in \mathbb{R}^n$.

(ii) Since f is locally injective, there exists an open ball $B := B(0; r)$ such that $f|_{\overline{B}} : \overline{B} \to \mathbb{R}^n$ is injective. Let

$$H(x, \lambda) := f\Big(\frac{1}{1+\lambda}x\Big) - f\Big(-\frac{\lambda}{1+\lambda}x\Big), \quad (x, \lambda) \in \overline{B} \times [0, 1].$$

Then the homotopy $H \in \mathcal{C}(\overline{B} \times [0,1]\,;\mathbb{R}^n)$ defined in this fashion is such that

$$H(\cdot,0) = f \quad \text{and} \quad H(\cdot,1) \text{ is an odd function.}$$

Besides,

$$0 \notin H(\partial B \times [0,1]),$$

since the equation

$$f\Big(\frac{1}{1+\lambda}x\Big) = f\Big(-\frac{\lambda}{1+\lambda}x\Big), \quad (x,\lambda) \in \overline{B} \times [0,1]\,,$$

and the assumed injectivity of $f|_{\overline{B}}$ together imply that $x = 0 \notin \partial B$.

(iii) Therefore, by *Borsuk's theorem* (which can be applied since $0 \in B$, $0 \notin H(\partial B \times [0,1])$, and B is symmetric with respect to 0),

$$\deg(H(\cdot,1), B, 0) \quad \text{is an odd number.}$$

But $\deg(f,B,0) = \deg(H(\cdot,1),B,0)$ by Theorem 9.15-6(a). Consequently,

$$\deg(f,B,0) \text{ is an odd number.}$$

(iv) Resorting to Theorem 9.15-4 as in the previous proof, we conclude that there exists $s > 0$ such that

$$B(0;s) \subset f(B).$$

Hence the assertion is established. □

Various applications of the Brouwer invariance of domain theorem in $\mathbb{R}^n$ are proposed in Problems 9.17-3–9.17-7.

Problems

9.17-1 This exercise constitutes a complement to Borsuk's theorem. Let Ω be a bounded open subset of $\mathbb{R}^n$ that is symmetric with respect to 0 but *does not contain* 0, and let $f \in \mathcal{C}(\overline{\Omega};\mathbb{R}^n)$ be an odd function such that $0 \notin f(\partial\Omega)$. Show that $\deg(f,\Omega,0)$ is even.

9.17-2 Show that, given n bounded measurable subsets $A_i, 1 \le i \le n$, of $\mathbb{R}^n$, there exists a hyperplane in $\mathbb{R}^n$, i.e., a subset of $\mathbb{R}^n$ of the form $\{y \in \mathbb{R}^n;\ y \cdot a = b\}$ for some vector $a \in \mathbb{R}^n$ and $b \in \mathbb{R}$, such that, for each $1 \le i \le n$,

$$\mu\,(A_i \cap \{y \in \mathbb{R}^n;\ y \cdot a > b\}) = \mu\,(A_i \cap \{y \in \mathbb{R}^n;\ y \cdot a < b\})$$

where μ denotes the n-dimensional Lebesgue measure.

Hint: Apply the Borsuk–Ulam theorem to the function

$$f = (f_i)_{i=1}^n : S := \{x \in \mathbb{R}^{n+1};\ |x| = 1\} \to \mathbb{R}^n,$$

whose components $f_i : S \to \mathbb{R}$, $1 \le i \le n$, are defined by

$$f_i(x) = \mu\,(A_i \cap \{y \in \mathbb{R}^n;\ y \cdot x' > x_{n+1}\}) \quad \text{for each } x = (x', x_{n+1}) \in S.$$

9.17-3 (1) Let Ω be a bounded open subset of $\mathbb{R}^n$ and let $f \in \mathcal{C}(\overline{\Omega}; \mathbb{R}^n)$ be such that $f|_\Omega : \Omega \to \mathbb{R}^n$ is injective. Using again the *invariance of domain theorem*, show that

$$f(\overline{\Omega}) = \overline{f(\Omega)}, \quad f(\Omega) \subset \operatorname{int} f(\overline{\Omega}), \quad f(\partial\Omega) \supset \partial(f(\overline{\Omega})).$$

(2) Assume in addition that $\operatorname{int} \overline{\Omega} = \Omega$ and that $f : \overline{\Omega} \to \mathbb{R}^n$ is injective. Using again the invariance of domain theorem, show that

$$f(\Omega) = \operatorname{int} f(\overline{\Omega}), \quad f(\partial\Omega) = \partial(f(\Omega)) = \partial(f(\overline{\Omega})).$$

9.17-4 Let $U \subset \mathbb{R}^m$ and $V \subset \mathbb{R}^n$ be open. Using the *invariance of domain theorem*, show that there is no homeomorphism from U onto V if $m \neq n$. Thus in particular, $\mathbb{R}^m$ *is not homeomorphic to* $\mathbb{R}^n$ *if* $m \neq n$.

9.17-5 (1) Let $f \in \mathcal{C}^1(\mathbb{R}^n; \mathbb{R}^n)$ be such that $\det \nabla f(x) \neq 0$ for all $x \in \mathbb{R}^n$ and $\lim_{|x| \to \infty} |f(x)| = \infty$. Show that $f(\mathbb{R}^n) = \mathbb{R}^n$ and that f is a $\mathcal{C}^1$-diffeomorphism of $\mathbb{R}^n$ onto $\mathbb{R}^n$.

(2) Show that, conversely, if a mapping f is a $\mathcal{C}^1$-diffeomorphism of $\mathbb{R}^n$ onto $\mathbb{R}^n$, then $\det \nabla f(x) \neq 0$ for all $x \in \mathbb{R}^n$ and $\lim_{|x| \to \infty} |f(x)| = \infty$.

9.17-6 Let $f \in \mathcal{C}(\mathbb{R}^n; \mathbb{R}^n)$ be a locally injective mapping such that $\lim_{|x| \to \infty} |f(x)| = \infty$. Show that $f(\mathbb{R}^n) = \mathbb{R}^n$.

9.17-7 Let Ω be a bounded open subset of $\mathbb{R}^n$, let $f \in \mathcal{C}(\overline{\Omega}; \mathbb{R}^n)$ be an injective mapping, and let $b \in f(\Omega)$. Show that either $\deg(f, \Omega, b) = 1$ or $\deg(f, \Omega, b) = -1$.

Remark If $f \in \mathcal{C}(\overline{\Omega}; \mathbb{R}^n) \cap \mathcal{C}^1(\Omega; \mathbb{R}^n)$, the proof is an immediate application of the formula given in Theorem 9.15-5(a). By contrast, if $f \in \mathcal{C}(\overline{\Omega}; \mathbb{R}^n)$, the proof is substantially more challenging. □

Hint: First, note that, by the *invariance of domain theorem*, there exists an open ball $B = B(b; r)$ such that $\overline{B} \subset f(\Omega)$; then, using Problem 9.15-2, show that it suffices to show that either $\deg(f, f^{-1}(B), b) = 1$ or $\deg(f, f^{-1}(B), b) = -1$. By the *Jordan–Brouwer separation theorem* (Problem 9.15-5), the set $\mathbb{R}^n - f^{-1}(\partial B)$ has one bounded connected component U. Applying *Leray's product formula* (Problem 9.15-4) to the composition $\tilde{f} \circ f^{-1} \in \mathcal{C}(\overline{B}; \mathbb{R}^n)$, where $\tilde{f} \in \mathcal{C}(\mathbb{R}^n; \mathbb{R}^n)$ is any continuous extension of f, show that $\deg(f^{-1}, B, U) \neq 0$, and finally, that $\deg(f, f^{-1}(B), b) \in \{-1, 1\}$.

9.17-8 This problem[89] provides in particular a useful *sufficient condition for the injectivity of a nonlinear mapping in* $\mathbb{R}^n$. Let Ω be a bounded open connected subset of $\mathbb{R}^n$ such that $\operatorname{int} \overline{\Omega} = \Omega$, let $f_0 \in \mathcal{C}(\overline{\Omega}; \mathbb{R}^n)$ be an injective mapping, and let $f \in \mathcal{C}(\overline{\Omega}; \mathbb{R}^n) \cap \mathcal{C}^1(\Omega; \mathbb{R}^n)$ be a mapping that satisfies

$$f(x) = f_0(x) \text{ for all } x \in \partial\Omega \quad \text{and} \quad \det \nabla f(x) > 0 \text{ for all } x \in \Omega.$$

---

[89] Adapted from Theorem 5.5-2 in CIARLET [1988]; a similar injectivity result, with slightly different assumptions, is found in:

G.H. MEISTERS; C. OLECH [1963]: Locally one-to-one mappings and a classical theorem on Schlicht functions, *Duke Mathematical Journal* **30**, 63–80.

If the function f is instead assumed to be in the Sobolev space $W^{1,p}(\Omega; \mathbb{R}^n)$ for some $p > n$ and Ω is a domain in $\mathbb{R}^n$ (so that $f \in \mathcal{C}(\overline{\Omega}; \mathbb{R}^n)$; cf. Theorem 6.6-1) and if f satisfies $\det \nabla f(x) > 0$ for almost all $x \in \Omega$, it can still be proved that $f : \overline{\Omega} \to \mathbb{R}^n$ is injective, but only *almost everywhere*, in the sense that $\operatorname{card} f^{-1}(b) = 1$ for almost all $b \in f(\Omega)$; see Theorems 1 and 2 in:

J. BALL [1981]: Global invertibility of Sobolev functions and the interpenetration of matter, *Proceedings of the Royal Society—Edinburgh* **88A**, 315–328.

(1) Using Problem 9.17-7, show that either $\deg(f, \Omega, b) = 1$ for all $b \in f_0(\Omega)$ or $\deg(f, \Omega, b) = -1$ for all $b \in f_0(\Omega)$, and that $\deg(f, \Omega, b) = 0$ if $b \notin f_0(\overline{\Omega})$.

(2) Show that $\operatorname{card} f^{-1}(b) = 1$ for all $b \in f_0(\Omega)$.

(3) Show that $f(\overline{\Omega}) = f_0(\overline{\Omega})$ and $f(\Omega) = f_0(\Omega)$ (use Problem 9.17-3).

(4) Show that $f : \overline{\Omega} \to f(\overline{\Omega})$ is a homeomorphism and that $f|_\Omega : \Omega \to f(\Omega)$ is a $\mathcal{C}^1$-diffeomorphism (use the invariance of domain theorem in Banach spaces; cf. Theorem 7.14-2).

BIBLIOGRAPHICAL NOTES

The list of books and handbook articles mentioned in these bibliographical notes is simply intended to provide a selection of titles that may usefully complement the text; otherwise it is by no means intended to be exhaustive. References to original papers are also found in the footnotes interspersed throughout the chapters.

Chapter 1: Real analysis and theory of functions

For detailed proofs, complements, and additional references, we refer the reader to such classic texts as DIEUDONNÉ [1960], ROYDEN [1963], HEWITT & STROMBERG [1965], or RUDIN [1966], where the core topics of *real analysis* are treated at length. More recent treatments include SCHWARTZ [1970, 1991], FOLLAND [1984], LANG [1993], DIBENEDETTO [2002], KRANTZ [2004], JOST [2005], KNAPP [2005a, 2005b], or AMANN & ESCHER [2005, 2008, 2009]. The book of LI [2011] contains many challenging *problems in real analysis.*

More specialized and in-depth treatments of *set theory* are found in COHEN [1966], BOURBAKI [1970], or HALMOS [1982]; of *general topology* in KELLEY [1955], TAYLOR [1965], CHOQUET [1966], or BOURBAKI [1966a, 1966b]; and of the *Lebesgue measure and integral* in HALMOS [1950], MUNROE [1953], SCHWARTZ [1993a, 1993b], RANA [2002], or BENEDETTO & CZAJA [2009].

The *theory of functions* is analyzed in depth in STEIN [1970], EVANS & GARIEPY [1992], CAROTHERS [2000], or STEIN & SHAKARCHI [2005]. A thorough treatment of *domains* in $\mathbb{R}^n$ and of Green's formulas is given in NEČAS [1967].

Chapters 2–5: Normed vector spaces; Banach spaces; inner-product spaces and Hilbert spaces; the "great theorems" of linear functional analysis

The contents of these chapters, which cover the basic results of *linear functional analysis*, can be usefully complemented by such classic texts as BANACH [1932] (the monograph that laid the foundations of modern linear functional analysis), RIESZ & NAGY [1955], TAYLOR [1958] (later revised and expanded as TAYLOR & LAY [1980]), the monumental treatise of DUNFORD & SCHWARTZ [1958, 1963, 1971], GOFFMAN & PEDRICK [1965], KATO [1966], YOSIDA [1966] (a classic among the classics), RUDIN [1973], SCHECHTER [1971], HALMOS [1974], DISTEL [1975], KREYSZIG [1978], KESAVAN [1989], CONWAY [1990], ZEIDLER [1995a, 1995b], DEBNATH & MIKUSIŃSKI [1999], AUBIN [2000], LAX [2002], HA [2007], KESAVAN [2009], ODEN & DEMKOWICZ [2010], STEIN & SHAKARCHI [2011], or BREZIS [2011] (the

English translation and expanded version of BREZIS [1983], the highly successful original French text).

Functional analysis of *spectral problems* (not considered here, save for compact self-adjoint operators) is treated in DUNFORD & SCHWARTZ [1971], FRIEDRICHS [1981], or DAUTRAY & LIONS [2000c].

Treatments that combine linear functional analysis with *interpolation theory, approximation theory,* or *numerical analysis,* are found in DAVIS [1963], CHENEY [1966], DAVIS & RABINOWITZ [1975], CROUZEIX & MIGNOT [1983], CHATELIN [1983], ATKINSON & HAN [2009], or MHASKAR & PAI [2007].

Thorough accounts of *matrix theory,* which is nothing but linear functional analysis in finite-dimensional spaces, are found in GANTMACHER [1959], VARGA [1962], HOUSEHOLDER [1964], STRANG [1976, 2009], HORN & JOHNSON [1985, 1991], CIARLET [1987], SERRE [2010].

Readers interested in the *history of functional analysis* and in *autobiographies, biographies,* and *photos* of functional analysts quoted in the text may consult the following books, all a delight to read or simply to browse through: REID [1970, 1976], BOURBAKI [1974], ULAM [1976], WESTFALL [1980], DIEUDONNÉ [1981], MAUDLIN [1981], ALBERS & ALEXANDERSON [1985], HALMOS [1985, 1987], PÓLYA [1987], CURIEN & SCHMIDT [1990], RUDIN [1997], BATTERSON [2000], SCHWARTZ [2001], MAZ'YA & SHAPOSHNIKOVA [1998], PIETSCH [2007], JAKIMOWICZ & MIRANOVICZ [2011], or GRAY [2012] (this list provides only a short sample of such books).

Chapter 6: Linear partial differential equations

For more details about the topologies of the spaces $\mathcal{D}(\Omega)$ and $\mathcal{D}'(\Omega)$, see YOSIDA [1966], VO-KHAC [1972a, 1972b], HÖRMANDER [1983, Chapters 1–7], DUISTERMAAT & KOLK [2010], and of course, the celebrated treatise of SCHWARTZ [1966], who created and formalized the theory of *distributions.* An illuminating introduction to this theory is given in SCHWARTZ [1965].

Detailed studies of the *Sobolev spaces* are found in LIONS & MAGENES [1972, Chapter 1] and DAUTRAY & LIONS [2000b, Chapter 4] in the Hilbertian case, and LIONS [1965, Chapters 1–3], NEČAS [1967, Chapter 2], ADAMS [1975], ATTOUCH, BUTTAZZO, & MICHAILLE [2006], TARTAR [2007], or BREZIS [2011, Chapter 9], in the general case.

Since there is a very large number of texts devoted to *linear partial differential equations,* our limited aim here is simply to quote a small selection of texts whose content and approach are (at least in part) similar to those of this chapter (references to the more specific topics treated at the end of this chapter, viz., the Poincaré, Saint-Venant, and Donati lemmas, and Pfaff systems, have been already provided in the footnotes).

More specifically, BUTTAZZO, GIAQUITA, & HILDEBRANDT [1998], CHIPOT [2009], or CIORANESCU, DONATO, & ROQUE [2012] constitute excellent introductions. At a more advanced level, the reader should consult such classic texts as NEČAS [1967], KINDERLEHRER & STAMPACCHIA [1980], GILBARG & TRUDINGER [1988], TAYLOR [1996a, 1996b], STAKGOLD [1998], ATTOUCH, BUTTAZZO, & MICHAILLE [2006], SAUVIGNY [2006a, 2006b], EVANS [2010], DIBENEDETTO [2010], and especially, the treatise of DAUTRAY & LIONS [2000a, 2000b, 2000c, 2000d, 2000e, 2000f], which treats in details an astonishingly wide number of applications.

Singularities are treated at length in GRISVARD [1992]. The mathematical analysis of *variational problems with "small" parameters,* such as *singular perturbation problems* or *homogenization problems*, has been essentially initiated by LIONS [1973]. More recent treatments are found in CIORANESCU & DONATO [1999], CIORANESCU & SAINT JEAN PAULIN [1999], or CIARLET [1997, 2000] for applications to *linearized plate and shell theories. Asymptotic analyses* of specific elliptic problems are treated in CHIPOT [2002], TARTAR [2009], or GHERGU & RĂDULESCU [2008].

The variational formulation of problems arising in *linearized elasticity* (including those modeled by variational inequalities) are treated at length in DUVAUT & LIONS [1972], FICHERA [1972a, 1972b], and NEČAS & HLAVÁČEK [1981].

The *approximation* of the solutions of problems modeled by variational equations or inequalities is thoroughly analyzed in CIARLET [1978, 1991], GLOWINSKI, LIONS, & TRÉMOLIÈRES [1981], RAVIART & THOMAS [1983], GLOWINSKI [1984], GIRAULT & RAVIART [1986], BREZZI & FORTIN [1991], BABUŠKA & OSBORN [1991], ROBERTS & THOMAS [1991], or BRENNER & SCOTT [2002] (the list is far from being exhaustive).

Chapter 7: Differential calculus in normed vector spaces

For further reading and complements, see DIEUDONNÉ [1960, Chapter 8], SCHWARTZ [1992] (some parts of this chapter were inspired by this beautiful text), LANG [1993, Chapter 13], or ABRAHAM, MARSDEN, & RATIU [1988, Chapter 2] (somewhat surprisingly, there are not so many texts in English that treat differential calculus in normed vector spaces).

Extensive treatments of *Newton's method* and, more generally, of the solution of *systems of nonlinear equations*, are found in ORTEGA & RHEINBOLDT [2000], DEUFLHARD [2004], or DEDIEU [2006].

Excellent texts on the *maximum principle* are PROTTER & WEINBERGER [1967], FRAENKEL [2000], and PUCCI & SERRIN [2007].

Applications of *Lagrange interpolation in* $\mathbb{R}^n$ to finite element methods (and also of Hermite interpolation in $\mathbb{R}^n$, not considered here) are treated at length in CIARLET [1978, 1991].

Optimization in $\mathbb{R}^n$ is briefly touched upon in Sections 7.12, 7.15, and 7.16 (the content of which is based on excerpts from CIARLET [1987], reused here with the kind permission of Dunod, Paris, current publisher of the original French edition). Otherwise, it is the subject of numerous texts; we only mention here LUENBERGER [1969], HESTENES [1975], CIARLET [1987], and HIRIART-URRUTY & LEMARÉCHAL [1993a, 1993b].

Chapter 8: Differential geometry in $\mathbb{R}^n$

For the most part, the content of this chapter closely follows that of Chapters 1 and 2 of CIARLET [2005] (this material has been adapted here with the kind permission of Springer, Dordrecht), where applications to *three-dimensional elasticity in curvilinear coordinates* and to *shell theory* are also given in Chapters 3 and 4; further applications to shell theory are found in CIARLET [2000].

Exhaustive treatments of *tensor analysis* are found in BOOTHBY [1975], MARSDEN & HUGHES [1999, Chapter 1], ABRAHAN, MARSDEN, & RATIU [1988], SIMMONDS [1994], or LEBEDEV & CLOUD [2003]. A gentle introduction to the subject is provided in ANTMAN [2005, Chapter 11, Sections 1–3].

For detailed treatments of *Riemannian geometry*, see classic texts such as CHOQUET-BRUHAT, DE WITT-MORETTE, & DILLARD-BLEICK [1982], MARSDEN & HUGHES [1999], BERGER [2003], GALLOT, HULIN, & LAFONTAINE [2004], and especially, SCHLICHTKRULL [2012].

More generally, useful complements to the text are found in classic texts such as STOKER [1969], KLINGENBERG [1973], DO CARMO [1976, 1994], BERGER & GOSTIAUX [1987], or SPIVAK [1999], as well as in KÜHNEL [2002], PRESSLEY [2005], or O'NEILL [2006].

Chapter 9: The "great theorems" of nonlinear functional analysis

While there are numerous texts that cover the essentials of *linear functional analysis*, there are comparatively few texts that comprehensively cover *nonlinear functional analysis*. Among these, SCHWARTZ [1969] and NIRENBERG [1974] stand as landmark classics. More recent texts include BERGER [1977], DEIMLING [1985], STRUWE [1990], KAVIAN [1993], AUBIN [1993, 2000], DENKOWSKI, MIGÓRSKI, & PAPAGEORGIOU [2003], and KESAVAN [2004]. Special mention must be made of the monumental treatise of ZEIDLER [1985, 1986, 1990a, 1990b], also an invaluable source of historical comments.

By contrast, there is a wide array of *specialized* texts: on the *calculus of variations* and *variational methods in general*, we mention EKELAND & TEMAM [1976], GOLDSTINE [1980] (for a scholarly historical perspective), STRUWE [1990], GIUSTI [2003], KESAVAN [2006] (for minimization problems of "domain-dependent" functionals), GIAQUINTA & HILDEBRANDT [2006a, 2006b], VAN BRUNT [2006], and especially, the scholarly treatment of DACAROGNA [2010].

On *nonlinear partial differential equations in general*, we mention LIONS [1969] (a masterpiece even to this day, unfortunately never translated into English), STRUWE [1990], TAYLOR [1996c], GILBARG & TRUDINGER [1998], CHIPOT [2000] (a nice introductory text), MOTREANU & RĂDULESCU [2003], SAUVIGNY [2006a, 2006b], GHERGU & RĂDULESCU [2008, 2012], and EVANS [2010]; on *Gamma-convergence*, we mention ATTOUCH [1984], DAL MASO [1993], and BRAIDES [2002]; on *monotone operators*, we mention BREZIS [1973].

On *nonlinear three-dimensional elasticity*, we mention MARSDEN & HUGHES [1999], VALENT [1988], CIARLET [1988], and ANTMAN [2005] (Section 9.7, which briefly touches upon this subject, is based on excerpts from Chapter 7 of CIARLET [1988], reused here with the kind permission of the publishers, North-Holland, Amsterdam); on *nonlinear plate theory*, we mention CIARLET & RABIER [1980] and LEWINSKI & TELEGA [2000]; on the *Navier–Stokes equations for incompressible fluids*, we mention TEMAM [1977, 1995], GIRAULT & RAVIART [1986], CONSTANTIN & FOIAS [1988], LIONS [1996], FOIAS, MANLEY, ROSA, & TEMAM [2001], GLOWINSKI [2003], and TARTAR [2006]; on the *minimal surface equation*, we mention EKELAND & TEMAM [1974], NITSCHE [1975], and GIUSTI [1984].

On *Brouwer's topological degree* and related topics, we mention RADO & REICHELDERFER [1955], MILNOR [1965], MAWHIN [1979], FONSECA & GANGBO [1995], and the forthcoming book of DINCA & MAWHIN [2013]; a scholarly account of the genesis of the Brouwer degree, Brouwer's theorem, and the invariance of domain theorem is given in DIEUDONNÉ [1989, Part 2].

BIBLIOGRAPHY

R. Abraham; J.E. Marsden; T. Ratiu [1988]: *Manifolds, Tensor Analysis, and Applications, Second Edition*, Springer, New York (First Edition: 1983, Addison-Wesley).

R.A. Adams [1975]: *Sobolev Spaces*, Academic Press, New York.

N.I. Akhiezer; I.M. Glazman [1961]: *Theory of Linear Operators in Hilbert Spaces, Volume* 1, Ungar, New York.

J.L. Akian [2003]: A simple proof of the ellipticity of Koiter's model, *Analysis and Applications* **1**, 1–16.

D.J. Albers; G.L. Alexanderson, editors [1985]: *Mathematical People—Profiles and Interieurs*, Birkhäuser, Boston.

H. Amann [1976]: Fixed point equations and nonlinear eigenvalue problems in ordered Banach spaces, *SIAM Review* **18**, 620–709.

H. Amann; J. Escher [2005]: *Analysis* I, Birkhäuser, Boston (translation of the original German edition, *Analysis* I, Birkhäuser, Basel, 1998).

H. Amann; J. Escher [2008]: *Analysis* II, Birkhäuser, Boston (translation of the original German edition, *Analysis* II, Birkhäuser, Basel, 1999).

H. Amann; J. Escher [2009]: *Analysis* III, Birkhäuser, Boston (translation of the original German edition, *Analysis* III, Birkhäuser, Basel, 2001).

C. Amrouche; P.G. Ciarlet; L. Gratie; S. Kesavan [2006]: On the characterization of matrix fields as linearized strain tensor fields, *Journal de Mathématiques Pures et Appliquées* **86**, 116–132.

C. Amrouche; V. Girault [1994]: Decomposition of vector spaces and application to the Stokes problem in arbitrary dimension, *Czechoslovak Mathematical Journal* **44**, 109–140.

S.S. Antman [1970]: Existence of solutions of the equilibrium equations for nonlinearly elastic rings and arches, *Indiana University Mathematics Journal* **20**, 281–302.

S.S. Antman [1983]: Regular and singular problems for large elastic deformations of tubes, wedges, and cylinders, *Archive for Rational Mechanics and Analysis* **82**, 1–52.

S.S. Antman [2005]: *Nonlinear Problems of Elasticity*, Springer, Berlin (First Edition: 1995).

D.N. Arnold; R.S. Falk; R. Winther [2006]: Finite element exterior calculus, homological techniques, and applications, in *Acta Numerica, Volume* 15 (A. Iserles, editor), pp. 1–155, Cambridge University Press, Cambridge, UK.

N. Aronszajn; K.T. Smith [1957]: Characterization of positive reproducing kernels. Applications to Green's functions, *American Journal of Mathematics* **79**, 611–622.

C. Arzelà [1883]: Un' osservazione intorno alle serie di funzioni, *Rendiconti delle Sessioni dell' Accademia Reale delle Scienze dell' Istituto di Bologna*, 142–159.

C. Ascoli [1883]: Le curve limiti di una varietà data di curve, *Atti della Accademia Nazionale dei Lincei, Classe di Scienze Fisiche, Matematiche e Naturali* **18**, 521–586.

K.E. Atkinson; W. Han [2009]: *Theoretical Numerical Analysis: A Functional Analysis Framework, Third Edition*, Springer, New York (First Edition: 2001).

H. Attouch [1984]: *Variational Convergence for Functions and Operators*, Pitman, Boston.

H. Attouch; G. Buttazzo; G. Michaille [2006]: *Variational Analysis in Sobolev and BV Spaces: Applications to PDEs and Optimization*, SIAM, Philadelphia.

J.P. Aubin [1993]: *Optima and Equilibria—An Introduction to Nonlinear Analysis*, Springer, Berlin.

J.P. Aubin [2000]: *Applied Functional Analysis, Second Edition*, Wiley-Interscience, New York (First Edition: 1979).

I. Babuška [1971]: Error bound for finite element method, *Numerische Mathematik* **16**, 322–333.

I. Babuška; A.K. Aziz [1976]: On the angle condition in the finite element method, *SIAM Journal on Numerical Analysis* **13**, 214–226.

I. Babuška; J. Osborn [1991]: Eigenvalue problems, in *Handbook of Numerical Analysis, Volume* II (P.G. Ciarlet & J.L. Lions, editors), pp. 641–784, North-Holland, Amsterdam.

R. Baire [1899]: Sur les fonctions de variables réelles, *Annali di Matematica Pura ed Applicata* **3**, 1–123.

J. Ball [1977]: Convexity conditions and existence theorems in nonlinear elasticity, *Archive for Rational Mechanics and Analysis* **63**, 337–403.

J. Ball [1981]: Global invertibility of Sobolev functions and the interpenetration of matter, *Proceedings of the Royal Society, Edinburgh* **88A**, 315–328.

J.M. Ball; R.J. Knops; J.E. Marsden [1978]: Two examples in nonlinear elasticity, in *Proceedings – Conference in Nonlinear Analysis, Besançon*, pp. 41–49, Lecture Notes in Mathematics, Volume 466, Springer, Berlin.

S. Banach [1922]: Sur les opérations dans les ensembles abstraits et leurs applications aux équations intégrales, *Fundamenta Mathematicae* **3**, 133–181.

S. Banach [1932]: *Théorie des Opérations Linéaires*, Monografje Matematyczne, Volume 1, Warsaw.

S. Banach; S. Saks [1930]: Sur la convergence forte dans le champ L^p, *Studia Mathematica* **2**, 51–57.

S. Banach; H. Steinhaus [1927]: Sur le principe de la condensation de singularités, *Fundamenta Mathematicae* **9**, 50–61.

S. Batterson [2000]: *Stephen Smale: The Mathematician Who Broke the Dimension Barrier*, American Mathematical Society, Providence, RI.

R. Beals; R. Wong [2010]: *Special Functions: A Graduate Text*, Cambridge University Press, Cambridge, UK.

P.R. Beesack; E. Hughes; M. Ortel [1979]: Rotund complex linear spaces, *Proceedings of the American Mathematical Society* **75**, 42–44.

J.J. Benedetto; W. Czaja [2009]: *Integration and Modern Analysis*, Birkhäuser, Boston.

A. Ben-Israel; T.N.E. Greville [2003]: *Generalized Inverses: Theory and Applications*, Second Edition, Springer.

M.S. Berger [1967]: On the von Kármán equations and the buckling of a thin elastic plate. I. The clamped plate, *Communications on Pure and Applied Mathematics* **20**, 687–719.

M.S. Berger [1977]: *Nonlinearity and Functional Analysis*, Academic Press, New York.

M. Berger [2003]: *A Panoramic View of Riemannian Geometry*, Springer, Berlin.

M. Berger; B. Gostiaux [1987]: *Géométrie Différentielle: Variétés, Courbes et Surfaces*, Presses Universitaires de France, Paris.

S. Bergman; M. Schiffer [1948]: Kernel functions in the theory of partial differential equations of elliptic type, *Duke Mathematical Journal* **15**, 535–566.

A. BERMAN; R.J. PLEMMONS [1994]: *Nonnegative Matrices in the Mathematical Sciences*, Classics in Applied Mathematics, Vol. 9, SIAM, Philadelphia.

M. BERNADOU; P.G. CIARLET [1976]: Sur l'ellipticité du modèle linéaire de coques de W.T. Koiter, in *Computing Methods in Applied Sciences and Engineering* (R. GLOWINSKI & J.L. LIONS, editors), pp. 89–136, Lecture Notes in Economics and Mathematical Systems, **134**, Springer, Heidelberg.

M. BERNADOU; P.G. CIARLET; B. MIARA [1994]: Existence theorems for two-dimensional linear shell theories, *Journal of Elasticity* **34**, 111–138.

J.M.E. BERNARD [2011]: Density results in Sobolev spaces whose elements vanish on a part of the boundary, *Chinese Annals of Mathematics, Series B*, **32**, 823–846.

S.N. BERNSTEIN [1912]: Démonstration du théorème de Weierstrass fondée sur le calcul de probabilités, *Communications of the Kharkov Mathematical Society* **13**, 1–2.

S.N. BERNSTEIN [1932]: Complément à l'article de E. Voronovskaya "Détermination de la forme asymptotique de l'approximation des fonctions par les polynômes de M. Bernstein," *Doklady Akademii Nauk SSSR* **4**, 86–92.

G. BIRKHOFF [1946]: Tres observaciones sobre el algebra lineal, *Universidad Nacional de Tucumán Revista A* **5**, 147–151.

E. BISHOP; R.R. PHELPS [1961]: A proof that every Banach space is subreflexive, *Bulletin of the American Mathematical Society* **67**, 97–98.

A. BLOUZA; H. LE DRET: [1999]: Existence and uniqueness for the linear Koiter model for shells with little regularity, *Quarterly of Applied Mathematics* **57**, 317–337.

A.B. BOGHOSSIAN; P.D. JOHNSON, JR. [1990]: A pointwise condition for an infinitely differentiable function of several variables to be a polynomial, *Journal of Mathematical Analysis and Applications* **151**, 17–19.

H. BOHMAN [1952]: On approximation of continuous and of analytic functions, *Arkiv för Mathematik* **2**, 43–56.

H.F. BOHNENBLUST; A. SOBCZYK [1938]: Extensions of functionals on complex linear spaces. *Bulletin of the American Mathematical Society* **44**, 91–93.

O. BOLZA [1946]: *Lectures on the Calculus of Variations*, Chelsea Publishing Company, New York.

O. BONNET [1848]: Mémoire sur la théorie générale des surfaces, *Journal de l'Ecole Polytechnique* **19**, 1–146.

W.M. BOOTHBY [1975]: *An Introduction to Differentiable Manifolds and Riemannian Geometry*, Academic Press, New York.

W. BORCHERS; H. SOHR [1990]: On the equations $\operatorname{rot} v = g$ and $\operatorname{div} u = f$ with zero boundary conditions, *Hokkaido Mathematical Journal* **19**, 67–87.

K. BORSUK [1933]: Drei Sätze über die n-dimensionale euklidische Sphäre, *Fundamenta Mathematicae* **21**, 177–190.

N. BOURBAKI [1966a]: *Éléments de Mathématique. Topologie Générale; Chapitres* 1 *à* 4, Hermann, Paris (English translation: *Elements of Mathematics, General Topology: Chapters* 1–4, Springer, New York, 1998).

N. BOURBAKI [1966b]: *Eléments de Mathématique. Topologie Générale: Chapitres* 5 *à* 10, Hermann, Paris (English translation: *Elements of Mathematics, General Topology: Chapters* 5–10, Addison-Wesley, Reading, MA, 1966).

N. BOURBAKI [1970]: *Éléments de Mathématique. Théorie des Ensembles*, Hermann, Paris (English translation: *Theory of Sets*, Springer, New York, 2004).

N. BOURBAKI [1974]: *Éléments d'Histoire des Mathématiques*, Hermann, Paris (English translation: *Elements of the History of Mathematics*, Springer, New York, 1998).

J. BOURGAIN [1977]: On dentability and the Bishop-Phelps property, *Israel Journal of Mathematics* **28**, 268–271.

R.E. BRADLEY; C.E. SANDIFER [2009]: *Cauchy's Cours d'Analyse—An Annotated Translation*, Springer, Heidelberg.

A. BRAIDES [2002]: *Γ-Convergence for Beginners*, Oxford University Press, Oxford, UK.

J.H. BRAMBLE; S.R. HILBERT [1970]: Estimation of linear functionals on Sobolev spaces with application to Fourier transforms and spline interpolation, *SIAM Journal on Numerical Analysis* **7**, 112–124.

J. BRANDTS; S. KOROTOV; M. KŘÍŽEK [2011]: Generalization of the Zlámal condition for simplicial finite elements in $\mathbb{R}^d$, *Applied Mathematics* **56**, 417–424.

S.C. BRENNER; R. SCOTT [2002]: *The Mathematical Theory of Finite Element Methods*, Springer, New York.

H. BREZIS [1971]: Problèmes unilatéraux, *Journal de Mathématiques Pures et Appliquées* **9**, 1–168.

H. BREZIS [1973]: *Opérateurs Maximaux Monotones*, North-Holland, Amsterdam.

H. BREZIS [1983]: *Analyse Fonctionnelle. Théorie et Applications*, Masson, Paris.

H. BREZIS [2011]: *Functional Analysis, Sobolev Spaces and Partial Differential Equations*, Springer, New York.

H. BREZIS; M. SIBONY [1971]: Equivalence de deux inéquations variationnelles, *Archive for Rational Mechanics and Analysis* **41**, 254–265.

H. BREZIS; G. STAMPACCHIA [1968]: Sur la régularité de la solution d'inéquations elliptiques, *Bulletin de la Société Mathématique de France* **96**, 153–180.

F. BREZZI [1974]: On the existence, uniqueness and approximation of saddle point problems arising from Lagrange multipliers, *Revue Française d'Automatique, Informatique, et Recherche Opérationnelle - Série Rouge* **8**, 129–151.

F. BREZZI; M. FORTIN [1991]: *Mixed and Hybrid Finite Element Methods*, Springer, New York.

L.E.J. BROUWER [1911]: Beweis des Jordanschen Satzes für den n-dimensionalen Raum, *Mathematische Annalen* **71**, 314–319 and 598.

L.E.J. BROUWER [1912]: Über Abbildungen von Mannigfaltigkeiten, *Mathematische Annalen* **71**, 97–115.

L.E.J. BROUWER [1912]: Beweis der Invarianz des n-dimensionalen Gebiets, *Mathematische Annalen* **71**, 305–315.

F.E. BROWDER [1963]: Nonlinear elliptic boundary value problems, *Bulletin of the American Mathematical Society* **69**, 862–874.

F.E. BROWDER [1965]: Existence and uniqueness theorems for solutions of nonlinear boundary value problems, in *Proceedings of Symposia in Applied Mathematics, Volume* XVII: *Applications of Nonlinear Partial Differential Equations in Mathematical Physics*, pp. 24–49, American Mathematical Society, Providence, RI.

B. VAN BRUNT [2006]: *The Calculus of Variations*, Springer, New York.

L. BRUTMAN [1997]: Lebesgue functions for polynomial interpolations – a survey, *Annals of Numerical Mathematics* **4**, 111–127.

V. BUNYAKOVSKIĬ: [1859]: Sur quelques inégalités concernant les intégrales aux différences finies, *Mémoires de l'Académie des Sciences de Saint-Peterbourg, 7ème Série, Tome* 1, No. 9, 1–18.

B. BUTTAZZO; M. GIAQUINTA; S. HILDEBRANDT [1998]: *One-dimensional Variational Problems: An Introduction*, Clarendon Press, Oxford.

G. CANTOR [1899]: *Beiträge zur Begründung der transfiniten Mengenlehre*, Georg Olms Verlag (English translation: *Contributions to the Founding of Transfinite Numbers*, Dover, New York, 1955).

C. CARATHÉODORY [1907]: Über den Variabilitätsbereich der Fourier'schen Konstanten von positiven harmonischen Funktionen, *Rendiconti del Circolo Matematico di Palermo* **32**, 193–217.

C. CARATHÉODORY [1965]: *Calculus of Variations and Partial Differential Equations of the First Order*, Holden Day, San Francisco.

L. CARLESON [1966]: On convergence and growth of partial sums of Fourier series, *Acta Mathematica* **116**, 135–157.

M.P. DO CARMO [1976]: *Differential Geometry of Curves and Surfaces*, Prentice-Hall, Englewood Cliffs, NJ.

M.P. DO CARMO [1994]: *Differential Forms and Applications*, Universitext, Springer, Berlin (English translation of: *Formas Diferenciais e Apliçöes*, Instituto da Matematica, Pura e Aplicada, Rio de Janeiro, 1971).

N.L. CAROTHERS [2000]: *Real Analysis*, Cambridge University Press.

E. CARTAN [1927]: Sur la possibilité de plonger un espace riemannien donné dans un espace euclidien, *Annales de la Société Polonaise de Mathématiques* **6**, 1–7.

E. CARTAN [1928]: *Leçons sur la Géométrie des Espaces de Riemann*, Gauthier-Villars, Paris.

A.L. CAUCHY [1821]: *Cours d'Analyse de l'École Royale Polytechnique*, de Bure, Paris.

E. ČECH [1937]: On bicompact spaces, *Annals of Mathematics* **38**, 823–844.

E. CESÀRO [1906]: Sulle formole del Volterra, fondamentali nella teoria delle distorsioni elastiche, *Rendiconti Napoli* **12**, 311–321.

F. CHATELIN [1983]: *Spectral Approximation of Linear Operators*, Academic Press, New York.

W. CHEN; J. JOST [2002]: A Riemannian version of Korn's inequality, *Calculus of Variations* **14**, 517–530.

W.W. CHENEY [1966]: *Introduction to Approximation Theory*, McGraw-Hill, New York.

M. CHIPOT [2000]: *Elements of Nonlinear Analysis*, Birkhäuser, Basel.

M. CHIPOT [2002]: *ℓ Goes to Plus Infinity*, Birkhäuser, Basel.

M. CHIPOT [2009]: *Elliptic Equations: An Introductory Course*, Birkhäuser, Basel.

G. CHOQUET [1966]: *Topology*, Academic Press, New York.

Y. CHOQUET-BRUHAT; C. DE WITT-MORETTE; M. DILLARD-BLEICK [1982]: *Analysis, Manifolds and Physics, Second Edition*, North-Holland, Amsterdam (First Edition: 1977).

E.B. CHRISTOFFEL [1869]: Über die Transformation der homogenen Differentialausdrücke zweiten Grades, *Journal für die Reine und Angewandte Mathematik* **70**, 46–70.

P.G. CIARLET [1975]: *Lectures on the Finite Element Method*, Tata Institute of Fundamental Research, Bombay.

P.G. CIARLET [1978]: *The Finite Element Method for Elliptic Problems*, North-Holland, Amsterdam (reprinted in 2002 as SIAM Classics in Applied Mathematics, Volume 40, SIAM, Philadelphia).

P.G. CIARLET [1978]: Interpolation error estimates for the reduced Hsieh-Clough-Tocher triangle, *Mathematics of Computation* **32**, 335–344.

P.G. CIARLET [1980]: A justification of the von Kármán equations, *Archive for Rational Mechanics and Analysis* **73**, 349–389.

P.G. CIARLET [1987]: *Introduction to Numerical Linear Algebra and Optimisation*, with the assistance of B. MIARA and J. M. THOMAS for the Exercises, Cambridge University Press, Cambridge, UK (translation of the original French edition, *Introduction à l'Analyse Numérique Matricielle et à l'Optimisation*, Masson, Paris, 1982, republished with a new presentation by Dunod, Paris, in 2007, and of *Exercices d'Analyse Numérique Matricielle et d'Optimisation, avec Solutions*, by P.G. CIARLET, B. MIARA, and J. M. THOMAS, Masson, Paris, 1991, republished with a new presentation by Dunod, Paris, in 2001).

P.G. Ciarlet [1988]: *Mathematical Elasticity, Volume* I: *Three-Dimensional Elasticity*, North-Holland, Amsterdam.

P.G. Ciarlet [1991]: Basic error estimates for elliptic problems, in *Handbook of Numerical Analysis, Volume* II (P.G. Ciarlet & J.L. Lions, editors), pp. 17–351, North-Holland, Amsterdam.

P.G. Ciarlet [1997]: *Mathematical Elasticity, Volume* II: *Theory of Plates*, North-Holland, Amsterdam.

P.G. Ciarlet [2000]: *Mathematical Elasticity, Volume* III: *Theory of Shells*, North-Holland, Amsterdam.

P.G. Ciarlet [2003]: The continuity of a surface as a function of its two fundamental forms, *Journal de Mathématiques Pures et Appliquées* **82**, 253–274.

P.G. Ciarlet [2005]: *An Introduction to Differential Geometry with Applications to Elasticity*, Springer, Dordrecht.

P.G. Ciarlet; P. Ciarlet, Jr. [2005]: Another approach to linearized elasticity and a new proof of Korn's inequality, *Mathematical Models and Methods in Applied Sciences* **15**, 259–271.

P.G. Ciarlet; P. Destuynder [1979]: A justification of a nonlinear model in plate theory, *Computer Methods in Applied Mechanics and Engineering* **17/18**, 227–258.

P.G. Ciarlet; G. Geymonat [1982]: Sur les lois de comportement en élasticité non-linéaire compressible, *Comptes Rendus de l'Académie des Sciences de Paris, Série* II, **295**, 423–426.

P.G. Ciarlet; G. Geymonat; F. Krasucki [2012]: A new duality approach to elasticity, *Mathematical Models and Methods in Applied Sciences* **22**, 1150003.

P.G. Ciarlet; L. Gratie; O. Iosifescu; C. Mardare; C. Vallée [2007]: Another approach to the fundamental theorem of Riemannian geometry in $\mathbb{R}^3$, by way of rotation fields, *Journal de Mathématiques Pures et Appliquées* **87**, 237–252.

P.G. Ciarlet; L. Gratie; C. Mardare [2005]: A nonlinear Korn inequality on a surface, *Journal de Mathématiques Pures et Appliquées* **85**, 2–16.

P.G. Ciarlet; L. Gratie; C. Mardare [2008]: A new approach to the fundamental theorem of surface theory, *Archive for Rational Mechanics and Analysis* **188**, 457–473.

P.G. Ciarlet; O. Iosifescu [2009]: A new approach to the fundamental theorem of surface theory, by means of the Darboux-Vallée-Fortunée compatibility relation, *Journal de Mathématiques Pures et Appliquées* **91**, 384–401.

P.G. Ciarlet; F. Larsonneur [2002]: On the recovery of a surface with prescribed first and second fundamental forms, *Journal de Mathématiques Pures et Appliquées* **81**, 167–185.

P.G. Ciarlet; F. Laurent [2003]: Continuity of a deformation as a function of its Cauchy-Green tensor. *Archive for Rational Mechanics and Analysis* **167**, 255–269.

P.G. Ciarlet; V. Lods [1996]: On the ellipticity of linear membrane shell equations, *Journal de Mathématiques Pures et Appliquées* **75**, 107–124.

P.G. Ciarlet; C. Mardare [2003]: On rigid and infinitesimal rigid displacements in shell theory, *Journal de Mathématiques Pures et Appliquées* **83**, 1–15.

P.G. Ciarlet; C. Mardare [2003]: On rigid and infinitesimal rigid displacements in three-dimensional elasticity, *Mathematical Models and Methods in Applied Sciences* **13**, 1589–1598.

P.G. Ciarlet; C. Mardare [2004]: Continuity of a deformation in H^1 as a function of its Cauchy-Green tensor in L^1, *Journal of Nonlinear Science* **14**, 415–427.

P.G. Ciarlet; C. Mardare [2004]: Recovery of a manifold with boundary and its continuity as a function of its metric tensor, *Journal de Mathématiques Pures et Appliquées* **83**, 811–843.

P.G. Ciarlet; C. Mardare [2005]: Recovery of a surface with boundary and its continuity as a function of its two fundamental forms, *Analysis and Applications* **3**, 99–117.

P.G. Ciarlet; C. Mardare [2012]: The Newton-Kantorovich theorem, *Analysis and Applications* **10**, 249–269.

P.G. Ciarlet; S. Mardare [2001]: On Korn's inequalities in curvilinear coordinates, *Mathematical Models and Methods in Applied Sciences* **11**, 1379–1391.

P.G. Ciarlet; J. Nečas [1987]: Injectivity and self-contact in nonlinear elasticity, *Archive for Rational Mechanics and Analysis* **97**, 171–188.

P.G. Ciarlet; P. Rabier [1980]: *Les Equations de von Kármán*, Lecture Notes in Mathematics, Volume 826, Springer, Berlin.

P.G. Ciarlet, P.A. Raviart [1972]: General Lagrange and Hermite interpolation in $\mathbb{R}^n$ with applications to finite element methods, *Archive for Rational Mechanics and Analysis* **46**, 177–199.

P.G. Ciarlet; E. Sanchez-Palencia [1996]: An existence and uniqueness theorem for the two-dimensional linear membrane shell equations, *Journal de Mathématiques Pures et Appliquées* **75**, 51–67.

P.G. Ciarlet; M.H. Schultz; R.S. Varga [1969]: Numerical methods of high-order accuracy for nonlinear boundary value problems V: Monotone operator theory, *Numerische Mathematik* **13**, 51–79.

P.G. Ciarlet; C. Wagschal [1971]: Multipoint Taylor formulas and applications to the finite element method, *Numerische Mathematik* **17**, 84–100.

D. Cioranescu; P. Donato [1999]: *An Introduction to Homogenization*, Oxford Lecture Series in Mathematics and Its Applications, Volume 17, Oxford University Press, Oxford, UK.

D. Cioranescu; P. Donato; M.P. Roque [2012]: *Introduction to Classical and Variational Partial Differential Equations*, The University of the Philippines Press, Quezon City.

D. Cioranescu; J. Saint Jean Paulin [1999]: *Homogenization of Reticulated Structures*, Applied Mathematical Sciences, Volume 136, Springer, Berlin.

J.A. Clarkson [1936]: Uniformly convex spaces, *Transactions of the American Mathematical Society* **40**, 396–414.

C. Coatmélec [1966]: Approximation et interpolation des fonctions différentiables de plusieurs variables, *Annales Scientifiques de l'Ecole Normale Supérieure* **83**, 271–341.

D. Codazzi [1868–1869]: Sulle coordinate curvilinee d'una superficie dello spazio, *Annali di Mathematica Pura e Applicata* **2**, 101–119.

E.A. Coddington; N. Levinson [1955]: *Theory of Ordinary Differential Equations*, McGraw Hill, New York.

P.J. Cohen [1963]: The independence of the continuum hypothesis, *Proceedings of the National Academy of Sciences, USA* **50**, 1143–1148.

P.J. Cohen [1964]: The independence of the continuum hypothesis, *Proceedings of the National Academy of Sciences, USA* **51**, 105–110.

P.J. Cohen [1966]: *Set Theory and the Continuum Hypothesis*, Benjamin, New York.

B.D. Coleman; W. Noll [1959]: On the thermostatics of continuous media, *Archive for Rational Mechanics and Analysis* **4**, 97–128.

P. Constantin; C. Foias [1988]: *Navier-Stokes Equations*, University of Chicago Press, Chicago, IL.

J. Conway [1990]: *A Course in Functional Analysis, Second Edition*, Springer, New York (First Edition: 1985).

E. Corominas; F.S. Balaguer [1954]: Conditions for an infinitely differentiable function to be a polynomial (title in Spanish), *Revista Matemática Hispano-Americana* **14**, 26–43.

E. Cosserat; F. Cosserat [1896]: Sur la théorie de l'élasticité. Premier mémoire, *Annales de la Faculté des Sciences de l'Université de Toulouse* **10**, 1–116.

R. Courant [1920]: Über die Eigenwerte bei den Differentialgleichungen der Mathematischen Physik, *Mathematische Zeitschrift* **7**, 1–57.

M. Crouzeix; A.L. Mignot [1983]: *Analyse Numérique des Equations Différentielles*, Masson, Paris.

G. Csato; B. Dacorogna; O. Kneuss [2011]: *The Pullback Equation*, Birkhäuser, Basel.

H. Curien; M. Schmidt, editors [1990]: *Hommes de Science*, Hermann, Paris.

B. Dacorogna [1982]: Minimal hypersurfaces in parametric form with nonconvex integrands, *Indiana University Mathematics Journal* **31**, 531–552.

B. Dacorogna [2010]: *Direct Methods in the Calculus of Variations, Second Edition*, Springer, Berlin (First Edition: 1989).

G. Dal Maso [1993]: *An Introduction to Γ-Convergence*, Birkhäuser, Boston.

R. Dautray; J.L. Lions [2000a]: *Mathematical Analysis and Numerical Methods for Science and Technology, Volume 1 : Physical Origins and Classical Methods*, Springer, Heidelberg.[90]

R. Dautray; J.L. Lions [2000b]: *Mathematical Analysis and Numerical Methods for Science and Technology, Volume 2 : Functional and Variational Methods*, Springer, Heidelberg.

R. Dautray; J.L. Lions [2000c]: *Mathematical Analysis and Numerical Methods for Science and Technology, Volume 3 : Spectral Theory and Applications*, Springer, Heidelberg.

R. Dautray; J.L. Lions [2000d]: *Mathematical Analysis and Numerical Methods for Science and Technology, Volume 4 : Integral Equations and Numerical Methods*, Springer, Heidelberg.

R. Dautray; J.L. Lions [2000e]: *Mathematical Analysis and Numerical Methods for Science and Technology, Volume 5 : Evolution Problems* I, Springer, Heidelberg.

R. Dautray; J.L. Lions [2000f]: *Mathematical Analysis and Numerical Methods for Science and Technology, Volume 6 : Evolution Problems* II, Springer, Heidelberg.

C. Davis [1971]: The Toeplitz-Hausdorff theorem explained, *Canadian Mathematical Bulletin* **14**, 245–246.

P.J. Davis [1963]: *Interpolation and Approximation*, Dover, New York.

P.J. Davis; P. Rabinowitz [1975]: *Methods of Numerical Integration*, Academic Press, New York.

L. Debnath; P. Mikusiński [1999]: *Hilbert Spaces with Applications, Second Edition*, Academic Press, New York (First Edition: 1990).

J.P. Dedieu [2006]: *Points Fixes, Zéros et la Méthode de Newton*, Springer, Berlin.

E. De Giorgi [1975]: Sulla convergenza di alcune successioni di integrali del tipo dell'area, *Rendiconti Mathematica Roma* **8**, 227–294.

E. De Giorgi [1977]: Γ-convergenza e G-convergenza, *Bolletina Unione Mathematica Italiana* **5**, 213–220.

E. De Giorgi; G. Dal Maso [1983]: *Γ-Convergence and Calculus of Variations*, Lecture Notes in Mathematics, Volume 979, Springer, Berlin.

K. Deimling [1985]: *Nonlinear Functional Analysis*, Springer, Berlin.

L. Demkowicz [2000]: Babuška ⇔ Brezzi??, Technical Report, Texas Institute for Computational and Applied Mathematics, TICAM Seminar (October 31, 2000).

Z. Denkowski; S. Migórski; N.S. Papageorgiou [2003]: *An Introduction to Nonlinear Analysis: Applications*, Kluwer, Boston.

[90]These six volumes are translated from *Analyse Mathématique et Calcul Numérique pour les Sciences et les Techniques*, Masson, Paris et Commissariat à l'Energie Atomique, Paris, 1984–1985.

P. DEUFLHARD [2004]: *Newton Methods for Nonlinear Problems—Affine Invariance and Adaptive Algorithms*, Springer, Berlin.

R. DEVORE, G.G. LORENTZ [1993]: *Constructive Approximation*, Springer, Heidelberg.

E. DIBENEDETTO [2002]: *Real Analysis*, Birkhäuser, Boston.

E. DIBENEDETTO [2010]: *Partial Differential Equations, Second Edition*, Birkhäuser, Boston (First Edition: 1995, Springer, New York).

J. DIESTEL [1975]: *Geometry of Banach Spaces: Selected Topics*, Springer, Berlin.

J. DIEUDONNÉ [1950]: Deux exemples singuliers d'équations différentielles, *Acta Scientiarum Mathematicarum B (Szeged)* **12**, 38–40.

J. DIEUDONNÉ [1960]: *Foundations of Modern Analysis*, Academic Press, New York.

J. DIEUDONNÉ [1981]: *History of Functional Analysis*, North-Holland, Amsterdam.

J. DIEUDONNÉ [1989]: *A History of Algebraic and Differential Topology*, 1900–1960, Birkhäuser, Boston.

G. DINCA [2001]: A Fredholm-type result for a couple of nonlinear operators, *Comptes Rendus de l'Académie des Sciences de Paris, Série* 1, **333**, 4015–4019.

G. DINCA [2004]: Duality mappings on infinite dimensional reflexive and smooth Banach spaces are not compact, *Bulletin de l'Académie Royale de Belgique, Classes des Sciences* **6**, 33–40.

G. DINCA; P. JEBELEAN; J. MAWHIN [2001]: Variational and topological methods for Dirichlet problems with p-Laplacian, *Portugaliae Mathematica* **58**, 339–378.

G. DINCA; J. MAWHIN [2013]: *Brouwer Degree and Applications*, to appear.

U. DINI [1878]: *Analisi Infinitesimale. Lezioni dettate nella Reale Università di Pisa, Anno Accademico* 1877–1878.

U. DINI [1878]: *Fondamenti per la Teoria delle Funzioni di Variabili Reali*, T. Nistri, Pisa.

J. DIXMIER [1953]: Sur les bases orthonormales dans les espaces préhilbertiens, *Acta Scientiarum Mathematicarum Szeged* **15**, 29–30.

L. DONATI [1890]: Illustrazione al teorema del Menabrea, *Memorie della Accademia delle Scienze dell'Istituto di Bologna* **10**, 267–274.

L. DONATI [1894]: Ulteriori osservazioni intorno al teorema del Menabrea, *Memorie della Accademia delle Scienze dell'Istituto di Bologna* **4**, 449–474.

P. DU BOIS-RAYMOND [1876]: Untersuchungen über die Convergenz und Divergenz der Fourierschen Darstellungsformeln, *Abhandlungen der Mathematisch-Physikalischen Klasse der Königlich Bayerischen Akademie der Wissenschaften* **12**, 1–103.

R.M. DUDLEY [1964]: On sequential convergence, *Transactions of the American Mathematical Society* **112**, 483–507.

J. DUISTERMAAT; J.A. KOLK [2010]: *Distributions: Theory and Applications*, Springer, New York.

N. DUNFORD; J. SCHWARTZ [1958]: *Linear Operators, Part* I: *General Theory*, Interscience, New York (Reprinting: Wiley Classics Library, 1988).

N. DUNFORD; J. SCHWARTZ [1963]: *Linear Operators, Part* II: *Spectral Theory—Self Adjoint Operators in Hilbert Spaces*, Interscience, New York (Reprinting: Wiley Classics Library, 1988).

N. DUNFORD; J. SCHWARTZ [1971]: *Linear Operators, Part* III: *Spectral Operators*, Interscience, New York (Reprinting: Wiley Classics Library, 1988).

G. DUVAUT; J.L. LIONS [1976]: *Inequalities in Mechanics and Physics*, Springer, Berlin (translation of the original French edition, *Les Inéquations en Mécanique et en Physique*, Dunod, Paris, 1972).

W.F. EBERLEIN [1947]: Weak compactness in Banach spaces I, *Proceedings of the National Academy of Sciences, USA* **33**, 51–53.

A. EISINBERG; G. FEDELE; G. FRANZÈ [2004]: Lebesgue constant for Lagrange interpolation on equidistant nodes, *Analysis in Theory and Applications* **20**, 323–331.

I. EKELAND [1974]: On the variational principle, *Journal of Mathematical Analysis and Applications* **47**, 324–353.

I. EKELAND [1979]: Nonconvex minimization problems, *Bulletin of the American Mathematical Society* **1**, 443–473.

I. EKELAND; R. TÉMAM [1976]: *Convex Analysis and Variational Problems*, North-Holland, Amsterdam (reprinted in 1999 as SIAM Classics in Applied Mathematics, Volume 28; translation of the original French edition, *Analyse Convexe et Problèmes Variationnels*, Dunod, Paris, 1974).

L. EULER [1775]: On representations of a spherical surface on the plane, *Proceedings of the Saint Petersburg Academy of Sciences*.

L.C. EVANS [2010]: *Partial Differential Equations, Second Edition*, American Mathematical Society, Providence, RI (First Edition: 1998).

L.C. EVANS; R.F. GARIEPY [1992]: *Measure Theory and Fine Properties of Functions*, Studies in Advanced Mathematics, CRC Press, Boca Raton, FL.

Ky FAN [1953]: Minimax theorems, *Proceedings of the National Academy of Sciences* **39**, 42–47.

J. FARKAŠ [1901]: Theorie der einfachen Ungleichungen, *Journal für die Reine und Angewandte Mathematik* **124**, 1–27.

H. FEDERER [1969]: *Geometric Measure Theory*, Springer, New York.

L. FEJÉR [1900]: Sur les fonctions bornées et intégrables, *Comptes Rendus de l'Académie des Sciences, Paris* **131**, 984–987.

W. FENCHEL [1949]: On conjugate convex functions, *Canadian Journal of Mathematics* **1**, 73–77.

G. FICHERA [1964]: Problemi elastostatici con vincoli unilaterali: il problema de Signorini con ambigue condizioni al contorno, *Memorie dell'Accademia Nazionale dei Lincei* **8**, 91–140.

G. FICHERA [1972a]: Existence theorems in elasticity, in *Handbuch der Physik* VIa/2 (S. FLÜGGE & C. TRUESDELL, editors), pp. 347–389, Springer, Berlin.

G. FICHERA [1972b]: Boundary value problems of elasticity with unilateral constraints, in *Handbuch der Physik* VIa/2 (S. FLÜGGE & C. TRUESDELL, editors), pp. 391–424, Springer, Berlin.

E. FISCHER [1905]: Über quadratische Formen mit reellen Koeffizienten, *Monatshefte für Mathematik und Physik* **16**, 234–249.

E. FISCHER [1907]: Sur la convergence en moyenne, *Comptes Rendus de l'Académie des Sciences* **144**, 1022–1024.

S.R. FOGUEL [1958]: On a theorem of A.E. Taylor, *Proceedings of the American Mathematical Society* **9**, 325.

C. FOIAS; O. MANLEY; R. ROSA; R. TEMAM [2001]: *Navier–Stokes Equations and Turbulence*, Cambridge University Press, Cambridge, UK.

G.B. FOLLAND [1984]: *Real Analysis*, Wiley, New York.

I. FONSECA; W. GANGBO [1995]: *Degree Theory in Analysis and Applications*, Clarendon Press, Oxford, UK.

L.E. FRAENKEL [2000]: *An Introduction to Maximum Principle and Symmetry in Elliptic Problems*, Cambridge University Press, Cambridge, UK.

S.P. FRANKLIN [1965]: Spaces in which sequences suffice, *Fundamenta Mathematicae* **57**, 107–115.

S.P. FRANKLIN [1967]: Spaces in which sequences suffice, *Fundamenta Mathematicae* **61**, 51–56.

T.G. FREEMAN [2002]: *Portraits of the Earth. A Mathematician Looks at Maps*, American Mathematical Society, Providence.

K.O. FRIEDRICHS [1947]: On the boundary-value problems of the theory of elasticity and Korn's inequality, *Annals of Mathematics* **48**, 441–471.

K.O. FRIEDRICHS [1981]: *Spectral Theory of Operators in Hilbert Spaces*, Springer, Berlin.

G. FRIESECKE; R.D. JAMES; M.G. MORA; S. MÜLLER [2003]: Derivation of nonlinear bending theory for shells from three-dimensional nonlinear elasticity by Gamma-convergence, *Comptes Rendus de l'Académie des Sciences de Paris, Série* 1, **336**, 697–702.

G. FRIESECKE; R.D. JAMES; S. MÜLLER [2002]: A theorem on geometric rigidity and the derivation of nonlinear plate theory from three dimensional elasticity, *Communications on Pure and Applied Mathematics* **55**, 1461–1506.

G. FRIESECKE; R.D. JAMES; S. MÜLLER [2006]: A hierarchy of plate models derived from nonlinear elasticity by Gamma-convergence, *Archive for Rational Mechanics and Analysis* **180**, 183–236.

G. FROBENIUS [1912]: Über Matrizen aus nicht negativen Elementen, *Sitzungsberichte Preußische Akademie der Wissenschaft*, Berlin, 456–477.

B. GALERKIN [1915]: *Rods and Plates*, Vestnik Inženerov **19** (in Russian).

S. GALLOT; D. HULIN; J. LAFONTAINE [2004]: *Riemannian Geometry, Third Edition*, Springer, Berlin (First Edition: 1987).

F.R. GANTMACHER [1959]: *The Theory of Matrices, Volumes* 1 *and* 2, Chelsea, New York.

R. GÂTEAUX [1919]: Fonctions d'une infinité de variables indépendantes, *Bulletin de la Société Mathématique de France* **47**, 70–96.

C.F. GAUß [1809]: *Theoria Motus Corporum Coelestium in Sectionibus Conicis Solum Ambientium*, Perthes und Besser, Hamburg.

C.F. GAUß [1822]: Anwendung der Wahrscheinlichkeitsrechnung auf eine Aufgabe der practischen Geometrie, *Astronomische Nachrichten* **1**, 81–86.

C.F. GAUß [1827]: Disquisitiones generales circa superficies curvas, *Commentationes Societatis Regiae Scientiarum Gottingensis Recentiores* **6**, 99–146.

C.F. GAUß [1828]: Disquisitiones generales circas superficies curvas, *Commentationes societatis regiae scientiarum Gottingensis recentiores* **6**, Göttingen.

G. GEYMONAT; F. KRASUCKI [2005]: Some remarks on the compatibility conditions in elasticity, *Accademia Nazionale delle Scienze detta dei* XL. *Rendiconti. Serie* V. *Memorie di Matematica e Applicazioni. Parte* I, **29**, 175–181.

G. GEYMONAT; F. KRASUCKI [2006]: Beltrami's solutions of general equilibrium equations in continuum mechanics, *Comptes Rendus de l'Académie des Sciences de Paris, Série* 1, **342**, 359–363.

G. GEYMONAT; G. GILARDI [1998]: Contre-exemple à l'inégalité de Korn et au lemme de Lions dans des domaines irréguliers, in *Equations aux Dérivées Partielles et Applications. Articles Dédiés à Jacques-Louis Lions*, pp. 541–548, Gauthier-Villars, Paris.

G. GEYMONAT; P. SUQUET [1986]: Functional spaces for Norton-Hoff materials, *Mathematical Methods in the Applied Sciences* **8**, 206–222.

M. GHERGU; V.D. RĂDULESCU [2008]: *Singular Elliptic Problems: Bifurcation and Asymptotic Analysis*, Clarendon Press, Oxford, UK.

M. GHERGU; V.D. RĂDULESCU [2012]: *Nonlinear PDEs—Mathematical Models in Biology, Chemistry and Population Genetics*, Springer, Heidelberg.

M. GIAQUINTA; S. HILDEBRANDT [2006a]: *Calculus of Variations* I: *The Lagrangian Formalism*, Springer, New York.

M. GIAQUINTA; S. HILDEBRANDT [2006b]: *Calculus of Variations* II: *The Hamiltonian Formalism*, Springer, New York.

D. GILBARG; N.S. TRUDINGER [1998]: *Elliptic Partial Differential Equations, Revised Second Edition*, Springer, Berlin (First Edition: 1977).

V. GIRAULT; P.A. RAVIART [1979]: *Finite Element Approximation of the Navier–Stokes Equations*, Lecture Notes in Mathematics, Volume 749, Springer, Berlin.

V. GIRAULT; P.A. RAVIART [1986]: *Finite Element Methods for Navier–Stokes Equations*, Springer, Berlin.

E. GIUSTI [1984]: *Minimal Surfaces and Functions of Bounded Variations*, Birkhäuser, Boston.

E. GIUSTI [2003]: *Direct Methods in the Calculus of Variations*, World Scientific, Singapore.

R. GLOWINSKI [1984]: *Numerical Methods for Nonlinear Variational Problems*, Springer, New York.

R. GLOWINSKI [2003]: Finite element methods for incompressible viscous flows, in *Handbook of Numerical Analysis, Volume* IX (P.G. CIARLET & J.L. LIONS, editors), pp. 3–1176, North-Holland, Amsterdam.

R. GLOWINSKI; H. LANCHON [1973]: Torsion élasto-plastique d'une barre cylindrique de section multi-connexe, *Journal de Mécanique* **12**, 151–171.

R. GLOWINSKI; J.L. LIONS; R. TRÉMOLIÈRES [1981]: *Numerical Analysis of Variational Inequalities*, North-Holland, Amsterdam (translation of the original French edition, *Analyse Numérique des Inéquations Variationnelles*, Dunod, Paris, 1976).

J. GOBERT [1962]: Une inégalité fondamentale de la théorie de l'élasticité, *Bulletin de la Société Royale des Sciences de Liège* **31**, 182–191.

K. GÖDEL [1940]: *The Consistency of the Axiom of Choice and of the Generalized Continuum Hypothesis with the Axioms of Set Theory*, Princeton University Press, Princeton, NJ.

C. GOFFMAN; G. PEDRICK [1965]: *First Course in Functional Analysis*, Prentice-Hall, Englewood Cliffs, NJ.

H.H. GOLDSTINE [1980]: *A History of the Calculus of Variations from the 17th to the 19th Century*, Springer, New York.

W.B. GRAGG; R.A. TAPIA [1974]: Optimal error bounds for the Newton–Kantorovich theorem, *SIAM Journal on Numerical Analysis* **11**, 10–13.

J.P. GRAM [1883]: Über die Entwicklung reeller Funktionen in Reihen mittelst der Methode der kleinsten Quadrate, *Journal für die Reine und Angewandte Mathematik* **94**, 41–73.

J. GRAY [2012]: *Henry Poincaré: A Scientific Biography*, Princeton University Press, Princeton, NJ.

P. GRISVARD [1992]: *Singularities in Boundary Value Problems*, Masson, Paris.

W. GROMES [1981]: Ein einfacher Beweis des Satzes von Borsuk, *Mathematische Zeitschrift* **178**, 399–400.

M.E. GURTIN [1972]: The linear theory of elasticity, in *Handbuch der Physik, Volume* VIa/2 ((S. FLÜGGE & C. TRUESDELL, editors), pp. 1–295, Springer, Berlin.

M.E. GURTIN [1981]: *Topics in Finite Elasticity*, CBMS-NSF Regional Conference Series in Applied Mathematics, Volume 35, SIAM, Philadelphia.

Dzung Minh HA [2007]: *Functional Analysis: A Gentle Introduction*, Matrix Editions, Ithaca, NY.

A. HAAR [1918]: Die Minkowskische Geometrie und die Annäherung an stetige Funktionen, *Mathematische Annalen* **78**, 294–311.

J. HADAMARD [1902]: Sur les problèmes aux dérivées partielles et leur signification physique, *Princeton University Bulletin* **13**, 49–52.

H. HAHN [1927]: Über lineare Gleichungssysteme in linearen Räumen, *Journal de Crelle* **157**, 214–229.

P.R. HALMOS [1950]: *Measure Theory*, Van Nostrand, Princeton, NJ.

P.R. HALMOS [1970]: How to write mathematics, *L'Enseignement Mathématique* **16**, 123–152.

P.R. HALMOS [1974]: *A Hilbert Space Problem Book, Second Edition*, Springer, New York (First Edition: 1960).

P.R. HALMOS [1985]: *I Want to Be a Mathematician*, Springer, New York.

P.R. HALMOS [1987]: *I Have a Photographic Memory*, American Mathematical Society, Providence, RI.

G. HAMEL [1905]: Eine Basis aller Zahlen und die unstetigen Lösungen der Funktionalgleichung $f(x+y) = f(x) + f(y)$, *Mathematische Annalen* **60**, 459–462.

G.H. HARDY [1916]: Weierstraß's non-differentiable function, *Transactions, American Mathematical Society* **17**, 301–325.

G.H. HARDY [1925]: Notes on some points in the integral calculus. LX. An inequality between integrals, *Messengers of Mathematics* **54**, 150–156.

P. HARTMAN [2002]: *Ordinary Differential Equations, Second Edition*, SIAM, Philadelphia (First Edition: 1964, John Wiley & Sons, New York).

P. HARTMAN; G. STAMPACCHIA [1966]: On some nonlinear elliptic differential functional equations, *Acta Mathematica* **115**, 271–310.

P. HARTMAN; A. WINTNER [1950]: On the embedding problem in differential geometry, *American Journal of Mathematics* **72**, 553–564.

P. HARTMAN; A. WINTNER [1950]: On the fundamental equations of differential geometry, *American Journal of Mathematics* **72**, 757–774.

F. HAUSDORFF [1919]: Der Wertvorrat einer Bilinearform, *Mathematische Zeitschrift* **3**, 314–316.

E. HEINZ [1959]: An elementary analytic theory of the degree of mapping in n-dimensional space, *Journal of Mathematics and Mechanics* **8**, 231–247.

E. HELLINGER; O. TOEPLITZ [1910]: Grundlagen für eine Theorie der unendlichen Matrizen, *Mathematische Annalen* **69**, 281–330.

M. HÉNON [1976]: A two-dimensional mapping with a strange attractor, *Communications in Mathematics and Physics* **50**, 69–77.

C. HERMITE [1878]: Sur la formule d'interpolation de Lagrange, *Journal für die reine und angewandte Mathematik* **84**, 70–79.

M.R. HESTENES [1975]: *Optimization Theory—The Finite Dimensional Case*, John Wiley, New York.

E. HEWITT; K. STROMBERG [1965]: *Real and Abstract Analysis—A Modern Treatment of the Theory of Functions of a Real Variable*, Springer, New York.

E. HILLE; C. SZEGÖ; J.D. TAMARKIN [1937]: On some generalizations of a theorem of A. Markoff, *Duke Mathematical Journal* **8**, 729–739.

J.B. HIRIART-URRUTY; C. LEMARÉCHAL [1993a]: *Convex Analysis and Minimization Algorithms* I: *Fundamentals*, Springer, Berlin.

J.B. HIRIART-URRUTY; C. LEMARÉCHAL [1993b]: *Convex Analysis and Minimization Algorithms* II: *Advanced Theory and Bundle Methods*, Springer, Berlin.

O. HÖLDER [1889]: *Über einen Mittelwertsatz*, Göttinger Nachrichten, 38–47.

E. HOPF [1927]: Elementare Bemerkungen über die Lösungen partieller Differentialgleichungen zweiter Ordnung vom elliptischen Typus, in *Sitzungsberichte der Preußischen Akademie der Wissenschaften, Berlin*, 147–152.

C.O. HORGAN [1995]: Korn's inequalities and their applications in continuum mechanics, SIAM Review **37**, 491–511.

L. HÖRMANDER [1955]: On the theory of general partial differential operators, *Acta Mathematica* **94**, 161–248.

L. HÖRMANDER [1983]: *The Analysis of Partial Differential Operators, Volume* 1, Springer, New York.

R.A. Horn; C.R. Johnson [1985]: *Matrix Analysis*, Cambridge University Press, Cambridge, UK.

R.A. Horn; C.R. Johnson [1991]: *Topics in Matrix Analysis*, Cambridge University Press, Cambridge, UK.

A.S. Householder [1964]: *The Theory of Matrices in Numerical Analysis*, Blaisdell, New York.

J.L.W.V. Jensen [1906]: Sur les fonctions convexes et les inégalités entre les valeurs moyennes, *Acta Mathematica* **30**, 175–193.

H. Jacobowith [1982]: *The Gauß-Codazzi equations, Tensor (N.S.)* **39**, 15–22.

E. Jakimowicz; A. Miranovicz, editors [2011]: *Stefan Banach: Remarkable Life, Brilliant Mathematics*, American Mathematical Society, Providence, RI.

R.C. James [1951]: A non-reflexive Banach space isometric with its second conjugate space, *Proceedings of the National Academy of Sciences, USA* **37**, 174–177.

R.C. James [1964]: Characterizations of reflexivity, *Studia Mathematica* **23**, 205–216.

R.C. James [1972]: Reflexivity and the sup of linear functionals, *Israel Journal of Mathematics* **13**, 289–301.

P. Jamet [1976]: Estimation d'erreur pour des éléments finis droits presque dégénérés, *Revue Française d'Automatique, Informatique, Recherche Opérationnelle, Série Rouge: Analyse Numérique* **10**, 43–61.

M. Janet [1926]: Sur la possibilité de plonger un espace riemannien donné dans un espace euclidien, *Annales de la Société Polonaise de Mathématiques* **5**, 38–43.

D. Jerison; C.E. Kenig [1995]: The inhomogeneous Dirichlet problem in Lipschitz domains, *Journal of Functional Analysis* **130**, 161–219.

J. Jost [2005]: *Postmodern Analysis*, Springer, Berlin.

Y. Kannai [1981]: An elementary proof of the no-retraction theorem, *American Mathematical Monthly* **88**, 264–268.

L.V. Kantorovich [1948]: Functional analysis and applied mathematics, *Uspehi Matematiceskii Nauk (New Series)* **3**, 89–185 (in Russian).

L.V. Kantorovich; G.P. Akilov [1959]: *Functional Analysis in Normed Vector Spaces*, Fizmatgiz, Moscow (in Russian) (English translation: Pergamon, New York, 1964).

S. Karlin [1959]: Positive operators, *Journal of Mathematics and Mechanics* **8**, 907–937.

T. von Kármán [1910]: Festigkeitsprobleme im Maschinenbau, in *Encyclopädie der Mathematischen Wissenschaften, Volume* IV/4, pp. 311–385, Leipzig.

T. Kato [1966]: *Perturbation Theory for Linear Operators*, Springer, Berlin (Corrected Printing of the Second Edition: 1980).

O. Kavian [1993]: *Introduction à la Théorie des Points Critiques et Applications aux Problèmes Elliptiques*, Springer, Paris.

J. Kelley [1955]: *General Topology*, Van Nostrand, Princeton, NJ.

O.D. Kellogg [1929]: *Foundations of Potential Theory*, Springer, Berlin.

S. Kesavan [1989]: *Topics in Functional Analysis and Applications*, Wiley, New Delhi.

S. Kesavan [2004]: *Nonlinear Functional Analysis—A First Course*, Hindustan Book Agency, Gurgaon.

S. Kesavan [2005]: On Poincaré's and J.L. Lions' lemmas, *Comptes Rendus de l'Académie des Sciences de Paris, Série* I, **340**, 27–30.

S. Kesavan [2006]: *Symmetrization & Applications*, World Scientific, Singapore.

S. Kesavan [2009]: *Functional Analysis*, Hindustan Book Agency, Gurgaon.

D. KINDERLEHRER; G. STAMPACCHIA [1980]: *An Introduction to Variational Inequalities and Their Applications*, Academic Press, New York (reprinted as Classics in Applied Mathematics, Volume 31, SIAM, Philadelphia, 2002).

J. KISYNSKI [1959]: Convergence du type L, *Colloquium Mathematicum* **7**, 205–211.

W. KLINGENBERG [1973]: *Eine Vorlesung über Differentialgeometrie*, Springer, Berlin (English translation: *A Course in Differential Geometry*, Springer, Berlin, 1978).

A.W. KNAPP [2005a]: *Basic Real Analysis*, Birkhäuser, Boston.

A.W. KNAPP [2005b]: *Advanced Real Analysis*, Birkhäuser, Boston.

V.I. KONDRACHOV [1945]: Certain properties of functions in the spaces L^p, *Doklady Akademii Nauk SSSR* **48**, 535–538.

V.A. KONDRAT'EV; O.A. OLEINIK [1988]: Boundary-value problems for the system of elasticity theory in unbounded domains. Korn's inequalities, *Uspehi Mathematiceskii Nauk* **43**, 55–98 (in Russian) [English translation: *Russian Mathematical Surveys* **43** (1988), 65–119].

A. KORN [1906]: Die Eigenschwingungen eines elastischen Körpers mit ruhender Oberfläche, *Sitzungsberichte der Mathematisch-physikalischen Klasse der Königlich bayerischen Akademie der Wissenschaften zu München* **36**, 351–402.

A. KORN [1908]: Solution générale du problème d'équilibre dans la théorie de l'élasticité, dans le cas où les efforts sont donnés à la surface, *Annales de la Faculté des Sciences de Toulouse* **10**, 165–269.

A. KORN [1909]: Über einige Ungleichungen, welche in der Theorie der elastischen und elektrischen Schwingungen eine Rolle spielen, *Bulletin International de l'Académie des Sciences de Cracovie* **9**, 705–724.

P.P. KOROVKIN [1959]: *Linear Operators and Approximation Theory*, Fitzmatgiz, Moscow (in Russian) [English translation, Hindustan Publishing Corporation, Delhi, 1960].

P.P. KOROVKIN [1959]: On convergence of linear positive operators in the space of continuous functions, *Doklady Akademii Nauk SSR* **90**, 961–964 (in Russian).

I. KRA; S.R. SIMANCA [2012]: On circulant matrices, *Notices of the American Mathematical Society* **59**, 368–377.

S.G. KRANTZ [2004]: *Real Analysis and Foundations, Second Edition*, Studies in Advanced Mathematics, Chapman & Hall/CRC, Boca Raton, FL (First Edition: 1991).

M.A. KRASNOSELSKII [1960]: Fixed points of cone-compressive or cone-extending operators, *Soviet Mathematics Doklady* **1**, 1285–1288.

M. KREIN; M. RUTMAN [1948]: Linear operators leaving invariant a cone in a Banach space, *Uspehi Mathematiceskii Nauk* **3**, 3–95 [in Russian; English translation: *American Mathematical Society Translations* 1950, No. 26].

E. KREYSZIG [1978]: *Introductory Functional Analysis with Applications*, John Wiley, New York (reprinted in the *Wiley Classics Library Edition*, 1989).

B. KRIPKE [1967]: One more reason why sequences are not enough, *American Mathematical Monthly* **74**, 563–565.

A. KUFNER; L. MALIGRANDA; L.E. PERSSON [2007]: *The Hardy Inequality: About Its History and Some Related Results*, Vydavatelský Servis, Pilsen.

H.W. KUHN; A.W. TUCKER [1951]: Nonlinear programming, in *Proceedings of the Second Berkeley Symposium on Mathematical Statistics and Probability* (J. NEYMAN, editor), pp. 481–492, University of California Press, Berkeley.

W. KÜHNEL [2002]: *Differentialgeometrie*, Fried. Vieweg & Sohn, Wiesbaden (English translation: *Differential Geometry: Curves-Surfaces-Manifolds*, American Mathematical Society, Providence, RI, 2002).

O.A. LADYZHENSKAYA [1969]: *The Mathematical Theory of Viscous Flows, Second Edition*, Gordon and Breach, New York.

J.L. LAGRANGE [1760]: Essai d'une nouvelle méthode pour déterminer les maxima et les minima des formules intégrales indéfinies, *Miscellanea Taurinensia* **325**, 173–199.

J.L. LAGRANGE [1773]: Solutions analytiques de quelques problèmes sur les pyramides triangulaires, *Mémoire de l'Académie Royale de Berlin.*

J.L. LAGRANGE [1812]: Leçons élémentaires de mathématiques données à l'Ecole Normale en 1795, *Journal de l'Ecole Polytechnique*, VIIe et VIIIe cahiers, t-II.

S. LANG [1993]: *Real and Functional Analysis, Third Edition*, Springer, New York.

P.S. LAPLACE [1820]: *Théorie Analytique des Probabilités, Troisième Edition, Premier Supplément: Sur l'Application du Calcul des Probabilités à la Philosophie Naturelle*, Courcier, Paris.

M. LAVRENTIEV [1926]: Sur quelques problèmes du calcul des variations, *Annales de Mathématiques Pures et Appliquées* **4**, 7–18.

D.F. LAWDEN [1989]: *Elliptic Functions and Applications*, Applied Mathematical Sciences Series, Volume 98, Springer, Heidelberg.

P.D. LAX [2002]: *Functional Analysis*, Wiley-Interscience, New York.

P.D. LAX; A.M. MILGRAM [1954]: Parabolic equations, in *Contributions to the Theory of Partial Differential Equations, Annals of Mathematics Studies, No.* 33, pp. 167–190, Princeton University Press Princeton, NJ.

L.P. LEBEDEV; M.J. CLOUD [2003]: *Tensor Analysis*, World Scientific, Singapore.

H. LEBESGUE [1901]: Sur une généralisation de l'intégrale définie, *Comptes Rendus des Séances de l'Académie des Sciences* **132**, 1025–1027.

H. LEBESGUE [1909]: Sur les intégrales singulières, *Annales de la Faculté des Sciences de l'Université de Toulouse* **1**, 25–117.

H. LE DRET; A. RAOULT [1995]: The nonlinear membrane model as variational limit of nonlinear three-dimensional elasticity, *Journal de Mathématiques Pures et Appliquées* **74**, 549–578.

H. LE DRET; A. RAOULT [1996]: The membrane shell model in nonlinear elasticity: A variational asymptotic derivation, *Journal of Nonlinear Science* **6**, 59–94.

A.M. LEGENDRE [1805]: *Nouvelle Méthode pour la Détermination des Orbites des Comètes*, Chez Didot, Paris.

J. LERAY [1933]: Essai sur le mouvement plan d'un liquide visqueux que limitent des parois, *Journal de Mathématiques Pures et Appliquées* **13**, 331–418.

J. LERAY [1933]: Sur le mouvement d'un liquide visqueux emplissant l'espace, *Acta Mathematica* **63**, 193–248.

J. LERAY [1935]: Topologie des espaces abstraits de M. Banach, *Comptes Rendus de l'Académie des Sciences de Paris* **200**, 1082–1084.

J. LERAY [1950]: La théorie des points fixes et ses applications en analyse, in *Proceedings—International Congress of Mathematicians, Volume* 2, pp. 202–208, Cambridge.

J. LERAY; J.L. LIONS [1965]: Quelques résultats de Visik sur les problèmes elliptiques non linéaires par les méthodes de Minty-Browder, *Bulletin de la Société Mathématique de France* **93**, 97–107.

J. LERAY; J. SCHAUDER [1934]: Topologie et équations fonctionnelles, *Annales Scientifiques de l'Ecole Normale Supérieure* **51**, 45–78.

T. LEWIŃSKI; J. TELEGA [2000]: *Plates, Laminates and Shells—Asymptotic Analysis and Homogenization*, World Scientific, Singapore.

H. LEWY; G. STAMPACCHIA [1969]: On the regularity of the solution of a variational inequality, *Communications on Pure and Applied Mathematics* **22**, 153–188.

Ta-Tsien LI [2011]: *Problems and Solutions in Mathematics, Second Edition*, World Scientific, Singapore.

X. LI; R.N. MOHAPATRA [1993]: On the convergence of Lagrange interpolation with equidistant nodes, *Proceedings of the American Mathematical Society* **118**, 1205–1212.

H. LIEBMANN [1899]: Eine neue Eigenschaft der Kugel, *Nachrichten von der Gesellschaft der Wissenschaften zu Göttingen, Mathematisch-Physikalische Klasse*, 45–55.

Minghua LIN [2012]: The AM-GM inequality and CBS inequality are equivalent, *The Mathematical Intelligencer* **34**, 6.

J.L. LIONS [1961]: *Equations Différentielles Opérationnelles et Problèmes aux Limites*, Springer, Berlin.

J.L. LIONS [1965]: *Problèmes aux Limites dans les Equations aux Dérivées Partielles*, Presses de l'Université de Montréal, Montréal, Que.

J.L. LIONS [1969]: *Quelques Méthodes de Résolution des Problèmes aux Limites Non Linéaires*, Dunod, Paris.

J.L. LIONS [1973]: *Perturbations Singulières dans les Problèmes aux Limites et en Contrôle Optimal*, Lecture Notes in Mathematics, Volume 323, Springer, Berlin.

J.L. LIONS; E. MAGENES [1972]: *Non-Homogeneous Boundary Value Problems and Applications, Volume* 1, Springer, Heidelberg (translation of the original French edition, *Problèmes aux Limites non Homogènes et Applications, Volume* 1, Dunod, Paris, 1968).

J.L. LIONS; G. STAMPACCHIA [1967]: Variational inequalities, *Communications on Pure and Applied Mathematics* **20**, 493–519.

P.L. LIONS [1984]: The concentration-compactness principle in the calculus of variations. The locally compact case – Part 1, *Annales de l'Institut Henri Poincaré – Analyse Non Linéaire* **1**, 109–145.

P.L. LIONS [1984]: The concentration-compactness principle in the calculus of variations. The locally compact case – Part 2, *Annales de l'Institut Henri Poincaré – Analyse Non Linéaire* **1**, 223–283.

P.L. LIONS [1985]: The concentration-compactness principle in the calculus of variations. The limit case – Part 1, *Revista Matematica Iberoamericana* **1.1**, 145–201.

P.L. LIONS [1985]: The concentration-compactness principle in the calculus of variations. The limit case – Part 2, *Revista Matematica Iberoamericana* **1.2**, 45–121.

P.L. LIONS [1996]: *Mathematical Topics in Fluid Mechanics, Volume* 1 : *Incompressible Models*, Clarendon Press, Oxford, UK.

J. LIOUVILLE [1850]: Extension au cas des trois dimensions de la question du tracé géographique, Note VI in the Appendix to G. MONGE: *Application de l'Analyse à la Géométrie, Cinquième Edition*, Bachelier, Paris.

G.G. LORENTZ [1986]: *Bernstein Polynomials*, Chelsea, New York.

S. LOZINSKI [1948]: On a class of linear operators, *Doklady Akademii Nauk SSSR* **61**, 193–196 (in Russian).

D.G. LUENBERGER [1969]: *Optimization by Vector Space Methods*, John Wiley, New York.

N. LUSIN [1913]: Sur la convergence des séries trigonométriques de Fourier, *Comptes Rendus de l'Académie des Sciences de Paris* **156**, 1655–1658.

C.R. MACCLUER [2000]: The many proofs and applications of Perron's theorem, *SIAM Review* **42**, 487–498.

E.J. MACSHANE [1934]: Extension of range of functions, *Bulletin of the American Mathematical Society* **40**, 837–842.

E. MAGENES; G. STAMPACCHIA [1958]: I problemi al contorno per le equazioni differenziali di tipo ellitico, *Annali della Scuola Normale Superiore di Pisa* **12**, 247–358.

G. MAINARDI [1856]: Su la teoria generale delle superficie, *Giornale dell' Istituto Lombardo* **9**, 385–404.

L. MALIGRANDA [2012]: The AM-GM inequality is equivalent to the Bernoulli inequality, *The Mathematical Intelligencer* **34**, 1–2.

D.H. MALING [1992]: *Coordinate Systems and Map Projections*, Second Edition, Pergamon Press, Oxford.

B. MANIÀ [1934]: Sopra un esempio di Lavrentieff, *Bolletone dell Unione Mathematica Italiana* **13**, 147–153.

C. MARDARE [2003]: On the recovery of a manifold with prescribed metric tensor, *Analysis and Applications* **1**, 433–453.

S. MARDARE [2003]: Inequality of Korn's type on compact surfaces without boundary, *Chinese Annals of Mathematics, Series B*, **24**, 191–204.

S. MARDARE [2005]: On Pfaff systems with L^p coefficients and their applications in differential geometry, *Journal de Mathématiques Pures et Appliquées* **84**, 1659–1692.

S. MARDARE [2007]: On systems of first order linear partial differential equations with L^p coefficients, *Advances in Differential Equations* **12**, 301–360.

S. MARDARE [2008]: On Poincaré and De Rham's theorems, *Revue Roumaine de Mathématiques Pures et Appliquées* **53**, 523–541.

I. MAREK [1970]: Frobenius theory of positive operators: Comparison theorems and applications, *SIAM Journal on Applied Mathematics* **19**, 607–628.

K. MARGUERRE [1939]: Zur Theorie der gekrümmten Platte großer Formänderung, in *Proceedings, Fifth International Congress for Applied Mechanics*, pp. 93–101, John Wiley & Sons, New York, 1939.

A.A. MARKOFF [1889]: Sur une question posée par Mendeleieff, *Izvestia Akademii Nauk SSSR* **62**, 1–24.

J.E. MARSDEN; T.J.R. HUGHES [1999]: *Mathematical Foundations of Elasticity*, Prentice-Hall, Englewood Cliffs, NJ (First Edition: 1983).

R.D. MAUDLIN, editor [1981]: *The Scottish Book—Mathematics from the Scottish Café*, Birkhäuser, Basel.

J. MAWHIN [1979]: *Topological Degree Methods in Nonlinear Boundary Value Problems*, American Mathematical Society, Providence, RI.

J. MAWHIN [1999]: Leray-Schauder degree: A half century of extensions and applications, *Topological Methods in Nonlinear Analysis* **14**, 195–228.

S. MAZUR [1933]: Über konvexe Mengen in linearen normierten Räumen, *Studia Mathematica* **5**, 70–84.

S. MAZUR; S. ULAM [1932]: Sur les transformations isométriques d'espaces vectoriels normés, *Comptes Rendus de l'Académie des Sciences de Paris* **194**, 946–948.

V. MAZ'YA; T. SHAPOSHNIKOVA [1998]: *Jacques Hadamard, a Universal Mathematician*, American Mathematical Society, Providence, RI.

A. MCINTOSCH [1978]: The Toeplitz-Hausdorff theorem and ellipticity conditions, *The American Mathematical Monthly* **85**, 475–477.

W.H. MEEKS III; J. PÉREZ [2011]: The classical theory of minimal surfaces, *Bulletin of the American Mathematical Society* **48**, 325–407.

G.H. MEISTERS; C. OLECH [1963]: Locally one-to-one mappings and a classical theorem on Schlicht functions, *Duke Mathematical Journal* **30**, 63–80.

H.N. MHASKAR; D.V. PAI [2007]: *Fundamentals of Approximation Theory, Revised Edition*, Alpha Science, Oxford, UK (First Edition: 2000).

D.P. MILMAN [1938]: On some criteria for the regularity of spaces of type (B), *Doklady Akademii Nauk SSSR* **20**, 243–246.

J. MILNOR [1965]: *Topology from the Differentiable Viewpoint*, Princeton University Press, Princeton, NJ.

J. MILNOR [1978]: Analytic proofs of the "hairy ball theorem" and the Brouwer fixed point theorem, *The American Mathematical Monthly* **85**, 521–524.

H. MINKOWSKI [1896]: *Geometrie der Zahlen*, Leipzig.

G.J. MINTY [1962]: Monotone (nonlinear) operators in Hilbert space, *Duke Mathematical Journal* **29**, 341–346.

G.J. MINTY [1963]: On a monotonicity method for the solution of nonlinear equations in Banach spaces, *Proceedings of the National Academy of Sciences USA* **50**, 1038–1041.

E.H. MOORE [1920]: On the reciprocal of the general algebraic matrix, *Bulletin of the American Mathematical Society* **26**, 394–395.

J.J. MOREAU [1970]: Inf-convolution, sous-additivité, convexité des fonctions numériques, *Journal de Mathématiques Pures et Appliquées* **49**, 109–154.

J.J. MOREAU [1979]: Duality characterization of strain tensor distributions in an arbitrary open set, *Journal of Mathematical Analysis and Applications* **72**, 760–770.

C.B. MORREY, JR. [1952]: Quasi-convexity and the lower semicontinuity of multiple integrals, *Pacific Journal of Mathematics* **2**, 25–53.

C.B. MORREY, JR. [1966]: *Multiple Integrals in the Calculus of Variations*, Springer, Berlin.

P.P. MOSOLOV; V.P. MJASNIKOV [1971]: A proof of Korn's inequality, *Soviet Mathematics Doklady* **12**, 1618–1622.

D. MOTREANU; V. RĂDULESCU [2003]: *Variational and Non-Variational Methods in Nonlinear Analysis and Boundary Value Problems*, Kluwer, Dordrecht.

M.E. MUNROE [1953]: *Introduction to Measure and Integration*, Addison-Wesley, Reading, MA.

C. MÜNTZ [1914]: Über den Approximationssatz von Weierstraß, in *H.A. Schwarz Festschrift*, pp. 303–312, Mathematische Abhandlungen, Springer, Berlin.

F. MURAT [1978]: Compacité par compensation, *Annali Scuola Normale Superiore de Pisa, Serie* IV, **5**, 489–507.

F. MURAT [1987]: A survey on compensated compactness, in *Contributions to Modern Calculus of Variations* (L. CESARI, editor), pp. 145–183, Longman, Harlow.

L. NACHBIN [1969]: *Topology on Spaces of Holomorphic Mappings*, Springer, Berlin.

M. NAGUMO [1951]: A theory of degree of mapping based on infinitesimal analysis, *American Journal of Mathematics* **73**, 485–496.

J. NASH [1954]: $\mathcal{C}^1$ isometric imbeddings, *Annals of Mathematics* **60**, 383–396.

C.L.M.H. NAVIER [1823]: Mémoire sur les lois du mouvement des fluides, *Mémoires de l'Académie Royale des Sciences de Paris* **6**, 389–416.

J. NEČAS [1962]: Sur une méthode pour résoudre les équations aux dérivées partielles du type elliptique, voisine de la variationnelle, *Annali della Scuola Normale Superiore di Pisa, Classe di Scienze, Serie* III, **16**, 305–326.

J. NEČAS [1965]: *Equations aux Dérivées Partielles*, Presses de l'Université de Montréal, Montréal.

J. NEČAS [1967]: *Les Méthodes Directes en Théorie des Equations Elliptiques*, Masson, Paris and Academia, Praha (English translation: *Direct Methods in the Theory of Elliptic Equations*, Springer, Heidelberg, 2012).

J. NEČAS; I. HLAVÁČEK [1981]: *Mathematical Theory of Elastic and Elasto-Plastic Bodies: An Introduction*, Elsevier, Amsterdam.

P.M. NEUMANN [2011]: *The Mathematical Writings of Évariste Galois*, European Mathematical Society, Zürich.

R.A. NICOLAIDES [1972]: On a class of finite elements generated by Lagrange interpolation, *SIAM Journal on Numerical Analysis* **9**, 435–445.

L. NIRENBERG [1974]: *Topics in Nonlinear Functional Analysis*, Lecture Notes, Courant Institute, New York University, NY (Second Edition: American Mathematical Society, Providence, RI, 1994).

J.A. NITSCHE [1981]: On Korn's second inequality, *RAIRO Analyse Numérique* **15**, 237–248.

J.C.C. NITSCHE [1975]: *Vorlesungen über Minimalflächen*, Springer, Berlin.

B. O'NEILL [2006]: *Elementary Differential Geometry, Revised Second Edition*, Elsevier/Academic Press, Burlington (First Edition: 1966).

J.T. ODEN; L.F. DEMKOWICZ [2010]: *Applied Functional Analysis, Second Edition*, Chapman & Hall, Boca Raton, FL (First Edition: 1996).

J.M. ORTEGA [1968]: The Newton-Kantorovich theorem, *The American Mathematical Monthly* **75**, 658–660.

J.M. ORTEGA; W.C. RHEINBOLDT [2000]: *Iterative Solution of Nonlinear Equations in Several Variables*, SIAM, Philadelphia.

A.M. OSTROWSKI [1954]: On the linear iteration procedures for symmetric matrices, *Rendiconti Lincei – Matematica e Applicazioni* **14**, 140–163.

C. PADOVANI [2000]: On the derivative of some tensor-valued functions, *Journal of Elasticity* **58**, 257–268.

R.S. PALAIS; S. SMALE [1964]: A generalized Morse theory, *Bulletin of the American Mathematical Society* **70**, 165–171.

R. PENROSE [1955]: A generalized inverse for matrices, *Proceedings of the Cambridge Philosophical Society* **51**, 406–413.

O. PERRON [1907]: Grundlagen für eine Theorie des Jacobischen Kettenbruchalgorithmus, *Mathematische Annalen* **64**, 11–76.

O. PERRON [1923]: Eine neue Behandlung der Randwertaufgabe für $\Delta u = 0$, *Mathematische Zeitschrift* **18**, 42–54.

B.J. PETTIS [1939]: A proof that every uniformly convex space is reflexive, *Duke Mathematical Journal* **5**, 249–253.

R. PHELPS [1960]: Uniqueness of Hahn–Banach extensions and unique best approximation, *Transactions of the American Mathematical Society* **95**, 238–255.

E. PICARD [1893]: Sur l'application des méthodes d'approximations successives à l'étude de certaines équations différentielles ordinaires, *Journal de Mathématiques Pures et Appliquées* **9**, 217–271.

A. PIETSCH [2007]: *History of Banach Spaces and Linear Operators*, Birkhäuser, Boston.

R.B. PLATTE; L.N. TREFETHEN; A.B.J. KUIJLAARS [2011]: Impossibility of fast stable approximation of analytic functions from equispaced samples, *SIAM Review* **53**, 308–318.

G. PÓLYA [1933]: Über die Konvergenz von Quadraturverfahren, *Mathematische Zeitschrift* **37**, 264–286.

G. PÓLYA [1987]: *The Pólya Picture Album: Encounters of a Mathematician*, Birkhäuser, Boston.

A. PRESSLEY [2005]: *Elementary Differential Geometry*, Springer, London.

M. PROTTER; H. WEINBERGER [1967]: *Maximum Principles in Differential Equations*, Prentice-Hall, Englewood Cliffs, NJ.

P. PUCCI; J. SERRIN [2007]: *The Maximum Principle*, Birkhäuser, Basel.

P. RABIER [1979]: Résultats d'existence dans des modèles non linéaires de plaques, *Comptes Rendus de l'Académie des Sciences de Paris, Série A*, **289**, 515–518.

R. RADO [1956]: Note on generalized inverses of matrices, *Proceedings of the Cambridge Philosophical Society* **52**, 600–601.

T. RADO [1930]: The problem of the least area and the problem of Plateau, *Mathematische Zeitschrift* **32**, 763–796.

T. RADO; P.V. REICHELDERFER [1955]: *Continuous Transformations in Analysis*, Springer, Berlin.

I.K. RANA [2002]: *An Introduction to Measure and Integration, Second Edition*, Graduate Studies in Mathematics, Volume 45, American Mathematical Society, Providence, RI.

P.A. RAVIART; J.M. THOMAS [1983]: *Introduction à l'Analyse Numérique des Equations aux Dérivées Partielles*, Masson, Paris.

E. REICH [1949]: On the convergence of the classical iterative method of solving linear simultaneous equations, *Annals of Mathematical Statistics* **20**, 448–451.

C. REID [1970]: *Hilbert—With an Appreciation of Hilbert's Mathematical Work by Hermann Weyl*, Springer, New York.

C. REID [1976]: *Courant in Göttingen and New York—The Story of an Improbable Mathematician*, Springer, New York.

F. RELLICH [1930]: Ein Satz über mittlere Konvergenz, *Nachrichten von der Gesellschaft der Wissenschaften zu Göttingen*, 30–35.

Y.G. RESHETNYAK [1967]: Liouville's theory on conformal mappings under minimal regularity assumptions, *Siberian Mathematical Journal* **8**, 69–85.

G. de RHAM [1955]: *Variétés Différentiables*, Hermann, Paris.

W.C. RHEINBOLDT [1968]: A unified convergence theory for a class of iterative processes, *SIAM Journal on Numerical Analysis* **5**, 42–63.

F. RIESZ [1907]: Sur les systèmes orthogonaux de fonctions, *Comptes Rendus de l'Académie des Sciences* **144**, 615–619.

F. RIESZ [1907]: Sur une espèce de géométrie analytique des systèmes de fonctions sommables, *Comptes Rendus de l'Académie des Sciences de Paris* **144**, 1409–1411.

F. RIESZ; B. SZ.-NAGY [1955]: *Leçons d'Analyse Fonctionnelle, Troisième Edition*, Gauthier-Villars, Paris, and Akadémiai Kiadó, Budapest (English translation: *Functional Analysis*, Dover, New York, 1990).

J.E. ROBERTS; J.M. THOMAS [1991]: Mixed and hybrid methods, in *Handbook of Numerical Analysis, Volume* II (P.G. CIARLET & J.L. LIONS, editors), pp. 523–639, North-Holland, Amsterdam.

H.L. ROYDEN [1963]: *Real Analysis*, MacMillan, New York (Third Edition: 1988).

W. RUDIN [1966]: *Real and Complex Analysis*, McGraw-Hill, New York (Third Edition: 1987).

W. RUDIN [1973]: *Functional Analysis*, McGraw-Hill, New York (Second Edition: 1991).

W. RUDIN [1997]: *The Way I Remember It*, American Mathematical Society, Providence, RI.

H. SAMELSON [2001]: Differential forms, the early days; or the stories of Deahna's theorem and of Volterra's theorem, *American Mathematical Monthly* **108**, 552–530.

A. SARD [1942]: The measure of the critical values of differential maps, *Bulletin of the American Mathematical Society* **48**, 883–890.

F. SAUVIGNY [2006a]: *Partial Differential Equations* 1 : *Foundations and Integral Representations*, Springer, Berlin.

F. SAUVIGNY [2006b]: *Partial Differential Equations* 2 : *Functional Analytic Methods*, Springer, Berlin.

G.M. Scarpello; D. Ritelli [2002]: A historical outline of the theorem of implicit functions, *Divulgaciones Matemáticas* **10**, 171–180.

H. Schäfer [1955]: Über die Methode der a priori Schranken, *Mathematische Annalen* **129**, 415–416.

J. Schauder [1930]: Der Fixpunktsatz in Funktionalräumen, *Studia Mathematica* **2**, 171–180.

J. Schauder [1934]: Über lineare elliptische Differentialgleichungen zweiter Ordnung, *Mathematische Zeitschrift* **38**, 257–282.

M. Schechter [1971]: *Principles of Functional Analysis, First Edition*, Graduate Studies in Mathematics, Volume 36, American Mathematical Society, Providence, RI (Second Edition: 2002).

H. Schlichtkrull [2012]: *Differential Manifolds*, Lecture Notes for Geometry 2, available online at www.math.ku.dk/~jakobsen/geom2/manusgeom2.pdf.

E. Schmidt [1907]: Zur Theorie der linearen und nichtlinearen Integralgleichungen. 1. Teil: Entwicklung willkürlicher Funktionen nach Systemen vorgeschriebener, *Mathematische Annalen* **63**, 433–476.

J. Schwartz [1969]: *Nonlinear Functional Analysis*, Gordon and Breach, New York.

L. Schwartz [1965]: *Méthodes Mathématiques pour les Sciences Physiques*, Hermann, Paris (English translation: *Mathematics for the Physical Sciences*, Dover, New York, 2008).

L. Schwartz [1966]: *Théorie des Distributions*, Hermann, Paris.

L. Schwartz [1970]: *Analyse: Deuxième Partie: Topologie Générale et Analyse Fonctionnelle*, Hermann, Paris.

L. Schwartz [1991]: *Analyse* I: *Théorie des Ensembles et Topologie*, Hermann, Paris.

L. Schwartz [1992]: *Analyse* II: *Calcul Différentiel et Equations Différentielles*, Hermann, Paris.

L. Schwartz [1993a]: *Analyse* III: *Calcul Intégral*, Hermann, Paris.

L. Schwartz [1993b]: *Analyse* IV: *Applications de la Théorie de la Mesure*, Hermann, Paris.

L. Schwartz [2001]: *A Mathematician Grappling with His Century*, Birkhäuser, Basel (translation of the original French edition, *Un Mathématicien aux Prises avec le Siècle*, Odile Jacob, Paris, 1997).

H.A. Schwarz [1885]: Über ein Flächen kleinsten Flächeninhalts betreffendes Problem der Variationsrechnung, *Acta Societatis Scientiarum Fennicae* **15**, 315–362.

D. Serre [2010]: *Matrices, Second Edition*, Springer, Heidelberg (translated from the original French edition, *Matrices*, Springer, New York, 2002).

R.T. Shield [1973]: The rotation associated with large strains, *SIAM Journal on Applied Mathematics* **25**, 483–491.

A. Signorini: Sopra alcune questioni di elastostatica, *Atti della Società Italiana per il Progresso della Scienza* (1933).

J.G. Simmonds [1994]: *A Brief on Tensor Analysis, Second Edition,* Springer, Berlin (First Edition: 1982).

M. Sion [1958]: On general mini-max theorems, *Pacific Journal of Mathematics* **8**, 171–176.

S. Slicaru [1998]: On the ellipticity of the middle surface of a shell and its application to the asymptotic analysis of "membrane shells," *Journal of Elasticity* **46**, 33–42.

K.T. Smith [1983]: *Primer of Modern Analysis, Second Edition*, Springer, New York (First Edition: 1971, Bogden & Quigley, Tarrytown-on-Hudson, NY).

S.J. Smith [2006]: Lebesgue constants in polynomial interpolation, *Annales Mathematicae et Informaticae* **33**, 109–123.

V.L. Šmulian [1940]: Über lineare topologische Räume, *Mathematiceskii Sbornik, N.S.* **49**, 425–448.

J.P. SNYDER [1993]: *Flattening the Earth: Two Thousand Years of Map Projection*, University of Chicago Press, Chicago.

S.L. SOBOLEV [1938]: On a theorem of functional analysis, *Matematicheskii Sbornik* **46**, 471–496.

S.L. SOBOLEV [1950]: *Applications of Functional Analysis in Mathematical Physics*, Leningrad (in Russian; English translation: American Mathematical Society, Providence, RI, 1963).

V.A. SOLONNIKOV [1982]: On the Stokes equations in domains with non-smooth boundaries and on viscous incompressible flow with a free surface, in *Nonlinear Partial Differential Equations and Their Applications* (H. BREZIS & J.L. LIONS, editors), pp. 340–423, Pitman, Boston.

G.A. SOUKHOMLINOFF [1938]: Über Fortsetzung von linearen Funktionalen in linearen komplexen Räumen und linearen Quaternionräumen, *Mathematiceskii Sbornik* **3**, 353–358.

M. SPIVAK [1999]: *A Comprehensive Introduction to Differential Geometry, Volumes* I *to* V*, Third Edition*, Publish or Perish, Berkeley, CA.

I. STAKGOLD [1998]: *Green's Functions and Boundary Value Problems, Second Edition*, John Wiley, New York (First Edition: 1979).

G. STAMPACCHIA [1964]: Formes bilinéaires coercitives sur les ensembles convexes, *Comptes Rendus de l'Académie des Sciences de Paris Série A*, **258**, 4413–4416.

G. STAMPACCHIA [1965]: *Equations Elliptiques du Second Ordre à Coefficients Discontinus*, Presses de l'Université de Montréal, Montréal, Que.

G. STAMPACCHIA [1965]: Le problème de Dirichlet pour les équations elliptiques du second ordre à coefficients discontinus, *Annales de l'Institut Fourier (Grenoble)* **15**, 189–258.

E.M. STEIN [1970]: *Singular Integrals and Differentiability Properties of Functions*, Princeton University Press, Princeton, NJ.

E.M. STEIN; R. SHAKARCHI [2005]: *Real Analysis: Measure Theory, Integration and Hilbert Spaces*, Princeton Lectures on Analysis, Volume III, Princeton University Press, Princeton, NJ.

E.M. STEIN; R. SHAKARCHI [2011]: *Functional Analysis: Introduction to Further Topics in Analysis*, Princeton University Press, Princeton, NJ.

H. STEINLEIN [1979]: Two results of J. Dugundji about extensions of maps and retractions, *Proceedings of the American Mathematical Society* **77**, 298–290.

R.A. STEPHENSON [1980]: On the uniqueness of the square-root of a symmetric, positive-definite tensor, *Journal of Elasticity* **10**, 213–214.

G.W. STEWART [1969]: On the continuity of the generalized inverse, *SIAM Journal on Applied Mathematics* **17**, 33–45.

J.J. STOKER [1969]: *Differential Geometry*, John Wiley, New York.

G.G. STOKES [1845]: On the theories of the internal friction of fluids in motion, *Transactions of the Cambridge Philosophical Society* **8**, 287–305.

M.H. STONE [1948]: The generalized Weierstrass approximation theorem, *Mathematics Magazine* **21**, 167–183 and 237–254.

G. STRANG [1976]: *Linear Algebra and Its Applications*, Academic Press, New York.

G. STRANG [2009]: *Introduction to Linear Algebra, Fourth Edition*, Wellesley Cambridge Press, UK.

M. STRUWE [1990]: *Variational Methods—Applications to Nonlinear Partial Differential Equations and Hamiltonian Systems*, Springer, Berlin.

A. STUBHAUG [2000]: *Niels Henrik Abel and his Times—Called Too Soon by Flames Afar*, Springer, Heidelberg (translated from the Norwegian).

R.H. SZCZARBA [1970]: On isometric immersions of Riemannian manifolds in Euclidean space, *Boletim da Sociedade Brasileira de Matemática* **1**, 31–45.

G. SZEGŐ [1975]: *Orthogonal Polynomials, Fourth Edition*, American Mathematical Society, Providence, RI (First Edition: 1939).

M. SZOPOS [2005]: On the recovery and continuity of a submanifold with boundary, *Analysis and Applications* **3**, 119–143.

L. TARTAR [1978]: *Topics in Nonlinear Analysis*, Publications Mathématiques d'Orsay No. 78.13, Université de Paris-Sud, Orsay.

L. TARTAR [1979]: Compensated compactness and partial differential equations, in *Nonlinear Analysis and Mechanics, Heriot-Watt Symposium, Volume* IV (R. J. KNOPS, editor), pp. 136–212, Pitman, Boston.

L. TARTAR [1983]: The compensated compactness method applied to systems of conservation laws, in *Systems of Nonlinear Partial Differential Equations* (J.M. BALL, editor), pp. 263–285, Reidel, Dordrecht.

L. TARTAR [2006]: *An Introduction to Navier–Stokes Equation and Oceanography*, Springer, Berlin.

L. TARTAR [2007]: *An Introduction to Sobolev Spaces and Interpolation Spaces*, Springer, Berlin.

L. TARTAR [2009]: *The General Theory of Homogenization: A Personalized Introduction*, Springer, Berlin.

A.E. TAYLOR [1939]: The extension of linear functionals, *Duke Mathematical Journal* **5**, 538–547.

A.E. TAYLOR [1958]: *Introduction to Functional Analysis*, John Wiley, New York.

A.E. TAYLOR [1965]: *General Theory of Functions and Integration*, Blaisdell, Waltham.

A.E. TAYLOR; D.C. LAY [1980]: *Introduction to Functional Analysis, Second Edition*, John Wiley, New York.

M.E. TAYLOR [1996a]: *Partial Differential Equations* I: *Basic Theory*, Springer, New York.

M.E. TAYLOR [1996b]: *Partial Differential Equations* II: *Qualitative Studies of Linear Equations*, Springer, New York.

M.E. TAYLOR [1996c]: *Partial Differential Equations* III: *Nonlinear Equations*, Springer, New York.

R. TEMAM [1971]: Solutions généralisées de certaines équations du type hypersurfaces minima, *Archive for Rational Mechanics and Analysis* **44**, 121–156.

R. TEMAM [1977]: *Navier–Stokes Equations*, North-Holland, Amsterdam.

R. TEMAM [1995]: *Navier–Stokes Equations and Nonlinear Functional Analysis, Second Edition*, SIAM, Philadelphia.

K. TENENBLAT [1971]: On isometric immersions of Riemannian manifolds, *Boletim da Sociedade Brasileira de Matemática* **2**, 23–36.

T.Y. THOMAS [1934]: Systems of total differential equations defined over simply connected domains, *Annals of Mathematics* **35**, 730–734.

T.W. TING [1974]: St. Venant's compatibility conditions, *Tensors, N.S.* **28**, 5–12.

K. TINTAREV; K.-H. FIESELER [2007]: *Concentration Compactness. Functional-Analytic Grounds and Applications*, Imperial College Press, London.

O. TOEPLITZ [1918]: Das algebraische Analogon zu einem Satze von Fejér, *Mathematische Zeitschrift* **2**, 187–197.

L. TONELLI [1920]: La semicontinuità nel calcolo delle variazioni, *Rendiconti del Circolo Matematico di Palermo* **44**, 167–249.

A. TYCHONOFF [1930]: Über die topologische Erweiterung von Räumen, *Mathematische Annalen* **102**, 544–561.

S.M. ULAM [1976]: *Adventures of a Mathematician*, reprinted and expanded by University of California Press, Berkeley, 1991.

H. UZAWA [1958]: Iterative methods for concave programming, in *Studies in Linear and Nonlinear Programming* (K.J. ARROW, L. HURWICZ, & H. UZAWA, editors), pp. 154–165, Stanford University Press, Stanford, CA.

M.M. VAINBERG [1952]: Some questions of differential calculus in linear spaces, *Uspehi Matematicheskii Nauk (New Series)* **7**, 55–102 (in Russian).

T. VALENT [1988]: *Boundary Value Problems of Finite Elasticity—Local Theorems on Existence, Uniqueness, and Analytic Dependence on Data*, Springer, New York.

C.J. DE LA VALLÉE POUSSIN [1910]: Sur les polynômes d'approximation et la représentation approchée d'un angle, *Académie Royale de Belgique, Bulletins de la Classe des Sciences* **12**.

C. VALLÉE; D. FORTUNÉ [1976]: Compatibility equations in shell theory, *International Journal of Engineering Science* **34**, 495–499.

R.S. VARGA [1962]: *Matrix Iterative Analysis*, Prentice-Hall, Englewood Cliffs, NJ.

K. VO-KHAC [1972a]: *Distributions—Analyse de Fourier—Opérateurs aux Dérivées Partielles, Volume* 1, Vuibert, Paris.

K. VO-KHAC [1972b]: *Distributions—Analyse de Fourier—Opérateurs aux Dérivées Partielles, Volume* 2, Vuibert, Paris.

V. VOLTERRA [1907]: Sur l'équilibre des corps élastiques multiplement connexes, *Annales de l'Ecole Normale* **24**, 401–517.

E.V. VORONOVSKAJA [1932]: Détermination de la forme asymptotique de l'approximation des fonctions par les polynômes de M. Bernstein, *Doklady Akademii Nauk SSSR* **4**, 79–85.

K. WEIERSTRAß [1872]: Über continuirliche Functionen eines reellen Arguments, die für keinen Werth des letzteren einen bestimmten Differentialquotienten besitzen, *Königliche Akademie der Wissenschaften*.

K. WEIERSTRAß [1885]: Über die analytische Darstellbarkeit sogenannter willkürlicher Funktionen einer reellen Veränderlichen, *Sitzungsberichte der Akademie zu Berlin*, 633–639 and 789–805.

J. WEINGARTEN [1861]: Über eine Klasse auf einander abwickelbarer Flächen, *Journal für Reine und Angewandte Mathematik* **59**, 382–393.

R.S. WESTFALL [1980]: *Never at Rest: A Biography of Isaac Newton*, Cambridge University Press, Cambridge, UK.

H. WEYL [1940]: The method of orthogonal projection in potential theory, *Duke Mathematical Journal* **7**, 414–444.

R. WHITLEY [1967]: An elementary proof of the Eberlein-Šmulian theorem, *Mathematische Annalen* **172**, 116–118.

H. WHITNEY [1934]: Analytic extensions of differentiable functions defined in closed sets, *Transactions of the American Mathematical Society* **36**, 63–89.

R. WONG [2010]: *Lecture Notes on Applied Analysis*, World Scientific, Singapore.

Q. YANG; J.P. SNYDER; W.R. TOBLER [2000]: *Map Projection Transformation—Principle and Applications*, Taylor and Francis, London.

W.H. YOUNG [1910]: *The Fundamental Theorems of the Differentiable Calculus*, Cambridge University Press, Cambridge, UK.

K. YOSIDA [1966]: *Functional Analysis, First Edition*, Springer, Berlin (Reprint of the Sixth Edition: 1980).

G. ZAMPIERI [1992]: Diffeomorphisms with Banach space domains, *Nonlinear Analysis, Theory, Methods & Applications* **19**, 923–932.

F. ZARANTONELLO [1960]: Solving functional equations by contractive averaging, *Mathematics Research Center Report No.* 160, University of wisconsin Madison, Madison, WI.

E. ZEIDLER [1985]: *Nonlinear Functional Analysis and Its Applications, Volume* III: *Variational Methods and Optimization*, Springer, New York.

E. ZEIDLER [1986]: *Nonlinear Functional Analysis and Its Applications, Volume* I: *Fixed-Point Theorems*, Springer, Berlin.

E. ZEIDLER [1990a]: *Nonlinear Functional Analysis and Its Applications, Volume* IIa: *Linear Monotone Operators*, Springer, New York.

E. ZEIDLER [1990b]: *Nonlinear Functional Analysis and Its Applications, Volume* IIb: *Fixed-Point Theorems*, Springer, New York.

E. ZEIDLER [1995a]: *Applied Functional Analysis: Main Principles and Their Applications*, Springer, New York.

E. ZEIDLER [1995b]: *Applied Functional Analysis: Applications of Mathematical Physics*, Springer, New York.

E. ZERMELO [1904]: Beweis dass jede Menge wohlgeordnet werden kann, *Mathematische Annalen* **LIX**, 514–516.

M. ZLÁMAL [1968]: On the finite element method, *Numerische Mathematik* **12**, 394–409.

MAIN NOTATIONS

Sets, mappings, sequences

$\varnothing$: empty set.

$A \subset B$: A is contained in B.

$A \subsetneqq B$: A is strictly contained in B.

$A \cup B$: union of A and B.

$A \cap B$: intersection of A and B.

$A \times B$: product of A and B.

$\bigcup_{i \in I} A_i$: union of sets of a family $(A_i)_{i \in I}$.

$\bigsqcup_{i \in I} A_i$: disjoint union of sets of a family $(A_i)_{i \in I}$.

$\bigcap_{i \in I} A_i$: intersection of sets of a family $(A_i)_{i \in I}$.

$\prod_{i \in I} A_i$: product of sets of a family $(A_i)_{i \in I}$.

$X - A = \{y \in X;\ y \notin A\}$: complement of a subset $A \subset X$.

$\mathbb{N} = \{0, 1, 2, \ldots\}$: set of natural integers.

$\mathbb{Z} = \{\ldots, -2, -1, 0, 1, 2, \ldots\}$: set of integers.

$\mathbb{Q}$: set of rational numbers.

$\mathbb{R}$: set of real numbers.

$\{-\infty\} \cup \mathbb{R} \cup \{\infty\}$: set of extended real numbers.

$\mathbb{C}$: set of complex numbers.

$\mathbb{K} = \mathbb{R}$ or $\mathbb{C}$: set of scalars.

$\overline{z}$: complex conjugate of $z \in \mathbb{C}$.

$\operatorname{Re} z$ and $\operatorname{Im} z$: real and imaginary parts of $z \in \mathbb{C}$.

δ_{ij}, or δ_i^j, or δ^{ij}: Kronecker's symbol ($\delta_{ij} = 1$ if $i = j$ and $\delta_{ij} = 0$ if $i \neq j$).

$\mathfrak{S}_n$: set of all permutations of $\{1, 2, \ldots, n\}$.

$\overline{A}$: closure of a set A.

$\mathring{A}$ or $\operatorname{int} A$: interior of a set A.

∂A: boundary of a set A.

$\operatorname{card} A$: cardinal number of a set A.

$f : X \to Y$, or $f : x \in X \to f(x) \in Y$: mapping, or function, of X into Y.

$g \circ f$: composition of f and g.

$f|_A$: restriction of a mapping f to a set A.

$f(\cdot, b)$: partial mapping $x \to f(x, b)$.

$f(A) = \{y \in Y;\ y = f(x) \text{ for some } x \in X\}$: image of a subset $A \subset X$ under a mapping $f : X \to Y$ (also denoted $\operatorname{Im}(A)$ if A is linear).

$f^{-1}(B) = \{x \in X;\ f(x) \in B\}$: inverse image of a subset $B \subset Y$ under the mapping $f : X \to Y$.

f^{-1}: inverse mapping of a bijective mapping.

$\operatorname{supp} f = \overline{\{x \in X;\ f(x) \neq 0\}}$: support of a function $f : X \to \mathbb{R}$.

id, or id_X: identity mapping of a set X.

$\operatorname{sgn} \alpha = 1$ if $\alpha > 0$, $\operatorname{sgn} \alpha = 0$ if $\alpha = 0$, $\operatorname{sgn} \alpha = -1$ if $\alpha < 0$.

$\deg(f, \Omega, b)$: Brouwer's topological degree of a mapping $f \in \mathcal{C}(\overline{\Omega}; \mathbb{R}^n)$ at a point $b \notin f(\partial\Omega)$ (here, Ω is a bounded open subset of $\mathbb{R}^n$).

$(x_k)_{k=\ell}^{\infty}$, or (x_k) if $\ell = 0$ or $\ell = 1$: sequence of elements $x_\ell, x_{\ell+1}, \ldots, x_k, \ldots$.

$(x_{\sigma(k)})_{k=1}^{\infty}$: subsequence of $(x_k)_{k=1}^{\infty}$ (where σ denotes any strictly increasing mapping of the set $\{1, 2, \ldots\}$ into itself).

$x = \lim_{k\to\infty} x_k$, or $x_k \xrightarrow[k\to\infty]{} x$, or $x_k \to x$ as $k \to \infty$: the sequence (x_k) converges, and its limit is x.

$\liminf_{k\to\infty} x_k$, $\limsup_{k\to\infty} x_k$: limit inferior, limit superior, of a sequence (x_k) of numbers in the set$\{-\infty\} \cup \mathbb{R} \cup \{\infty\}$.

When no confusion should arise, the symbol "$k \to \infty$" is sometimes omitted for notational brevity (e.g., $x = \lim x_k$, $x = \limsup x_k$, $x_k \to x$, etc.).

$x \to a^+$: the real numbers $x > a$ converge to $a \in \mathbb{R}$.

$x \to a^-$: the real numbers $x < a$ converge to $a \in \mathbb{R}$.

dx, or meas, or dx-meas: n-dimensional Lebesgue measure.

Vector spaces

$X = Y \oplus Z$: X is the direct sum of its subspaces Y and Z.

$[a, b] = \{ta + (1-t)b;\ 0 \le t \le 1\}$: closed segment with end-points a and b.

$]a, b[= \{ta + (1-t)b;\ 0 < t < 1\}$: open segment with end-points a and b.

I: identity mapping of a vector space.

$\operatorname{Ker} A = \{x \in X;\ Ax = 0\}$: kernel of the linear operator $A : X \to Y$.

$\operatorname{Im} A = \{y \in Y;\ y = Ax \text{ for some } x \in X\}$: image of the space X under the linear operator $A : X \to Y$ (also denoted $A(X)$).

$(X, \|\cdot\|)$: vector space X equipped with the norm $\|\cdot\|$.

$\|\cdot\|_X$: norm in a vector space X.

$\|\cdot\|_p$: norm in the space $\ell^p, 1 \le p \le \infty$.

$(X, (\cdot,\cdot))$: Hilbert space X equipped with the inner product $(\cdot,\cdot)$.

$|\cdot|$: Euclidean norm in $\mathbb{R}^n$.

$|\cdot|$: operator norm of a matrix subordinate to the Euclidean norm.

$B(a; x) = \{x \in X;\ \|x - a\| < r\}$: open ball of radius r centered at a.

$\operatorname{co} A$: convex hull of a set A.

$\overline{\operatorname{co}} A$: closed convex hull of a set A.

$\mathcal{L}(X; Y)$: space of all continuous linear mappings from a normed vector space X into a normed vector space Y.

$\mathcal{L}(X) = \mathcal{L}(X; X)$.

$X' = \mathcal{L}(X; \mathbb{K})$: dual (space) of a normed vector space X over $\mathbb{K}$.

${}_{X'}\langle x', x\rangle_X = x'(x)$ for any $x' \in X'$ and $x \in X$.

$X'' = \mathcal{L}(X'; \mathbb{K})$: bidual (space) of a normed vector space over $\mathbb{K}$.

$A' \in \mathcal{L}(Y'; X)$: dual (operator) of a linear operator $A \in \mathcal{L}(X; Y)$.

$A^* \in \mathcal{L}(Y; X)$: adjoint (operator) of $A \in \mathcal{L}(X; Y)$ when X and Y are Hilbert spaces.

$\mathcal{L}_k(X_1, X_2, \dots, X_k; Y)$, or $\mathcal{L}_k(X; Y)$ if $X := X_1 = X_2 = \cdots = X_k$: space of all continuous k-linear mappings from a product $X_1 \times X_2 \times \cdots \times X_k$ of normed vector spaces into a normed vector space Y, $k \ge 2$.

X/Y: quotient of a vector space X by a vector subspace Y of X.

$X \hookrightarrow Y$: X is contained in Y with a continuous injection.

$X \Subset Y$: X is contained in Y with a compact injection.

$A^{\perp} = \{y \in X;\ (y, x) = 0 \text{ for all } x \in A\}$: orthogonal complement of a subset A of a Hilbert space $(X, (\cdot, \cdot))$.

$x_k \to x$, or $x = \lim x_k$: strong convergence in $(X; \|\cdot\|)$, i.e., $\lim \|x_k - x\|_x = 0$.

$x_k \rightharpoonup x$: weak convergence in X, i.e., $\lim x'(x_k) = x'(x)$ for all $x' \in X'$.

$x'_k \overset{*}{\rightharpoonup} x'$: weak $*$ convergence in X', i.e., $\lim x'_k(x) = x'(x)$ for all $x \in X$.

Some function spaces

$\mathcal{P}_n$: space of all real polynomials of degree $\leq n$.

$\mathcal{P} := \bigcup_{n=0}^{\infty} \mathcal{P}_n$: space of all real polynomials.

$\mathcal{P}_n\,[a, b] = \{p|_{[a,b]}; p \in \mathcal{P}_n\}$.

$\mathcal{P}\,[a, b] = \{p|_{[a,b]}; p \in \mathcal{P}\}$.

$\mathcal{P}([a, b]\,; \mathbb{C}) := \{p|_{[a,b]};\ p$ is a polynomial with complex coefficients$\}$.

$\mathcal{Q}_n\,[0, 2\pi]$: space of all real 2π-periodic trigonometric polynomials of degree $\leq n$.

$\mathcal{Q}_n\,[0, 2\pi] = \bigcup_{n=0}^{\infty} \mathcal{Q}_n\,[0, 2\pi]$: space of all real 2π-periodic trigonometric polynomials.

$\mathcal{Q}_n([0, 2\pi]\,; \mathbb{C})$: space of all complex 2π-periodic polynomials of degree $\leq n$.

$\mathcal{Q}([0, 2\pi]\,; \mathbb{C} = \bigcup_{n=0}^{\infty} \mathcal{Q}_n([0, 2\pi]\,; \mathbb{C})$: space of all complex 2π-periodic polynomials.

$\mathcal{C}(X; Y)$: set of all continuous mappings from a topological space X into a topological space Y.

$\mathcal{C}(X) = \mathcal{C}(X; \mathbb{R})$.

$\mathcal{C}\,[a, b] = \mathcal{C}([a, b]\,; \mathbb{R})$.

$\mathcal{C}_{\text{per}}\,[0, 2\pi] = \{g \in \mathcal{C}\,[0, 2\pi]\,;\ g(0) = g(2\pi)\}$.

$\mathcal{C}^m(\Omega; Y)$: space of all m times continuously differentiable mappings from an open subset Ω of a normed vector space into a normed vector space Y, $1 \leq m \leq \infty$.

$\mathcal{C}^m(\Omega) = \mathcal{C}^m(\Omega; \mathbb{R})$.

$\mathcal{C}^m(\overline{\Omega})$, where Ω is a bounded open subset of $\mathbb{R}^n$, and $1 \leq m \leq \infty$: space of all functions $v \in \mathcal{C}^m(\Omega)$ such that, for each multi-index $\boldsymbol{\alpha}$ with $|\boldsymbol{\alpha}| \leq m$, there exists a function $v^{\boldsymbol{\alpha}} \in \mathcal{C}^0(\overline{\Omega})$ such that $v^{\boldsymbol{\alpha}}|_{\Omega} = \partial^{\boldsymbol{\alpha}} v$.

$\|v\|_{\mathcal{C}^m(\overline{\Omega})} = \max_{|\boldsymbol{\alpha}| \leq m} \sup_{x \in \overline{\Omega}} |v^{\boldsymbol{\alpha}}(x)|$.

$\mathcal{C}^m\,[a, b] = \{f|_{[a,b]};\ f \in \mathcal{C}^m(\mathbb{R})\}$.

In what follows, Ω is an open subset of $\mathbb{R}^n$, or a domain in $\mathbb{R}^n$.

$\mathcal{D}(\Omega) = \{v \in \mathcal{C}^{\infty}(\Omega);\ \operatorname{supp} v$ is a compact subset of $\Omega\}$.

$\mathcal{D}'(\Omega)$: space of distributions on Ω.

$L^p(\Omega)$, resp. $L^p(\Omega;\mathbb{C})$, $1 \le p \le \infty$: space of equivalence classes of dx-almost everywhere equal functions, resp. complex-valued functions, v that satisfy $\|v\|_{0,p,\Omega} < \infty$.

$\|v\|_{0,\infty,\Omega} = \inf\{\alpha \ge 0;\ \mathrm{d}x\text{-meas}\,\{x \in \Omega;\ |v(x)| \ge \alpha\}\}$ if $p = \infty$.

$\|v\|_{0,p,\Omega} = \{\int_\Omega |v(x)|^p\}^{1/p} < \infty$ if $1 \le p < \infty$.

$\|v\|_{0,\Omega} = \|v\|_{0,2,\Omega}$.

$L^p(a,b) = L^p(\Omega)$ with $\Omega =]a,b[\subset \mathbb{R}$.

$L^p(\Gamma)$, $1 \le p < \infty$, where $\Gamma = \partial\Omega$: space of equivalence classes of dΓ-almost everywhere equal functions that satisfy $\int_\Gamma |v|^p \,\mathrm{d}\Gamma < \infty$.

$\|v\|_{L^p(\Gamma)} = \{\int_\Gamma |v|^p \,\mathrm{d}a\}^{1/p} < \infty$, $1 \le p < \infty$.

$W^{m,p}(\Omega) = \{v \in L^p(\Omega);\ \partial^\alpha v \in L^p(\Omega)$ for all $|\alpha| \le m\}$, $1 \le m, 1 \le p \le \infty$.

$W_0^{m,p}(\Omega)$: closure of $\mathcal{D}(\Omega)$ in $W^{m,p}(\Omega)$, $1 \le m, 1 \le p < \infty$.

$\|v\|_{m,p,\Omega} = \{\int_\Omega \sum_{|\alpha| \le m} |\partial^\alpha v|^p \,\mathrm{d}x\}^{1/p}$, $1 \le m, 1 \le p < \infty$.

$\|v\|_{m,\infty,\Omega} = \max_{|\alpha| \le m} |\partial^\alpha v|_{0,\infty,\Omega}$.

$|v|_{m,p,\Omega} = \{\int_\Omega \sum_{|\alpha| = m} |\partial^\alpha v|^p \,\mathrm{d}x\}^{1/p}$, $1 \le m, 1 \le p < \infty$.

$|v|_{m,\infty,\Omega} = \max_{|\alpha| = m} |\partial^\alpha v|_{0,\infty,\Omega}$, $1 \le m$.

$H^m(\Omega) = W^{m,2}(\Omega)$, $1 \le m$.

$H_0^m(\Omega) = W_0^{m,2}(\Omega)$, $1 \le m$.

$\|v\|_{m,\Omega} = \|v\|_{m,2,\Omega}$, $1 \le m$.

$|v|_{m,\Omega} = |v|_{m,2,\Omega}$, $1 \le m$.

$\mathrm{tr} \in \mathcal{L}(W^{1,p}(\Omega), L^q(\Gamma)$: trace operator from the Sobolev space $W^{1,p}(\Omega)$, $1 \le p < \infty$, into the space $L^q(\Gamma)$ ($\mathrm{tr}\,\boldsymbol{A}$ also denotes the trace of a matrix $\boldsymbol{A}$).

If $V(\Omega)$ denotes a space of real-valued functions defined over Ω, $\boldsymbol{V}(\Omega)$, resp. $\mathbb{V}(\Omega)$, denotes any space of vector-valued, resp. symmetric tensor-valued, mappings whose components, resp. elements, are in $V(\Omega)$; for instance:

$\boldsymbol{W}^{1,p}(\Omega) = \{\boldsymbol{v} = (v_i);\ v_i \in W^{1,p}(\Omega)\}$,

$\mathbb{L}^2(\Omega) = \{\boldsymbol{\sigma} = (\sigma_{ij});\ \sigma_{ij} = \sigma_{ji} \in L^2(\Omega\}$, etc.,

and the associated norms or seminorms are denoted by the same symbols; for instance,

$\|\boldsymbol{v}\|_{1,p,\Omega} = \left(\sum_i \|v_i\|_{1,p,\Omega}^p\right)^{1/p}$ for each $\boldsymbol{v} = (v_i) \in \boldsymbol{W}^{1,p}(\Omega)$,

$\|\boldsymbol{\sigma}\|_{0,\Omega} = \left(\sum_{i,j} \|\boldsymbol{\sigma}_{ij}\|_{0,\Omega}^2\right)^{1/2}$ for each $\boldsymbol{\sigma} = (\sigma_{ij}) \in \mathbb{L}^2(\Omega)$, etc.

Differential calculus

In what follows, X and Y are normed vector spaces, Ω is an open subset of X, and f is a mapping from Ω into Y.

$f'(a) \in \mathcal{L}(X;Y)$: Fréchet derivative of f at $a \in \Omega$.

$\dfrac{\mathrm{d}f}{\mathrm{d}x}(a) = f'(a)$ if $X = \mathbb{R}$.

$\partial_j f(a) \in \mathcal{L}(X_j;Y)$, or $\dfrac{\partial f}{\partial x_j}(a)$: jth partial derivative of f at a, when $X = \prod_{j=1}^n X_j$.

$\nabla f(a)$ or $\operatorname{grad} f(a) = \left(\dfrac{\partial f}{\partial x_i}(a)\right)_{i=1}^n \in \mathbb{R}^n$: gradient at $a \in \Omega$ of a function $f : \Omega \subset \mathbb{R}^n \to \mathbb{R}$ at $a \in \Omega$.

$\operatorname{div} \boldsymbol{v}(a) = \sum_{j=1}^n \partial_j v_j(a)$: divergence at $a \in \Omega$ of a vector-valued function $\boldsymbol{v} = (v_j) : \Omega \subset \mathbb{R}^n \to \mathbb{R}^n$.

$\mathbf{div}\, \boldsymbol{\sigma}(a) = \left(\sum_{j=1}^n \partial_j \sigma_{ij}(a)\right)_{i=1}^n$: divergence at $a \in \Omega$ of a matrix-valued function $\boldsymbol{\sigma} = (\sigma_{ij}) : \Omega \subset \mathbb{R}^n \to \mathbb{M}^n$.

$\mathbf{curl}\, \boldsymbol{h}(a) = (\partial_j h_i(a) - \partial_i h_j(a))_{1 \le i < j \le n}$: curl at $a \in \Omega$ of a vector-valued function $\boldsymbol{h} = (h_i) : \Omega \subset \mathbb{R}^n \to \mathbb{R}^n$.

$f''(a) \in \mathcal{L}_2(X;Y)$: second derivative of f at $a \in \Omega$.

$\partial_{ij} f(a) = \dfrac{\partial^2 f}{\partial x_i \partial x_j}(a) \in \mathbb{R}$: second-order partial derivative of $f : \Omega \subset \mathbb{R}^n \to \mathbb{R}$ at $a \in \Omega$.

$f^{(m)}(a) \in \mathcal{L}_m(X;Y)$: mth derivative of a mapping f at $a \in \Omega$.

$f^{(m)}(a)h^m = f^{(m)}(a)(h_1, h_2, \ldots, h_m) \in Y$ when $h_i = h$, $1 \le i \le m$.

$\partial^{\boldsymbol{\alpha}} v(a) = \dfrac{\partial^{|\boldsymbol{\alpha}|} v}{\partial x_1^{\alpha_1} \cdots \partial x_n^{\alpha_n}}$, $|\boldsymbol{\alpha}| = \alpha_1 + \cdots + \alpha_n$: multi-index notation for partial derivatives of functions $v : \Omega \subset \mathbb{R}^n \to \mathbb{R}$, with $\boldsymbol{\alpha} = (\alpha_1, \ldots, \alpha_n) \in \mathbb{N}^n$.

The same notations $\partial_j f$, $\partial f/\partial x_j$, $\partial^2 f/\partial x_i \partial x_j$, or $\partial^{\boldsymbol{\alpha}} v$, also denote partial derivatives in the sense of distributions.

Vectors, matrices, tensors

When viewed as a matrix, a vector in $\mathbb{R}^n$ is identified with a column vector, i.e., an $n \times 1$ matrix.

$\boldsymbol{u}^T = (u_1 u_2 \cdots u_n)$: transpose of a vector $\boldsymbol{u}$ (a row vector, i.e., a $1 \times n$ matrix).

$\boldsymbol{u} \cdot \boldsymbol{v} = \boldsymbol{u}^T \boldsymbol{v}$: Euclidean inner product in $\mathbb{R}^n$.

$|\boldsymbol{u}| = \sqrt{\boldsymbol{u} \cdot \boldsymbol{u}}$: Euclidean norm in $\mathbb{R}^n$.

$\boldsymbol{u} \otimes \boldsymbol{v} = \boldsymbol{u}\boldsymbol{v}^T = (u_i v_j)$: tensor product in $\mathbb{R}^n$.

$\boldsymbol{u} \wedge \boldsymbol{v} = \varepsilon_{ijk} u_j v_k \boldsymbol{e}_i$: vector product in $\mathbb{R}^3$, where the orientation tensor (ε_{ijk}) is the tensor of order 3 defined by

$\varepsilon_{ijk} = 1$ if $\{i,j,k\}$ is an even permutation of $\{1,2,3\}$, $\varepsilon_{ijk} = -1$ if $\{i,j,k\}$ is an odd permutation of $\{1,2,3\}$, and $\varepsilon_{ijk} = 0$ if at least two indices are equal.

$\mathbb{M}^n$: space of all real $n \times n$ matrices.

$\mathbb{M}^{m \times n}$: set of all real $m \times n$ matrices (m rows, n columns).

$\mathbb{U}^n$: set of all real $n \times n$ invertible matrices.

$\mathbb{M}^n_+ = \{\boldsymbol{F} \in \mathbb{M}^n;\ \det \boldsymbol{F} > 0\}$.

$\mathbb{O}^n = \{\boldsymbol{P} \in \mathbb{M}^n;\ \boldsymbol{P}\boldsymbol{P}^T = \boldsymbol{P}^T\boldsymbol{P} = \boldsymbol{I}\}$: set of all real $n \times n$ orthogonal matrices.

$\mathbb{O}^n_+ = \{\boldsymbol{P} \in \mathbb{O}^n;\ \det \boldsymbol{P} = 1\}$: set of all real $n \times n$ proper orthogonal matrices.

$\mathbb{S}^n = \{\boldsymbol{B} \in \mathbb{M}^n : \boldsymbol{B} = \boldsymbol{B}^T\}$: set of all real $n \times n$ symmetric matrices.

$\mathbb{S}^n_>$: set of all real $n \times n$ symmetric, positive-definite, matrices.

Given a matrix $\boldsymbol{A} \in \mathbb{M}^{m \times n}$, $(\boldsymbol{A})_{ij}$ denotes its element at the ith row and jth column.

The notation $\boldsymbol{A} = (a_{ij})$ means that $a_{ij} = (\boldsymbol{A})_{ij}$; equivalently,

$$\boldsymbol{A} = (a_{ij}) = \begin{pmatrix} a_{11} & a_{12} & \cdots & a_{1n} \\ a_{21} & a_{22} & \cdots & a_{2n} \\ \vdots & \vdots & \vdots & \\ a_{m1} & a_{m2} & \cdots & a_{mn} \end{pmatrix}.$$

$\boldsymbol{I} = (\delta_{ij})$: unit matrix.

$\boldsymbol{A}^T$: transpose of a matrix $\boldsymbol{A}$.

$\boldsymbol{A}^{-1}$: inverse of a matrix $\boldsymbol{A}$.

$\boldsymbol{A}^{-T} = (\boldsymbol{A}^{-1})^T = (\boldsymbol{A}^T)^{-1}$.

$\boldsymbol{A}^{1/2} \in \mathbb{S}^n_>$: square root of a matrix $\boldsymbol{A} \in \mathbb{S}^n_>$.

$\mathbf{Diag}\,\mu_i$, or $\mathbf{Diag}(\mu_1, \mu_2, \ldots, \mu_n)$: diagonal matrix whose diagonal elements are $\mu_1, \mu_2, \ldots, \mu_n$ (in this order).

$\operatorname{tr} \boldsymbol{A}$: trace of a matrix $\boldsymbol{A}$ (tr also denotes the trace operator in Sobolev spaces).

$\det \boldsymbol{A}$: determinant of a matrix $\boldsymbol{A}$.

$\lambda_i = \lambda_i(\boldsymbol{A})$, $1 \le i \le n$: eigenvalues of a matrix $\boldsymbol{A} \in \mathbb{M}^n$.

$|\boldsymbol{A}| = \sup_{\boldsymbol{v} \neq \boldsymbol{0}} (|\boldsymbol{A}\boldsymbol{v}| \,/\, |\boldsymbol{v}|)$: operator norm of a matrix $\boldsymbol{A}$ subordinate to the Euclidean norm.

$\boldsymbol{A} : \boldsymbol{B} = \operatorname{tr} \boldsymbol{A}^T \boldsymbol{B}$: matrix inner product in $\mathbb{M}^{m \times n}$.

$\|\boldsymbol{A}\|_F = \{\boldsymbol{A} : \boldsymbol{A}\}^{1/2}$: Frobenius norm of a matrix $\boldsymbol{A} \in \mathbb{M}^{m \times n}$.

$\mathbf{Cof}\,\boldsymbol{A} \in \mathbb{M}^n$: cofactor matrix of a matrix $\boldsymbol{A} \in \mathbb{M}^n$.

Differential geometry in $\mathbb{R}^n$

Latin indices or exponents vary in the set $\{1, 2, \ldots, n\}$; Greek indices or exponents vary in the set $\{1, 2\}$; the repeated index or exponent summation convention is used.

$\mathbb{E}^n$: n-dimensional Euclidean space.

In what follows, Ω is an open subset of $\mathbb{R}^n$ and $\boldsymbol{\Theta} = (\Theta_i) : \Omega \to \mathbb{R}^n$ is a smooth enough immersion.

$\boldsymbol{g}_i = \partial_i \boldsymbol{\Theta}$, $\quad g_{ij} = \boldsymbol{g}_i \cdot \boldsymbol{g}_j$, $\quad (g^{ij}) = (g_{ij})^{-1}$, $\quad \boldsymbol{g}^i = g^{ij} \boldsymbol{g}_j$.

$(g_{ij}) : \Omega \to \mathbb{S}^n_>$: metric tensor.

$\Gamma_{ijq} = \dfrac{1}{2}(\partial_j g_{iq} + \partial_i g_{jq} - \partial_q g_{ij}) = \partial_i \boldsymbol{g}_j \cdot \boldsymbol{g}_q$: Christoffel symbols of the first kind.

$\Gamma^p_{ij} = a^{pq} \Gamma_{ijq} = \boldsymbol{g}^p \cdot \partial_i \boldsymbol{g}_j$: Christoffel symbols of the second kind.

$v_{i\|j} = \partial_j v_i - \Gamma^p_{ij} v_p$: covariant derivative of a vector field $v_i \boldsymbol{g}^i : \Omega \to \mathbb{E}^n$.

$R_{qijk} = \partial_j \Gamma_{ikq} - \partial_k \Gamma_{ijq} + \Gamma^p_{ij} \Gamma_{kqp} - \Gamma^p_{ik} \Gamma_{jqp}$: covariant components of the Riemann curvature tensor.

In what follows, ω is an open subset of $\mathbb{R}^2$ and $\boldsymbol{\theta} = (\theta_\alpha) : \omega \to \mathbb{R}^3$ is a smooth enough immersion.

$\boldsymbol{a}_\alpha = \partial_\alpha \boldsymbol{\theta}$; $\quad \boldsymbol{a}_3 = \boldsymbol{a}^3 = \dfrac{\boldsymbol{a}_1 \wedge \boldsymbol{a}_2}{|\boldsymbol{a}_1 \wedge \boldsymbol{a}_2|}$, $\quad a_{\alpha\beta} = \boldsymbol{a}_\alpha \cdot \boldsymbol{a}_\beta$, $\quad (a^{\alpha\beta}) = (a_{\alpha\beta})^{-1}$, $\quad \boldsymbol{a}^\alpha = a^{\alpha\beta} \boldsymbol{a}_\beta$.

$b_{\alpha\beta} = \partial_\alpha \boldsymbol{a}_\beta \cdot \boldsymbol{a}_3 = -\boldsymbol{a}_\beta \cdot \partial_\alpha \boldsymbol{a}_3$, $\quad b^\beta_\alpha = a^{\beta\sigma} b_{\alpha\sigma}$.

$(a_{\alpha\beta}) : \omega \to \mathbb{S}^2_>$: first fundamental form of the surface $\boldsymbol{\theta}(\omega)$.

$(b_{\alpha\beta}) : \omega \to \mathbb{S}^2$: second fundamental form of the surface $\boldsymbol{\theta}(\omega)$.

$\Gamma_{\alpha\beta\tau} = \dfrac{1}{2}(\partial_\beta a_{\alpha\tau} + \partial_\alpha a_{\beta\tau} - \partial_\tau a_{\alpha\beta}) = \partial_\alpha \boldsymbol{a}_\beta \cdot \boldsymbol{a}_\tau$: Christoffel symbols of the first kind.

$\Gamma^\sigma_{\alpha\beta} = a^{\sigma\tau} \Gamma_{\alpha\beta\tau} = \boldsymbol{a}^\sigma \cdot \partial_\alpha \boldsymbol{a}_\beta$: Christoffel symbols of the second kind.

$\eta_{\alpha|\beta} = \partial_\beta \eta_\alpha - \Gamma^\sigma_{\alpha\beta} \eta_\sigma$: covariant derivative of a tangent vector field $\eta_\alpha \boldsymbol{a}^\alpha : \omega \to \mathbb{E}^3$.

$R_{\tau\alpha\beta\sigma} = \partial_\beta \Gamma_{\alpha\sigma\tau} - \partial_\sigma \Gamma_{\alpha\beta\tau} + \Gamma^\mu_{\alpha\beta} \Gamma_{\sigma\tau\mu} - \Gamma^\mu_{\alpha\sigma} \Gamma_{\beta\tau\mu}$: covariant components of the Riemann curvature tensor of the surface $\boldsymbol{\theta}(\omega)$.

INDEX